ALGEBRA

Lines

Slope of the line through $P_1 = (x_1, y_1)$ and $P_2 = (x_2, y_2)$:

$$m = \frac{y_2 - y_1}{x_2 - x_1}$$

Slope-intercept equation of line with slope m and y-intercept b:

$$y = mx + b$$

Point-slope equation of line through $P_1 = (x_1, y_1)$ with slope m:

$$y - y_1 = m(x - x_1)$$

Point-point equation of line through $P_1 = (x_1, y_1)$ and $P_2 = (x_2, y_2)$:

$$y - y_1 = m(x - x_1) \quad \text{where } m = \frac{y_2 - y_1}{x_2 - x_1}$$

Lines of slope m_1 and m_2 are parallel if and only if $m_1 = m_2$.
Lines of slope m_1 and m_2 are perpendicular if and only if $m_1 = -\frac{1}{m_2}$.

Circles

Equation of the circle with center (a, b) and radius r:

$$(x - a)^2 + (y - b)^2 = r^2$$

Distance and Midpoint Formulas

Distance between $P_1 = (x_1, y_1)$ and $P_2 = (x_2, y_2)$:

$$d = \sqrt{(x_2 - x_1)^2 + (y_2 - y_1)^2}$$

Midpoint of $\overline{P_1 P_2}$: $\left(\dfrac{x_1 + x_2}{2}, \dfrac{y_1 + y_2}{2} \right)$

Laws of Exponents

$$x^m x^n = x^{m+n} \qquad \frac{x^m}{x^n} = x^{m-n} \qquad (x^m)^n = x^{mn}$$

$$x^{-n} = \frac{1}{x^n} \qquad (xy)^n = x^n y^n \qquad \left(\frac{x}{y} \right)^n = \frac{x^n}{y^n}$$

$$x^{1/n} = \sqrt[n]{x} \qquad \sqrt[n]{xy} = \sqrt[n]{x}\,\sqrt[n]{y} \qquad \sqrt[n]{\frac{x}{y}} = \frac{\sqrt[n]{x}}{\sqrt[n]{y}}$$

$$x^{m/n} = \sqrt[n]{x^m} = \left(\sqrt[n]{x} \right)^m$$

Special Factorizations

$$x^2 - y^2 = (x + y)(x - y)$$
$$x^3 + y^3 = (x + y)(x^2 - xy + y^2)$$
$$x^3 - y^3 = (x - y)(x^2 + xy + y^2)$$

Binomial Theorem

$$(x + y)^2 = x^2 + 2xy + y^2$$
$$(x - y)^2 = x^2 - 2xy + y^2$$
$$(x + y)^3 = x^3 + 3x^2 y + 3xy^2 + y^3$$
$$(x - y)^3 = x^3 - 3x^2 y + 3xy^2 - y^3$$

$$(x + y)^n = x^n + nx^{n-1}y + \frac{n(n-1)}{2}x^{n-2}y^2$$
$$+ \cdots + \binom{n}{k} x^{n-k} y^k + \cdots + nxy^{n-1} + y^n$$

where $\dbinom{n}{k} = \dfrac{n(n-1) \cdots (n - k + 1)}{1 \cdot 2 \cdot 3 \cdots \cdot k}$

Quadratic Formula

If $ax^2 + bx + c = 0$, then $x = \dfrac{-b \pm \sqrt{b^2 - 4ac}}{2a}$.

Inequalities and Absolute Value

If $a < b$ and $b < c$, then $a < c$.
If $a < b$, then $a + c < b + c$.
If $a < b$ and $c > 0$, then $ca < cb$.
If $a < b$ and $c < 0$, then $ca > cb$.
$|x| = x \quad \text{if } x \geq 0$
$|x| = -x \quad \text{if } x \leq 0$

$|x| < a$ means $-a < x < a.$

$|x - c| < a$ means $c - a < x < c + a.$

GEOMETRY

Formulas for area A, circumference C, and volume V

Triangle	Circle	Sector of Circle	Sphere	Cylinder	Cone	Cone with arbitrary base
$A = \frac{1}{2}bh$ $= \frac{1}{2}ab \sin \theta$	$A = \pi r^2$ $C = 2\pi r$	$A = \frac{1}{2}r^2 \theta$ $s = r\theta$ (θ in radians)	$V = \frac{4}{3}\pi r^3$ $A = 4\pi r^2$	$V = \pi r^2 h$	$V = \frac{1}{3}\pi r^2 h$ $A = \pi r \sqrt{r^2 + h^2}$	$V = \frac{1}{3}Ah$ where A is the area of the base

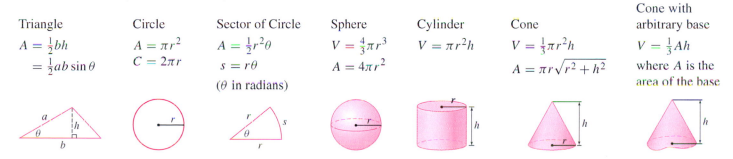

Pythagorean Theorem: For a right triangle with hypotenuse of length c and legs of lengths a and b, $c^2 = a^2 + b^2$.

TRIGONOMETRY

Angle Measurement

π radians $= 180°$

$1° = \dfrac{\pi}{180}$ rad $\qquad$ 1 rad $= \dfrac{180°}{\pi}$

$s = r\theta \quad$ (θ in radians)

Right Triangle Definitions

$\sin\theta = \dfrac{\text{opp}}{\text{hyp}} \qquad\qquad \cos\theta = \dfrac{\text{adj}}{\text{hyp}}$

$\tan\theta = \dfrac{\sin\theta}{\cos\theta} = \dfrac{\text{opp}}{\text{adj}} \qquad \cot\theta = \dfrac{\cos\theta}{\sin\theta} = \dfrac{\text{adj}}{\text{opp}}$

$\sec\theta = \dfrac{1}{\cos\theta} = \dfrac{\text{hyp}}{\text{adj}} \qquad \csc\theta = \dfrac{1}{\sin\theta} = \dfrac{\text{hyp}}{\text{opp}}$

Trigonometric Functions

$\sin\theta = \dfrac{y}{r} \qquad\qquad \csc\theta = \dfrac{r}{y}$

$\cos\theta = \dfrac{x}{r} \qquad\qquad \sec\theta = \dfrac{r}{x}$

$\tan\theta = \dfrac{y}{x} \qquad\qquad \cot\theta = \dfrac{x}{y}$

$\displaystyle\lim_{\theta\to 0}\dfrac{\sin\theta}{\theta} = 1 \qquad \lim_{\theta\to 0}\dfrac{1-\cos\theta}{\theta} = 0$

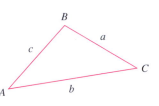

$P = (r\cos\theta, r\sin\theta)$

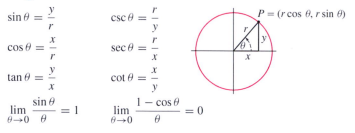

Fundamental Identities

$\sin^2\theta + \cos^2\theta = 1 \qquad\qquad \sin(-\theta) = -\sin\theta$

$1 + \tan^2\theta = \sec^2\theta \qquad\qquad \cos(-\theta) = \cos\theta$

$1 + \cot^2\theta = \csc^2\theta \qquad\qquad \tan(-\theta) = -\tan\theta$

$\sin\left(\dfrac{\pi}{2} - \theta\right) = \cos\theta \qquad \sin(\theta + 2\pi) = \sin\theta$

$\cos\left(\dfrac{\pi}{2} - \theta\right) = \sin\theta \qquad \cos(\theta + 2\pi) = \cos\theta$

$\tan\left(\dfrac{\pi}{2} - \theta\right) = \cot\theta \qquad \tan(\theta + \pi) = \tan\theta$

The Law of Sines

$\dfrac{\sin A}{a} = \dfrac{\sin B}{b} = \dfrac{\sin C}{c}$

The Law of Cosines

$a^2 = b^2 + c^2 - 2bc\cos A$

Addition and Subtraction Formulas

$\sin(x + y) = \sin x \cos y + \cos x \sin y$

$\sin(x - y) = \sin x \cos y - \cos x \sin y$

$\cos(x + y) = \cos x \cos y - \sin x \sin y$

$\cos(x - y) = \cos x \cos y + \sin x \sin y$

$\tan(x + y) = \dfrac{\tan x + \tan y}{1 - \tan x \tan y}$

$\tan(x - y) = \dfrac{\tan x - \tan y}{1 + \tan x \tan y}$

Double-Angle Formulas

$\sin 2x = 2\sin x \cos x$

$\cos 2x = \cos^2 x - \sin^2 x = 2\cos^2 x - 1 = 1 - 2\sin^2 x$

$\tan 2x = \dfrac{2\tan x}{1 - \tan^2 x}$

$\sin^2 x = \dfrac{1 - \cos 2x}{2} \qquad\qquad \cos^2 x = \dfrac{1 + \cos 2x}{2}$

Graphs of Trigonometric Functions

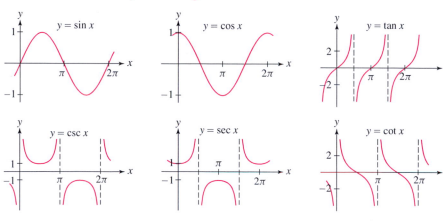

2 UCSD POLICY ON INTEGRITY OF SCHOLARSHIP

[Enacted 5/23/78, Amended 3/2/82, 5/28/85, 1/27/87, 5/22/90, 5/28/91, 4/26/94, 11/22/94, 4/23/96, 11/25/97, 5/27/03, Effective 9/25/03, 4/25/06, 5/26/09, 3/1/11, 1/31/12, 6/3/14, Effective 9/29/14]

Integrity of scholarship is essential for an academic community. The University expects that both faculty and students will honor this principle and in so doing protect the validity of University intellectual work. For students, this means that all academic work will be done by the individual to whom it is assigned, without unauthorized aid of any kind. Instructors, for their part, will exercise care in planning and supervising academic work, so that honest effort will be upheld.

The UCSD Policy on Integrity of Scholarship (herein the "Policy") states the general rules and procedures associated with student integrity of scholarship. This Policy applies to undergraduate and graduate students enrolled at UCSD and/or enrolled in a UCSD course. A separate policy exists governing integrity of research. Medical students are governed by policies specified in the Handbook for School of Medicine Advisors and Students, as formulated by the School of Medicine Committee on Educational Policy. Pharmacy students are governed by the Skaggs School of Pharmacy and Pharmaceutical Sciences (SSPPS) Policy on Integrity of Scholarship as formulated by the SSPPS faculty. In this Policy, the term "in writing" is defined as communications delivered either on paper or electronically via email.

I)   **Instructors' Responsibility**

The Instructor shall state in writing how graded assignments and exams will contribute to the final grade in the course. If there are any course-specific rules required by the Instructor for maintaining academic integrity, the instructor shall also inform students of these in writing.

2   UCSD POLICY ON INTEGRITY OF SCHOLARSHIP

II)   **Students' Responsibility**

Students are expected to complete the course in compliance with the instructor's standards. No student shall engage in an activity that involves attempting to receive a grade by means other than honest effort; for example:

- No student shall knowingly procure, provide, or accept any unauthorized material that contains questions or answers to any examination or assignment that is being, or will be, administered.

- No student shall complete, in part or in total, any examination or assignment for another person.

- No student shall knowingly allow any examination or assignment to be completed, in part or in whole, for himself or herself by another person.

- No student shall plagiarize or copy the work of another person and submit it as his or her own work.

- No student shall employ aids excluded by the instructor in undertaking course work or in completing any exam or assignment.

- No student shall alter graded class assignments or examinations and then resubmit them for regrading.

- No student shall submit substantially the same material in more than one course without prior authorization.

III)   **The Instructional Assistant's (IA) Responsibilities**

A student acting in the capacity of an Instructional Assistant (IA), a category including but not limited to teaching assistants, readers, and tutors, has a special responsibility to safeguard integrity of scholarship. In this role the student functions as an apprentice instructor, under the tutelage of the responsible instructor. An IA shall equitably grade student work in the manner agreed upon with the course instructor. An IA shall not provide a student with any information or collaboration that would aid the student in completing the course in a dishonest manner (e.g., providing access to unauthorized material related to tests, exams, and homework).

IV)   **Responsibility for Resolution of Cases of Violation of the Policy**

The responsibility for maintaining the standards of academic honesty rests with two University authorities: the faculty and the administration. Under the Standing Orders of the Regents, authority over courses and curricula is delegated to the faculty through the Academic Senate. The University of California's Policy on Student Conduct and Discipline authorizes UCSD administration to discipline students for academic misconduct. All cases in which the Student is found responsible for violating the Policy will result in both an academic and an administrative sanction.

2   UCSD POLICY ON INTEGRITY OF SCHOLARSHIP

A)   Academic Responsibilities and Sanctions

The Instructor shall report the alleged violation to the Academic Integrity Office, shall participate in the process according to the Policy, and when the case is resolved, shall determine the Student's grade in the course. Any violation of the Policy by the Student may be considered grounds for failure in the course, although less serious consequences may be incurred in less serious circumstances. An Instructor shall not assign an academic sanction for academic dishonesty unless he or she has submitted a report of an alleged violation of the Policy and the Student has either admitted responsibility for, or has been found responsible for, violating the Policy.

B)   Administrative Authority and Sanction

The appropriate administrative authority shall impose an administrative sanction in accordance with guidelines authorized by the Educational Policy Committee. For undergraduates, the appropriate administrative authority is the Council of Deans of Student Affairs. For graduate students, the appropriate administrative authority is the Assistant Dean of Graduate Studies. For non-matriculated students enrolled in a UCSD course through Summer Session, the appropriate administrative authority is the Director of Summer Session. For non-matriculated students enrolled in a UCSD course through University Extension, the appropriate administrative authority is the Student Affairs Manager. Administrative sanctions range in severity from administrative probation to dismissal from the University. Students found responsible for multiple cases of academic dishonesty shall be subject to dismissal from the University. Sanctioning guidelines can be found at http://academicintegrity.ucsd.edu.

C)   The Academic Integrity Office (AI Office)

The AI Office is the initial contact for the Instructor and the administrative manager for the processing of cases of Policy violations. The AI Office may extend any timelines in the Policy when practical exigencies so dictate, in which case all involved parties will normally be notified in writing. The AI Office shall maintain a record of all cases and shall report annually to the Academic Senate Educational Policy Committee, to the Council of Provosts, and to the Executive Vice Chancellor for Academic Affairs on the number, nature, and type of cases; the pattern of decision- making; the severity and type of academic and administrative sanctions; and other relevant matters as directed by the Educational Policy Committee.

V)   **Procedures for Resolution of Cases of Violations of the Policy**

The procedure for resolution of cases of violating the Policy is divided into three phases: A, Reporting Phase; B, Decision and Resolution Phase; C, Appeals Phase.

A)   **The Reporting Phase**

When an Instructor has reason to believe that a Student has violated the Policy, the Instructor should proceed in one of two ways:

2   UCSD POLICY ON INTEGRITY OF SCHOLARSHIP

I)   Meet with the Student to discuss the suspected violation. If the Instructor decides that there is evidence of a Policy violation, he or she must submit a formal charge describing the suspected violation to the AI Office.

II)   Submit a formal charge to the AI Office describing the alleged violation.

All alleged cases of academic dishonesty must be reported. To file a charge of violating the Policy with the AI Office, an Instructor must submit in writing the following information: the Student's name, the Student's PID, the course name and number, the date of the alleged incident, and a description of the incident. Upon receiving the charge, the AI Office will initiate the resolution process, as described in Section B below.

If the Instructor has submitted a formal charge of violating the Policy, he or she will refrain from assigning a course grade for the Student until the charge has been resolved. If the course concludes before the charge is resolved, the Instructor will assign an "X" code for the course in eGrades, which indicates that the X is due to a "Pending Charge of Academic Dishonesty".

If there is insufficient time to submit a charge of violating the Policy before grades are due (e.g., suspected violation occurred during the final exam), then the Instructor may assign an X code for the course before a charge is filed with the AI Office. In this case, the Instructor must:

i)   assign an X code in eGrades, which indicates that the X is due to a "Pending Charge of Academic Dishonesty". eGrades will automatically notify the AI Office that a formal charge is forthcoming, and

ii)   file a formal charge to the AI Office as soon as possible, normally within fifteen (15) business days of the grades due date.

If, after reporting a charge to the AI Office, the Instructor decides to withdraw the charge, the Instructor shall notify the AI Office in writing of his or her decision. The Instructor shall determine the grade for the course based on the student's academic work. The AI Office shall notify the Student, the appropriate administrative authority, the department chair/program director, and Academic Records that the charge against the Student has been withdrawn by the Instructor. All notations of the charge shall be removed from the Student's academic record. If an X has been assigned, the Instructor shall assign a grade for the course in eGrades once Academic Records has removed all notations of the charge. The charge may be reinstated in accordance with this Policy should new evidence become available.

An instructor may not withdraw a charge if a student has accepted responsibility of violating the Policy or has been found responsible for violating the Policy.

2   UCSD POLICY ON INTEGRITY OF SCHOLARSHIP

B)   **The Decision and Resolution Phase**

Once the Instructor has reported a charge of violating the Policy to the AI Office, the AI Office shall immediately notify the appropriate administrative authority in writing, with a copy to the Instructor and to Academic Records that the Student is charged with violating the Policy. Within two (2) business days, the administrative authority shall notify the Student in writing of the charge and copy the AI Office and the Instructor.

If Students from two or more undergraduate colleges are allegedly involved in the same incident, the AI Office will direct the case to the chair of the Council of Deans of Student Affairs. The chair will then appoint one of the Deans to proceed with the case for all Students, regardless of undergraduate college. If the charge involves both undergraduate and graduate Students, the chair of the Council of Deans of Student Affairs and the Assistant Dean of Graduate Studies shall consult and agree on how to proceed with the case. If the charges involve non-matriculated students enrolled through Summer Session or Extension, please refer to Section IV.B. on the appropriate administrative authority to be consulted.

1)   The Student's deadline for responding to charge(s) of violating the Policy

Within ten (10) business days of the date of notification by the appropriate administrative authority, the Student must respond to the administrative authority acknowledging receipt of the charge and arranging to meet (either in person or via telephone) with the administrative authority to discuss the charge(s) and possible administrative sanctions. The administrative authority shall review the charge(s) with the Student and may advise and assist the Student regarding possible administrative sanctions and the process for resolution of the charge(s) of violating the Policy. Within twelve (12) business days of the date of notification by the administrative authority, the Student must report to the administrative authority his or her decision either to accept the charge of violating the Policy or to contest the charge and request an Academic Integrity Review.

If the Student does not meet with or notify the Dean of his or her decision by the end of the twelfth (12) business day following the date of notification by the Dean, he or she will be presumed to have decided to accept the charge(s) of violating the policy.

2)   Decision I: Student accepts charge(s) of violating the Policy

a)   Administrative Sanction

Administrative sanctions range from administrative probation to dismissal from the University, depending on the severity of the case, any previously recorded offenses, and any mitigating circumstances.

For undergraduate students, the appropriate Dean shall make a recommendation of the administrative sanction(s) to the Council of Deans of Student Affairs. The Council of Deans of Student Affairs shall determine the administrative sanction(s) and shall notify the AI Office of the decision within thirty (30) calendar days from the date of the AI Office notification of the charge.

For graduate students, the Assistant Dean of Graduate Studies shall determine the administrative sanction(s) and shall notify the AI Office of the administrative decision within thirty (30) calendar days from the date of the AI Office's notification of the charge.

For non-matriculated student enrolled in a UCSD course through Summer Session or UCSD Extension, the appropriate administrative authority identified in Section IV.B. shall determine the administrative sanction(s) and shall notify the AI Office of the administrative decision within thirty (30) calendar days from the date of the AI Office's notification of the charge.

A record of the administrative sanction(s) shall be maintained by the AI Office, the appropriate administrative authority, the Council of Deans of Student Affairs (for undergraduates), and Academic Records.

b)   Academic Sanction

Within ten (10) business days of being notified of the administrative sanction(s), the AI Office shall notify the Instructor, the department chair/program director, and Academic Records of the administrative sanction(s). Academic Records shall update or remove all notations of the charge from the Student's academic record and direct the Instructor to assign a grade for the course in eGrades.

The Instructor shall determine the grade for the course. If an X has been assigned, the Instructor shall assign a grade for the course by submitting the grade in eGrades. If the outcome is not determined within the calendar year, Academic Records shall direct the Instructor to submit a clerical error form to assign a grade and Academic Records will post the assigned grade. Upon notification from Academic Records that the final grade has been recorded, the AI Office will notify the Instructor, the appropriate administrative authority, and the department chair/program director of the resolution of the case with a report of both the administrative and academic sanctions.

2   UCSD POLICY ON INTEGRITY OF SCHOLARSHIP

3)   Decision II: The Student contests the charge of violating the Policy and requests an Academic Integrity Review

If the Student contests the charge of violating the Policy (Decision II), he or she must submit to the appropriate administrative authority a written request for an Academic Integrity Review with an explanation of why the charge is contested.

a)   This request must be received by the appropriate administrative authority within twelve (12) business days of the date of the notification of the charge.

b)   Within two (2) business days of receiving the Student's written request for an Academic Integrity Review, the administrative authority shall transmit the written request to the AI Office along with any additional relevant documentation.

c)   Within two (2) business days of receiving the administrative authority's request, the AI Office shall notify the Student, the Instructor, and the administrative authority in writing that the request for an Academic Integrity Review was received.

4)   The Academic Integrity Review (AIR)

The purpose of an Academic Integrity Review is to explore and investigate the incident giving rise to the charge and to reach an informed, evidence-based conclusion as to whether the Policy was violated.

5)   Composition of the Academic Integrity Review Board and the Review Panel

The composition of the Academic Integrity Review Board (AIRB) shall be as follows:

a)   Twenty-five (25) faculty members appointed by the Academic Senate Committee on Committees.

b)   At least six (6) graduate students appointed by the Graduate Student Association in collaboration with the Assistant Dean of Graduate Studies.

c) At least twelve (12) upper division undergraduate students, two from each college, appointed by the college Dean.

For each AIR request, the AI Office shall select from the AIRB five (5) members (the "Review Panel"), which shall normally be composed of three faculty members, one graduate student, and one undergraduate student. The AI Office shall also select a college Dean or the Assistant Dean of Graduate Studies, who is not the Dean or Assistant Dean of the Student, to serve as the Presiding Officer. The Presiding Officer shall conduct the review and advise the Review Panel on procedure, but shall not vote. In the event that a five-member Review Panel is not available (e.g., during the summer months or due to unforeseen circumstances, a recusal or challenge of a Review Panel member, or last minute absences), the Student shall be given the option of electing to proceed with a reduced Review Panel. If the Student elects to proceed with a reduced Review Panel, the Presiding Officer, or the AI Office when appropriate, may agree to proceed with not less than two (2) faculty members and one (1) student (either undergraduate or graduate).

A Review Panel member may recuse himself or herself or the Student may challenge the participation of a Review Panel member only when a reasonable person would recognize a conflict of interest or an inability of the Review Panel member to be unbiased; for example, when there is a personal or authoritative relationship between the Student and a Review Panel member. The Presiding Officer shall make the final determination on challenges to Review Panel composition. In the event that the AIR cannot proceed due to Review Panel composition, the Presiding Officer shall call for a continuance until such time as an appropriate Review Panel can be constituted.

2   UCSD POLICY ON INTEGRITY OF SCHOLARSHIP

6)   Notice of the Academic Integrity Review

As soon as possible, and normally no longer than one quarter after receipt of the request for an AIR, the AI Office shall schedule a review of the case by a Review Panel. The AI Office shall normally provide at least ten (10) business days' notice to the Student and the Instructor of the time, date, and location of the AIR, although exceptions can be made if both the Student and the Instructor agree.

The notice shall include a statement that the UCSD Policy on Integrity of Scholarship is alleged to have been violated and a statement that an AIR has been scheduled. If the time and place of the AIR are not known, the notice shall include a statement indicating that a subsequent notice will be sent specifying same. In the event that the time or place is adjusted after the original notice is sent, an email notifying the parties to this effect shall be deemed sufficient notice.

Objections to the time and date of the AIR will be ruled on by the Presiding Officer no later than five (5) business days before the AIR. Academic Integrity Reviews shall not normally be rescheduled to accommodate the Student's work, class, or personal conflicts unless undue hardship would otherwise be experienced by the Student. Academic Integrity Reviews shall not normally be rescheduled to accommodate the availability of Relevant Parties.

7)   The Review Packet

Once an AIR has been requested by the Student, the relevant documents will be collected, including the facts of the charge by the Instructor and the Student's dispute of the facts of the charge. The Student or the Instructor may also submit to the AI Office additional documents relevant to the charge, or the names and contact information of any additional people (e.g., classmates, teaching assistants) who have knowledge relevant to the charge (Relevant Parties). All documents must be submitted to the AI Office within ten (10) business days of the receipt of the review request by the AI Office.

The AI Office will make available to the Presiding Officer, the Instructor, and the Student a copy of the documents relevant to the charge (the Review Packet) no later than five (5) business days before the date of the AIR.

Newly available documents not included in the Review Packet can be presented at the AIR subject to the approval of the Presiding Officer. In such circumstances, the Presiding Officer should provide the Review Panel, the Student, and the Instructor with adequate time to review the new information.

2   UCSD POLICY ON INTEGRITY OF SCHOLARSHIP

8)   Parties Attending the AIR

A Relevant Party is one with direct and material understanding of the case.

Normally, the Instructor bringing the charge forward and the Student requesting the AIR must be present for the AIR. However, in lieu of attending the AIR, the Instructor and/or the Student may forfeit in-person participation and provide a written statement.

The Student's absence from or silence during the AIR shall not imply acceptance of responsibility. The University will normally conduct a single AIR to address the charges made against multiple Students in the same incident unless the Students would experience substantial prejudice as a result of a joint AIR. The appropriate administrative authority with whom the Students meet to request an AIR will, in consultation with the AI Office, hear and decide on prejudice concerns.

Recognizing their formal role in the University instruction, in cases where an Instructional Assistant (IA) is involved, the IA may also be present for the entire AIR rather than partially as a Relevant Party.

The Student may be accompanied by an Associated Students Student Advocate in the AIR. The Student should present his or her own case, but the Advocate may assist the Student with questioning and procedural issues. The Advocate may not normally appear at the AIR in lieu of the Student, but in the event that the Advocate is present but the Student is not, the AIR may continue at the discretion of the Presiding Officer, questions may be asked of the Advocate, and the Advocate may address procedural issues on behalf of the Student. Prior to the AIR, the AI Office shall be available to advise the Instructor of the procedures and options for presentation of the case, but the Instructor may be accompanied in the AIR only by a faculty colleague acting under the same restrictions as a Student Advocate.

The Instructor and the Student shall have the right to present Relevant Parties and question all Relevant Parties present at the AIR. In lieu of Relevant Parties attending the AIR, the Instructor and/or the Student may submit written statements from Relevant Parties as part of the Review Packet. Normally, Relevant Parties are present at the AIR only for the time they are presenting their statements and being questioned by the Instructor, the Student, and the Review Panel.

2   **UCSD POLICY ON INTEGRITY OF SCHOLARSHIP**

9)   The Academic Review Process

The Review Panel shall hold an AIR and decide based on the preponderance of evidence presented at the AIR whether or not the Student is responsible for violating the Policy. Academic Integrity Reviews are fundamentally educative and investigative in nature, and thus the rules of evidence used in legal proceedings do not apply.

The Presiding Officer shall conduct the AIR in such a manner as to ensure fairness to the Student and to the Instructor, to maintain order and decorum, to facilitate presentation of evidence, and to provide an opportunity for questions to be asked by the Review Panel.

No AIR shall be undertaken without a reliable recording. The Presiding Officer shall provide for either a reliable audio recording of the AIR or keep written minutes summarizing the AIR. Any recording shall be retained as part of the permanent record by the Student's administrative authority. Transcripts of the AIR will not be made by the University, but if either the Instructor or the Student makes a transcript at his or her own expense, copies should be provided to the other party for the cost of the copy or ten cents per page, whichever is less. Procedures for such record keeping are covered by the UCSD Student Records Policy as implemented by PPM 160-2.

No other recording or broadcasting devices shall be allowed in the AIR.

The final determination of the case shall rest with the Review Panel. The Instructor and the Student, along with any other parties to the AIR, will be excused before the Review Panel begins its deliberations. Review Panel deliberations shall always be confidential and conducted in private with only the Review Panel members and the Presiding Officer present. The responsibility of the Review Panel is only to determine whether the Student violated the Policy, although the Review Panel can make recommendations regarding administrative sanctions to be considered by the appropriate administrative authority. In AIRs where there is more than one Student charged, the Review Panel must make a separate determination for each Student.

Within five (5) business days from the date on which the AIR is completed, the Presiding Officer shall forward via email the Review Panel's determination to the appropriate administrative authority, with copies to the AI Office, the department chair/program director, the Instructor, and the Student.

10)   Determination of Sanctions

If the student is found responsible for violating the Policy, sanctions shall be determined as follows:

2   **UCSD POLICY ON INTEGRITY OF SCHOLARSHIP**

a) Administrative Sanction

If an undergraduate Student is found responsible for violating the Policy, the appropriate Dean shall make a recommendation of the administrative sanction(s) to the Council of Deans of Student Affairs. The Council of Deans of Student Affairs shall determine the administrative sanction(s) and shall inform the Student and the AI Office in writing within ten (10) business days after the receipt of the notice of the Review Panel's determination.

If a graduate Student or non-matriculated Student enrolled in a UCSD course through Summer Session or University Extension is found responsible for violating the Policy, the appropriate administrative authority identified in Section IV.B. shall determine the administrative sanction(s) and shall inform the Student and the AI Office in writing within ten (10) business days after the receipt of the notice of the Review Panel's determination.

A record of the administrative sanction(s) shall be maintained by the AIC, the appropriate Dean or administrative authority, the Council of Deans of Student Affairs (for undergraduates), and Academic Records.

2   UCSD POLICY ON INTEGRITY OF SCHOLARSHIP

b)   Academic Sanction

The AI Office shall notify the Instructor, the department chair/program director, and Academic Records of the administrative sanction(s) and shall direct the Instructor to assign a grade for the course in eGrades if an X has been assigned.

Within ten (10) business days after receiving the official notice from the AI Office, the Instructor shall determine the grade for the course. Academic Records shall update or remove all notations of the charge from the Student's academic record and direct the Instructor to assign a grade for the course in eGrades. If the outcome is not determined within the calendar year, Academic Records shall direct the Instructor to submit a clerical error form to assign a grade and Academic Records will post the assigned grade. Upon notification from Academic Records that the final grade has been recorded, the AI Office will notify the Instructor, the appropriate administrative authority, and the department chair/program director of the resolution of the case with a report of both the administrative and academic sanctions and that the case is closed.

If the Review Panel finds the evidence insufficient to sustain the charge of violating the Policy, the administrative authority and the Instructor shall dismiss the matter without further action against the Student, who shall be permitted either to complete the course without prejudice or to withdraw from it. The AI Office shall notify the Student of his or her options and, within five (5) business days of the date of the letter, the Student shall notify the AI Office of his or her decision. If the Student does not notify the AI Office within this timeframe, it shall be assumed that the Student is electing to complete the course without prejudice. The AI Office shall then notify the Instructor and Academic Records of the Student's decision. If the Student withdraws from the course, the course shall not be listed on his or her transcript.

C)   **The Appeals Phase**

The Student may appeal the determination of the Review Panel, the academic sanction determined by the Instructor, and/or the administrative sanction(s) determined by the appropriate administrative authority.

2   UCSD POLICY ON INTEGRITY OF SCHOLARSHIP

1)   Appeal of the Determination of the Review Panel:

An undergraduate student may appeal the Review Panel's determination by submitting a written appeal to the AI Office, within five (5) business days of formal notification of the determination of the Review Panel. The AI Office will forward the student's appeal to the appropriate Provost. Council of Provosts will consider the appeal within ten (10) business days from the date the appeal was received. The decision of the Council of Provosts regarding the student's appeal shall be sent to the student in writing and copied to the Student's Dean, the AI Office, and Academic Records.

A graduate student may appeal the Review Panel's determination by submitting a written appeal to the AI Office within five (5) business days of formal notification of the determination of the Review Panel. The AI Office will forward the student's appeal to the Dean of Graduate Studies. The Dean of Graduate Studies will consider the appeal within ten (10) business days from the date the appeal was received. The decision of the Dean of Graduate Studies shall be sent directly to the Student in writing and copied to the Assistant Dean of Graduate Studies, the AI Office, and Academic Records.

A non-matriculated student enrolled in a UCSD course through Summer Session or UCSD Extension may appeal the Review Panel's determination by submitting a written appeal to the AI Office within five (5) business days of formal notification of the determination of the Review Panel. The AI Office will forward the student's appeal to the Council of Provosts for students enrolled through Summer Session and the Dean of Extension for students enrolled through UCSD Extension. The designated authority will consider the appeal within ten (10) business days from the date the appeal was received. The decision of the designated authority shall be sent directly to the Student in writing and copied to the appropriate administrative authority, the AI Office, and Academic Records.

The basis for appeal of the Review Board's determination shall be: (i) that standards of procedural fairness were violated, e.g., that the Student did not have sufficient opportunity to present his or her side of the case; or (ii) that there exists newly discovered important evidence that has substantial bearing on the determination of the Review Panel. If the appeal is sustained, the case shall be referred back to the AI Office to schedule a new AIR before a new Review Panel. Except for such appeals, the determination of the Review Panel shall be final.

2)   Appeal of the Academic Sanction:

Appeals must be submitted to the Educational Policy Committee within five (5) business days of receiving notice from the AI Office of the academic sanction assigned. If the case was reviewed by a Review Panel, the Committee shall receive the determination of the Review Panel and accept its determination as to the facts of the case. The Educational Policy Committee shall consider the appeal in accordance with its established procedures.

2 UCSD POLICY ON INTEGRITY OF SCHOLARSHIP

3) Appeal of Administrative Sanction:

An appeal of the administrative sanction(s) shall be submitted by an undergraduate student to the Council of Provosts with a copy to the AI Office within five (5) business days of receiving notice of the administrative sanction. The Council of Provosts shall evaluate the Student's appeal and make a final decision within ten (10) business days of receiving the appeal. The decision of the Council of Provosts shall be sent by the Chair of the Council of Provosts to the Student in writing and copied to the Dean, the AI Office, and Academic Records. [EC 2/10/17]

An appeal by a graduate student shall be directed to the Dean of Graduate Studies with a copy to the AI Office within five (5) business days of receiving notice from the AIC of the administrative sanction. The Dean of Graduate Studies shall evaluate the Student's appeal and make a decision within ten (10) business days of receiving the appeal. The decision of the Dean of Graduate Studies shall be sent to the Student in writing and copied to the Assistant Dean of Graduate Studies, the AI Office, and Academic Records.

An appeal by a non-matriculated student enrolled in a UCSD course shall be directed to the Council of Provosts for students enrolled through Summer Session and the Dean of Extension for students enrolled through UCSD Extension, with a copy to the AI Office within five (5) business days of receiving notice from the AIC of the administrative sanction. The designated authority shall evaluate the Student's appeal and make a decision within ten (10) business days of receiving the appeal. The decision of the designated authority shall be sent to the Student in writing and copied to the appropriate administrative authority, the AI Office, and Academic Records.

A decision of the Council of Provosts, the Dean of Graduate Studies, or Dean of Extension regarding an appeal is final.

2   UCSD POLICY ON INTEGRITY OF SCHOLARSHIP

VI)   **Policy Regarding Student Academic Records**

- Until a charge of violating the Policy has been resolved, the student's transcript will show a blank grade for the course. Academic Records will note in attached text to the course (i.e., not on the student's transcript) that the hold is for a "Pending Charge of Academic Dishonesty".

- Once a charge is filed with the AI Office, the student shall not drop or withdraw from the course. If the Student drops the course before the charge of violating the Policy has been resolved, he or she will be administratively reenrolled in the course by Academic Records. If a student drops or withdraws from a course before a charge is filed with the AI Office, the resolution process will proceed as described in the Policy but no academic sanction will be applied.

- The "Pending Charge of Academic Dishonesty" notation shall not be removed by Academic Records until notification from the AI Office that the case has been resolved or that the Instructor has withdrawn the charge.

- If a passing grade is assigned to a student found responsible for violating the Policy and a conflict arises because of the Student's enrollment in a duplicate, cross listed, or equivalent course taken after the charge has been resolved, Academic Records shall ensure that the grade given in the course with the Academic Dishonesty charge is not removed from the GPA. All other academic regulations pertaining to duplicate course enrollment will be enforced.

- If the student has been found responsible for violating the Policy, the grade assigned by the Instructor will be counted in the Student's GPA even if the course is retaken. Academic Records will permanently note in text attached to the course (i.e., not on the Student's transcript) that the grade was given as a result of "Academic Dishonesty".

- If the student withdraws from UCSD before the final resolution of the case, the following policy shall govern. If the student is found responsible for violating the Policy and the Instructor assigns the student a final grade in the course, this grade shall be permanently entered on the transcript. If the administrative sanction is dismissal, the transcript shall bear a notation that readmission is contingent upon approval from the Chancellor. Any administrative penalty less severe than dismissal shall be imposed if and when the student returns to the University.

- If a case of alleged Policy violation is also the subject of an administrative inquiry under the Policy on Integrity of Research, then the Executive Vice Chancellor for Academic Affairs, in consultation with the Review Panel, may make such modifications in procedure as are necessary to coordinate the two inquiries.

- If the administrative sanction is suspension or dismissal, the fact that the student was suspended or dismissed for violating the Policy must be posted on the academic transcript for the duration of the sanction.

VII)   **Review of this Policy**

The Educational Policy Committee shall periodically review this Policy and propose changes as it deems necessary.

2   **UCSD POLICY ON INTEGRITY OF SCHOLARSHIP**

VIII) **Academic Dishonesty in Independent Exams**

In cases where academic dishonesty is reported in independent exams (exams held outside of coursework), such as placement exams and qualifying exams, the procedures described above shall apply, with exception to the language regarding administration of a grade as an academic sanction in section V.B.10.b and appeals of the academic sanction (section V.C.2).

A)   Academic Sanctions

The academic sanction will be determined by the faculty member or faculty committee with ultimate responsibility for evaluating the exam. The sanction will establish the following:

1)   Evaluation of exam results. This may include granting the student full, partial, or no credit for the exam.

2)   Provision to allow or deny the student the ability to repeat the exam.

The responsible party shall report the academic sanction to the Student and the AI Office, which shall notify the appropriate administrative offices of the sanction (per Section V.B.10.b).

B)   Appeals of the Academic Sanctions

Appeals of academic sanctions must be submitted to the Educational Policy Committee following the timelines specified in the section V.C.2 of the Policy.

Single Variable

CALCULUS

THIRD EDITION

Early Transcendentals

JON ROGAWSKI
University of California, Los Angeles

COLIN ADAMS
Williams College

W. H. FREEMAN
& COMPANY

A Macmillan Education Imprint

TO JULIE —Jon
TO ALEXA AND COLTON —Colin

Publisher: Terri Ward
Developmental Editors: Tony Palermino, Katrina Wilhelm
Marketing Manager: Cara LeClair
Market Development Manager: Shannon Howard
Executive Media Editor: Laura Judge
Associate Editor: Marie Dripchak
Editorial Assistant: Victoria Garvey
Director of Editing, Design, and Media Production: Tracey Kuehn
Managing Editor: Lisa Kinne
Project Editor: Kerry O'Shaughnessy
Production Manager: Paul Rohloff
Cover and Text Designer: Blake Logan
Illustration Coordinator: Janice Donnola
Illustrations: Network Graphics and Techsetters, Inc.
Photo Editors: Eileen Liang, Christine Buese
Photo Researcher: Eileen Liang
Composition: John Rogosich/Techsetters, Inc.
Printing and Binding: King Printing Co., Inc.
Cover photo: ayzek/Shutterstock

Library of Congress Preassigned Control Number: 2014957105

Hardcover:
ISBN-13: 978-1-4641-7174-1
ISBN-10: 1-4641-7174-2

Loose-leaf:
ISBN-13: 978-1-4641-9376-7
ISBN-10: 1-4641-9376-2

Printed in the United States of America
Fourth printing

W. H. Freeman and Company, 41 Madison Avenue, New York, NY 10010
Houndmills, Basingstoke RG21 6XS, England
www.macmillanhighered.com

ABOUT THE AUTHORS

COLIN ADAMS

Colin Adams is the Thomas T. Read professor of Mathematics at Williams College, where he has taught since 1985. Colin received his undergraduate degree from MIT and his PhD from the University of Wisconsin. His research is in the area of knot theory and low-dimensional topology. He has held various grants to support his research, and written numerous research articles.

Colin is the author or co-author of *The Knot Book, How to Ace Calculus: The Streetwise Guide, How to Ace the Rest of Calculus: The Streetwise Guide, Riot at the Calc Exam and Other Mathematically Bent Stories, Why Knot?, Introduction to Topology: Pure and Applied*, and *Zombies & Calculus*. He co-wrote and appears in the videos "The Great Pi vs. E Debate" and "Derivative vs. Integral: the Final Smackdown."

He is a recipient of the Haimo National Distinguished Teaching Award from the Mathematical Association of America (MAA) in 1998, an MAA Polya Lecturer for 1998-2000, a Sigma Xi Distinguished Lecturer for 2000-2002, and the recipient of the Robert Foster Cherry Teaching Award in 2003.

Colin has two children and one slightly crazy dog, who is great at providing the entertainment.

JON ROGAWSKI

As a successful teacher for more than 30 years, Jon Rogawski listened and learned much from his own students. These valuable lessons made an impact on his thinking, his writing, and his shaping of a calculus text.

Jon Rogawski received his undergraduate and master's degrees in mathematics simultaneously from Yale University, and he earned his PhD in mathematics from Princeton University, where he studied under Robert Langlands. Before joining the Department of Mathematics at UCLA in 1986, where he was a full professor, he held teaching and visiting positions at the Institute for Advanced Study, the University of Bonn, and the University of Paris at Jussieu and Orsay.

Jon's areas of interest were number theory, automorphic forms, and harmonic analysis on semisimple groups. He published numerous research articles in leading mathematics journals, including the research monograph *Automorphic Representations of Unitary Groups in Three Variables* (Princeton University Press). He was the recipient of a Sloan Fellowship and an editor of the Pacific Journal of Mathematics and the Transactions of the AMS.

Sadly, Jon Rogawski passed away in September 2011. Jon's commitment to presenting the beauty of calculus and the important role it plays in students' understanding of the wider world is the legacy that lives on in each new edition of *Calculus*.

CONTENTS | SINGLE VARIABLE CALCULUS

Early Transcendentals

Additional content can be accessed online via **LaunchPad**:

ADDITIONAL PROOFS

- L'Hôpital's Rule
- Error Bounds for Numerical Integration
- Comparison Test for Improper Integrals

ADDITIONAL CONTENT

- Second Order Differential Equations
- Complex Numbers

PREFACE

ABOUT *CALCULUS*

On Teaching Mathematics

I consider myself very lucky to have a career as a teacher and practitioner of mathematics. When I was young, I decided I wanted to be a writer. I loved telling stories. But I was also good at math, and, once in college, it didn't take me long to become enamored with it. I loved the fact that success in mathematics does not depend on your presentation skills or your interpersonal relationships. You are either right or you are wrong and there is little subjective evaluation involved. And I loved the satisfaction of coming up with a solution. That intensified when I started solving problems that were open research questions that had previously remained unsolved.

So, I became a professor of mathematics. And I soon realized that teaching mathematics is about telling a story. The goal is to explain to students in an intriguing manner, at the right pace, and in as clear a way as possible, how mathematics works and what it can do for you. I find mathematics immensely beautiful. I want students to feel that way, too.

On Writing a Calculus Text

I had always thought I might write a calculus text. But that is a daunting task. These days, calculus books average over a thousand pages. And I would need to convince myself that I had something to offer that was different enough from what already appears in the existing books. Then, I was approached about writing the third edition of Jon Rogawski's calculus book. Here was a book for which I already had great respect. Jon's vision of what a calculus book should be fit very closely with my own. Jon believed that as math teachers, how we say it is as important as what we say. Although he insisted on rigor at all times, he also wanted a book that was written in plain English, a book that could be read and that would entice students to read further and learn more. Moreover, Jon strived to create a text in which exposition, graphics, and layout would work together to enhance all facets of a student's calculus experience.

In writing his book, Jon paid special attention to certain aspects of the text:

1. Clear, accessible exposition that anticipates and addresses student difficulties.

2. Layout and figures that communicate the flow of ideas.

3. Highlighted features that emphasize concepts and mathematical reasoning: Conceptual Insight, Graphical Insight, Assumptions Matter, Reminder, and Historical Perspective.

4. A rich collection of examples and exercises of graduated difficulty that teach basic skills, problem-solving techniques, reinforce conceptual understanding, and motivate calculus through interesting applications. Each section also contains exercises that develop additional insights and challenge students to further develop their skills.

Coming into the project of creating the third edition, I was somewhat apprehensive. Here was an already excellent book that had attained the goals set for it by its author. First and foremost, I wanted to be sure that I did it no harm. On the other hand, I have been teaching calculus now for 30 years, and in that time, I have come to some conclusions about what does and does not work well for students.

As a mathematician, I want to make sure that the theorems, proofs, arguments and development are correct. There is no place in mathematics for sloppiness of any kind. As a teacher, I want the material to be accessible. The book should not be written at the mathematical level of the instructor. Students should be able to use the book to learn the material, with the help of their instructor. Working from the high standard that Jon set, I have tried hard to maintain the level of quality of the previous edition while making the changes that I believe will bring the book to the next level.

Placement of Taylor Polynomials

Taylor polynomials appear in Chapter 8, before infinite series in Chapter 10. The goal here is to present Taylor polynomials as a natural extension of linear approximation. When teaching infinite series, the primary focus is on convergence, a topic that many students find challenging. By the time we have covered the basic convergence tests and studied the convergence of power series, students are ready to tackle the issues involved in representing a function by its Taylor series. They can then rely on their previous work with Taylor polynomials and the error bound from Chapter 8. However, the section on Taylor polynomials is written so that you can cover this topic together with the materials on infinite series if this order is preferred.

Careful, Precise Development

W. H. Freeman is committed to high quality and precise textbooks and supplements. From this project's inception and throughout its development and production, quality and precision have been given significant priority. We have in place unparalleled procedures to ensure the accuracy of the text:

- Exercises and Examples
- Exposition
- Figures
- Editing
- Composition

Together, these procedures far exceed prior industry standards to safeguard the quality and precision of a calculus textbook.

New to the Third Edition

There are a variety of changes that have been implemented in this edition. Following are some of the most important.

MORE FOCUS ON CONCEPTS The emphasis has been shifted to focus less on the memorization of specific formulas, and more on understanding the underlying concepts. Memorization can never be completely avoided, but it is in no way the crux of calculus. Students will remember how to apply a procedure or technique if they see the logical progression that generates it. And they then understand the underlying concepts rather than seeing the topic as a black box in which you insert numbers. Specific examples include:

- (**Section 1.2**) Removed the general formula for the completion of a square and instead, emphasized the method so students need not memorize the formula.
- (**Section 7.2**) Changed the methods for evaluating trigonometric integrals to focus on techniques to apply rather than formulas to memorize.
- (**Chapter 9**) Discouraged the memorization of solutions of specific types of differential equations and instead, encouraged the use of methods of solution.
- (**Section 12.2**) Decreased number of formulas for parametrizing a line from two to one, as the second can easily be derived from the first.
- (**Section 12.6**) De-emphasized the memorization of the various formulas for quadric surfaces. Instead, moved the focus to slicing with planes to find curves and using those to determine the shape of the surface. These methods will be useful regardless of the type of surface it is.
- (**Section 14.4**) Decreased the number of essential formulas for linear approximation of functions of two variables from four to two, providing the background to derive the others from these.

CHANGES IN NOTATION There are numerous notational changes. Some were made to bring the notation more into line with standard usage in mathematics and other fields in which mathematics is applied. Some were implemented to make it easier for students to remember the meaning of the notation. Some were made to help make the corresponding concepts that are represented more transparent. Specific examples include:

- **(Section 4.6)** Presented a new notation for graphing that gives the signs of the first and second derivative and then simple symbols (slanted up and down arrows and up and down u's) to help the student keep track of when the graph is increasing or decreasing and concave up or concave down over the given interval.
- **(Section 7.1)** Simplified the notation for integration by parts and provided a visual method for remembering it.
- **(Chapter 10)** Changed names of the various tests for convergence/divergence of infinite series to evoke the usage of the test and thereby make it easier for students to remember them.
- **(Chapters 13–17)** Rather than using $c(t)$ for a path, we consistently switched to the vector-valued function $\mathbf{r}(t)$. This also allowed us to replace $d\mathbf{s}$ with $d\mathbf{r}$ as a differential, which means there is less likely to be confusion with $d\mathbf{s}$, dS and $d\mathbf{S}$.

MORE EXPLANATIONS OF DERIVATIONS Occasionally, in the previous edition, a result was given and verified, without motivating where the derivation came from. I believe it is important for students to understand how someone might come up with a particular result, thereby helping them to picture how they might themselves one day be able to derive results.

- **(Section 8.3)** Developed the center of mass formulas by first discussing the one-dimensional case of a seesaw.
- **(Section 14.4)** Developed the equation of the tangent plane in a manner that makes geometric sense.
- **(Section 14.5)** Included a proof of the fact the gradient of a function f of three variables is orthogonal to the surfaces that are the level sets of f.
- **(Section 14.8)** Gave an intuitive explanation for why the Method of Lagrange Multipliers works.

REORDERING AND ADDING TOPICS There were some specific rearrangements among the sections and additions. These include:

- A subsection on piecewise-defined functions has been added to Section 1.3.
- The section on implicit differentiation in Chapter 3 (previously Section 3.10) has been moved up to become Section 3.8 and has absorbed the previous Section 3.8 (inverse functions) so that implicit differentiation can be applied to derive the various derivatives as necessary.
- The section on indefinite integrals (previously Section 4.9) has been moved from Chapter 4 (Applications of the Derivative) to Chapter 5 (The Integral). This is a more natural placement for it.
- A new section on choosing from amongst the various methods of integration has been added to Chapter 7.
- A subsection on choosing the appropriate convergence/divergence test has been added to Section 10.5.
- An explanation of how to find indefinite limits using power series has been added to Section 10.6.
- The definitions of divergence and curl have been moved from Chapter 17 to Section 16.1. This allows us to utilize them at an appropriate earlier point in the text.
- A list all of the different types of integrals that have been introduced in Chapter 16 has been added to Section 16.5.
- A subsection on the Vector Form of Green's Theorem has been added to Section 17.1.

NEW EXAMPLES, FIGURES, AND EXERCISES Numerous examples and accompanying figures have been added to clarify concepts. A variety of exercises have also been added throughout the text, particularly where new applications are available or further conceptual development is advantageous. Figures marked with a **DF** icon have been made dynamic and can be accessed via *LaunchPad*. A selection of these figures also includes brief tutorial videos explaining the concepts at work.

SUPPLEMENTS

For Instructors

Instructor's Solutions Manual
Contains worked-out solutions to all exercises in the text.

Test Bank
Computerized (CD-ROM), ISBN:1-3190-0939-5
Includes a comprehensive set of multiple-choice test items.

Instructor's Resource Manual
Provides sample course outlines, suggested class time, key points, lecture material, discussion topics, class activities, worksheets, projects, and questions to accompany the Dynamic Figures.

For Students

Student Solutions Manual
Single Variable ISBN: 1-4641-7188-2
Multivariable ISBN: 1-4641-7189-0
Contains worked-out solutions to all odd-numbered exercises in the text.

Software Manuals
Maple™ and Mathematica® software manuals serve as basic introductions to popular mathematical software options.

ONLINE HOMEWORK OPTIONS

Our new course space, LaunchPad, combines an interactive e-Book with high-quality multimedia content and ready-made assessment options, including LearningCurve adaptive quizzing. Pre-built, curated units are easy to assign or adapt with your own material, such as readings, videos, quizzes, discussion groups, and more. LaunchPad includes a gradebook that provides a clear window on performance for your whole class, for individual students, and for individual assignments. While a streamlined interface helps students focus on what's due next, social commenting tools let them engage, make connections, and learn from each other. Use LaunchPad on its own or integrate it with your school's learning management system so your class is always on the same page. Contact your rep to make sure you have access.

Assets integrated into LaunchPad include:

Interactive e-Book: Every LaunchPad e-Book comes with powerful study tools for students, video and multimedia content, and easy customization for instructors. Students can search, highlight, and bookmark, making it easier to study and access key content. And instructors can make sure their class gets just the book they want to deliver: customize and rearrange chapters, add and share notes and discussions, and link to quizzes, activities, and other resources.

LearningCurve provides students and instructors with powerful adaptive quizzing, a game-like format, direct links to the e-Book, and instant feedback. The quizzing system features questions tailored specifically to the text and adapts to students' responses, providing material at different difficulty levels and topics based on student performance.

Dynamic Figures: Over 250 figures from the text have been recreated in a new interactive format for students and instructors to manipulate and explore, making the visual aspects and dimensions of calculus concepts easier to grasp. Brief tutorial videos accompany selected figures and explain the concepts at work.

CalcClips: These whiteboard tutorials provide animated and narrated step-by-step solutions to exercises that are based on key problems in the text.

SolutionMaster offers an easy-to-use Web-based version of the instructor's solutions, allowing instructors to generate a solution file for any set of homework exercises.

WebAssign Premium integrates the book's exercises into the world's most popular and trusted online homework system, making it easy to assign algorithmically generated homework and quizzes. Algorithmic exercises offer the instructor optional algorithmic solutions. WebAssign Premium also offers access to resources, including Dynamic Figures, CalcClips whiteboard tutorials, and a "Show My Work" feature. In addition, WebAssign Premium is available with a fully customizable e-Book option.

W. H. Freeman offers thousands of algorithmically generated questions (with full solutions) through this free, open-source online homework system created at the University of Rochester. Adopters also have access to a shared national library test bank with thousands of additional questions, including 2,500 problem sets matched to the book's table of contents.

FEATURES

Conceptual Insights

encourage students to develop a conceptual understanding of calculus by explaining important ideas clearly but informally.

CONCEPTUAL INSIGHT Leibniz notation is widely used for several reasons. First, it reminds us that the derivative df/dx, although not itself a ratio, is in fact a *limit* of ratios $\Delta f/\Delta x$. Second, the notation specifies the independent variable. This is useful when variables other than x are used. For example, if the independent variable is t, we write df/dt. Third, we often think of d/dx as an "operator" that performs differentiation on functions. In other words, we apply the operator d/dx to f to obtain the derivative df/dx. We will see other advantages of Leibniz notation when we discuss the Chain Rule in Section 3.7.

Ch. 3, p. 123

Graphical Insights enhance

students' visual understanding by making the crucial connections between graphical properties and the underlying concepts.

GRAPHICAL INSIGHT Can we visualize the rate represented by $f''(x)$? The second derivative is the rate at which $f'(x)$ is changing, so $f''(x)$ is large if the slopes of the tangent lines change rapidly, as in Figure 3(A). Similarly, $f''(x)$ is small if the slopes of the tangent lines change slowly—in this case, the curve is relatively flat, as in Figure 3(B). If f is a linear function [Figure 3(C)], then the tangent line does not change at all and $f''(x) = 0$. Thus, $f''(x)$ measures the "bending" or concavity of the graph.

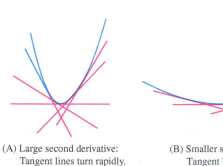

(A) Large second derivative:
   Tangent lines turn rapidly.

(B) Smaller second derivative:
   Tangent lines turn slowly.

(C) Second derivative is zero:
   Tangent line does not change.

DF FIGURE 3

Ch. 3, p. 153

Reminders

are margin notes that link the current discussion to important concepts introduced earlier in the text to give students a quick review and make connections with related ideas.

■ **EXAMPLE 3** Evaluate $\int \sin^2 x \, dx$.

Solution We could apply the reduction formula Eq. (5) from the last section. However, instead, we apply a method that does not rely on knowing that formula. We utilize the trigonometric identity called the double angle formula $\sin^2 x = \frac{1}{2}(1 - \cos 2x)$. Then

$$\int \sin^2 x \, dx = \int \frac{1}{2}(1 - \cos 2x) \, dx = \frac{x}{2} - \frac{\sin 2x}{4} + C$$ ■

←··· **REMINDER** *Useful Identities:*

$$\sin^2 x = \frac{1}{2}(1 - \cos 2x)$$

$$\cos^2 x = \frac{1}{2}(1 + \cos 2x)$$

$$\sin 2x = 2 \sin x \cos x$$

$$\cos 2x = \cos^2 x - \sin^2 x$$

Using the trigonometric identities in the margin, we can also integrate $\cos^2 x$, obtaining the following:

$$\int \sin^2 x \, dx = \frac{x}{2} - \frac{\sin 2x}{4} + C = \frac{x}{2} - \frac{1}{2} \sin x \cos x + C \qquad \boxed{1}$$

$$\int \cos^2 x \, dx = \frac{x}{2} + \frac{\sin 2x}{4} + C = \frac{x}{2} + \frac{1}{2} \sin x \cos x + C \qquad \boxed{2}$$

Ch. 7, p. 380

Caution Notes

warn students of common pitfalls they may encounter in understanding the material.

CAUTION *When using L'Hôpital's Rule, be sure to take the derivative of the numerator and denominator separately:*

$$\lim_{x \to a} \frac{f(x)}{g(x)} = \lim_{x \to a} \frac{f'(x)}{g'(x)}$$

Do not differentiate the quotient function $y = f(x)/g(x)$.

■ **EXAMPLE 1** Use L'Hôpital's Rule to evaluate $\displaystyle\lim_{x \to 2} \frac{x^3 - 8}{x^4 + 2x - 20}$.

Solution Let $f(x) = x^3 - 8$ and $g(x) = x^4 + 2x - 20$. Both f and g are differentiable and $f(x)/g(x)$ is indeterminate of type 0/0 at $a = 2$ because $f(2) = g(2) = 0$:

• Numerator: $f(2) = 2^3 - 1 = 0$
• Denominator: $g(2) = 2^4 + 2(2) - 20 = 0$

Furthermore, $g'(x) = 4x^3 + 2$ is nonzero near $x = 2$, so L'Hôpital's Rule applies. We may replace the numerator and denominator by their derivatives to obtain

$$\underbrace{\lim_{x \to 2} \frac{x^3 - 8}{x^4 + 2x - 2} = \lim_{x \to 2} \frac{(x^3 - 8)'}{(x^4 + 2x - 2)'}}_{\text{L'Hôpital's Rule}} = \lim_{x \to 2} \frac{3x^2}{4x^3 + 2} = \frac{3(2^2)}{4(2^3) + 2} = \frac{12}{34} = \frac{6}{17}$$ ■

Ch. 4, p. 224

Historical Perspectives

are brief vignettes that place key discoveries and conceptual advances in their historical context. They give students a glimpse into some of the accomplishments of great mathematicians and an appreciation for their significance.

HISTORICAL PERSPECTIVE

(Mechanics Magazine London, 1824)

Geometric series were used as early as the third century BCE by Archimedes in a brilliant argument for determining the area S of a "parabolic segment" (shaded region in Figure 3). Given two points A and C on a parabola, there is a point B between A and C where the tangent line is parallel to $\overline{AC}$ (apparently, Archimedes was aware of the Mean Value Theorem more than 2000 years before the invention of calculus). Let T be the area of triangle $\triangle ABC$. Archimedes proved that if D is chosen in a similar fashion relative to $\overline{AB}$ and E is chosen relative to $\overline{BC}$, then

$$\frac{1}{4}T = \text{Area}(\triangle ADB) + \text{Area}(\triangle BEC) \qquad \boxed{6}$$

This construction of triangles can be continued. The next step would be to construct the four triangles on the segments $\overline{AD}, \overline{DB}, \overline{BE}, \overline{EC}$, of total area $\frac{1}{4}^2 T$. Then construct eight triangles of total area $\frac{1}{4}^3 T$, etc. In this way, we obtain infinitely many triangles that completely fill up the parabolic segment. By the formula for the sum of a geometric series, we get

$$S = T + \frac{1}{4}T + \frac{1}{16}T + \cdots = T \sum_{n=0}^{\infty} \frac{1}{4^n} = \frac{4}{3}T$$

For this and many other achievements, Archi-

medes is ranked together with Newton and Gauss as one of the greatest scientists of all time.

The modern study of infinite series began in the seventeenth century with Newton, Leibniz, and their contemporaries. The divergence of $\displaystyle\sum_{n=1}^{\infty} 1/n$ (called the **harmonic series**) was known to the medieval scholar Nicole d'Oresme (1323–1382), but his proof was lost for centuries, and the result was rediscovered on more than one occasion. It was also known that the sum of the reciprocal squares $\displaystyle\sum_{n=1}^{\infty} 1/n^2$ converges, and in the 1640s, the Italian Pietro Mengoli put forward the challenge of finding its sum. Despite the efforts of the best mathematicians of the day, including Leibniz and the Bernoulli brothers Jakob and Johann, the problem resisted solution for nearly a century. In 1735 the great master Leonhard Euler (at the time, 28 years old) astonished his contemporaries by proving that

$$\frac{1}{1^2} + \frac{1}{2^2} + \frac{1}{3^2} + \frac{1}{4^2} + \frac{1}{5^2} + \frac{1}{6^2} + \cdots = \frac{\pi^2}{6}$$

This formula, surprising in itself, plays a role in a variety of mathematical fields. A theorem from number theory states that two whole numbers, chosen randomly, have no common factor with probability $6/\pi^2 \approx 0.6$ (the reciprocal of Euler's result). On the other hand, Euler's result and its generalizations appear in the field of statistical mechanics.

Ch. 10, p. 530

Assumptions Matter uses short explanations and well-chosen counterexamples to help students appreciate why hypotheses are needed in theorems.

■ **EXAMPLE 3** Assumptions Matter Show that the Product Law cannot be applied to $\lim_{x \to 0} f(x)g(x)$ if $f(x) = x$ and $g(x) = x^{-1}$.

Solution For all $x \neq 0$, we have $f(x)g(x) = x \cdot x^{-1} = 1$, so the limit of the product exists:

$$\lim_{x \to 0} f(x)g(x) = \lim_{x \to 0} 1 = 1$$

However, $\lim_{x \to 0} x^{-1}$ does not exist because $g(x) = x^{-1}$ approaches ∞ as $x \to 0^+$ and it approaches $-\infty$ as $x \to 0^-$. Therefore, the Product Law cannot be applied and its conclusion does not hold:

$$\left(\lim_{x \to 0} f(x) \right) \left(\lim_{x \to 0} g(x) \right) = \left(\lim_{x \to 0} x \right) \underbrace{\left(\lim_{x \to 0} x^{-1} \right)}_{\text{Does not exist}}$$

■

Ch. 2, p. 58

Section Summaries summarize a section's key points in a concise and useful way and emphasize for students what is most important in each section.

Section Exercise Sets offer a comprehensive set of exercises closely coordinated with the text. These exercises vary in difficulty from routine, to moderate, to more challenging. Also included are icons indicating problems that require the student to give a written response or require the use of technology T.

Chapter Review Exercises offer a comprehensive set of exercises closely coordinated with the chapter material to provide additional problems for self-study or assignments.

ACKNOWLEDGMENTS

Colin Adams and W. H. Freeman and Company are grateful to the many instructors from across the United States and Canada who have offered comments that assisted in the development and refinement of this book. These contributions included class testing, manuscript reviewing, problems reviewing, and participating in surveys about the book and general course needs.

ALABAMA Tammy Potter, *Gadsden State Community College*; David Dempsey, *Jacksonville State University*; Edwin Smith, *Jacksonville State University*; Jeff Dodd, *Jacksonville State University*; Douglas Bailer, *Northeast Alabama Community College*; Michael Hicks, *Shelton State Community College*; Patricia C. Eiland, *Troy University, Montgomery Campus*; Chadia Affane Aji, *Tuskegee University*; James L. Wang, *The University of Alabama*; Stephen Brick, *University of South Alabama*; Joerg Feldvoss, *University of South Alabama* **ALASKA** Mark A. Fitch, *University of Alaska Anchorage*; Kamal Narang, *University of Alaska Anchorage*; Alexei Rybkin, *University of Alaska Fairbanks*; Martin Getz, *University of Alaska Fairbanks* **ARIZONA** Stefania Tracogna, *Arizona State University*; Bruno Welfert, *Arizona State University*; Light Bryant, *Arizona Western College*; Daniel Russow, *Arizona Western College*; Jennifer Jameson, *Coconino College*; George Cole, *Mesa Community College*; David Schultz, *Mesa Community College*; Michael Bezusko, *Pima Community College, Desert Vista Campus*; Garry Carpenter, *Pima Community College, Northwest Campus*; Paul Flasch, *Pima County Community College*; Jessica Knapp, *Pima Community College, Northwest Campus*; Roger Werbylo, *Pima County Community College*; Katie Louchart, *Northern Arizona University*; Janet McShane, *Northern Arizona University*; Donna M. Krawczyk, *The University of Arizona* **ARKANSAS** Deborah Parker, *Arkansas Northeastern College*;

J. Michael Hall, *Arkansas State University*; Kevin Cornelius, *Ouachita Baptist University*; Hyungkoo Mark Park, *Southern Arkansas University*; Katherine Pinzon, *University of Arkansas at Fort Smith*; Denise LeGrand, *University of Arkansas at Little Rock*; John Annulis, *University of Arkansas at Monticello*; Erin Haller, *University of Arkansas, Fayetteville*; Shannon Dingman, *University of Arkansas, Fayetteville*; Daniel J. Arrigo, *University of Central Arkansas* **CALIFORNIA** Michael S. Gagliardo, *California Lutheran University*; Harvey Greenwald, *California Polytechnic State University, San Luis Obispo*; Charles Hale, *California Polytechnic State University*; John Hagen, *California Polytechnic State University, San Luis Obispo*; Donald Hartig, *California Polytechnic State University, San Luis Obispo*; Colleen Margarita Kirk, *California Polytechnic State University, San Luis Obispo*; Lawrence Sze, *California Polytechnic State University, San Luis Obispo*; Raymond Terry, *California Polytechnic State University, San Luis Obispo*; James R. McKinney, *California State Polytechnic University, Pomona*; Robin Wilson, *California State Polytechnic University, Pomona*; Charles Lam, *California State University, Bakersfield*; David McKay, *California State University, Long Beach*; Melvin Lax, *California State University, Long Beach*; Wallace A. Etterbeek, *California State University, Sacramento*; Mohamed Allali, *Chapman University*; George Rhys, *College of the Canyons*; Janice Hector, *DeAnza College*; Isabelle Saber, *Glendale Community College*;

Peter Stathis, *Glendale Community College*; Douglas B. Lloyd, *Golden West College*; Thomas Scardina, *Golden West College*; Kristin Hartford, *Long Beach City College*; Eduardo Arismendi-Pardi, *Orange Coast College*; Mitchell Alves, *Orange Coast College*; Yenkanh Vu, *Orange Coast College*; Yan Tian, *Palomar College*; Donna E. Nordstrom, *Pasadena City College*; Don L. Hancock, *Pepperdine University*; Kevin Iga, *Pepperdine University*; Adolfo J. Rumbos, *Pomona College*; Virginia May, *Sacramento City College*; Carlos de la Lama, *San Diego City College*; Matthias Beck, *San Francisco State University*; Arek Goetz, *San Francisco State University*; Nick Bykov, *San Joaquin Delta College*; Eleanor Lang Kendrick, *San Jose City College*; Elizabeth Hodes, *Santa Barbara City College*; William Konya, *Santa Monica College*; John Kennedy, *Santa Monica College*; Peter Lee, *Santa Monica College*; Richard Salome, *Scotts Valley High School*; Norman Feldman, *Sonoma State University*; Elaine McDonald, *Sonoma State University*; John D. Eggers, *University of California, San Diego*; Adam Bowers, *University of California, San Diego*; Bruno Nachtergaele, *University of California, Davis*; Boumediene Hamzi, *University of California, Davis*; Olga Radko, *University of California, Los Angeles*; Richard Leborne, *University of California, San Diego*; Peter Stevenhagen, *University of California, San Diego*; Jeffrey Stopple, *University of California, Santa Barbara*; Guofang Wei, *University of California, Santa Barbara*; Rick A. Simon, *University of La Verne*; Alexander E. Koonce, *University of Redlands*; Mohamad A. Alwash, *West Los Angeles College*; Calder Daenzer, *University of California, Berkeley*; Jude Thaddeus Socrates, *Pasadena City College*; Cheuk Ying Lam, *California State University Bakersfield*; Borislava Gutarts, *California State University, Los Angeles*; Daniel Rogalski, *University of California, San Diego*; Don Hartig, *California Polytechnic State University*; Anne Voth, *Palomar College*; Jay Wiestling, *Palomar College*; Lindsey Bramlett-Smith, *Santa Barbara City College*; Dennis Morrow, *College of the Canyons*; Sydney Shanks, *College of the Canyons*; Bob Tolar, *College of the Canyons*; Gene W. Majors, *Fullerton College*; Robert Diaz, *Fullerton College*; Gregory Nguyen, *Fullerton College*; Paul Sjoberg, *Fullerton College*; Deborah Ritchie, *Moorpark College*; Maya Rahnamaie, *Moorpark College*; Kathy Fink, *Moorpark College*; Christine Cole, *Moorpark College*; K. Di Passero, *Moorpark College*; Sid Kolpas, *Glendale Community College*; Miriam Castrconde, *Irvine Valley College*; Ilkner Erbas-White, *Irvine Valley College*; Corey Manchester, *Grossmont College*; Donald Murray, *Santa Monica College*; Barbara McGee, *Cuesta College*; Marie Larsen, *Cuesta College*; Joe Vasta, *Cuesta College*; Mike Kinter, *Cuesta College*; Mark Turner, *Cuesta College*; G. Lewis, *Cuesta College*; Daniel Kleinfelter, *College of the Desert*; Esmeralda Medrano, *Citrus College*; James Swatzel, *Citrus College*; Mark Littrell, *Rio Hondo College*; Rich Zucker, *Irvine Valley College*; Cindy Torigison, *Palomar College*; Craig Chamberline, *Palomar College*; Lindsey Lang, *Diablo Valley College*; Sam Needham, *Diablo Valley College*; Dan Bach, *Diablo Valley College*; Ted Nirgiotis, *Diablo Valley College*; Monte Collazo, *Diablo Valley College*; Tina Levy, *Diablo Valley College*; Mona Panchal, *East Los Angeles College*; Ron Sandvick, *San Diego Mesa College*; Larry Handa, *West Valley College*; Frederick Utter, *Santa Rosa Junior College*; Farshod Mosh, *DeAnza College*; Doli Bambhania, *DeAnza College*; Charles Klein, *DeAnza College*; Tammi Marshall, *Cauyamaca College*; Inwon Leu, *Cauyamaca College*; Michael Moretti, *Bakersfield College*; Janet Tarjan, *Bakersfield College*; Hoat Le, *San Diego City College*; Richard Fielding, *Southwestern College*; Shannon Gracey, *Southwestern College*; Janet Mazzarella, *Southwestern College*; Christina Soderlund, *California Lutheran University*; Rudy Gonzalez, *Citrus College*; Robert Crise, *Crafton Hills College*; Joseph Kazimir, *East Los Angeles College*; Randall Rogers, *Fullerton College*; Peter Bouzar, *Golden West College*; Linda Ternes, *Golden West College*; Hsiao-Ling Liu, *Los Angeles Trade Tech Community College*; Yu-Chung Chang-Hou, *Pasadena City College*; Guillermo Alvarez, *San Diego City College*; Ken Kuniyuki, *San Diego Mesa College*; Laleh Howard, *San Diego Mesa College*; Sharareh Masooman, *Santa Barbara City College*; Jared Hersh, *Santa Barbara City College*; Betty Wong, *Santa Monica College*; Brian Rodas, *Santa Monica College*; Veasna Chiek, *Riverside City College* **COLORADO** Tony Weathers, *Adams State College*; Erica Johnson, *Arapahoe Community College*; Karen Walters, *Arapahoe Community College*; Joshua D. Laison, *Colorado College*; G. Gustave Greivel, *Colorado School of Mines*; Holly Eklund, *Colorado School of the Mines*; Mike Nicholas, *Colorado School of the Mines*; Jim Thomas, *Colorado State University*; Eleanor Storey, *Front Range Community College*; Larry Johnson, *Metropolitan State College of Denver*; Carol Kuper, *Morgan Community College*; Larry A. Pontaski, *Pueblo Community College*; Terry Chen Reeves, *Red Rocks Community College*; Debra S. Carney, *Colorado School of the Mines*; Louis A. Talman, *Metropolitan State College of Denver*; Mary A. Nelson, *University of Colorado at Boulder*; J. Kyle Pula, *University of Denver*; Jon Von Stroh, *University of Denver*; Sharon Butz, *University of Denver*; Daniel Daly, *University of Denver*; Tracy Lawrence, *Arapahoe Community College*; Shawna Mahan, *University of Colorado Denver*; Adam Norris, *University of Colorado at Boulder*; Anca Radulescu, *University of Colorado at Boulder*; Mike Kawai, *University of Colorado Denver*; Janet Barnett, *Colorado State University–Pueblo*; Byron Hurley, *Colorado State University–Pueblo*; Jonathan Portiz, *Colorado State University–Pueblo*; Bill Emerson, *Metropolitan State College of Denver*; Suzanne Caulk, *Regis University*; Anton Dzhamay, *University of Northern Colorado* **CONNECTICUT** Jeffrey McGowan, *Central Connecticut State University*; Ivan Gotchev, *Central Connecticut State University*; Charles Waiveris, *Central Connecticut State University*; Christopher Hammond, *Connecticut College*; Anthony Y. Aidoo, *Eastern Connecticut State University*; Kim Ward, *Eastern Connecticut State University*; Joan W. Weiss, *Fairfield University*; Theresa M. Sandifer, *Southern Connecticut State University*; Cristian Rios, *Trinity College*; Melanie Stein, *Trinity College*; Steven Orszag, *Yale University* **DELAWARE** Patrick F. Mwerinde, *University of Delaware* **DISTRICT OF COLUMBIA** Jeffrey Hakim, *American University*; Joshua M. Lansky, *American University*; James A. Nickerson, *Gallaudet University* **FLORIDA** Gregory Spradlin, *Embry-Riddle University at Daytona Beach*; Daniela Popova, *Florida Atlantic University*; Abbas Zadegan, *Florida International University*; Gerardo Aladro, *Florida International University*; Gregory Henderson, *Hillsborough Community College*; Pam Crawford, *Jacksonville University*; Penny Morris, *Polk Community College*; George Schultz, *St. Petersburg College*; Jimmy Chang, *St. Petersburg College*; Carolyn Kistner, *St. Petersburg College*; Aida Kadic-Galeb, *The University of Tampa*; Constance Schober, *University of Central Florida*; S. Roy Choudhury, *University of Central Florida*; Kurt Overhiser, *Valencia Community College*; Jiongmin Yong, *University of Central Florida*; Giray Okten, *The Florida State University*; Frederick Hoffman, *Florida Atlantic University*; Thomas Beatty, *Florida Gulf Coast University*; Witny Librun, *Palm Beach Community College North*; Joe Castillo, *Broward County College*; Joann Lewin, *Edison College*; Donald Ransford, *Edison College*; Scott Berthiaume, *Edison College*; Alexander Ambrioso, *Hillsborough Community College*; Jane Golden, *Hillsborough Community College*; Susan Hiatt, *Polk Community College–Lakeland Campus*; Li Zhou, *Polk Community College–Winter Haven Campus*; Heather Edwards, *Seminole Community College*; Benjamin Landon, *Daytona State College*; Tony Malaret, *Seminole Community College*; Lane Vosbury, *Seminole Community College*; William Rickman, *Seminole Community College*; Cheryl Cantwell, *Seminole Community College*; Michael Schramm, *Indian River State College*; Janette Campbell, *Palm Beach Community College–Lake Worth*; Kwai-Lee Chui, *University of Florida*; Shu-Jen Huang, *University of Florida* **GEORGIA** Christian Barrientos, *Clayton State University*; Thomas T. Morley, *Georgia Institute of Technology*; Doron Lubinsky, *Georgia Institute of Technology*; Ralph Wildy, *Georgia Military College*; Shahram Nazari, *Georgia Perimeter College*; Alice Eiko Pierce, *Georgia Perimeter College, Clarkson Campus*; Susan Nelson, *Georgia Perimeter College, Clarkson Campus*; Laurene Fausett, *Georgia Southern University*; Scott N. Kersey, *Georgia Southern University*; Jimmy L. Solomon, *Georgia Southern University*; Allen G. Fuller, *Gordon College*; Marwan Zabdawi, *Gordon College*; Carolyn A. Yackel, *Mercer University*; Blane Hollingsworth, *Middle Georgia State College*; Shahryar Heydari, *Piedmont College*; Dan Kannan, *The University of Georgia*; June

Jones, *Middle Georgia State College*; Abdelkrim Brania, *Morehouse College*; Ying Wang, *Augusta State University*; James M. Benedict, *Augusta State University*; Kouong Law, *Georgia Perimeter College*; Rob Williams, *Georgia Perimeter College*; Alvina Atkinson, *Georgia Gwinnett College*; Amy Erickson, *Georgia Gwinnett College* **HAWAII** Shuguang Li, *University of Hawaii at Hilo*; Raina B. Ivanova, *University of Hawaii at Hilo* **IDAHO** Uwe Kaiser, *Boise State University*; Charles Kerr, *Boise State University*; Zach Teitler, *Boise State University*; Otis Kenny, *Boise State University*; Alex Feldman, *Boise State University*; Doug Bullock, *Boise State University*; Brian Dietel, *Lewis-Clark State College*; Ed Korntved, *Northwest Nazarene University*; Cynthia Piez, *University of Idaho* **ILLINOIS** Chris Morin, *Blackburn College*; Alberto L. Delgado, *Bradley University*; John Haverhals, *Bradley University*; Herbert E. Kasube, *Bradley University*; Marvin Doubet, *Lake Forest College*; Marvin A. Gordon, *Lake Forest Graduate School of Management*; Richard J. Maher, *Loyola University Chicago*; Joseph H. Mayne, *Loyola University Chicago*; Marian Gidea, *Northeastern Illinois University*; John M. Alongi, *Northwestern University*; Miguel Angel Lerma, *Northwestern University*; Mehmet Dik, *Rockford College*; Tammy Voepel, *Southern Illinois University Edwardsville*; Rahim G. Karimpour, *Southern Illinois University*; Thomas Smith, *University of Chicago*; Laura DeMarco, *University of Illinois*; Evangelos Kobotis, *University of Illinois at Chicago*; Jennifer McNeilly, *University of Illinois at Urbana-Champaign*; Timur Oikhberg, *University of Illinois at Urbana-Champaign*; Manouchehr Azad, *Harper College*; Minhua Liu, *Harper College*; Mary Hill, *College of DuPage*; Arthur N. DiVito, *Harold Washington College* **INDIANA** Vania Mascioni, *Ball State University*; Julie A. Killingbeck, *Ball State University*; Kathie Freed, *Butler University*; Zhixin Wu, *DePauw University*; John P. Boardman, *Franklin College*; Robert N. Talbert, *Franklin College*; Robin Symonds, *Indiana University Kokomo*; Henry L. Wyzinski, *Indiana University Northwest*; Melvin Royer, *Indiana Wesleyan University*; Gail P. Greene, *Indiana Wesleyan University*; David L. Finn, *Rose-Hulman Institute of Technology*; Chong Keat Arthur Lim, *University of Notre Dame* **IOWA** Nasser Dastrange, *Buena Vista University*; Mark A. Mills, *Central College*; Karen Ernst, *Hawkeye Community College*; Richard Mason, *Indian Hills Community College*; Robert S. Keller, *Loras College*; Eric Robert Westlund, *Luther College*; Weimin Han, *The University of Iowa* **KANSAS** Timothy W. Flood, *Pittsburg State University*; Sarah Cook, *Washburn University*; Kevin E. Charlwood, *Washburn University*; Conrad Uwe, *Cowley County Community College*; David N. Yetter, *Kansas State University* **KENTUCKY** Alex M. McAllister, *Center College*; Sandy Spears, *Jefferson Community & Technical College*; Leanne Faulkner, *Kentucky Wesleyan College*; Donald O. Clayton, *Madisonville Community College*; Thomas Riedel, *University of Louisville*; Manabendra Das, *University of Louisville*; Lee Larson, *University of Louisville*; Jens E. Harlander, *Western Kentucky University*; Philip McCartney, *Northern Kentucky University*; Andy Long, *Northern Kentucky University*; Omer Yayenie, *Murray State University*; Donald Krug, *Northern Kentucky University* **LOUISIANA** William Forrest, *Baton Rouge Community College*; Paul Wayne Britt, *Louisiana State University*; Galen Turner, *Louisiana Tech University*; Randall Wills, *Southeastern Louisiana University*; Kent Neuerburg, *Southeastern Louisiana University*; Guoli Ding, *Louisiana State University*; Julia Ledet, *Louisiana State University*; Brent Strunk, *University of Louisiana at Monroe* **MAINE** Andrew Knightly, *The University of Maine*; Sergey Lvin, *The University of Maine*; Joel W. Irish, *University of Southern Maine*; Laurie Woodman, *University of Southern Maine*; David M. Bradley, *The University of Maine*; William O. Bray, *The University of Maine* **MARYLAND** Leonid Stern, *Towson University*; Jacob Kogan, *University of Maryland Baltimore County*; Mark E. Williams, *University of Maryland Eastern Shore*; Austin A. Lobo, *Washington College*; Supawan Lertskrai, *Harford Community College*; Fary Sami, *Harford Community College*; Andrew Bulleri, *Howard Community College* **MASSACHUSETTS** Sean McGrath, *Algonquin Regional High School*; Norton Starr, *Amherst College*; Renato Mirollo, *Boston College*; Emma Previato, *Boston University*; Laura K Gross, *Bridgewater State University*; Richard H. Stout, *Gordon College*; Matthew P. Leingang, *Harvard University*; Suellen Robinson, *North Shore Community College*; Walter Stone, *North Shore Community College*; Barbara Loud, *Regis College*; Andrew B. Perry, *Springfield College*; Tawanda Gwena, *Tufts University*; Gary Simundza, *Wentworth Institute of Technology*; Mikhail Chkhenkeli, *Western New England College*; David Daniels, *Western New England College*; Alan Gorfin, *Western New England College*; Saeed Ghahramani, *Western New England College*; Julian Fleron, *Westfield State College*; Maria Fung, *Worchester State University*; Brigitte Servatius, *Worcester Polytechnic Institute*; John Goulet, *Worcester Polytechnic Institute*; Alexander Martsinkovsky, *Northeastern University*; Marie Clote, *Boston College*; Alexander Kastner, *Williams College*; Margaret Peard, *Williams College*; Mihai Stoiciu, *Williams College* **MICHIGAN** Mark E. Bollman, *Albion College*; Jim Chesla, *Grand Rapids Community College*; Jeanne Wald, *Michigan State University*; Allan A. Struthers, *Michigan Technological University*; Debra Pharo, *Northwestern Michigan College*; Anna Maria Spagnuolo, *Oakland University*; Diana Faoro, *Romeo Senior High School*; Andrew Strowe, *University of Michigan–Dearborn*; Daniel Stephen Drucker, *Wayne State University*; Christopher Cartwright, *Lawrence Technological University*; Jay Treiman, *Western Michigan University* **MINNESOTA** Bruce Bordwell, *Anoka-Ramsey Community College*; Robert Dobrow, *Carleton College*; Jessie K. Lenarz, *Concordia College–Moorhead Minnesota*; Bill Tomhave, *Concordia College*; David L. Frank, *University of Minnesota*; Steven I. Sperber, *University of Minnesota*; Jeffrey T. McLean, *University of St. Thomas*; Chehrzad Shakiban, *University of St. Thomas*; Melissa Loe, *University of St. Thomas*; Nick Christopher Fiala, *St. Cloud State University*; Victor Padron, *Normandale Community College*; Mark Ahrens, *Normandale Community College*; Gerry Naughton, *Century Community College*; Carrie Naughton, *Inver Hills Community College* **MISSISSIPPI** Vivien G. Miller, *Mississippi State University*; Ted Dobson, *Mississippi State University*; Len Miller, *Mississippi State University*; Tristan Denley, *The University of Mississippi* **MISSOURI** Robert Robertson, *Drury University*; Gregory A. Mitchell, *Metropolitan Community College–Penn Valley*; Charles N. Curtis, *Missouri Southern State University*; Vivek Narayanan, *Moberly Area Community College*; Russell Blyth, *Saint Louis University*; Julianne Rainbolt, *Saint Louis University*; Blake Thornton, *Saint Louis University*; Kevin W. Hopkins, *Southwest Baptist University*; Joe Howe, *St. Charles Community College*; Wanda Long, *St. Charles Community College*; Andrew Stephan, *St. Charles Community College* **MONTANA** Kelly Cline, *Carroll College*; Veronica Baker, *Montana State University, Bozeman*; Richard C. Swanson, *Montana State University*; Thomas Hayes-McGoff, *Montana State University*; Nikolaus Vonessen, *The University of Montana* **NEBRASKA** Edward G. Reinke Jr., *Concordia University*; Judith Downey, *University of Nebraska at Omaha* **NEVADA** Jennifer Gorman, *College of Southern Nevada*; Jonathan Pearsall, *College of Southern Nevada*; Rohan Dalpatadu, *University of Nevada, Las Vegas*; Paul Aizley, *University of Nevada, Las Vegas* **NEW HAMPSHIRE** Richard Jardine, *Keene State College*; Michael Cullinane, *Keene State College*; Roberta Kieronski, *University of New Hampshire at Manchester*; Erik Van Erp, *Dartmouth College* **NEW JERSEY** Paul S. Rossi, *College of Saint Elizabeth*; Mark Galit, *Essex County College*; Katarzyna Potocka, *Ramapo College of New Jersey*; Nora S. Thornber, *Raritan Valley Community College*; Abdulkadir Hassen, *Rowan University*; Olcay Ilicasu, *Rowan University*; Avraham Soffer, *Rutgers, The State University of New Jersey*; Chengwen Wang, *Rutgers, The State University of New Jersey*; Shabnam Beheshti, *Rutgers University, The State University of New Jersey*; Stephen J. Greenfield, *Rutgers, The State University of New Jersey*; John T. Saccoman, *Seton Hall University*; Lawrence E. Levine, *Stevens Institute of Technology*; Jana Gevertz, *The College of New Jersey*; Barry Burd, *Drew University*; Penny Luczak, *Camden County College*; John Climent, *Cecil Community College*; Kristyanna Erickson, *Cecil Community College*; Eric Compton, *Brookdale Community College*; John Atsu-Swanzy, *Atlantic Cape Community College* **NEW MEXICO** Kevin Leith, *Central New Mexico Community College*; David Blankenbaker, *Central New Mexico Community College*; Joseph Lakey, *New Mexico State University*; Kees Onneweer, *University of New Mexico*; Jurg Bolli,

The University of New Mexico   **NEW YORK**   Robert C. Williams, *Alfred University*; Timmy G. Bremer, *Broome Community College State University of New York*; Joaquin O. Carbonara, *Buffalo State College*; Robin Sue Sanders, *Buffalo State College*; Daniel Cunningham, *Buffalo State College*; Rose Marie Castner, *Canisius College*; Sharon L. Sullivan, *Catawba College*; Fabio Nironi, *Columbia University*; Camil Muscalu, *Cornell University*; Maria S. Terrell, *Cornell University*; Margaret Mulligan, *Dominican College of Blauvelt*; Robert Andersen, *Farmingdale State University of New York*; Leonard Nissim, *Fordham University*; Jennifer Roche, *Hobart and William Smith Colleges*; James E. Carpenter, *Iona College*; Peter Shenkin, *John Jay College of Criminal Justice/CUNY*; Gordon Crandall, *LaGuardia Community College/CUNY*; Gilbert Traub, *Maritime College, State University of New York*; Paul E. Seeburger, *Monroe Community College Brighton Campus*; Abraham S. Mantell, *Nassau Community College*; Daniel D. Birmajer, *Nazareth College*; Sybil G. Shaver, *Pace University*; Margaret Kiehl, *Rensselaer Polytechnic Institute*; Carl V. Lutzer, *Rochester Institute of Technology*; Michael A. Radin, *Rochester Institute of Technology*; Hossein Shahmohamad, *Rochester Institute of Technology*; Thomas Rousseau, *Siena College*; Jason Hofstein, *Siena College*; Leon E. Gerber, *St. Johns University*; Christopher Bishop, *Stony Brook University*; James Fulton, *Suffolk County Community College*; John G. Michaels, *SUNY Brockport*; Howard J. Skogman, *SUNY Brockport*; Cristina Bacuta, *SUNY Cortland*; Jean Harper, *SUNY Fredonia*; David Hobby, *SUNY New Paltz*; Kelly Black, *Union College*; Thomas W. Cusick, *University at Buffalo/The State University of New York*; Gino Biondini, *University at Buffalo/The State University of New York*; Robert Koehler, *University at Buffalo/The State University of New York*; Donald Larson, *University of Rochester*; Robert Thompson, *Hunter College*; Ed Grossman, *The City College of New York*   **NORTH CAROLINA**   Jeffrey Clark, *Elon University*; William L. Burgin, *Gaston College*; Manouchehr H. Misaghian, *Johnson C. Smith University*; Legunchim L. Emmanwori, *North Carolina A&T State University*; Drew Pasteur, *North Carolina State University*; Demetrio Labate, *North Carolina State University*; Mohammad Kazemi, *The University of North Carolina at Charlotte*; Richard Carmichael, *Wake Forest University*; Gretchen Wilke Whipple, *Warren Wilson College*; John Russell Taylor, *University of North Carolina at Charlotte*; Mark Ellis, *Piedmont Community College*   **NORTH DAKOTA**   Jim Coykendall, *North Dakota State University*; Anthony J. Bevelacqua, *The University of North Dakota*; Richard P. Millspaugh, *The University of North Dakota*; Thomas Gilsdorf, *The University of North Dakota*; Michele Iiams, *The University of North Dakota*; Mohammad Khavanin, *University of North Dakota*   **OHIO**   Christopher Butler, *Case Western Reserve University*; Pamela Pierce, *The College of Wooster*; Barbara H. Margolius, *Cleveland State University*; Tzu-Yi Alan Yang, *Columbus State Community College*; Greg S. Goodhart, *Columbus State Community College*; Kelly C. Stady, *Cuyahoga Community College*; Brian T. Van Pelt, *Cuyahoga Community College*; David Robert Ericson, *Miami University*; Frederick S. Gass, *Miami University*; Thomas Stacklin, *Ohio Dominican University*; Vitaly Bergelson, *The Ohio State University*; Robert Knight, *Ohio University*; John R. Pather, *Ohio University, Eastern Campus*; Teresa Contenza, *Otterbein College*; Ali Hajjafar, *The University of Akron*; Jianping Zhu, *The University of Akron*; Ian Clough, *University of Cincinnati Clermont College*; Atif Abueida, *University of Dayton*; Judith McCrory, *The University at Findlay*; Thomas Smotzer, *Youngstown State University*; Angela Spalsbury, *Youngstown State University*; James Osterburg, *The University of Cincinnati*; Mihaela A. Poplicher, *University of Cincinnati*; Frederick Thulin, *University of Illinois at Chicago*; Weimin Han, *The Ohio State University*; Crichton Ogle, *The Ohio State University*; Jackie Miller, *The Ohio State University*; Walter Mackey, *Owens Community College*; Jonathan Baker, *Columbus State Community College*   **OKLAHOMA**   Christopher Francisco, *Oklahoma State University*; Michael McClendon, *University of Central Oklahoma*; Teri Jo Murphy, *The University of Oklahoma*; Kimberly Adams, *University of Tulsa*; Shirley Pomeranz, *University of Tulsa*   **OREGON**   Lorna TenEyck, *Chemeketa Community College*; Angela Martinek, *Linn-Benton Community College*; Filix Maisch, *Oregon State University*; Tevian Dray, *Oregon State University*; Mark Ferguson,

Chemekata Community College; Andrew Flight, *Portland State University*; Austina Fong, *Portland State University*; Jeanette R. Palmiter, *Portland State University*   **PENNSYLVANIA**   John B. Polhill, *Bloomsburg University of Pennsylvania*; Russell C. Walker, *Carnegie Mellon University*; Jon A. Beal, *Clarion University of Pennsylvania*; Kathleen Kane, *Community College of Allegheny County*; David A. Santos, *Community College of Philadelphia*; David S. Richeson, *Dickinson College*; Christine Marie Cedzo, *Gannon University*; Monica Pierri-Galvao, *Gannon University*; John H. Ellison, *Grove City College*; Gary L. Thompson, *Grove City College*; Dale McIntyre, *Grove City College*; Dennis Benchoff, *Harrisburg Area Community College*; William A. Drumin, *King's College*; Denise Reboli, *King's College*; Chawne Kimber, *Lafayette College*; Elizabeth McMahon, *Lafayette College*; Lorenzo Traldi, *Lafayette College*; David L. Johnson, *Lehigh University*; Matthew Hyatt, *Lehigh University*; Zia Uddin, *Lock Haven University of Pennsylvania*; Donna A. Dietz, *Mansfield University of Pennsylvania*; Samuel Wilcock, *Messiah College*; Richard R. Kern, *Montgomery County Community College*; Michael Fraboni, *Moravian College*; Neena T. Chopra, *The Pennsylvania State University*; Boris A. Datskovsky, *Temple University*; Dennis M. DeTurck, *University of Pennsylvania*; Jacob Burbea, *University of Pittsburgh*; Mohammed Yahdi, *Ursinus College*; Timothy Feeman, *Villanova University*; Douglas Norton, *Villanova University*; Robert Styer, *Villanova University*; Michael J. Fisher, *West Chester University of Pennsylvania*; Peter Brooksbank, *Bucknell University*; Emily Dryden, *Bucknell University*; Larry Friesen, *Butler County Community College*; Lisa Angelo, *Bucks County College*; Elaine Fitt, *Bucks County College*; Pauline Chow, *Harrisburg Area Community College*; Diane Benner, *Harrisburg Area Community College*; Emily B. Dryden, *Bucknell University*; Erica Chauvet, *Waynesburg University*   **RHODE ISLAND**   Thomas F. Banchoff, *Brown University*; Yajni Warnapala-Yehiya, *Roger Williams University*; Carol Gibbons, *Salve Regina University*; Joe Allen, *Community College of Rhode Island*; Michael Latina, *Community College of Rhode Island*   **SOUTH CAROLINA**   Stanley O. Perrine, *Charleston Southern University*; Joan Hoffacker, *Clemson University*; Constance C. Edwards, *Coastal Carolina University*; Thomas L. Fitzkee, *Francis Marion University*; Richard West, *Francis Marion University*; John Harris, *Furman University*; Douglas B. Meade, *University of South Carolina*; George Androulakis, *University of South Carolina*; Art Mark, *University of South Carolina Aiken*; Sherry Biggers, *Clemson University*; Mary Zachary Krohn, *Clemson University*; Andrew Incognito, *Coastal Carolina University*; Deanna Caveny, *College of Charleston*   **SOUTH DAKOTA**   Dan Kemp, *South Dakota State University*   **TENNESSEE**   Andrew Miller, *Belmont University*; Arthur A. Yanushka, *Christian Brothers University*; Laurie Plunk Dishman, *Cumberland University*; Maria Siopsis, *Maryville College*; Beth Long, *Pellissippi State Technical Community College*; Judith Fethe, *Pellissippi State Technical Community College*; Andrzej Gutek, *Tennessee Technological University*; Sabine Le Borne, *Tennessee Technological University*; Richard Le Borne, *Tennessee Technological University*; Maria F. Bothelho, *University of Memphis*; Roberto Triggiani, *University of Memphis*; Jim Conant, *The University of Tennessee*; Pavlos Tzermias, *The University of Tennessee*; Luis Renato Abib Finotti, *University of Tennessee, Knoxville*; Jennifer Fowler, *University of Tennessee, Knoxville*; Jo Ann W. Staples, *Vanderbilt University*; Dave Vinson, *Pellissippi State Community College*; Jonathan Lamb, *Pellissippi State Community College*   **TEXAS**   Sally Haas, *Angelina College*; Karl Havlak, *Angelo State University*; Michael Huff, *Austin Community College*; John M. Davis, *Baylor University*; Scott Wilde, *Baylor University and The University of Texas at Arlington*; Rob Eby, *Blinn College*; Tim Sever, *Houston Community College–Central*; Ernest Lowery, *Houston Community College–Northwest*; Brian Loft, *Sam Houston State University*; Jianzhong Wang, *Sam Houston State University*; Shirley Davis, *South Plains College*; Todd M. Steckler, *South Texas College*; Mary E. Wagner-Krankel, *St. Mary's University*; Elise Z. Price, *Tarrant County College, Southeast Campus*; David Price, *Tarrant County College, Southeast Campus*; Runchang Lin, *Texas A&M University*; Michael Stecher, *Texas A&M University*; Philip B. Yasskin, *Texas A&M University*; Brock Williams, *Texas*

Tech University*; I. Wayne Lewis, *Texas Tech University*; Robert E. Byerly, *Texas Tech University*; Ellina Grigorieva, *Texas Woman's University*; Abraham Haje, *Tomball College*; Scott Chapman, *Trinity University*; Elias Y. Deeba, *University of Houston Downtown*; Jianping Zhu, *The University of Texas at Arlington*; Tuncay Aktosun, *The University of Texas at Arlington*; John E. Gilbert, *The University of Texas at Austin*; Jorge R. Viramontes-Olivias, *The University of Texas at El Paso*; Fengxin Chen, *University of Texas at San Antonio*; Melanie Ledwig, *The Victoria College*; Gary L. Walls, *West Texas A&M University*; William Heierman, *Wharton County Junior College*; Lisa Rezac, *University of St. Thomas*; Raymond J. Cannon, *Baylor University*; Kathryn Flores, *McMurry University*; Jacqueline A. Jensen, *Sam Houston State University*; James Galloway, *Collin County College*; Raja Khoury, *Collin County College*; Annette Benbow, *Tarrant County College–Northwest*; Greta Harland, *Tarrant County College–Northeast*; Doug Smith, *Tarrant County College–Northeast*; Marcus McGuff, *Austin Community College*; Clarence McGuff, *Austin Community College*; Steve Rodi, *Austin Community College*; Vicki Payne, *Austin Community College*; Anne Pradera, *Austin Community College*; Christy Babu, *Laredo Community College*; Deborah Hewitt, *McLennan Community College*; W. Duncan, *McLennan Community College*; Hugh Griffith, *Mt. San Antonio College* **UTAH** Ruth Trygstad, *Salt Lake City Community College* **VIRGINIA** Verne E. Leininger, *Bridgewater College*; Brian Bradie, *Christopher Newport University*; Hongwei Chen, *Christopher Newport University*; John J. Avioli, *Christopher Newport University*; James H. Martin, *Christopher Newport University*; David Walnut, *George Mason University*; Mike Shirazi, *Germanna Community College*; Julie Clark, *Hollins University*; Ramon A. Mata-Toledo, *James Madison University*; Adrian Riskin, *Mary Baldwin College*; Josephine Letts, *Ocean Lakes High School*; Przemyslaw Bogacki, *Old Dominion University*; Deborah Denvir, *Randolph-Macon Woman's College*; Linda Powers, *Virginia Tech*; Gregory Dresden, *Washington and Lee University*; Jacob A. Siehler, *Washington and Lee University*; Yuan-Jen Chiang, *University of Mary Washington*; Nicholas Hamblet, *University of Virginia*; Bernard Fulgham, *University of Virginia*; Manouchehr "Mike" Mohajeri, *University of Virginia*; Lester Frank Caudill, *University of Richmond* **VERMONT** David Dorman, *Middlebury College*; Rachel Repstad, *Vermont Technical College* **WASHINGTON** Ricardo Chavez, *Bellevue College*; Jennifer Laveglia, *Bellevue Community College*; David Whittaker, *Cascadia Community College*; Sharon Saxton, *Cascadia Community College*; Aaron Montgomery, *Central Washington University*; Patrick Averbeck, *Edmonds Community College*; Tana Knudson, *Heritage University*; Kelly Brooks, *Pierce College*; Shana P. Calaway, *Shoreline Community College*; Abel Gage, *Skagit Valley College*; Scott MacDonald, *Tacoma Community College*; Jason Preszler, *University of Puget Sound*; Martha A. Gady, *Whitworth College*; Wayne L. Neidhardt, *Edmonds Community College*; Simrat Ghuman, *Bellevue College*; Jeff Eldridge, *Edmonds Community College*; Kris Kissel, *Green River Community College*; Laura Moore-Mueller, *Green River Community College*; David Stacy, *Bellevue College*; Eric Schultz, *Walla Walla Community College*; Julianne Sachs, *Walla Walla Community College* **WEST VIRGINIA** David Cusick, *Marshall University*; Ralph Oberste-Vorth, *Marshall University*; Suda Kunyosying, *Shepard University*; Nicholas Martin, *Shepherd University*; Rajeev Rajaram, *Shepherd University*; Xiaohong Zhang, *West Virginia State University*; Sam B. Nadler, *West Virginia University* **WYOMING** Claudia Stewart, *Casper College*; Pete Wildman, *Casper College*; Charles Newberg, *Western Wyoming Community College*; Lynne Ipina, *University of Wyoming*; John Spitler, *University of Wyoming* **WISCONSIN** Erik R. Tou, *Carthage College*; Paul Bankston, *Marquette University*; Jane Nichols, *Milwaukee School of Engineering*; Yvonne Yaz, *Milwaukee School of Engineering*; Simei Tong, *University of Wisconsin–Eau Claire*; Terry Nyman, *University of Wisconsin–Fox Valley*; Robert L. Wilson, *University of Wisconsin–Madison*; Dietrich A. Uhlenbrock, *University of Wisconsin–Madison*; Paul Milewski, *University of Wisconsin–Madison*; Donald Solomon, *University of Wisconsin–Milwaukee*; Kandasamy Muthuvel, *University of Wisconsin–Oshkosh*; Sheryl Wills, *University of Wisconsin–Platteville*; Kathy A. Tomlinson, *University of Wisconsin–River Falls*; Cynthia L. McCabe, *University of Wisconsin–Stevens Point*; Matthew Welz, *University of Wisconsin–Stevens Point*; Joy Becker, *University of Wisconsin-Stout*; Jeganathan Sriskandarajah , *Madison Area Tech College*; Wayne Sigelko, *Madison Area Tech College* **CANADA** Don St. Jean, *George Brown College*; Robert Dawson, *St. Mary's University*; Len Bos, *University of Calgary*; Tony Ware, *University of Calgary*; Peter David Papez, *University of Calgary*; John O'Conner, *Grant MacEwan University*; Michael P. Lamoureux, *University of Calgary*; Yousry Elsabrouty, *University of Calgary*; Darja Kalajdzievska, *University of Manitoba*; Andrew Skelton, *University of Guelph*; Douglas Farenick, *University of Regina*

The creation of this third edition could not have happened without the help of many people. First, I want to thank the individuals whom I have worked with at W. H. Freeman. Terri Ward and Ruth Baruth convinced me that I should take on this project, and I am grateful to them for their support and their confidence in my ability to tackle it. Throughout this process, Terri has been a huge help. I can always count on her to keep this train on track. Katrina Wilhelm has also been an amazing resource. She brings calm competence and organizational skills that constantly impress me. Tony Palermino has provided expert editorial help throughout the process. He is incredibly knowledgeable about all aspects of mathematics textbooks and has an eye for the details that make a book work. Kerry O'Shaughnessy kept the production process moving forward in a timely manner without ever resorting to threats. John Rogosich was the superb compositor. Patti Brecht handled the copyediting in an expert manner. My thanks are also due to W. H. Freeman's superb production team: Janice Donnola, Eileen Liang, Blake Logan, Paul Rohloff, and to Ron Weickart at Network Graphics for his skilled and creative execution of the art program.

Many faculty gave critical feedback on the second edition, and their names appear above. I am deeply grateful to them. I do want to particularly thank all of the advisory board members who gave me feedback month after month. Maria Shea Terrell continually sent me excellent unsolicited feedback until I asked to have her on the board. Then it became solicited. The accuracy reviewers, John Alongi, CK Cheung, Kwai-Lee Chui, John Davis, John Eggers, Stephen Greenfield, Roger Lipsett, Vivek Narayanan, and Olga Radko, helped to bring the final version into the form in which it now appears. You think you have found the errors, but you have not.

I also want to thank my colleagues in the Mathematics and Statistics Department at Williams College. I have always known I am incredibly lucky to be a member of this department. There are so many interesting projects and clever pedagogical ideas coming out of the department that it motivates me just because I am trying to keep up.

I would further like to thank my students. Their enthusiasm is what makes teaching fun. I enjoy coming to work every day, and they are what make it such a pleasure.

Finally, I want to thank my two children, Alexa and Colton. They are the ones who keep me grounded, who remind me what works and what doesn't in the real world. This book is dedicated to them.

Colin Adams

1 PRECALCULUS REVIEW

Functions that yield the amount of seismic activity as a function of time help scientists to predict volcanic eruptions and earthquakes.

(Douglas Peebles/Science Source)

Calculus builds on the foundation of algebra, analytic geometry, and trigonometry. In this chapter, therefore, we review some concepts, facts, and formulas from precalculus that are used throughout the text. In the last section, we discuss ways in which technology can be used to enhance your visual understanding of functions and their properties.

1.1 Real Numbers, Functions, and Graphs

We begin with a short discussion of real numbers. This gives us the opportunity to recall some basic properties and standard notation.

A **real number** is a number represented by a decimal or decimal expansion. There are three types of decimal expansions: finite, repeating, and infinite but nonrepeating. For example,

$$\frac{3}{8} = 0.375, \qquad \frac{1}{7} = 0.142857142857\ldots = 0.\overline{142857}$$

$$\pi = 3.141592653589793\ldots$$

The number $\frac{3}{8}$ is represented by a finite decimal, whereas $\frac{1}{7}$ is represented by a *repeating* or *periodic* decimal. The bar over 142857 indicates that this sequence repeats indefinitely. The decimal expansion of π is infinite but nonrepeating.

The set of all real numbers is denoted by a boldface **R**. When there is no risk of confusion, we refer to a real number simply as a *number*. We also use the standard symbol $\in$ for the phrase belongs to. Thus,

$$a \in \mathbf{R} \qquad \text{reads} \qquad a \text{ belongs to } \mathbf{R}$$

The set of integers is commonly denoted by the letter **Z** (this choice comes from the German word *Zahl*, meaning number). Thus, $\mathbf{Z} = \{\ldots, -2, -1, 0, 1, 2, \ldots\}$. A **whole number** is a nonnegative integer that is, one of the numbers $0, 1, 2, \ldots$.

Additional properties of real numbers are discussed in Appendix B.

A real number is called **rational** if it can be represented by a fraction p/q, where p and q are integers with $q \neq 0$. The set of rational numbers is denoted **Q** (for quotient). Numbers that are not rational, such as π and $\sqrt{2}$, are called **irrational**.

We can tell whether a number is rational from its decimal expansion: Rational numbers have finite or repeating decimal expansions, and irrational numbers have infinite, nonrepeating decimal expansions. Furthermore, the decimal expansion of a number is unique, apart from the following exception: Every finite decimal is equal to an infinite decimal in which the digit 9 repeats. For example,

$$1 = 0.999\ldots, \qquad \frac{3}{8} = 0.375 = 0.374999\ldots, \qquad \frac{47}{20} = 2.35 = 2.34999\ldots$$

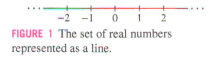

FIGURE 1 The set of real numbers represented as a line.

We visualize real numbers as points on a line (Figure 1). For this reason, real numbers are often referred to as **points**. The point corresponding to 0 is called the **origin**.

The **absolute value** of a real number a, denoted $|a|$, is defined by (Figure 2)

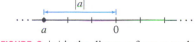

FIGURE 2 $|a|$ is the distance from a to the origin.

$$|a| = distance\ from\ the\ origin = \begin{cases} a & \text{if } a \geq 0 \\ -a & \text{if } a < 0 \end{cases}$$

For example, $|1.2| = 1.2$ and $|-8.35| = 8.35$. The absolute value satisfies

$$|a| = |-a|, \qquad |ab| = |a|\,|b|$$

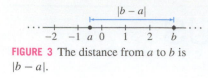

FIGURE 3 The distance from a to b is $|b - a|$.

The **distance** between two real numbers a and b is $|b - a|$, which is the length of the line segment joining a and b (Figure 3).

Two real numbers a and b are close to each other if $|b - a|$ is small, and this is the case if their decimal expansions agree to many places. More precisely, *if the decimal expansions of a and b agree to k places (to the right of the decimal point), then the distance $|b - a|$ is at most 10^{-k}.* Thus, the distance between $a = 3.1415$ and $b = 3.1478$ is at most 10^{-2} because a and b agree to two places. In fact, the distance is exactly $|3.1478 - 3.1415| = 0.0063$.

Beware that $|a + b|$ is not equal to $|a| + |b|$ unless a and b have the same sign or at least one of a and b is zero. If they have opposite signs, cancellation occurs in the sum $a + b$, and $|a + b| < |a| + |b|$. For example, $|2 + 5| = |2| + |5|$ but $|-2 + 5| = 3$, which is less than $|-2| + |5| = 7$. In any case, $|a + b|$ is never larger than $|a| + |b|$ and this gives us the simple but important **triangle inequality**:

$$|a + b| \le |a| + |b| \qquad \boxed{1}$$

We use standard notation for intervals. Given real numbers $a < b$, there are four intervals with endpoints a and b (Figure 4). They all have length $b - a$ but differ according to which endpoints are included.

FIGURE 4 The four intervals with endpoints a and b.

Closed interval $[a, b]$ (endpoints included)   Open interval (a, b) (endpoints excluded)   Half-open interval $[a, b)$ Half-open interval $(a, b]$

The **closed interval** $[a, b]$ is the set of all real numbers x such that $a \le x \le b$:

$$[a, b] = \{x \in \mathbf{R} : a \le x \le b\}$$

The notation $(2, 3)$ could mean the open interval $\{x : 2 < x < 3\}$ or it could mean the point in the xy-plane with $x = 2$ and $y = 3$. In general, the meaning will be apparent from the context.

We usually write this more simply as $\{x : a \le x \le b\}$, it being understood that x belongs to $\mathbf{R}$. The **open** and **half-open intervals** are the sets

$$\underbrace{(a, b) = \{x : a < x < b\}}_{\text{Open interval (endpoints excluded)}}, \qquad \underbrace{[a, b) = \{x : a \le x < b\}}_{\text{Half-open interval}}, \qquad \underbrace{(a, b] = \{x : a < x \le b\}}_{\text{Half-open interval}}$$

The infinite interval $(-\infty, \infty)$ is the entire real line $\mathbf{R}$. A half-infinite interval is closed if it contains its finite endpoint and is open otherwise (Figure 5):

$$[a, \infty) = \{x : a \le x\}, \qquad (-\infty, b] = \{x : x \le b\}$$

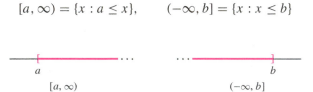

$[a, \infty)$ $(-\infty, b]$

FIGURE 5 Closed half-infinite intervals.

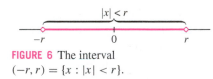

FIGURE 6 The interval $(-r, r) = \{x : |x| < r\}$.

Open and closed intervals may be described by inequalities. For example, the interval $(-r, r)$ is described by the inequality $|x| < r$ (Figure 6):

$$|x| < r \quad \Leftrightarrow \quad -r < x < r \quad \Leftrightarrow \quad x \in (-r, r) \qquad \boxed{2}$$

More generally, for an interval symmetric about the value c (Figure 7),

$$|x - c| < r \quad \Leftrightarrow \quad c - r < x < c + r \quad \Leftrightarrow \quad x \in (c - r, c + r) \qquad \boxed{3}$$

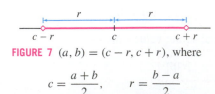

FIGURE 7 $(a, b) = (c - r, c + r)$, where
$$c = \frac{a + b}{2}, \qquad r = \frac{b - a}{2}$$

Closed intervals are similar, with $<$ replaced by $\le$. We refer to r as the **radius** and to c as the **midpoint** or **center**. The intervals (a, b) and $[a, b]$ have midpoint $c = \frac{1}{2}(a + b)$ and radius $r = \frac{1}{2}(b - a)$ (Figure 7).

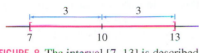

FIGURE 8 The interval $[7, 13]$ is described by $|x - 10| \le 3$.

■ **EXAMPLE 1** Describe $[7, 13]$ using inequalities.

Solution The midpoint of the interval $[7, 13]$ is $c = \frac{1}{2}(7 + 13) = 10$ and its radius is $r = \frac{1}{2}(13 - 7) = 3$ (Figure 8). Therefore,

$$[7, 13] = \left\{ x \in \mathbf{R} : |x - 10| \le 3 \right\}$$ ■

■ **EXAMPLE 2** Describe the set $S = \left\{ x : \left| \frac{1}{2}x - 3 \right| > 4 \right\}$ in terms of intervals.

Solution It is easier to consider the opposite inequality $\left| \frac{1}{2}x - 3 \right| \le 4$ first. By (2),

In Example 2 we use the notation $\cup$ to denote union : The union $A \cup B$ of sets A and B consists of all elements that belong to either A or B (or to both).

$$\left| \frac{1}{2}x - 3 \right| \le 4 \quad \Leftrightarrow \quad -4 \le \frac{1}{2}x - 3 \le 4$$

$$-1 \le \frac{1}{2}x \le 7 \qquad \text{(add 3)}$$

$$-2 \le x \le 14 \qquad \text{(multiply by 2)}$$

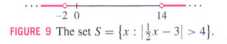

FIGURE 9 The set $S = \left\{ x : \left| \frac{1}{2}x - 3 \right| > 4 \right\}$.

Thus, $\left| \frac{1}{2}x - 3 \right| \le 4$ is satisfied when x belongs to $[-2, 14]$. The set S is the *complement*, consisting of all numbers x *not in* $[-2, 14]$. We can describe S as the union of two intervals: $S = (-\infty, -2) \cup (14, \infty)$ (Figure 9). ■

Graphing

The term Cartesian refers to the French philosopher and mathematician René Descartes (1596 1650), whose Latin name was Cartesius. He is credited (along with Pierre de Fermat) with the invention of analytic geometry. In his great work La Géométrie, *Descartes used the letters x, y, z for unknowns and a, b, c for constants, a convention that has been followed ever since.*

Graphing is a basic tool in calculus, as it is in algebra and trigonometry. Recall that rect-angular (or Cartesian) coordinates in the plane are defined by choosing two perpendicular axes, the x-axis and the y-axis. To a pair of numbers (a, b) we associate the point P located at the intersection of the line perpendicular to the x-axis at a and the line perpendicular to the y-axis at b [Figure 10(A)]. The numbers a and b are the x- and y-**coordinates** of P. The x-coordinate is sometimes called the abscissa and the y-coordinate the ordinate. The **origin** is the point with coordinates $(0, 0)$.

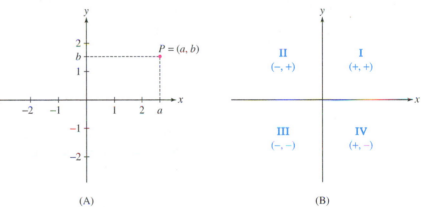

(A) (B)

FIGURE 10 Rectangular coordinate system.

The axes divide the plane into four quadrants labeled I IV, determined by the signs of the coordinates [Figure 10(B)]. For example, quadrant III consists of points (x, y) such that $x < 0$ and $y < 0$.

The distance d between two points $P_1 = (x_1, y_1)$ and $P_2 = (x_2, y_2)$ is computed using the Pythagorean Theorem. In Figure 11, we see that $\overline{P_1 P_2}$ is the hypotenuse of a right triangle with sides $a = |x_2 - x_1|$ and $b = |y_2 - y_1|$. Therefore,

$$d^2 = a^2 + b^2 = (x_2 - x_1)^2 + (y_2 - y_1)^2$$

We obtain the distance formula by taking square roots.

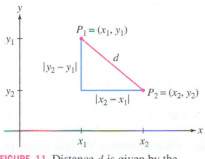

FIGURE 11 Distance d is given by the distance formula.

> **Distance Formula** The distance between $P_1 = (x_1, y_1)$ and $P_2 = (x_2, y_2)$ is equal to
>
> $$d = \sqrt{(x_2 - x_1)^2 + (y_2 - y_1)^2}$$

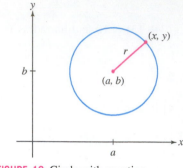

FIGURE 12 Circle with equation $(x-a)^2 + (y-b)^2 = r^2$.

Once we have the distance formula, we can derive the equation of a circle of radius r and center (a, b) (Figure 12). A point (x, y) lies on this circle if the distance from (x, y) to (a, b) is r:

$$\sqrt{(x-a)^2 + (y-b)^2} = r$$

Squaring both sides, we obtain the standard equation of the circle:

$$(x-a)^2 + (y-b)^2 = r^2$$

We now review some definitions and notation concerning functions.

> **DEFINITION**   A **function** f from a set D to a set Y is a rule that assigns, to each element x in D, a unique element $y = f(x)$ in Y. We write
>
> $$f : D \to Y$$

The set D, called the **domain** of f, is the set of "allowable inputs." For $x \in D$, $f(x)$ is called the **value** of f at x (Figure 13). The **range** R of f is the subset of Y consisting of all values $f(x)$:

$$R = \{y \in Y : f(x) = y \text{ for some } x \in D\}$$

A function $f : D \to Y$ is also called a "map." The sets D and Y can be arbitrary. For example, we can define a map from the set of living people to the set of whole numbers by mapping each person to his or her year of birth. The range of this map is the set of years in which a living person was born. In multivariable calculus, the domain might be a set of points in the two-dimensional plane and the range a set of numbers, points, or vectors.

Informally, we think of f as a "machine" that produces an output y for every input x in the domain D (Figure 14).

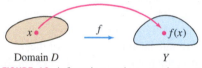

FIGURE 13 A function assigns an element $f(x)$ in Y to each $x \in D$.

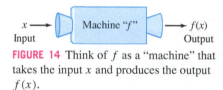

FIGURE 14 Think of f as a "machine" that takes the input x and produces the output $f(x)$.

The first part of this text deals with *numerical* functions f, where both the domain and the range are sets of real numbers. We refer to such a function as f and its value at x as $f(x)$. The letter x is used often to denote the **independent variable** that can take on any value in the domain D. We write $y = f(x)$ and refer to y as the **dependent variable** (because its value depends on the choice of x).

When f is defined by a formula, its natural domain is the set of real numbers x for which the formula is meaningful. For example, the function $f(x) = \sqrt{9 - x}$ has domain $D = \{x : x \le 9\}$ because $\sqrt{9 - x}$ is defined if $9 - x \ge 0$. Here are some other examples of domains and ranges:

$f(x)$	**Domain** D	**Range** R
x^2	$\mathbf{R}$	$\{y : y \ge 0\}$
$\cos x$	$\mathbf{R}$	$\{y : -1 \le y \le 1\}$
$\dfrac{1}{x+1}$	$\{x : x \ne -1\}$	$\{y : y \ne 0\}$

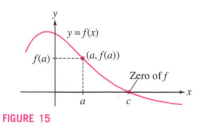

FIGURE 15

The **graph** of a function $y = f(x)$ is obtained by plotting the points $(a, f(a))$ for a in the domain D (Figure 15). If you start at $x = a$ on the x-axis, move up to the graph and then over to the y-axis, you arrive at the value $f(a)$. The absolute value $|f(a)|$ is the distance from the graph to the x-axis.

A **zero** or **root** of a function f is a number c such that $f(c) = 0$. The zeros are the values of x where the graph intersects the x-axis.

In Chapter 4, we will use calculus to sketch and analyze graphs. At this stage, to sketch a graph by hand, we can make a table of function values, plot the corresponding points (including any zeros), and connect them by a smooth curve.

■ **EXAMPLE 3** Find the roots and sketch the graph of $f(x) = x^3 - 2x$.

Solution First, we solve

$$x^3 - 2x = x(x^2 - 2) = 0$$

The roots of f are $x = 0$ and $x = \pm\sqrt{2}$. To sketch the graph, we plot the roots and a few values listed in Table 1 and join them by a curve (Figure 16).   ■

TABLE 1

x	$x^3 - 2x$
-2	-4
-1	1
0	0
1	-1
2	4

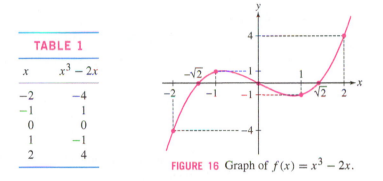

FIGURE 16 Graph of $f(x) = x^3 - 2x$.

Functions arising in applications are not always given by formulas. For example, data collected from observation or experiment define functions for which there may be no exact formula. Such functions can be displayed either graphically or by a table of values. Figure 17 and Table 2 display data collected by biologist Julian Huxley (1887–1975) in a study of the antler weight W of male red deer as a function of age t. We will see that many of the tools from calculus can be applied to functions constructed from data in this way.

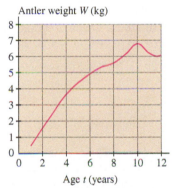

FIGURE 17 Male red deer shed their antlers every winter and regrow them in the spring. This graph shows average antler weight as a function of age.

TABLE 2

t (years)	W (kg)	t (years)	W (kg)
1	0.48	7	5.34
2	1.59	8	5.62
3	2.66	9	6.18
4	3.68	10	6.81
5	4.35	11	6.21
6	4.92	12	6.1

We can graph not just functions but, more generally, any equation relating y and x. Figure 18 shows the graph of the equation $4y^2 - x^3 = 3$; it consists of all pairs (x, y) satisfying the equation. This curve is not the graph of a function because some x-values are associated with two y-values. For example, $x = 1$ is associated with $y = \pm 1$. A curve is the graph of a function if and only if it passes the **Vertical Line Test**; that is, every vertical line $x = a$ intersects the curve in at most one point.

We are often interested in whether a function is increasing or decreasing. Roughly speaking, a function f is increasing if its graph goes up as we move to the right and is decreasing if its graph goes down [Figures 19(A) and (B)]. More precisely, we define the notion of increase/decrease on an open interval.

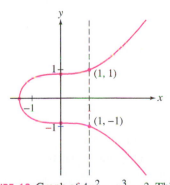

FIGURE 18 Graph of $4y^2 - x^3 = 3$. This graph fails the Vertical Line Test, so it is not the graph of a function.

A function f is:

- **increasing** on (a, b) if $f(x_1) < f(x_2)$ for all $x_1, x_2 \in (a, b)$ such that $x_1 < x_2$.
- **decreasing** on (a, b) if $f(x_1) > f(x_2)$ for all $x_1, x_2 \in (a, b)$ such that $x_1 < x_2$.

We say that f is **monotonic** if it is either increasing or decreasing. In Figure 19(C), the function is not monotonic because it is neither increasing nor decreasing for all x.

A function f is called **nondecreasing** if $f(x_1) \leq f(x_2)$ for $x_1 < x_2$ (defined by $\leq$ rather than a strict inequality $<$). **Nonincreasing** functions are defined similarly. Function (D) in Figure 19 is nondecreasing, but it is not increasing on the intervals where the graph is horizontal. Function (E) is increasing everywhere even though it levels off momentarily.

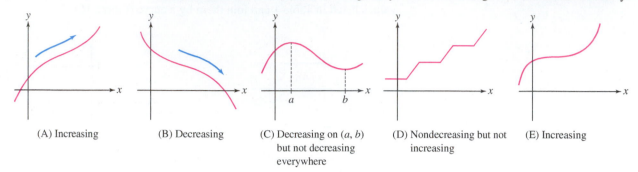

(A) Increasing

(B) Decreasing

(C) Decreasing on (a, b) but not decreasing everywhere

(D) Nondecreasing but not increasing

(E) Increasing

FIGURE 19

Another important property is **parity**, which refers to whether a function is even or odd:

> - f is **even** if $\quad f(-x) = f(x)$
> - f is **odd** if $\quad f(-x) = -f(x)$

The graphs of functions with even or odd parity have a special symmetry:

- **Even function:** Graph is symmetric about the y-axis. This means that if $P = (a, b)$ lies on the graph, then so does $Q = (-a, b)$ [Figure 20(A)].
- **Odd function:** Graph is symmetric with respect to the origin. This means that if $P = (a, b)$ lies on the graph, then so does $Q = (-a, -b)$ [Figure 20(B)].

Many functions are neither even nor odd [Figure 20(C)].

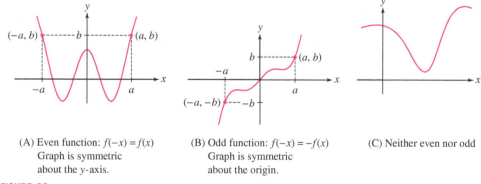

(A) Even function: $f(-x) = f(x)$
Graph is symmetric about the y-axis.

(B) Odd function: $f(-x) = -f(x)$
Graph is symmetric about the origin.

(C) Neither even nor odd

FIGURE 20

■ **EXAMPLE 4** Determine whether the function is even, odd, or neither.

(a) $f(x) = x^4$ **(b)** $g(x) = x^{-1}$ **(c)** $h(x) = x^2 + x$

Solution

(a) $f(-x) = (-x)^4 = x^4$. Thus, $f(x) = f(-x)$, and f is even.

(b) $g(-x) = (-x)^{-1} = -x^{-1}$. Thus, $g(-x) = -g(x)$, and g is odd.

(c) $h(-x) = (-x)^2 + (-x) = x^2 - x$. We see that $h(-x)$ is not equal to $h(x)$ or to $-h(x) = -x^2 - x$. Therefore, h is neither even nor odd. ■

■ **EXAMPLE 5** **Using Symmetry** Sketch the graph of $f(x) = \dfrac{1}{x^2 + 1}$.

Solution The function f is positive [$f(x) > 0$] and even [$f(-x) = f(x)$]. Therefore, the graph lies above the x-axis and is symmetric with respect to the y-axis. Furthermore,

f is decreasing for $x \geq 0$ (because a larger value of x makes the denominator larger). We use this information and a short table of values (Table 3) to sketch the graph (Figure 21). Note that the graph approaches the x-axis as we move to the right or left because $f(x)$ gets closer to 0 as $|x|$ increases. ∎

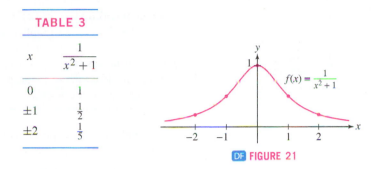

TABLE 3

x	$\dfrac{1}{x^2+1}$
0	1
± 1	$\dfrac{1}{2}$
± 2	$\dfrac{1}{5}$

$f(x) = \dfrac{1}{x^2+1}$

DF FIGURE 21

Two important ways of modifying a graph are **translation** (or **shifting**) and **scaling**. Translation consists of moving the graph horizontally or vertically:

> *Remember that $f(x)+c$ and $f(x+c)$ are different. The graph of $y = f(x)+c$ is a vertical translation and $y = f(x+c)$ a horizontal translation of the graph of $y = f(x)$.*

DEFINITION Translation (Shifting)

- **Vertical translation** $y = f(x) + c$: Shifts the graph by $|c|$ units *vertically*, upward if $c > 0$ and downward if $c < 0$.
- **Horizontal translation** $y = f(x + c)$: Shifts the graph by $|c|$ units *horizontally*, to the right if $c < 0$ and c units to the left if $c > 0$.

Figure 22 shows the effect of translating the graph of $f(x) = 1/(x^2 + 1)$ vertically and horizontally.

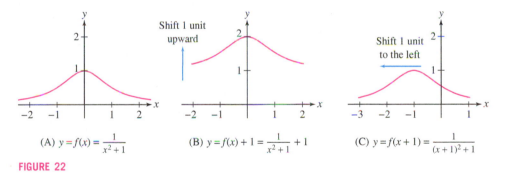

(A) $y = f(x) = \dfrac{1}{x^2+1}$ (B) $y = f(x) + 1 = \dfrac{1}{x^2+1} + 1$ (C) $y = f(x+1) = \dfrac{1}{(x+1)^2+1}$

FIGURE 22

■ **EXAMPLE 6** Figure 23(A) is the graph of $f(x) = x^2$, and Figure 23(B) is a horizontal and vertical shift of (A). What is the equation of graph (B)?

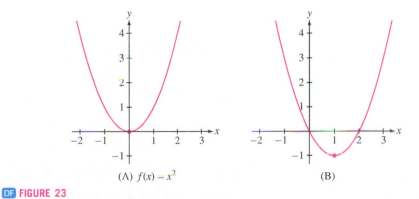

(A) $f(x) = x^2$ (B)

DF FIGURE 23

Solution Graph (B) is obtained by shifting graph (A) 1 unit to the right and 1 unit down. We can see this by observing that the point $(0, 0)$ on the graph of f is shifted to $(1, -1)$. Therefore, (B) is the graph of $g(x) = (x - 1)^2 - 1$. ∎

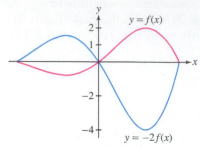

FIGURE 24 Negative vertical scale factor $k = -2$.

Scaling (also called **dilation**) consists of compressing or expanding the graph in the vertical or horizontal directions:

DEFINITION Scaling

- **Vertical scaling** $y = kf(x)$: If $k > 1$, the graph is expanded vertically by the factor k. If $0 < k < 1$, the graph is compressed vertically. When the scale factor k is negative ($k < 0$), the graph is also reflected across the x-axis (Figure 24).
- **Horizontal scaling** $y = f(kx)$: If $k > 1$, the graph is compressed in the horizontal direction. If $0 < k < 1$, the graph is expanded. If $k < 0$, then the graph is also reflected across the y-axis.

The amplitude of a function is half the difference between its greatest value and its least value, if it has both a greatest value and least value. Thus, vertical scaling changes the amplitude by the factor $|k|$.

■ **EXAMPLE 7** Sketch the graphs of $f(x) = \sin(\pi x)$ and its dilates $f(3x)$ and $3f(x)$.

Solution The graph of $f(x) = \sin(\pi x)$ is a sine curve with period 2. It completes one cycle over every interval of length 2 see Figure 25(A). It has amplitude 1.

- The graph of $f(3x) = \sin(3\pi x)$ is a compressed version of $y = f(x)$, completing three cycles instead of one over intervals of length 2 [Figure 25(B)]. It also has amplitude 1.
- The graph of $y = 3f(x) = 3\sin(\pi x)$ differs from $y = f(x)$ only in amplitude: It is expanded in the vertical direction by a factor of 3 [Figure 25(C)], so its amplitude is 3.   ■

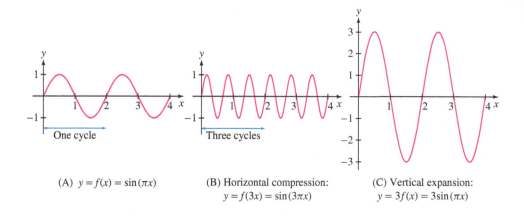

(A) $y = f(x) = \sin(\pi x)$

(B) Horizontal compression: $y = f(3x) = \sin(3\pi x)$

(C) Vertical expansion: $y = 3f(x) = 3\sin(\pi x)$

DF **FIGURE 25** Horizontal and vertical scaling of $f(x) = \sin(\pi x)$.

1.1 SUMMARY

- Absolute value: $|a| = \begin{cases} a & \text{if } a \geq 0 \\ -a & \text{if } a < 0 \end{cases}$

- Triangle inequality: $|a + b| \leq |a| + |b|$
- Four intervals with endpoints a and b:

$$(a, b), \qquad [a, b], \qquad [a, b), \qquad (a, b]$$

- Writing open and closed intervals using inequalities:

$$(a, b) = \{x : |x - c| < r\}, \qquad [a, b] = \{x : |x - c| \leq r\}$$

where $c = \frac{1}{2}(a + b)$ is the midpoint and $r = \frac{1}{2}(b - a)$ is the radius.

- Distance d between (x_1, y_1) and (x_2, y_2):

$$d = \sqrt{(x_2 - x_1)^2 + (y_2 - y_1)^2}$$

- Equation of circle of radius r with center (a, b):

$$(x - a)^2 + (y - b)^2 = r^2$$

- A *zero* or *root* of a function f is a number c such that $f(c) = 0$.
- Vertical Line Test: A curve in the plane is the graph of a function if and only if each vertical line $x = a$ intersects the curve in at most one point.

Increasing:	$f(x_1) < f(x_2)$ if $x_1 < x_2$
Nondecreasing:	$f(x_1) \le f(x_2)$ if $x_1 < x_2$
Decreasing:	$f(x_1) > f(x_2)$ if $x_1 < x_2$
Nonincreasing:	$f(x_1) \ge f(x_2)$ if $x_1 < x_2$

- Even function: $f(-x) = f(x)$ (graph is symmetric about the y-axis).
- Odd function: $f(-x) = -f(x)$ (graph is symmetric about the origin).
- Four ways to transform the graph of f:

$f(x) + c$	Shifts graph vertically $|c|$ units (upward if $c > 0$, downward if $c < 0$)
$f(x + c)$	Shifts graph horizontally $|c|$ units (to the right if $c < 0$, to the left if $c > 0$)
$kf(x)$	Scales graph vertically by factor k; if $k < 0$, graph is reflected across x-axis
$f(kx)$	Scales graph horizontally by factor k (compresses if $k > 1$); if $k < 0$, graph is reflected across y-axis

1.1 EXERCISES

Preliminary Questions

1. Give an example of numbers a and b such that $a < b$ and $|a| > |b|$.

2. Which numbers satisfy $|a| = a$? Which satisfy $|a| = -a$? What about $|-a| = a$?

3. Give an example of numbers a and b such that $|a + b| < |a| + |b|$.

4. Are there numbers a and b such that $|a + b| > |a| + |b|$?

5. What are the coordinates of the point lying at the intersection of the lines $x = 9$ and $y = -4$?

6. In which quadrant do the following points lie?
(a) $(1, 4)$ (b) $(-3, 2)$ (c) $(4, -3)$ (d) $(-4, -1)$

7. What is the radius of the circle with equation $(x - 7)^2 + (y - 8)^2 = 9$?

8. The equation $f(x) = 5$ has a solution if (choose one):
(a) 5 belongs to the domain of f.
(b) 5 belongs to the range of f.

9. What kind of symmetry does the graph have if $f(-x) = -f(x)$?

10. Is there a function that is both even and odd?

Exercises

1. Use a calculator to find a rational number r such that $|r - \pi^2| < 10^{-4}$.

2. Which of (a)–(f) are true for $a = -3$ and $b = 2$?
(a) $a < b$
(b) $|a| < |b|$
(c) $ab > 0$
(d) $3a < 3b$
(e) $-4a < -4b$
(f) $\dfrac{1}{a} < \dfrac{1}{b}$

In Exercises 3–8, express the interval in terms of an inequality involving absolute value.

3. $[-2, 2]$
4. $(-4, 4)$
5. $(0, 4)$
6. $[-4, 0]$
7. $[1, 5]$
8. $(-2, 8)$

In Exercises 9–12, write the inequality in the form $a < x < b$.

9. $|x| < 8$
10. $|x - 12| < 8$
11. $|2x + 1| < 5$
12. $|3x - 4| < 2$

In Exercises 13–18, express the set of numbers x satisfying the given condition as an interval.

13. $|x| < 4$
14. $|x| \le 9$
15. $|x - 4| < 2$
16. $|x + 7| < 2$
17. $|4x - 1| \le 8$
18. $|3x + 5| < 1$

In Exercises 19 22, describe the set as a union of ₋nite or in₋nite intervals.

19. $\{x : |x - 4| > 2\}$

20. $\{x : |2x + 4| > 3\}$

21. $\{x : |x^2 - 1| > 2\}$

22. $\{x : |x^2 + 2x| > 2\}$

23. Match (a) (f) with (i) (vi).

(a) $a > 3$

(b) $|a - 5| < \dfrac{1}{3}$

(c) $\left|a - \dfrac{1}{3}\right| < 5$

(d) $|a| > 5$

(e) $|a - 4| < 3$

(f) $1 \le a \le 5$

(i) a lies to the right of 3.

(ii) a lies between 1 and 7.

(iii) The distance from a to 5 is less than $\frac{1}{3}$.

(iv) The distance from a to 3 is at most 2.

(v) a is less than 5 units from $\frac{1}{3}$.

(vi) a lies either to the left of -5 or to the right of 5.

24. Describe $\left\{x : \dfrac{x}{x+1} < 0\right\}$ as an interval. *Hint: Consider the sign* of x and $x + 1$ individually.

25. Describe $\{x : x^2 + 2x < 3\}$ as an interval. *Hint: Plot* $y = x^2 + 2x - 3$.

26. Describe the set of real numbers satisfying $|x - 3| = |x - 2| + 1$ as a half-in₋nite interval.

27. Show that if $a > b$, and $a, b \neq 0$, then $b^{-1} > a^{-1}$, provided that a and b have the same sign. What happens if $a > 0$ and $b < 0$?

28. Which x satis₋es both $|x - 3| < 2$ and $|x - 5| < 1$?

29. Show that if $|a - 5| < \frac{1}{2}$ and $|b - 8| < \frac{1}{2}$, then $|(a + b) - 13| < 1$. *Hint: Use the triangle inequality* $(|a + b| \le |a| + |b|)$.

30. Suppose that $|x - 4| \le 1$.

(a) What is the maximum possible value of $|x + 4|$?

(b) Show that $|x^2 - 16| \le 9$.

31. Suppose that $|a - 6| \le 2$ and $|b| \le 3$.

(a) What is the largest possible value of $|a + b|$?

(b) What is the smallest possible value of $|a + b|$?

32. Prove that $|x| - |y| \le |x - y|$. *Hint: Apply the triangle inequality* to y and $x - y$.

33. Express $r_1 = 0.\overline{27}$ as a fraction. *Hint:* $100r_1 - r_1$ is an integer. Then express $r_2 = 0.2666\ldots$ as a fraction.

34. Represent $1/7$ and $4/27$ as repeating decimals.

35. The text states: *If the decimal expansions of numbers a and b agree to k places, then $|a - b| \le 10^{-k}$.* Show that the converse is false: For all k there are numbers a and b whose decimal expansions *do not agree at all* but $|a - b| \le 10^{-k}$.

36. Plot each pair of points and compute the distance between them:

(a) $(1, 4)$ and $(3, 2)$

(b) $(2, 1)$ and $(2, 4)$

(c) $(0, 0)$ and $(-2, 3)$

(d) $(-3, -3)$ and $(-2, 3)$

37. Find the equation of the circle with center $(2, 4)$:

(a) with radius $r = 3$.

(b) that passes through $(1, -1)$.

38. Find all points in the xy-plane with integer coordinates located at a distance 5 from the origin. Then ₋nd all points with integer coordinates located at a distance 5 from $(2, 3)$.

39. Determine the domain and range of the function

$$f : \{r, s, t, u\} \to \{A, B, C, D, E\}$$

de₋ned by $f(r) = A$, $f(s) = B$, $f(t) = B$, $f(u) = E$.

40. Give an example of a function whose domain D has three elements and whose range R has two elements. Does a function exist whose domain D has two elements and whose range R has three elements?

In Exercises 41 48, ₋nd the domain and range of the function.

41. $f(x) = -x$

42. $g(t) = t^4$

43. $f(x) = x^3$

44. $g(t) = \sqrt{2 - t}$

45. $f(x) = |x|$

46. $h(s) = \dfrac{1}{s}$

47. $f(x) = \dfrac{1}{x^2}$

48. $g(t) = \cos \dfrac{1}{t}$

In Exercises 49 52, determine where f is increasing.

49. $f(x) = |x + 1|$

50. $f(x) = x^3$

51. $f(x) = x^4$

52. $f(x) = \dfrac{1}{x^4 + x^2 + 1}$

In Exercises 53 58, ₋nd the zeros of f and sketch its graph by plotting points. Use symmetry and increase/decrease information where appropriate.

53. $f(x) = x^2 - 4$

54. $f(x) = 2x^2 - 4$

55. $f(x) = x^3 - 4x$

56. $f(x) = x^3$

57. $f(x) = 2 - x^3$

58. $f(x) = \dfrac{1}{(x - 1)^2 + 1}$

59. Which of the curves in Figure 26 is the graph of a function?

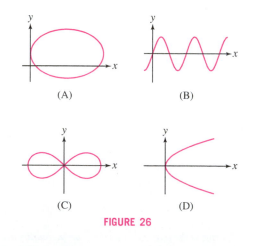

FIGURE 26

60. Determine whether the function is even, odd, or neither.

(a) $f(x) = x^5$

(b) $g(t) = t^3 - t^2$

(c) $F(t) = \dfrac{1}{t^4 + t^2}$

61. Determine whether the function is even, odd, or neither.

(a) $f(t) = \dfrac{1}{t^4 + t + 1} - \dfrac{1}{t^4 - t + 1}$ (b) $g(t) = 2^t - 2^{-t}$

(c) $G(\theta) = \sin\theta + \cos\theta$ (d) $H(\theta) = \sin(\theta^2)$

62. Write $f(x) = 2x^4 - 5x^3 + 12x^2 - 3x + 4$ as the sum of an even and an odd function.

63. Show that $f(x) = \ln\left(\dfrac{1-x}{1+x}\right)$ is an odd function.

64. State whether the function is increasing, decreasing, or neither.

(a) Surface area of a sphere as a function of its radius

(b) Temperature at a point on the equator as a function of time

(c) Price of an airline ticket as a function of the price of oil

(d) Pressure of the gas in a piston as a function of volume

In Exercises 65–70, let f be the function shown in Figure 27.

65. Find the domain and range of f.

66. Sketch the graphs of $y = f(x + 2)$ and $y = f(x) + 2$.

67. Sketch the graphs of $y = f(2x)$, $y = f\left(\frac{1}{2}x\right)$, and $y = 2f(x)$.

68. Sketch the graphs of $y = f(-x)$ and $y = -f(-x)$.

69. Extend the graph of f to $[-4, 4]$ so that it is an even function.

70. Extend the graph of f to $[-4, 4]$ so that it is an odd function.

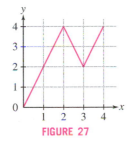

FIGURE 27

71. Suppose that f has domain $[4, 8]$ and range $[2, 6]$. Find the domain and range of:

(a) $y = f(x) + 3$ (b) $y = f(x + 3)$

(c) $y = f(3x)$ (d) $y = 3f(x)$

72. Let $f(x) = x^2$. Sketch the graph over $[-2, 2]$ of:

(a) $y = f(x + 1)$ (b) $y = f(x) + 1$

(c) $y = f(5x)$ (d) $y = 5f(x)$

73. Suppose that the graph of $f(x) = \sin x$ is compressed horizontally by a factor of 2 and then shifted 5 units to the right.

(a) What is the equation for the new graph?

(b) What is the equation if you first shift by 5 and then compress by 2?

(c) [GU] Verify your answers by plotting your equations.

74. Figure 28 shows the graph of $f(x) = |x| + 1$. Match the functions (a)–(e) with their graphs (i)–(v).

(a) $y = f(x - 1)$ (b) $y = -f(x)$ (c) $y = -f(x) + 2$

(d) $y = f(x - 1) - 2$ (e) $y = f(x + 1)$

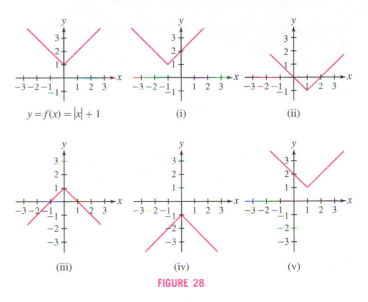

$y = f(x) = |x| + 1$

FIGURE 28

75. Sketch the graph of $y = f(2x)$ and $y = f\left(\frac{1}{2}x\right)$, where $f(x) = |x| + 1$ (Figure 28).

76. Find the function f whose graph is obtained by shifting the parabola $y = x^2$ by 3 units to the right and 4 units down, as in Figure 29.

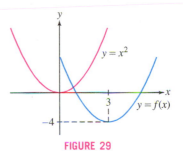

FIGURE 29

77. Define $f(x)$ to be the larger of x and $2 - x$. Sketch the graph of f. What are its domain and range? Express $f(x)$ in terms of the absolute value function.

78. For each curve in Figure 30, state whether it is symmetric with respect to the y-axis, the origin, both, or neither.

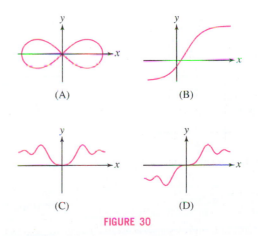

FIGURE 30

79. Show that the sum of two even functions is even and the sum of two odd functions is odd.

80. Suppose that f and g are both odd. Which of the following functions are even? Which are odd?

(a) $y = f(x)g(x)$

(b) $y = f(x)^3$

(c) $y = f(x) - g(x)$

(d) $y = \dfrac{f(x)}{g(x)}$

81. Prove that the only function whose graph is symmetric with respect to both the y-axis and the origin is the function $f(x) = 0$.

Further Insights and Challenges

82. Prove the triangle inequality ($|a + b| \le |a| + |b|$) by adding the two inequalities

$$-|a| \le a \le |a|, \qquad -|b| \le b \le |b|$$

83. Show that a fraction $r = a/b$ in lowest terms has a *finite* decimal expansion if and only if

$$b = 2^n 5^m \quad \text{for some } n, m \ge 0.$$

Hint: Observe that r has a finite decimal expansion when $10^N r$ is an integer for some $N \ge 0$ (and hence b divides 10^N).

84. Let $p = p_1 \ldots p_s$ be an integer with digits $p_1, \ldots, p_s$. Show that

$$\frac{p}{10^s - 1} = 0.\overline{p_1 \ldots p_s}$$

Use this to find the decimal expansion of $r = \frac{2}{11}$. Note that

$$r = \frac{2}{11} = \frac{18}{10^2 - 1}$$

85. ✏️   A function f is symmetric with respect to the vertical line $x = a$ if $f(a - x) = f(a + x)$.

(a) Draw the graph of a function that is symmetric with respect to $x = 2$.

(b) Show that if f is symmetric with respect to $x = a$, then $g(x) = f(x + a)$ is even.

86. ✏️   Formulate a condition for f to be symmetric with respect to the point $(a, 0)$ on the x-axis.

1.2 Linear and Quadratic Functions

Linear functions are the simplest of all functions, and their graphs (lines) are the simplest of all curves. However, linear functions and lines play an enormously important role in calculus. For this reason, you should be thoroughly familiar with the basic properties of linear functions and the different ways of writing an equation of a line.

Let's recall that a **linear function** is a function of the form

$$\boxed{f(x) = mx + b \quad (m \text{ and } b \text{ constants})}$$

The graph of f is a line of slope m, and since $f(0) = b$, the graph intersects the y-axis at the point $(0, b)$ (Figure 1). The number b is called the y-intercept.

> The **slope-intercept form** of the line with slope m and y-intercept b is given by
>
> $$\boxed{y = mx + b}$$

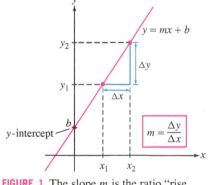

FIGURE 1 The slope m is the ratio "rise over run."

We use the symbols Δx and Δy to denote the *change* (or *increment*) in x and $y = f(x)$ over an interval $[x_1, x_2]$ (Figure 1):

$$\boxed{\Delta x = x_2 - x_1, \qquad \Delta y = y_2 - y_1 = f(x_2) - f(x_1)}$$

The slope m of a line is equal to the ratio

$$m = \frac{\Delta y}{\Delta x} = \frac{\text{vertical change}}{\text{horizontal change}} = \frac{\text{rise}}{\text{run}}$$

This follows from the formula $y = mx + b$:

$$\frac{\Delta y}{\Delta x} = \frac{y_2 - y_1}{x_2 - x_1} = \frac{(mx_2 + b) - (mx_1 + b)}{x_2 - x_1} = \frac{m(x_2 - x_1)}{x_2 - x_1} = m$$

The slope m measures the *rate of change* of y with respect to x. In fact, by writing

$$\Delta y = m\,\Delta x$$

we see that a 1-unit increase in x (i.e., $\Delta x = 1$) produces an m-unit change Δy in y. For example, if $m = 5$, then y increases by 5 units per unit increase in x. The rate-of-change interpretation of the slope is fundamental in calculus. We discuss it in greater detail in Section 2.1.

Graphically, the slope m measures the steepness of the line $y = mx + b$. Figure 2(A) shows lines through a point of varying slope m. Note the following properties:

- **Steepness:** The larger the absolute value $|m|$, the steeper the line.
- **Positive slope:** If $m > 0$, the line slants upward from left to right.
- **Negative slope:** If $m < 0$, the line slants downward from left to right.
- $f(x) = mx + b$ is increasing if $m > 0$ and decreasing if $m < 0$.
- The **horizontal line** $y = b$ has slope $m = 0$ [Figure 2(B)].
- A **vertical line** has equation $x = c$, where c is a constant. The slope of a vertical line is undefined. It is not possible to write the equation of a vertical line in slope-intercept form $y = mx + b$. A vertical line is not the graph of a function [Figure 2(B)].

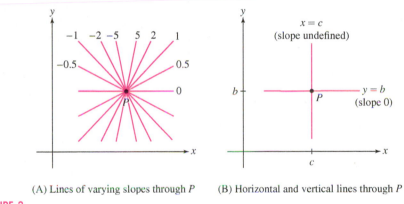

(A) Lines of varying slopes through P (B) Horizontal and vertical lines through P

FIGURE 2

Scale is especially important in applications because the steepness of a graph depends on the choice of units for the x- and y-axes. We can create very different *subjective* impressions by changing the scale. Figure 3 shows the growth of company profits over a 4-year period. The two plots convey the same information, but the left-hand plot makes the growth look more dramatic.

CAUTION Graphs are often plotted using different scales for the x- and y-axes. This is necessary to keep the sizes of graphs within reasonable bounds. However, when the scales are different, lines do not appear with their true slopes.

FIGURE 3 Growth of company profits.

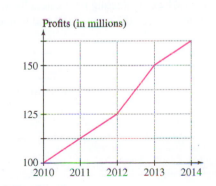

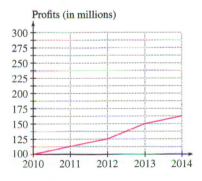

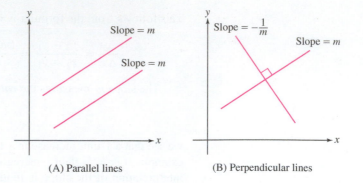

FIGURE 4 Parallel and perpendicular lines.

(A) Parallel lines   (B) Perpendicular lines

Next, we recall the relation between the slopes of parallel and perpendicular lines (Figure 4):

- Lines of slopes m_1 and m_2 are **parallel** if and only if $m_1 = m_2$.
- Lines of slopes m_1 and m_2 are **perpendicular** if and only if

$$m_1 = -\frac{1}{m_2} \qquad (\text{or } m_1 m_2 = -1)$$

CONCEPTUAL INSIGHT The increments over an interval $[x_1, x_2]$:

$$\Delta x = x_2 - x_1, \qquad \Delta y = f(x_2) - f(x_1)$$

are de,ned for any function f (linear or not), but the ratio $\Delta y / \Delta x$ may depend on the interval (Figure 5). The characteristic property of a linear function $f(x) = mx + b$ is that $\Delta y / \Delta x$ has the same value m for every interval. In other words, y has a constant rate of change with respect to x. We can use this property to test if two quantities are related by a linear equation.

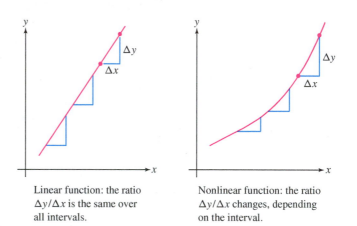

Linear function: the ratio $\Delta y / \Delta x$ is the same over all intervals.

Nonlinear function: the ratio $\Delta y / \Delta x$ changes, depending on the interval.

DF **FIGURE 5**

■ **EXAMPLE 1** **Testing for a Linear Relationship** Do the data in Table 1 suggest a linear relation between the pressure P and temperature T of a gas?

TABLE 1

Temperature (°C)	Pressure (kPa)
40	1365.80
45	1385.40
55	1424.60
70	1483.40
80	1522.60

Solution We calculate $\Delta P / \Delta T$ at successive data points and check whether this ratio is constant:

(T_1, P_1)	(T_2, P_2)	$\dfrac{\Delta P}{\Delta T}$
(40, 1365.80)	(45, 1385.40)	$\dfrac{1385.40 - 1365.80}{45 - 40} = 3.92$
(45, 1385.40)	(55, 1424.60)	$\dfrac{1424.60 - 1385.40}{55 - 45} = 3.92$
(55, 1424.60)	(70, 1483.40)	$\dfrac{1483.40 - 1424.60}{70 - 55} = 3.92$
(70, 1483.40)	(80, 1522.60)	$\dfrac{1522.60 - 1483.40}{80 - 70} = 3.92$

Real experimental data are unlikely to reveal perfect linearity, even if the data points do essentially lie on a line. The method of linear regression is used to find the linear function that best fits the data.

Because $\Delta P / \Delta T$ has the constant value 3.92, the data points lie on a line with slope $m = 3.92$. This is confirmed in the plot in Figure 6. ■

As mentioned above, it is important to be familiar with the standard ways of writing the equation of a line. The general **linear equation** is

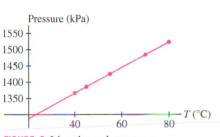

Pressure (kPa)

FIGURE 6 Line through pressure-temperature data points.

$$ax + by = c \qquad \boxed{1}$$

where a and b are not *both* zero. For $b = 0$, we obtain the vertical line $ax = c$. When $b \neq 0$, we can rewrite Eq. (1) in slope-intercept form. For example, $-6x + 2y = 3$ can be rewritten as $y = 3x + \frac{3}{2}$.

Two other forms we will use frequently are the **point-slope** and **point-point** forms. Given a point $P = (a, b)$ and a slope m, the equation of the line through P with slope m is $y - b = m(x - a)$. Similarly, the line through two distinct points $P = (a_1, b_1)$ and $Q = (a_2, b_2)$ has slope (Figure 7)

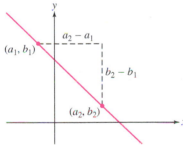

$$m = \frac{b_2 - b_1}{a_2 - a_1}$$

Therefore, we can write its equation as $y - b_1 = m(x - a_1)$.

DF FIGURE 7 Slope of the line between $P = (a_1, b_1)$ and $Q = (a_2, b_2)$ is $m = \dfrac{b_2 - b_1}{a_2 - a_1}$.

If $a = 0$, point-slope form becomes slope-intercept form $y = mx + b$.

Additional Equations for Lines

1. Point-slope form of the line through $P = (a, b)$ with slope m:

$$y - b = m(x - a)$$

2. Point-point form of the line through $P = (a_1, b_1)$ and $Q = (a_2, b_2)$:

$$y - b_1 = m(x - a_1), \qquad \text{where } m = \frac{b_2 - b_1}{a_2 - a_1}$$

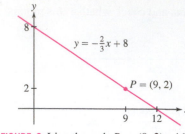

FIGURE 8 Line through $P = (9, 2)$ with slope $m = -\frac{2}{3}$.

■ **EXAMPLE 2 Line of Given Slope Through a Given Point** Find the equation of the line through $(9, 2)$ with slope $-\frac{2}{3}$.

Solution In point-slope form:

$$y - 2 = -\frac{2}{3}(x - 9)$$

In slope-intercept form: $y = -\frac{2}{3}(x - 9) + 2$ or $y = -\frac{2}{3}x + 8$. See Figure 8.   ■

■ **EXAMPLE 3 Line Through Two Points** Find the equation of the line through $(2, 1)$ and $(9, 5)$.

Solution The line has slope

$$m = \frac{5 - 1}{9 - 2} = \frac{4}{7}$$

Because $(2, 1)$ lies on the line, its equation in point-slope form is $y - 1 = \frac{4}{7}(x - 2)$.   ■

A **quadratic function** is a function de ned by a quadratic polynomial

$$f(x) = ax^2 + bx + c \quad (a, b, c, \text{ constants with } a \neq 0)$$

The graph of f is a **parabola** (Figure 9). The parabola opens upward if the leading coef cient a is positive and downward if a is negative. The **discriminant** of $f(x)$ is the quantity

$$D = b^2 - 4ac$$

The roots of f are given by the **quadratic formula** (see Exercise 58):

$$\text{Roots of } f = \frac{-b \pm \sqrt{b^2 - 4ac}}{2a} = \frac{-b \pm \sqrt{D}}{2a}$$

The sign of D determines whether or not f has real roots (Figure 9). If $D > 0$, then f has two real roots, and if $D = 0$, it has one real root (a double root). If $D < 0$, then $\sqrt{D}$ is imaginary and f has no real roots.

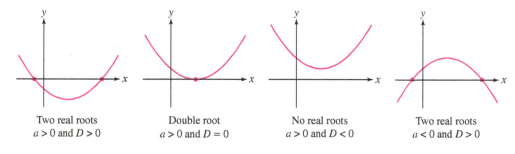

FIGURE 9 Graphs of quadratic functions $f(x) = ax^2 + bx + c$.

Two real roots
$a > 0$ and $D > 0$

Double root
$a > 0$ and $D = 0$

No real roots
$a > 0$ and $D < 0$

Two real roots
$a < 0$ and $D > 0$

When f has two real roots r_1 and r_2, then $f(x)$ factors as

$$f(x) = a(x - r_1)(x - r_2)$$

For example, $f(x) = 2x^2 - 3x + 1$ has discriminant $D = b^2 - 4ac = 9 - 8 = 1 > 0$, and by the quadratic formula, its roots are $(3 \pm 1)/4$ or 1 and $\frac{1}{2}$. Therefore,

$$f(x) = 2x^2 - 3x + 1 = 2(x - 1)\left(x - \frac{1}{2}\right)$$

The technique of **completing the square** consists of writing a quadratic polynomial as a multiple of a square plus a constant. Then

$$x^2 + bx + c = x^2 + bx + \left(\frac{b}{2}\right)^2 - \left(\frac{b}{2}\right)^2 + c = \left(x + \frac{b}{2}\right)^2 - \left(\frac{b}{2}\right)^2 + c$$

If there is a constant a multiplying the x^2 term, we factor that out rst, as demonstrated in the following example.

Cuneiform texts written on clay tablets show that the method of completing the square was known to ancient Babylonian mathematicians who lived some 4000 years ago.

Ignoring air resistance, a basketball follows a parabolic path (Figure 10).

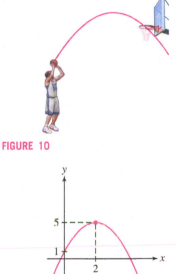

FIGURE 10

DF FIGURE 11 Graph of $f(x) = -x^2 + 4x + 1$.

■ **EXAMPLE 4   Completing the Square**   Complete the square for the quadratic polynomial $f(x) = 4x^2 - 12x + 3$.

Solution   First factor out the leading coefficient:

$$4x^2 - 12x + 3 = 4\left(x^2 - 3x + \frac{3}{4}\right)$$

Then complete the square for the term $x^2 - 3x$:

$$x^2 - 3x = x^2 - 3x + \left(\frac{3}{2}\right)^2 - \left(\frac{3}{2}\right)^2 = \left(x - \frac{3}{2}\right)^2 - \frac{9}{4}$$

Therefore,

$$4x^2 - 12x + 3 = 4\left(\left(x - \frac{3}{2}\right)^2 - \frac{9}{4} + \frac{3}{4}\right) = 4\left(x - \frac{3}{2}\right)^2 - 6 \qquad ■$$

The method of completing the square can be used to find the **minimum** or **maximum** value of a quadratic function.

■ **EXAMPLE 5   Finding the Maximum of a Quadratic Function**   Complete the square and find the maximum value of $f(x) = -x^2 + 4x + 1$.

Solution   We have

This term is ≤ 0

$$f(x) = -(x^2 - 4x - 1) = -(x^2 - 4x + 4 - 4 - 1) = -((x - 2)^2 - 5) = \overbrace{-(x - 2)^2} + 5$$

Thus, $f(x) \leq 5$ for all x, and the maximum value of f is $f(2) = 5$ (Figure 11). ■

1.2 SUMMARY

- A linear function is a function of the form $f(x) = mx + b$.
- The general equation of a line is $ax + by = c$. The line $y = c$ is horizontal and $x = c$ is vertical.
- Three convenient ways of writing the equation of a nonvertical line:

    Slope-intercept form: $y = mx + b$ (slope m and y-intercept b)

    Point-slope form: $y - b = m(x - a)$ [slope m, passes through (a, b)]

    Point-point form: The line through two points $P = (a_1, b_1)$ and $Q = (a_2, b_2)$ has slope $m = \dfrac{b_2 - b_1}{a_2 - a_1}$ and equation $y - b_1 = m(x - a_1)$.

- Two lines of slopes m_1 and m_2 are parallel if and only if $m_1 = m_2$, and they are perpendicular if and only if $m_1 = -1/m_2$.
- Quadratic function: $f(x) = ax^2 + bx + c$. The roots are $x = (-b \pm \sqrt{D})/2a$, where $D = b^2 - 4ac$ is the discriminant. The roots are real and distinct if $D > 0$, there is a double root if $D = 0$, and there are no real roots if $D < 0$.
- Completing the square consists of writing a quadratic function as a multiple of a square plus a constant.

1.2 EXERCISES

Preliminary Questions

1. What is the slope of the line $y = -4x - 9$?

2. Are the lines $y = 2x + 1$ and $y = -2x - 4$ perpendicular?

3. When is the line $ax + by = c$ parallel to the y-axis? To the x-axis?

4. Suppose $y = 3x + 2$. What is Δy if x increases by 3?

5. What is the minimum of $f(x) = (x + 3)^2 - 4$?

6. What is the result of completing the square for $f(x) = x^2 + 1$?

Exercises

In Exercises 1 4, nd the slope, the y-intercept, and the x-intercept of the line with the given equation.

1. $y = 3x + 12$

2. $y = 4 - x$

3. $4x + 9y = 3$

4. $y - 3 = \frac{1}{2}(x - 6)$

In Exercises 5 8, nd the slope of the line.

5. $y = 3x + 2$

6. $y = 3(x - 9) + 2$

7. $3x + 4y = 12$

8. $3x + 4y = -8$

In Exercises 9 20, nd the equation of the line with the given description.

9. Slope 3, y-intercept 8

10. Slope -2, y-intercept 3

11. Slope 3, passes through $(7, 9)$

12. Slope -5, passes through $(0, 0)$

13. Horizontal, passes through $(0, -2)$

14. Passes through $(-1, 4)$ and $(2, 7)$

15. Parallel to $y = 3x - 4$, passes through $(1, 1)$

16. Passes through $(1, 4)$ and $(12, -3)$

17. Perpendicular to $3x + 5y = 9$, passes through $(2, 3)$

18. Vertical, passes through $(-4, 9)$

19. Horizontal, passes through $(8, 4)$

20. Slope 3, x-intercept 6

21. Find the equation of the perpendicular bisector of the segment joining $(1, 2)$ and $(5, 4)$ (Figure 12). *Hint:* The midpoint Q of the segment joining (a, b) and (c, d) is $\left(\dfrac{a + c}{2}, \dfrac{b + d}{2} \right)$.

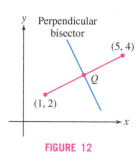

FIGURE 12

22. Intercept-Intercept Form   Show that if $a, b \neq 0$, then the line with x-intercept $x = a$ and y-intercept $y = b$ has equation (Figure 13)

$$\frac{x}{a} + \frac{y}{b} = 1$$

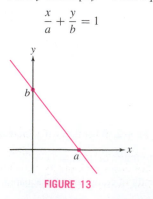

FIGURE 13

23. Find an equation of the line with x-intercept $x = 4$ and y-intercept $y = 3$.

24. Find y such that $(3, y)$ lies on the line of slope $m = 2$ through $(1, 4)$.

25. Determine whether there exists a constant c such that the line $x + cy = 1$:

(a) has slope 4.   **(b)** passes through $(3, 1)$.

(c) is horizontal.   **(d)** is vertical.

26. Assume that the number N of concert tickets that can be sold at a price of P dollars per ticket is a linear function $N(P)$ for $10 \leq P \leq 40$. Determine $N(P)$ (called the demand function) if $N(10) = 500$ and $N(40) = 0$. What is the decrease ΔN in the number of tickets sold if the price is increased by $\Delta P = 5$ dollars?

27. Suppose that the number of a certain type of computer that can be sold when its price is P (in dollars) is given by a linear function $N(P)$. Determine $N(P)$ if $N(1000) = 10,000$ and $N(1500) = 7,500$. What is the change ΔN in the number of computers sold if the price is increased by $\Delta P = 100$ dollars?

28. Suppose that the demand for Colin's kidney pies is linear in the price P. Determine the demand function N as a function of P giving the number of pies sold when the price is P if he can sell 100 pies when the price is \$5.00 and he can sell 40 pies when the price is \$10.00. Determine the revenue ($N \times P$) for prices $P = 5, 6, 7, 8, 9, 10$ and then choose a price to maximize the revenue.

29. Materials expand when heated. Consider a metal rod of length L_0 at temperature T_0. If the temperature is changed by an amount ΔT, then the rod's length approximately changes by $\Delta L = \alpha L_0 \Delta T$, where α is the thermal expansion coef cient and ΔT is not an extreme temperature change. For steel, $\alpha = 1.24 \times 10^{-5} \,^\circ\text{C}^{-1}$.

(a) A steel rod has length $L_0 = 40$ cm at $T_0 = 40^\circ$C. Find its length at $T = 90^\circ$C.

(b) Find its length at $T = 50^\circ$C if its length at $T_0 = 100^\circ$C is 65 cm.

(c) Express length L as a function of T if $L_0 = 65$ cm at $T_0 = 100^\circ$C.

30. Do the points $(0.5, 1)$, $(1, 1.2)$, $(2, 2)$ lie on a line?

31. Find b such that $(2, -1)$, $(3, 2)$, and $(b, 5)$ lie on a line.

32. Find an expression for the velocity v as a linear function of t that matches the following data:

t (s)	0	2	4	6
v (m/s)	39.2	58.6	78	97.4

33. The period T of a pendulum is measured for pendulums of several different lengths L. Based on the following data, does T appear to be a linear function of L?

L (cm)	20	30	40	50
T (s)	0.9	1.1	1.27	1.42

34. Show that f is linear of slope m if and only if

$$f(x + h) - f(x) = mh \quad \text{(for all } x \text{ and } h)$$

That is to say, prove the following two statements:

(a) f is linear of slope m implies that $f(x + h) - f(x) = mh$ (for all x and h).

(b) $f(x+h) - f(x) = mh$ (for all x and h) implies that f is linear of slope m.

35. Find the roots of the quadratic polynomials:

(a) $f(x) = 4x^2 - 3x - 1$

(b) $f(x) = x^2 - 2x - 1$

In Exercises 36–43, complete the square and find the minimum or maximum value of the quadratic function.

36. $y = x^2 + 2x + 5$

37. $y = x^2 - 6x + 9$

38. $y = -9x^2 + x$

39. $y = x^2 + 6x + 2$

40. $y = 2x^2 - 4x - 7$

41. $y = -4x^2 + 3x + 8$

42. $y = 3x^2 + 12x - 5$

43. $y = 4x - 12x^2$

44. Sketch the graph of $y = x^2 - 6x + 8$ by plotting the roots and the minimum point.

45. Sketch the graph of $y = x^2 + 4x + 6$ by plotting the minimum point, the y-intercept, and one other point.

46. If the alleles A and B of the cystic fibrosis gene occur in a population with frequencies p and $1 - p$ (where p is a fraction between 0 and 1), then the frequency of heterozygous carriers (carriers with both alleles) is $2p(1 - p)$. Which value of p gives the largest frequency of heterozygous carriers?

47. For which values of c does $f(x) = x^2 + cx + 1$ have a double root? No real roots?

48. Let f be a quadratic function and c a constant. Which of the following statements is correct? Explain graphically.

(a) There is a unique value of c such that $y = f(x) - c$ has a double root.

(b) There is a unique value of c such that $y = f(x - c)$ has a double root.

49. Prove that $x + \frac{1}{x} \geq 2$ for all $x > 0$. *Hint:* Consider $(x^{1/2} - x^{-1/2})^2$.

50. Let $a, b > 0$. Show that the *geometric mean* $\sqrt{ab}$ is not larger than the *arithmetic mean* $(a + b)/2$. *Hint:* Use a variation of the hint given in Exercise 49.

51. If objects of weights x and w_1 are suspended from the balance in Figure 14(A), the cross-beam is horizontal if $bx = aw_1$. If the lengths a and b are known, we may use this equation to determine an unknown weight x by selecting w_1 such that the cross-beam is horizontal. If a and b are not known precisely, we might proceed as follows. First balance x by w_1 on the left as in (A). Then switch places and balance x by w_2 on the right as in (B). The average $\bar{x} = \frac{1}{2}(w_1 + w_2)$ gives an estimate for x. Show that $\bar{x}$ is greater than or equal to the true weight x.

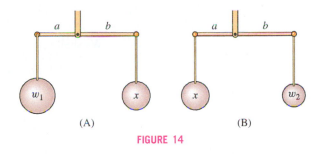

(A)   (B)

FIGURE 14

52. Find numbers x and y with sum 10 and product 24. *Hint:* Find a quadratic polynomial satisfied by x.

53. Find a pair of numbers whose sum and product are both equal to 8.

54. Show that the parabola $y = x^2$ consists of all points P such that $d_1 = d_2$, where d_1 is the distance from P to $\left(0, \frac{1}{4}\right)$ and d_2 is the distance from P to the line $y = -\frac{1}{4}$ (Figure 15).

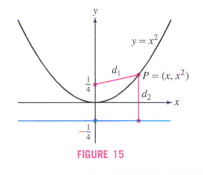

FIGURE 15

Further Insights and Challenges

55. Show that if f and g are linear, then so is $f + g$. Is the same true of fg?

56. Show that if f and g are linear functions such that $f(0) = g(0)$ and $f(1) = g(1)$, then $f = g$.

57. Show that $\Delta y / \Delta x$ for the function $f(x) = x^2$ over the interval $[x_1, x_2]$ is not a constant, but depends on the interval. Determine the exact dependence of $\Delta y / \Delta x$ on x_1 and x_2.

58. Complete the square and use the result to derive the quadratic formula for the roots of $ax^2 + bx + c = 0$.

59. Let $a, c \neq 0$. Show that the roots of

$$ax^2 + bx + c = 0 \quad \text{and} \quad cx^2 + bx + a = 0$$

are reciprocals of each other.

60. Show, by completing the square, that the parabola

$$y = ax^2 + bx + c$$

is congruent to $y = ax^2$ by a vertical and horizontal translation.

61. Prove **Viète's Formulas**: The quadratic polynomial with α and β as roots is $x^2 + bx + c$, where $b = -\alpha - \beta$ and $c = \alpha\beta$.

1.3 The Basic Classes of Functions

It would be impossible (and useless) to describe all possible functions f. Since the values of a function can be assigned arbitrarily, a function chosen at random would likely be so complicated that we could neither graph it nor describe it in any reasonable way. However, calculus makes no attempt to deal with all functions. The techniques of calculus, powerful and general as they are, apply only to functions that are sufficiently well-behaved (we will see what well-behaved means when we study the derivative in Chapter 3). Fortunately, such functions are adequate for a vast range of applications.

Most of the functions considered in this text are constructed from the following familiar classes of well-behaved functions:

polynomials   rational functions   algebraic functions

exponential functions   trigonometric functions

logarithmic functions   inverse trigonometric functions

We shall refer to these as the **basic functions**.

- **Polynomials:** For any real number m, $f(x) = x^m$ is called the **power function** with exponent m. Power functions include $f(x) = x^3$, $f(x) = x^{-7}$ and $f(x) = x^\pi$. The base is the variable and the exponent is a constant. For now, we are interested in power functions with exponents that are positive integers. A **polynomial** is a sum of multiples of power functions with exponents that are positive integers or zero (making the term a constant in that case) (Figure 1):

$$f(x) = x^5 - 5x^3 + 4x, \qquad g(t) = 7t^6 + t^3 - 3t - 1, \qquad h(x) = x^9$$

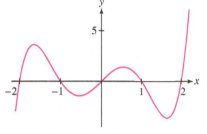

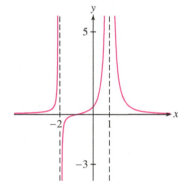

FIGURE 1 The polynomial $y = x^5 - 5x^3 + 4x$.

Thus, the function $f(x) = x + x^{-1}$ is not a polynomial because it includes a power x^{-1} with a negative exponent. The general polynomial P in the variable x may be written

$$P(x) = a_n x^n + a_{n-1} x^{n-1} + \cdots + a_1 x + a_0$$

The numbers $a_0, a_1, \ldots, a_n$ are called **coefficients**.

The **degree** of P is n (assuming that $a_n \neq 0$).

The coefficient a_n is called the **leading coefficient**.

The domain of P is **R**.

- A **rational function** is a *quotient* of two polynomials (Figure 2):

$$f(x) = \frac{P(x)}{Q(x)} \qquad [P(x) \text{ and } Q(x) \text{ polynomials}]$$

FIGURE 2 The rational function $f(x) = \dfrac{x+1}{x^3 - 3x + 2}$.

The domain of f is the set of numbers x such that $Q(x) \neq 0$. For example,

$$f(x) = \frac{1}{x^2} \qquad \text{domain } \{x : x \neq 0\}$$

$$h(t) = \frac{7t^6 + t^3 - 3t - 1}{t^2 - 1} \qquad \text{domain } \{t : t \neq \pm 1\}$$

Every polynomial is also a rational function [with $Q(x) = 1$].

- An **algebraic function** is produced by taking sums, products, and quotients of *roots* of polynomials and rational functions (Figure 3):

$$f(x) = \sqrt{1 + 3x^2 - x^4}, \qquad g(t) = (\sqrt{t} - 2)^{-2}, \qquad h(z) = \frac{z + z^{-5/3}}{5z^3 - \sqrt{z}}$$

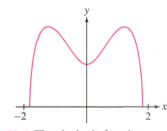

FIGURE 3 The algebraic function $f(x) = \sqrt{1 + 3x^2 - x^4}$.

A number x belongs to the domain of f if each term in the formula is defined and the result does not involve division by zero. For example, $g(t)$ is defined if $t \geq 0$ and $\sqrt{t} \neq 2$, so the domain of g is $D = \{t : t \geq 0 \text{ and } t \neq 4\}$. More generally, algebraic functions are defined by polynomial equations between x and y. In this case, we say that y is **implicitly defined** as a function of x. For example, the equation $y^4 + 2x^2 y + x^4 = 1$ defines y implicitly as a function of x.

- **Exponential functions:** The function $f(x) = b^x$, where $b > 0$, is called the exponential function with base b. Some examples are

$$f(x) = 2^x, \qquad g(t) = 10^t, \qquad h(x) = \left(\frac{1}{3}\right)^x, \qquad p(t) = (\sqrt{5})^t$$

Any function that is not algebraic is called *transcendental*. Exponential and trigonometric functions are examples, as are the Bessel and gamma functions that appear in engineering and statistics. The term *transcendental* goes back to the 1670s, when it was used by Gottfried Wilhelm Leibniz (1646–1716) to describe functions of this type.

Exponential functions and their *inverses*, the **logarithmic functions**, are treated in greater detail in Section 1.6.

- **Trigonometric functions** are functions built from $\sin x$ and $\cos x$. These functions and their inverses are discussed in the next two sections.

Constructing New Functions

Given functions f and g, we can construct new functions by forming the sum, difference, product, and quotient functions:

$$(f + g)(x) = f(x) + g(x), \qquad (f - g)(x) = f(x) - g(x)$$

$$(fg)(x) = f(x)\, g(x), \qquad \left(\frac{f}{g}\right)(x) = \frac{f(x)}{g(x)} \quad \text{(where } g(x) \neq 0\text{)}$$

For example, if $f(x) = x^2$ and $g(x) = \sin x$, then

$$(f + g)(x) = x^2 + \sin x, \qquad (f - g)(x) = x^2 - \sin x$$

$$(fg)(x) = x^2 \sin x, \qquad \left(\frac{f}{g}\right)(x) = \frac{x^2}{\sin x}$$

We can also multiply functions by constants. A function of the form

$$h(x) = c_1 f(x) + c_2 g(x) \quad (c_1, c_2 \text{ constants})$$

is called a **linear combination** of f and g.

Composition is another important way of constructing new functions. The composition of f and g is the function $f \circ g$ defined by $(f \circ g)(x) = f(g(x))$. The domain of $f \circ g$ is the set of values of x in the domain of g such that $g(x)$ lies in the domain of f.

■ **EXAMPLE 1** Compute the composite functions $f \circ g$ and $g \circ f$ and discuss their domains, where

$$f(x) = \sqrt{x}, \qquad g(x) = 1 - x$$

Example 1 shows that the composition of functions is not commutative: The functions $f \circ g$ and $g \circ f$ may be (and usually are) different.

Solution We have

$$(f \circ g)(x) = f(g(x)) = f(1 - x) = \sqrt{1 - x}$$

The square root $\sqrt{1 - x}$ is defined if $1 - x \geq 0$ or $x \leq 1$, so the domain of $f \circ g$ is $\{x : x \leq 1\}$. On the other hand,

$$(g \circ f)(x) = g(f(x)) = g(\sqrt{x}) = 1 - \sqrt{x}$$

The domain of $g \circ f$ is $\{x : x \geq 0\}$. ■

Elementary Functions

Inverse functions are discussed in Section 1.5.

As noted above, we can produce new functions by applying the operations of addition, subtraction, multiplication, division, and composition. It is convenient to refer to a function constructed in this way from the basic functions listed above as an **elementary function**. The following functions are elementary:

$$f(x) = \sqrt{2x + \sin x}, \qquad f(x) = 10^{\sqrt{x}}, \qquad f(x) = \frac{1 + x^{-1}}{1 + \cos x}$$

Piecewise-Defined Functions

We can also create new functions by piecing together functions defined over limited domains, obtaining **piecewise-defined functions**. One example we have already seen is the absolute value function defined by

$$|x| = \begin{cases} -x & \text{when } x < 0 \\ x & \text{when } x \geq 0 \end{cases}$$

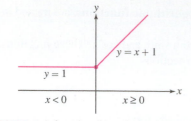

FIGURE 4 A function defined piecewise.

■ **EXAMPLE 2** Given the function f, determine its domain, range, and whether or not it is increasing or decreasing for different values of x.

$$f(x) = \begin{cases} 1 & \text{when } x < 0 \\ x + 1 & \text{when } x \geq 0 \end{cases}$$

Solution The function f appears in Figure 4. It is defined for all values of x so the domain is all real numbers. However, for $x < 0$ the range is just the single value of 1, and for $x \geq 0$ the range is all $x \geq 1$. Hence, the range of the function is $\{x : x \geq 1\}$. The function is neither increasing nor decreasing for $x < 0$; however, the function is increasing for $x \geq 0$.   ■

1.3 SUMMARY

- For m a real number, $f(x) = x^m$ is called the *power function* with exponent m. A polynomial P is a sum of multiples of x^m, where m is a whole number:

$$P(x) = a_n x^n + a_{n-1} x^{n-1} + \cdots + a_1 x + a_0$$

This polynomial has degree n (assuming that $a_n \neq 0$) and a_n is called the leading coefficient.
- A rational function is a quotient P/Q of two polynomials (defined when $Q(x) \neq 0$).
- An algebraic function is produced by taking sums, products, and nth roots of polynomials and rational functions.
- Exponential function: $f(x) = b^x$, where $b > 0$ (b is called the base).
- The composite function $f \circ g$ is defined by $(f \circ g)(x) = f(g(x))$. The domain of $f \circ g$ is the set of x in the domain of g such that $g(x)$ belongs to the domain of f.
- The elementary functions are obtained by taking products, sums, differences, quotients, and compositions of the basic functions, which include polynomials, rational functions, algebraic functions, exponential functions, trigonometric functions, logarithmic functions, and inverse trigonometric functions.
- A piecewise-defined function is obtained by defining a function over two or more distinct domains.

1.3 EXERCISES

Preliminary Questions

1. Give an example of a rational function.

2. Is $y = |x|$ a polynomial function? What about $y = |x^2 + 1|$?

3. What is unusual about the domain of the composite function $f \circ g$ for the functions $f(x) = x^{1/2}$ and $g(x) = -1 - |x|$?

4. Is $f(x) = \left(\frac{1}{2}\right)^x$ increasing or decreasing?

5. Give an example of a transcendental function.

Exercises

In Exercises 1–12, determine the domain of the function.

1. $f(x) = x^{1/4}$

2. $g(t) = t^{2/3}$

3. $f(x) = x^3 + 3x - 4$

4. $h(z) = z^3 + z^{-3}$

5. $g(t) = \dfrac{1}{t + 2}$

6. $f(x) = \dfrac{1}{x^2 + 4}$

7. $G(u) = \dfrac{1}{u^2 - 4}$

8. $f(x) = \dfrac{\sqrt{x}}{x^2 - 9}$

9. $f(x) = x^{-4} + (x - 1)^{-3}$

10. $F(s) = \sin\left(\dfrac{s}{s + 1}\right)$

11. $g(y) = 10^{\sqrt{y} + y^{-1}}$

12. $f(x) = \dfrac{x + x^{-1}}{(x - 3)(x + 4)}$

In Exercises 13–24, identify each of the following functions as polynomial, rational, algebraic, or transcendental.

13. $f(x) = 4x^3 + 9x^2 - 8$

14. $f(x) = x^{-4}$

15. $f(x) = \sqrt{x}$

16. $f(x) = \sqrt{1 - x^2}$

17. $f(x) = \dfrac{x^2}{x + \sin x}$

18. $f(x) = 2^x$

19. $f(x) = \dfrac{2x^3 + 3x}{9 - 7x^2}$

20. $f(x) = \dfrac{3x - 9x^{-1/2}}{9 - 7x^2}$

21. $f(x) = \sin(x^2)$

22. $f(x) = \dfrac{x}{\sqrt{x} + 1}$

23. $f(x) = x^2 + 3x^{-1}$

24. $f(x) = \sin(3^x)$

25. Is $f(x) = 2^{x^2}$ a transcendental function?

26. Show that $f(x) = x^2 + 3x^{-1}$ and $g(x) = 3x^3 - 9x + x^{-2}$ are rational functions that is, quotients of polynomials.

In Exercises 27 34, calculate the composite functions $f \circ g$ and $g \circ f$, and determine their domains.

27. $f(x) = \sqrt{x}$, $g(x) = x + 1$

28. $f(x) = \dfrac{1}{x}$, $g(x) = x^{-4}$

29. $f(x) = 2^x$, $g(x) = x^2$

30. $f(x) = |x|$, $g(\theta) = \sin\theta$

31. $f(\theta) = \cos\theta$, $g(x) = x^3 + x^2$

32. $f(x) = \dfrac{1}{x^2 + 1}$, $g(x) = x^{-2}$

33. $f(t) = \dfrac{1}{\sqrt{t}}$, $g(t) = -t^2$

34. $f(t) = \sqrt{t}$, $g(t) = 1 - t^3$

In Exercises 35 38, draw the graphs of each of the piecewise-defined functions.

35.
$$f(x) = \begin{cases} 3 & \text{when } x < 0 \\ x^2 + 3 & \text{when } x \geq 0 \end{cases}$$

36.
$$f(x) = \begin{cases} x + 1 & \text{when } x < 0 \\ 1 - x & \text{when } x \geq 0 \end{cases}$$

37.
$$f(x) = \begin{cases} x^2 & \text{when } x < 0 \\ -x^2 & \text{when } x \geq 0 \end{cases}$$

38.
$$f(x) = \begin{cases} 2x - 2 & \text{when } x < 0 \\ x & \text{when } x \geq 0 \end{cases}$$

39. The population (in millions) of a country as a function of time t (years) is $P(t) = 30 \cdot 2^{0.1t}$. Show that the population doubles every 10 years. Show more generally that for any positive constants a and k, the function $g(t) = a2^{kt}$ doubles after $1/k$ years.

40. Find all values of c such that $f(x) = \dfrac{x + 1}{x^2 + 2cx + 4}$ has domain **R**.

Further Insights and Challenges

In Exercises 41 47, we define the first difference δf of a function f by $\delta f(x) = f(x + 1) - f(x)$.

41. Show that if $f(x) = x^2$, then $\delta f(x) = 2x + 1$. Calculate δf for $f(x) = x$ and $f(x) = x^3$.

42. Show that $\delta(10^x) = 9 \cdot 10^x$ and, more generally, that $\delta(b^x) = (b - 1)b^x$.

43. Show that for any two functions f and g, $\delta(f + g) = \delta f + \delta g$ and $\delta(cf) = c\delta(f)$, where c is any constant.

44. Suppose we can find a function P such that $\delta P(x) = (x + 1)^k$ and $P(0) = 0$. Prove that $P(1) = 1^k$, $P(2) = 1^k + 2^k$, and, more generally, for every whole number n,

$$P(n) = 1^k + 2^k + \cdots + n^k \qquad \boxed{1}$$

45. First show that

$$P(x) = \dfrac{x(x + 1)}{2}$$

satisfies $\delta P = (x + 1)$. Then apply Exercise 44 to conclude that

$$1 + 2 + 3 + \cdots + n = \dfrac{n(n + 1)}{2}$$

46. Calculate $\delta(x^3)$, $\delta(x^2)$, and $\delta(x)$. Then find a polynomial P of degree 3 such that $\delta P = (x + 1)^2$ and $P(0) = 0$. Conclude that $P(n) = 1^2 + 2^2 + \cdots + n^2$.

47. This exercise combined with Exercise 44 shows that for all whole numbers k, there exists a polynomial P satisfying Eq. (1). The solution requires the Binomial Theorem and proof by induction (see Appendix C).

(a) Show that $\delta(x^{k+1}) = (k + 1)x^k + \cdots$, where the dots indicate terms involving smaller powers of x.

(b) Show by induction that there exists a polynomial of degree $k + 1$ with leading coefficient $1/(k + 1)$:

$$P(x) = \dfrac{1}{k + 1}x^{k+1} + \cdots$$

such that $\delta P = (x + 1)^k$ and $P(0) = 0$.

1.4 Trigonometric Functions

We begin our trigonometric review by recalling the two systems of angle measurement: **radians** and **degrees**. They are best described using the relationship between angles and rotation. As is customary, we often use the lowercase Greek letter θ (theta) to denote angles and rotations.

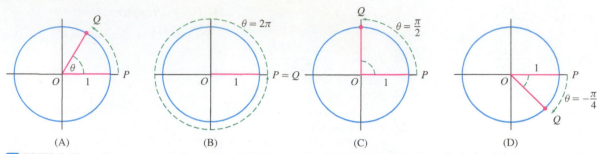

DF **FIGURE 1** The radian measure θ of a counterclockwise rotation is the length along the unit circle of the arc traversed by P as it rotates into Q.

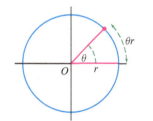

FIGURE 2 On a circle of radius r, the arc traversed by a counterclockwise rotation of θ radians has length θr.

TABLE 1

Rotation through	Radian measure
Two full circles	4π
Full circle	2π
Half circle	π
Quarter circle	$2\pi/4 = \pi/2$
One-sixth circle	$2\pi/6 = \pi/3$

Radians	Degrees
0	0°
$\dfrac{\pi}{6}$	30°
$\dfrac{\pi}{4}$	45°
$\dfrac{\pi}{3}$	60°
$\dfrac{\pi}{2}$	90°

Radian measurement is usually the better choice for mathematical purposes, but there are good practical reasons for using degrees. The number 360 has many divisors ($360 = 8 \cdot 9 \cdot 5$), and consequently, many fractional parts of the circle can be expressed as an integer number of degrees. For example, one-fifth of the circle is 72°, two-ninths is 80°, three-eighths is 135°, etc.

Figure 1(A) shows a unit circle with radius $\overline{OP}$ rotating counterclockwise into radius $\overline{OQ}$. *The radian measure of this rotation is the length θ of the circular arc traversed by P as it rotates into Q.* On a circle of radius r, the arc traversed by a counterclockwise rotation of θ radians has length θr (Figure 2).

The unit circle has circumference 2π. Therefore, a rotation through a full circle has radian measure $\theta = 2\pi$ [Figure 1(B)]. The radian measure of a rotation through one-quarter of a circle is $\theta = 2\pi/4 = \pi/2$ [Figure 1(C)] and, in general, the rotation through one-nth of a circle has radian measure $2\pi/n$ (Table 1). A negative rotation (with $\theta < 0$) is a rotation in the *clockwise* direction [Figure 1(D)].

The radian measure of an angle such as $\angle POQ$ in Figure 1(A) is defined as the radian measure of a rotation that carries $\overline{OP}$ to $\overline{OQ}$. Notice, however, that the radian measure of an angle is not unique. The rotations through θ and $\theta + 2\pi$ both carry $\overline{OP}$ to $\overline{OQ}$. Therefore, θ and $\theta + 2\pi$ represent the same angle even though the rotation through $\theta + 2\pi$ takes an extra trip around the circle. In general, *two radian measures represent the same angle if the corresponding rotations differ by an integer multiple of 2π*. For example, $\pi/4$, $9\pi/4$, and $-15\pi/4$ all represent the same angle because they differ by multiples of 2π:

$$\frac{\pi}{4} = \frac{9\pi}{4} - 2\pi = -\frac{15\pi}{4} + 4\pi$$

Every angle has a unique radian measure satisfying $0 \le \theta < 2\pi$. With this choice, the angle θ subtends an arc of length θr on a circle of radius r (Figure 2).

Degrees are defined by dividing the circle (not necessarily the unit circle) into 360 equal parts. A degree is $\frac{1}{360}$ of a circle. A rotation through θ degrees (denoted $\theta°$) is a rotation through the fraction $\theta/360$ of the complete circle. For example, a rotation through 90° is a rotation through the fraction $\frac{90}{360}$, or $\frac{1}{4}$, of a circle.

As with radians, the degree measure of an angle is not unique. Two degree measures represent that same angle if they differ by an integer multiple of 360. For example, the angles $-45°$ and $675°$ coincide because $675 = -45 + 2(360)$. Every angle has a unique degree measure θ with $0 \le \theta < 360$.

To convert between radians and degrees, remember that 2π radians is equal to $360°$. Therefore, 1 radian equals $360/2\pi$ or $180/\pi$ degrees.

- To convert from radians to degrees, multiply by $180/\pi$.
- To convert from degrees to radians, multiply by $\pi/180$.

■ **EXAMPLE 1** Convert   **(a)** 55° to radians   and   **(b)** 0.5 radians to degrees.

Solution

(a) $55° \times \dfrac{\pi}{180°} \approx 0.9599$ radians   **(b)** 0.5 radians $\times \dfrac{180°}{\pi} \approx 28.648°$ ■

Convention *Unless otherwise stated, we always measure angles in radians.*

The trigonometric functions sine and cosine can be defined in terms of right triangles. Let θ be an acute angle in a right triangle, and let us label the sides as in Figure 3. Then

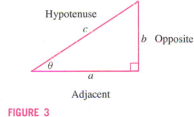

Hypotenuse
c

b Opposite

θ

a

Adjacent

FIGURE 3

$$\sin\theta = \frac{b}{c} = \frac{\text{opposite}}{\text{hypotenuse}}, \qquad \cos\theta = \frac{a}{c} = \frac{\text{adjacent}}{\text{hypotenuse}}$$

A disadvantage of this definition is that it makes sense only if θ lies between 0 and $\pi/2$ (because an angle in a right triangle cannot exceed $\pi/2$). However, sine and cosine can be defined for all angles in terms of the unit circle. Let $P = (x, y)$ be the point on the unit circle corresponding to the angle θ as in Figures 4(A) and (B), and define

$$\cos\theta = x\text{-coordinate of } P, \qquad \sin\theta = y\text{-coordinate of } P$$

This agrees with the right-triangle definition when $0 < \theta < \frac{\pi}{2}$. On the circle of radius r (centered at the origin), the point corresponding to the angle θ has coordinates

$$(r\cos\theta, r\sin\theta)$$

Furthermore, we see from Figure 4(C) that $f(x) = \sin\theta$ is an odd function and $f(x) = \cos\theta$ is an even function:

$$\sin(-\theta) = -\sin\theta, \qquad \cos(-\theta) = \cos\theta$$

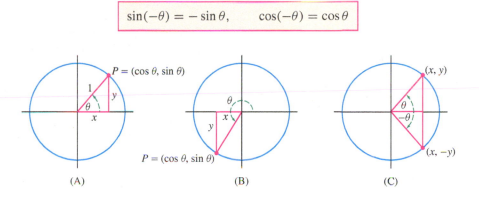

FIGURE 4 The unit circle definition of sine and cosine is valid for all angles θ.

(A)   (B)   (C)

Although we use a calculator to evaluate sine and cosine for general angles, the standard values listed in Figure 5 and Table 2 appear often and should be memorized.

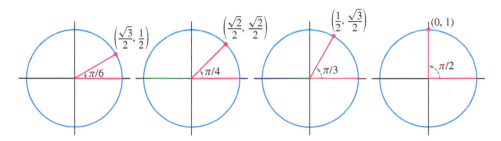

FIGURE 5 Four standard angles: The x- and y-coordinates of the points are $\cos\theta$ and $\sin\theta$.

TABLE 2

θ	0	$\dfrac{\pi}{6}$	$\dfrac{\pi}{4}$	$\dfrac{\pi}{3}$	$\dfrac{\pi}{2}$	$\dfrac{2\pi}{3}$	$\dfrac{3\pi}{4}$	$\dfrac{5\pi}{6}$	π
$\sin\theta$	0	$\dfrac{1}{2}$	$\dfrac{\sqrt{2}}{2}$	$\dfrac{\sqrt{3}}{2}$	1	$\dfrac{\sqrt{3}}{2}$	$\dfrac{\sqrt{2}}{2}$	$\dfrac{1}{2}$	0
$\cos\theta$	1	$\dfrac{\sqrt{3}}{2}$	$\dfrac{\sqrt{2}}{2}$	$\dfrac{1}{2}$	0	$-\dfrac{1}{2}$	$-\dfrac{\sqrt{2}}{2}$	$-\dfrac{\sqrt{3}}{2}$	-1

The graph of $y = \sin\theta$ is the familiar sine wave shown in Figure 6. Observe how the graph is generated by the y-coordinate of the point $P = (\cos\theta, \sin\theta)$ moving around the unit circle.

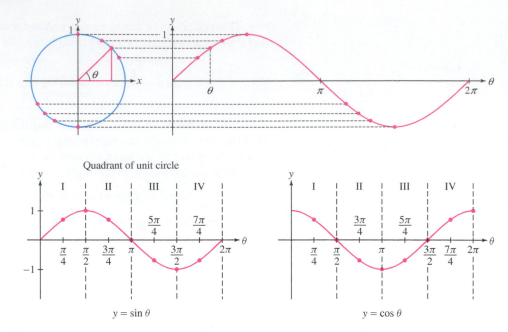

DF FIGURE 6 The graph of $y = \sin\theta$ is generated as the point $P = (\cos\theta, \sin\theta)$ moves around the unit circle.

FIGURE 7 Graphs of $y = \sin\theta$ and $y = \cos\theta$ over one *period* of length 2π.

The graph of $y = \cos\theta$ has the same shape but is shifted to the left $\pi/2$ units (Figure 7). The signs of $\sin\theta$ and $\cos\theta$ vary as $P = (\cos\theta, \sin\theta)$ changes quadrant.

A function f is called **periodic** with period T if $f(x + T) = f(x)$ (for all x) and T is the smallest positive number with this property. The sine and cosine functions are periodic with period $T = 2\pi$ (Figure 8) because the radian measures x and $x + 2\pi k$ correspond to the same point on the unit circle for any integer k:

We often write $\sin x$ and $\cos x$, using x instead of θ. Depending on the application, we may think of x as an angle or simply as a real number.

$$\sin x = \sin(x + 2\pi k), \qquad \cos x = \cos(x + 2\pi k)$$

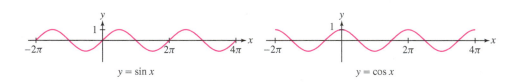

FIGURE 8 Sine and cosine have period 2π.

There are four other standard trigonometric functions, each defined in terms of $\sin x$ and $\cos x$ or as ratios of sides in a right triangle (Figure 9):

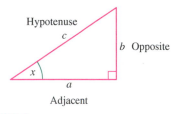

FIGURE 9

Tangent: $\tan x = \dfrac{\sin x}{\cos x} = \dfrac{b}{a}$,	Cotangent: $\cot x = \dfrac{\cos x}{\sin x} = \dfrac{a}{b}$
Secant: $\sec x = \dfrac{1}{\cos x} = \dfrac{c}{a}$,	Cosecant: $\csc x = \dfrac{1}{\sin x} = \dfrac{c}{b}$

These functions are periodic (Figure 10): $y = \tan x$ and $y = \cot x$ have period π; $y = \sec x$ and $y = \csc x$ have period 2π (see Exercise 57).

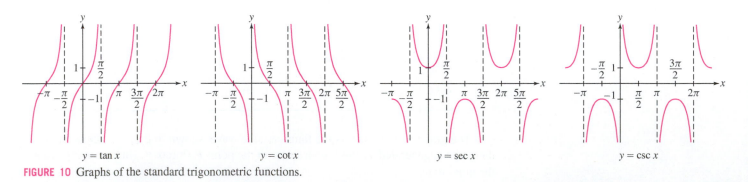

FIGURE 10 Graphs of the standard trigonometric functions.

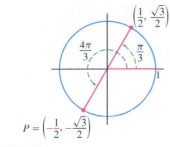

FIGURE 11

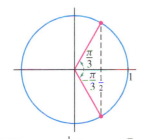

FIGURE 12 $\cos x = \frac{1}{2}$ for $x = \pm\frac{\pi}{3}$

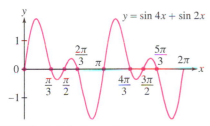

DF **FIGURE 13** $\sin\theta_2 = -\sin\theta_1$ when $\theta_2 = -\theta_1$ or $\theta_2 = \theta_1 + \pi$.

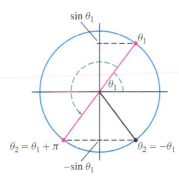

FIGURE 14 Solutions of $\sin 4x + \sin 2x = 0$.

■ **EXAMPLE 2** **Computing Values of Trigonometric Functions** Find the values of the six trigonometric functions at $x = 4\pi/3$.

Solution The point P on the unit circle corresponding to the angle $x = 4\pi/3$ lies opposite the point with angle $\pi/3$ (Figure 11). Thus, we see that (refer to Table 2)

$$\sin\frac{4\pi}{3} = -\sin\frac{\pi}{3} = -\frac{\sqrt{3}}{2}, \qquad \cos\frac{4\pi}{3} = -\cos\frac{\pi}{3} = -\frac{1}{2}$$

The remaining values are

$$\tan\frac{4\pi}{3} = \frac{\sin 4\pi/3}{\cos 4\pi/3} = \frac{-\sqrt{3}/2}{-1/2} = \sqrt{3}, \quad \cot\frac{4\pi}{3} = \frac{\cos 4\pi/3}{\sin 4\pi/3} = \frac{\sqrt{3}}{3}$$

$$\sec\frac{4\pi}{3} = \frac{1}{\cos 4\pi/3} = \frac{1}{-1/2} = -2, \quad \csc\frac{4\pi}{3} = \frac{1}{\sin 4\pi/3} = \frac{-2\sqrt{3}}{3} \qquad ■$$

■ **EXAMPLE 3** Find the angles x such that $\sec x = 2$.

Solution Because $\sec x = 1/\cos x$, we must solve $\cos x = \frac{1}{2}$. From Figure 12 we see that $x = \pi/3$ and $x = -\pi/3$ are solutions. We may add any integer multiple of 2π, so the general solution is $x = \pm\pi/3 + 2\pi k$ for any integer k. ■

■ **EXAMPLE 4** **Trigonometric Equation** Solve $\sin 4x + \sin 2x = 0$ for $x \in [0, 2\pi)$.

Solution We must find the angles x such that $\sin 4x = -\sin 2x$. First, let's determine when angles θ_1 and θ_2 satisfy $\sin\theta_2 = -\sin\theta_1$. Figure 13 shows that this occurs if $\theta_2 = -\theta_1$ or $\theta_2 = \theta_1 + \pi$. Because the sine function is periodic with period 2π,

$$\sin\theta_2 = -\sin\theta_1 \quad \Leftrightarrow \quad \theta_2 = -\theta_1 + 2\pi k \quad \text{or} \quad \theta_2 = \theta_1 + \pi + 2\pi k$$

where k is an integer. Taking $\theta_2 = 4x$ and $\theta_1 = 2x$, we see that

$$\sin 4x = -\sin 2x \quad \Leftrightarrow \quad 4x = -2x + 2\pi k \quad \text{or} \quad 4x = 2x + \pi + 2\pi k$$

The first equation gives $6x = 2\pi k$ or $x = (\pi/3)k$ and the second equation gives $2x = \pi + 2\pi k$ or $x = \pi/2 + \pi k$. We obtain eight solutions in $[0, 2\pi)$ (Figure 14):

$$x = 0, \quad \frac{\pi}{3}, \quad \frac{2\pi}{3}, \quad \pi, \quad \frac{4\pi}{3}, \quad \frac{5\pi}{3} \quad \text{and} \quad x = \frac{\pi}{2}, \quad \frac{3\pi}{2} \qquad ■$$

■ **EXAMPLE 5** Sketch the graph of $f(x) = 3\cos\left(2\left(x + \frac{\pi}{2}\right)\right)$ over $[0, 2\pi]$.

Solution The graph is obtained by scaling and shifting the graph of $y = \cos x$ in three steps (Figure 15):

- Compress horizontally by a factor of 2: $\qquad y = \cos 2x$
- Shift to the left $\pi/2$ units: $\qquad\qquad y = \cos\left(2\left(x + \frac{\pi}{2}\right)\right)$

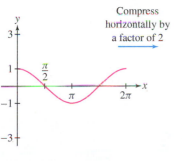

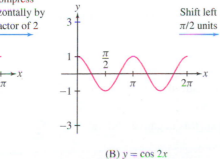

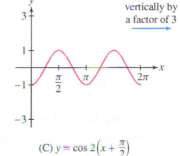

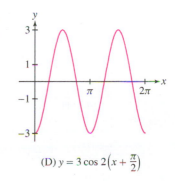

(A) $y = \cos x$ (B) $y = \cos 2x$ (periodic with period π)   (C) $y = \cos 2\left(x + \frac{\pi}{2}\right)$ (D) $y = 3\cos 2\left(x + \frac{\pi}{2}\right)$

DF **FIGURE 15**

CAUTION *To shift the graph of* $y = \cos 2x$ *to the left* $\pi/2$ *units, we must replace* x *by* $x + \frac{\pi}{2}$ *to obtain* $\cos\left(2\left(x + \frac{\pi}{2}\right)\right)$. *It is incorrect to take* $\cos\left(2x + \frac{\pi}{2}\right)$. *Note that to shift left (in the* $-x$ *direction), we add* $\pi/2$. *To shift right (in the* $+x$ *direction), we subtract* $\pi/2$, *counter to what you might expect.*

• Expand vertically by a factor of 3: $y = 3\cos\left(2\left(x + \frac{\pi}{2}\right)\right)$ ■

The vertical height taken on by such a function is the **amplitude**. So in this last example, the amplitude was 3.

Trigonometric Identities

A key feature of trigonometric functions is that they satisfy a large number of identities. First and foremost, sine and cosine satisfy a fundamental identity, which is equivalent to the Pythagorean Theorem:

$$\sin^2 x + \cos^2 x = 1$$ **1**

The expression $(\sin x)^k$ *is usually denoted* $\sin^k x$. *For example,* $\sin^2 x$ *is the square of* $\sin x$. *We use similar notation for the other trigonometric functions. However, we reserve* $\sin^{-1} x$ *for the inverse sine function discussed in the next section, rather than for* $\frac{1}{\sin x}$.

Equivalent versions are obtained by dividing Eq. (1) by $\cos^2 x$ or $\sin^2 x$:

$$\tan^2 x + 1 = \sec^2 x, \qquad 1 + \cot^2 x = \csc^2 x$$ **2**

Here is a list of some other commonly used identities. The identities for complementary angles are justified by Figure 16.

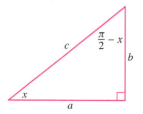

FIGURE 16 For complementary angles, the sine of one angle is equal to the cosine of the complementary angle.

Basic Trigonometric Identities

Complementary angles: $\sin\left(\frac{\pi}{2} - x\right) = \cos x, \quad \cos\left(\frac{\pi}{2} - x\right) = \sin x$

Addition formulas: $\sin(x + y) = \sin x \cos y + \cos x \sin y$

$\cos(x + y) = \cos x \cos y - \sin x \sin y$

Double-angle formulas: $\sin^2 x = \frac{1}{2}(1 - \cos 2x), \quad \cos^2 x = \frac{1}{2}(1 + \cos 2x)$

$\cos 2x = \cos^2 x - \sin^2 x, \quad \sin 2x = 2\sin x \cos x$

Shift formulas: $\sin\left(x + \frac{\pi}{2}\right) = \cos x, \quad \cos\left(x + \frac{\pi}{2}\right) = -\sin x$

■ **EXAMPLE 6** Suppose that $\cos\theta = \frac{2}{5}$. Calculate $\tan\theta$ in the following two cases:
(a) $0 < \theta < \frac{\pi}{2}$ and   **(b)** $\pi < \theta < 2\pi$.

Solution First, using the identity $\cos^2\theta + \sin^2\theta = 1$, we obtain

$$\sin\theta = \pm\sqrt{1 - \cos^2\theta} = \pm\sqrt{1 - \frac{4}{25}} = \pm\frac{\sqrt{21}}{5}$$

(a) If $0 < \theta < \frac{\pi}{2}$, then $\sin\theta$ is positive and we take the positive square root:

$$\tan\theta = \frac{\sin\theta}{\cos\theta} = \frac{\sqrt{21}/5}{2/5} = \frac{\sqrt{21}}{2}$$

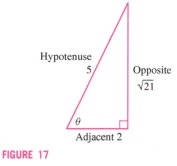

FIGURE 17

To visualize this computation, draw a right triangle with angle θ such that $\cos\theta = \frac{2}{5}$ as in Figure 17. The opposite side then has length $\sqrt{21} = \sqrt{5^2 - 2^2}$ by the Pythagorean Theorem.

(b) If $\pi < \theta < 2\pi$, then $\sin\theta$ is negative and $\tan\theta = -\frac{\sqrt{21}}{2}$.   ■

We conclude this section by quoting the **Law of Cosines** (Figure 18), which is a generalization of the Pythagorean Theorem (see Exercise 60).

FIGURE 18

> **THEOREM 1 Law of Cosines** If a triangle has sides a, b, and c, and θ is the angle opposite side c, then
>
> $$c^2 = a^2 + b^2 - 2ab \cos \theta$$

If $\theta = \pi/2$, then $\cos \theta = 0$ and the Law of Cosines reduces to the Pythagorean Theorem.

1.4 SUMMARY

- An angle of θ radians subtends an arc of length θr on a circle of radius r.
- To convert from radians to degrees, multiply by $180/\pi$.
- To convert from degrees to radians, multiply by $\pi/180$.
- Unless otherwise stated, all angles in this text are given in radians.
- The functions $f(x) = \cos \theta$ and $f(x) = \sin \theta$ are defined in terms of right triangles for acute angles and as coordinates of a point on the unit circle for general angles (Figure 19):

$$\sin \theta = \frac{b}{c} = \frac{\text{opposite}}{\text{hypotenuse}}, \qquad \cos \theta = \frac{a}{c} = \frac{\text{adjacent}}{\text{hypotenuse}}$$

FIGURE 19

- Basic properties of sine and cosine:

Periodicity: $\quad \sin(\theta + 2\pi) = \sin \theta, \quad \cos(\theta + 2\pi) = \cos \theta$

Parity: $\quad \sin(-\theta) = -\sin \theta, \quad \cos(-\theta) = \cos \theta$

Basic identity: $\quad \sin^2 \theta + \cos^2 \theta = 1$

- The four additional trigonometric functions:

$$\tan \theta = \frac{\sin \theta}{\cos \theta}, \qquad \cot \theta = \frac{\cos \theta}{\sin \theta}, \qquad \sec \theta = \frac{1}{\cos \theta}, \qquad \csc \theta = \frac{1}{\sin \theta}$$

1.4 EXERCISES

Preliminary Questions

1. How is it possible for two different rotations to define the same angle?

2. Give two different positive rotations that define the angle $\pi/4$.

3. Give a negative rotation that defines the angle $\pi/3$.

4. The definition of $\cos \theta$ using right triangles applies when (choose the correct answer):

(a) $0 < \theta < \dfrac{\pi}{2}$ (b) $0 < \theta < \pi$ (c) $0 < \theta < 2\pi$

5. What is the unit circle definition of $\sin \theta$?

6. How does the periodicity of $f(x) = \sin \theta$ and $f(x) = \cos \theta$ follow from the unit circle definition?

Exercises

1. Find the angle between 0 and 2π equivalent to $13\pi/4$.

2. Describe $\theta = \pi/6$ by an angle of negative radian measure.

3. Convert from radians to degrees:

(a) 1   (b) $\dfrac{\pi}{3}$ (c) $\dfrac{5}{12}$ (d) $-\dfrac{3\pi}{4}$

4. Convert from degrees to radians:

(a) $1°$ (b) $30°$ (c) $25°$ (d) $120°$

5. Find the lengths of the arcs subtended by the angles θ and ϕ radians in Figure 20.

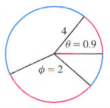

FIGURE 20 Circle of radius 4.

6. Calculate the values of the six standard trigonometric functions for the angle θ in Figure 21.

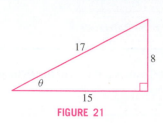

FIGURE 21

7. Fill in the remaining values of $(\cos\theta, \sin\theta)$ for the points in Figure 22.

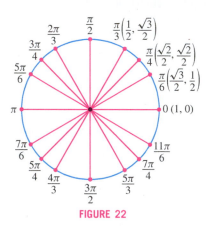

FIGURE 22

8. Find the values of the six standard trigonometric functions at $\theta = 11\pi/6$.

In Exercises 9 14, use Figure 22 to ₋nd all angles between 0 and 2π satisfying the given condition.

9. $\cos\theta = \dfrac{1}{2}$

10. $\tan\theta = 1$

11. $\tan\theta = -1$

12. $\csc\theta = 2$

13. $\sin x = \dfrac{\sqrt{3}}{2}$

14. $\sec t = 2$

15. Fill in the following table of values:

θ	$\dfrac{\pi}{6}$	$\dfrac{\pi}{4}$	$\dfrac{\pi}{3}$	$\dfrac{\pi}{2}$	$\dfrac{2\pi}{3}$	$\dfrac{3\pi}{4}$	$\dfrac{5\pi}{6}$
$\tan\theta$							
$\sec\theta$							

16. Complete the following table of signs:

θ	$\sin\theta$	$\cos\theta$	$\tan\theta$	$\cot\theta$	$\sec\theta$	$\csc\theta$
$0 < \theta < \dfrac{\pi}{2}$	+	+				
$\dfrac{\pi}{2} < \theta < \pi$						
$\pi < \theta < \dfrac{3\pi}{2}$						
$\dfrac{3\pi}{2} < \theta < 2\pi$						

17. Show that if $\tan\theta = c$ and $0 \le \theta < \pi/2$, then $\cos\theta = 1/\sqrt{1+c^2}$. *Hint:* Draw a right triangle whose opposite and adjacent sides have lengths c and 1.

18. Suppose that $\cos\theta = \frac{1}{3}$.

(a) Show that if $0 \le \theta < \pi/2$, then $\sin\theta = 2\sqrt{2}/3$ and $\tan\theta = 2\sqrt{2}$.

(b) Find $\sin\theta$ and $\tan\theta$ if $3\pi/2 \le \theta < 2\pi$.

In Exercises 19 24, assume that $0 \le \theta < \pi/2$.

19. Find $\sin\theta$ and $\tan\theta$ if $\cos\theta = \frac{5}{13}$.

20. Find $\cos\theta$ and $\tan\theta$ if $\sin\theta = \frac{3}{5}$.

21. Find $\sin\theta$, $\sec\theta$, and $\cot\theta$ if $\tan\theta = \frac{2}{7}$.

22. Find $\sin\theta$, $\cos\theta$, and $\sec\theta$ if $\cot\theta = 4$.

23. Find $\cos 2\theta$ if $\sin\theta = \frac{1}{5}$.

24. Find $\sin 2\theta$ and $\cos 2\theta$ if $\tan\theta = \sqrt{2}$.

25. Find $\cos\theta$ and $\tan\theta$ if $\sin\theta = 0.4$ and $\pi/2 \le \theta < \pi$.

26. Find $\cos\theta$ and $\sin\theta$ if $\tan\theta = 4$ and $\pi \le \theta < 3\pi/2$.

27. Find $\cos\theta$ if $\cot\theta = \frac{4}{3}$ and $\sin\theta < 0$.

28. Find $\tan\theta$ if $\sec\theta = \sqrt{5}$ and $\sin\theta < 0$.

29. Find the values of $\sin\theta$, $\cos\theta$, and $\tan\theta$ for the angles corresponding to the eight points on the unit circles in Figure 23(A) and (B).

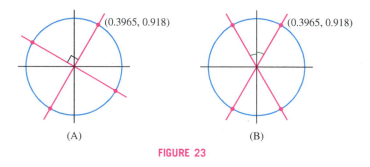

(A) (0.3965, 0.918)   **(B)** (0.3965, 0.918)

FIGURE 23

30. Refer to Figure 24(A). Express the functions $\sin\theta$, $\tan\theta$, and $\csc\theta$ in terms of c.

31. Refer to Figure 24(B). Compute $\cos\psi$, $\sin\psi$, $\cot\psi$, and $\csc\psi$.

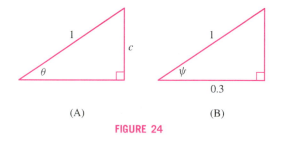

(A)   **(B)**

FIGURE 24

32. Express $\cos\left(\theta + \frac{\pi}{2}\right)$ and $\sin\left(\theta + \frac{\pi}{2}\right)$ in terms of $\cos\theta$ and $\sin\theta$. *Hint:* Find the relation between the coordinates (a, b) and (c, d) in Figure 25.

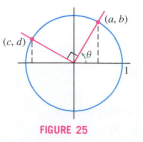

FIGURE 25

33. Use the addition formula to compute $\cos\left(\frac{\pi}{3} + \frac{\pi}{4}\right)$ exactly.

34. Use the addition formula to compute $\sin\left(\frac{\pi}{3} - \frac{\pi}{4}\right)$ exactly.

In Exercises 35–38, sketch the graph over $[0, 2\pi]$.

35. $f(\theta) = 2\sin 4\theta$

36. $f(\theta) = \cos\left(2\left(\theta - \frac{\pi}{2}\right)\right)$

37. $f(\theta) = \cos\left(2\theta - \frac{\pi}{2}\right)$

38. $f(\theta) = \sin\left(2\left(\theta - \frac{\pi}{2}\right) + \pi\right) + 2$

39. Determine a function that would have a graph as in Figure 26(A), stating the period and amplitude.

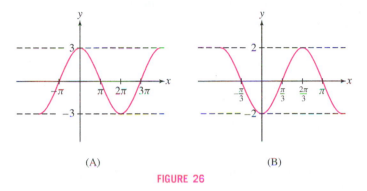

(A)   (B)

FIGURE 26

40. Determine a function that would have a graph as in Figure 26(B), stating the period and amplitude.

41. How many points lie on the intersection of the horizontal line $y = c$ and the graph of $y = \sin x$ for $0 \le x < 2\pi$? *Hint: The answer depends on c.*

42. How many points lie on the intersection of the horizontal line $y = c$ and the graph of $y = \tan x$ for $0 \le x < 2\pi$?

In Exercises 43–46, solve for $0 \le \theta < 2\pi$ (see Example 4).

43. $\sin 2\theta + \sin 3\theta = 0$

44. $\sin\theta = \sin 2\theta$

45. $\cos 4\theta + \cos 2\theta = 0$

46. $\sin\theta = \cos 2\theta$

In Exercises 47–56, derive the identity using the identities listed in this section.

47. $\cos 2\theta = 2\cos^2\theta - 1$

48. $\cos^2\frac{\theta}{2} = \frac{1 + \cos\theta}{2}$

49. $\sin\frac{\theta}{2} = \sqrt{\frac{1 - \cos\theta}{2}}$

50. $\sin(\theta + \pi) = -\sin\theta$

51. $\cos(\theta + \pi) = -\cos\theta$

52. $\tan x = \cot\left(\frac{\pi}{2} - x\right)$

53. $\tan(\pi - \theta) = -\tan\theta$

54. $\tan 2x = \frac{2\tan x}{1 - \tan^2 x}$

55. $\tan x = \frac{\sin 2x}{1 + \cos 2x}$

56. $\sin^2 x\cos^2 x = \frac{1 - \cos 4x}{8}$

57. Use Exercises 50 and 51 to show that $\tan\theta$ and $\cot\theta$ are periodic with period π.

58. Use the identity of Exercise 48 to show that $\cos\frac{\pi}{8}$ is equal to
$$\sqrt{\frac{1}{2} + \frac{\sqrt{2}}{4}}.$$

59. Use the Law of Cosines to find the distance from P to Q in Figure 27.

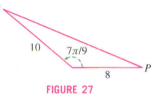

FIGURE 27

Further Insights and Challenges

60. Use Figure 28 to derive the Law of Cosines from the Pythagorean Theorem.

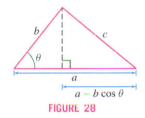

FIGURE 28

61. Use the addition formula to prove
$$\cos 3\theta = 4\cos^3\theta - 3\cos\theta$$

62. Use the addition formulas for sine and cosine to prove
$$\tan(a + b) = \frac{\tan a + \tan b}{1 - \tan a\tan b}$$
$$\cot(a - b) = \frac{\cot a\cot b + 1}{\cot b - \cot a}$$

63. Let θ be the angle between the line $y = mx + b$ and the x-axis [Figure 29(A)]. Prove that $m = \tan\theta$.

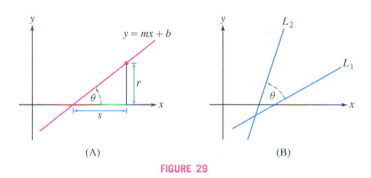

(A)   (B)

FIGURE 29

64. Let L_1 and L_2 be the lines of slope m_1 and m_2 [Figure 29(B)]. Show that the angle θ between L_1 and L_2 satisfies $\cot\theta = \frac{m_2 m_1 + 1}{m_2 - m_1}$.

65. Perpendicular Lines Use Exercise 64 to prove that two lines with nonzero slopes m_1 and m_2 are perpendicular if and only if $m_2 = -1/m_1$.

66. Apply the double-angle formula to prove:

(a) $\cos\frac{\pi}{8} = \frac{1}{2}\sqrt{2 + \sqrt{2}}$

(b) $\cos\frac{\pi}{16} = \frac{1}{2}\sqrt{2 + \sqrt{2 + \sqrt{2}}}$

Guess the values of $\cos\frac{\pi}{32}$ and of $\cos\frac{\pi}{2^n}$ for all n.

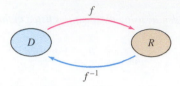

FIGURE 1 A function and its inverse.

In general, $f^{-1}(x) \neq \frac{1}{f(x)}$. The expression $f^{-1}(x)$ is simply a notation for the inverse function, and the '−1' does not represent an exponent.

1.5 Inverse Functions

Many important functions, such as logarithms, roots, and the arcsine, are defined as inverse functions. In this section, we review inverse functions and their graphs, and we discuss the inverse trigonometric functions.

The inverse of f, denoted f^{-1}, is the function that *reverses* the effect of f (Figure 1). For example, the inverse of $f(x) = x^3$ is the cube root function $f^{-1}(x) = x^{1/3}$. Given a table of function values for f, we obtain a table for f^{-1} by interchanging the x and y columns, assuming the resulting f^{-1} is a function:

Function				Inverse	
x	$f(x) = x^3$			x	$f^{-1}(x) = x^{1/3}$
−2	−8	(Interchange columns)		−8	−2
−1	−1	$\Longrightarrow$		−1	−1
0	0			0	0
1	1			1	1
2	8			8	2
3	27			27	3

If we apply both f and f^{-1} to a number x in either order, we get back x. For instance,

Apply f and then f^{-1}: $2 \xrightarrow{\text{(Apply } x^3)} 8 \xrightarrow{\text{(Apply } x^{1/3})} 2$

Apply f^{-1} and then f: $8 \xrightarrow{\text{(Apply } x^{1/3})} 2 \xrightarrow{\text{(Apply } x^3)} 8$

This property is used in the formal definition of the inverse function.

◄··· *REMINDER The domain is the set of numbers x such that $f(x)$ is defined (the set of allowable inputs), and the range is the set of all values $f(x)$ (the set of outputs).*

> **DEFINITION Inverse**   Let f have domain D and range R. If there is a function g with domain R such that
>
> $$g\big(f(x)\big) = x \quad \text{for } x \in D \qquad \text{and} \qquad f\big(g(x)\big) = x \quad \text{for } x \in R$$
>
> then f is said to be **invertible**. The function g is called the **inverse function** and is denoted f^{-1}.

■ **EXAMPLE 1**   Show that $f(x) = 2x - 18$ is invertible. What are the domain and range of f^{-1}?

Solution   We show that f is invertible by computing the inverse function in two steps.

Step 1. **Solve the equation $y = f(x)$ for x in terms of y.**

$$y = 2x - 18$$
$$y + 18 = 2x$$
$$x = \frac{1}{2}y + 9$$

This gives us the inverse as a function of the variable y: $f^{-1}(y) = \frac{1}{2}y + 9$.

Step 2. **Interchange variables.**

We usually prefer to write the inverse as a function of x, so we interchange the roles of x and y (Figure 2):

$$f^{-1}(x) = \frac{1}{2}x + 9$$

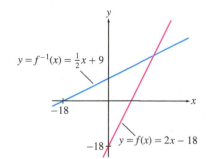

FIGURE 2

To check our calculation, let's verify that $f^{-1}(f(x)) = x$ and $f(f^{-1}(x)) = x$:

$$f^{-1}(f(x)) = f^{-1}(2x - 18) = \frac{1}{2}(2x - 18) + 9 = (x - 9) + 9 = x$$

$$f(f^{-1}(x)) = f\left(\frac{1}{2}x + 9\right) = 2\left(\frac{1}{2}x + 9\right) - 18 = (x + 18) - 18 = x$$

Because f^{-1} is a linear function, its domain and range are **R**. ∎

The inverse function, if it exists, is unique. However, some functions do not have an inverse. Consider $f(x) = x^2$. When we interchange the columns in a table of values (which should give us a table of values for f^{-1}), the resulting table does not define a function:

Function			Inverse (?)		
x	$f(x) = x^2$		x	$f^{-1}(x)$	
-2	4	(Interchange columns) $\Longrightarrow$	4	-2	$f^{-1}(1)$ has two values: 1 and -1.
-1	1		1	-1	
0	0		0	0	
1	1		1	1	
2	4		4	2	

The problem is that every positive number occurs twice as an output of $f(x) = x^2$. For example, 1 occurs twice as an *output* in the first table and therefore occurs twice as an *input* in the second table. So the second table gives us two possible values for $f^{-1}(1)$, namely $f^{-1}(1) = 1$ and $f^{-1}(1) = -1$. Neither value satisfies the inverse property. For instance, if we set $f^{-1}(1) = 1$, then $f^{-1}(f(-1)) = f^{-1}(1) = 1$, but an inverse would have to satisfy $f^{-1}(f(-1)) = -1$.

So when does a function f have an inverse? The answer is: If f is **one-to-one**, which means that f takes on each value at most once (Figure 3). Here is the formal definition:

Another standard term for one-to-one is injective.

DEFINITION **One-to-One Function** A function f is one-to-one on a domain D if, for every value c, the equation $f(x) = c$ has at most one solution for $x \in D$. Or, equivalently, if for all $a, b \in D$,

$$f(a) \neq f(b) \qquad \text{unless} \qquad a = b$$

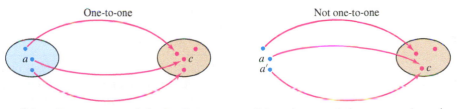

One-to-one   Not one-to-one

$f(x) = c$ has at most one solution for all c $f(x) = c$ has two solutions: $x = a$ and $x = a'$

FIGURE 3 A one-to-one function takes on each value at most once.

Think of a function as a device for labeling members of the range by members of the domain. When f is one-to-one, this labeling is unique and f^{-1} maps each number in the range back to its label.

When f is one-to-one on its domain D, the inverse function f^{-1} exists and its domain is equal to the range R of f (Figure 4). Indeed, for every $c \in R$, there is precisely one element $a \in D$ such that $f(a) = c$ and we may define $f^{-1}(c) = a$. With this definition, $f(f^{-1}(c)) = f(a) = c$ and $f^{-1}(f(a)) = f^{-1}(c) = a$. This proves the following theorem.

THEOREM 1 **Existence of Inverses** The inverse function f^{-1} exists if and only if f is one-to-one on its domain D. Furthermore,

- Domain of f = range of f^{-1}.
- Range of f = domain of f^{-1}.

Domain of f = range of f^{-1} Range of f = domain of f^{-1}

FIGURE 4 In passing from f to f^{-1}, the domain and range are interchanged.

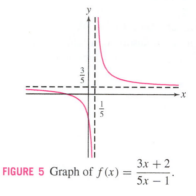

FIGURE 5 Graph of $f(x) = \dfrac{3x+2}{5x-1}$.

Often, it is impossible to find a formula for the inverse because we cannot solve for x explicitly in the equation $y = f(x)$. For example, the function $f(x) = x + e^x$ has an inverse, but we must make do without an explicit formula for it.

■ **EXAMPLE 2** Show that $f(x) = \dfrac{3x+2}{5x-1}$ is invertible. Determine the domain and range of f and f^{-1}.

Solution The domain of f is $D = \left\{ x : x \neq \frac{1}{5} \right\}$ (Figure 5). Assume that $x \in D$, and let's solve $y = f(x)$ for x in terms of y:

$$y = \frac{3x+2}{5x-1}$$

$$y(5x-1) = 3x+2$$

$$5xy - y = 3x+2$$

$$5xy - 3x = y+2 \qquad \text{(gather terms involving } x\text{)}$$

$$x(5y-3) = y+2 \qquad \text{(factor out } x \text{ in order to solve for } x\text{)} \qquad \boxed{1}$$

$$x = \frac{y+2}{5y-3} \qquad \text{(divide by } 5y-3\text{)} \qquad \boxed{2}$$

The last step is valid if $5y - 3 \neq 0$ that is, if $y \neq \frac{3}{5}$. But note that $y = \frac{3}{5}$ is not in the range of f. For if it were, Eq. (1) would yield the false equation $0 = \frac{3}{5} + 2$. Now Eq. (2) shows that for all $y \neq \frac{3}{5}$ there is a unique value x such that $f(x) = y$. Therefore, f is one-to-one on its domain. By Theorem 1, f is invertible. The range of f is $R = \left\{ x : x \neq \frac{3}{5} \right\}$ and

$$f^{-1}(x) = \frac{x+2}{5x-3}$$

The inverse function has domain R and range D.   ■

We can tell whether f is one-to-one from its graph. The horizontal line $y = c$ intersects the graph of f at points $(a, f(a))$, where $f(a) = c$ (Figure 6). There is at most one such point if $f(x) = c$ has at most one solution. This gives us the

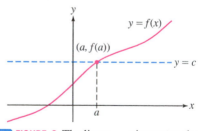

DF **FIGURE 6** The line $y = c$ intersects the graph at points where $f(a) = c$.

Horizontal Line Test A function f is one-to-one if and only if every horizontal line intersects the graph of f in at most one point.

In Figure 7, we see that $f(x) = x^3$ passes the Horizontal Line Test and therefore is one-to-one, whereas $f(x) = x^2$ fails the test and is not one-to-one.

■ **EXAMPLE 3** **Increasing Functions Are One-to-One** Show that increasing functions are one-to-one. Then show that $f(x) = x^5 + 4x + 3$ is one-to-one.

Solution An increasing function satisfies $f(a) < f(b)$ if $a < b$. Therefore, f cannot take on any value more than once, and thus f is one-to-one. Note similarly that decreasing functions are one-to-one.

Now observe that

- If n is odd and $c > 0$, then cx^n is increasing.
- A sum of increasing functions is increasing.

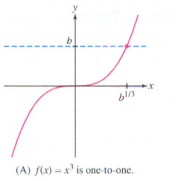
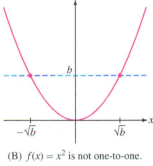

(A) $f(x) = x^3$ is one-to-one.　　(B) $f(x) = x^2$ is not one-to-one.

FIGURE 7

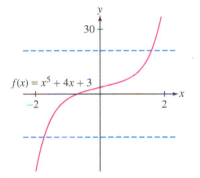

FIGURE 8 The increasing function $f(x) = x^5 + 4x + 3$ satisfies the Horizontal Line Test.

One-to-one for $x \geq 0$

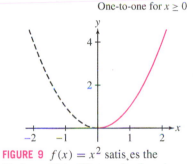

FIGURE 9 $f(x) = x^2$ satisfies the Horizontal Line Test on the domain $\{x : x \geq 0\}$.

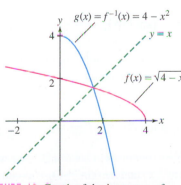

FIGURE 12 Graph of the inverse g of $f(x) = \sqrt{4 - x}$.

Thus, $x^5, 4x$, and hence the sum $x^5 + 4x$ are increasing. It follows that the function $f(x) = x^5 + 4x + 3$ is increasing and therefore one-to-one (Figure 8). However, determining an explicit formula for its inverse would be difficult. ∎

We can make a function one-to-one by restricting its domain suitably.

■ **EXAMPLE 4** **Restricting the Domain** Find a domain on which $f(x) = x^2$ is one-to-one and determine its inverse on this domain.

Solution The function $f(x) = x^2$ is one-to-one on the domain $D = \{x : x \geq 0\}$, for if $a^2 = b^2$ where a and b are both nonnegative, then $a = b$ (Figure 9). The inverse of f on D is the positive square root $f^{-1}(x) = \sqrt{x}$. Alternatively, we may restrict f to the domain $\{x : x \leq 0\}$, on which the inverse function is $f^{-1}(x) = -\sqrt{x}$. ∎

Next we describe the graph of the inverse function. The **reflection** of a point (a, b) through the line $y = x$ is defined to be the point (b, a) (Figure 10). Note that if the x- and y-axes are drawn to the same scale, then (a, b) and (b, a) are equidistant from the line $y = x$ and the segment joining them is perpendicular to $y = x$.

The graph of f^{-1} is the reflection of the graph of f through $y = x$ (Figure 11). To check this, note that (a, b) lies on the graph of f if $f(a) = b$. But $f(a) = b$ if and only if $f^{-1}(b) = a$, and in this case, (b, a) lies on the graph of f^{-1}.

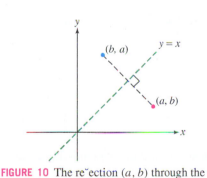

FIGURE 10 The reflection (a, b) through the line $y = x$ is the point (b, a).

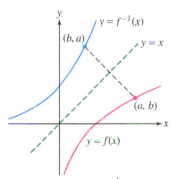

FIGURE 11 The graph of f^{-1} is the reflection of the graph of f through the line $y = x$.

■ **EXAMPLE 5** **Sketching the Graph of the Inverse** Sketch the graph of the inverse of $f(x) = \sqrt{4 - x}$.

Solution Let $g(x) = f^{-1}(x)$. Observe that f has domain $\{x : x \leq 4\}$ and range $\{x : x \geq 0\}$. We do not need a formula for $g(x)$ to draw its graph. We simply reflect the graph of f through the line $y = x$ as in Figure 12. If desired, however, we can easily solve $y = \sqrt{4 - x}$ to obtain $x = 4 - y^2$ and thus $g(x) = 4 - x^2$ with domain $\{x : x \geq 0\}$. ∎

Inverse Trigonometric Functions

We have seen that the inverse function f^{-1} exists if and only if f is one-to-one on its domain. Because the trigonometric functions are not one-to-one, we must restrict their domains to define their inverses.

First consider the sine function. Figure 13 shows that $f(\theta) = \sin\theta$ is one-to-one on $\left[-\frac{\pi}{2}, \frac{\pi}{2}\right]$. With this interval as domain, the inverse is called the **arcsine function** and is denoted $\theta = \sin^{-1} x$ or $\theta = \arcsin x$. By definition,

$$\theta = \sin^{-1} x \text{ is the unique angle in } \left[-\frac{\pi}{2}, \frac{\pi}{2}\right] \text{ such that } \sin\theta = x$$

Do not confuse the inverse $\sin^{-1} x$ with the reciprocal

$$(\sin x)^{-1} = \frac{1}{\sin x} = \csc x$$

The inverse functions $\sin^{-1} x$, $\cos^{-1} x$, ... are often denoted $\arcsin x$, $\arccos x$, etc.

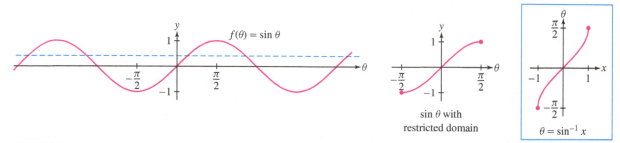

FIGURE 13

sin θ with restricted domain

$\theta = \sin^{-1} x$

The range of $f(x) = \sin x$ is $[-1, 1]$, so $f^{-1}(x) = \sin^{-1} x$ has domain $[-1, 1]$. A table of values for the arcsine (Table 1) is obtained by reversing the columns in a table of values for $\sin x$.

Summary of inverse relation between the sine and arcsine functions:

$\sin(\sin^{-1} x) = x$ for $-1 \le x \le 1$

$\sin^{-1}(\sin\theta) = \theta$ for $-\frac{\pi}{2} \le \theta \le \frac{\pi}{2}$

TABLE 1

x	-1	$-\frac{\sqrt{3}}{2}$	$-\frac{\sqrt{2}}{2}$	$-\frac{1}{2}$	0	$\frac{1}{2}$	$\frac{\sqrt{2}}{2}$	$\frac{\sqrt{3}}{2}$	1
$\theta = \sin^{-1} x$	$-\frac{\pi}{2}$	$-\frac{\pi}{3}$	$-\frac{\pi}{4}$	$-\frac{\pi}{6}$	0	$\frac{\pi}{6}$	$\frac{\pi}{4}$	$\frac{\pi}{3}$	$\frac{\pi}{2}$

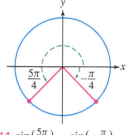

FIGURE 14 $\sin\left(\frac{5\pi}{4}\right) = \sin\left(-\frac{\pi}{4}\right)$.

■ **EXAMPLE 6** (a) Show that $\sin^{-1}\left(\sin\left(\frac{\pi}{4}\right)\right) = \frac{\pi}{4}$.

(b) Explain why $\sin^{-1}\left(\sin\left(\frac{5\pi}{4}\right)\right) \ne \frac{5\pi}{4}$.

Solution The equation $\sin^{-1}(\sin\theta) = \theta$ is valid only if θ lies in $\left[-\frac{\pi}{2}, \frac{\pi}{2}\right]$.

(a) Because $\frac{\pi}{4}$ lies in the required interval, $\sin^{-1}\left(\sin\left(\frac{\pi}{4}\right)\right) = \frac{\pi}{4}$.

(b) Let $\theta = \sin^{-1}\left(\sin\left(\frac{5\pi}{4}\right)\right)$. By definition, θ is the angle in $\left[-\frac{\pi}{2}, \frac{\pi}{2}\right]$ such that $\sin\theta = \sin\left(\frac{5\pi}{4}\right)$. By the identity $\sin\theta = \sin(\pi - \theta)$ (Figure 14),

$$\sin\left(\frac{5\pi}{4}\right) = \sin\left(\pi - \frac{5\pi}{4}\right) = \sin\left(-\frac{\pi}{4}\right)$$

The angle $-\frac{\pi}{4}$ lies in the required interval, so $\theta = \sin^{-1}\left(\sin\left(\frac{5\pi}{4}\right)\right) = -\frac{\pi}{4}$. ■

The cosine function is one-to-one on $[0, \pi]$ rather than $\left[-\frac{\pi}{2}, \frac{\pi}{2}\right]$ (Figure 15). With this domain, the inverse is called the **arccosine function** and is denoted $\theta = \cos^{-1} x$ or $\theta = \arccos x$. It has domain $[-1, 1]$. By definition,

$$\theta = \cos^{-1} x \text{ is the unique angle in } [0, \pi] \text{ such that } \cos\theta = x$$

Summary of inverse relation between the cosine and arccosine:

$\cos(\cos^{-1} x) = x$ for $-1 \le x \le 1$

$\cos^{-1}(\cos\theta) = \theta$ for $0 \le \theta \le \pi$

When we study the calculus of inverse trigonometric functions in Section 3.8, we will need to simplify composite expressions such as $\cos(\sin^{-1} x)$ and $\tan(\sin^{-1} x)$. This can be done in two ways: by referring to the appropriate right triangle or by using trigonometric identities.

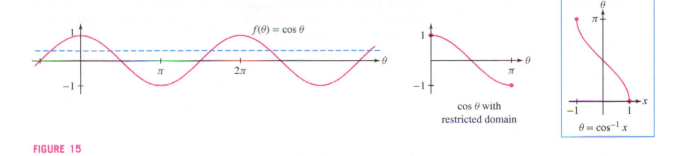

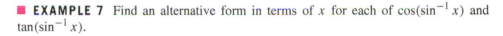

cos θ with
restricted domain

$\theta = \cos^{-1} x$

FIGURE 15

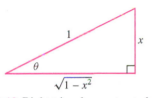

FIGURE 16 Right triangle constructed such that $\sin \theta = x$.

■ **EXAMPLE 7** Find an alternative form in terms of x for each of $\cos(\sin^{-1} x)$ and $\tan(\sin^{-1} x)$.

Solution This problem asks for the values of $\cos \theta$ and $\tan \theta$ at the angle $\theta = \sin^{-1} x$. Consider a right triangle with hypotenuse of length 1 and angle θ such that $\sin \theta = x$, as in Figure 16. By the Pythagorean Theorem, the adjacent side has length $\sqrt{1 - x^2}$. Now we can read off the values from Figure 16:

$$\cos(\sin^{-1} x) = \cos \theta = \frac{\text{adjacent}}{\text{hypotenuse}} = \sqrt{1 - x^2}$$

$$\tan(\sin^{-1} x) = \tan \theta = \frac{\text{opposite}}{\text{adjacent}} = \frac{x}{\sqrt{1 - x^2}}$$

Alternatively, we may argue using trigonometric identities. Because $\sin \theta = x$,

$$\cos(\sin^{-1} x) = \cos \theta = \sqrt{1 - \sin^2 \theta} = \sqrt{1 - x^2}$$

We are justified in taking the positive square root in either approach because $\theta = \sin^{-1} x$ lies in $\left[-\frac{\pi}{2}, \frac{\pi}{2}\right]$ and $\cos \theta$ is positive in this interval.   ■

We now address the remaining trigonometric functions. The function $f(\theta) = \tan \theta$ is one-to-one on $\left(-\frac{\pi}{2}, \frac{\pi}{2}\right)$, and $f(\theta) = \cot \theta$ is one-to-one on $(0, \pi)$ (see Figure 10 in Section 1.4). We define their inverses by restricting them to these domains:

> $\theta = \tan^{-1} x$ is the unique angle in $\left(-\dfrac{\pi}{2}, \dfrac{\pi}{2}\right)$ such that $\tan \theta = x$

> $\theta = \cot^{-1} x$ is the unique angle in $(0, \pi)$ such that $\cot \theta = x$

The range of both $f(\theta) = \tan \theta$ and $f(\theta) = \cot \theta$ is the set of all real numbers **R**. Therefore, $\theta = \tan^{-1} x$ and $\theta = \cot^{-1} x$ have domain **R** (Figure 17).

The function $f(\theta) = \sec \theta$ is not defined at $\theta = \frac{\pi}{2}$, but we see in Figure 18 that it is one-to-one on both $\left[0, \frac{\pi}{2}\right)$ and $\left(\frac{\pi}{2}, \pi\right]$. Similarly, $f(\theta) = \csc \theta$ is not defined at $\theta = 0$, but it is one-to-one on $\left[-\frac{\pi}{2}, 0\right)$ and $\left(0, \frac{\pi}{2}\right]$. We define the inverse functions as follows:

> $\theta = \sec^{-1} x$ is the unique angle in $\left[0, \dfrac{\pi}{2}\right) \cup \left(\dfrac{\pi}{2}, \pi\right]$ such that $\sec \theta = x$

> $\theta = \csc^{-1} x$ is the unique angle in $\left[-\dfrac{\pi}{2}, 0\right) \cup \left(0, \dfrac{\pi}{2}\right]$ such that $\csc \theta = x$

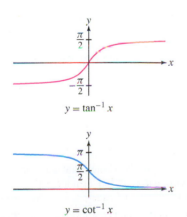

$y = \tan^{-1} x$

$y = \cot^{-1} x$

FIGURE 17

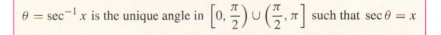

Figure 18 shows that the range of $f(\theta) = \sec \theta$ is the set of real numbers x such that $|x| \geq 1$. The same is true of $f(\theta) = \csc \theta$. It follows that both $\theta = \sec^{-1} x$ and $\theta = \csc^{-1} x$ have domain $\{x : |x| \geq 1\}$.

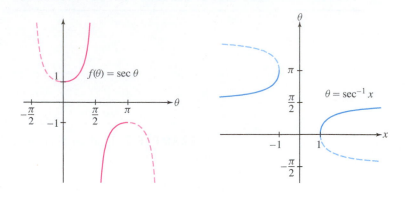

FIGURE 18 $f(\theta) = \sec \theta$ is one-to-one on the interval $[0, \pi]$ with $\frac{\pi}{2}$ removed.

1.5 SUMMARY

- A function f is *one-to-one* on a domain D if for every value c, the equation $f(x) = c$ has at most one solution for $x \in D$, or, equivalently, if for all $a, b \in D$, $f(a) \neq f(b)$ unless $a = b$.
- Let f have domain D and range R. The *inverse* f^{-1} (if it exists) is the unique function with domain R and range D satisfying $f(f^{-1}(x)) = x$ and $f^{-1}(f(x)) = x$.
- The inverse of f exists if and only if f is one-to-one on its domain.
- To find the inverse function, solve $y = f(x)$ for x in terms of y to obtain $x = g(y)$. The inverse is the function g.
- *Horizontal Line Test:* f is one-to-one if and only if every horizontal line intersects the graph of f in at most one point.
- The graph of f^{-1} is obtained by reflecting the graph of f through the line $y = x$.
- The *arcsine* and *arccosine* are defined for $-1 \leq x \leq 1$:

$$\theta = \sin^{-1} x \text{ is the unique angle in } \left[-\frac{\pi}{2}, \frac{\pi}{2}\right] \text{ such that } \sin \theta = x$$

$$\theta = \cos^{-1} x \text{ is the unique angle in } [0, \pi] \text{ such that } \cos \theta = x$$

- $\tan^{-1} x$ and $\cot^{-1} x$ are defined for all x:

$$\theta = \tan^{-1} x \text{ is the unique angle in } \left(-\frac{\pi}{2}, \frac{\pi}{2}\right) \text{ such that } \tan \theta = x$$

$$\theta = \cot^{-1} x \text{ is the unique angle in } (0, \pi) \text{ such that } \cot \theta = x$$

- $\sec^{-1} x$ and $\csc^{-1} x$ are defined for $|x| \geq 1$:

$$\theta = \sec^{-1} x \text{ is the unique angle in } \left[0, \frac{\pi}{2}\right) \cup \left(\frac{\pi}{2}, \pi\right] \text{ such that } \sec \theta = x$$

$$\theta = \csc^{-1} x \text{ is the unique angle in } \left[-\frac{\pi}{2}, 0\right) \cup \left(0, \frac{\pi}{2}\right] \text{ such that } \csc \theta = x$$

1.5 EXERCISES

Preliminary Questions

1. Which of the following satisfy $f^{-1}(x) = f(x)$?

(a) $f(x) = x$

(b) $f(x) = 1 - x$

(c) $f(x) = 1$

(d) $f(x) = \sqrt{x}$

(e) $f(x) = |x|$

(f) $f(x) = x^{-1}$

2. The function f maps teenagers in the United States to their last names. Explain why the inverse function f^{-1} does not exist.

3. The following fragment of a train schedule for the New Jersey Transit System defines a function f from towns to times. Is f one-to-one?

What is $f^{-1}(6{:}27)$?

Trenton	6:21
Hamilton Township	6:27
Princeton Junction	6:34
New Brunswick	6:38

4. A homework problem asks for a sketch of the graph of the *inverse* of $f(x) = x + \cos x$. Frank, after trying but failing to find a formula for $f^{-1}(x)$, says it's impossible to graph the inverse. Bianca hands in an accurate sketch without solving for f^{-1}. How did Bianca complete the problem?

5. Which of the following quantities is undefined?

(a) $\sin^{-1}\left(-\frac{1}{2}\right)$ **(b)** $\cos^{-1}(2)$

(c) $\csc^{-1}\left(\frac{1}{2}\right)$ **(d)** $\csc^{-1}(2)$

6. Give an example of an angle θ such that $\cos^{-1}(\cos\theta) \neq \theta$. Does this contradict the definition of inverse function?

Exercises

1. Show that $f(x) = 7x - 4$ is invertible and find its inverse.

2. Is $f(x) = x^2 + 2$ one-to-one? If not, describe a domain on which it is one-to-one.

3. What is the largest interval containing zero on which $f(x) = \sin x$ is one-to-one?

4. Show that $f(x) = \dfrac{x-2}{x+3}$ is invertible and find its inverse.

(a) What is the domain of f? The range of f^{-1}?
(b) What is the domain of f^{-1}? The range of f?

5. Verify that $f(x) = x^3 + 3$ and $g(x) = (x-3)^{1/3}$ are inverses by showing that $f(g(x)) = x$ and $g(f(x)) = x$.

6. Repeat Exercise 5 for $f(t) = \dfrac{t+1}{t-1}$ and $g(t) = \dfrac{t+1}{t-1}$.

7. The escape velocity from a planet of radius R is $v(R) = \sqrt{\dfrac{2GM}{R}}$, where G is the universal gravitational constant and M is the mass. Find the inverse of $v(R)$ expressing R in terms of v.

In Exercises 8 15, find a domain on which f is one-to-one and a formula for the inverse of f restricted to this domain. Sketch the graphs of f and f^{-1}.

8. $f(x) = 3x - 2$ **9.** $f(x) = 4 - x$

10. $f(x) = \dfrac{1}{x+1}$ **11.** $f(x) = \dfrac{1}{7x-3}$

12. $f(s) = \dfrac{1}{s^2}$ **13.** $f(x) = \dfrac{1}{\sqrt{x^2+1}}$

14. $f(z) = z^3$ **15.** $f(x) = \sqrt{x^3+9}$

16. For each function shown in Figure 19, sketch the graph of the inverse (restrict the function's domain if necessary).

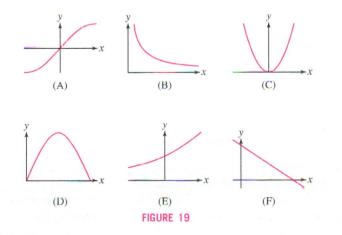

(A) (B) (C)

(D) (E) (F)

FIGURE 19

17. Which of the graphs in Figure 20 is the graph of a function satisfying $f^{-1} = f$?

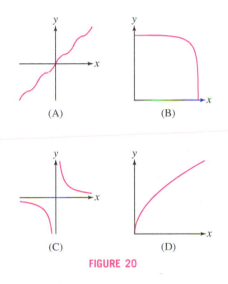

(A) (B)

(C) (D)

FIGURE 20

18. Let n be a nonzero integer. Find a domain on which $f(x) = (1 - x^n)^{1/n}$ coincides with its inverse. *Hint:* The answer depends on whether n is even or odd.

19. Let $f(x) = x^7 + x + 1$.

(a) Show that f^{-1} exists (but do not attempt to find it). *Hint:* Show that f is increasing.

(b) What is the domain of f^{-1}?

(c) Find $f^{-1}(3)$.

20. Show that $f(x) = (x^2 + 1)^{-1}$ is one-to-one on $(-\infty, 0]$, and find a formula for f^{-1} for this domain of f.

21. Let $f(x) = x^2 - 2x$. Determine a domain on which f^{-1} exists, and find a formula for f^{-1} for this domain of f.

22. Show that $f(x) = x + x^{-1}$ is one-to-one on $[1, \infty)$, and find the corresponding inverse f^{-1}. What is the domain of f^{-1}?

For each of the piecewise-defined functions in Exercises 23 26, determine whether or not the function is one-to-one, and if it is, determine its inverse function.

23.

$$f(x) = \begin{cases} x & \text{when } x < 0 \\ 2x & \text{when } x \geq 0 \end{cases}$$

24.

$$g(x) = \begin{cases} -x & \text{when } x < -1 \\ x & \text{when } x \geq -1 \end{cases}$$

25.

$$f(x) = \begin{cases} x^2 & \text{when } x < 0 \\ x & \text{when } x \geq 0 \end{cases}$$

26.

$$g(x) = \begin{cases} x & \text{when } x < 0 \\ x^2 & \text{when } x \geq 0 \end{cases}$$

In Exercises 27 32, evaluate without using a calculator.

27. $\cos^{-1} 1$

28. $\sin^{-1} \dfrac{1}{2}$

29. $\cot^{-1} 1$

30. $\sec^{-1} \dfrac{2}{\sqrt{3}}$

31. $\tan^{-1} \sqrt{3}$

32. $\sin^{-1}(-1)$

In Exercises 33 42, compute without using a calculator.

33. $\sin^{-1}\left(\sin \dfrac{\pi}{3}\right)$

34. $\sin^{-1}\left(\sin \dfrac{4\pi}{3}\right)$

35. $\cos^{-1}\left(\cos \dfrac{3\pi}{2}\right)$

36. $\sin^{-1}\left(\sin\left(-\dfrac{5\pi}{6}\right)\right)$

37. $\tan^{-1}\left(\tan \dfrac{3\pi}{4}\right)$

38. $\tan^{-1}(\tan \pi)$

39. $\sec^{-1}(\sec 3\pi)$

40. $\sec^{-1}\left(\sec \dfrac{3\pi}{2}\right)$

41. $\csc^{-1}(\csc(-\pi))$

42. $\cot^{-1}\left(\cot\left(-\dfrac{\pi}{4}\right)\right)$

In Exercises 43 46, simplify by referring to the appropriate triangle or trigonometric identity.

43. $\tan(\cos^{-1} x)$

44. $\cos(\tan^{-1} x)$

45. $\cot(\sec^{-1} x)$

46. $\cot(\sin^{-1} x)$

In Exercises 47 54, refer to the appropriate triangle or trigonometric identity to compute the given value.

47. $\cos\left(\sin^{-1} \dfrac{2}{3}\right)$

48. $\tan\left(\cos^{-1} \dfrac{2}{3}\right)$

49. $\tan(\sin^{-1} 0.8)$

50. $\cos(\cot^{-1} 1)$

51. $\cot(\csc^{-1} 2)$

52. $\tan(\sec^{-1}(-2))$

53. $\cot(\tan^{-1} 20)$

54. $\sin(\csc^{-1} 20)$

Further Insights and Challenges

55. Show that if f is odd and f^{-1} exists, then f^{-1} is odd. Show, on the other hand, that an even function does not have an inverse.

56. A cylindrical tank of radius R and length L lying horizontally as in Figure 21 is lled with oil to height h. Show that the volume $V(h)$ of oil in the tank as a function of height h is

$$V(h) = L\left(R^2 \cos^{-1}\left(1 - \frac{h}{R}\right) - (R - h)\sqrt{2hR - h^2}\right)$$

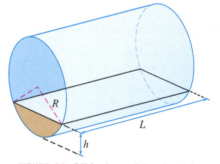

FIGURE 21 Oil in the tank has level h.

1.6 Exponential and Logarithmic Functions

An **exponential function** is a function of the form $f(x) = b^x$, where $b > 0$ and $b \neq 1$. The number b is called the **base**. Some examples are $f(x) = 2^x$, $g(x) = (1.4)^x$, and $h(x) = 10^x$. The case $b = 1$ is excluded because $f(x) = 1^x$ is a constant function. Calculators give good decimal approximations to values of exponential functions:

$$2^4 = 16, \quad 2^{-3} = 0.125, \quad (1.4)^{0.8} \approx 1.309, \quad 10^{4.6} \approx 39{,}810.717$$

Three properties of exponential functions should be singled out from the start (see Figure 1 for the case $b = 2$):

- *Exponential functions are positive*: $b^x > 0$ for all x.
- The *range* of $f(x) = b^x$ is the set of all positive real numbers.
- $f(x) = b^x$ is increasing if $b > 1$ and decreasing if $0 < b < 1$.

If $b > 1$, the exponential function $f(x) = b^x$ is not merely increasing but is, in a certain sense, rapidly increasing. Although the term rapid increase is perhaps subjective, the following precise statement is true: $f(x) = b^x$ increases more rapidly than the power function $g(x) = x^n$ for all n (we will prove this in Section 4.5). For example, Figure 2

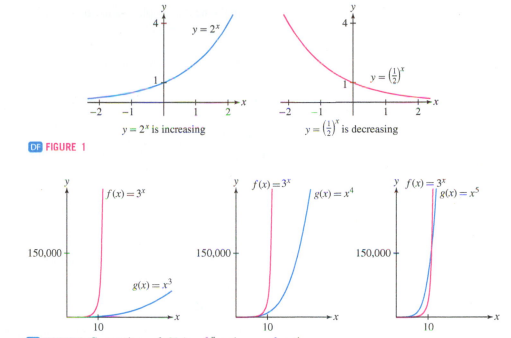

DF **FIGURE 1**

DF **FIGURE 2** Comparison of $f(x) = 3^x$ and power functions.

TABLE 1

x	x^5	3^x
1	1	3
5	3125	243
10	100,000	59,049
15	759,375	14,348,907
25	9,765,625	847,288,609,443

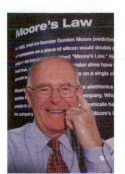

Gordon Moore (1929). Moore, who later became chairman of Intel Corporation, predicted that in the decades following 1965, the number of transistors per integrated circuit would grow exponentially. This prediction has held up for nearly ,ve decades and may well continue for several more years. Moore has said, Moore's Law is a term that got applied to a curve I plotted in the mid-sixties showing the increase in complexity of integrated circuits versus time. It's been expanded to include a lot more than that, and I'm happy to take credit for all of it. (*AP Photo/Paul Sakuma*)

Be sure you are familiar with the Laws of Exponents. They are used throughout calculus.

shows that $f(x) = 3^x$ eventually overtakes and increases faster than the power functions $g(x) = x^3$, $g(x) = x^4$, and $g(x) = x^5$. Table 1 compares $f(x) = 3^x$ and $g(x) = x^5$.

We now review the Laws of Exponents. The most important law is

$$b^x b^y = b^{x+y}$$

In other words, *under multiplication, the exponents add*, provided that the bases are the same. This law does not apply to a product such as $3^2 \cdot 5^4$.

Laws of Exponents ($b > 0$)

	Rule	Example
Exponent zero	$b^0 = 1$	
Products	$b^x b^y = b^{x+y}$	$2^5 \cdot 2^3 = 2^{5+3} = 2^8$
Quotients	$\dfrac{b^x}{b^y} = b^{x-y}$	$\dfrac{4^7}{4^2} = 4^{7-2} = 4^5$
Negative exponents	$b^{-x} = \dfrac{1}{b^x}$	$3^{-4} = \dfrac{1}{3^4} = \dfrac{1}{81}$
Power to a power	$\left(b^x\right)^y = b^{xy}$	$\left(3^2\right)^4 = 3^{2(4)} = 3^8$
Roots	$b^{1/n} = \sqrt[n]{b}$	$5^{1/2} = \sqrt{5}$

■ **EXAMPLE 1** Rewrite as a whole number or fraction:

(a) $16^{-1/2}$ **(b)** $27^{2/3}$ **(c)** $4^{16} \cdot 4^{-18}$ **(d)** $\dfrac{9^3}{3^7}$

Solution

(a) $16^{-1/2} = \dfrac{1}{16^{1/2}} = \dfrac{1}{\sqrt{16}} = \dfrac{1}{4}$ **(b)** $27^{2/3} = \left(27^{1/3}\right)^2 = 3^2 = 9$

(c) $4^{16} \cdot 4^{-18} = 4^{-2} = \dfrac{1}{4^2} = \dfrac{1}{16}$ **(d)** $\dfrac{9^3}{3^7} = \dfrac{\left(3^2\right)^3}{3^7} = \dfrac{3^6}{3^7} = 3^{-1} = \dfrac{1}{3}$ ■

In the next example, we use the fact that $f(x) = b^x$ is one-to-one. In other words, if $b^x = b^y$, then $x = y$.

■ **EXAMPLE 2** Solve for the unknown:

(a) $2^{3x+1} = 2^5$ **(b)** $b^3 = 5^6$ **(c)** $7^{t+1} = \left(\dfrac{1}{7}\right)^{2t}$

Solution

(a) If $2^{3x+1} = 2^5$, then $3x + 1 = 5$ and thus $x = \frac{4}{3}$.

(b) Raise both sides of $b^3 = 5^6$ to the $\frac{1}{3}$ power. By the power to a power rule,

$$b = \left(b^3\right)^{1/3} = \left(5^6\right)^{1/3} = 5^{6/3} = 5^2 = 25$$

(c) Since $\frac{1}{7} = 7^{-1}$, the right-hand side of the equation is $\left(\frac{1}{7}\right)^{2t} = (7^{-1})^{2t} = 7^{-2t}$. The equation becomes $7^{t+1} = 7^{-2t}$. Therefore, $t + 1 = -2t$, or $t = -\frac{1}{3}$. ■

The Number e

Although written references to the number π go back more than 4000 years, mathematicians first became aware of the special role played by e in the seventeenth century. The notation e was introduced by Leonhard Euler, who discovered many fundamental properties of this important number.

In Chapter 3, we will use calculus to study exponential functions. One of the surprising insights of calculus is that the most convenient or natural base for an exponential function is not $b = 10$ or $b = 2$, as one might think at first, but rather a certain irrational number, denoted by e, whose value is approximately $e \approx 2.718$. A calculator is used to evaluate specific values of $f(x) = e^x$. For example,

$$e^3 \approx 20.0855, \qquad e^{-1/4} \approx 0.7788$$

In calculus, when we speak of *the* exponential function, it is understood that the base is e. Another common notation for e^x is $\exp(x)$.

How is e defined? There are many different definitions, but they all rely on the calculus concept of a limit. We shall discuss one way of defining e in Section 3.2. Another definition is described in Example 4 of Section 1.7. For now, we mention the following two graphical descriptions:

• Using Figure 3(A): Among all exponential functions $y = b^x$, $b = e$ is the unique base for which the slope of the tangent line to the graph at $(0, 1)$ is equal to 1.

• Using Figure 3(B): The number e is the unique number such that the area of the region under the hyperbola $y = 1/x$ for $1 \leq x \leq e$ is equal to 1.

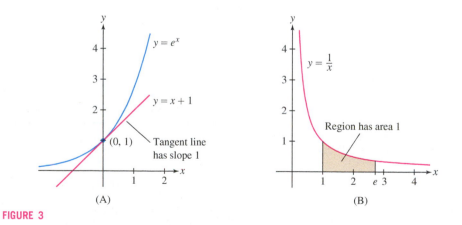

FIGURE 3

From these descriptions it is not clear why e is important. As we will learn, however, the exponential function $f(x) = e^x$ plays a fundamental role because it behaves in a particularly simple way with respect to the basic operations of calculus: differentiation and integration.

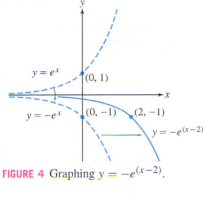

FIGURE 4 Graphing $y = -e^{(x-2)}$.

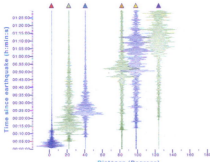

Seismograph of the 2010 Haiti earthquake, which registered 7.0 on the Richter scale. The Richter scale is based on the logarithm (to base 10) of the amplitude of seismic waves. Each whole-number increase in Richter magnitude corresponds to a tenfold increase in amplitude and approximately 31.6 times more energy. (*University of South Carolina and the IRIS Consortium*)

In this text, the natural logarithm is denoted $\ln x$. Other common notations are $\log x$ and $\mathrm{Log}\, x$.

■ EXAMPLE 3 Draw the graph of $y = -e^{(x-2)}$.

Solution Figure 3(A) shows the graph of $y = e^x$. The graph of $y = -e^x$ is simply the reflection of this graph over the x-axis, making all values negative instead of positive, as in Figure 4. The graph of $y = -e^{(x-2)}$ is obtained by translating this graph 2 units to the right. ■

Logarithms

Logarithmic functions are inverses of exponential functions. More precisely, if $b > 0$ and $b \neq 1$, then the *logarithm to the base b*, denoted $\log_b x$, is the inverse of $f(x) = b^x$. By definition, $y = \log_b x$ if $b^y = x$, so we have

$$b^{\log_b x} = x \quad \text{and} \quad \log_b(b^x) = x$$

In other words, $\log_b x$ is the number to which b must be raised in order to get x. For example,

$$\log_2(8) = 3 \quad \text{because} \quad 2^3 = 8$$

$$\log_{10}(1) = 0 \quad \text{because} \quad 10^0 = 1$$

$$\log_3\left(\frac{1}{9}\right) = -2 \quad \text{because} \quad 3^{-2} = \frac{1}{3^2} = \frac{1}{9}$$

The logarithm to the base e, denoted $\ln x$, plays a special role and is called the **natural logarithm**.

$$\ln x = \log_e x$$

We use a calculator to evaluate logarithms numerically. For example,

$$\ln 17 \approx 2.83321 \quad \text{because} \quad e^{2.83321} \approx 17$$

As in Figure 5, $f(x) = \ln x$ and $g(x) = e^x$ are inverse functions, so we have

$$e^{\ln x} = x \quad \text{and} \quad \ln(e^x) = x$$

Recall that the domain of $f(x) = b^x$ is **R** and its range is the set of positive real numbers $\{x : x > 0\}$. Since the domain and range are reversed in the inverse function,

- The *domain* of $f(x) = \log_b x$ is $\{x : x > 0\}$.
- The *range* of $f(x) = \log_b x$ is the set of all real numbers **R**.

If $b > 1$, then $\log_b x$ is positive for $x > 1$ and negative for $0 < x < 1$. Figure 5 illustrates these facts for the base $b = e$. Keep in mind that the logarithm of a negative number does not exist. For example, $\log_{10}(-2)$ does not exist because $10^y = -2$ has no solution.

For each law of exponents, there is a corresponding law for logarithms. The rule $b^{x+y} = b^x b^y$ corresponds to the rule

$$\log_b(xy) = \log_b x + \log_b y$$

In words: *The log of a product is the sum of the logs.* To verify this rule, observe that

$$b^{\log_b(xy)} = xy = b^{\log_b x} \cdot b^{\log_b y}$$

$$= b^{\log_b x + \log_b y}$$

The exponents $\log_b(xy)$ and $\log_b x + \log_b y$ are equal as claimed because $f(x) = b^x$ is one-to-one. The logarithm laws are collected in the following table.

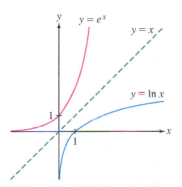

FIGURE 5 $y = \ln x$ is the inverse of $y = e^x$.

Laws of Logarithms

	Law	Example
Log of 1	$\log_b(1) = 0$	
Log of b	$\log_b(b) = 1$	
Products	$\log_b(xy) = \log_b x + \log_b y$	$\log_5(2 \cdot 3) = \log_5 2 + \log_5 3$
Quotients	$\log_b\left(\dfrac{x}{y}\right) = \log_b x - \log_b y$	$\log_2\left(\dfrac{3}{7}\right) = \log_2 3 - \log_2 7$
Reciprocals	$\log_b\left(\dfrac{1}{x}\right) = -\log_b x$	$\log_2\left(\dfrac{1}{7}\right) = -\log_2 7$
Powers (any n)	$\log_b(x^n) = n \log_b x$	$\log_{10}(8^2) = 2 \cdot \log_{10} 8$

We note also that all logarithmic functions are proportional. More precisely, the following **change-of-base** formula holds (see Exercise 51):

$$\log_b x = \frac{\log_a x}{\log_a b}, \qquad \log_b x = \frac{\ln x}{\ln b} \qquad \boxed{1}$$

■ **EXAMPLE 4** **Using the Logarithm Laws** Evaluate:

(a) $\log_6 9 + \log_6 4$ **(b)** $\ln\left(\dfrac{1}{\sqrt{e}}\right)$ **(c)** $10\log_b(b^3) - 4\log_b(\sqrt{b})$

Solution

(a) $\log_6 9 + \log_6 4 = \log_6(9 \cdot 4) = \log_6(36) = \log_6(6^2) = 2$

(b) $\ln\left(\dfrac{1}{\sqrt{e}}\right) = \ln(e^{-1/2}) = -\dfrac{1}{2}\ln(e) = -\dfrac{1}{2}$

(c) $10\log_b(b^3) - 4\log_b(\sqrt{b}) = 10(3) - 4\log_b(b^{1/2}) = 30 - 4\left(\dfrac{1}{2}\right) = 28$ ■

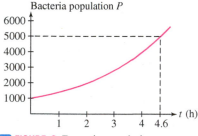
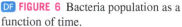

■ **EXAMPLE 5** **Solving an Exponential Equation** The bacteria population in a bottle at time t (in hours) has size $P(t) = 1000e^{0.35t}$. After how many hours will there be 5000 bacteria?

Solution We must solve $P(t) = 1000e^{0.35t} = 5000$ for t (Figure 6):

$$e^{0.35t} = \frac{5000}{1000} = 5$$

$$\ln(e^{0.35t}) = \ln 5 \qquad \text{(take logarithm of both sides)}$$

$$0.35t = \ln 5 \approx 1.609 \qquad [\text{because } \ln(e^a) = a]$$

$$t \approx \frac{1.609}{0.35} \approx 4.6 \text{ hours} \qquad ■$$

DF FIGURE 6 Bacteria population as a function of time.

FIGURE 7 The St. Louis Arch has the shape of an inverted hyperbolic cosine. (*Corbis*)

Hyperbolic Functions

The hyperbolic functions are certain special combinations of e^x and e^{-x} that play a role in engineering and physics (see Figure 7 for a real-life example). The hyperbolic sine and cosine, often called cinch and cosh, are de฀ned as follows:

$$\sinh x = \frac{e^x - e^{-x}}{2}, \qquad \cosh x = \frac{e^x + e^{-x}}{2}$$

As the terminology suggests, there are similarities between the hyperbolic and trigonometric functions. Here are some examples:

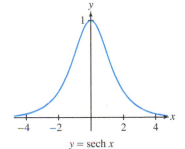

$y = \sinh x$

$y = \cosh x$

FIGURE 8 $y = \sinh x$ is an odd function; $y = \cosh x$ is an even function.

- **Parity:** The trigonometric functions and their hyperbolic analogs have the same parity. Thus, $f(x) = \sin x$ and $f(x) = \sinh x$ are both odd, and $f(x) = \cos x$ and $f(x) = \cosh x$ are both even (Figure 8):

$$\sinh(-x) = -\sinh x, \qquad \cosh(-x) = \cosh x$$

- **Identities:** The basic trigonometric identity $\sin^2 x + \cos^2 x = 1$ has a hyperbolic analog:

$$\boxed{\cosh^2 x - \sinh^2 x = 1} \qquad \boxed{2}$$

The addition formulas satisfied by $\sin x$ and $\cos x$ also have hyperbolic analogs:

$$\boxed{\begin{array}{l}\sinh(x + y) = \sinh x \cosh y + \cosh x \sinh y \\ \cosh(x + y) = \cosh x \cosh y + \sinh x \sinh y\end{array}}$$

- **Hyperbola instead of the circle:** Because of the identity $\cosh^2 t - \sinh^2 t = 1$, the point $(\cosh t, \sinh t)$ lies on the hyperbola $x^2 - y^2 = 1$, just as $(\cos t, \sin t)$ lies on the unit circle $x^2 + y^2 = 1$ (Figure 9).

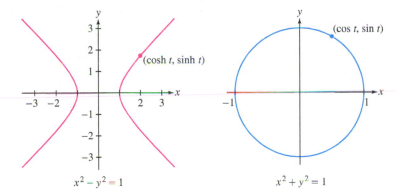

$x^2 - y^2 = 1$ $\qquad$ $x^2 + y^2 = 1$

FIGURE 9

- **Other hyperbolic functions:** The hyperbolic tangent, cotangent, secant, and cosecant functions (see Figures 10 and 11) are defined like their trigonometric counterparts:

$$\tanh x = \frac{\sinh x}{\cosh x} = \frac{e^x - e^{-x}}{e^x + e^{-x}}, \qquad \operatorname{sech} x = \frac{1}{\cosh x} = \frac{2}{e^x + e^{-x}}$$

$$\coth x = \frac{\cosh x}{\sinh x} = \frac{e^x + e^{-x}}{e^x - e^{-x}}, \qquad \operatorname{csch} x = \frac{1}{\sinh x} = \frac{2}{e^x - e^{-x}}$$

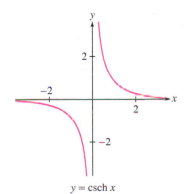

$y = \operatorname{sech} x$

$y = \operatorname{csch} x$

FIGURE 10 The hyperbolic secant and cosecant.

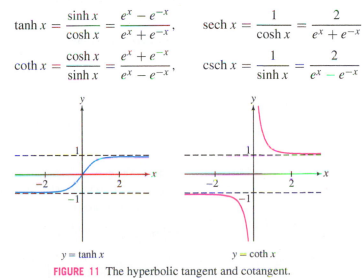

$y = \tanh x$ $\qquad$ $y = \coth x$

FIGURE 11 The hyperbolic tangent and cotangent.

■ **EXAMPLE 6** **Verifying the Basic Identity** Verify Eq. (2): $\cosh^2 x - \sinh^2 x = 1$.

Solution Because $\cosh x = \frac{1}{2}(e^x + e^{-x})$ and $\sinh x = \frac{1}{2}(e^x - e^{-x})$, we have

$$\cosh x + \sinh x = e^x, \qquad \cosh x - \sinh x = e^{-x}$$

We obtain Eq. (2) by multiplying these two equations together:

$$\cosh^2 x - \sinh^2 x = (\cosh x + \sinh x)(\cosh x - \sinh x) = e^x \cdot e^{-x} = 1 \quad \blacksquare$$

Inverse hyperbolic functions

Function	Domain
$y = \sinh^{-1} x$	all x
$y = \cosh^{-1} x$	$x \geq 1$
$y = \tanh^{-1} x$	$|x| < 1$
$y = \coth^{-1} x$	$|x| > 1$
$y = \operatorname{sech}^{-1} x$	$0 < x \leq 1$
$y = \operatorname{csch}^{-1} x$	$x \neq 0$

Inverse Hyperbolic Functions

Each of the hyperbolic functions, except $y = \cosh x$ and $y = \operatorname{sech} x$, is one-to-one on its domain and therefore has a well-defined inverse. The functions $y = \cosh x$ and $y = \operatorname{sech} x$ are one-to-one on the restricted domain $\{x : x \geq 0\}$. We let $y = \cosh^{-1} x$ and $y = \operatorname{sech}^{-1} x$ denote the corresponding inverses.

Einstein's Law of Velocity Addition

The inverse hyperbolic tangent plays a role in the Special Theory of Relativity, developed by Albert Einstein in 1905. One consequence of this theory is that no object can travel faster than the speed of light, $c \approx 3 \times 10^8$ m/s. Einstein realized that this contradicts a law stated by Galileo more than 250 years earlier, namely that *velocities add*. Imagine a train traveling at $u = 50$ m/s and a man walking down the aisle in the train at $v = 2$ m/s. According to Galileo, the man's velocity relative to the ground is $u + v = 52$ m/s. This agrees with our everyday experience. But now imagine an (unrealistic) rocket traveling away from the earth at $u = 2 \times 10^8$ m/s, and suppose that the rocket fires a missile with velocity $v = 1.5 \times 10^8$ m/s (relative to the rocket). If Galileo's Law were correct, the velocity of the missile relative to the earth would be $u + v = 3.5 \times 10^8$ m/s, which exceeds Einstein's maximum speed limit of $c \approx 3 \times 10^8$ m/s.

However, Einstein's theory replaces Galileo's Law with a new law stating that the *inverse hyperbolic tangents of velocities add*. More precisely, if u is the rocket's velocity relative to the earth and v is the missile's velocity relative to the rocket, then the velocity of the missile relative to the earth (Figure 12) is w, where

$$\tanh^{-1}\left(\frac{w}{c}\right) = \tanh^{-1}\left(\frac{u}{c}\right) + \tanh^{-1}\left(\frac{v}{c}\right) \qquad \boxed{3}$$

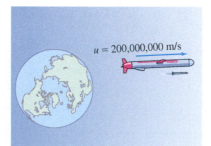

$u = 200{,}000{,}000$ m/s

FIGURE 12 What is the missile's velocity relative to the earth?

Einstein's Law of Velocity Addition [Eq. (3)] reduces to Galileo's Law, $w = u + v$, when u and v are small relative to the velocity of light c. See Exercise 52 for another way of expressing Eq. (3).

■ **EXAMPLE 7** A rocket travels away from the earth at a velocity of 2×10^8 m/s. A missile is fired from the rocket at a velocity of 1.5×10^8 m/s (relative to the rocket) away from the earth. Use Einstein's Law to find the velocity w of the missile relative to the earth.

Solution According to Eq. (3),

$$\tanh^{-1}\left(\frac{w}{c}\right) = \tanh^{-1}\left(\frac{2 \times 10^8}{3 \times 10^8}\right) + \tanh^{-1}\left(\frac{1.5 \times 10^8}{3 \times 10^8}\right) \approx 0.805 + 0.549 \approx 1.354$$

Therefore, $w/c \approx \tanh(1.354) \approx 0.875$, and $w \approx 0.875c \approx 2.6 \times 10^8$ m/s. This value obeys the Einstein speed limit of 3×10^8 m/s. ■

■ **EXAMPLE 8** **Low Velocities** A plane traveling at 300 m/s fires a missile forward at a velocity of 200 m/s. Calculate the missile's velocity w relative to the earth using both Einstein's Law and Galileo's Law.

Solution According to Einstein's Law,

$$\tanh^{-1}\left(\frac{w}{c}\right) = \tanh^{-1}\left(\frac{300}{c}\right) + \tanh^{-1}\left(\frac{200}{c}\right)$$

$$w = c \cdot \tanh\left(\tanh^{-1}\left(\frac{300}{c}\right) + \tanh^{-1}\left(\frac{200}{c}\right)\right) \approx 499.99999999967 \text{ m/s}$$

This is practically indistinguishable from the value $w = 300 + 200 = 500$ m/s obtained using Galileo's Law. ■

1.6 SUMMARY

- $f(x) = b^x$ is the *exponential function* with base b (where $b > 0$ and $b \neq 1$).
- $f(x) = b^x$ is increasing if $b > 1$ and decreasing if $b < 1$.
- Important exponent laws:

  (i) $b^x b^y = b^{x+y}$ **(ii)** $\dfrac{b^x}{b^y} = b^{x-y}$

  (iii) $b^{-x} = \dfrac{1}{b^x}$ **(iv)** $(b^x)^y = b^{xy}$

- The number $e \approx 2.718$.
- For $b > 0$ with $b \neq 1$, the *logarithmic function* $f(x) = \log_b x$ is the inverse of $f(x) = b^x$;

$$y = \log_b x \quad \Leftrightarrow \quad x = b^y$$

- The *natural logarithm* is the logarithm with base e and is denoted $\ln x$.
- $e^{\ln x} = x$ for $x > 0$ and $\ln(e^x) = x$ for all x.
- Important logarithm laws:

  (i) $\log_b(xy) = \log_b x + \log_b y$ **(ii)** $\log_b \left(\dfrac{x}{y} \right) = \log_b x - \log_b y$

  (iii) $\log_b(x^n) = n \log_b x$ **(iv)** $\log_b 1 = 0$ and $\log_b b = 1$

- The *hyperbolic sine and cosine*:

$$\sinh x = \frac{e^x - e^{-x}}{2} \quad \text{(odd function)}, \qquad \cosh x = \frac{e^x + e^{-x}}{2} \quad \text{(even function)}$$

  The remaining hyperbolic functions:

$$\tanh x = \frac{\sinh x}{\cosh x}, \qquad \coth x = \frac{\cosh x}{\sinh x}, \qquad \text{sech } x = \frac{1}{\cosh x}, \qquad \text{csch } x = \frac{1}{\sinh x}$$

- Basic identity: $\cosh^2 x - \sinh^2 x = 1$.
- The inverse hyperbolic functions and their domains:

$$f(x) = \sinh^{-1} x, \text{ for all } x \qquad f(x) = \coth^{-1} x, \text{ for } |x| > 1$$
$$f(x) = \cosh^{-1} x, \text{ for } x \geq 1 \qquad f(x) = \text{sech}^{-1} x, \text{ for } 0 < x \leq 1$$
$$f(x) = \tanh^{-1} x, \text{ for } |x| < 1 \qquad f(x) = \text{csch}^{-1} x, \text{ for } x \neq 0$$

1.6 EXERCISES

Preliminary Questions

1. Which of the following equations is incorrect?

(a) $3^2 \cdot 3^5 = 3^7$ **(b)** $(\sqrt{5})^{4/3} = 5^{2/3}$

(c) $3^2 \cdot 2^3 = 1$ **(d)** $(2^{-2})^{-2} = 16$

2. Compute $\log_{b^2}(b^4)$. **3.** When is $\ln x$ negative?

4. What is $\ln(-3)$? Explain.

5. Explain the phrase, The logarithm converts multiplication into addition.

6. What are the domain and range of $f(x) = \ln x$?

7. Which hyperbolic functions take on only positive values?

8. Which hyperbolic functions are increasing on their domains?

9. Describe three properties of hyperbolic functions that have trigonometric analogs.

Exercises

1. Rewrite as a whole number (without using a calculator):

(a) 7^0 **(b)** $10^2(2^{-2} + 5^{-2})$

(c) $\dfrac{(4^3)^5}{(4^5)^3}$ **(d)** $27^{4/3}$

(e) $8^{-1/3} \cdot 8^{5/3}$ **(f)** $3 \cdot 4^{1/4} - 12 \cdot 2^{-3/2}$

In Exercises 2–10, solve for the unknown variable.

2. $9^{2x} = 9^8$ **3.** $e^{2x} = e^{x+1}$

4. $e^{t^2} = e^{4t-3}$

5. $3^x = \left(\frac{1}{3} \right)^{x+1}$

6. $(\sqrt{5})^x = 125$

7. $4^{-x} = 2^{x+1}$

8. $b^4 = 10^{12}$

9. $k^{3/2} = 27$

10. $(b^2)^{x+1} = b^{-6}$

In Exercises 11–26, calculate without using a calculator.

11. $\log_3 27$ **12.** $\log_5 \frac{1}{25}$

13. $\ln 1$

14. $\log_5(5^4)$

15. $\log_2(2^{5/3})$

16. $\log_2(8^{5/3})$

17. $\log_{64} 4$

18. $\log_7(49^2)$

19. $\log_8 2 + \log_4 2$

20. $\log_{25} 30 + \log_{25} \frac{5}{6}$

21. $\log_4 48 - \log_4 12$

22. $\ln(\sqrt{e} \cdot e^{7/5})$

23. $\ln(e^3) + \ln(e^4)$

24. $\log_2 \frac{4}{3} + \log_2 24$

25. $7^{\log_7(29)}$

26. $8^{3\log_8(2)}$

27. Write as the natural log of a single expression:

(a) $2\ln 5 + 3\ln 4$

(b) $5\ln(x^{1/2}) + \ln(9x)$

28. Solve for x: $\ln(x^2 + 1) - 3\ln x = \ln(2)$.

In Exercises 29 34, solve for the unknown.

29. $7e^{5t} = 100$

30. $6e^{-4t} = 2$

31. $2^{x^2 - 2x} = 8$

32. $e^{2t+1} = 9e^{1-t}$

33. $\ln(x^4) - \ln(x^2) = 2$

34. $\log_3 y + 3\log_3(y^2) = 14$

35. Find the inverse of $y = e^{2x-3}$.

36. Find the inverse of $y = \ln(x^2 - 2)$ for $x > \sqrt{2}$.

37. Use a calculator to compute $\sinh x$ and $\cosh x$ for $x = -3, 0, 5$.

38. Compute $\sinh(\ln 5)$ and $\tanh(3\ln 5)$ without using a calculator.

39. Show, by producing a counterexample, that $\ln(ab)$ is not equal to $(\ln a)(\ln b)$.

40. For which values of x are $y = \sinh x$ and $y = \cosh x$ increasing and decreasing?

41. Show that $y = \tanh x$ is an odd function.

42. The population of a city (in millions) at time t (years) is $P(t) = 2.4e^{0.06t}$, where $t = 0$ is the year 2000.

(a) What is the population at time $t = 0$?

(b) When will the population double from its size at $t = 0$?

43. The **Gutenberg Richter Law** states that the number N of earthquakes per year worldwide of Richter magnitude at least M satisűes an approximate relation $\log_{10} N = a - M$ for some constant a. Find a, assuming that there is one earthquake of magnitude $M \geq 8$ per year. How many earthquakes of magnitude $M \geq 5$ occur per year?

44. The energy E (in joules) radiated as seismic waves from an earthquake of Richter magnitude M is given by the formula $\log_{10} E = 4.8 + 1.5M$.

(a) Express E as a function of M.

(b) Show that when M increases by 1, the energy increases by a factor of approximately 31.6.

45. Refer to the graphs to explain why the equation $\sinh x = t$ has a unique solution for every t and why $\cosh x = t$ has two solutions for every $t > 1$.

46. Compute $\cosh x$ and $\tanh x$, assuming that $\sinh x = 0.8$.

47. Prove the addition formula for $\cosh x$ given by $\cosh(x + y) = \cosh x \cosh y + \sinh x \sinh y$.

48. Use the addition formulas to prove

$$\sinh(2x) = 2\cosh x \sinh x$$

$$\cosh(2x) = \cosh^2 x + \sinh^2 x$$

49. A train moves along a track at velocity v. Bionica walks down the aisle of the train with velocity u in the direction of the train's motion. Compute the velocity w of Bionica relative to the ground using the laws of both Galileo and Einstein in the following cases:

(a) $v = 500$ m/s and $u = 10$ m/s. Is your calculator accurate enough to detect the difference between the two laws?

(b) $v = 10^7$ m/s and $u = 10^6$ m/s.

Further Insights and Challenges

50. Show that $\log_a b \log_b a = 1$ for all $a, b > 0$ such that $a \neq 1$ and $b \neq 1$.

51. Verify that for all x, the formula holds. $\log_b x = \dfrac{\log_a x}{\log_a b}$ for $a, b > 0$ such that $a \neq 1, b \neq 1$.

52. (a) Use the addition formulas for $\sinh x$ and $\cosh x$ to prove

$$\tanh(u + v) = \frac{\tanh u + \tanh v}{1 + \tanh u \tanh v}$$

(b) Use (a) to show that Einstein's Law of Velocity Addition [Eq. (3)] is equivalent to

$$w = \frac{u + v}{1 + \dfrac{uv}{c^2}}$$

53. Prove that every function f can be written as a sum $f(x) = f_+(x) + f_-(x)$ of an even function $f_+(x)$ and an odd function $f_-(x)$. Express $f(x) = 5e^x + 8e^{-x}$ in terms of $\cosh x$ and $\sinh x$. *Hint:* $y = f(x) + f(-x)$ is an even function, and $y = f(x) - f(-x)$ is an odd function.

1.7 Technology: Calculators and Computers

Computer technology has vastly extended our ability to calculate and visualize mathematical relationships. In applied settings, computers are indispensable for solving complex systems of equations and analyzing data, as in weather prediction and medical imaging. Mathematicians use computers to study complex structures such as the Mandelbrot Set (Figures 1 and 2). We take advantage of this technology to explore the ideas of calculus visually and numerically.

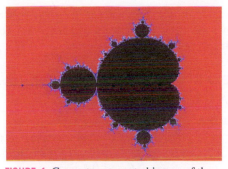

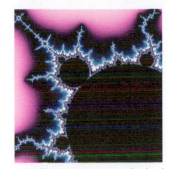

FIGURE 1 Computer-generated image of the Mandelbrot Set, which occurs in the mathematical theory of chaos and fractals. (*Scott Camazine/Science Source*)

FIGURE 2 Even greater complexity is revealed when we zoom in on a portion of the Mandelbrot Set. (*Laguna Design/Science Photo Library*)

When we plot a function with a graphing calculator or computer algebra system, the graph is contained within a **viewing rectangle**, the region determined by the range of x- and y-values in the plot. We write $[a, b] \times [c, d]$ to denote the rectangle where $a \leq x \leq b$ and $c \leq y \leq d$.

The appearance of the graph depends heavily on the choice of viewing rectangle. Different choices may convey very different impressions that are sometimes misleading. Compare the three viewing rectangles for the graph of $f(x) = 12 - x - x^2$ in Figure 3. Only (A) successfully displays the shape of the graph as a parabola. In (B), the graph is cut off, and no graph at all appears in (C). Keep in mind that the scales along the axes may change with the viewing rectangle. For example, the unit increment along the y-axis is larger in (B) than in (A), so the graph in (B) is steeper.

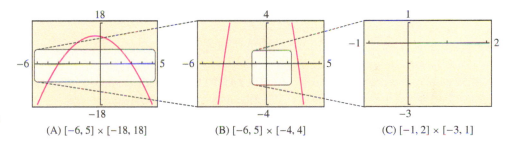

FIGURE 3 Viewing rectangles for the graph of $f(x) = 12 - x - x^2$.

(A) $[-6, 5] \times [-18, 18]$ (B) $[-6, 5] \times [-4, 4]$ (C) $[-1, 2] \times [-3, 1]$

There is no single correct viewing rectangle. The goal is to select the viewing rectangle that displays the properties you wish to investigate. This usually requires experimentation.

Technology is indispensable but also has its limitations. When shown the computer-generated results of a complex calculation, the Nobel prize winning physicist Eugene Wigner (1902–1995) is reported to have said: It is nice to know that the computer understands the problem, but I would like to understand it too.

■ **EXAMPLE 1** **How Many Roots and Where?** How many real roots does the function $f(x) = x^9 - 20x + 1$ have? Find their approximate locations.

Solution We experiment with several viewing rectangles (Figure 4). Our first attempt (A) displays a cut-off graph, so we try a viewing rectangle that includes a larger range of y-values. Plot (B) shows that the roots of f probably lie somewhere in the interval $[-3, 3]$, but it does not reveal how many real roots there are. Therefore, we try the viewing

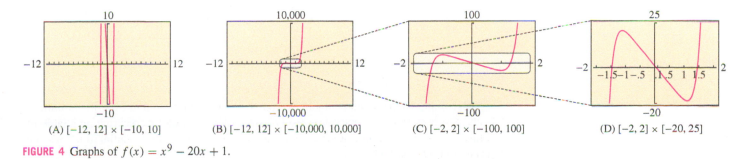

(A) $[-12, 12] \times [-10, 10]$ (B) $[-12, 12] \times [-10{,}000, 10{,}000]$ (C) $[-2, 2] \times [-100, 100]$ (D) $[-2, 2] \times [-20, 25]$

FIGURE 4 Graphs of $f(x) = x^9 - 20x + 1$.

rectangle in (C). Now we can see clearly that f has three roots. A further zoom in (D) shows that these roots are located near $-1.5, 0.1$, and 1.5. Further zooming would provide their locations with greater accuracy.   ■

■ **EXAMPLE 2   Does a Solution Exist?**   Does $\cos x = \tan x$ have a solution? Describe the set of all solutions.

Solution   The solutions of $\cos x = \tan x$ are the x-coordinates of the points where the graphs of $y = \cos x$ and $y = \tan x$ intersect. Figure 5(A) shows that there are two solutions in the interval $[0, 2\pi]$. By zooming in on the graph as in (B), we see that the first positive root lies between 0.6 and 0.7 and the second positive root lies between 2.4 and 2.5. Further zooming shows that the first root is approximately 0.67 [Figure 5(C)]. Continuing this process, we find that the first two roots are $x \approx 0.666$ and $x \approx 2.475$.

Since $f(x) = \cos x$ and $f(x) = \tan x$ are periodic, the picture repeats itself with period 2π. All solutions are obtained by adding multiples of 2π to the two solutions in $[0, 2\pi]$:

$$x \approx 0.666 + 2\pi k \qquad \text{and} \qquad x \approx 2.475 + 2\pi k \quad \text{(for any integer } k)$$ ■

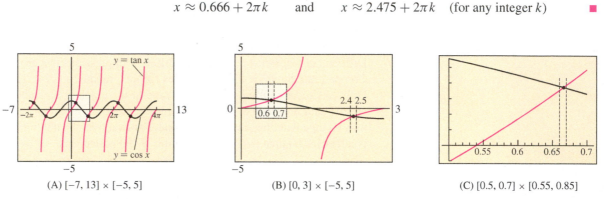

(A) $[-7, 13] \times [-5, 5]$ (B) $[0, 3] \times [-5, 5]$ (C) $[0.5, 0.7] \times [0.55, 0.85]$

FIGURE 5 Graphs of $y = \cos x$ and $y = \tan x$.

CAUTION When considering the graph of a function such as $y = \ln x$ (Figure 6), it may appear to approach a horizontal asymptote, but in fact, it does not. For any given horizontal line, the graph eventually rises above it. In this respect, graphing calculators and computer graphing systems must be used judiciously.

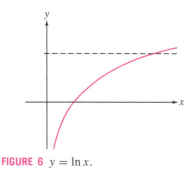

FIGURE 6 $y = \ln x$.

FIGURE 7 Graphs of $f(x) = \dfrac{1 - 3x}{x - 2}$.

■ **EXAMPLE 3   Functions with Asymptotes**   Plot the function $f(x) = \dfrac{1 - 3x}{x - 2}$ and describe its asymptotic behavior.

Solution   First, we plot f in the viewing rectangle $[-10, 20] \times [-5, 5]$ as in Figure 7(A). The vertical line $x = 2$ is called a **vertical asymptote**. Many graphing calculators display this line, but it is *not* part of the graph (and it can usually be eliminated by choosing a smaller range of y-values). We see that $f(x)$ tends to ∞ as x approaches 2 from the left, and to $-\infty$ as x approaches 2 from the right. To display the horizontal asymptotic behavior of f, we use the viewing rectangle $[-10, 20] \times [-10, 5]$ [Figure 7(B)]. Here, we see that the graph approaches the horizontal line $y = -3$, called a **horizontal asymptote** (which we have added as a dashed horizontal line in the figure).   ■

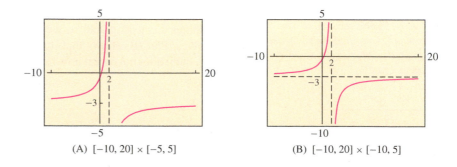

(A) $[-10, 20] \times [-5, 5]$ (B) $[-10, 20] \times [-10, 5]$

Calculators and computer algebra systems give us the freedom to experiment numerically. For instance, we can explore the behavior of a function by constructing a table of values. In the next example, we investigate a function related to exponential functions and compound interest (see Section 5.8).

■ **EXAMPLE 4** **Investigating the Behavior of a Function** How does $f(n) = (1 + 1/n)^n$ behave for large whole-number values of n? Does $f(n)$ tend to infinity as n gets larger?

Solution First, we make a table of values of $f(n)$ for larger and larger values of n. Table 1 suggests that $f(n)$ does not tend to infinity. Rather, as n grows larger, $f(n)$ appears to get closer to some value near 2.718 (a number resembling e). This is an example of limiting behavior that we will discuss in Chapter 2. Next, replace n by the variable x and plot the function $f(x) = (1 + 1/x)^x$. The graphs in Figure 8 confirm that $f(x)$ approaches a limit of approximately 2.7. We will prove that $f(n)$ approaches e as n tends to infinity in Section 5.9. ■

TABLE 1

n	$\left(1 + \dfrac{1}{n}\right)^n$
10	2.59374
10^2	2.70481
10^3	2.71692
10^4	2.71815
10^5	2.71827
10^6	2.71828

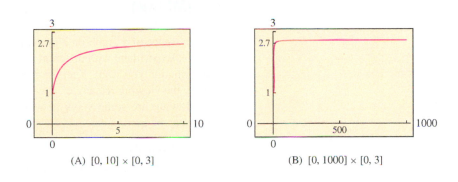

(A) $[0, 10] \times [0, 3]$ (B) $[0, 1000] \times [0, 3]$

FIGURE 8 Graphs of $f(x) = \left(1 + \dfrac{1}{x}\right)^x$.

■ **EXAMPLE 5** **Bird Flight: Finding a Minimum Graphically** According to one model of bird flight, the power consumed by a pigeon flying at velocity v (in meters per second) is $P(v) = 17v^{-1} + 10^{-3}v^3$ (in joules per second). Use a graph of P to find the velocity that minimizes power consumption.

Solution The velocity that minimizes power consumption corresponds to the lowest point on the graph of P. We plot P first in a large viewing rectangle (Figure 9). This figure reveals the general shape of the graph and shows that P takes on a minimum value for v somewhere between $v = 8$ and $v = 9$. In the viewing rectangle $[8, 9.2] \times [2.6, 2.65]$, we see that the minimum occurs at approximately $v = 8.65$ m/s. ■

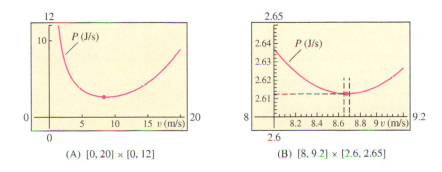

(A) $[0, 20] \times [0, 12]$ (B) $[8, 9.2] \times [2.6, 2.65]$

FIGURE 9 Power consumption $P(v)$ as a function of velocity v.

Local linearity is an important concept in calculus that is based on the idea that many functions are *nearly linear* over small intervals. Local linearity can be illustrated effectively with a graphing calculator.

■ **EXAMPLE 6** **Illustrating Local Linearity** Illustrate local linearity for the function $f(x) = x^{\sin x}$ at $x = 1$.

Solution First, we plot $f(x) = x^{\sin x}$ in the viewing window of Figure 10(A). The graph moves up and down and appears very wavy. However, as we zoom in, the graph straightens out. Figures (B)–(D) show the result of zooming in on the point $(1, f(1))$. When viewed up close, the graph looks like a straight line. This illustrates the local linearity of f at $x = 1$. ■

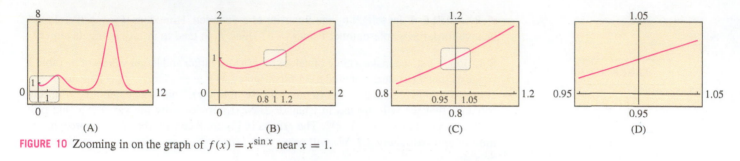

FIGURE 10 Zooming in on the graph of $f(x) = x^{\sin x}$ near $x = 1$.

1.7 SUMMARY

- The appearance of a graph on a graphing calculator depends on the choice of viewing rectangle. Experiment with different viewing rectangles until you find one that displays the information you want. Keep in mind that the scales along the axes may change as you vary the viewing rectangle.
- The following are some ways in which graphing calculators and computer algebra systems can be used in calculus:

  Visualizing the behavior of a function
  Finding solutions graphically or numerically
  Conducting numerical or graphical experiments
  Illustrating theoretical ideas (such as local linearity)

1.7 EXERCISES

Preliminary Questions

1. Is there a definite way of choosing the optimal viewing rectangle, or is it best to experiment until you find a viewing rectangle appropriate to the problem at hand?

2. Describe the calculator screen produced when the function $y = 3 + x^2$ is plotted with a viewing rectangle:
(a) $[-1, 1] \times [0, 2]$ **(b)** $[0, 1] \times [0, 4]$

3. According to the evidence in Example 4, it appears that $f(n) = (1 + 1/n)^n$ never takes on a value greater than 3 for $n > 0$. Does this evidence *prove* that $f(n) \leq 3$ for $n > 0$?

4. How can a graphing calculator be used to find the minimum value of a function?

Exercises

The exercises in this section should be done using a graphing calculator or computer algebra system.

1. Plot $f(x) = 2x^4 + 3x^3 - 14x^2 - 9x + 18$ in the appropriate viewing rectangles and determine its roots.

2. How many solutions does $x^3 - 4x + 8 = 0$ have?

3. How many *positive* solutions does $x^3 - 12x + 8 = 0$ have?

4. Does $\cos x + x = 0$ have a solution? A positive solution?

5. Find all the solutions of $\sin x = \sqrt{x}$ for $x > 0$.

6. How many solutions does $\cos x = x^2$ have?

7. Let $f(x) = (x - 100)^2 + 1000$. What will the display show if you graph f in the viewing rectangle $[-10, 10]$ by $[-10, 10]$? Find an appropriate viewing rectangle.

8. Plot $f(x) = \dfrac{8x + 1}{8x - 4}$ in an appropriate viewing rectangle. What are the vertical and horizontal asymptotes?

9. Plot the graph of $f(x) = x/(4 - x)$ in a viewing rectangle that clearly displays the vertical and horizontal asymptotes.

10. Illustrate local linearity for $f(x) = x^2$ by zooming in on the graph at $x = 0.5$ (see Example 6).

11. Plot $f(x) = \cos(x^2) \sin x$ for $0 \leq x \leq 2\pi$. Then illustrate local linearity at $x = 3.8$ by choosing appropriate viewing rectangles.

12. If P_0 dollars are deposited in a bank account paying 5% interest compounded monthly, then the account has value $P_0 \left(1 + \frac{0.05}{12}\right)^N$ after N months. Find, to the nearest integer N, the number of months after which the account value doubles.

In Exercises 13 18, investigate the behavior of the function as n or x grows large by making a table of function values and plotting a graph (see Example 4). Describe the behavior in words.

13. $f(n) = n^{1/n}$

14. $f(n) = \dfrac{4n + 1}{6n - 5}$

15. $f(n) = \left(1 + \dfrac{1}{n}\right)^{n^2}$

16. $f(x) = \left(\dfrac{x + 6}{x - 4}\right)^x$

17. $f(x) = \left(x \tan \dfrac{1}{x}\right)^x$

18. $f(x) = \left(x \tan \dfrac{1}{x}\right)^{x^2}$

19. The graph of $f(\theta) = A \cos \theta + B \sin \theta$ is a sinusoidal wave for any constants A and B. Confirm this for $(A, B) = (1, 1)$, $(1, 2)$, and $(3, 4)$ by plotting f.

20. Find the maximum value of f for the graphs produced in Exercise 19. Can you guess the formula for the maximum value in terms of A and B?

21. Find the intervals on which $f(x) = x(x + 2)(x - 3)$ is positive by plotting a graph.

22. Find the set of solutions to the inequality $(x^2 - 4)(x^2 - 1) < 0$ by plotting a graph.

Further Insights and Challenges

23. CAS Let $f_1(x) = x$ and define a sequence of functions by $f_{n+1}(x) = \frac{1}{2}(f_n(x) + x/f_n(x))$. For example, $f_2(x) = \frac{1}{2}(x + 1)$. Use a computer algebra system to compute $f_n(x)$ for $n = 3, 4, 5$ and plot $y = f_n(x)$ together with $y = \sqrt{x}$ for $x \geq 0$. What do you notice?

24. Set $P_0(x) = 1$ and $P_1(x) = x$. The **Chebyshev polynomials** (useful in approximation theory) are defined inductively by the formula $P_{n+1}(x) = 2x P_n(x) - P_{n-1}(x)$.

(a) Show that $P_2(x) = 2x^2 - 1$.

(b) Compute $P_n(x)$ for $3 \leq n \leq 6$ using a computer algebra system or by hand, and plot $y = P_n(x)$ over $[-1, 1]$.

(c) Check that your plots confirm two interesting properties: (a) $y = P_n(x)$ has n real roots in $[-1, 1]$ and (b) for $x \in [-1, 1]$, $P_n(x)$ lies between -1 and 1.

CHAPTER REVIEW EXERCISES

1. Express $(4, 10)$ as a set $\{x : |x - a| < c\}$ for suitable a and c.

2. Express as an interval:

(a) $\{x : |x - 5| < 4\}$ **(b)** $\{x : |5x + 3| \leq 2\}$

3. Express $\{x : 2 \leq |x - 1| \leq 6\}$ as a union of two intervals.

4. Give an example of numbers x, y such that $|x| + |y| = x - y$.

5. Describe the pairs of numbers x, y such that $|x + y| = x - y$.

6. Sketch the graph of $y = f(x + 2) - 1$, where $f(x) = x^2$ for $-2 \leq x \leq 2$.

In Exercises 7–10, let $f(x)$ be the function shown in Figure 1.

7. Sketch the graphs of $y = f(x) + 2$ and $y = f(x + 2)$.

8. Sketch the graphs of $y = \frac{1}{2} f(x)$ and $y = f(\frac{1}{2}x)$.

9. Continue the graph of f to the interval $[-4, 4]$ as an even function.

10. Continue the graph of f to the interval $[-4, 4]$ as an odd function.

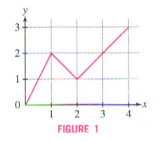

FIGURE 1

In Exercises 11–14, find the domain and range of the function.

11. $f(x) = \sqrt{x + 1}$ **12.** $f(x) = \dfrac{4}{x^4 + 1}$

13. $f(x) = \dfrac{2}{3 - x}$ **14.** $f(x) = \sqrt{x^2 - x + 5}$

15. Determine whether the function is increasing, decreasing, or neither:

(a) $f(x) = 3^{-x}$ **(b)** $f(x) = \dfrac{1}{x^2 + 1}$

(c) $g(t) = t^2 + t$ **(d)** $g(t) = t^3 + t$

16. Determine whether the function is even, odd, or neither:

(a) $f(x) = x^4 - 3x^2$

(b) $g(x) = \sin(x + 1)$

(c) $f(x) = 2^{-x^2}$

In Exercises 17–24, find the equation of the line.

17. Line passing through $(-1, 4)$ and $(2, 6)$

18. Line passing through $(-1, 4)$ and $(-1, 6)$

19. Line of slope 6 through $(9, 1)$

20. Line of slope $-\frac{3}{2}$ through $(4, -12)$

21. Line through $(2, 1)$ perpendicular to the line given by $y = 3x + 7$

22. Line through $(3, 4)$ perpendicular to the line given by $y = 4x - 2$

23. Line through $(2, 3)$ parallel to $y = 4 - x$

24. Horizontal line through $(-3, 5)$

25. Does the following table of market data suggest a linear relationship between price and number of homes sold during a one-year period? Explain.

Price (thousands of $)	180	195	220	240
No. of homes sold	127	118	103	91

26. Does the following table of revenue data for a computer manufacturer suggest a linear relation between revenue and time? Explain.

Year	2005	2009	2011	2014
Revenue (billions of $)	13	18	15	11

27. Suppose that a cell phone plan that is offered at a price of P dollars per month attracts C customers, where $C(P)$ is a linear demand function for $100 \leq P \leq 500$. If $C(100) = 1,000,000$ and $C(500) = 100,000$, determine the demand function C. What is the decrease in the number of customers for each increase of $100 in the price?

28. Suppose that Internet domain names are sold at a price of P per month for $2 \leq P \leq 100$. The number of customers C who buy the domain names is a linear function of the price. If 10,000 customers buy a domain name when the price is \$2 per month and 1000 customers buy when the price is \$100 per month, determine the demand function C. What is the decrease in the number of customers for every \$1 increase in the cost of the domain names?

29. Find the roots of $f(x) = x^4 - 4x^2$ and sketch its graph. On which intervals is f decreasing?

30. Let $h(z) = -2z^2 + 12z + 3$. Complete the square and find the maximum value of h.

31. Let $f(x)$ be the square of the distance from the point $(2, 1)$ to a point $(x, 3x + 2)$ on the line $y = 3x + 2$. Show that f is a quadratic function, and find its minimum value by completing the square.

32. Prove that $x^2 + 3x + 3 \geq 0$ for all x.

In Exercises 33 38, sketch the graph by hand.

33. $y = t^4$

34. $y = t^5$

35. $y = \sin \dfrac{\theta}{2}$

36. $y = 10^{-x}$

37. $y = x^{1/3}$

38. $y = \dfrac{1}{x^2}$

39. Show that the graph of $y = f\left(\frac{1}{3}x - b\right)$ is obtained by shifting the graph of $y = f\left(\frac{1}{3}x\right)$ to the right $3b$ units. Use this observation to sketch the graph of $y = \left|\frac{1}{3}x - 4\right|$.

40. Let $h(x) = \cos x$ and $g(x) = x^{-1}$. Compute the composite functions $h(g)$ and $g(h)$, and find their domains.

41. Find functions f and g such that the function

$$f(g(t)) = (12t + 9)^4$$

42. Sketch the points on the unit circle corresponding to the following three angles, and find the values of the six standard trigonometric functions at each angle:

(a) $\dfrac{2\pi}{3}$

(b) $\dfrac{7\pi}{4}$

(c) $\dfrac{7\pi}{6}$

43. What are the periods of these functions?

(a) $y = \sin 2\theta$

(b) $y = \sin \frac{\theta}{2}$

(c) $y = \sin 2\theta + \sin \frac{\theta}{2}$

44. Assume that $\sin \theta = \frac{4}{5}$, where $\pi/2 < \theta < \pi$. Find:

(a) $\tan \theta$

(b) $\sin 2\theta$

(c) $\csc \dfrac{\theta}{2}$

45. Give an example of values a, b such that

(a) $\cos(a + b) \neq \cos a + \cos b$

(b) $\cos \dfrac{a}{2} \neq \dfrac{\cos a}{2}$

46. Let $f(x) = \cos x$. Sketch the graph of $y = 2f\left(\frac{1}{3}x - \frac{\pi}{4}\right)$ for $0 \leq x \leq 6\pi$.

47. Solve $\sin 2x + \cos x = 0$ for $0 \leq x < 2\pi$.

48. How does $h(n) = n^2/2^n$ behave for large whole-number values of n? Does $h(n)$ tend to infinity?

49. [GU] Use a graphing calculator to determine whether the equation $\cos x = 5x^2 - 8x^4$ has any solutions.

50. [GU] Using a graphing calculator, find the number of real roots and estimate the largest root to two decimal places:

(a) $f(x) = 1.8x^4 - x^5 - x$

(b) $g(x) = 1.7x^4 - x^5 - x$

51. Match each quantity (a) (d) with (i), (ii), or (iii) if possible, or state that no match exists.

(a) $2^a 3^b$

(b) $\dfrac{2^a}{3^b}$

(c) $(2^a)^b$

(d) $2^{a-b} 3^{b-a}$

(i) 2^{ab}

(ii) 6^{a+b}

(iii) $\left(\frac{2}{3}\right)^{a-b}$

52. Match each quantity (a) (d) with (i), (ii), or (iii) if possible, or state that no match exists.

(a) $\ln \left(\dfrac{a}{b}\right)$

(b) $\dfrac{\ln a}{\ln b}$

(c) $e^{\ln a - \ln b}$

(d) $(\ln a)(\ln b)$

(i) $\ln a + \ln b$

(ii) $\ln a - \ln b$

(iii) $\dfrac{a}{b}$

53. Find the inverse of $f(x) = \sqrt{x^3 - 8}$ and determine its domain and range.

54. Find the inverse of $f(x) = \dfrac{x - 2}{x - 1}$ and determine its domain and range.

55. Find a domain on which $h(t) = (t - 3)^2$ is one-to-one and determine the inverse on this domain.

56. Show that $g(x) = \dfrac{x}{x - 1}$ is equal to its inverse on the domain $\{x : x \neq 1\}$.

57. Let

$$f(x) = \begin{cases} -x^2 & \text{when } x < 0 \\ x & \text{when } x \geq 0 \end{cases}$$

(a) Is f increasing?

(b) Does f have an inverse? If so, what is it?

58. Let

$$f(x) = \begin{cases} x - 1 & \text{when } x < 1 \\ \ln x & \text{when } x \geq 1 \end{cases}$$

(a) Is f increasing?

(b) Does f have an inverse? If so, what is it?

59. Suppose that g is the inverse of f. Match the functions (a) (d) with their inverses (i) (iv).

(a) $f(x) + 1$

(b) $f(x + 1)$

(c) $4f(x)$

(d) $f(4x)$

(i) $g(x)/4$

(ii) $g(x/4)$

(iii) $g(x - 1)$

(iv) $g(x) - 1$

60. [GU] Plot $f(x) = xe^{-x}$ and use the zoom feature to find two solutions of $f(x) = 0.3$.

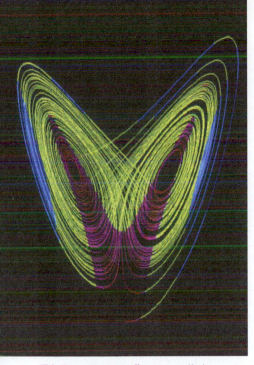

This "strange attractor" represents limit behavior that appeared first in weather models studied by meteorologist E. Lorenz in 1963. (*Scott Camazine/Science Source*)

2 LIMITS

Calculus is usually divided into two branches, differential and integral, partly for historical reasons. The subject grew out of efforts in the seventeenth century to solve two important geometric problems: finding tangent lines to curves (differential calculus) and computing areas under curves (integral calculus). However, calculus is a broad subject with no clear boundaries. It includes other topics, such as the theory of infinite series, and it has an extraordinarily wide range of applications. What makes these methods and applications part of calculus is that they all rely on the concept of a limit. We will see throughout the text how limits allow us to make computations and solve problems that cannot be solved using algebra alone.

This chapter introduces the limit concept and sets the stage for our study of the derivative in Chapter 3. The first section, intended as motivation, discusses how limits arise in the study of rates of change and tangent lines.

2.1 Limits, Rates of Change, and Tangent Lines

Limits are about how a function f behaves, as x approaches a number a. Does $f(x)$ get closer and closer to a number L? If so, we say L is the limit of $f(x)$ as x approaches a. If not, we say the limit does not exist. Limits play a key role throughout calculus. In this section, we discuss their role in understanding rates of change.

Rates of change become critical whenever we study the relationship between two changing quantities. Velocity is a familiar example (the rate of change of position with respect to time), but there are many others, such as

- Rate of change in the number of infected individuals with respect to time
- Rate of change in consumer price index with respect to time
- Rate of change of atmospheric temperature with respect to altitude

Roughly speaking, if y and x are related quantities, the rate of change should tell us how much y changes in response to a unit change in x. For example, if an automobile travels at a velocity of 80 km/h, then its position changes by 80 km for each unit change in time (the unit being 1 hour). If the trip lasts only half an hour, its position changes by 40 km, and in general, the change in position is $80t$ km, where t is the change in time (i.e., the time elapsed in hours). In other words,

$$\boxed{\text{Change in position} = \text{velocity} \times \text{change in time}}$$

However, this simple formula is not valid or even meaningful if the velocity is not constant. After all, if the automobile is accelerating or decelerating, which velocity would we use in the formula?

The problem of extending this formula to account for changing velocity lies at the heart of calculus. As we will learn, differential calculus uses the limit concept to define *instantaneous velocity*, and integral calculus enables us to compute the change in position in terms of instantaneous velocity. But these ideas are very general. They apply to all rates of change, making calculus an indispensable tool for modeling an amazing range of real-world phenomena.

In this section, we discuss velocity and other rates of change, emphasizing their graphical interpretation in terms of *tangent lines*. Although at this stage, we do not define precisely what a tangent line is—this will have to wait until Chapter 3—you can think of a tangent line as a line that *skims* a curve at a point, as in Figures 1(A) and (B) but not (C).

DF FIGURE 1 The line is tangent in (A) and (B) but not in (C).

(A)   (B)   (C)

(© Jeremy Pembrey/Alamy)

This statue of Isaac Newton in Cambridge University was described in *The Prelude*, a poem by William Wordsworth (1770–1850):

Newton with his prism and silent face,
The marble index of a mind for ever
Voyaging through strange seas of Thought,
    alone.

In linear motion, velocity may be positive or negative (indicating forward or backward direction of motion). Speed, by definition, is the absolute value of velocity and is always positive.

(NASA/JPL-Caltech/STScI)

HISTORICAL PERSPECTIVE

Philosophy is written in this grand book—I mean the universe—which stands continually open to our gaze, but it cannot be understood unless one first learns to comprehend the language … in which it is written. It is written in the language of mathematics.…

—GALILEO GALILEI, 1623

The scientific revolution of the sixteenth and seventeenth centuries reached its high point in the work of Isaac Newton (1643–1727), who was the first scientist to show that the physical world, despite its complexity and diversity, is governed by a small number of universal laws. One of Newton's great insights was that the universal laws are dynamical, describing how the world changes over time in response to forces, rather than how the world actually is at any given moment in time. These laws are expressed best in the language of calculus, which is the mathematics of change.

More than 50 years before the work of Newton, the astronomer Johannes Kepler (1571–1630) discovered his three laws of planetary motion, the most famous of which states that the path of a planet around the sun is an ellipse. Kepler arrived at these laws through a painstaking analysis of astronomical data, but he could not explain why they were true. According to Newton, the motion of any object—planet or pebble—is determined by the forces acting on it. The planets, if left undisturbed, would travel in straight lines. Since their paths are elliptical, some force—in this case, the gravitational force of the sun—must be acting to make them change direction continuously. In his magnum opus *Principia Mathematica*, published in 1687, Newton proved that Kepler's laws follow from Newton's own universal laws of motion and gravity.

For these discoveries, Newton gained widespread fame in his lifetime. His fame continued to increase after his death, assuming a nearly mythic dimension, and his ideas had a profound influence, not only in science but also in the arts and literature, as expressed in this epitaph by British poet Alexander Pope: "Nature and Nature's Laws lay hid in Night. God said, *Let Newton be!* and all was Light."

Velocity

When we speak of velocity, we usually mean *instantaneous* velocity, which indicates the speed and direction of an object at a particular moment. The idea of instantaneous velocity makes intuitive sense, but care is required to define it precisely.

Consider an object traveling in a straight line (linear motion). The **average velocity** over a given time interval has a straightforward definition as the ratio

$$\text{Average velocity} = \frac{\text{change in position}}{\text{length of time interval}}$$

For example, if an automobile travels 200 km in 4 hours, then its average velocity during this 4-hour period is $\frac{200}{4} = 50$ km/h. At any given moment, the automobile may be going faster or slower than the average.

We cannot define instantaneous velocity as a ratio because we would have to divide by the length of the time interval (which is zero). However, we should be able to estimate instantaneous velocity by computing average velocity over successively smaller time intervals. The guiding principle is: *Average velocity over a very small time interval is very close to instantaneous velocity*. To explore this idea further, we introduce some notation.

The Greek letter Δ (delta) is commonly used to denote the *change* in a function or variable. If $s(t)$ is the position of an object (distance from the origin) at time t and $[t_0, t_1]$ is a time interval, we set

$$\Delta s = s(t_1) - s(t_0) = \text{change in position}$$

$$\Delta t = \quad t_1 - t_0 \quad = \text{change in time (length of time interval)}$$

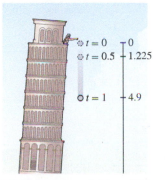

FIGURE 2 Distance traveled by a falling object after t seconds is $s(t) = 4.9t^2$ meters.

TABLE 1

Time intervals	Average velocity
[0.8, 0.81]	7.889
[0.8, 0.805]	7.8645
[0.8, 0.8001]	7.8405
[0.8, 0.80005]	7.84024
[0.8, 0.800001]	7.840005

There is nothing special about the particular time intervals in Table 1. We are looking for a trend, and we could have chosen any intervals $[0.8, t]$ for values of t approaching 0.8. We could also have chosen intervals $[t, 0.8]$ for $t < 0.8$.

The change in position Δs is also called the **displacement**, or **net change** in position. For $t_1 \neq t_0$,

$$\text{Average velocity over } [t_0, t_1] = \frac{\Delta s}{\Delta t} = \frac{s(t_1) - s(t_0)}{t_1 - t_0}$$

One type of motion we will study is the motion of an object falling to earth under the influence of gravity (assuming no air resistance). Galileo discovered that if the object is released at time $t = 0$ from a state of rest (Figure 2), then the distance traveled after t seconds is given by the formula

$$s(t) = 4.9t^2 \text{ m}$$

■ **EXAMPLE 1** A hammer falls off a scaffolding at time $t = 0$. Estimate the instantaneous velocity at $t = 0.8$ s.

Solution We use Galileo's formula $s(t) = 4.9t^2$ to compute the average velocity over the five short time intervals listed in Table 1. Consider the first interval $[t_0, t_1] = [0.8, 0.81]$:

$$\Delta s = s(0.81) - s(0.8) = 4.9(0.81)^2 - 4.9(0.8)^2 \approx 3.2149 - 3.1360 = 0.7889 \text{ m}$$

$$\Delta t = 0.81 - 0.8 = 0.01 \text{ s}$$

The average velocity over $[0.8, 0.81]$ is the ratio

$$\frac{\Delta s}{\Delta t} = \frac{s(0.81) - s(0.8)}{0.81 - 0.8} = \frac{0.07889}{0.01} = 7.889 \text{ m/s}$$

Table 1 shows the results of similar calculations for intervals of successively shorter lengths. It looks like these average velocities are getting closer to 7.84 m/s as the length of the time interval shrinks:

$$7.889, \quad 7.8645, \quad \textbf{7.84}05, \quad \textbf{7.84}024, \quad \textbf{7.84}0005$$

This suggests that 7.84 m/s is a good candidate for the instantaneous velocity at $t = 0.8$. ■

We express our conclusion in the previous example by saying the following:

*Average velocity **converges** to instantaneous velocity. Instantaneous velocity is the **limit** of average velocity as the length of the time interval shrinks to zero.*

We might write this as *instantaneous velocity* $= \lim\limits_{\Delta t \to 0}$ *(average velocity)*.

Graphical Interpretation of Velocity

The idea that average velocity converges to instantaneous velocity as we shorten the time interval has a vivid interpretation in terms of secant lines. The term **secant line** refers to a line through two points on a curve.

Consider the graph of position $s(t)$ for an object traveling in a straight line such as a bicycle (Figure 3). The ratio defining average velocity over $[t_0, t_1]$ is nothing more than the slope of the secant line through the points $(t_0, s(t_0))$ and $(t_1, s(t_1))$. For $t_1 \neq t_0$,

$$\text{Average velocity} = \text{slope of secant line} = \frac{\Delta s}{\Delta t} = \frac{s(t_1) - s(t_0)}{t_1 - t_0}$$

By interpreting average velocity as a slope, we can visualize what happens as the time interval gets smaller. Figure 4 shows the graph of position for the falling hammer of Example 1, where $s(t) = 4.9t^2$. As the time interval shrinks, the secant lines get closer to—*and seem to rotate into*—the tangent line at $t = 0.8$.

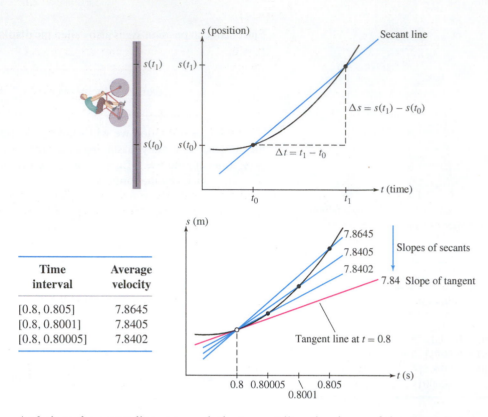

FIGURE 3 The average velocity over $[t_0, t_1]$ is equal to the slope of the secant line.

Time interval	Average velocity
[0.8, 0.805]	7.8645
[0.8, 0.8001]	7.8405
[0.8, 0.80005]	7.8402

DF **FIGURE 4** The secant lines "rotate into" the tangent line as the time interval shrinks. *Note:* The graph is not drawn to scale.

And since the secant lines approach the tangent line, the slopes of the secant lines get closer and closer to the slope of the tangent line. In other words, the statement

As the time interval shrinks to zero, the average velocity approaches the instantaneous velocity.

has the graphical interpretation

As the time interval shrinks to zero, the slope of the secant line approaches the slope of the tangent line.

We conclude the following:

Instantaneous velocity is equal to the slope of the tangent line to the graph of position as a function of time.

This conclusion and its generalization to other rates of change are of fundamental importance in differential calculus.

■ **EXAMPLE 2** Describe the motion and velocities of a shuttle train that runs on a straight track at the airport, ferrying passengers from one terminal to another according to the graph given in Figure 5.

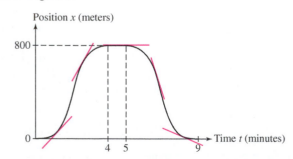

FIGURE 5 The position of a shuttle train traveling along a straight track.

Solution At time $t = 0$, the train is at position $x = 0$ on the track at the first terminal. It gradually begins moving, speeding up until around 2 min. It then begins slowing down, and reaches its destination 800 m down the track at $t = 4$ min. At this point, it stops. Notice that for t in the interval $(0, 4)$, the slopes of the tangent lines, which represent the velocity, are positive. So the instantaneous velocities are positive. The slopes of the tangent lines begin very small, increase, and then decrease to 0. In the interval [4, 5], the graph has a horizontal tangent line with slope 0. In this interval, the train is not moving, having instantaneous velocity 0. The shuttle train then moves in the reverse direction. The slopes of the tangent lines in the interval $(5, 9)$ are all negative. The speed increases and then decreases in this interval but the velocities are all negative. The train has returned to the original terminal at $x = 0$ on the track at time $t = 9$ min. ∎

Other Rates of Change

Velocity is only one of many examples of a rate of change. Our reasoning applies to any quantity y that depends on a variable x—say, $y = f(x)$. For any interval $[x_0, x_1]$, we set

$$\Delta f = f(x_1) - f(x_0), \qquad \Delta x = x_1 - x_0$$

Sometimes, we write Δy and $\Delta y/\Delta x$ instead of Δf and $\Delta f/\Delta x$.

For $x_1 \neq x_0$, the **average rate of change** of y with respect to x over $[x_0, x_1]$ is the ratio

$$\text{Average rate of change} = \frac{\Delta f}{\Delta x} = \underbrace{\frac{f(x_1) - f(x_0)}{x_1 - x_0}}_{\text{Slope of secant line}}$$

The word "instantaneous" is often dropped. When we use the term "rate of change," it is understood that the instantaneous rate is the intended not average rate.

The **instantaneous rate of change** at $x = x_0$ is the limit of the average rates of change. We estimate it by computing the average rate over smaller and smaller intervals.

In Example 1 above, we considered only right-hand intervals $[x_0, x_1]$. In the next example, we compute the average rate of change for intervals lying to both the left and the right of x_0.

■ **EXAMPLE 3** **Bacteria in a Petri Dish** Suppose the formula $B = 5\sqrt{T}$ yields a good approximation to the amount of bacteria, as measured by its weight, that grows in a particular petri dish (in milligrams) over one day as a function of the temperature T (in degrees centigrade) for temperatures in the range $0 \leq T \leq 25$. Estimate the instantaneous rate of change of B with respect to T when $T = 10°C$. What are the units of this rate?

Solution To estimate the instantaneous rate of change at $T = 10$, we compute the average rate for several intervals lying to the left and right of $T = 10$. For example, the average rate of change over [9.5, 10] is

$$\frac{B(10) - B(9.5)}{10 - 9.5} = \frac{5\sqrt{10} - 5\sqrt{9.5}}{0.5} \approx 0.8007$$

Tables 2 and 3 suggest that the instantaneous rate is approximately 0.790. This is the rate of increase in the weight of the bacteria with respect to temperature, so it has units of mg/°C, or *milligrams per degree centigrade*. The secant lines corresponding to the values in the tables are shown in Figures 6 and 7. ∎

TABLE 2 **Left-Hand Intervals**

Temperature interval	Average rate of change
[9.5, 10]	0.8007
[9.8, 10]	0.7946
[9.9, 10]	0.7926
[9.99, 10]	0.7908

TABLE 3 **Right-Hand Intervals**

Temperature interval	Average rate of change
[10, 10.5]	0.7809
[10, 10.2]	0.7867
[10, 10.1]	0.7886
[10, 10.01]	0.7904

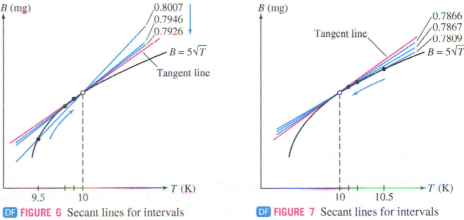

DF **FIGURE 6** Secant lines for intervals lying to the left of $T = 10$.

DF **FIGURE 7** Secant lines for intervals lying to the right of $T = 10$.

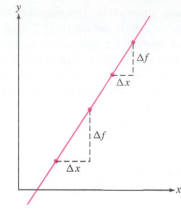

FIGURE 8 For a linear function $f(x) = mx + b$, the ratio $\Delta f/\Delta x$ is equal to the slope m for every interval.

To conclude this section, we recall an important point discussed in Section 1.2: For any linear function $f(x) = mx + b$, *the average rate of change over every interval is equal to the slope m* (Figure 8). We verify as follows:

$$\frac{\Delta f}{\Delta x} = \frac{f(x_1) - f(x_0)}{x_1 - x_0} = \frac{(mx_1 + b) - (mx_0 + b)}{x_1 - x_0} = \frac{m(x_1 - x_0)}{x_1 - x_0} = m$$

The instantaneous rate of change at $x = x_0$, which is the limit of these average rates, is also equal to m. This makes sense graphically because all secant lines and all tangent lines to the graph of f coincide with the graph itself.

2.1 SUMMARY

- The limit as x approaches a of a function f is a number L that $f(x)$ approaches, if such an L exists.
- The *average rate of change* of $y = f(x)$ over an interval $[x_0, x_1]$:

$$\text{Average rate of change} = \frac{\Delta f}{\Delta x} = \frac{f(x_1) - f(x_0)}{x_1 - x_0} \qquad (x_1 \neq x_0)$$

- The *instantaneous rate of change* is the limit of the average rates of change as the time interval shrinks.
- Graphical interpretation:

  - Average rate of change is the slope of the secant line through the points $(x_0, f(x_0))$ and $(x_1, f(x_1))$ on the graph of f.
  - Instantaneous rate of change is the slope of the tangent line at x_0.

- To estimate the instantaneous rate of change at $x = x_0$, compute the average rate of change over several intervals $[x_0, x_1]$ (or $[x_1, x_0]$) for x_1 close to x_0.
- The velocity of an object in linear motion is the rate of change of position $s(t)$.
- Linear function $f(x) = mx + b$: The average rate of change over every interval and the instantaneous rate of change at every point are equal to the slope m.

2.1 EXERCISES

Preliminary Questions

1. Average velocity is equal to the slope of a secant line through two points on a graph. Which graph?

2. Can instantaneous velocity be defined as a ratio? If not, how is instantaneous velocity computed?

3. What is the graphical interpretation of instantaneous velocity at a specific time $t = t_0$?

4. Find a graphical interpretation of the following statement: The average rate of change approaches the instantaneous rate of change as the interval $[x_0, x_1]$ shrinks to x_0.

5. The rate of change of atmospheric temperature with respect to altitude is equal to the slope of the tangent line to a graph. Which graph? What are possible units for this rate?

Exercises

1. A ball dropped from a state of rest at time $t = 0$ travels a distance $s(t) = 4.9t^2$ m in t seconds.

(a) How far does the ball travel during the time interval $[2, 2.5]$?

(b) Compute the average velocity over $[2, 2.5]$.

(c) Compute the average velocity for the time intervals in the table and estimate the ball's instantaneous velocity at $t = 2$.

Interval	[2, 2.01]	[2, 2.005]	[2, 2.001]	[2, 2.00001]
Average velocity				

2. A wrench dropped from a state of rest at time $t = 0$ travels a distance $s(t) = 4.9t^2$ m in t seconds. Estimate the instantaneous velocity at $t = 3$.

3. Let $B = 5\sqrt{T}$ as in Example 3. Estimate the instantaneous rate of change of B with respect to T when $T = 20°$C.

4. Compute $\Delta y/\Delta x$ for the interval $[2, 5]$, where $y = 4x - 9$. What is the instantaneous rate of change of y with respect to x at $x = 2$?

In Exercises 5–6, a stone is tossed vertically into the air from ground level with an initial velocity of 15 m/s. Its height at time t is $h(t) = 15t - 4.9t^2$ m.

5. Compute the stone's average velocity over the time interval [0.5, 2.5] and indicate the corresponding secant line on a sketch of the graph of h.

6. Compute the stone's average velocity over the time intervals [1, 1.01], [1, 1.001], [1, 1.0001] and [0.99, 1], [0.999, 1], [0.9999, 1], and then estimate the instantaneous velocity at $t = 1$.

7. With an initial deposit of $100, and an interest rate of 8%, the balance in a bank account after t years is $f(t) = 100(1.08)^t$ dollars.

(a) What are the units of the rate of change of $f(t)$?

(b) Find the average rate of change over [0, 0.5] and [0, 1].

(c) Estimate the instantaneous rate of change at $t = 0.5$ by computing the average rate of change over intervals to the left and right of $t = 0.5$.

8. The position of a particle at time t is $s(t) = t^3 + t$. Compute the average velocity over the time interval [1, 4] and estimate the instantaneous velocity at $t = 1$.

9. 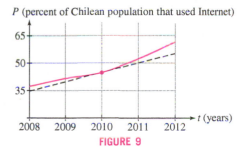 Figure 9 shows the estimated percentage P of the Chilean population that uses the Internet, based on data from the United Nations Statistics Division.

(a) Estimate the rate of change of P at $t = 2010$.

(b) Does the rate of change increase or decrease as t increases? Explain graphically.

(c) Let R be the average rate of change over [2008, 2012]. Compute R.

(d) Is the rate of change at $t = 2012$ greater than or less than the average rate R? What about the rate at $t = 2008$? Explain graphically.

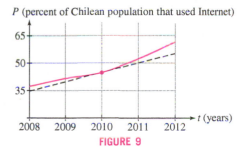

P (percent of Chilean population that used Internet)

FIGURE 9

10. The **atmospheric temperature** T (in degrees Celsius) at altitude h meters above a certain point on Earth can be approximated by $T = 15 - 0.0065h$ for $h \leq 12{,}000$ m. What are the average and instantaneous rates of change of T with respect to h? Why are they the same? Sketch the graph of T for $h \leq 12{,}000$.

In Exercises 11--18, estimate the instantaneous rate of change at the point indicated.

11. $P(x) = 3x^2 - 5$; $x = 2$

12. $f(t) = 12t - 7$; $t = -4$

13. $y(x) = \dfrac{1}{x+2}$; $x = 2$

14. $y(t) = \sqrt{3t+1}$; $t = 1$

15. $f(x) = e^x$; $x = 0$

16. $f(x) = e^x$; $x = e$

17. $f(x) = \ln x$; $x = 3$

18. $f(x) = \tan x$; $x = \dfrac{\pi}{4}$

19. The height (in centimeters) at time t (in seconds) of a small mass oscillating at the end of a spring is $h(t) = 8\cos(12\pi t)$.

(a) Calculate the mass's average velocity over the time intervals [0, 0.1] and [3, 3.5].

(b) Estimate its instantaneous velocity at $t = 3$.

20. 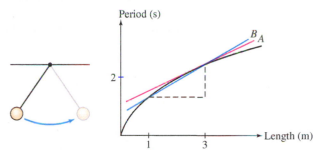 Assume that the period T (in seconds) of a pendulum (the time required for a complete back-and-forth cycle) is $T = \frac{3}{2}\sqrt{L}$, where L is the pendulum's length (in meters).

(a) What are the units for the rate of change of T with respect to L? Explain what this rate measures.

(b) Which quantities are represented by the slopes of lines A and B in Figure 10?

(c) Estimate the instantaneous rate of change of T with respect to L when $L = 3$ m.

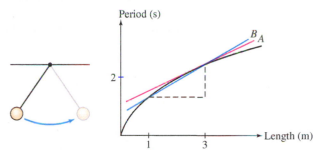

FIGURE 10 The period T is the time required for a pendulum to swing back and forth.

21. The number $P(t)$ of *E. coli* cells at time t (hours) in a petri dish is plotted in Figure 11.

(a) Calculate the average rate of change of $P(t)$ over the time interval [1, 3] and draw the corresponding secant line.

(b) Estimate the slope m of the line in Figure 11. What does m represent?

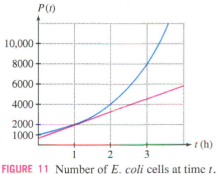

FIGURE 11 Number of *E. coli* cells at time t.

22. The graphs in Figure 12 represent the positions of moving particles as functions of time.

(a) Do the instantaneous velocities at times t_1, t_2, t_3 in (A) form an increasing or a decreasing sequence?

(b) Is the particle speeding up or slowing down in (A)?

(c) Is the particle speeding up or slowing down in (B)?

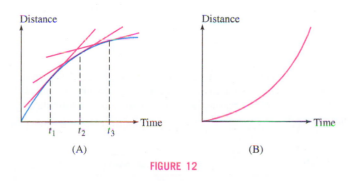

FIGURE 12

23. GU   An advertising campaign boosted sales of Crunchy Crust frozen pizza to a peak level of S_0 dollars per month. A marketing study showed that after t months, monthly sales declined to

$$S(t) = S_0 g(t), \quad \text{where } g(t) = \frac{1}{\sqrt{1+t}}$$

Do sales decline more slowly or more rapidly as time increases? Answer by referring to a sketch of the graph of g together with several tangent lines.

24. The fraction of a city's population infected by a flu virus is plotted as a function of time (in weeks) in Figure 13.

(a) Which quantities are represented by the slopes of lines A and B? Estimate these slopes.

(b) Is the flu spreading more rapidly at $t = 1, 2$, or 3?

(c) Is the flu spreading more rapidly at $t = 4, 5$, or 6?

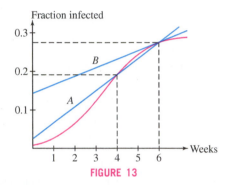

FIGURE 13

25. The graphs in Figure 14 represent the positions s of moving particles as functions of time t. Match each graph with a description:

(a) Speeding up

(b) Speeding up and then slowing down

(c) Slowing down

(d) Slowing down and then speeding up

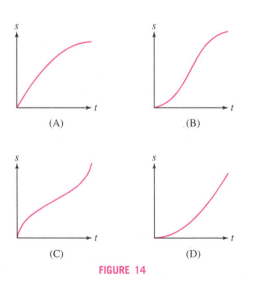

FIGURE 14

26. An epidemiologist finds that the percentage $N(t)$ of susceptible children who are sick on day t during the first 3 weeks of a measles outbreak is given, to a reasonable approximation, by the formula (Figure 15)

$$N(t) = \frac{100t^2}{t^3 + 5t^2 - 100t + 380}$$

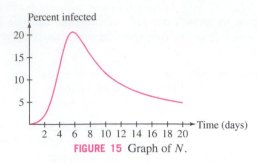

FIGURE 15   Graph of N.

(a) Draw the secant line whose slope is the average rate of change in infected children over the intervals $[4, 6]$ and $[12, 14]$. Then compute these average rates (in units of percent per day).

(b) Is the rate of decline greater at $t = 8$ or $t = 16$?

(c) Estimate the rate of change of $N(t)$ on day 12.

27. The fungus *Fusarium exosporium* infects a field of flax plants through the roots and causes the plants to wilt. Eventually, the entire field is infected. The percentage $f(t)$ of infected plants as a function of time t (in days) since planting is shown in Figure 16.

(a) What are the units of the rate of change of $f(t)$ with respect to t? What does this rate measure?

(b) Use the graph to rank (from smallest to largest) the average infection rates over the intervals $[0, 12]$, $[20, 32]$, and $[40, 52]$.

(c) Use the following table to compute the average rates of infection over the intervals $[30, 40]$, $[40, 50]$, $[30, 50]$:

Days	0	10	20	30	40	50	60
Percent infected	0	18	56	82	91	96	98

(d) Draw the tangent line at $t = 40$ and estimate its slope.

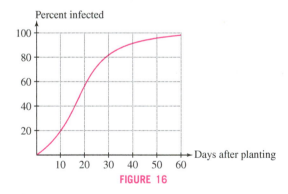

FIGURE 16

28. ✎   Let $B = 5\sqrt{T}$ as in Example 3. Is the rate of change of B with respect to T greater at low temperatures or high temperatures? Explain in terms of the graph.

29. ✎   If an object in linear motion (but with changing velocity) covers Δs meters in Δt seconds, then its average velocity is $v_0 = \Delta s / \Delta t$ m/s. Show that it would cover the same distance if it traveled at constant velocity v_0 over the same time interval. This justifies our calling $\Delta s / \Delta t$ the *average velocity*.

30. ✎   Sketch the graph of $f(x) = x(1 - x)$ over $[0, 1]$. Refer to the graph and, without making any computations, find:

(a) the average rate of change over $[0, 1]$.

(b) the (instantaneous) rate of change at $x = \frac{1}{2}$.

(c) the values of x at which the rate of change is positive.

31. 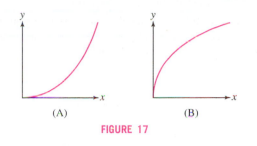 Which graph in Figure 17 has the following property: For all x, the average rate of change over $[0, x]$ is greater than the instantaneous rate of change at x. Explain.

(A) (B)

FIGURE 17

Further Insights and Challenges

32. The height of a projectile fired in the air vertically with initial velocity 25 m/s is

$$h(t) = 25t - 4.9t^2 \text{ m}$$

(a) Compute $h(1)$. Show that $h(t) - h(1)$ can be factored with $(t - 1)$ as a factor.

(b) Using part (a), show that the average velocity over the interval $[1, t]$ is $20.1 - 4.9t$.

(c) Use this formula to find the average velocity over several intervals $[1, t]$ with t close to 1. Then estimate the instantaneous velocity at time $t = 1$.

33. Let $Q(t) = t^2$. As in the previous exercise, find a formula for the average rate of change of Q over the interval $[1, t]$ and use it to estimate the instantaneous rate of change at $t = 1$. Repeat for the interval $[2, t]$ and estimate the rate of change at $t = 2$.

34. Show that the average rate of change of $f(x) = x^3$ over $[1, x]$ is equal to

$$x^2 + x + 1$$

Use this to estimate the instantaneous rate of change of $f(x)$ at $x = 1$.

35. Find a formula for the average rate of change of $f(x) = x^3$ over $[2, x]$ and use it to estimate the instantaneous rate of change at $x = 2$.

36. Let $T = \frac{3}{2}\sqrt{L}$ as in Exercise 20. The numbers in the second column of Table 4 are increasing, and those in the last column are decreasing. Explain why in terms of the graph of T as a function of L. Also, explain graphically why the instantaneous rate of change at $L = 3$ lies between 0.43299 and 0.43303.

TABLE 4   Average Rates of Change of T with Respect to L

Interval	Average rate of change	Interval	Average rate of change
[3, 3.2]	0.42603	[2.8, 3]	0.44048
[3, 3.1]	0.42946	[2.9, 3]	0.43668
[3, 3.001]	0.43298	[2.999, 3]	0.43305
[3, 3.0005]	0.43299	[2.9995, 3]	0.43303

2.2 Limits: A Numerical and Graphical Approach

The goal in this section is to define limits and study them using numerical and graphical techniques. We begin with the following question: *How do the values of a function f behave when x approaches a number c, whether or not $f(c)$ is defined?*

To explore this question, we'll experiment with the function

$$f(x) = \frac{\sin x}{x} \quad (x \text{ in radians})$$

The undefined expression 0/0 is referred to as an "indeterminate form."

Notice that $f(0)$ is not defined. In fact, when we set $x = 0$ in

$$f(x) = \frac{\sin x}{x}$$

we obtain the undefined expression 0/0 because $\sin 0 = 0$. Nevertheless, we can compute $f(x)$ for values of x *close* to 0. When we do this, a clear trend emerges.

To describe the trend, we use the phrase "x approaches 0" or "x tends to 0" to indicate that x takes on values (both positive and negative) that get closer and closer to 0. The notation for this is $x \to 0$, and more specifically we write

- $x \to 0^+$ if x approaches 0 from the right (on the number line).
- $x \to 0^-$ if x approaches 0 from the left (on the number line).

Now consider the values listed in Table 1. The table gives the unmistakable impression that $f(x)$ gets closer and closer to 1 as $x \to 0^+$ and as $x \to 0^-$.

This conclusion is supported by the graph of f in Figure 1. The point $(0, 1)$ is missing from the graph because $f(x)$ is not defined at $x = 0$, but the graph approaches

TABLE 1

x	$\dfrac{\sin x}{x}$	x	$\dfrac{\sin x}{x}$
1	0.841470985	−1	0.841470985
0.5	0.958851077	−0.5	0.958851077
0.1	0.998334166	−0.1	0.998334166
0.05	0.999583385	−0.05	0.999583385
0.01	0.999983333	−0.01	0.999983333
0.005	0.999995833	−0.005	0.999995833
0.001	0.999999833	−0.001	0.999999833
$x \to 0^+$	$f(x) \to 1$	$x \to 0^-$	$f(x) \to 1$

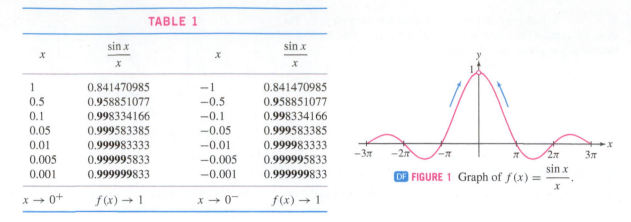

DF FIGURE 1 Graph of $f(x) = \dfrac{\sin x}{x}$.

this missing point as x approaches 0 from the left and right. We say that the *limit* of $f(x)$ as $x \to 0$ is equal to 1, and we write

$$\lim_{x \to 0} f(x) = 1$$

We also say that $f(x)$ *approaches* or *converges* to 1 as $x \to 0$.

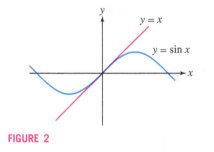

FIGURE 2

The fact that $\frac{\sin x}{x}$ approaches 1 for x approaching 0 can be interpreted to mean that the two functions $y = \sin x$ and $y = x$ behave similarly as x approaches 0. Their graphs in Figure 2 make it clear this is the case. The closer we get to $x = 0$, the closer the two graphs become.

CONCEPTUAL INSIGHT   The numerical and graphical evidence may convince us that $f(x) = \dfrac{\sin x}{x}$ converges to 1 as $x \to 0$, but since $f(0)$ yields the undefined expression 0/0, could we not arrive at this conclusion more simply by saying that 0/0 is equal to 1? The answer is no. *Algebra does not allow us to divide by* 0 *under any circumstances*, and it is not correct to say that 0/0 equals 1 or any other number.

What we have learned, however, is that a function f may approach a limit as $x \to c$ even if the formula for $f(c)$ produces the undefined expression 0/0. The limit of $f(x) = \dfrac{\sin x}{x}$ turns out to be 1. We will encounter other examples where the formula for $f(c)$ produces 0/0 but the limit is a number other than 1 (or the limit does not exist).

Definition of a Limit

The concept of a limit was not fully clarified until the nineteenth century. The French mathematician Augustin-Louis Cauchy (1789–1857, pronounced Koh-shee) gave the following verbal definition: "When the values successively attributed to the same variable approach a fixed value indefinitely, in such a way as to end up differing from it by as little as one could wish, this last value is called the limit of all the others. So, for example, an irrational number is the limit of the various fractions which provide values that approximate it more and more closely." (Translated by J. Grabiner)

To define limits, let us recall that the distance between two numbers a and b is the absolute value $|a - b|$. So we can express the idea that $f(x)$ is close to L by saying that $|f(x) - L|$ is small.

DEFINITION   Limit   Assume that $f(x)$ is defined for all x in an open interval containing c, but not necessarily at c itself. We say that

the limit of $f(x)$ as x approaches c is equal to the number L

if $|f(x) - L|$ can be made arbitrarily small by taking x sufficiently close (but not equal) to c. In this case, we write

$$\lim_{x \to c} f(x) = L$$

We also say that $f(x)$ *approaches* or *converges* to L as $x \to c$ [and we write $f(x) \to L$].

In other words, as x approaches c, $f(x)$ approaches L. See Figure 3 for the graphical interpretation. If the values of $f(x)$ do not converge to any number L as $x \to c$, we say that $\lim_{x \to c} f(x)$ *does not exist*. It is important to note that the value $f(c)$ itself, which may or may not be defined, plays no role in the limit. All that matters are the values of $f(x)$ for x close to c. Furthermore, if $f(x)$ approaches a limit as $x \to c$, then the limiting value L is unique.

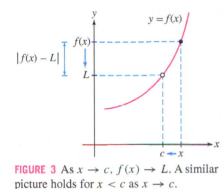

FIGURE 3 As $x \to c$, $f(x) \to L$. A similar picture holds for $x < c$ as $x \to c$.

■ **EXAMPLE 1** Use the definition above to verify the following limits:

(a) $\lim_{x \to 7} 5 = 5$

(b) $\lim_{x \to 4} (3x + 1) = 13$

Solution

(a) Let $f(x) = 5$. To show that $\lim_{x \to 7} f(x) = 5$, we must show that $|f(x) - 5|$ becomes arbitrarily small when x is sufficiently close (but not equal) to 7. But observe that $|f(x) - 5| = |5 - 5| = 0$ *for all* x, so what we are required to show is automatic (and it is not necessary to take x close to 7).

(b) Let $f(x) = 3x + 1$. To show that $\lim_{x \to 4} (3x + 1) = 13$, we must show that $|f(x) - 13|$ becomes arbitrarily small when x is sufficiently close (but not equal) to 4. We have

$$|f(x) - 13| = |(3x + 1) - 13| = |3x - 12| = 3|x - 4|$$

Because $|f(x) - 13|$ is a multiple of $|x - 4|$, we can make $|f(x) - 13|$ arbitrarily small by taking x sufficiently close to 4. ■

Reasoning as in Example 1 but with arbitrary constants, we obtain the following simple but important results:

THEOREM 1 For any constants k and c, (a) $\lim_{x \to c} k = k$, (b) $\lim_{x \to c} x = c$.

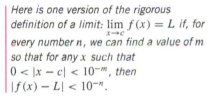

Here is one version of the rigorous definition of a limit: $\lim_{x \to c} f(x) = L$ if, for every number n, we can find a value of m so that for any x such that $0 < |x - c| < 10^{-m}$, then $|f(x) - L| < 10^{-n}$.

To deal with more complicated limits and, especially, to provide mathematically rigorous proofs, a more precise version of the above limit definition is needed. This more precise version is discussed in Section 2.9, where inequalities are used to pin down the exact meaning of the phrases "arbitrarily small" and "sufficiently close."

Graphical and Numerical Investigation

Our goal in the rest of this section is to develop a better intuitive understanding of limits by investigating them graphically and numerically.

Graphical Investigation Use a graphing utility to produce a graph of f. The graph should give a visual impression of whether or not a limit exists. It can often be used to estimate the value of the limit.

Numerical Investigation We write $x \to c^-$ to indicate that x approaches c through values less than c (i.e., from the left), and we write $x \to c^+$ to indicate that x approaches c through values greater than c (i.e., from the right). To investigate $\lim_{x \to c} f(x)$,

(i) Make a table of values of $f(x)$ for x close to but less than c—that is, as $x \to c^-$.

(ii) Make a second table of values of $f(x)$ for x close to but greater than c—that is, as $x \to c^+$.

(iii) If both tables indicate convergence to the same number L, we take L to be an estimate for the limit.

Keep in mind that graphical and numerical investigations provide evidence for a limit, but they do not prove that the limit exists or has a given value. This is done using the Limit Laws established in the following sections.

The tables should contain enough values to reveal a clear trend of convergence to a value L. If $f(x)$ approaches a limit, the successive values of $f(x)$ will generally agree to more and more decimal places as x is taken closer to c. If no pattern emerges, then the limit may not exist.

■ **EXAMPLE 2** Investigate $\lim\limits_{x \to 9} \dfrac{x-9}{\sqrt{x}-3}$ graphically and numerically.

Solution The function $f(x) = \dfrac{x-9}{\sqrt{x}-3}$ is undefined at $x = 9$ because the formula for $f(9)$ leads to the undefined expression $0/0$. Therefore, the graph in Figure 4 has a gap at $x = 9$. However, the graph suggests that $f(x)$ approaches 6 as $x \to 9$.

For numerical evidence, we consider a table of values of $f(x)$ for x approaching 9 from both the left and the right. Table 2 supports our impression that

$$\lim_{x \to 9} \frac{x-9}{\sqrt{x}-3} = 6$$

We will see shortly that, in fact, we can write $\lim\limits_{x \to 9} \dfrac{x-9}{\sqrt{x}-3} = \lim\limits_{x \to 9} \dfrac{\left(\sqrt{x}-3\right)\left(\sqrt{x}+3\right)}{\sqrt{x}-3} =$ $\lim\limits_{x \to 9} (\sqrt{x}+3) = 6$. Thus, the algebraic solution will confirm our numerical and geometric conclusion. ■

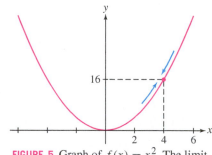

DF **FIGURE 4** Graph of $f(x) = \dfrac{x-9}{\sqrt{x}-3}$.

		TABLE 2		
$x \to 9^-$	$\dfrac{x-9}{\sqrt{x}-3}$	$x \to 9^+$	$\dfrac{x-9}{\sqrt{x}-3}$
8.9	**5.98**329	9.1	**6.0**1662
8.99	**5.99**833	9.01	**6.00**1666
8.999	**5.999**83	9.001	**6.000**167
8.9999	**5.9999**833	9.0001	**6.0000**167

■ **EXAMPLE 3** **Limit Equals Value of the Function** Investigate $\lim\limits_{x \to 4} x^2$.

Solution Figure 5 and Table 3 both suggest that $\lim\limits_{x \to 4} x^2 = 16$. But $f(x) = x^2$ is defined at $x = 4$ and $f(4) = 16$, so in this case, *the limit is equal to the function value*. This pleasant conclusion is valid whenever f is a *continuous* function, a concept treated in Section 2.4. ■

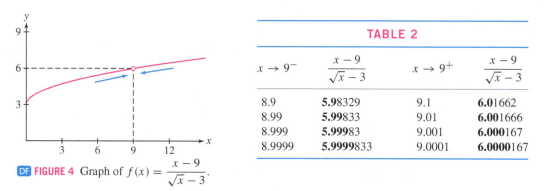

FIGURE 5 Graph of $f(x) = x^2$. The limit is equal to the value of the function $f(4) = 16$.

		TABLE 3		
$x \to 4^-$	x^2	$x \to 4^+$	x^2
3.9	15.21	4.1	16.81
3.99	**15.9**201	4.01	**16.0**801
3.999	**15.99**2001	4.001	**16.00**8001
3.9999	**15.999**20001	4.0001	**16.000**80001

■ **EXAMPLE 4** **Defining Property of e** Verify numerically that $\lim\limits_{h \to 0} \dfrac{e^h - 1}{h} = 1$.

Solution The function $f(h) = (e^h - 1)/h$ is undefined at $h = 0$, but both Figure 6 and Table 4 suggest that $\lim\limits_{h \to 0} (e^h - 1)/h = 1$. We will ultimately see that the fact this limit is equal to 1 reflects the fact that the tangent line to the graph of $y = e^x$ at $x = 0$ has slope 1, as in Figure 3 from Section 1.6. ■

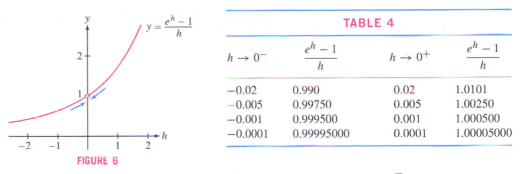

FIGURE 6

TABLE 4			
$h \to 0^-$	$\dfrac{e^h - 1}{h}$	$h \to 0^+$	$\dfrac{e^h - 1}{h}$
-0.02	0.990	0.02	1.0101
-0.005	0.99750	0.005	1.00250
-0.001	0.999500	0.001	1.000500
-0.0001	0.99995000	0.0001	1.00005000

CAUTION *Numerical investigations are often suggestive, but may be misleading in some cases. If, in Example 5, we had chosen to evaluate $f(x) = \sin \dfrac{\pi}{x}$ at the values $x = 0.1, 0.01, 0.001, \dots$, we might have concluded incorrectly that $f(x)$ approaches the limit 0 as $x \to 0$. The problem is that $f(10^{-n}) = \sin(10^n \pi) = 0$ for every whole number n, but $f(x)$ itself does not approach any limit.*

■ **EXAMPLE 5** **A Limit That Does Not Exist** Investigate $\lim\limits_{x \to 0} \sin \dfrac{\pi}{x}$ graphically and numerically.

Solution The function $f(x) = \sin \dfrac{\pi}{x}$ is not defined at $x = 0$, but Figure 7 suggests that it oscillates between $+1$ and -1 infinitely often as $x \to 0$. It appears, therefore, that $\lim\limits_{x \to 0} \sin \dfrac{\pi}{x}$ does not exist. This impression is confirmed by Table 5, which shows that the values of $f(x)$ bounce around and do not tend toward any limit L as $x \to 0$. ■

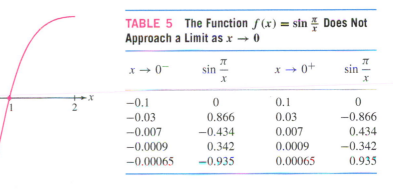

FIGURE 7 Graph of $f(x) = \sin \dfrac{\pi}{x}$.

TABLE 5	The Function $f(x) = \sin \dfrac{\pi}{x}$ Does Not Approach a Limit as $x \to 0$		
$x \to 0^-$	$\sin \dfrac{\pi}{x}$	$x \to 0^+$	$\sin \dfrac{\pi}{x}$
-0.1	0	0.1	0
-0.03	0.866	0.03	-0.866
-0.007	-0.434	0.007	0.434
-0.0009	0.342	0.0009	-0.342
-0.00065	-0.935	0.00065	0.935

One-Sided Limits

The limits discussed so far are *two-sided*. To show that $\lim\limits_{x \to c} f(x) = L$, it is necessary to check that $f(x)$ converges to L as x approaches c through values both larger and smaller than c. In some instances, $f(x)$ may approach L from one side of c without necessarily approaching it from the other side, or $f(x)$ may be defined on only one side of c. For this reason, we define the one-sided limits

$$\lim_{x \to c^-} f(x) \quad \text{(left-hand limit)}, \qquad \lim_{x \to c^+} f(x) \quad \text{(right-hand limit)}$$

The limit itself exists if both one-sided limits exist and are equal.

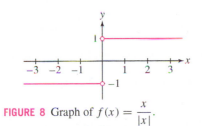

FIGURE 8 Graph of $f(x) = \dfrac{x}{|x|}$.

■ **EXAMPLE 6** **Left- and Right-Hand Limits Not Equal** Investigate the one-sided limits of $f(x) = \dfrac{x}{|x|}$ as $x \to 0$. Does $\lim\limits_{x \to 0} f(x)$ exist?

Solution Figure 8 shows what is going on. For $x < 0$,

$$f(x) = \frac{x}{|x|} = \frac{x}{-x} = -1$$

Therefore, the left-hand limit is $\lim\limits_{x \to 0^-} f(x) = -1$. But for $x > 0$,

$$f(x) = \frac{x}{|x|} = \frac{x}{x} = 1$$

Therefore, $\lim\limits_{x \to 0^+} f(x) = 1$. These one-sided limits are not equal, so $\lim\limits_{x \to 0} f(x)$ does not exist. ■

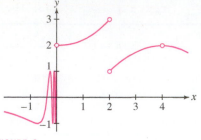

FIGURE 9

■ EXAMPLE 7 The function f in Figure 9 is not defined at $c = 0, 2, 4$. Investigate the one- and two-sided limits at these points.

Solution

- $c = 0$: The left-hand limit $\lim_{x \to 0^-} f(x)$ does not seem to exist because $f(x)$ appears to oscillate infinitely often to the left of $x = 0$. On the other hand, $\lim_{x \to 0^+} f(x) = 2$.
- $c = 2$: The one-sided limits exist but are not equal:

$$\lim_{x \to 2^-} f(x) = 3 \qquad \text{and} \qquad \lim_{x \to 2^+} f(x) = 1$$

Therefore, $\lim_{x \to 2} f(x)$ does not exist.

- $c = 4$: The one-sided limits exist and both have the value 2. Therefore, the two-sided limit exists and $\lim_{x \to 4} f(x) = 2$.   ■

Infinite Limits

For some functions, $f(x)$ tends to ∞ or $-\infty$ as x approaches a value c. If so, $\lim_{x \to c} f(x)$ does not exist, but we say that $f(x)$ has an *infinite limit*. More precisely, we write

- $\lim_{x \to c} f(x) = \infty$ if $f(x)$ increases without bound as $x \to c$.
- $\lim_{x \to c} f(x) = -\infty$ if $f(x)$ decreases without bound as $x \to c$.

Here, "decrease without bound" means that $f(x)$ becomes negative and $|f(x)| \to \infty$. One-sided infinite limits are defined similarly. When using this notation, keep in mind that ∞ and $-\infty$ are not numbers.

When $f(x)$ approaches ∞ or $-\infty$ as x approaches c from one or both sides, the line $x = c$ is called a **vertical asymptote**. In Figure 10, the line $x = 2$ is a vertical asymptote in (A), and $x = 0$ is a vertical asymptote in both (B) and (C).

In the next example, the notation $x \to c^{\pm}$ is used to indicate that the left- and right-hand limits are to be considered separately.

■ EXAMPLE 8 |GU| Investigate the one-sided limits graphically:

(a) $\lim_{x \to 2^{\pm}} \dfrac{1}{x - 2}$ **(b)** $\lim_{x \to 0^{\pm}} \dfrac{1}{x^2}$ **(c)** $\lim_{x \to 0^+} \ln x$

Solution

(a) Figure 10(A) suggests that

$$\lim_{x \to 2^-} \frac{1}{x - 2} = -\infty, \qquad \lim_{x \to 2^+} \frac{1}{x - 2} = \infty$$

The vertical line $x = 2$ is a vertical asymptote. Why are the one-sided limits different? Because $f(x) = \dfrac{1}{x - 2}$ is negative for $x < 2$ (so the limit from the left is $-\infty$) and $f(x)$ is positive for $x > 2$ (so the limit from the right is ∞).

(b) Figure 10(B) suggests that $\lim_{x \to 0} \dfrac{1}{x^2} = \infty$. Indeed, $f(x) = \dfrac{1}{x^2}$ is positive for all $x \neq 0$ and becomes arbitrarily large as $x \to 0$ from either side. The line $x = 0$ is a vertical asymptote.

(c) Figure 10(C) suggests that $\lim_{x \to 0^+} \ln x = -\infty$ because $f(x) = \ln x$ is negative for $0 < x < 1$ and tends to $-\infty$ as $x \to 0^+$. The line $x = 0$ is a vertical asymptote.   ■

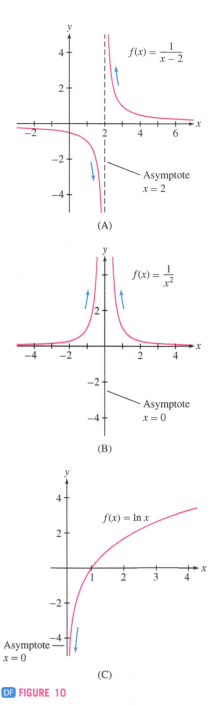

(A)

(B)

(C)

DF FIGURE 10

CONCEPTUAL INSIGHT You should not think of an infinite limit as a true limit. The notation $\lim_{x\to c} f(x) = \infty$ is merely a shorthand way of saying that $f(x)$ increases beyond all bounds as x approaches c. The limit itself does not exist. We must be careful when using this notation because ∞ and $-\infty$ *are not numbers*, and contradictions can arise if we try to manipulate them as numbers. For example, if ∞ were a number, it would be larger than any finite number, and presumably, $\infty + 1 = \infty$. But then

$$\infty + 1 = \infty$$

$$(\infty + 1) - \infty = \infty - \infty$$

$$1 = 0 \qquad \text{(contradiction!)}$$

To avoid errors, keep in mind the ∞ is not a number but rather a convenient shorthand notation.

2.2 SUMMARY

- By definition, $\lim_{x\to c} f(x) = L$ if $|f(x) - L|$ can be made arbitrarily small by taking x sufficiently close (but not equal) to c. We say that

    - *The limit of $f(x)$ as x approaches c is L,* or

    - *$f(x)$ approaches (or converges) to L as x approaches c.*

- If $f(x)$ approaches a limit as $x \to c$, then the value of the limit L is unique.
- If $f(x)$ does not approach a limit as $x \to c$, we say that $\lim_{x\to c} f(x)$ does not exist.
- The limit may exist even if $f(c)$ is not defined.
- *One-sided limits*:

    - $\lim_{x\to c^-} f(x) = L$ if $f(x)$ converges to L as x approaches c through values less than c.

    - $\lim_{x\to c^+} f(x) = L$ if $f(x)$ converges to L as x approaches c through values greater than c.

- The limit exists if and only if both one-sided limits exist and are equal.
- *Infinite limits*: $\lim_{x\to c} f(x) = \infty$ if $f(x)$ increases beyond bound as x approaches c, and $\lim_{x\to c} f(x) = -\infty$ if $f(x)$ becomes arbitrarily large (in absolute value) but negative as x approaches c.
- In the case of a one- or two-sided infinite limit, the vertical line $x = c$ is called a *vertical asymptote*.

2.2 EXERCISES

Preliminary Questions

1. What is the limit of $f(x) = 1$ as $x \to \pi$?

2. What is the limit of $g(t) = t$ as $t \to \pi$?

3. Is $\lim_{x\to 10} 20$ equal to 10 or 20?

4. Can $f(x)$ approach a limit as $x \to c$ if $f(c)$ is undefined? If so, give an example.

5. What does the following table suggest about $\lim_{x\to 1^-} f(x)$ and $\lim_{x\to 1^+} f(x)$?

x	0.9	0.99	0.999	1.1	1.01	1.001
$f(x)$	7	25	4317	3.0126	3.0047	3.00011

6. Can you tell whether $\lim_{x\to 5} f(x)$ exists from a plot of f for $x > 5$? Explain.

7. If you know in advance that $\lim_{x\to 5} f(x)$ exists, can you determine its value from a plot of f for all $x > 5$?

Exercises

In Exercises 1–4, fill in the tables and guess the value of the limit.

1. $\lim\limits_{x \to 1} f(x)$, where $f(x) = \dfrac{x^3 - 1}{x^2 - 1}$.

x	$f(x)$	x	$f(x)$
1.002		0.998	
1.001		0.999	
1.0005		0.9995	
1.00001		0.99999	

2. $\lim\limits_{t \to 0} h(t)$, where $h(t) = \dfrac{\cos t - 1}{t^2}$. Note that h is even; that is, $h(t) = h(-t)$.

t	± 0.002	± 0.0001	± 0.00005	± 0.00001
$h(t)$				

3. $\lim\limits_{y \to 2} f(y)$, where $f(y) = \dfrac{y^2 - y - 2}{y^2 + y - 6}$.

y	$f(y)$	y	$f(y)$
2.002		1.998	
2.001		1.999	
2.0001		1.9999	

4. $\lim\limits_{x \to 0^+} f(x)$, where $f(x) = x \ln x$.

x	1	0.5	0.1	0.05	0.01	0.005	0.001
$f(x)$							

5. Determine $\lim\limits_{x \to 0.5} f(x)$ for f as in Figure 11.

6. Determine $\lim\limits_{x \to 0.5} g(x)$ for g as in Figure 12.

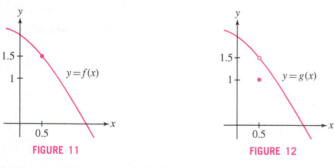

FIGURE 11   **FIGURE 12**

In Exercises 7–8, evaluate the limit.

7. $\lim\limits_{x \to 21} x$

8. $\lim\limits_{x \to 4.2} \sqrt{3}$

In Exercises 9–16, verify each limit using the limit definition. For example, in Exercise 9, show that $|3x - 12|$ can be made as small as desired by taking x close to 4.

9. $\lim\limits_{x \to 4} 3x = 12$

10. $\lim\limits_{x \to 5} 3 = 3$

11. $\lim\limits_{x \to 3} (5x + 2) = 17$

12. $\lim\limits_{x \to 2} (7x - 4) = 10$

13. $\lim\limits_{x \to 0} x^2 = 0$

14. $\lim\limits_{x \to 0} (3x^2 - 9) = -9$

15. $\lim\limits_{x \to 0} (4x^2 + 2x + 5) = 5$

16. $\lim\limits_{x \to 0} (x^3 + 12) = 12$

In Exercises 17–38, estimate the limit numerically or state that the limit does not exist. If infinite, state whether the one-sided limits are ∞ or $-\infty$.

17. $\lim\limits_{x \to 1} \dfrac{\sqrt{x} - 1}{x - 1}$

18. $\lim\limits_{x \to -4} \dfrac{2x^2 - 32}{x + 4}$

19. $\lim\limits_{x \to 2} \dfrac{x^2 + x - 6}{x^2 - x - 2}$

20. $\lim\limits_{x \to 3} \dfrac{x^3 - 2x^2 - 9}{x^2 - 2x - 3}$

21. $\lim\limits_{x \to 0} \dfrac{\sin 2x}{x}$

22. $\lim\limits_{x \to 0} \dfrac{\sin 5x}{x}$

23. $\lim\limits_{\theta \to 0} \dfrac{\cos \theta - 1}{\theta}$

24. $\lim\limits_{x \to 0} \dfrac{\sin x}{x^2}$

25. $\lim\limits_{x \to 4} \dfrac{1}{(x - 4)^3}$

26. $\lim\limits_{x \to 1^-} \dfrac{3 - x}{x - 1}$

27. $\lim\limits_{x \to -3} \dfrac{x + 3}{x^2 + x - 6}$

28. $\lim\limits_{x \to -2^-} \dfrac{x + 1}{x + 2}$

29. $\lim\limits_{x \to 3^+} \dfrac{x - 4}{x^2 - 9}$

30. $\lim\limits_{h \to 0} \dfrac{3^h - 1}{h}$

31. $\lim\limits_{h \to 0} \sin h \cos \dfrac{1}{h}$

32. $\lim\limits_{h \to 0} \cos \dfrac{1}{h}$

33. $\lim\limits_{x \to 0} |x|^x$

34. $\lim\limits_{x \to 1^+} \dfrac{\sec^{-1} x}{\sqrt{x - 1}}$

35. $\lim\limits_{t \to e} \dfrac{t - e}{\ln t - 1}$

36. $\lim\limits_{r \to 0} (1 + r)^{1/r}$

37. $\lim\limits_{x \to 1^-} \dfrac{\tan^{-1} x}{\cos^{-1} x}$

38. $\lim\limits_{x \to 0} \dfrac{\tan^{-1} x - x}{\sin^{-1} x - x}$

39. The **greatest integer function**, also known as the **floor function**, is defined by $\lfloor x \rfloor = n$, where n is the unique integer such that $n \le x < n + 1$. Sketch the graph of $y = \lfloor x \rfloor$. Calculate for c an integer:

(a) $\lim\limits_{x \to c^-} \lfloor x \rfloor$ **(b)** $\lim\limits_{x \to c^+} \lfloor x \rfloor$ **(c)** $\lim\limits_{x \to 2.6} \lfloor x \rfloor$

40. Determine the one-sided limits at $c = 1, 2$, and 4 of the function g shown in Figure 13, and state whether the limit exists at these points.

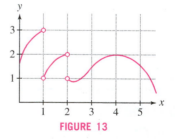

FIGURE 13

In Exercises 41–48, determine the one-sided limits numerically or graphically. If infinite, state whether the one-sided limits are ∞ or $-\infty$, and describe the corresponding vertical asymptote. In Exercise 48, $f(x) = \lfloor x \rfloor$ is the greatest integer function defined in Exercise 39.

41. $\lim\limits_{x \to 0^\pm} \dfrac{\sin x}{|x|}$

42. $\lim\limits_{x \to 0^\pm} |x|^{1/x}$

43. $\lim\limits_{x \to 0^\pm} \dfrac{x - \sin |x|}{x^3}$

44. $\lim\limits_{x \to 4^\pm} \dfrac{x + 1}{x - 4}$

45. $\displaystyle\lim_{x\to-2\pm}\frac{4x^2+7}{x^3+8}$

46. $\displaystyle\lim_{x\to-3\pm}\frac{x^2}{x^2-9}$

47. $\displaystyle\lim_{x\to1\pm}\frac{x^5+x-2}{x^2+x-2}$

48. $\displaystyle\lim_{x\to2\pm}\cos\left(\frac{\pi}{2}(x-\lfloor x\rfloor)\right)$

49. Determine the one-sided limits at $c=2$ and $c=4$ of the function f in Figure 14. What are the vertical asymptotes of f?

50. Determine the infinite one- and two-sided limits in Figure 15.

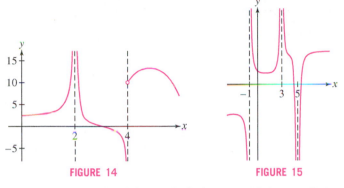

FIGURE 14   FIGURE 15

In Exercises 51–54, sketch the graph of a function with the given limits.

51. $\displaystyle\lim_{x\to1}f(x)=2,\quad \lim_{x\to3^-}f(x)=0,\quad \lim_{x\to3^+}f(x)=4$

52. $\displaystyle\lim_{x\to1}f(x)=\infty,\quad \lim_{x\to3^-}f(x)=0,\quad \lim_{x\to3^+}f(x)=-\infty$

53. $\displaystyle\lim_{x\to2^+}f(x)=f(2)=3,\quad \lim_{x\to2^-}f(x)=-1,$
$\displaystyle\lim_{x\to4}f(x)=2\neq f(4)$

54. $\displaystyle\lim_{x\to1^+}f(x)=\infty,\quad \lim_{x\to1^-}f(x)=3,\quad \lim_{x\to4}f(x)=-\infty$

55. Determine the one-sided limits of the function f in Figure 16, at the points $c=1,3,5,6$.

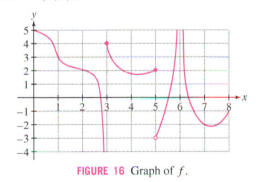

FIGURE 16 Graph of f.

56. Does either of the two oscillating functions in Figure 17 appear to approach a limit as $x\to0$?

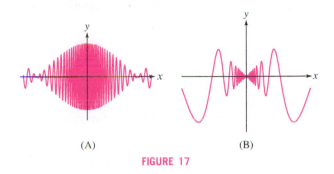

(A)   (B)

FIGURE 17

GU *In Exercises 57–62, plot the function and use the graph to estimate the value of the limit.*

57. $\displaystyle\lim_{\theta\to0}\frac{\sin5\theta}{\sin2\theta}$

58. $\displaystyle\lim_{x\to0}\frac{12^x-1}{4^x-1}$

59. $\displaystyle\lim_{x\to0}\frac{2^x-\cos x}{x}$

60. $\displaystyle\lim_{\theta\to0}\frac{\sin^2 4\theta}{\cos\theta-1}$

61. $\displaystyle\lim_{\theta\to0}\frac{\cos7\theta-\cos5\theta}{\theta^2}$

62. $\displaystyle\lim_{\theta\to0}\frac{\sin^2 2\theta-\theta\sin4\theta}{\theta^4}$

63. Let n be a positive integer. For which n are the two infinite one-sided limits $\displaystyle\lim_{x\to0\pm}1/x^n$ equal?

64. Let $\displaystyle L(n)=\lim_{x\to1}\left(\frac{n}{1-x^n}-\frac{1}{1-x}\right)$ for n a positive integer. Investigate $L(n)$ numerically for several values of n, and then guess the value of $L(n)$ in general.

65. GU   In some cases, numerical investigations can be misleading. Plot $f(x)=\cos\frac{\pi}{x}$.

(a) Does $\displaystyle\lim_{x\to0}f(x)$ exist?

(b) Show, by evaluating $f(x)$ at $x=\pm\frac{1}{2},\pm\frac{1}{4},\pm\frac{1}{6},\ldots$, that you might be able to trick your friends into believing that the limit exists and is equal to $L=1$.

(c) Which sequence of evaluations might trick them into believing that the limit is $L=-1$.

Further Insights and Challenges

66. Light waves of frequency λ passing through a slit of width a produce a **Fraunhofer diffraction pattern** of light and dark fringes (Figure 18). The intensity as a function of the angle θ is

$$I(\theta)=I_m\left(\frac{\sin(R\sin\theta)}{R\sin\theta}\right)^2$$

where $R=\pi a/\lambda$ and I_m is a constant. Show that the intensity function is not defined at $\theta=0$. Then choose any two values for R and check numerically that $I(\theta)$ approaches I_m as $\theta\to0$.

67. Investigate $\displaystyle\lim_{\theta\to0}\frac{\sin n\theta}{\theta}$ numerically for several positive integer values of n. Then guess the value in general.

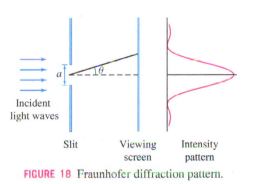

Incident light waves

Slit   Viewing screen   Intensity pattern

FIGURE 18 Fraunhofer diffraction pattern.

68. Show numerically that $\lim\limits_{x \to 0} \dfrac{b^x - 1}{x}$ for $b = 3$ and $b = 5$ appears to equal $\ln 3$ and $\ln 5$, respectively, where $\ln x$ is the natural logarithm. Then make a conjecture (guess) for the value in general and test your conjecture for two additional values of b.

69. Investigate $\lim\limits_{x \to 1} \dfrac{x^n - 1}{x^m - 1}$ for (m, n) equal to $(2, 1)$, $(1, 2)$, $(2, 3)$, and $(3, 2)$. Then guess the value of the limit in general and check your guess for two additional pairs.

70. Find by numerical experimentation the positive integers k such that $\lim\limits_{x \to 0} \dfrac{\sin(\sin^2 x)}{x^k}$ exists.

71. 📖 **GU**   Plot the graph of $f(x) = \dfrac{2^x - 8}{x - 3}$.

(a) Zoom in on the graph to estimate $L = \lim\limits_{x \to 3} f(x)$.

(b) Explain why

$$f(2.99999) \le L \le f(3.00001)$$

Use this to determine L to three decimal places.

72. **GU**   The function $f(x) = \dfrac{2^{1/x} - 2^{-1/x}}{2^{1/x} + 2^{-1/x}}$ is defined for $x \ne 0$.

(a) Investigate $\lim\limits_{x \to 0+} f(x)$ and $\lim\limits_{x \to 0-} f(x)$ numerically.

(b) Plot the graph of f and describe its behavior near $x = 0$.

2.3 Basic Limit Laws

In Section 2.2, we relied on graphical and numerical approaches to investigate limits and estimate their values. In the next four sections, we go beyond this intuitive approach and develop tools for computing limits in a precise way. The next theorem provides our first set of tools.

The proof of Theorem 1 is discussed in Section 2.9 and Appendix D. To illustrate the underlying idea, consider two numbers such as 2.99 and 5.001. Observe that 2.99 is close to 3 and 5.0001 is close to 5, so certainly the sum 2.99 + 5.0001 is close to 3 + 5 and the product 2.99 × 5.0001 is close to 3 × 5. In the same way, if $f(x)$ approaches L and $g(x)$ approaches M as $x \to c$, then $f(x) + g(x)$ approaches the sum $L + M$, and $f(x)g(x)$ approaches the product LM. The other laws are similar.

THEOREM 1 Basic Limit Laws   If $\lim\limits_{x \to c} f(x)$ and $\lim\limits_{x \to c} g(x)$ exist, then

(i) Sum Law: $\lim\limits_{x \to c} \big(f(x) + g(x)\big)$ exists and

$$\lim_{x \to c} \big(f(x) + g(x)\big) = \lim_{x \to c} f(x) + \lim_{x \to c} g(x)$$

(ii) Constant Multiple Law: For any number k, $\lim\limits_{x \to c} kf(x)$ exists and

$$\lim_{x \to c} kf(x) = k \lim_{x \to c} f(x)$$

(iii) Product Law: $\lim\limits_{x \to c} f(x)g(x)$ exists and

$$\lim_{x \to c} f(x)g(x) = \left(\lim_{x \to c} f(x) \right) \left(\lim_{x \to c} g(x) \right)$$

(iv) Quotient Law: If $\lim\limits_{x \to c} g(x) \ne 0$, then $\lim\limits_{x \to c} \dfrac{f(x)}{g(x)}$ exists and

$$\lim_{x \to c} \frac{f(x)}{g(x)} = \frac{\lim\limits_{x \to c} f(x)}{\lim\limits_{x \to c} g(x)}$$

(v) Powers and Roots: If n is a positive integer, then

$$\lim_{x \to c} [f(x)]^n = \left(\lim_{x \to c} f(x) \right)^n, \qquad \lim_{x \to c} \sqrt[n]{f(x)} = \sqrt[n]{\lim_{x \to c} f(x)}$$

In the second limit, assume that $\lim\limits_{x \to c} f(x) \ge 0$ if n is even.

If p, q are integers with $q \ne 0$, then $\lim\limits_{x \to c} [f(x)]^{p/q}$ exists and

$$\lim_{x \to c} [f(x)]^{p/q} = \left(\lim_{x \to c} f(x) \right)^{p/q}$$

Assume that $\lim\limits_{x \to c} f(x) \ge 0$ if q is even, and that $\lim\limits_{x \to c} f(x) \ne 0$ if $p/q < 0$.

Before proceeding to the examples, we make some useful remarks.

• The Sum and Product Laws are valid for any number of functions. For example,

$$\lim_{x \to c} \big(f_1(x) + f_2(x) + f_3(x)\big) = \lim_{x \to c} f_1(x) + \lim_{x \to c} f_2(x) + \lim_{x \to c} f_3(x)$$

- The Sum Law has a counterpart for differences:

$$\lim_{x \to c} \left(f(x) - g(x) \right) = \lim_{x \to c} f(x) - \lim_{x \to c} g(x)$$

This follows from the Sum and Constant Multiple Laws (with $k = -1$):

$$\lim_{x \to c} \left(f(x) - g(x) \right) = \lim_{x \to c} f(x) + \lim_{x \to c} \left(- g(x) \right) = \lim_{x \to c} f(x) - \lim_{x \to c} g(x)$$

- Recall two basic limits from Theorem 1 in Section 2.2:

$$\lim_{x \to c} k = k, \qquad \lim_{x \to c} x = c$$

Applying Law (v) to $f(x) = x$, we obtain

$$\boxed{\lim_{x \to c} x^{p/q} = c^{p/q}} \qquad \qquad \boxed{1}$$

for integers $p, q, q \neq 0$. Assume that $c \geq 0$ if q is even and that $c \neq 0$ if $p/q < 0$.

■ EXAMPLE 1   Use the Basic Limit Laws to evaluate:

(a) $\displaystyle \lim_{x \to 2} x^3$ **(b)** $\displaystyle \lim_{x \to 2} (x^3 + 5x + 7)$ **(c)** $\displaystyle \lim_{x \to 2} \sqrt{x^3 + 5x + 7}$

Solution

(a) By Eq. (1), $\displaystyle \lim_{x \to 2} x^3 = 2^3 = 8$.

(b)

$$\lim_{x \to 2} (x^3 + 5x + 7) = \lim_{x \to 2} x^3 + \lim_{x \to 2} 5x + \lim_{x \to 2} 7 \qquad \text{(Sum Law)}$$

$$= \lim_{x \to 2} x^3 + 5 \lim_{x \to 2} x + \lim_{x \to 2} 7 \qquad \text{(Constant Multiple Law)}$$

$$= 8 + 5(2) + 7 = 25$$

(c) By Law (v) for roots and (b),

$$\lim_{x \to 2} \sqrt{x^3 + 5x + 7} = \sqrt{\lim_{x \to 2} (x^3 + 5x + 7)} = \sqrt{25} = 5 \qquad \qquad ■$$

You may have noticed that each of the limits in Examples 1 and 2 could have been evaluated by a simple substitution. For example, set $t = -1$ to evaluate

$$\lim_{t \to -1} \frac{t + 6}{2t^4} = \frac{-1 + 6}{2(-1)^4} = \frac{5}{2}$$

*Substitution is valid when the function is **continuous**, a concept we shall study in the next section.*

■ EXAMPLE 2   Evaluate   **(a)** $\displaystyle \lim_{t \to -1} \frac{t + 6}{2t^4}$ and   **(b)** $\displaystyle \lim_{t \to 3} t^{-1/4}(t + 5)^{1/3}$.

Solution

(a) Use the Quotient, Sum, and Constant Multiple Laws:

$$\lim_{t \to -1} \frac{t + 6}{2t^4} = \frac{\lim_{t \to -1} (t + 6)}{\lim_{t \to -1} 2t^4} = \frac{\lim_{t \to -1} t + \lim_{t \to -1} 6}{2 \lim_{t \to -1} t^4} = \frac{-1 + 6}{2(-1)^4} = \frac{5}{2}$$

(b) Use the Product, Powers, and Sum Laws:

$$\lim_{t \to 3} t^{-1/4}(t + 5)^{1/3} = \left(\lim_{t \to 3} t^{-1/4} \right) \left(\lim_{t \to 3} \sqrt[3]{t + 5} \right) = \left(3^{-1/4} \right) \left(\sqrt[3]{\lim_{t \to 3} t + 5} \right)$$

$$= 3^{-1/4} \sqrt[3]{3 + 5} = 3^{-1/4}(2) = \frac{2}{3^{1/4}} \qquad \qquad ■$$

The next example reminds us that the Basic Limit Laws apply only when the limits of both $f(x)$ and $g(x)$ exist.

■ **EXAMPLE 3 Assumptions Matter**   Show that the Product Law cannot be applied to $\lim\limits_{x\to 0} f(x)g(x)$ if $f(x) = x$ and $g(x) = x^{-1}$.

Solution   For all $x \neq 0$, we have $f(x)g(x) = x \cdot x^{-1} = 1$, so the limit of the product exists:

$$\lim_{x\to 0} f(x)g(x) = \lim_{x\to 0} 1 = 1$$

However, $\lim\limits_{x\to 0} x^{-1}$ does not exist because $g(x) = x^{-1}$ approaches ∞ as $x \to 0^+$ and it approaches $-\infty$ as $x \to 0^-$. Therefore, the Product Law cannot be applied and its conclusion does not hold:

$$\left(\lim_{x\to 0} f(x) \right) \left(\lim_{x\to 0} g(x) \right) = \left(\lim_{x\to 0} x \right) \underbrace{\left(\lim_{x\to 0} x^{-1} \right)}_{\text{Does not exist}}$$

■

2.3 SUMMARY

- The Basic Limit Laws: If $\lim\limits_{x\to c} f(x)$ and $\lim\limits_{x\to c} g(x)$ both exist, then

  (i) $\lim\limits_{x\to c} \big(f(x) + g(x)\big) = \lim\limits_{x\to c} f(x) + \lim\limits_{x\to c} g(x)$

  (ii) $\lim\limits_{x\to c} kf(x) = k \lim\limits_{x\to c} f(x)$

  (iii) $\lim\limits_{x\to c} f(x)\,g(x) = \left(\lim\limits_{x\to c} f(x) \right)\left(\lim\limits_{x\to c} g(x) \right)$

  (iv) If $\lim\limits_{x\to c} g(x) \neq 0$, then $\lim\limits_{x\to c} \dfrac{f(x)}{g(x)} = \dfrac{\lim\limits_{x\to c} f(x)}{\lim\limits_{x\to c} g(x)}$

  (v) If p, q are integers with $q \neq 0$,

  $$\lim_{x\to c} [f(x)]^{p/q} = \left(\lim_{x\to c} f(x) \right)^{p/q}$$

  For n a positive integer,

  $$\lim_{x\to c} [f(x)]^n = \left(\lim_{x\to c} f(x) \right)^n, \qquad \lim_{x\to c} \sqrt[n]{f(x)} = \sqrt[n]{\lim_{x\to c} f(x)}$$

- If $\lim\limits_{x\to c} f(x)$ or $\lim\limits_{x\to c} g(x)$ does not exist, then the Basic Limit Laws cannot be applied.

2.3 EXERCISES

Preliminary Questions

1. State the Sum Law and Quotient Law.

2. Which of the following is a verbal version of the Product Law (assuming the limits exist)?

(a) The product of two functions has a limit.

(b) The limit of the product is the product of the limits.

(c) The product of a limit is a product of functions.

(d) A limit produces a product of functions.

3. Which statement is correct? The Quotient Law does not hold if:

(a) the limit of the denominator is zero.

(b) the limit of the numerator is zero.

Exercises

In Exercises 1–24, evaluate the limit using the Basic Limit Laws and the limits $\lim\limits_{x\to c} x^{p/q} = c^{p/q}$ *and* $\lim\limits_{x\to c} k = k$.

1. $\lim\limits_{x\to 9} x$

2. $\lim\limits_{x\to -3} 14$

3. $\lim\limits_{x\to \frac{1}{2}} x^4$

4. $\lim\limits_{z\to 27} z^{2/3}$

5. $\lim\limits_{t\to 2} t^{-1}$

6. $\lim\limits_{x\to 5} x^{-2}$

7. $\lim\limits_{x\to 0.2} (3x + 4)$

8. $\lim\limits_{x\to \frac{1}{3}} (3x^3 + 2x^2)$

9. $\lim\limits_{x\to -1} (3x^4 - 2x^3 + 4x)$

10. $\lim\limits_{x\to 8} (3x^{2/3} - 16x^{-1})$

11. $\lim\limits_{x\to 2} (x+1)(3x^2-9)$

12. $\lim\limits_{x\to \frac{1}{2}} (4x+1)(6x-1)$

13. $\lim\limits_{t\to 4} \dfrac{3t-14}{t+1}$

14. $\lim\limits_{z\to 9} \dfrac{\sqrt{z}}{z-2}$

15. $\lim\limits_{y\to \frac{1}{4}} (16y+1)(2y^{1/2}+1)$

16. $\lim\limits_{x\to 2} x(x+1)(x+2)$

17. $\lim\limits_{y\to 4} \dfrac{1}{\sqrt{6y+1}}$

18. $\lim\limits_{w\to 7} \dfrac{\sqrt{w+2}+1}{\sqrt{w-3}-1}$

19. $\lim\limits_{x\to -1} \dfrac{x}{x^3+4x}$

20. $\lim\limits_{t\to -1} \dfrac{t^2+1}{(t^3+2)(t^4+1)}$

21. $\lim\limits_{t\to 25} \dfrac{3\sqrt{t}-\frac{1}{5}t}{(t-20)^2}$

22. $\lim\limits_{y\to \frac{1}{3}} (18y^2-4)^4$

23. $\lim\limits_{t\to \frac{3}{2}} (4t^2+8t-5)^{3/2}$

24. $\lim\limits_{t\to 7} \dfrac{(t+2)^{1/2}}{(t+1)^{2/3}}$

25. Use the Quotient Law to prove that if $\lim\limits_{x\to c} f(x)$ exists and is nonzero, then
$$\lim\limits_{x\to c} \frac{1}{f(x)} = \frac{1}{\lim\limits_{x\to c} f(x)}$$

26. Assuming that $\lim\limits_{x\to 6} f(x) = 4$, compute:

(a) $\lim\limits_{x\to 6} f(x)^2$

(b) $\lim\limits_{x\to 6} \dfrac{1}{f(x)}$

(c) $\lim\limits_{x\to 6} x\sqrt{f(x)}$

In Exercises 27–30, evaluate the limit assuming that $\lim\limits_{x\to -4} f(x) = 3$ and $\lim\limits_{x\to -4} g(x) = 1$.

27. $\lim\limits_{x\to -4} f(x)g(x)$

28. $\lim\limits_{x\to -4} (2f(x)+3g(x))$

29. $\lim\limits_{x\to -4} \dfrac{g(x)}{x^2}$

30. $\lim\limits_{x\to -4} \dfrac{f(x)+1}{3g(x)-9}$

31. Can the Quotient Law be applied to evaluate $\lim\limits_{x\to 0} \dfrac{\sin x}{x}$? Explain.

32. Show that the Product Law cannot be used to evaluate the limit $\lim\limits_{x\to \pi/2} \left(x-\frac{\pi}{2}\right) \tan x$.

33. Give an example where $\lim\limits_{x\to 0} (f(x)+g(x))$ exists but neither $\lim\limits_{x\to 0} f(x)$ nor $\lim\limits_{x\to 0} g(x)$ exists.

34. Give an example where $\lim\limits_{x\to 0} (f(x)\cdot g(x))$ exists but neither $\lim\limits_{x\to 0} f(x)$ nor $\lim\limits_{x\to 0} g(x)$ exists.

35. Give an example where $\lim\limits_{x\to 0} \dfrac{f(x)}{g(x)}$ exists but neither $\lim\limits_{x\to 0} f(x)$ nor $\lim\limits_{x\to 0} g(x)$ exists.

Further Insights and Challenges

36. Show that if both $\lim\limits_{x\to c} f(x)\,g(x)$ and $\lim\limits_{x\to c} g(x)$ exist and $\lim\limits_{x\to c} g(x) \neq 0$, then $\lim\limits_{x\to c} f(x)$ exists. *Hint:* Write $f(x) = \dfrac{f(x)\,g(x)}{g(x)}$.

37. Suppose that $\lim\limits_{t\to 3} tg(t) = 12$. Show that $\lim\limits_{t\to 3} g(t)$ exists and equals 4.

38. Prove that if $\lim\limits_{t\to 3} \dfrac{h(t)}{t} = 5$, then $\lim\limits_{t\to 3} h(t) = 15$.

39. Assuming that $\lim\limits_{x\to 0} \dfrac{f(x)}{x} = 1$, which of the following statements is necessarily true? Why?

(a) $f(0) = 0$

(b) $\lim\limits_{x\to 0} f(x) = 0$

40. Prove that if $\lim\limits_{x\to c} f(x) = L \neq 0$ and $\lim\limits_{x\to c} g(x) = 0$, then the limit $\lim\limits_{x\to c} \dfrac{f(x)}{g(x)}$ does not exist.

41. Suppose that $\lim\limits_{h\to 0} g(h) = L$.

(a) Explain why $\lim\limits_{h\to 0} g(ah) = L$ for any constant $a \neq 0$.

(b) If we assume instead that $\lim\limits_{h\to 1} g(h) = L$, is it still necessarily true that $\lim\limits_{h\to 1} g(ah) = L$?

(c) Illustrate (a) and (b) with the function $f(x) = x^2$.

42. Assume that $L(a) = \lim\limits_{x\to 0} \dfrac{a^x-1}{x}$ exists for all $a > 0$. Assume also that $\lim\limits_{x\to 0} a^x = 1$.

(a) Prove that $L(ab) = L(a) + L(b)$ for $a, b > 0$. *Hint:* $(ab)^x - 1 = a^x b^x - a^x + a^x - 1 = a^x(b^x - 1) + (a^x - 1)$. This shows that $L(a)$ "behaves" like a logarithm, in the sense that $\ln(ab) = \ln a + \ln b$. We will see that $L(a) = \ln a$ in Section 3.9.

(b) Verify numerically that $L(12) = L(3) + L(4)$.

2.4 Limits and Continuity

In everyday speech, the word "continuous" means having no breaks or interruptions. In calculus, continuity is used to describe functions whose graphs have no breaks. If we imagine the graph of a function f as a wavy metal wire, then f is continuous if its graph consists of a single piece of wire as in Figure 1.

Most physical phenomena are continuous in nature. Our position and velocity vary continuously with time. Temperature varies continuously with time. Viscosity of a fluid varies continuously with temperature, assuming we do not freeze or boil the liquid. Ultimately, when we determine the rate of change of a function as we do in the next chapter, we will need the function to be continuous.

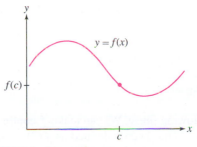

FIGURE 1 f is continuous at $x = c$.

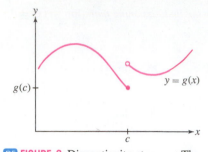

DF FIGURE 2 Discontinuity at $x = c$: The left- and right-hand limits as $x \to c$ are not equal.

A break in the wire as in Figure 2 is called a **discontinuity**. Observe in Figure 2 that the break in the graph occurs because the left- and right-hand limits as x approaches c are not equal and thus $\lim_{x \to c} g(x)$ does not exist. By contrast, in Figure 1, $\lim_{x \to c} f(x)$ exists and is equal to the function value $f(c)$. This suggests the following definition of continuity in terms of limits.

DEFINITION Continuity at a Point Assume that $f(x)$ is defined on an open interval containing $x = c$. Then f is **continuous** at $x = c$ if

$$\lim_{x \to c} f(x) = f(c)$$

If the limit does not exist, or if it exists but is not equal to $f(c)$, we say that f has a **discontinuity** (or is **discontinuous**) at $x = c$.

Note that for f to be continuous at c, three conditions must hold:

1. $f(c)$ is defined.   **2.** $\lim_{x \to c} f(x)$ exists.   **3.** They are equal.

A function f may be continuous at some points and discontinuous at others. If f is continuous at all points in an interval I, then f is said to be continuous on I. If I is an interval $[a, b]$ or $[a, b)$ that includes a as a left endpoint, we require that $\lim_{x \to a+} f(x) = f(a)$. Similarly, we require that $\lim_{x \to b-} f(x) = f(b)$ if I includes b as a right endpoint. If f is continuous at all points in its domain, then f is simply called continuous.

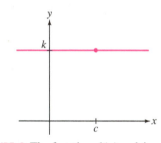

FIGURE 3 The function $f(x) = k$ is continuous.

■ **EXAMPLE 1** Show that the following functions are continuous:

(a) $f(x) = k$ (k any constant)   **(b)** $g(x) = x^n$ (n a whole number)

Solution

(a) We have $\lim_{x \to c} f(x) = \lim_{x \to c} k = k$ and $f(c) = k$. The limit exists and is equal to the function value for all c, so f is continuous (Figure 3).

(b) By Eq. (1) in Section 2.3, $\lim_{x \to c} g(x) = \lim_{x \to c} x^n = c^n$ for all c. Also $g(c) = c^n$, so again, the limit exists and is equal to the function value. Therefore, g is continuous. (Figure 4 illustrates the case $n = 1$.) ■

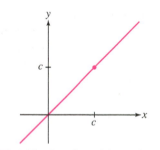

FIGURE 4 The function $g(x) = x$ is continuous.

Examples of Discontinuities

To understand continuity better, let's consider some ways in which a function can fail to be continuous. Keep in mind that continuity at a point $x = c$ requires that:

1. $f(c)$ is defined.   **2.** $\lim_{x \to c} f(x)$ exists.   **3.** They are equal.

If $\lim_{x \to c} f(x)$ exists but is not equal to $f(c)$, we say that f has a **removable discontinuity** at $x = c$. The function in Figure 5(A) has a removable discontinuity at $c = 2$ because

$$f(2) = 10 \quad \text{but} \quad \underbrace{\lim_{x \to 2} f(x) = 5}$$

$$\text{Limit exists but is not equal to function value}$$

Removable discontinuities are "mild" in the following sense: We can make f continuous at $x = c$ by redefining $f(c)$. In Figure 5(B), $f(2)$ has been redefined as $f(2) = 5$, and this makes f continuous at $x = 2$.

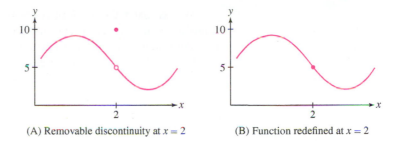

FIGURE 5 Removable discontinuity: The discontinuity can be removed by redefining $f(2)$.

(A) Removable discontinuity at $x = 2$

(B) Function redefined at $x = 2$

A "worse" type of discontinuity is a **jump discontinuity**, which occurs if the one-sided limits $\lim\limits_{x \to c^-} f(x)$ and $\lim\limits_{x \to c^+} f(x)$ exist but are not equal. Figure 6 shows two functions with jump discontinuities at $c = 2$. Unlike the removable case, we cannot make f continuous by redefining $f(c)$.

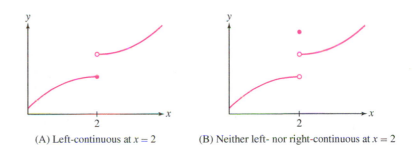

DF **FIGURE 6** Jump discontinuities.

(A) Left-continuous at $x = 2$

(B) Neither left- nor right-continuous at $x = 2$

In connection with jump discontinuities, it is convenient to define *one-sided continuity*.

DEFINITION **One-Sided Continuity**   A function f is called:

- **Left-continuous** at $x = c$ if $\lim\limits_{x \to c^-} f(x) = f(c)$
- **Right-continuous** at $x = c$ if $\lim\limits_{x \to c^+} f(x) = f(c)$

In Figure 6 above, the function in (A) is left-continuous but the function in (B) is neither left- nor right-continuous. The next example explores one-sided continuity using a piecewise-defined function—that is, a function defined by different formulas on different intervals.

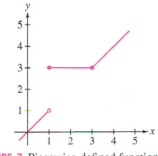

FIGURE 7 Piecewise-defined function F in Example 2.

■ **EXAMPLE 2** **Piecewise-Defined Function**   Discuss the continuity of

$$F(x) = \begin{cases} x & \text{for } x < 1 \\ 3 & \text{for } 1 \leq x \leq 3 \\ x & \text{for } x > 3 \end{cases}$$

Solution The functions $f(x) = x$ and $g(x) = 3$ are continuous, so F is also continuous, except possibly at the transition points $x = 1$ and $x = 3$, where the formula for $F(x)$ changes (Figure 7).

- At $x = 1$, the one-sided limits exist but are not equal:

$$\lim_{x \to 1^-} F(x) = \lim_{x \to 1^-} x = 1, \qquad \lim_{x \to 1^+} F(x) = \lim_{x \to 1^+} 3 = 3$$

Thus, F has a jump discontinuity at $x = 1$. However, the right-hand limit is equal to the function value $F(1) = 3$, so F is *right-continuous* at $x = 1$.

- At $x = 3$, the left- and right-hand limits exist and both are equal to $F(3)$, so F is *continuous* at $x = 3$:

$$\lim_{x \to 3^-} F(x) = \lim_{x \to 3^-} 3 = 3, \qquad \lim_{x \to 3^+} F(x) = \lim_{x \to 3^+} x = 3 \qquad ■$$

CAUTION *Piecewise-defined functions may OR may not be continuous at points where they are pieced together.*

We say that f has an **infinite discontinuity** at $x = c$ if one or both of the one-sided limits are infinite [even if $f(x)$ itself is not defined at $x = c$]. Figure 8 illustrates three types of infinite discontinuities occurring at $x = 2$. Notice that $x = 2$ does not belong to the domain of the function in cases (A) and (B).

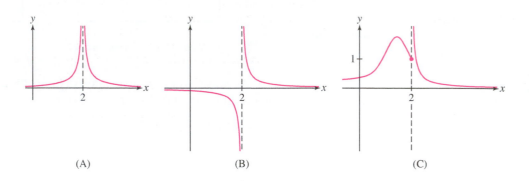

(A)   (B)   (C)

FIGURE 8 Functions with an infinite discontinuity at $x = 2$.

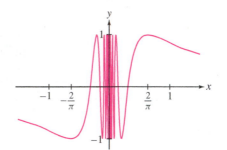

DF FIGURE 9 Graph of $y = \sin \frac{1}{x}$. The discontinuity at $x = 0$ is not a jump, removable, or infinite discontinuity.

Finally, we note that some functions have more "severe" types of discontinuity than those discussed above. For example, $f(x) = \sin \frac{1}{x}$ oscillates infinitely often between $+1$ and -1 as $x \to 0$ (Figure 9). Neither the left- nor the right-hand limit exists at $x = 0$, so this discontinuity is not a jump discontinuity. See Exercises 88 and 89 for even stranger examples. Although of interest from a theoretical point of view, these discontinuities rarely arise in practice.

Building Continuous Functions

Having studied some examples of discontinuities, we focus again on continuous functions. How can we show that a function is continuous? One way is to use the **Laws of Continuity**, which state, roughly speaking, that a function is continuous if it is built out of functions that are known to be continuous.

> **THEOREM 1   Basic Laws of Continuity**   If f and g are continuous at $x = c$, then the following functions are also continuous at $x = c$:
> (i) $f + g$ and $f - g$ (iii) fg
> (ii) kf for any constant k (iv) f/g if $g(c) \neq 0$

Proof   These laws follow directly from the corresponding Basic Limit Laws (Theorem 1, Section 2.3). We illustrate by proving the first part of (i) in detail. The remaining laws are proved similarly. By definition, we must show that $\lim\limits_{x \to c} (f(x) + g(x)) = f(c) + g(c)$. Because f and g are both continuous at $x = c$, we have

$$\lim_{x \to c} f(x) = f(c), \qquad \lim_{x \to c} g(x) = g(c)$$

The Sum Law for limits yields the desired result:

$$\lim_{x \to c} (f(x) + g(x)) = \lim_{x \to c} f(x) + \lim_{x \to c} g(x) = f(c) + g(c) \qquad \blacksquare$$

In Section 2.3, we noted that the Basic Limit Laws for Sums and Products are valid for an arbitrary number of functions. The same is true for continuity; that is, if $f_1, \dots, f_n$ are continuous, then so are the functions

$$f_1 + f_2 + \cdots + f_n, \qquad f_1 \cdot f_2 \cdots f_n$$

When a function f is defined and continuous for all values of x, we say that f is continuous on the real line.

The basic functions are continuous on their domains. Recall (Section 1.3) that the term *basic function* refers to polynomials, rational functions, nth-root and algebraic functions, trigonometric functions and their inverses, and exponential and logarithmic functions.

> **THEOREM 2 Continuity of Polynomial and Rational Functions** Let P and Q be polynomials. Then:
>
> - P and Q are continuous on the real line.
> - P/Q is continuous on its domain [at all values $x = c$ such that $Q(c) \neq 0$].

Proof The function $f(x) = x^m$ is continuous for all whole numbers m by Example 1. By Continuity Law (ii), $f(x) = ax^m$ is continuous for every constant a. A polynomial

$$P(x) = a_n x^n + a_{n-1} x^{n-1} + \cdots + a_1 x + a_0$$

is a sum of continuous functions, so it too is continuous. By Continuity Law (iv), a quotient function P/Q is continuous at $x = c$, provided that $Q(c) \neq 0$. ∎

This result shows, for example, that $f(x) = 3x^4 - 2x^3 + 8x$ is continuous for all x and that

$$g(x) = \frac{x+3}{x^2 - 1}$$

is continuous for $x \neq \pm 1$. Note that if n is a positive integer, then $f(x) = x^{-n}$ is continuous for $x \neq 0$ because $f(x) = x^{-n} = 1/x^n$ is a rational function.

The continuity of the nth-root, sine, cosine, exponential, and logarithmic functions should not be surprising because their graphs have no visible breaks (Figure 10). However, complete proofs of continuity are somewhat technical and are omitted.

> **THEOREM 3 Continuity of Some Basic Functions**
>
> - $y = x^{1/n}$ is continuous on its domain for n a natural number.
> - $y = \sin x$ and $y = \cos x$ are continuous on the real line.
> - $y = b^x$ is continuous on the real line (for $b > 0$, $b \neq 1$).
> - $y = \log_b x$ is continuous for $x > 0$ (for $b > 0$, $b \neq 1$).

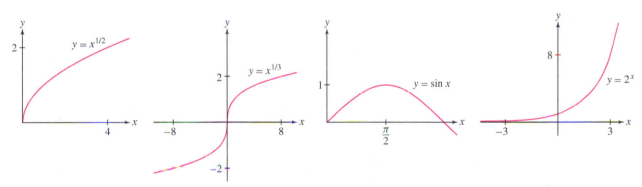

FIGURE 10 As the graphs suggest, these functions are continuous on their domains.

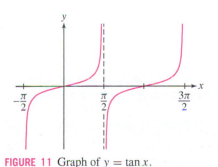

FIGURE 11 Graph of $y = \tan x$.

Because $f(x) = \sin x$ and $f(x) = \cos x$ are continuous, the Continuity Law (iv) for Quotients implies that the other standard trigonometric functions are continuous on their domains, consisting of the values of x where their denominators are nonzero:

$$\tan x = \frac{\sin x}{\cos x}, \qquad \cot x = \frac{\cos x}{\sin x}, \qquad \sec x = \frac{1}{\cos x}, \qquad \csc x = \frac{1}{\sin x}$$

They have infinite discontinuities at points where their denominators are zero. For example, $f(x) = \tan x$ has infinite discontinuities at the points (Figure 11)

$$x = \pm\frac{\pi}{2}, \quad \pm\frac{3\pi}{2}, \quad \pm\frac{5\pi}{2}, \ldots$$

The next theorem states that the inverse f^{-1} of a continuous function f is continuous. This is to be expected because the graph of f^{-1} is the reflection of the graph of f through the line $y = x$. If the graph of f has "no breaks," the same ought to be true of the graph of f^{-1} (see the proof of Theorem 6 in Appendix D).

> **THEOREM 4** **Continuity of the Inverse Function**   If f is continuous on an interval I with range R, and if f^{-1} exists, then f^{-1} is continuous with domain R.

One consequence of this theorem is that the logarithms, nth roots, and inverse trigonometric functions $f(x) = \sin^{-1} x$, $f(x) = \cos^{-1} x$, $f(x) = \tan^{-1} x$, and so on are all continuous on their domains.

Finally, it is important to know that a composition of continuous functions is again continuous. The following theorem is proved in Appendix D.

> **THEOREM 5** **Continuity of Composite Functions**   If g is continuous at $x = c$, and f is continuous at $x = g(c)$, then the composite function $F(x) = f(g(x))$ is continuous at $x = c$.

For example, $F(x) = (x^2 + 9)^{1/3}$ is continuous because it is the composite of the continuous functions $f(x) = x^{1/3}$ and $g(x) = x^2 + 9$. Similarly, $F(x) = \cos(x^{-1})$ is continuous for all $x \neq 0$, and $F(x) = 2^{\sin x}$ is continuous for all x.

More generally, an **elementary function** is a function that is constructed out of basic functions using the operations of addition, subtraction, multiplication, division, and composition. Since the basic functions are continuous (on their domains), an elementary function is also continuous on its domain by the Laws of Continuity. An example of an elementary function is

$$F(x) = \tan^{-1}\left(\frac{x^2 + \cos(2^x + 9)}{x - 8}\right)$$

This function is continuous on its domain $\{x : x \neq 8\}$.

Substitution: Evaluating Limits Using Continuity

It is easy to evaluate a limit when the function in question is known to be continuous. In this case, by definition, the limit is equal to the function value:

$$\lim_{x \to c} f(x) = f(c)$$

We call this the **Substitution Method** because the limit is evaluated by "plugging in" $x = c$.

■ **EXAMPLE 3** Evaluate   **(a)** $\displaystyle\lim_{y \to \frac{\pi}{3}} \sin y$ and   **(b)** $\displaystyle\lim_{x \to -1} \frac{3^x}{\sqrt{x + 5}}$.

Solution

(a) We can use substitution because $f(y) = \sin y$ is continuous.

$$\lim_{y \to \frac{\pi}{3}} \sin y = \sin\frac{\pi}{3} = \frac{\sqrt{3}}{2}$$

(b) The function $f(x) = 3^x/\sqrt{x + 5}$ is continuous at $x = -1$ because the numerator and denominator are both continuous at $x = -1$ and the denominator $\sqrt{x + 5}$ is nonzero at $x = -1$. Therefore, we can use substitution:

$$\lim_{x \to -1} \frac{3^x}{\sqrt{x + 5}} = \frac{3^{-1}}{\sqrt{-1 + 5}} = \frac{1}{6}$$ ■

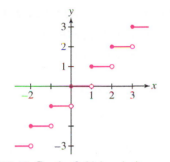

FIGURE 12 Graph of $f(x) = \lfloor x \rfloor$.

The **greatest integer function** $f(x) = \lfloor x \rfloor$ is the function defined by $\lfloor x \rfloor = n$, where n is the unique integer such that $n \leq x < n + 1$ (Figure 12). For example, $\lfloor 4.7 \rfloor = 4$ and $\lfloor -2.3 \rfloor = -3$.

■ **EXAMPLE 4** Assumptions Matter   Can we evaluate $\lim_{x \to 2} \lfloor x \rfloor$ using substitution?

Solution Substitution cannot be applied because $f(x) = \lfloor x \rfloor$ is not continuous at $x = 2$. Although $f(2) = 2$, $\lim_{x \to 2} \lfloor x \rfloor$ does not exist because the one-sided limits are not equal:

$$\lim_{x \to 2^+} \lfloor x \rfloor = 2 \qquad \text{and} \qquad \lim_{x \to 2^-} \lfloor x \rfloor = 1 \qquad ■$$

CONCEPTUAL INSIGHT Real-World Modeling by Continuous Functions   Continuous functions are used often to represent physical quantities such as velocity, temperature, and voltage. This reflects our everyday experience that change in the physical world tends to occur continuously rather than through abrupt transitions. However, mathematical models are at best approximations to reality, and it is important to be aware of their limitations.

In Figure 13, atmospheric temperature is represented as a continuous function of altitude. This is justified for large-scale objects such as the earth's atmosphere because the reading on a thermometer appears to vary continuously as altitude changes. However, temperature is a measure of the average kinetic energy of molecules. At the microscopic level, it would not be meaningful to treat temperature as a quantity that varies continuously from point to point.

Similarly, the size $P(t)$ of a population is usually treated as a continuous function of time t. Strictly speaking, $P(t)$ is a whole number that changes by ± 1 when an individual is born or dies, so it is not continuous, but if the population is large, the effect of an individual birth or death is small, and it is both reasonable and convenient to treat P as a continuous function.

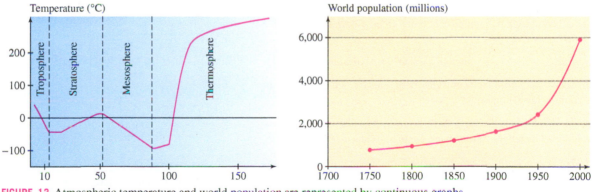

FIGURE 13 Atmospheric temperature and world population are represented by continuous graphs.

2.4 SUMMARY

- Definition: f is *continuous* at $x = c$ if $\lim_{x \to c} f(x) = f(c)$. This means that $f(c)$ exists, $\lim_{x \to c} f(x)$ exists, and they are equal.
- If $\lim_{x \to c} f(x)$ does not exist, or if it exists but does not equal $f(c)$, then f is *discontinuous* at $x = c$.
- If f is continuous at all points in its domain, f is simply called *continuous*.
- *Right-continuous* at $x = c$: $\lim_{x \to c^+} f(x) = f(c)$.
- *Left-continuous* at $x = c$: $\lim_{x \to c^-} f(x) = f(c)$.

- Three common types of discontinuities: *removable discontinuity* $\left[\lim_{x \to c} f(x)\right.$ exists but does not equal $f(c)\left.\right]$, *jump discontinuity* (the one-sided limits both exist but are not equal), and *infinite discontinuity* (the limit is infinite as x approaches c from one or both sides).
- Laws of Continuity: Sums, products, multiples, inverses, and composites of continuous functions are again continuous. The same holds for a quotient f/g at points where $g(x) \neq 0$.
- Basic functions: Polynomials, rational functions, nth-root and algebraic functions, trigonometric functions and their inverses, exponential and logarithmic functions. Basic functions are continuous on their domains.
- Substitution Method: If f is known to be continuous at $x = c$, then the value of the limit $\lim_{x \to c} f(x)$ is $f(c)$.

2.4 EXERCISES

Preliminary Questions

1. Which property of $f(x) = x^3$ allows us to conclude that $\lim_{x \to 2} x^3 = 8$?

2. What can be said about $f(3)$ if f is continuous and $\lim_{x \to 3} f(x) = \frac{1}{2}$?

3. Suppose that $f(x) < 0$ if x is positive and $f(x) > 1$ if x is negative. Can f be continuous at $x = 0$?

4. Is it possible to determine $f(7)$ if $f(x) = 3$ for all $x < 7$ and f is right-continuous at $x = 7$? What if f is left-continuous?

5. Are the following true or false? If false, then draw or give a counterexample, and state a correct version.

(a) f is continuous at $x = a$ if the left- and right-hand limits of $f(x)$ as $x \to a$ exist and are equal.

(b) f is continuous at $x = a$ if the left- and right-hand limits of $f(x)$ as $x \to a$ exist and equal $f(a)$.

(c) If the left- and right-hand limits of $f(x)$ as $x \to a$ exist, then f has a removable discontinuity at $x = a$.

(d) If f and g are continuous at $x = a$, then $f + g$ is continuous at $x = a$.

(e) If f and g are continuous at $x = a$, then f/g is continuous at $x = a$.

Exercises

1. Referring to Figure 14, state whether f is left- or right-continuous (or neither) at each point of discontinuity. Does f have any removable discontinuities?

Exercises 2–4 refer to the function g whose graph appears in Figure 15.

2. State whether g is left- or right-continuous (or neither) at each of its points of discontinuity.

3. At which point c does g have a removable discontinuity? How should $g(c)$ be redefined to make g continuous at $x = c$?

4. Find the point c_1 at which g has a jump discontinuity but is left-continuous. How should $g(c_1)$ be redefined to make g right-continuous at $x = c_1$?

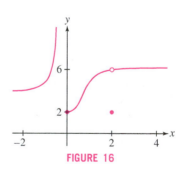

FIGURE 16

6. Suppose that $f(x) = 2$ for $x < 3$ and $f(x) = -4$ for $x > 3$.

(a) What is $f(3)$ if f is left-continuous at $x = 3$?

(b) What is $f(3)$ if f is right-continuous at $x = 3$?

In Exercises 7–16, use the Laws of Continuity and Theorems 2 and 3 to show that the function is continuous.

7. $f(x) = x + \sin x$

8. $f(x) = x \sin x$

9. $f(x) = 3x + 4 \sin x$

10. $f(x) = 3x^3 + 8x^2 - 20x$

11. $f(x) = \dfrac{1}{x^2 + 1}$

12. $f(x) = \dfrac{x^2 - \cos x}{3 + \cos x}$

13. $f(x) = \cos(x^2)$

14. $f(x) = \tan^{-1}(4^x)$

15. $f(x) = e^x \cos 3x$

16. $f(x) = \ln(x^4 + 1)$

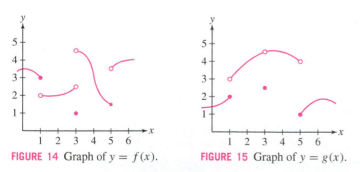

FIGURE 14 Graph of $y = f(x)$.   **FIGURE 15** Graph of $y = g(x)$.

5. In Figure 16, determine the one-sided limits at the points of discontinuity. Which discontinuity is removable and how should f be redefined to make it continuous at this point?

In Exercises 17–34, determine the points of discontinuity. State the type of discontinuity (removable, jump, infinite, or none of these) and whether the function is left- or right-continuous.

17. $f(x) = \dfrac{1}{x}$

18. $f(x) = |x|$

19. $f(x) = \dfrac{x - 2}{|x - 1|}$

20. $f(x) = \lfloor x \rfloor$

21. $f(x) = \left\lfloor \dfrac{x}{2} \right\rfloor$

22. $g(t) = \dfrac{1}{t^2 - 1}$

23. $f(x) = \dfrac{x + 1}{4x - 2}$

24. $h(z) = \dfrac{1 - 2z}{z^2 - z - 6}$

25. $f(x) = 3x^{2/3} - 9x^3$

26. $g(t) = 3t^{-2/3} - 9t^3$

27. $f(x) = \begin{cases} \dfrac{x - 2}{|x - 2|} & x \neq 2 \\ -1 & x = 2 \end{cases}$

28. $f(x) = \begin{cases} \cos \dfrac{1}{x} & x \neq 0 \\ 1 & x = 0 \end{cases}$

29. $g(t) = \tan 2t$

30. $f(x) = \csc(x^2)$

31. $f(x) = \tan(\sin x)$

32. $f(x) = \cos(\pi \lfloor x \rfloor)$

33. $f(x) = \dfrac{1}{e^x - e^{-x}}$

34. $f(x) = \ln|x - 4|$

In Exercises 35–48, determine the domain of the function and prove that it is continuous on its domain using the Laws of Continuity and the facts quoted in this section.

35. $f(x) = 2\sin x + 3\cos x$

36. $f(x) = \sqrt{x^2 + 9}$

37. $f(x) = \sqrt{x} \sin x$

38. $f(x) = \dfrac{x^2}{x + x^{1/4}}$

39. $f(x) = x^{2/3} 2^x$

40. $f(x) = x^{1/3} + x^{3/4}$

41. $f(x) = x^{-4/3}$

42. $f(x) = \ln(9 - x^2)$

43. $f(x) = \tan^2 x$

44. $f(x) = \cos(2^x)$

45. $f(x) = (x^4 + 1)^{3/2}$

46. $f(x) = e^{-x^2}$

47. $f(x) = \dfrac{\cos(x^2)}{x^2 - 1}$

48. $f(x) = 9^{\tan x}$

49. Show that the function
$$f(x) = \begin{cases} x^2 + 3 & \text{for } x < 1 \\ 10 - x & \text{for } 1 \leq x \leq 2 \\ 6x - x^2 & \text{for } x > 2 \end{cases}$$

is continuous for $x \neq 1, 2$. Then compute the right- and left-hand limits at $x = 1, 2$, and determine whether f is left-continuous, right-continuous, or continuous at these points (Figure 17).

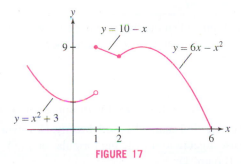

FIGURE 17

50. Sawtooth Function Draw the graph of $f(x) = x - \lfloor x \rfloor$. At which points is f discontinuous? Is it left- or right-continuous at those points?

In Exercises 51–54, sketch the graph of f. At each point of discontinuity, state whether f is left- or right-continuous.

51. $f(x) = \begin{cases} x^2 & \text{for } x \leq 1 \\ 2 - x & \text{for } x > 1 \end{cases}$

52. $f(x) = \begin{cases} x + 1 & \text{for } x < 1 \\ \dfrac{1}{x} & \text{for } x \geq 1 \end{cases}$

53. $f(x) = \begin{cases} \dfrac{x^2 - 3x + 2}{|x - 2|} & x \neq 2 \\ 0 & x = 2 \end{cases}$

54. $f(x) = \begin{cases} x^3 + 1 & \text{for } -\infty < x \leq 0 \\ -x + 1 & \text{for } 0 < x < 2 \\ -x^2 + 10x - 15 & \text{for } x \geq 2 \end{cases}$

55. Show that the function
$$f(x) = \begin{cases} \dfrac{x^2 - 16}{x - 4} & x \neq 4 \\ 10 & x = 4 \end{cases}$$

has a removable discontinuity at $x = 4$.

56. GU Define $f(x) = x \sin \dfrac{1}{x} + 2$ for $x \neq 0$. Plot f. How should $f(0)$ be defined so that f is continuous at $x = 0$?

In Exercises 57–59, find the value of the constant (a, b, or c) that makes the function continuous.

57. $f(x) = \begin{cases} x^2 - c & \text{for } x < 5 \\ 4x + 2c & \text{for } x \geq 5 \end{cases}$

58. $f(x) = \begin{cases} 2x + 9x^{-1} & \text{for } x \leq 3 \\ -4x + c & \text{for } x > 3 \end{cases}$

59. $f(x) = \begin{cases} x^{-1} & \text{for } x < -1 \\ ax + b & \text{for } -1 \leq x \leq \frac{1}{2} \\ x^{-1} & \text{for } x > \frac{1}{2} \end{cases}$

60. Define
$$g(x) = \begin{cases} x + 3 & \text{for } x < -1 \\ cx & \text{for } -1 \leq x \leq 2 \\ x + 2 & \text{for } x > 2 \end{cases}$$

Find a value of c such that g is

(a) left-continuous   **(b)** right-continuous

In each case, sketch the graph of g.

61. Define $g(t) = \tan^{-1}\left(\dfrac{1}{t - 1}\right)$ for $t \neq 1$. Answer the following questions, using a plot if necessary.

(a) Can $g(1)$ be defined so that g is continuous at $t = 1$?

(b) How should $g(1)$ be defined so that g is left-continuous at $t = 1$?

62. Each of the following statements is *false*. For each statement, sketch the graph of a function that provides a counterexample.

(a) If $\lim\limits_{x \to a} f(x)$ exists, then f is continuous at $x = a$.

(b) If f has a jump discontinuity at $x = a$, then $f(a)$ is equal to either $\lim\limits_{x \to a^-} f(x)$ or $\lim\limits_{x \to a^+} f(x)$.

In Exercises 63–66, draw the graph of a function on [0, 5] with the given properties.

63. f is not continuous at $x = 1$, but $\lim\limits_{x \to 1^+} f(x)$ and $\lim\limits_{x \to 1^-} f(x)$ exist and are equal.

64. f is left-continuous but not continuous at $x = 2$, and right-continuous but not continuous at $x = 3$.

65. f has a removable discontinuity at $x = 1$, a jump discontinuity at $x = 2$, and

$$\lim_{x \to 3^-} f(x) = -\infty, \qquad \lim_{x \to 3^+} f(x) = 2$$

66. f is right- but not left-continuous at $x = 1$, left- but not right-continuous at $x = 2$, and neither left- nor right-continuous at $x = 3$.

In Exercises 67–80, evaluate using substitution.

67. $\lim\limits_{x \to -1} (2x^3 - 4)$

68. $\lim\limits_{x \to 2} (5x - 12x^{-2})$

69. $\lim\limits_{x \to 3} \dfrac{x + 2}{x^2 + 2x}$

70. $\lim\limits_{x \to \pi} \sin\left(\dfrac{x}{2} - \pi\right)$

71. $\lim\limits_{x \to \frac{\pi}{4}} \tan(3x)$

72. $\lim\limits_{x \to \pi} \dfrac{1}{\cos x}$

73. $\lim\limits_{x \to 4} x^{-5/2}$

74. $\lim\limits_{x \to 2} \sqrt{x^3 + 4x}$

75. $\lim\limits_{x \to -1} (1 - 8x^3)^{3/2}$

76. $\lim\limits_{x \to 2} \left(\dfrac{7x + 2}{4 - x}\right)^{2/3}$

77. $\lim\limits_{x \to 3} 10^{x^2 - 2x}$

78. $\lim\limits_{x \to -\frac{\pi}{2}} 3^{\sin x}$

79. $\lim\limits_{x \to 4^-} \sin^{-1}\left(\dfrac{x}{4}\right)$

80. $\lim\limits_{x \to 0} \tan^{-1}(e^x)$

81. Suppose that f and g are discontinuous at $x = c$. Does it follow that $f + g$ is discontinuous at $x = c$? If not, give a counterexample. Does this contradict Theorem 1(i)?

82. Prove that $f(x) = |x|$ is continuous for all x. *Hint:* To prove continuity at $x = 0$, consider the one-sided limits.

83. Use the result of Exercise 82 to prove that if g is continuous, then $f(x) = |g(x)|$ is also continuous.

84. Which of the following quantities would be represented by continuous functions of time and which would have one or more discontinuities?

(a) Velocity of an airplane during a flight

(b) Temperature in a room under ordinary conditions

(c) Value of a bank account with interest paid yearly

(d) The salary of a teacher

(e) The population of the world

85. In 2009, the federal income tax T on income of x dollars (up to \$82,250) was determined by the formula

$$T(x) = \begin{cases} 0.10x & \text{for } 0 \le x < 8350 \\ 0.15x - 417.50 & \text{for } 8350 \le x < 33{,}950 \\ 0.25x - 3812.50 & \text{for } 33{,}950 \le x < 82{,}250 \end{cases}$$

Sketch the graph of T. Does T have any discontinuities? Explain why, if T had a jump discontinuity, it might be advantageous in some situations to earn *less* money.

Further Insights and Challenges

86. If f has a removable discontinuity at $x = c$, then it is possible to redefine $f(c)$ so that f is continuous at $x = c$. Can this be done in more than one way?

87. Give an example of functions f and g such that $f(g(x))$ is continuous but g has at least one discontinuity.

88. Continuous at Only One Point Show that the following function is continuous only at $x = 0$:

$$f(x) = \begin{cases} x & \text{for } x \text{ rational} \\ -x & \text{for } x \text{ irrational} \end{cases}$$

89. Show that f is a discontinuous function for all x, where $f(x)$ is defined as follows:

$$f(x) = \begin{cases} 1 & \text{for } x \text{ rational} \\ -1 & \text{for } x \text{ irrational} \end{cases}$$

Show that f^2 is continuous for all x.

2.5 Evaluating Limits Algebraically

Substitution can be used to evaluate limits when the function in question is known to be continuous. For example, $f(x) = x^{-2}$ is continuous at $x = 3$, and therefore,

$$\lim_{x \to 3} x^{-2} = 3^{-2} = \frac{1}{9}$$

When we study derivatives in Chapter 3, we will be faced with limits $\lim\limits_{x \to c} f(x)$, where $f(c)$ is not defined. In such cases, substitution cannot be used directly. However, many of these limits can be evaluated if we use algebra to rewrite the formula for $f(x)$.

To illustrate, consider the limit (Figure 1):

$$\lim_{x \to 4} \frac{x^2 - 16}{x - 4}$$

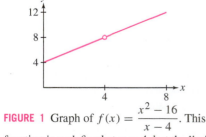

FIGURE 1 Graph of $f(x) = \dfrac{x^2 - 16}{x - 4}$. This function is undefined at $x = 4$, but the limit as $x \to 4$ exists.

The function $f(x) = \dfrac{x^2 - 16}{x - 4}$ is not defined at $x = 4$ because the formula for $f(4)$ produces the undefined expression $0/0$. However, the numerator of $f(x)$ factors:

$$\frac{x^2 - 16}{x - 4} = \frac{(x + 4)(x - 4)}{x - 4} = x + 4 \quad \text{(valid for } x \neq 4)$$

This shows that f coincides with the *continuous* function $y = x + 4$ for all $x \neq 4$. Since the limit depends only on the values of $f(x)$ for $x \neq 4$, we have

$$\lim_{x \to 4} \frac{x^2 - 16}{x - 4} = \underbrace{\lim_{x \to 4} (x + 4) = 8}_{\text{Evaluate by substitution}}$$

Other indeterminate forms are 1^∞, ∞^0, and 0^0. These are treated in Section 4.5.

> We say that $f(x)$ has an **indeterminate form** (or is **indeterminate**) at $x = c$ if the formula for $f(c)$ yields an undefined expression of the type $\frac{0}{0}$, $\frac{\infty}{\infty}$, $\infty \cdot 0$, $\infty - \infty$.

Indeterminate forms are warning signs that tell us more work needs to be done to evaluate the limit. Our strategy, when this occurs, is to *transform $f(x)$ algebraically, if possible, into a new expression that is defined and continuous at $x = c$, and then evaluate the limit by substitution ("plugging in")*. As you study the following examples, notice that the critical step is to cancel a common factor from the numerator and denominator at the appropriate moment, thereby removing the indeterminacy.

■ **EXAMPLE 1** Calculate $\displaystyle\lim_{x \to 3} \frac{x^2 - 4x + 3}{x^2 + x - 12}$.

Solution The function has the indeterminate form $0/0$ at $x = 3$ because

Numerator at $x = 3$: $x^2 - 4x + 3 = 3^2 - 4(3) + 3 = 0$

Denominator at $x = 3$: $x^2 + x - 12 = 3^2 + 3 - 12 = 0$

Step 1. **Transform algebraically and cancel.**

$$\frac{x^2 - 4x + 3}{x^2 + x - 12} = \underbrace{\frac{(x - 3)(x - 1)}{(x - 3)(x + 4)}}_{\text{Cancel common factor}} = \underbrace{\frac{x - 1}{x + 4}}_{\text{Continuous at } x = 3} \quad \text{(if } x \neq 3) \qquad \boxed{1}$$

Step 2. **Substitute (evaluate using continuity).**
Because the expression on the right in Eq. (1) is *continuous* at $x = 3$,

$$\lim_{x \to 3} \frac{x^2 - 4x + 3}{x^2 + x - 12} = \underbrace{\lim_{x \to 3} \frac{x - 1}{x + 4} = \frac{2}{7}}_{\text{Evaluate by substitution}}$$

■ **EXAMPLE 2** **The Form $\dfrac{\infty}{\infty}$** Calculate $\displaystyle\lim_{x \to \frac{\pi}{2}} \frac{\tan x}{\sec x}$.

Solution As we see in Figure 2, both $f(x) = \tan x$ and $f(x) = \sec x$ have infinite discontinuities at $x = \frac{\pi}{2}$, so this limit has the indeterminate form ∞/∞ at $x = \frac{\pi}{2}$.

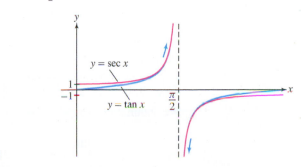

DF **FIGURE 2**

Step 1. **Transform algebraically and cancel.**

$$\frac{\tan x}{\sec x} = \frac{(\sin x)\left(\dfrac{1}{\cos x}\right)}{\dfrac{1}{\cos x}} = \sin x \quad (\text{if } \cos x \neq 0)$$

Step 2. **Substitute (evaluate using continuity).**
Because $f(x) = \sin x$ is continuous,

$$\lim_{x \to \frac{\pi}{2}} \frac{\tan x}{\sec x} = \lim_{x \to \frac{\pi}{2}} \sin x = \sin \frac{\pi}{2} = 1 \qquad \blacksquare$$

The next example illustrates the algebraic technique of "multiplying by the conjugate," which can be used to treat some indeterminate forms involving square roots.

■ **EXAMPLE 3** **Multiplying by the Conjugate** Evaluate $\displaystyle\lim_{x \to 4} \frac{\sqrt{x} - 2}{x - 4}$.

Solution We check that $f(x) = \dfrac{\sqrt{x} - 2}{x - 4}$ has the indeterminate form $0/0$ at $x = 4$:

Numerator at $x = 4$: $\sqrt{x} - 2 = \sqrt{4} - 2 = 0$

Denominator at $x = 4$: $x - 4 = 4 - 4 = 0$

Step 1. **Multiply by the conjugate and cancel.**

$$\left(\frac{\sqrt{x} - 2}{x - 4}\right)\left(\frac{\sqrt{x} + 2}{\sqrt{x} + 2}\right) = \frac{x - 4}{(x - 4)(\sqrt{x} + 2)} = \frac{1}{\sqrt{x} + 2} \quad (\text{if } x \neq 4)$$

Step 2. **Substitute (evaluate using continuity).**
Because $f(x) = 1/(\sqrt{x} + 2)$ is continuous at $x = 4$,

$$\lim_{x \to 4} \frac{\sqrt{x} - 2}{x - 4} = \lim_{x \to 4} \frac{1}{\sqrt{x} + 2} = \frac{1}{4} \qquad \blacksquare$$

Note, in Step 1, that the conjugate of $\sqrt{x} - 2$ is $\sqrt{x} + 2$, so $(\sqrt{x} - 2)(\sqrt{x} + 2) = x - 4$.

■ **EXAMPLE 4** Evaluate $\displaystyle\lim_{h \to 5} \frac{h - 5}{\sqrt{h + 4} - 3}$.

Solution We note that $f(h) = \dfrac{h - 5}{\sqrt{h + 4} - 3}$ yields $0/0$ at $h = 5$:

Numerator at $h = 5$: $h - 5 = 5 - 5 = 0$

Denominator at $h = 5$: $\sqrt{h + 4} - 3 = \sqrt{5 + 4} - 3 = 0$

The conjugate of $\sqrt{h + 4} - 3$ is $\sqrt{h + 4} + 3$, and

$$\frac{h - 5}{\sqrt{h + 4} - 3} = \left(\frac{h - 5}{\sqrt{h + 4} - 3}\right)\left(\frac{\sqrt{h + 4} + 3}{\sqrt{h + 4} + 3}\right) = \frac{(h - 5)(\sqrt{h + 4} + 3)}{(\sqrt{h + 4} - 3)(\sqrt{h + 4} + 3)}$$

The denominator is equal to

$$\left(\sqrt{h + 4} - 3\right)\left(\sqrt{h + 4} + 3\right) = \left(\sqrt{h + 4}\right)^2 - 9 = h - 5$$

Thus, for $h \neq 5$,

$$f(h) = \frac{h - 5}{\sqrt{h + 4} - 3} = \frac{(h - 5)(\sqrt{h + 4} + 3)}{h - 5} = \sqrt{h + 4} + 3$$

We obtain

$$\lim_{h \to 5} \frac{h - 5}{\sqrt{h + 4} - 3} = \lim_{h \to 5} \left(\sqrt{h + 4} + 3\right) = \sqrt{9} + 3 = 6 \qquad \blacksquare$$

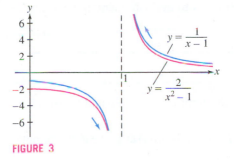

FIGURE 3

■ **EXAMPLE 5** **The Form** $\infty - \infty$ Calculate $\lim\limits_{x \to 1} \left(\dfrac{1}{x-1} - \dfrac{2}{x^2-1} \right)$.

Solution As we see in Figure 3, $y = \dfrac{1}{x-1}$ and $y = \dfrac{2}{x^2-1}$ both have infinite discontinuities at $x = 1$, so this limit has the indeterminate form $\infty - \infty$.

Step 1. Transform algebraically and cancel.

Combine the fractions and simplify (for $x \neq 1$):

$$\frac{1}{x-1} - \frac{2}{x^2-1} = \frac{x+1}{x^2-1} - \frac{2}{x^2-1} = \frac{x-1}{x^2-1} = \frac{x-1}{(x-1)(x+1)} = \frac{1}{x+1}$$

Step 2. Substitute (evaluate using continuity).

$$\lim_{x \to 1} \left(\frac{1}{x-1} - \frac{2}{x^2-1} \right) = \lim_{x \to 1} \frac{1}{x+1} = \frac{1}{1+1} = \frac{1}{2}$$ ■

> **CAUTION** *The form* $\frac{a}{0}$ *with nonzero* a *is NOT an indeterminate form. In this case, the limit does not exist, and we have a vertical asymptote.*

In the next example, the function has the undefined form $a/0$ with a nonzero. This is *not* an indeterminate form (it is not of the form $0/0$, ∞/∞, etc.).

■ **EXAMPLE 6** **Infinite But Not Indeterminate** Evaluate $\lim\limits_{x \to 2} \dfrac{x^2-x+5}{x-2}$.

Solution The function $f(x) = \dfrac{x^2-x+5}{x-2}$ is undefined at $x = 2$ because the formula for $f(2)$ yields $7/0$:

Numerator at $x = 2$: $x^2 - x + 5 = 2^2 - 2 + 5 = 7$

Denominator at $x = 2$: $x - 2 = 2 - 2 = 0$

But $f(x)$ *is not indeterminate* at $x = 2$ because $7/0$ is not an indeterminate form. As in Figure 4, the one-sided limits are infinite:

$$\lim_{x \to 2^-} \frac{x^2-x+5}{x-2} = -\infty, \qquad \lim_{x \to 2^+} \frac{x^2-x+5}{x-2} = \infty$$

The limit itself does not exist since the numerator approaches a fixed nonzero value and the denominator approaches 0. ■

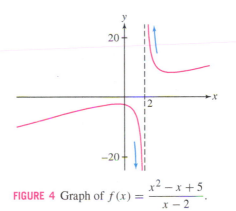

FIGURE 4 Graph of $f(x) = \dfrac{x^2-x+5}{x-2}$.

As preparation for the derivative in Chapter 3, we evaluate a limit involving a symbolic constant.

■ **EXAMPLE 7** **Symbolic Constant** Calculate $\lim\limits_{h \to 0} \dfrac{(h+a)^2 - a^2}{h}$, where a is a constant.

Solution We have the indeterminate form $0/0$ at $h = 0$ because

Numerator at $h = 0$: $(h+a)^2 - a^2 = (0+a)^2 - a^2 = 0$

Denominator at $h = 0$: $h = 0$

Expand the numerator and simplify (for $h \neq 0$):

$$\frac{(h+a)^2 - a^2}{h} = \frac{(h^2 + 2ah + a^2) - a^2}{h} = \frac{h^2 + 2ah}{h} = \frac{h(h+2a)}{h} = h + 2a$$

The function $y = h + 2a$ is continuous (for any constant a), so

$$\lim_{h \to 0} \frac{(h+a)^2 - a^2}{h} = \lim_{h \to 0} (h + 2a) = 2a$$ ■

Given an expression that yields an indeterminate form, it will not always be the case that algebraic manipulation yields the limit. For instance, consider $\lim\limits_{x \to 0} \dfrac{\sin x}{x}$. Although this is an indeterminate form of type $0/0$, algebraic manipulation does not provide an answer. In the next section, we use the Squeeze Theorem to evaluate this limit. It can also be evaluated by L'Hôpital's Rule, which will allow us to evaluate many indeterminate forms and which we will discuss in Section 4.5.

2.5 SUMMARY

- When f is known to be continuous at $x = c$, the limit can be evaluated by substitution: $\lim\limits_{x \to c} f(x) = f(c)$.
- We say that $f(x)$ is *indeterminate* (or has an *indeterminate form*) at $x = c$ if the formula for $f(c)$ yields an undefined expression of the type

$$\frac{0}{0}, \quad \frac{\infty}{\infty}, \quad \infty \cdot 0, \quad \infty - \infty$$

- If $f(x)$ is indeterminate at $x = c$, try to transform $f(x)$ algebraically into a new expression that is defined and continuous at $x = c$. Then evaluate by substitution.

2.5 EXERCISES

Preliminary Questions

1. Which of the following is indeterminate at $x = 1$?

$$\frac{x^2 + 1}{x - 1}, \qquad \frac{x^2 - 1}{x + 2}, \qquad \frac{x^2 - 1}{\sqrt{x + 3} - 2}, \qquad \frac{x^2 + 1}{\sqrt{x + 3} - 2}$$

2. Give counterexamples to show that these statements are false:

(a) If $f(c)$ is indeterminate, then the right- and left-hand limits as $x \to c$ are not equal.

(b) If $\lim\limits_{x \to c} f(x)$ exists, then $f(c)$ is not indeterminate.

(c) If $f(x)$ is undefined at $x = c$, then $f(x)$ has an indeterminate form at $x = c$.

3. The method for evaluating limits discussed in this section is sometimes called "simplify and plug in." Explain how it actually relies on the property of continuity.

Exercises

In Exercises 1–4, show that the limit leads to an indeterminate form. Then carry out the two-step procedure: Transform the function algebraically and evaluate using continuity.

1. $\lim\limits_{x \to 6} \dfrac{x^2 - 36}{x - 6}$

2. $\lim\limits_{h \to 3} \dfrac{9 - h^2}{h - 3}$

3. $\lim\limits_{x \to -1} \dfrac{x^2 + 2x + 1}{x + 1}$

4. $\lim\limits_{t \to 9} \dfrac{2t - 18}{5t - 45}$

In Exercises 5–34, evaluate the limit, if it exists. If not, determine whether the one-sided limits exist (finite or infinite).

5. $\lim\limits_{x \to 7} \dfrac{x - 7}{x^2 - 49}$

6. $\lim\limits_{x \to 8} \dfrac{x^2 - 64}{x - 9}$

7. $\lim\limits_{x \to -2} \dfrac{x^2 + 3x + 2}{x + 2}$

8. $\lim\limits_{x \to 8} \dfrac{x^3 - 64x}{x - 8}$

9. $\lim\limits_{x \to 5} \dfrac{2x^2 - 9x - 5}{x^2 - 25}$

10. $\lim\limits_{h \to 0} \dfrac{(1 + h)^3 - 1}{h}$

11. $\lim\limits_{x \to -\frac{1}{2}} \dfrac{2x + 1}{2x^2 + 3x + 1}$

12. $\lim\limits_{x \to 3} \dfrac{x^2 - x}{x^2 - 9}$

13. $\lim\limits_{x \to 2} \dfrac{3x^2 - 4x - 4}{2x^2 - 8}$

14. $\lim\limits_{h \to 0} \dfrac{(3 + h)^3 - 27}{h}$

15. $\lim\limits_{t \to 0} \dfrac{4^{2t} - 1}{4^t - 1}$

16. $\lim\limits_{h \to 4} \dfrac{(h + 2)^2 - 9h}{h - 4}$

17. $\lim\limits_{x \to 16} \dfrac{\sqrt{x} - 4}{x - 16}$

18. $\lim\limits_{t \to -2} \dfrac{2t + 4}{12 - 3t^2}$

19. $\lim\limits_{h \to 0} \dfrac{\dfrac{1}{(h + 2)^2} - \dfrac{1}{4}}{h}$

20. $\lim\limits_{y \to 3} \dfrac{y^2 + y - 12}{y^3 - 10y + 3}$

21. $\lim\limits_{h \to 0} \dfrac{\sqrt{2 + h} - 2}{h}$

22. $\lim\limits_{x \to 8} \dfrac{\sqrt{x - 4} - 2}{x - 8}$

23. $\lim\limits_{x \to 4} \dfrac{x - 4}{\sqrt{x} - \sqrt{8 - x}}$

24. $\lim\limits_{x \to 4} \dfrac{\sqrt{5 - x} - 1}{2 - \sqrt{x}}$

25. $\lim\limits_{x \to 4} \left(\dfrac{1}{\sqrt{x} - 2} - \dfrac{4}{x - 4} \right)$

26. $\lim\limits_{x \to 0+} \left(\dfrac{1}{\sqrt{x}} - \dfrac{1}{\sqrt{x^2 + x}} \right)$

27. $\lim\limits_{x \to 0} \dfrac{\cot x}{\csc x}$

28. $\lim\limits_{\theta \to \frac{\pi}{2}} \dfrac{\cot \theta}{\csc \theta}$

29. $\lim\limits_{x \to 1} \left(\dfrac{1}{1 - x} - \dfrac{2}{1 - x^2} \right)$

30. $\lim\limits_{x \to \frac{\pi}{4}} \dfrac{\sin x - \cos x}{\tan x - 1}$

31. $\lim\limits_{t \to 2} \dfrac{2^{2t} + 2^t - 20}{2^t - 4}$

32. $\lim\limits_{\theta \to \frac{\pi}{2}} \left(\sec \theta - \tan \theta \right)$

33. $\lim\limits_{\theta \to \frac{\pi}{4}} \left(\dfrac{1}{\tan \theta - 1} - \dfrac{2}{\tan^2 \theta - 1} \right)$

34. $\lim\limits_{x \to \frac{\pi}{3}} \dfrac{2 \cos^2 x + 3 \cos x - 2}{2 \cos x - 1}$

35. [GU] Use a plot of $f(x) = \dfrac{x-4}{\sqrt{x} - \sqrt{8-x}}$ to estimate $\lim\limits_{x\to 4} f(x)$ to two decimal places. Compare with the answer obtained algebraically in Exercise 23.

36. [GU] Use a plot of $f(x) = \dfrac{1}{\sqrt{x}-2} - \dfrac{4}{x-4}$ to estimate $\lim\limits_{x\to 4} f(x)$ numerically. Compare with the answer obtained algebraically in Exercise 25.

In Exercises 37–42, evaluate using the identity

$$a^3 - b^3 = (a-b)(a^2 + ab + b^2)$$

37. $\lim\limits_{x\to 2} \dfrac{x^3 - 8}{x-2}$

38. $\lim\limits_{x\to 3} \dfrac{x^3 - 27}{x^2 - 9}$

39. $\lim\limits_{x\to 1} \dfrac{x^2 - 5x + 4}{x^3 - 1}$

40. $\lim\limits_{x\to -2} \dfrac{x^3 + 8}{x^2 + 6x + 8}$

41. $\lim\limits_{x\to 1} \dfrac{x^4 - 1}{x^3 - 1}$

42. $\lim\limits_{x\to 27} \dfrac{x - 27}{x^{1/3} - 3}$

43. Evaluate $\lim\limits_{h\to 0} \dfrac{\sqrt[4]{1+h} - 1}{h}$. *Hint:* Set $x = \sqrt[4]{1+h}$, express h as a function of x, and rewrite as a limit as $x \to 1$.

44. Evaluate $\lim\limits_{h\to 0} \dfrac{\sqrt[3]{1+h} - 1}{\sqrt[2]{1+h} - 1}$. *Hint:* Set $x = \sqrt[6]{1+h}$, express h as a function of x, and rewrite as a limit as $x \to 1$.

In Exercises 45–54, evaluate in terms of the constant a.

45. $\lim\limits_{x\to 0} (2a + x)$

46. $\lim\limits_{h\to -2} (4ah + 7a)$

47. $\lim\limits_{t\to -1} (4t - 2at + 3a)$

48. $\lim\limits_{h\to 0} \dfrac{(3a+h)^2 - 9a^2}{h}$

49. $\lim\limits_{h\to 0} \dfrac{2(a+h)^2 - 2a^2}{h}$

50. $\lim\limits_{x\to a} \dfrac{(x+a)^2 - 4x^2}{x-a}$

51. $\lim\limits_{x\to a} \dfrac{\sqrt{x} - \sqrt{a}}{x-a}$

52. $\lim\limits_{h\to 0} \dfrac{\sqrt{a+2h} - \sqrt{a}}{h}$

53. $\lim\limits_{x\to 0} \dfrac{(x+a)^3 - a^3}{x}$

54. $\lim\limits_{h\to a} \dfrac{\frac{1}{h} - \frac{1}{a}}{h - a}$

Further Insights and Challenges

In Exercises 55–58, find all values of c such that the limit exists.

55. $\lim\limits_{x\to c} \dfrac{x^2 - 5x - 6}{x - c}$

56. $\lim\limits_{x\to 1} \dfrac{x^2 + 3x + c}{x - 1}$

57. $\lim\limits_{x\to 1} \left(\dfrac{1}{x-1} - \dfrac{c}{x^3 - 1} \right)$

58. $\lim\limits_{x\to 0} \dfrac{1 + cx^2 - \sqrt{1+x^2}}{x^4}$

59. For which sign, $+$ or $-$, does the following limit exist?

$$\lim\limits_{x\to 0} \left(\dfrac{1}{x} \pm \dfrac{1}{x(x-1)} \right)$$

2.6 Trigonometric Limits

In our study of the derivative, we will need to evaluate certain limits involving transcendental functions such as sine and cosine. The algebraic techniques of the previous section are often ineffective for such functions, and other tools are required. In this section, we discuss one such tool—the Squeeze Theorem—and use it to evaluate the trigonometric limits needed in Section 3.6.

The Squeeze Theorem

Consider a function f that is "trapped" between two functions l, for lower bound, and u, for upper bound, on an interval I. In other words,

$$l(x) \le f(x) \le u(x) \quad \text{for all } x \in I$$

Thus, the graph of f lies between the graphs of l and u (Figure 1).

The Squeeze Theorem applies when f is not just trapped but **squeezed** at a point $x = c$ (Figure 2). By this we mean that for all $x \ne c$ in some open interval containing c,

$$l(x) \le f(x) \le u(x) \quad \text{and} \quad \lim\limits_{x\to c} l(x) = \lim\limits_{x\to c} u(x) = L$$

We do not require that $f(x)$ be defined at $x = c$, but it is clear graphically that $f(x)$ must approach the limit L, as stated in the next theorem. See Appendix D for a proof.

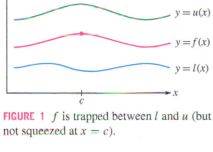

FIGURE 1 f is trapped between l and u (but not squeezed at $x = c$).

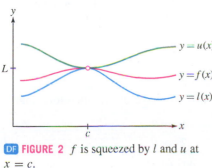

[DF] **FIGURE 2** f is squeezed by l and u at $x = c$.

THEOREM 1 Squeeze Theorem Assume that for $x \ne c$ (in some open interval containing c),

$$l(x) \le f(x) \le u(x) \quad \text{and} \quad \lim\limits_{x\to c} l(x) = \lim\limits_{x\to c} u(x) = L$$

Then $\lim\limits_{x\to c} f(x)$ exists and $\lim\limits_{x\to c} f(x) = L$.

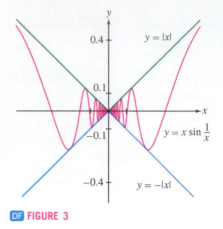

DF FIGURE 3

■ **EXAMPLE 1** Show that $\lim\limits_{x \to 0} x \sin \frac{1}{x} = 0$.

Solution Although $f(x) = x \sin \frac{1}{x}$ is a product of two functions, we cannot use the Product Law because $\lim\limits_{x \to 0} \sin \frac{1}{x}$ does not exist. However, the sine function takes on values between 1 and -1, and therefore $\left| \sin \frac{1}{x} \right| \leq 1$ for all $x \neq 0$. Multiplying by $|x|$, we obtain $\left| x \sin \frac{1}{x} \right| \leq |x|$ and conclude that (Figure 3)

$$-|x| \leq x \sin \frac{1}{x} \leq |x|$$

Because

$$\lim_{x \to 0} |x| = 0 \qquad \text{and} \qquad \lim_{x \to 0} (-|x|) = -\lim_{x \to 0} |x| = 0$$

we can apply the Squeeze Theorem to conclude that $\lim\limits_{x \to 0} x \sin \frac{1}{x} = 0$.   ■

In Section 2.2, we found numerical and graphical evidence suggesting that the limit $\lim\limits_{\theta \to 0} \dfrac{\sin \theta}{\theta}$ is equal to 1. The Squeeze Theorem will allow us to prove this fact.

THEOREM 2 Important Trigonometric Limits

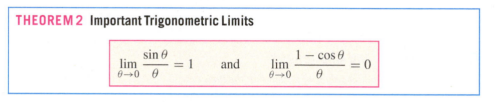

$$\lim_{\theta \to 0} \frac{\sin \theta}{\theta} = 1 \qquad \text{and} \qquad \lim_{\theta \to 0} \frac{1 - \cos \theta}{\theta} = 0$$

Note that both $\frac{\sin \theta}{\theta}$ and $\frac{\cos \theta - 1}{\theta}$ are indeterminate at $\theta = 0$, so Theorem 2 cannot be proved by substitution.

To apply the Squeeze Theorem, we must find functions that squeeze $\dfrac{\sin \theta}{\theta}$ at $\theta = 0$. These are provided by the next theorem (Figure 4).

DF FIGURE 4 Graph illustrating the inequalities of Theorem 3.

THEOREM 3

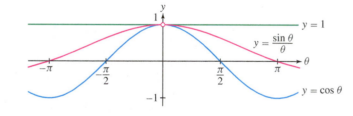

$$\cos \theta \leq \frac{\sin \theta}{\theta} \leq 1 \qquad \text{for} \qquad -\frac{\pi}{2} < \theta < \frac{\pi}{2}, \qquad \theta \neq 0, \qquad \boxed{1}$$

◀⋯ *REMINDER Let's recall why a sector of angle θ in a circle of radius r has area $\frac{1}{2}r^2\theta$. A sector of angle θ represents a fraction $\frac{\theta}{2\pi}$ of the entire circle. The circle has area πr^2, so the sector has area $\left(\frac{\theta}{2\pi}\right)\pi r^2 = \frac{1}{2}r^2\theta$. In the unit circle $(r = 1)$, the sector has area $\frac{1}{2}\theta$.*

NOTE Our proof of Theorem 3 uses the formula $\frac{1}{2}\theta$ for the area of a sector, but this formula is based on the formula πr^2 for the area of a circle, a complete proof of which requires integral calculus.

Proof Assume first that $0 < \theta < \frac{\pi}{2}$. Our proof is based on the following relation between the areas in Figure 5:

$$\text{Area of } \triangle OAB < \text{area of sector } BOA < \text{area of } \triangle OAC \qquad \boxed{2}$$

Let's compute these three areas. First, $\triangle OAB$ has base 1 and height $\sin \theta$, so its area is $\frac{1}{2}\sin \theta$. Next, recall that a sector of angle θ has area $\frac{1}{2}\theta$. Finally, to compute the area of $\triangle OAC$, we observe that

$$\tan \theta = \frac{\text{opposite side}}{\text{adjacent side}} = \frac{AC}{OA} = \frac{AC}{1} = AC$$

Thus, $\triangle OAC$ has base 1, height $\tan \theta$, and area $\frac{1}{2}\tan \theta$. We have shown, therefore, that

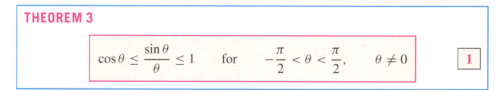

$$\underbrace{\frac{1}{2}\sin \theta}_{\text{Area } \triangle OAB} \leq \underbrace{\frac{1}{2}\theta}_{\text{Area of sector}} \leq \underbrace{\frac{1}{2}\frac{\sin \theta}{\cos \theta}}_{\text{Area } \triangle OAC} \qquad \boxed{3}$$

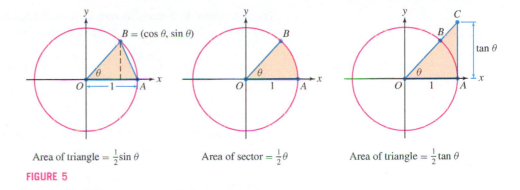

$$\text{Area of triangle} = \tfrac{1}{2}\sin\theta \qquad \text{Area of sector} = \tfrac{1}{2}\theta \qquad \text{Area of triangle} = \tfrac{1}{2}\tan\theta$$

FIGURE 5

The first inequality yields $\sin\theta \le \theta$, and because $\theta > 0$, we obtain

$$\frac{\sin\theta}{\theta} \le 1 \qquad \boxed{4}$$

Next, multiply the second inequality in (3) by $\dfrac{2\cos\theta}{\theta}$ to obtain

$$\cos\theta \le \frac{\sin\theta}{\theta} \qquad \boxed{5}$$

The combination of (4) and (5) gives us (1) when $0 < \theta < \frac{\pi}{2}$. However, the functions in (1) do not change value when θ is replaced by $-\theta$ because both $f(x) = \cos\theta$ and $f(x) = \dfrac{\sin\theta}{\theta}$ are even functions. Indeed, $\cos(-\theta) = \cos\theta$ and

$$\frac{\sin(-\theta)}{-\theta} = \frac{-\sin\theta}{-\theta} = \frac{\sin\theta}{\theta}$$

Therefore, (1) holds for $-\frac{\pi}{2} < \theta < 0$ as well. This completes the proof of Theorem 3. ∎

Proof of Theorem 2 According to Theorem 3,

$$\cos\theta \le \frac{\sin\theta}{\theta} \le 1$$

Since $\lim_{\theta \to 0} \cos\theta = \cos 0 = 1$ and $\lim_{\theta \to 0} 1 = 1$, the Squeeze Theorem yields $\lim_{\theta \to 0} \dfrac{\sin\theta}{\theta} = 1$, as required. It then follows that

$$\lim_{\theta \to 0} \frac{1 - \cos\theta}{\theta} = \lim_{\theta \to 0} \left(\frac{1 + \cos\theta}{1 + \cos\theta}\right) \frac{1 - \cos\theta}{\theta} = \lim_{\theta \to 0} \frac{1 - \cos^2\theta}{(1 + \cos\theta)\theta}$$

$$= \lim_{\theta \to 0} \frac{1}{1 + \cos\theta} \frac{\sin^2\theta}{\theta} = \lim_{x \to 0} \frac{\sin\theta}{1 + \cos\theta} \frac{\sin\theta}{\theta} = \frac{0}{2} \cdot 1 = 0 \qquad ∎$$

In the next example, we evaluate another trigonometric limit. The key idea is to rewrite the function of h in terms of the new variable $\theta = 4h$.

h	$\dfrac{\sin 4h}{h}$
± 1.0	-0.75680
± 0.5	1.81859
± 0.2	3.58678
± 0.1	$\mathbf{3.8}9\,418$
± 0.05	$\mathbf{3.9}7\,339$
± 0.01	$\mathbf{3.99}\,893$
± 0.005	$\mathbf{3.999}\,73$

■ **EXAMPLE 2** **Evaluating a Limit by Changing Variables** Investigate $\lim_{h \to 0} \dfrac{\sin 4h}{h}$ numerically and then evaluate it exactly.

Solution The table of values at the left suggests that the limit is equal to 4. To evaluate the limit exactly, we rewrite it in terms of the limit of $\dfrac{\sin\theta}{\theta}$ so that Theorem 2 can be applied. Thus, we set $\theta = 4h$ and write

$$\frac{\sin 4h}{h} = 4\left(\frac{\sin 4h}{4h}\right) = 4\frac{\sin\theta}{\theta}$$

The new variable θ tends to zero as $h \to 0$ because θ is a multiple of h. Therefore, we may change the limit as $h \to 0$ into a limit as $\theta \to 0$ to obtain

$$\lim_{h \to 0} \frac{\sin 4h}{h} = \lim_{\theta \to 0} 4\frac{\sin \theta}{\theta} = 4\left(\lim_{\theta \to 0} \frac{\sin \theta}{\theta}\right) = 4(1) = 4 \qquad \boxed{6}$$

∎

Note that the change of variables demonstrates that $\dfrac{\sin(kx)}{kx}$ approaches 1 as $x \to 0$. We can use this to our advantage in the next example.

∎ **EXAMPLE 3** Find $\displaystyle\lim_{x \to 0} \frac{\tan(3x)}{\tan(2x)}$.

Solution

$$\lim_{x \to 0} \frac{\tan(3x)}{\tan(2x)} = \lim_{x \to 0} \frac{\sin(3x)}{\cos(3x)} \cdot \frac{\cos(2x)}{\sin(2x)} = \lim_{x \to 0} \frac{\sin(3x)}{\cos(3x)} \cdot \frac{\cos(2x)}{\sin(2x)} \cdot \frac{x}{x} \qquad \boxed{7}$$

$$= \lim_{x \to 0} \frac{3}{2}\left(\frac{\sin(3x)}{3x}\right)\left(\frac{2x}{\sin(2x)}\right)\frac{\cos(2x)}{\cos(3x)} = \frac{3}{2} \cdot 1 \cdot 1 \cdot \frac{1}{1} = \frac{3}{2} \qquad \boxed{8}$$

2.6 SUMMARY

- We say that f is *squeezed* at $x = c$ if there exist functions l and u such that $l(x) \leq f(x) \leq u(x)$ for all $x \neq c$ in an open interval I containing c, and

$$\lim_{x \to c} l(x) = \lim_{x \to c} u(x) = L$$

The Squeeze Theorem states that in this case, $\displaystyle\lim_{x \to c} f(x) = L$.

- Two important trigonometric limits:

$$\lim_{\theta \to 0} \frac{\sin \theta}{\theta} = 1 \qquad \text{and} \qquad \lim_{\theta \to 0} \frac{1 - \cos \theta}{\theta} = 0$$

2.6 EXERCISES

Preliminary Questions

1. Assume that $-x^4 \leq f(x) \leq x^2$. What is $\displaystyle\lim_{x \to 0} f(x)$? Is there enough information to evaluate $\displaystyle\lim_{x \to \frac{1}{2}} f(x)$? Explain.

2. State the Squeeze Theorem carefully.

3. If you want to evaluate $\displaystyle\lim_{h \to 0} \frac{\sin 5h}{3h}$, it is a good idea to rewrite the limit in terms of the variable (choose one):

(a) $\theta = 5h$ **(b)** $\theta = 3h$ **(c)** $\theta = \dfrac{5h}{3}$

Exercises

1. State precisely the hypothesis and conclusions of the Squeeze Theorem for the situation in Figure 6.

2. In Figure 7, is f squeezed by u and l at $x = 3$? At $x = 2$?

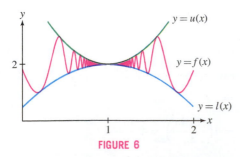

FIGURE 6

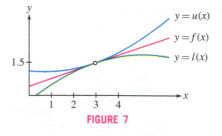

FIGURE 7

3. What does the Squeeze Theorem say about $\lim\limits_{x \to 7} f(x)$ if the limits $\lim\limits_{x \to 7} l(x) = \lim\limits_{x \to 7} u(x) = 6$ and f, u, and l are related as in Figure 8? The inequality $f(x) \le u(x)$ is not satisfied for all x. Does this affect the validity of your conclusion?

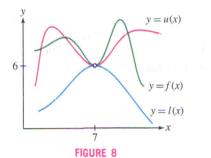

FIGURE 8

4. Determine $\lim\limits_{x \to 0} f(x)$ assuming that $\cos x \le f(x) \le 1$.

5. State whether the inequality provides sufficient information to determine $\lim\limits_{x \to 1} f(x)$, and if so, find the limit.

(a) $4x - 5 \le f(x) \le x^2$

(b) $2x - 1 \le f(x) \le x^2$

(c) $4x - x^2 \le f(x) \le x^2 + 2$

6. [GU] Plot the graphs of $u(x) = 1 + \left|x - \frac{\pi}{2}\right|$ and $l(x) = \sin x$ on the same set of axes. What can you say about $\lim\limits_{x \to \frac{\pi}{2}} f(x)$ if f is squeezed by l and u at $x = \frac{\pi}{2}$?

In Exercises 7–16, evaluate using the Squeeze Theorem.

7. $\lim\limits_{x \to 0} x^2 \cos \dfrac{1}{x}$

8. $\lim\limits_{x \to 0} x \sin \dfrac{1}{x^2}$

9. $\lim\limits_{x \to 1} (x - 1) \sin \dfrac{\pi}{x - 1}$

10. $\lim\limits_{x \to 3} (x^2 - 9) \dfrac{x - 3}{|x - 3|}$

11. $\lim\limits_{t \to 0} (2^t - 1) \cos \dfrac{1}{t}$

12. $\lim\limits_{x \to 0^+} \sqrt{x}\, e^{\cos(\pi/x)}$

13. $\lim\limits_{t \to 2} (t^2 - 4) \cos \dfrac{1}{t - 2}$

14. $\lim\limits_{x \to 0} \tan x \cos \left(\sin \dfrac{1}{x}\right)$

15. $\lim\limits_{\theta \to \frac{\pi}{2}} \cos \theta \cos(\tan \theta)$

16. $\lim\limits_{t \to 0^+} \sin t \tan^{-1}(\ln t)$

In Exercises 17–26, evaluate using Theorem 2 as necessary.

17. $\lim\limits_{x \to 0} \dfrac{\tan x}{x}$

18. $\lim\limits_{x \to 0} \dfrac{\sin x \sec x}{x}$

19. $\lim\limits_{t \to 0} \dfrac{\sqrt{t^3 + 9}\, \sin t}{t}$

20. $\lim\limits_{t \to 0} \dfrac{\sin^2 t}{t}$

21. $\lim\limits_{x \to 0} \dfrac{x^2}{\sin^2 x}$

22. $\lim\limits_{t \to \frac{\pi}{2}} \dfrac{1 - \cos t}{t}$

23. $\lim\limits_{\theta \to 0} \dfrac{\sec \theta - 1}{\theta}$

24. $\lim\limits_{\theta \to 0} \dfrac{1 - \cos \theta}{\sin \theta}$

25. $\lim\limits_{t \to \frac{\pi}{4}} \dfrac{\sin t}{t}$

26. $\lim\limits_{t \to 0} \dfrac{\cos t - \cos^2 t}{t}$

27. Let $L = \lim\limits_{x \to 0} \dfrac{\sin 14x}{x}$.

(a) Show, by letting $\theta = 14x$, that $L = \lim\limits_{\theta \to 0} 14 \dfrac{\sin \theta}{\theta}$.

(b) Compute L.

28. Evaluate $\lim\limits_{h \to 0} \dfrac{\sin 9h}{\sin 7h}$. *Hint:* $\dfrac{\sin 9h}{\sin 7h} = \left(\dfrac{9}{7}\right)\left(\dfrac{\sin 9h}{9h}\right)\left(\dfrac{7h}{\sin 7h}\right)$.

In Exercises 29–48, evaluate the limit.

29. $\lim\limits_{h \to 0} \dfrac{\sin 9h}{h}$

30. $\lim\limits_{h \to 0} \dfrac{\sin 4h}{4h}$

31. $\lim\limits_{h \to 0} \dfrac{\sin h}{5h}$

32. $\lim\limits_{x \to \frac{\pi}{6}} \dfrac{x}{\sin 3x}$

33. $\lim\limits_{\theta \to 0} \dfrac{\sin 7\theta}{\sin 3\theta}$

34. $\lim\limits_{x \to 0} \dfrac{\tan 4x}{9x}$

35. $\lim\limits_{x \to 0} x \csc 25x$

36. $\lim\limits_{t \to 0} \dfrac{\tan 4t}{t \sec t}$

37. $\lim\limits_{h \to 0} \dfrac{\sin 2h \sin 3h}{h^2}$

38. $\lim\limits_{z \to 0} \dfrac{\sin(z/3)}{\sin z}$

39. $\lim\limits_{\theta \to 0} \dfrac{\sin(-3\theta)}{\sin(4\theta)}$

40. $\lim\limits_{x \to 0} \dfrac{\tan 4x}{\tan 9x}$

41. $\lim\limits_{t \to 0} \dfrac{\csc 8t}{\csc 4t}$

42. $\lim\limits_{x \to 0} \dfrac{\sin 5x \sin 2x}{\sin 3x \sin 5x}$

43. $\lim\limits_{x \to 0} \dfrac{\sin 3x \sin 2x}{x \sin 5x}$

44. $\lim\limits_{h \to 0} \dfrac{1 - \cos 2h}{h}$

45. $\lim\limits_{h \to 0} \dfrac{\sin(2h)(1 - \cos h)}{h^2}$

46. $\lim\limits_{t \to 0} \dfrac{1 - \cos 2t}{\sin^2 3t}$

47. $\lim\limits_{\theta \to 0} \dfrac{\cos 2\theta - \cos \theta}{\theta}$

48. $\lim\limits_{h \to \frac{\pi}{2}} \dfrac{1 - \cos 3h}{h}$

49. Calculate $\lim\limits_{x \to 0^-} \dfrac{\sin x}{|x|}$.

50. Use the identity $\sin 3\theta = 3 \sin \theta - 4 \sin^3 \theta$ to evaluate the limit $\lim\limits_{\theta \to 0} \dfrac{\sin 3\theta - 3 \sin \theta}{\theta^3}$.

51. Prove the following result:

$$\lim\limits_{\theta \to 0} \csc \theta - \cot \theta = 0 \qquad \boxed{9}$$

52. [GU] Investigate $\lim\limits_{h \to 0} \dfrac{1 - \cos h}{h^2}$ numerically (and graphically if you have a graphing utility). Then prove that the limit is equal to $\frac{1}{2}$. *Hint:* See the proof of Theorem 2.

In Exercises 53–55, evaluate using the result of Exercise 52.

53. $\lim\limits_{h \to 0} \dfrac{\cos 3h - 1}{h^2}$

54. $\lim\limits_{h \to 0} \dfrac{\cos 3h - 1}{\cos 2h - 1}$

55. $\lim\limits_{t \to 0} \dfrac{\sqrt{1 - \cos t}}{t}$

56. Use the Squeeze Theorem to prove that if $\lim\limits_{x \to c} |f(x)| = 0$, then $\lim\limits_{x \to c} f(x) = 0$.

Further Insights and Challenges

57. Use the result of Exercise 52 to prove that for $m \neq 0$,

$$\lim_{x \to 0} \frac{\cos mx - 1}{x^2} = -\frac{m^2}{2}$$

58. Using a diagram of the unit circle and the Pythagorean Theorem, show that

$$\sin^2 \theta \leq (1 - \cos \theta)^2 + \sin^2 \theta \leq \theta^2$$

Conclude that $\sin^2 \theta \leq 2(1 - \cos \theta) \leq \theta^2$ and use this to give an alternative proof of Eq. (9) in Exercise 51. Then give an alternative proof of the result in Exercise 52.

59. (a) Investigate $\displaystyle\lim_{x \to c} \frac{\sin x - \sin c}{x - c}$ numerically for the five values $c = 0, \frac{\pi}{6}, \frac{\pi}{4}, \frac{\pi}{3}, \frac{\pi}{2}$.

(b) Can you guess the answer for general c?

(c) Check numerically that your answer to (b) works for two other values of c.

(NASA)

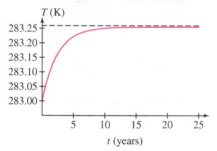

FIGURE 1 The earth's average temperature (according to a simple climate model) in response to an 0.25% increase in solar radiation. According to this model, $\lim_{t \to \infty} T(t) = 283.255$.

2.7 Limits at Infinity

So far we have considered limits as x approaches a number c. It is also important to consider limits where x approaches ∞ or $-\infty$, which we refer to as **limits at infinity**. In applications, limits at infinity arise naturally when we describe the "long-term" behavior of a system as in Figure 1.

The notation $x \to \infty$ indicates that x increases without bound, and $x \to -\infty$ indicates that x decreases (through negative values) without bound. We write

- $\displaystyle\lim_{x \to \infty} f(x) = L$ if $f(x)$ gets closer and closer to L as $x \to \infty$.
- $\displaystyle\lim_{x \to -\infty} f(x) = L$ if $f(x)$ gets closer and closer to L as $x \to -\infty$.

As before, "closer and closer" means that $|f(x) - L|$ becomes arbitrarily small. In either case, the line $y = L$ is called a **horizontal asymptote**. We use the notation $x \to \pm\infty$ to indicate that we are considering both infinite limits, as $x \to \infty$ and as $x \to -\infty$.

Infinite limits describe the **asymptotic behavior** of a function, which is determined by the behavior of the graph as we move out indefinitely to the right or the left.

■ **EXAMPLE 1** Discuss the asymptotic behavior in Figure 2.

Solution The function g approaches $L = 7$ as we move to the right and it approaches $L = 3$ as we move to left, so

$$\lim_{x \to \infty} g(x) = 7, \qquad \lim_{x \to -\infty} g(x) = 3$$

Accordingly, the lines $y = 7$ and $y = 3$ are horizontal asymptotes of g.   ■

A function may approach an infinite limit as $x \to \pm\infty$. We write

$$\lim_{x \to \infty} f(x) = \infty \qquad \text{or} \qquad \lim_{x \to -\infty} f(x) = \infty$$

if $f(x)$ becomes arbitrarily large as $x \to \infty$ or $-\infty$. Similar notation is used if $f(x)$ approaches $-\infty$ as $x \to \pm\infty$. For example, we see in Figure 3(A) that

$$\lim_{x \to \infty} e^x = \infty \qquad \text{and} \qquad \lim_{x \to -\infty} e^x = 0$$

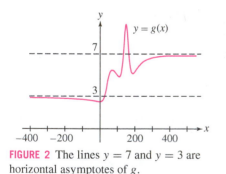

FIGURE 2 The lines $y = 7$ and $y = 3$ are horizontal asymptotes of g.

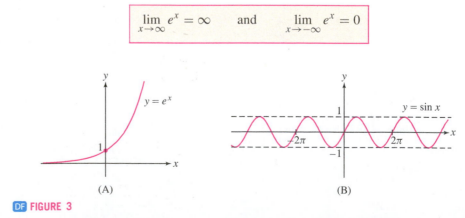

(A)   (B)

DF **FIGURE 3**

When a limit equals either ∞ or $-\infty$, the limit does not exist, since the function is not approaching a finite number. But the fact that the limit is ∞ or $-\infty$ is useful information nonetheless, and we record that fact, rather than just saying the limit does not exist.

However, limits at infinity do not always exist. For example, $f(x) = \sin x$ oscillates indefinitely [Figure 3(B)], so

$$\lim_{x \to \infty} \sin x \quad \text{and} \quad \lim_{x \to -\infty} \sin x$$

do not exist.

The limits at infinity of the power functions $f(x) = x^n$ are easily determined. If $n > 0$, then x^n certainly increases without bound as $x \to \infty$, so (Figure 4)

$$\lim_{x \to \infty} x^n = \infty \quad \text{and} \quad \lim_{x \to \infty} x^{-n} = \lim_{x \to \infty} \frac{1}{x^n} = 0$$

To describe the limits as $x \to -\infty$, assume that n is a whole number so that x^n is defined for $x < 0$. If n is even, then x^n becomes large and positive as $x \to -\infty$, and if n is odd, it becomes large and negative. In summary,

CAUTION $\lim_{x \to -\infty} x^{1/2}$ does not exist, since the square root of a negative number is not a real number.

THEOREM 1   For all $n > 0$,

$$\lim_{x \to \infty} x^n = \infty \quad \text{and} \quad \lim_{x \to \infty} x^{-n} = \lim_{x \to \infty} \frac{1}{x^n} = 0$$

If n is a positive whole number,

$$\lim_{x \to -\infty} x^n = \begin{cases} \infty & \text{if } n \text{ is even} \\ -\infty & \text{if } n \text{ is odd} \end{cases} \quad \text{and} \quad \lim_{x \to -\infty} x^{-n} = \lim_{x \to -\infty} \frac{1}{x^n} = 0$$

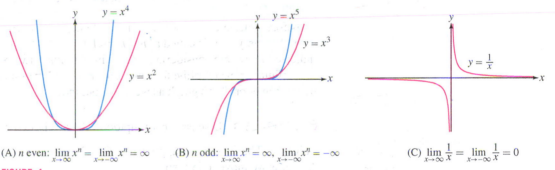

(A) n even: $\lim_{x \to \infty} x^n = \lim_{x \to -\infty} x^n = \infty$ (B) n odd: $\lim_{x \to \infty} x^n = \infty$, $\lim_{x \to -\infty} x^n = -\infty$ (C) $\lim_{x \to \infty} \frac{1}{x} = \lim_{x \to -\infty} \frac{1}{x} = 0$

FIGURE 4

Note also that if p and q are positive integers, then

$$\lim_{x \to -\infty} x^{p/q} = \begin{cases} -\infty & \text{if } q \text{ is odd} \\ \text{undefined} & \text{if } q \text{ is even} \end{cases}$$

The Basic Limit Laws (Theorem 1 in Section 2.3) are valid for limits at infinity. For example, the Sum and Constant Multiple Laws yield

$$\lim_{x \to \infty} \left(3 - 4x^{-3} + 5x^{-5} \right) = \lim_{x \to \infty} 3 - 4 \lim_{x \to \infty} x^{-3} + 5 \lim_{x \to \infty} x^{-5}$$

$$= 3 + 0 + 0 = 3$$

■ **EXAMPLE 2**   Calculate $\lim_{x \to \infty} \dfrac{20x^2 - 3x}{3x^5 - 4x^2 + 5}$.

Solution It would be nice if we could apply the Quotient Law directly, but this law is valid only if the denominator has a finite, nonzero limit. Our limit has the indeterminate form ∞/∞ because

$$\lim_{x \to \infty} (20x^2 - 3x) = \infty \quad \text{and} \quad \lim_{x \to \infty} (3x^5 - 4x^2 + 5) = \infty$$

The way around this difficulty is to divide the numerator and denominator by x^5 (the highest power of x in the denominator):

$$\frac{20x^2 - 3x}{3x^5 - 4x^2 + 5} = \frac{x^{-5}(20x^2 - 3x)}{x^{-5}(3x^5 - 4x^2 + 5)} = \frac{20x^{-3} - 3x^{-4}}{3 - 4x^{-3} + 5x^{-5}}$$

Now we can use the Quotient Law:

$$\lim_{x \to \infty} \frac{20x^2 - 3x}{3x^5 - 4x^2 + 5} = \frac{\lim\limits_{x \to \infty} (20x^{-3} - 3x^{-4})}{\lim\limits_{x \to \infty} (3 - 4x^{-3} + 5x^{-5})} = \frac{0}{3} = 0$$

In general, if

$$f(x) = \frac{a_n x^n + a_{n-1} x^{n-1} + \cdots + a_0}{b_m x^m + b_{m-1} x^{m-1} + \cdots + b_0}$$

where $a_n \neq 0$ and $b_m \neq 0$, divide the numerator and denominator by x^m:

$$f(x) = \frac{a_n x^{n-m} + a_{n-1} x^{n-1-m} + \cdots + a_0 x^{-m}}{b_m + b_{m-1} x^{-1} + \cdots + b_0 x^{-m}}$$

$$= x^{n-m} \left(\frac{a_n + a_{n-1} x^{-1} + \cdots + a_0 x^{-n}}{b_m + b_{m-1} x^{-1} + \cdots + b_0 x^{-m}} \right)$$

The quotient in parentheses approaches the finite limit a_n/b_m because

$$\lim_{x \to \infty} (a_n + a_{n-1} x^{-1} + \cdots + a_0 x^{-n}) = a_n$$

$$\lim_{x \to \infty} (b_m + b_{m-1} x^{-1} + \cdots + b_0 x^{-m}) = b_m$$

This also holds true for $x \to -\infty$, and therefore,

$$\lim_{x \to \pm\infty} f(x) = \lim_{x \to \pm\infty} x^{n-m} \lim_{x \to \pm\infty} \frac{a_n + a_{n-1} x^{-1} + \cdots + a_0 x^{-n}}{b_m + b_{m-1} x^{-1} + \cdots + b_0 x^{-m}} = \frac{a_n}{b_m} \lim_{x \to \pm\infty} x^{n-m}$$

> **THEOREM 2 Limits at Infinity of a Rational Function** The asymptotic behavior of a rational function depends only on the leading terms of its numerator and denominator. If $a_n, b_m \neq 0$, then
>
> $$\lim_{x \to \pm\infty} \frac{a_n x^n + a_{n-1} x^{n-1} + \cdots + a_0}{b_m x^m + b_{m-1} x^{m-1} + \cdots + b_0} = \frac{a_n}{b_m} \lim_{x \to \pm\infty} x^{n-m}$$

Here are some examples:

- $n = m$:

$$\lim_{x \to \infty} \frac{3x^4 - 7x + 9}{7x^4 - 4} = \frac{3}{7} \lim_{x \to \infty} x^0 = \frac{3}{7}$$

- $n < m$:

$$\lim_{x \to \infty} \frac{3x^3 - 7x + 9}{7x^4 - 4} = \frac{3}{7} \lim_{x \to \infty} x^{-1} = 0$$

- $n > m$, $n - m$ odd:

$$\lim_{x \to -\infty} \frac{3x^8 - 7x + 9}{7x^3 - 4} = \frac{3}{7} \lim_{x \to -\infty} x^5 = -\infty$$

- $n > m$, $n - m$ even:

$$\lim_{x \to -\infty} \frac{3x^7 - 7x + 9}{7x^3 - 4} = \frac{3}{7} \lim_{x \to -\infty} x^4 = \infty$$

Our method can be adapted to noninteger exponents and algebraic functions.

■ **EXAMPLE 3** Calculate the limits   **(a)** $\lim\limits_{x \to \infty} \dfrac{3x^{7/2} + 7x^{-1/2}}{x^2 - x^{1/2}}$ **(b)** $\lim\limits_{x \to \infty} \dfrac{x^2}{\sqrt{x^3 + 1}}$.

Solution

The Quotient Law is valid if $\lim\limits_{x \to c} f(x) = \infty$ and $\lim\limits_{x \to c} g(x) = L$, where $L \neq 0$:

$$\lim_{x \to c} \frac{f(x)}{g(x)} = \frac{\lim\limits_{x \to c} f(x)}{\lim\limits_{x \to c} g(x)} = \begin{cases} \infty & \text{if } L > 0 \\ -\infty & \text{if } L < 0 \end{cases}$$

A similar result holds when $\lim\limits_{x \to c} f(x) = -\infty$.

(a) As before, divide the numerator and denominator by x^2, which is the highest power of x occurring in the denominator (this means multiply by x^{-2}):

$$\frac{3x^{7/2} + 7x^{-1/2}}{x^2 - x^{1/2}} = \left(\frac{x^{-2}}{x^{-2}} \right) \frac{3x^{7/2} + 7x^{-1/2}}{x^2 - x^{1/2}} = \frac{3x^{3/2} + 7x^{-5/2}}{1 - x^{-3/2}}$$

$$\lim_{x \to \infty} \frac{3x^{7/2} + 7x^{-1/2}}{x^2 - x^{1/2}} = \frac{\lim\limits_{x \to \infty} (3x^{3/2} + 7x^{-5/2})}{\lim\limits_{x \to \infty} (1 - x^{-3/2})} = \frac{\infty}{1} = \infty$$

(b) The key is to observe that the denominator of $\dfrac{x^2}{\sqrt{x^3 + 1}}$ "behaves" like $x^{3/2}$:

$$\sqrt{x^3 + 1} = \sqrt{x^3(1 + x^{-3})} = x^{3/2}\sqrt{1 + x^{-3}} \qquad (\text{for } x > 0)$$

This suggests that we divide the numerator and denominator by $x^{3/2}$:

$$\frac{x^2}{\sqrt{x^3 + 1}} = \left(\frac{x^{-3/2}}{x^{-3/2}}\right)\frac{x^2}{x^{3/2}\sqrt{1 + x^{-3}}} = \frac{x^{1/2}}{\sqrt{1 + x^{-3}}}$$

Then apply the Quotient Law:

$$\lim_{x \to \infty}\frac{x^2}{\sqrt{x^3 + 1}} = \lim_{x \to \infty}\frac{x^{1/2}}{\sqrt{1 + x^{-3}}} = \frac{\lim\limits_{x \to \infty} x^{1/2}}{\lim\limits_{x \to \infty}\sqrt{1 + x^{-3}}}$$

$$= \frac{\infty}{1} = \infty \qquad\blacksquare$$

■ **EXAMPLE 4** Calculate the limits at infinity of $f(x) = \dfrac{12x + 25}{\sqrt{16x^2 + 100x + 500}}$.

Solution Divide the numerator and denominator by x (multiply by x^{-1}), but notice the difference between x positive and x negative. For $x > 0$,

$$x^{-1}\sqrt{16x^2 + 100x + 500} = \sqrt{x^{-2}}\sqrt{16x^2 + 100x + 500} = \sqrt{16 + \frac{100}{x} + \frac{500}{x^2}}$$

$$\lim_{x \to \infty}\frac{12x + 25}{\sqrt{16x^2 + 100x + 500}} = \frac{\lim\limits_{x \to \infty}\left(12 + \frac{25}{x}\right)}{\lim\limits_{x \to \infty}\sqrt{16 + \frac{100}{x} + \frac{500}{x^2}}} = \frac{12}{\sqrt{16}} = 3$$

However, if $x < 0$, then $x = -\sqrt{x^2}$ and

$$x^{-1}\sqrt{16x^2 + 100x + 500} = -\sqrt{x^{-2}}\sqrt{16x^2 + 100x + 500} = -\sqrt{16 + \frac{100}{x} + \frac{500}{x^2}}$$

So the limit as $x \to -\infty$ is -3 instead of 3 (Figure 5):

$$\lim_{x \to -\infty}\frac{12x + 25}{\sqrt{16x^2 + 100x + 500}} = \frac{\lim\limits_{x \to -\infty}\left(12 + \frac{25}{x}\right)}{-\lim\limits_{x \to -\infty}\sqrt{16 + \frac{100}{x} + \frac{500}{x^2}}} = \frac{12}{-\sqrt{16}} = -3 \qquad\blacksquare$$

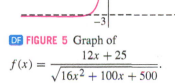

DF FIGURE 5 Graph of
$$f(x) = \frac{12x + 25}{\sqrt{16x^2 + 100x + 500}}.$$

2.7 SUMMARY

• *Limits at infinity*:

    − $\lim\limits_{x \to \infty} f(x) = L$ if $|f(x) - L|$ becomes arbitrarily small as x increases without bound.

    − $\lim\limits_{x \to -\infty} f(x) = L$ if $|f(x) - L|$ becomes arbitrarily small as x decreases without bound.

    − $\lim\limits_{x \to \infty} e^x = \infty$ and $\lim\limits_{x \to -\infty} e^x = 0$

• A horizontal line $y = L$ is a *horizontal asymptote* if

$$\lim_{x \to \infty} f(x) = L \qquad \text{and/or} \qquad \lim_{x \to -\infty} f(x) = L$$

A function can have 0, 1 or 2 horizontal asymptotes.

- If $n > 0$, then $\lim\limits_{x \to \infty} x^n = \infty$ and $\lim\limits_{x \to \pm\infty} x^{-n} = 0$. If $n > 0$ is a whole number, then

$$\lim_{x \to -\infty} x^n = \begin{cases} \infty & \text{if } n \text{ is even} \\ -\infty & \text{if } n \text{ is odd} \end{cases} \quad \text{and} \quad \lim_{x \to -\infty} x^{-n} = 0$$

- If $f(x) = \dfrac{a_n x^n + a_{n-1} x^{n-1} + \cdots + a_0}{b_m x^m + b_{m-1} x^{m-1} + \cdots + b_0}$ with $a_n, b_m \neq 0$, then

$$\lim_{x \to \pm\infty} f(x) = \frac{a_n}{b_m} \lim_{x \to \pm\infty} x^{n-m}$$

2.7 EXERCISES

Preliminary Questions

1. Assume that

$$\lim_{x \to \infty} f(x) = L \quad \text{and} \quad \lim_{x \to L} g(x) = \infty$$

Which of the following statements are correct?

(a) $x = L$ is a vertical asymptote of g.
(b) $y = L$ is a horizontal asymptote of g.
(c) $x = L$ is a vertical asymptote of f.
(d) $y = L$ is a horizontal asymptote of f.

2. What are the following limits?

(a) $\lim\limits_{x \to \infty} x^3$ (b) $\lim\limits_{x \to -\infty} x^3$ (c) $\lim\limits_{x \to -\infty} x^4$

3. Sketch the graph of a function that approaches a limit as $x \to \infty$ but does not approach a limit (either finite or infinite) as $x \to -\infty$.

4. What is the sign of a if $f(x) = ax^3 + x + 1$ satisfies $\lim\limits_{x \to -\infty} f(x) = \infty$?

5. What is the sign of the coefficient multiplying x^7 if f is a polynomial of degree 7 such that $\lim\limits_{x \to -\infty} f(x) = \infty$?

6. Explain why $\lim\limits_{x \to \infty} \sin \frac{1}{x}$ exists but $\lim\limits_{x \to 0} \sin \frac{1}{x}$ does not exist. What is $\lim\limits_{x \to \infty} \sin \frac{1}{x}$?

Exercises

1. What are the horizontal asymptotes of the function in Figure 6?

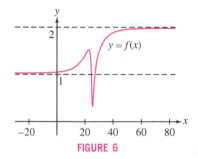

FIGURE 6

2. Sketch the graph of a function f that has both $y = -1$ and $y = 5$ as horizontal asymptotes.

3. Sketch the graph of a function f with a single horizontal asymptote $y = 3$.

4. Sketch the graphs of two functions f and g that have both $y = -2$ and $y = 4$ as horizontal asymptotes but $\lim\limits_{x \to \infty} f(x) \neq \lim\limits_{x \to \infty} g(x)$.

5. $\boxed{\text{GU}}$ Investigate the asymptotic behavior of $f(x) = \dfrac{x^3}{x^3 + x}$ numerically and graphically:

(a) Make a table of values of $f(x)$ for $x = \pm 50, \pm 100, \pm 500, \pm 1000$.

(b) Plot the graph of f.

(c) What are the horizontal asymptotes of f?

6. $\boxed{\text{GU}}$ Investigate $\lim\limits_{x \to \pm\infty} \dfrac{12x + 1}{\sqrt{4x^2 + 9}}$ numerically and graphically:

(a) Make a table of values of $f(x) = \dfrac{12x + 1}{\sqrt{4x^2 + 9}}$ for $x = \pm 100, \pm 500, \pm 1000, \pm 10,000$.

(b) Plot the graph of f.

(c) What are the horizontal asymptotes of f?

In Exercises 7–16, evaluate the limit.

7. $\lim\limits_{x \to \infty} \dfrac{x}{x + 9}$ **8.** $\lim\limits_{x \to \infty} \dfrac{3x^2 + 20x}{4x^2 + 9}$

9. $\lim\limits_{x \to \infty} \dfrac{3x^2 + 20x}{2x^4 + 3x^3 - 29}$ **10.** $\lim\limits_{x \to \infty} \dfrac{4}{x + 5}$

11. $\lim\limits_{x \to \infty} \dfrac{7x - 9}{4x + 3}$ **12.** $\lim\limits_{x \to \infty} \dfrac{9x^2 - 2}{6 - 29x}$

13. $\lim\limits_{x \to -\infty} \dfrac{7x^2 - 9}{4x + 3}$ **14.** $\lim\limits_{x \to -\infty} \dfrac{5x - 9}{4x^3 + 2x + 7}$

15. $\lim\limits_{x \to -\infty} \dfrac{3x^3 - 10}{x + 4}$ **16.** $\lim\limits_{x \to -\infty} \dfrac{2x^5 + 3x^4 - 31x}{8x^4 - 31x^2 + 12}$

In Exercises 17–22, find the horizontal asymptotes.

17. $f(x) = \dfrac{2x^2 - 3x}{8x^2 + 8}$ **18.** $f(x) = \dfrac{8x^3 - x^2}{7 + 11x - 4x^4}$

19. $f(x) = \dfrac{\sqrt{36x^2 + 7}}{9x + 4}$ **20.** $f(x) = \dfrac{\sqrt{36x^4 + 7}}{9x^2 + 4}$

21. $f(t) = \dfrac{e^t}{1 + e^{-t}}$ **22.** $f(t) = \dfrac{t^{1/3}}{(64t^2 + 9)^{1/6}}$

In Exercises 23–30, evaluate the limit.

23. $\lim\limits_{x \to \infty} \dfrac{\sqrt{9x^4 + 3x + 2}}{4x^3 + 1}$ **24.** $\lim\limits_{x \to \infty} \dfrac{\sqrt{x^3 + 20x}}{10x - 2}$

25. $\lim\limits_{x \to -\infty} \dfrac{8x^2 + 7x^{1/3}}{\sqrt{16x^4 + 6}}$

26. $\lim\limits_{x \to -\infty} \dfrac{4x - 3}{\sqrt{25x^2 + 4x}}$

27. $\lim\limits_{t \to \infty} \dfrac{t^{4/3} + t^{1/3}}{(4t^{2/3} + 1)^2}$

28. $\lim\limits_{t \to \infty} \dfrac{t^{4/3} - 9t^{1/3}}{(8t^4 + 2)^{1/3}}$

29. $\lim\limits_{x \to -\infty} \dfrac{|x| + x}{x + 1}$

30. $\lim\limits_{t \to -\infty} \dfrac{4 + 6e^{2t}}{5 - 9e^{3t}}$

31.  Determine $\lim\limits_{x \to \infty} \tan^{-1} x$. Explain geometrically.

32. Show that $\lim\limits_{x \to \infty} (\sqrt{x^2 + 1} - x) = 0$. *Hint:* Observe that

$$\sqrt{x^2 + 1} - x = \dfrac{1}{\sqrt{x^2 + 1} + x}$$

33. According to the **Michaelis–Menten equation** (Figure 7), when an enzyme is combined with a substrate of concentration s (in millimolars), the reaction rate (in micromolars/min) is

$$R(s) = \dfrac{As}{K + s} \qquad (A, K \text{ constants})$$

(a) Show, by computing $\lim\limits_{s \to \infty} R(s)$, that A is the limiting reaction rate as the concentration s approaches ∞.

(b) Show that the reaction rate $R(s)$ attains one-half of the limiting value A when $s = K$.

(c) For a certain reaction, $K = 1.25$ mM and $A = 0.1$. For which concentration s is $R(s)$ equal to 75% of its limiting value?

Leonor Michaelis
1875–1949
(Rockefeller Archive Center)

Maud Menten
1879–1960
(University Archives, University of Pittsburgh)

FIGURE 7 Canadian-born biochemist Maud Menten is best known for her fundamental work on enzyme kinetics with German scientist Leonor Michaelis. She was also an accomplished painter, clarinetist, mountain climber, and master of numerous languages.

34. Suppose that the average temperature of Earth is $T(t) = 283 + 3(1 - e^{-0.03t})$ kelvins, where t is the number of years since 2000.

(a) Calculate the long-term average $L = \lim\limits_{t \to \infty} T(t)$.

(b) At what time is $T(t)$ within one-half a degree of its limiting value?

In Exercises 35–42, calculate the limit.

35. $\lim\limits_{x \to \infty} (\sqrt{4x^4 + 9x} - 2x^2)$

36. $\lim\limits_{x \to \infty} (\sqrt{9x^3 + x} - x^{3/2})$

37. $\lim\limits_{x \to \infty} (2\sqrt{x} - \sqrt{x + 2})$

38. $\lim\limits_{x \to \infty} \left(\dfrac{1}{x} - \dfrac{1}{x + 2} \right)$

39. $\lim\limits_{x \to \infty} (\ln(3x + 1) - \ln(2x + 1))$

40. $\lim\limits_{x \to \infty} \left(\ln(\sqrt{5x^2 + 2}) - \ln x \right)$

41. $\lim\limits_{x \to \infty} \tan^{-1} \left(\dfrac{x^2 + 9}{9 - x} \right)$

42. $\lim\limits_{x \to \infty} \tan^{-1} \left(\dfrac{1 + x}{1 - x} \right)$

43.  Let $P(n)$ be the perimeter of an n-gon inscribed in a unit circle (Figure 8).

(a) Explain, intuitively, why $P(n)$ approaches 2π as $n \to \infty$.

(b) Show that $P(n) = 2n \sin\left(\dfrac{\pi}{n} \right)$.

(c) Combine (a) and (b) to conclude that $\lim\limits_{n \to \infty} \dfrac{n}{\pi} \sin\left(\dfrac{\pi}{n} \right) = 1$.

(d) Use this to give another argument that $\lim\limits_{\theta \to 0} \dfrac{\sin \theta}{\theta} = 1$.

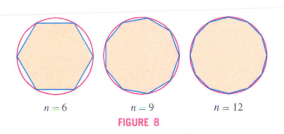

$n = 6$ $n = 9$ $n = 12$

FIGURE 8

44. Physicists have observed that Einstein's theory of **special relativity** reduces to Newtonian mechanics in the limit as $c \to \infty$, where c is the speed of light. This is illustrated by a stone tossed up vertically from ground level so that it returns to Earth 1 s later. Using Newton's Laws, we find that the stone's maximum height is $h = g/8$ meters ($g = 9.8$ m/s^2). According to special relativity, the stone's mass depends on its velocity divided by c, and the maximum height is

$$h(c) = c\sqrt{c^2/g^2 + 1/4} - c^2/g$$

Prove that $\lim\limits_{c \to \infty} h(c) = g/8$.

Further Insights and Challenges

45. Every limit as $x \to \infty$ can be rewritten as a one-sided limit as $t \to 0^+$, where $t = x^{-1}$. Setting $g(t) = f(t^{-1})$, we have

$$\lim\limits_{x \to \infty} f(x) = \lim\limits_{t \to 0^+} g(t)$$

Show that $\lim\limits_{x \to \infty} \dfrac{3x^2 - x}{2x^2 + 5} = \lim\limits_{t \to 0^+} \dfrac{3 - t}{2 + 5t^2}$, and evaluate using the Quotient Law.

46. Rewrite the following as one-sided limits as in Exercise 45 and evaluate.

(a) $\lim\limits_{x \to \infty} \dfrac{3 - 12x^3}{4x^3 + 3x + 1}$

(b) $\lim\limits_{x \to \infty} e^{1/x}$

(c) $\lim\limits_{x \to \infty} x \sin \dfrac{1}{x}$

(d) $\lim\limits_{x \to \infty} \ln\left(\dfrac{x + 1}{x - 1} \right)$

47. Let $G(b) = \lim\limits_{x \to \infty} (1 + b^x)^{1/x}$ for $b \geq 0$. Investigate $G(b)$ numerically and graphically for $b = 0.2, 0.8, 2, 3, 5$ (and additional values if necessary). Then make a conjecture for the value of $G(b)$ as a function of b. Draw a graph of $y = G(b)$. Does G appear to be continuous? We will evaluate $G(b)$ using L'Hôpital's Rule in Section 4.5 (see Exercise 69 there).

2.8 Intermediate Value Theorem

The **Intermediate Value Theorem (IVT)** says, roughly speaking, that *a continuous function cannot skip values*. Consider a plane that takes off and climbs from 0 to 10,000 m in 20 min. The plane must reach every altitude between 0 and 10,000 m during this 20-min interval. Thus, at some moment, the plane's altitude must have been exactly 8371 m. Of course, this assumes that the plane's motion is continuous, so its altitude cannot jump abruptly from, say, 8000 to 9000 m.

To state this conclusion formally, let $A(t)$ be the plane's altitude at time t. The IVT asserts that for every altitude M between 0 and 10,000, there is a time t_0 between 0 and 20 such that $A(t_0) = M$. In other words, the graph of A must intersect the horizontal line $y = M$ [Figure 1(A)].

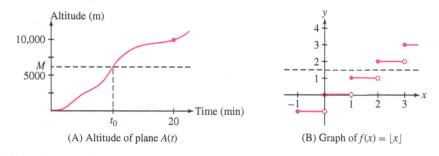

(A) Altitude of plane $A(t)$ (B) Graph of $f(x) = \lfloor x \rfloor$

DF FIGURE 1

By contrast, a discontinuous function can skip values. The greatest integer function $f(x) = \lfloor x \rfloor$ in Figure 1(B) satisfies $\lfloor 1 \rfloor = 1$ and $\lfloor 2 \rfloor = 2$, but it does not take on the value 1.5 (or any other value between 1 and 2).

> **THEOREM 1 Intermediate Value Theorem** If f is continuous on a closed interval $[a, b]$, then for every value M between $f(a)$ and $f(b)$, there exists at least one value $c \in (a, b)$ such that $f(c) = M$.

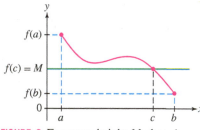

FIGURE 2 For every height M, there is a c in (a, b) such that $f(c) = M$.

Graphically, as in Figure 2, the result appears obvious. For a continuous function, every horizontal line at height M between $f(a)$ and $f(b)$ is forced to hit the continuous graph and therefore there must be at least one value c in (a, b) such that $f(c) = M$. The proof appears in Appendix B.

■ **EXAMPLE 1** Prove that the equation $\sin x = 0.3$ has at least one solution.

Solution We may apply the IVT because $f(x) = \sin x$ is continuous. We choose an interval where we suspect that a solution exists. The desired value 0.3 lies between the two values of the function

$$\sin 0 = 0 \quad \text{and} \quad \sin \frac{\pi}{2} = 1$$

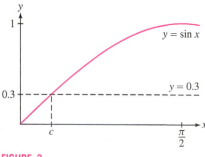

FIGURE 3

so the interval $\left[0, \frac{\pi}{2}\right]$ will work (Figure 3). The IVT tells us that $\sin x = 0.3$ has at least one solution in $\left(0, \frac{\pi}{2}\right)$.   ■

The IVT can be used to show the existence of zeros of functions. If f is continuous and takes on both positive and negative values—say, $f(a) < 0$ and $f(b) > 0$—then the IVT guarantees that $f(c) = 0$ for some c between a and b. This is extremely useful in a case where we cannot explicitly solve for the zero but would like to know that it is there.

A zero or root of a function is a value c such that $f(c) = 0$. Sometimes the word "root" is reserved to refer specifically to the zero of a polynomial.

> **COROLLARY 2 Existence of Zeros** If f is continuous on $[a, b]$ and if $f(a)$ and $f(b)$ are nonzero and have opposite signs, then f has a zero in (a, b).

We can locate zeros of functions to arbitrary accuracy using the **Bisection Method**. The idea is to find an interval $[a, b]$ such that the function has opposite signs at the endpoints. Then Corollary 2 tells us that there is a zero on this interval. To find its location more precisely, we cut the interval into two equal subintervals. Then, check the signs at the end points of each of these intervals to see which one Corollary 2 tells us has a zero. (But keep in mind that there may be more than one zero, so both could contain a zero). Next we repeat the process on this smaller interval. Eventually, we narrow down on the zero. This is illustrated in the next example.

■ **EXAMPLE 2** The Bisection Method Show that $f(x) = \cos^2 x - 2 \sin \frac{x}{4}$ has a zero in $(0, 2)$. Then locate the zero more accurately using the Bisection Method.

Solution Using a calculator, we find that $f(0)$ and $f(2)$ have opposite signs:

$$f(0) = 1 > 0, \qquad f(2) \approx -0.786 < 0$$

Corollary 2 guarantees that $f(x) = 0$ has a solution in $(0, 2)$ (Figure 4).

To locate a zero more accurately, divide $[0, 2]$ into two intervals $[0, 1]$ and $[1, 2]$. At least one of these intervals must contain a zero of f. To determine which, evaluate f at the midpoint $m = 1$. A calculator gives $f(1) \approx -0.203 < 0$, and since $f(0) = 1$, we see that

$$f(x) \text{ takes on opposite signs at the endpoints of } [0, 1]$$

Therefore, $(0, 1)$ must contain a zero. We discard $[1, 2]$ because both $f(1)$ and $f(2)$ are negative.

The Bisection Method consists of continuing this process until we narrow down the location of the zero to any desired accuracy. In the following table, the process is carried out three times:

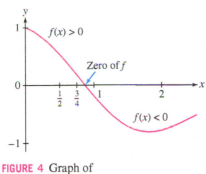

FIGURE 4 Graph of $f(x) = \cos^2 x - 2 \sin \frac{x}{4}$.

Computer algebra systems have built-in commands for finding roots of a function or solving an equation numerically. These systems use a variety of methods, including more sophisticated versions of the Bisection Method. Notice that to use the Bisection Method, we must first find an interval containing a root.

Interval	Midpoint of interval	Function values	Conclusion
$[0, 1]$	$\frac{1}{2}$	$f\left(\frac{1}{2}\right) \approx 0.521$ $f(1) \approx -0.203$	Zero lies in $\left(\frac{1}{2}, 1\right)$
$\left[\frac{1}{2}, 1\right]$	$\frac{3}{4}$	$f\left(\frac{3}{4}\right) \approx 0.163$ $f(1) \approx -0.203$	Zero lies in $\left(\frac{3}{4}, 1\right)$
$\left[\frac{3}{4}, 1\right]$	$\frac{7}{8}$	$f\left(\frac{7}{8}\right) \approx -0.0231$ $f\left(\frac{3}{4}\right) \approx 0.163$	Zero lies in $\left(\frac{3}{4}, \frac{7}{8}\right)$

We conclude that f has a zero c satisfying $0.75 < c < 0.875$. ■

CONCEPTUAL INSIGHT The IVT seems to state the obvious, namely that a continuous function cannot skip values. Yet its proof (given in Appendix B) is subtle because it depends on the *completeness* property of real numbers. To highlight the subtlety, observe that the IVT is *false* for functions defined only on the *rational numbers*. For example, $f(x) = x^2$ is continuous, but it does not have the intermediate value property if we restrict its domain to the rational numbers. Indeed, $f(0) = 0$ and $f(2) = 4$, but $f(c) = 2$ *has no solution* for c rational. The solution $c = \sqrt{2}$ is "missing" from the set of rational numbers because it is irrational. No doubt the IVT was always regarded as obvious, but it was not possible to give a correct proof until the completeness property was clarified in the second half of the nineteenth century.

2.8 SUMMARY

- The Intermediate Value Theorem (IVT) says that a continuous function cannot *skip* values.
- More precisely, if f is continuous on $[a, b]$ with $f(a) \neq f(b)$, and if M is a number between $f(a)$ and $f(b)$, then $f(c) = M$ for some $c \in (a, b)$.

- Existence of zeros: If f is continuous on $[a, b]$ and if $f(a)$ and $f(b)$ take opposite signs (one is positive and the other negative), then $f(c) = 0$ for some $c \in (a, b)$.
- Bisection Method: Assume f is continuous and that $f(a)$ and $f(b)$ have opposite signs, so that f has a zero in (a, b). Then f has a zero in $[a, m]$ or $[m, b]$, where $m = (a + b)/2$ is the midpoint of $[a, b]$. A zero lies in (a, m) if $f(a)$ and $f(m)$ have opposite signs and in (m, b) if $f(m)$ and $f(b)$ have opposite signs. Continuing the process, we can locate a zero with arbitrary accuracy.

2.8 EXERCISES

Preliminary Questions

1. Prove that $f(x) = x^2$ takes on the value 0.5 in the interval $[0, 1]$.

2. The temperature in Vancouver was 8°C at 6 AM and rose to 20°C at noon. Which assumption about temperature allows us to conclude that the temperature was 15°C at some moment of time between 6 AM and noon?

3. What is the graphical interpretation of the IVT?

4. Show that the following statement is false by drawing a graph that provides a counterexample:

If f is continuous and has a root in $[a, b]$, then $f(a)$ and $f(b)$ have opposite signs.

5. Assume that f is continuous on $[1, 5]$ and that $f(1) = 20$, $f(5) = 100$. Determine whether each of the following statements is always true, never true, or sometimes true.

(a) $f(c) = 3$ has a solution with $c \in [1, 5]$.
(b) $f(c) = 75$ has a solution with $c \in [1, 5]$.
(c) $f(c) = 50$ has no solution with $c \in [1, 5]$.
(d) $f(c) = 30$ has exactly one solution with $c \in [1, 5]$.

Exercises

1. Use the IVT to show that $f(x) = x^3 + x$ takes on the value 9 for some x in $[1, 2]$.

2. Show that $g(t) = \dfrac{t}{t + 1}$ takes on the value 0.499 for some t in $[0, 1]$.

3. Show that $g(t) = t^2 \tan t$ takes on the value $\frac{1}{2}$ for some t in $\left[0, \frac{\pi}{4}\right]$.

4. Show that $f(x) = \dfrac{x^2}{x^7 + 1}$ takes on the value 0.4.

5. Show that $\cos x = x$ has a solution in the interval $[0, 1]$. *Hint:* Show that $f(x) = x - \cos x$ has a zero in $[0, 1]$.

6. Use the IVT to find an interval of length $\frac{1}{2}$ containing a root of $f(x) = x^3 + 2x + 1$.

In Exercises 7–16, prove using the IVT.

7. $\sqrt{c} + \sqrt{c + 2} = 3$ has a solution.

8. For all integers n, $\sin nx = \cos x$ for some $x \in [0, \pi]$.

9. $\sqrt{2}$ exists. *Hint:* Consider $f(x) = x^2$.

10. A positive number c has an nth root for all positive integers n.

11. For all positive integers k, $\cos x = x^k$ has a solution.

12. $2^x = bx$ has a solution if $b > 2$.

13. $2^x + 3^x = 4^x$ has a solution.

14. $\cos x = \cos^{-1} x$ has a solution in $(0, 1)$.

15. $e^x + \ln x = 0$ has a solution.

16. $\tan^{-1} x = \cos^{-1} x$ has a solution.

17. Use the Intermediate Value Theorem to show that the equation $x^6 - 8x^4 + 10x^2 - 1 = 0$ has at least six distinct solutions.

In Exercises 18–20, determine whether or not the IVT applies to show that the given function takes on all values between $f(a)$ and $f(b)$ for $x \in (a, b)$. If it does not apply, determine any values between $f(a)$ and $f(b)$ that the function does not take on for $x \in (a, b)$.

18.
$$f(x) = \begin{cases} x & \text{for } x < 0 \\ x^2 & \text{for } x \geq 0 \end{cases}$$

for the interval $[-1, 1]$.

19.
$$g(x) = \begin{cases} -x & \text{for } x < 0 \\ x^3 + 1 & \text{for } x \geq 0 \end{cases}$$

for the interval $[-1, 1]$.

20.
$$j(x) = \begin{cases} -x^2 & \text{for } x < 0 \\ 1 & \text{for } x = 0 \\ x & \text{for } x > 0 \end{cases}$$

for the interval $[-2, 2]$.

21. Carry out three steps of the Bisection Method for $f(x) = 2^x - x^3$ as follows:

(a) Show that f has a zero in $[1, 1.5]$.
(b) Show that f has a zero in $[1.25, 1.5]$.
(c) Determine whether $[1.25, 1.375]$ or $[1.375, 1.5]$ contains a zero.

22. Figure 5 shows that $f(x) = x^3 - 8x - 1$ has a root in the interval $[2.75, 3]$. Apply the Bisection Method twice to find an interval of length $\frac{1}{16}$ containing this root.

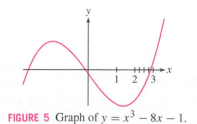

FIGURE 5 Graph of $y = x^3 - 8x - 1$.

23. Find an interval of length $\frac{1}{4}$ in $[1, 2]$ containing a root of the equation $x^7 + 3x - 10 = 0$.

24. Show that $\tan^3 \theta - 8 \tan^2 \theta + 17 \tan \theta - 8 = 0$ has a root in $[0.5, 0.6]$. Apply the Bisection Method twice to find an interval of length 0.025 containing this root.

In Exercises 25–28, draw the graph of a function f on $[0, 4]$ with the given property.

25. Jump discontinuity at $x = 2$ and does not satisfy the conclusion of the IVT

26. Jump discontinuity at $x = 2$ and satisfies the conclusion of the IVT on $[0, 4]$

27. Infinite one-sided limits at $x = 2$ and does not satisfy the conclusion of the IVT

28. Infinite one-sided limits at $x = 2$ and satisfies the conclusion of the IVT on $[0, 4]$

29. Can Corollary 2 be applied to $f(x) = x^{-1}$ on $[-1, 1]$? Does f have any roots?

30. $x^6 - 8x^4 + 10x^2 - 1 = 0$ has at least six distinct solutions.

Further Insights and Challenges

31. Take any map and draw a circle on it anywhere (Figure 6). Prove that at any moment in time there exists a pair of diametrically opposite points A and B on that circle corresponding to locations where the temperatures at that moment are equal. *Hint:* Let θ be an angular coordinate along the circle and let $f(\theta)$ be the difference in temperatures at the locations corresponding to θ and $\theta + \pi$.

FIGURE 6 $f(\theta)$ is the difference between the temperatures at A and B.

32. Show that if f is continuous and $0 \leq f(x) \leq 1$ for $0 \leq x \leq 1$, then $f(c) = c$ for some c in $[0, 1]$ (Figure 7).

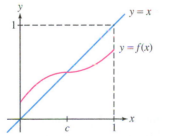

FIGURE 7 A function satisfying $0 \leq f(x) \leq 1$ for $0 \leq x \leq 1$.

33. Use the IVT to show that if f is continuous and one-to-one on an interval $[a, b]$, then f is either an increasing or a decreasing function.

34. **Ham Sandwich Theorem** Figure 8(A) shows a slice of ham. Prove that for any angle θ ($0 \leq \theta \leq \pi$), it is possible to cut the slice in half with a cut of incline θ. *Hint:* The lines of inclination θ are given by the equations $y = (\tan \theta)x + b$, where b varies from $-\infty$ to ∞. Each such line divides the slice into two pieces (one of which may be empty). Let $A(b)$ be the amount of ham to the left of the line minus the amount to the right, and let A be the total area of the ham. Show that $A(b) = -A$ if b is sufficiently large and $A(b) = A$ if b is sufficiently negative. Then use the IVT. This works if $\theta \neq 0$ or $\frac{\pi}{2}$. If $\theta = 0$, define $A(b)$ as the amount of ham above the line $y = b$ minus the amount below. How can you modify the argument to work when $\theta = \frac{\pi}{2}$ (in which case $\tan \theta = \infty$)?

35. Figure 8(B) shows a slice of ham on a piece of bread. Prove that it is possible to slice this open-faced sandwich so that each part has equal amounts of ham and bread. *Hint:* By Exercise 34, for all $0 \leq \theta \leq \pi$ there is a line $L(\theta)$ of incline θ (which we assume is unique) that divides the ham into two equal pieces. Let $B(\theta)$ denote the amount of bread to the left of (or above) $L(\theta)$ minus the amount to the right (or below). Notice that $L(\pi)$ and $L(0)$ are the same line, but $B(\pi) = -B(0)$ since left and right get interchanged as the angle moves from 0 to π. Assume that B is continuous and apply the IVT. (By a further extension of this argument, one can prove the full "Ham Sandwich Theorem," which states that if you allow the knife to cut at a slant, then it is possible to cut a sandwich consisting of a slice of ham and two slices of bread so that all three layers are divided in half.)

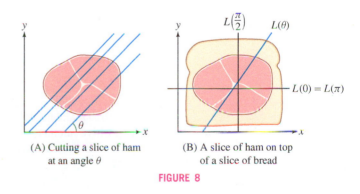

(A) Cutting a slice of ham at an angle θ

(B) A slice of ham on top of a slice of bread

FIGURE 8

2.9 The Formal Definition of a Limit

In this section, we reexamine the definition of a limit in order to state it in a more rigorous and precise fashion. Why is this necessary? In Section 2.2, we defined limits by saying that $\lim_{x \to c} f(x) = L$ if $|f(x) - L|$ becomes arbitrarily small when x is sufficiently close (but not equal) to c. The problem with this definition lies in the phrases "arbitrarily small" and "sufficiently close." We must find a way to specify just how close is sufficiently close.

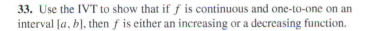

The Size of the Gap

A "rigorous proof" in mathematics is a proof based on a complete chain of logic without any gaps or ambiguity. The formal definition of a limit is a key ingredient of rigorous proofs in calculus. A few such proofs are included in Appendix D. More complete developments can be found in textbooks on the branch of mathematics called "analysis."

Recall that the distance from $f(x)$ to L is $|f(x) - L|$. It is convenient to refer to the quantity $|f(x) - L|$ as the *gap* between the value $f(x)$ and the limit L.

Let's reexamine the trigonometric limit

$$\lim_{x \to 0} \frac{\sin x}{x} = 1 \qquad \boxed{1}$$

In this example, $f(x) = \dfrac{\sin x}{x}$ and $L = 1$, so Eq. (1) tells us that the gap $|f(x) - 1|$ gets arbitrarily small when x is sufficiently close, but not equal, to 0 [Figure 1(A)].

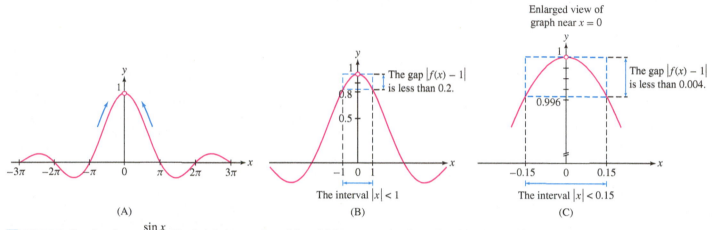

(A)   (B)   (C)

DF FIGURE 1 Graphs of $y = \dfrac{\sin x}{x}$. To shrink the gap from 0.2 to 0.004, we require that x lie within 0.15 of 0.

Suppose we want the gap $|f(x) - 1|$ to be less than 0.2. How close to 0 must x be? Figure 1(B) shows that $f(x)$ lies within 0.2 of $L = 1$ for all values of x in the interval $[-1, 1]$. In other words, the following statement is true:

$$\text{If } 0 < |x| < 1, \quad \text{then} \quad \left| \frac{\sin x}{x} - 1 \right| < 0.2$$

If we insist instead that the gap be smaller than 0.004, we can check by zooming in on the graph, as in Figure 1(C), that

$$\text{If } 0 < |x| < 0.15, \quad \text{then} \quad \left| \frac{\sin x}{x} - 1 \right| < 0.004$$

It would seem that this process can be continued: By zooming in on the graph, we can find a small interval around $c = 0$, where the gap $|f(x) - 1|$ is smaller than any prescribed positive number.

To express this in a precise fashion, we follow time-honored tradition in using the Greek letters ϵ (epsilon) and δ (delta) to denote small numbers specifying the sizes of the gap and the quantity $|x - c|$, respectively. In our case, $c = 0$ and $|x - c| = |x|$. The precise meaning of Eq. (1) is that for every choice of $\epsilon > 0$, there exists some δ (depending on ϵ) such that

$$\text{If } 0 < |x| < \delta, \quad \text{then} \quad \left| \frac{\sin x}{x} - 1 \right| < \epsilon$$

The number δ pins down just how close is "sufficiently close" for a given ϵ. With this motivation, we are ready to state the formal definition of the limit.

The formal definition of a limit is often called the ϵ-δ definition. The tradition of using the symbols ϵ and δ originated in the writings of Augustin-Louis Cauchy on calculus and analysis in the 1820s.

FORMAL DEFINITION OF A LIMIT Suppose that $f(x)$ is defined for all x in an open interval containing c (but not necessarily at $x = c$). Then

$$\lim_{x \to c} f(x) = L$$

if for all $\epsilon > 0$, there exists $\delta > 0$ such that

$$\text{if } 0 < |x - c| < \delta, \quad \text{then } |f(x) - L| < \epsilon$$

If the symbols ϵ and δ seem to make this definition too abstract, keep in mind that we can take $\epsilon = 10^{-n}$ and $\delta = 10^{-m}$. Thus, $\lim_{x \to c} f(x) = L$ if, for any n, there exist $m > 0$ such that if $0 < |x - c| < 10^{-m}$, then $|f(x) - L| < 10^{-n}$.

The condition $0 < |x - c| < \delta$ in this definition excludes $x = c$. In other words, the limit depends only on values of $f(x)$ near c but not on $f(c)$ itself. As we have seen in previous sections, the limit may exist even when $f(c)$ is not defined.

■ **EXAMPLE 1** Let $f(x) = 8x + 3$.

(a) Prove that $\lim_{x \to 3} f(x) = 27$ using the formal definition of the limit.

(b) Find values of δ that work for $\epsilon = 0.2$ and 0.001.

Solution

(a) We break the proof into two steps.

Step 1. **Relate the gap to $|x - c|$.**
We must find a relation between two absolute values: $|f(x) - L|$ for $L = 27$ and $|x - c|$ for $c = 3$. Observe that

$$\underbrace{|f(x) - 27|}_{\text{Size of gap}} = |(8x + 3) - 27| = |8x - 24| = 8|x - 3|$$

Thus, the gap is 8 times as large as $|x - 3|$.

Step 2. **Choose δ (in terms of ϵ).**
We can now see how to make the gap small: If $|x - 3| < \frac{\epsilon}{8}$, then the gap is less than $8\left(\frac{\epsilon}{8}\right) = \epsilon$. Therefore, for any $\epsilon > 0$, we choose $\delta = \frac{\epsilon}{8}$. With this choice, the following statement holds:

$$\text{If } 0 < |x - 3| < \delta, \quad \text{then } |f(x) - 27| < \epsilon, \quad \text{where } \delta = \frac{\epsilon}{8}$$

Since we have specified δ for all $\epsilon > 0$, we have fulfilled the requirements of the formal definition, thus proving rigorously that $\lim_{x \to 3} (8x + 3) = 27$.

(b) For the particular choice $\epsilon = 0.2$, we may take $\delta = \frac{\epsilon}{8} = \frac{0.2}{8} = 0.025$:

$$\text{If } 0 < |x - 3| < 0.025, \quad \text{then } |f(x) - 27| < 0.2$$

This statement is illustrated in Figure 2. But note that *any positive δ smaller than* 0.025 *will also work*. For example, the following statement is also true, although it places an unnecessary restriction on x:

$$\text{If } 0 < |x - 3| < 0.019, \quad \text{then } |f(x) - 27| < 0.2$$

Similarly, to make the gap less than $\epsilon = 0.001$, we may take

$$\delta = \frac{\epsilon}{8} = \frac{0.001}{8} = 0.000125 \qquad ■$$

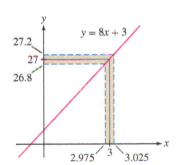

FIGURE 2 To make the gap less than 0.2, we may take $\delta = 0.025$ (not drawn to scale).

The difficulty in applying the limit definition lies in trying to relate $|f(x) - L|$ to $|x - c|$. The next two examples illustrate how this can be done in special cases.

■ **EXAMPLE 2** Prove that $\lim_{x \to 2} x^2 = 4$.

Solution Let $f(x) = x^2$.

Step 1. **Relate the gap to $|x - c|$.**
In this case, we must relate the gap $|f(x) - 4| = |x^2 - 4|$ to the quantity $|x - 2|$ (Figure 3). This is more difficult than in the previous example because the gap is not a constant multiple of $|x - 2|$. To proceed, consider the factorization

$$|x^2 - 4| = |x + 2|\,|x - 2|$$

Because we are going to require that $|x - 2|$ be small, we may as well assume from the outset that $|x - 2| < 1$, which means that $1 < x < 3$. In this case, $|x + 2|$ is less than 5 and the gap satisfies

$$\text{If } |x - 2| < 1, \quad \text{then } |x^2 - 4| = |x + 2|\,|x - 2| < 5\,|x - 2| \qquad \boxed{2}$$

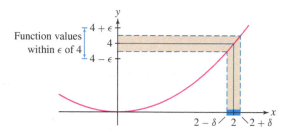

DF FIGURE 3 Graph of $f(x) = x^2$. We may choose δ so that $f(x)$ lies within ϵ of 4 for all x in $[2 - \delta, 2 + \delta]$.

Step 2. **Choose δ (in terms of ϵ).**
We see from Eq. (2) that if $|x - 2|$ is smaller than both $\frac{\epsilon}{5}$ and 1, then the gap satisfies

$$|x^2 - 4| < 5|x - 2| < 5\left(\frac{\epsilon}{5}\right) = \epsilon$$

Therefore, the following statement holds for all $\epsilon > 0$:

> If $0 < |x - 2| < \delta$, $\quad$ then $|x^2 - 4| < \epsilon$, $\quad$ where δ is the smaller of $\frac{\epsilon}{5}$ and 1

We have specified δ for all $\epsilon > 0$, so we have fulfilled the requirements of the formal limit definition, thus proving that $\lim_{x \to 2} x^2 = 4$. ■

■ **EXAMPLE 3** Prove that $\lim_{x \to 3} \dfrac{1}{x} = \dfrac{1}{3}$.

Solution

Step 1. **Relate the gap to $|x - c|$.**
The gap is equal to

$$\left|\frac{1}{x} - \frac{1}{3}\right| = \left|\frac{3 - x}{3x}\right| = |x - 3|\left|\frac{1}{3x}\right|$$

◀┄ **REMINDER** If $a > b > 0$, then $\frac{1}{a} < \frac{1}{b}$. Thus, if $3x > 6$, then $\frac{1}{3x} < \frac{1}{6}$.

Because we are going to require that $|x - 3|$ be small, we may as well assume from the outset that $|x - 3| < 1$, or equivalently, $2 < x < 4$. Now observe that if $x > 2$, then $3x > 6$ and $\frac{1}{3x} < \frac{1}{6}$, so the following inequality is valid if $|x - 3| < 1$:

$$\left|f(x) - \frac{1}{3}\right| = \left|\frac{3 - x}{3x}\right| = \left|\frac{1}{3x}\right| |x - 3| < \frac{1}{6}|x - 3| \qquad \boxed{3}$$

Step 2. **Choose δ (in terms of ϵ).**
By Eq. (3), if $|x - 3| < 1$ and $|x - 3| < 6\epsilon$, then

$$\left|\frac{1}{x} - \frac{1}{3}\right| < \frac{1}{6}|x - 3| < \frac{1}{6}(6\epsilon) = \epsilon$$

Therefore, given any $\epsilon > 0$, we let δ be the smaller of the numbers 6ϵ and 1. Then we have:

$$\text{If } 0 < |x - 3| < \delta, \quad \text{then} \quad \left| \frac{1}{x} - \frac{1}{3} \right| < \epsilon, \quad \text{where } \delta \text{ is the smaller of } 6\epsilon \text{ and } 1$$

Again, we have fulfilled the requirements of the formal limit definition, thus proving rigorously that $\lim\limits_{x \to 3} \frac{1}{x} = \frac{1}{3}$. ∎

GRAPHICAL INSIGHT Keep the graphical interpretation of limits in mind. In Figure 4(A), $f(x)$ approaches L as $x \to c$ because for any $\epsilon > 0$, we can make the gap less than ϵ by taking δ sufficiently small. By contrast, the function in Figure 4(B) has a jump discontinuity at $x = c$. The gap cannot be made small, no matter how small δ is taken. Therefore, the limit does not exist.

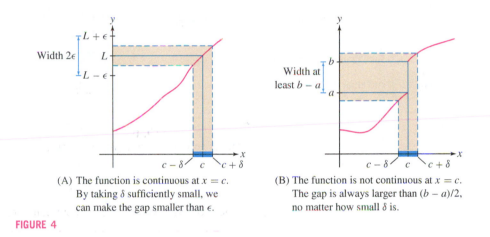

(A) The function is continuous at $x = c$. By taking δ sufficiently small, we can make the gap smaller than ϵ.

(B) The function is not continuous at $x = c$. The gap is always larger than $(b - a)/2$, no matter how small δ is.

FIGURE 4

Proving Limit Theorems

In practice, the formal definition of the limit is rarely used to evaluate limits. Most limits are evaluated using the Basic Limit Laws or other techniques such as the Squeeze Theorem. However, the formal definition allows us to prove these laws in a rigorous fashion and thereby ensure that calculus is built on a solid foundation. We illustrate by proving the Sum Law. Other proofs are given in Appendix D.

Proof of the Sum Law Assume that

$$\lim_{x \to c} f(x) = L \qquad \text{and} \qquad \lim_{x \to c} g(x) = M$$

We must prove that $\lim\limits_{x \to c} (f(x) + g(x)) = L + M$.

◀·· REMINDER *The Triangle Inequality [Eq. (1) in Section 1.1] states*

$$|a + b| \leq |a| + |b|$$

for all a and b.

Apply the Triangle Inequality (see margin) with $a = f(x) - L$ and $b = g(x) - M$:

$$|(f(x) + g(x)) - (L + M)| \leq |f(x) - L| + |g(x) - M| \qquad \boxed{4}$$

Each term on the right in (4) can be made small by the limit definition. More precisely, given $\epsilon > 0$, we can choose δ such that $|f(x) - L| < \frac{\epsilon}{2}$ and $|g(x) - M| < \frac{\epsilon}{2}$ if $0 < |x - c| < \delta$ (in principle, we might choose different δ's for f and g, but we may then use the smaller of the two δ's). Thus, Eq. (4) gives

$$\text{If } 0 < |x - c| < \delta, \quad \text{then } |f(x) + g(x) - (L + M)| < \frac{\epsilon}{2} + \frac{\epsilon}{2} = \epsilon \qquad \boxed{5}$$

This proves that

$$\lim_{x \to c} (f(x) + g(x)) = L + M = \lim_{x \to c} f(x) + \lim_{x \to c} g(x) \qquad \blacksquare$$

2.9 SUMMARY

- Informally speaking, the statement $\lim_{x \to c} f(x) = L$ means that the gap $|f(x) - L|$ tends to 0 as x approaches c.
- The *formal definition* (called the ϵ-δ definition): $\lim_{x \to c} f(x) = L$ if, for all $\epsilon > 0$, there exists a $\delta > 0$ such that

$$\text{if } 0 < |x - c| < \delta, \quad \text{then } |f(x) - L| < \epsilon$$

2.9 EXERCISES

Preliminary Questions

1. Given that $\lim_{x \to 0} \cos x = 1$, which of the following statements is true?

(a) If $|\cos x - 1|$ is very small, then x is close to 0.

(b) There is an $\epsilon > 0$ such that if if $0 < |\cos x - 1| < \epsilon$, then $|x| < 10^{-5}$.

(c) There is a $\delta > 0$ such that if $0 < |x| < \delta$, then $|\cos x - 1| < 10^{-5}$.

(d) There is a $\delta > 0$ such that if $0 < |x - 1| < \delta$, then $|\cos x| < 10^{-5}$.

2. Suppose it is known that for a given ϵ and δ, if $0 < |x - 3| < \delta$, then $|f(x) - 2| < \epsilon$. Which of the following statements must also be true?

(a) If $0 < |x - 3| < 2\delta$, then $|f(x) - 2| < \epsilon$.

(b) If $0 < |x - 3| < \delta$, then $|f(x) - 2| < 2\epsilon$.

(c) If $0 < |x - 3| < \dfrac{\delta}{2}$, then $|f(x) - 2| < \dfrac{\epsilon}{2}$.

(d) If $0 < |x - 3| < \dfrac{\delta}{2}$, then $|f(x) - 2| < \epsilon$.

Exercises

1. Based on the information conveyed in Figure 5(A), find values of L, ϵ, and $\delta > 0$ such that the following statement holds: If $|x| < \delta$, then $|f(x) - L| < \epsilon$.

2. Based on the information conveyed in Figure 5(B), find values of c, L, ϵ, and $\delta > 0$ such that the following statement holds: If $0 < |x - c| < \delta$, then $|f(x) - L| < \epsilon$.

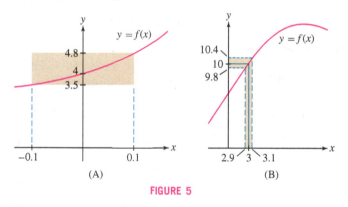

(A) (B)

FIGURE 5

3. Consider $\lim_{x \to 4} f(x)$, where $f(x) = 8x + 3$.

(a) Show that $|f(x) - 35| = 8|x - 4|$.

(b) Show that for any $\epsilon > 0$, if $0 < |x - 4| < \delta$, then $|f(x) - 35| < \epsilon$, where $\delta = \frac{\epsilon}{8}$. Explain how this proves rigorously that $\lim_{x \to 4} f(x) = 35$.

4. Consider $\lim_{x \to 2} f(x)$, where $f(x) = 4x - 1$.

(a) Show that if $0 < |x - 2| < \delta$, then $|f(x) - 7| < 4\delta$.

(b) Find a δ such that

$$\text{If } 0 < |x - 2| < \delta, \quad \text{then } |f(x) - 7| < 0.01$$

(c) Prove rigorously that $\lim_{x \to 2} f(x) = 7$.

5. Consider $\lim_{x \to 2} x^2 = 4$ (refer to Example 2).

(a) Show that if $0 < |x - 2| < 0.01$, then $|x^2 - 4| < 0.05$.

(b) Show that if $0 < |x - 2| < 0.0002$, then $|x^2 - 4| < 0.0009$.

(c) Find a value of δ such that if $0 < |x - 2| < \delta$, then $|x^2 - 4|$ is less than 10^{-4}.

6. With regard to the limit $\lim_{x \to 5} x^2 = 25$,

(a) show that if $4 < x < 6$, then $|x^2 - 25| < 11|x - 5|$. *Hint:* Write $|x^2 - 25| = |x + 5| \cdot |x - 5|$.

(b) find a δ such that if $0 < |x - 5| < \delta$, then $|x^2 - 25| < 10^{-3}$.

(c) give a rigorous proof of the limit by showing that if $0 < |x - 5| < \delta$, then $|x^2 - 25| < \epsilon$, where δ is the smaller of $\frac{\epsilon}{11}$ and 1.

7. Refer to Example 3 to find a value of $\delta > 0$ such that

$$\text{If } 0 < |x - 3| < \delta, \quad \text{then } \left| \frac{1}{x} - \frac{1}{3} \right| < 10^{-4}$$

8. Use Figure 6 to find a value of $\delta > 0$ such that the following statement holds: If $0 < |x - 2| < \delta$, then $\left| 1/x^2 - \frac{1}{4} \right| < \epsilon$ for $\epsilon = 0.03$. Then find a value of δ that works for $\epsilon = 0.01$.

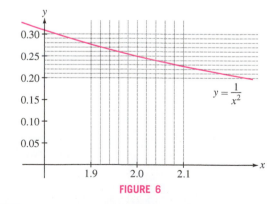

FIGURE 6

9. [GU] Plot $f(x) = \sqrt{2x - 1}$ together with the horizontal lines $y = 2.9$ and $y = 3.1$. Use this plot to find a value of $\delta > 0$ such that if $0 < |x - 5| < \delta$, then $|\sqrt{2x - 1} - 3| < 0.1$.

10. [GU] Plot $f(x) = \tan x$ together with the horizontal lines $y = 0.99$ and $y = 1.01$. Use this plot to find a value of $\delta > 0$ such that if $0 < \left| x - \frac{\pi}{4} \right| < \delta$, then $|\tan x - 1| < 0.01$.

11. [GU] The number e has the following property: $\lim\limits_{x \to 0} \dfrac{e^x - 1}{x} = 1$.

Use a plot of $f(x) = \dfrac{e^x - 1}{x}$ to find a value of $\delta > 0$ such that if $0 < |x - 1| < \delta$, then $|f(x) - 1| < 0.01$.

12. [GU] Let $f(x) = \dfrac{4}{x^2 + 1}$ and $\epsilon = 0.5$. Using a plot of f, find a value of $\delta > 0$ such that if $0 < \left| x - \frac{1}{2} \right| < \delta$, then $\left| f(x) - \frac{16}{5} \right| < \epsilon$. Repeat for $\epsilon = 0.2$ and 0.1.

13. Consider $\lim\limits_{x \to 2} \dfrac{1}{x}$.

(a) Show that if $|x - 2| < 1$, then

$$\left| \frac{1}{x} - \frac{1}{2} \right| < \frac{1}{2} |x - 2|$$

(b) Let δ be the smaller of 1 and 2ϵ. Prove the following:

If $0 < |x - 2| < \delta$, then $\left| \dfrac{1}{x} - \dfrac{1}{2} \right| < \epsilon$

(c) Find a $\delta > 0$ such that if $0 < |x - 2| < \delta$, then $\left| \frac{1}{x} - \frac{1}{2} \right| < 0.01$.

(d) Prove rigorously that $\lim\limits_{x \to 2} \dfrac{1}{x} = \dfrac{1}{2}$.

14. Consider $\lim\limits_{x \to 1} \sqrt{x + 3}$.

(a) Show that if $|x - 1| < 4$, then $|\sqrt{x + 3} - 2| < \frac{1}{2} |x - 1|$. *Hint:* Multiply the inequality by $|\sqrt{x + 3} + 2|$ and observe that $|\sqrt{x + 3} + 2| > 2$.

(b) Find $\delta > 0$ such that if $0 < |x - 1| < \delta$, then $|\sqrt{x + 3} - 2| < 10^{-4}$.

(c) Prove rigorously that the limit is equal to 2.

15. [pencil icon] Let $f(x) = \sin x$. Using a calculator, we find

$$f\left(\frac{\pi}{4} - 0.1 \right) \approx 0.633, \quad f\left(\frac{\pi}{4} \right) \approx 0.707, \quad f\left(\frac{\pi}{4} + 0.1 \right) \approx 0.774$$

Use these values and the fact that f is increasing on $\left[0, \frac{\pi}{2} \right]$ to justify the statement

If $0 < \left| x - \frac{\pi}{4} \right| < 0.1$, then $\left| f(x) - f\left(\frac{\pi}{4} \right) \right| < 0.08$

Then draw a figure like Figure 3 to illustrate this statement.

16. Adapt the argument in Example 1 to prove rigorously that $\lim\limits_{x \to c} (ax + b) = ac + b$, where a, b, c are arbitrary.

17. Adapt the argument in Example 2 to prove rigorously that $\lim\limits_{x \to c} x^2 = c^2$ for all c.

18. Adapt the argument in Example 3 to prove rigorously that $\lim\limits_{x \to c} x^{-1} = \frac{1}{c}$ for all $c \neq 0$.

In Exercises 19–24, use the formal definition of the limit to prove the statement rigorously.

19. $\lim\limits_{x \to 4} \sqrt{x} = 2$

20. $\lim\limits_{x \to 1} (3x^2 + x) = 4$

21. $\lim\limits_{x \to 1} x^3 = 1$

22. $\lim\limits_{x \to 0} (x^2 + x^3) = 0$

23. $\lim\limits_{x \to 2} x^{-2} = \dfrac{1}{4}$

24. $\lim\limits_{x \to 0} x \sin \dfrac{1}{x} = 0$

25. Let $f(x) = \dfrac{x}{|x|}$. Prove rigorously that $\lim\limits_{x \to 0} f(x)$ does not exist. *Hint:* Show that for any L, there always exists some x such that $|x| < \delta$ but $|f(x) - L| \geq \frac{1}{2}$, no matter how small δ is taken.

26. Prove rigorously that $\lim\limits_{x \to 0} |x| = 0$.

27. Let $f(x) = \min(x, x^2)$, where $\min(a, b)$ is the minimum of a and b. Prove rigorously that $\lim\limits_{x \to 1} f(x) = 1$.

28. Prove rigorously that $\lim\limits_{x \to 0} \sin \frac{1}{x}$ does not exist.

29. First, use the identity

$$\sin x + \sin y = 2 \sin \left(\frac{x + y}{2} \right) \cos \left(\frac{x - y}{2} \right)$$

to verify the relation

$$\sin(a + h) - \sin a = h \frac{\sin(h/2)}{h/2} \cos \left(a + \frac{h}{2} \right) \qquad \boxed{6}$$

Then use the inequality $\left| \dfrac{\sin x}{x} \right| \leq 1$ for $x \neq 0$ to show that

$|\sin(a + h) - \sin a| < |h|$ for all a. Finally, prove rigorously that $\lim\limits_{x \to a} \sin x = \sin a$.

Further Insights and Challenges

30. Uniqueness of the Limit Prove that a function converges to at most one limiting value. In other words, use the limit definition to prove that if $\lim\limits_{x \to c} f(x) = L_1$ and $\lim\limits_{x \to c} f(x) = L_2$, then $L_1 = L_2$.

In Exercises 31–33, prove the statement using the formal limit definition.

31. The Constant Multiple Law [Theorem 1, part (ii) in Section 2.3].

32. The Squeeze Theorem (Theorem 1 in Section 2.6).

33. The Product Law [Theorem 1, part (iii) in Section 2.3]. *Hint:* Use the identity

$$f(x)g(x) - LM = (f(x) - L)g(x) + L(g(x) - M)$$

34. Let $f(x) = 1$ if x is rational and $f(x) = 0$ if x is irrational. Prove that $\lim\limits_{x \to c} f(x)$ does not exist for any c. *Hint:* There exist rational and irrational numbers arbitrarily close to any c.

35. [pencil icon] Here is a function with strange continuity properties:

$$f(x) = \begin{cases} \dfrac{1}{q} & \text{if } x \text{ is the rational number } p/q \text{ in lowest terms} \\ 0 & \text{if } x \text{ is an irrational number} \end{cases}$$

(a) Show that f is discontinuous at c if c is rational. *Hint:* There exist irrational numbers arbitrarily close to c.

(b) Show that f is continuous at c if c is irrational. *Hint:* Let I be the interval $\{x : |x - c| < 1\}$. Show that for any $Q > 0$, I contains at most finitely many fractions p/q with $q < Q$. Conclude that there is a δ such that all fractions in $\{x : |x - c| < \delta\}$ have a denominator larger than Q.

36. Write a formal definition of the following:

$$\lim_{x \to \infty} f(x) = L$$

37. Write a formal definition of the following:

$$\lim_{x \to a} f(x) = \infty$$

CHAPTER REVIEW EXERCISES

1. The position of a particle at time t (s) is $s(t) = \sqrt{t^2 + 1}$ m. Compute its average velocity over $[2, 5]$ and estimate its instantaneous velocity at $t = 2$.

2. The "wellhead" price p of natural gas in the United States (in dollars per 1000 ft^3) on the first day of each month in 2012 is listed in the table below.

J	F	M	A	M	J
2.89	2.46	2.25	1.89	1.94	2.54

J	A	S	O	N	D
2.59	2.86	2.71	3.03	3.35	3.35

Compute the average rate of change of p (in dollars per 1000 ft^3 per month) over the quarterly periods January–March, April–June, and July–September.

3. For a positive integer n, let $P(n)$ be the number of *partitions* of n, that is, the number of ways of writing n as a sum of one or more positive integers. For example, $P(4) = 5$ since the number 4 can be partitioned in five different ways: 4, $3 + 1$, $2 + 2$, $2 + 1 + 1$, and $1 + 1 + 1 + 1$. Suppose P is a continuous function whose values at positive integers are known. Use Figure 1 to estimate the rate of change of P at $n = 12$.

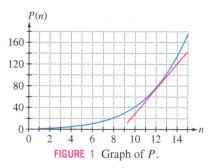

FIGURE 1 Graph of P.

4. The average velocity v (m/s) of an oxygen molecule in the air at temperature T (°C) is $v = 25.7\sqrt{273.15 + T}$. What is the average speed at $T = 25°$ (room temperature)? Estimate the rate of change of average velocity with respect to temperature at $T = 25°$. What are the units of this rate?

In Exercises 5–10, estimate the limit numerically to two decimal places or state that the limit does not exist.

5. $\displaystyle\lim_{x \to 0} \frac{1 - \cos^3(x)}{x^2}$

6. $\displaystyle\lim_{x \to 1} x^{1/(x-1)}$

7. $\displaystyle\lim_{x \to 2} \frac{x^x - 4}{x^2 - 4}$

8. $\displaystyle\lim_{x \to 2} \frac{x - 2}{\ln(3x - 5)}$

9. $\displaystyle\lim_{x \to 1} \left(\frac{7}{1 - x^7} - \frac{3}{1 - x^3} \right)$

10. $\displaystyle\lim_{x \to 2} \frac{3^x - 9}{5^x - 25}$

In Exercises 11–50, evaluate the limit if it exists. If not, determine whether the one-sided limits exist (finite or infinite).

11. $\displaystyle\lim_{x \to 4} (3 + x^{1/2})$

12. $\displaystyle\lim_{x \to 1} \frac{5 - x^2}{4x + 7}$

13. $\displaystyle\lim_{x \to -2} \frac{4}{x^3}$

14. $\displaystyle\lim_{x \to -1} \frac{3x^2 + 4x + 1}{x + 1}$

15. $\displaystyle\lim_{t \to 9} \frac{\sqrt{t} - 3}{t - 9}$

16. $\displaystyle\lim_{x \to 3} \frac{\sqrt{x + 1} - 2}{x - 3}$

17. $\displaystyle\lim_{x \to 1} \frac{x^3 - x}{x - 1}$

18. $\displaystyle\lim_{h \to 0} \frac{2(a + h)^2 - 2a^2}{h}$

19. $\displaystyle\lim_{t \to 9} \frac{t - 6}{\sqrt{t} - 3}$

20. $\displaystyle\lim_{s \to 0} \frac{1 - \sqrt{s^2 + 1}}{s^2}$

21. $\displaystyle\lim_{x \to -1+} \frac{1}{x + 1}$

22. $\displaystyle\lim_{y \to \frac{1}{3}} \frac{3y^2 + 5y - 2}{6y^2 - 5y + 1}$

23. $\displaystyle\lim_{x \to 1} \frac{x^3 - 2x}{x - 1}$

24. $\displaystyle\lim_{a \to b} \frac{a^2 - 3ab + 2b^2}{a - b}$

25. $\displaystyle\lim_{x \to 0} \frac{e^{3x} - e^x}{e^x - 1}$

26. $\displaystyle\lim_{\theta \to 0} \frac{\sin 5\theta}{\theta}$

27. $\displaystyle\lim_{x \to 1.5} \left\lfloor \frac{1}{x} \right\rfloor$

28. $\displaystyle\lim_{\theta \to \frac{\pi}{4}} \sec \theta$

29. $\displaystyle\lim_{z \to -3} \frac{z + 3}{z^2 + 4z + 3}$

30. $\displaystyle\lim_{x \to 1} \frac{x^3 - ax^2 + ax - 1}{x - 1}$

31. $\displaystyle\lim_{x \to b} \frac{x^3 - b^3}{x - b}$

32. $\displaystyle\lim_{x \to 0} \frac{\sin 4x}{\sin 3x}$

33. $\displaystyle\lim_{x \to 0} \left(\frac{1}{3x} - \frac{1}{x(x + 3)} \right)$

34. $\displaystyle\lim_{\theta \to \frac{1}{4}} 3^{\tan(\pi\theta)}$

35. $\displaystyle\lim_{x \to 0-} \frac{\lfloor x \rfloor}{x}$

36. $\displaystyle\lim_{x \to 0+} \frac{\lfloor x \rfloor}{x}$

37. $\displaystyle\lim_{\theta \to \frac{\pi}{2}} \theta \sec \theta$

38. $\displaystyle\lim_{y \to 2} \ln \left(\sin \frac{\pi}{y} \right)$

39. $\displaystyle\lim_{\theta \to 0} \frac{\cos \theta - 2}{\theta}$

40. $\displaystyle\lim_{x \to 4.3} \frac{1}{x - \lfloor x \rfloor}$

41. $\displaystyle\lim_{x \to 2-} \frac{x - 3}{x - 2}$

42. $\displaystyle\lim_{t \to 0} \frac{\sin^2 t}{t^3}$

43. $\displaystyle\lim_{x \to 1+} \left(\frac{1}{\sqrt{x - 1}} - \frac{1}{\sqrt{x^2 - 1}} \right)$

44. $\displaystyle\lim_{t \to e} \sqrt{t}(\ln t - 1)$

45. $\lim\limits_{x\to\frac{\pi}{2}} \tan x$

46. $\lim\limits_{t\to 0} \cos \dfrac{1}{t}$

47. $\lim\limits_{t\to 0^+} \sqrt{t} \cos \dfrac{1}{t}$

48. $\lim\limits_{x\to 5^+} \dfrac{x^2-24}{x^2-25}$

49. $\lim\limits_{x\to 0} \dfrac{\cos x - 1}{\sin x}$

50. $\lim\limits_{\theta\to 0} \dfrac{\tan\theta - \sin\theta}{\sin^3\theta}$

51. Find the left- and right-hand limits of the function f in Figure 2 at $x = 0, 2, 4$. State whether f is left- or right-continuous (or both) at these points.

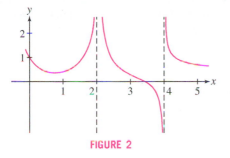

FIGURE 2

52. Sketch the graph of a function f such that

(a) $\lim\limits_{x\to 2^-} f(x) = 1$, $\lim\limits_{x\to 2^+} f(x) = 3$

(b) $\lim\limits_{x\to 4} f(x)$ exists but does not equal $f(4)$.

53. Graph h and describe the discontinuity:

$$h(x) = \begin{cases} e^x & \text{for } x \le 0 \\ \ln x & \text{for } x > 0 \end{cases}$$

Is h left- or right-continuous?

54. Sketch the graph of a function g such that

$$\lim\limits_{x\to -3^-} g(x) = \infty, \qquad \lim\limits_{x\to -3^+} g(x) = -\infty, \qquad \lim\limits_{x\to 4} g(x) = \infty$$

55. Find the points of discontinuity of

$$g(x) = \begin{cases} \cos\left(\dfrac{\pi x}{2}\right) & \text{for } |x| < 1 \\[2mm] |x-1| & \text{for } |x| \ge 1 \end{cases}$$

Determine the type of discontinuity and whether g is left- or right-continuous.

56. Show that $f(x) = xe^{\sin x}$ is continuous on its domain.

57. Find a constant b such that h is continuous at $x = 2$, where

$$h(x) = \begin{cases} x+1 & \text{for } |x| < 2 \\ b - x^2 & \text{for } |x| \ge 2 \end{cases}$$

With this choice of b, find all points of discontinuity.

In Exercises 58–63, find the horizontal asymptotes of the function by computing the limits at infinity.

58. $f(x) = \dfrac{9x^2-4}{2x^2-x}$

59. $f(x) = \dfrac{x^2-3x^4}{x-1}$

60. $f(u) = \dfrac{8u-3}{\sqrt{16u^2+6}}$

61. $f(u) = \dfrac{2u^2-1}{\sqrt{6+u^4}}$

62. $f(x) = \dfrac{3x^{2/3}+9x^{3/7}}{7x^{4/5}-4x^{-1/3}}$

63. $f(t) = \dfrac{t^{1/3}-t^{-1/3}}{(t-t^{-1})^{1/3}}$

64. Calculate (a)–(d), assuming that

$$\lim\limits_{x\to 3} f(x) = 6, \qquad \lim\limits_{x\to 3} g(x) = 4$$

(a) $\lim\limits_{x\to 3} (f(x) - 2g(x))$

(b) $\lim\limits_{x\to 3} x^2 f(x)$

(c) $\lim\limits_{x\to 3} \dfrac{f(x)}{g(x)+x}$

(d) $\lim\limits_{x\to 3} (2g(x)^3 - g(x)^{3/2})$

65. Assume that the following limits exist:

$$A = \lim\limits_{x\to a} f(x), \qquad B = \lim\limits_{x\to a} g(x), \qquad L = \lim\limits_{x\to a} \dfrac{f(x)}{g(x)}$$

Prove that if $L = 1$, then $A = B$. *Hint:* You cannot use the Quotient Law if $B = 0$, so apply the Product Law to L and B instead.

66. $\boxed{\text{GU}}$ Define $g(t) = (1 + 2^{1/t})^{-1}$ for $t \ne 0$. How should $g(0)$ be defined to make g left-continuous at $t = 0$?

67. In the notation of Exercise 65, give an example where L exists but neither A nor B exists.

68. True or false?

(a) If $\lim\limits_{x\to 3} f(x)$ exists, then $\lim\limits_{x\to 3} f(x) = f(3)$.

(b) If $\lim\limits_{x\to 0} \dfrac{f(x)}{x} = 1$, then $f(0) = 0$.

(c) If $\lim\limits_{x\to -7} f(x) = 8$, then $\lim\limits_{x\to -7} \dfrac{1}{f(x)} = \dfrac{1}{8}$.

(d) If $\lim\limits_{x\to 5^+} f(x) = 4$ and $\lim\limits_{x\to 5^-} f(x) = 8$, then $\lim\limits_{x\to 5} f(x) = 6$.

(e) If $\lim\limits_{x\to 0} \dfrac{f(x)}{x} = 1$, then $\lim\limits_{x\to 0} f(x) = 0$.

(f) If $\lim\limits_{x\to 5} f(x) = 2$, then $\lim\limits_{x\to 5} f(x)^3 = 8$.

69. Let $f(x) = \left\lfloor \dfrac{1}{x} \right\rfloor$, where $\lfloor x \rfloor$ is the greatest integer function. Show that for $x \ne 0$,

$$\frac{1}{x} - 1 < \left\lfloor \frac{1}{x} \right\rfloor \le \frac{1}{x}$$

Then use the Squeeze Theorem to prove that

$$\lim\limits_{x\to 0} x \left\lfloor \frac{1}{x} \right\rfloor = 1$$

Hint: Treat the one-sided limits separately.

70. Let r_1 and r_2 be the roots of $f(x) = ax^2 - 2x + 20$. Observe that f "approaches" the linear function $L(x) = -2x + 20$ as $a \to 0$. Because $r = 10$ is the unique root of L, we might expect one of the roots of f to approach 10 as $a \to 0$ (Figure 3). Prove that the roots can be labeled so that $\lim\limits_{a\to 0} r_1 = 10$ and $\lim\limits_{a\to 0} r_2 = \infty$.

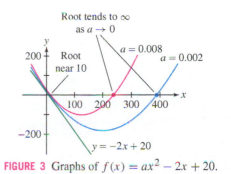

FIGURE 3 Graphs of $f(x) = ax^2 - 2x + 20$.

71. Use the IVT to prove that the curves $y = x^2$ and $y = \cos x$ intersect.

72. Use the IVT to prove that $f(x) = x^3 - \dfrac{x^2 + 2}{\cos x + 2}$ has a root in the interval $[0, 2]$.

73. Use the IVT to show that $e^{-x^2} = x$ has a solution on $(0, 1)$.

74. Use the Bisection Method to locate a solution of $x^2 - 7 = 0$ to two decimal places.

75. Give an example of a (discontinuous) function that does not satisfy the conclusion of the IVT on $[-1, 1]$. Then show that the function

$$f(x) = \begin{cases} \sin \dfrac{1}{x} & x \neq 0 \\ 0 & x = 0 \end{cases}$$

satisfies the conclusion of the IVT on every interval $[-a, a]$.

76. Let $f(x) = \dfrac{1}{x + 2}$.

(a) Show that if $|x - 2| < 1$, then $\left| f(x) - \frac{1}{4} \right| < \dfrac{|x - 2|}{12}$. *Hint:* Observe that if $|x - 2| < 1$, then $|4(x + 2)| > 12$.

(b) Find $\delta > 0$ such that if $|x - 2| < \delta$, then $\left| f(x) - \frac{1}{4} \right| < 0.01$.

(c) Prove rigorously that $\lim_{x \to 2} f(x) = \frac{1}{4}$.

77. [GU] Plot the function $f(x) = x^{1/3}$. Use the zoom feature to find a $\delta > 0$ such that if $|x - 8| < \delta$, then $|x^{1/3} - 2| < 0.05$.

78. Use the fact that $f(x) = 2^x$ is increasing to find a value of δ such that $|2^x - 8| < 0.001$ if $|x - 2| < \delta$. *Hint:* Find c_1 and c_2 such that $7.999 < f(c_1) < f(c_2) < 8.001$.

79. Prove rigorously that $\lim_{x \to -1} (4 + 8x) = -4$.

80. Prove rigorously that $\lim_{x \to 3} (x^2 - x) = 6$.

The velocity at any given moment in free fall is given by evaluating the derivative of the position function at that time. (*Germanskydiver/Shutterstock*)

3 DIFFERENTIATION

Differential calculus is the study of the derivative, and differentiation is the process of computing derivatives. What is a derivative? There are three equally important answers: A derivative is a rate of change, it is the slope of a tangent line, and (more formally), it is the limit of a difference quotient, as we will explain shortly. In this chapter, we explore all three facets of the derivative and develop the basic rules of differentiation. When you master these techniques, you will possess one of the most useful and flexible tools that mathematics has to offer.

3.1 Definition of the Derivative

We begin with two questions: What is the precise definition of a tangent line? And how can we compute its slope? To answer these questions, let's return to the relationship between tangent and secant lines first mentioned in Section 2.1.

The secant line through distinct points $P = (a, f(a))$ and $Q = (x, f(x))$ on the graph of a function f has slope [Figure 1(A)]

$$\frac{\Delta f}{\Delta x} = \frac{f(x) - f(a)}{x - a}$$

where

$$\Delta f = f(x) - f(a) \qquad \text{and} \qquad \Delta x = x - a$$

The expression $\dfrac{f(x) - f(a)}{x - a}$ is called the **difference quotient**.

····· **REMINDER** A **secant line** is any line through two points on a curve or graph.

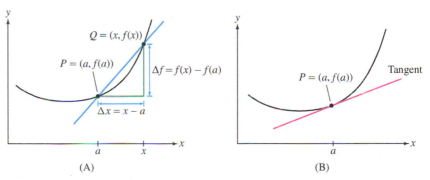

FIGURE 1 The secant line has slope $\Delta f / \Delta x$. Our goal is to compute the slope of the tangent line at $(a, f(a))$.

Now observe what happens as Q approaches P or, equivalently, as x approaches a. Figure 2 suggests that the secant lines get progressively closer to the tangent line. If we imagine Q moving toward P, then the secant line appears to rotate into the tangent line as in (D). Therefore, we may expect the slopes of the secant lines to approach the slope of the tangent line.

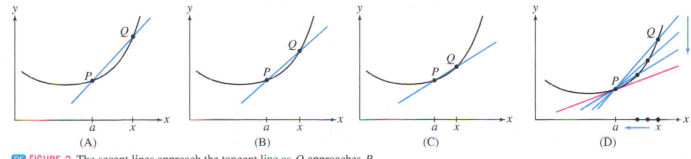

FIGURE 2 The secant lines approach the tangent line as Q approaches P.

113

Based on this intuition, we define the **derivative** $f'(a)$ (which is read "f prime of a") at a point $x = a$ as the limit

$$\underbrace{f'(a)}_{\substack{\text{Slope of the} \\ \text{tangent line}}} = \underbrace{\lim_{x \to a} \frac{f(x) - f(a)}{x - a}}_{\text{Limit of slopes of secant lines}}$$

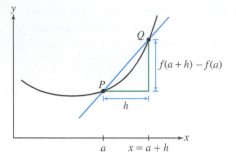

FIGURE 3 The difference quotient can be written in terms of h.

There is another way of writing the difference quotient using a new variable h:

$$h = x - a$$

We have $x = a + h$ and, for $x \neq a$ (Figure 3),

$$\frac{f(x) - f(a)}{x - a} = \frac{f(a + h) - f(a)}{h}$$

The variable h approaches 0 as $x \to a$, so we can rewrite the derivative as

$$f'(a) = \lim_{h \to 0} \frac{f(a + h) - f(a)}{h}$$

Each way of writing the derivative is useful. The version using h is often more convenient in computations.

DEFINITION The Derivative   The derivative of f at a point a is the limit of the difference quotients (if it exists):

$$f'(a) = \lim_{h \to 0} \frac{f(a + h) - f(a)}{h} \qquad \boxed{1}$$

When the limit exists, we say that f is **differentiable** at a. An equivalent definition of the derivative at a point a is

$$f'(a) = \lim_{x \to a} \frac{f(x) - f(a)}{x - a} \qquad \boxed{2}$$

We can now define the tangent line in a precise way, as the line of slope $f'(a)$ through $P = (a, f(a))$.

◄·· **REMINDER** *The equation of the line through* $P = (a, b)$ *of slope m in point-slope form:*

$$y - b = m(x - a)$$

DEFINITION Tangent Line   Assume that f is differentiable at a. The tangent line to the graph of $y = f(x)$ at $P = (a, f(a))$ is the line through P of slope $f'(a)$. The equation of the tangent line in point-slope form is

$$y - f(a) = f'(a)(x - a) \qquad \boxed{3}$$

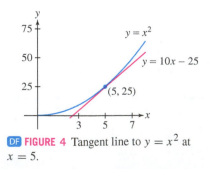

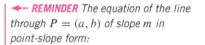

DF FIGURE 4 Tangent line to $y = x^2$ at $x = 5$.

■ **EXAMPLE 1 Equation of a Tangent Line**   Find an equation of the tangent line to the graph of $f(x) = x^2$ at $x = 5$.

Solution   First, we must compute $f'(5)$. We are free to use either Eq. (1) or Eq. (2). Using Eq. (2), we have

$$f'(5) = \lim_{x \to 5} \frac{f(x) - f(5)}{x - 5} = \lim_{x \to 5} \frac{x^2 - 25}{x - 5} = \lim_{x \to 5} \frac{(x - 5)(x + 5)}{x - 5}$$

$$= \lim_{x \to 5} (x + 5) = 10$$

Next, we apply Eq. (3) with $a = 5$. Because $f(5) = 25$, an equation of the tangent line is $y - 25 = 10(x - 5)$, or in slope-intercept form, $y = 10x - 25$ (Figure 4).

If instead, we had used Eq. (1) to find $f'(5)$, we obtain the same slope:

$$f'(5) = \lim_{h \to 0} \frac{f(5+h) - f(5)}{h} = \lim_{h \to 0} \frac{(5+h)^2 - 5^2}{h}$$

$$= \lim_{h \to 0} \frac{(25 + 10h + h^2) - 25}{h} = \lim_{h \to 0} (10 + h) = 10 \qquad \blacksquare$$

In the next two examples, we perform the differentiation (the process of computing the derivative) using Eq. (1). For clarity, we break up the computations into three steps.

Isaac Newton referred to calculus as the "method of fluxions" (from the Latin word for "flow"), but the term "differential calculus," introduced in its Latin form "calculus differentialis" by Gottfried Wilhelm Leibniz, eventually won out and was adopted universally.

■ **EXAMPLE 2**   Compute $f'(3)$, where $f(x) = x^2 - 8x$.

Solution   Using Eq. (1), we write the difference quotient at $a = 3$ as

$$\frac{f(a+h) - f(a)}{h} = \frac{f(3+h) - f(3)}{h} \qquad (h \neq 0)$$

Step 1. **Write out the numerator of the difference quotient.**

$$f(3+h) - f(3) = \big((3+h)^2 - 8(3+h)\big) - \big(3^2 - 8(3)\big)$$

$$= \big((9 + 6h + h^2) - (24 + 8h)\big) - (9 - 24)$$

$$= h^2 - 2h$$

Step 2. **Divide by h and simplify.**

$$\frac{f(3+h) - f(3)}{h} = \frac{h^2 - 2h}{h} = \underbrace{\frac{h(h-2)}{h}}_{\text{Cancel } h} = h - 2$$

Step 3. **Compute the limit.**

$$f'(3) = \lim_{h \to 0} \frac{f(3+h) - f(3)}{h} = \lim_{h \to 0} (h - 2) = -2 \qquad \blacksquare$$

■ **EXAMPLE 3**   Sketch the graph of $f(x) = \dfrac{1}{x}$ and the tangent line at $x = 2$.

(a) Based on the sketch, do you expect $f'(2)$ to be positive or negative?

(b) Find an equation of the tangent line at $x = 2$.

Solution   The graph and tangent line at $x = 2$ are shown in Figure 5.

(a) We see that the tangent line has negative slope, so $f'(2)$ must be negative.

(b) We compute $f'(2)$ in three steps as before.

Step 1. **Write out the numerator of the difference quotient.**

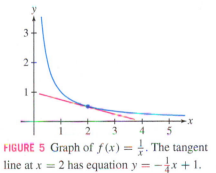

FIGURE 5 Graph of $f(x) = \frac{1}{x}$. The tangent line at $x = 2$ has equation $y = -\frac{1}{4}x + 1$.

$$f(2+h) - f(2) = \frac{1}{2+h} - \frac{1}{2} = \frac{2}{2(2+h)} - \frac{2+h}{2(2+h)} = -\frac{h}{2(2+h)}$$

Step 2. **Divide by h and simplify.**

$$\frac{f(2+h) - f(2)}{h} = \frac{1}{h} \cdot \left(-\frac{h}{2(2+h)}\right) = -\frac{1}{2(2+h)}$$

Step 3. **Compute the limit.**

$$f'(2) = \lim_{h \to 0} \frac{f(2+h) - f(2)}{h} = \lim_{h \to 0} \frac{-1}{2(2+h)} = -\frac{1}{4}$$

The function value is $f(2) = \frac{1}{2}$, so the tangent line passes through $\left(2, \frac{1}{2}\right)$ and has equation

$$y - \frac{1}{2} = -\frac{1}{4}(x - 2)$$

In slope-intercept form, $y = -\frac{1}{4}x + 1$. ∎

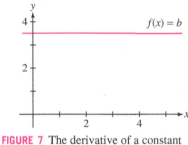

FIGURE 6 The derivative of $f(x) = mx + b$ is $f'(a) = m$ for all a.

The graph of a **linear function** $f(x) = mx + b$ (where m and b are constants) is a line of slope m. The tangent line at any point coincides with the line itself (Figure 6), so we should expect that $f'(a) = m$ for all a. Let's check this by computing the derivative:

$$f'(a) = \lim_{h \to 0} \frac{f(a+h) - f(a)}{h} = \lim_{h \to 0} \frac{(m(a+h) + b) - (ma + b)}{h}$$

$$= \lim_{h \to 0} \frac{mh}{h} = \lim_{h \to 0} m = m$$

If $m = 0$, then $f(x) = b$ is constant and $f'(a) = 0$ (Figure 7). In summary,

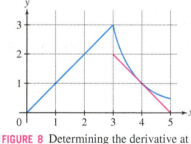

FIGURE 7 The derivative of a constant function $f(x) = b$ is $f'(a) = 0$ for all a.

THEOREM 1 Derivative of Linear and Constant Functions

- If $f(x) = mx + b$ is a linear function, then $f'(a) = m$ for all a.
- If $f(x) = b$ is a constant function, then $f'(a) = 0$ for all a.

■ **EXAMPLE 4** Find the derivative of $f(x) = 9x - 5$ at $x = 2$ and $x = 5$.

Solution We have $f'(a) = 9$ for all a. Hence, $f'(2) = f'(5) = 9$. ∎

■ **EXAMPLE 5** Determine the derivative of the function f whose graph appears in Figure 8 at $x = 2, 3, 4$.

Solution At $x = 2$, the graph is a line with slope given by

$$m = \frac{f(3) - f(0)}{3 - 0} = \frac{3 - 0}{3 - 0} = 1$$

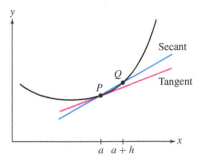

FIGURE 8 Determining the derivative at $x = 2, 3, 4$.

Hence, $f'(2) = 1$.

At $x = 3$, there is no well-defined tangent line, since the limit of the slopes of the secant lines as x approaches 3 from the left is different from the limit of the slopes of the secant lines as x approaches 3 from the right. So the derivative $f'(3)$ does not exist.

At $x = 4$, the tangent line passes through the two points $(3, 2)$ and $(5, 0)$. Hence, it has slope $m = \frac{2-0}{3-5} = -1$. Therefore, $f'(4) = -1$. ∎

Estimating the Derivative

Approximations to the derivative are useful in situations where we cannot evaluate $f'(a)$ exactly. Since the derivative is the limit of difference quotients, the difference quotient should give a good numerical approximation when h is sufficiently small:

$$\boxed{f'(a) \approx \frac{f(a+h) - f(a)}{h} \qquad \text{if } h \text{ is small}}$$

FIGURE 9 When h is small, the secant line has nearly the same slope as the tangent line.

Graphically, this says that for small h, the slope of the secant line is nearly equal to the slope of the tangent line (Figure 9).

■ **EXAMPLE 6** Estimate the derivative of $f(x) = \sin x$ at $x = \frac{\pi}{6}$.

Solution We calculate the difference quotient for several small values of h:

$$\frac{\sin\left(\frac{\pi}{6} + h\right) - \sin\frac{\pi}{6}}{h} = \frac{\sin\left(\frac{\pi}{6} + h\right) - 0.5}{h}$$

Table 1 suggests that the limit has a decimal expansion beginning 0.866. In other words, $f'\left(\frac{\pi}{6}\right) \approx 0.866$. ■

TABLE 1   Values of the Difference Quotient for Small h

$h > 0$	$\dfrac{\sin\left(\frac{\pi}{6} + h\right) - 0.5}{h}$	$h < 0$	$\dfrac{\sin\left(\frac{\pi}{6} + h\right) - 0.5}{h}$
0.01	0.863511	−0.01	0.868511
0.001	0.865775	−0.001	0.866275
0.0001	0.866000	−0.0001	0.866050
0.00001	0.8660229	−0.00001	0.8660279

In the next example, we use graphical reasoning to determine the accuracy of the estimates obtained in Example 6.

■ **EXAMPLE 7** GU **Determining Accuracy Graphically** Let $f(x) = \sin x$. Show that the approximation $f'\left(\frac{\pi}{6}\right) \approx 0.8660$ is accurate to four decimal places.

Solution Observe in Figure 10 that the position of the secant line relative to the tangent line depends on whether h is positive or negative. When $h > 0$, the slope of the secant line is *smaller* than the slope of the tangent line, but it is *larger* when $h < 0$. This tells us that the difference quotients in the second column of Table 1 are smaller than $f'\left(\frac{\pi}{6}\right)$ and those in the fourth column are greater than $f'\left(\frac{\pi}{6}\right)$. From the last line in Table 1 we may conclude that

$$0.866022 \le f'\left(\frac{\pi}{6}\right) \le 0.866028$$

It follows that the estimate $f'\left(\frac{\pi}{6}\right) \approx 0.8660$ is accurate to four decimal places. In Section 3.6, we will see that the exact value is $f'\left(\frac{\pi}{6}\right) = \cos\left(\frac{\pi}{6}\right) = \frac{\sqrt{3}}{2} \approx 0.8660254$, just about midway between 0.866022 and 0.866028. ■

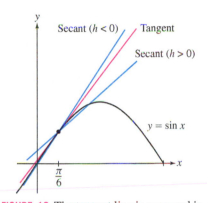

FIGURE 10 The tangent line is squeezed in *between* the secant lines with $h > 0$ and $h < 0$.

This technique of estimating an unknown quantity by showing that it lies between two known values ("squeezing it") is used frequently in calculus.

CONCEPTUAL INSIGHT Are Limits Really Necessary?   It is natural to ask whether limits are really necessary. The tangent line is easy to visualize. Is there perhaps a better or simpler way to find its equation? History gives one answer: The methods of calculus based on limits have stood the test of time and are used more widely today than ever before.

History aside, we can see directly why limits play such a crucial role. The slope of a line can be computed if the coordinates of *two* points $P = (x_1, y_1)$ and $Q = (x_2, y_2)$ on the line are known:

$$\text{Slope of line} = \frac{y_2 - y_1}{x_2 - x_1}$$

This formula cannot be applied to the tangent line because we know only that it passes through the single point $P = (a, f(a))$. Limits provide an ingenious way around this obstacle. We choose a point $Q = (a + h, f(a + h))$ on the graph near P and form the secant line. The slope of this secant line is just an approximation to the slope of the tangent line:

$$\text{Slope of secant line} = \frac{f(a + h) - f(a)}{h} \approx \text{slope of tangent line}$$

But this approximation improves as $h \to 0$, and by taking the limit, we convert our approximations into the exact slope.

3.1 SUMMARY

- The *difference quotient*:

$$\frac{f(a+h) - f(a)}{h}$$

The difference quotient is the slope of the secant line through the points $P = (a, f(a))$ and $Q = (a + h, f(a + h))$ on the graph of f.
- The *derivative* $f'(a)$ is defined by the following equivalent limits:

$$f'(a) = \lim_{h \to 0} \frac{f(a+h) - f(a)}{h} = \lim_{x \to a} \frac{f(x) - f(a)}{x - a}$$

If the limit exists, we say that f is *differentiable* at $x = a$.
- By definition, the tangent line at $P = (a, f(a))$ is the line through P with slope $f'(a)$ [assuming that $f'(a)$ exists].
- Equation of the tangent line in point-slope form:

$$y - f(a) = f'(a)(x - a)$$

- To calculate $f'(a)$ using the limit definition:

    Step 1. Write out the numerator of the difference quotient.

    Step 2. Divide by h and simplify.

    Step 3. Compute the derivative by taking the limit.

- For small values of h, we have the estimate $f'(a) \approx \dfrac{f(a+h) - f(a)}{h}$.

3.1 EXERCISES

Preliminary Questions

1. Which of the lines in Figure 11 are tangent to the curve?

FIGURE 11

2. What are the two ways of writing the difference quotient?

3. Find a and h such that $\dfrac{f(a+h) - f(a)}{h}$ is equal to the slope of the secant line between $(3, f(3))$ and $(5, f(5))$.

4. Which derivative is approximated by $\dfrac{\tan(\frac{\pi}{4} + 0.0001) - 1}{0.0001}$?

5. What do the following quantities represent in terms of the graph of $f(x) = \sin x$?

(a) $\sin 1.3 - \sin 0.9$ (b) $\dfrac{\sin 1.3 - \sin 0.9}{0.4}$ (c) $f'(0.9)$

Exercises

1. Let $f(x) = 5x^2$. Show that $f(3 + h) = 5h^2 + 30h + 45$. Then show that

$$\frac{f(3+h) - f(3)}{h} = 5h + 30$$

and compute $f'(3)$ by taking the limit as $h \to 0$.

2. Let $f(x) = 2x^2 - 3x - 5$. Show that the secant line through $(2, f(2))$ and $(2 + h, f(2 + h))$ has slope $2h + 5$. Then use this formula to compute the slope of:

(a) The secant line through $(2, f(2))$ and $(3, f(3))$

(b) The tangent line at $x = 2$ (by taking a limit)

In Exercises 3–8, compute $f'(a)$ in two ways, using Eq. (1) and Eq. (2).

3. $f(x) = x^2 + 9x$, $a = 0$

4. $f(x) = x^2 + 9x$, $a = 2$

5. $f(x) = 3x^2 + 4x + 2$, $a = -1$

6. $f(x) = x^3$, $a = 2$

7. $f(x) = x^3 + 2x$, $a = 1$

8. $f(x) = \frac{1}{x}$, $a = 2$

In Exercises 9–12, refer to Figure 12.

9. Find the slope of the secant line through $(2, f(2))$ and $(2.5, f(2.5))$. Is it larger or smaller than $f'(2)$? Explain.

10. Estimate $\dfrac{f(2+h) - f(2)}{h}$ for $h = -0.5$. What does this quantity represent? Is it larger or smaller than $f'(2)$? Explain.

11. Estimate $f'(1)$ and $f'(2)$.

12. Find a value of h for which $\dfrac{f(2+h) - f(2)}{h} = 0$.

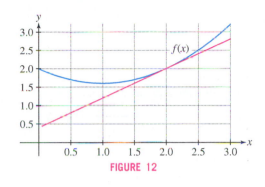

FIGURE 12

In Exercises 13–16, refer to Figure 13.

13. Determine $f'(a)$ for $a = 1, 2, 4, 7$.

14. For which values of x is $f'(x) < 0$?

15. Which is larger, $f'(5.5)$ or $f'(6.5)$?

16. Show that $f'(3)$ does not exist.

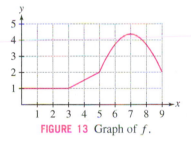

FIGURE 13 Graph of f.

In Exercises 17–20, use the limit definition to calculate the derivative of the linear function.

17. $f(x) = 7x - 9$

18. $f(x) = 12$

19. $g(t) = 8 - 3t$

20. $k(z) = 14z + 12$

21. Find an equation of the tangent line at $x = 3$, assuming that $f(3) = 5$ and $f'(3) = 2$.

22. Find $f(3)$ and $f'(3)$, assuming that the tangent line to $y = f(x)$ at $a = 3$ has equation $y = 5x + 2$.

23. Describe the tangent line at an arbitrary point on the "curve" $y = 2x + 8$.

24. Suppose that $f(2+h) - f(2) = 3h^2 + 5h$. Calculate:

(a) The slope of the secant line through $(2, f(2))$ and $(6, f(6))$

(b) $f'(2)$

25. Let $f(x) = \dfrac{1}{x}$. Does $f(-2+h)$ equal $\dfrac{1}{-2+h}$ or $\dfrac{1}{-2} + \dfrac{1}{h}$?
Compute the difference quotient at $a = -2$ with $h = 0.5$.

26. Let $f(x) = \sqrt{x}$. Does $f(5+h)$ equal $\sqrt{5+h}$ or $\sqrt{5} + \sqrt{h}$?
Compute the difference quotient at $a = 5$ with $h = 1$.

27. Let $f(x) = 1/\sqrt{x}$. Compute $f'(5)$ by showing that

$$\frac{f(5+h) - f(5)}{h} = -\frac{1}{\sqrt{5}\sqrt{5+h}(\sqrt{5+h} + \sqrt{5})}$$

28. Find an equation of the tangent line to the graph of $f(x) = 1/\sqrt{x}$ at $x = 9$.

In Exercises 29–46, use the limit definition to compute $f'(a)$ and find an equation of the tangent line.

29. $f(x) = 2x^2 + 10x$, $a = 3$

30. $f(x) = 4 - x^2$, $a = -1$

31. $f(t) = t - 2t^2$, $a = 3$

32. $f(x) = 8x^3$, $a = 1$

33. $f(x) = x^3 + x$, $a = 0$

34. $f(t) = 2t^3 + 4t$, $a = 4$

35. $f(x) = x^{-1}$, $a = 8$

36. $f(x) = x + x^{-1}$, $a = 4$

37. $f(x) = \dfrac{1}{x+3}$, $a = -2$

38. $f(t) = \dfrac{2}{1-t}$, $a = -1$

39. $f(x) = \sqrt{x+4}$, $a = 1$

40. $f(t) = \sqrt{3t+5}$, $a = -1$

41. $f(x) = \dfrac{1}{\sqrt{x}}$, $a = 4$

42. $f(x) = \dfrac{1}{\sqrt{2x+1}}$, $a = 4$

43. $f(t) = \sqrt{t^2+1}$, $a = 3$

44. $f(x) = x^{-2}$, $a = -1$

45. $f(x) = \dfrac{1}{x^2+1}$, $a = 0$

46. $f(t) = t^{-3}$, $a = 1$

47. Figure 14 displays data collected by the biologist Julian Huxley (1887–1975) on the average antler weight W of male red deer as a function of age t. Estimate the derivative at $t = 4$. For which values of t is the slope of the tangent line equal to zero? For which values is it negative?

FIGURE 14

48. Figure 15(A) shows the graph of $f(x) = \sqrt{x}$. The close-up in Figure 15(B) shows that the graph is nearly a straight line near $x = 16$. Estimate the slope of this line and take it as an estimate for $f'(16)$. Then compute $f'(16)$ and compare with your estimate.

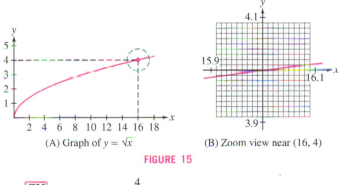

(A) Graph of $y = \sqrt{x}$ **(B)** Zoom view near $(16, 4)$

FIGURE 15

49. [GU] Let $f(x) = \dfrac{4}{1 + 2^x}$.

(a) Plot f over $[-2, 2]$. Then zoom in near $x = 0$ until the graph appears straight, and estimate the slope $f'(0)$.

(b) Use (a) to find an approximate equation to the tangent line at $x = 0$. Plot this line and $y = f(x)$ on the same set of axes.

50. [GU] Let $f(x) = \cot x$. Estimate $f'\left(\frac{\pi}{2}\right)$ graphically by zooming in on a plot of f near $x = \frac{\pi}{2}$.

51. Determine the intervals along the x-axis on which the derivative in Figure 16 is positive.

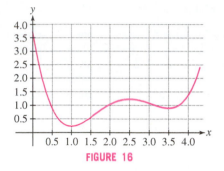

FIGURE 16

52. Sketch the graph of $f(x) = \sin x$ on $[0, \pi]$ and guess the value of $f'\left(\frac{\pi}{2}\right)$. Then calculate the difference quotient at $x = \frac{\pi}{2}$ for two small positive and negative values of h. Are these calculations consistent with your guess?

In Exercises 53–58, each limit represents a derivative $f'(a)$. Find $f(x)$ and a.

53. $\lim_{h \to 0} \dfrac{(5+h)^3 - 125}{h}$

54. $\lim_{x \to 5} \dfrac{x^3 - 125}{x - 5}$

55. $\lim_{h \to 0} \dfrac{\sin\left(\frac{\pi}{6} + h\right) - 0.5}{h}$

56. $\lim_{x \to \frac{1}{4}} \dfrac{x^{-1} - 4}{x - \frac{1}{4}}$

57. $\lim_{h \to 0} \dfrac{5^{2+h} - 25}{h}$

58. $\lim_{h \to 0} \dfrac{5^h - 1}{h}$

59. Apply the method of Example 7 to $f(x) = \sin x$ to determine $f'\left(\frac{\pi}{4}\right)$ accurately to four decimal places.

60. Apply the method of Example 7 to $f(x) = \cos x$ to determine $f'\left(\frac{\pi}{5}\right)$ accurately to four decimal places. Use a graph of f to explain how the method works in this case.

61. For each graph in Figure 17, determine whether $f'(1)$ is larger or smaller than the slope of the secant line between $x = 1$ and $x = 1 + h$ for $h > 0$. Explain.

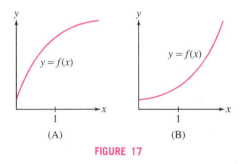

FIGURE 17

62. Refer to the graph of $f(x) = 2^x$ in Figure 18.
(a) Explain graphically why, for $h > 0$,

$$\frac{f(-h) - f(0)}{-h} \le f'(0) \le \frac{f(h) - f(0)}{h}$$

(b) Use (a) to show that $0.69314 \le f'(0) \le 0.69315$.

(c) Similarly, compute $f'(x)$ to four decimal places for $x = 1, 2, 3, 4$.
(d) Now compute the ratios $f'(x)/f'(0)$ for $x = 1, 2, 3, 4$. Can you guess an approximate formula for $f'(x)$?

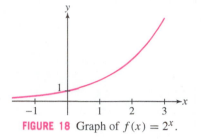

FIGURE 18 Graph of $f(x) = 2^x$.

63. [GU] Sketch the graph of $f(x) = x^{5/2}$ on $[0, 6]$.
(a) Use the sketch to justify the inequalities for $h > 0$:

$$\frac{f(4) - f(4 - h)}{h} \le f'(4) \le \frac{f(4 + h) - f(4)}{h}$$

(b) Use (a) to compute $f'(4)$ to four decimal places.
(c) Use a graphing utility to plot $y = f(x)$ and the tangent line at $x = 4$, utilizing your estimate for $f'(4)$.

64. [GU] Verify that $P = \left(1, \frac{1}{2}\right)$ lies on the graphs of both $f(x) = 1/(1 + x^2)$ and $L(x) = \frac{1}{2} + m(x - 1)$ for every slope m. Plot $y = f(x)$ and $y = L(x)$ on the same axes for several values of m until you find a value of m for which $y = L(x)$ appears tangent to the graph of f. What is your estimate for $f'(1)$?

65. [GU] Use a plot of $f(x) = x^x$ to estimate the value c such that $f'(c) = 0$. Find c to sufficient accuracy so that

$$\left| \frac{f(c + h) - f(c)}{h} \right| \le 0.006 \quad \text{for} \quad h = \pm 0.001$$

66. [GU] Plot $f(x) = x^x$ and $y = 2x + a$ on the same set of axes for several values of a until the line becomes tangent to the graph. Then estimate the value c such that $f'(c) = 2$.

*In Exercises 67–73, estimate derivatives using the **symmetric difference quotient** (SDQ), defined as the average of the difference quotients at h and $-h$:*

$$\frac{1}{2}\left(\frac{f(a + h) - f(a)}{h} + \frac{f(a - h) - f(a)}{-h} \right)$$

$$= \frac{f(a + h) - f(a - h)}{2h} \qquad \boxed{4}$$

The SDQ usually gives a better approximation to the derivative than the difference quotient.

67. The vapor pressure of water at temperature T (in kelvins) is the atmospheric pressure P at which no net evaporation takes place. Use the following table to estimate $P'(T)$ for $T = 303, 313, 323, 333, 343$ by computing the SDQ given by Eq. (4) with $h = 10$.

T (K)	293	303	313	323	333	343	353
P (atm)	0.0278	0.0482	0.0808	0.1311	0.2067	0.3173	0.4754

68. Use the SDQ with $h = 1$ year to estimate $P'(T)$ in the years 2005, 2007, 2009, 2011, where $P(T)$ is the U.S. ethanol production (Figure 19). Express your answer in the correct units.

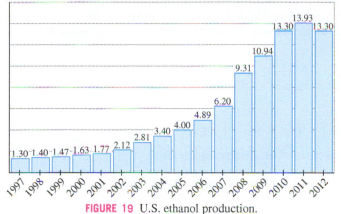

P (billions of gallons)

13.93
13.30   13.30
10.94
9.31
6.20
4.89
4.00
3.40
2.81
1.30 1.40 1.47 1.63 1.77 2.12

1997 1998 1999 2000 2001 2002 2003 2004 2005 2006 2007 2008 2009 2010 2011 2012

FIGURE 19 U.S. ethanol production.

In Exercises 69–70, traffic speed S along a certain road (in kilometers per hour) varies as a function of traffic density q (number of cars per kilometer of road). Use the following data to answer the questions:

q (density)	60	70	80	90	100
S (speed)	72.5	67.5	63.5	60	56

69. Estimate $S'(80)$.

70. Explain why $V = qS$, called *traffic volume*, is equal to the number of cars passing a point per hour. Use the data to estimate $V'(80)$.

Exercises 71–73: The current (in amperes) at time t (in seconds) flowing in the circuit in Figure 20 is given by Kirchhoff's Law:

$$i(t) = Cv'(t) + R^{-1}v(t)$$

where $v(t)$ is the voltage (in volts, V), C the capacitance (in farads, F), and R the resistance (in ohms, Ω).

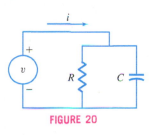

i

+
v
−
R   C

FIGURE 20

71. Calculate the current at $t = 3$ if

$$v(t) = 0.5t + 4 \text{ V}$$

where $C = 0.01$ F and $R = 100\ \Omega$.

72. Use the following data to estimate $v'(10)$ (by an SDQ). Then estimate $i(10)$, assuming $C = 0.03$ and $R = 1000$.

t	9.8	9.9	10	10.1	10.2
v(t)	256.52	257.32	258.11	258.9	259.69

73. Assume that $R = 200\ \Omega$ but C is unknown. Use the following data to estimate $v'(4)$ (by an SDQ) and deduce an approximate value for the capacitance C.

t	3.8	3.9	4	4.1	4.2
v(t)	388.8	404.2	420	436.2	452.8
i(t)	32.34	33.22	34.1	34.98	35.86

Further Insights and Challenges

74. The SDQ usually approximates the derivative much more closely than does the ordinary difference quotient. Let $f(x) = 2^x$ and $a = 0$. Compute the SDQ with $h = 0.001$ and the ordinary difference quotients with $h = \pm 0.001$. Compare with the actual value, which is $f'(0) = \ln 2$.

75. Explain how the symmetric difference quotient defined by Eq. (4) can be interpreted as the slope of a secant line.

76. Which of the two functions in Figure 21 satisfies the inequality

$$\frac{f(a+h) - f(a-h)}{2h} \le \frac{f(a+h) - f(a)}{h}$$

for $h > 0$? Explain in terms of secant lines.

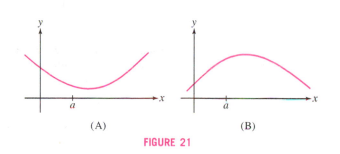

(A)   (B)

FIGURE 21

77. Show that if f is a quadratic polynomial, then the SDQ at $x = a$ (for any $h \ne 0$) is *equal* to $f'(a)$. Explain the graphical meaning of this result.

78. Let $f(x) = x^{-2}$. Compute $f'(1)$ by taking the limit of the SDQs (with $a = 1$) as $h \to 0$.

3.2 The Derivative as a Function

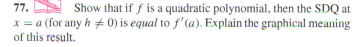

In the previous section, we computed the derivative $f'(a)$ for specific values of a. It is also useful to view the derivative as a function f' whose value at $x = a$ is $f'(a)$. The function f' is still defined as a limit, but the fixed number a is replaced by the variable x:

$$f'(x) = \lim_{h \to 0} \frac{f(x+h) - f(x)}{h} \qquad \boxed{1}$$

If $y = f(x)$, we also write y' or $y'(x)$ for $f'(x)$.

Often, the domain of f' is clear from the context. If so, we usually do not mention the domain explicitly.

The domain of f' consists of all values of x in the domain of f for which the limit in Eq. (1) exists. We say that f is **differentiable** on (a, b) if $f'(x)$ exists for all x in (a, b). When $f'(x)$ exists for all x in the interval or intervals on which $f(x)$ is defined, we say simply that f is differentiable.

■ **EXAMPLE 1** Prove that $f(x) = x^3 - 12x$ is differentiable. Compute $f'(x)$ and find an equation of the tangent line at $x = -3$.

Solution We compute $f'(x)$ in three steps as in the previous section.

Step 1. **Write out the numerator of the difference quotient.**

$$
\begin{aligned}
f(x + h) - f(x) &= \big((x + h)^3 - 12(x + h)\big) - \big(x^3 - 12x\big) \\
&= (x^3 + 3x^2 h + 3xh^2 + h^3 - 12x - 12h) - (x^3 - 12x) \\
&= 3x^2 h + 3xh^2 + h^3 - 12h \\
&= h(3x^2 + 3xh + h^2 - 12) \qquad \text{(factor out } h\text{)}
\end{aligned}
$$

Step 2. **Divide by h and simplify.**

$$
\frac{f(x + h) - f(x)}{h} = \frac{h(3x^2 + 3xh + h^2 - 12)}{h} = 3x^2 + 3xh + h^2 - 12 \quad (h \neq 0)
$$

Step 3. **Compute the limit.**

$$
f'(x) = \lim_{h \to 0} \frac{f(x + h) - f(x)}{h} = \lim_{h \to 0} (3x^2 + 3xh + h^2 - 12) = 3x^2 - 12
$$

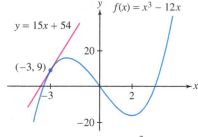

In this limit, x is treated as a constant because it does not change as $h \to 0$. We see that the limit exists for all x, so f is differentiable and $f'(x) = 3x^2 - 12$.

Now evaluate:

$$
f'(-3) = 3(-3)^2 - 12 = 15
$$

This is the slope of the tangent line. Since $f(-3) = 9$, the line passes through $(-3, 9)$. So an equation of the tangent line at $x = -3$ is $y - 9 = 15(x + 3)$ or $y = 15x + 54$ (Figure 1). ■

FIGURE 1 Graph of $f(x) = x^3 - 12x$.

■ **EXAMPLE 2** Prove that $y = x^{-2}$ is differentiable and calculate y'.

Solution The domain of $f(x) = x^{-2}$ is $\{x : x \neq 0\}$, so assume that $x \neq 0$. We compute $f'(x)$ directly, without the separate steps of the previous example:

$$
\begin{aligned}
y' &= \lim_{h \to 0} \frac{f(x + h) - f(x)}{h} = \lim_{h \to 0} \frac{\dfrac{1}{(x + h)^2} - \dfrac{1}{x^2}}{h} \\[2em]
&= \lim_{h \to 0} \frac{\dfrac{x^2 - (x + h)^2}{x^2(x + h)^2}}{h} = \lim_{h \to 0} \frac{1}{h}\left(\frac{x^2 - (x + h)^2}{x^2(x + h)^2}\right) \\[2em]
&= \lim_{h \to 0} \frac{1}{h}\left(\frac{-h(2x + h)}{x^2(x + h)^2}\right) = \lim_{h \to 0} -\frac{2x + h}{x^2(x + h)^2} \qquad \text{(cancel } h\text{)} \\[2em]
&= -\frac{2x + 0}{x^2(x + 0)^2} = -\frac{2x}{x^4} = -2x^{-3}
\end{aligned}
$$

The limit exists for all $x \neq 0$, so y is differentiable and $y' = -2x^{-3}$. ■

FIGURE 2 Gottfried Wilhelm von Leibniz (1646–1716), German philosopher and scientist. Newton and Leibniz (pronounced "Libe-nitz") are often regarded as the inventors of calculus (working independently). It is more accurate to credit them with developing calculus into a general and fundamental discipline, because many particular results of calculus had been discovered previously by other mathematicians. (*The Granger Collection, NYC. All rights reserved.*)

Leibniz Notation

The "prime" notation y' and $f'(x)$ was introduced by the French mathematician Joseph Louis Lagrange (1736–1813). There is another standard notation for the derivative that we owe to Leibniz (Figure 2):

$$\frac{df}{dx} \quad \text{or} \quad \frac{dy}{dx}$$

In Example 2, we showed that the derivative of $y = x^{-2}$ is $y' = -2x^{-3}$. In Leibniz notation, we would write

$$\frac{dy}{dx} = -2x^{-3} \quad \text{or} \quad \frac{d}{dx}x^{-2} = -2x^{-3}$$

To specify the value of the derivative for a fixed value of x, say, $x = 4$, we write

$$\frac{df}{dx}\bigg|_{x=4} \quad \text{or} \quad \frac{dy}{dx}\bigg|_{x=4}$$

You should not think of dy/dx as the fraction "dy divided by dx." The expressions dy and dx are called **differentials**. They play a role in some situations (in linear approximation and in more advanced calculus). At this stage, we treat them merely as symbols with no independent meaning.

CONCEPTUAL INSIGHT Leibniz notation is widely used for several reasons. First, it reminds us that the derivative df/dx, although not itself a ratio, is in fact a *limit* of ratios $\Delta f/\Delta x$. Second, the notation specifies the independent variable. This is useful when variables other than x are used. For example, if the independent variable is t, we write df/dt. Third, we often think of d/dx as an "operator" that performs differentiation on functions. In other words, we apply the operator d/dx to f to obtain the derivative df/dx. We will see other advantages of Leibniz notation when we discuss the Chain Rule in Section 3.7.

A main goal of this chapter is to develop the basic rules of differentiation. These rules enable us to find derivatives without computing limits. For example, in Theorem 1 of the previous section, we stated the following fact in slightly different notation. We will prove it again directly.

THEOREM 1 **The Constant Rule** For any constant c,

$$\frac{d}{dx}(c) = 0$$

Proof If $f(x) = c$, then

$$f'(x) = \lim_{h \to 0} \frac{f(x+h) - f(x)}{h} = \lim_{h \to 0} \frac{c - c}{h} = \lim_{h \to 0} 0 = 0 \qquad \boxed{2}$$

■

Here is another useful rule that is straightforward to prove:

THEOREM 2 **The x Rule**

$$\frac{d}{dx}(x) = 1$$

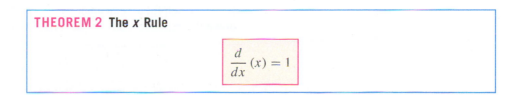

Proof If $f(x) = x$, then

$$f'(x) = \lim_{h \to 0} \frac{f(x+h) - f(x)}{h} = \lim_{h \to 0} \frac{(x+h) - x}{h} = \lim_{h \to 0} \frac{h}{h} = 1 \qquad \boxed{3}$$

■

This next theorem will prove to be incredibly useful for differentiating polynomials. It includes the last rule as a special case.

The Power Rule is valid for all exponents. We prove it here for a positive integer n (see Exercise 95 for a negative integer n and the marginal note on p. 176 for arbitrary n).

THEOREM 3 The Power Rule   For all exponents n,

$$\boxed{\frac{d}{dx} x^n = nx^{n-1}}$$

Proof We prove this only for the case that n is a positive integer. Let $f(x) = x^n$. Then

$$f'(a) = \lim_{x \to a} \frac{x^n - a^n}{x - a}$$

To simplify the difference quotient, we need to generalize the following identities:

$$x^2 - a^2 = (x - a)(x + a)$$

$$x^3 - a^3 = (x - a)(x^2 + xa + a^2)$$

$$x^4 - a^4 = (x - a)(x^3 + x^2 a + xa^2 + a^3)$$

The generalization is

$$x^n - a^n = (x - a)(x^{n-1} + x^{n-2}a + x^{n-3}a^2 + \cdots + xa^{n-2} + a^{n-1}) \qquad \boxed{4}$$

To verify Eq. (4), observe that the right-hand side is equal to

$$x(x^{n-1} + x^{n-2}a + x^{n-3}a^2 + \cdots + xa^{n-2} + a^{n-1})$$
$$- a(x^{n-1} + x^{n-2}a + x^{n-3}a^2 + \cdots + xa^{n-2} + a^{n-1})$$

When we carry out the multiplications, all terms cancel except the first and the last, so only $x^n - a^n$ remains, as required.

Equation (4) gives us

$$\frac{x^n - a^n}{x - a} = \underbrace{x^{n-1} + x^{n-2}a + x^{n-3}a^2 + \cdots + xa^{n-2} + a^{n-1}}_{n \text{ terms}} \quad (x \neq a) \qquad \boxed{5}$$

Therefore,

$$f'(a) = \lim_{x \to a} (x^{n-1} + x^{n-2}a + x^{n-3}a^2 + \cdots + xa^{n-2} + a^{n-1})$$

$$= a^{n-1} + a^{n-2}a + a^{n-3}a^2 + \cdots + aa^{n-2} + a^{n-1} \quad (n \text{ terms})$$

$$= na^{n-1}$$

This proves that $f'(a) = na^{n-1}$, which we may also write as $f'(x) = nx^{n-1}$.   ■

We make a few remarks before proceeding:

• It may be helpful to remember the Power Rule in words: To differentiate x^n, "bring down the exponent and subtract one (from the exponent)."

$$\frac{d}{dx} x^{\text{exponent}} = (\text{exponent})\, x^{\text{exponent}-1}$$

CAUTION *The Power Rule applies only to the power functions* $y = x^n$. *It does not apply to exponential functions such as* $y = 2^x$. *The derivative of* $y = 2^x$ **is not** $x2^{x-1}$. *We will study the derivatives of exponential functions later in this section.*

- The Power Rule is valid for all exponents, whether negative, fractional, or irrational:

$$\frac{d}{dx}x^{-3/5} = -\frac{3}{5}x^{-8/5}, \qquad \frac{d}{dx}x^{\sqrt{2}} = \sqrt{2}x^{\sqrt{2}-1}$$

- The Power Rule can be applied with any variable, not just x. For example,

$$\frac{d}{dz}z^2 = 2z, \qquad \frac{d}{dt}t^{20} = 20t^{19}, \qquad \frac{d}{dr}r^{1/2} = \frac{1}{2}r^{-1/2}$$

Next, we state the Linearity Rules for derivatives, which are analogous to the linearity laws for limits.

THEOREM 4 **Linearity Rules** Assume that f and g are differentiable. Then

Sum and Difference Rules: $f + g$ and $f - g$ are differentiable, and

$$\boxed{(f + g)' = f' + g', \qquad (f - g)' = f' - g'}$$

Constant Multiple Rule: For any constant c, cf is differentiable and

$$\boxed{(cf)' = cf'}$$

Proof To prove the Sum Rule, we use the definition

$$(f + g)'(x) = \lim_{h \to 0} \frac{(f(x + h) + g(x + h)) - (f(x) + g(x))}{h}$$

This difference quotient is equal to a sum ($h \neq 0$):

$$\frac{(f(x + h) + g(x + h)) - (f(x) + g(x))}{h} = \frac{f(x + h) - f(x)}{h} + \frac{g(x + h) - g(x)}{h}$$

Therefore, by the Sum Law for limits,

$$(f + g)'(x) = \lim_{h \to 0} \frac{f(x + h) - f(x)}{h} + \lim_{h \to 0} \frac{g(x + h) - g(x)}{h}$$

$$= f'(x) + g'(x)$$

as claimed. The Difference and Constant Multiple Rules are proved similarly. ∎

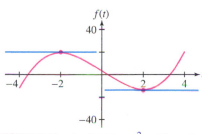

FIGURE 3 Graph of $f(t) = t^3 - 12t + 4$. Tangent lines at $t = \pm 2$ are horizontal.

■ **EXAMPLE 3** Find the points on the graph of $f(t) = t^3 - 12t + 4$ where the tangent line is horizontal (Figure 3).

Solution We calculate the derivative:

$$\frac{df}{dt} = \frac{d}{dt}(t^3 - 12t + 4)$$

$$= \frac{d}{dt}t^3 - \frac{d}{dt}(12t) + \frac{d}{dt}4 \quad \text{(Sum and Difference Rules)}$$

$$= \frac{d}{dt}t^3 - 12\frac{d}{dt}t + 0 \qquad \text{(Constant Multiple Rule and Constant Rule)}$$

$$= 3t^2 - 12 \qquad \text{(Power Rule)}$$

Note in the second line that the derivative of the constant 4 is zero. The tangent line is horizontal at points where the slope $f'(t)$ is zero, so we solve

$$f'(t) = 3t^2 - 12 = 0 \quad \Rightarrow \quad t = \pm 2$$

Now $f(2) = -12$ and $f(-2) = 20$. Hence, the tangent lines are horizontal at $(2, -12)$ and $(-2, 20)$. ∎

■ **EXAMPLE 4** Calculate $\dfrac{dg}{dt}\Big|_{t=1}$, where $g(t) = t^{-3} + 2\sqrt{t} - t^{-4/5}$.

Solution We differentiate term-by-term using the Power Rule without justifying the intermediate steps. Writing $\sqrt{t}$ as $t^{1/2}$, we have

$$\frac{dg}{dt} = \frac{d}{dt}\left(t^{-3} + 2t^{1/2} - t^{-4/5}\right) = -3t^{-4} + 2\left(\frac{1}{2}\right)t^{-1/2} - \left(-\frac{4}{5}\right)t^{-9/5}$$

$$= -3t^{-4} + t^{-1/2} + \frac{4}{5}t^{-9/5}$$

$$\frac{dg}{dt}\Big|_{t=1} = -3 + 1 + \frac{4}{5} = -\frac{6}{5}$$ ■

The Derivative and Behavior of the Graph

The derivative f' gives us important information about the graph of f. For example, the sign of $f'(x)$ tells us whether the tangent line has positive or negative slope. When the tangent line has positive slope, it slopes upward and the graph must be increasing. When the tangent line has negative slope, it slopes downward and the graph must be decreasing. The magnitude of $f'(x)$ reveals how steep the slope is.

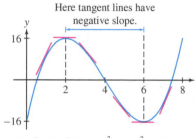

Here tangent lines have negative slope.

(A) Graph of $f(x) = x^3 - 12x^2 + 36x - 16$

■ **EXAMPLE 5** **Graphical Insight** How is the graph of $f(x) = x^3 - 12x^2 + 36x - 16$ related to the derivative $f'(x) = 3x^2 - 24x + 36$?

Solution The derivative $f'(x) = 3x^2 - 24x + 36 = 3(x - 6)(x - 2)$ is negative for $2 < x < 6$ and positive elsewhere [Figure 4(B)]. The following table summarizes this sign information [Figure 4(A)]:

Property of $f'(x)$	Property of the Graph of f
$f'(x) < 0$ for $2 < x < 6$	Tangent has negative slope for $2 < x < 6$ (graph is decreasing).
$f'(2) = f'(6) = 0$	Tangent is horizontal at $x = 2$ and $x = 6$.
$f'(x) > 0$ for $x < 2$ and $x > 6$	Tangent has positive slope for $x < 2$ and $x > 6$ (graph is increasing).

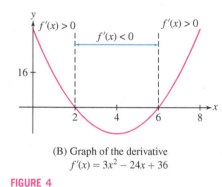

(B) Graph of the derivative $f'(x) = 3x^2 - 24x + 36$

FIGURE 4

Note also that $f'(x) \to \infty$ as $|x|$ becomes large. This corresponds to the fact that the tangent lines to the graph of f get steeper as $|x|$ grows large. ■

■ **EXAMPLE 6** **Identifying the Derivative** The graph of f is shown in Figure 5(A). Which graph, (B) or (C), is the graph of f'?

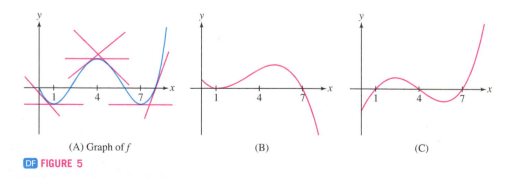

(A) Graph of f (B)   (C)

DF **FIGURE 5**

Slope of Tangent Line	Where
Negative	$(0, 1)$ and $(4, 7)$
Zero	$x = 1, 4, 7$
Positive	$(1, 4)$ and $(7, \infty)$

Solution In Figure 5(A) we see that on the intervals $(0, 1)$ and $(4, 7)$, the graph is decreasing, and therefore the tangent lines to the graph have negative slope . Thus, $f'(x)$ is negative on these intervals. Similarly, on the intervals $(1, 4)$ and $(7, \infty)$, the graph is increasing, and therefore the tangent lines have positive slope and $f'(x)$ is positive (see the table in the margin). Only (C) has these properties, so (C) is the graph of f'. ■

The Derivative of $f(x) = e^x$

The number e was introduced informally in Section 1.6. Now that we have the derivative in our arsenal, we can define e as follows: e is the unique number for which the exponential function $f(x) = e^x$ is its own derivative. To justify this definition, we must prove that a number with this property exists.

In some ways, the number e is "complicated": It is irrational and it cannot be defined without using limits. However, the elegant formula $\frac{d}{dx}e^x = e^x$ shows that e is "simple" from the point of view of calculus and that $f(x) = e^x$ is simpler than the seemingly more natural exponential functions $f(x) = 2^x$ and $f(x) = 10^x$.

THEOREM 5 **The Number e** There is a unique positive real number e with the property

$$\frac{d}{dx}e^x = e^x \qquad \boxed{6}$$

The number e is irrational, with approximate value $e \approx 2.718$.

Proof We shall take for granted a few plausible facts whose proofs are somewhat technical. The first fact is that $f(x) = b^x$ is differentiable for all $b > 0$. Assuming this, let us compute its derivative:

$$\frac{f(x+h) - f(x)}{h} = \frac{b^{x+h} - b^x}{h} = \frac{b^x b^h - b^x}{h} = \frac{b^x(b^h - 1)}{h}$$

$$f'(x) = \lim_{h \to 0} \frac{f(x+h) - f(x)}{h} = \lim_{h \to 0} \frac{b^x(b^h - 1)}{h}$$

$$= b^x \lim_{h \to 0}\left(\frac{b^h - 1}{h}\right)$$

Notice that we took the factor b^x outside the limit. This is legitimate because b^x does not depend on h. Denote the value of the limit on the right by $m(b)$:

$$m(b) = \lim_{h \to 0}\left(\frac{b^h - 1}{h}\right) \qquad \boxed{7}$$

What we have shown, then, is that *the derivative of $f(x) = b^x$ is proportional to b^x*:

$$\frac{d}{dx}b^x = m(b)\, b^x \qquad \boxed{8}$$

Before continuing, let's investigate $m(b)$ numerically using Eq. (7).

■ **EXAMPLE 7** Estimate $m(b)$ numerically for $b = 2$, 2.5, 3, and 10.

Solution We create a table of values of difference quotients to estimate $m(b)$.

h	$\dfrac{2^h - 1}{h}$	$\dfrac{(2.5)^h - 1}{h}$	$\dfrac{3^h - 1}{h}$	$\dfrac{10^h - 1}{h}$
0.01	0.69556	0.92050	1.10467	2.32930
0.001	0.69339	0.91671	1.09921	2.30524
0.0001	0.69317	0.91633	1.09867	2.30285
0.00001	0.69315	0.916295	1.09861	2.30261
	$\boxed{m(2) \approx 0.69}$	$\boxed{m(2.5) \approx 0.92}$	$\boxed{m(3) \approx 1.10}$	$\boxed{m(10) \approx 2.30}$

In many books, e^x is denoted $\exp(x)$. Whenever we refer to the exponential function without specifying the base, the reference is to $f(x) = e^x$. The number e has been computed to an accuracy of more than 100 billion digits. To 20 places,

$$e = 2.71828182845904523536\ldots$$

Since $m(2.5) \approx 0.92$ and $m(3) \approx 1.10$, there must exist a number b *between* 2.5 and 3 such that $m(b) = 1$. This follows from the Intermediate Value Theorem [if we assume the fact that $m(b)$ is a continuous function of b]. If we also use the fact that $m(b)$ is an increasing function of b, we may conclude that there is precisely one number b such that $m(b) = 1$. This is the number e. ■

Using infinite series (see Exercise 93 in Section 10.7), we can show that e is irrational and we can compute its value to any desired degree of accuracy. For most purposes, the approximation $e \approx 2.718$ is adequate.

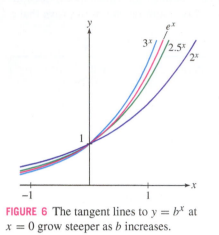

FIGURE 6 The tangent lines to $y = b^x$ at $x = 0$ grow steeper as b increases.

GRAPHICAL INSIGHT   The graph of $f(x) = b^x$ passes through $(0, 1)$ because $b^0 = 1$ (Figure 6). The number $m(b)$ is simply the slope of the tangent line at $(0, 1)$:

$$\frac{d}{dx}b^x\bigg|_{x=0} = m(b) \cdot b^0 = m(b)$$

These tangent lines become steeper as b increases, and $b = e$ is the unique value for which the tangent line has slope 1. In Section 3.9, we will show more generally that $m(b) = \ln b$, the natural logarithm of b, and therefore $\frac{d}{dx}b^x = b^x \ln b$.

■ **EXAMPLE 8**   Find the tangent line to the graph of $f(x) = 3e^x - 5x^2$ at $x = 2$.

Solution   We compute both $f'(2)$ and $f(2)$:

$$f'(x) = \frac{d}{dx}(3e^x - 5x^2) = 3\frac{d}{dx}e^x - 5\frac{d}{dx}x^2 = 3e^x - 10x$$

$$f'(2) = 3e^2 - 10(2) \approx 2.17$$

$$f(2) = 3e^2 - 5(2^2) \approx 2.17$$

Since $f'(2)$ is the slope and the tangent line passes through $(2, f(2))$, an equation of the tangent line is $y - f(2) = f'(2)(x - 2)$. Using these approximate values, we write the equation as (Figure 7)

$$y - 2.17 = 2.17(x - 2) \quad \text{or} \quad y = 2.17x - 2.17 \qquad ■$$

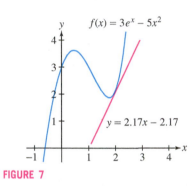

FIGURE 7

CONCEPTUAL INSIGHT   What precisely do we mean by b^x? We have taken for granted that b^x is meaningful for all real numbers x, but we never specified how b^x is defined when x is irrational. If n is a whole number, b^n is simply the product $b \cdot b \cdots b$ (n times), and for any rational number $x = m/n$,

$$b^x = b^{m/n} = \left(b^{1/n}\right)^m = \left(\sqrt[n]{b}\right)^m$$

When x is irrational, this definition does not apply and b^x cannot be defined directly in terms of roots and powers of b. However, it makes sense to view $b^{m/n}$ as an approximation to b^x when m/n is a rational number close to x. For example, $3^{\sqrt{2}}$ should be approximately equal to $3^{1.4142} \approx 4.729$ because 1.4142 is a good rational approximation to $\sqrt{2}$. Formally, then, we may define b^x as a limit over rational numbers m/n approaching x:

$$b^x = \lim_{m/n \to x} b^{m/n}$$

We can show that this limit exists and that the function $f(x) = b^x$ thus defined is not only continuous but also differentiable (see Exercise 80 in Section 5.8).

Differentiability, Continuity, and Local Linearity

In the rest of this section, we examine the concept of **differentiability** more closely. We begin by proving that a differentiable function is necessarily continuous. In particular, a differentiable function cannot have any jumps. Figure 8 shows why: Although the secant lines from the right approach the line L (which is tangent to the right half of the graph), the secant lines from the left approach the vertical (and their slopes tend to ∞).

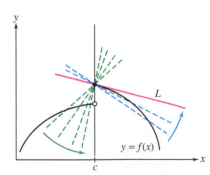

DF FIGURE 8 Secant lines at a jump discontinuity.

> **THEOREM 6** **Differentiability Implies Continuity** If f is differentiable at $x = c$, then f is continuous at $x = c$.

Proof By definition, if f is differentiable at $x = c$, then the following limit exists:

$$f'(c) = \lim_{x \to c} \frac{f(x) - f(c)}{x - c}$$

We must prove that $\lim\limits_{x \to c} f(x) = f(c)$, because this is the definition of continuity at $x = c$. To relate the two limits, consider the equation (valid for $x \neq c$)

$$f(x) - f(c) = (x - c) \frac{f(x) - f(c)}{x - c}$$

Both factors on the right approach a limit as $x \to c$, so

$$\lim_{x \to c} \left(f(x) - f(c) \right) = \lim_{x \to c} \left((x - c) \frac{f(x) - f(c)}{x - c} \right)$$

$$= \left(\lim_{x \to c} (x - c) \right) \left(\lim_{x \to c} \frac{f(x) - f(c)}{x - c} \right)$$

$$= 0 \cdot f'(c) = 0$$

by the Product Law for limits. The Sum Law now yields the desired conclusion:

$$\lim_{x \to c} f(x) = \lim_{x \to c} (f(x) - f(c)) + \lim_{x \to c} f(c) = 0 + f(c) = f(c) \qquad \blacksquare$$

Most of the functions encountered in this text are differentiable, but exceptions exist, as the next example shows.

All differentiable functions are continuous by Theorem 6, but Example 9 shows that the converse is false. A continuous function is not necessarily differentiable.

■ **EXAMPLE 9** **Continuous But Not Differentiable** Show that $f(x) = |x|$ is continuous but not differentiable at $x = 0$.

Solution The function f is continuous at $x = 0$ because $\lim\limits_{x \to 0} |x| = 0 = f(0)$. On the other hand,

$$f'(0) = \lim_{h \to 0} \frac{f(0 + h) - f(0)}{h} = \lim_{h \to 0} \frac{|0 + h| - |0|}{h} = \lim_{h \to 0} \frac{|h|}{h}$$

This limit does not exist [and hence f is not differentiable at $x = 0$] because

$$\frac{|h|}{h} = \begin{cases} 1 & \text{if } h > 0 \\ -1 & \text{if } h < 0 \end{cases}$$

and thus the one-sided limits are not equal:

$$\lim_{h \to 0+} \frac{|h|}{h} = 1 \qquad \text{and} \qquad \lim_{h \to 0-} \frac{|h|}{h} = -1 \qquad \blacksquare$$

> **GRAPHICAL INSIGHT** Differentiability has an important graphical interpretation in terms of local linearity. We say that f is **locally linear** at $x = a$ if the graph looks more and more like a straight line as we zoom in on the point $(a, f(a))$. In this context, the adjective *linear* means "resembling a line," and *local* indicates that we are concerned only with the behavior of the graph near $(a, f(a))$. The graph of a locally linear function may be very wavy or *nonlinear*, as in Figure 9. But as soon as we zoom in on a sufficiently small piece of the graph, it begins to appear straight.
>
> Not only does the graph look like a line as we zoom in on a point, but as Figure 9 suggests, the "zoom line" is the tangent line. Thus, the relation between differentiability and local linearity can be expressed as follows:
>
> If $f'(a)$ exists, then f is locally linear at $x = a$: As we zoom in on the point $(a, f(a))$, the graph becomes nearly indistinguishable from its tangent line.

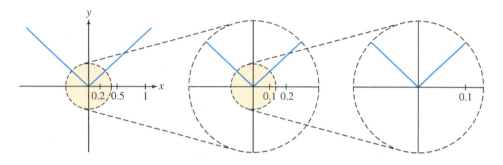

DF **FIGURE 9** Local linearity: The graph looks more and more like the tangent line as we zoom in on a point.

Local linearity gives us a graphical way to understand why $f(x) = |x|$ is not differentiable at $x = 0$ (as shown in Example 9). Figure 10 shows that the graph of $f(x) = |x|$ has a corner at $x = 0$, and this corner *does not disappear*, no matter how closely we zoom in on the origin. Since the graph does not straighten out under zooming, f is not locally linear at $x = 0$, and we cannot expect $f'(0)$ to exist.

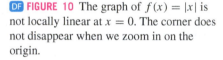

DF **FIGURE 10** The graph of $f(x) = |x|$ is not locally linear at $x = 0$. The corner does not disappear when we zoom in on the origin.

Another way that a continuous function can fail to be differentiable is if the tangent line exists but is vertical (in which case the slope of the tangent line is undefined).

■ **EXAMPLE 10** **Vertical Tangents** Show that $f(x) = x^{1/3}$ is not differentiable at $x = 0$.

Solution The limit defining $f'(0)$ is infinite:

$$\lim_{h \to 0} \frac{f(h) - f(0)}{h} = \lim_{h \to 0} \frac{h^{1/3} - 0}{h} = \lim_{h \to 0} \frac{h^{1/3}}{h} = \lim_{h \to 0} \frac{1}{h^{2/3}} = \infty$$

Therefore, $f'(0)$ does not exist (Figure 11). ■

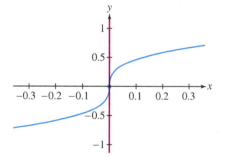

FIGURE 11 The tangent line to the graph of $f(x) = x^{1/3}$ at the origin is the (vertical) y-axis. The derivative $f'(0)$ does not exist.

As a final remark, we mention that there are more complicated ways in which a continuous function can fail to be differentiable. Figure 12 shows the graph of $f(x) = x \sin \frac{1}{x}$. If we define $f(0) = 0$, then f is continuous but not differentiable at $x = 0$. The secant lines keep oscillating and never settle down to a limiting position (see Exercise 97).

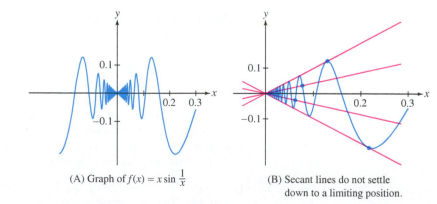

(A) Graph of $f(x) = x \sin \frac{1}{x}$

(B) Secant lines do not settle down to a limiting position.

FIGURE 12

3.2 SUMMARY

- The derivative f' is the function whose value at $x = a$ is the derivative $f'(a)$.
- We have several different notations for the derivative of $y = f(x)$:

$$y', \quad y'(x), \quad f'(x), \quad \frac{dy}{dx}, \quad \frac{df}{dx}$$

The value of the derivative at $x = a$ is written

$$y'(a), \quad f'(a), \quad \frac{dy}{dx}\bigg|_{x=a}, \quad \frac{df}{dx}\bigg|_{x=a}$$

THEOREM 7 **Rules for Taking a Derivative**

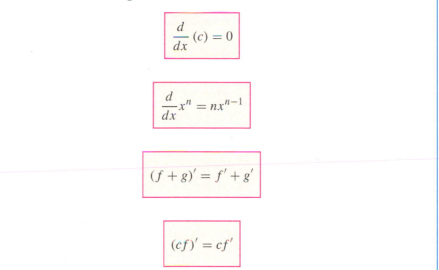

$$\frac{d}{dx}(c) = 0$$

$$\frac{d}{dx}x^n = nx^{n-1}$$

$$(f + g)' = f' + g'$$

$$(cf)' = cf'$$

- The derivative of $f(x) = b^x$ is proportional to b^x:

$$\frac{d}{dx}b^x = m(b)b^x, \qquad \text{where} \quad m(b) = \lim_{h \to 0} \frac{b^h - 1}{h}$$

- The number $e \approx 2.718$ is defined by the property $m(e) = 1$, so that

$$\frac{d}{dx}e^x = e^x$$

- Differentiability implies continuity: If f is differentiable at $x = a$, then f is continuous at $x = a$. However, there exist continuous functions that are not differentiable.
- If $f'(a)$ exists, then f is locally linear in the following sense: As we zoom in on the point $(a, f(a))$, the graph becomes nearly indistinguishable from its tangent line.

3.2 EXERCISES

Preliminary Questions

1. What is the slope of the tangent line through the point $(2, f(2))$ if $f'(x) = x^3$?

2. Evaluate $(f - g)'(1)$ and $(3f + 2g)'(1)$, assuming that $f'(1) = 3$ and $g'(1) = 5$.

3. To which of the following does the Power Rule apply?

(a) $f(x) = x^2$ (b) $f(x) = 2^e$

(c) $f(x) = x^e$ (d) $f(x) = e^x$

(e) $f(x) = x^x$ (f) $f(x) = x^{-4/5}$

4. Choose (a) or (b). The derivative does not exist if the tangent line is: (a) horizontal (b) vertical.

5. Which property distinguishes $f(x) = e^x$ from all other exponential functions $g(x) = b^x$?

Exercises

In Exercises 1–6, compute $f'(x)$ using the limit definition.

1. $f(x) = 3x - 7$

2. $f(x) = x^2 + 3x$

3. $f(x) = x^3$

4. $f(x) = 1 - x^{-1}$

5. $f(x) = x - \sqrt{x}$

6. $f(x) = x^{-1/2}$

In Exercises 7–14, use the Power Rule to compute the derivative.

7. $\left. \dfrac{d}{dx} x^4 \right|_{x=-2}$

8. $\left. \dfrac{d}{dt} t^{-3} \right|_{t=4}$

9. $\left. \dfrac{d}{dt} t^{2/3} \right|_{t=8}$

10. $\left. \dfrac{d}{dt} t^{-2/5} \right|_{t=1}$

11. $\dfrac{d}{dx} x^{0.35}$

12. $\dfrac{d}{dx} x^{14/3}$

13. $\dfrac{d}{dt} t^{\sqrt{17}}$

14. $\dfrac{d}{dt} t^{-\pi^2}$

In Exercises 15–18, compute $f'(x)$ and find an equation of the tangent line to the graph at $x = a$.

15. $f(x) = x^4, \quad a = 2$

16. $f(x) = x^{-2}, \quad a = 5$

17. $f(x) = 5x - 32\sqrt{x}, \quad a = 4$

18. $f(x) = \sqrt[3]{x}, \quad a = 8$

19. Calculate:

(a) $\dfrac{d}{dx} 12e^x$ **(b)** $\dfrac{d}{dt}(25t - 8e^t)$ **(c)** $\dfrac{d}{dt} e^{t-3}$

Hint for (c): Write e^{t-3} as $e^{-3}e^t$.

20. Find an equation of the tangent line to $y = 24e^x$ at $x = 2$.

In Exercises 21–32, calculate the derivative.

21. $f(x) = 2x^3 - 3x^2 + 5$

22. $f(x) = 2x^3 - 3x^2 + 2x$

23. $f(x) = 4x^{5/3} - 3x^{-2} - 12$

24. $f(x) = x^{5/4} + 4x^{-3/2} + 11x$

25. $g(z) = 7z^{-5/14} + z^{-5} + 9$

26. $h(t) = 6\sqrt{t} + \dfrac{1}{\sqrt{t}}$

27. $f(s) = \sqrt[4]{s} + \sqrt[3]{s}$

28. $W(y) = 6y^4 + 7y^{2/3}$

29. $g(x) = e^2$

30. $f(x) = 3e^x - x^3$

31. $h(t) = 5e^{t-3}$

32. $f(x) = 9 - 12x^{1/3} + 8e^x$

In Exercises 33–36, calculate the derivative by expanding or simplifying the function.

33. $P(s) = (4s - 3)^2$

34. $Q(r) = (1 - 2r)(3r + 5)$

35. $g(x) = \dfrac{x^2 + 4x^{1/2}}{x^2}$

36. $s(t) = \dfrac{1 - 2t}{t^{1/2}}$

In Exercises 37–42, calculate the derivative indicated.

37. $\left. \dfrac{dT}{dC} \right|_{C=8}, \quad T = 3C^{2/3}$

38. $\left. \dfrac{dP}{dV} \right|_{V=-2}, \quad P = \dfrac{7}{V}$

39. $\left. \dfrac{ds}{dz} \right|_{z=2}, \quad s = 4z - 16z^2$

40. $\left. \dfrac{dR}{dW} \right|_{W=1}, \quad R = W^\pi$

41. $\left. \dfrac{dr}{dt} \right|_{t=4}, \quad r = t - e^t$

42. $\left. \dfrac{dp}{dh} \right|_{h=4}, \quad p = 7e^{h-2}$

43. Match the functions in graphs (A)–(D) with their derivatives (I)–(III) in Figure 13. Note that two of the functions have the same derivative. Explain why.

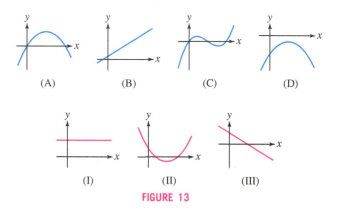

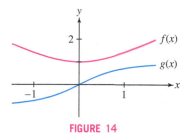

FIGURE 13

44. Of the two functions f and g in Figure 14, which is the derivative of the other? Justify your answer.

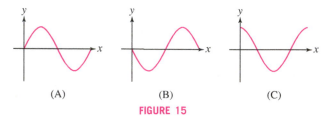

FIGURE 14

45. Assign the labels $y = f(x)$, $y = g(x)$, and $y = h(x)$ to the graphs in Figure 15 in such a way that $f'(x) = g(x)$ and $g'(x) = h(x)$.

(A) **(B)** **(C)**

FIGURE 15

46. According to the *peak oil theory*, first proposed in 1956 by geophysicist M. Hubbert, the total amount of crude oil $Q(t)$ produced worldwide up to time t has a graph like that in Figure 16.

(a) Sketch the derivative $Q'(t)$ for $1900 \le t \le 2150$. What does $Q'(t)$ represent?

(b) In which year (approximately) does $Q'(t)$ take on its maximum value?

(c) What is $L = \lim\limits_{t \to \infty} Q(t)$? And what is its interpretation?

(d) What is the value of $\lim\limits_{t \to \infty} Q'(t)$?

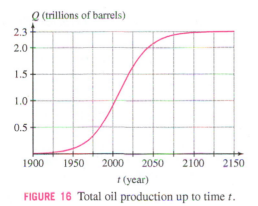

FIGURE 16 Total oil production up to time t.

47. ✏️ Use the table of values of f to determine which of (A) or (B) in Figure 17 is the graph of f'. Explain.

x	0	0.5	1	1.5	2	2.5	3	3.5	4
$f(x)$	10	55	98	139	177	210	237	257	268

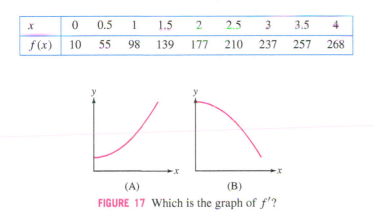

FIGURE 17 Which is the graph of f'?

48. Let R be a variable and r a constant. Compute the derivatives:

(a) $\dfrac{d}{dR} R$ (b) $\dfrac{d}{dR} r$ (c) $\dfrac{d}{dR} r^2 R^3$

49. Compute the derivatives, where c is a constant.

(a) $\dfrac{d}{dt} ct^3$ (b) $\dfrac{d}{dz} (5z + 4cz^2)$

(c) $\dfrac{d}{dy} (9c^2y^3 - 24c)$

50. Find the points on the graph of $f(x) = 12x - x^3$ where the tangent line is horizontal.

51. Find the points on the graph of $y = x^2 + 3x - 7$ at which the slope of the tangent line is equal to 4.

52. Find the values of x where $y = x^3$ and $y = x^2 + 5x$ have parallel tangent lines.

53. Determine a and b such that $p(x) = x^2 + ax + b$ satisfies $p(1) = 0$ and $p'(1) = 4$.

54. Find all values of x such that the tangent line to $y = 4x^2 + 11x + 2$ is steeper than the tangent line to $y = x^3$.

55. Let $f(x) = x^3 - 3x + 1$. Show that $f'(x) \geq -3$ for all x and that, for every $m > -3$, there are precisely two points where $f'(x) = m$. Indicate the position of these points and the corresponding tangent lines for one value of m in a sketch of the graph of f.

56. Show that the tangent lines to $y = \frac{1}{3}x^3 - x^2$ at $x = a$ and at $x = b$ are parallel if $a = b$ or $a + b = 2$.

57. Compute the derivative of $f(x) = x^{3/2}$ using the limit definition. *Hint:* Show that

$$\frac{f(x+h) - f(x)}{h} = \frac{(x+h)^3 - x^3}{h} \left(\frac{1}{\sqrt{(x+h)^3} + \sqrt{x^3}} \right)$$

58. Use the limit definition of $m(b)$ to approximate $m(4)$. Then estimate the slope of the tangent line to $y = 4^x$ at $x = 0$ and $x = 2$.

59. Let $f(x) = xe^x$. Use the limit definition to compute $f'(0)$, and find the equation of the tangent line at $x = 0$.

60. The average speed (in meters per second) of a gas molecule is

$$v_{avg} = \sqrt{\frac{8RT}{\pi M}}$$

where T is the temperature (in kelvins), M is the molar mass (in kilograms per mole), and $R = 8.31$. Calculate dv_{avg}/dT at $T = 300$ K for oxygen, which has a molar mass of 0.032 kg/mol.

61. Biologists have observed that the pulse rate P (in beats per minute) in animals is related to body mass (in kilograms) by the approximate formula $P = 200m^{-1/4}$. This is one of many *allometric scaling laws* prevalent in biology. Is P an increasing or decreasing function of m? Find an equation of the tangent line at the points on the graph in Figure 18 that represent goat ($m = 33$) and man ($m = 68$).

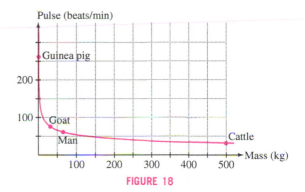

FIGURE 18

62. Some studies suggest that kidney mass K in mammals (in kilograms) is related to body mass m (in kilograms) by the approximate formula $K = 0.007m^{0.85}$. Calculate dK/dm at $m = 68$. Then calculate the derivative with respect to m of the relative kidney-to-mass ratio K/m at $m = 68$.

63. The Clausius–Clapeyron Law relates the *vapor pressure* of water P (in atmospheres) to the temperature T (in kelvins):

$$\frac{dP}{dT} = k\frac{P}{T^2}$$

where k is a constant. Estimate dP/dT for $T = 303, 313, 323, 333, 343$ using the data and the approximation

$$\frac{dP}{dT} \approx \frac{P(T+10) - P(T-10)}{20}$$

T (K)	293	303	313	323	333	343	353
P (atm)	0.0278	0.0482	0.0808	0.1311	0.2067	0.3173	0.4754

Do your estimates seem to confirm the Clausius–Clapeyron Law? What is the approximate value of k?

64. Let L be the tangent line to the hyperbola $xy = 1$ at $x = a$, where $a > 0$. Show that the area of the triangle bounded by L and the coordinate axes does not depend on a.

65. In the setting of Exercise 64, show that the point of tangency is the midpoint of the segment of L lying in the first quadrant.

66. Match functions (A)–(C) with their derivatives (I)–(III) in Figure 19.

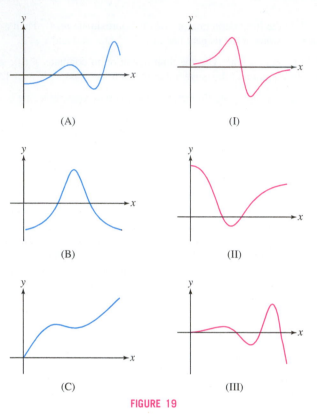

FIGURE 19

67. Make a rough sketch of the graph of the derivative of the function in Figure 20(A).

68. Graph the derivative of the function in Figure 20(B), omitting points where the derivative is not defined.

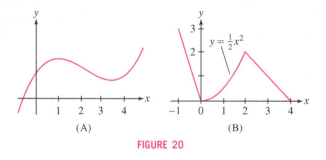

FIGURE 20

69. Sketch the graph of $f(x) = x\,|x|$. Then show that $f'(0)$ exists.

70. Determine the values of x at which the function in Figure 21 is:
(a) discontinuous and (b) nondifferentiable.

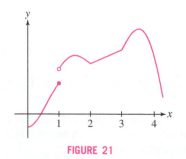

FIGURE 21

In Exercises 71–76, find the points c (if any) such that $f'(c)$ does not exist.

71. $f(x) = |x - 1|$

72. $f(x) = \lfloor x \rfloor$

73. $f(x) = x^{2/3}$

74. $f(x) = x^{3/2}$

75. $f(x) = |x^2 - 1|$

76. $f(x) = |x - 1|^2$

GU *In Exercises 77–82, zoom in on a plot of f at the point $(a, f(a))$ and state whether or not f appears to be differentiable at $x = a$. If it is nondifferentiable, state whether the tangent line appears to be vertical or does not exist.*

77. $f(x) = (x - 1)|x|, \quad a = 0$

78. $f(x) = (x - 3)^{5/3}, \quad a = 3$

79. $f(x) = (x - 3)^{1/3}, \quad a = 3$

80. $f(x) = \sin(x^{1/3}), \quad a = 0$ **81.** $f(x) = |\sin x|, \quad a = 0$

82. $f(x) = |x - \sin x|, \quad a = 0$

83. Find the coordinates of the point P in Figure 22 at which the tangent line passes through $(5, 0)$.

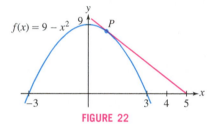

FIGURE 22

84. **GU** Plot the derivative f' of $f(x) = 2x^3 - 10x^{-1}$ for $x > 0$ (set the bounds of the viewing box appropriately) and observe that $f'(x) > 0$. What does the positivity of $f'(x)$ tell us about the graph of f itself? Plot f and confirm this conclusion.

Exercises 85–88 refer to Figure 23. Length QR is called the subtangent *at P, and length RT is called the* subnormal.

85. Calculate the subtangent of

$$f(x) = x^2 + 3x \quad \text{at } x = 2$$

86. Show that the subtangent of $f(x) = e^x$ is everywhere equal to 1.

87. Prove in general that the subnormal at P is $|f'(x)f(x)|$.

88. Show that $\overline{PQ}$ has length $|f(x)|\sqrt{1 + f'(x)^{-2}}$.

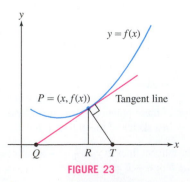

FIGURE 23

89. Prove the following theorem of Apollonius of Perga (the Greek mathematician born in 262 BCE who gave the parabola, ellipse, and hyperbola their names): The subtangent of the parabola $y = x^2$ at $x = a$ is equal to $a/2$.

90. Show that the subtangent to $y = x^3$ at $x = a$ is equal to $\frac{1}{3}a$.

91. Formulate and prove a generalization of Exercise 90 for $y = x^n$.

Further Insights and Challenges

92. Two small arches have the shape of parabolas. The first is given by $f(x) = 1 - x^2$ for $-1 \le x \le 1$ and the second by $g(x) = 4 - (x-4)^2$ for $2 \le x \le 6$. A board is placed on top of these arches so it rests on both (Figure 24). What is the slope of the board? *Hint:* Find the tangent line to $y = f(x)$ that intersects $y = g(x)$ in exactly one point.

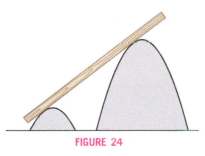

FIGURE 24

93. A vase is formed by rotating $y = x^2$ around the y-axis. If we drop in a marble, it will either touch the bottom point of the vase or be suspended above the bottom by touching the sides (Figure 25). How small must the marble be to touch the bottom?

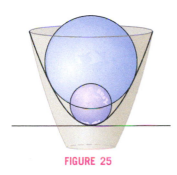

FIGURE 25

94. Let f be a differentiable function, and set the function $g(x) = f(x + c)$, where c is a constant. Use the limit definition to show that $g'(x) = f'(x + c)$. Explain this result graphically, recalling that the graph of g is obtained by shifting the graph of f c units to the left (if $c > 0$) or right (if $c < 0$).

95. Negative Exponents Let n be a whole number. Use the Power Rule for x^n to calculate the derivative of $f(x) = x^{-n}$ by showing that

$$\frac{f(x+h) - f(x)}{h} = \frac{-1}{x^n(x+h)^n} \frac{(x+h)^n - x^n}{h}$$

96. Verify the Power Rule for the exponent $1/n$, where n is a positive integer, using the following trick: Rewrite the difference quotient for $y = x^{1/n}$ at $x = b$ in terms of

$$u = (b+h)^{1/n} \quad \text{and} \quad a = b^{1/n}$$

97. Infinitely Rapid Oscillations Define

$$f(x) = \begin{cases} x \sin \dfrac{1}{x} & x \ne 0 \\ 0 & x = 0 \end{cases}$$

Show that f is continuous at $x = 0$ but $f'(0)$ does not exist (see Figure 12).

98. For which value of λ does the equation $e^x = \lambda x$ have a unique solution? For which values of λ does it have at least one solution? For intuition, plot $y = e^x$ and the line $y = \lambda x$.

3.3 Product and Quotient Rules

This section covers the **Product Rule** and **Quotient Rule** for computing derivatives. These two rules, together with the Chain Rule and implicit differentiation (covered in later sections), make up an extremely effective "differentiation toolkit."

←·· REMINDER *The product function fg is defined by $(fg)(x) = f(x)g(x)$.*

THEOREM 1 **Product Rule** If f and g are differentiable functions, then fg is differentiable and

$$(fg)'(x) = f'(x)g(x) + f(x)g'(x)$$

It may be helpful to remember the Product Rule in words: The derivative of a product is equal to *the derivative of the first function times the second function plus the first function times the derivative of the second function*:

$$(\text{First})' \cdot \text{Second} + \text{First} \cdot (\text{Second})'$$

We prove the Product Rule after presenting three examples.

■ **EXAMPLE 1** Find the derivative of $h(x) = x^2(9x + 2)$.

Solution This function is a product:

$$h(x) = \overbrace{x^2}^{\text{First}} \overbrace{(9x + 2)}^{\text{Second}}$$

By the Product Rule (in Leibniz notation),

$$h'(x) = \overbrace{\frac{d}{dx}(x^2)}^{(\text{First})'} \overbrace{(9x + 2)}^{\text{Second}} + \overbrace{(x^2)}^{\text{First}} \overbrace{\frac{d}{dx}(9x + 2)}^{(\text{Second})'}$$

$$= (2x)(9x + 2) + (x^2)(9) = 27x^2 + 4x \qquad ■$$

■ **EXAMPLE 2** Find the derivative of $y = (2 + x^{-1})(x^{3/2} + 1)$.

Solution Use the Product Rule:

> Note how the prime notation is used in the solution to Example 2. We write $(x^{3/2} + 1)'$ to denote the derivative of $f(x) = x^{3/2} + 1$, etc.

$$y' = \overbrace{(2 + x^{-1})'(x^{3/2} + 1) + (2 + x^{-1})(x^{3/2} + 1)'}^{(\text{First})' \cdot \text{Second} + \text{First} \cdot (\text{Second})'}$$

$$= (-x^{-2})(x^{3/2} + 1) + (2 + x^{-1})\left(\frac{3}{2}x^{1/2}\right) \quad \text{(compute the derivatives)}$$

$$= -x^{-1/2} - x^{-2} + 3x^{1/2} + \tfrac{3}{2}x^{-1/2} = \tfrac{1}{2}x^{-1/2} - x^{-2} + 3x^{1/2} \quad \text{(simplify)} \qquad ■$$

In the previous two examples, we could have avoided the Product Rule by expanding the function. Thus, the result of Example 2 can be obtained as follows:

$$y = (2 + x^{-1})(x^{3/2} + 1) = 2x^{3/2} + 2 + x^{1/2} + x^{-1}$$

$$y' = \frac{d}{dx}(2x^{3/2} + 2 + x^{1/2} + x^{-1}) = 3x^{1/2} + \tfrac{1}{2}x^{-1/2} - x^{-2}$$

In the next example, the function cannot be expanded, so we must use the Product Rule (or go back to the limit definition of the derivative).

■ **EXAMPLE 3** Calculate $\dfrac{d}{dt}\left(t^2 e^t\right)$.

Solution Use the Product Rule and the formula $\dfrac{d}{dt}e^t = e^t$:

$$\frac{d}{dt}\left(t^2 e^t\right) = \overbrace{\left(\frac{d}{dt}t^2\right)e^t + t^2\left(\frac{d}{dt}e^t\right)}^{(\text{First})' \cdot \text{Second} + \text{First} \cdot (\text{Second})'} = 2te^t + t^2(e^t) = (2t + t^2)e^t \qquad ■$$

Proof of the Product Rule According to the limit definition of the derivative,

$$(fg)'(x) = \lim_{h \to 0} \frac{f(x + h)g(x + h) - f(x)g(x)}{h}$$

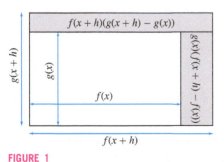

FIGURE 1

We can interpret the numerator as the area of the shaded region in Figure 1: the area of the larger rectangle $f(x + h)g(x + h)$ minus the area of the smaller rectangle $f(x)g(x)$. This shaded region is the union of two rectangular strips, so we obtain the following identity [which we can also obtain algebraically by adding and subtracting the term $f(x + h)g(x)$ from the left-hand side and then manipulating the result algebraically]:

$$f(x + h)g(x + h) - f(x)g(x) = \big(f(x + h) - f(x)\big)g(x) + f(x + h)\big(g(x + h) - g(x)\big)$$

Now use this identity to write $(fg)'(x)$ as a sum of two limits:

$$(fg)'(x) = \underbrace{\lim_{h \to 0} \frac{f(x + h) - f(x)}{h}g(x)}_{\text{Show that this equals } f'(x)g(x).} + \underbrace{\lim_{h \to 0} f(x + h)\frac{g(x + h) - g(x)}{h}}_{\text{Show that this equals } f(x)g'(x).} \qquad \boxed{1}$$

The use of the Sum Law is valid, provided that each limit on the right exists. To check that the first limit exists and to evaluate it, we note that f is differentiable. Thus,

$$\lim_{h \to 0} \frac{f(x+h) - f(x)}{h} g(x) = \lim_{h \to 0} \frac{f(x+h) - f(x)}{h} \lim_{h \to 0} g(x)$$

$$= f'(x) g(x) \qquad \boxed{2}$$

The second limit is similar, but using the facts f is continuous (because it is differentiable) and g is differentiable:

$$\lim_{h \to 0} f(x+h) \frac{g(x+h) - g(x)}{h} = \lim_{h \to 0} f(x+h) \lim_{h \to 0} \frac{g(x+h) - g(x)}{h} = f(x) g'(x)$$

$$\boxed{3}$$

Using Eq. (2) and Eq. (3) in Eq. (1), we conclude that fg is differentiable and that $(fg)'(x) = f'(x)g(x) + f(x)g'(x)$ as claimed. ∎

CONCEPTUAL INSIGHT The Product Rule was first stated by the 29-year-old Leibniz in 1675, the year he developed some of his major ideas on calculus. To document his process of discovery for posterity, he recorded his thoughts and struggles, the moments of inspiration as well as the mistakes. In a manuscript dated November 11, 1675, Leibniz suggests *incorrectly* that $(fg)'$ equals $f'g'$. He then catches his error by taking $f(x) = g(x) = x$ and noticing that

$$(fg)'(x) = (x^2)' = 2x \qquad \text{is *not* equal to} \qquad f'(x)g'(x) = 1 \cdot 1 = 1$$

Ten days later, on November 21, Leibniz writes down the correct Product Rule and comments, "*Now this is a really noteworthy theorem.*"

With the benefit of hindsight, we can point out that Leibniz might have avoided his error if he had paid attention to units. Suppose $f(t)$ and $g(t)$ represent distances in meters, where t is time in seconds. Then $(fg)'$ has units of m^2/s. This cannot equal $f'g'$, which has units of (m/s)(m/s) = m^2/s^2.

The next theorem states the rule for differentiating quotients. Note, in particular, that $(f/g)'$ is *not* equal to the quotient f'/g'.

← **REMINDER** The quotient function f/g is defined by

$$\left(\frac{f}{g} \right)(x) = \frac{f(x)}{g(x)}$$

THEOREM 2 Quotient Rule If f and g are differentiable functions, then f/g is differentiable for all x such that $g(x) \neq 0$, and

$$\left(\frac{f}{g} \right)'(x) = \frac{g(x)f'(x) - f(x)g'(x)}{g(x)^2}$$

The numerator in the Quotient Rule is equal to *the bottom times the derivative of the top minus the top times the derivative of the bottom*:

$$\frac{\text{Bottom} \cdot (\text{Top})' - \text{Top} \cdot (\text{Bottom})'}{\text{Bottom}^2}$$

CAUTION It is easy to mistakenly reverse the order of the two terms in the numerator of the quotient rule. Thinking of the original function as Hi/Lo, the derivative is "Lo D Hi minus Hi D Lo, draw the line and square below."

Proof of the Quotient Rule Let $Q(x) = f(x)/g(x)$. Our goal is to find the formula for $Q'(x)$. Then $f(x) = Q(x) \cdot g(x)$. Differentiating both sides, utilizing the Product Rule for the right side, we obtain $f'(x) = Q'(x) \cdot g(x) + Q(x) \cdot g'(x)$. Solving for $Q'(x)$, we obtain

$$Q'(x) = \frac{f'(x) - Q(x) \cdot g'(x)}{g(x)} = \frac{f'(x) - \frac{f(x)}{g(x)} \cdot g'(x)}{g(x)} = \frac{f'(x)g(x) - f(x)g'(x)}{g(x)^2}$$

as we wanted to show. ∎

An alternative proof appears in Exercises 58–60.

■ EXAMPLE 4 Compute the derivative of $f(x) = \dfrac{x}{1+x^2}$.

Solution Apply the Quotient Rule:

$$f'(x) = \frac{\overbrace{(1+x^2)}^{\text{Bottom}}\,\overbrace{(x)'}^{\text{(Top)}'} - \overbrace{(x)}^{\text{Top}}\,\overbrace{(1+x^2)'}^{\text{(Bottom)}'}}{(1+x^2)^2} = \frac{(1+x^2)(1) - (x)(2x)}{(1+x^2)^2}$$

$$= \frac{1+x^2-2x^2}{(1+x^2)^2} = \frac{1-x^2}{(1+x^2)^2} \qquad ■$$

> *Note that it is not always the simplest method to apply the quotient rule. If we want to differentiate the function*
> $f(x) = \dfrac{\sqrt{x} - x^3}{x}$, *it is easier to rewrite it as* $f(x) = x^{-1/2} - x^2$ *and then differentiate it directly.*

■ EXAMPLE 5 Calculate $\dfrac{d}{dt}\left(\dfrac{e^t}{e^t + t}\right)$.

Solution Use the Quotient Rule and the formula $(e^t)' = e^t$:

$$\frac{d}{dt}\left(\frac{e^t}{e^t + t}\right) = \frac{(e^t + t)(e^t)' - e^t(e^t + t)'}{(e^t + t)^2} = \frac{(e^t + t)e^t - e^t(e^t + 1)}{(e^t + t)^2} = \frac{(t-1)e^t}{(e^t + t)^2} \qquad ■$$

■ EXAMPLE 6 Find the tangent line to the graph of $f(x) = \dfrac{3x^2 + x - 2}{4x^3 + 1}$ at $x = 1$.

Solution

$$f'(x) = \frac{d}{dx}\left(\frac{3x^2 + x - 2}{4x^3 + 1}\right) = \frac{\overbrace{(4x^3 + 1)}^{\text{Bottom}}\,\overbrace{(3x^2 + x - 2)'}^{\text{(Top)}'} - \overbrace{(3x^2 + x - 2)}^{\text{Top}}\,\overbrace{(4x^3 + 1)'}^{\text{(Bottom)}'}}{(4x^3 + 1)^2}$$

$$= \frac{(4x^3 + 1)(6x + 1) - (3x^2 + x - 2)(12x^2)}{(4x^3 + 1)^2}$$

$$= \frac{(24x^4 + 4x^3 + 6x + 1) - (36x^4 + 12x^3 - 24x^2)}{(4x^3 + 1)^2}$$

$$= \frac{-12x^4 - 8x^3 + 24x^2 + 6x + 1}{(4x^3 + 1)^2}$$

At $x = 1$,

$$f(1) = \frac{3 + 1 - 2}{4 + 1} = \frac{2}{5}$$

$$f'(1) = \frac{-12 - 8 + 24 + 6 + 1}{5^2} = \frac{11}{25}$$

An equation of the tangent line at $\left(1, \frac{2}{5}\right)$ is

$$y - \frac{2}{5} = \frac{11}{25}(x - 1) \qquad \text{or} \qquad y = \frac{11}{25}x - \frac{1}{25} \qquad ■$$

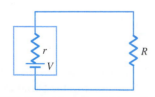

FIGURE 2 Apparatus of resistance R attached to a battery of voltage V.

■ EXAMPLE 7 **Power Delivered by a Battery** The power that a battery supplies to an apparatus such as a laptop depends on the *internal resistance* of the battery. For a battery of voltage V and internal resistance r, the total power delivered to an apparatus of resistance R (Figure 2) is

$$P = \frac{V^2 R}{(R + r)^2}$$

(a) Calculate dP/dR, assuming that V and r are constants.

(b) Where, in the graph of P versus R, is the tangent line horizontal?

Solution

(a) Because V is a constant, we obtain (using the Quotient Rule)

$$\frac{dP}{dR} = V^2 \frac{d}{dR}\left(\frac{R}{(R + r)^2}\right) = V^2 \frac{(R + r)^2 \frac{d}{dR}R - R \frac{d}{dR}(R + r)^2}{(R + r)^4}$$

$$\boxed{4}$$

We have $\frac{d}{dR}R = 1$, and $\frac{d}{dR}r = 0$ because r is a constant. Thus,

$$\frac{d}{dR}(R+r)^2 = \frac{d}{dR}(R^2 + 2rR + r^2)$$

$$= \frac{d}{dR}R^2 + 2r\frac{d}{dR}R + \frac{d}{dR}r^2$$

$$= 2R + 2r + 0 = 2(R+r) \qquad \boxed{5}$$

Using Eq. (5) in Eq. (4), we obtain

$$\frac{dP}{dR} = V^2\frac{(R+r)^2 - 2R(R+r)}{(R+r)^4} = V^2\frac{(R+r) - 2R}{(R+r)^3} = V^2\frac{r-R}{(R+r)^3} \qquad \boxed{6}$$

(b) The tangent line is horizontal when the derivative is zero. We see from Eq. (6) that the derivative is zero when $r - R = 0$—that is, when $R = r$. ∎

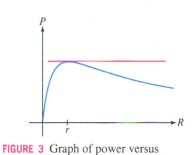

FIGURE 3 Graph of power versus resistance:

$$P = \frac{V^2R}{(R+r)^2}$$

> **GRAPHICAL INSIGHT**   Figure 3 shows that the point where the tangent line is horizontal is the *maximum point* on the graph. This proves an important result in circuit design: Maximum power is delivered when the resistance of the load (apparatus) is equal to the internal resistance of the battery.

3.3 SUMMARY

- Two basic rules of differentiation:

    Product Rule: $\qquad (fg)' = f'g + fg'$

    Quotient Rule: $\qquad \left(\dfrac{f}{g}\right)' = \dfrac{gf' - fg'}{g^2}$

- Remember: The derivative of fg is *not* equal to $f'g'$. Similarly, the derivative of f/g is *not* equal to f'/g'.

3.3 EXERCISES

Preliminary Questions

1. Are the following statements true or false? If false, state the correct version.

(a) fg denotes the function whose value at x is $f(g(x))$.

(b) f/g denotes the function whose value at x is $f(x)/g(x)$.

(c) The derivative of the product is the product of the derivatives.

(d) $\left.\dfrac{d}{dx}(fg)\right|_{x=4} = f(4)g'(4) - g(4)f'(4)$

(e) $\left.\dfrac{d}{dx}(fg)\right|_{x=0} = f'(0)g(0) + f(0)g'(0)$

2. Find $(f/g)'(1)$ if $f(1) = f'(1) = g(1) = 2$ and $g'(1) = 4$.

3. Find $g(1)$ if $f(1) = 0$, $f'(1) = 2$, and $(fg)'(1) = 10$.

Exercises

In Exercises 1–6, use the Product Rule to calculate the derivative.

1. $f(x) = x^3(2x^2 + 1)$

2. $f(x) = (3x - 5)(2x^2 - 3)$

3. $f(x) = x^2 e^x$

4. $f(x) = (2x - 9)(4e^x + 1)$

5. $\left.\dfrac{dh}{ds}\right|_{s=4}, \quad h(s) = (s^{-1/2} + 2s)(7 - s^{-1})$

6. $\left.\dfrac{dy}{dt}\right|_{t=2}, \quad y = (t - 8t^{-1})(e^t + t^2)$

In Exercises 7–12, use the Quotient Rule to calculate the derivative.

7. $f(x) = \dfrac{x}{x-2}$

8. $f(x) = \dfrac{x+4}{x^2+x+1}$

9. $\left.\dfrac{dg}{dt}\right|_{t=-2}, \quad g(t) = \dfrac{t^2+1}{t^2-1}$

10. $\left.\dfrac{dw}{dz}\right|_{z=9}, \quad w = \dfrac{z^2}{\sqrt{z}+z}$

11. $g(x) = \dfrac{1}{1+e^x}$

12. $f(x) = \dfrac{e^x}{x^2+1}$

In Exercises 13–16, calculate the derivative in two ways. First use the Product or Quotient Rule; then rewrite the function algebraically and apply the Power Rule directly.

13. $f(t) = (2t + 1)(t^2 - 2)$ **14.** $f(x) = x^2(3 + x^{-1})$

15. $h(t) = \dfrac{t^2 - 1}{t - 1}$

16. $g(x) = \dfrac{x^3 + 2x^2 + 3x^{-1}}{x}$

In Exercises 17–38, calculate the derivative.

17. $f(x) = (x^3 + 5)(x^3 + x + 1)$

18. $f(x) = (4e^x - x^2)(x^3 + 1)$

19. $\left.\dfrac{dy}{dx}\right|_{x=3}$, $y = \dfrac{1}{x + 10}$ **20.** $\left.\dfrac{dz}{dx}\right|_{x=-2}$, $z = \dfrac{x}{3x^2 + 1}$

21. $f(x) = (\sqrt{x} + 1)(\sqrt{x} - 1)$ **22.** $f(x) = \dfrac{9x^{5/2} - 2}{x}$

23. $\left.\dfrac{dy}{dx}\right|_{x=2}$, $y = \dfrac{x^4 - 4}{x^2 - 5}$ **24.** $f(x) = \dfrac{x^4 + e^x}{x + 1}$

25. $\left.\dfrac{dz}{dx}\right|_{x=1}$, $z = \dfrac{1}{x^3 + 1}$ **26.** $f(x) = \dfrac{3x^3 - x^2 + 2}{\sqrt{x}}$

27. $h(t) = \dfrac{t}{(t + 1)(t^2 + 1)}$

28. $f(x) = x^{3/2}(2x^4 - 3x + x^{-1/2})$

29. $f(t) = 3^{1/2} \cdot 5^{1/2}$ **30.** $h(x) = \pi^2(x - 1)$

31. $f(x) = (x + 3)(x - 1)(x - 5)$

32. $f(x) = e^x(x^2 + 1)(x + 4)$

33. $f(x) = \dfrac{e^x}{x + 1}$ **34.** $g(x) = \dfrac{e^{x+1} + e^x}{e + 1}$

35. $g(z) = \left(\dfrac{z^2 - 4}{z - 1}\right)\left(\dfrac{z^2 - 1}{z + 2}\right)$ *Hint: Simplify first.*

36. $\dfrac{d}{dx}\left((ax + b)(abx^2 + 1)\right)$ $(a, b$ constants$)$

37. $\dfrac{d}{dt}\left(\dfrac{xt - 4}{t^2 - x}\right)$ $(x$ constant$)$

38. $\dfrac{d}{dx}\left(\dfrac{ax + b}{cx + d}\right)$ $(a, b, c, d$ constants$)$

In Exercises 39–42, calculate the derivative using the values:

$f(4)$	$f'(4)$	$g(4)$	$g'(4)$
10	-2	5	-1

39. $(fg)'(4)$ and $(f/g)'(4)$

40. $F'(4)$, where $F(x) = x^2 f(x)$

41. $G'(4)$, where $G(x) = (g(x))^2$

42. $H'(4)$, where $H(x) = \dfrac{x}{g(x)f(x)}$

43. Calculate $F'(0)$, where

$$F(x) = \dfrac{x^9 + x^8 + 4x^5 - 7x}{x^4 - 3x^2 + 2x + 1}$$

Hint: Do not calculate $F'(x)$. Instead, write $F(x) = f(x)/g(x)$ and express $F'(0)$ directly in terms of $f(0), f'(0), g(0), g'(0)$.

44. Proceed as in Exercise 43 to calculate $F'(0)$, where

$$F(x) = \left(1 + x + x^{4/3} + x^{5/3}\right)\dfrac{3x^5 + 5x^4 + 5x + 1}{8x^9 - 7x^4 + 1}$$

45. Use the Product Rule to calculate $\dfrac{d}{dx}e^{2x}$.

46. $\boxed{\text{GU}}$ Plot the derivative of $f(x) = x/(x^2 + 1)$ over $[-4, 4]$. Use the graph to determine the intervals on which $f'(x) > 0$ and $f'(x) < 0$. Then plot f and describe how the sign of $f'(x)$ is reflected in the graph of f.

47. $\boxed{\text{GU}}$ Plot $f(x) = x/(x^2 - 1)$ (in a suitably bounded viewing box). Use the plot to determine whether $f'(x)$ is positive or negative on its domain $\{x : x \neq \pm 1\}$. Then compute $f'(x)$ and confirm your conclusion algebraically.

48. Let $P = V^2 R/(R + r)^2$ as in Example 7. Calculate dP/dr, assuming that r is variable and R is constant.

49. Find $a > 0$ such that the tangent line to the graph of

$$f(x) = x^2 e^{-x} \quad \text{at } x = a$$

passes through the origin (Figure 4).

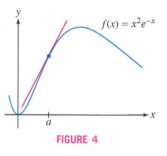

$f(x) = x^2 e^{-x}$

FIGURE 4

50. Current I (amperes), voltage V (volts), and resistance R (ohms) in a circuit are related by Ohm's Law, $I = V/R$.

(a) Calculate $\left.\dfrac{dI}{dR}\right|_{R=6}$ if V is constant with value $V = 24$.

(b) Calculate $\left.\dfrac{dV}{dR}\right|_{R=6}$ if I is constant with value $I = 4$.

51. The revenue per month earned by the Couture clothing chain at time t is $R(t) = N(t)S(t)$, where $N(t)$ is the number of stores and $S(t)$ is average revenue per store per month. Couture embarks on a two-part campaign: (A) to build new stores at a rate of 5 stores per month, and (B) to use advertising to increase average revenue per store at a rate of $10,000 per month. Assume that $N(0) = 50$ and $S(0) = \$150,000$.

(a) Show that total revenue will increase at the rate

$$\dfrac{dR}{dt} = 5S(t) + 10{,}000N(t)$$

Note that the two terms in the Product Rule correspond to the separate effects of increasing the number of stores on the one hand, and the average revenue per store on the other.

(b) Calculate $\left.\dfrac{dR}{dt}\right|_{t=0}$.

(c) If Couture can implement only one leg (A or B) of its expansion at $t = 0$, which choice will grow revenue most rapidly?

52. The **tip speed ratio** of a turbine (Figure 5) is the ratio $R = T/W$, where T is the speed of the tip of a blade and W is the speed of the wind. (Engineers have found empirically that a turbine with n blades extracts maximum power from the wind when $R = 2\pi/n$.) Calculate dR/dt (t in minutes) if $W = 35$ km/h and W decreases at a rate of 4 km/h per minute, and the tip speed has constant value $T = 150$ km/h.

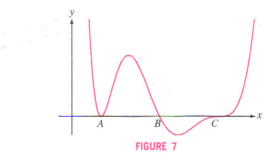

FIGURE 5 Turbines on a wind farm

(*Brian A. Jackson/iStockphoto.com*)

53. The curve $y = 1/(x^2 + 1)$ is called the *witch of Agnesi* (Figure 6) after the Italian mathematician Maria Agnesi (1718–1799), who wrote one of the first books on calculus. This strange name is the result of a mistranslation of the Italian word *la versiera*, meaning "that which turns." Find equations of the tangent lines at $x = \pm 1$.

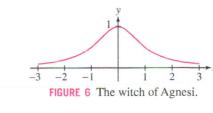

FIGURE 6 The witch of Agnesi.

54. Let $f(x) = g(x) = x$. Show that $(f/g)' \neq f'/g'$.

55. Use the Product Rule to show that $(f^2)' = 2ff'$.

56. Show that $(f^3)' = 3f^2 f'$.

Further Insights and Challenges

57. Let f, g, h be differentiable functions. Show that $(fgh)'(x)$ is equal to

$$f'(x)g(x)h(x) + f(x)g'(x)h(x) + f(x)g(x)h'(x)$$

Hint: Write fgh as $f(gh)$.

58. Prove the Quotient Rule using the limit definition of the derivative.

59. Derivative of the Reciprocal Use the limit definition to prove

$$\frac{d}{dx}\left(\frac{1}{f(x)}\right) = -\frac{f'(x)}{f^2(x)} \qquad \boxed{7}$$

Hint: Show that the difference quotient for $1/f(x)$ is equal to

$$\frac{f(x) - f(x+h)}{hf(x)f(x+h)}$$

60. Prove the Quotient Rule using Eq. (7) and the Product Rule.

61. Use the limit definition of the derivative to prove the following special case of the Product Rule:

$$\frac{d}{dx}(xf(x)) = f(x) + xf'(x)$$

62. Carry out Maria Agnesi's proof of the Quotient Rule from her book on calculus, published in 1748: Assume that f, g, and $h = f/g$ are differentiable. Compute the derivative of $hg = f$ using the Product Rule, and solve for h'.

63. The Power Rule Revisited If you are familiar with *proof by induction*, use induction to prove the Power Rule for all whole numbers n. Show that the Power Rule holds for $n = 1$; then write x^n as $x \cdot x^{n-1}$ and use the Product Rule.

Exercises 64 and 65: A basic fact of algebra states that c is a root of a polynomial f if and only if $f(x) = (x - c)g(x)$ for some polynomial g. We say that c is a **multiple root** *if $f(x) = (x - c)^2 h(x)$, where h is a polynomial.*

64. Show that c is a multiple root of f if and only if c is a root of both f and f'.

65. Use Exercise 64 to determine whether $c = -1$ is a multiple root:

(a) $x^5 + 2x^4 - 4x^3 - 8x^2 - x + 2$

(b) $x^4 + x^3 - 5x^2 - 3x + 2$

66. Figure 7 is the graph of a polynomial with roots at A, B, and C. Which of these is a multiple root? Explain your reasoning using Exercise 64.

FIGURE 7

67. According to Eq. (8) in Section 3.2, $\frac{d}{dx}b^x = m(b)b^x$. Use the Product Rule to show that $m(ab) = m(a) + m(b)$.

3.4 Rates of Change

Recall the notation for the average rate of change of a function $y = f(x)$ over an interval $[x_0, x_1]$:

$$\Delta y = \text{change in } y = f(x_1) - f(x_0)$$

$$\Delta x = \text{change in } x = x_1 - x_0$$

$$\boxed{\text{Average Rate of Change} = \frac{\Delta y}{\Delta x} = \frac{f(x_1) - f(x_0)}{x_1 - x_0}}$$

We usually omit the word "instantaneous" and refer to the derivative simply as the rate of change. This is shorter and also more accurate when applied to general rates, because the term "instantaneous" would seem to refer only to rates with respect to time.

In our prior discussion in Section 2.1, limits and derivatives had not yet been introduced. Now that we have them at our disposal, we can define the **instantaneous** rate of change of y with respect to x at $x = x_0$:

$$\boxed{\text{Instantaneous Rate of Change} = f'(x_0) = \lim_{\Delta x \to 0} \frac{\Delta y}{\Delta x} = \lim_{x_1 \to x_0} \frac{f(x_1) - f(x_0)}{x_1 - x_0}}$$

Keep in mind the geometric interpretations: The average rate of change is the slope of the secant line (Figure 1), and the instantaneous rate of change is the slope of the tangent line (Figure 2).

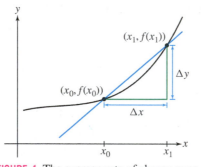

FIGURE 1 The average rate of change over $[x_0, x_1]$ is the slope of the secant line.

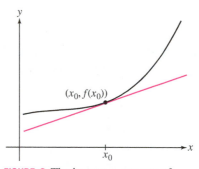

FIGURE 2 The instantaneous rate of change at x_0 is the slope of the tangent line.

Leibniz notation dy/dx is particularly convenient because it specifies that we are considering the rate of change of y with respect to the independent variable x. The rate dy/dx is measured in units of y per unit of x. For example, the rate of change of temperature with respect to time has units such as degrees per minute, whereas the rate of change of temperature with respect to altitude has units such as degrees per kilometer.

TABLE 1 Data from Mars Pathfinder Mission, July 1997

Time	Temperature (°C)
5:42	−74.7
6:11	−71.6
6:40	−67.2
7:09	−63.7
7:38	−59.5
8:07	−53
8:36	−47.7
9:05	−44.3
9:34	−42

■ **EXAMPLE 1** Table 1 contains data on the temperature T on the surface of Mars at Martian time t, collected by the NASA Pathfinder space probe.

(a) Calculate the average rate of change of temperature T from 6:11 AM to 9:05 AM.

(b) Use Figure 3 to estimate the rate of change at $t = 12$:28 PM.

Solution

(a) The time interval [6:11, 9:05] has length 2 h, 54 min, or $\Delta t = 2.9$ h. According to Table 1, the change in temperature over this time interval is

$$\Delta T = -44.3 - (-71.6) = 27.3°C$$

The average rate of change is the ratio

$$\frac{\Delta T}{\Delta t} = \frac{27.3}{2.9} \approx 9.4°C/h$$

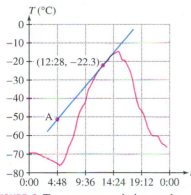

T (°C)

FIGURE 3 Temperature variation on the surface of Mars on July 6, 1997.

By Eq. (1), dA/dr is equal to the circumference $2\pi r$. We can explain this intuitively as follows: Up to a small error, the area ΔA of the band of width Δr in Figure 4 is equal to the circumference $2\pi r$ times the width Δr. Therefore, $\Delta A \approx 2\pi r \, \Delta r$ and

$$\frac{dA}{dr} = \lim_{\Delta r \to 0} \frac{\Delta A}{\Delta r} = 2\pi r$$

(b) The rate of change is the derivative dT/dt, which is equal to the slope of the tangent line through the point $(12{:}28, -22.3)$ in Figure 3. To estimate the slope, we must choose a second point on the tangent line. Let's use the point labeled A, whose coordinates are approximately $(4{:}48, -51)$. The time interval from 4:48 AM to 12:28 PM has length 7 h, 40 min, or $\Delta t \approx 7.67$ h, and

$$\frac{dT}{dt} = \text{slope of tangent line} \approx \frac{-22.3 - (-51)}{7.67} \approx 3.7°\text{C/h}$$ ■

■ **EXAMPLE 2** Let $A = \pi r^2$ be the area of a circle of radius r.

(a) Compute dA/dr at $r = 2$ and $r = 5$.

(b) Why is dA/dr larger at $r = 5$?

Solution The rate of change of area with respect to radius is the derivative

$$\frac{dA}{dr} = \frac{d}{dr}(\pi r^2) = 2\pi r \qquad \boxed{1}$$

(a) We have

$$\left.\frac{dA}{dr}\right|_{r=2} = 2\pi(2) \approx 12.57 \qquad \text{and} \qquad \left.\frac{dA}{dr}\right|_{r=5} = 2\pi(5) \approx 31.42$$

(b) The derivative dA/dr measures how the area of the circle changes when r increases. Figure 4 shows that when the radius increases by Δr, the area increases by a band of thickness Δr. The area of the band is greater at $r = 5$ than at $r = 2$. Therefore, the derivative is larger (and the tangent line is steeper) at $r = 5$. In general, for a fixed Δr, the change in area ΔA is greater when r is larger. ■

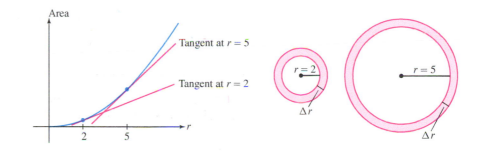

DF **FIGURE 4** The pink bands represent the change in area when r is increased by Δr.

The Effect of a 1-Unit Change

For small values of h, the difference quotient is close to the derivative itself:

$$f'(x_0) \approx \frac{f(x_0 + h) - f(x_0)}{h} \qquad \boxed{2}$$

This approximation generally improves as h gets smaller, but in some applications, the approximation is already useful with $h = 1$. Setting $h = 1$ in Eq. (2) gives

$$\boxed{f'(x_0) \approx f(x_0 + 1) - f(x_0)} \qquad \boxed{3}$$

In other words, $f'(x_0)$ is approximately equal to the change in f caused by a 1-unit change in x when $x = x_0$.

■ **EXAMPLE 3** **Stopping Distance** For speeds s between 30 and 75 mph, the stopping distance of an automobile after the brakes are applied is approximately $F(s) = 1.1s + 0.05s^2$ ft. For $s = 60$ mph:

(a) Estimate the change in stopping distance if the speed is increased by 1 mph.

(b) Compare your estimate with the actual increase in stopping distance.

Solution

(a) We have

$$F'(s) = \frac{d}{ds}(1.1s + 0.05s^2) = 1.1 + 0.1s \text{ ft/mph}$$

$$F'(60) = 1.1 + 6 = 7.1 \text{ ft/mph}$$

Using Eq. (3), we estimate

$$\underbrace{F(61) - F(60)}_{\text{Change in stopping distance}} \approx F'(60) = 7.1 \text{ ft}$$

Thus, when you increase your speed from 60 to 61 mph, your stopping distance increases by roughly 7 ft.

(b) The actual change in stopping distance is $F(61) - F(60) = 253.15 - 246 = 7.15$, so the estimate in (a) is fairly accurate. ■

Marginal Cost in Economics

Although $C(x)$ is meaningful only when x is a whole number, economists often treat $C(x)$ as a differentiable function of x so that the techniques of calculus can be applied.

Let $C(x)$ denote the dollar cost (including labor and parts) of producing x units of a particular product. The number x of units manufactured is called the **production level**. To study the relation between costs and production, economists define the **marginal cost** at production level x_0 as the cost of producing 1 additional unit:

$$\text{Marginal cost} = C(x_0 + 1) - C(x_0)$$

In this setting, Eq. (3) usually gives a good approximation, so we take $C'(x_0)$ as an estimate of the marginal cost.

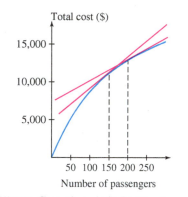

FIGURE 5 Cost of an air flight. The slopes of the tangent lines are decreasing, so marginal cost is decreasing.

■ **EXAMPLE 4 Cost of an Air Flight** Company data suggest that the total dollar cost of a certain flight is approximately $C(x) = 0.0005x^3 - 0.38x^2 + 120x$, where x is the number of passengers (Figure 5).

(a) Estimate the marginal cost of an additional passenger if the flight already has 150 passengers.

(b) Compare your estimate with the actual cost of an additional passenger.

(c) Is it more expensive to add a passenger when $x = 150$ or when $x = 200$?

Solution The derivative is $C'(x) = 0.0015x^2 - 0.76x + 120$.

(a) We estimate the marginal cost at $x = 150$ by the derivative

$$C'(150) = 0.0015(150)^2 - 0.76(150) + 120 = 39.75$$

Thus, it costs approximately \$39.75 to add one additional passenger.

(b) The actual cost of adding one additional passenger is

$$C(151) - C(150) \approx 11{,}177.10 - 11{,}137.50 = 39.60$$

Our estimate of \$39.75 is close enough for practical purposes.

(c) The marginal cost at $x = 200$ is approximately

$$C'(200) = 0.0015(200)^2 - 0.76(200) + 120 = 28$$

Since $39.75 > 28$, it is more expensive to add a passenger when $x = 150$. ■

In his famous textbook Lectures on Physics, *Nobel laureate Richard Feynman (1918–1988) uses a dialogue to make a point about instantaneous velocity:*

Policeman: "My friend, you were going 75 miles an hour."

Driver: "That's impossible, sir, I was traveling for only seven minutes."

Linear Motion

Recall that *linear motion* is motion along a straight line. This includes horizontal motion along a straight highway and vertical motion of a falling object. Let $s(t)$ denote the position or distance from the origin at time t. Velocity is the rate of change of position with respect to time:

$$v(t) = \text{velocity} = \frac{ds}{dt}$$

The *sign* of $v(t)$ indicates the direction of motion. For example, if $s(t)$ is the height above ground, then $v(t) > 0$ indicates that the object is rising. **Speed** is defined as the absolute value of velocity $|v(t)|$.

Figure 6 shows the position of a car as a function of time. Remember that the height of the graph represents the car's distance from the point of origin. The slope of the tangent line is the velocity. Here are some facts we can glean from the graph:

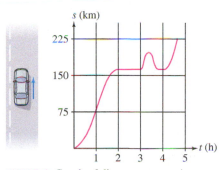

FIGURE 6 Graph of distance versus time.

- **Speeding up or slowing down?** The tangent lines get steeper in the interval $[0, 1]$, so *the car was speeding up during the first hour*. They get flatter in the interval $[1, 2]$, so the car slowed down in the second hour.
- **Standing still** The graph is horizontal over $[2, 3]$ (perhaps the driver stopped at a restaurant for an hour).
- **Returning to the same spot** The graph rises and falls in the interval $[3, 4]$, indicating that the driver returned to the restaurant (perhaps she left her cell phone there).
- **Average velocity** The graph rises more over $[0, 2]$ than over $[3, 5]$, so the average velocity was greater over the first 2 hours than over the last two hours.

■ **EXAMPLE 5** A truck enters the off-ramp of a highway at $t = 0$. Its position after t seconds is $s(t) = 25t - 0.3t^3$ m for $0 \le t \le 5$.

(a) How fast is the truck going at the moment it enters the off-ramp?

(b) Is the truck speeding up or slowing down?

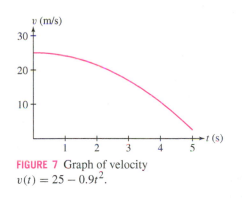

FIGURE 7 Graph of velocity $v(t) = 25 - 0.9t^2$.

Solution The truck's velocity at time t is $v(t) = \frac{d}{dt}(25t - 0.3t^3) = 25 - 0.9t^2$.

(a) The truck enters the off-ramp with velocity $v(0) = 25$ m/s.

(b) Since $v(t) = 25 - 0.9t^2$ is decreasing (Figure 7), the truck is slowing down. ■

Motion Under the Influence of Gravity

Galileo discovered that the height $s(t)$ and velocity $v(t)$ at time t (seconds) of an object tossed vertically in the air near the earth's surface are given by the formulas

Galileo's formulas are valid only when air resistance is negligible. We assume this to be the case in all examples.

$$s(t) = s_0 + v_0 t - \frac{1}{2}gt^2, \qquad v(t) = \frac{ds}{dt} = v_0 - gt \qquad \boxed{4}$$

The constants s_0 and v_0 are the *initial values*:

- $s_0 = s(0)$, the position at time $t = 0$.
- $v_0 = v(0)$, the velocity at $t = 0$.
- $-g$ is the acceleration due to gravity on the surface of the earth (negative because the up direction is positive), where

$$g \approx 9.8 \text{ m/s}^2 \qquad \text{or} \qquad g \approx 32 \text{ ft/s}^2$$

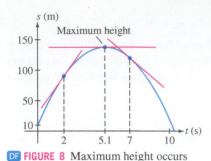

DF **FIGURE 8** Maximum height occurs when $s'(t) = v(t) = 0$, where the tangent line is horizontal.

Galileo's formulas:

$$s(t) = s_0 + v_0 t - \frac{1}{2} g t^2$$

$$v(t) = \frac{ds}{dt} = v_0 - gt$$

A simple observation enables us to find the object's maximum height. Since velocity is positive as the object rises and negative as it falls back to Earth, the object reaches its maximum height at the moment of transition, when it is no longer rising and has not yet begun to fall. At that moment, its velocity is zero. In other words, *the maximum height is attained when $v(t) = 0$*. At this moment, the tangent line to the graph of s is horizontal (Figure 8).

■ **EXAMPLE 6 Finding the Maximum Height** A stone is shot with a slingshot vertically upward with an initial velocity of 50 m/s from an initial height of 10 m.

(a) Find the velocity at $t = 2$ and at $t = 7$. Explain the change in sign.

(b) What is the stone's maximum height and when does it reach that height?

Solution Apply Eq. (4) with $s_0 = 10$, $v_0 = 50$, and $g = 9.8$:

$$s(t) = 10 + 50t - 4.9t^2, \qquad v(t) = 50 - 9.8t$$

(a) Therefore,

$$v(2) = 50 - 9.8(2) = 30.4 \text{ m/s}, \qquad v(7) = 50 - 9.8(7) = -18.6 \text{ m/s}$$

At $t = 2$, the stone is rising and its velocity $v(2)$ is positive (Figure 8). At $t = 7$, the stone is already on the way down and its velocity $v(7)$ is negative.

(b) Maximum height is attained when the velocity is zero, so we solve

$$v(t) = 50 - 9.8t = 0 \quad \Rightarrow \quad t = \frac{50}{9.8} \approx 5.1$$

The stone reaches maximum height at $t = 5.1$ s. Its maximum height is

$$s(5.1) = 10 + 50(5.1) - 4.9(5.1)^2 \approx 137.6 \text{ m} \qquad ■$$

In the previous example, we specified the initial values of position and velocity. In the next example, the goal is to determine initial velocity.

How important are units? In September 1999 the $125 million Mars Climate Orbiter spacecraft burned up in the Martian atmosphere before completing its scientific mission. According to Arthur Stephenson, NASA chairman of the Mars Climate Orbiter Mission Failure Investigation Board, 1999, "The 'root cause' of the loss of the spacecraft was the failed translation of English units into metric units in a segment of ground-based, navigation-related mission software."

■ **EXAMPLE 7 Finding Initial Conditions** What initial velocity v_0 is required for a bullet, fired vertically from ground level, to reach a maximum height of 2 km?

Solution We need a formula for maximum height as a function of initial velocity v_0. The initial height is $s_0 = 0$, so the bullet's height is $s(t) = v_0 t - \frac{1}{2} g t^2$ by Galileo's formula. Maximum height is attained when the velocity is zero:

$$v(t) = v_0 - gt = 0 \quad \Rightarrow \quad t = \frac{v_0}{g}$$

The maximum height is the value of $s(t)$ at $t = v_0/g$:

$$s\left(\frac{v_0}{g}\right) = v_0\left(\frac{v_0}{g}\right) - \frac{1}{2}g\left(\frac{v_0}{g}\right)^2 = \frac{v_0^2}{g} - \frac{1}{2}\frac{v_0^2}{g} = \frac{v_0^2}{2g}$$

In Eq. (5), distance must be in meters because our value of g has units of m/s².

Now we can solve for v_0 using the value $g = 9.8$ m/s² (note that 2 km = 2000 m).

$$\text{Maximum height} = \frac{v_0^2}{2g} = \frac{v_0^2}{2(9.8)} = 2000 \text{ m} \qquad \boxed{5}$$

This yields $v_0 = \sqrt{(2)(9.8)2000} \approx 198$ m/s. In reality, the initial velocity would have to be considerably greater to overcome air resistance. ■

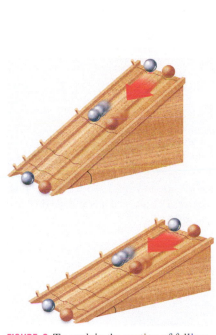

HISTORICAL PERSPECTIVE

Galileo Galilei (1564–1642) discovered the laws of motion for falling objects on the earth's surface around 1600. This paved the way for Newton's general laws of motion. How did Galileo arrive at his formulas? The motion of a falling object is too rapid to measure directly, without modern photographic or electronic apparatus. To get around this difficulty, Galileo experimented with balls rolling down an incline (Figure 9). For a sufficiently flat incline, he was able to measure the motion with a water clock and found that the velocity of the rolling ball is proportional to time. He then reasoned that motion in free-fall is just a faster version of motion down an incline and deduced the formula $v(t) = -gt$ for falling objects (assuming zero initial velocity).

Prior to Galileo, it had been assumed incorrectly that heavy objects fall more rapidly than lighter ones. Galileo realized that this was not true (as long as air resistance is negligible), and indeed, the formula $v(t) = -gt$ shows that the velocity depends on time but not on the weight of the object. Interestingly, 300 years later, another great physicist, Albert Einstein, was deeply puzzled by Galileo's discovery that all objects fall at the same rate regardless of their weight. He called this the Principle of Equivalence and sought to understand why it was true. In 1916, after a decade of intensive work, Einstein developed the General Theory of Relativity, which finally gave a full explanation of the Principle of Equivalence in terms of the geometry of space and time.

FIGURE 9 To explain the motion of falling objects, Galileo studied the motion of balls on an inclined plane. (*Dorling Kindersley/Getty Images*)

3.4 SUMMARY

- The (instantaneous) rate of change of $y = f(x)$ with respect to x at $x = x_0$ is defined as the derivative

$$f'(x_0) = \lim_{\Delta x \to 0} \frac{\Delta y}{\Delta x} = \lim_{x_1 \to x_0} \frac{f(x_1) - f(x_0)}{x_1 - x_0}$$

- The rate dy/dx is measured in *units of y per unit of x*.
- For linear motion, velocity $v(t)$ is the rate of change of position $s(t)$ with respect to time—that is, $v(t) = s'(t)$.
- In some applications, $f'(x_0)$ provides a good estimate of the change in f due to a 1-unit increase in x when $x = x_0$:

$$f'(x_0) \approx f(x_0 + 1) - f(x_0)$$

- Marginal cost is the cost of producing one additional unit. If $C(x)$ is the cost of producing x units, then the marginal cost at production level x_0 is $C(x_0 + 1) - C(x_0)$. The derivative $C'(x_0)$ is often a good estimate for marginal cost.
- Galileo's formulas for an object rising or falling under the influence of gravity near the Earth's surface ignoring air resistance (s_0 = initial position, v_0 = initial velocity):

$$s(t) = s_0 + v_0 t - \frac{1}{2}gt^2, \qquad v(t) = v_0 - gt$$

where $g \approx 9.8$ m/s^2, or $g \approx 32$ ft/s^2. Maximum height is attained when $v(t) = 0$.

3.4 EXERCISES

Preliminary Questions

1. Which units might be used for each rate of change?

(a) Pressure (in atmospheres) in a water tank with respect to depth
(b) The rate of a chemical reaction (change in concentration with respect to time with concentration in moles per liter)

2. Two trains travel from New Orleans to Memphis in 4 h. The first train travels at a constant velocity of 90 mph, but the velocity of the second train varies. What was the second train's average velocity during the trip?

3. Estimate $f(26)$, assuming that

$$f(25) = 43, \qquad f'(25) = 0.75$$

4. The population $P(t)$ of Freedonia in 2009 was $P(2009) = 5$ million.

(a) What is the meaning of $P'(2009)$?
(b) Estimate $P(2010)$ if $P'(2009) = 0.2$.

Exercises

In Exercises 1–8, find the rate of change.

1. Area of a square with respect to its side s when $s = 5$

2. Volume of a cube with respect to its side s when $s = 5$

3. Cube root $\sqrt[3]{x}$ with respect to x when $x = 1, 8, 27$

4. The reciprocal $1/x$ with respect to x when $x = 1, 2, 3$

5. The diameter of a circle with respect to radius

6. Surface area A of a sphere with respect to radius r $(A = 4\pi r^2)$

7. Volume V of a cylinder with respect to radius if the height is equal to the radius

8. Speed of sound v (in m/s) with respect to air temperature T (in kelvins), where $v = 20\sqrt{T}$

In Exercises 9–11, refer to Figure 10, the graph of distance s from the origin as a function of time for a car trip.

9. Find the average velocity over each interval.
(a) $[0, 0.5]$ (b) $[0.5, 1]$ (c) $[1, 1.5]$ (d) $[1, 2]$

10. At what time is velocity at a maximum?

11. Match the descriptions (i)–(iii) with the intervals (a)–(c).
(i) Velocity increasing
(ii) Velocity decreasing
(iii) Velocity negative
(a) $[0, 0.5]$ (b) $[2.5, 3]$ (c) $[1.5, 2]$

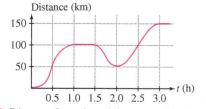

FIGURE 10 Distance from the origin versus time for a car trip.

12. Use the data from Table 1 in Example 1 to calculate the average rate of change of Martian temperature T with respect to time t over the interval from 8:36 AM to 9:34 AM.

13. Use Figure 3 from Example 1 to estimate the instantaneous rate of change of Martian temperature with respect to time (in degrees Celsius per hour) at $t = 4$ AM.

14. The temperature (in degrees Celsius) of an object at time t (in minutes) is $T(t) = \frac{3}{8}t^2 - 15t + 180$ for $0 \le t \le 20$. At what rate is the object cooling at $t = 10$? (Give correct units.)

15. The velocity (in centimeters per second) of blood molecules flowing through a capillary of radius 0.008 cm is $v = 6.4 \times 10^{-8} - 0.001r^2$, where r is the distance from the molecule to the center of the capillary. Find the rate of change of velocity with respect to r when $r = 0.004$ cm.

16. Figure 11 displays the voltage V across a capacitor as a function of time while the capacitor is being charged. Estimate the rate of change of voltage at $t = 20$ s. Indicate the values in your calculation and include proper units. Does voltage change more quickly or more slowly as time goes on? Explain in terms of tangent lines.

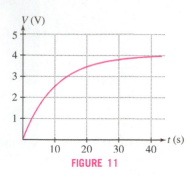

FIGURE 11

17. Use Figure 12 to estimate dT/dh at $h = 30$ and 70, where T is atmospheric temperature (in degrees Celsius) and h is altitude (in kilometers). Where is dT/dh equal to zero?

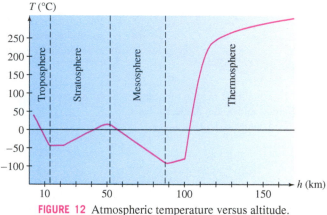

FIGURE 12 Atmospheric temperature versus altitude.

18. The earth exerts a gravitational force of $F(r) = (2.99 \times 10^{16})/r^2$ newtons on an object with a mass of 75 kg located r meters from the center of the earth. Find the rate of change of force with respect to distance r at the surface of the earth.

19. Calculate the rate of change of escape velocity $v_{esc} = (2.82 \times 10^7)r^{-1/2}$ m/s with respect to distance r from the center of the earth.

20. The power delivered by a battery to an apparatus of resistance R (in ohms) is $P = 2.25R/(R + 0.5)^2$ watts (W). Find the rate of change of power with respect to resistance for $R = 3\ \Omega$ and $R = 5\ \Omega$.

21. The position of a particle moving in a straight line during a 5-s trip is $s(t) = t^2 - t + 10$ cm. Find a time t at which the instantaneous velocity is equal to the average velocity for the entire trip beginning at $t = 0$.

22. The height (in meters) of a helicopter at time t (in minutes) is $s(t) = 600t - 3t^3$ for $0 \le t \le 12$.
(a) Plot s and velocity v as functions of time.
(b) Find the velocity at $t = 8$ and $t = 10$.
(c) Find the maximum height of the helicopter.

23. A particle moving along a line has position $s(t) = t^4 - 18t^2$ m at time t seconds. At which times does the particle pass through the origin? At which times is the particle instantaneously motionless (i.e., it has zero velocity)?

24. **GU** Plot the position of the particle in Exercise 23. What is the farthest distance to the left of the origin attained by the particle?

25. A bullet is fired in the air vertically from ground level with an initial velocity 200 m/s. Find the bullet's maximum velocity and maximum height.

26. Find the velocity of an air conditioner accidentally dropped from a height of 300 m at the moment it hits the ground.

27. A ball tossed in the air vertically from ground level returns to earth 4 s later. Find the initial velocity and maximum height of the ball.

28. Olivia is gazing out a window from the tenth floor of a building when a bucket (dropped by a window washer) passes by. She notes that it hits the ground 1.5 s later. Determine the floor from which the bucket was dropped if each floor is 5 m high and the window is in the middle of the tenth floor. Neglect air friction.

29. Show that for an object falling according to Galileo's formula, the average velocity over any time interval $[t_1, t_2]$ is equal to the average of the instantaneous velocities at t_1 and t_2.

30. An object falls under the influence of gravity near the earth's surface. Which of the following statements is true? Explain.

(a) Distance traveled increases by equal amounts in equal time intervals.

(b) Velocity increases by equal amounts in equal time intervals.

(c) The derivative of velocity increases with time.

31. By Faraday's Law, if a conducting wire of length ℓ meters moves at velocity v m/s perpendicular to a magnetic field of strength B (in teslas), a voltage of size $V = -B\ell v$ is induced in the wire. Assume that $B = 2$ and $\ell = 0.5$.

(a) Calculate dV/dv.

(b) Find the rate of change of V with respect to time t if $v(t) = 4t + 9$.

32. The voltage V, current I, and resistance R in a circuit are related by Ohm's Law: $V = IR$, where the units are volts, amperes, and ohms. Assume that voltage is constant with $V = 12$ volts (V). Calculate (specifying units):

(a) The average rate of change of I with respect to R for the interval from $R = 8$ to $R = 8.1$

(b) The rate of change of I with respect to R when $R = 8$

(c) The rate of change of R with respect to I when $I = 1.5$

33. Ethan finds that with h hours of tutoring, he is able to answer correctly $S(h)$ percent of the problems on a math exam. Which would you expect to be larger: $S'(3)$ or $S'(30)$? Explain.

34. Suppose $\theta(t)$ measures the angle between a clock's minute and hour hands. What is $\theta'(t)$ at 3 o'clock?

35. To determine drug dosages, doctors estimate a person's body surface area (BSA) (in meters squared) using the formula BSA $= \sqrt{hm}/60$, where h is the height in centimeters and m the mass in kilograms. Calculate the rate of change of BSA with respect to mass for a person of constant height $h = 180$. What is this rate at $m = 70$ and $m = 80$? Express your result in the correct units. Does BSA increase more rapidly with respect to mass at lower or higher body mass?

36. The atmospheric CO_2 levels at Mauna Loa, Hawaii, at time t (in parts per million by volume) is recorded by the Scripps Institute of Oceanography. Reading across, the average annual value $A(t)$ for the given years are:

1960	1964	1968	1972	1976	1980	1984
316.91	319.20	323.05	327.45	332.15	338.69	344.42

1988	1992	1996	2000	2004	2008	2012
351.48	356.47	362.64	369.48	377.38	385.34	393.87

(a) Estimate $A'(t)$ in 1962, 1970, 1978, 1986, 1994, 2002, and 2010.

(b) In which of the years from (a) did the approximation to $A'(t)$ take on its largest and smallest values?

(c) In which of these years does the approximation suggest that the CO_2 level was the most constant?

37. The tangent lines to the graph of $f(x) = x^2$ grow steeper as x increases. At what rate do the slopes of the tangent lines increase?

38. Figure 13 shows the height y of a mass oscillating at the end of a spring, through one cycle of the oscillation. Sketch the graph of velocity as a function of time.

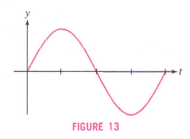

FIGURE 13

In Exercises 39–46, use Eq. (3) to estimate the unit change.

39. Estimate $\sqrt{2} - \sqrt{1}$ and $\sqrt{101} - \sqrt{100}$. Compare your estimates with the actual values.

40. Estimate $f(4) - f(3)$ if $f'(x) = 2^{-x}$. Then estimate $f(4)$, assuming that $f(3) = 12$.

41. Let $F(s) = 1.1s + 0.05s^2$ be the stopping distance as in Example 3. Calculate $F(65)$ and estimate the increase in stopping distance if speed is increased from 65 to 66 mph. Compare your estimate with the actual increase.

42. According to Kleiber's Law, the metabolic rate P (in kilocalories per day) and body mass m (in kilograms) of an animal are related by a *three-quarter-power law* $P = 73.3m^{3/4}$. Estimate the increase in metabolic rate when body mass increases from 60 to 61 kg.

43. The dollar cost of producing x bagels is given by the function $C(x) = 300 + 0.25x - 0.5(x/1000)^3$. Determine the cost of producing 2000 bagels and estimate the cost of the 2001st bagel. Compare your estimate with the actual cost of the 2001st bagel.

44. Suppose the dollar cost of producing x video cameras is $C(x) = 500x - 0.003x^2 + 10^{-8}x^3$.

(a) Estimate the marginal cost at production level $x = 5000$ and compare it with the actual cost $C(5001) - C(5000)$.

(b) Compare the marginal cost at $x = 5000$ with the average cost per camera, defined as $C(x)/x$.

45. Demand for a commodity generally decreases as the price is raised. Suppose that the demand for oil (per capita per year) is $D(p) = 900/p$ barrels, where p is the dollar price per barrel. Find the demand when $p = \$40$. Estimate the decrease in demand if p rises to $\$41$ and the increase if p declines to $\$39$.

46. The reproduction rate f of the fruit fly *Drosophila melanogaster*, grown in bottles in a laboratory, decreases with the number p of flies in the bottle. A researcher has found the number of offspring per female per day to be approximately $f(p) = (34 - 0.612p)p^{-0.658}$.

(a) Calculate $f(15)$ and $f'(15)$.

(b) Estimate the decrease in daily offspring per female when p is increased from 15 to 16. Is this estimate larger or smaller than the actual value $f(16) - f(15)$?

(c) GU Plot f for $5 \leq p \leq 25$ and verify that $f(p)$ is a decreasing function of p. Do you expect $f'(p)$ to be positive or negative? Plot f' and confirm your expectation.

47. According to Stevens' Law in psychology, the perceived magnitude of a stimulus is proportional (approximately) to a power of the actual intensity I of the stimulus. Experiments show that the *perceived brightness B* of a light satisfies $B = kI^{2/3}$, where I is the light intensity, whereas the *perceived heaviness H* of a weight W satisfies $H = kW^{3/2}$ (k is a constant that is different in the two cases). Compute dB/dI and dH/dW and state whether they are increasing or decreasing functions. Then explain the following statements:

(a) A 1-unit increase in light intensity is felt more strongly when I is small than when I is large.

(b) Adding another pound to a load W is felt more strongly when W is large than when W is small.

48. Let $M(t)$ be the mass (in kilograms) of a plant as a function of time (in years). Recent studies by Niklas and Enquist have suggested that a remarkably wide range of plants (from algae and grass to palm trees) obey a *three-quarter-power growth law*—that is,

$$\frac{dM}{dt} = CM^{3/4} \quad \text{for some constant } C$$

(a) If a tree has a growth rate of 6 kg/year when $M = 100$ kg, what is its growth rate when $M = 125$ kg?

(b) If $M = 0.5$ kg, how much more mass must the plant acquire to double its growth rate?

Further Insights and Challenges

*Exercises 49–51: The **Lorenz curve** $y = F(r)$ is used by economists to study income distribution in a given country (see Figure 14). By definition, $F(r)$ is the fraction of the total income that goes to the bottom rth part of the population, where $0 \le r \le 1$. For example, if $F(0.4) = 0.245$, then the bottom 40% of households receive 24.5% of the total income. Note that $F(0) = 0$ and $F(1) = 1$.*

49. Our goal is to find an interpretation for $F'(r)$. The average income for a group of households is the total income going to the group divided by the number of households in the group. The national average income is $A = T/N$, where N is the total number of households and T is the total income earned by the entire population.

(a) Show that the average income among households in the bottom rth part is equal to $(F(r)/r)A$.

(b) Show more generally that the average income of households belonging to an interval $[r, r + \Delta r]$ is equal to

$$\left(\frac{F(r + \Delta r) - F(r)}{\Delta r} \right) A$$

(c) Let $0 \le r \le 1$. A household belongs to the $100r$th percentile if its income is greater than or equal to the income of $100r$ % of all households. Pass to the limit as $\Delta r \to 0$ in (b) to derive the following interpretation: A household in the $100r$th percentile has income $F'(r)A$. In particular, a household in the $100r$th percentile receives more than the national average if $F'(r) > 1$ and less if $F'(r) < 1$.

(d) For the Lorenz curves L_1 and L_2 in Figure 14(B), what percentage of households have above-average income?

50. The following table provides values of $F(r)$ for the United States in 2010. Assume that the national average income was $A = \$66,000$.

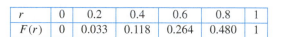

r	0	0.2	0.4	0.6	0.8	1
$F(r)$	0	0.033	0.118	0.264	0.480	1

(a) What was the average income in the lowest 40% of households?

(b) Show that the average income of the households belonging to the interval [0.4, 0.6] was $48,180.

(c) Estimate $F'(0.5)$. Estimate the income of households in the 50th percentile? Was it greater or less than the national average?

51. Use Exercise 49(c) to prove:

(a) $F'(r)$ is an increasing function of r.

(b) Income is distributed equally (all households have the same income) if and only if $F(r) = r$ for $0 \le r \le 1$.

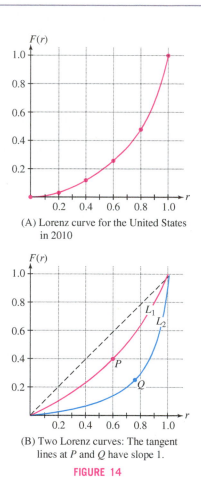

(A) Lorenz curve for the United States in 2010

(B) Two Lorenz curves: The tangent lines at P and Q have slope 1.

FIGURE 14

52. CAS Studies of Internet usage show that Web site popularity is described quite well by Zipf's Law, according to which the nth most popular Web site receives roughly the fraction $1/n$ of all visits. Suppose that on a particular day, the nth most popular site had approximately $V(n) = 10^6/n$ visitors (for $n \le 15,000$).

(a) Verify that the top 50 Web sites received nearly 45% of the visits. *Hint:* Let $T(N)$ denote the sum of $V(n)$ for $1 \le n \le N$. Use a computer algebra system to compute $T(50)$ and $T(15,000)$.

(b) Verify, by numerical experimentation, that when Eq. (3) is used to estimate $V(n + 1) - V(n)$, the error in the estimate decreases as n grows larger. Find (again, by experimentation) an N such that the error is at most 10 for $n \ge N$.

(c) Using Eq. (3), show that for $n \ge 100$, the nth Web site received at most 100 more visitors than the $(n + 1)$st Web site.

In Exercises 53 and 54, the average cost per unit at production level x is defined as $C_{avg}(x) = C(x)/x$, where $C(x)$ is the cost of producing x units. Average cost is a measure of the efficiency of the production process.

53. Show that $C_{avg}(x)$ is equal to the slope of the line through the origin and the point $(x, C(x))$ on the graph of $y = C(x)$. Using this interpretation, determine whether average cost or marginal cost is greater at points A, B, C, D in Figure 15.

54. The cost in dollars of producing alarm clocks is given by $C(x) = 50x^3 - 750x^2 + 3740x + 3750$, where x is in units of 1000.

(a) Calculate the average cost at $x = 4, 6, 8$, and 10.

(b) Use the graphical interpretation of average cost to find the production level x_0 at which average cost is lowest. What is the relation between average cost and marginal cost at x_0 (see Figure 16)?

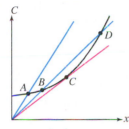

FIGURE 15 Graph of $y = C(x)$.

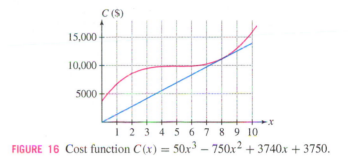

FIGURE 16 Cost function $C(x) = 50x^3 - 750x^2 + 3740x + 3750$.

3.5 Higher Derivatives

Higher derivatives are obtained by repeatedly differentiating a function $y = f(x)$. If f' is differentiable, then the **second derivative**, denoted f'' or y'', is the derivative

$$f''(x) = \frac{d}{dx}\left(f'(x)\right)$$

The second derivative is the rate of change of $f'(x)$. The next example highlights the difference between the first and second derivatives.

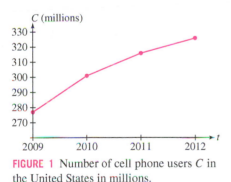

FIGURE 1 Number of cell phone users C in the United States in millions.

■ **EXAMPLE 1** Figure 1 and Table 1 show the number of cell phone subscribers $C(t)$ in the United States in year t. Discuss $C'(t)$ and $C''(t)$.

TABLE 1 Number of Cell Phone Subscribers in the United States

Year	2009	2010	2011	2012
Number in millions	277	301	316	326
Yearly increase		24	15	10

Solution We will show that $C'(t)$ is positive but $C''(t)$ is negative. According to Table 1, the number of cell phone users each year was greater than the previous year, so the rate of change $C'(t)$ is certainly positive. However, the amount of increase declined from 24 million in 2010 to 10 million in 2012. So although $C'(t)$ is positive, $C'(t)$ decreases from one year to the next, and therefore its rate of change $C''(t)$ is negative. Figure 1 supports this conclusion: The slopes of the segments in the graph are decreasing. ■

• dy/dx has units of y per unit of x.
• d^2y/dx^2 has units of dy/dx per unit of x or units of y per unit of x squared.

The process of differentiation can be continued, provided that the derivatives exist. The third derivative, denoted $f'''(x)$ or $f^{(3)}(x)$, is the derivative of $f''(x)$. More generally, the nth derivative $f^{(n)}(x)$ is the derivative of the $(n-1)$st derivative. We call $f(x)$ the zeroeth derivative and $f'(x)$ the first derivative. In Leibniz notation, we write

$$\frac{df}{dx}, \quad \frac{d^2f}{dx^2}, \quad \frac{d^3f}{dx^3}, \quad \frac{d^4f}{dx^4}, \cdots$$

■ **EXAMPLE 2** Calculate $f'''(-1)$ for $f(x) = 3x^5 - 2x^2 + 7x^{-2}$.

Solution We must calculate the first three derivatives:

$$f'(x) = \frac{d}{dx}\left(3x^5 - 2x^2 + 7x^{-2}\right) = 15x^4 - 4x - 14x^{-3}$$

$$f''(x) = \frac{d}{dx}\left(15x^4 - 4x - 14x^{-3}\right) = 60x^3 - 4 + 42x^{-4}$$

$$f'''(x) = \frac{d}{dx}\left(60x^3 - 4 + 42x^{-4}\right) = 180x^2 - 168x^{-5}$$

At $x = -1$, $f'''(-1) = 180 + 168 = 348$. ■

Polynomials have a special property: Their higher derivatives are eventually the zero function. More precisely, if f is a polynomial of degree k, then $f^{(n)}(x)$ is zero for $n > k$. Table 2 illustrates this property for $f(x) = x^5$. By contrast, the higher derivatives of a nonpolynomial function are never the zero function (see Exercise 87, Section 5.3).

TABLE 2   Derivatives of x^5

$f(x)$	$f'(x)$	$f''(x)$	$f'''(x)$	$f^{(4)}(x)$	$f^{(5)}(x)$	$f^{(6)}(x)$
x^5	$5x^4$	$20x^3$	$60x^2$	$120x$	120	0

■ **EXAMPLE 3** Calculate the first four derivatives of $y = x^{-1}$. Then find the pattern and determine a general formula for $y^{(n)}$.

Solution By the Power Rule,

$$y'(x) = -x^{-2}, \quad y'' = 2x^{-3}, \quad y''' = -2(3)x^{-4}, \quad y^{(4)} = 2(3)(4)x^{-5}$$

We see that $y^{(n)}(x)$ is equal to $\pm n!\, x^{-n-1}$. Now observe that the sign alternates. Since the odd-order derivatives occur with a minus sign, the sign of $y^{(n)}(x)$ is $(-1)^n$. In general, therefore, $y^{(n)}(x) = (-1)^n n!\, x^{-n-1}$. ■

◄··· REMINDER *n-factorial is the number*

$$n! = n(n-1)(n-2)\cdots(2)(1)$$

Thus,

$$1! = 1, \quad 2! = (2)(1) = 2$$

$$3! = (3)(2)(1) = 6$$

By convention, we set $0! = 1$.

It is not always possible to find a simple formula for the higher derivatives of a function. In most cases, they become increasingly complicated.

■ **EXAMPLE 4** Calculate the first three derivatives of $f(x) = xe^x$. Then determine a general formula for $f^{(n)}(x)$.

Solution Use the Product Rule:

$$f'(x) = \frac{d}{dx}(xe^x) = e^x + xe^x = (1+x)e^x$$

$$f''(x) = \frac{d}{dx}\left((1+x)e^x\right) = e^x + (1+x)e^x = (2+x)e^x$$

$$f'''(x) = \frac{d}{dx}\left((2+x)e^x\right) = e^x + (2+x)e^x = (3+x)e^x$$

We see that $f^n(x) = e^x + f^{n-1}(x)$, which leads to the general formula

$$f^{(n)}(x) = (n+x)e^x$$

■

One familiar second derivative is acceleration. An object in linear motion with position $s(t)$ at time t has velocity $v(t) = s'(t)$ and acceleration $a(t) = v'(t) = s''(t)$. Thus, acceleration is the rate at which velocity changes and is measured in units of velocity per unit of time or "distance per time squared" such as m/s^2.

■ **EXAMPLE 5** **Acceleration Due to Gravity** Find the acceleration $a(t)$ of a ball tossed vertically in the air from ground level with an initial velocity of 12 m/s. How does $a(t)$ describe the change in the ball's velocity as it rises and falls?

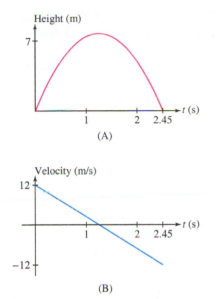

Height (m)

(A)

Velocity (m/s)

(B)

FIGURE 2 Height and velocity of a ball tossed vertically with initial velocity 12 m/s.

Solution The ball's height at time t is $s(t) = s_0 + v_0 t - 4.9t^2$ m by Galileo's formula [Figure 2(A)]. In our case, $s_0 = 0$ and $v_0 = 12$, so $s(t) = 12t - 4.9t^2$ m. Therefore, $v(t) = s'(t) = 12 - 9.8t$ m/s and the ball's acceleration is

$$a(t) = s''(t) = \frac{d}{dt}(12 - 9.8t) = -9.8 \text{ m/s}^2$$

As expected, the acceleration is constant with value $-g = -9.8$ m/s^2. As the ball rises and falls, its velocity decreases from 12 to -12 m/s at the constant rate $-g$ [Figure 2(B)]. ∎

GRAPHICAL INSIGHT   Can we visualize the rate represented by $f''(x)$? The second derivative is the rate at which $f'(x)$ is changing, so $f''(x)$ is large if the slopes of the tangent lines change rapidly, as in Figure 3(A). Similarly, $f''(x)$ is small if the slopes of the tangent lines change slowly—in this case, the curve is relatively flat, as in Figure 3(B). If f is a linear function [Figure 3(C)], then the tangent line does not change at all and $f''(x) = 0$. Thus, $f''(x)$ measures the "bending" or concavity of the graph.

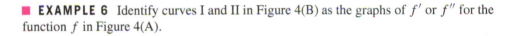

(A) Large second derivative: Tangent lines turn rapidly.

(B) Smaller second derivative: Tangent lines turn slowly.

(C) Second derivative is zero: Tangent line does not change.

DF FIGURE 3

■ **EXAMPLE 6**   Identify curves I and II in Figure 4(B) as the graphs of f' or f'' for the function f in Figure 4(A).

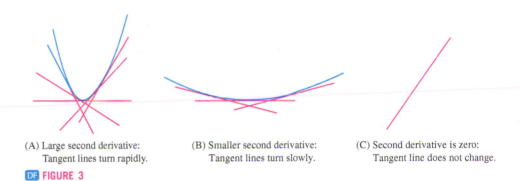

Slopes of tangent lines increasing

(A) Graph of f

(B) Graph of first two derivatives

DF FIGURE 4

Solution The slopes of the tangent lines to the graph of f are *increasing* on the interval $[a, b]$. Therefore, f' is an increasing function and its graph must be II. Since $f''(x)$ is the rate of change of $f'(x)$, $f''(x)$ is positive and its graph must be I. ∎

3.5 SUMMARY

• The higher derivatives f', f'', f''', ... are defined by successive differentiation:

$$f''(x) = \frac{d}{dx}f'(x) = \frac{d^2 f}{dx^2}, \qquad f'''(x) = \frac{d}{dx}f''(x) = \frac{d^3 f}{dx^3}, \dots$$

The nth derivative is denoted $f^{(n)}(x) = \frac{d^n f}{dx^n}$.

- The second derivative plays an important role: It is the rate at which $f'(x)$ changes. Graphically, $f''(x)$ measures how fast the tangent lines change direction and thus measures the "bending" of the graph.
- If $s(t)$ is the position of an object at time t, then $s'(t)$ is velocity and $s''(t)$ is acceleration.

3.5 EXERCISES

Preliminary Questions

1. On September 4, 2003, the *Wall Street Journal* printed the headline "Stocks Go Higher, Though the Pace of Their Gains Slows." Rephrase this headline as a statement about the first and second derivatives of stock prices and sketch a possible graph.

2. True or false? The third derivative of position with respect to time

is zero for an object falling to Earth under the influence of gravity. Explain.

3. Which type of polynomial satisfies $f'''(x) = 0$ for all x?

4. What is the millionth derivative of $f(x) = e^x$?

Exercises

In Exercises 1–16, calculate y'' and y'''.

1. $y = 14x^2$

2. $y = 7 - 2x$

3. $y = x^4 - 25x^2 + 2x$

4. $y = 4t^3 - 9t^2 + 7$

5. $y = \frac{4}{3}\pi r^3$

6. $y = \sqrt{x}$

7. $y = 20t^{4/5} - 6t^{2/3}$

8. $y = x^{-9/5}$

9. $y = z - \frac{4}{z}$

10. $y = 5t^{-3} + 7t^{-8/3}$

11. $y = \theta^2(2\theta + 7)$

12. $y = (x^2 + x)(x^3 + 1)$

13. $y = \frac{x - 4}{x}$

14. $y = \frac{1}{1 - x}$

15. $y = x^5 e^x$

16. $y = \frac{e^x}{x}$

In Exercises 17–26, calculate the derivative indicated.

17. $f^{(4)}(1)$, $f(x) = x^4$

18. $g'''(-1)$, $g(t) = -4t^{-5}$

19. $\left.\dfrac{d^2 y}{dt^2}\right|_{t=1}$, $y = 4t^{-3} + 3t^2$

20. $\left.\dfrac{d^4 f}{dt^4}\right|_{t=1}$, $f(t) = 6t^9 - 2t^5$

21. $\left.\dfrac{d^4 x}{dt^4}\right|_{t=16}$, $x = t^{-3/4}$

22. $f'''(4)$, $f(t) = 2t^2 - t$

23. $f'''(-3)$, $f(x) = 4e^x - x^3$

24. $f''(1)$, $f(t) = \dfrac{t}{t + 1}$

25. $h''(1)$, $h(w) = \sqrt{w}e^w$

26. $g''(0)$, $g(s) = \dfrac{e^s}{s + 1}$

27. Calculate $y^{(k)}(0)$ for $0 \le k \le 5$, where $y = x^4 + ax^3 + bx^2 + cx + d$ (with a, b, c, d the constants).

28. Which of the following satisfy $f^{(k)}(x) = 0$ for all $k \ge 6$?

(a) $f(x) = 7x^4 + 4 + x^{-1}$

(b) $f(x) = x^3 - 2$

(c) $f(x) = \sqrt{x}$

(d) $f(x) = 1 - x^6$

(e) $f(x) = x^{9/5}$

(f) $f(x) = 2x^2 + 3x^5$

29. Use the result in Example 3 to find $\dfrac{d^6}{dx^6} x^{-1}$.

30. Calculate the first five derivatives of $f(x) = \sqrt{x}$.

(a) Show that $f^{(n)}(x)$ is a multiple of $x^{-n+1/2}$.

(b) Show that $f^{(n)}(x)$ alternates in sign as $(-1)^{n-1}$ for $n \ge 1$.

(c) Find a formula for $f^{(n)}(x)$ for $n \ge 2$. *Hint:* Verify that the coefficient is $\pm 1 \cdot 3 \cdot 5 \cdots \dfrac{2n - 3}{2^n}$.

In Exercises 31–36, find a general formula for $f^{(n)}(x)$.

31. $f(x) = x^{-2}$

32. $f(x) = (x + 2)^{-1}$

33. $f(x) = x^{-1/2}$

34. $f(x) = x^{-3/2}$

35. $f(x) = xe^{-x}$

36. $f(x) = x^2 e^x$

37. (a) Find the acceleration at time $t = 5$ min of a helicopter whose height is $s(t) = 300t - 4t^3$ m.

(b) Plot the acceleration s'' for $0 \le t \le 6$. How does this graph show that the helicopter is slowing down during this time interval?

38. Find an equation of the tangent to the graph of $y = f'(x)$ at $x = 3$, where $f(x) = x^4$.

39. Figure 5 shows f, f', and f''. Determine which is which.

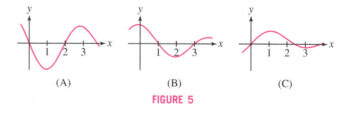

(A)   (B)   (C)

FIGURE 5

40. The second derivative f'' is shown in Figure 6. Which of (A) or (B) is the graph of f and which is f'?

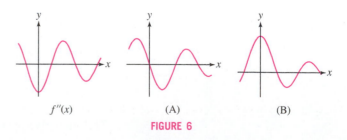

$f''(x)$ (A)   (B)

FIGURE 6

41. Figure 7 shows the graph of the position s of an object as a function of time t. Determine the intervals on which the acceleration is positive.

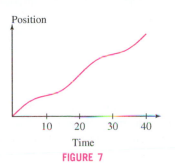

FIGURE 7

42. Find a polynomial $f(x)$ that satisfies the equation $xf''(x) + f(x) = x^2$.

43. Find all values of n such that $y = x^n$ satisfies

$$x^2y'' - 2xy' = 4y$$

44. Which of the following descriptions could *not* apply to Figure 8? Explain.

(a) Graph of acceleration when velocity is constant

(b) Graph of velocity when acceleration is constant

(c) Graph of position when acceleration is zero

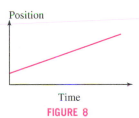

FIGURE 8

45. According to one model that takes into account air resistance, the acceleration $a(t)$ (in m/s^2) of a skydiver of mass m in free-fall satisfies

$$a(t) = -9.8 + \frac{k}{m}v(t)^2$$

where $v(t)$ is velocity (negative since the skydiver is falling) and k is a constant. Suppose that $m = 75$ kg and $k = 14$ kg/m.

(a) What is the skydiver's velocity when $a(t) = -4.9$?

(b) What is the skydiver's velocity when $a(t) = 0$? This velocity is the skydiver's terminal velocity.

46. According to one model that attempts to account for air resistance, the distance $s(t)$ (in meters) traveled by a falling raindrop satisfies

$$\frac{d^2s}{dt^2} = g - \frac{0.0005}{D}\left(\frac{ds}{dt}\right)^2$$

where D is the raindrop diameter and $g = 9.8$ m/s^2. Terminal velocity v_{term} is defined as the velocity at which the drop has zero acceleration (one can show that velocity approaches v_{term} as time proceeds).

(a) Show that $v_{\text{term}} = \sqrt{2000gD}$.

(b) Find v_{term} for drops of diameter 10^{-3} m and 10^{-4} m.

(c) In this model, do raindrops accelerate more rapidly at higher or lower velocities?

47. A servomotor controls the vertical movement of a drill bit that will drill a pattern of holes in sheet metal. The maximum vertical speed of the drill bit is 4 in./s, and while drilling the hole, it must move no more than 2.6 in./s to avoid warping the metal. During a cycle, the bit begins and ends at rest, quickly approaches the sheet metal, and quickly returns to its initial position after the hole is drilled. Sketch possible graphs of the drill bit's vertical velocity and acceleration. Label the point where the bit enters the sheet metal.

In Exercises 48 and 49, refer to the following. In a 1997 study, Boardman and Lave related the traffic speed S on a two-lane road to traffic density Q (number of cars per mile of road) by the formula

$$S = 2882Q^{-1} - 0.052Q + 31.73$$

for $60 \le Q \le 400$ (Figure 9).

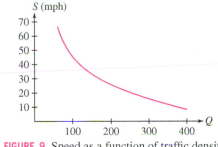

FIGURE 9 Speed as a function of traffic density.

48. Calculate dS/dQ and d^2S/dQ^2.

49. (a) Explain intuitively why we should expect that $dS/dQ < 0$.

(b) Show that $d^2S/dQ^2 > 0$. Then use the fact that $dS/dQ < 0$ and $d^2S/dQ^2 > 0$ to justify the following statement: *A 1-unit increase in traffic density slows down traffic more when Q is small than when Q is large.*

(c) **GU** Plot dS/dQ. Which property of this graph shows that $d^2S/dQ^2 > 0$?

50. *CAS* Use a computer algebra system to compute $f^{(k)}(x)$ for $k = 1, 2, 3$ for the following functions:

(a) $f(x) = (1 + x^3)^{5/3}$

(b) $f(x) = \dfrac{1 - x^4}{1 - 5x - 6x^2}$

51. *CAS* Let $f(x) = \dfrac{x+2}{x-1}$. Use a computer algebra system to compute the $f^{(k)}(x)$ for $1 \le k \le 4$. Can you find a general formula for $f^{(k)}(x)$?

Further Insights and Challenges

52. Find the 100th derivative of

$$p(x) = (x + x^5 + x^7)^{10}(1 + x^2)^{11}(x^3 + x^5 + x^7)$$

53. What is $p^{(99)}(x)$ for $p(x)$ as in Exercise 52?

54. Use the Product Rule twice to find a formula for $(fg)''$ in terms of f and g and their first and second derivatives.

55. Use the Product Rule to find a formula for $(fg)'''$ and compare your result with the expansion of $(a + b)^3$. Then try to guess the general formula for $(fg)^{(n)}$.

56. Compute

$$\Delta f(x) = \lim_{h \to 0} \frac{f(x+h) + f(x-h) - 2f(x)}{h^2}$$

for the following functions:

(a) $f(x) = x$ **(b)** $f(x) = x^2$ **(c)** $f(x) = x^3$

Based on these examples, what do you think the limit Δf represents?

3.6 Trigonometric Functions

We can use the rules developed so far to differentiate functions involving powers of x, but we cannot yet handle the trigonometric functions. What is missing are the formulas for the derivatives of $\sin x$ and $\cos x$. Fortunately, their derivatives are simple—each is the derivative of the other up to a sign.

Recall our convention: *Angles are measured in radians, unless otherwise specified.*

> *CAUTION* In Theorem 1, we are differentiating with respect to x measured in radians. The derivatives of sine and cosine with respect to degrees involve an extra, unwieldy factor of $\pi/180$ (see Example 8 in Section 3.7).

> **THEOREM 1 Derivative of Sine and Cosine**   The functions $y = \sin x$ and $y = \cos x$ are differentiable and
>
> $$\frac{d}{dx} \sin x = \cos x \qquad \text{and} \qquad \frac{d}{dx} \cos x = -\sin x$$

Proof   We must go back to the definition of the derivative:

$$\frac{d}{dx} \sin x = \lim_{h \to 0} \frac{\sin(x+h) - \sin x}{h} \qquad \boxed{1}$$

> ◀⋯ *REMINDER* Addition formula for $\sin x$:
>
> $$\sin(x+h) = \sin x \cos h + \cos x \sin h$$

We cannot cancel the h by rewriting the difference quotient, but we can use the addition formula (see marginal note) to write the numerator as a sum of two terms:

$$\sin(x+h) - \sin x = \sin x \cos h + \cos x \sin h - \sin x \qquad \text{(addition formula)}$$

$$= (\sin x \cos h - \sin x) + \cos x \sin h$$

$$= \sin x (\cos h - 1) + \cos x \sin h$$

This gives us

$$\frac{\sin(x+h) - \sin x}{h} = \frac{\sin x \,(\cos h - 1)}{h} + \frac{\cos x \,\sin h}{h}$$

$$\frac{d \sin x}{dx} = \lim_{h \to 0} \frac{\sin(x+h) - \sin x}{h} = \lim_{h \to 0} \frac{\sin x \,(\cos h - 1)}{h} + \lim_{h \to 0} \frac{\cos x \,\sin h}{h}$$

$$= (\sin x) \underbrace{\lim_{h \to 0} \frac{\cos h - 1}{h}}_{\text{This equals 0.}} + (\cos x) \underbrace{\lim_{h \to 0} \frac{\sin h}{h}}_{\text{This equals 1.}} \qquad \boxed{2}$$

Here, we can take $\sin x$ and $\cos x$ outside the limits in Eq. (2) because they do not depend on h. The two limits are given by Theorem 2 in Section 2.6:

$$\lim_{h \to 0} \frac{\cos h - 1}{h} = 0 \qquad \text{and} \qquad \lim_{h \to 0} \frac{\sin h}{h} = 1$$

Therefore, Eq. (2) reduces to the formula $\frac{d}{dx} \sin x = \cos x$, as desired. The formula $\frac{d}{dx} \cos x = -\sin x$ is proved similarly (see Exercise 53). ∎

■ **EXAMPLE 1**   Calculate $f''(x)$, where $f(x) = x \cos x$.

Solution   By the Product Rule,

$$f'(x) = x' \cos x + x (\cos x)' = \cos x - x \sin x$$

$$f''(x) = (\cos x - x \sin x)' = -\sin x - \big(x'(\sin x) + x(\sin x)'\big) = -2 \sin x - x \cos x \;\blacksquare$$

GRAPHICAL INSIGHT The formula $(\sin x)' = \cos x$ is made plausible when we compare the graphs in Figure 1. The tangent lines to the graph of $y = \sin x$ have positive slope on the interval $\left(-\frac{\pi}{2}, \frac{\pi}{2}\right)$, and on this interval, the derivative $y' = \cos x$ is positive. Similarly, the tangent lines have negative slope on the interval $\left(\frac{\pi}{2}, \frac{3\pi}{2}\right)$, where $y' = \cos x$ is negative. The tangent lines are horizontal at $x = -\frac{\pi}{2}, \frac{\pi}{2}, \frac{3\pi}{2}$, where $\cos x = 0$.

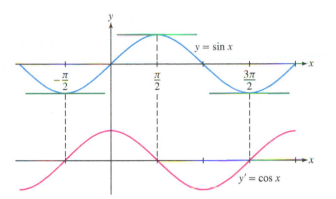

DF **FIGURE 1** Compare the graphs of $y = \sin x$ and its derivative $y' = \cos x$.

◄·· *REMINDER The standard trigonometric functions are defined in Section 1.4.*

The derivatives of the other **standard trigonometric functions** can be computed using the Quotient Rule. We derive the formula for $(\tan x)'$ in Example 2 and leave the remaining formulas for the exercises (Exercises 35–37).

> **THEOREM 2 Derivatives of Standard Trigonometric Functions**
>
> $$\frac{d}{dx}\tan x = \sec^2 x, \qquad \frac{d}{dx}\sec x = \sec x \tan x$$
>
> $$\frac{d}{dx}\cot x = -\csc^2 x, \qquad \frac{d}{dx}\csc x = -\csc x \cot x$$

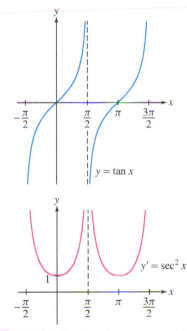

DF **FIGURE 2** Graphs of $y = \tan x$ and its derivative $y' = \sec^2 x$.

■ **EXAMPLE 2** Verify the formula $\dfrac{d}{dx}(\tan x) = \sec^2 x$ (Figure 2).

Solution Use the Quotient Rule and the identity $\cos^2 x + \sin^2 x = 1$:

$$\frac{d}{dx}(\tan x) = \left(\frac{\sin x}{\cos x}\right)' = \frac{\cos x \cdot (\sin x)' - \sin x \cdot (\cos x)'}{\cos^2 x}$$

$$= \frac{\cos x \cos x - \sin x (-\sin x)}{\cos^2 x}$$

$$= \frac{\cos^2 x + \sin^2 x}{\cos^2 x} = \frac{1}{\cos^2 x} = \sec^2 x \qquad ■$$

■ **EXAMPLE 3** Find the tangent line to the graph of $y = \tan \theta \sec \theta$ at $\theta = \frac{\pi}{4}$.

Solution By the Product Rule,

$$y' = (\tan \theta)' \sec \theta + \tan \theta (\sec \theta)' = \sec^2 \theta \, \sec \theta + \tan \theta (\sec \theta \tan \theta)$$

$$= \sec^3 \theta + \tan^2 \theta \sec \theta$$

Now use the values $\sec \frac{\pi}{4} = \sqrt{2}$ and $\tan \frac{\pi}{4} = 1$ to compute

$$y\left(\frac{\pi}{4}\right) = \tan\left(\frac{\pi}{4}\right) \sec\left(\frac{\pi}{4}\right) = \sqrt{2}$$

$$y'\left(\frac{\pi}{4}\right) = \sec^3\left(\frac{\pi}{4}\right) + \tan^2\left(\frac{\pi}{4}\right) \sec\left(\frac{\pi}{4}\right) = 2\sqrt{2} + \sqrt{2} = 3\sqrt{2}$$

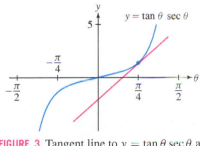

FIGURE 3 Tangent line to $y = \tan \theta \sec \theta$ at $\theta = \frac{\pi}{4}$.

An equation of the tangent line (Figure 3) is $y - \sqrt{2} = 3\sqrt{2}\left(\theta - \frac{\pi}{4}\right)$. ■

3.6 SUMMARY

- Basic trigonometric derivatives:

$$\frac{d}{dx} \sin x = \cos x, \qquad \frac{d}{dx} \cos x = -\sin x$$

- Additional formulas:

$$\frac{d}{dx} \tan x = \sec^2 x, \qquad \frac{d}{dx} \sec x = \sec x \tan x$$

$$\frac{d}{dx} \cot x = -\csc^2 x, \qquad \frac{d}{dx} \csc x = -\csc x \cot x$$

3.6 EXERCISES

Preliminary Questions

1. Determine the sign ($+$ or $-$) that yields the correct formula for the following:

(a) $\dfrac{d}{dx}(\sin x + \cos x) = \pm \sin x \pm \cos x$

(b) $\dfrac{d}{dx} \sec x = \pm \sec x \tan x$

(c) $\dfrac{d}{dx} \cot x = \pm \csc^2 x$

2. Which of the following functions can be differentiated using the rules we have covered so far?

(a) $y = 3 \cos x \cot x$ **(b)** $y = \cos(x^2)$ **(c)** $y = e^x \sin x$

3. Compute $\dfrac{d}{dx}(\sin^2 x + \cos^2 x)$ without using the derivative formulas for $\sin x$ and $\cos x$.

4. How is the addition formula used in deriving the formula $(\sin x)' = \cos x$?

Exercises

In Exercises 1–4, find an equation of the tangent line at the point indicated.

1. $y = \sin x, \quad x = \frac{\pi}{4}$

2. $y = \cos x, \quad x = \frac{\pi}{3}$

3. $y = \tan x, \quad x = \frac{\pi}{4}$

4. $y = \sec x, \quad x = \frac{\pi}{6}$

In Exercises 5–24, compute the derivative.

5. $f(x) = \sin x \cos x$

6. $f(x) = x^2 \cos x$

7. $f(x) = \sin^2 x$

8. $f(x) = 9 \sec x + 12 \cot x$

9. $H(t) = \sin t \sec^2 t$

10. $h(t) = 9 \csc t + t \cot t$

11. $f(\theta) = \tan \theta \sec \theta$

12. $k(\theta) = \theta^2 \sin^2 \theta$

13. $f(x) = (2x^4 - 4x^{-1}) \sec x$

14. $f(z) = z \tan z$

15. $y = \dfrac{\sec \theta}{\theta}$

16. $G(z) = \dfrac{1}{\tan z - \cot z}$

17. $R(y) = \dfrac{3 \cos y - 4}{\sin y}$

18. $f(x) = \dfrac{x}{\sin x + 2}$

19. $f(x) = \dfrac{1 + \tan x}{1 - \tan x}$

20. $f(\theta) = \theta \tan \theta \sec \theta$

21. $f(x) = e^x \sin x$

22. $h(t) = e^t \csc t$

23. $f(\theta) = e^\theta (5 \sin \theta - 4 \tan \theta)$

24. $f(x) = x e^x \cos x$

In Exercises 25–34, find an equation of the tangent line at the point specified.

25. $y = x^3 + \cos x, \quad x = 0$

26. $y = \tan \theta, \quad \theta = \frac{\pi}{6}$

27. $y = \dfrac{\sin t}{1 + \cos t}, \quad t = \frac{\pi}{3}$

28. $y = \sin x + 3 \cos x, \quad x = 0$

29. $y = 2(\sin \theta + \cos \theta), \quad \theta = \frac{\pi}{3}$

30. $y = \csc x - \cot x, \quad x = \frac{\pi}{4}$

31. $y = e^x \cos x, \quad x = 0$

32. $y = e^x \cos^2 x, \quad x = \frac{\pi}{4}$

33. $y = e^t (1 - \cos t), \quad t = \frac{\pi}{2}$

34. $y = e^\theta \sec \theta, \quad \theta = \frac{\pi}{4}$

In Exercises 35–37, use Theorem 1 to verify the formula.

35. $\dfrac{d}{dx} \cot x = -\csc^2 x$

36. $\dfrac{d}{dx} \sec x = \sec x \tan x$

37. $\dfrac{d}{dx} \csc x = -\csc x \cot x$

38. Show that both $y = \sin x$ and $y = \cos x$ satisfy $y'' = -y$.

In Exercises 39–42, calculate the higher derivative.

39. $f''(\theta), \quad f(\theta) = \theta \sin \theta$

40. $\dfrac{d^2}{dt^2} \cos^2 t$

41. $y'', \quad y''', \quad y = \tan x$

42. $y'', \quad y''', \quad y = e^t \sin t$

43. Calculate the first five derivatives of $f(x) = \cos x$. Then determine $f^{(8)}(x)$ and $f^{(37)}(x)$.

44. Find $y^{(157)}$, where $y = \sin x$.

45. Find the values of x between 0 and 2π where the tangent line to the graph of $y = \sin x \cos x$ is horizontal.

46. [GU] Plot the graph $f(\theta) = \sec\theta + \csc\theta$ over $[0, 2\pi]$ and determine the number of solutions to $f'(\theta) = 0$ in this interval graphically. Then compute $f'(\theta)$ and find the solutions.

47. [GU] Let $g(t) = t - \sin t$.

(a) Plot the graph of g with a graphing utility for $0 \le t \le 4\pi$.

(b) Show that the slope of the tangent line is nonnegative. Verify this on your graph.

(c) For which values of t in the given range is the tangent line horizontal?

48. CAS Let $f(x) = (\sin x)/x$ for $x \neq 0$ and $f(0) = 1$.

(a) Plot f on $[-3\pi, 3\pi]$.

(b) Show that $f'(c) = 0$ if $c = \tan c$. Use the numerical root finder on a computer algebra system to find a good approximation to the smallest *positive* value c_0 such that $f'(c_0) = 0$.

(c) Verify that the horizontal line $y = f(c_0)$ is tangent to the graph of $y = f(x)$ at $x = c_0$ by plotting them on the same set of axes.

49. Show that no tangent line to the graph of $f(x) = \tan x$ has zero slope. What is the least slope of a tangent line? Justify by sketching the graph of $f'(x) = (\tan x)'$.

50. The height at time t (in seconds) of a mass, oscillating at the end of a spring, is $s(t) = 300 + 40 \sin t$ cm. Find the velocity and acceleration at $t = \frac{\pi}{3}$ s.

51. The horizontal range R of a projectile launched from ground level at an angle θ and initial velocity v_0 m/s is $R = (v_0^2/9.8) \sin\theta \cos\theta$. Calculate $dR/d\theta$. If $\theta = 7\pi/24$, will the range increase or decrease if the angle is increased slightly? Base your answer on the sign of the derivative.

52. Show that if $\frac{\pi}{2} < \theta < \pi$, then the distance along the x-axis between θ and the point where the tangent line intersects the x-axis is equal to $|\tan\theta|$ (Figure 4).

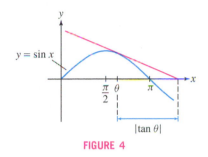

FIGURE 4

Further Insights and Challenges

53. Use the limit definition of the derivative and the addition law for the cosine function to prove that $(\cos x)' = -\sin x$.

54. Use the addition formula for the tangent

$$\tan(x + h) = \frac{\tan x + \tan h}{1 + \tan x \tan h}$$

to compute $(\tan x)'$ directly as a limit of the difference quotients. You will also need to show that $\lim_{h\to 0} \dfrac{\tan h}{h} = 1$.

55. Verify the following identity and use it to give another proof of the formula $(\sin x)' = \cos x$:

$$\sin(x + h) - \sin x = 2\cos\left(x + \tfrac{1}{2}h\right)\sin\left(\tfrac{1}{2}h\right)$$

Hint: Use the addition formula to prove that $\sin(a + b) - \sin(a - b) = 2\cos a \sin b$.

56. Show that a nonzero polynomial function $y = f(x)$ *cannot* satisfy the equation $y'' = -y$. Use this to prove that neither $f(x) = \sin x$ nor $f(x) = \cos x$ is a polynomial. Can you think of another way to reach this conclusion by considering limits as $x \to \infty$?

57. Let $f(x) = x \sin x$ and $g(x) = x \cos x$.

(a) Show that $f'(x) = g(x) + \sin x$ and $g'(x) = -f(x) + \cos x$.

(b) Verify that $f''(x) = -f(x) + 2\cos x$ and $g''(x) = -g(x) - 2\sin x$.

(c) By further experimentation, try to find formulas for all higher derivatives of f and g. *Hint:* The kth derivative depends on whether $k = 4n, 4n + 1, 4n + 2$, or $4n + 3$.

58. Figure 5 shows the geometry behind the derivative formula $(\sin\theta)' = \cos\theta$. Segments $\overline{BA}$ and $\overline{BD}$ are parallel to the x- and y-axes. Let $\Delta\sin\theta = \sin(\theta + h) - \sin\theta$. Verify the following statements:

(a) $\Delta\sin\theta = BC$

(b) $\angle BDA = \theta$ *Hint:* $\overline{OA} \perp AD$.

(c) $BD = (\cos\theta)AD$

Now explain the following intuitive argument: If h is small, then $BC \approx BD$ and $AD \approx h$, so $\Delta\sin\theta \approx (\cos\theta)h$ and $(\sin\theta)' = \cos\theta$.

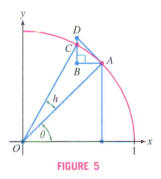

FIGURE 5

3.7 The Chain Rule

The **Chain Rule** is used to differentiate composite functions such as $y = \cos(x^3)$ and $y = \sqrt{x^4 + 1}$.

Recall that a *composite function* is obtained by "plugging" one function into another. The composite of f and g, denoted $f \circ g$, is defined by

$$(f \circ g)(x) = f\big(g(x)\big)$$

For convenience, we call f the *outside* function and g the *inside* function. Often, we write the composite function as $f(u)$, where $u = g(x)$. For example, $y = \cos(x^3)$ is the function $y = \cos u$, where $u = x^3$.

In verbal form, the Chain Rule says

$$\big(f(g(x))\big)' = \text{outside}'(\text{inside}) \cdot \text{inside}'$$

In words, it is " the derivative of the outside function at the inside function times the derivative of the inside function." A proof of the Chain Rule is given at the end of this section.

THEOREM 1 Chain Rule If f and g are differentiable, then the composite function $(f \circ g)(x) = f(g(x))$ is differentiable and

$$\big(f(g(x))\big)' = f'\big(g(x)\big)\, g'(x)$$

■ **EXAMPLE 1** Calculate the derivative of $y = \cos(x^3)$.

Solution As noted above, $y = \cos(x^3)$ is a composite $f(g(x))$, where

$$f(u) = \cos u, \qquad u = g(x) = x^3$$

$$f'(u) = -\sin u, \qquad g'(x) = 3x^2$$

Since $u = x^3$, $f'(g(x)) = f'(u) = f'(x^3) = -\sin(x^3)$. So, by the Chain Rule,

$$\frac{d}{dx}\cos(x^3) = \underbrace{-\sin(x^3)}_{f'(g(x))}\ \underbrace{(3x^2)}_{g'(x)} = -3x^2\sin(x^3) \qquad \blacksquare$$

■ **EXAMPLE 2** Calculate the derivative of $y = \sqrt{x^4 + 1}$.

Solution The function $y = \sqrt{x^4 + 1}$ is a composite $f(g(x))$, where

$$f(u) = u^{1/2}, \qquad u = g(x) = x^4 + 1$$

$$f'(u) = \frac{1}{2}u^{-1/2}, \qquad g'(x) = 4x^3$$

Note that $f'(g(x)) = \frac{1}{2}(x^4 + 1)^{-1/2}$, so by the Chain Rule,

$$\frac{d}{dx}\sqrt{x^4 + 1} = \underbrace{\frac{1}{2}(x^4 + 1)^{-1/2}}_{f'(g(x))}\ \underbrace{(4x^3)}_{g'(x)} = \frac{4x^3}{2\sqrt{x^4 + 1}} \qquad \blacksquare$$

■ **EXAMPLE 3** Calculate $\dfrac{dy}{dx}$ for $y = \tan\left(\dfrac{x}{x + 1}\right)$.

Solution The outside function is $f(u) = \tan u$. Because $f'(u) = \sec^2 u$, the Chain Rule gives us

$$\frac{d}{dx}\tan\left(\frac{x}{x + 1}\right) = \sec^2\left(\frac{x}{x + 1}\right)\ \underbrace{\frac{d}{dx}\left(\frac{x}{x + 1}\right)}_{\substack{\text{Derivative of} \\ \text{inside function}}}$$

Now, by the Quotient Rule,

$$\frac{d}{dx}\left(\frac{x}{x + 1}\right) = \frac{(x + 1)\dfrac{d}{dx}x - x\dfrac{d}{dx}(x + 1)}{(x + 1)^2} = \frac{1}{(x + 1)^2}$$

We obtain

$$\frac{d}{dx}\tan\left(\frac{x}{x + 1}\right) = \sec^2\left(\frac{x}{x + 1}\right)\frac{1}{(x + 1)^2} = \frac{\sec^2\left(\dfrac{x}{x + 1}\right)}{(x + 1)^2} \qquad \blacksquare$$

■ **EXAMPLE 4** The daily natural gas usage in a household follows a pattern given by $G(t) = 300(1 + \cos\frac{2\pi t}{365})$ cubic feet per day, where $t = 1$ corresponds to January 1. Assuming it is not Leap Year, find the rate of change of usage on March 1, June 1, September 1 and December 1.

Solution By the Chain Rule,

$$G'(t) = -300 \sin\left(\frac{2\pi t}{365}\right) \cdot \frac{2\pi}{365}$$

On March 1, $t = 60$, so $G'(60) = -300 \sin\left(\frac{2\pi(60)}{365}\right) \cdot \frac{2\pi}{365} \approx -4.4349$.

On June 1, $t = 152$, so $G'(152) = -300 \sin\left(\frac{2\pi(152)}{365}\right) \cdot \frac{2\pi}{365} \approx -2.5885$.

On Sept. 1, $t = 244$, so $G'(244) = -300 \sin\left(\frac{2\pi(244)}{365}\right) \cdot \frac{2\pi}{365} \approx 4.5017$.

On Dec. 1, $t = 335$, so $G'(335) = -300 \sin\left(\frac{2\pi(335)}{365}\right) \cdot \frac{2\pi}{365} \approx 2.5500$.

It is not surprising that the derivative is negative in the spring, as usage drops as temperatures rise. The derivative is nearer 0 in the summer, as the weather remains warm, and usage does not change. Usage increases substantially in the fall and is still increasing in December. ∎

It is instructive to write the Chain Rule in Leibniz notation. Let

$$y = f(u) = f(g(x))$$

Then, by the Chain Rule,

$$\frac{dy}{dx} = f'(u)\, g'(x) = \frac{df}{du}\frac{du}{dx}$$

or

$$\boxed{\frac{dy}{dx} = \frac{dy}{du}\frac{du}{dx}}$$

Christiaan Huygens (1629–1695), one of the greatest scientists of his age, was Leibniz's teacher in mathematics and physics. He admired Isaac Newton greatly but did not accept Newton's theory of gravitation. He referred to it as the "improbable principle of attraction," because it did not explain how two masses separated by a distance could influence each other. *(© Bettmann/CORBIS)*

CONCEPTUAL INSIGHT In Leibniz notation, it appears as if we are multiplying fractions and the Chain Rule is simply a matter of "canceling the du." Since the symbolic expressions dy/du and du/dx are not fractions, this does not make sense literally, but it does suggest that derivatives behave *as if they were fractions* (this is reasonable because a derivative is a *limit* of fractions, namely of the difference quotients). Leibniz's form also emphasizes a key aspect of the Chain Rule: *Rates of change multiply*. To illustrate, suppose that (thanks to your knowledge of calculus) your salary increases twice as fast as your friend's. If your friend's salary increases \$4000 per year, your salary will increase at the rate of 2×4000 or \$8000 per year. In terms of derivatives,

$$\frac{d(\text{your salary})}{dt} = \frac{d(\text{your salary})}{d(\text{friend's salary})} \times \frac{d(\text{friend's salary})}{dt}$$

$$\$8000/\text{year} = \qquad 2 \qquad \times \quad \$4000/\text{year}$$

■ **EXAMPLE 5** Imagine a sphere whose radius r increases at a rate of 3 cm/s. At what rate is the volume V of the sphere increasing when $r = 10$ cm?

Solution Because we are asked to determine the rate at which V is increasing, we must find dV/dt. What we are given is the rate dr/dt, namely $dr/dt = 3$ cm/s. The Chain Rule allows us to express dV/dt in terms of dV/dr and dr/dt:

$$\underbrace{\frac{dV}{dt}}_{\substack{\text{Rate of change of volume} \\ \text{with respect to time}}} = \underbrace{\frac{dV}{dr}}_{\substack{\text{Rate of change of volume} \\ \text{with respect to radius}}} \times \underbrace{\frac{dr}{dt}}_{\substack{\text{Rate of change of radius} \\ \text{with respect to time}}}$$

To compute dV/dr, we use the formula for the volume of a sphere, $V = \frac{4}{3}\pi r^3$:

$$\frac{dV}{dr} = \frac{d}{dr}\left(\frac{4}{3}\pi r^3\right) = 4\pi r^2$$

Because $dr/dt = 3$, we obtain

$$\frac{dV}{dt} = \frac{dV}{dr}\frac{dr}{dt} = 4\pi r^2(3) = 12\pi r^2$$

For $r = 10$,

$$\left.\frac{dV}{dt}\right|_{r=10} = (12\pi)10^2 = 1200\pi \approx 3770 \text{ cm}^3/\text{s} \qquad \blacksquare$$

We now discuss some important special cases of the Chain Rule.

THEOREM 2 General Power and Exponential Rules   If g is differentiable, then

- $\dfrac{d}{dx}(g(x))^n = n(g(x))^{n-1}g'(x)$ (for any number n)

- $\dfrac{d}{dx}e^{g(x)} = e^{g(x)}g'(x)$

Proof   Let $f(u) = u^n$. Then $(g(x))^n = f(g(x))$, and the Chain Rule yields

$$\frac{d}{dx}(g(x))^n = f'(g(x))g'(x) = n(g(x))^{n-1}g'(x)$$

On the other hand, $e^{g(x)} = h(g(x))$, where $h(u) = e^u$. We obtain

$$\frac{d}{dx}e^{g(x)} = h'(g(x))g'(x) = e^{g(x)}g'(x) \qquad \blacksquare$$

■ **EXAMPLE 6 General Power and Exponential Rules**   Find the derivatives of
(a) $y = (x^2 + 7x + 2)^{-1/3}$ and   **(b)** $y = e^{\cos t}$.

Solution   Apply $\dfrac{d}{dx}g(x)^n = ng(x)^{n-1}g'(x)$ in (A) and $\dfrac{d}{dx}e^{g(x)} = e^{g(x)}g'(x)$ in (B).

(a)
$$\frac{d}{dx}(x^2 + 7x + 2)^{-1/3} = -\frac{1}{3}(x^2 + 7x + 2)^{-4/3}\frac{d}{dx}(x^2 + 7x + 2)$$

$$= -\frac{1}{3}(x^2 + 7x + 2)^{-4/3}(2x + 7)$$

(b)
$$\frac{d}{dt}e^{\cos t} = e^{\cos t}\frac{d}{dt}(\cos t) = -(\sin t)e^{\cos t} \qquad \blacksquare$$

■ **EXAMPLE 7 Using the Chain Rule Twice**   Calculate $\dfrac{d}{dx}\sqrt{1 + \sqrt{x^2 + 1}}$.

Solution   First apply the Chain Rule with inside function $u = 1 + \sqrt{x^2 + 1}$:

$$\frac{d}{dx}\left(1 + (x^2 + 1)^{1/2}\right)^{1/2} = \frac{1}{2}\left(1 + (x^2 + 1)^{1/2}\right)^{-1/2}\frac{d}{dx}\left(1 + (x^2 + 1)^{1/2}\right)$$

Then apply the Chain Rule again to the remaining derivative:

$$\frac{d}{dx}\left(1 + (x^2 + 1)^{1/2}\right)^{1/2} = \frac{1}{2}(1 + (x^2 + 1)^{1/2})^{-1/2}\left(\frac{1}{2}(x^2 + 1)^{-1/2}(2x)\right)$$

$$= \frac{1}{2}x(x^2 + 1)^{-1/2}\left(1 + (x^2 + 1)^{1/2}\right)^{-1/2} \qquad \blacksquare$$

According to our convention, $\sin x$ denotes the sine of x radians, and with this convention, the formula $(\sin x)' = \cos x$ holds. In the next example, we derive a formula for the derivative of the sine function when x is measured in degrees.

■ **EXAMPLE 8** **Trigonometric Derivatives in Degrees** Calculate the derivative of the sine function as a function of degrees rather than radians.

Solution To solve this problem, it is convenient to use an underline to indicate a function of degrees rather than radians. For example,

$$\underline{\sin} \, x = \text{sine of } x \text{ degrees}$$

The functions $f(x) = \sin x$ and $g(x) = \underline{\sin} \, x$ are different, but they are related by

$$\underline{\sin} \, x = \sin\left(\frac{\pi}{180}x\right)$$

because x degrees corresponds to $\frac{\pi}{180}x$ radians. By Theorem 1,

$$\frac{d}{dx}\underline{\sin} \, x = \frac{d}{dx}\sin\left(\frac{\pi}{180}x\right) = \left(\frac{\pi}{180}\right)\cos\left(\frac{\pi}{180}x\right) = \left(\frac{\pi}{180}\right)\underline{\cos} \, x \qquad ■$$

> *A similar calculation shows that the factor $\frac{\pi}{180}$ appears in the formulas for the derivatives of the other standard trigonometric functions with respect to degrees. For example,*
>
> $$\frac{d}{dx}\underline{\tan} \, x = \left(\frac{\pi}{180}\right)\underline{\sec}^2 x$$

Proof of the Chain Rule The difference quotient for the composite $f \circ g$ is

$$\frac{f(g(x+h)) - f(g(x))}{h} \qquad (h \neq 0)$$

Our goal is to show that $(f \circ g)'(x)$ is the product of $f'(g(x))$ and $g'(x)$, so it makes sense to write the difference quotient as a product:

$$\frac{f(g(x+h)) - f(g(x))}{h} = \frac{f(g(x+h)) - f(g(x))}{g(x+h) - g(x)} \times \frac{g(x+h) - g(x)}{h} \qquad \boxed{1}$$

This is legitimate only if the denominator $g(x+h) - g(x)$ is nonzero. Therefore, to continue our proof, we make the extra assumption that $g(x+h) - g(x) \neq 0$ for all h close to but not equal to 0. This assumption is not necessary, but without it, the argument is more technical (see Exercise 105).

Under our assumption, we may use Eq. (1) to write $(f \circ g)'(x)$ as a product of two limits:

$$(f \circ g)'(x) = \underbrace{\lim_{h \to 0} \frac{f(g(x+h)) - f(g(x))}{g(x+h) - g(x)}}_{\text{Show that this equals } f'(g(x)).} \times \underbrace{\lim_{h \to 0} \frac{g(x+h) - g(x)}{h}}_{\text{This is } g'(x).}$$

The second limit on the right is $g'(x)$. The Chain Rule will follow if we show that the first limit equals $f'(g(x))$. To verify this, set

$$k = g(x+h) - g(x)$$

Then $g(x+h) = g(x) + k$ and

$$\frac{f(g(x+h)) - f(g(x))}{g(x+h) - g(x)} = \frac{f(g(x) + k) - f(g(x))}{k}$$

The function g is continuous because it is differentiable. Therefore, $g(x+h)$ tends to $g(x)$ and $k = g(x+h) - g(x)$ tends to zero as $h \to 0$. Thus, we may rewrite the limit in terms of k to obtain the desired result:

$$\lim_{h \to 0} \frac{f(g(x+h)) - f(g(x))}{g(x+h) - g(x)} = \lim_{k \to 0} \frac{f(g(x) + k) - f(g(x))}{k} = f'(g(x)) \qquad ■$$

3.7 SUMMARY

- The Chain Rule expresses $(f \circ g)'$ in terms of f' and g':

$$(f(g(x)))' = f'(g(x)) \, g'(x)$$

- In Leibniz notation: $\dfrac{dy}{dx} = \dfrac{dy}{du} \dfrac{du}{dx}$, where $y = f(u)$ and $u = g(x)$

- General Power Rule: $\dfrac{d}{dx}(g(x))^n = n(g(x))^{n-1} g'(x)$

- General Exponential Rule: $\dfrac{d}{dx} e^{g(x)} = e^{g(x)} g'(x)$

3.7 EXERCISES

Preliminary Questions

1. Identify the outside and inside functions for each of these composite functions.

(a) $y = \sqrt{4x + 9x^2}$ (b) $y = \tan(x^2 + 1)$

(c) $y = \sec^5 x$ (d) $y = (1 + e^x)^4$

2. Which of the following can be differentiated easily *without* using the Chain Rule?

(a) $y = \tan(7x^2 + 2)$ (b) $y = \dfrac{x}{x + 1}$

(c) $y = \sqrt{x} \cdot \sec x$ (d) $y = \sqrt{x \cos x}$

(e) $y = xe^x$ (f) $y = e^{\sin x}$

3. Which is the derivative of $f(5x)$?

(a) $5f'(x)$ (b) $5f'(5x)$ (c) $f'(5x)$

4. Suppose that $f'(4) = g(4) = g'(4) = 1$. Do we have enough information to compute $F'(4)$, where $F(x) = f(g(x))$? If not, what is missing?

Exercises

In Exercises 1–4, fill in a table of the following type:

$f(g(x))$	$f'(u)$	$f'(g(x))$	$g'(x)$	$(f \circ g)'(x)$

1. $f(u) = u^{3/2}, \quad g(x) = x^4 + 1$

2. $f(u) = u^3, \quad g(x) = 3x + 5$

3. $f(u) = \tan u, \quad g(x) = x^4$

4. $f(u) = u^4 + u, \quad g(x) = \cos x$

In Exercises 5 and 6, write the function as a composite $f(g(x))$ and compute the derivative using the Chain Rule.

5. $y = (x + \sin x)^4$ **6.** $y = \cos(x^3)$

7. Calculate $\dfrac{d}{dx} \cos u$ for the following choices of $u(x)$:

(a) $u(x) = 9 - x^2$ (b) $u(x) = x^{-1}$ (c) $u(x) = \tan x$

8. Calculate $\dfrac{d}{dx} f(x^2 + 1)$ for the following choices of $f(u)$:

(a) $f(u) = \sin u$ (b) $f(u) = 3u^{3/2}$ (c) $f(u) = u^2 - u$

9. Compute $\dfrac{df}{dx}$ if $\dfrac{df}{du} = 2$ and $\dfrac{du}{dx} = 6$.

10. Compute $\dfrac{df}{dx}\Big|_{x=2}$ if $f(u) = u^2$, $u(2) = -5$, and $u'(2) = -5$.

In Exercises 11–22, use the General Power Rule, Exponential Rule, or the Chain Rule to compute the derivative.

11. $y = (x^4 + 5)^3$ **12.** $y = (8x^4 + 5)^3$

13. $y = \sqrt{7x - 3}$ **14.** $y = (4 - 2x - 3x^2)^5$

15. $y = (x^2 + 9x)^{-2}$ **16.** $y = (x^3 + 3x + 9)^{-4/3}$

17. $y = \cos^4 \theta$ **18.** $y = \cos(9\theta + 41)$

19. $y = (2\cos\theta + 5\sin\theta)^9$ **20.** $y = \sqrt{9 + x + \sin x}$

21. $y = e^{x-12}$ **22.** $y = e^{8x+9}$

In Exercises 23–26, compute the derivative of $f \circ g$.

23. $f(u) = \sin u, \quad g(x) = 2x + 1$

24. $f(u) = 2u + 1, \quad g(x) = \sin x$

25. $f(u) = e^u, \quad g(x) = x + x^{-1}$

26. $f(u) = \dfrac{u}{u - 1}, \quad g(x) = \csc x$

In Exercises 27 and 28, find the derivatives of $f(g(x))$ and $g(f(x))$.

27. $f(u) = \cos u, \quad u = g(x) = x^2 + 1$

28. $f(u) = u^3, \quad u = g(x) = \dfrac{1}{x + 1}$

In Exercises 29–42, use the Chain Rule to find the derivative.

29. $y = \sin(x^2)$ **30.** $y = \sin^2 x$

31. $y = \sqrt{t^2 + 9}$ **32.** $y = (t^2 + 3t + 1)^{-5/2}$

33. $y = (x^4 - x^3 - 1)^{2/3}$

34. $y = (\sqrt{x+1} - 1)^{3/2}$

35. $y = \left(\dfrac{x+1}{x-1}\right)^4$

36. $y = \cos^3(12\theta)$

37. $y = \sec\dfrac{1}{x}$

38. $y = \tan(\theta^2 - 4\theta)$

39. $y = \tan(\theta + \cos\theta)$

40. $y = e^{2x^2}$

41. $y = e^{2-9t^2}$

42. $y = \cos^3(e^{4\theta})$

In Exercises 43–72, find the derivative using the appropriate rule or combination of rules.

43. $y = \tan(x^2 + 4x)$

44. $y = \sin(x^2 + 4x)$

45. $y = x\cos(1 - 3x)$

46. $y = \sin(x^2)\cos(x^2)$

47. $y = (4t + 9)^{1/2}$

48. $y = (z + 1)^4(2z - 1)^3$

49. $y = (x^3 + \cos x)^{-4}$

50. $y = \sin(\cos(\sin x))$

51. $y = \sqrt{\sin x \cos x}$

52. $y = (9 - (5 - 2x^4)^7)^3$

53. $y = (\cos 6x + \sin x^2)^{1/2}$

54. $y = \dfrac{(x+1)^{1/2}}{x+2}$

55. $y = \tan^3 x + \tan(x^3)$

56. $y = \sqrt{4 - 3\cos x}$

57. $y = \sqrt{\dfrac{z+1}{z-1}}$

58. $y = (\cos^3 x + 3\cos x + 7)^9$

59. $y = \dfrac{\cos(1 + x)}{1 + \cos x}$

60. $y = \sec(\sqrt{t^2 - 9})$

61. $y = \cot^7(x^5)$

62. $y = \dfrac{\cos(1/x)}{1 + x^2}$

63. $y = \left(1 + \cot^5(x^4 + 1)\right)^9$

64. $y = 4e^{-x} + 7e^{-2x}$

65. $y = (2e^{3x} + 3e^{-2x})^4$

66. $y = \cos(te^{-2t})$

67. $y = e^{(x^2 + 2x + 3)^2}$

68. $y = e^{e^x}$

69. $y = \sqrt{1 + \sqrt{1 + \sqrt{x}}}$

70. $y = \sqrt{\sqrt{x+1} + 1}$

71. $y = (kx + b)^{-1/3}$; k and b any constants

72. $y = \dfrac{1}{\sqrt{kt^4 + b}}$; k, b constants, not both zero

In Exercises 73–76, compute the higher derivative.

73. $\dfrac{d^2}{dx^2}\sin(x^2)$

74. $\dfrac{d^2}{dx^2}(x^2 + 9)^5$

75. $\dfrac{d^3}{dx^3}(9 - x)^8$

76. $\dfrac{d^3}{dx^3}\sin(2x)$

77. The average molecular velocity v of a gas in a certain container is given by $v(T) = 29\sqrt{T}$ m/s, where T is the temperature in kelvins. The temperature is related to the pressure (in atmospheres) by $T = 200P$. Find $\left.\dfrac{dv}{dP}\right|_{P=1.5}$.

78. The power P in a circuit is $P = Ri^2$, where R is the resistance and i is the current. Find dP/dt at $t = \frac{1}{3}$ if $R = 1000\ \Omega$ and i varies according to $i = \sin(4\pi t)$ (time in seconds).

79. An expanding sphere has radius $r = 0.4t$ cm at time t (in seconds). Let V be the sphere's volume. Find dV/dt when (a) $r = 3$ and (b) $t = 3$.

80. A 2005 study by the Fisheries Research Services in Aberdeen, Scotland, suggests that the average length of the species *Clupea harengus* (Atlantic herring) as a function of age t (in years) can be modeled by $L(t) = 32(1 - e^{-0.37t})$ cm for $0 \le t \le 13$. See Figure 1.

(a) How fast is the average length changing at age $t = 6$ years?

(b) At what age is the average length changing at a rate of 5 cm/year?

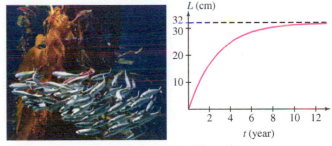

FIGURE 1 Average length of the species *Clupea harengus*.
(© A & J Visage/Alamy)

81. A 1999 study by Starkey and Scarnecchia developed the following model for the average weight (in kilograms) at age t (in years) of channel catfish in the Lower Yellowstone River (Figure 2):

$$W(t) = (3.46293 - 3.32173e^{-0.03456t})^{3.4026}$$

Find the rate at which average weight is changing at age $t = 10$.

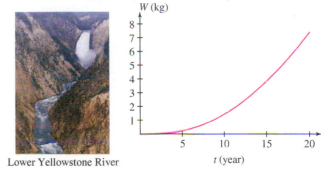

Lower Yellowstone River
FIGURE 2 Average weight of channel catfish at age t.
(© Doug James/Dreamstime.com)

82. The functions in Exercises 80 and 81 are examples of the **von Bertalanffy growth function**

$$M(t) = \left(a + (b - a)e^{kmt}\right)^{1/m} \qquad (m \ne 0)$$

introduced in the 1930s by Austrian-born biologist Karl Ludwig von Bertalanffy. Calculate $M'(0)$ in terms of the constants a, b, k and m.

83. With notation as in Example 8, calculate

(a) $\left.\dfrac{d}{d\theta}\sin\theta\right|_{\theta=60°}$

(b) $\left.\dfrac{d}{d\theta}(\theta + \tan\theta)\right|_{\theta=45°}$

84. Assume that

$$f(0) = 2, \quad f'(0) = 3, \quad h(0) = -1, \quad h'(0) = 7$$

Calculate the derivatives of the following functions at $x = 0$:

(a) $(f(x))^3$

(b) $f(7x)$

(c) $f(4x)h(5x)$

85. Compute the derivative of $h(\sin x)$ at $x = \frac{\pi}{6}$, assuming that $h'(0.5) = 10$.

86. Let $F(x) = f(g(x))$, where the graphs of f and g are shown in Figure 3. Estimate $g'(2)$ and $f'(g(2))$ and compute $F'(2)$.

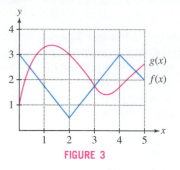

FIGURE 3

In Exercises 87–90, use the table of values to calculate the derivative of the function at the given point.

x	1	4	6
$f(x)$	4	0	6
$f'(x)$	5	7	4
$g(x)$	4	1	6
$g'(x)$	5	$\frac{1}{2}$	3

87. $f(g(x))$, $x = 6$

88. $e^{f(x)}$, $x = 4$

89. $g(\sqrt{x})$, $x = 16$

90. $f(2x + g(x))$, $x = 1$

91. The price (in dollars) of a computer component is $P = 2C - 18C^{-1}$, where C is the manufacturer's cost to produce it. Assume that cost at time t (in years) is $C = 9 + 3t^{-1}$. Determine the rate of change of price with respect to time at $t = 3$.

92. ☐ GU Plot the "astroid" $y = (4 - x^{2/3})^{3/2}$ for $0 \le x \le 8$. Show that the part of every tangent line in the first quadrant has a constant length 8.

93. According to the U.S. standard atmospheric model, developed by the National Oceanic and Atmospheric Administration for use in aircraft and rocket design, atmospheric temperature T (in degrees Celsius), pressure P (kPa = 1000 pascals), and altitude h (in meters) are related by these formulas (valid in the troposphere $h \le 11{,}000$):

$$T = 15.04 - 0.000649h, \qquad P = 101.29 + \left(\frac{T + 273.1}{288.08}\right)^{5.256}$$

Use the Chain Rule to calculate dP/dh. Then estimate the change in P (in pascals, Pa) per additional meter of altitude when $h = 3000$.

94. Climate scientists use the **Stefan–Boltzmann Law** $R = \sigma T^4$ to estimate the change in the earth's average temperature T (in kelvins) caused by a change in the radiation R (in joules per square meter per second) that the earth receives from the sun. Here, $\sigma = 5.67 \times 10^{-8}$ $Js^{-1}m^{-2}K^{-4}$. Calculate dR/dt, assuming that $T = 283$ and $\frac{dT}{dt} = 0.05$ K/year. What are the units of the derivative?

95. In the setting of Exercise 94, calculate the yearly rate of change of T if $T = 283$ K and R increases at a rate of 0.5 $Js^{-1}m^{-2}$ per year.

96. ⌐CAS⌐ Use a computer algebra system to compute $f^{(k)}(x)$ for $k = 1, 2, 3$ for the following functions:

(a) $f(x) = \cot(x^2)$ **(b)** $f(x) = \sqrt{x^3 + 1}$

97. Use the Chain Rule to express the second derivative of $f \circ g$ in terms of the first and second derivatives of f and g.

98. Compute the second derivative of $\sin(g(x))$ at $x = 2$, assuming that $g(2) = \frac{\pi}{4}$, $g'(2) = 5$, and $g''(2) = 3$.

Further Insights and Challenges

99. Show that if f, g, and h are differentiable, then

$$[f(g(h(x)))]' = f'(g(h(x)))g'(h(x))h'(x)$$

100. 📖 Show that differentiation reverses parity: If f is even, then f' is odd, and if f is odd, then f' is even. *Hint:* Differentiate $f(-x)$.

101. (a) 📖 Sketch a graph of any even function f and explain graphically why f' is odd.

(b) Suppose that f' is even. Is f necessarily odd? *Hint:* Check whether this is true for linear functions.

102. Power Rule for Fractional Exponents Let $f(u) = u^q$ and $g(x) = x^{p/q}$. Assume that g is differentiable.

(a) Show that $f(g(x)) = x^p$ (recall the Laws of Exponents).

(b) Apply the Chain Rule and the Power Rule for whole-number exponents to show that $f'(g(x))\,g'(x) = px^{p-1}$.

(c) Then derive the Power Rule for $x^{p/q}$.

103. Prove that for all whole numbers $n \ge 1$,

$$\frac{d^n}{dx^n}\sin x = \sin\left(x + \frac{n\pi}{2}\right)$$

Hint: Use the identity $\cos x = \sin\left(x + \frac{\pi}{2}\right)$.

104. A Discontinuous Derivative Use the limit definition to show that $g'(0)$ exists but $g'(0) \ne \lim_{x \to 0} g'(x)$, where

$$g(x) = \begin{cases} x^2 \sin \dfrac{1}{x} & x \ne 0 \\ 0 & x = 0 \end{cases}$$

105. Chain Rule This exercise proves the Chain Rule without the special assumption made in the text. For any number b, define a new function

$$F(u) = \frac{f(u) - f(b)}{u - b} \qquad \text{for all } u \ne b$$

(a) Show that if we define $F(b) = f'(b)$, then F is continuous at $u = b$.

(b) Take $b = g(a)$. Show that if $x \ne a$, then for all u,

$$\frac{f(u) - f(g(a))}{x - a} = F(u)\frac{u - g(a)}{x - a} \qquad \boxed{2}$$

Note that both sides are zero if $u = g(a)$.

(c) Substitute $u = g(x)$ in Eq. (2) to obtain

$$\frac{f(g(x)) - f(g(a))}{x - a} = F(g(x))\frac{g(x) - g(a)}{x - a}$$

Derive the Chain Rule by computing the limit of both sides as $x \to a$.

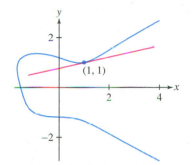

FIGURE 1 Graph of the implicitly defined function $y^4 + xy = x^3 - x + 2$.

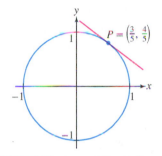

DF FIGURE 2 The tangent line to the unit circle $x^2 + y^2 = 1$ at P has slope $-\frac{3}{4}$.

3.8 Implicit Differentiation

We have developed the basic techniques for calculating a derivative dy/dx when y is given in terms of x by a formula—such as $y = x^3 + 1$. But suppose that y is determined instead by an equation such as

$$y^4 + xy = x^3 - x + 2 \qquad \boxed{1}$$

In this case, we say that y is defined *implicitly*. How can we find the slope of the tangent line at a point on the graph (Figure 1)? Although it may be difficult or even impossible to solve for y explicitly as a function of x, we can find dy/dx using the method of **implicit differentiation**.

To illustrate, consider the equation of the unit circle (Figure 2):

$$x^2 + y^2 = 1$$

Compute dy/dx by taking the derivative of both sides of the equation:

$$\frac{d}{dx}\left(x^2 + y^2\right) = \frac{d}{dx}(1)$$

$$\frac{d}{dx}\left(x^2\right) + \frac{d}{dx}\left(y^2\right) = 0$$

$$2x + \frac{d}{dx}\left(y^2\right) = 0 \qquad \boxed{2}$$

How do we handle the term $\frac{d}{dx}\left(y^2\right)$? We use the Chain Rule. Think of y as a function $y = f(x)$. Then $y^2 = (f(x))^2$ and by the Chain Rule,

$$\frac{d}{dx}y^2 = \frac{d}{dx}(f(x))^2 = 2f(x)\frac{df}{dx} = 2y\frac{dy}{dx}$$

Equation (2) becomes $2x + 2y\frac{dy}{dx} = 0$, and we can solve for $\frac{dy}{dx}$ if $y \neq 0$:

$$\boxed{\frac{dy}{dx} = -\frac{x}{y}} \qquad \boxed{3}$$

■ **EXAMPLE 1** Use Eq. (3) to find the slope of the tangent line at the point $P = \left(\frac{3}{5}, \frac{4}{5}\right)$ on the unit circle.

Solution Set $x = \frac{3}{5}$ and $y = \frac{4}{5}$ in Eq. (3):

$$\frac{dy}{dx}\bigg|_P = -\frac{x}{y} = -\frac{\frac{3}{5}}{\frac{4}{5}} = -\frac{3}{4} \qquad ■$$

In this particular example, we could have computed dy/dx directly, without implicit differentiation. The upper semicircle is the graph of $y = \sqrt{1 - x^2}$ and

$$\frac{dy}{dx} = \frac{d}{dx}\sqrt{1 - x^2} = \frac{1}{2}(1 - x^2)^{-1/2}\frac{d}{dx}(1 - x^2) = -\frac{x}{\sqrt{1 - x^2}}$$

This formula expresses dy/dx in terms of x alone, whereas Eq. (3) expresses dy/dx in terms of both x and y, as is typical when we use implicit differentiation. The two formulas agree because $y = \sqrt{1 - x^2}$.

Before presenting additional examples, let's examine again how the factor dy/dx arises when we differentiate an expression involving y with respect to x. It would not appear if we were differentiating with respect to y. Thus,

Notice what happens if we insist on applying the Chain Rule to $\frac{d}{dy}\sin y$. The extra factor appears, but it is equal to 1:

$$\frac{d}{dy}\sin y = (\cos y)\frac{dy}{dy} = \cos y$$

$$\frac{d}{dy}\sin y = \cos y \qquad \text{but} \qquad \frac{d}{dx}\sin y = (\cos y)\frac{dy}{dx}$$

$$\frac{d}{dy}y^4 = 4y^3 \qquad \text{but} \qquad \frac{d}{dx}y^4 = 4y^3\frac{dy}{dx}$$

Similarly, the Product Rule applied to xy yields

$$\frac{d}{dx}(xy) = \frac{dx}{dx}y + x\frac{dy}{dx} = y + x\frac{dy}{dx}$$

The Quotient Rule applied to t^2/y yields

$$\frac{d}{dt}\left(\frac{t^2}{y}\right) = \frac{y\frac{d}{dt}t^2 - t^2\frac{dy}{dt}}{y^2} = \frac{2ty - t^2\frac{dy}{dt}}{y^2}$$

■ **EXAMPLE 2** Find an equation of the tangent line at the point $P = (1, 1)$ on the curve (Figure 1)

$$y^4 + xy = x^3 - x + 2$$

Solution We break up the calculation into two steps.

Step 1. **Differentiate both sides of the equation with respect to x.**
Note that each occurrence of y in the original equation generates an additional $\frac{dy}{dx}$ upon differentiation.

$$\frac{d}{dx}y^4 + \frac{d}{dx}(xy) = \frac{d}{dx}\left(x^3 - x + 2\right)$$

$$4y^3\frac{dy}{dx} + \left(y + x\frac{dy}{dx}\right) = 3x^2 - 1 \qquad \boxed{4}$$

Step 2. **Solve for $\dfrac{dy}{dx}$.**

Move the terms involving dy/dx in Eq. (4) to the left and place the remaining terms on the right:

$$4y^3\frac{dy}{dx} + x\frac{dy}{dx} = 3x^2 - 1 - y$$

Then factor out dy/dx and divide:

$$\left(4y^3 + x\right)\frac{dy}{dx} = 3x^2 - 1 - y$$

$$\frac{dy}{dx} = \frac{3x^2 - 1 - y}{4y^3 + x} \qquad \boxed{5}$$

To find the derivative at $P = (1, 1)$, apply Eq. (5) with $x = 1$ and $y = 1$:

$$\frac{dy}{dx}\bigg|_{(1,1)} = \frac{3 \cdot 1^2 - 1 - 1}{4 \cdot 1^3 + 1} = \frac{1}{5}$$

An equation of the tangent line is $y - 1 = \frac{1}{5}(x - 1)$ or $y = \frac{1}{5}x + \frac{4}{5}$. ■

CONCEPTUAL INSIGHT The graph of an equation does not always define a function because there may be more than one y-value for a given value of x. Implicit differentiation works because the graph is generally made up of several pieces called **branches**, each of which does define a function (a proof of this fact relies on the Implicit Function Theorem from advanced calculus). For example, the branches of the unit circle $x^2 + y^2 = 1$ are the graphs of the functions $y = \sqrt{1 - x^2}$ and $y = -\sqrt{1 - x^2}$. Similarly, the graph in Figure 3 has an upper and a lower branch. In most examples, the branches are differentiable except at certain exceptional points where the tangent line may be vertical.

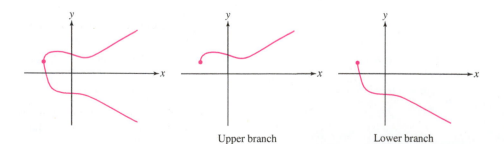

FIGURE 3 Each branch of the graph of $y^4 + xy = x^3 - x + 2$ defines a function of x.

Upper branch   Lower branch

■ **EXAMPLE 3** Find the slope of the tangent line at the point $P = (1, 1)$ on the graph of $e^{x-y} = 2x^2 - y^2$.

Solution We follow the steps of the previous example, this time writing y' for dy/dx:

$$\frac{d}{dx}e^{x-y} = \frac{d}{dx}(2x^2 - y^2)$$

$$e^{x-y}(1 - y') = 4x - 2yy' \qquad \text{(Chain Rule applied to } e^{x-y})$$

$$e^{x-y} - e^{x-y}y' = 4x - 2yy'$$

$$(2y - e^{x-y})y' = 4x - e^{x-y} \qquad \text{(place all } y'\text{-terms on left)}$$

$$y' = \frac{4x - e^{x-y}}{2y - e^{x-y}}$$

The slope of the tangent line at $P = (1, 1)$ is (Figure 4)

$$\left.\frac{dy}{dx}\right|_{(1,1)} = \frac{4(1) - e^{1-1}}{2(1) - e^{1-1}} = \frac{4 - 1}{2 - 1} = 3 \qquad ■$$

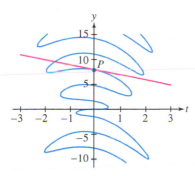

FIGURE 4 Graph of $e^{x-y} = 2x^2 - y^2$.

■ **EXAMPLE 4** **Shortcut to Derivative at a Specific Point** Calculate $\left.\dfrac{dy}{dt}\right|_P$ at the point $P = \left(0, \dfrac{5\pi}{2}\right)$ on the curve (Figure 5):

$$y\cos(y + t + t^2) = t^3$$

Solution As before, differentiate both sides of the equation (we write y' for dy/dt):

$$\frac{d}{dt}y\cos(y + t + t^2) = \frac{d}{dt}t^3$$

$$y'\cos(y + t + t^2) - y\sin(y + t + t^2)(y' + 1 + 2t) = 3t^2 \qquad \boxed{6}$$

We could continue to solve for y', but that is not necessary. Instead, we can substitute $t = 0$, $y = \frac{5\pi}{2}$ directly in Eq. (6) to obtain

$$y'\cos\left(\frac{5\pi}{2} + 0 + 0^2\right) - \left(\frac{5\pi}{2}\right)\sin\left(\frac{5\pi}{2} + 0 + 0^2\right)(y' + 1 + 0) = 0$$

$$0 - \left(\frac{5\pi}{2}\right)(1)(y' + 1) = 0$$

This gives us $y' + 1 = 0$ or $y' = -1$. ■

DF **FIGURE 5** Graph of $y\cos(y + t + t^2) = t^3$. The tangent line at $P = \left(0, \frac{5\pi}{2}\right)$ has slope -1.

Derivatives of Inverse Trigonometric Functions

We now apply implicit differentiation to determine the derivatives of the inverse trigonometric functions. An interesting feature of these functions is that their derivatives are not trigonometric. Rather, they involve quadratic expressions and their square roots. Keep in mind the restricted domains of these functions.

THEOREM 1 **Derivatives of Arcsine and Arccosine**

$$\frac{d}{dx}(\sin^{-1}x) = \frac{1}{\sqrt{1 - x^2}}, \qquad \frac{d}{dx}(\cos^{-1}x) = -\frac{1}{\sqrt{1 - x^2}} \qquad \boxed{7}$$

Proof If $y = \sin^{-1}x$, our goal is to find $\frac{dy}{dx}$. By applying sine to both sides, we have

$$\sin y = x$$

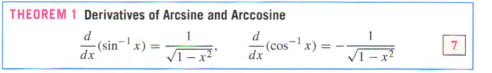

◄·· *REMINDER In Example 7 of Section 1.5, we used the right triangle in Figure 6 in the computation:*

$$\cos(\sin^{-1} x) = \cos y = \frac{\text{adjacent}}{\text{hypotenuse}}$$

$$= \sqrt{1 - x^2}$$

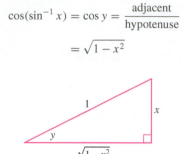

FIGURE 6 Right triangle constructed so that $\sin y = x$.

Differentiating both sides of the equation, treating x as itself and y as a function of x, we obtain

$$\cos y \frac{dy}{dx} = 1$$

$$\frac{dy}{dx} = \frac{1}{\cos y}$$

In order to determine an algebraic expression in x for $\cos y$, we construct a right triangle as in Figure 6 such that $\sin y = x$. We choose y to be its angle, and take its hypotenuse to be of length 1 and its opposite edge to have length x. Then, by the Pythagorean Theorem, its adjacent side must have length $\sqrt{1 - x^2}$. We can therefore read off the triangle that $\cos y = \frac{\sqrt{1-x^2}}{1} = \sqrt{1 - x^2}$. Thus,

$$\frac{dy}{dx} = \frac{1}{\cos y} = \frac{1}{\sqrt{1 - x^2}}$$

The fact that the domain of the inverse sine function is from $-\pi/2$ to $\pi/2$, over which cosine is nonnegative, allows us to take the positive square root rather than the negative square root.

The computation of $\frac{d}{dx}(\cos^{-1} x)$ is similar (see Exercise 45 or the next example). ∎

■ **EXAMPLE 5   Complementary Angles**   The derivatives of $\sin^{-1} x$ and $\cos^{-1} x$ are equal up to a minus sign. Explain this by proving that

$$\sin^{-1} x + \cos^{-1} x = \frac{\pi}{2}$$

Solution   In Figure 7, we have $\theta = \sin^{-1} x$ and $\psi = \cos^{-1} x$. These angles are complementary, so $\theta + \psi = \frac{\pi}{2}$ as claimed. Therefore,

$$\sin^{-1} x = \frac{\pi}{2} - \cos^{-1} x$$

$$\frac{d}{dx}\sin^{-1} x = \frac{d}{dx}\left(\frac{\pi}{2} - \cos^{-1} x\right) = -\frac{d}{dx}\cos^{-1} x \qquad ∎$$

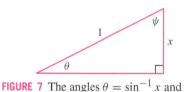

FIGURE 7 The angles $\theta = \sin^{-1} x$ and $\psi = \cos^{-1} x$ are complementary and thus sum to $\pi/2$.

■ **EXAMPLE 6**   Calculate $f'\left(\frac{1}{2}\right)$, where $f(x) = \arcsin(x^2)$.

Solution   Recall that $\arcsin x$ is another notation for $\sin^{-1} x$. By the Chain Rule,

$$\frac{d}{dx}\arcsin(x^2) = \frac{d}{dx}(\sin^{-1}(x^2)) = \frac{1}{\sqrt{1 - (x^2)^2}}\frac{d}{dx}(x^2) = \frac{2x}{\sqrt{1 - x^4}}$$

$$f'\left(\frac{1}{2}\right) = \frac{2\left(\frac{1}{2}\right)}{\sqrt{1 - \left(\frac{1}{2}\right)^4}} = \frac{1}{\sqrt{\frac{15}{16}}} = \frac{4}{\sqrt{15}} \qquad ∎$$

The proofs of the formulas in Theorem 2 are similar to the proof of Theorem 1. See Exercises 46–48.

THEOREM 2   Derivatives of Inverse Trigonometric Functions

$$\frac{d}{dx}\tan^{-1} x = \frac{1}{x^2 + 1}, \qquad \frac{d}{dx}\cot^{-1} x = -\frac{1}{x^2 + 1}$$

$$\frac{d}{dx}\sec^{-1} x = \frac{1}{|x|\sqrt{x^2 - 1}}, \qquad \frac{d}{dx}\csc^{-1} x = -\frac{1}{|x|\sqrt{x^2 - 1}}$$

■ **EXAMPLE 7** Calculate $\dfrac{d}{dx}(\csc^{-1}(e^x + 1))\Big|_{x=0}$.

Solution Apply the Chain Rule using the formula $\dfrac{d}{du}\csc^{-1} u = -\dfrac{1}{|u|\sqrt{u^2 - 1}}$:

$$\frac{d}{dx}\csc^{-1}(e^x + 1) = -\frac{1}{|e^x + 1|\sqrt{(e^x + 1)^2 - 1}}\ \frac{d}{dx}(e^x + 1)$$

$$= -\frac{e^x}{(e^x + 1)\sqrt{e^{2x} + 2e^x}}$$

We have replaced $|e^x + 1|$ by $e^x + 1$ because this quantity is positive. Now we have

$$\frac{d}{dx}\csc^{-1}(e^x + 1)\Big|_{x=0} = -\frac{e^0}{(e^0 + 1)\sqrt{e^0 + 2e^0}} = -\frac{1}{2\sqrt{3}} \qquad ■$$

Finding Higher Order Derivatives Implicitly

We may need to find a higher order derivative of a function that is defined implicitly, as in the next example.

■ **EXAMPLE 8** Find a formula for $\dfrac{d^2 y}{dx^2}$ if y is defined implicitly as a function of x by $x^2 + 4y^2 = 7$.

Solution We differentiate with respect to x, writing y' for $\dfrac{dy}{dx}$.

$$2x + 8yy' = 0$$

Solving for y', we obtain

$$y' = \frac{-x}{4y}$$

Differentiating again with respect to x, we obtain

$$y'' = \frac{4y(-1) - (-x)(4y')}{16y^2} = \frac{-y + xy'}{4y^2}$$

Substituting in the fact that $y' = \frac{-x}{4y}$ yields

$$y'' = \frac{-y + x(-x/4y)}{4y^2} = \frac{-4y^2 - x^2}{16y^3} \qquad ■$$

3.8 SUMMARY

- Implicit differentiation is used to compute dy/dx when x and y are related by an equation.

  Step 1. Take the derivative of both sides of the equation with respect to x.

  Step 2. Solve for dy/dx by collecting the terms involving dy/dx on one side and the remaining terms on the other side of the equation.

- Remember to include the factor dy/dx when differentiating expressions involving y with respect to x. For instance,

$$\frac{d}{dx}\sin y = (\cos y)\frac{dy}{dx}$$

• Derivative formulas:

$$\frac{d}{dx} \sin^{-1} x = \frac{1}{\sqrt{1-x^2}}, \qquad \frac{d}{dx} \cos^{-1} x = -\frac{1}{\sqrt{1-x^2}}$$

$$\frac{d}{dx} \tan^{-1} x = \frac{1}{x^2+1}, \qquad \frac{d}{dx} \cot^{-1} x = -\frac{1}{x^2+1}$$

$$\frac{d}{dx} \sec^{-1} x = \frac{1}{|x|\sqrt{x^2-1}}, \qquad \frac{d}{dx} \csc^{-1} x = -\frac{1}{|x|\sqrt{x^2-1}}$$

3.8 EXERCISES

Preliminary Questions

1. Which differentiation rule is used to show $\frac{d}{dx} \sin y = \cos y \frac{dy}{dx}$?

2. One of (a)–(c) is incorrect. Find and correct the mistake.

(a) $\frac{d}{dy} \sin(y^2) = 2y \cos(y^2)$ **(b)** $\frac{d}{dx} \sin(x^2) = 2x \cos(x^2)$

(c) $\frac{d}{dx} \sin(y^2) = 2y \cos(y^2)$

3. On an exam, Jason was asked to differentiate the equation

$$x^2 + 2xy + y^3 = 7$$

Find the errors in Jason's answer: $2x + 2xy' + 3y^2 = 0$.

4. Which of (a) or (b) is equal to $\frac{d}{dx} (x \sin t)$?

(a) $(x \cos t) \frac{dt}{dx}$ **(b)** $(x \cos t) \frac{dt}{dx} + \sin t$

5. Determine which inverse trigonometric function g has the derivative

$$g'(x) = \frac{1}{x^2+1}$$

6. What does the following identity tell us about the derivatives of $\sin^{-1} x$ and $\cos^{-1} x$?

$$\sin^{-1} x + \cos^{-1} x = \frac{\pi}{2}$$

Exercises

1. Show that if you differentiate both sides of $x^2 + 2y^3 = 6$, the result is $2x + 6y^2 \frac{dy}{dx} = 0$. Then solve for dy/dx and evaluate it at the point $(2, 1)$.

2. Show that if you differentiate both sides of $xy + 4x + 2y = 1$, the result is $(x + 2) \frac{dy}{dx} + y + 4 = 0$. Then solve for dy/dx and evaluate it at the point $(1, -1)$.

In Exercises 3–8, differentiate the expression with respect to x, assuming that $y = f(x)$.

3. $x^2 y^3$ **4.** $\dfrac{x^3}{y^2}$ **5.** $(x^2 + y^2)^{3/2}$

6. $\tan(xy)$ **7.** $\dfrac{y}{y+1}$ **8.** $e^{y/x}$

In Exercises 9–26, calculate the derivative of the other variable with respect to x.

9. $3y^3 + x^2 = 5$ **10.** $y^4 - 2y = 4x^3 + x$

11. $x^2 y + 2x^3 y = x + y$ **12.** $xy^2 + x^2 y^5 - x^3 = 3$

13. $x^3 R^5 = 1$ **14.** $x^4 + z^4 = 1$

15. $\dfrac{y}{x} + \dfrac{x}{y} = 2y$ **16.** $\sqrt{x + s} = \dfrac{1}{x} + \dfrac{1}{s}$

17. $y^{-2/3} + x^{3/2} = 1$ **18.** $x^{1/2} + y^{2/3} = -4y$

19. $y + \dfrac{1}{y} = x^2 + x$ **20.** $\sin(xt) = t$

21. $\sin(x + y) = x + \cos y$ **22.** $\tan(x^2 y) = (x + y)^3$

23. $xe^y = 2xy + y^3$ **24.** $e^{xy} = \sin(y^2)$

25. $e^x + e^y = x - y$ **26.** $e(x^2 + y^2) = x + 4$

In Exercises 27–30, compute the derivative at the point indicated without using a calculator.

27. $y = \sin^{-1} x, \quad x = \frac{3}{5}$ **28.** $y = \tan^{-1} x, \quad x = \frac{1}{2}$

29. $y = \sec^{-1} x, \quad x = 4$ **30.** $y = \arccos(4x), \quad x = \frac{1}{5}$

In Exercises 31–44, find the derivative.

31. $y = \sin^{-1}(7x)$ **32.** $y = \arctan\left(\dfrac{x}{3}\right)$

33. $y = \cos^{-1}(x^2)$ **34.** $y = \sec^{-1}(t + 1)$

35. $y = x \tan^{-1} x$ **36.** $y = e^{\cos^{-1} x}$

37. $y = \arcsin(e^x)$ **38.** $y = \csc^{-1}(x^{-1})$

39. $y = \sqrt{1 - t^2} + \sin^{-1} t$ **40.** $y = \tan^{-1}\left(\dfrac{1+t}{1-t}\right)$

41. $y = (\tan^{-1} x)^3$ **42.** $y = \dfrac{\cos^{-1} x}{\sin^{-1} x}$

43. $y = \cos^{-1} t^{-1} - \sec^{-1} t$ **44.** $y = \cos^{-1}(x + \sin^{-1} x)$

45. Use Figure 8 to prove that $(\cos^{-1} x)' = -\dfrac{1}{\sqrt{1-x^2}}$.

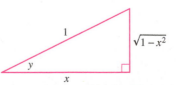

FIGURE 8 Right triangle with $y = \cos^{-1} x$.

46. Show that $(\tan^{-1} x)' = \cos^2(\tan^{-1} x)$ and then use Figure 9 to prove that $(\tan^{-1} x)' = (x^2 + 1)^{-1}$.

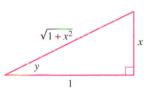

FIGURE 9 Right triangle with $y = \tan^{-1} x$.

47. Let $y = \sec^{-1} x$. Show that $\tan y = \sqrt{x^2 - 1}$ if $x \geq 1$ and that $\tan y = -\sqrt{x^2 - 1}$ if $x \leq -1$. *Hint:* $\tan y \geq 0$ on $\left(0, \frac{\pi}{2}\right)$ and $\tan y \leq 0$ on $\left(\frac{\pi}{2}, \pi\right)$.

48. Use Exercise 47 to verify the formula

$$(\sec^{-1} x)' = \frac{1}{|x|\sqrt{x^2 - 1}}$$

49. Show that $x + yx^{-1} = 1$ and $y = x - x^2$ define the same curve [except that $(0, 0)$ is not a solution of the first equation] and that implicit differentiation yields $y' = yx^{-1} - x$ and $y' = 1 - 2x$. Explain why these formulas produce the same values for the derivative.

50. Use the method of Example 4 to compute $\frac{dy}{dx}\big|_P$ at $P = (2, 1)$ on the curve $y^2 x^3 + y^3 x^4 - 10x + y = 5$.

In Exercises 51 and 52, find dy/dx at the given point.

51. $(x + 2)^2 - 6(2y + 3)^2 = 3$, $(1, -1)$

52. $\sin^2(3y) = x + y$, $\left(\dfrac{2 - \pi}{4}, \dfrac{\pi}{4}\right)$

In Exercises 53–60, find an equation of the tangent line at the given point.

53. $xy + x^2 y^2 = 6$, $(2, 1)$ **54.** $x^{2/3} + y^{2/3} = 2$, $(1, 1)$

55. $x^2 + \sin y = xy^2 + 1$, $(1, 0)$

56. $\sin(x - y) = x \cos\left(y + \frac{\pi}{4}\right)$, $\left(\frac{\pi}{4}, \frac{\pi}{4}\right)$

57. $2x^{1/2} + 4y^{-1/2} = xy$, $(1, 4)$ **58.** $x^2 e^y + ye^x = 4$, $(2, 0)$

59. $e^{2x - y} = \dfrac{x^2}{y}$, $(2, 4)$

60. $y^2 e^{x^2 - 16} - xy^{-1} = 2$, $(4, 2)$

61. Find the points on the graph of $y^2 = x^3 - 3x + 1$ (Figure 10) where the tangent line is horizontal.

(a) First show that $2yy' = 3x^2 - 3$, where $y' = dy/dx$.

(b) Do not solve for y'. Rather, set $y' = 0$ and solve for x. This yields two values of x where the slope may be zero.

(c) Show that the positive value of x does not correspond to a point on the graph.

(d) The negative value corresponds to the two points on the graph where the tangent line is horizontal. Find their coordinates.

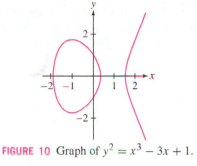

FIGURE 10 Graph of $y^2 = x^3 - 3x + 1$.

62. Show, by differentiating the equation, that if the tangent line at a point (x, y) on the curve $x^2 y - 2x + 8y = 2$ is horizontal, then $xy = 1$. Then substitute $y = x^{-1}$ in $x^2 y - 2x + 8y = 2$ to show that the tangent line is horizontal at the points $\left(2, \frac{1}{2}\right)$ and $\left(-4, -\frac{1}{4}\right)$.

63. Find all points on the graph of $3x^2 + 4y^2 + 3xy = 24$ where the tangent line is horizontal (Figure 11).

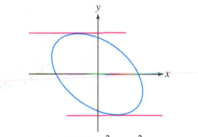

FIGURE 11 Graph of $3x^2 + 4y^2 + 3xy = 24$.

64. Show that no point on the graph of $x^2 - 3xy + y^2 = 1$ has a horizontal tangent line.

65. Figure 1 shows the graph of $y^4 + xy = x^3 - x + 2$. Find dy/dx at the two points on the graph with x-coordinate 0 and find an equation of the tangent line at $(1, 1)$.

66. Folium of Descartes The curve $x^3 + y^3 = 3xy$ (Figure 12) was first discussed in 1638 by the French philosopher-mathematician René Descartes, who called it the folium (meaning "leaf"). Descartes's scientific colleague Gilles de Roberval called it the jasmine flower. Both men believed incorrectly that the leaf shape in the first quadrant was repeated in each quadrant, giving the appearance of petals of a flower. Find an equation of the tangent line at the point $\left(\frac{2}{3}, \frac{4}{3}\right)$.

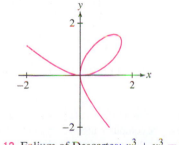

FIGURE 12 Folium of Descartes: $x^3 + y^3 = 3xy$.

67. Find a point on the folium $x^3 + y^3 = 3xy$ other than the origin at which the tangent line is horizontal.

68. [GU] Plot $x^3 + y^3 = 3xy + b$ for several values of b and describe how the graph changes as $b \to 0$. Then compute dy/dx at the point $(b^{1/3}, 0)$. How does this value change as $b \to \infty$? Do your plots confirm this conclusion?

69. Find the x-coordinates of the points where the tangent line is horizontal on the *trident curve* $xy = x^3 - 5x^2 + 2x - 1$, so named by Isaac Newton in his treatise on curves published in 1710 (Figure 13). *Hint:* $2x^3 - 5x^2 + 1 = (2x - 1)(x^2 - 2x - 1)$.

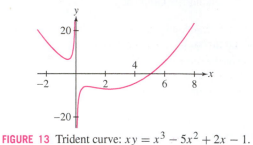

FIGURE 13 Trident curve: $xy = x^3 - 5x^2 + 2x - 1$.

70. Find an equation of the tangent line at each of the four points on the curve $(x^2 + y^2 - 4x)^2 = 2(x^2 + y^2)$ where $x = 1$. This curve (Figure 14) is an example of a *limaçon of Pascal*, named after the father of the French philosopher Blaise Pascal, who first described it in 1650.

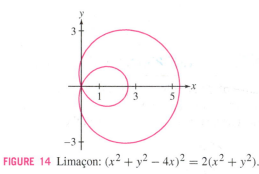

FIGURE 14 Limaçon: $(x^2 + y^2 - 4x)^2 = 2(x^2 + y^2)$.

71. Find the derivative at the points where $x = 1$ on the folium $(x^2 + y^2)^2 = \frac{25}{4}xy^2$. See Figure 15.

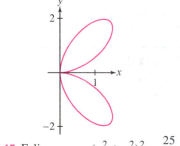

FIGURE 15 Folium curve: $(x^2 + y^2)^2 = \dfrac{25}{4}xy^2$.

72. [CAS] Plot $(x^2 + y^2)^2 = 12(x^2 - y^2) + 2$ for $-4 \le x \le 4$, $-4 \le y \le 4$ using a computer algebra system. How many horizontal tangent lines does the curve appear to have? Find the points where these occur.

73. Calculate dx/dy for the equation $y^4 + 1 = y^2 + x^2$ and find the points on the graph where the tangent line is vertical.

74. Show that the tangent lines at $x = 1 \pm \sqrt{2}$ to the *conchoid* with equation $(x - 1)^2(x^2 + y^2) = 2x^2$ are vertical (Figure 16).

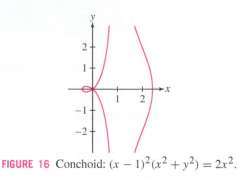

FIGURE 16 Conchoid: $(x - 1)^2(x^2 + y^2) = 2x^2$.

75. [CAS] Use a computer algebra system to plot $y^2 = x^3 - 4x$ for $-4 \le x \le 4$, $-4 \le y \le 4$. Show that if $dx/dy = 0$, then $y = 0$. Conclude that the tangent line is vertical at the points where the curve intersects the x-axis. Does your plot confirm this conclusion?

76. Show that for all points P on the graph in Figure 17, the segments $\overline{OP}$ and $\overline{PR}$ have equal length.

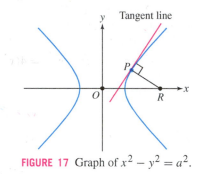

FIGURE 17 Graph of $x^2 - y^2 = a^2$.

In Exercises 77–80, use implicit differentiation to calculate higher derivatives.

77. Consider the equation $y^3 - \frac{3}{2}x^2 = 1$.

(a) Show that $y' = x/y^2$ and differentiate again to show that

$$y'' = \frac{y^2 - 2xyy'}{y^4}$$

(b) Express y'' in terms of x and y using part (a).

78. Use the method of the previous exercise to show that $y'' = -y^{-3}$ on the circle $x^2 + y^2 = 1$.

79. Calculate y'' at the point $(1, 1)$ on the curve $xy^2 + y - 2 = 0$ by the following steps:

(a) Find y' by implicit differentiation and calculate y' at the point $(1, 1)$.

(b) Differentiate the expression for y' found in (a). Then compute y'' at $(1, 1)$ by substituting $x = 1$, $y = 1$, and the value of y' found in (a).

80. Use the method of the previous exercise to compute y'' at the point $(1, 1)$ on the curve $x^3 + y^3 = 3x + y - 2$.

In Exercises 81–83, x and y are functions of a variable t and use implicit differentiation to relate dy/dt and dx/dt.

81. Differentiate $xy = 1$ with respect to t and derive the relation $\dfrac{dy}{dt} = -\dfrac{y}{x}\dfrac{dx}{dt}$.

82. Differentiate

$$x^3 + 3xy^2 = 1$$

with respect to t and express dy/dt in terms of dx/dt, as in Exercise 81.

83. Calculate dy/dt in terms of dx/dt.

(a) $x^3 - y^3 = 1$
(b) $y^4 + 2xy + x^2 = 0$

84. 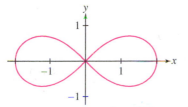 The volume V and pressure P of gas in a piston (which vary in time t) satisfy $PV^{3/2} = C$, where C is a constant. Prove that

$$\frac{dP/dt}{dV/dt} = -\frac{3}{2}\frac{P}{V}$$

The ratio of the derivatives is negative. Could you have predicted this from the relation $PV^{3/2} = C$?

Further Insights and Challenges

85. Show that if P lies on the intersection of the two curves $x^2 - y^2 = c$ and $xy = d$ (c, d constants), then the tangents to the curves at P are perpendicular.

86. The *lemniscate curve* $(x^2 + y^2)^2 = 4(x^2 - y^2)$ was discovered by Jacob Bernoulli in 1694, who noted that it is "shaped like a figure 8, or a knot, or the bow of a ribbon." Find the coordinates of the four points at which the tangent line is horizontal (Figure 18).

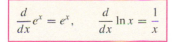

FIGURE 18 Lemniscate curve: $(x^2 + y^2)^2 = 4(x^2 - y^2)$.

87. Divide the curve in Figure 19 into five branches, each of which is the graph of a function. Sketch the branches.

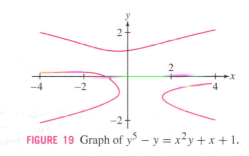

FIGURE 19 Graph of $y^5 - y = x^2 y + x + 1$.

3.9 Derivatives of General Exponential and Logarithmic Functions

◄··· **REMINDER** $\ln x$ is the natural logarithm; that is, $\ln x = \log_e x$.

We would like to know the derivative of $f(x) = \ln x$. Implicit differentiation makes this straightforward to determine the derivative of $f(x) = \ln x$. Letting $y = \ln x$ and assuming $x > 0$ so it is in the domain of $\ln x$, our goal is to find dy/dx. But then we have

$$e^y = e^{\ln x}$$

$$e^y = x$$

Implicitly differentiating, we obtain

$$e^y \frac{dy}{dx} = 1$$

$$\frac{dy}{dx} = \frac{1}{e^y} = \frac{1}{x}$$

THEOREM 1 **Derivative of the Natural Logarithm**

$$\frac{d}{dx}\ln x = \frac{1}{x} \qquad \text{for } x > 0 \qquad \boxed{1}$$

The two most important calculus facts about exponentials and logs are

$$\frac{d}{dx}e^x = e^x, \qquad \frac{d}{dx}\ln x = \frac{1}{x}$$

■ **EXAMPLE 1** Differentiate: **(a)** $y = x \ln x$ and **(b)** $y = (\ln x)^2$.

Solution

(a) Use the Product Rule:

$$\frac{d}{dx}(x \ln x) = x \cdot (\ln x)' + (x)' \cdot \ln x$$

$$= x \cdot \frac{1}{x} + \ln x = 1 + \ln x$$

(b) Use the General Power Rule:

$$\frac{d}{dx}(\ln x)^2 = 2 \ln x \cdot \frac{d}{dx} \ln x = \frac{2 \ln x}{x}$$

∎

We obtain a useful formula for the derivative of $\ln(f(x))$ by applying the Chain Rule with $u = f(x)$:

$$\frac{d}{dx} \ln(f(x)) = \frac{d}{du} \ln(u) \frac{du}{dx} = \frac{1}{u} \cdot u' = \frac{1}{f(x)} f'(x)$$

$$\boxed{\frac{d}{dx} \ln(f(x)) = \frac{f'(x)}{f(x)}}$$

2

> In Section 3.2, we proved the Power Rule for whole-number exponents. We can now prove it for all exponents n by writing x^n as an exponential and using the Chain Rule. For $x > 0$,
>
> $$x^n = (e^{\ln x})^n = e^{n \ln x}$$
> $$\frac{d}{dx} x^n = \frac{d}{dx} e^{n \ln x} = \left(\frac{d}{dx} n \ln x\right) e^{n \ln x}$$
> $$= \left(\frac{n}{x}\right) x^n = nx^{n-1}$$

■ **EXAMPLE 2** Differentiate: **(a)** $y = \ln(x^3 + 1)$ and   **(b)** $y = \ln(\sqrt{\sin x})$.

Solution   Use Eq. (2):

(a) $\dfrac{d}{dx} \ln(x^3 + 1) = \dfrac{(x^3 + 1)'}{x^3 + 1} = \dfrac{3x^2}{x^3 + 1}$

(b) The algebra is simpler if we write $\ln(\sqrt{\sin x}) = \ln\left((\sin x)^{1/2}\right) = \frac{1}{2} \ln(\sin x)$:

$$\frac{d}{dx} \ln(\sqrt{\sin x}) = \frac{1}{2} \frac{d}{dx} \ln(\sin x)$$

$$= \frac{1}{2} \frac{(\sin x)'}{\sin x} = \frac{1}{2} \frac{\cos x}{\sin x} = \frac{1}{2} \cot x$$

■

> ◀┅ **REMINDER** According to Eq. (1) in Section 1.6, we have the "change-of-base" formulas:
>
> $$\boxed{\log_b x = \frac{\log_a x}{\log_a b}, \qquad \log_b x = \frac{\ln x}{\ln b}}$$
>
> It follows, as in Example 3, that for any base $b > 0$, $b \neq 1$:
>
> $$\frac{d}{dx} \log_b x = \frac{1}{(\ln b)x}$$

■ **EXAMPLE 3**   **Logarithm to Another Base**   Calculate $\dfrac{d}{dx} \log_{10} x$.

Solution   By the change-of-base formula (see margin), $\log_{10} x = \frac{\ln x}{\ln 10}$. Therefore,

$$\frac{d}{dx} \log_{10} x = \frac{d}{dx} \left(\frac{\ln x}{\ln 10}\right) = \frac{1}{\ln 10} \frac{d}{dx} \ln x = \frac{1}{(\ln 10)x}$$

■

The Derivative of $f(x) = b^x$

In Section 3.2, we proved that for any base $b > 0$,

$$\frac{d}{dx} b^x = m(b) \, b^x, \qquad \text{where} \quad m(b) = \lim_{h \to 0} \frac{b^h - 1}{h}$$

but we were not able to identify the factor $m(b)$ [other than to say that e is the unique number for which $m(e) = 1$]. Now we can use implicit differentiation to prove that $m(b) = \ln b$.

If $y = b^x$, then our goal is to find $\frac{dy}{dx}$. Taking the natural log of both sides yields

$$\ln y = \ln b^x = x \ln b$$

Then we implicitly differentiate each side of the equation with respect to x, treating x as itself and treating y as a function of x. This yields

$$\frac{1}{y} \frac{dy}{dx} = \ln b$$

Thus, since $y = b^x$, we have $\frac{dy}{dx} = y \ln b = b^x \ln b$. Therefore, we have the following theorem.

THEOREM 2 Derivative of $f(x) = b^x$

$$\frac{d}{dx}b^x = (\ln b)b^x \qquad \text{for } b > 0 \qquad \boxed{3}$$

For example, $(10^x)' = (\ln 10)10^x$.

■ **EXAMPLE 4** Differentiate: **(a)** $f(x) = 4^{3x}$ and **(b)** $f(x) = 5^{x^2}$.

Solution

(a) The function $f(x) = 4^{3x}$ is a composite of 4^u and $u = 3x$:

$$\frac{d}{dx}4^{3x} = \left(\frac{d}{du}4^u\right)\frac{du}{dx} = (\ln 4)4^u(3x)' = (\ln 4)4^{3x}(3) = (3\ln 4)4^{3x}$$

(b) The function $f(x) = 5^{x^2}$ is a composite of 5^u and $u = x^2$:

$$\frac{d}{dx}5^{x^2} = \left(\frac{d}{du}5^u\right)\frac{du}{dx} = (\ln 5)5^u(x^2)' = (\ln 5)5^{x^2}(2x) = (2\ln 5)x\,5^{x^2}$$ ■

Logarithmic Differentiation

The next example illustrates **logarithmic differentiation**. This technique saves work when the function is a product or quotient with several factors.

Logarithmic differentiation makes what would be a painful and tedious derivative to take, involving multiple applications of the product and quotient rules, into a relatively easy procedure.

■ **EXAMPLE 5** Find the derivative of

$$f(x) = \frac{(x+1)^2(2x^2-3)}{\sqrt{x^2+1}}$$

Solution In logarithmic differentiation, we differentiate $\ln(f(x))$ rather than $f(x)$ itself. First, we take the natural log of both sides of the equation:

$$\ln f(x) = \ln\left(\frac{(x+1)^2(2x^2-3)}{\sqrt{x^2+1}}\right)$$

Then we expand the right-hand side using the logarithm rules:

$$\ln(f(x)) = \ln\left((x+1)^2\right) + \ln\left(2x^2-3\right) - \ln\left(\sqrt{x^2+1}\right)$$

$$= 2\ln(x+1) + \ln\left(2x^2-3\right) - \frac{1}{2}\ln(x^2+1)$$

Next use Eq. (2):

$$\frac{f'(x)}{f(x)} = \frac{d}{dx}\ln(f(x)) = 2\frac{d}{dx}\ln(x+1) + \frac{d}{dx}\ln(2x^2-3) - \frac{1}{2}\frac{d}{dx}\ln(x^2+1)$$

$$\frac{f'(x)}{f(x)} = \frac{2}{x+1} + \frac{4x}{2x^2-3} - \frac{1}{2}\frac{2x}{x^2+1}$$

Finally, multiply through by $f(x)$:

$$f'(x) = \left(\frac{2}{x+1} + \frac{4x}{2x^2-3} - \frac{x}{x^2+1}\right)\left(\frac{(x+1)^2(2x^2-3)}{\sqrt{x^2+1}}\right)$$ ■

Logarithmic differentiation also allows us to take the derivative of functions of the form $y = f(x)^{g(x)}$, where both the base and the exponent depend on x.

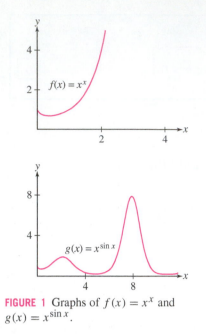

FIGURE 1 Graphs of $f(x) = x^x$ and $g(x) = x^{\sin x}$.

■ **EXAMPLE 6** Differentiate (for $x > 0$): **(a)** $f(x) = x^x$ and **(b)** $g(x) = x^{\sin x}$.

Solution The two problems are similar (Figure 1). We illustrate two different methods.

(a) Method 1: Use the identity $x = e^{\ln x}$ to rewrite $f(x)$ as an exponential:

$$f(x) = x^x = (e^{\ln x})^x = e^{x \ln x}$$

$$f'(x) = (x \ln x)' e^{x \ln x} = (1 + \ln x) e^{x \ln x} = (1 + \ln x) x^x$$

(b) Method 2: Apply Eq. (2) to $\ln(g(x))$. Since $\ln(g(x)) = \ln(x^{\sin x}) = (\sin x) \ln x$,

$$\frac{g'(x)}{g(x)} = \frac{d}{dx} \ln(g(x)) = \frac{d}{dx}\big((\sin x) \ln x\big) = \frac{\sin x}{x} + (\cos x) \ln x$$

$$g'(x) = \left(\frac{\sin x}{x} + (\cos x) \ln x\right) g(x) = \left(\frac{\sin x}{x} + (\cos x) \ln x\right) x^{\sin x} \qquad ■$$

Derivatives of Hyperbolic Functions

Recall from Section 1.6 that the hyperbolic functions are special combinations of e^x and e^{-x}. The formulas for their derivatives are similar to those for the corresponding trigonometric functions, differing at most by a sign.

Consider the hyperbolic sine and cosine:

$$\sinh x = \frac{e^x - e^{-x}}{2}, \qquad \cosh x = \frac{e^x + e^{-x}}{2}$$

Their derivatives are

$$\frac{d}{dx} \sinh x = \cosh x, \qquad \frac{d}{dx} \cosh x = \sinh x$$

We can check this directly. For example,

$$\frac{d}{dx}\left(\frac{e^x - e^{-x}}{2}\right) = \left(\frac{e^x - e^{-x}}{2}\right)' = \frac{e^x + e^{-x}}{2} = \cosh x$$

Note the resemblance to the formulas $\frac{d}{dx} \sin x = \cos x$, $\frac{d}{dx} \cos x = -\sin x$. The derivatives of the other hyperbolic functions, which are computed in a similar fashion, also differ from their trigonometric counterparts by a sign at most.

◀·· **REMINDER**

$$\tanh x = \frac{\sinh x}{\cosh x} = \frac{e^x - e^{-x}}{e^x + e^{-x}}$$

$$\operatorname{sech} x = \frac{1}{\cosh x} = \frac{2}{e^x + e^{-x}}$$

$$\coth x = \frac{\cosh x}{\sinh x} = \frac{e^x + e^{-x}}{e^x - e^{-x}}$$

$$\operatorname{csch} x = \frac{1}{\sinh x} = \frac{2}{e^x - e^{-x}}$$

◀·· **REMINDER** Hyperbolic sine and cosine satisfy the basic identity (Section 1.6)

$$\cosh^2 x - \sinh^2 x = 1$$

Derivatives of Hyperbolic and Trigonometric Functions

$$\frac{d}{dx} \tanh x = \operatorname{sech}^2 x, \qquad \frac{d}{dx} \tan x = \sec^2 x$$

$$\frac{d}{dx} \coth x = -\operatorname{csch}^2 x, \qquad \frac{d}{dx} \cot x = -\csc^2 x$$

$$\frac{d}{dx} \operatorname{sech} x = -\operatorname{sech} x \tanh x, \qquad \frac{d}{dx} \sec x = \sec x \tan x$$

$$\frac{d}{dx} \operatorname{csch} x = -\operatorname{csch} x \coth x, \qquad \frac{d}{dx} \csc x = -\csc x \cot x$$

■ **EXAMPLE 7** Verify $\dfrac{d}{dx} \coth x = -\operatorname{csch}^2 x$.

Solution By the Quotient Rule and the identity $\cosh^2 x - \sinh^2 x = 1$,

$$\frac{d}{dx} \coth x = \left(\frac{\cosh x}{\sinh x}\right)' = \frac{(\sinh x)(\cosh x)' - (\cosh x)(\sinh x)'}{\sinh^2 x}$$

$$= \frac{\sinh^2 x - \cosh^2 x}{\sinh^2 x} = \frac{-1}{\sinh^2 x} = -\operatorname{csch}^2 x \qquad ■$$

■ **EXAMPLE 8** Calculate: (a) $\dfrac{d}{dx}\cosh(3x^2+1)$ and (b) $\dfrac{d}{dx}\sinh x \tanh x$.

Solution

(a) By the Chain Rule, $\dfrac{d}{dx}\cosh(3x^2+1)=6x\sinh(3x^2+1)$.

(b) By the Product Rule,

$$\frac{d}{dx}(\sinh x \tanh x)=\sinh x \,\text{sech}^2 x + \tanh x \cosh x = \text{sech}\, x \tanh x + \sinh x$$ ■

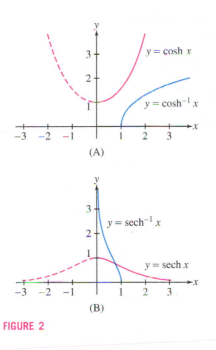

FIGURE 2

←·· **REMINDER** *The derivatives of* $\cosh^{-1}x$ *and* $\text{sech}^{-1}x$ *are undefined at the endpoint* $x=1$ *of their domains.*

Inverse Hyperbolic Functions

Recall that a function f with domain D has an inverse if it is one-to-one on D. Each of the hyperbolic functions except $f(x)=\cosh x$ and $f(x)=\text{sech}\, x$ is one-to-one on its domain and therefore has a well-defined inverse. The functions $f(x)=\cosh x$ and $f(x)=\text{sech}\, x$ are one-to-one on the restricted domain $\{x : x \ge 0\}$. We let $f(x)=\cosh^{-1}x$ and $f(x)=\text{sech}^{-1}x$ denote the corresponding inverses (Figure 2). In reading the following table, keep in mind that the domain of the inverse is equal to the range of the function.

Inverse Hyperbolic Functions and Their Derivatives

Function	Domain	Derivative
$y=\sinh^{-1}x$	all x	$\dfrac{d}{dx}\sinh^{-1}x=\dfrac{1}{\sqrt{x^2+1}}$
$y=\cosh^{-1}x$	$x\ge 1$	$\dfrac{d}{dx}\cosh^{-1}x=\dfrac{1}{\sqrt{x^2-1}}$
$y=\tanh^{-1}x$	$|x|<1$	$\dfrac{d}{dx}\tanh^{-1}x=\dfrac{1}{1-x^2}$
$y=\coth^{-1}x$	$|x|>1$	$\dfrac{d}{dx}\coth^{-1}x=\dfrac{1}{1-x^2}$
$y=\text{sech}^{-1}x$	$0<x\le 1$	$\dfrac{d}{dx}\text{sech}^{-1}x=-\dfrac{1}{x\sqrt{1-x^2}}$
$y=\text{csch}^{-1}x$	$x\ne 0$	$\dfrac{d}{dx}\text{csch}^{-1}x=-\dfrac{1}{|x|\sqrt{x^2+1}}$

■ **EXAMPLE 9** Verify $\dfrac{d}{dx}\tanh^{-1}x=\dfrac{1}{1-x^2}$.

Solution Let $y=\tanh^{-1}x$. Then $\tanh y = x$. Differentiating implicitly yields

$$\text{sech}^2 y\,\frac{dy}{dx}=1$$

$$\frac{dy}{dx}=\frac{1}{\text{sech}^2 y}$$

Therefore,

$$\frac{d}{dx}\tanh^{-1}x=\frac{1}{\text{sech}^2(\tanh^{-1}x)}$$

To compute $\text{sech}^2(y)$,

$$\cosh^2 y - \sinh^2 y = 1 \qquad \text{(basic identity)}$$

$$1-\tanh^2 y = \text{sech}^2 y \qquad \text{(divide by } \cosh^2 t)$$

$$1-x^2 = \text{sech}^2(y) \qquad \text{(because } x=\tanh y)$$

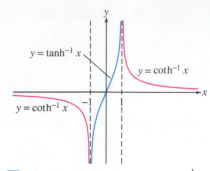

The functions $y = \tanh^{-1} x$ and $y = \coth^{-1} x$ have disjoint domains.

This gives the desired result:

$$\frac{d}{dx}\tanh^{-1} x = \frac{1}{\text{sech}^2 y} = \frac{1}{1 - x^2}$$ ∎

The functions $y = \tanh^{-1} x$ and $y = \coth^{-1} x$ both have derivative $1/(1 - x^2)$. Note, however, that their domains are disjoint (Figure 3).

3.9 SUMMARY

- Derivative formulas:

$$\frac{d}{dx}e^x = e^x, \qquad \frac{d}{dx}\ln x = \frac{1}{x}, \qquad \frac{d}{dx}b^x = (\ln b)b^x, \qquad \frac{d}{dx}\log_b x = \frac{1}{(\ln b)x}$$

- Use logarithmic differentiation on functions that are products or quotients of several factors, and on functions of the form $y = f(x)^{g(x)}$.
- Hyperbolic functions:

$$\frac{d}{dx}\sinh x = \cosh x, \qquad\qquad \frac{d}{dx}\cosh x = \sinh x$$

$$\frac{d}{dx}\tanh x = \text{sech}^2 x, \qquad\qquad \frac{d}{dx}\coth x = -\text{csch}^2 x$$

$$\frac{d}{dx}\text{sech}\, x = -\text{sech}\, x \tanh x, \qquad \frac{d}{dx}\text{csch}\, x = -\text{csch}\, x \coth x$$

- Inverse hyperbolic functions:

$$\frac{d}{dx}\sinh^{-1} x = \frac{1}{\sqrt{x^2 + 1}}, \qquad\qquad \frac{d}{dx}\cosh^{-1} x = \frac{1}{\sqrt{x^2 - 1}} \quad (x > 1)$$

$$\frac{d}{dx}\tanh^{-1} x = \frac{1}{1 - x^2} \quad (|x| < 1), \qquad \frac{d}{dx}\coth^{-1} x = \frac{1}{1 - x^2} \quad (|x| > 1)$$

$$\frac{d}{dx}\text{sech}^{-1} x = \frac{-1}{x\sqrt{1 - x^2}} \quad (0 < x < 1), \qquad \frac{d}{dx}\text{csch}^{-1} x = -\frac{1}{|x|\sqrt{x^2 + 1}} \quad (x \neq 0)$$

3.9 EXERCISES

Preliminary Questions

1. What is the slope of the tangent line to $y = 4^x$ at $x = 0$?

2. What is the rate of change of $y = \ln x$ at $x = 10$?

3. What is $b > 0$ if the tangent line to $y = b^x$ at $x = 0$ has slope 2?

4. What is b if $(\log_b x)' = \dfrac{1}{3x}$?

5. What are $y^{(100)}$ and $y^{(101)}$ for $y = \cosh x$?

Exercises

In Exercises 1–20, find the derivative.

1. $y = x \ln x$

2. $y = t \ln t - t$

3. $y = 2^{x^3}$

4. $y = \ln(x^5)$

5. $y = \ln(9x^2 - 8)$

6. $y = \ln(t5^t)$

7. $y = (\ln x)^2$

8. $y = x^2 \ln x$

9. $y = e^{(\ln x)^2}$

10. $y = \dfrac{\ln x}{x}$

11. $y = \ln(\ln x)$

12. $y = \ln(\cot x)$

13. $y = \left(\ln(\ln x)\right)^3$

14. $y = \ln\left((\ln x)^3\right)$

15. $y = \ln\left((x + 1)(2x + 9)\right)$

16. $y = \ln\left(\dfrac{x + 1}{x^3 + 1}\right)$

17. $y = 11^x$

18. $y = 7^{4x - x^2}$

19. $y = \dfrac{2^x - 3^{-x}}{x}$

20. $y = 16^{\sin x}$

In Exercises 21–24, compute the derivative.

21. $f'(x), \quad f(x) = \log_2 x$

22. $f'(3), \quad f(x) = \log_5 x$

23. $\dfrac{d}{dt}\log_3(\sin t)$

24. $\dfrac{d}{dt}\log_{10}(t + 2^t)$

In Exercises 25–36, find an equation of the tangent line at the point indicated.

25. $f(x) = 6^x$, $x = 2$

26. $y = (\sqrt{2})^x$, $x = 8$

27. $s(t) = 3^{9t}$, $t = 2$

28. $y = \pi^{5x-2}$, $x = 1$

29. $f(x) = 5^{x^2-2x}$, $x = 1$

30. $s(t) = \ln t$, $t = 5$

31. $s(t) = \ln(8 - 4t)$, $t = 1$

32. $f(x) = \ln(x^2)$, $x = 4$

33. $R(z) = \log_5(2z^2 + 7)$, $z = 3$

34. $y = \ln(\sin x)$, $x = \dfrac{\pi}{4}$

35. $f(w) = \log_2 w$, $w = \frac{1}{8}$

36. $y = \log_2(1 + 4x^{-1})$, $x = 4$

In Exercises 37–44, find the derivative using logarithmic differentiation as in Example 5.

37. $y = (x + 5)(x + 9)$

38. $y = (3x + 5)(4x + 9)$

39. $y = (x - 1)(x - 12)(x + 7)$

40. $y = \dfrac{x(x + 1)^3}{(3x - 1)^2}$

41. $y = \dfrac{x(x^2 + 1)}{\sqrt{x + 1}}$

42. $y = (2x + 1)(4x^2)\sqrt{x - 9}$

43. $y = \sqrt{\dfrac{x(x + 2)}{(2x + 1)(3x + 2)}}$

44. $y = (x^3 + 1)(x^4 + 2)(x^5 + 3)^2$

In Exercises 45–50, find the derivative using either method of Example 6.

45. $f(x) = x^{3x}$

46. $f(x) = x^{3^x}$

47. $f(x) = x^{e^x}$

48. $f(x) = x^{x^2}$

49. $f(x) = x^{\cos x}$

50. $f(x) = e^{x^x}$

In Exercises 51–74, calculate the derivative.

51. $y = \sinh(9x)$

52. $y = \sinh(x^2)$

53. $y = \cosh^2(9 - 3t)$

54. $y = \tanh(t^2 + 1)$

55. $y = \sqrt{\cosh x + 1}$

56. $y = \sinh x \tanh x$

57. $y = \dfrac{\coth t}{1 + \tanh t}$

58. $y = (\ln(\cosh x))^5$

59. $y = \sinh(\ln x)$

60. $y = e^{\coth x}$

61. $y = \tanh(e^x)$

62. $y = \sinh(\cosh^3 x)$

63. $y = \operatorname{sech}(\sqrt{x})$

64. $y = \ln(\coth x)$

65. $y = \operatorname{sech} x \coth x$

66. $y = x^{\sinh x}$

67. $y = \cosh^{-1}(3x)$

68. $y = \tanh^{-1}(e^x + x^2)$

69. $y = (\sinh^{-1}(x^2))^3$

70. $y = (\operatorname{csch}^{-1} 3x)^4$

71. $y = e^{\cosh^{-1} x}$

72. $y = \sinh^{-1}(\sqrt{x^2 + 1})$

73. $y = \tanh^{-1}(\ln t)$

74. $y = \ln(\tanh^{-1} x)$

In Exercises 75–77, prove the formula.

75. $\dfrac{d}{dx}(\coth x) = -\operatorname{csch}^2 x$

76. $\dfrac{d}{dt} \sinh^{-1} t = \dfrac{1}{\sqrt{t^2 + 1}}$

77. $\dfrac{d}{dt} \cosh^{-1} t = \dfrac{1}{\sqrt{t^2 - 1}}$ for $t > 1$

78. Use the formula $(\ln f(x))' = f'(x)/f(x)$ to show that $\ln x$ and $\ln(2x)$ have the same derivative. Is there a simpler explanation of this result?

79. According to one simplified model, the purchasing power of a dollar in the year $2000 + t$ is equal to $P(t) = 0.68(1.04)^{-t}$ (in 1983 dollars). Calculate the predicted rate of decline in purchasing power (in cents per year) in the year 2020.

80. The energy E (in joules) radiated as seismic waves by an earthquake of Richter magnitude M satisfies $\log_{10} E = 4.8 + 1.5M$.

(a) Show that when M increases by 1, the energy increases by a factor of approximately 31.5.

(b) Calculate dE/dM.

81. Show that for any constants M, k, and a, the function

$$y(t) = \frac{1}{2}M\left(1 + \tanh\left(\frac{k(t - a)}{2}\right)\right)$$

satisfies the **logistic equation**: $\dfrac{y'}{y} = k\left(1 - \dfrac{y}{M}\right)$.

82. Show that $V(x) = 2\ln(\tanh(x/2))$ satisfies the **Poisson–Boltzmann** equation $V''(x) = \sinh(V(x))$, which is used to describe electrostatic forces in certain molecules.

83. The Palermo Technical Impact Hazard Scale P is used to quantify the risk associated with the impact of an asteroid colliding with the earth:

$$P = \log_{10}\left(\frac{p_i E^{0.8}}{0.03T}\right)$$

where p_i is the probability of impact, T is the number of years until impact, and E is the energy of impact (in megatons of TNT). The risk is greater than a random event of similar magnitude if $P > 0$.

(a) Calculate dP/dT, assuming that $p_i = 2 \times 10^{-5}$ and $E = 2$ megatons.

(b) Use the derivative to estimate the change in P if T increases from 8 to 9 years.

Further Insights and Challenges

84. (a) Show that if f and g are differentiable, then

$$\frac{d}{dx}\ln(f(x)g(x)) = \frac{f'(x)}{f(x)} + \frac{g'(x)}{g(x)} \qquad \boxed{4}$$

(b) Give a new proof of the Product Rule by observing that the left-hand side of Eq. (4) is equal to $\dfrac{(f(x)g(x))'}{f(x)g(x)}$.

85. Use the formula $\log_b x = \dfrac{\log_a x}{\log_a b}$ for $a, b > 0$ to verify the formula

$$\frac{d}{dx}\log_b x = \frac{1}{(\ln b)x}$$

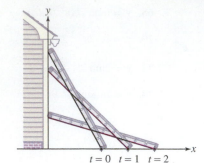

FIGURE 1 Positions of a ladder at times $t = 0, 1, 2$.

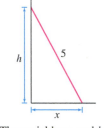

FIGURE 2 The variables x and h.

t	x	h	dh/dt
0	1.5	4.77	−0.25
1	2.3	4.44	−0.41
2	3.1	3.92	−0.63
3	3.9	3.13	−1.00

This table of values confirms that the top of the ladder is speeding up.

3.10 Related Rates

In *related-rate* problems, the goal is to calculate an unknown rate of change in terms of other rates of change that are known. The "sliding ladder problem" is a good example: A ladder leans against a wall as the bottom is pulled away at constant velocity. *How fast does the top of the ladder move?* What is interesting and perhaps surprising is that the top and bottom travel at different speeds. Figure 1 shows this clearly: The bottom travels the same distance over each time interval, but the top travels farther during the second time interval than the first. In other words, the top is speeding up while the bottom moves at a constant speed. In the next example, we use calculus to find the velocity of the ladder's top.

■ **EXAMPLE 1** **Sliding Ladder Problem** A 5-m ladder leans against a wall. The bottom of the ladder is 1.5 m from the wall at time $t = 0$ and slides away from the wall at a rate of 0.8 m/s. Find the velocity of the top of the ladder at time $t = 1$.

Solution The first step in any related-rate problem is to choose variables for the relevant quantities. Since we are considering how the top and bottom of the ladder change position, we use variables (Figure 2):

- $x = x(t)$ distance from the bottom of the ladder to the wall
- $h = h(t)$ height of the ladder's top

Both x and h are functions of time. The velocity of the bottom is $dx/dt = 0.8$ m/s. The unknown velocity of the top is dh/dt, and the initial distance from the bottom to the wall is $x(0) = 1.5$, so we can restate the problem as

> Compute $\dfrac{dh}{dt}$ at $t = 1$ given that $\dfrac{dx}{dt} = 0.8$ m/s and $x(0) = 1.5$ m

To solve this problem, we need an equation relating x and h (Figure 2). This is provided by the Pythagorean Theorem:

$$x^2 + h^2 = 5^2$$

To calculate dh/dt, we differentiate both sides of this equation *with respect to t*:

$$\frac{d}{dt}x^2 + \frac{d}{dt}h^2 = \frac{d}{dt}5^2$$

$$2x\frac{dx}{dt} + 2h\frac{dh}{dt} = 0$$

Therefore, $\dfrac{dh}{dt} = -\dfrac{x}{h}\dfrac{dx}{dt}$, and because $\dfrac{dx}{dt} = 0.8$ m/s, the velocity of the top is

$$\boxed{\frac{dh}{dt} = -0.8\frac{x}{h} \text{ m/s}} \qquad \boxed{1}$$

To apply this formula, we must find x and h at time $t = 1$. Since the bottom slides away at 0.8 m/s and $x(0) = 1.5$, we have $x(1) = 2.3$ and $h(1) = \sqrt{5^2 - 2.3^2} \approx 4.44$. We obtain (note that the answer is negative because the ladder top is falling)

$$\frac{dh}{dt}\bigg|_{t=1} = -0.8\frac{x(1)}{h(1)} \approx -0.8\frac{2.3}{4.44} \approx -0.41 \text{ m/s} \qquad ■$$

CONCEPTUAL INSIGHT A puzzling feature of Eq. (1) is that the velocity dh/dt, which is equal to $-0.8x/h$, becomes infinite as $h \to 0$ (as the top of the ladder gets close to the ground). Since this is impossible, our mathematical model must break down as $h \to 0$. In fact, one can show that the ladder's top loses contact with the wall before reaching the bottom and from that moment on, the formula is no longer valid.

In the next examples, we divide the solution into three steps that can be followed when working the exercises.

■ **EXAMPLE 2** Filling a Rectangular Tank Water pours into a fish tank at a rate of 0.3 m³/min. How fast is the water level rising if the base of the tank is a rectangle of dimensions 2 × 3 m?

Solution To solve a related-rate problem, it is useful to draw a diagram if possible. Figure 3 illustrates our problem.

Step 1. Identify variables and restate the problem.
First, we must recognize that the rate at which water pours into the tank is the derivative of water volume with respect to time. Therefore, let V be the volume and h the height of the water at time t. Then

$$\frac{dV}{dt} = \text{rate at which water is added to the tank}$$

$$\frac{dh}{dt} = \text{rate at which the water level is rising}$$

Now we can restate our problem in terms of derivatives:

$$\text{Compute } \frac{dh}{dt} \quad \text{given that} \quad \frac{dV}{dt} = 0.3 \text{ m}^3/\text{min}$$

Step 2. Find an equation relating the variables and differentiate with respect to time.
We need a relation between V and h. We have $V = 6h$ since the tank's base has area 6 m². Therefore,

$$\frac{dV}{dt} = 6\frac{dh}{dt} \quad \Rightarrow \quad \frac{dh}{dt} = \frac{1}{6}\frac{dV}{dt}$$

Step 3. Use the data to find the unknown derivative.
Because $dV/dt = 0.3$, the water level rises at the rate

$$\frac{dh}{dt} = \frac{1}{6}\frac{dV}{dt} = \frac{1}{6}(0.3) = 0.05 \text{ m/min}$$

Note that dh/dt has units of meters per minute because h and t are in meters and minutes, respectively. ■

The set-up in the next example is similar but more complicated because the water tank has the shape of a circular cone. We use similar triangles to derive a relation between the volume and height of the water. We also need the formula $V = \frac{1}{3}\pi hr^2$ for the volume of a circular cone of height h and radius r.

■ **EXAMPLE 3** 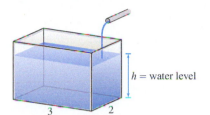 Filling a Conical Tank Water pours into a conical tank of height 10 m and radius 4 m at a rate of 6 m³/min.

(a) At what rate is the water level rising when the level is 5 m high?

(b) As time passes, what happens to the rate at which the water level rises?

Solution

(a) *Step 1.* **Assign variables and restate the problem.**
As in the previous example, let V and h be the volume and height of the water in the tank at time t. Our problem, in terms of derivatives, is

$$\text{Compute } \frac{dh}{dt} \text{ at } h = 5 \quad \text{given that} \quad \frac{dV}{dt} = 6 \text{ m}^3/\text{min}$$

FIGURE 3 $V = $ water volume at time t.

$h = $ water level

It is helpful to choose variables that are related to or traditionally associated with the quantity represented, such as V for volume, θ for an angle, h or y for height, and r for radius.

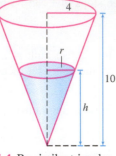

FIGURE 4 By similar triangles,

$$\frac{r}{h} = \frac{4}{10}$$

CAUTION *A common mistake is substituting the particular value h = 5 in Eq. (2). Do not set h = 5 until the end of the problem, after the derivatives have been computed. This applies to all related-rate problems.*

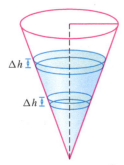

FIGURE 5 When h is larger, it takes more water to raise the level by an amount Δh.

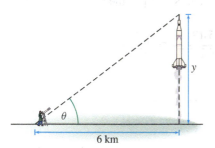

FIGURE 6 Tracking a rocket through a telescope.

Step 2. Find an equation relating the variables and differentiate with respect to time.
When the water level is h, the volume of water in the cone is $V = \frac{1}{3}\pi h r^2$, where r is the radius of the cone at height h. In order to use this relation, we eliminate the variable r. Using similar triangles in Figure 4, we see that

$$\frac{r}{h} = \frac{4}{10}$$

or

$$r = 0.4 h$$

Therefore,

$$V = \frac{1}{3}\pi h(0.4 h)^2 = \left(\frac{0.16}{3}\right)\pi h^3$$

$$\frac{dV}{dt} = (0.16)\pi h^2 \frac{dh}{dt} \qquad \boxed{2}$$

Step 3. Use the data to find the unknown derivative.
We are given $\dfrac{dV}{dt} = 6$. Using this in Eq. (2), we obtain

$$(0.16)\pi h^2 \frac{dh}{dt} = 6$$

$$\frac{dh}{dt} = \frac{6}{(0.16)\pi h^2} \qquad \boxed{3}$$

When $h = 5$, the level is rising at a rate of $\dfrac{dh}{dt} = \dfrac{6}{(0.16)\pi 5^2} \approx 0.48$ m/min.

(b) Eq. (3) shows that dh/dt is inversely proportional to h^2. As h increases, the water level rises more slowly. This is reasonable if you consider that a thin slice of the cone of width Δh has more volume when h is large, so more water is needed to raise the level when h is large (Figure 5). ∎

■ **EXAMPLE 4** **Tracking a Rocket** A spy uses a telescope to track a rocket launched vertically from a launching pad 6 km away, as in Figure 6. At a certain moment, the angle θ between the telescope and the ground is equal to $\frac{\pi}{3}$ and is changing at a rate of 0.9 radians rad/min. What is the rocket's velocity at that moment?

Solution

Step 1. Assign variables and restate the problem.
Let y be the height of the rocket at time t. Our goal is to compute the rocket's velocity dy/dt when $\theta = \frac{\pi}{3}$ so we can restate the problem as follows:

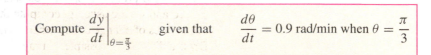

> Compute $\dfrac{dy}{dt}\bigg|_{\theta=\frac{\pi}{3}}$ given that $\dfrac{d\theta}{dt} = 0.9$ rad/min when $\theta = \dfrac{\pi}{3}$

Step 2. Find an equation relating the variables and differentiate.
We need a relation between θ and y. As we see in Figure 6,

$$\tan\theta = \frac{y}{6}$$

Now differentiate with respect to time:

$$\sec^2\theta \frac{d\theta}{dt} = \frac{1}{6}\frac{dy}{dt}$$

$$\frac{dy}{dt} = \frac{6}{\cos^2\theta}\frac{d\theta}{dt} \qquad \boxed{4}$$

Step 3. **Use the given data to find the unknown derivative.**
At the given moment, $\theta = \frac{\pi}{3}$ and $d\theta/dt = 0.9$, so Eq. (4) yields

$$\frac{dy}{dt} = \frac{6}{\cos^2(\pi/3)}(0.9) = \frac{6}{(0.5)^2}(0.9) = 21.6 \text{ km/min}$$

The rocket's velocity at this moment is 21.6 km/min, or approximately 1296 km/h. ∎

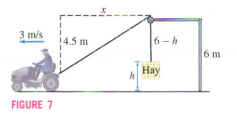

FIGURE 7

■ **EXAMPLE 5** Farmer John's tractor, traveling at 3 m/s, pulls a rope attached to a bale of hay through a pulley. With dimensions as indicated in Figure 7, how fast is the bale rising when the tractor is 5 m from the bale?

Solution

Step 1. **Assign variables and restate the problem.** Let x be the horizontal distance from the tractor to the bale of hay, and let h be the height above ground of the top of the bale. The tractor is 5 m from the bale when $x = 5$, so we can restate the problem as follows:

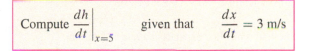

$$\text{Compute } \frac{dh}{dt}\bigg|_{x=5} \quad \text{given that} \quad \frac{dx}{dt} = 3 \text{ m/s}$$

Step 2. **Find an equation relating the variables and differentiate.**
Let L be the total length of the rope. From Figure 7 (using the Pythagorean Theorem),

$$L = \sqrt{x^2 + 4.5^2} + (6 - h)$$

Although the length L is not given, it is a constant, and therefore $dL/dt = 0$. Thus,

$$\frac{dL}{dt} = \frac{d}{dt}\left(\sqrt{x^2 + 4.5^2} + (6 - h)\right) = \frac{x\frac{dx}{dt}}{\sqrt{x^2 + 4.5^2}} - \frac{dh}{dt} = 0 \qquad \boxed{5}$$

Step 3. **Use the given data to find the unknown derivative.**
Apply Eq. (5) with $x = 5$ and $dx/dt = 3$. The bale is rising at the rate

$$\frac{dh}{dt} = \frac{x\frac{dx}{dt}}{\sqrt{x^2 + 4.5^2}} = \frac{(5)(3)}{\sqrt{5^2 + 4.5^2}} \approx 2.23 \text{ m/s} \qquad ∎$$

3.10 SUMMARY

- Related-rate problems present us with situations in which two or more variables are related and we are asked to compute the rate of change of one of the variables in terms of the rates of change of the other variable(s).
- Draw a diagram if possible. It may also be useful to break the solution into three steps:

Step 1. Identify relevant variables and restate the problem.

Step 2. Find an equation that relates the variables and differentiate with respect to time.

This gives us an equation relating the known and unknown derivatives. Remember not to substitute the specific values for the variables until after you have computed all derivatives.

Step 3. Use the given data to find the unknown derivative.

- Two facts from geometry arise often in related-rate problems: the Pythagorean Theorem and Theorem of Similar Triangles (ratios of corresponding sides are equal).

3.10 EXERCISES

Preliminary Questions

1. Assign variables and restate the following problem in terms of known and unknown derivatives (but do not solve it): How fast is the volume of a cube increasing if its side increases at a rate of 0.5 cm/s?

2. What is the relation between dV/dt and dr/dt if $V = \left(\frac{4}{3}\right)\pi r^3$?

In Questions 3 and 4, water pours into a cylindrical glass of radius 4 cm. Let V and h denote the volume and water level respectively, at time t.

3. Restate this question in terms of dV/dt and dh/dt: How fast is the water level rising if water pours in at a rate of 2 cm³/min?

4. Restate this question in terms of dV/dt and dh/dt: At what rate is water pouring in if the water level rises at a rate of 1 cm/min?

Exercises

In Exercises 1 and 2, consider a rectangular bathtub whose base is 18 ft².

1. How fast is the water level rising if water is filling the tub at a rate of 0.7 ft³/min?

2. At what rate is water pouring into the tub if the water level rises at a rate of 0.8 ft/min?

3. The radius of a circular oil slick expands at a rate of 2 m/min.

(a) How fast is the area of the oil slick increasing when the radius is 25 m?

(b) If the radius is 0 at time $t = 0$, how fast is the area increasing after 3 min?

4. At what rate is the diagonal of a cube increasing if its edges are increasing at a rate of 2 cm/s?

In Exercises 5–8, assume that the radius r of a sphere is expanding at a rate of 30 cm/min. The volume of a sphere is $V = \frac{4}{3}\pi r^3$ and its surface area is $4\pi r^2$. Determine the given rate.

5. Volume with respect to time when $r = 15$ cm

6. Volume with respect to time at $t = 2$ min, assuming that $r = 0$ at $t = 0$

7. Surface area with respect to time when $r = 40$ cm

8. Surface area with respect to time at $t = 2$ min, assuming that $r = 10$ at $t = 0$

In Exercises 9–12, refer to a 5-m ladder sliding down a wall, as in Figures 1 and 2. The variable h is the height of the ladder's top at time t, and x is the distance from the wall to the ladder's bottom.

9. Assume the bottom slides away from the wall at a rate of 0.8 m/s. Find the velocity of the top of the ladder at $t = 2$ s if the bottom is 1.5 m from the wall at $t = 0$ s.

10. Suppose that the top is sliding down the wall at a rate of 1.2 m/s. Calculate dx/dt when $h = 3$ m.

11. Suppose that $h(0) = 4$ and the top slides down the wall at a rate of 1.2 m/s. Calculate x and dx/dt at $t = 2$ s.

12. What is the relation between h and x at the moment when the top and bottom of the ladder move at the same speed?

13. A conical tank has height 3 m and radius 2 m at the top. Water flows in at a rate of 2 m³/min. How fast is the water level rising when it is 2 m?

14. Follow the same set-up as in Exercise 13, but assume that the water level is rising at a rate of 0.3 m/min when it is 2 m. At what rate is water flowing in?

15. The radius r and height h of a circular cone change at a rate of 2 cm/s. How fast is the volume of the cone increasing when $r = 10$ and $h = 20$?

16. A road perpendicular to a highway leads to a farmhouse located 2 km away (Figure 8). An automobile travels past the farmhouse at a speed of 80 km/h. How fast is the distance between the automobile and the farmhouse increasing when the automobile is 6 km past the intersection of the highway and the road?

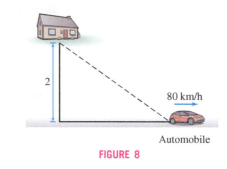

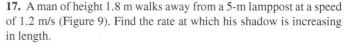

Automobile

FIGURE 8

17. A man of height 1.8 m walks away from a 5-m lamppost at a speed of 1.2 m/s (Figure 9). Find the rate at which his shadow is increasing in length.

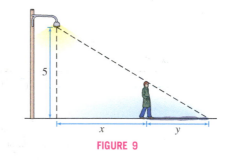

FIGURE 9

18. As Claudia walks away from a 264-cm lamppost, the tip of her shadow moves twice as fast as she does. What is Claudia's height?

19. At a given moment, a plane passes directly above a radar station at an altitude of 6 km.

(a) The plane's speed is 800 km/h. How fast is the distance between the plane and the station changing half a minute later?

(b) How fast is the distance between the plane and the station changing when the plane passes directly above the station?

20. In the setting of Exercise 19, let θ be the angle that the line through the radar station and the plane makes with the horizontal. How fast is θ changing 12 min after the plane passes over the radar station?

21. A hot air balloon rising vertically is tracked by an observer located 4 km from the lift-off point. At a certain moment, the angle between the observer's line of sight and the horizontal is $\frac{\pi}{5}$, and it is changing at a rate of 0.2 rad/min. How fast is the balloon rising at this moment?

22. A laser pointer is placed on a platform that rotates at a rate of 20 revolutions per minute. The beam hits a wall 8 m away, producing a dot of light that moves horizontally along the wall. Let θ be the angle between the beam and the line through the searchlight perpendicular to the wall (Figure 10). How fast is this dot moving when $\theta = \frac{\pi}{6}$?

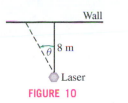

Wall

8 m

θ

Laser

FIGURE 10

23. A rocket travels vertically at a speed of 1200 km/h. The rocket is tracked through a telescope by an observer located 16 km from the launching pad. Find the rate at which the angle between the telescope and the ground is increasing 3 min after lift-off.

24. Using a telescope, you track a rocket that was launched 4 km away, recording the angle θ between the telescope and the ground at half-second intervals. Estimate the velocity of the rocket if $\theta(10) = 0.205$ and $\theta(10.5) = 0.225$.

25. A police car traveling south toward Sioux Falls at 160 km/h pursues a truck traveling east away from Sioux Falls, Iowa, at 140 km/h (Figure 11). At time $t = 0$, the police car is 20 km north and the truck is 30 km east of Sioux Falls. Calculate the rate at which the distance between the vehicles is changing:

(a) At time $t = 0$ **(b)** 5 min later

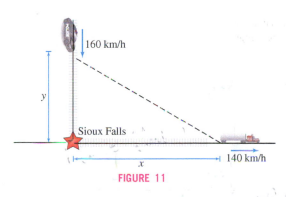

160 km/h

y

Sioux Falls

x

140 km/h

FIGURE 11

26. A car travels down a highway at 25 m/s. An observer stands 150 m from the highway.

(a) How fast is the distance from the observer to the car increasing when the car passes in front of the observer? Explain your answer without making any calculations.

(b) How fast is the distance increasing 20 s later?

27. In the setting of Example 5, at a certain moment, the tractor's speed is 3 m/s and the bale is rising at 2 m/s. How far is the tractor from the bale at this moment?

28. Placido pulls a rope attached to a wagon through a pulley at a rate of q m/s. With dimensions as in Figure 12:

(a) Find a formula for the speed of the wagon in terms of q and the variable x in the figure.

(b) Find the speed of the wagon when $x = 0.6$ if $q = 0.5$ m/s.

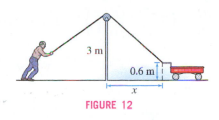

3 m

0.6 m

x

FIGURE 12

29. Julian is jogging around a circular track of radius 50 m. In a coordinate system with its origin at the center of the track, Julian's x-coordinate is changing at a rate of -1.25 m/s when his coordinates are (40, 30). Find dy/dt at this moment.

30. A particle moves counterclockwise around the ellipse with equation $9x^2 + 16y^2 = 25$ (Figure 13).

(a) In which of the four quadrants is $dx/dt > 0$? Explain.

(b) Find a relation between dx/dt and dy/dt.

(c) At what rate is the x-coordinate changing when the particle passes the point (1, 1) if its y-coordinate is increasing at a rate of 6 m/s?

(d) Find dy/dt when the particle is at the top and bottom of the ellipse.

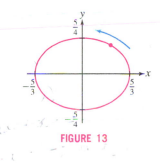

y

$\frac{5}{4}$

$-\frac{5}{3}$

$\frac{5}{3}$

x

$-\frac{5}{4}$

FIGURE 13

In Exercises 31 and 32, assume that the pressure P (in kilopascals) and volume V (in cubic centimeters) of an expanding gas are related by $PV^b = C$, where b and C are constants (this holds in an adiabatic expansion, without heat gain or loss).

31. Find dP/dt if $b = 1.2$, $P = 8$ kPa, $V = 100$ cm^2, and $dV/dt = 20$ cm^3/min.

32. Find b if $P = 25$ kPa, $dP/dt = 12$ kPa/min, $V = 100$ cm^2, and $dV/dt = 20$ cm^3/min.

33. The base x of the right triangle in Figure 14 increases at a rate of 5 cm/s, while the height remains constant at $h = 20$. How fast is the angle θ changing when $x = 20$?

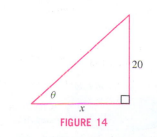

20

θ

x

FIGURE 14

34. Two parallel paths 15 m apart run east–west through the woods. Brooke jogs east on one path at 10 km/h, while Jamail walks west on the other path at 6 km/h. If they pass each other at time $t = 0$, how far apart are they 3 s later, and how fast is the distance between them changing at that moment?

35. A particle travels along a curve $y = f(x)$ as in Figure 15. Let $L(t)$ be the particle's distance from the origin.

(a) Show that $\dfrac{dL}{dt} = \left(\dfrac{x + f(x)f'(x)}{\sqrt{x^2 + f(x)^2}} \right) \dfrac{dx}{dt}$ if the particle's location at time t is $P = (x, f(x))$.

(b) Calculate $L'(t)$ when $x = 1$ and $x = 2$ if $f(x) = \sqrt{3x^2 - 8x + 9}$ and $dx/dt = 4$.

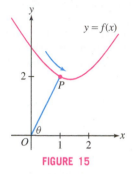

FIGURE 15

36. Let θ be the angle in Figure 15, where $P = (x, f(x))$. In the setting of the previous exercise, show that

$$\frac{d\theta}{dt} = \left(\frac{xf'(x) - f(x)}{x^2 + f(x)^2} \right) \frac{dx}{dt}$$

Hint: Differentiate $\tan \theta = f(x)/x$ and observe that $\cos \theta = x/\sqrt{x^2 + f(x)^2}$.

Exercises 37 and 38 refer to the baseball diamond (a square of side 90 ft) in Figure 16.

37. A baseball player runs from home plate toward first base at 20 ft/s. How fast is the player's distance from second base changing when the player is halfway to first base?

38. Player 1 runs to first base at a speed of 20 ft/s, while Player 2 runs from second base to third base at a speed of 15 ft/s. Let s be the distance between the two players. How fast is s changing when Player 1 is 30 ft from home plate and Player 2 is 60 ft from second base?

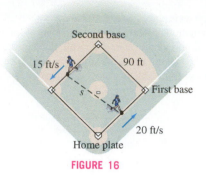

FIGURE 16

39. The conical watering pail in Figure 17 has a grid of holes. Water flows out through the holes at a rate of kA m^3/min, where k is a constant and A is the surface area of the part of the cone in contact with the water. This surface area is $A = \pi r \sqrt{h^2 + r^2}$ and the volume is $V = \frac{1}{3}\pi r^2 h$. Calculate the rate dh/dt at which the water level changes at $h = 0.3$ m, assuming that $k = 0.25$ m.

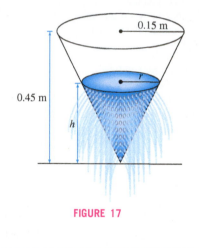

FIGURE 17

Further Insights and Challenges

40. 📖   A bowl contains water that evaporates at a rate proportional to the surface area of water exposed to the air (Figure 18). Let $A(h)$ be the cross-sectional area of the bowl at height h.

(a) Explain why $V(h + \Delta h) - V(h) \approx A(h)\Delta h$ if Δh is small.

(b) Use (a) to argue that $\dfrac{dV}{dh} = A(h)$.

(c) Show that the water level h decreases at a constant rate.

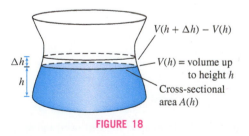

FIGURE 18

41. A roller coaster has the shape of the graph in Figure 19. Show that when the roller coaster passes the point $(x, f(x))$, the vertical velocity of the roller coaster is equal to $f'(x)$ times its horizontal velocity.

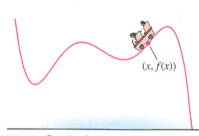

FIGURE 19 Graph of f as a roller coaster track.

42. Two trains leave a station at $t = 0$ and travel with constant velocity v along straight tracks that make an angle θ.

(a) Show that the trains are separating from each other at a rate $v\sqrt{2 - 2\cos\theta}$.

(b) What does this formula give for $\theta = \pi$?

43. As the wheel of radius r cm in Figure 20 rotates, the rod of length L attached at point P drives a piston back and forth in a straight line. Let x be the distance from the origin to point Q at the end of the rod, as shown in the figure.

(a) Use the Pythagorean Theorem to show that

$$L^2 = (x - r \cos \theta)^2 + r^2 \sin^2 \theta \qquad \boxed{6}$$

(b) Differentiate Eq. (6) with respect to t to prove that

$$2(x - r \cos \theta)\left(\frac{dx}{dt} + r \sin \theta \frac{d\theta}{dt}\right) + 2r^2 \sin \theta \cos \theta \frac{d\theta}{dt} = 0$$

(c) Calculate the speed of the piston when $\theta = \frac{\pi}{2}$, assuming that $r = 10$ cm, $L = 30$ cm, and the wheel rotates at 4 revolutions per minute.

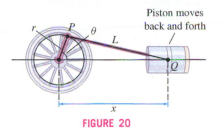

FIGURE 20

44. A spectator seated 300 m away from the center of a circular track of radius 100 m watches an athlete run laps at a speed of 5 m/s. How fast is the distance between the spectator and athlete changing when the runner is approaching the spectator and the distance between them is 250 m? *Hint:* The diagram for this problem is similar to Figure 20, with $r = 100$ and $x = 300$.

45. A cylindrical tank of radius R and length L lying horizontally as in Figure 21 is filled with oil to height h.

(a) Show that the volume $V(h)$ of oil in the tank is

$$V(h) = L\left(R^2 \cos^{-1}\left(1 - \frac{h}{R}\right) - (R - h)\sqrt{2hR - h^2}\right)$$

(b) Show that $\frac{dV}{dh} = 2L\sqrt{h(2R - h)}$.

(c) Suppose that $R = 1.5$ m and $L = 10$ m and that the tank is filled at a constant rate of 0.6 m³/min. How fast is the height h increasing when $h = 0.5$?

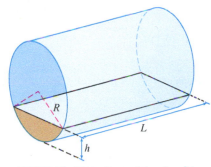

FIGURE 21 Oil in the tank has level h.

CHAPTER REVIEW EXERCISES

In Exercises 1–4, refer to the function f whose graph is shown in Figure 1.

1. Compute the average rate of change of $f(x)$ over $[0, 2]$. What is the graphical interpretation of this average rate?

2. For which value of h is $\dfrac{f(0.7 + h) - f(0.7)}{h}$ equal to the slope of the secant line between the points where $x = 0.7$ and $x = 1.1$?

3. Estimate $\dfrac{f(0.7 + h) - f(0.7)}{h}$ for $h = 0.3$. Is this number larger or smaller than $f'(0.7)$?

4. Estimate $f'(0.7)$ and $f'(1.1)$.

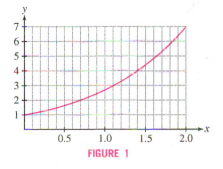

FIGURE 1

In Exercises 5–8, compute $f'(a)$ using the limit definition and find an equation of the tangent line to the graph of f at $x = a$.

5. $f(x) = x^2 - x, \quad a = 1$

6. $f(x) = 5 - 3x, \quad a = 2$

7. $f(x) = x^{-1}, \quad a = 4$

8. $f(x) = x^3, \quad a = -2$

In Exercises 9–12, compute dy/dx using the limit definition.

9. $y = 4 - x^2$

10. $y = \sqrt{2x + 1}$

11. $y = \dfrac{1}{2 - x}$

12. $y = \dfrac{1}{(x - 1)^2}$

In Exercises 13–16, express the limit as a derivative.

13. $\lim\limits_{h \to 0} \dfrac{\sqrt{1 + h} - 1}{h}$

14. $\lim\limits_{x \to -1} \dfrac{x^3 + 1}{x + 1}$

15. $\lim\limits_{t \to \pi} \dfrac{\sin t \cos t}{t - \pi}$

16. $\lim\limits_{\theta \to \pi} \dfrac{\cos \theta - \sin \theta + 1}{\theta - \pi}$

17. Find $f(4)$ and $f'(4)$ if the tangent line to the graph of f at $x = 4$ has equation $y = 3x - 14$.

18. Each graph in Figure 2 shows the graph of a function f and its derivative f'. Determine which is the function and which is the derivative.

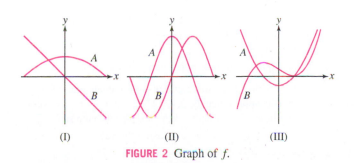

FIGURE 2 Graph of f.

19. Is (A), (B), or (C) the graph of the derivative of the function f shown in Figure 3?

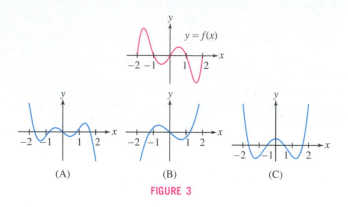

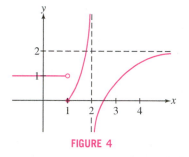

(A)   (B)   (C)

FIGURE 3

20. Sketch the graph of f' if the graph of f appears as in Figure 4.

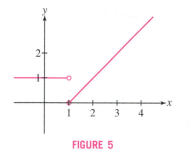

FIGURE 4

21. Sketch the graph of a continuous function f if the graph of f' appears as in Figure 5 and $f(0) = 0$.

FIGURE 5

22. Let $N(t)$ be the percentage of a state population infected with a flu virus on week t of an epidemic. What percentage is likely to be infected in week 4 if $N(3) = 8$ and $N'(3) = 1.2$?

23. A girl's height $h(t)$ (in centimeters) is measured at time t (in years) for $0 \le t \le 14$:

   52,   75.1,   87.5,   96.7,   104.5,   111.8,   118.7,   125.2,
   131.5,   137.5,   143.3,   149.2,   155.3,   160.8,   164.7

(a) What is the average growth rate over the 14-year period?

(b) Is the average growth rate larger over the first half or the second half of this period?

(c) Estimate $h'(t)$ (in centimeters per year) for $t = 3, 8$.

24. A planet's period P (number of days to complete one revolution around the sun) is approximately $0.199A^{3/2}$, where A is the average distance (in millions of kilometers) from the planet to the sun.

(a) Calculate P and dP/dA for Earth using the value $A = 150$.

(b) Estimate the increase in P if A is increased to 152.

In Exercises 25 and 26, use the following table of values for the number $A(t)$ of automobiles (in millions) manufactured in the United States in year t.

t	1970	1971	1972	1973	1974	1975	1976
$A(t)$	6.55	8.58	8.83	9.67	7.32	6.72	8.50

25. What is the interpretation of $A'(t)$? Estimate $A'(1971)$. Does $A'(1974)$ appear to be positive or negative?

26. Given the data, which of (A)–(C) in Figure 6 could be the graph of the derivative A'? Explain.

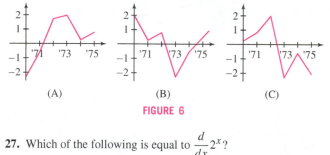

(A)   (B)   (C)

FIGURE 6

27. Which of the following is equal to $\dfrac{d}{dx}2^x$?

(a) 2^x **(b)** $(\ln 2)2^x$ **(c)** $x2^{x-1}$ **(d)** $\dfrac{1}{\ln 2}2^x$

28. Use the Chain Rule to show that if g is the inverse of f, then $g'(x) = 1/f'(g(x))$ for all x in the domain of g such that $f(g(x)) \ne 0$. Use this to obtain another method for finding the derivative of $\ln x$ using the derivative of e^x.

In Exercises 29–80, compute the derivative.

29. $y = 3x^5 - 7x^2 + 4$

30. $y = 4x^{-3/2}$

31. $y = t^{-7.3}$

32. $y = 4x^2 - x^{-2}$

33. $y = \dfrac{x+1}{x^2+1}$

34. $y = \dfrac{3t-2}{4t-9}$

35. $y = (x^4 - 9x)^6$

36. $y = (3t^2 + 20t^{-3})^6$

37. $y = (2 + 9x^2)^{3/2}$

38. $y = (x+1)^3(x+4)^4$

39. $y = \dfrac{z}{\sqrt{1-z}}$

40. $y = \left(1 + \dfrac{1}{x}\right)^3$

41. $y = \dfrac{x^4 + \sqrt{x}}{x^2}$

42. $y = \dfrac{1}{(1-x)\sqrt{2-x}}$

43. $y = \sqrt{x + \sqrt{x + \sqrt{x}}}$

44. $h(z) = \left(z + (z+1)^{1/2}\right)^{-3/2}$

45. $y = \tan(t^{-3})$

46. $y = 4\cos(2 - 3x)$

47. $y = \sin(2x)\cos^2 x$

48. $y = \sin\left(\dfrac{4}{\theta}\right)$

49. $y = \dfrac{t}{1 + \sec t}$

50. $y = z\csc(9z + 1)$

51. $y = \dfrac{8}{1 + \cot\theta}$

52. $y = \tan(\cos x)$

53. $y = \tan(\sqrt{1 + \csc\theta})$

54. $y = \cos(\cos(\cos(\theta)))$

55. $f(x) = 9e^{-4x}$

56. $f(x) = \dfrac{e^{-x}}{x}$

57. $g(t) = e^{4t-t^2}$

58. $g(t) = t^2 e^{1/t}$

59. $f(x) = \ln(4x^2 + 1)$

60. $f(x) = \ln(e^x - 4x)$

61. $G(s) = (\ln(s))^2$

62. $G(s) = \ln(s^2)$

63. $f(\theta) = \ln(\sin\theta)$

64. $f(\theta) = \sin(\ln\theta)$

65. $h(z) = \sec(z + \ln z)$

66. $f(x) = e^{\sin^2 x}$

67. $f(x) = 7^{-2x}$

68. $h(y) = \dfrac{1 + e^y}{1 - e^y}$

69. $g(x) = \tan^{-1}(\ln x)$

70. $G(s) = \cos^{-1}(s^{-1})$

71. $f(x) = \ln(\csc^{-1} x)$

72. $f(x) = e^{\sec^{-1} x}$

73. $R(s) = s^{\ln s}$

74. $f(x) = (\cos^2 x)^{\cos x}$

75. $G(t) = (\sin^2 t)^t$

76. $h(t) = t^{(t^t)}$

77. $g(t) = \sinh(t^2)$

78. $h(y) = y\tanh(4y)$

79. $g(x) = \tanh^{-1}(e^x)$

80. $g(t) = \sqrt{t^2 - 1}\,\sinh^{-1} t$

81. For which values of α is $f(x) = |x|^\alpha$ differentiable at $x = 0$?

82. Find $f'(2)$ if $f(g(x)) = e^{x^2}$, $g(1) = 2$, and $g'(1) = 4$.

In Exercises 83 and 84, let $f(x) = xe^{-x}$.

83. Show that f has an inverse on $[1, \infty)$. Let g be this inverse. Find the domain and range of g and compute $g'(2e^{-2})$.

84. Show that $f(x) = c$ has two solutions if $0 < c < e^{-1}$.

In Exercises 85–90, use the following table of values to calculate the derivative of the given function at $x = 2$:

x	$f(x)$	$g(x)$	$f'(x)$	$g'(x)$
2	5	4	-3	9
4	3	2	-2	3

85. $S(x) = 3f(x) - 2g(x)$

86. $H(x) = f(x)g(x)$

87. $R(x) = \dfrac{f(x)}{g(x)}$

88. $G(x) = f(g(x))$

89. $F(x) = f(g(2x))$

90. $K(x) = f(x^2)$

91. Find the points on the graph of $f(x) = x^3 - 3x^2 + x + 4$ where the tangent line has slope 10.

92. Find the points on the graph of $x^{2/3} + y^{2/3} = 1$ where the tangent line has slope 1.

93. Find a such that the tangent lines to $y = x^3 - 2x^2 + x + 1$ at $x = a$ and $x = a + 1$ are parallel.

94. ✏ Use the table to compute the average rate of change of Candidate A's percentage of votes over the intervals from day 20 to day 15, day 15 to day 10, and day 10 to day 5. If this trend continues over the last 5 days before the election, will Candidate A win?

Days Before Election	20	15	10	5
Candidate A	44.8%	46.8%	48.3%	49.3%
Candidate B	55.2%	53.2%	51.7%	50.7%

In Exercises 95–100, calculate y''.

95. $y = 12x^3 - 5x^2 + 3x$

96. $y = x^{-2/5}$

97. $y = \sqrt{2x + 3}$

98. $y = \dfrac{4x}{x + 1}$

99. $y = \tan(x^2)$

100. $y = \sin^2(4x + 9)$

In Exercises 101–106, compute $\dfrac{dy}{dx}$.

101. $x^3 - y^3 = 4$

102. $4x^2 - 9y^2 = 36$

103. $y = xy^2 + 2x^2$

104. $\dfrac{y}{x} = x + y$

105. $y = \sin(x + y)$

106. $\tan(x + y) = xy$

107. In Figure 7, label the graphs f, f', and f''.

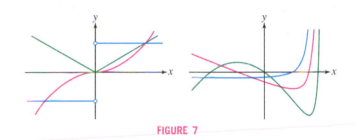

FIGURE 7

108. Let $f(x) = x^2 \sin(x^{-1})$ for $x \neq 0$ and $f(0) = 0$. Show that $f'(x)$ exists for all x (including $x = 0$) but that f' is not continuous at $x = 0$ (Figure 8).

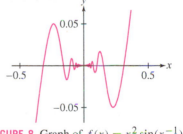

FIGURE 8 Graph of $f(x) = x^2 \sin(x^{-1})$.

In Exercises 109–114, use logarithmic differentiation to find the derivative.

109. $y = \dfrac{(x + 1)^3}{(4x - 2)^2}$

110. $y = \dfrac{(x + 1)(x + 2)^2}{(x + 3)(x + 4)}$

111. $y = e^{(x-1)^2} e^{(x-3)^2}$

112. $y = \dfrac{e^x \sin^{-1} x}{\ln x}$

113. $y = \dfrac{e^{3x}(x - 2)^2}{(x + 1)^2}$

114. $y = x^{\sqrt{x}}(x^{\ln x})$

*Exercises 115–117: Let q be the number of units of a product (cell phones, barrels of oil, etc.) that can be sold at the price p. The **price elasticity of demand** E is defined as the percentage rate of change of q with respect to p. In terms of derivatives,*

$$E = \frac{p}{q}\frac{dq}{dp} = \lim_{\Delta p \to 0} \frac{(100\Delta q)/q}{(100\Delta p)/p}$$

115. Show that the total revenue $R = pq$ satisfies $\dfrac{dR}{dp} = q(1 + E)$.

116. 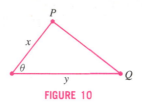 A commercial bakery can sell q chocolate cakes per week at price $\$p$, where $q = 50p(10 - p)$ for $5 < p < 10$.

(a) Show that $E(p) = \dfrac{2p - 10}{p - 10}$.

(b) Show, by computing $E(8)$, that if $p = \$8$, then a 1% increase in price reduces demand by approximately 3%.

117. The monthly demand (in thousands) for flights between Chicago and St. Louis at the price p is $q = 40 - 0.2p$. Calculate the price elasticity of demand when $p = \$150$ and estimate the percentage increase in number of additional passengers if the ticket price is lowered by 1%.

118. How fast does the water level rise in the tank in Figure 9 when the water level is $h = 4$ m and water pours in at 20 m^3/min?

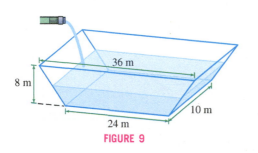

FIGURE 9

119. The minute hand of a clock is 8 cm long, and the hour hand is 5 cm long. How fast is the distance between the tips of the hands changing at 3 o'clock?

120. Chloe and Bao are in motorboats at the center of a lake. At time $t = 0$, Chloe begins traveling south at a speed of 50 km/h. One minute later, Bao takes off, heading east at a speed of 40 km/h. At what rate is the distance between them increasing at $t = 12$ min?

121. A bead slides down the curve $xy = 10$. Find the bead's horizontal velocity at time $t = 2$ s if its height at time t seconds is $y = 400 - 16t^2$ cm.

122. In Figure 10, x is increasing at 2 cm/s, y is increasing at 3 cm/s, and θ is decreasing such that the area of the triangle has the constant value 4 cm^2.

(a) How fast is θ decreasing when $x = 4$, $y = 4$?

(b) How fast is the distance between P and Q changing when $x = 4$, $y = 4$?

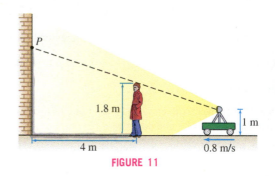

FIGURE 10

123. A light moving at 0.8 m/s approaches a man standing 4 m from a wall (Figure 11). The light is 1 m above the ground. How fast is the tip P of the man's shadow moving when the light is 7 m from the wall?

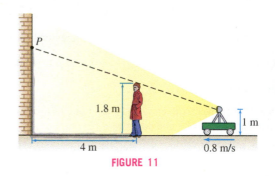

FIGURE 11

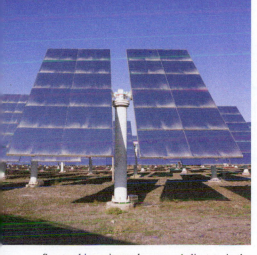

Sun-tracking mirrors, known as heliostats, in the Tabernas Desert in Spain, use the Principle of Least Distance (see Section 4.7) to concentrate the Sun's light and generate energy. (*Thomas Dressler/Gallo Images/Getty Images*)

4 APPLICATIONS OF THE DERIVATIVE

This chapter puts the derivative to work. The first and second derivatives are used to analyze functions and their graphs and to solve optimization problems (finding minimum and maximum values of a function). Newton's Method in Section 4.8 employs the derivative to approximate solutions of equations.

4.1 Linear Approximation and Applications

In some situations we are interested in determining the "effect of a small change." For example:

- How does a small change in angle affect the distance of a basketball shot? (Exercise 39)
- How are revenues at the box office affected by a small change in ticket prices? (Exercise 29)
- The cube root of 27 is 3. How much larger is the cube root of 27.2? (Exercise 7)

In each case, we have a function f and we're interested in the change

$$\Delta f = f(a + \Delta x) - f(a)$$

where Δx is small. The **Linear Approximation** uses the derivative to estimate Δf without computing it exactly. By definition, the derivative is the limit

$$f'(a) = \lim_{\Delta x \to 0} \frac{f(a + \Delta x) - f(a)}{\Delta x} = \lim_{\Delta x \to 0} \frac{\Delta f}{\Delta x}$$

So when Δx is small, we have $\Delta f / \Delta x \approx f'(a)$, and thus,

$$\Delta f \approx f'(a) \Delta x$$

◄·· **REMINDER** *The notation $\approx$ means "approximately equal to." The accuracy of the Linear Approximation is discussed at the end of this section.*

Linear Approximation of Δf   If f is differentiable at $x = a$ and Δx is small, then

$$\Delta f \approx f'(a) \Delta x \qquad \boxed{1}$$

where $\Delta f = f(a + \Delta x) - f(a)$. Therefore,

$$f(a + \Delta x) \approx f(a) + f'(a) \Delta x \qquad \boxed{2}$$

Keep in mind the different roles played by Δf and $f'(a) \Delta x$. The quantity of interest is the *actual change* Δf. We estimate it by $f'(a) \Delta x$. The Linear Approximation tells us that up to a small error, Δf is directly proportional to Δx when Δx is small.

GRAPHICAL INSIGHT   The Linear Approximation is sometimes called the **tangent line approximation**. Why? Observe in Figure 1 that Δf is the vertical change in the graph from $x = a$ to $x = a + \Delta x$. For a straight line, the vertical change is equal to the slope times the horizontal change Δx, and since the tangent line has slope $f'(a)$, its vertical change is $f'(a) \Delta x$. So the Linear Approximation approximates Δf by the vertical change in the tangent line. When Δx is small, the two quantities are nearly equal.

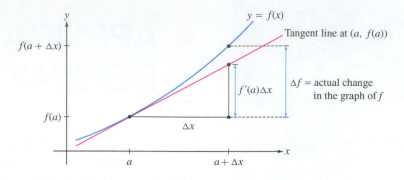

DF **FIGURE 1** Graphical meaning of the Linear Approximation $\Delta f \approx f'(a)\Delta x$.

The Linear Approximation

$$\Delta f \approx f'(a)\Delta x$$

where $\Delta f = f(a + \Delta x) - f(a)$ yields the approximation for the value of the function at $a + \Delta x$ to be

$$f(a + \Delta x) \approx f(a) + f'(a)\Delta x$$

■ **EXAMPLE 1** Use the Linear Approximation to estimate $\frac{1}{10.2} - \frac{1}{10}$, and hence $\frac{1}{10.2}$. How accurate is your estimate?

Solution We apply the Linear Approximation to $f(x) = \frac{1}{x}$ with $a = 10$ and $\Delta x = 0.2$:

$$\Delta f = f(10.2) - f(10) = \frac{1}{10.2} - \frac{1}{10}$$

We have $f'(x) = -x^{-2}$ and $f'(10) = -0.01$, so Δf is approximated by

$$f'(10)\Delta x = (-0.01)(0.2) = -0.002$$

In other words,

$$\frac{1}{10.2} - \frac{1}{10} \approx -0.002$$

Then, we have

$$\frac{1}{10.2} \approx \frac{1}{10} + (-0.002) \approx 0.098$$

The error in the Linear Approximation is the quantity

$$\text{Error} = \left| \Delta f - f'(a)\Delta x \right|$$

A calculator gives the value $\frac{1}{10.2} - \frac{1}{10} \approx -0.00196$ and thus our error is less than 10^{-4}:

$$\text{Error} \approx \left| -0.00196 - (-0.002) \right| = 0.00004 < 10^{-4} \qquad ■$$

Differential Notation The Linear Approximation to $y = f(x)$ is often written using the "differentials" dx and dy. In this notation, dx is used interchangeably with Δx to represent the change in x, and dy is the corresponding vertical change in the tangent line:

$$\boxed{dy = f'(a)dx} \qquad \boxed{3}$$

Let $\Delta y = f(a + \Delta x) - f(a)$. Then the Linear Approximation says

$$\boxed{\Delta y \approx dy} \qquad \boxed{4}$$

This is simply another way of writing $\Delta f \approx f'(a)\Delta x$.

■ **EXAMPLE 2** **Differential Notation** How much larger is $\sqrt[3]{8.1}$ than $\sqrt[3]{8} = 2$?

Solution We are interested in $\sqrt[3]{8.1} - \sqrt[3]{8}$, so we apply the Linear Approximation to $f(x) = x^{1/3}$ with $a = 8$ and change $\Delta x = dx = 0.1$.

Step 1. **Write out Δy.**

$$\Delta y = f(a + \Delta x) - f(a) = \sqrt[3]{8 + 0.1} - \sqrt[3]{8} = \sqrt[3]{8.1} - 2$$

Step 2. **Compute dy.**

$$f'(x) = \frac{1}{3}x^{-2/3} \qquad \text{and} \qquad f'(8) = \left(\frac{1}{3}\right)8^{-2/3} = \left(\frac{1}{3}\right)\left(\frac{1}{4}\right) = \frac{1}{12}$$

Therefore, $dy = f'(8)\,dx = \frac{1}{12}(0.1) \approx 0.0083$.

Step 3. **Use the Linear Approximation.**

$$\Delta y \approx dy \quad \Rightarrow \quad \sqrt[3]{8.1} - 2 \approx 0.0083$$

Thus, $\sqrt[3]{8.1}$ is larger than $\sqrt[3]{8}$ by approximately the amount 0.0083, and $\sqrt[3]{8.1} \approx 2.0083$.

∎

FIGURE 2 Cable position transducer (manufactured by Space Age Control, Inc.). In one application, a transducer was used to compare the changes in throttle position on a Formula 1 race car with the shifting actions of the driver. (*Space Age Control Inc., photo by Tom Anderson*)

When engineers need to monitor the change in position of an object with great accuracy, they may use a cable position transducer (Figure 2). This device detects and records the movement of a metal cable attached to the object. Its accuracy is affected by changes in temperature because heat causes the cable to stretch. The Linear Approximation can be used to estimate these effects.

■ **EXAMPLE 3** **Thermal Expansion** A thin metal cable has length $L = 12$ cm when the temperature is $T = 21°C$. Estimate the change in length when T rises to 24°C, assuming that

$$\frac{dL}{dT} = kL \qquad \boxed{5}$$

where $k = 1.7 \times 10^{-5}\,°C^{-1}$ (k is called the coefficient of thermal expansion).

Solution How does the Linear Approximation apply here? We will use the differential dL to estimate the actual change in length ΔL when T increases from 21° to 24°—that is, when $dT = \Delta T = 3°$. By Eq. (3), the differential dL is

$$dL = \left(\frac{dL}{dT}\right) dT$$

By Eq. (5), since $L = 12$,

$$\frac{dL}{dT}\bigg|_{L=12} = kL = (1.7 \times 10^{-5})(12) \approx 2 \times 10^{-4}\ \text{cm/°C}$$

The Linear Approximation $\Delta L \approx dL$ tells us that the change in length is approximately

$$\Delta L \approx \underbrace{\left(\frac{dL}{dT}\right) dT}_{dL} \approx (2 \times 10^{-4})(3) = 6 \times 10^{-4}\ \text{cm}$$

∎

Suppose that we measure the *diameter D* of a circle and use this result to compute the *area* of the circle. If our measurement of D is inexact, the area computation will also be inexact. What is the effect of the measurement error on the resulting area computation? This can be estimated using the Linear Approximation, as in the next example.

■ **EXAMPLE 4** **Effect of an Inexact Measurement** The Cheezy Pizza Parlor claims that its pizzas are circular with diameter 50 cm (Figure 3).

(a) What is the area of the pizza?

(b) Estimate the quantity of pizza lost or gained if the diameter is off by at most 1.2 cm.

Solution First, we need a formula for the area A of a circle in terms of its diameter D. Since the radius is $r = D/2$, the area is

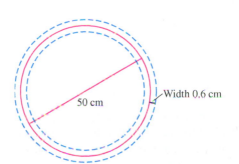

FIGURE 3 The border of the actual pizza lies between the dashed circles.

In this example, we interpret ΔA as the possible error in the computation of $A(D)$. This should not be confused with the error in the Linear Approximation. This latter error refers to the accuracy in using $A'(D)\,\Delta D$ to approximate ΔA.

$$A(D) = \pi r^2 = \pi \left(\frac{D}{2}\right)^2 = \frac{\pi}{4} D^2$$

(a) If $D = 50$ cm, then the pizza has area $A(50) = \left(\frac{\pi}{4}\right)(50)^2 \approx 1963.5\ \text{cm}^2$.

(b) If the actual diameter is equal to $50 + \Delta D$, then the loss or gain in pizza area is $\Delta A = A(50 + \Delta D) - A(50)$. Observe that $A'(D) = \frac{\pi}{2} D$ and $A'(50) = 25\pi \approx 78.5$ cm, so the Linear Approximation yields

$$\Delta A = A(50 + \Delta D) - A(50) \approx A'(D)\Delta D \approx (78.5)\,\Delta D$$

Because ΔD is at most ± 1.2 cm, the loss or gain in pizza is no more than around

$$\Delta A \approx \pm(78.5)(1.2) \approx \pm 94.2 \text{ cm}^2$$

This is a loss or gain of approximately 4.8%.   ■

Linearization

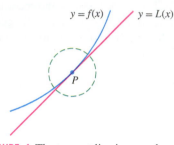

To approximate the function f itself rather than the change Δf, we use the linearization $L(x)$ "centered at $x = a$," defined by

$$L(x) = f'(a)(x - a) + f(a)$$

Notice that $y = L(x)$ is the equation of the tangent line at $x = a$ (Figure 4). For values of x close to a, $L(x)$ provides a good approximation to $f(x)$.

FIGURE 4 The tangent line is a good approximation in a small neighborhood of $P = (a, f(a))$.

> **Approximating f by Its Linearization**   If f is differentiable at $x = a$ and x is close to a, then
>
> $$f(x) \approx L(x) = f(a) + f'(a)(x - a)$$

CONCEPTUAL INSIGHT   Keep in mind that the linearization and the Linear Approximation are two ways of saying the same thing. Indeed, when we apply the linearization with $x = a + \Delta x$ and rearrange, we obtain the Linear Approximation:

$$f(x) \approx f(a) + f'(a)(x - a)$$
$$f(a + \Delta x) \approx f(a) + f'(a)\,\Delta x \qquad (\text{since } \Delta x = x - a)$$
$$f(a + \Delta x) - f(a) \approx f'(a)\Delta x$$

■ **EXAMPLE 5**   Compute the linearization of $f(x) = \sqrt{x}e^{x-1}$ at $a = 1$.

Solution   By the Product Rule:

$$f'(x) = \frac{1}{2}x^{-1/2}e^{x-1} + x^{1/2}e^{x-1} = \left(\frac{1}{2}x^{-1/2} + x^{1/2}\right)e^{x-1}$$

$$f(1) = \sqrt{1}e^0 = 1, \qquad f'(1) = \left(\frac{1}{2} + 1\right)e^0 = \frac{3}{2}$$

Therefore, the linearization at $a = 1$ is

$$L(x) = f(1) + f'(1)(x - 1) = 1 + \frac{3}{2}(x - 1) = \frac{3}{2}x - \frac{1}{2}$$ ■

The linearization can be used to approximate function values. The following table compares values of the linearization to values obtained from a calculator for the function $f(x) = \sqrt{x}e^{x-1}$ in the previous example. Note that the error is large for $x = 2.5$, as expected, because 2.5 is not close to the center $a = 1$ (Figure 5).

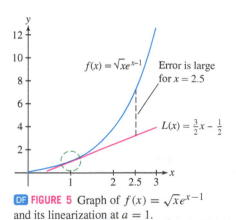

DF **FIGURE 5** Graph of $f(x) = \sqrt{x}e^{x-1}$ and its linearization at $a = 1$.

x	$\sqrt{x}e^{x-1}$	Linearization at $a = 1$: $L(x) = \frac{3}{2}x - \frac{1}{2}$	Calculator	Error
1.1	$\sqrt{1.1}e^{0.1}$	$L(1.1) = \frac{3}{2}(1.1) - \frac{1}{2} = 1.15$	1.15911	10^{-2}
0.999	$\sqrt{0.999}e^{-0.001}$	$L(0.999) = \frac{3}{2}(0.999) - \frac{1}{2} = 0.9985$	0.998501	10^{-6}
2.5	$\sqrt{2.5}e^{1.5}$	$L(2.5) = \frac{3}{2}(2.5) - \frac{1}{2} = 3.25$	7.086	3.84

In the next example, we compute the **percentage error**, which is often more important than the error itself. By definition,

$$\text{Percentage error} = \left| \frac{\text{error}}{\text{actual value}} \right| \times 100\%$$

■ **EXAMPLE 6** Estimate $\tan\left(\frac{\pi}{4} + 0.02\right)$ and compute the percentage error.

Solution We find the linearization of $f(x) = \tan x$ at $a = \frac{\pi}{4}$:

$$f\left(\frac{\pi}{4}\right) = \tan\left(\frac{\pi}{4}\right) = 1, \qquad f'\left(\frac{\pi}{4}\right) = \sec^2\left(\frac{\pi}{4}\right) = \left(\sqrt{2}\right)^2 = 2$$

$$L(x) = f\left(\frac{\pi}{4}\right) + f'\left(\frac{\pi}{4}\right)\left(x - \frac{\pi}{4}\right) = 1 + 2\left(x - \frac{\pi}{4}\right)$$

At $x = \frac{\pi}{4} + 0.02$, the linearization yields the estimate

$$\tan\left(\frac{\pi}{4} + 0.02\right) \approx L\left(\frac{\pi}{4} + 0.02\right) = 1 + 2(0.02) = 1.04$$

A calculator gives $\tan\left(\frac{\pi}{4} + 0.02\right) \approx 1.0408$, so

$$\text{Percentage error} \approx \left| \frac{1.0408 - 1.04}{1.0408} \right| \times 100 \approx 0.08\% \qquad ■$$

The Size of the Error

The examples in this section may have convinced you that the Linear Approximation yields a good approximation to Δf when Δx is small, but if we want to rely on the Linear Approximation, we need to know more about the size of the error:

$$E = \text{Error} = \left| \Delta f - f'(a)\Delta x \right|$$

Remember that the error E is simply the vertical gap between the graph and the tangent line (Figure 6). In Section 8.4, we will prove the following **Error Bound**:

$$\boxed{E \le \frac{1}{2} K \left(\Delta x\right)^2} \qquad \boxed{6}$$

where K is the maximum value of $|f''(x)|$ on the interval from a to $a + \Delta x$.

The Error Bound tells us two important things. First, it says that the error is small when the second derivative (and hence K) is small. This makes sense, because $f''(x)$ measures how quickly the tangent lines change direction. When $|f''(x)|$ is smaller, the graph is flatter and the Linear Approximation is more accurate over a larger interval around $x = a$ (compare the graphs in Figure 7).

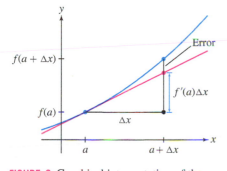

FIGURE 6 Graphical interpretation of the error in the Linear Approximation.

Error Bound:

$$E \le \frac{1}{2} K \left(\Delta x\right)^2$$

where K is the max value of $|f''(x)|$ on the interval $[a, a + \Delta x]$.

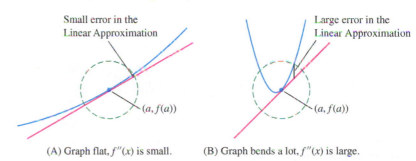

Small error in the Linear Approximation

Large error in the Linear Approximation

$(a, f(a))$

$(a, f(a))$

(A) Graph flat, $f''(x)$ is small.   (B) Graph bends a lot, $f''(x)$ is large.

FIGURE 7 The accuracy of the Linear Approximation depends on how much the curve bends.

Second, the Error Bound tells us that the error is of *order 2* in Δx, meaning that E is no larger than a constant times $(\Delta x)^2$. So if Δx is small, say, $\Delta x = 10^{-n}$, then E has a substantially smaller order of magnitude, since $(\Delta x)^2 = 10^{-2n}$. In particular, $E/\Delta x$ tends to zero (because $E/\Delta x < K \Delta x$), so the Error Bound tells us that the graph becomes nearly indistinguishable from its tangent line as we zoom in on the graph around $x = a$. This is a precise version of the "local linearity" property discussed in Section 3.2.

4.1 SUMMARY

- Let $\Delta f = f(a + \Delta x) - f(a)$. The *Linear Approximation* is the estimate

$$\Delta f \approx f'(a)\Delta x \quad \text{(for } \Delta x \text{ small)} \quad \text{and} \quad f(a + \Delta x) \approx f(a) + f'(a)\Delta x$$

- Differential notation: $dx = \Delta x$ is the change in x, $dy = f'(a)dx$, $\Delta y = f(a + \Delta x) - f(a)$. In this notation, the Linear Approximation reads

$$\Delta y \approx dy \qquad \text{(for } dx \text{ small)}$$

- The *linearization* of $f(x)$ at $x = a$ is the function

$$L(x) = f(a) + f'(a)(x - a)$$

- The Linear Approximation is equivalent to the approximation

$$f(x) \approx L(x) \qquad \text{(for } x \text{ close to } a\text{)}$$

- The error in the Linear Approximation is the quantity

$$\text{Error} = \left| \Delta f - f'(a)\Delta x \right|$$

In many cases, the percentage error is more important than the error itself:

$$\text{Percentage error} = \left| \frac{\text{error}}{\text{actual value}} \right| \times 100\%$$

4.1 EXERCISES

Preliminary Questions

1. True or False? The Linear Approximation says that the vertical change in the graph is approximately equal to the vertical change in the tangent line.

2. Estimate $g(1.2) - g(1)$ if $g'(1) = 4$.

3. Estimate $f(2.1)$ if $f(2) = 1$ and $f'(2) = 3$.

4. Complete the following sentence: The Linear Approximation shows that up to a small error, the change in output Δf is directly proportional to ....

Exercises

In Exercises 1–6, use Eq. (1) to estimate $\Delta f = f(3.02) - f(3)$.

1. $f(x) = x^2$

2. $f(x) = x^4$

3. $f(x) = x^{-1}$

4. $f(x) = \dfrac{1}{x + 1}$

5. $f(x) = \sqrt{x + 6}$

6. $f(x) = \tan\dfrac{\pi x}{3}$

7. The cube root of 27 is 3. How much larger is the cube root of 27.2? Estimate using the Linear Approximation.

8. Estimate $\ln(e^3 + 0.1) - \ln(e^3)$ using differentials.

In Exercises 9–12, use Eq. (2) to estimate Δf. Use a calculator to compute both the error and the percentage error.

9. $f(x) = \sqrt{1 + x}$, $\quad a = 3$, $\quad \Delta x = 0.2$

10. $f(x) = 2x^2 - x$, $\quad a = 5$, $\quad \Delta x = -0.4$

11. $f(x) = \dfrac{1}{1 + x^2}$, $\quad a = 3$, $\quad \Delta x = 0.5$

12. $f(x) = \ln(x^2 + 1)$, $\quad a = 1$, $\quad \Delta x = 0.1$

In Exercises 13–16, estimate Δy using differentials [Eq. (4)].

13. $y = \cos x$, $\quad a = \frac{\pi}{6}$, $\quad dx = 0.014$

14. $y = \tan^2 x$, $\quad a = \frac{\pi}{4}$, $\quad dx = -0.02$

15. $y = \dfrac{10 - x^2}{2 + x^2}$, $\quad a = 1$, $\quad dx = 0.01$

16. $y = x^{1/3}e^{x-1}$, $\quad a = 1$, $\quad dx = 0.1$

In Exercises 17–24, estimate using the Linear Approximation and find the error using a calculator.

17. $\sqrt{26} - \sqrt{25}$

18. $16.5^{1/4} - 16^{1/4}$

19. $\dfrac{1}{\sqrt{101}} - \dfrac{1}{10}$

20. $\dfrac{1}{\sqrt{98}} - \dfrac{1}{10}$

21. $9^{1/3} - 2$

22. $\tan^{-1}(1.05) - \frac{\pi}{4}$

23. $e^{-0.1} - 1$

24. $\ln(0.97)$

25. Estimate $f(4.03)$ for $f(x)$ as in Figure 8.

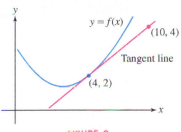

y = f(x)

(10, 4)

Tangent line

(4, 2)

FIGURE 8

26. At a certain moment, an object in linear motion has velocity 100 m/s. Estimate the distance traveled over the next quarter-second, and explain how this is an application of the Linear Approximation.

27. Which is larger: $\sqrt{2.1} - \sqrt{2}$ or $\sqrt{9.1} - \sqrt{9}$? Explain using the Linear Approximation.

28. Estimate $\sin 61° - \sin 60°$ using the Linear Approximation. *Hint:* Express $\Delta\theta$ in radians.

29. Box office revenue at a multiplex cinema in Paris is $R(p) = 3600p - 10p^3$ euros per showing when the ticket price is p euros. Calculate $R(p)$ for $p = 9$ and use the Linear Approximation to estimate ΔR if p is raised or lowered by 0.5 euros.

30. The *stopping distance* for an automobile is $F(s) = 1.1s + 0.054s^2$ ft, where s is the speed in mph. Use the Linear Approximation to estimate the change in stopping distance per additional mph when $s = 35$ and when $s = 55$.

31. A thin silver wire has length $L = 18$ cm when the temperature is $T = 30°C$. Estimate ΔL when T decreases to $25°C$ if the coefficient of thermal expansion is $k = 1.9 \times 10^{-5}°C^{-1}$ (see Example 3).

32. At a certain moment, the temperature in a snake cage satisfies $dT/dt = 0.008°C/s$. Estimate the rise in temperature over the next 10 s.

33. The atmospheric pressure at altitude h (kilometers) for $11 \le h \le 25$ is approximately

$$P(h) = 128e^{-0.157h} \text{ kilopascals}$$

(a) Estimate ΔP at $h = 20$ when $\Delta h = 0.5$.

(b) Compute the actual change, and compute the percentage error in the Linear Approximation.

34. The resistance R of a copper wire at temperature $T = 20°C$ is $R = 15\ \Omega$. Estimate the resistance at $T = 22°C$, assuming that $dR/dT\big|_{T=20} = 0.06\ \Omega/°C$.

35. Newton's Law of Gravitation shows that if a person weighs w pounds on the surface of the earth, then his or her weight at distance x from the center of the earth is

$$W(x) = \frac{wR^2}{x^2} \qquad \text{(for } x \ge R)$$

where $R = 3960$ miles is the radius of the earth (Figure 9).

(a) Show that the weight lost at altitude h miles above the earth's surface is approximately $\Delta W \approx -(0.0005w)h$. *Hint:* Use the Linear Approximation with $dx = h$.

(b) Estimate the weight lost by a 200-lb football player flying in a jet at an altitude of 7 miles.

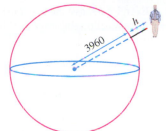

FIGURE 9 The distance to the center of the earth is $3960 + h$ miles.

36. Using Exercise 35(a), estimate the altitude at which a 130-lb pilot would weigh 129.5 lb.

37. A stone tossed vertically into the air with initial velocity v cm/s reaches a maximum height of $h = v^2/1960$ cm.

(a) Estimate Δh if $v = 700$ cm/s and $\Delta v = 1$ cm/s.

(b) Estimate Δh if $v = 1000$ cm/s and $\Delta v = 1$ cm/s.

(c) In general, does a 1 cm/s increase in v lead to a greater change in h at low or high initial velocities? Explain.

38. The side s of a square carpet is measured at 6 m. Estimate the maximum error in the area A of the carpet if s is accurate to within 2 cm.

In Exercises 39 and 40, use the following fact derived from Newton's Laws: An object released at an angle θ with initial velocity v ft/s travels a horizontal distance

$$s = \frac{1}{32}v^2 \sin 2\theta \text{ ft (Figure 10)}$$

39. A player located 18.1 ft from the basket launches a successful jump shot from a height of 10 ft (level with the rim of the basket), at an angle $\theta = 34°$ and initial velocity $v = 25$ ft/s.

(a) Show that $\Delta s \approx 0.255\Delta\theta$ ft for a small change of $\Delta\theta$.

(b) Is it likely that the shot would have been successful if the angle had been off by $2°$?

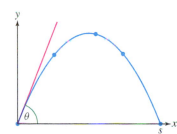

FIGURE 10 Trajectory of an object released at an angle θ.

40. Estimate Δs if $\theta = 34°$, $v = 25$ ft/s, and $\Delta v = 2$.

41. The radius of a spherical ball is measured at $r = 25$ cm. Estimate the maximum error in the volume and surface area if r is accurate to within 0.5 cm.

42. The dosage D of diphenhydramine for a dog of body mass w kg is $D = 4.7w^{2/3}$ mg. Estimate the maximum allowable error in w for a cocker spaniel of mass $w = 10$ kg if the percentage error in D must be less than 3%.

43. The volume (in liters) and pressure P (in atmospheres) of a certain gas satisfy $PV = 24$. A measurement yields $V = 4$ with a possible error of ± 0.3 L. Compute P and estimate the maximum error in this computation.

44. In the notation of Exercise 43, assume that a measurement yields $V = 4$. Estimate the maximum allowable error in V if P must have an error of less than 0.2 atm.

In Exercises 45–54, find the linearization at $x = a$.

45. $f(x) = x^4$, $a = 1$

46. $f(x) = \dfrac{1}{x}$, $a = 2$

47. $f(\theta) = \sin^2 \theta$, $a = \frac{\pi}{4}$

48. $g(x) = \dfrac{x^2}{x - 3}$, $a = 4$

49. $y = (1 + x)^{-1/2}$, $a = 0$

50. $y = (1 + x)^{-1/2}$, $a = 3$

51. $y = (1 + x^2)^{-1/2}$, $a = 0$

52. $y = \tan^{-1} x$, $a = 1$

53. $y = e^{\sqrt{x}}$, $a = 1$

54. $y = e^x \ln x$, $a = 1$

55. What is $f(2)$ if the linearization of $f(x)$ at $a = 2$ is $L(x) = 2x + 4$?

56. Compute the linearization of $f(x) = 3x - 4$ at $a = 0$ and $a = 2$. Prove more generally that a linear function coincides with its linearization at $x = a$ for all a.

57. Estimate $\sqrt{16.2}$ using the linearization $L(x)$ of $f(x) = \sqrt{x}$ at $a = 16$. Plot f and L on the same set of axes and determine whether the estimate is too large or too small.

58. GU   Estimate $1/\sqrt{15}$ using a suitable linearization of $f(x) = 1/\sqrt{x}$. Plot f and L on the same set of axes and determine whether the estimate is too large or too small. Use a calculator to compute the percentage error.

In Exercises 59–67, approximate using linearization and use a calculator to compute the percentage error.

59. $\dfrac{1}{\sqrt{17}}$

60. $\dfrac{1}{101}$

61. $\dfrac{1}{(10.03)^2}$

62. $(17)^{1/4}$

63. $(64.1)^{1/3}$

64. $(1.2)^{5/3}$

65. $\cos^{-1}(0.52)$

66. $\ln 1.07$

67. $e^{-0.012}$

68. GU   Compute the linearization $L(x)$ of $f(x) = x^2 - x^{3/2}$ at $a = 4$. Then plot $f - L$ and find an interval I around $a = 4$ such that $|f(x) - L(x)| \le 0.1$ for $x \in I$.

69. Show that the Linear Approximation to $f(x) = \sqrt{x}$ at $x = 9$ yields the estimate $\sqrt{9 + h} - 3 \approx \frac{1}{6} h$. Set $K = 0.01$ and show that $|f''(x)| \le K$ for $x \ge 9$. Then verify numerically that the error E satisfies Eq. (6) for $h = 10^{-n}$, for $1 \le n \le 4$.

70. GU   The Linear Approximation to $f(x) = \tan x$ at $x = \frac{\pi}{4}$ yields the estimate $\tan\left(\frac{\pi}{4} + h\right) - 1 \approx 2h$. Set $K = 6.2$ and show, using a plot, that $|f''(x)| \le K$ for $x \in [\frac{\pi}{4}, \frac{\pi}{4} + 0.1]$. Then verify numerically that the error E satisfies Eq. (6) for $h = 10^{-n}$, for $1 \le n \le 4$.

Further Insights and Challenges

71. Compute dy/dx at the point $P = (2, 1)$ on the curve $y^3 + 3xy = 7$ and show that the linearization at P is $L(x) = -\frac{1}{3} x + \frac{5}{3}$. Use $L(x)$ to estimate the y-coordinate of the point on the curve where $x = 2.1$.

72. Apply the method of Exercise 71 to $P = (0.5, 1)$ on $y^5 + y - 2x = 1$ to estimate the y-coordinate of the point on the curve where $x = 0.55$.

73. Apply the method of Exercise 71 to $P = (-1, 2)$ on $y^4 + 7xy = 2$ to estimate the solution of $y^4 - 7.7y = 2$ near $y = 2$.

74. Show that for any real number k, $(1 + \Delta x)^k \approx 1 + k \Delta x$ for small Δx. Estimate $(1.02)^{0.7}$ and $(1.02)^{-0.3}$.

75. Let $\Delta f = f(5 + h) - f(5)$, where $f(x) = x^2$. Verify directly that $E = |\Delta f - f'(5)h|$ satisfies (6) with $K = 2$.

76. Let $\Delta f = f(1 + h) - f(1)$, where $f(x) = x^{-1}$. Show directly that $E = |\Delta f - f'(1)h|$ is equal to $h^2/(1 + h)$. Then prove that $E \le 2h^2$ if $-\frac{1}{2} \le h \le \frac{1}{2}$. *Hint:* In this case, $\frac{1}{2} \le 1 + h \le \frac{3}{2}$.

4.2 Extreme Values

In many applications, it is important to find the minimum or maximum value of a function f. For example, a physician needs to know the maximum drug concentration in a patient's bloodstream when a drug is administered. This amounts to finding the highest point on the graph of C, the concentration at time t (Figure 1).

We refer to the maximum and minimum values (max and min for short) as **extreme values** or **extrema** (singular: extremum) and to the process of finding them as **optimization**. Sometimes, we are interested in finding the min or max for x in a particular interval I, rather than on the entire domain of f.

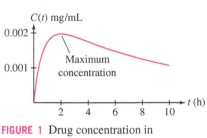

FIGURE 1 Drug concentration in bloodstream (see Exercise 74).

Often, we drop the word "absolute" and speak simply of the min or max on an interval I. When no interval is mentioned, it is understood that we refer to the extreme values on the entire domain of the function.

DEFINITION **Extreme Values on an Interval**   Let f be a function on an interval I and let $a \in I$. We say that $f(a)$ is the

- **Absolute minimum** of f on I if $f(a) \le f(x)$ for all $x \in I$.
- **Absolute maximum** of f on I if $f(a) \ge f(x)$ for all $x \in I$.

Does every function have a minimum or maximum value? Clearly not, as we see by taking $f(x) = x$. Indeed, $f(x) = x$ increases without bound as $x \to \infty$ and decreases without bound as $x \to -\infty$. In fact, extreme values do not always exist even if we restrict ourselves to an interval I. Figure 2 illustrates what can go wrong if I is open or f has a discontinuity.

- **Discontinuity:** (A) shows a discontinuous function with no maximum value. The values of $f(x)$ get arbitrarily close to 3 from below, but 3 is not the maximum value because $f(x)$ never actually takes on the value 3.
- **Open interval:** In (B), $g(x)$ is defined on the *open* interval (a, b). It has no max because it tends to ∞ on the right, and it has no min because it tends to 10 on the left without ever reaching this value.

Fortunately, our next theorem guarantees that extreme values exist when the function is continuous and I is closed [Figure 2(C)].

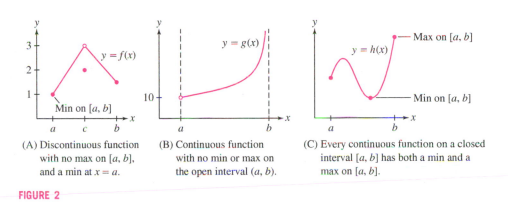

(A) Discontinuous function with no max on $[a, b]$, and a min at $x = a$.

(B) Continuous function with no min or max on the open interval (a, b).

(C) Every continuous function on a closed interval $[a, b]$ has both a min and a max on $[a, b]$.

FIGURE 2

REMINDER *A closed, bounded interval is an interval $I = [a, b]$ (endpoints included), where a and b are both finite. Often, we drop the word "bounded" and refer to I more simply as a closed interval. An open interval (a, b) (endpoints not included) may have one or two infinite endpoints.*

THEOREM 1 Existence of Extrema on a Closed Interval   A continuous function f on a closed (bounded) interval $I = [a, b]$ takes on both a minimum and a maximum value on I.

CONCEPTUAL INSIGHT   Why does Theorem 1 require a closed interval? Think of the graph of a continuous function as a string. If the interval is closed, the string is pinned down at the two endpoints and cannot fly off to infinity (or approach a min/max without reaching it) as in Figure 2(B). Intuitively, therefore, it must have a highest and lowest point. However, a rigorous proof of Theorem 1 relies on the *completeness property* of the real numbers (see Appendix D).

Local Extrema and Critical Points

We focus now on the problem of finding extreme values. A key concept is that of a local minimum or maximum.

When we get to the top of a hill in an otherwise flat region, our altitude is at a local maximum, but we are still far from the point of absolute maximum altitude, which is located at the peak of Mt. Everest. That's the difference between local and absolute extrema.

Adapted from V. M. Tikhomirov, Stories About Maxima and Minima, 1990.

DEFINITION Local Extrema   We say that $f(c)$ is a

- **Local minimum** occurring at $x = c$ if $f(c)$ is the minimum value of f on some open interval (in the domain of f) containing c.
- **Local maximum** occurring at $x = c$ if $f(c)$ is the maximum value of f on some open interval (in the domain of f) containing c.

A local max occurs at $x = c$ if $(c, f(c))$ is the highest point on the graph within some small box [Figure 3(A)]. Thus, $f(c)$ is greater than or equal to all other *nearby* values, but it does not have to be the absolute maximum value of f. Local minima are similar. Figure 3(B) illustrates the difference between local and absolute extrema: $f(a)$ is the absolute max on $[a, b]$ but is not a local max because $f(x)$ takes on larger values to the left of $x = a$.

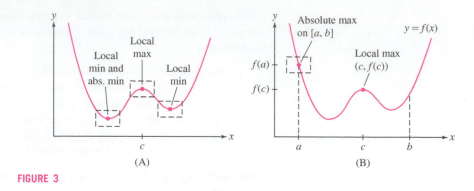

FIGURE 3

How do we find the local extrema? The crucial observation is that *the tangent line at a local min or max is horizontal* [Figure 4(A)]. In other words, if $f(c)$ is a local min or max, then $f'(c) = 0$. However, this assumes that f is differentiable. Otherwise, the tangent line may not exist, as in Figure 4(B). To take both possibilities into account, we define the notion of a critical point.

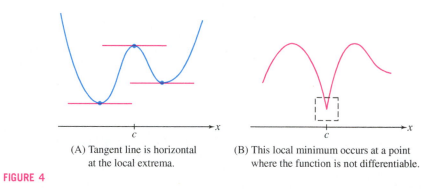

(A) Tangent line is horizontal
at the local extrema.

(B) This local minimum occurs at a point
where the function is not differentiable.

FIGURE 4

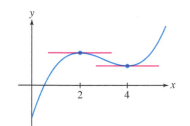

FIGURE 5 Graph of
$f(x) = x^3 - 9x^2 + 24x - 10$.

> **DEFINITION Critical Points**   A number c in the domain of f is called a **critical point** if *either* $f'(c) = 0$ or $f'(c)$ does not exist.

■ **EXAMPLE 1**   Find the critical points of $f(x) = x^3 - 9x^2 + 24x - 10$.

Solution   The function f is differentiable everywhere (Figure 5), so the critical points are the solutions of $f'(x) = 0$:

$$f'(x) = 3x^2 - 18x + 24 = 3(x^2 - 6x + 8)$$
$$= 3(x - 2)(x - 4) = 0$$

The critical points are the roots $c = 2$ and $c = 4$.   ■

■ **EXAMPLE 2**   Nondifferentiable Function   Find the critical points of $f(x) = |x|$.

Solution   As we see in Figure 6, $f'(x) = -1$ for $x < 0$ and $f'(x) = 1$ for $x > 0$. Therefore, $f'(x) = 0$ has no solutions with $x \neq 0$. However, $f'(0)$ does not exist. Thus, $c = 0$ is a critical point.   ■

The next theorem tells us that we can find local extrema by solving for the critical points. It is one of the most important results in calculus.

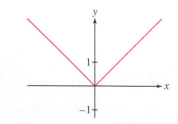

FIGURE 6 Graph of $f(x) = |x|$.

> **THEOREM 2 Fermat's Theorem on Local Extrema**   If $f(c)$ is a local min or max, then c is a critical point of f.

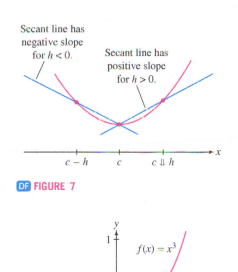

Secant line has negative slope for $h < 0$.

Secant line has positive slope for $h > 0$.

$c - h$ c $c \Downarrow h$ x

DF FIGURE 7

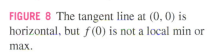

$f(x) = x^3$

FIGURE 8 The tangent line at $(0, 0)$ is horizontal, but $f(0)$ is not a local min or max.

In this section, we restrict our attention to closed intervals because in this case extreme values are guaranteed to exist (Theorem 1). Optimization on open intervals is discussed in Section 4.7.

Proof Suppose that $f(c)$ is a local minimum (the case of a local maximum is similar). If $f'(c)$ does not exist, then c is a critical point and there is nothing more to prove. So assume that $f'(c)$ exists. We must then prove that $f'(c) = 0$.

Because $f(c)$ is a local minimum, we have $f(c + h) \geq f(c)$ for all sufficiently small $h \neq 0$. Equivalently, $f(c + h) - f(c) \geq 0$. Now divide this inequality by h:

$$\frac{f(c + h) - f(c)}{h} \geq 0 \qquad \text{if } h > 0 \qquad \boxed{1}$$

$$\frac{f(c + h) - f(c)}{h} \leq 0 \qquad \text{if } h < 0 \qquad \boxed{2}$$

Figure 7 shows the graphical interpretation of these inequalities. Taking the one-sided limits of both sides of (1) and (2), we obtain

$$f'(c) = \lim_{h \to 0^+} \frac{f(c + h) - f(c)}{h} \geq \lim_{h \to 0^+} 0 = 0$$

$$f'(c) = \lim_{h \to 0^-} \frac{f(c + h) - f(c)}{h} \leq \lim_{h \to 0^-} 0 = 0$$

Thus, $f'(c) \geq 0$ and $f'(c) \leq 0$. The only possibility is $f'(c) = 0$ as claimed. ■

CONCEPTUAL INSIGHT Fermat's Theorem *does not claim* that all critical points yield local extrema. "False positives" may exist—that is, we might have $f'(c) = 0$ without $f(c)$ being a local min or max. For example, $f(x) = x^3$ has derivative $f'(x) = 3x^2$ and $f'(0) = 0$, but $f(0)$ is neither a local min nor max (Figure 8). The origin is a point of inflection (studied in Section 4.4), where the tangent line crosses the graph.

Optimizing on a Closed Interval

Finally, we have all the tools needed for optimizing a continuous function on a closed interval. Theorem 1 guarantees that the extreme values exist, and the next theorem tells us where to find them, namely among the critical points or endpoints of the interval.

> **THEOREM 3 Extreme Values on a Closed Interval** Assume that f is continuous on $[a, b]$ and let $f(c)$ be the minimum or maximum value on $[a, b]$. Then c is either a critical point or one of the endpoints a or b.

Proof If c is one of the endpoints a or b, there is nothing to prove. If not, then c belongs to the open interval (a, b). In this case, $f(c)$ is also a local min or max because it is the min or max on (a, b). By Fermat's Theorem, c is a critical point. ■

■ **EXAMPLE 3** Find the extrema of $f(x) = 2x^3 - 15x^2 + 24x + 7$ on $[0, 6]$.

Solution The extreme values occur at critical points or endpoints by Theorem 3, so we can break up the problem neatly into two steps.

Step 1. **Find the critical points.**

The function f is differentiable, so we find the critical points by solving

$$f'(x) = 6x^2 - 30x + 24 = 6(x - 1)(x - 4) = 0$$

The critical points are $c = 1$ and 4.

Step 2. **Compare values of $f(x)$ at the critical points and endpoints.**

x-value	Value of $f(x)$	
1 (critical point)	$f(1) = 18$	
4 (critical point)	$f(4) = -9$	min
0 (endpoint)	$f(0) = 7$	
6 (endpoint)	$f(6) = 43$	max

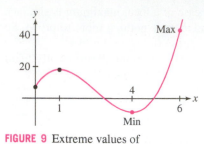

FIGURE 9 Extreme values of $f(x) = 2x^3 - 15x^2 + 24x + 7$ on $[0, 6]$.

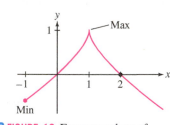

DF **FIGURE 10** Extreme values of $f(x) = 1 - (x - 1)^{2/3}$ on $[-1, 2]$.

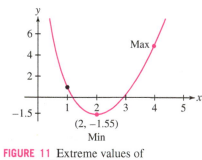

FIGURE 11 Extreme values of $f(x) = x^2 - 8 \ln x$ on $[1, 4]$.

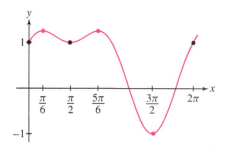

DF **FIGURE 12** f attains a max at $\frac{\pi}{6}$ and $\frac{5\pi}{6}$ and a min at $\frac{3\pi}{2}$.

The maximum value of $f(x)$ on $[0, 6]$ is the largest of the values in this table, namely $f(6) = 43$. Similarly, the minimum is $f(4) = -9$. See Figure 9. ∎

■ **EXAMPLE 4   Function with a Cusp**   Find the maximum of $f(x) = 1 - (x - 1)^{2/3}$ on $[-1, 2]$.

Solution   First, find the critical points:

$$f'(x) = -\frac{2}{3}(x - 1)^{-1/3} = -\frac{2}{3(x - 1)^{1/3}}$$

The equation $f'(x) = 0$ has no solutions because $f'(x)$ is never zero. However, $f'(x)$ does not exist at $x = 1$, so $c = 1$ is a critical point (Figure 10).

Next, compare values of $f(x)$ at the critical points and endpoints:

x-value	Value of $f(x)$	
1 (critical point)	$f(1) = 1$	max
-1 (endpoint)	$f(-1) \approx -0.59$	min
2 (endpoint)	$f(2) = 0$	

∎

■ **EXAMPLE 5   Logarithmic Example**   Find the extreme values of the function $f(x) = x^2 - 8 \ln x$ on $[1, 4]$.

Solution   First, solve for the critical points:

$$f'(x) = 2x - \frac{8}{x} = 0 \quad \Rightarrow \quad 2x = \frac{8}{x} \quad \Rightarrow \quad x = \pm 2$$

The only critical point in the interval $[1, 4]$ is $c = 2$. Next, compare the values of $f(x)$ at the critical points and endpoints (Figure 11):

x-value	Value of $f(x)$	
2 (critical point)	$f(2) \approx -1.55$	min
1 (endpoint)	$f(1) = 1$	
4 (endpoint)	$f(4) \approx 4.9$	max

We see that the min on $[1, 4]$ is $f(2) \approx -1.55$ and the max is $f(4) \approx 4.9$. ∎

■ **EXAMPLE 6   Trigonometric Function**   Find the minimum and maximum of the function $f(x) = \sin x + \cos^2 x$ on $[0, 2\pi]$ (Figure 12).

Solution   First, solve for the critical points:

$$f'(x) = \cos x - 2 \sin x \cos x = \cos x(1 - 2 \sin x) = 0 \quad \Rightarrow \quad \cos x = 0 \text{ or } \sin x = \frac{1}{2}$$

$$\cos x = 0 \quad \Rightarrow \quad x = \frac{\pi}{2}, \frac{3\pi}{2} \quad \text{and} \quad \sin x = \frac{1}{2} \quad \Rightarrow \quad x = \frac{\pi}{6}, \frac{5\pi}{6}$$

Then compare the values of $f(x)$ at the critical points and endpoints:

x-value	Value of $f(x)$	
$\frac{\pi}{2}$ (critical point)	$f\left(\frac{\pi}{2}\right) = 1 + 0^2 = 1$	
$\frac{3\pi}{2}$ (critical point)	$f\left(\frac{3\pi}{2}\right) = -1 + 0^2 = -1$	min
$\frac{\pi}{6}$ (critical point)	$f\left(\frac{\pi}{6}\right) = \frac{1}{2} + \left(\frac{\sqrt{3}}{2}\right)^2 = \frac{5}{4}$	max
$\frac{5\pi}{6}$ (critical point)	$f\left(\frac{5\pi}{6}\right) = \frac{1}{2} + \left(-\frac{\sqrt{3}}{2}\right)^2 = \frac{5}{4}$	max
0 and 2π (endpoints)	$f(0) = f(2\pi) = 1$	

∎

Rolle's Theorem

As an application of our optimization methods, we prove Rolle's Theorem: If f is differentiable and takes on the same value at two different points a and b, then somewhere between these two points the derivative is zero. Graphically, if the secant line between $x = a$ and $x = b$ is horizontal, then at least one tangent line between a and b is also horizontal (Figure 13).

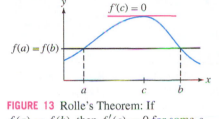

FIGURE 13 Rolle's Theorem: If $f(a) = f(b)$, then $f'(c) = 0$ for some c between a and b.

> **THEOREM 4 Rolle's Theorem** Assume that f is continuous on $[a, b]$ and differentiable on (a, b). If $f(a) = f(b)$, then there exists a number c between a and b such that $f'(c) = 0$.

Proof Since f is continuous and $[a, b]$ is closed, f has a min and a max in $[a, b]$. Where do they occur? If either the min or the max occurs at a point c in the *open* interval (a, b), then $f(c)$ is a local extreme value and $f'(c) = 0$ by Fermat's Theorem (Theorem 2). Otherwise, both the min and the max occur at the endpoints. However, $f(a) = f(b)$, so in this case, the min and max coincide and f is a constant function with zero derivative. Then, $f'(c) = 0$ for all c in (a, b). ∎

■ **EXAMPLE 7** **Illustrating Rolle's Theorem** Verify Rolle's Theorem for

$$f(x) = x^4 - x^2 \qquad \text{on} \qquad [-2, 2]$$

Solution The hypotheses of Rolle's Theorem are satisfied because f is differentiable (and therefore continuous) everywhere, and $f(2) = f(-2)$:

$$f(2) = 2^4 - 2^2 = 12, \qquad f(-2) = (-2)^4 - (-2)^2 = 12$$

We must verify that $f'(c) = 0$ has a solution in $(-2, 2)$, so we solve $f'(x) = 4x^3 - 2x = 2x(2x^2 - 1) = 0$. The solutions are $c = 0$ and $c = \pm 1/\sqrt{2} \approx \pm 0.707$. They all lie in $(-2, 2)$, so Rolle's Theorem is satisfied with three values of c. ■

■ **EXAMPLE 8** **Using Rolle's Theorem** Show that $f(x) = x^3 + 9x - 4$ has precisely one real root.

Solution First, we note that $f(0) = -4$ is negative and $f(1) = 6$ is positive. By the Intermediate Value Theorem (Section 2.8), f has *at least* one root a in $[0, 1]$. If f had a second root b, then $f(a) = f(b) = 0$ and Rolle's Theorem would imply that $f'(c) = 0$ for some $c \in (a, b)$. This is not possible because $f'(x) = 3x^2 + 9 \geq 9$, so $f'(c) = 0$ has no solutions. We conclude that a is *the only* real root of f (Figure 14). ■

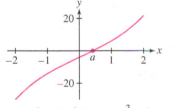

FIGURE 14 Graph of $f(x) = x^3 + 9x - 4$. This function has one real root.

We can hardly expect a more general method.... This method never fails and could be extended to a number of beautiful problems; with its aid we have found the centers of gravity of figures bounded by straight lines or curves, as well as those of solids, and a number of other results which we may treat elsewhere if we have the time to do so.

—From Fermat's *On Maxima and Minima and on Tangents*

4.2 SUMMARY

- The *extreme values* of f on an interval I are the minimum and maximum values of f for $x \in I$ (also called *absolute extrema* on I).
- Basic Theorem: If f is continuous on a closed interval $[a, b]$, then f has both a min and a max on $[a, b]$.
- $f(c)$ is a *local minimum* if $f(x) \geq f(c)$ for all x in some open interval around c. Local maxima are defined similarly.
- $x = c$ is a *critical point* of f if either $f'(c) = 0$ or $f'(c)$ does not exist.
- Fermat's Theorem: If $f(c)$ is a local min or max, then c is a critical point.
- To find the extreme values of a continuous function f on a closed interval $[a, b]$:

  Step 1. Find the critical points of f in $[a, b]$.

  Step 2. Calculate $f(x)$ at the critical points in $[a, b]$ and at the endpoints.

  The min and max on $[a, b]$ are the smallest and largest among the values computed in Step 2.
- Rolle's Theorem: If f is continuous on $[a, b]$ and differentiable on (a, b), and if $f(a) = f(b)$, then there exists c between a and b such that $f'(c) = 0$.

HISTORICAL PERSPECTIVE

Pierre de Fermat
(1601–1665)
(© Bettmann/Corbis)

René Descartes
(1596–1650)
(© Stapleton Collection/Corbis)

Sometime in the 1630s, in the decade before Isaac Newton was born, the French mathematician Pierre de Fermat invented a general method for finding extreme values. Fermat said, in essence, that if you want to find extrema, you must set the derivative equal to zero and solve for the critical points, just as we have done in this section. He also described a general method for finding tangent lines that is not essentially different from our method of derivatives. For this reason, Fermat is often regarded as an inventor of calculus, together with Newton and Leibniz.

At around the same time, René Descartes (1596-1650) developed a different but less effective approach to finding tangent lines. Descartes, after whom Cartesian coordinates are named, was a profound thinker—the leading philosopher and scientist of his time in Europe. He is regarded today as the father of modern philosophy and the founder (along with Fermat) of analytic geometry. A dispute developed when Descartes learned through an intermediary that Fermat had criticized his work on optics. Sensitive and stubborn, Descartes retaliated by attacking Fermat's method of finding tangents and only after some

third-party refereeing did he admit that Fermat was correct. He wrote:

...Seeing the last method that you use for finding tangents to curved lines, I can reply to it in no other way than to say that it is very good and that, if you had explained it in this manner at the outset, I would have not contradicted it at all.

However, in subsequent private correspondence, Descartes was less generous, referring at one point to some of Fermat's work as "*le galimatias le plus ridicule*"—meaning the most ridiculous gibberish. Today Fermat is recognized as one of the greatest mathematicians of his age who made far-reaching contributions in several areas of mathematics.

4.2 EXERCISES

Preliminary Questions

1. What is the definition of a critical point?

In Questions 2 and 3, choose the correct conclusion.

2. If f is not continuous on $[0, 1]$, then

(a) f has no extreme values on $[0, 1]$.

(b) f might not have any extreme values on $[0, 1]$.

3. If f is continuous but has no critical points in $[0, 1]$, then

(a) f has no min or max on $[0, 1]$.

(b) Either $f(0)$ or $f(1)$ is the minimum value on $[0, 1]$.

4. Fermat's Theorem *does not* claim that if $f'(c) = 0$, then $f(c)$ is a local extreme value (this is false). What *does* Fermat's Theorem assert?

Exercises

1. The following questions refer to Figure 15.

(a) How many critical points does f have on $[0, 8]$?

(b) What is the maximum value of f on $[0, 8]$?

(c) What are the local maximum values of f?

(d) Find a closed interval on which both the minimum and maximum values of f occur at critical points.

(e) Find an interval on which the minimum value occurs at an endpoint.

2. State whether $f(x) = x^{-1}$ (Figure 16) has a minimum or maximum value on the following intervals:

(a) $(0, 2)$ **(b)** $(1, 2)$ **(c)** $[1, 2]$

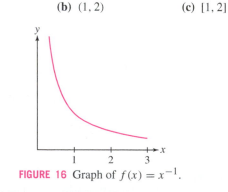

FIGURE 16 Graph of $f(x) = x^{-1}$.

In Exercises 3–20, find all critical points of the function.

3. $f(x) = x^2 - 2x + 4$ **4.** $f(x) = 7x - 2$

5. $f(x) = x^3 - \frac{9}{2}x^2 - 54x + 2$ **6.** $f(t) = 8t^3 - t^2$

FIGURE 15

7. $f(x) = x^{-1} - x^{-2}$

8. $g(z) = \dfrac{1}{z-1} - \dfrac{1}{z}$

9. $f(x) = \dfrac{x}{x^2 + 1}$

10. $f(x) = \dfrac{x^2}{x^2 - 4x + 8}$

11. $f(t) = t - 4\sqrt{t+1}$

12. $f(t) = 4t - \sqrt{t^2 + 1}$

13. $f(x) = xe^{2x}$

14. $f(x) = x + |2x + 1|$

15. $g(\theta) = \sin^2 \theta$

16. $R(\theta) = \cos \theta + \sin^2 \theta$

17. $f(x) = x \ln x$

18. $f(x) = x^2 \sqrt{1 - x^2}$

19. $f(x) = \sin^{-1} x - 2x$

20. $f(x) = \sec^{-1} x - \ln x$

21. Let $f(x) = x^2 - 4x + 1$.
(a) Find the critical point c of f and compute $f(c)$.
(b) Compute the value of $f(x)$ at the endpoints of the interval $[0, 4]$.
(c) Determine the min and max of f on $[0, 4]$.
(d) Find the extreme values of f on $[0, 1]$.

22. Find the extreme values of $f(x) = 2x^3 - 9x^2 + 12x$ on $[0, 3]$ and $[0, 2]$.

23. Find the critical points of $f(x) = \sin x + \cos x$ and determine the extreme values on $\left[0, \frac{\pi}{2}\right]$.

24. Compute the critical points of $h(t) = (t^2 - 1)^{1/3}$. Check that your answer is consistent with Figure 17. Then find the extreme values of h on $[0, 1]$ and on $[0, 2]$.

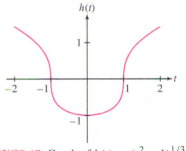

$h(t)$

FIGURE 17 Graph of $h(t) = (t^2 - 1)^{1/3}$.

25. $\boxed{GU}$ Plot $f(x) = 2\sqrt{x} - x$ on $[0, 4]$ and determine the maximum value graphically. Then verify your answer using calculus.

26. $\boxed{GU}$ Plot $f(x) = \ln x - 5 \sin x$ on $[0.1, 2]$ and approximate both the critical points and the extreme values.

27. $\mathcal{CAS}$ Approximate the critical points of $g(x) = x \cos^{-1} x$ and estimate the maximum value of g.

28. $\mathcal{CAS}$ Approximate the critical points of $g(x) = 5e^x - \tan x$ in $\left(-\frac{\pi}{2}, \frac{\pi}{2}\right)$.

In Exercises 29–58, find the minimum and maximum value of the function on the given interval by comparing values at the critical points and endpoints.

29. $y = 2x^2 + 4x + 5$, $[-2, 2]$

30. $y = 2x^2 + 4x + 5$, $[0, 2]$

31. $y = 6t - t^2$, $[0, 5]$

32. $y = 6t - t^2$, $[4, 6]$

33. $y = x^3 - 6x^2 + 8$, $[1, 6]$

34. $y = x^3 + x^2 - x$, $[-2, 2]$

35. $y = 2t^3 + 3t^2$, $[1, 2]$

36. $y = x^3 - 12x^2 + 21x$, $[0, 2]$

37. $y = z^5 - 80z$, $[-3, 3]$

38. $y = 2x^5 + 5x^2$, $[-2, 2]$

39. $y = \dfrac{x^2 + 1}{x - 4}$, $[5, 6]$

40. $y = \dfrac{1 - x}{x^2 + 3x}$, $[1, 4]$

41. $y = x - \dfrac{4x}{x + 1}$, $[0, 3]$

42. $y = 2\sqrt{x^2 + 1} - x$, $[0, 2]$

43. $y = (2 + x)\sqrt{2 + (2 - x)^2}$, $[0, 2]$

44. $y = \sqrt{1 + x^2} - 2x$, $[0, 1]$

45. $y = \sqrt{x + x^2} - 2\sqrt{x}$, $[0, 4]$ **46.** $y = (t - t^2)^{1/3}$, $[-1, 2]$

47. $y = \sin x \cos x$, $\left[0, \frac{\pi}{2}\right]$ **48.** $y = x + \sin x$, $[0, 2\pi]$

49. $y = \sqrt{2}\,\theta - \sec \theta$, $\left[0, \frac{\pi}{3}\right]$

50. $y = \cos \theta + \sin \theta$, $[0, 2\pi]$

51. $y = \theta - 2 \sin \theta$, $[0, 2\pi]$

52. $y = 4 \sin^3 \theta - 3 \cos^2 \theta$, $[0, 2\pi]$

53. $y = \tan x - 2x$, $[0, 1]$ **54.** $y = xe^{-x}$, $[0, 2]$

55. $y = \dfrac{\ln x}{x}$, $[1, 3]$ **56.** $y = 5 \tan^{-1} x - x$, $[1, 5]$

57. $y = 3e^x - e^{2x}$, $\left[-\frac{1}{2}, 1\right]$ **58.** $y = x^3 - 24 \ln x$, $\left[\frac{1}{2}, 3\right]$

59. Let $f(\theta) = 2 \sin 2\theta + \sin 4\theta$.
(a) Show that θ is a critical point if $\cos 4\theta = -\cos 2\theta$.
(b) Show, using a unit circle, that $\cos \theta_1 = -\cos \theta_2$ if and only if $\theta_1 = \pi \pm \theta_2 + 2\pi k$ for an integer k.
(c) Show that $\cos 4\theta = -\cos 2\theta$ if and only if $\theta = \frac{\pi}{2} + \pi k$ or $\theta = \frac{\pi}{6} + \left(\frac{\pi}{3}\right)k$.
(d) Find the six critical points of f on $[0, 2\pi]$ and find the extreme values of f on this interval.
(e) $\boxed{GU}$ Check your results against a graph of f.

60. $\boxed{GU}$ Find the critical points of $f(x) = 2 \cos 3x + 3 \cos 2x$ in $[0, 2\pi]$. Check your answer against a graph of f.

In Exercises 61–64, find the critical points and the extreme values on $[0, 4]$. In Exercises 63 and 64, refer to Figure 18.

61. $y = |x - 2|$

62. $y = |3x - 9|$

63. $y = |x^2 + 4x - 12|$

64. $y = |\cos x|$

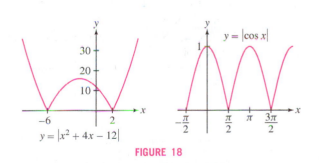

$y = |x^2 + 4x - 12|$

$y = |\cos x|$

FIGURE 18

In Exercises 65–68, verify Rolle's Theorem for the given interval by checking $f(a) = f(b)$ and then finding a value c in (a, b) such that $f'(c) = 0$.

65. $f(x) = x + x^{-1}$, $\left[\frac{1}{2}, 2\right]$ **66.** $f(x) = \sin x$, $\left[\frac{\pi}{4}, \frac{3\pi}{4}\right]$

67. $f(x) = \dfrac{x^2}{8x - 15}$, $[3, 5]$

68. $f(x) = \sin^2 x - \cos^2 x$, $\left[\frac{\pi}{4}, \frac{3\pi}{4}\right]$

69. Prove that $f(x) = x^5 + 2x^3 + 4x - 12$ has precisely one real root.

70. Prove that $f(x) = x^3 + 3x^2 + 6x$ has precisely one real root.

71. Prove that $f(x) = x^4 + 5x^3 + 4x$ has no root c satisfying $c > 0$. *Hint:* Note that $x = 0$ is a root and apply Rolle's Theorem.

72. Prove that $c = 4$ is the largest root of $f(x) = x^4 - 8x^2 - 128$.

73. The position of a mass oscillating at the end of a spring is $s(t) = A \sin \omega t$, where A is the amplitude and ω is the angular frequency. Show that the speed $|v(t)|$ is at a maximum when the acceleration $a(t)$ is zero and that $|a(t)|$ is at a maximum when $v(t)$ is zero.

74. The concentration $C(t)$ (in milligrams per cubic centimeter) of a drug in a patient's bloodstream after t hours is

$$C(t) = \frac{0.016t}{t^2 + 4t + 4}$$

Find the maximum concentration in the time interval $[0, 8]$ and the time at which it occurs.

75. $\mathcal{CAS}$ **Antibiotic Levels** A study shows that the concentration $C(t)$ (in micrograms per milliliter) of antibiotic in a patient's blood serum after t hours is $C(t) = 120(e^{-0.2t} - e^{-bt})$, where $b \geq 1$ is a constant that depends on the particular combination of antibiotic agents used. Solve numerically for the value of b (to two decimal places) for which maximum concentration occurs at $t = 1$ h. You may assume that the maximum occurs at a critical point as suggested by Figure 19.

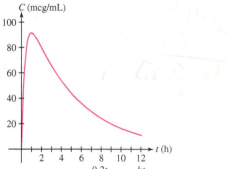

FIGURE 19 Graph of $C(t) = 120(e^{-0.2t} - e^{-bt})$ with b chosen so that the maximum occurs at $t = 1$ h.

76. $\mathcal{CAS}$ In the notation of Exercise 75, find the value of b (to two decimal places) for which the maximum value of C is equal to 100 mcg/ml.

77. In 1919 physicist Alfred Betz argued that the maximum efficiency of a wind turbine is around 59%. If wind enters a turbine with speed v_1 and exits with speed v_2, then the power extracted is the difference in kinetic energy per unit time:

$$P = \frac{1}{2}mv_1^2 - \frac{1}{2}mv_2^2 \quad \text{watts}$$

where m is the mass of wind flowing through the rotor per unit time (Figure 20). Betz assumed that $m = \rho A(v_1 + v_2)/2$, where ρ is the density of air and A is the area swept out by the rotor. Wind flowing undisturbed through the same area A would have mass per unit time $\rho A v_1$ and power $P_0 = \frac{1}{2}\rho A v_1^3$. The fraction of power extracted by the turbine is $F = P/P_0$.

(a) Show that F depends only on the ratio $r = v_2/v_1$ and is equal to $F(r) = \frac{1}{2}(1 - r^2)(1 + r)$, where $0 \leq r \leq 1$.

(b) Show that the maximum value of F, called the **Betz Limit**, is $16/27 \approx 0.59$.

(c) 🖊️   Explain why Betz's formula for F is not meaningful for r close to zero. *Hint:* How much wind would pass through the turbine if v_2 were zero? Is this realistic?

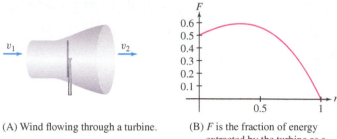

(A) Wind flowing through a turbine.   (B) F is the fraction of energy extracted by the turbine as a function of $r = v_2/v_1$.

FIGURE 20

78. GU   The **Bohr radius** a_0 of the hydrogen atom is the value of r that minimizes the energy

$$E(r) = \frac{\hbar^2}{2mr^2} - \frac{e^2}{4\pi \epsilon_0 r}$$

where $\hbar$, m, e, and ϵ_0 are physical constants. Show that $a_0 = 4\pi \epsilon_0 \hbar^2/(me^2)$. Assume that the minimum occurs at a critical point, as suggested by Figure 21.

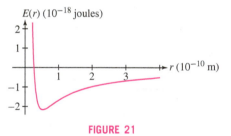

FIGURE 21

79. The response of a circuit or other oscillatory system to an input of frequency ω ("omega") is described by the function

$$\phi(\omega) = \frac{1}{\sqrt{(\omega_0^2 - \omega^2)^2 + 4D^2\omega^2}}$$

Both ω_0 (the natural frequency of the system) and D (the damping factor) are positive constants. The graph of ϕ is called a **resonance curve**, and the positive frequency $\omega_r > 0$, where ϕ takes its maximum value, if it exists, is called the **resonant frequency**. Show that $\omega_r = \sqrt{\omega_0^2 - 2D^2}$ if $0 < D < \omega_0/\sqrt{2}$ and that no resonant frequency exists otherwise (Figure 22).

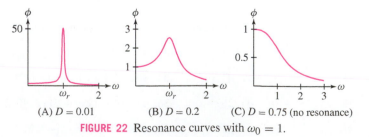

(A) $D = 0.01$ (B) $D = 0.2$ (C) $D = 0.75$ (no resonance)

FIGURE 22 Resonance curves with $\omega_0 = 1$.

80. Bees build honeycomb structures out of cells with a hexagonal base and three rhombus-shaped faces on top, as in Figure 23. We can show that the surface area of this cell is

$$A(\theta) = 6hs + \frac{3}{2}s^2(\sqrt{3}\csc\theta - \cot\theta)$$

with h, s, and θ as indicated in the figure. Remarkably, bees "know" which angle θ minimizes the surface area (and therefore requires the least amount of wax).

(a) Show that $\theta \approx 54.7°$ (assume h and s are constant). *Hint:* Find the critical point of $A(\theta)$ for $0 < \theta < \pi/2$.

(b) GU Confirm, by graphing $f(\theta) = \sqrt{3}\csc\theta - \cot\theta$, that the critical point indeed minimizes the surface area.

FIGURE 23 A cell in a honeycomb constructed by bees.

81. Find the maximum of $y = x^a - x^b$ on $[0, 1]$, where $0 < a < b$. In particular, find the maximum of $y = x^5 - x^{10}$ on $[0, 1]$.

In Exercises 82–84, plot the function using a graphing utility and find its critical points and extreme values on $[-5, 5]$.

82. GU $y = \dfrac{1}{1 + |x - 1|}$

83. GU $y = \dfrac{1}{1 + |x - 1|} + \dfrac{1}{1 + |x - 4|}$

84. GU $y = \dfrac{x}{|x^2 - 1| + |x^2 - 4|}$

85. (a) Use implicit differentiation to find the critical points on the curve $27x^2 = (x^2 + y^2)^3$.

(b) GU Plot the curve and the horizontal tangent lines on the same set of axes.

86. Sketch the graph of a continuous function on $(0, 4)$ with a minimum value but no maximum value.

87. Sketch the graph of a continuous function on $(0, 4)$ having a local minimum but no absolute minimum.

88. Sketch the graph of a function on $[0, 4]$ having

(a) Two local maxima and one local minimum.

(b) An absolute minimum that occurs at an endpoint, and an absolute maximum that occurs at a critical point.

89. Sketch the graph of a function f on $[0, 4]$ with a discontinuity such that f has an absolute minimum but no absolute maximum.

90. A rainbow is produced by light rays that enter a raindrop (assumed spherical) and exit after being reflected internally as in Figure 24. The angle between the incoming and reflected rays is $\theta = 4r - 2i$, where the angle of incidence i and refraction r are related by Snell's Law $\sin i = n \sin r$ with $n \approx 1.33$ (the index of refraction for air and water).

(a) Use Snell's Law to show that $\dfrac{dr}{di} = \dfrac{\cos i}{n \cos r}$.

(b) Show that the maximum value $\theta_{\max}$ of θ occurs when i satisfies $\cos i = \sqrt{\dfrac{n^2 - 1}{3}}$. *Hint:* Show that $\dfrac{d\theta}{di} = 0$ if $\cos i = \dfrac{n}{2}\cos r$. Then use Snell's Law to eliminate r.

(c) Show that $\theta_{\max} \approx 42.53°$.

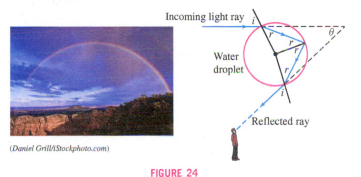

(Daniel Grill/iStockphoto.com)

FIGURE 24

Further Insights and Challenges

91. Show that the extreme values of $f(x) = a \sin x + b \cos x$ are $\pm\sqrt{a^2 + b^2}$.

92. Show, by considering its minimum, that $f(x) = x^2 - 2x + 3$ takes on only positive values. More generally, find the conditions on r and s under which the quadratic function $f(x) = x^2 + rx + s$ takes on only positive values. Give examples of r and s for which f takes on both positive and negative values.

93. Show that if the quadratic polynomial $f(x) = x^2 + rx + s$ takes on both positive and negative values, then its minimum value occurs at the midpoint between the two roots.

94. Generalize Exercise 93: Show that if the horizontal line $y = c$ intersects the graph of $f(x) = x^2 + rx + s$ at two points $(x_1, f(x_1))$ and $(x_2, f(x_2))$, then f takes its minimum value at the midpoint $M = \dfrac{x_1 + x_2}{2}$ (Figure 25).

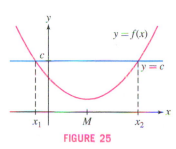

FIGURE 25

95. A cubic polynomial may have a local min and max, or it may have neither (Figure 26). Find conditions on the coefficients a and b of

$$f(x) = \frac{1}{3}x^3 + \frac{1}{2}ax^2 + bx + c$$

that ensure f has neither a local min nor a local max. *Hint:* Apply Exercise 92 to $f'(x)$.

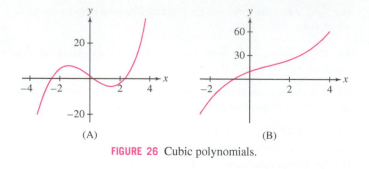

FIGURE 26 Cubic polynomials.

96. Find the min and max of

$$f(x) = x^p(1-x)^q \quad \text{on } [0, 1]$$

where $p, q > 0$.

97. Prove that if f is continuous and $f(a)$ and $f(b)$ are local minima where $a < b$, then there exists a value c between a and b such that $f(c)$ is a local maximum. (*Hint:* Apply Theorem 1 to the interval $[a, b]$.) Show that continuity is a necessary hypothesis by sketching the graph of a function (necessarily discontinuous) with two local minima but no local maximum.

4.3 The Mean Value Theorem and Monotonicity

We have taken for granted that if $f'(x)$ is positive, the function f is increasing, and if $f'(x)$ is negative, f is decreasing. In this section, we prove this rigorously using an important result called the Mean Value Theorem (MVT). Then we develop a method for "testing" critical points—that is, for determining whether they correspond to local minima or maxima.

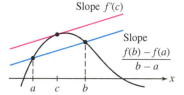
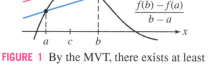

FIGURE 1 By the MVT, there exists at least one tangent line parallel to the secant line.

The MVT says that a secant line between two points $(a, f(a))$ and $(b, f(b))$ on a graph is parallel to at least one tangent line in the interval (a, b) (Figure 1). Since the secant line between $(a, f(a))$ and $(b, f(b)$ has slope $\dfrac{f(b) - f(a)}{b - a}$ and since two lines are parallel if they have the same slope, the MVT is claiming that there exists a point c between a and b such that

$$\underbrace{f'(c)}_{\text{Slope of tangent line}} = \underbrace{\frac{f(b) - f(a)}{b - a}}_{\text{Slope of secant line}}$$

> **THEOREM 1 The Mean Value Theorem** Assume that f is continuous on the closed interval $[a, b]$ and differentiable on (a, b). Then there exists at least one value c in (a, b) such that
>
> $$f'(c) = \frac{f(b) - f(a)}{b - a}$$

Rolle's Theorem (Section 4.2) is the special case of the MVT in which $f(a) = f(b)$. In this case, the conclusion is that $f'(c) = 0$.

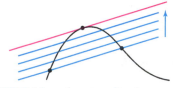

FIGURE 2 Move the secant line in a parallel fashion until it becomes tangent to the curve.

GRAPHICAL INSIGHT Imagine what happens when a secant line is moved parallel to itself. Eventually, it becomes a tangent line, as shown in Figure 2. This is the idea behind the MVT. We present a formal proof at the end of this section.

CONCEPTUAL INSIGHT The conclusion of the MVT can be rewritten as

$$f(b) - f(a) = f'(c)(b - a)$$

We can think of this as a variation on the Linear Approximation, which says

$$f(b) - f(a) \approx f'(a)(b - a)$$

The MVT turns this approximation into an equality by replacing $f'(a)$ with $f'(c)$ for a suitable choice of c in (a, b).

■ **EXAMPLE 1** Verify the MVT with $f(x) = \sqrt{x}$, $a = 1$, and $b = 9$.

Solution First, compute the slope of the secant line (Figure 3):

$$\frac{f(b) - f(a)}{b - a} = \frac{\sqrt{9} - \sqrt{1}}{9 - 1} = \frac{3 - 1}{9 - 1} = \frac{1}{4}$$

We must find c such that $f'(c) = 1/4$. The derivative is $f'(x) = \frac{1}{2}x^{-1/2}$, and

$$f'(c) = \frac{1}{2\sqrt{c}} = \frac{1}{4} \quad \Rightarrow \quad 2\sqrt{c} = 4 \quad \Rightarrow \quad c = 4$$

The value $c = 4$ lies in $(1, 9)$ and satisfies $f'(4) = \frac{1}{4}$. This verifies the MVT. ■

As a first application, we prove that a function with zero derivative is constant.

COROLLARY If f is differentiable and $f'(x) = 0$ for all $x \in (a, b)$, then f is constant on (a, b). In other words, $f(x) = C$ for some constant C.

Proof If a_1 and b_1 are any two distinct points in (a, b), then, by the MVT, there exists c between a_1 and b_1 such that

$$f(b_1) - f(a_1) = f'(c)(b_1 - a_1) = 0 \qquad \text{(since } f'(c) = 0)$$

Thus, $f(b_1) = f(a_1)$. This says that $f(x)$ is constant on (a, b). ■

We say that f is "nondecreasing" if

$$f(x_1) \leq f(x_2) \quad \text{for} \quad x_1 \leq x_2$$

"Nonincreasing" is defined similarly. In Theorem 2, if we assume that $f'(x) \geq 0$ (instead of > 0), then f is nondecreasing on (a, b). If $f'(x) \leq 0$, then f is nonincreasing on (a, b).

Increasing/Decreasing Behavior of Functions

We prove now that the sign of the derivative determines whether a function f is increasing or decreasing. Recall that f is

- **Increasing on (a, b)** if $f(x_1) < f(x_2)$ for all $x_1, x_2 \in (a, b)$ such that $x_1 < x_2$.
- **Decreasing on (a, b)** if $f(x_1) > f(x_2)$ for all $x_1, x_2 \in (a, b)$ such that $x_1 < x_2$.

We say that f is **monotonic** on (a, b) if it is either increasing or decreasing on (a, b).

THEOREM 2 The Sign of the Derivative Let f be a differentiable function on an open interval (a, b).

- If $f'(x) > 0$ for $x \in (a, b)$, then f is increasing on (a, b).
- If $f'(x) < 0$ for $x \in (a, b)$, then f is decreasing on (a, b).

Proof Suppose first that $f'(x) > 0$ for all $x \in (a, b)$. The MVT tells us that for any two points $x_1 < x_2$ in (a, b), there exists c between x_1 and x_2 such that

$$f(x_2) - f(x_1) = f'(c)(x_2 - x_1) > 0$$

The inequality holds because $f'(c)$ and $(x_2 - x_1)$ are both positive. Therefore, $f(x_2) > f(x_1)$, as required. The case $f'(x) < 0$ is similar. ■

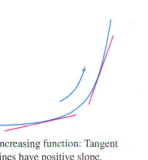

Increasing function: Tangent lines have positive slope.

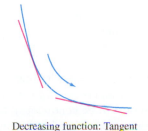

Decreasing function: Tangent lines have negative slope.

FIGURE 4

FIGURE 3 The tangent line at $c = 4$ is parallel to the secant line.

GRAPHICAL INSIGHT Theorem 2 confirms our graphical intuition (Figure 4):

- $f'(x) > 0 \quad \Rightarrow \quad$ Tangent lines have positive slope $\quad \Rightarrow \quad f$ increasing
- $f'(x) < 0 \quad \Rightarrow \quad$ Tangent lines have negative slope $\quad \Rightarrow \quad f$ decreasing

■ **EXAMPLE 2** Show that $f(x) = \ln x$ is increasing.

Solution The derivative $f'(x) = x^{-1}$ is positive on the domain $\{x : x > 0\}$, so $f(x) = \ln x$ is increasing. Observe, however, that $f'(x) = x^{-1}$ is decreasing, so the graph of f grows flatter as $x \to \infty$ (Figure 5). ■

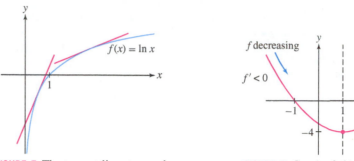

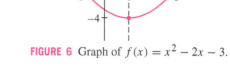

DF **FIGURE 5** The tangent lines to $y = \ln x$ get flatter as $x \to \infty$.

FIGURE 6 Graph of $f(x) = x^2 - 2x - 3$.

■ **EXAMPLE 3** Find the intervals on which $f(x) = x^2 - 2x - 3$ is monotonic.

Solution The derivative $f'(x) = 2x - 2 = 2(x - 1)$ is positive for $x > 1$ and negative for $x < 1$. By Theorem 2, f is decreasing on the interval $(-\infty, 1)$ and increasing on the interval $(1, \infty)$, as confirmed in Figure 6. ■

Testing Critical Points

There is a useful test for determining whether a critical point yields a min or max (or neither) based on the *sign change* of the derivative $f'(x)$.

To explain the term "sign change," suppose that a function g satisfies $g(c) = 0$. We say that $g(x)$ *changes from positive to negative* at $x = c$ if $g(x) > 0$ to the left of c and $g(x) < 0$ to the right of c for x within a small open interval around c (Figure 7). A sign change from negative to positive is defined similarly. Observe in Figure 7 that $g(5) = 0$ but $g(x)$ does not change sign at $x = 5$.

Now suppose that $f'(c) = 0$ and that $f'(x)$ changes sign at $x = c$, say, from $+$ to $-$. Then f is increasing to the left of c and decreasing to the right, so $f(c)$ is a local maximum. Similarly, if $f'(x)$ changes sign from $-$ to $+$, then $f(c)$ is a local minimum. See Figure 8(A).

Figure 8(B) illustrates a case where $f'(c) = 0$ but $f'(x)$ does not change sign. In this case, $f'(x) > 0$ for all x near but not equal to c, so f is increasing and has neither a local min nor a local max at c. The same analysis holds true when $f'(c)$ does not exist.

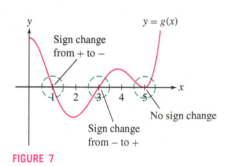

FIGURE 7

THEOREM 3 **First Derivative Test for Critical Points**   Let c be a critical point of f. Then

- $f'(x)$ changes from $+$ to $-$ at c $\Rightarrow$ $f(c)$ is a local maximum.
- $f'(x)$ changes from $-$ to $+$ at c $\Rightarrow$ $f(c)$ is a local minimum.

To carry out the First Derivative Test, we make a useful observation: $f'(x)$ can change sign at a critical point, but *it cannot change sign on the interval between two consecutive critical points* (this can be proved even if f' is not assumed to be continuous). So we can determine the sign of $f'(x)$ on an interval between consecutive critical points by evaluating $f'(x)$ at an any *test point* x_0 inside the interval. The sign of $f'(x_0)$ is the sign of $f'(x)$ on the entire interval.

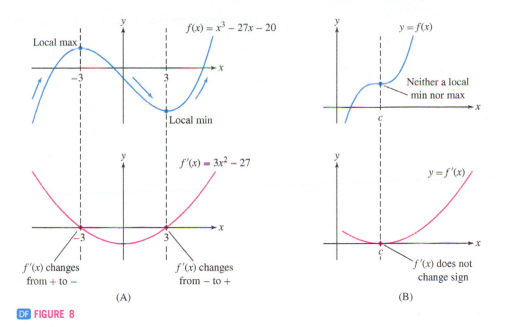

DF FIGURE 8

■ **EXAMPLE 4** Analyze the critical points of $f(x) = x^3 - 27x - 20$.

Solution Our analysis will confirm the picture in Figure 8(A).

Step 1. **Find the critical points.**
The roots of $f'(x) = 3x^2 - 27 = 3(x^2 - 9) = 0$ are $c = \pm 3$.

Step 2. **Find the sign of $f'(x)$ on the intervals between the critical points.**
The critical points $c = \pm 3$ divide the real line into three intervals:

$$(-\infty, -3), \qquad (-3, 3), \qquad (3, \infty)$$

To determine the sign of $f'(x)$ on these intervals, we choose a test point inside each interval and evaluate. For example, in $(-\infty, -3)$ we choose $x = -4$. Because $f'(-4) = 21 > 0$, $f'(x)$ is positive on the entire interval $(-3, \infty)$. Similarly,

We chose the test points $-4, 0$, and 4 arbitrarily. To find the sign of $f'(x)$ on $(-\infty, -3)$, we could just as well have computed $f'(-5)$ or any other value of f' in the interval $(-\infty, -3)$.

$$f'(-4) = 21 > 0 \quad \Rightarrow \quad f'(x) > 0 \quad \text{for all } x \in (-\infty, -3)$$

$$f'(0) = -27 < 0 \quad \Rightarrow \quad f'(x) < 0 \quad \text{for all } x \in (-3, 3)$$

$$f'(4) = 21 > 0 \quad \Rightarrow \quad f'(x) > 0 \quad \text{for all } x \in (3, \infty)$$

This information is displayed in the following sign diagram:

Behavior of $f(x)$ ↗ ↘ ↗

Sign of $f'(x)$ +   −   +

−3   0   3

Step 3. **Use the First Derivative Test.**

- $c = -3$: $f'(x)$ changes from $+$ to $-$ $\Rightarrow$ $f(-3) = 34$ is a local maximum value.
- $c = 3$: $f'(x)$ changes from $-$ to $+$ $\Rightarrow$ $f(3) = -74$ is a local minimum value.   ■

■ **EXAMPLE 5** Analyze the critical points and the increase/decrease behavior of $f(x) = \cos^2 x + \sin x$ in $(0, \pi)$.

Solution First, find the critical points:

$$f'(x) = -2\cos x \sin x + \cos x = (\cos x)(1 - 2\sin x) = 0 \quad \Rightarrow \quad \cos x = 0 \text{ or } \sin x = \frac{1}{2}$$

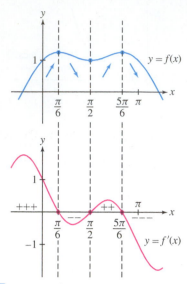

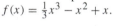

DF **FIGURE 9** Graph of $f(x) = \cos^2 x + \sin x$ and its derivative.

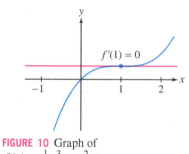

FIGURE 10 Graph of $f(x) = \frac{1}{3}x^3 - x^2 + x$.

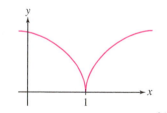

FIGURE 11 Graph of $f(x) = (1 - x)^{2/3}$.

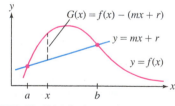

FIGURE 12 $G(x)$ is the vertical distance between the graph and the secant line.

The critical points are $\frac{\pi}{6}$, $\frac{\pi}{2}$, and $\frac{5\pi}{6}$. They divide $(0, \pi)$ into four intervals:

$$\left(0, \frac{\pi}{6}\right), \qquad \left(\frac{\pi}{6}, \frac{\pi}{2}\right), \qquad \left(\frac{\pi}{2}, \frac{5\pi}{6}\right), \qquad \left(\frac{5\pi}{6}, \pi\right)$$

We determine the sign of $f'(x)$ by evaluating $f'(x)$ at a test point inside each interval. Since $\frac{\pi}{6} \approx 0.52$, $\frac{\pi}{2} \approx 1.57$, $\frac{5\pi}{6} \approx 2.62$, and $\pi \approx 3.14$, we can use the following test points:

Interval	Test Value	Sign of $f'(x)$	Behavior of $f(x)$
$\left(0, \frac{\pi}{6}\right)$	$f'(0.5) \approx 0.04$	$+$	↗
$\left(\frac{\pi}{6}, \frac{\pi}{2}\right)$	$f'(1) \approx -0.37$	$-$	↘
$\left(\frac{\pi}{2}, \frac{5\pi}{6}\right)$	$f'(2) \approx 0.34$	$+$	↗
$\left(\frac{5\pi}{6}, \pi\right)$	$f'(3) \approx -0.71$	$-$	↘

Now apply the First Derivative Test:

- Local max at $c = \frac{\pi}{6}$ and $c = \frac{5\pi}{6}$ because $f'(x)$ changes from $+$ to $-$.

- Local min at $c = \frac{\pi}{2}$ because $f'(x)$ changes from $-$ to $+$.

The behavior of $f(x)$ and $f'(x)$ is reflected in the graphs in Figure 9. ■

■ **EXAMPLE 6** **A Critical Point Without a Sign Transition** Analyze the critical points of $f(x) = \frac{1}{3}x^3 - x^2 + x$.

Solution The derivative is $f'(x) = x^2 - 2x + 1 = (x - 1)^2$, so $c = 1$ is the only critical point. However, $(x - 1)^2 \geq 0$, so $f'(x)$ does not change sign at $c = 1$, and therefore $f(1)$ is neither a local min nor a local max. See Figure 10. ■

■ **EXAMPLE 7** **A Critical Point Where $f'(x)$ Is Undefined** Analyze the critical points of $f(x) = (1 - x)^{2/3}$.

Solution The derivative is $f'(x) = -\frac{2}{3}(1 - x)^{-1/3} = \frac{-2}{3(1-x)^{1/3}}$. The only critical point occurs at $c = 1$, when $f'(x)$ is undefined. For $x < 1$, $f'(x)$ is negative. For $x > 1$, $f'(x)$ is positive. So $f'(x)$ changes sign as we pass through $c = 1$, and by the First Derivative Test, $f(c)$ is a local minimum. See Figure 11. ■

Proof of the MVT Let $m = \dfrac{f(b) - f(a)}{b - a}$ be the slope of the secant line joining $(a, f(a))$ and $(b, f(b))$. The secant line has equation $y = mx + r$ for some r (Figure 12). The value of r is not important, but you can check that $r = f(a) - ma$. Now consider the function

$$G(x) = f(x) - (mx + r)$$

As indicated in Figure 12, $G(x)$ is the vertical distance between the graph and the secant line at x (it is negative at points where the graph of f lies below the secant line). This distance is zero at the endpoints, and therefore, $G(a) = G(b) = 0$. By Rolle's Theorem (Section 4.2), there exists a point c in (a, b) such that $G'(c) = 0$. But $G'(x) = f'(x) - m$, so $G'(c) = f'(c) - m = 0$, and $f'(c) = m$ as desired. ■

4.3 SUMMARY

- The Mean Value Theorem (MVT): If f is continuous on $[a, b]$ and differentiable on (a, b), then there exists at least one value c in (a, b) such that

$$f'(c) = \frac{f(b) - f(a)}{b - a}$$

This conclusion can also be written

$$f(b) - f(a) = f'(c)(b - a)$$

- Important corollary of the MVT: If $f'(x) = 0$ for all $x \in (a, b)$, then f is constant on (a, b).
- The *sign* of $f'(x)$ determines whether f is increasing or decreasing:

$$f'(x) > 0 \text{ for } x \in (a, b) \quad \Rightarrow \quad f \text{ is increasing on } (a, b)$$
$$f'(x) < 0 \text{ for } x \in (a, b) \quad \Rightarrow \quad f \text{ is decreasing on } (a, b)$$

- The sign of $f'(x)$ can change only at the critical points, so f is *monotonic* (increasing or decreasing) on the intervals between the critical points.
- To find the sign of $f'(x)$ on the interval between two critical points, calculate the sign of $f'(x_0)$ at any test point x_0 in that interval.
- First Derivative Test: If f is differentiable and c is a critical point, then

Sign Change of $f'(x)$ at c	Type of Critical Point
From + to −	Local maximum
From − to +	Local minimum

4.3 EXERCISES

Preliminary Questions

1. For which value of m is the following statement correct? If $f(2) = 3$ and $f(4) = 9$, and f is differentiable, then f has a tangent line of slope m.

2. Assume f is differentiable. Which of the following statements does *not* follow from the MVT?

(a) If f has a secant line of slope 0, then f has a tangent line of slope 0.

(b) If $f(5) < f(9)$, then $f'(c) > 0$ for some $c \in (5, 9)$.

(c) If f has a tangent line of slope 0, then f has a secant line of slope 0.

(d) If $f'(x) > 0$ for all x, then every secant line has positive slope.

3. Can a function with the real numbers as its domain that takes on only negative values have a positive derivative? If so, sketch an example.

4. For f with derivative as in Figure 13:

(a) Is $f(c)$ a local minimum or maximum?

(b) Is f a decreasing function?

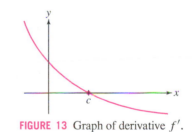

FIGURE 13 Graph of derivative f'.

Exercises

In Exercises 1–8, find a point c satisfying the conclusion of the MVT for the given function and interval.

1. $y = x^{-1}$, [2, 8]

2. $y = \sqrt{x}$, [9, 25]

3. $y = \cos x - \sin x$, [0, 2π]

4. $y = \dfrac{x}{x+2}$, [1, 4]

5. $y = x^3$, [−4, 5]

6. $y = x \ln x$, [1, 2]

7. $y = e^{-2x}$, [0, 3]

8. $y = e^x - x$, [−1, 1]

In Exercises 9–12, find a point c satisfying the conclusion of the MVT for the given function and interval. Then draw the graph of the function, the secant line between the endpoints of the graph and the tangent line at (c, f(c)), to see that the secant and tangent lines are, in fact, parallel.

9. $y = x^2$, [0, 1]

10. $y = x^{2/3}$, [0, 8]

11. $y = e^x$, [0, 1]

12. $y = \sqrt{x}$, [0, 3]

13. GU Let $f(x) = x^5 + x^2$. The secant line between (0, 0) and (1, 2) has slope 2 (check this), so by the MVT, $f'(c) = 2$ for some $c \in (0, 1)$. Plot f and the secant line on the same axes. Then plot $y = 2x + b$ for different values of b until the line becomes tangent to the graph of f. Zoom in on the point of tangency to estimate the x-coordinate c of the point of tangency.

14. GU Plot the derivative of $f(x) = 3x^5 - 5x^3$. Describe its sign changes and use this to determine the local extreme values of f. Then graph f to confirm your conclusions.

15. Determine the intervals on which $f'(x)$ is positive and negative, assuming that Figure 14 is the graph of f.

16. Determine the intervals on which f is increasing or decreasing, assuming that Figure 14 is the graph of f'.

17. State whether $f(2)$ and $f(4)$ are local minima or local maxima, assuming that Figure 14 is the graph of f'.

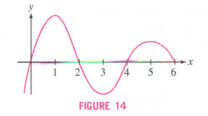

FIGURE 14

18. Figure 15 shows the graph of the derivative f' of a function f. Find the critical points of f and determine whether they are local minima, local maxima, or neither.

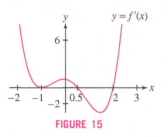

FIGURE 15

In Exercises 19–22, sketch the graph of a function f whose derivative f' has the given description.

19. $f'(x) > 0$ for $x > 3$ and $f'(x) < 0$ for $x < 3$

20. $f'(x) > 0$ for $x < 1$ and $f'(x) < 0$ for $x > 1$

21. $f'(x)$ is negative on $(1, 3)$ and positive everywhere else.

22. $f'(x)$ makes the sign transitions $+, -, +, -$.

In Exercises 23–26, find all critical points of f and use the First Derivative Test to determine whether they are local minima or maxima.

23. $f(x) = 4 + 6x - x^2$

24. $f(x) = x^3 - 12x - 4$

25. $f(x) = \dfrac{x^2}{x+1}$

26. $f(x) = x^3 + x^{-3}$

In Exercises 27–58, find the critical points and the intervals on which the function is increasing or decreasing. Use the First Derivative Test to determine whether the critical point yields a local min or max (or neither).

27. $y = -x^2 + 7x - 17$

28. $y = 5x^2 + 6x - 4$

29. $y = x^3 - 12x^2$

30. $y = x(x-2)^3$

31. $y = 3x^4 + 8x^3 - 6x^2 - 24x$

32. $y = x^2 + (10-x)^2$

33. $y = \frac{1}{3}x^3 + \frac{3}{2}x^2 + 2x + 4$

34. $y = x^4 + x^3$

35. $y = x^5 + x^3 + 1$

36. $y = x^5 + x^3 + x$

37. $y = x^4 - 4x^{3/2}$ $(x > 0)$

38. $y = x^{5/2} - x^2$ $(x > 0)$

39. $y = x + x^{-1}$ $(x > 0)$

40. $y = x^{-2} - 4x^{-1}$ $(x > 0)$

41. $y = \dfrac{1}{x^2+1}$

42. $y = \dfrac{2x+1}{x^2+1}$

43. $y = \dfrac{x^3}{x^2+1}$

44. $y = \dfrac{x^3}{x^2-3}$

45. $y = \theta + \sin\theta + \cos\theta$

46. $y = \sin\theta + \sqrt{3}\cos\theta$

47. $y = \sin^2\theta + \sin\theta$

48. $y = \theta - 2\cos\theta$, $[0, 2\pi]$

49. $y = x + e^{-x}$

50. $y = \dfrac{e^x}{x}$ $(x > 0)$

51. $y = e^{-x}\cos x$, $\left[-\frac{\pi}{2}, \frac{\pi}{2}\right]$

52. $y = x^2 e^x$

53. $y = \tan^{-1}x - \frac{1}{2}x$

54. $y = (x^2 - 2x)e^x$

55. $y = x - \ln x$ $(x > 0)$

56. $y = \dfrac{\ln x}{x}$ $(x > 0)$

57. $y = x^{1/3}$

58. $y = x^{2/3} - x^2$

59. Find the minimum value of $f(x) = x^x$ for $x > 0$.

60. Show that $f(x) = x^2 + bx + c$ is decreasing on $\left(-\infty, -\frac{b}{2}\right)$ and increasing on $\left(-\frac{b}{2}, \infty\right)$.

61. Show that $f(x) = x^3 - 2x^2 + 2x$ is an increasing function. *Hint:* Find the minimum value of f'.

62. Find conditions on a and b that ensure $f(x) = x^3 + ax + b$ is increasing on $(-\infty, \infty)$.

63. ⏐GU⏐ Let $h(x) = \dfrac{x(x^2 - 1)}{x^2 + 1}$ and suppose that $f'(x) = h(x)$. Plot h and use the plot to describe the local extrema and the increasing/decreasing behavior of f. Sketch a plausible graph for f itself.

64. Sam made two statements that Deborah found dubious.

(a) "The average velocity for my trip was 70 mph; at no point in time did my speedometer read 70 mph."

(b) "A policeman clocked me going 70 mph, but my speedometer never read 65 mph."

In each case, which theorem did Deborah apply to prove Sam's statement false: the Intermediate Value Theorem or the Mean Value Theorem? Explain.

65. Determine where $f(x) = (1{,}000 - x)^2 + x^2$ is decreasing. Use this to decide which is larger: $800^2 + 200^2$ or $600^2 + 400^2$.

66. Show that $f(x) = 1 - |x|$ satisfies the conclusion of the MVT on $[a, b]$ if both a and b are positive or negative, but not if $a < 0$ and $b > 0$.

67. Which values of c satisfy the conclusion of the MVT on the interval $[a, b]$ if f is a linear function?

68. Show that if f is any quadratic polynomial, then the midpoint $c = \dfrac{a+b}{2}$ satisfies the conclusion of the MVT on $[a, b]$ for any a and b.

69. Suppose that $f(0) = 2$ and $f'(x) \le 3$ for $x > 0$. Apply the MVT to the interval $[0, 4]$ to prove that $f(4) \le 14$. Prove more generally that $f(x) \le 2 + 3x$ for all $x > 0$.

70. Show that if $f(2) = -2$ and $f'(x) \ge 5$ for $x > 2$, then $f(4) \ge 8$.

71. Show that if $f(2) = 5$ and $f'(x) \ge 10$ for $x > 2$, then $f(x) \ge 10x - 15$ for all $x > 2$.

Further Insights and Challenges

72. Show that a cubic function $f(x) = x^3 + ax^2 + bx + c$ is increasing on $(-\infty, \infty)$ if $b > a^2/3$.

73. Prove that if $f(0) = g(0)$ and $f'(x) \le g'(x)$ for $x \ge 0$, then $f(x) \le g(x)$ for all $x \ge 0$. *Hint:* Show that the function given by $y = f(x) - g(x)$ is nonincreasing.

74. Use Exercise 73 to prove that $x \le \tan x$ for $0 \le x < \frac{\pi}{2}$.

75. Use Exercise 73 and the inequality $\sin x \le x$ for $x \ge 0$ (established in Theorem 3 of Section 2.6) to prove the following assertions for all $x \ge 0$ (each assertion follows from the previous one):

(a) $\cos x \ge 1 - \frac{1}{2}x^2$

(b) $\sin x \geq x - \frac{1}{6}x^3$

(c) $\cos x \leq 1 - \frac{1}{2}x^2 + \frac{1}{24}x^4$
Can you guess the next inequality in the series?

76. Let $f(x) = e^{-x}$. Use the method of Exercise 75 to prove the following inequalities for $x \geq 0$:

(a) $e^{-x} \geq 1 - x$

(b) $e^{-x} \leq 1 - x + \frac{1}{2}x^2$

(c) $e^{-x} \geq 1 - x + \frac{1}{2}x^2 - \frac{1}{6}x^3$
Can you guess the next inequality in the series?

77. Assume that f'' exists and $f''(x) = 0$ for all x. Prove that $f(x) = mx + b$, where $m = f'(0)$ and $b = f(0)$.

78. Define $f(x) = x^3 \sin\left(\frac{1}{x}\right)$ for $x \neq 0$ and $f(0) = 0$.
(a) Show that f' is continuous at $x = 0$ and that $x = 0$ is a critical point of f.
(b) GU Examine the graphs of f and f'. Can the First Derivative Test be applied?

(c) Show that $f(0)$ is neither a local min nor a local max.

79. Suppose that $f(x)$ satisfies the following equation (an example of a **differential equation**):

$$f''(x) = -f(x) \qquad \boxed{1}$$

(a) Show that $f(x)^2 + f'(x)^2 = f(0)^2 + f'(0)^2$ for all x. *Hint:* Show that the function on the left has zero derivative.

(b) Verify that $\sin x$ and $\cos x$ satisfy Eq. (1), and deduce that $\sin^2 x + \cos^2 x = 1$.

80. Suppose that functions f and g satisfy Eq. (1) and have the same initial values—that is, $f(0) = g(0)$ and $f'(0) = g'(0)$. Prove that $f(x) = g(x)$ for all x. *Hint:* Apply Exercise 79(a) to $f - g$.

81. Use Exercise 80 to prove $f(x) = \sin x$ is the unique solution of Eq. (1) such that $f(0) = 0$ and $f'(0) = 1$; and $g(x) = \cos x$ is the unique solution such that $g(0) = 1$ and $g'(0) = 0$. This result can be used to develop all the properties of the trigonometric functions "analytically"—that is, without reference to triangles.

4.4 The Shape of a Graph

In the previous section, we studied the increasing/decreasing behavior of a function, as determined by the sign of the derivative. Another important property is concavity, which refers to the way the graph bends. Informally, a curve is *concave up* if it bends up and *concave down* if it bends down (Figure 1).

Concave up   Concave down

FIGURE 1

To analyze concavity in a precise fashion, let's examine how concavity is related to tangent lines and derivatives. Observe in Figure 2 that when f is concave up, f' is increasing (the slopes of the tangent lines increase as we move to the right). Similarly, when f is concave down, f' is decreasing. This suggests the following definition.

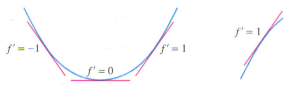

Concave up: Slopes of tangent lines are increasing.

Concave down: Slopes of tangent lines are decreasing.

FIGURE 2

> **DEFINITION Concavity** Let f be a differentiable function on an open interval (a, b). Then
>
> - f is **concave up** on (a, b) if f' is increasing on (a, b).
> - f is **concave down** on (a, b) if f' is decreasing on (a, b).

■ **EXAMPLE 1 Concavity and Stock Prices** The stocks of two companies, Arenot Industries (AI) and Blurbenthal Business Associates (BBA), went up in value, and both currently sell for $75 (Figure 3). However, one is clearly a better investment than the other. Explain in terms of concavity.

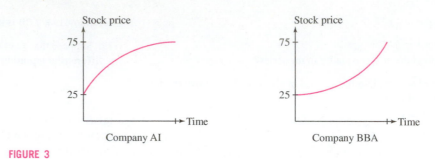

FIGURE 3

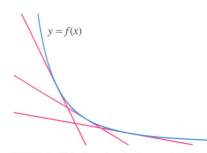

$y = f(x)$

FIGURE 4 This function is decreasing. Its derivative is negative but increasing.

Solution The graph of Stock AI is concave down, so its growth rate (first derivative) is declining as time goes on. The graph of Stock BBA is concave up, so its growth rate is increasing. If these trends continue, Stock BBA is the better investment. ∎

> **GRAPHICAL INSIGHT** Keep in mind that a function can decrease while its derivative increases. In Figure 4, the derivative f' is increasing. Although the tangent lines are getting less steep, their slopes are becoming *less negative*.

The concavity of a function is determined by the *sign* of its second derivative. Indeed, if $f''(x) > 0$, then f' is increasing and hence f is concave up. Similarly, if $f''(x) < 0$, then f' is decreasing and f is concave down.

> **THEOREM 1 Test for Concavity** Assume that $f''(x)$ exists for all $x \in (a, b)$.
> - If $f''(x) > 0$ for all $x \in (a, b)$, then f is concave up on (a, b).
> - If $f''(x) < 0$ for all $x \in (a, b)$, then f is concave down on (a, b).

| **CAUTION** A critical point c is just a single number, whereas a point of inflection $(c, f(c))$ is a point in the xy-plane.

Of special interest are the points on the graph where the concavity changes. We say that $P = (c, f(c))$ is a **point of inflection** of f if the concavity changes from up to down or from down to up at $x = c$. Figure 5 shows a curve made up of two arcs—one is concave down and one is concave up (the word "arc" refers to a piece of a curve). The point P where the arcs are joined is a point of inflection. We will denote points of inflection in graphs by a solid square ■.

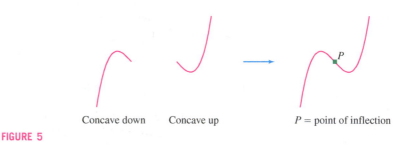

FIGURE 5

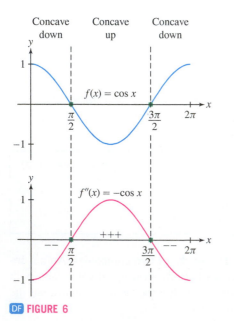

DF FIGURE 6

According to Theorem 1, the concavity of f is determined by the sign of $f''(x)$. Therefore, a point of inflection is a point where $f''(x)$ changes sign.

> **THEOREM 2 Test for Inflection Points** If $f''(c) = 0$ or $f''(c)$ does not exist and $f''(x)$ changes sign at $x = c$, then f has a point of inflection at $x = c$.

■ **EXAMPLE 2** Find the points of inflection of $f(x) = \cos x$ on $[0, 2\pi]$.

Solution We have $f''(x) = -\cos x$, and $f''(x) = 0$ for $x = \frac{\pi}{2}, \frac{3\pi}{2}$. Figure 6 shows that $f''(x)$ changes sign at $x = \frac{\pi}{2}$ and $\frac{3\pi}{2}$, so f has a point of inflection at both points. ∎

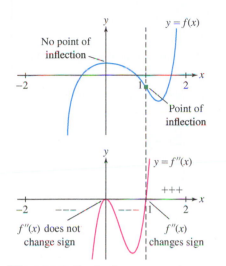

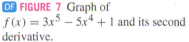

FIGURE 7 Graph of $f(x) = 3x^5 - 5x^4 + 1$ and its second derivative.

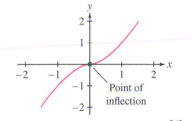

FIGURE 8 The concavity of $f(x) = x^{5/3}$ changes at $x = 0$ even though $f''(0)$ does not exist.

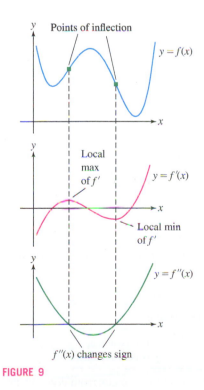

FIGURE 9

■ **EXAMPLE 3** **Points of Inflection and Intervals of Concavity** Find the points of inflection and the intervals on which $f(x) = 3x^5 - 5x^4 + 1$ is concave up and concave down.

Solution The first derivative is $f'(x) = 15x^4 - 20x^3$ and

$$f''(x) = 60x^3 - 60x^2 = 60x^2(x - 1)$$

The zeros of $f''(x) = 60x^2(x - 1)$ are $x = 0$ and $x = 1$. They divide the x-axis into three intervals: $(-\infty, 0)$, $(0, 1)$, and $(1, \infty)$. We determine the sign of $f''(x)$ and the concavity of f by computing "test values" within each interval (Figure 7):

Interval	Test Value	Sign of $f''(x)$	Behavior of $f(x)$
$(-\infty, 0)$	$f''(-1) = -120$	$-$	Concave down
$(0, 1)$	$f''(\frac{1}{2}) = -\frac{15}{2}$	$-$	Concave down
$(1, \infty)$	$f''(2) = 240$	$+$	Concave up

We can read off the points of inflection from this table:

- $c = 0$: no point of inflection, because $f''(x)$ does not change sign at 0.
- $c = 1$: point of inflection, because $f''(x)$ changes sign at 1. ■

Usually, we find the inflection points by solving $f''(x) = 0$. However, an inflection point can also occur at a point $(c, f(c))$, where $f''(c)$ does not exist.

■ **EXAMPLE 4** **A Case Where the Second Derivative Does Not Exist** Find the points of inflection of $f(x) = x^{5/3}$.

Solution In this case, $f'(x) = \frac{5}{3}x^{2/3}$ and $f''(x) = \frac{10}{9}x^{-1/3}$. Although $f''(0)$ does not exist, $f''(x)$ does change sign at $x = 0$:

$$f''(x) = \frac{10}{9x^{1/3}} \begin{cases} > 0 & \text{for } x > 0 \\ < 0 & \text{for } x < 0 \end{cases}$$

Therefore, the concavity of f changes at $x = 0$, and $(0, 0)$ is a point of inflection (Figure 8). ■

GRAPHICAL INSIGHT Points of inflection are easy to spot on the graph of the first derivative f'. If $f''(c) = 0$ and $f''(x)$ changes sign at $x = c$, then the increasing/decreasing behavior of f' changes at $x = c$. Thus, *inflection points of f occur where f' has a local min or max* (Figure 9).

Second Derivative Test for Critical Points

There is a simple test for critical points based on concavity. Suppose that $f'(c) = 0$. As we see in Figure 10, $f(c)$ is a local max if f is concave down, and it is a local min if f is concave up. Concavity is determined by the sign of $f''(x)$, so we obtain the following Second Derivative Test. (See Exercise 67 for a detailed proof.)

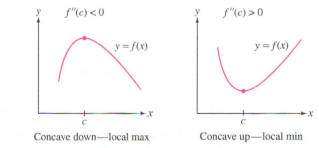

FIGURE 10 Concavity determines the type of the critical point.

Mnemonic Device:

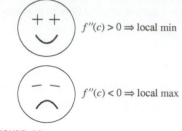

$f''(c) > 0 \Rightarrow$ local min

$f''(c) < 0 \Rightarrow$ local max

FIGURE 11

> **THEOREM 3 Second Derivative Test** Let c be a critical point of $f(x)$. If $f''(c)$ exists, then
>
> - $f''(c) > 0 \quad \Rightarrow \quad f(c)$ is a local minimum.
> - $f''(c) < 0 \quad \Rightarrow \quad f(c)$ is a local maximum.
> - $f''(c) = 0 \quad \Rightarrow \quad$ inconclusive: $f(c)$ may be a local min, a local max, or neither.

The mnemonic device appearing in Figure 11 provides an easy way to remember the test.

■ **EXAMPLE 5** Analyze the critical points of $f(x) = (2x - x^2)e^x$.

Solution First, solve

$$f'(x) = e^x(2 - 2x) + (2x - x^2)e^x = (2 - x^2)e^x = 0$$

The critical points are $c = \pm\sqrt{2}$ (Figure 12). Next, determine the sign of the second derivative at the critical points:

$$f''(x) = (-2x)e^x + (2 - x^2)e^x = (2 - 2x - x^2)e^x$$

$$f''(-\sqrt{2}) = \left(2 - 2(-\sqrt{2}) - (-\sqrt{2})^2\right)e^{-\sqrt{2}} = 2\sqrt{2}e^{-\sqrt{2}} \qquad > 0 \quad \text{(local min)}$$

$$f''(\sqrt{2}) = \left(2 - 2\sqrt{2} - (\sqrt{2})^2\right)e^{\sqrt{2}} = -2\sqrt{2}e^{\sqrt{2}} \qquad < 0 \quad \text{(local max)}$$

By the Second Derivative Test, f has a local min at $c = -\sqrt{2}$ and a local max at $c = \sqrt{2}$ (Figure 12). ■

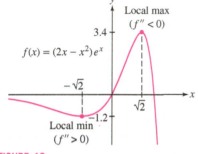

FIGURE 12

■ **EXAMPLE 6** *Second Derivative Test Inconclusive* Analyze the critical points of $f(x) = x^5 - 5x^4$.

Solution The first two derivatives are

$$f'(x) = 5x^4 - 20x^3 = 5x^3(x - 4)$$

$$f''(x) = 20x^3 - 60x^2$$

The critical points are $c = 0, 4$, and the Second Derivative Test yields

$$f''(0) = 0 \qquad \Rightarrow \quad \text{Second Derivative Test fails}$$

$$f''(4) = 320 > 0 \quad \Rightarrow \quad f(4) \text{ is a local min}$$

The Second Derivative Test fails at $c = 0$, so we fall back on the First Derivative Test. Choosing test points to the left and right of $c = 0$, we find

$$f'(-1) = 5 + 20 = 25 > 0 \qquad \Rightarrow \quad f'(x) \text{ is positive on } (-\infty, 0)$$

$$f'(1) = 5 - 20 = -15 < 0 \qquad \Rightarrow \quad f'(x) \text{ is negative on } (0, 4)$$

Since $f'(x)$ changes from $+$ to $-$ at $c = 0$, $f(0)$ is a local max (Figure 13). ■

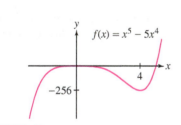

FIGURE 13

4.4 SUMMARY

- A differentiable function f is *concave up* on (a, b) if f' is increasing and *concave down* if f' is decreasing on (a, b).
- The signs of the first two derivatives provide the following information:

First Derivative		Second Derivative	
$f' > 0 \quad \Rightarrow$	f is increasing	$f'' > 0 \quad \Rightarrow$	f is concave up
$f' < 0 \quad \Rightarrow$	f is decreasing	$f'' < 0 \quad \Rightarrow$	f is concave down

- A *point of inflection* is a point $(c, f(c))$ where the concavity changes from concave up to concave down, or vice versa.

- If $f''(c) = 0$ or does not exist and $f''(x)$ changes sign at c, then $(c, f(c))$ is a point of inflection.
- Second Derivative Test: If $f'(c) = 0$ and $f''(c)$ exists, then
  - $f(c)$ is a local maximum value if $f''(c) < 0$.
  - $f(c)$ is a local minimum value if $f''(c) > 0$.
  - The test fails if $f''(c) = 0$.

If this test fails, use the First Derivative Test.

4.4 EXERCISES

Preliminary Questions

1. If f is concave up, then f' is (choose one):

(a) increasing.   **(b)** decreasing.

2. What conclusion can you draw if $f'(c) = 0$ and $f''(c) < 0$?

3. True or False? If $f(c)$ is a local min, then $f''(c)$ must be positive.

4. True or False? If $f''(x)$ changes from $+$ to $-$ at $x = c$, then f has a point of inflection at $x = c$.

Exercises

1. Match the graphs in Figure 14 with the description:

(a) $f''(x) < 0$ for all x.

(b) $f''(x)$ goes from $+$ to $-$.

(c) $f''(x) > 0$ for all x.

(d) $f''(x)$ goes from $-$ to $+$.

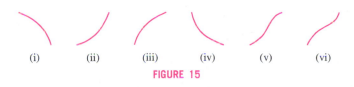

(A)   (B)   (C)   (D)

FIGURE 14

2. Match each statement with a graph in Figure 15 that represents company profits as a function of time.

(a) The outlook is great: The growth rate keeps increasing.

(b) We're losing money, but not as quickly as before.

(c) We're losing money, and it's getting worse as time goes on.

(d) We're doing well, but our growth rate is leveling off.

(e) Business had been cooling off, but now it's picking up.

(f) Business had been picking up, but now it's cooling off.

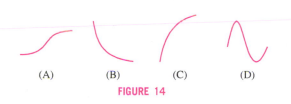

(i)   (ii)   (iii)   (iv)   (v)   (vi)

FIGURE 15

In Exercises 3–18, determine the intervals on which the function is concave up or down and find the points of inflection.

3. $y = x^2 - 4x + 3$

4. $y = t^3 - 6t^2 + 4$

5. $y = 10x^3 - x^5$

6. $y = 5x^2 + x^4$

7. $y = \theta - 2\sin\theta$, $[0, 2\pi]$

8. $y = \theta + \sin^2\theta$, $[0, \pi]$

9. $y = x(x - 8\sqrt{x})$ $(x \geq 0)$

10. $y = x^{7/2} - 35x^2$

11. $y = (x - 2)(1 - x^3)$

12. $y = x^{7/5}$

13. $y = \dfrac{1}{x^2 + 3}$

14. $y = \dfrac{x}{x^2 + 9}$

15. $y = xe^{-3x}$

16. $y = (x^2 - 7)e^x$

17. $y = 2x^2 + \ln x$ $(x > 0)$

18. $y = x - \ln x$ $(x > 0)$

19. The position of an ambulance in kilometers on a straight road over a period of 4 h is given by the graph in Figure 16.

(a) Describe the motion of the ambulance.

(b) Explain what the fact that this graph is concave up tells us about the speed of the ambulance.

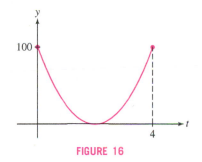

FIGURE 16

20. The position of a bicyclist on a straight road in kilometers over a period of 4 h is given by the graph in Figure 17, where inflection points occur when $t = 0.5$ and $t = 2$.

(a) Describe the motion of the bicyclist.

(b) Explain what the concavity of the graph over various intervals tells us about the speed of the bicyclist.

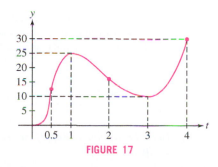

FIGURE 17

21. The growth of a sunflower during the first 100 days after sprouting is modeled well by the *logistic curve* $y = h(t)$ shown in Figure 18. Estimate the growth rate at the point of inflection and explain its significance. Then make a rough sketch of the first and second derivatives of h.

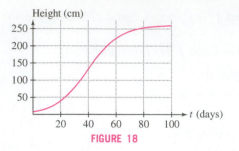

FIGURE 18

22. Assume that Figure 19 is the graph of f. Where do the points of inflection of f occur, and on which interval is f concave down?

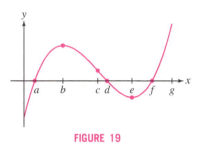

FIGURE 19

23. Repeat Exercise 22 but assume that Figure 19 is the graph of the *derivative* f'.

24. Repeat Exercise 22 but assume that Figure 19 is the graph of the *second derivative* f''.

25. Figure 20 shows the *derivative* f' on $[0, 1.2]$. Locate the points of inflection of f and the points where the local minima and maxima occur. Determine the intervals on which f has the following properties:

(a) Increasing   **(b)** Decreasing

(c) Concave up   **(d)** Concave down

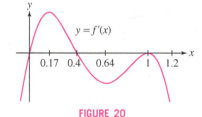

FIGURE 20

26. Leticia has been selling solar-powered laptop chargers through her website, with monthly sales as recorded below. In a report to investors, she states, "Sales reached a point of inflection when I started using pay-per-click advertising." In which month did that occur? Explain.

Month	1	2	3	4	5	6	7	8
Sales	2	30	50	60	90	150	230	340

In Exercises 27–40, find the critical points and apply the Second Derivative Test (or state that it fails).

27. $f(x) = x^3 - 12x^2 + 45x$ **28.** $f(x) = x^4 - 8x^2 + 1$

29. $f(x) = 3x^4 - 8x^3 + 6x^2$ **30.** $f(x) = x^5 - x^3$

31. $f(x) = \dfrac{x^2 - 8x}{x + 1}$ **32.** $f(x) = \dfrac{1}{x^2 - x + 2}$

33. $y = 6x^{3/2} - 4x^{1/2}$ **34.** $y = 9x^{7/3} - 21x^{1/2}$

35. $f(x) = \sin^2 x + \cos x,$ $[0, \pi]$

36. $y = \dfrac{1}{\sin x + 4},$ $[0, 2\pi]$

37. $f(x) = xe^{-x^2}$ **38.** $f(x) = e^{-x} - 4e^{-2x}$

39. $f(x) = x^3 \ln x$ $(x > 0)$

40. $f(x) = \ln x + \ln(4 - x^2),$ $(0, 2)$

In Exercises 41–56, find the intervals on which f is concave up or down, the points of inflection, the critical points, and the local minima and maxima.

41. $f(x) = x^3 - 2x^2 + x$ **42.** $f(x) = x^2(x - 4)$

43. $f(t) = t^2 - t^3$ **44.** $f(x) = 2x^4 - 3x^2 + 2$

45. $f(x) = x^2 - 8x^{1/2}$ $(x \geq 0)$

46. $f(x) = x^{3/2} - 4x^{-1/2}$ $(x > 0)$

47. $f(x) = \dfrac{x}{x^2 + 27}$ **48.** $f(x) = \dfrac{1}{x^4 + 1}$

49. $f(x) = x^{5/3} - x$ **50.** $f(x) = (x - 1)^{3/5}$

51. $f(\theta) = \theta + \sin \theta,$ $[0, 2\pi]$ **52.** $f(x) = \cos^2 x,$ $[0, \pi]$

53. $f(x) = \tan x,$ $\left(-\frac{\pi}{2}, \frac{\pi}{2}\right)$

54. $f(x) = e^{-x} \cos x,$ $\left[-\frac{\pi}{2}, \frac{3\pi}{2}\right]$

55. $y = (x^2 - 2)e^{-x}$ $(x > 0)$ **56.** $y = \ln(x^2 + 2x + 5)$

57. Sketch the graph of an increasing function such that $f''(x)$ changes from $+$ to $-$ at $x = 2$ and from $-$ to $+$ at $x = 4$. Do the same for a decreasing function.

In Exercises 58–60, sketch the graph of a function f satisfying all of the given conditions.

58. $f'(x) > 0$ and $f''(x) < 0$ for all x.

59. **(i)** $f'(x) > 0$ for all x, and

(ii) $f''(x) < 0$ for $x < 0$ and $f''(x) > 0$ for $x > 0$.

60. **(i)** $f'(x) < 0$ for $x < 0$ and $f'(x) > 0$ for $x > 0$, and

(ii) $f''(x) < 0$ for $|x| > 2$, and $f''(x) > 0$ for $|x| < 2$.

61. An infectious flu spreads slowly at the beginning of an epidemic. The infection process accelerates until a majority of the susceptible individuals are infected, at which point the process slows down.

(a) If $R(t)$ is the number of individuals infected at time t, describe the concavity of the graph of R near the beginning and end of the epidemic.

(b) Describe the status of the epidemic on the day that R has a point of inflection.

62. Water is pumped into a sphere at a constant rate (Figure 21). Let $h(t)$ be the water level at time t. Sketch the graph of h (approximately, but with the correct concavity). Where does the point of inflection occur?

63. Water is pumped into a sphere of radius R at a variable rate in such a way that the water level rises at a constant rate (Figure 21). Let $V(t)$ be the volume of water in the tank at time t. Sketch the graph V (approximately, but with the correct concavity). Where does the point of inflection occur?

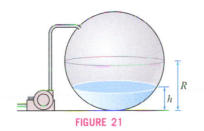

FIGURE 21

Original

Sigmoidal correction

(both: Library of Congress Prints and Photographs Division)

FIGURE 22

64. (Continuation of Exercise 63) If the sphere has radius R, the volume of water is $V = \pi\left(Rh^2 - \frac{1}{3}h^3\right)$, where h is the water level. Assume the level rises at a constant rate of 1 (i.e., $h = t$).

(a) Find the inflection point of V. Does this agree with your conclusion in Exercise 63?

(b) [GU] Plot V for $R = 1$.

65. Image Processing The intensity of a pixel in a digital image is measured by a number u between 0 and 1. Often, images can be enhanced by rescaling intensities (Figure 22), where pixels of intensity u are displayed with intensity $g(u)$ for a suitable function g. One common choice is the **sigmoidal correction**, defined for constants a, b by

$$g(u) = \frac{f(u) - f(0)}{f(1) - f(0)}, \qquad \text{where} \quad f(u) = \left(1 + e^{b(a-u)}\right)^{-1}$$

Figure 23 shows that $g(u)$ reduces the intensity of low-intensity pixels [where $g(u) < u$] and increases the intensity of high-intensity pixels.

(a) Verify that $f'(u) > 0$ and use this to show that $g(u)$ increases from 0 to 1 for $0 \le u \le 1$.

(b) Where does $g(u)$ have a point of inflection?

FIGURE 23 Sigmoidal correction with $a = 0.47$, $b = 12$.

66. Use graphical reasoning to determine whether the following statements are true or false. If false, modify the statement to make it correct.

(a) If f is increasing, then f^{-1} is decreasing.

(b) If f is decreasing, then f^{-1} is decreasing.

(c) If f is concave up, then f^{-1} is concave up.

(d) If f is concave down, then f^{-1} is concave up.

Further Insights and Challenges

In Exercises 67–69, assume that f is differentiable.

67. Proof of the Second Derivative Test Let c be a critical point such that $f''(c) > 0$ [the case $f''(c) < 0$ is similar].

(a) Show that $f''(c) = \lim_{h \to 0} \dfrac{f'(c + h)}{h}$.

(b) Use (a) to show that there exists an open interval (a, b) containing c such that $f'(x) < 0$ if $a < x < c$ and $f'(x) > 0$ if $c < x < b$. Conclude that $f(c)$ is a local minimum.

68. Prove that if f'' exists and $f''(x) > 0$ for all x, then the graph of f "sits above" its tangent lines.

(a) For any c, set $G(x) = f(x) - f'(c)(x - c) - f(c)$. It is sufficient to prove that $G(x) \ge 0$ for all c. Explain why with a sketch.

(b) Show that $G(c) = G'(c) = 0$ and $G''(x) > 0$ for all x. Conclude that $G'(x) < 0$ for $x < c$ and $G'(x) > 0$ for $x > c$. Then deduce, using the MVT, that $G(x) > G(c)$ for $x \ne c$.

69. Assume that f'' exists and let c be a point of inflection of f.

(a) Use the method of Exercise 68 to prove that the tangent line at $x = c$ crosses the graph (Figure 24). *Hint:* Show that $G(x)$ changes sign at $x = c$.

(b) [GU] Verify this conclusion for $f(x) = \dfrac{x}{3x^2 + 1}$ by graphing f and the tangent line at each inflection point on the same set of axes.

FIGURE 24 Tangent line crosses graph at point of inflection.

70. Let $C(x)$ be the cost of producing x units of a certain good. Assume that the graph of C is concave up.

(a) Show that the average cost $A(x) = C(x)/x$ is minimized at the production level x_0 such that average cost equals marginal cost—that is, $A(x_0) = C'(x_0)$.

(b) Show that the line through $(0, 0)$ and $(x_0, C(x_0))$ is tangent to the graph of C.

71. Let f be a polynomial of degree $n \geq 2$. Show that f has at least one point of inflection if n is odd. Then give an example to show that f need not have a point of inflection if n is even.

72. Critical and Inflection Points If $f'(c) = 0$ and $f(c)$ is neither a local min nor a local max, must $x = c$ be a point of inflection? This is true for "reasonable" functions (including the functions studied in this text), but it is not true in general. Let

$$f(x) = \begin{cases} x^2 \sin \frac{1}{x} & \text{for } x \neq 0 \\ 0 & \text{for } x = 0 \end{cases}$$

(a) Use the limit definition of the derivative to show that $f'(0)$ exists and $f'(0) = 0$.

(b) Show that $f(0)$ is neither a local min nor a local max.

(c) Show that $f'(x)$ changes sign infinitely often near $x = 0$. Conclude that $x = 0$ is not a point of inflection.

4.5 L'Hôpital's Rule

L'Hôpital's Rule is named for the French mathematician Guillaume François Antoine Marquis de L'Hôpital (1661–1704), who wrote the first textbook on calculus in 1696. The name L'Hôpital is pronounced "Lo-pee-tal."

L'Hôpital's Rule is a valuable tool for computing certain limits that are otherwise difficult to evaluate, and also for determining "asymptotic behavior" (limits at infinity). We will use it for graph sketching in the next section.

Consider the limit of a quotient

$$\lim_{x \to a} \frac{f(x)}{g(x)}$$

Roughly speaking, L'Hôpital's Rule states that *when $f(x)/g(x)$ has an indeterminate form of type $0/0$ or ∞/∞ at $x = a$, then we can replace $f(x)/g(x)$ by the quotient of the derivatives $f'(x)/g'(x)$.*

THEOREM 1 L'Hôpital's Rule Assume that f and g are differentiable on an open interval containing a and that

$$f(a) = g(a) = 0$$

Also assume that $g'(x) \neq 0$ (except possibly at a). Then

$$\lim_{x \to a} \frac{f(x)}{g(x)} = \lim_{x \to a} \frac{f'(x)}{g'(x)}$$

if the limit on the right exists or is infinite (∞ or $-\infty$). This conclusion also holds if f and g are differentiable for x near (but not equal to) a and

$$\lim_{x \to a} f(x) = \pm\infty \qquad \text{and} \qquad \lim_{x \to a} g(x) = \pm\infty$$

Furthermore, this rule is valid for one-sided limits.

CAUTION When using L'Hôpital's Rule, be sure to take the derivative of the numerator and denominator separately:

$$\lim_{x \to a} \frac{f(x)}{g(x)} = \lim_{x \to a} \frac{f'(x)}{g'(x)}$$

Do not differentiate the quotient function $y = f(x)/g(x)$.

■ **EXAMPLE 1** Use L'Hôpital's Rule to evaluate $\displaystyle\lim_{x \to 2} \frac{x^3 - 8}{x^4 + 2x - 20}$.

Solution Let $f(x) = x^3 - 8$ and $g(x) = x^4 + 2x - 20$. Both f and g are differentiable and $f(x)/g(x)$ is indeterminate of type $0/0$ at $a = 2$ because $f(2) = g(2) = 0$:

- Numerator: $f(2) = 2^3 - 1 = 0$
- Denominator: $g(2) = 2^4 + 2(2) - 20 = 0$

Furthermore, $g'(x) = 4x^3 + 2$ is nonzero near $x = 2$, so L'Hôpital's Rule applies. We may replace the numerator and denominator by their derivatives to obtain

$$\underbrace{\lim_{x \to 2} \frac{x^3 - 8}{x^4 + 2x - 2} = \lim_{x \to 2} \frac{(x^3 - 8)'}{(x^4 + 2x - 2)'}}_{\text{L'Hôpital's Rule}} = \lim_{x \to 2} \frac{3x^2}{4x^3 + 2} = \frac{3(2^2)}{4(2^3) + 2} = \frac{12}{34} = \frac{6}{17} \quad ■$$

■ **EXAMPLE 2** Evaluate $\lim\limits_{x \to 2} \dfrac{4 - x^2}{\sin \pi x}$.

Solution The quotient is indeterminate of type $0/0$ at $x = 2$:

- Numerator: $4 - x^2 = 4 - 2^2 = 0$
- Denominator: $\sin \pi x = \sin 2\pi = 0$

The other hypotheses [that f and g are differentiable and $g'(x) \neq 0$ for x near $a = 2$] are also satisfied, so we may apply L'Hôpital's Rule:

$$\underbrace{\lim_{x \to 2} \frac{4 - x^2}{\sin \pi x} = \lim_{x \to 2} \frac{(4 - x^2)'}{(\sin \pi x)'}}_{\text{L'Hôpital's Rule}} = \lim_{x \to 2} \frac{-2x}{\pi \cos \pi x} = \frac{-2(2)}{\pi \cos 2\pi} = \frac{-4}{\pi}$$ ■

■ **EXAMPLE 3** Evaluate $\lim\limits_{x \to \pi/2} \dfrac{\cos^2 x}{1 - \sin x}$.

Solution Again, the quotient is indeterminate of type $0/0$ at $x = \frac{\pi}{2}$:

$$\cos^2\left(\frac{\pi}{2}\right) = 0, \qquad 1 - \sin\frac{\pi}{2} = 1 - 1 = 0$$

The other hypotheses are satisfied, so we may apply L'Hôpital's Rule:

$$\underbrace{\lim_{x \to \pi/2} \frac{\cos^2 x}{1 - \sin x} = \lim_{x \to \pi/2} \frac{(\cos^2 x)'}{(1 - \sin x)'}}_{\text{L'Hôpital's Rule}} = \lim_{x \to \pi/2} \frac{-2\cos x \, \sin x}{-\cos x} = \underbrace{\lim_{x \to \pi/2} (2 \sin x) = 2}_{\text{simplified}}$$

Note that the quotient $\dfrac{-2\cos x \, \sin x}{-\cos x}$ is still indeterminate at $x = \pi/2$. We removed this indeterminacy by cancelling the factor $-\cos x$. ■

■ **EXAMPLE 4** **The Form $0 \cdot \infty$** Evaluate $\lim\limits_{x \to 0^+} x \ln x$.

Solution This limit is one-sided because $f(x) = x \ln x$ is not defined for $x \le 0$. Furthermore, as $x \to 0^+$,

- x approaches 0.
- $\ln x$ approaches $-\infty$.

So $f(x)$ presents an indeterminate form of type $0 \cdot \infty$. To apply L'Hôpital's Rule, we rewrite our function as $f(x) = (\ln x)/x^{-1}$ so that $f(x)$ presents an indeterminate form of type $-\infty/\infty$. Then L'Hôpital's Rule applies:

$$\lim_{x \to 0^+} x \ln x = \underbrace{\lim_{x \to 0^+} \frac{\ln x}{x^{-1}} = \lim_{x \to 0^+} \frac{(\ln x)'}{(x^{-1})'}}_{\text{L'Hôpital's Rule}} = \lim_{x \to 0^+} \left(\frac{x^{-1}}{-x^{-2}}\right) = \underbrace{\lim_{x \to 0^+} (-x) = 0}_{\text{simplified}}$$ ■

■ **EXAMPLE 5** **Using L'Hôpital's Rule Twice** Evaluate $\lim\limits_{x \to 0} \dfrac{e^x - x - 1}{\cos x - 1}$.

Solution For $x = 0$, we have

$$e^x - x - 1 = e^0 - 0 - 1 = 0, \qquad \cos x - 1 = \cos 0 - 1 = 0$$

A first application of L'Hôpital's Rule gives

$$\lim_{x \to 0} \frac{e^x - x - 1}{\cos x - 1} = \lim_{x \to 0} \frac{(e^x - x - 1)'}{(\cos x - 1)'} = \lim_{x \to 0} \left(\frac{e^x - 1}{-\sin x}\right) = \lim_{x \to 0} \frac{1 - e^x}{\sin x}$$

This limit is again indeterminate of type $0/0$, so we apply L'Hôpital's Rule a second time:

$$\lim_{x \to 0} \frac{1 - e^x}{\sin x} = \lim_{x \to 0} \frac{-e^x}{\cos x} = \frac{-e^0}{\cos 0} = -1$$ ■

■ **EXAMPLE 6** **Assumptions Matter** Can L'Hôpital's Rule be applied to $\lim\limits_{x\to 1}\dfrac{x^2+1}{2x+1}$?

Solution The answer is no. The function does *not* have an indeterminate form because

$$\left.\frac{x^2+1}{2x+1}\right|_{x=1}=\frac{1^2+1}{2\cdot 1+1}=\frac{2}{3}$$

However, the limit can be evaluated directly by substitution: $\lim\limits_{x\to 1}\dfrac{x^2+1}{2x+1}=\dfrac{2}{3}$. An incorrect application of L'Hôpital's Rule gives the wrong answer:

$$\lim_{x\to 1}\frac{(x^2+1)'}{(2x+1)'}=\lim_{x\to 1}\frac{2x}{2}=1\quad\text{(not equal to original limit)}\qquad\blacksquare$$

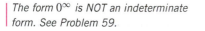

■ **EXAMPLE 7** **The Form $\infty-\infty$** Evaluate $\lim\limits_{x\to 0}\left(\dfrac{1}{\sin x}-\dfrac{1}{x}\right)$.

Solution Both $1/\sin x$ and $1/x$ become infinite at $x=0$, so we have an indeterminate form of type $\infty-\infty$. We must rewrite the function as

$$\frac{1}{\sin x}-\frac{1}{x}=\frac{x-\sin x}{x\sin x}$$

to obtain an indeterminate form of type $0/0$. L'Hôpital's Rule yields (see Figure 1)

$$\lim_{x\to 0}\left(\frac{1}{\sin x}-\frac{1}{x}\right)=\underbrace{\lim_{x\to 0}\frac{x-\sin x}{x\sin x}=\lim_{x\to 0}\frac{1-\cos x}{x\cos x+\sin x}}_{\text{L'Hôpital's Rule}}$$

$$=\underbrace{\lim_{x\to 0}\frac{\sin x}{-x\sin x+2\cos x}}_{\text{L'Hôpital's Rule again}}=\frac{0}{2}=0\qquad\blacksquare$$

FIGURE 1 The graph confirms that $y=\dfrac{1}{\sin x}-\dfrac{1}{x}$ approaches 0 as $x\to 0$.

Limits of functions of the form $f(x)^{g(x)}$ can lead to the indeterminate forms 0^0, 1^∞, or ∞^0. These are indeterminate as the limit can take on a variety of values, depending on the relative speed with which the base and exponent approach their limits. In such cases, we take the logarithm and then apply L'Hôpital's Rule.

The form 0^∞ is NOT an indeterminate form. See Problem 59.

■ **EXAMPLE 8** **The Form 0^0** Evaluate $\lim\limits_{x\to 0+}x^x$.

Solution First, set $y=x^x$ so that $\ln y=x\ln x$.
Then we take the limit of $\ln y$:

$$\lim_{x\to 0+}\ln y=\lim_{x\to 0+}x\ln x=\lim_{x\to 0+}\frac{\ln x}{x^{-1}}=0\qquad\text{(by Example 4)}$$

Since $f(x)=e^x$ is continuous, we can exponentiate to obtain the desired limit (see Figure 2):

$$\lim_{x\to 0+}x^x=\lim_{x\to 0+}e^{\ln(x^x)}=e^{\lim_{x\to 0+}\ln(x^x)}=e^{\lim_{x\to 0+}y}=e^0=1\qquad\blacksquare$$

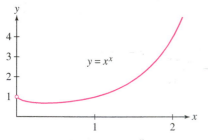

FIGURE 2 The function $y=x^x$ approaches 1 as $x\to 0+$.

■ **EXAMPLE 9** **The Form 1^∞** Find $\lim\limits_{x\to 0}(1+4x)^{1/2x}$.

Solution This has the indeterminate form 1^∞. Thus, we take $y=(1+4x)^{1/2x}$ and therefore $\ln y=\dfrac{\ln(1+4x)}{2x}$.
Then

$$\lim_{x\to 0}\ln y=\underbrace{\lim_{x\to 0}\frac{\ln(1+4x)}{2x}=\lim_{x\to 0}\frac{\frac{4}{1+4x}}{2}}_{\text{L'Hôpital's Rule}}=2$$

So

$$\lim_{x\to 0}(1+4x)^{1/2x}=\lim_{x\to 0}e^{\ln y}=e^{\lim_{x\to 0}\ln y}=e^2.\qquad\blacksquare$$

Comparing Growth of Functions

Sometimes, we are interested in determining which of two functions, f and g, grows faster. For example, there are two standard computer algorithms for sorting data (alphabetizing, ordering according to rank, etc.): **Quick Sort** and **Bubble Sort**. The average time required to sort a list of size n has order of magnitude $n \ln n$ for Quick Sort and n^2 for Bubble Sort. Which algorithm is faster when the size n is large? Although n is a whole number, this problem amounts to comparing the growth of $f(x) = x \ln x$ and $g(x) = x^2$ as $x \to \infty$.

We say that $f(x)$ grows *faster* than $g(x)$ if

$$\lim_{x \to \infty} \frac{f(x)}{g(x)} = \infty \qquad \text{or, equivalently,} \qquad \lim_{x \to \infty} \frac{g(x)}{f(x)} = 0$$

To indicate that $f(x)$ grows faster than $g(x)$, we use the notation $f(x) \ll g(x)$. For example, $x^2 \ll x$ because

$$\lim_{x \to \infty} \frac{x^2}{x} = \lim_{x \to \infty} x = \infty$$

To compare the growth of functions, we need a version of L'Hôpital's Rule that applies to limits at infinity.

THEOREM 2 L'Hôpital's Rule for Limits at Infinity Assume that f and g are differentiable in an interval (b, ∞) and that $g'(x) \neq 0$ for $x > b$. If $\lim\limits_{x \to \infty} f(x)$ and $\lim\limits_{x \to \infty} g(x)$ exist and either both are zero or both are infinite, then

$$\boxed{\lim_{x \to \infty} \frac{f(x)}{g(x)} = \lim_{x \to \infty} \frac{f'(x)}{g'(x)}}$$

provided that the limit on the right exists. A similar result holds for limits as $x \to -\infty$.

■ **EXAMPLE 10 The Form $\dfrac{\infty}{\infty}$** Which of $f(x) = x^2$ and $g(x) = x \ln x$ grows faster as $x \to \infty$?

Solution Both $f(x)$ and $g(x)$ approach infinity as $x \to \infty$, so L'Hôpital's Rule applies to the quotient:

$$\lim_{x \to \infty} \frac{f(x)}{g(x)} = \lim_{x \to \infty} \frac{x^2}{x \ln x} = \underbrace{\lim_{x \to \infty} \frac{x}{\ln x} = \lim_{x \to \infty} \frac{1}{x^{-1}}}_{\text{L'Hôpital's Rule}} = \lim_{x \to \infty} x = \infty$$

We conclude that $x \ln x \ll x^2$ (Figure 3). ■

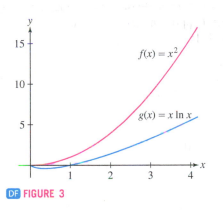

FIGURE 3

Note that this example implies that Quick Sort is a much faster sorting algorithm than Bubble Sort for large n.

■ **EXAMPLE 11** Jonathan is interested in comparing two computer algorithms whose average run times are approximately $(\ln n)^2$ and $\sqrt{n}$. Which algorithm takes less time for large values of n?

Solution Replace n by the continuous variable x and apply L'Hôpital's Rule twice:

$$\underbrace{\lim_{x \to \infty} \frac{\sqrt{x}}{(\ln x)^2} = \lim_{x \to \infty} \frac{\frac{1}{2} x^{-1/2}}{2x^{-1} \ln x}}_{\text{L'Hôpital's Rule}} = \underbrace{\lim_{x \to \infty} \frac{x^{1/2}}{4 \ln x}}_{\text{simplified}} = \underbrace{\lim_{x \to \infty} \frac{\frac{1}{2} x^{-1/2}}{4x^{-1}}}_{\text{L'Hôpital's Rule again}} = \underbrace{\lim_{x \to \infty} \frac{x^{1/2}}{8} = \infty}_{\text{simplified}}$$

This shows that $(\ln x)^2 \ll \sqrt{x}$. We conclude that the algorithm whose average time is proportional to $(\ln n)^2$ takes less time for large n. ■

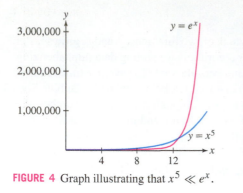

FIGURE 4 Graph illustrating that $x^5 \ll e^x$.

In Section 1.6, we asserted that exponential functions increase more rapidly than the power functions. We now prove this by showing that $x^n \ll e^x$ for every exponent n (Figure 4).

THEOREM 3 Growth of $f(x) = e^x$

$$x^n \ll e^x \qquad \text{for every exponent } n$$

In other words, $\displaystyle\lim_{x\to\infty} \frac{e^x}{x^n} = \infty$ for all k.

Proof The theorem is true for $n = 0$ since $\displaystyle\lim_{x\to\infty} e^x = \infty$. We use L'Hôpital's Rule repeatedly to prove that e^x/x^n tends to ∞ for $n = 1, 2, 3 \dots$ For example,

$$\lim_{x\to\infty} \frac{e^x}{x} = \lim_{x\to\infty} \frac{e^x}{1} = \lim_{x\to\infty} e^x = \infty$$

Then, having proved that $e^x/x \to \infty$, we use L'Hôpital's Rule again:

$$\lim_{x\to\infty} \frac{e^x}{x^2} = \lim_{x\to\infty} \frac{e^x}{2x} = \frac{1}{2}\lim_{x\to\infty} \frac{e^x}{x} = \infty$$

Proceeding in this way, we prove the result for all whole numbers n. A more formal proof would use the principle of induction. Finally, if k is any exponent, choose any whole number n such that $n > k$. Then $e^x/x^n < e^x/x^k$ for $x > 1$, so e^x/x^k must also tend to infinity as $x \to \infty$. ■

Proof of L'Hôpital's Rule

A full proof of L'Hôpital's Rule, without simplifying assumptions, is presented in a supplement on the text's Companion Web Site.

We prove L'Hôpital's Rule here only in the first case of Theorem 1—namely, in the case that $f(a) = g(a) = 0$. We also assume that f' and g' are continuous at $x = a$ and that $g'(a) \neq 0$. Then $g(x) \neq g(a)$ for x near but not equal to a, and

$$\frac{f(x)}{g(x)} = \frac{f(x) - f(a)}{g(x) - g(a)} = \frac{\dfrac{f(x) - f(a)}{x - a}}{\dfrac{g(x) - g(a)}{x - a}}$$

By the Quotient Law for Limits and the definition of the derivative,

$$\lim_{x\to a} \frac{f(x)}{g(x)} = \frac{\displaystyle\lim_{x\to a} \frac{f(x) - f(a)}{x - a}}{\displaystyle\lim_{x\to a} \frac{g(x) - g(a)}{x - a}} = \frac{f'(a)}{g'(a)} = \lim_{x\to a} \frac{f'(x)}{g'(x)} \qquad ■$$

4.5 SUMMARY

- L'Hôpital's Rule: Assume that f and g are differentiable near a and that

$$f(a) = g(a) = 0$$

Assume also that $g'(x) \neq 0$ (except possibly at a). Then

$$\lim_{x\to a} \frac{f(x)}{g(x)} = \lim_{x\to a} \frac{f'(x)}{g'(x)}$$

provided that the limit on the right exists or is infinite (∞ or $-\infty$).
- L'Hôpital's Rule applies to indeterminate forms $0/0$ and $\pm\infty/\infty$. It often also applies to $0 \cdot \infty$ and $\infty - \infty$, once they are rewritten appropriately.
- L'Hôpital's Rule also applies to limits as $x \to \infty$ or $x \to -\infty$.

- Limits involving the indeterminate forms 0^0, 1^∞, or ∞^0 can often be evaluated by first taking the logarithm and then applying L'Hôpital's Rule.
- In comparing the growth rates of functions, we say that $f(x)$ grows faster than $g(x)$, and we write $f \ll g$, if

$$\lim_{x \to \infty} \frac{f(x)}{g(x)} = \infty$$

4.5 EXERCISES

Preliminary Questions

1. What is wrong with applying L'Hôpital's Rule to $\lim_{x \to 0} \frac{x^2 - 2x}{3x - 2}$?

2. Does L'Hôpital's Rule apply to $\lim_{x \to a} f(x)g(x)$ if $f(x)$ and $g(x)$ both approach ∞ as $x \to a$?

Exercises

In Exercises 1–10, use L'Hôpital's Rule to evaluate the limit, or state that L'Hôpital's Rule does not apply.

1. $\lim_{x \to 3} \frac{2x^2 - 5x - 3}{x - 4}$

2. $\lim_{x \to -5} \frac{x^2 - 25}{5 - 4x - x^2}$

3. $\lim_{x \to 4} \frac{x^3 - 64}{x^2 + 16}$

4. $\lim_{x \to -1} \frac{x^4 + 2x + 1}{x^5 - 2x - 1}$

5. $\lim_{x \to 9} \frac{x^{1/2} + x - 6}{x^{3/2} - 27}$

6. $\lim_{x \to 3} \frac{\sqrt{x + 1} - 2}{x^3 - 7x - 6}$

7. $\lim_{x \to 0} \frac{\sin 4x}{x^2 + 3x + 1}$

8. $\lim_{x \to 0} \frac{x^3}{\sin x - x}$

9. $\lim_{x \to 0} \frac{\cos 2x - 1}{\sin 5x}$

10. $\lim_{x \to 0} \frac{\cos x - \sin^2 x}{\sin x}$

In Exercises 11–16, show that L'Hôpital's Rule is applicable to the limit as $x \to \pm\infty$ and evaluate.

11. $\lim_{x \to \infty} \frac{9x + 4}{3 - 2x}$

12. $\lim_{x \to -\infty} x \sin \frac{1}{x}$

13. $\lim_{x \to \infty} \frac{\ln x}{x^{1/2}}$

14. $\lim_{x \to \infty} \frac{x}{e^x}$

15. $\lim_{x \to -\infty} \frac{\ln(x^4 + 1)}{x}$

16. $\lim_{x \to \infty} \frac{x^2}{e^x}$

In Exercises 17–54, evaluate the limit.

17. $\lim_{x \to 1} \frac{\sqrt{8 + x} - 3x^{1/3}}{x^2 - 3x + 2}$

18. $\lim_{x \to 4} \left[\frac{1}{\sqrt{x} - 2} - \frac{4}{x - 4} \right]$

19. $\lim_{x \to -\infty} \frac{3x - 2}{1 - 5x}$

20. $\lim_{x \to \infty} \frac{x^{2/3} + 3x}{x^{5/3} - x}$

21. $\lim_{x \to -\infty} \frac{7x^2 + 4x}{9 - 3x^2}$

22. $\lim_{x \to \infty} \frac{3x^3 + 4x^2}{4x^3 - 7}$

23. $\lim_{x \to 1} \frac{(1 + 3x)^{1/2} - 2}{(1 + 7x)^{1/3} - 2}$

24. $\lim_{x \to 8} \frac{x^{5/3} - 2x - 16}{x^{1/3} - 2}$

25. $\lim_{x \to 0} \frac{\sin 2x}{\sin 7x}$

26. $\lim_{x \to 0} \frac{\tan 4x}{\tan 5x}$

27. $\lim_{x \to 0} \frac{\tan x}{x}$

28. $\lim_{x \to 0} \left(\cot x - \frac{1}{x} \right)$

29. $\lim_{x \to 0} \frac{\sin x - x \cos x}{x - \sin x}$

30. $\lim_{x \to \pi/2} \left(x - \frac{\pi}{2} \right) \tan x$

31. $\lim_{x \to 0} \frac{\cos(x + \frac{\pi}{2})}{\sin x}$

32. $\lim_{x \to 0} \frac{x^2}{1 - \cos x}$

33. $\lim_{x \to \pi/2} \frac{\cos x}{\sin(2x)}$

34. $\lim_{x \to 0} \left(\frac{1}{x^2} - \csc^2 x \right)$

35. $\lim_{x \to \pi/2} (\sec x - \tan x)$

36. $\lim_{x \to 2} \frac{e^{x^2} - e^4}{x - 2}$

37. $\lim_{x \to 1} \tan \left(\frac{\pi x}{2} \right) \ln x$

38. $\lim_{x \to 1} \frac{x(\ln x - 1) + 1}{(x - 1) \ln x}$

39. $\lim_{x \to 0} \frac{e^x - 1}{\sin x}$

40. $\lim_{x \to 1} \frac{e^x - e}{\ln x}$

41. $\lim_{x \to 0} \frac{e^{2x} - 1 - x}{x^2}$

42. $\lim_{x \to \infty} \frac{e^{2x} - 1 - x}{x^2}$

43. $\lim_{t \to 0+} (\sin t)(\ln t)$

44. $\lim_{x \to \infty} e^{-x}(x^3 - x^2 + 9)$

45. $\lim_{x \to 0} \frac{a^x - 1}{x}$ $(a > 0)$

46. $\lim_{x \to \infty} x^{1/x^2}$

47. $\lim_{x \to 1} (1 + \ln x)^{1/(x-1)}$

48. $\lim_{x \to 0+} x^{\sin x}$

49. $\lim_{x \to 0} (\cos x)^{3/x^2}$

50. $\lim_{x \to \infty} \left(\frac{x}{x + 1} \right)^x$

51. $\lim_{x \to 0} \frac{\sin^{-1} x}{x}$

52. $\lim_{x \to 0} \frac{\tan^{-1} x}{\sin^{-1} x}$

53. $\lim_{x \to 1} \frac{\tan^{-1} x - \frac{\pi}{4}}{\tan \frac{\pi}{4} x - 1}$

54. $\lim_{x \to 0+} \ln x \tan^{-1} x$

55. Evaluate $\lim_{x \to \pi/2} \frac{\cos mx}{\cos nx}$, where $m, n \neq 0$ are integers.

56. Evaluate $\lim_{x \to 1} \frac{x^m - 1}{x^n - 1}$ for any numbers $m, n \neq 0$.

57. Prove the following limit formula for e:

$$e = \lim_{x \to 0} (1 + x)^{1/x}$$

Then find a value of x such that $|(1 + x)^{1/x} - e| \leq 0.001$.

58. Prove the following limit formula for e:

$$e = \lim_{x \to \infty} \left(1 + \frac{1}{x}\right)^x$$

59. Show that 0^∞ is not an indeterminate form by showing that if f is a positive function and $\lim_{x \to 0} f(x) = 0$ and $\lim_{x \to 0} g(x) = \infty$, then $\lim_{x \to 0} (f(x))^{g(x)} = 0$.

60. $\boxed{\text{GU}}$ Can L'Hôpital's Rule be applied to $\lim_{x \to 0+} x^{\sin(1/x)}$? Does a graphical or numerical investigation suggest that the limit exists?

61. Let $f(x) = x^{1/x}$ for $x > 0$.
(a) Calculate $\lim_{x \to 0+} f(x)$ and $\lim_{x \to \infty} f(x)$.
(b) Find the maximum value of f and determine the intervals on which f is increasing or decreasing.

62. (a) Use the results of Exercise 61 to prove that $x^{1/x} = c$ has a unique solution if $0 < c \leq 1$ or $c = e^{1/e}$, two solutions if $1 < c < e^{1/e}$, and no solutions if $c > e^{1/e}$.
(b) $\boxed{\text{GU}}$ Plot the graph of $f(x) = x^{1/x}$ and verify that it confirms the conclusions of (a).

63. Determine whether $f \ll g$ or $g \ll f$ (or neither) for the functions $f(x) = \log_{10} x$ and $g(x) = \ln x$.

64. Show that $(\ln x)^3 \ll x^{1/3}$ and $(\ln x)^4 \ll x^{1/10}$.

65. Just as exponential functions are distinguished by their rapid rate of increase, the logarithm functions grow particularly slowly. Show that $\ln x \ll x^a$ for all $a > 0$.

66. Show that $(\ln x)^N \ll x^a$ for all N and all $a > 0$.

67. Determine whether $\sqrt{x} \ll e^{\sqrt{\ln x}}$ or $e^{\sqrt{\ln x}} \ll \sqrt{x}$. *Hint:* Use the substitution $u = \ln x$ instead of L'Hôpital's Rule.

68. Show that $\lim_{x \to \infty} x^n e^{-x} = 0$ for all whole numbers $n > 0$.

69. Assumptions Matter   Suppose $f(x) = x(2 + \sin x)$ and let $g(x) = x^2 + 1$.

(a) Show directly that $\lim_{x \to \infty} f(x)/g(x) = 0$.
(b) Show that $\lim_{x \to \infty} f(x) = \lim_{x \to \infty} g(x) = \infty$, but $\lim_{x \to \infty} f'(x)/g'(x)$ does not exist.
Do (a) and (b) contradict L'Hôpital's Rule? Explain.

70. Let $H(b) = \lim_{x \to \infty} \frac{\ln(1 + b^x)}{x}$ for $b > 0$.
(a) Show that $H(b) = \ln b$ if $b \geq 1$.
(b) Determine $H(b)$ for $0 < b \leq 1$.

71. Let $G(b) = \lim_{x \to \infty} (1 + b^x)^{1/x}$.
(a) Use the result of Exercise 70 to evaluate $G(b)$ for all $b > 0$.
(b) $\boxed{\text{GU}}$ Verify your result graphically by plotting $y = (1 + b^x)^{1/x}$ together with the horizontal line $y = G(b)$ for the values $b = 0.25$, $0.5, 2, 3$.

72. Show that $\lim_{t \to \infty} t^k e^{-t^2} = 0$ for all k. *Hint:* Compare with $\lim_{t \to \infty} t^k e^{-t} = 0$.

In Exercises 73–75, let

$$f(x) = \begin{cases} e^{-1/x^2} & \text{for } x \neq 0 \\ 0 & \text{for } x = 0 \end{cases}$$

These exercises show that f has an unusual property: All of its derivatives at $x = 0$ exist and are equal to zero.

73. Show that $\lim_{x \to 0} \frac{f(x)}{x^k} = 0$ for all k. *Hint:* Let $t = x^{-1}$ and apply the result of Exercise 72.

74. Show that $f'(0)$ exists and is equal to zero. Also, verify that $f''(0)$ exists and is equal to zero.

75. Show that for $k \geq 1$ and $x \neq 0$,

$$f^{(k)}(x) = \frac{P(x)e^{-1/x^2}}{x^r}$$

for some polynomial $P(x)$ and some exponent $r \geq 1$. Use the result of Exercise 73 to show that $f^{(k)}(0)$ exists and is equal to zero for all $k \geq 1$.

Further Insights and Challenges

76. Show that L'Hôpital's Rule applies to $\lim_{x \to \infty} \dfrac{x}{\sqrt{x^2 + 1}}$ but that it does not help. Then evaluate the limit directly.

77. The Second Derivative Test for critical points fails if $f''(c) = 0$. This exercise develops a **Higher Derivative Test** based on the sign of the first nonzero derivative. Suppose that

$$f'(c) = f''(c) = \cdots = f^{(n-1)}(c) = 0, \quad \text{but} \quad f^{(n)}(c) \neq 0$$

(a) Show, by applying L'Hôpital's Rule n times, that

$$\lim_{x \to c} \frac{f(x) - f(c)}{(x - c)^n} = \frac{1}{n!} f^{(n)}(c)$$

where $n! = n(n - 1)(n - 2) \cdots (2)(1)$.
(b) Use (a) to show that if n is even, then $f(c)$ is a local minimum if $f^{(n)}(c) > 0$ and is a local maximum if $f^{(n)}(c) < 0$. *Hint:* If n is even, then $(x - c)^n > 0$ for $x \neq a$, so $f(x) - f(c)$ must be positive for x near c if $f^{(n)}(c) > 0$.
(c) Use (a) to show that if n is odd, then $f(c)$ is neither a local minimum nor a local maximum.

78. When a spring with natural frequency $\lambda/2\pi$ is driven with a sinusoidal force $\sin(\omega t)$ with $\omega \neq \lambda$, it oscillates according to

$$y(t) = \frac{1}{\lambda^2 - \omega^2} \left(\lambda \sin(\omega t) - \omega \sin(\lambda t)\right)$$

Let $y_0(t) = \lim_{\omega \to \lambda} y(t)$.

(a) Use L'Hôpital's Rule to determine $y_0(t)$.

(b) Show that $y_0(t)$ ceases to be periodic and that its amplitude $|y_0(t)|$ tends to ∞ as $t \to \infty$ (the system is said to be in **resonance**; eventually, the spring is stretched beyond its structural tolerance).

(c) CAS Plot y for $\lambda = 1$ and $\omega = 0.8, 0.9, 0.99,$ and 0.999. Do the graphs confirm your conclusion in (b)?

79. $\boxed{\quad}$ We expended a lot of effort to evaluate $\lim_{x \to 0} \dfrac{\sin x}{x}$ in Chapter 2. Show that we could have evaluated it easily using L'Hôpital's Rule. Then explain why this method would involve *circular reasoning*.

80. By a fact from algebra, if f, g are polynomials such that $f(a) = g(a) = 0$, then there are polynomials f_1, g_1 such that

$$f(x) = (x - a)f_1(x), \qquad g(x) = (x - a)g_1(x)$$

Use this to verify L'Hôpital's Rule directly for $\lim\limits_{x \to a} f(x)/g(x)$.

81. Patience Required Use L'Hôpital's Rule to evaluate and check your answers numerically:

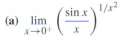

(a) $\lim\limits_{x \to 0^+} \left(\dfrac{\sin x}{x} \right)^{1/x^2}$ (b) $\lim\limits_{x \to 0} \left(\dfrac{1}{\sin^2 x} - \dfrac{1}{x^2} \right)$

82. In the following cases, check that $x = c$ is a critical point and use Exercise 77 to determine whether $f(c)$ is a local minimum or a local maximum.

(a) $f(x) = x^5 - 6x^4 + 14x^3 - 16x^2 + 9x + 12$ $(c = 1)$

(b) $f(x) = x^6 - x^3$ $(c = 0)$

4.6 Graph Sketching and Asymptotes

In this section, our goal is to sketch graphs using the information provided by the first two derivatives f' and f''. We will see that a useful sketch can be produced without plotting a large number of points. Although nowadays almost all graphs are produced by computer (including, of course, the graphs in this textbook), sketching graphs by hand is a useful way of solidifying your understanding of the basic concepts in this chapter.

Most graphs are made up of smaller *arcs* that have one of the four basic shapes, corresponding to the four possible sign combinations of f' and f'' (Figure 1). Since f' and f'' can each have sign $+$ or $-$, the sign combinations are

$$++ \qquad +- \qquad -+ \qquad --$$

In this notation, the first sign refers to f' and the second sign to f''. For instance, $-+$ indicates that $f'(x) < 0$ and $f''(x) > 0$. We use a slanted arrow over the first sign to indicate whether the function is increasing or decreasing, and an upturned or downturned ∪ over the second sign to indicate the concavity.

In graph sketching, we focus on the **transition points**, where the basic shape changes due to a sign change in either f' (local min or max) or f'' (point of inflection). In this section, local extrema are indicated by solid dots, and points of inflection are indicated by green solid squares (Figure 2).

In graph sketching, we must also pay attention to **asymptotic behavior**—that is, to the behavior of $f(x)$ as x approaches either $\pm\infty$ or a vertical asymptote.

The next three examples treat polynomials. Recall from Section 2.7 that the limits at infinity of a polynomial

$$f(x) = a_n x^n + a_{n-1}x^{n-1} + \cdots + a_1 x + a_0$$

(assuming that $a_n \neq 0$) are determined by

$$\lim_{x \to \infty} f(x) = a_n \lim_{x \to \infty} x^n$$

In general, then, the graph of a polynomial "wiggles" up and down a finite number of times and then tends to positive or negative infinity. Typical examples appear in Figure 3.

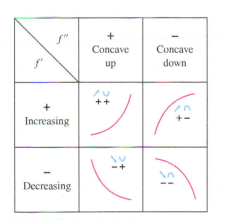

FIGURE 1 The four basic shapes.

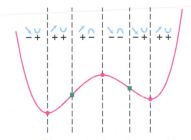

FIGURE 2 The graph of f with transition points and sign combinations of f' and f''.

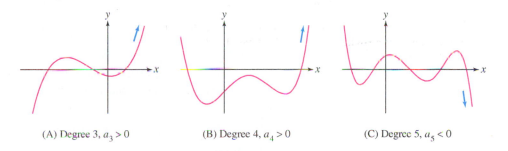

(A) Degree 3, $a_3 > 0$ (B) Degree 4, $a_4 > 0$ (C) Degree 5, $a_5 < 0$

FIGURE 3 Graphs of polynomials.

■ **EXAMPLE 1 Quadratic Polynomial** Sketch the graph of $f(x) = x^2 - 4x + 3$.

Solution Note that $f(x) = (x - 1)(x - 3)$ so the graph intersects the x-axis at $x = 1$ and $x = 3$. We have $f'(x) = 2x - 4 = 2(x - 2)$. We can see directly that $f'(x)$ is negative for $x < 2$ and positive for $x > 2$, but let's confirm this using test values, as in previous sections:

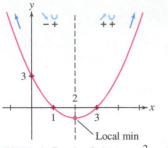

FIGURE 4 Graph of $f(x) = x^2 - 4x + 3$.

Interval	Test Value	Sign of f'
$(-\infty, 2)$	$f'(1) = -2$	$-$
$(2, \infty)$	$f'(3) = 2$	$+$

Furthermore, $f''(x) = 2$ is positive, so the graph is everywhere concave up. To sketch the graph, plot the local minimum $(2, -1)$, the y-intercept, and the roots $x = 1, 3$. Since the leading term of f is x^2, $f(x)$ tends to ∞ as $x \to \pm\infty$. This asymptotic behavior is noted by the arrows in Figure 4. ∎

■ **EXAMPLE 2** **Cubic Polynomial** Sketch the graph of $f(x) = \frac{1}{3}x^3 - \frac{1}{2}x^2 - 2x + 3$.

Solution

Step 1. **Determine the signs of f' and f''.**
First, solve for the critical points:

$$f'(x) = x^2 - x - 2 = (x + 1)(x - 2) = 0$$

The critical points $c = -1, 2$ divide the x-axis into three intervals $(-\infty, -1)$, $(-1, 2)$, and $(2, \infty)$, on which we determine the sign of f' by computing test values:

Interval	Test Value	Sign of f'
$(-\infty, -1)$	$f'(-2) = 4$	$+$
$(-1, 2)$	$f'(0) = -2$	$-$
$(2, \infty)$	$f'(3) = 4$	$+$

Next, solve $f''(x) = 2x - 1 = 0$. The solution is $c = \frac{1}{2}$ and we have

Interval	Test Value	Sign of f''
$\left(-\infty, \frac{1}{2}\right)$	$f''(0) = -1$	$-$
$\left(\frac{1}{2}, \infty\right)$	$f''(1) = 1$	$+$

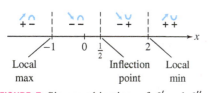

FIGURE 5 Sign combinations of f' and f''.

Local max   Inflection point   Local min

Step 2. **Note transition points and sign combinations.**
This step merges the information about f' and f'' in a sign diagram (Figure 5). There are three transition points:

- $c = -1$: local max since f' changes from $+$ to $-$ at $c = -1$.
- $c = \frac{1}{2}$: corresponds to a point of inflection since f'' changes sign at $c = \frac{1}{2}$.
- $c = 2$: local min since f' changes from $-$ to $+$ at $c = 2$.

In Figure 6(A), we plot the transition points and, for added accuracy, the y-intercept $f(0)$, using the values

$$f(-1) = \frac{25}{6}, \qquad f\left(\frac{1}{2}\right) = \frac{23}{12}, \qquad f(0) = 3, \qquad f(2) = -\frac{1}{3}$$

Step 3. **Draw arcs of appropriate shape and asymptotic behavior.**
The leading term of $f(x)$ is $\frac{1}{3}x^3$. Therefore, $\lim_{x \to \infty} f(x) = \infty$ and $\lim_{x \to -\infty} f(x) = -\infty$.
To create the sketch, it remains only to connect the transition points by arcs of the appropriate concavity and asymptotic behavior, as in Figure 6(B) and (C). ∎

■ **EXAMPLE 3** Sketch the graph of $f(x) = 3x^4 - 8x^3 + 6x^2 + 1$.

Solution

Step 1. **Determine the signs of f' and f''.**
First, solve for the transition points:

$$f'(x) = 12x^3 - 24x^2 + 12x = 12x(x - 1)^2 = 0 \quad \Rightarrow \quad x = 0, 1$$

$$f''(x) = 36x^2 - 48x + 12 = 12(x - 1)(3x - 1) = 0 \quad \Rightarrow \quad x = \frac{1}{3}, 1$$

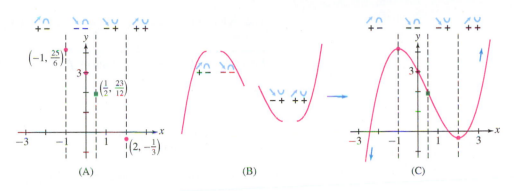

FIGURE 6 Graph of
$f(x) = \frac{1}{3}x^3 - \frac{1}{2}x^2 - 2x + 3.$

(A)   (B)   (C)

The signs of f' and f'' are recorded in the following tables:

Interval	Test Value	Sign of f'
$(-\infty, 0)$	$f'(-1) = -48$	$-$
$(0, 1)$	$f'(\frac{1}{2}) = \frac{3}{2}$	$+$
$(1, \infty)$	$f'(2) = 24$	$+$

Interval	Test Value	Sign of f''
$\left(-\infty, \frac{1}{3}\right)$	$f''(0) = 12$	$+$
$\left(\frac{1}{3}, 1\right)$	$f''(\frac{1}{2}) = -3$	$-$
$(1, \infty)$	$f''(2) = 60$	$+$

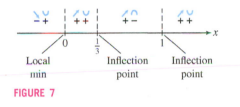

FIGURE 7

Step 2. Note transition points and sign combinations.
The transition points $c = 0, \frac{1}{3}, 1$ divide the x-axis into four intervals (Figure 7). The type of sign change determines the nature of the transition point:

- $c = 0$: local min since f' changes from $-$ to $+$ at $c = 0$.
- $c = \frac{1}{3}$: corresponds to a point of inflection since f'' changes sign at $c = \frac{1}{3}$.
- $c = 1$: neither a local min nor a local max since f' does not change sign, but it is a point of inflection since $f''(x)$ changes sign at $c = 1$.

We plot the transition points $c = 0, \frac{1}{3}, 1$ in Figure 8(A) using function values $f(0) = 1$, $f\left(\frac{1}{3}\right) = \frac{38}{27}$, and $f(1) = 2$.

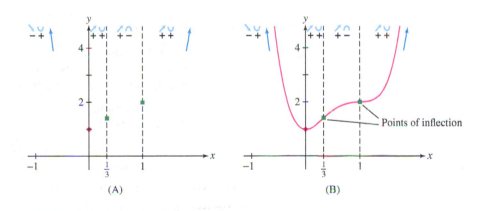

FIGURE 8 $f(x) = 3x^4 - 8x^3 + 6x^2 + 1.$

(A)   (B)

Step 3. Draw arcs of appropriate shape and asymptotic behavior.
Before drawing the arcs, we note that $f(x)$ has leading term $3x^4$, so $f(x)$ tends to ∞ as $x \to \infty$ and as $x \to -\infty$. We obtain Figure 8(B). ■

■ **EXAMPLE 4** **Trigonometric Function** Sketch $f(x) = \cos x + \frac{1}{2}x$ over $[0, \pi]$.

Solution First, we find the transition points for x in $[0, \pi]$:

$$f'(x) = -\sin x + \frac{1}{2} = 0 \quad \Rightarrow \quad x = \frac{\pi}{6}, \frac{5\pi}{6}$$

$$f''(x) = -\cos x = 0 \quad \Rightarrow \quad x = \frac{\pi}{2}$$

The sign combinations are shown in the following tables:

Interval	Test Value	Sign of f'
$(0, \frac{\pi}{6})$	$f'(\frac{\pi}{12}) \approx 0.24$	$+$
$(\frac{\pi}{6}, \frac{5\pi}{6})$	$f'(\frac{\pi}{2}) = -\frac{1}{2}$	$-$
$(\frac{5\pi}{6}, \pi)$	$f'(\frac{11\pi}{12}) \approx 0.24$	$+$

Interval	Test Value	Sign of f''
$(0, \frac{\pi}{2})$	$f''(\frac{\pi}{4}) = -\frac{\sqrt{2}}{2}$	$-$
$(\frac{\pi}{2}, \pi)$	$f''(\frac{3\pi}{4}) = \frac{\sqrt{2}}{2}$	$+$

We record the sign changes and transition points in Figure 9 and sketch the graph using the values

$$f(0) = 1, \quad f\left(\frac{\pi}{6}\right) \approx 1.13, \quad f\left(\frac{\pi}{2}\right) \approx 0.79, \quad f\left(\frac{5\pi}{6}\right) \approx 0.44, \quad f(\pi) \approx 0.57 \blacksquare$$

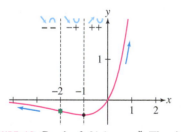

FIGURE 9 $f(x) = \cos x + \frac{1}{2}x$.

■ **EXAMPLE 5 A Function Involving e^x** Sketch the graph of $f(x) = xe^x$.

Solution As usual, we solve for the transition points and determine the signs:

$$f'(x) = xe^x + e^x = (x + 1)e^x = 0 \quad \Rightarrow \quad x = -1$$

$$f''(x) = (x + 1)e^x + e^x = (x + 2)e^x = 0 \quad \Rightarrow \quad x = -2$$

Interval	Test Value	Sign of f'
$(-\infty, -1)$	$f'(-2) = -e^{-2}$	$-$
$(-1, \infty)$	$f'(0) = e^0$	$+$

Interval	Test Value	Sign of f''
$(-\infty, -2)$	$f''(-3) = -e^{-3}$	$-$
$(-2, \infty)$	$f''(0) = 2e^0$	$+$

The sign change of f' shows that $f(-1)$ is a local min. The sign change of f'' shows that f has a point of inflection at $x = -2$, where the graph changes from concave down to concave up.

The last pieces of information we need are the limits at infinity. Both x and e^x tend to ∞ as $x \to \infty$, so $\lim\limits_{x \to \infty} xe^x = \infty$. On the other hand, the limit as $x \to -\infty$ is indeterminate of type $\infty \cdot 0$ because x tends to $-\infty$ and e^x tends to zero. Therefore, we write $xe^x = x/e^{-x}$ and apply L'Hôpital's Rule:

$$\lim_{x \to -\infty} xe^x = \lim_{x \to -\infty} \frac{x}{e^{-x}} = \lim_{x \to -\infty} \frac{1}{-e^{-x}} = -\lim_{x \to -\infty} e^x = 0$$

Figure 10 shows the graph with its local minimum and point of inflection, drawn with the correct concavity and asymptotic behavior. ■

FIGURE 10 Graph of $f(x) = xe^x$. The sign combinations $--$, $-+$, $++$ indicate the signs of f' and f''.

The next two examples deal with horizontal and vertical asymptotes.

■ **EXAMPLE 6** Sketch the graph of $f(x) = \dfrac{3x + 2}{2x - 4}$.

Solution The function f is not defined for all x. This plays a role in our analysis so we add a Step 0 to our procedure.

Step 0. **Determine the domain of f.**

Since $f(x)$ is not defined for $x = 2$, the domain of f consists of the two intervals $(-\infty, 2)$ and $(2, \infty)$. We must analyze f on these intervals separately.

Step 1. **Determine the signs of f' and f''.**

Calculation shows that

$$f'(x) = -\frac{4}{(x - 2)^2}, \qquad f''(x) = \frac{8}{(x - 2)^3}$$

Although $f'(x)$ is not defined at $x = 2$, we do not call it a critical point because $x = 2$ is not in the domain of f. In fact, $f'(x)$ is negative for $x \neq 2$, so f is decreasing and has no critical points.

On the other hand, $f''(x) > 0$ for $x > 2$ and $f''(x) < 0$ for $x < 2$. Although $f''(x)$ changes sign at $x = 2$, we do not call $x = 2$ a point of inflection because it is not in the domain of f.

Step 2. Note transition points and sign combinations.

There are no transition points in the domain of f.

$(-\infty, 2)$	$f'(x) < 0$ and $f''(x) < 0$
$(2, \infty)$	$f'(x) < 0$ and $f''(x) > 0$

Step 3. Draw arcs of appropriate shape and asymptotic behavior.

The following limits show that $y = \frac{3}{2}$ is a horizontal asymptote:

$$\lim_{x \to \pm\infty} \frac{3x + 2}{2x - 4} = \lim_{x \to \pm\infty} \frac{3 + 2x^{-1}}{2 - 4x^{-1}} = \frac{3}{2}$$

The line $x = 2$ is a vertical asymptote because $f(x)$ has infinite one-sided limits

$$\lim_{x \to 2^-} \frac{3x + 2}{2x - 4} = -\infty, \qquad \lim_{x \to 2^+} \frac{3x + 2}{2x - 4} = \infty$$

To verify this, note that for x near 2, the denominator $2x - 4$ is small and negative if $x < 2$ and small and positive if $x > 2$, whereas the numerator $3x + 2$ is positive.

Figure 11(A) summarizes the asymptotic behavior. What does the graph look like to the right of $x = 2$? It is decreasing and concave up since $f' < 0$ and $f'' > 0$, and it approaches the asymptotes. The only possibility is the right-hand curve in Figure 11(B). To the left of $x = 2$, the graph is decreasing, is concave down, and approaches the asymptotes. The x-intercept is $x = -\frac{2}{3}$ because $f\left(-\frac{2}{3}\right) = 0$ and the y-intercept is $y = f(0) = -\frac{1}{2}$. ∎

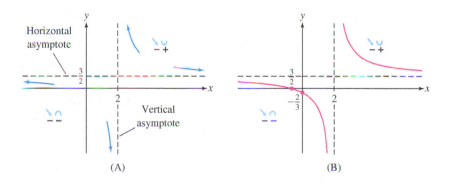

FIGURE 11 Graph of $y = \dfrac{3x + 2}{2x - 4}$.

(A)

(B)

■ **EXAMPLE 7** Sketch the graph of $f(x) = \dfrac{1}{x^2 - 1}$.

Solution The function f is defined for $x \neq \pm 1$. By calculation,

$$f'(x) = -\frac{2x}{(x^2 - 1)^2}, \qquad f''(x) = \frac{6x^2 + 2}{(x^2 - 1)^3}$$

For $x \neq \pm 1$, the denominator of $f'(x)$ is positive. Therefore, $f'(x)$ and x have opposite signs:

• $f'(x) > 0$ for $x < 0$, $f'(x) < 0$ for $x > 0$, $x = 0$ is a local max

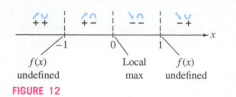

FIGURE 12

In this example,

$$f(x) = \frac{1}{x^2 - 1}$$

$$f'(x) = -\frac{2x}{(x^2 - 1)^2}$$

$$f''(x) = \frac{6x^2 + 2}{(x^2 - 1)^3}$$

The sign of $f''(x)$ is equal to the sign of $x^2 - 1$ because $6x^2 + 2$ is positive:

- $f''(x) > 0$ for $x < -1$ or $x > 1$ and $f''(x) < 0$ for $-1 < x < 1$

Figure 12 summarizes the sign information.

The x-axis, $y = 0$, is a horizontal asymptote because

$$\lim_{x \to \infty} \frac{1}{x^2 - 1} = 0 \quad \text{and} \quad \lim_{x \to -\infty} \frac{1}{x^2 - 1} = 0$$

The lines $x = \pm 1$ are vertical asymptotes. To determine the one-sided limits, note that $f(x) < 0$ for $-1 < x < 1$ and $f(x) > 0$ for $|x| > 1$. Therefore, as $x \to \pm 1$, $f(x)$ approaches $-\infty$ from within the interval $(-1, 1)$, and it approaches ∞ from outside $(-1, 1)$ (Figure 13). We display the sketch in Figure 14.

Vertical Asymptote	Left-Hand Limit	Right-Hand Limit
$x = -1$	$\lim\limits_{x \to -1^-} \dfrac{1}{x^2 - 1} = \infty$	$\lim\limits_{x \to -1^+} \dfrac{1}{x^2 - 1} = -\infty$
$x = 1$	$\lim\limits_{x \to 1^-} \dfrac{1}{x^2 - 1} = -\infty$	$\lim\limits_{x \to 1^+} \dfrac{1}{x^2 - 1} = \infty$

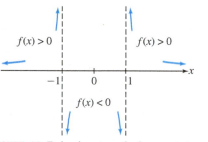

FIGURE 13 Behavior at vertical asymptotes.

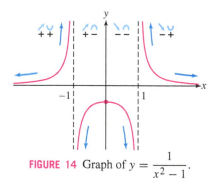

FIGURE 14 Graph of $y = \dfrac{1}{x^2 - 1}$.

4.6 SUMMARY

- Most graphs are made up of arcs that have one of the four basic shapes (Figure 15):

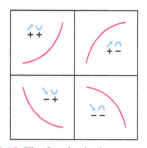

FIGURE 15 The four basic shapes.

	Sign Combination	Curve Type
++	$f' > 0, f'' > 0$	Increasing and concave up
+−	$f' > 0, f'' < 0$	Increasing and concave down
−+	$f' < 0, f'' > 0$	Decreasing and concave up
−−	$f' < 0, f'' < 0$	Decreasing and concave down

- A *transition point* is a point in the domain of f at which either f' changes sign (local min or max) or f'' changes sign (point of inflection).
- It is convenient to break up the curve-sketching process into steps:

Step 0. Determine the domain of f.

Step 1. Determine the signs of f' and f''.

Step 2. Note transition points and sign combinations.

Step 3. Determine the asymptotic behavior of $f(x)$.

Step 4. Draw arcs of appropriate shape and asymptotic behavior.

4.6 EXERCISES

Preliminary Questions

1. Sketch an arc where f' and f'' have the sign combination ++. Do the same for $-+$.

2. If the sign combination of f' and f'' changes from ++ to $+-$ at $x = c$, then (choose the correct answer):

(a) $f(c)$ is a local min.   **(b)** $f(c)$ is a local max.

(c) $(c, f(c))$ is a point of inflection.

3. The second derivative of the function $f(x) = (x-4)^{-1}$ is $f''(x) = 2(x-4)^{-3}$. Although $f''(x)$ changes sign at $x = 4$, f does not have a point of inflection at $x = 4$. Why not?

Exercises

1. Determine the sign combinations of f' and f'' for each interval A–G in Figure 16.

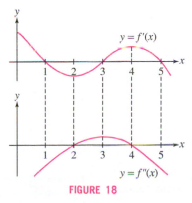

FIGURE 16

2. State the sign change at each transition point A–G in Figure 17. Example: $f'(x)$ goes from $+$ to $-$ at A.

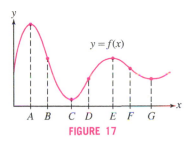

FIGURE 17

In Exercises 3–6, draw the graph of a function for which f' and f'' take on the given sign combinations.

3. ++, +−, −−   **4.** +−, −−, −+

5. −+, −−, −+   **6.** −+, ++, +−

7. Sketch the graph of a function that could have the graphs of f' and f'' appearing in Figure 18.

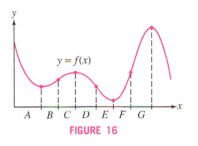

FIGURE 18

8. Sketch the graph of a function that could have the graphs of f' and f'' appearing in Figure 19.

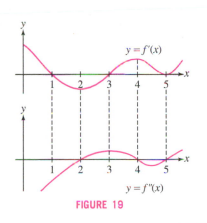

FIGURE 19

9. Sketch the graph of $y = x^2 - 5x + 4$.

10. Sketch the graph of $y = 12 - 5x - 2x^2$.

11. Sketch the graph of $f(x) = x^3 - 3x^2 + 2$. Include the zeros of f, which are $x = 1$ and $1 \pm \sqrt{3}$ (approximately -0.73, 2.73).

12. Show that $f(x) = x^3 - 3x^2 + 6x$ has a point of inflection but no local extreme values. Sketch the graph.

13. Extend the sketch of the graph of $f(x) = \cos x + \frac{1}{2}x$ in Example 4 to the interval $[0, 5\pi]$.

14. Sketch the graphs of $y = x^{2/3}$ and $y = x^{4/3}$.

In Exercises 15–36, find the transition points, intervals of increase/decrease, concavity, and asymptotic behavior. Then sketch the graph, with this information indicated.

15. $y = x^3 + 24x^2$ **16.** $y = x^3 - 3x + 5$

17. $y = x^2 - 4x^3$ **18.** $y = \frac{1}{3}x^3 + x^2 + 3x$

19. $y = 4 - 2x^2 + \frac{1}{6}x^4$ **20.** $y = 7x^4 - 6x^2 + 1$

21. $y = x^5 + 5x$ **22.** $y = x^5 - 15x^3$

23. $y = x^4 - 3x^3 + 4x$ **24.** $y = x^2(x-4)^2$

25. $y = x^7 - 14x^6$ **26.** $y = x^6 - 9x^4$

27. $y = x - 4\sqrt{x}$ **28.** $y = \sqrt{x} + \sqrt{16 - x}$

29. $y = x(8 - x)^{1/3}$ **30.** $y = (x^2 - 4x)^{1/3}$

31. $y = xe^{-x^2}$ **32.** $y = (2x^2 - 1)e^{-x^2}$

33. $y = x - 2\ln x$ **34.** $y = x(4 - x) - 3\ln x$

35. $y = x^2 - 2\ln x$ **36.** $y = x - 2\ln(x^2 + 1)$

37. Sketch the graph of $f(x) = 18(x - 3)(x - 1)^{2/3}$ using the formulas

$$f'(x) = \frac{30(x - \frac{9}{5})}{(x - 1)^{1/3}}, \qquad f''(x) = \frac{20(x - \frac{3}{5})}{(x - 1)^{4/3}}$$

38. Sketch the graph of $f(x) = \dfrac{x}{x^2 + 1}$ using the formulas

$$f'(x) = \frac{1 - x^2}{(1 + x^2)^2}, \qquad f''(x) = \frac{2x(x^2 - 3)}{(x^2 + 1)^3}$$

⌐ᴀ⌐ *In Exercises 39–42, sketch the graph of the function, indicating all transition points. If necessary, use a graphing utility or computer algebra system to locate the transition points numerically.*

39. $y = x^2 - 10\ln(x^2 + 1)$ **40.** $y = e^{-x/2}\ln x$

41. $y = x^4 - 4x^2 + x + 1$

42. $y = 2\sqrt{x} - \sin x, \quad 0 \leq x \leq 2\pi$

In Exercises 43–48, sketch the graph over the given interval, with all transition points indicated.

43. $y = x + \sin x, \quad [0, 2\pi]$

44. $y = \sin x + \cos x, \quad [0, 2\pi]$

45. $y = 2\sin x - \cos^2 x, \quad [0, 2\pi]$

46. $y = \sin x + \tfrac{1}{2}x, \quad [0, 2\pi]$

47. $y = \sin x + \sqrt{3}\cos x, \quad [0, \pi]$

48. $y = \sin x - \tfrac{1}{2}\sin 2x, \quad [0, \pi]$

49. ✎ Are all sign transitions possible? Explain with a sketch why the transitions $++ \to -+$ and $-- \to +-$ do not occur if the function is differentiable. (See Exercise 78 for a proof.)

50. Suppose that f is twice differentiable satisfying (i) $f(0) = 1$, (ii) $f'(x) > 0$ for all $x \neq 0$, and (iii) $f''(x) < 0$ for $x < 0$ and $f''(x) > 0$ for $x > 0$. Let $g(x) = f(x^2)$.
(a) Sketch a possible graph of f.
(b) Prove that g has no points of inflection and a unique local extreme value at $x = 0$. Sketch a possible graph of g.

51. Which of the graphs in Figure 20 *cannot* be the graph of a polynomial? Explain.

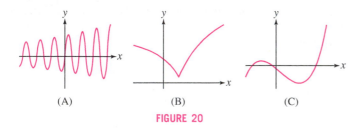

(A)   (B)   (C)

FIGURE 20

52. Which curve in Figure 21 is the graph of $f(x) = \dfrac{2x^4 - 1}{1 + x^4}$? Explain on the basis of horizontal asymptotes.

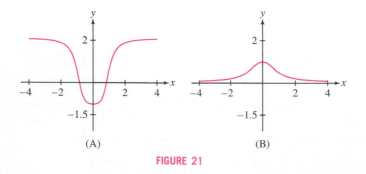

(A)   (B)

FIGURE 21

53. Match the graphs in Figure 22 with the two functions $y = \dfrac{3x}{x^2 - 1}$ and $y = \dfrac{3x^2}{x^2 - 1}$. Explain.

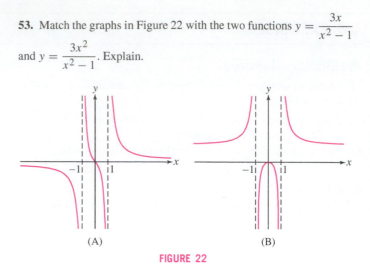

(A)   (B)

FIGURE 22

54. Match the functions below with their graphs in Figure 23.

(a) $y = \dfrac{1}{x^2 - 1}$ **(b)** $y = \dfrac{x^2}{x^2 + 1}$

(c) $y = \dfrac{1}{x^2 + 1}$ **(d)** $y = \dfrac{x}{x^2 - 1}$

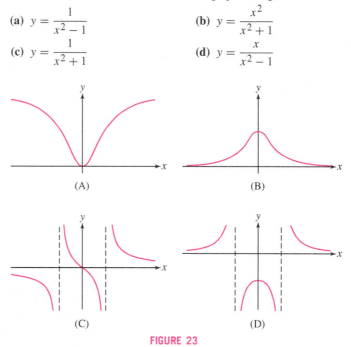

(A)   (B)

(C)   (D)

FIGURE 23

In Exercises 55–72, sketch the graph of the function. Indicate the transition points and asymptotes.

55. $y = \dfrac{1}{3x - 1}$ **56.** $y = \dfrac{x - 2}{x - 3}$

57. $y = \dfrac{x + 3}{x - 2}$ **58.** $y = x + \dfrac{1}{x}$

59. $y = \dfrac{1}{x} + \dfrac{1}{x - 1}$ **60.** $y = \dfrac{1}{x} - \dfrac{1}{x - 1}$

61. $y = \dfrac{1}{x(x - 2)}$ **62.** $y = \dfrac{x}{x^2 - 9}$

63. $y = \dfrac{1}{x^2 - 6x + 8}$ **64.** $y = \dfrac{x^3 + 1}{x}$

65. $y = 1 - \dfrac{3}{x} + \dfrac{4}{x^3}$ **66.** $y = \dfrac{1}{x^2} + \dfrac{1}{(x - 2)^2}$

67. $y = \dfrac{1}{x^2} - \dfrac{1}{(x - 2)^2}$ **68.** $y = \dfrac{4}{x^2 - 9}$

69. $y = \dfrac{1}{(x^2 + 1)^2}$

70. $y = \dfrac{x^2}{(x^2 - 1)(x^2 + 1)}$

71. $y = \dfrac{1}{\sqrt{x^2 + 1}}$

72. $y = \dfrac{x}{\sqrt{x^2 + 1}}$

Further Insights and Challenges

*In Exercises 73–77, we explore functions whose graphs approach a nonhorizontal line as $x \to \infty$. A line $y = ax + b$ is called a **slant asymptote** if*

$$\lim_{x \to \infty} (f(x) - (ax + b)) = 0$$

or

$$\lim_{x \to -\infty} (f(x) - (ax + b)) = 0$$

73. Let $f(x) = \dfrac{x^2}{x - 1}$ (Figure 24). Verify the following:

(a) $f(0)$ is a local max and $f(2)$ a local min.

(b) f is concave down on $(-\infty, 1)$ and concave up on $(1, \infty)$.

(c) $\lim\limits_{x \to 1-} f(x) = -\infty$ and $\lim\limits_{x \to 1+} f(x) = \infty$.

(d) $y = x + 1$ is a slant asymptote of f as $x \to \pm\infty$.

(e) The slant asymptote lies above the graph of f for $x < 1$ and below the graph for $x > 1$.

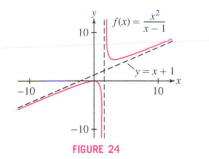

$f(x) = \dfrac{x^2}{x - 1}$

$y = x + 1$

FIGURE 24

74. If $f(x) = P(x)/Q(x)$, where P and Q are polynomials of degrees $m + 1$ and m, then by long division, we can write

$$f(x) = (ax + b) + P_1(x)/Q(x)$$

where P_1 is a polynomial of degree $< m$. Show that $y = ax + b$ is the slant asymptote of $f(x)$. Use this procedure to find the slant asymptotes of the following functions:

(a) $y = \dfrac{x^2}{x + 2}$

(b) $y = \dfrac{x^3 + x}{x^2 + x + 1}$

75. Sketch the graph of

$$f(x) = \dfrac{x^2}{x + 1}$$

Proceed as in the previous exercise to find the slant asymptote.

76. Show that $y = 3x$ is a slant asymptote for $f(x) = 3x + x^{-2}$. Determine whether $f(x)$ approaches the slant asymptote from above or below and make a sketch of the graph.

77. Sketch the graph of $f(x) = \dfrac{1 - x^2}{2 - x}$.

78. Assume that f' and f'' exist for all x and let c be a critical point of f. Show that $f(x)$ cannot make a transition from $++$ to $-+$ at $x = c$. *Hint:* Apply the MVT to $f'(x)$.

79. Assume that f'' exists and $f''(x) > 0$ for all x. Show that $f(x)$ cannot be negative for all x. *Hint:* Show that $f'(b) \neq 0$ for some b and use the result of Exercise 68 in Section 4.4.

4.7 Applied Optimization

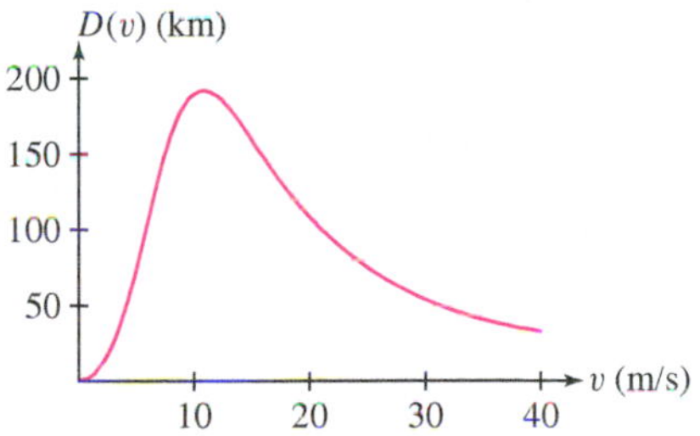

Optimization plays a role in a wide range of disciplines, including the physical sciences, economics, and biology. For example, scientists have studied how migrating birds choose an optimal velocity v that maximizes the distance they can travel without stopping, given the energy that can be stored as body fat (Figure 1).

In many optimization problems, the first step is to write down the **objective function**. This is the function whose minimum or maximum we seek. Once we find the objective function, we can apply the techniques developed in this chapter. Our first examples require optimization on a closed interval $[a, b]$. Let's recall the steps for finding extrema developed in Section 4.2:

(i) Find the critical points of f in $[a, b]$.

(ii) Evaluate $f(x)$ at the critical points and the endpoints a and b.

(iii) The largest and smallest values are the extreme values of f on $[a, b]$.

FIGURE 1 Physiology and aerodynamics are applied to obtain a plausible formula for bird migration distance D as a function of velocity v. The optimal velocity corresponds to the maximum point on the graph (see Exercise 67).

■ **EXAMPLE 1** A piece of wire of length L is bent into the shape of a rectangle (Figure 2). Which dimensions produce the rectangle of maximum area?

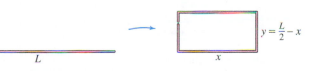

L

$y = \dfrac{L}{2} - x$

x

FIGURE 2

An equation relating two or more variables in an optimization problem is called a "constraint equation." In Example 1, the constraint equation is

$$2x + 2y = L$$

Solution The rectangle has area $A = xy$, where x and y are the lengths of the sides. Since A depends on two variables x and y, we cannot find the maximum until we eliminate one of the variables. We can do this because the variables are related: The rectangle has perimeter $L = 2x + 2y$, so $y = \frac{1}{2}L - x$. This allows us to rewrite the area in terms of x alone to obtain the objective function

$$A(x) = x\left(\frac{1}{2}L - x\right) = \frac{1}{2}Lx - x^2$$

On which interval does the optimization take place? The sides of the rectangle are non-negative, so we require both $x \geq 0$ and $\frac{1}{2}L - x \geq 0$. Thus, $0 \leq x \leq \frac{1}{2}L$. Our problem is to maximize $A(x)$ on the closed interval $\left[0, \frac{1}{2}L\right]$.

We solve $A'(x) = \frac{1}{2}L - 2x = 0$ to obtain the critical point $x = \frac{1}{4}L$ and compare:

Endpoints: $\qquad A(0) = 0$

$$A\left(\frac{1}{2}L\right) = \frac{1}{2}L\left(\frac{1}{2}L - \frac{1}{2}L\right) = 0$$

Critical point: $\quad A\left(\frac{1}{4}L\right) = \left(\frac{1}{4}L\right)\left(\frac{1}{2}L - \frac{1}{4}L\right) = \frac{1}{16}L^2$

The largest value occurs for $x = \frac{1}{4}L$, and in this case, $y = \frac{1}{2}L - \frac{1}{4}L = \frac{1}{4}L$. The rectangle of maximum area is the square of sides $x = y = \frac{1}{4}L$. ■

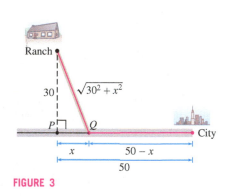

Ranch

30 $\qquad \sqrt{30^2 + x^2}$

P $\qquad Q$ $\qquad$ • City

$x \qquad 50 - x$

50

FIGURE 3

■ **EXAMPLE 2** Minimizing Travel Time Your task is to build a road joining a ranch to a highway that enables drivers to reach the city in the shortest time (Figure 3). How should this be done if the speed limit is 60 km/h on the road and 110 km/h on the highway? The perpendicular distance from the ranch to the highway is 30 km, and the city is 50 km down the highway.

Solution This problem is more complicated than the previous one, so we'll analyze it in three steps. You can follow these steps to solve other optimization problems.

Step 1. Choose variables.
We need to determine the point Q where the road will join the highway. So let x be the distance from Q to the point P where the perpendicular joins the highway.

Step 2. Find the objective function and the interval.
Our objective function is the time $T(x)$ of the trip as a function of x. To find a formula for $T(x)$, recall that distance traveled at constant velocity v is $d = vt$, and the *time* required to travel a distance d is $t = d/v$. The road has length $\sqrt{30^2 + x^2}$ by the Pythagorean Theorem, so at velocity $v = 60$ km/h, it takes

$$\frac{\sqrt{30^2 + x^2}}{60} \text{ hours to travel from the ranch to } Q$$

The strip of highway from Q to the city has length $50 - x$. At velocity $v = 110$ km/h, it takes

$$\frac{50 - x}{110} \text{ hours to travel from } Q \text{ to the city}$$

The total number of hours for the trip is

$$T(x) = \frac{\sqrt{30^2 + x^2}}{60} + \frac{50 - x}{110}$$

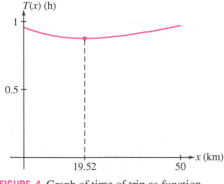

$T(x)$ (h)

1

0.5

$19.52 \qquad 50 \qquad x$ (km)

FIGURE 4 Graph of time of trip as function of x.

Our interval is $0 \leq x \leq 50$ because the road joins the highway somewhere between P and the city. So our task is to minimize T on $[0, 50]$ (Figure 4).

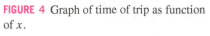

Step 3. **Optimize.**

Solve for the critical points:

$$T'(x) = \frac{x}{60\sqrt{30^2 + x^2}} - \frac{1}{110} = 0$$

$$110x = 60\sqrt{30^2 + x^2} \quad \Rightarrow \quad 11x = 6\sqrt{30^2 + x^2} \quad \Rightarrow$$

$$121x^2 = 36(30^2 + x^2) \quad \Rightarrow \quad 85x^2 = 32,400 \quad \Rightarrow \quad x = \sqrt{32,400/85} \approx 19.52$$

To find the minimum value of T, we compare the values of $T(x)$ at the critical point and the endpoints of $[0, 50]$:

$$T(0) \approx 0.95 \text{ h}, \qquad T(19.52) \approx 0.87 \text{ h}, \qquad T(50) \approx 0.97 \text{ h}$$

We conclude that the travel time is minimized if the road joins the highway at a distance $x \approx 19.52$ km along the highway from P. ∎

■ **EXAMPLE 3** **Optimal Price** All units in a 30-unit apartment building are rented out when the monthly rent is set at $r = \$1000$/month. A survey reveals that for each \$40 increase in rent, one additional apartment becomes vacant. Suppose that each occupied unit costs \$120/month in maintenance. Which rent r maximizes monthly profit?

Solution

Step 1. **Choose variables.**

Our goal is to maximize the total monthly profit P. Let r be the monthly rent and let $N(r)$ be the number of occupied units when the rent is set at r.

Step 2. **Find the objective function and the interval.**

Since one unit becomes vacant with each \$40 increase in rent above \$1000, we find that $(r - 1000)/40$ units are vacant when $r > 1000$. Therefore,

$$N(r) = 30 - \frac{1}{40}(r - 1000) = 55 - \frac{1}{40}r$$

Total monthly profit is equal to the number of occupied units times the profit per unit, which is $r - 120$ (because each unit costs \$120 in maintenance), so

$$P(r) = N(r)(r - 120) = \left(55 - \frac{1}{40}r\right)(r - 120) = -6600 + 58r - \frac{1}{40}r^2$$

Which interval of r-values should we consider? There is no reason to lower the rent below $r = 1000$ because all units are already occupied when $r = 1000$. On the other hand, $N(r) = 0$ for $r = 40 \cdot 55 = 2200$. Therefore, zero units are occupied when $r = 2200$ and it makes sense to take $1000 \le r \le 2200$.

Step 3. **Optimize.**

Solve for the critical points:

$$P'(r) = 58 - \frac{1}{20}r = 0 \quad \Rightarrow \quad r = 1160$$

and compare values at the critical point and the endpoints:

$$P(1000) = 26,400, \qquad P(1160) = 27,040, \qquad P(2200) = 0$$

We conclude that the profit is maximized when the rent is set at $r = \$1160$. In this case, 26 units are occupied. Note that if the maximum profit had occurred at a price that gave us a fractional number of units occupied, we could not have achieved that maximum. Instead, we would have taken the price corresponding to rounding the fractional number up or down to the integer number of units that maximized our profit. ∎

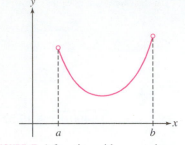

FIGURE 5 A function with no maximum over the open interval (a, b).

Open Versus Closed Intervals

When we have to optimize over an open interval, there is no guarantee that a min or max exists as for example in Figure 5 (unlike the case of closed intervals). However, if a min or max does exist, then it must occur at a critical point (because it is also a local min or max). Often, we can show that a min or max exists by examining $f(x)$ near the endpoints of the open interval. If $f(x)$ tends to infinity at the endpoints (as in Figure 7), then a minimum occurs at a critical point somewhere in the interval.

■ **EXAMPLE 4** Design a cylindrical can of volume 900 cm³ so that it uses the least amount of metal (Figure 6). In other words, minimize the surface area of the can (including its top and bottom).

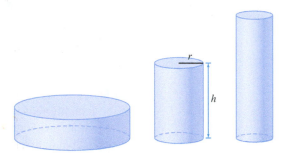

DF FIGURE 6 Cylinders with the same volume but different surface areas.

Solution

Step 1. Choose variables.

We must specify the can's radius and height. Therefore, let r be the radius and h the height. Let A be the surface area of the can.

Step 2. Find the objective function and the interval.

We compute A as a function of r and h:

$$A = \underbrace{\pi r^2}_{\text{Top}} + \underbrace{\pi r^2}_{\text{Bottom}} + \underbrace{2\pi rh}_{\text{Side}} = 2\pi r^2 + 2\pi rh$$

The can's volume is $V = \pi r^2 h$. Since we require that $V = 900$ cm³, we have the constraint equation $\pi r^2 h = 900$. Thus, $h = (900/\pi)r^{-2}$ and

$$A(r) = 2\pi r^2 + 2\pi r\left(\frac{900}{\pi r^2}\right) = 2\pi r^2 + \frac{1800}{r}$$

The radius r can take on any positive value, so we minimize $A(r)$ on $(0, \infty)$.

Step 3. Optimize the function.

Observe that $A(r)$ tends to infinity as r approaches the endpoints of $(0, \infty)$:

- $A(r) \to \infty$ as $r \to \infty$ (because of the r^2 term).
- $A(r) \to \infty$ as $r \to 0$ (because of the $1/r$ term).

Therefore, $A(r)$ must take on a minimum value at a critical point in $(0, \infty)$ (Figure 7). We solve in the usual way:

$$\frac{dA}{dr} = 4\pi r - \frac{1800}{r^2} = 0 \quad \Rightarrow \quad r^3 = \frac{450}{\pi} \quad \Rightarrow \quad r = \left(\frac{450}{\pi}\right)^{1/3} \approx 5.23 \text{ cm}$$

We also need to calculate the height:

$$h = \frac{900}{\pi r^2} = 2\left(\frac{450}{\pi}\right)r^{-2} = 2\left(\frac{450}{\pi}\right)\left(\frac{450}{\pi}\right)^{-2/3} = 2\left(\frac{450}{\pi}\right)^{1/3} \approx 10.46 \text{ cm}$$

Notice that the optimal dimensions satisfy $h = 2r$. In other words, the optimal can is as tall as it is wide. ■

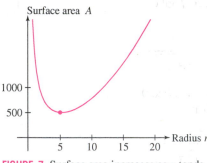

FIGURE 7 Surface area increases as r tends to 0 or ∞. The minimum value exists.

In the case of a single critical point, as we have here, a second method for proving that the point corresponds to a minimum is to apply the First Derivative Test. Since $A'(r) < 0$ for $r < \left(\frac{450}{\pi}\right)^{1/3}$ and $A'(r) > 0$ for $r > \left(\frac{450}{\pi}\right)^{1/3}$, the critical point must be a local minimum and as the only critical point, the global minimum. A third method would be to apply the Second Derivative Test to show this is a local minimum and therefore, as the only extreme point, the global minimum.

■ **EXAMPLE 5** Optimization Problem with No Solution Is it possible to design a cylinder of volume 900 cm^3 with the largest possible surface area?

Solution The answer is no. In the previous example, we showed that a cylinder of volume 900 and radius r has surface area

$$A(r) = 2\pi r^2 + \frac{1800}{r}$$

This function has no maximum value because it tends to infinity as $r \to 0$ or $r \to \infty$ (Figure 7). This means that a cylinder of fixed volume has a large surface area if it is either very fat and short (r large) or very tall and skinny (r small). ■

The **Principle of Least Distance** states that a light beam reflected in a mirror travels along the shortest path. More precisely, a beam traveling from A to B, as in Figure 8, is reflected at the point P for which the path APB has minimum length. In the next example, we show that this minimum occurs when *the angle of incidence is equal to the angle of reflection*, that is, $\theta_1 = \theta_2$.

*The Principle of Least Distance is also called **Heron's Principle** after the mathematician Heron of Alexandria (c. 100 CE). See Exercise 79 for an elementary proof that does not use calculus and would have been known to Heron. Exercise 54 develops Snell's Law, a more general optical law based on the Principle of Least Time.*

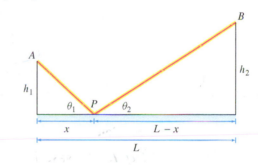

DF **FIGURE 8** Reflection of a light beam in a mirror.

■ **EXAMPLE 6** Show that if P is the point for which the path APB in Figure 8 has minimal length, then $\theta_1 = \theta_2$.

Solution By the Pythagorean Theorem, the path APB has length

$$f(x) = AP + PB = \sqrt{x^2 + h_1^2} + \sqrt{(L-x)^2 + h_2^2}$$

with x, h_1, and h_2 as in the figure. The function f tends to infinity as x approaches $\pm\infty$ (i.e., as P moves arbitrarily far to the right or left), so f takes on its minimum value at a critical point x such that (see Figure 9)

$$f'(x) = \frac{x}{\sqrt{x^2 + h_1^2}} - \frac{L-x}{\sqrt{(L-x)^2 + h_2^2}} = 0 \qquad \boxed{1}$$

It is not necessary to solve for x because our goal is not to find the critical point, but rather to show that $\theta_1 = \theta_2$. To do this, we rewrite Eq. (1) as

$$\underbrace{\frac{x}{\sqrt{x^2 + h_1^2}}}_{\cos\theta_1} = \underbrace{\frac{L-x}{\sqrt{(L-x)^2 + h_2^2}}}_{\cos\theta_2}$$

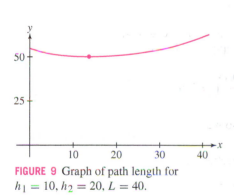

FIGURE 9 Graph of path length for $h_1 = 10$, $h_2 = 20$, $L = 40$.

Referring to Figure 8, we see that this equation says $\cos\theta_1 = \cos\theta_2$, and since θ_1 and θ_2 lie between 0 and $\frac{\pi}{2}$, we conclude that $\theta_1 = \theta_2$ as claimed. ■

CONCEPTUAL INSIGHT The examples in this section were selected because they lead to optimization problems where the min or max occurs at a critical point. Often, the critical point represents the best compromise between "competing factors." In Example 3, we maximized profit by finding the best compromise between raising the rent and keeping the apartment units occupied. In Example 4, our solution minimizes surface area by finding the best compromise between height and width. In daily life, however, we often encounter endpoint rather than critical point solutions. For example, to run 10 m in minimal time, you should run as fast as you can—the solution is not a critical point but rather an endpoint (your maximum speed).

4.7 SUMMARY

- There are usually three main steps in solving an applied optimization problem:

Step 1. Choose variables.
Determine which quantities are relevant, often by drawing a diagram, and assign appropriate variables.

Step 2. Find the objective function and the interval.
Restate as an optimization problem for a function f over an interval. If f depends on more than one variable, use a *constraint equation* to write f as a function of just one variable.

Step 3. Optimize the objective function.

- If the interval is open, f does not necessarily take on a minimum or maximum value. But if it does, these must occur at critical points within the interval. To determine if a min or max exists, analyze the behavior of f as x approaches the endpoints of the interval.

4.7 EXERCISES

Preliminary Questions

1. The problem is to find the right triangle of perimeter 10 whose area is as large as possible. What is the constraint equation relating the base b and height h of the triangle?

2. Describe a way of showing that a continuous function on an open interval (a, b) has a minimum value.

3. Is there a rectangle of area 100 of largest perimeter? Explain.

Exercises

1. Find the dimensions x and y of the rectangle of maximum area that can be formed using 3 m of wire.

(a) What is the constraint equation relating x and y?

(b) Find a formula for the area in terms of x alone.

(c) What is the interval of optimization? Is it open or closed?

(d) Solve the optimization problem.

2. Wire of length 12 m is divided into two pieces and each piece is bent into a square. How should this be done in order to minimize the sum of the areas of the two squares?

(a) Express the sum of the areas of the squares in terms of the lengths x and y of the two pieces.

(b) What is the constraint equation relating x and y?

(c) What is the interval of optimization? Is it open or closed?

(d) Solve the optimization problem.

3. A rectangular bird sanctuary is being created with one side along a straight riverbank. The remaining three sides are to be enclosed with a protective fence. If there are 12 km of fence available, find the dimension of the rectangle to maximize the area of the sanctuary.

4. The rectangular bird sanctuary with one side along a straight river is to be constructed so that it contains 8 km^2 of area. Find the dimensions of the rectangle to minimize the amount of fence necessary to enclose the remaining three sides.

5. Find two positive real numbers such that the sum of the first number squared and the second number is 48 and their product is a maximum.

6. Find two positive real numbers such that they sum to 108 and the product of the first times the square of the second is a maximum.

7. A wire of length 12 m is divided into two pieces and the pieces are bent into a square and a circle. How should this be done in order to minimize the sum of their areas?

8. Find the positive number x such that the sum of x and its reciprocal is as small as possible. Does this problem require optimization over an open interval or a closed interval?

9. Find two positive real numbers such that they add to 40 and their product is as large as possible.

10. Find two positive real numbers x and y such that they add to 120 and $x^2 y$ is as large as possible.

11. Find two positive real numbers x and y such that their product is 800 and $x + 2y$ is as small as possible.

12. A flexible tube of length 4 m is bent into an L-shape. Where should the bend be made to minimize the distance between the two ends?

13. Find the dimensions of the box with square base with:

(a) Volume 12 and the minimal surface area.

(b) Surface area 20 and maximal volume.

14. A jewelry box with a square base is to be built with copper plated sides, nickel plated bottom and top, and a volume of 40 cm^3. If nickel plating costs \$2 per cm^2 and copper plating costs \$1 per cm^2, find the dimensions of the box to minimize the cost of the materials.

15. A rancher will use 600 m of fencing to build a corral in the shape of a semicircle on top of a rectangle (Figure 10). Find the dimensions that maximize the area of the corral.

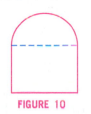

FIGURE 10

16. What is the maximum area of a rectangle inscribed in a right triangle with legs of length 3 and 4 as in Figure 11. The sides of the rectangle are parallel to the legs of the triangle.

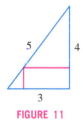

FIGURE 11

17. Find the dimensions of the rectangle of maximum area that can be inscribed in a circle of radius $r = 4$ (Figure 12).

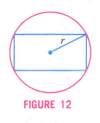

FIGURE 12

18. Find the dimensions x and y of the rectangle inscribed in a circle of radius r that maximizes the quantity xy^2.

19. In the article "Do Dogs Know Calculus?" the author Timothy Pennings explained how he noticed that when he threw a ball diagonally into Lake Michigan along a straight shoreline, his dog Elvis seemed to pick the optimal point in which to enter the water so as to minimize his time to reach the ball, as in Figure 13. He timed the dog and found Elvis could run at 6.4 m/s on the sand and swim at 0.91 m/s. If Tim stood at point A and threw the ball to a point B in the water, which was a perpendicular distance 10 m from point C on the shore, where C is a distance 15 m from where he stood, at what distance x from point C did Elvis enter the water if the dog effectively minimized his time to reach the ball?

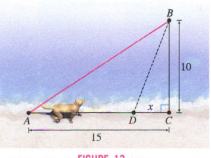

FIGURE 13

20. A four-wheel-drive vehicle is transporting an injured hiker to the hospital from a point that is 30 km from the nearest point on a straight road. The hospital is 50 km down that road from that nearest point. If the vehicle can drive at 30 kph over the terrain and at 120 kph on the road, how far down the road should the vehicle aim to reach the road to minimize the time it takes to reach the hospital?

21. Find the point on the line $y = x$ closest to the point $(1, 0)$. *Hint:* It is equivalent and easier to minimize the *square* of the distance.

22. Find the point P on the parabola $y = x^2$ closest to the point $(3, 0)$ (Figure 14).

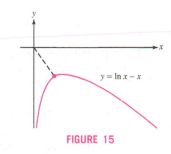

FIGURE 14

23. **CAS** Find a good numerical approximation to the coordinates of the point on the graph of $y = \ln x - x$ closest to the origin (Figure 15).

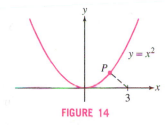

FIGURE 15

24. **Problem of Tartaglia (1500–1557)** Among all positive numbers a, b whose sum is 8, find those for which the product of the two numbers and their difference is largest.

25. Find the angle θ that maximizes the area of the isosceles triangle whose legs have length ℓ (Figure 16), using the fact the area is given by $A = \frac{1}{2}\ell^2 \sin\theta$.

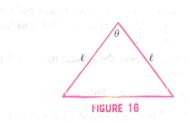

FIGURE 16

26. A right circular cone (Figure 17) has volume $V = \frac{\pi}{3}r^2 h$ and surface area $S = \pi r\sqrt{r^2 + h^2}$. Find the dimensions of the cone with surface area 1 and maximal volume.

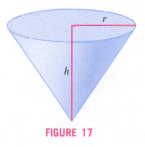

FIGURE 17

27. Find the area of the largest isosceles triangle that can be inscribed in a circle of radius 1 (Figure 18).

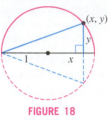

FIGURE 18

28. Find the radius and height of a cylindrical can of total surface area A whose volume is as large as possible. Does there exist a cylinder of surface area A and minimal total volume?

29. A poster of area 6000 cm^2 has blank margins of width 10 cm on the top and bottom and 6 cm on the sides. Find the dimensions that maximize the printed area.

30. According to postal regulations, a carton is classified as "oversized" if the sum of its height and girth (perimeter of its base) exceeds 108 in. Find the dimensions of a carton with a square base that is not oversized and has maximum volume.

31. Kepler's Wine Barrel Problem   In his work *Nova stereometria doliorum vinariorum* (New Solid Geometry of a Wine Barrel), published in 1615, astronomer Johannes Kepler stated and solved the following problem: Find the dimensions of the cylinder of largest volume that can be inscribed in a sphere of radius R. *Hint:* Show that an inscribed cylinder has volume $2\pi x(R^2 - x^2)$, where x is one-half the height of the cylinder.

32. Find the angle θ that maximizes the area of the trapezoid with a base of length 4 and sides of length 2, as in Figure 19.

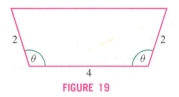

FIGURE 19

33. A landscape architect wishes to enclose a rectangular garden of area 1000 m^2 on one side by a brick wall costing \$90/m and on the other three sides by a metal fence costing \$30/m. Which dimensions minimize the total cost?

34. The amount of light reaching a point at a distance r from a light source A of intensity I_A is I_A/r^2. Suppose that a second light source B of intensity $I_B = 4I_A$ is located 10 m from A. Find the point on the segment joining A and B where the total amount of light is at a minimum.

35. Find the maximum area of a rectangle inscribed in the region bounded by the graph of $y = \dfrac{4-x}{2+x}$ and the axes (Figure 20).

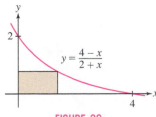

FIGURE 20

36. Find the maximum area of a triangle formed by the axes and a tangent line to the graph of $y = (x+1)^{-2}$ with $x > 0$.

37. Find the maximum area of a rectangle circumscribed around a rectangle of sides L and H. *Hint:* Express the area in terms of the angle θ (Figure 21).

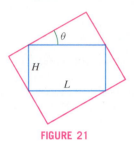

FIGURE 21

38. A contractor is engaged to build steps up the slope of a hill that has the shape of the graph of $y = \dfrac{x^2(120 - x)}{6400}$ for $0 \le x \le 80$ with x in meters (Figure 22). What is the maximum vertical rise of a stair if each stair has a horizontal length of $\frac{1}{3}$ m.

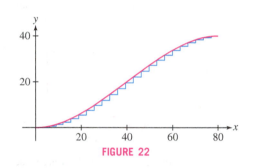

FIGURE 22

39. Find the equation of the line through $P = (4, 12)$ such that the triangle bounded by this line and the axes in the first quadrant has minimal area.

40. Let $P = (a, b)$ lie in the first quadrant. Find the slope of the line through P such that the triangle bounded by this line and the axes in the first quadrant has minimal area. Then show that P is the midpoint of the hypotenuse of this triangle.

41. Archimedes's Problem   A spherical cap (Figure 23) of radius r and height h has volume $V = \pi h^2(r - \frac{1}{3}h)$ and surface area $S = 2\pi rh$. Prove that the hemisphere encloses the largest volume among all spherical caps of fixed surface area S.

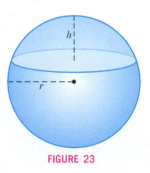

FIGURE 23

42. Find the isosceles triangle of smallest area (Figure 24) that circumscribes a circle of radius 1 (from Thomas Simpson's *The Doctrine and Application of Fluxions*, a calculus text that appeared in 1750).

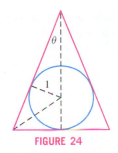

FIGURE 24

43. A box of volume $72 \, \text{m}^3$ with a square bottom and no top is constructed out of two different materials. The cost of the bottom is $40/m^2$ and the cost of the sides is $30/m^2$. Find the dimensions of the box that minimize total cost.

44. Find the dimensions of a cylinder of volume $1 \, \text{m}^3$ of minimal cost if the top and bottom are made of material that costs twice as much as the material for the side.

45. Your task is to design a rectangular industrial warehouse consisting of three separate spaces of equal size as in Figure 25. The wall materials cost $500 per linear meter and your company allocates $2,400,000 for that part of the project involving the walls.

(a) Which dimensions maximize the area of the warehouse?

(b) What is the area of each compartment in this case?

FIGURE 25

46. Suppose, in the previous exercise, that the warehouse consists of n separate spaces of equal size. Find a formula in terms of n for the maximum possible area of the warehouse.

47. According to a model developed by economists E. Heady and J. Pesek, if fertilizer made from N pounds of nitrogen and P pounds of phosphate is used on an acre of farmland, then the yield of corn (in bushels per acre) is

$$Y = 7.5 + 0.6N + 0.7P - 0.001N^2 - 0.002P^2 + 0.001NP$$

A farmer intends to spend $30/acre on fertilizer. If nitrogen costs 25 cents/lb and phosphate costs 20 cents/lb, which combination of N and P produces the highest yield of corn?

48. Experiments show that the quantities x of corn and y of soybean required to produce a hog of weight Q satisfy $Q = 0.5x^{1/2}y^{1/4}$. The unit of x, y, and Q is the cwt, an agricultural unit equal to 100 lb. Find the values of x and y that minimize the cost of a hog of weight $Q = 2.5$ cwt if corn costs $3/cwt and soy costs $7/cwt.

49. All units in a 100-unit apartment building are rented out when the monthly rent is set at $r = \$900$/month. Suppose that one unit becomes vacant with each $10 increase in rent and that each occupied unit costs $80/month in maintenance. Which rent r maximizes monthly profit?

50. An 8-billion-bushel corn crop brings a price of $2.40/bushel. A commodity broker uses the rule of thumb: If the crop is reduced by x percent, then the price increases by $10x$ cents. Which crop size results in maximum revenue and what is the price per bushel? *Hint:* Revenue is equal to price times crop size.

51. The monthly output of a Spanish light bulb factory is $P = 2LK^2$ (in millions), where L is the cost of labor and K is the cost of equipment (in millions of euros). The company needs to produce 1.7 million units per month. Which values of L and K would minimize the total cost $L + K$?

52. The rectangular plot in Figure 26 has size 100 m $\times$ 200 m. Pipe is to be laid from A to a point P on side BC and from there to C. The cost of laying pipe along the side of the plot is $45/m and the cost through the plot is $80/m (since it is underground).

(a) Let $f(x)$ be the total cost, where x is the distance from P to B. Determine $f(x)$, but note that f is discontinuous at $x = 0$ (when $x = 0$, the cost of the entire pipe is $45/m).

(b) What is the most economical way to lay the pipe? What if the cost along the sides is $65/m?

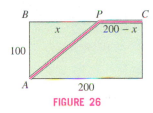

FIGURE 26

53. Brandon is on one side of a river that is 50 m wide and wants to reach a point 200 m downstream on the opposite side as quickly as possible by swimming diagonally across the river and then running the rest of the way. Find the best route if Brandon can swim at 1.5 m/s and run at 4 m/s.

54. Snell's Law When a light beam travels from a point A above a swimming pool to a point B below the water (Figure 27), it chooses the path that takes the *least time*. Let v_1 be the velocity of light in air and v_2 the velocity in water (it is known that $v_1 > v_2$). Prove Snell's Law of Refraction:

$$\frac{\sin \theta_1}{v_1} = \frac{\sin \theta_2}{v_2}$$

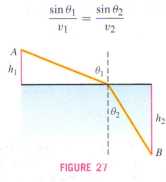

FIGURE 27

55. Vascular Branching A small blood vessel of radius r branches off at an angle θ from a larger vessel of radius R to supply blood along a path from A to B. According to Poiseuille's Law, the total resistance to blood flow is proportional to

$$T = \left(\frac{a - b \cot \theta}{R^4} + \frac{b \csc \theta}{r^4} \right)$$

where a and b are as in Figure 28. Show that the total resistance is minimized when $\cos \theta = (r/R)^4$.

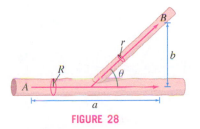

FIGURE 28

In Exercises 56–58, a box (with no top) is to be constructed from a piece of cardboard with sides of length A and B by cutting out squares of length h from the corners and folding up the sides (Figure 29).

56. Find the value of h that maximizes the volume of the box if $A = 15$ and $B = 24$. What are the dimensions of this box?

57. Which values of A and B maximize the volume of the box if $h = 10$ cm and $AB = 900$ cm^2.

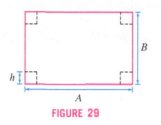

FIGURE 29

58. Which value of h maximizes the volume of the box if $A = B$?

59. Given n numbers $x_1, \ldots, x_n$, find the value of x minimizing the sum of the squares:

$$(x - x_1)^2 + (x - x_2)^2 + \cdots + (x - x_n)^2$$

First solve for $n = 2, 3$ and then try it for arbitrary n.

60. A billboard of height b is mounted on the side of a building with its bottom edge at a distance h from the street as in Figure 30. At what distance x should an observer stand from the wall to maximize the angle of observation θ?

61. Solve Exercise 60 again using geometry rather than calculus. There is a unique circle passing through points B and C that is tangent to the street. Let R be the point of tangency. Note that the two angles labeled ψ in Figure 30 are equal because they subtend equal arcs on the circle.

(a) Show that the maximum value of θ is $\theta = \psi$. *Hint:* Show that $\psi = \theta + \angle PBA$, where A is the intersection of the circle with PC.

(b) Prove that this agrees with the answer to Exercise 60.

(c) Show that $\angle QRB = \angle RCQ$ for the maximal angle ψ.

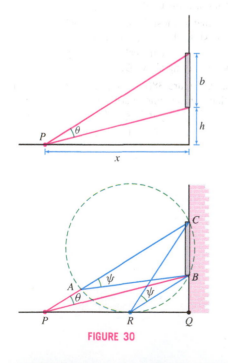

FIGURE 30

62. Optimal Delivery Schedule   A gas station sells Q gallons of gasoline per year, which is delivered N times per year in equal shipments of Q/N gallons. The cost of each delivery is d dollars and the yearly storage costs are sQT, where T is the length of time (a fraction of a year) between shipments and s is a constant. Show that costs are minimized for $N = \sqrt{sQ/d}$. *(Hint: $T = 1/N$.)* Find the optimal number of deliveries if $Q = 2$ million gal, $d = \$8000$, and $s = 30$ cents/gal-year. Your answer should be a whole number, so compare costs for the two integer values of N nearest the optimal value.

63. Victor Klee's Endpoint Maximum Problem   Given 40 m of straight fence, your goal is to build a rectangular enclosure using 80 additional meters of fence that encompasses the greatest area. Let $A(x)$ be the area of the enclosure, with x as in Figure 31.

(a) Find the maximum value of $A(x)$.

(b) Which interval of x values is relevant to our problem? Find the maximum value of $A(x)$ on this interval.

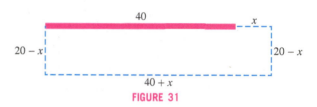

FIGURE 31

64. Let (a, b) be a fixed point in the first quadrant and let $S(d)$ be the sum of the distances from $(d, 0)$ to the points $(0, 0)$, (a, b), and $(a, -b)$.

(a) Find the value of d for which $S(d)$ is minimal. The answer depends on whether $b < \sqrt{3}a$ or $b \geq \sqrt{3}a$. *Hint:* Show that $d = 0$ when $b \geq \sqrt{3}a$.

(b) ⬚GU⬚ Let $a = 1$. Plot S for $b = 0.5, \sqrt{3}, 3$ and describe the position of the minimum.

65. The force F (in Newtons) required to move a box of mass m kg in motion by pulling on an attached rope (Figure 32) is

$$F(\theta) = \frac{fmg}{\cos\theta + f\sin\theta}$$

where θ is the angle between the rope and the horizontal, f is the coefficient of static friction, and $g = 9.8$ m/s^2. Find the angle θ that minimizes the required force F, assuming $f = 0.4$. *Hint:* Find the maximum value of $\cos\theta + f\sin\theta$.

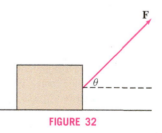

FIGURE 32

66. In the setting of Exercise 65, show that for any f the minimal force required is proportional to $1/\sqrt{1 + f^2}$.

67. Bird Migration   Ornithologists have found that the power (in joules per second) consumed by a certain pigeon flying at velocity v m/s is described well by the function $P(v) = 17v^{-1} + 10^{-3}v^3$ joules/s. Assume that the pigeon can store 5×10^4 joules of usable energy as body fat.

(a) Show that at velocity v, a pigeon can fly a total distance of $D(v) = (5 \times 10^4)v/P(v)$ if it uses all of its stored energy.

(b) Find the velocity v_p that *minimizes* P.

(c) Migrating birds are smart enough to fly at the velocity that maximizes distance traveled rather than minimizes power consumption. Show that the velocity v_d which maximizes $D(v)$ satisfies $P'(v_d) = P(v_d)/v_d$. Show that v_d is obtained graphically as the velocity coordinate of the point where a line through the origin is tangent to the graph of P (Figure 33).

(d) Find v_d and the maximum distance $D(v_d)$.

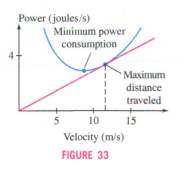

FIGURE 33

68. The problem is to put a "roof" of side s on an attic room of height h and width b. Find the smallest length s for which this is possible if $b = 27$ and $h = 8$ (Figure 34).

69. Redo Exercise 68 for arbitrary b and h.

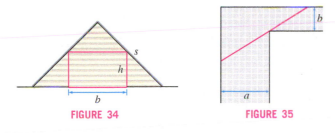

FIGURE 34   **FIGURE 35**

70. Find the maximum length of a pole that can be carried horizontally around a corner joining corridors of widths $a = 24$ and $b = 3$ (Figure 35).

71. Redo Exercise 70 for arbitrary widths a and b.

72. Find the minimum length ℓ of a beam that can clear a fence of height h and touch a wall located b ft behind the fence (Figure 36).

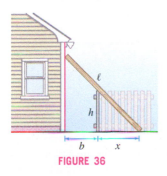

FIGURE 36

73. A basketball player stands d feet from the basket. Let h and α be as in Figure 37. Using physics, one can show that if the player releases the ball at an angle θ, then the initial velocity required to make the ball go through the basket satisfies

$$v^2 = \frac{16d}{\cos^2 \theta (\tan \theta - \tan \alpha)}$$

(a) Explain why this formula is meaningful only for $\alpha < \theta < \frac{\pi}{2}$. Why does v approach infinity at the endpoints of this interval?

(b) GU   Take $\alpha = \frac{\pi}{6}$ and plot v^2 as a function of θ for $\frac{\pi}{6} < \theta < \frac{\pi}{2}$. Verify that the minimum occurs at $\theta = \frac{\pi}{3}$.

(c) Set $F(\theta) = \cos^2 \theta (\tan \theta - \tan \alpha)$. Explain why v is minimized for θ such that $F(\theta)$ is maximized.

(d) Verify that $F'(\theta) = \cos(\alpha - 2\theta) \sec \alpha$ (you will need to use the addition formula for cosine) and show that the maximum value of F on $\left[\alpha, \frac{\pi}{2}\right]$ occurs at $\theta_0 = \frac{\alpha}{2} + \frac{\pi}{4}$.

(e) For a given α, the optimal angle for shooting the basket is θ_0 because it minimizes v^2 and therefore minimizes the energy required to make the shot (energy is proportional to v^2). Show that the velocity v_{opt} at the optimal angle θ_0 satisfies

$$v_{\text{opt}}^2 = \frac{32d \cos \alpha}{1 - \sin \alpha} = \frac{32 d^2}{-h + \sqrt{d^2 + h^2}}$$

(f) GU   Show with a graph that for fixed d (say, $d = 15$ ft, the distance of a free throw), v_{opt}^2 is an increasing function of h. Use this to explain why taller players have an advantage and why it can help to jump while shooting.

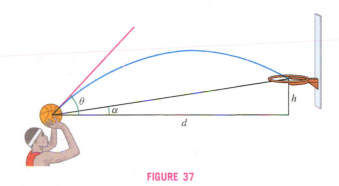

FIGURE 37

74. Three towns A, B, and C are to be joined by an underground fiber cable as illustrated in Figure 38(A). Assume that C is located directly below the midpoint of $\overline{AB}$. Find the junction point P that minimizes the total amount of cable used.

(a) First show that P must lie directly above C. *Hint:* Use the result of Example 6 to show that if the junction is placed at point Q in Figure 38(B), then we can reduce the cable length by moving Q horizontally over to the point P lying above C.

(b) With x as in Figure 38(A), let $f(x)$ be the total length of cable used. Show that f has a unique critical point c. Compute c and show that $0 \le c \le L$ if and only if $D \le 2\sqrt{3} L$.

(c) Find the minimum of f on $[0, L]$ in two cases: $D = 2, L = 4$ and $D = 8, L = 2$.

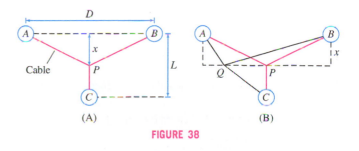

(A)   **(B)**

FIGURE 38

Further Insights and Challenges

75. Tom and Ali drive along a highway represented by the graph of f in Figure 39. During the trip, Ali views a billboard represented by the segment $\overline{BC}$ along the y-axis. Let Q be the y-intercept of the tangent line to $y = f(x)$. Show that θ is maximized at the value of x for which the angles $\angle QPB$ and $\angle QCP$ are equal. This generalizes Exercise 61 (c) [which corresponds to the case $f(x) = 0$]. *Hints:*

(a) Show that $d\theta/dx$ is equal to

$$(b - c) \cdot \frac{(x^2 + (xf'(x))^2) - (b - (f(x) - xf'(x)))(c - (f(x) - xf'(x)))}{(x^2 + (b - f(x))^2)(x^2 + (c - f(x))^2)}$$

(b) Show that the y-coordinate of Q is $f(x) - xf'(x)$.

(c) Show that the condition $d\theta/dx = 0$ is equivalent to

$$PQ^2 = BQ \cdot CQ$$

(d) Conclude that $\triangle QPB$ and $\triangle QCP$ are similar triangles.

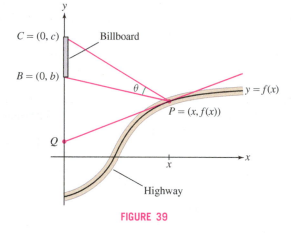

FIGURE 39

Seismic Prospecting *Exercises 76–78 are concerned with determining the thickness d of a layer of soil that lies on top of a rock formation. Geologists send two sound pulses from point A to point D separated by a distance s. The first pulse travels directly from A to D along the surface of the earth. The second pulse travels down to the rock formation, then along its surface, and then back up to D (path ABCD), as in Figure 40. The pulse travels with velocity v_1 in the soil and v_2 in the rock.*

76. (a) Show that the time required for the first pulse to travel from A to D is $t_1 = s/v_1$.

(b) Show that the time required for the second pulse is

$$t_2 = \frac{2d}{v_1} \sec \theta + \frac{s - 2d \tan \theta}{v_2}$$

provided that

$$\tan \theta \le \frac{s}{2d} \qquad \boxed{2}$$

(*Note:* If this inequality is not satisfied, then point B does not lie to the left of C.)

(c) Show that t_2 is minimized when $\sin \theta = v_1/v_2$.

77. In this exercise, assume that $v_2/v_1 \ge \sqrt{1 + 4(d/s)^2}$.

(a) Show that inequality (2) holds if $\sin \theta = v_1/v_2$.

(b) Show that the minimal time for the second pulse is

$$t_2 = \frac{2d}{v_1}(1 - k^2)^{1/2} + \frac{s}{v_2}$$

where $k = v_1/v_2$.

(c) Conclude that $\dfrac{t_2}{t_1} = \dfrac{2d(1 - k^2)^{1/2}}{s} + k$.

78. Continue with the assumption of the previous exercise.

(a) Find the thickness of the soil layer, assuming that $v_1 = 0.7v_2$, $t_2/t_1 = 1.3$, and $s = 400$ m.

(b) The times t_1 and t_2 are measured experimentally. The equation in Exercise 77(c) shows that t_2/t_1 is a linear function of $1/s$. What might you conclude if experiments were formed for several values of s and the points $(1/s, t_2/t_1)$ did *not* lie on a straight line?

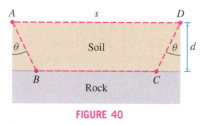

FIGURE 40

79. In this exercise, we use Figure 41 to prove Heron's principle of Example 6 without calculus. By definition, C is the reflection of B across the line $\overline{MN}$ (so that $\overline{BC}$ is perpendicular to $\overline{MN}$ and $BN = CN$). Let P be the intersection of $\overline{AC}$ and $\overline{MN}$. Use geometry to justify the following:

(a) $\triangle PNB$ and $\triangle PNC$ are congruent and $\theta_1 = \theta_2$.

(b) The paths APB and APC have equal length.

(c) Similarly AQB and AQC have equal length.

(d) The path APC is shorter than AQC for all $Q \ne P$.

Conclude that the shortest path AQB occurs for $Q = P$.

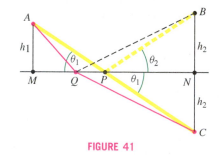

FIGURE 41

80. A jewelry designer plans to incorporate a component made of gold in the shape of a frustum of a cone of height 1 cm and fixed lower radius r (Figure 42). The upper radius x can take on any value between 0 and r. Note that $x = 0$ and $x = r$ correspond to a cone and cylinder, respectively. As a function of x, the surface area (not including the top and bottom) is $S(x) = \pi s(r + x)$, where s is the *slant height* as indicated in the figure. Which value of x yields the least expensive design [the minimum value of $S(x)$ for $0 \le x \le r$]?

(a) Show that $S(x) = \pi(r + x)\sqrt{1 + (r - x)^2}$.

(b) Show that if $r < \sqrt{2}$, then S is an increasing function. Conclude that the cone ($x = 0$) has minimal area in this case.

(c) Assume that $r > \sqrt{2}$. Show that S has two critical points $x_1 < x_2$ in $(0, r)$, and that $S(x_1)$ is a local maximum, and $S(x_2)$ is a local minimum.

(d) Conclude that the minimum occurs at $x = 0$ or x_2.

(e) Find the minimum in the cases $r = 1.5$ and $r = 2$.

(f) Challenge: Let $c = \sqrt{(5 + 3\sqrt{3})/4} \approx 1.597$. Prove that the minimum occurs at $x = 0$ (cone) if $\sqrt{2} < r < c$, but the minimum occurs at $x = x_2$ if $r > c$.

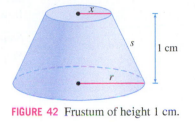

FIGURE 42 Frustum of height 1 cm.

4.8 Newton's Method

◄·· REMINDER A "zero" or "root" of a function f is a solution of the equation $f(x) = 0$.

Newton's Method is a procedure for finding numerical approximations to zeros of functions. Numerical approximations are important because it is often impossible to find the zeros exactly. For example, the polynomial $f(x) = x^5 - x - 1$ has one real root c (see Figure 1), but we can prove, using an advanced branch of mathematics called *Galois Theory*, that there is no algebraic formula for this root. Newton's Method shows that $c \approx 1.1673$, and with enough computation, we can compute c to any desired degree of accuracy.

In Newton's Method, we begin by choosing a number x_0, which we believe is close to a root of the equation $f(x) = 0$. This starting value x_0 is called the **initial guess**. Newton's Method then produces a sequence $x_0, x_1, x_2, \ldots$ of successive approximations that, in favorable situations, converge to a root.

Figure 2 illustrates the procedure. Given an initial guess x_0, we draw the tangent line to the graph at $(x_0, f(x_0))$. The approximation x_1 is defined as the x-coordinate of the point where the tangent line intersects the x-axis. To produce the second approximation x_2 (also called the second iterate), we apply this procedure to x_1.

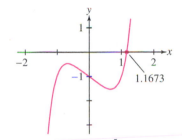

FIGURE 1 Graph of $y = x^5 - x - 1$. The value 1.1673 is a good numerical approximation to the root.

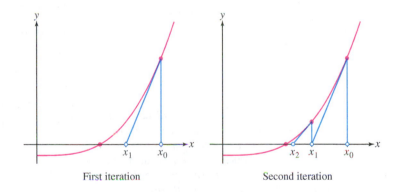

First iteration   Second iteration

DF FIGURE 2 The sequence produced by iteration converges to a root.

Let's derive a formula for x_1. The tangent line at $(x_0, f(x_0))$ has equation

$$y = f(x_0) + f'(x_0)(x - x_0)$$

The tangent line crosses the x-axis at x_1, where

$$y = f(x_0) + f'(x_0)(x_1 - x_0) = 0$$

If $f'(x_0) \neq 0$, we can solve for x_1 to obtain $x_1 - x_0 = -f(x_0)/f'(x_0)$, or

$$x_1 = x_0 - \frac{f(x_0)}{f'(x_0)}$$

The second iterate x_2 is obtained by applying this formula to x_1 instead of x_0:

$$x_2 = x_1 - \frac{f(x_1)}{f'(x_1)}$$

and so on. Notice in Figure 2 that x_1 is closer to the root than x_0 and that x_2 is closer still. This is typical: The successive approximations usually converge to the actual root. However, there are cases where Newton's Method fails (see Figure 4).

Newton's Method is an example of an iterative procedure. To "iterate" means to repeat, and in Newton's Method, we use Eq. (1) repeatedly to produce the sequence of approximations.

Newton's Method   To approximate a root of $f(x) = 0$:

Step 1. Choose an initial guess x_0 (close to the desired root if possible).

Step 2. Generate successive approximations $x_1, x_2, \ldots$, where

$$x_{n+1} = x_n - \frac{f(x_n)}{f'(x_n)} \qquad \boxed{1}$$

■ **EXAMPLE 1** Approximating $\sqrt{5}$ Calculate the first three approximations x_1, x_2, x_3 to a root of $f(x) = x^2 - 5$ using the initial guess $x_0 = 2$.

Solution We have $f'(x) = 2x$. Therefore,

$$x_1 = x_0 - \frac{f(x_0)}{f'(x_0)} = x_0 - \frac{x_0^2 - 5}{2x_0}$$

We compute the successive approximations as follows:

$$x_1 = x_0 - \frac{f(x_0)}{f'(x_0)} = 2 - \frac{2^2 - 5}{2 \cdot 2} \qquad = 2.25$$

$$x_2 = x_1 - \frac{f(x_1)}{f'(x_1)} = 2.25 - \frac{2.25^2 - 5}{2 \cdot 2.25} \qquad \approx 2.23611$$

$$x_3 = x_2 - \frac{f(x_2)}{f'(x_2)} = 2.23611 - \frac{2.23611^2 - 5}{2 \cdot 2.23611} \approx \mathbf{2.23606797789}$$

This sequence provides successive approximations to a root of $x^2 - 5 = 0$, namely

$$\sqrt{5} = \mathbf{2.236067977}499789696\ldots$$

Observe that x_3 is accurate to within an error of less than 10^{-9}. This is impressive accuracy for just three iterations of Newton's Method. ■

How Many Iterations Are Required?

How many iterations of Newton's Method are required to approximate a root to within a given accuracy? There is no definitive answer, but in practice, it is usually safe to assume that if x_n and x_{n+1} agree to m decimal places, then the approximation x_n is correct to these m places.

■ **EXAMPLE 2** $\boxed{\text{GU}}$ Let c be the smallest positive solution of $\sin 3x = \cos x$.

(a) Use a computer-generated graph to choose an initial guess x_0 for c.

(b) Use Newton's Method to approximate c to within an error of at most 10^{-6}.

Solution

(a) A solution of $\sin 3x = \cos x$ is a zero of the function $f(x) = \sin 3x - \cos x$. Figure 3 shows that the smallest zero is approximately halfway between 0 and $\frac{\pi}{4}$. Because $\frac{\pi}{4} \approx 0.785$, a good initial guess is $x_0 = 0.4$.

(b) Since $f'(x) = 3\cos 3x + \sin x$, Eq. (1) yields the formula

$$x_{n+1} = x_n - \frac{\sin 3x_n - \cos x_n}{3\cos 3x_n + \sin x_n}$$

With $x_0 = 0.4$ as the initial guess, the first four iterates are

$$x_1 \approx \mathbf{0.3925}647447$$

$$x_2 \approx \mathbf{0.3926990}382$$

$$x_3 \approx \mathbf{0.3926990816987}196$$

$$x_4 \approx \mathbf{0.3926990816987}241$$

Stopping here, we can be fairly confident that x_4 approximates the smallest positive root c to at least 12 places. In fact, $c = \frac{\pi}{8}$ and x_4 is accurate to 16 places. ■

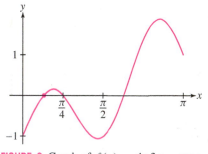

FIGURE 3 Graph of $f(x) = \sin 3x - \cos x$.

There is no single "correct" initial guess. In Example 2, we chose $x_0 = 0.4$, but another possible choice is $x_0 = 0$, leading to the sequence

$$x_1 \approx 0.3333333333$$

$$x_2 \approx 0.3864547725$$

$$x_3 \approx 0.3926082513$$

$$x_4 \approx 0.3926990816$$

You can check, however, that $x_0 = 1$ yields a sequence converging to $\frac{\pi}{4}$, which is the second positive solution of $\sin 3x = \cos x$.

Which Root Does Newton's Method Compute?

Sometimes, Newton's Method computes no root at all. In Figure 4, the iterates diverge to infinity. In practice, however, Newton's Method usually converges quickly, and if a particular choice of x_0 does not lead to a root, the best strategy is to try a different initial guess, consulting a graph if possible. If $f(x) = 0$ has more than one root, different initial guesses x_0 may lead to different roots.

■ **EXAMPLE 3** Figure 5 shows that $f(x) = x^4 - 6x^2 + x + 5$ has four real roots.

(a) Show that with $x_0 = 0$, Newton's Method converges to the root near -2.

(b) Show that with $x_0 = -1$, Newton's Method converges to the root near -1.

Solution We have $f'(x) = 4x^3 - 12x + 1$ and

$$x_{n+1} = x_n - \frac{x_n^4 - 6x_n^2 + x_n + 5}{4x_n^3 - 12x_n + 1} = \frac{3x_n^4 - 6x_n^2 - 5}{4x_n^3 - 12x_n + 1}$$

(a) On the basis of Table 1, we can be confident that when $x_0 = 0$, Newton's Method converges to a root near -2.3. Notice in Figure 5 that this is not the closest root to x_0.

(b) Table 2 suggests that with $x_0 = -1$, Newton's Method converges to the root near -0.9. ■

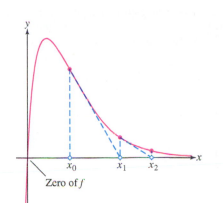

DF FIGURE 4 Function has only one zero but the sequence of Newton iterates goes off to infinity.

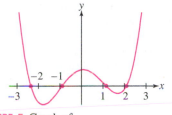

FIGURE 5 Graph of $f(x) = x^4 - 6x^2 + x + 5$.

TABLE 1	
x_0	0
x_1	-5
x_2	-3.9179954
x_3	-3.1669480
x_4	-2.6871270
x_5	-2.4363303
x_6	-2.3572979
x_7	-2.3495000

TABLE 2	
x_0	-1
x_1	-0.8888888888
x_2	-0.8882866140
x_3	-0.88828656234358
x_4	-0.888286562343575

4.8 SUMMARY

- Newton's Method: To find a sequence of numerical approximations to a root of f, begin with an initial guess x_0. Then construct the sequence $x_0, x_1, x_2, \ldots$ using the formula

$$x_{n+1} = x_n - \frac{f(x_n)}{f'(x_n)}$$

You should choose the initial guess x_0 as close as possible to a root, possibly by referring to a graph. In favorable cases, the sequence converges rapidly to a root.

- If x_n and x_{n+1} agree to m decimal places, it is usually safe to assume that x_n agrees with a root to m decimal places.

4.8 EXERCISES

Preliminary Questions

1. How many iterations of Newton's Method are required to compute a root if f is a linear function?

2. What happens in Newton's Method if your initial guess happens to be a zero of f?

3. What happens in Newton's Method if your initial guess happens to be a local min or max of f?

4. Is the following a reasonable description of Newton's Method: "A root of the equation of the tangent line to the graph of f is used as an approximation to a root of f itself"? Explain.

Exercises

In this exercise set, all approximations should be carried out using Newton's Method.

In Exercises 1–6, apply Newton's Method to f and initial guess x_0 to calculate x_1, x_2, x_3.

1. $f(x) = x^2 - 6, \quad x_0 = 2$

2. $f(x) = x^2 - 3x + 1, \quad x_0 = 3$

3. $f(x) = x^3 - 10, \quad x_0 = 2$

4. $f(x) = x^3 + x + 1, \quad x_0 = -1$

5. $f(x) = \cos x - 4x, \quad x_0 = 1$

6. $f(x) = 1 - x \sin x, \quad x_0 = 7$

7. Use Figure 6 to choose an initial guess x_0 to the unique real root of $x^3 + 2x + 5 = 0$ and compute the first three Newton iterates.

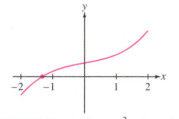

FIGURE 6 Graph of $y = x^3 + 2x + 5$.

8. Approximate a solution of $\sin x = \cos 2x$ in the interval $\left[0, \frac{\pi}{2}\right]$ to three decimal places. Then find the exact solution and compare with your approximation.

9. Approximate both solutions of $e^x = 5x$ to three decimal places (Figure 7).

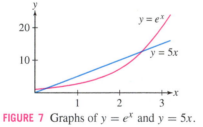

FIGURE 7 Graphs of $y = e^x$ and $y = 5x$.

10. The first positive solution of $\sin x = 0$ is $x = \pi$. Use Newton's Method to calculate π to four decimal places.

In Exercises 11–14, approximate to three decimal places using Newton's Method and compare with the value from a calculator.

11. $\sqrt{11}$ **12.** $5^{1/3}$ **13.** $2^{7/3}$ **14.** $3^{-1/4}$

15. Approximate the largest positive root of $f(x) = x^4 - 6x^2 + x + 5$ to within an error of at most 10^{-4}. Refer to Figure 5.

GU *In Exercises 16–19, approximate the root specified to three decimal places using Newton's Method. Use a plot to choose an initial guess.*

16. Largest positive root of $f(x) = x^3 - 5x + 1$

17. Negative root of $f(x) = x^5 - 20x + 10$

18. Positive solution of $\sin \theta = 0.8\theta$

19. Solution of $\ln(x + 4) = x$

20. Let x_1, x_2 be the estimates to a root obtained by applying Newton's Method with $x_0 = 1$ to the function graphed in Figure 8. Estimate the numerical values of x_1 and x_2, and draw the tangent lines used to obtain them.

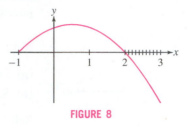

FIGURE 8

21. GU Find the smallest positive value of x at which $y = x$ and $y = \tan x$ intersect. *Hint:* Draw a plot.

22. In 1535 the mathematician Antonio Fior challenged his rival Niccolo Tartaglia to solve this problem: A tree stands 12 *braccia* high; it is broken into two parts at such a point that the height of the part left standing is the cube root of the length of the part cut away. What is the height of the part left standing? Show that this is equivalent to solving $x^3 + x = 12$ and finding the height to three decimal places. Tartaglia, who had discovered the secret of solving the cubic equation, was able to determine the exact answer:

$$x = \left(\sqrt[3]{\sqrt{2919} + 54} - \sqrt[3]{\sqrt{2919} - 54} \right) \Big/ \sqrt[3]{9}$$

23. Find (to two decimal places) the coordinates of the point P in Figure 9 where the tangent line to $y = \cos x$ passes through the origin.

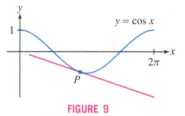

FIGURE 9

Newton's Method is often used to determine interest rates in financial calculations. In Exercises 24–26, r denotes a yearly interest rate expressed as a decimal (rather than as a percent).

24. If P dollars are deposited every month in an account earning interest at the yearly rate r, then the value S of the account after N years is

$$S = P \left(\frac{b^{12N+1} - b}{b - 1} \right), \qquad \text{where } b = 1 + \frac{r}{12}$$

You have decided to deposit $P = \$100$ per month.

(a) Determine S after 5 years if $r = 0.07$ (i.e., 7%).

(b) Show that to save \$10,000 after 5 years, you must earn interest at a rate r determined by the equation $b^{61} - 101b + 100 = 0$. Use Newton's Method to solve for b. Then find r. Note that $b = 1$ is a root, but you want the root satisfying $b > 1$.

25. If you borrow L dollars for N years at a yearly interest rate r, your monthly payment of P dollars is calculated using the equation

$$L = P \left(\frac{1 - b^{-12N}}{b - 1} \right), \qquad \text{where } b = 1 + \frac{r}{12}$$

(a) Find P if $L = \$5000$, $N = 3$, and $r = 0.08$ (8%).

(b) You are offered a loan of $L = \$5000$ to be paid back over 3 years with monthly payments of $P = \$200$. Use Newton's Method to compute b and find the implied interest rate r of this loan. *Hint:* Show that

$$(L/P)b^{12N+1} - (1 + L/P)b^{12N} + 1 = 0$$

26. If you deposit P dollars in a retirement fund every year for N years with the intention of then withdrawing Q dollars per year for M years, you must earn interest at a rate r satisfying

$$P(b^N - 1) = Q(1 - b^{-M}), \quad \text{where } b = 1 + r$$

Assume that $\$2000$ is deposited each year for 30 years and the goal is to withdraw $\$10,000$ per year for 25 years. Use Newton's Method to compute b and then find r. Note that $b = 1$ is a root, but you want the root satisfying $b > 1$.

27. There is no simple formula for the position at time t of a planet P in its orbit (an ellipse) around the sun. Introduce the auxiliary circle and angle θ in Figure 10 (note that P determines θ because it is the central angle of point B on the circle). Let $a = OA$ and $e = OS/OA$ (the eccentricity of the orbit).

(a) Show that sector BSA has area $(a^2/2)(\theta - e\sin\theta)$.

(b) By Kepler's Second Law, the area of sector BSA is proportional to the time t elapsed since the planet passed point A, and because the circle has area πa^2, BSA has area $(\pi a^2)(t/T)$, where T is the period of the orbit. Deduce **Kepler's Equation**:

$$\frac{2\pi t}{T} = \theta - e\sin\theta$$

(c) The eccentricity of Mercury's orbit is approximately $e = 0.2$. Use Newton's Method to find θ after a quarter of Mercury's year has elapsed ($t = T/4$). Convert θ to degrees. Has Mercury covered more than a quarter of its orbit at $t = T/4$?

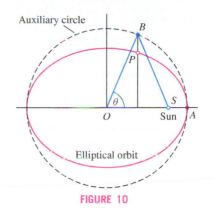

Auxiliary circle

Elliptical orbit

FIGURE 10

28. The roots of $f(x) = \frac{1}{3}x^3 - 4x + 1$ to three decimal places are -3.583, 0.251, and 3.332 (Figure 11). Determine the root to which Newton's Method converges for the initial choices $x_0 = 1.85$, 1.7, and 1.55. The answer shows that a small change in x_0 can have a significant effect on the outcome of Newton's Method.

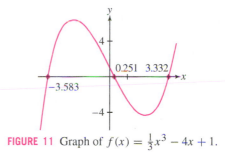

FIGURE 11 Graph of $f(x) = \frac{1}{3}x^3 - 4x + 1$.

29. What happens when you apply Newton's Method to find a zero of $f(x) = x^{1/3}$? Note that $x = 0$ is the only zero.

30. What happens when you apply Newton's Method to the equation $x^3 - 20x = 0$ with the unlucky initial guess $x_0 = 2$?

Further Insights and Challenges

31. Newton's Method can be used to compute reciprocals without performing division. Let $c > 0$ and set $f(x) = x^{-1} - c$.

(a) Show that $x - (f(x)/f'(x)) = 2x - cx^2$.

(b) Calculate the first three iterates of Newton's Method with $c = 10.3$ and the two initial guesses $x_0 = 0.1$ and $x_0 = 0.5$.

(c) Explain graphically why $x_0 = 0.5$ does not yield a sequence converging to $1/10.3$.

In Exercises 32 and 33, consider a metal rod of length L fastened at both ends. If you cut the rod and weld on an additional segment of length m, leaving the ends fixed, the rod will bow up into a circular arc of radius R (unknown), as indicated in Figure 12.

32. Let h be the maximum vertical displacement of the rod.

(a) Show that $L = 2R\sin\theta$ and conclude that

$$h = \frac{L(1 - \cos\theta)}{2\sin\theta}$$

(b) Show that $L + m = 2R\theta$ and then prove

$$\frac{\sin\theta}{\theta} = \frac{L}{L+m} \qquad \boxed{2}$$

33. Let $L = 3$ and $m = 1$. Apply Newton's Method to Eq. (2) to estimate θ, and use this to estimate h.

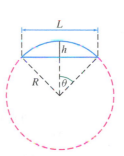

FIGURE 12 The bold circular arc has length $L + m$.

34. Quadratic Convergence to Square Roots Let $f(x) = x^2 - c$ and let $e_n = x_n - \sqrt{c}$ be the error in x_n.

(a) Show that $x_{n+1} = \frac{1}{2}(x_n + c/x_n)$ and $e_{n+1} = e_n^2/2x_n$.

(b) Show that if $x_0 > \sqrt{c}$, then $x_n > \sqrt{c}$ for all n. Explain graphically.

(c) Show that if $x_0 > \sqrt{c}$, then $e_{n+1} \le e_n^2/(2\sqrt{c})$.

*In Exercises 35–37, a flexible chain of length L is suspended between two poles of equal height separated by a distance $2M$ (Figure 13). By Newton's laws, the chain describes a **catenary***

$$y = a\cosh\left(\frac{x}{a}\right)$$

where a is the number such that

$$L = 2a \sinh\left(\frac{M}{a}\right)$$

The sag s is the vertical distance from the highest to the lowest point on the chain.

35. Suppose that $L = 120$ and $M = 50$.

(a) Use Newton's Method to find a value of a (to two decimal places) satisfying $L = 2a \sinh(M/a)$.

(b) Compute the sag s.

36. Assume that M is fixed.

(a) Calculate $\frac{ds}{da}$. Note that $s = a \cosh\left(\frac{M}{a}\right) - a$.

(b) Calculate $\frac{da}{dL}$ by implicit differentiation using the relation $L = 2a \sinh\left(\frac{M}{a}\right)$.

(c) Use (a) and (b) and the Chain Rule to show that

$$\frac{ds}{dL} = \frac{ds}{da}\frac{da}{dL} = \frac{\cosh(M/a) - (M/a)\sinh(M/a) - 1}{2\sinh(M/a) - (2M/a)\cosh(M/a)} \quad \boxed{3}$$

37. Suppose that $L = 160$ and $M = 50$.

(a) Use Newton's Method to find a value of a (to two decimal places) satisfying $L = 2a \sinh(M/a)$.

(b) Use Eq. (3) and the Linear Approximation to estimate the increase in sag Δs for changes in length $\Delta L = 1$ and $\Delta L = 5$.

(c) CAS Compute $s(161) - s(160)$ and $s(165) - s(160)$ directly and compare with your estimates in (b).

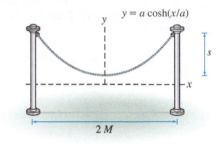

FIGURE 13 Chain hanging between two poles.

CHAPTER REVIEW EXERCISES

In Exercises 1–6, estimate using the Linear Approximation or linearization, and use a calculator to estimate the error.

1. $8.1^{1/3} - 2$

2. $\frac{1}{\sqrt{4.1}} - \frac{1}{2}$

3. $625^{1/4} - 624^{1/4}$

4. $\sqrt{101}$

5. $\frac{1}{1.02}$

6. $\sqrt[5]{33}$

In Exercises 7–12, find the linearization at the point indicated.

7. $y = \sqrt{x}, \quad a = 25$

8. $v(t) = 32t - 4t^2, \quad a = 2$

9. $A(r) = \frac{4}{3}\pi r^3, \quad a = 3$

10. $V(h) = 4h(2-h)(4-2h), \quad a = 1$

11. $P(x) = e^{-x^2/2}, \quad a = 1$

12. $f(x) = \ln(x + e), \quad a = e$

In Exercises 13–18, use the Linear Approximation.

13. The position of an object in linear motion at time t is $s(t) = 0.4t^2 + (t+1)^{-1}$. Estimate the distance traveled over the time interval $[4, 4.2]$.

14. A bond that pays $10,000 in 6 years is offered for sale at a price P. The percentage yield Y of the bond is

$$Y = 100\left(\left(\frac{10{,}000}{P}\right)^{1/6} - 1\right)$$

Verify that if $P = \$7500$, then $Y = 4.91\%$. Estimate the drop in yield if the price rises to $7700.

15. When a bus pass from Albuquerque to Los Alamos is priced at p dollars, a bus company takes in a monthly revenue of $R(p) = 1.5p - 0.01p^2$ (in thousands of dollars).

(a) Estimate ΔR if the price rises from $50 to $53.

(b) If $p = 80$, how will revenue be affected by a small increase in price? Explain using the Linear Approximation.

16. A store sells 80 MP4 players per week when the players are priced at $P = \$75$. Estimate the number N sold if P is raised to $80, assuming that $dN/dP = -4$. Estimate N if the price is lowered to $69.

17. The circumference of a sphere is measured at $C = 100$ cm. Estimate the maximum percentage error in V if the error in C is at most 3 cm.

18. Show that $\sqrt{a^2 + b} \approx a + \frac{b}{2a}$ if b is small. Use this to estimate $\sqrt{26}$ and find the error using a calculator.

19. Use the Intermediate Value Theorem to prove that $\sin x - \cos x = 3x$ has a solution, and use Rolle's Theorem to show that this solution is unique.

20. Show that $f(x) = 2x^3 + 2x + \sin x + 1$ has precisely one real root.

21. Verify the MVT for $f(x) = \ln x$ on $[1, 4]$.

22. Suppose that $f(1) = 5$ and $f'(x) \geq 2$ for $x \geq 1$. Use the MVT to show that $f(8) \geq 19$.

23. Use the MVT to prove that if $f'(x) \leq 2$ for $x > 0$ and $f(0) = 4$, then $f(x) \leq 2x + 4$ for all $x \geq 0$.

24. A function f has derivative $f'(x) = \frac{1}{x^4 + 1}$. Where on the interval $[1, 4]$ does f take on its maximum value?

In Exercises 25–30, find the critical points and determine whether they are minima, maxima, or neither.

25. $f(x) = x^3 - 4x^2 + 4x$

26. $s(t) = t^4 - 8t^2$

27. $f(x) = x^2(x+2)^3$

28. $f(x) = x^{2/3}(1 - x)$

29. $g(\theta) = \sin^2\theta + \theta$

30. $h(\theta) = 2\cos 2\theta + \cos 4\theta$

In Exercises 31–38, find the extreme values on the interval.

31. $f(x) = x(10 - x), \quad [-1, 3]$

32. $f(x) = 6x^4 - 4x^6, \quad [-2, 2]$

33. $g(\theta) = \sin^2\theta - \cos\theta, \quad [0, 2\pi]$

34. $R(t) = \frac{t}{t^2 + t + 1}, \quad [0, 3]$

35. $f(x) = x^{2/3} - 2x^{1/3}$, $[-1, 3]$

36. $f(x) = 4x - \tan^2 x$, $[-\frac{\pi}{4}, \frac{\pi}{3}]$

37. $f(x) = x - 12 \ln x$, $[5, 40]$

38. $f(x) = e^x - 20x - 1$, $[0, 5]$

39. Find the critical points and extreme values of $f(x) = |x - 1| + |2x - 6|$ in $[0, 8]$.

40. Match the description of f with the graph of its *derivative* f' in Figure 1.

(a) f is increasing and concave up.

(b) f is decreasing and concave up.

(c) f is increasing and concave down.

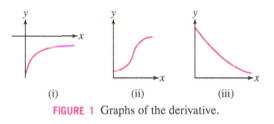

(i)   (ii)   (iii)

FIGURE 1 Graphs of the derivative.

In Exercises 41–46, find the points of inflection.

41. $y = x^3 - 4x^2 + 4x$

42. $y = x - 2 \cos x$

43. $y = \dfrac{x^2}{x^2 + 4}$

44. $y = \dfrac{x}{(x^2 - 4)^{1/3}}$

45. $f(x) = (x^2 - x)e^{-x}$

46. $f(x) = x(\ln x)^2$

In Exercises 47–56, sketch the graph, noting the transition points and asymptotic behavior.

47. $y = 12x - 3x^2$

48. $y = 8x^2 - x^4$

49. $y = x^3 - 2x^2 + 3$

50. $y = 4x - x^{3/2}$

51. $y = \dfrac{x}{x^3 + 1}$

52. $y = \dfrac{x}{(x^2 - 4)^{2/3}}$

53. $y = \dfrac{1}{|x + 2| + 1}$

54. $y = \sqrt{2 - x^3}$

55. $y = \sqrt{3} \sin x - \cos x$ on $[0, 2\pi]$

56. $y = 2x - \tan x$ on $[0, 2\pi]$

57. Draw a curve $y = f(x)$ for which f' and f'' have signs as indicated in Figure 2.

FIGURE 2

58. Find the dimensions of a cylindrical can with a bottom but no top of volume 4 m³ that uses the least amount of metal.

59. A rectangular open-topped box of height h with a square base of side b has volume $V = 4$ m³. Two of the side faces are made of material costing \$40/m². The remaining sides cost \$20/m². Which values of b and h minimize the cost of the box?

60. The corn yield on a certain farm is

$$Y = -0.118x^2 + 8.5x + 12.9 \quad \text{(bushels per acre)}$$

where x is the number of corn plants per acre (in thousands). Assume that corn seed costs \$1.25 (per thousand seeds) and that corn can be sold for \$1.50/bushel. Let $P(x)$ be the profit (revenue minus the cost of seeds) at planting level x.

(a) Compute $P(x_0)$ for the value x_0 that maximizes yield Y.

(b) Find the maximum value of $P(x)$. Does maximum yield lead to maximum profit?

61. Let $N(t)$ be the size of a tumor (in units of 10^6 cells) at time t (in days). According to the **Gompertz Model**, $dN/dt = N(a - b \ln N)$ where a, b are positive constants. Show that the maximum value of N is $e^{\frac{a}{b}}$ and that the tumor increases most rapidly when $N = e^{\frac{a}{b} - 1}$.

62. A truck gets 10 miles per gallon (mpg) of diesel fuel traveling along an interstate highway at 50 mph. This mileage decreases by 0.15 mpg for each mile per hour increase above 50 mph.

(a) If the truck driver is paid \$30/h and diesel fuel costs $P = \$3$/gal, which speed v between 50 and 70 mph will minimize the cost of a trip along the highway? Notice that the actual cost depends on the length of the trip, but the optimal speed does not.

(b) $\boxed{\text{GU}}$ Plot cost as a function of v (choose the length arbitrarily) and verify your answer to part (a).

(c) $\boxed{\text{GU}}$ Do you expect the optimal speed v to increase or decrease if fuel costs go down to $P = \$2$/gal? Plot the graphs of cost as a function of v for $P = 2$ and $P = 3$ on the same axis and verify your conclusion.

63. Find the maximum volume of a right-circular cone placed upside-down in a right-circular cone of radius $R = 3$ and height $H = 4$ as in Figure 3. A cone of radius r and height h has volume $\frac{1}{3}\pi r^2 h$.

64. Redo Exercise 63 for arbitrary R and H.

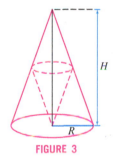

FIGURE 3

65. Show that the maximum area of a parallelogram $ADEF$ that is inscribed in a triangle ABC, as in Figure 4, is equal to one-half the area of $\triangle ABC$.

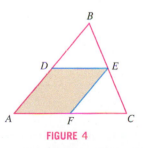

FIGURE 4

66. A box of volume 8 m³ with a square top and bottom is constructed out of two types of metal. The metal for the top and bottom costs \$50/m² and the metal for the sides costs \$30/m². Find the dimensions of the box that minimize total cost.

67. Let f be a function whose graph does not pass through the x-axis and let $Q = (a, 0)$. Let $P = (x_0, f(x_0))$ be the point on the graph closest to Q (Figure 5). Prove that $\overline{PQ}$ is perpendicular to the tangent line to the graph of x_0. *Hint:* Find the minimum value of the *square* of the distance from $(x, f(x))$ to $(a, 0)$.

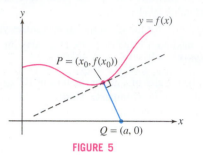

FIGURE 5

68. Take a circular piece of paper of radius R, remove a sector of angle θ (Figure 6), and fold the remaining piece into a cone-shaped cup. Which angle θ produces the cup of largest volume?

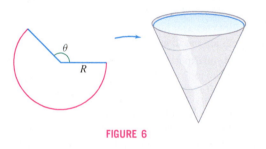

FIGURE 6

69. Use Newton's Method to estimate $\sqrt[3]{25}$ to four decimal places.

70. Use Newton's Method to find a root of $f(x) = x^2 - x - 1$ to four decimal places.

71. Find the local extrema of $f(x) = \dfrac{e^{2x} + 1}{e^{x+1}}$.

72. Find the points of inflection of $f(x) = \ln(x^2 + 1)$ and, at each point, determine whether the concavity changes from up to down or from down to up.

In Exercises 73–76, find the local extrema and points of inflection, and sketch the graph. Use L'Hôpital's Rule to determine the limits as $x \to 0+$ or $x \to \pm\infty$ if necessary.

73. $y = x \ln x$ $(x > 0)$

74. $y = e^{x - x^2}$

75. $y = x(\ln x)^2$ $(x > 0)$

76. $y = \tan^{-1}\left(\dfrac{x^2}{4}\right)$

77. 📖 Explain why L'Hôpital's Rule gives no information about $\displaystyle\lim_{x\to\infty} \dfrac{2x - \sin x}{3x + \cos 2x}$. Evaluate the limit by another method.

78. Let f be a differentiable function with inverse g which is also differentiable. Assume that $f(0) = 0$ and $f'(0) \neq 0$.

(a) Use the fact that $f(g(x)) = x$ and the Chain Rule to show that
$$g'(x) = \frac{1}{f'(g(x))}.$$

(b) Prove that
$$\lim_{x\to 0} \frac{f(x)}{g(x)} = f'(0)^2$$

In Exercises 79–90, verify that L'Hôpital's Rule applies and evaluate the limit.

79. $\displaystyle\lim_{x\to 3} \dfrac{4x - 12}{x^2 - 5x + 6}$

80. $\displaystyle\lim_{x\to -2} \dfrac{x^3 + 2x^2 - x - 2}{x^4 + 2x^3 - 4x - 8}$

81. $\displaystyle\lim_{x\to 0+} x^{1/2} \ln x$

82. $\displaystyle\lim_{t\to\infty} \dfrac{\ln(e^t + 1)}{t}$

83. $\displaystyle\lim_{\theta\to 0} \dfrac{2\sin\theta - \sin 2\theta}{\sin\theta - \theta\cos\theta}$

84. $\displaystyle\lim_{x\to 0} \dfrac{\sqrt{4 + x} - 2\sqrt[8]{1 + x}}{x^2}$

85. $\displaystyle\lim_{t\to\infty} \dfrac{\ln(t + 2)}{\log_2 t}$

86. $\displaystyle\lim_{x\to 0} \left(\dfrac{e^x}{e^x - 1} - \dfrac{1}{x}\right)$

87. $\displaystyle\lim_{y\to 0} \dfrac{\sin^{-1} y - y}{y^3}$

88. $\displaystyle\lim_{x\to 1} \dfrac{\sqrt{1 - x^2}}{\cos^{-1} x}$

89. $\displaystyle\lim_{x\to 0} \dfrac{\sinh(x^2)}{\cosh x - 1}$

90. $\displaystyle\lim_{x\to 0} \dfrac{\tanh x - \sinh x}{\sin x - x}$

91. Let $f(x) = e^{-Ax^2/2}$, where $A > 0$ is a constant. Given any n numbers $a_1, a_2, \ldots, a_n$, set
$$\Phi(x) = f(x - a_1)f(x - a_2)\cdots f(x - a_n)$$

(a) Assume $n = 2$ and prove that Φ attains its maximum value at the average $x = \frac{1}{2}(a_1 + a_2)$. *Hint:* Calculate $\Phi'(x)$ using logarithmic differentiation.

(b) Show that for any n, Φ attains its maximum value at $x = \frac{1}{n}(a_1 + a_2 + \cdots + a_n)$. This fact is related to the role of $f(x)$ (whose graph is a bell-shaped curve) in statistics.

5 THE INTEGRAL

Carbon dating, which relies on the exponential decay of C^{14} relative to C^{12}, allows for the determination of the age of these cave paintings.
(© John Mitchell/Alamy)

The basic problem in integral calculus is finding the area under a curve. You may wonder why calculus deals with two seemingly unrelated topics: tangent lines on the one hand and areas on the other. One reason is that both are computed using limits. A deeper connection is revealed by the Fundamental Theorem of Calculus, discussed in Sections 5.4 and 5.5. This theorem expresses the "inverse" relationship between integration and differentiation. It plays a truly fundamental role in nearly all applications of calculus, both theoretical and practical.

5.1 Approximating and Computing Area

Why might we be interested in the area under a graph? Consider an object moving in a straight line with *constant velocity* v (assumed positive). The distance traveled over a time interval $[t_1, t_2]$ is equal to $v\Delta t$, where $\Delta t = (t_2 - t_1)$ is the time elapsed. This is the well-known formula

$$\text{Distance traveled} = \overbrace{\text{velocity} \times \text{time elapsed}}^{v\Delta t}$$

1

Because v is constant, the graph of velocity is a horizontal line (Figure 1) and $v\Delta t$ is equal to the area of the rectangular region under the graph of velocity over $[t_1, t_2]$. So we can write Eq. (1) as

$$\text{Distance traveled} = \text{area under the graph of velocity over } [t_1, t_2]$$

2

There is, however, an important difference between these two equations: Eq. (1) makes sense only if velocity v is constant, whereas Eq. (2) is correct *even if the velocity changes with time* (we will prove this in Section 5.6). Thus, the advantage of expressing distance traveled as an area is that it enables us to deal with much more general types of motion.

To see why Eq. (2) might be true in general, let's consider the case where velocity changes over time but is constant on intervals. In other words, we assume that the object's velocity changes abruptly from one interval to the next as in Figure 2. The distance traveled over each time interval is equal to the area of the rectangle above that interval, so the total distance traveled is the sum of the areas of the rectangles. In Figure 2,

$$\text{Distance traveled over } [0, 8] \text{ s} = \underbrace{10 + 15 + 30 + 10}_{\text{Sum of areas of rectangles}} = 65 \text{ m}$$

Our strategy when velocity changes continuously (Figure 3) is to *approximate* the area under the graph by sums of areas of rectangles and then pass to a limit. This idea leads to the concept of an integral.

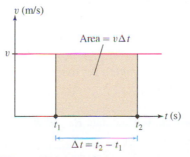

FIGURE 1 The rectangle has area $v\Delta t$, which is equal to the distance traveled.

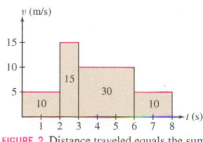

FIGURE 2 Distance traveled equals the sum of the areas of the rectangles.

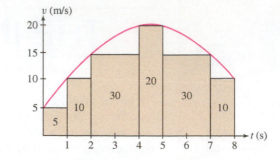

FIGURE 3 Distance traveled is equal to the area under the graph. It is *approximated* by the sum of the areas of the rectangles.

Approximating Area by Rectangles

Recall the two-step procedure for finding the slope of the tangent line (the derivative): First approximate the slope using secant lines and then compute the limit of these approximations. In Integral Calculus, there are also two steps:

- *First, approximate the area under the graph using rectangles.*
- *Then compute the exact area (the integral) as the limit of these approximations.*

Our goal is to compute the area under the graph of a function f. In this section, we assume that f is continuous and *positive*, so that the graph of f lies above the x-axis (Figure 4). The first step is to approximate the area using rectangles.

To begin, choose a whole number N and divide $[a, b]$ into N subintervals of equal width, as in Figure 4(A). The full interval $[a, b]$ has width $b - a$, so each subinterval has width $\Delta x = (b - a)/N$. The right endpoints of the subintervals are

$$x_1 = a + \Delta x, \quad x_2 = a + 2\Delta x, \quad \ldots, \quad x_{N-1} = a + (N - 1)\Delta x, \quad x_N = a + N\Delta x$$

Note that the last right endpoint is $x_N = b$ because $a + N\Delta x = a + N((b - a)/N) = b$. Next, as in Figure 4(B), construct, above each subinterval, a rectangle whose height is the value of $f(x)$ at the *right endpoint* of the subinterval.

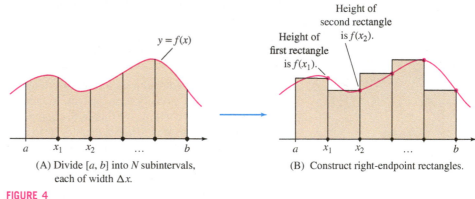

(A) Divide $[a, b]$ into N subintervals, each of width Δx.

(B) Construct right-endpoint rectangles.

FIGURE 4

The sum of the areas of these rectangles provides an *approximation* to the area under the graph. The first rectangle has base Δx and height $f(x_1)$, so its area is $f(x_1)\Delta x$. Similarly, the second rectangle has height $f(x_2)$ and area $f(x_2)\Delta x$, etc. The sum of the areas of the rectangles is denoted R_N and is called the **Nth right-endpoint approximation**:

$$R_N = f(x_1)\Delta x + f(x_2)\Delta x + \cdots + f(x_N)\Delta x$$

Factoring out Δx, we obtain the formula

To summarize,

$$a = left\ endpoint\ of\ interval\ [a, b]$$

$$b = right\ endpoint\ of\ interval\ [a, b]$$

$$N = number\ of\ subintervals\ in\ [a, b]$$

$$\Delta x = \frac{b - a}{N}$$

$$R_N = \Delta x \left(f(x_1) + f(x_2) + \cdots + f(x_N) \right)$$

In words: R_N is equal to Δx times the sum of the function values at the right endpoints of the subintervals.

■ **EXAMPLE 1**   Calculate R_4 and R_6 for $f(x) = x^2$ on the interval $[1, 3]$.

Solution

Step 1. **Determine Δx and the right endpoints.**

To calculate R_4, divide $[1, 3]$ into four subintervals of width $\Delta x = \frac{3-1}{4} = \frac{1}{2}$. The right endpoints are the numbers $x_j = a + j\Delta x = 1 + j(\frac{1}{2})$ for $j = 1, 2, 3, 4$. They are spaced at intervals of $\frac{1}{2}$ beginning at $\frac{3}{2}$, so, as we see in Figure 5(A), the right endpoints are $\frac{3}{2}, \frac{4}{2}, \frac{5}{2}, \frac{6}{2}$.

Step 2. Calculate Δx times the sum of function values.

R_4 is Δx times the sum of the function values at the right endpoints:

$$R_4 = \frac{1}{2}\left(f\left(\frac{3}{2}\right) + f\left(\frac{4}{2}\right) + f\left(\frac{5}{2}\right) + f\left(\frac{6}{2}\right) \right)$$

$$= \frac{1}{2}\left(\left(\frac{3}{2}\right)^2 + \left(\frac{4}{2}\right)^2 + \left(\frac{5}{2}\right)^2 + \left(\frac{6}{2}\right)^2 \right) = \frac{43}{4} = 10.75$$

R_6 is similar: $\Delta x = \frac{3-1}{6} = \frac{1}{3}$, and the right endpoints are spaced at intervals of $\frac{1}{3}$ beginning at $\frac{4}{3}$ and ending at 3, as in Figure 5(B). Thus,

$$R_6 = \frac{1}{3}\left(f\left(\frac{4}{3}\right) + f\left(\frac{5}{3}\right) + f\left(\frac{6}{3}\right) + f\left(\frac{7}{3}\right) + f\left(\frac{8}{3}\right) + f\left(\frac{9}{3}\right) \right)$$

$$= \frac{1}{3}\left(\frac{16}{9} + \frac{25}{9} + \frac{36}{9} + \frac{49}{9} + \frac{64}{9} + \frac{81}{9} \right) = \frac{271}{27} \approx 10.037 \qquad \blacksquare$$

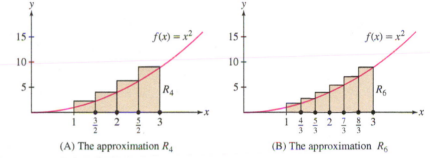

(A) The approximation R_4 (B) The approximation R_6

DF FIGURE 5

Summation Notation

Summation notation is a standard notation for writing sums in compact form. The sum of numbers $a_m, \ldots, a_n$ $(m \le n)$ is denoted

$$\sum_{j=m}^{n} a_j = a_m + a_{m+1} + \cdots + a_n$$

The Greek letter $\sum$ (capital sigma) stands for "sum," and the notation $\displaystyle\sum_{j=m}^{n}$ tells us to start the summation at $j = m$ and end it at $j = n$. For example,

$$\sum_{j=1}^{5} j^2 = 1^2 + 2^2 + 3^2 + 4^2 + 5^2 = 55$$

In this summation, the jth term is $a_j = j^2$. We refer to j^2 as the **general term**. The letter j is called the **summation index**. It is also referred to as a **dummy variable** because any other letter can be used instead. For example,

$$\sum_{k=4}^{6} (k^2 - 2k) = \overbrace{(4^2 - 2(4))}^{k=4} + \overbrace{(5^2 - 2(5))}^{k=5} + \overbrace{(6^2 - 2(6))}^{k=6} = 47$$

$$\sum_{m=7}^{9} 1 = 1 + 1 + 1 = 3 \qquad \text{(because } a_7 = a_8 = a_9 = 1\text{)}$$

The usual commutative, associative, and distributive laws of addition give us the following rules for manipulating summations.

Linearity of Summations

- $$\sum_{j=m}^{n} (a_j + b_j) = \sum_{j=m}^{n} a_j + \sum_{j=m}^{n} b_j$$

- $$\sum_{j=m}^{n} C a_j = C \sum_{j=m}^{n} a_j \qquad (C \text{ any constant})$$

- $$\sum_{j=1}^{n} C = nC \qquad (C \text{ any constant and } n \geq 1)$$

For example,

$$\sum_{j=3}^{5} (j^2 + j) = (3^2 + 3) + (4^2 + 4) + (5^2 + 5)$$

is equal to

$$\sum_{j=3}^{5} j^2 + \sum_{j=3}^{5} j = \left(3^2 + 4^2 + 5^2\right) + (3 + 4 + 5)$$

Linearity can be used to write a single summation as a sum of several summations. For example,

$$\sum_{k=0}^{100} (7k^2 - 4k + 9) = \sum_{k=0}^{100} 7k^2 + \sum_{k=0}^{100} (-4k) + \sum_{k=0}^{100} 9$$

$$= 7 \sum_{k=0}^{100} k^2 - 4 \sum_{k=0}^{100} k + 9 \sum_{k=0}^{100} 1$$

It is convenient to use summation notation when working with area approximations. For example, R_N is a sum with general term $f(x_j)$:

$$R_N = \Delta x \big[f(x_1) + f(x_2) + \cdots + f(x_N) \big]$$

The summation extends from $j = 1$ to $j = N$, so we can write R_N concisely as

$$R_N = \Delta x \sum_{j=1}^{N} f(x_j)$$

We shall make use of two other rectangular approximations to area: the left-endpoint and the midpoint approximations. Divide $[a, b]$ into N subintervals as before. In the **left-endpoint approximation** L_N, the heights of the rectangles are the values of $f(x)$ at the left endpoints [Figure 6(A)]. These left endpoints are

◀⋯ REMINDER

$$\Delta x = \frac{b - a}{N}$$

$$x_0 = a, \ x_1 = a + \Delta x, \ x_2 = a + 2\Delta x, \ \ldots, \ x_{N-1} = a + (N-1)\Delta x$$

and the sum of the areas of the left-endpoint rectangles is

$$L_N = \Delta x \Big(f(x_0) + f(x_1) + f(x_2) + \cdots + f(x_{N-1}) \Big)$$

Note that both R_N and L_N have general term $f(x_j)$, but the sum for L_N runs from $j = 0$ to $j = N - 1$ rather than from $j = 1$ to $j = N$:

$$L_N = \Delta x \sum_{j=0}^{N-1} f(x_j)$$

In the **midpoint approximation** M_N, the heights of the rectangles are the values of $f(x)$ at the midpoints of the subintervals rather than at the endpoints. As we see in Figure 6(B), the midpoints are

$$\frac{x_0 + x_1}{2}, \quad \frac{x_1 + x_2}{2}, \quad \ldots, \quad \frac{x_{N-1} + x_N}{2}$$

The sum of the areas of the midpoint rectangles is

$$M_N = \Delta x \left(f\left(\frac{x_0 + x_1}{2}\right) + f\left(\frac{x_1 + x_2}{2}\right) + \cdots + f\left(\frac{x_{N-1} + x_N}{2}\right) \right)$$

In summation notation,

$$M_N = \Delta x \sum_{j=0}^{N-1} f\left(\frac{x_j + x_{j+1}}{2}\right)$$

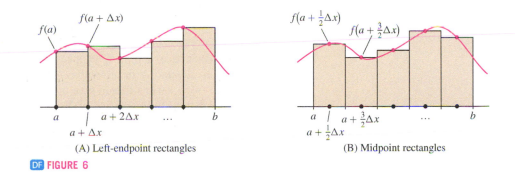

(A) Left-endpoint rectangles   (B) Midpoint rectangles

DF **FIGURE 6**

■ **EXAMPLE 2** Calculate R_6, L_6, and M_6 for $f(x) = x^{-1}$ on $[2, 4]$.

Solution In this case, $\Delta x = (b - a)/N = (4 - 2)/6 = \frac{1}{3}$. The general term in the summation for R_6 and L_6 is

$$f(x_j) = f(a + j\Delta x) = f\left(2 + j\left(\frac{1}{3}\right)\right) = \frac{1}{2 + \frac{1}{3}j} = \frac{3}{6 + j}$$

Therefore (Figure 7),

$$R_6 = \frac{1}{3} \sum_{j=1}^{6} f(x_j) = \frac{1}{3} \sum_{j=1}^{6} \frac{3}{6 + j}$$

$$= \frac{1}{3}\left(\frac{3}{7} + \frac{3}{8} + \frac{3}{9} + \frac{3}{10} + \frac{3}{11} + \frac{3}{12}\right) \approx 0.653$$

In L_6, the sum begins at $j = 0$ and ends at $j = 5$:

$$L_6 = \frac{1}{3} \sum_{j=0}^{5} \frac{3}{6 + j} = \frac{1}{3}\left(\frac{3}{6} + \frac{3}{7} + \frac{3}{8} + \frac{3}{9} + \frac{3}{10} + \frac{3}{11}\right) \approx 0.737$$

The general term in M_6 is

$$f\left(\frac{x_j + x_{j+1}}{2}\right)$$

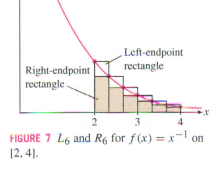

FIGURE 7 L_6 and R_6 for $f(x) = x^{-1}$ on $[2, 4]$.

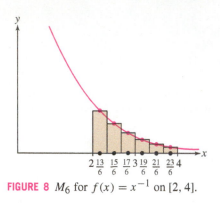

FIGURE 8 M_6 for $f(x) = x^{-1}$ on $[2, 4]$.

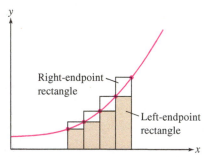

FIGURE 9 When f is increasing, the left-endpoint rectangles lie below the graph and right-endpoint rectangles lie above it.

In this case, the midpoints are $\frac{13}{6}, \frac{15}{6}, \frac{17}{6}, \frac{19}{6}, \frac{21}{6}$ and $\frac{23}{6}$. Summing up from $j = 0$ to 5, we obtain (Figure 8)

$$M_6 = \frac{1}{3}\left(f\left(\frac{13}{6}\right) + f\left(\frac{15}{6}\right) + f\left(\frac{17}{6}\right) + f\left(\frac{19}{6}\right) + f\left(\frac{21}{6}\right) + f\left(\frac{23}{6}\right)\right)$$

$$= \frac{1}{3}\left(\frac{6}{13} + \frac{6}{15} + \frac{6}{17} + \frac{6}{19} + \frac{6}{21} + \frac{6}{23}\right) \approx 0.692 \qquad \blacksquare$$

GRAPHICAL INSIGHT   Monotonic Functions   Observe in Figure 7 that the left-endpoint rectangles for $f(x) = x^{-1}$ extend above the graph and the right-endpoint rectangles lie below it. The exact area A must lie between R_6 and L_6, and so, according to the previous example, $0.65 \leq A \leq 0.74$. More generally, *when f is monotonic (increasing or decreasing), the exact area lies between R_N and L_N* (Figure 9):

- f increasing $\Rightarrow L_N \leq$ area under graph $\leq R_N$
- f decreasing $\Rightarrow R_N \leq$ area under graph $\leq L_N$

Notice that M_6 lies between R_6 and L_6. This is always the case for a monotonic function. (See Problem 93.)

Computing Area as the Limit of Approximations

Figure 10 shows several right-endpoint approximations. Notice that the *error* in computing the area, corresponding to the yellow region above the graph, gets smaller as the number of rectangles increases. In fact, it appears that *we can make the error as small as we please by taking the number N of rectangles large enough.* If so, it makes sense to consider the limit as $N \to \infty$, as this should give us the exact area under the curve. The next theorem guarantees that the limit exists (see Theorem 8 in Appendix D for a proof and Exercise 89 for a special case).

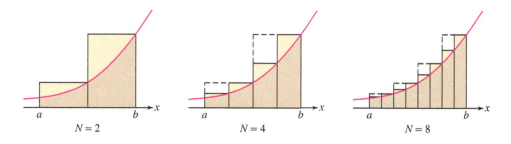

FIGURE 10 The error decreases as we use more rectangles.

In Theorem 1, it is not assumed that $f(x) \geq 0$. If $f(x)$ takes on negative values, the limit L no longer represents area under the graph, but we can interpret it as a "signed area," discussed in the next section.

THEOREM 1   If f is continuous on $[a, b]$, then the endpoint and midpoint approximations approach one and the same limit as $N \to \infty$. In other words, there is a value L such that

$$\lim_{N\to\infty} R_N = \lim_{N\to\infty} L_N = \lim_{N\to\infty} M_N = L$$

If $f(x) \geq 0$ on $[a, b]$, we define the area under the graph over $[a, b]$ to be L.

CONCEPTUAL INSIGHT   In calculus, limits are used to define basic quantities that otherwise would not have a precise meaning. Theorem 1 allows us to *define* area as a limit L in much the same way that we define the slope of a tangent line as the limit of slopes of secant lines.

The next three examples illustrate Theorem 1 using formulas for **power sums**. The kth power sum is defined as the sum of the kth powers of the first N integers. We shall use the power sum formulas for $k = 1, 2, 3$.

A method for proving power sum formulas is developed in Exercises 44–47 of Section 1.3. Formulas (3)–(5) can also be verified using the method of induction.

Power Sums

$$\sum_{j=1}^{N} j = 1 + 2 + \cdots + N = \frac{N(N+1)}{2} = \frac{N^2}{2} + \frac{N}{2} \qquad \boxed{3}$$

$$\sum_{j=1}^{N} j^2 = 1^2 + 2^2 + \cdots + N^2 = \frac{N(N+1)(2N+1)}{6} = \frac{N^3}{3} + \frac{N^2}{2} + \frac{N}{6} \qquad \boxed{4}$$

$$\sum_{j=1}^{N} j^3 = 1^3 + 2^3 + \cdots + N^3 = \frac{N^2(N+1)^2}{4} = \frac{N^4}{4} + \frac{N^3}{2} + \frac{N^2}{4} \qquad \boxed{5}$$

For example, by Eq. (4),

$$\sum_{j=1}^{6} j^2 = 1^2 + 2^2 + 3^2 + 4^2 + 5^2 + 6^2 = \underbrace{\frac{6^3}{3} + \frac{6^2}{2} + \frac{6}{6}}_{\frac{N^3}{3} + \frac{N^2}{2} + \frac{N}{6} \text{ for } N=6} = 91$$

As a first illustration, we compute the area of a right triangle "the hard way."

■ **EXAMPLE 3** Find the area A under the graph of $f(x) = x$ over $[0, 4]$ in three ways:

(a) Using geometry   **(b)** $\lim\limits_{N \to \infty} R_N$ **(c)** $\lim\limits_{N \to \infty} L_N$

Solution The region under the graph is a right triangle with base $b = 4$ and height $h = 4$ (Figure 11).

(a) By geometry, $A = \frac{1}{2}bh = \left(\frac{1}{2}\right)(4)(4) = 8$.

(b) We compute this area again as a limit. Since $\Delta x = (b - a)/N = 4/N$ and $f(x) = x$,

◀··· **REMINDER**

$$R_N = \Delta x \sum_{j=1}^{N} f(x_j)$$

$$L_N = \Delta x \sum_{j=0}^{N-1} f(x_j)$$

$$\Delta x = \frac{b - a}{N}$$

$$x_j = a + j\Delta x$$

$$f(x_j) = f(a + j\Delta x) = f\left(0 + j\left(\frac{4}{N}\right)\right) = \frac{4j}{N}$$

$$R_N = \Delta x \sum_{j=1}^{N} f(x_j) = \frac{4}{N} \sum_{j=1}^{N} \frac{4j}{N} = \frac{16}{N^2} \sum_{j=1}^{N} j$$

In the last equality, we factored out $4/N$ from the sum. This is valid because $4/N$ is a constant that does not depend on j. Now use formula (3):

$$R_N = \frac{16}{N^2} \sum_{j=1}^{N} j = \frac{16}{N^2} \underbrace{\left(\frac{N(N+1)}{2}\right)}_{\text{Formula for power sum}} = \frac{8}{N^2}\left(N^2 + N\right) = 8 + \frac{8}{N}$$

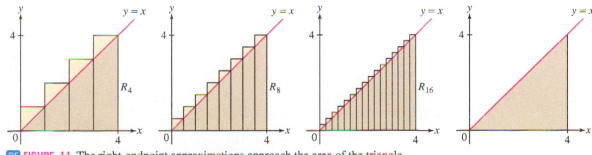

DF FIGURE 11 The right-endpoint approximations approach the area of the triangle.

The second term $8/N$ tends to zero as N approaches ∞, so

$$A = \lim_{N \to \infty} R_N = \lim_{N \to \infty} \left(8 + \frac{8}{N}\right) = 8$$

As expected, this limit yields the same value as the formula $\frac{1}{2}bh$.

(c) The left-endpoint approximation is similar, but the sum begins at $j = 0$ and ends at $j = N - 1$:

In Eq. (6), we apply the formula

$$\sum_{j=1}^{N} j = \frac{N(N+1)}{2}$$

with $N - 1$ in place of N:

$$\sum_{j=1}^{N-1} j = \frac{(N-1)N}{2}$$

$$L_N = \frac{16}{N^2} \sum_{j=0}^{N-1} j = \frac{16}{N^2} \sum_{j=1}^{N-1} j = \frac{16}{N^2}\left(\frac{(N-1)N}{2}\right) = 8 - \frac{8}{N} \qquad \boxed{6}$$

Note in the second step that we replaced the sum beginning at $j = 0$ with a sum beginning at $j = 1$. This is valid because the term for $j = 0$ is zero and may be dropped. Again, we find that $A = \lim_{N \to \infty} L_N = \lim_{N \to \infty} (8 - 8/N) = 8$. ∎

In the next example, we compute the area under a curved graph. Unlike the previous example, it is not possible to compute this area directly using geometry.

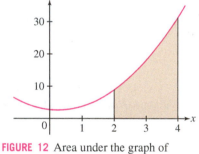

FIGURE 12 Area under the graph of $f(x) = 2x^2 - x + 3$ over $[2, 4]$.

■ **EXAMPLE 4** Let A be the area under the graph of $f(x) = 2x^2 - x + 3$ over $[2, 4]$ (Figure 12). Compute A as the limit $\lim_{N \to \infty} R_N$.

Solution

Step 1. **Express R_N in terms of power sums.**
In this case, $\Delta x = (4 - 2)/N = 2/N$ and

$$R_N = \Delta x \sum_{j=1}^{N} f(x_j) = \Delta x \sum_{j=1}^{N} f(a + j\Delta x) = \frac{2}{N} \sum_{j=1}^{N} f\left(2 + \frac{2j}{N}\right)$$

Let's use algebra to simplify the general term. Since $f(x) = 2x^2 - x + 3$,

$$f\left(2 + \frac{2j}{N}\right) = 2\left(2 + \frac{2j}{N}\right)^2 - \left(2 + \frac{2j}{N}\right) + 3$$

$$= 2\left(4 + \frac{8j}{N} + \frac{4j^2}{N^2}\right) - \left(2 + \frac{2j}{N}\right) + 3 = \frac{8}{N^2}j^2 + \frac{14}{N}j + 9$$

Now we can express R_N in terms of power sums:

$$R_N = \frac{2}{N} \sum_{j=1}^{N} \left(\frac{8}{N^2}j^2 + \frac{14}{N}j + 9\right) = \frac{2}{N} \sum_{j=1}^{N} \frac{8}{N^2}j^2 + \frac{2}{N} \sum_{j=1}^{N} \frac{14}{N}j + \frac{2}{N} \sum_{j=1}^{N} 9$$

$$= \frac{16}{N^3} \sum_{j=1}^{N} j^2 + \frac{28}{N^2} \sum_{j=1}^{N} j + \frac{18}{N} \sum_{j=1}^{N} 1 \qquad \boxed{7}$$

Step 2. **Use the formulas for the power sums.**
Using formulas (3) and (4) for the power sums in Eq. (7), we obtain

$$R_N = \frac{16}{N^3}\left(\frac{N^3}{3} + \frac{N^2}{2} + \frac{N}{6}\right) + \frac{28}{N^2}\left(\frac{N^2}{2} + \frac{N}{2}\right) + \frac{18}{N}(N)$$

$$= \left(\frac{16}{3} + \frac{8}{N} + \frac{8}{3N^2}\right) + \left(14 + \frac{14}{N}\right) + 18$$

$$= \frac{112}{3} + \frac{22}{N} + \frac{8}{3N^2}$$

Step 3. **Calculate the limit.**

$$A = \lim_{N \to \infty} R_N = \lim_{N \to \infty} \left(\frac{112}{3} + \frac{22}{N} + \frac{8}{3N^2} \right) = \frac{112}{3} \qquad \blacksquare$$

■ **EXAMPLE 5** Prove that for all $b > 0$, the area A under the graph of $f(x) = x^2$ over $[0, b]$ is equal to $b^3/3$, as indicated in Figure 13.

Solution We'll compute with R_N. We have $\Delta x = (b - 0)/N = b/N$ and

$$R_N = \Delta x \sum_{j=1}^{N} f(x_j) = \Delta x \sum_{j=1}^{N} f(0 + j\Delta x) = \frac{b}{N} \sum_{j=1}^{N} \left(0 + j\frac{b}{N} \right)^2 = \frac{b}{N} \sum_{j=1}^{N} \left(j^2 \frac{b^2}{N^2} \right)$$

$$= \frac{b^3}{N^3} \sum_{j=1}^{N} j^2$$

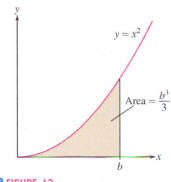

DF FIGURE 13

◀··· **REMINDER** By Eq. (4),

$$\sum_{j=1}^{N} j^2 = \frac{N^3}{3} + \frac{N^2}{2} + \frac{N}{6}$$

By the formula for the power sum recalled in the margin,

$$R_N = \frac{b^3}{N^3} \left(\frac{N^3}{3} + \frac{N^2}{2} + \frac{N}{6} \right) = \frac{b^3}{3} + \frac{b^3}{2N} + \frac{b^3}{6N^2}$$

$$A = \lim_{N \to \infty} R_N = \lim_{N \to \infty} \left(\frac{b^3}{3} + \frac{b^3}{2N} + \frac{b^3}{6N^2} \right) = \frac{b^3}{3} \qquad \blacksquare$$

The area under the graph of any polynomial can be calculated using power sum formulas as in the examples above. For other functions, the limit defining the area may be difficult or impossible to evaluate directly. Consider $f(x) = \sin x$ on the interval $\left[\frac{\pi}{4}, \frac{3\pi}{4} \right]$. In this case (Figure 14), $\Delta x = (3\pi/4 - \pi/4)/N = \pi/(2N)$ and the area A is

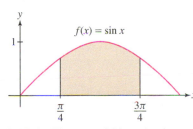

FIGURE 14 The area of this region is more difficult to compute as a limit of endpoint approximations.

$$A = \lim_{N \to \infty} R_N = \lim_{N \to \infty} \Delta x \sum_{j=1}^{N} f(a + j\Delta x) = \lim_{N \to \infty} \frac{\pi}{2N} \sum_{j=1}^{N} \sin \left(\frac{\pi}{4} + \frac{\pi j}{2N} \right)$$

With some work, we can show that the limit is equal to $A = \sqrt{2}$. However, in Section 5.4, we will see that it is much easier to apply the Fundamental Theorem of Calculus, which reduces area computations to the problem of finding antiderivatives.

HISTORICAL PERSPECTIVE

Jacob Bernoulli (1654–1705)

We used the formulas for the kth power sums for $k = 1, 2, 3$. Do similar formulas exist for all powers k? This problem was studied in the seventeenth century and eventually solved around 1690 by the great Swiss mathematician Jacob Bernoulli. Of this discovery, he wrote

With the help of [these formulas] it took me less than half of a quarter of an hour to find that the 10th powers of the first 1000 numbers being added together will yield the sum

9140992424142424342424191924242500

Bernoulli's formula has the general form

$$\sum_{j=1}^{n} j^k = \frac{1}{k+1} n^{k+1} + \frac{1}{2} n^k + \frac{k}{12} n^{k-1} + \cdots$$

The dots indicate terms involving smaller powers of n whose coefficients are expressed in terms of the so-called Bernoulli numbers. For example,

$$\sum_{j=1}^{n} j^4 = \frac{1}{5} n^5 + \frac{1}{2} n^4 + \frac{1}{3} n^3 - \frac{1}{30} n$$

These formulas are available on most computer algebra systems.

5.1 SUMMARY

Power Sums

$$\sum_{j=1}^{N} j = \frac{N(N+1)}{2} = \frac{N^2}{2} + \frac{N}{2}$$

$$\sum_{j=1}^{N} j^2 = \frac{N(N+1)(2N+1)}{6} = \frac{N^3}{3} + \frac{N^2}{2} + \frac{N}{6}$$

$$\sum_{j=1}^{N} j^3 = \frac{N^2(N+1)^2}{4} = \frac{N^4}{4} + \frac{N^3}{2} + \frac{N^2}{4}$$

- Approximations to the area under the graph of f over the interval $[a, b]$

$$\left(\Delta x = \frac{b-a}{N}, \; x_j = a + j\Delta x \right):$$

$$R_N = \Delta x \sum_{j=1}^{N} f(x_j) = \Delta x \big(f(x_1) + f(x_2) + \cdots + f(x_N) \big)$$

$$L_N = \Delta x \sum_{j=0}^{N-1} f(x_j) = \Delta x \big(f(x_0) + f(x_1) + \cdots + f(x_{N-1}) \big)$$

$$M_N = \Delta x \sum_{j=0}^{N-1} f\left(\frac{x_j + x_{j+1}}{2} \right)$$

$$= \Delta x \left(f\left(\frac{x_0 + x_1}{2} \right) + \cdots + f\left(\frac{x_{N-1} + x_N}{2} \right) \right)$$

- If f is continuous on $[a, b]$, then the endpoint and midpoint approximations approach one and the same limit L:

$$\lim_{N \to \infty} R_N = \lim_{N \to \infty} L_N = \lim_{N \to \infty} M_N = L$$

- If $f(x) \geq 0$ on $[a, b]$, we take L as the definition of the area under the graph of $y = f(x)$ over $[a, b]$.

5.1 EXERCISES

Preliminary Questions

1. What are the right and left endpoints if $[2, 5]$ is divided into six subintervals?

2. The interval $[1, 5]$ is divided into eight subintervals.
(a) What is the left endpoint of the last subinterval?
(b) What are the right endpoints of the first two subintervals?

3. Which of the following pairs of sums are *not* equal?

(a) $\displaystyle\sum_{i=1}^{4} i, \quad \sum_{\ell=1}^{4} \ell$

(b) $\displaystyle\sum_{j=1}^{4} j^2, \quad \sum_{k=2}^{5} k^2$

(c) $\displaystyle\sum_{j=1}^{4} j, \quad \sum_{i=2}^{5} (i-1)$

(d) $\displaystyle\sum_{i=1}^{4} i(i+1), \quad \sum_{j=2}^{5} (j-1)j$

4. Explain: $\displaystyle\sum_{j=1}^{100} j = \sum_{j=0}^{100} j$ but $\displaystyle\sum_{j=1}^{100} 1$ is not equal to $\displaystyle\sum_{j=0}^{100} 1$.

5. Explain why $L_{100} \geq R_{100}$ for $f(x) = x^{-2}$ on $[3, 7]$.

Exercises

1. Figure 15 shows the velocity of an object over a 3-minute (min) interval. Determine the distance traveled over the intervals $[0, 3]$ and $[1, 2.5]$ (remember to convert from kilometers per hour to kilometers per minute).

2. An ostrich (Figure 16) runs with velocity 20 km/h for 2 minutes (min), 12 km/h for 3 min, and 40 km/h for another minute. Compute the total distance traveled and indicate with a graph how this quantity can be interpreted as an area.

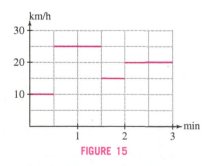

FIGURE 15

FIGURE 16 Ostriches can reach speeds as high as 70 km/h. (© Daryl Balfour/Gallo Images/Alamy)

3. A rainstorm hit Portland, Maine, in October 1996, resulting in record rainfall. The rainfall rate $R(t)$ on October 21 is recorded, in centimeters per hour, in the following table, where t is the number of hours since midnight. Compute the total rainfall during this 24-hour period and indicate on a graph how this quantity can be interpreted as an area.

t (h)	0–2	2–4	4–9	9–12	12–20	20–24
$R(t)$ (cm/h)	0.5	0.3	1.0	2.5	1.5	0.6

4. The velocity of an object is $v(t) = 12t$ m/s. Use Eq. (2) and geometry to find the distance traveled over the time intervals $[0, 2]$ and $[2, 5]$.

5. Compute R_5 and L_5 over $[0, 1]$ using the following values:

x	0	0.2	0.4	0.6	0.8	1
$f(x)$	50	48	46	44	42	40

6. Compute R_6, L_6, and M_3 to estimate the distance traveled over $[0, 3]$ if the velocity at half-second intervals is as follows:

t (s)	0	0.5	1	1.5	2	2.5	3
v (m/s)	0	12	18	25	20	14	20

7. Let $f(x) = 2x + 3$.

(a) Compute R_6 and L_6 over $[0, 3]$.

(b) Use geometry to find the exact area A and compute the errors $|A - R_6|$ and $|A - L_6|$ in the approximations.

8. Repeat Exercise 7 for $f(x) = 20 - 3x$ over $[2, 4]$.

9. Calculate R_3 and L_3 for $f(x) = x^2 - x + 4$ over $[1, 4]$. Then sketch the graph of f and the rectangles that make up each approximation. Is the area under the graph larger or smaller than R_3? Is it larger or smaller than L_3?

10. Let $f(x) = \sqrt{x^2 + 1}$ and $\Delta x = \frac{1}{3}$. Sketch the graph of f and draw the right-endpoint rectangles whose area is represented by the sum
$$\sum_{i=1}^{6} f(1 + i\Delta x)\Delta x.$$

11. Estimate $R_3, M_3,$ and L_6 over $[0, 1.5]$ for the function in Figure 17.

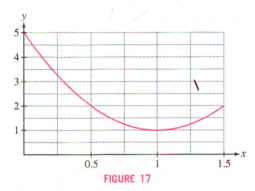

FIGURE 17

12. Calculate the area of the shaded rectangles in Figure 18. Which approximation do these rectangles represent?

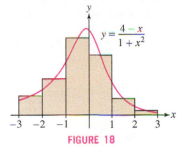

$y = \dfrac{4 - x}{1 + x^2}$

FIGURE 18

13. Let $f(x) = x^2$.

(a) Sketch the function over the interval $[0, 2]$ and the rectangles corresponding to L_4. Calculate the area contained within them.

(b) Sketch the function over the interval $[0, 2]$ again but with the rectangles corresponding to R_4. Calculate the area contained within them.

(c) Make a conclusion about the area under the curve $f(x) = x^2$ over the interval $[0, 2]$.

14. Let $f(x) = \sqrt{x}$.

(a) Sketch the function over the interval $[0, 4]$ and the rectangles corresponding to L_4. Calculate the area contained within them.

(b) Sketch the function over the interval $[0, 4]$ again but with the rectangles corresponding to R_4. Calculate the area contained within them.

(c) Make a conclusion about the area under the curve $f(x) = \sqrt{x}$ over the interval $[0, 4]$.

In Exercises 15–22, calculate the approximation for the given function and interval.

15. R_3, $\quad f(x) = 7 - x$, $\quad [3, 5]$

16. L_6, $\quad f(x) = \sqrt{6x + 2}$, $\quad [1, 3]$

17. M_6, $\quad f(x) = 4x + 3$, $\quad [5, 8]$

18. R_5, $\quad f(x) = x^2 + x$, $\quad [-1, 1]$

19. M_5, $\quad f(x) = \ln x$, $\quad [1, 3]$

20. M_4, $\quad f(x) = \sqrt{x}$, $\quad [3, 5]$

21. L_4, $\quad f(x) = \cos^2 x$, $\quad \left[\frac{\pi}{6}, \frac{\pi}{2}\right]$

22. L_6, $\quad f(x) = x^2 + 3|x|$, $\quad [-2, 1]$

In Exercises 23–28, write the sum in summation notation.

23. $4^7 + 5^7 + 6^7 + 7^7 + 8^7$

24. $(2^2 + 2) + (3^2 + 3) + (4^2 + 4) + (5^2 + 5)$

25. $(2^2 + 2) + (2^3 + 2) + (2^4 + 2) + (2^5 + 2)$

26. $\sqrt{1 + 1^3} + \sqrt{2 + 2^3} + \cdots + \sqrt{n + n^3}$

27. $\dfrac{1}{2 \cdot 3} + \dfrac{2}{3 \cdot 4} + \cdots + \dfrac{n}{(n + 1)(n + 2)}$

28. $e^{\pi} + e^{\pi/2} + e^{\pi/3} + \cdots + e^{\pi/n}$

29. Calculate the sums:

(a) $\displaystyle\sum_{i=1}^{5} 9$ 　　**(b)** $\displaystyle\sum_{i=0}^{5} 4$ 　　**(c)** $\displaystyle\sum_{k=2}^{4} k^3$

30. Calculate the sums:

(a) $\displaystyle\sum_{j=3}^{4} \sin\left(j\frac{\pi}{2}\right)$ 　　**(b)** $\displaystyle\sum_{k=3}^{5} \frac{1}{k-1}$ 　　**(c)** $\displaystyle\sum_{j=0}^{2} 3^{j-1}$

31. Let $b_1 = 4$, $b_2 = 1$, $b_3 = 2$, and $b_4 = -4$. Calculate:

(a) $\displaystyle\sum_{i=2}^{4} b_i$ (b) $\displaystyle\sum_{j=1}^{2}(2^{b_j} - b_j)$ (c) $\displaystyle\sum_{k=1}^{3} k b_k$

32. Assume that $a_1 = -5$, $\displaystyle\sum_{i=1}^{10} a_i = 20$, and $\displaystyle\sum_{i=1}^{10} b_i = 7$. Calculate:

(a) $\displaystyle\sum_{i=1}^{10}(4a_i + 3)$ (b) $\displaystyle\sum_{i=2}^{10} a_i$ (c) $\displaystyle\sum_{i=1}^{10}(2a_i - 3b_i)$

33. Calculate $\displaystyle\sum_{j=101}^{200} j$. *Hint:* Write as a difference of two sums and use formula (3).

34. Calculate $\displaystyle\sum_{j=1}^{30}(2j+1)^2$. *Hint:* Expand and use formulas (3)–(4).

In Exercises 35–42, use linearity and formulas (3)–(5) to rewrite and evaluate the sums.

35. $\displaystyle\sum_{j=1}^{20} 8j^3$

36. $\displaystyle\sum_{k=1}^{30}(4k-3)$

37. $\displaystyle\sum_{n=51}^{150} n^2$

38. $\displaystyle\sum_{k=101}^{200} k^3$

39. $\displaystyle\sum_{j=0}^{50} j(j-1)$

40. $\displaystyle\sum_{j=2}^{30}\left(6j + \frac{4j^2}{3}\right)$

41. $\displaystyle\sum_{m=1}^{30}(4-m)^3$

42. $\displaystyle\sum_{m=1}^{20}\left(5 + \frac{3m}{2}\right)^2$

In Exercises 43–46, use formulas (3)–(5) to evaluate the limit.

43. $\displaystyle\lim_{N\to\infty}\sum_{i=1}^{N}\frac{i}{N^2}$

44. $\displaystyle\lim_{N\to\infty}\sum_{j=1}^{N}\frac{j^3}{N^4}$

45. $\displaystyle\lim_{N\to\infty}\sum_{i=1}^{N}\frac{i^2 - i + 1}{N^3}$

46. $\displaystyle\lim_{N\to\infty}\sum_{i=1}^{N}\left(\frac{i^3}{N^4} - \frac{20}{N}\right)$

In Exercises 47–52, calculate the limit for the given function and interval. Verify your answer by using geometry.

47. $\displaystyle\lim_{N\to\infty} R_N$, $f(x) = 9x$, $[0, 2]$

48. $\displaystyle\lim_{N\to\infty} R_N$, $f(x) = 3x + 6$, $[1, 4]$

49. $\displaystyle\lim_{N\to\infty} L_N$, $f(x) = \frac{1}{2}x + 2$, $[0, 4]$

50. $\displaystyle\lim_{N\to\infty} L_N$, $f(x) = 4x - 2$, $[1, 3]$

51. $\displaystyle\lim_{N\to\infty} M_N$, $f(x) = x$, $[0, 2]$

52. $\displaystyle\lim_{N\to\infty} M_N$, $f(x) = 12 - 4x$, $[2, 6]$

53. Show, for $f(x) = 3x^2 + 4x$ over $[0, 2]$, that

$$R_N = \frac{2}{N}\sum_{j=1}^{N}\left(\frac{12j^2}{N^2} + \frac{8j}{N}\right)$$

Then evaluate $\displaystyle\lim_{N\to\infty} R_N$.

54. Show, for $f(x) = 3x^3 - x^2$ over $[1, 5]$, that

$$R_N = \frac{4}{N}\sum_{j=1}^{N}\left(\frac{192j^3}{N^3} + \frac{128j^2}{N^2} + \frac{28j}{N} + 2\right)$$

Then evaluate $\displaystyle\lim_{N\to\infty} R_N$.

In Exercises 55–62, find a formula for R_N and compute the area under the graph as a limit.

55. $f(x) = x^2$, $[0, 1]$ **56.** $f(x) = x^2$, $[-1, 5]$

57. $f(x) = 6x^2 - 4$, $[2, 5]$ **58.** $f(x) = x^2 + 7x$, $[6, 11]$

59. $f(x) = x^3 - x$, $[0, 2]$

60. $f(x) = 2x^3 + x^2$, $[-2, 2]$

61. $f(x) = 2x + 1$, $[a, b]$ (a, b constants with $a < b$)

62. $f(x) = x^2$, $[a, b]$ (a, b constants with $a < b$)

In Exercises 63–66, describe the area represented by the limits.

63. $\displaystyle\lim_{N\to\infty}\frac{1}{N}\sum_{j=1}^{N}\left(\frac{j}{N}\right)^4$

64. $\displaystyle\lim_{N\to\infty}\frac{3}{N}\sum_{j=1}^{N}\left(2 + \frac{3j}{N}\right)^4$

65. $\displaystyle\lim_{N\to\infty}\frac{5}{N}\sum_{j=0}^{N-1} e^{-2+5j/N}$

66. $\displaystyle\lim_{N\to\infty}\frac{\pi}{2N}\sum_{j=1}^{N}\sin\left(\frac{\pi}{3} - \frac{\pi}{4N} + \frac{j\pi}{2N}\right)$

In Exercises 67–72, express the area under the graph as a limit using the approximation indicated (in summation notation), but do not evaluate.

67. R_N, $f(x) = \sin x$ over $[0, \pi]$

68. R_N, $f(x) = x^{-1}$ over $[1, 7]$

69. L_N, $f(x) = \sqrt{2x + 1}$ over $[7, 11]$

70. L_N, $f(x) = \cos x$ over $\left[\frac{\pi}{8}, \frac{\pi}{4}\right]$

71. M_N, $f(x) = \tan x$ over $\left[\frac{1}{2}, 1\right]$

72. M_N, $f(x) = x^{-2}$ over $[3, 5]$

73. Evaluate $\displaystyle\lim_{N\to\infty}\frac{1}{N}\sum_{j=1}^{N}\sqrt{1 - \left(\frac{j}{N}\right)^2}$ by interpreting it as the area of part of a familiar geometric figure.

In Exercises 74–76, let $f(x) = x^2$ and let R_N, L_N, and M_N be the approximations for the interval $[0, 1]$.

74. ✏️ Show that $R_N = \dfrac{1}{3} + \dfrac{1}{2N} + \dfrac{1}{6N^2}$. Interpret the quantity $\dfrac{1}{2N} + \dfrac{1}{6N^2}$ as the area of a region.

75. Show that

$$L_N = \frac{1}{3} - \frac{1}{2N} + \frac{1}{6N^2}, \qquad M_N = \frac{1}{3} - \frac{1}{12N^2}$$

Then rank the three approximations R_N, L_N, and M_N in order of increasing accuracy (use Exercise 74).

76. For each of R_N, L_N, and M_N, find the smallest integer N for which the error is less than 0.001.

In Exercises 77–82, use the Graphical Insight on page 264 to obtain bounds on the area.

77. Let A be the area under $f(x) = \sqrt{x}$ over $[0, 1]$. Prove that $0.51 \leq A \leq 0.77$ by computing R_4 and L_4. Explain your reasoning.

78. Use R_5 and L_5 to show that the area A under $y = x^{-2}$ over $[10, 13]$ satisfies $0.0218 \leq A \leq 0.0244$.

79. Use R_4 and L_4 to show that the area A under the graph of $y = \sin x$ over $\left[0, \frac{\pi}{2}\right]$ satisfies $0.79 \leq A \leq 1.19$.

80. Show that the area A under $f(x) = x^{-1}$ over $[1, 8]$ satisfies

$$\tfrac{1}{2} + \tfrac{1}{3} + \tfrac{1}{4} + \tfrac{1}{5} + \tfrac{1}{6} + \tfrac{1}{7} + \tfrac{1}{8} \leq A \leq 1 + \tfrac{1}{2} + \tfrac{1}{3} + \tfrac{1}{4} + \tfrac{1}{5} + \tfrac{1}{6} + \tfrac{1}{7}$$

81. *CAS* Show that the area A under $y = x^{1/4}$ over $[0, 1]$ satisfies $L_N \leq A \leq R_N$ for all N. Use a computer algebra system to calculate L_N and R_N for $N = 100$ and 200, and determine A to two decimal places.

82. *CAS* Show that the area A under $y = 4/(x^2 + 1)$ over $[0, 1]$ satisfies $R_N \leq A \leq L_N$ for all N. Determine A to at least three decimal places using a computer algebra system. Can you guess the exact value of A?

83. In this exercise, we evaluate the area A under the graph of $y = e^x$ over $[0, 1]$ [Figure 19(A)] using the formula for a geometric sum (valid for $r \neq 1$):

$$1 + r + r^2 + \cdots + r^{N-1} = \sum_{j=0}^{N-1} r^j = \frac{r^N - 1}{r - 1} \qquad \boxed{8}$$

(a) Show that $L_N = \dfrac{1}{N} \displaystyle\sum_{j=0}^{N-1} e^{j/N}$.

(b) Apply Eq. (8) with $r = e^{1/N}$ to prove $L_N = \dfrac{e - 1}{N(e^{1/N} - 1)}$.

(c) Compute $A = \displaystyle\lim_{N \to \infty} L_N$ using L'Hôpital's Rule.

84. Use the result of Exercise 83 to show that the area B under the graph of $f(x) = \ln x$ over $[1, e]$ is equal to 1. *Hint:* Relate B in Figure 19(B) to the area A computed in Exercise 83.

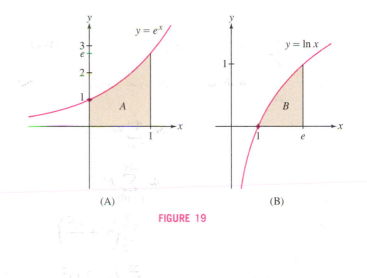

(A)   (B)

FIGURE 19

Further Insights and Challenges

85. Although the accuracy of R_N generally improves as N increases, this need not be true for small values of N. Draw the graph of a positive continuous function f on an interval such that R_1 is closer than R_2 to the exact area under the graph. Can such a function be monotonic?

86. Draw the graph of a positive continuous function on an interval such that R_2 and L_2 are both smaller than the exact area under the graph. Can such a function be monotonic?

87. Explain graphically: *The endpoint approximations are less accurate when $f'(x)$ is large.*

88. Prove that for any function f on $[a, b]$,

$$R_N - L_N = \frac{b - a}{N}(f(b) - f(a)) \qquad \boxed{9}$$

89. In this exercise, we prove that $\displaystyle\lim_{N \to \infty} R_N$ and $\displaystyle\lim_{N \to \infty} L_N$ exist and are equal if f is increasing [the case of f decreasing is similar]. We use the concept of a least upper bound discussed in Appendix B.

(a) Explain with a graph why $L_N \leq R_M$ for all $N, M \geq 1$.

(b) By (a), the sequence $\{L_N\}$ is bounded, so it has a least upper bound L. By definition, L is the smallest number such that $L_N \leq L$ for all N. Show that $L \leq R_M$ for all M.

(c) According to (b), $L_N \leq L \leq R_N$ for all N. Use Eq. (9) to show that $\displaystyle\lim_{N \to \infty} L_N = L$ and $\displaystyle\lim_{N \to \infty} R_N = L$.

90. Use Eq. (9) to show that if f is positive and monotonic, then the area A under its graph over $[a, b]$ satisfies

$$|R_N - A| \leq \frac{b - a}{N}|f(b) - f(a)| \qquad \boxed{10}$$

In Exercises 91–92, use Eq. (10) to find a value of N such that $|R_N - A| < 10^{-4}$ for the given function and interval.

91. $f(x) = \sqrt{x}$, $[1, 4]$ **92.** $f(x) = \sqrt{9 - x^2}$, $[0, 3]$

93. Prove that if f is positive and monotonic, then M_N lies between R_N and L_N and is closer to the actual area under the graph than both R_N and L_N. *Hint:* In the case that f is increasing, Figure 20 shows that the part of the error in R_N due to the ith rectangle is the sum of the areas $A + B + D$, and for M_N it is $|B - E|$. On the other hand, $A \geq E$.

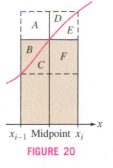

FIGURE 20

5.2 The Definite Integral

In the previous section, we saw that if f is continuous on an interval $[a, b]$, then the endpoint and midpoint approximations approach a common limit L as $N \to \infty$:

$$L = \lim_{N \to \infty} R_N = \lim_{N \to \infty} L_N = \lim_{N \to \infty} M_N \qquad \boxed{1}$$

When $f(x) \geq 0$, L is the area under the graph of f. In a moment, we will state formally that L is the *definite integral* of f over $[a, b]$. Before doing so, we introduce more general approximations called **Riemann sums**.

Recall that R_N, L_N, and M_N use rectangles of equal width Δx, whose heights are the values of $f(x)$ at the endpoints or midpoints of the subintervals. In Riemann sum approximations, we relax these requirements: The rectangles need not have equal width, and the height may be *any* value of $f(x)$ within the subinterval.

To specify a Riemann sum, we choose a partition and a set of sample points:

- **Partition** P of size N: a choice of points that divides $[a, b]$ into N subintervals.

$$P : a = x_0 < x_1 < x_2 < \cdots < x_N = b$$

- **Sample points** $C = \{c_1, \ldots, c_N\}$: c_i belongs to the subinterval $[x_{i-1}, x_i]$ for all $i = 1, \ldots, N$.

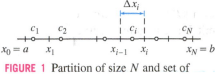

FIGURE 1 Partition of size N and set of sample points.

See Figures 1 and 2(A). The length of the ith subinterval $[x_{i-1}, x_i]$ is

$$\boxed{\Delta x_i = x_i - x_{i-1}}$$

The **norm** of P, denoted $\|P\|$, is the maximum of the lengths Δx_i.

Given P and C, we construct the rectangle of height $f(c_i)$ and base Δx_i over each subinterval $[x_{i-1}, x_i]$, as in Figure 2(B). This rectangle has area $f(c_i)\Delta x_i$ if $f(c_i) \geq 0$. If $f(c_i) < 0$, the rectangle extends below the x-axis, and $f(c_i)\Delta x_i$ is the negative of its area. The Riemann sum is the sum

Keep in mind that R_N, L_N, and M_N are particular examples of Riemann sums in which $\Delta x_i = (b - a)/N$ for all i, and the sample points c_i are either endpoints or midpoints.

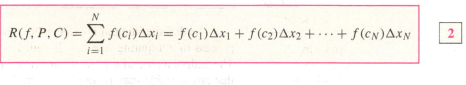

$$R(f, P, C) = \sum_{i=1}^{N} f(c_i)\Delta x_i = f(c_1)\Delta x_1 + f(c_2)\Delta x_2 + \cdots + f(c_N)\Delta x_N \qquad \boxed{2}$$

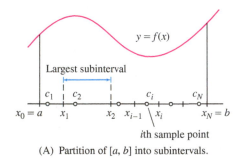

$y = f(x)$

Largest subinterval

(A) Partition of $[a, b]$ into subintervals.

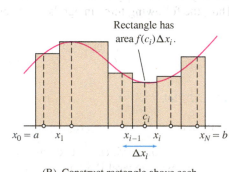

Rectangle has area $f(c_i)\Delta x_i$.

(B) Construct rectangle above each subinterval of height $f(c_i)$.

(C) Rectangles corresponding to a Riemann sum with $\|P\|$ small (a large number of rectangles).

FIGURE 2 Construction of $R(f, P, C)$. In this case, Δx_2 is the norm of the partition.

■ **EXAMPLE 1** Calculate the Riemann sum $R(f, P, C)$, where $f(x) = 8 + 12\sin x - 4x$ on $[0, 4]$,

$$P : x_0 = 0 < x_1 = 1 < x_2 = 1.8 < x_3 = 2.9 < x_4 = 4$$

$$C = \{0.4, 1.2, 2, 3.5\}$$

What is the norm $\|P\|$?

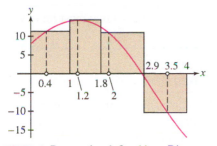

FIGURE 3 Rectangles defined by a Riemann sum for $f(x) = 8 + 12 \sin x - 4x$.

Solution The widths of the subintervals in the partition (Figure 3) are

$$\Delta x_1 = x_1 - x_0 = 1 - 0 = 1, \qquad \Delta x_2 = x_2 - x_1 = 1.8 - 1 = 0.8$$

$$\Delta x_3 = x_3 - x_2 = 2.9 - 1.8 = 1.1, \qquad \Delta x_4 = x_4 - x_3 = 4 - 2.9 = 1.1$$

The norm of the partition is $\|P\| = 1.1$ since the two longest subintervals have width 1.1. Using a calculator, we obtain

$$R(f, P, C) = f(0.4)\Delta x_1 + f(1.2)\Delta x_2 + f(2)\Delta x_3 + f(3.5)\Delta x_4$$

$$\approx 11.07(1) + 14.38(0.8) + 10.91(1.1) - 10.2(1.1) \approx 23.35 \qquad \blacksquare$$

Note in Figure 2(C) that as the norm $\|P\|$ tends to zero (meaning that the rectangles get thinner), the number of rectangles N tends to ∞ and they approximate the area under the graph more closely. This leads to the following definition: f is **integrable** over $[a, b]$ if *all* of the Riemann sums (not just the endpoint and midpoint approximations) approach one and the same limit L as $\|P\|$ tends to zero. Formally, we write

$$L = \lim_{\|P\| \to 0} R(f, P, C) = \lim_{\|P\| \to 0} \sum_{i=1}^{N} f(c_i)\Delta x_i \qquad \boxed{3}$$

if $|R(f, P, C) - L|$ gets arbitrarily small as the norm $\|P\|$ tends to zero, no matter how we choose the partition and sample points. The limit L is called the **definite integral** of f over $[a, b]$.

The notation $\int f(x)\,dx$ was introduced by Leibniz in 1686. The symbol $\int$ is an elongated S standing for "summation." The differential dx corresponds to the length Δx_i along the x-axis.

> **DEFINITION Definite Integral** The definite integral of f over $[a, b]$, denoted by the integral sign, is the limit of Riemann sums:
>
> $$\int_a^b f(x)\,dx = \lim_{\|P\| \to 0} R(f, P, C) = \lim_{\|P\| \to 0} \sum_{i=1}^{N} f(c_i)\Delta x_i$$
>
> When this limit exists, we say that f is integrable over $[a, b]$.

The definite integral is often called, more simply, the *integral* of f over $[a, b]$. The process of computing integrals is called **integration** and $f(x)$ is called the **integrand**. The endpoints a and b of $[a, b]$ are called the **limits of integration**. Finally, we remark that any variable may be used as a variable of integration (this is a "dummy" variable). Thus, the following three integrals all denote the same quantity:

$$\int_a^b \sin x\,dx, \qquad \int_a^b \sin t\,dt, \qquad \int_a^b \sin u\,du$$

One of the greatest mathematicians of the nineteenth century and perhaps second only to his teacher C. F. Gauss, Riemann transformed the fields of geometry, analysis, and number theory. Albert Einstein based his General Theory of Relativity on Riemann's geometry. The "Riemann Hypothesis" dealing with prime numbers is one of the great unsolved problems in present-day mathematics. The Clay Foundation has offered a $1 million prize for its solution (http://claymath.org/millenium-problems/riemann-hypothesis).

> **CONCEPTUAL INSIGHT** Keep in mind that a Riemann sum $R(f, P, C)$ is nothing more than an approximation to area based on rectangles, and that $\int_a^b f(x)\,dx$ is the number we obtain in the limit as we take thinner and thinner rectangles.
>
> However, general Riemann sums (with arbitrary partitions and sample points) are rarely used for computations. In practice, we use particular approximations such as M_N, or the Fundamental Theorem of Calculus, as we'll learn shortly. If so, why bother introducing Riemann sums? The answer is that Riemann sums play a theoretical rather than a computational role. They are useful in proofs and for dealing rigorously with certain discontinuous functions. In later sections, Riemann sums are used to show that volumes and other quantities can be expressed as definite integrals.

The next theorem assures us that continuous functions (and even functions with finitely many jump discontinuities) are integrable (see Appendix D for a proof). In practice, we rely on this theorem rather than attempting to prove directly that a given function is integrable.

Georg Friedrich Riemann (1826–1866)

THEOREM 1   If f is continuous on $[a, b]$, or if f is continuous with at most finitely many jump discontinuities, then f is integrable over $[a, b]$.

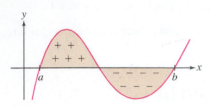

FIGURE 4 Signed area is the area above the x-axis minus the area below the x-axis.

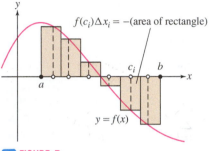

DF FIGURE 5

Interpretation of the Definite Integral as Signed Area

When $f(x) \geq 0$, the definite integral defines the area under the graph. To interpret the integral when $f(x)$ takes on both positive and negative values, we define the notion of **signed area**, where regions below the x-axis are considered to have "negative area" (Figure 4); that is,

$$\boxed{\text{Signed area of a region} = (\text{area above } x\text{-axis}) - (\text{area below } x\text{-axis})}$$

Now observe that a Riemann sum is equal to the signed area of the corresponding rectangles:

$$R(f, C, P) = f(c_1)\Delta x_1 + f(c_2)\Delta x_2 + \cdots + f(c_N)\Delta x_N$$

Indeed, if $f(c_i) < 0$, then the corresponding rectangle lies below the x-axis and has signed area $f(c_i)\Delta x_i$ (Figure 5). The limit of the Riemann sums is the signed area of the region between the graph and the x-axis:

$$\boxed{\int_a^b f(x)\, dx = \text{signed area of region between the graph and } x\text{-axis over } [a, b]}$$

■ **EXAMPLE 2   Signed Area**   Calculate

$$\int_0^5 (3 - x)\, dx \qquad \text{and} \qquad \int_0^5 |3 - x|\, dx$$

Solution   The region between $y = 3 - x$ and the x-axis consists of two triangles of areas $\frac{9}{2}$ and 2 [Figure 6(A)]. However, the second triangle lies below the x-axis, so it has signed area -2. In the graph of $y = |3 - x|$, both triangles lie above the x-axis [Figure 6(B)]. Therefore,

$$\int_0^5 (3 - x)\, dx = \frac{9}{2} - 2 = \frac{5}{2} \qquad \int_0^5 |3 - x|\, dx = \frac{9}{2} + 2 = \frac{13}{2} \qquad ■$$

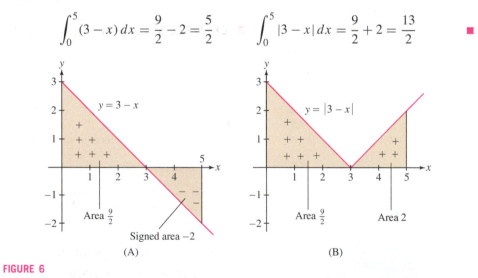

FIGURE 6

Properties of the Definite Integral

In the rest of this section, we discuss some basic properties of definite integrals. First, we note that the integral of a constant function $f(x) = C$ over $[a, b]$ is the signed area $C(b - a)$ of a rectangle (as we can see from Figure 7).

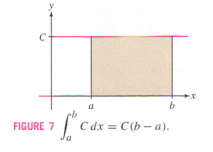

FIGURE 7 $\int_a^b C \, dx = C(b - a)$.

THEOREM 2 Integral of a Constant For any constant C,

$$\int_a^b C \, dx = C(b - a)$$

| | 4 |

Next, we state the linearity properties of the definite integral.

THEOREM 3 Linearity of the Definite Integral If f and g are integrable over $[a, b]$, then $f + g$ and Cf are integrable (for any constant C), and

- $\int_a^b \big(f(x) + g(x)\big) \, dx = \int_a^b f(x) \, dx + \int_a^b g(x) \, dx$

- $\int_a^b Cf(x) \, dx = C \int_a^b f(x) \, dx$

Proof These properties follow from the corresponding linearity properties of sums and limits. For example, Riemann sums are additive:

$$R(f + g, P, C) = \sum_{i=1}^{N} \big(f(c_i) + g(c_i)\big)\Delta x_i = \sum_{i=1}^{N} f(c_i)\Delta x_i + \sum_{i=1}^{N} g(c_i)\Delta x_i$$

$$= R(f, P, C) + R(g, P, C)$$

By the additivity of limits,

$$\int_a^b (f(x) + g(x)) \, dx = \lim_{\|P\| \to 0} R(f + g, P, C)$$

$$= \lim_{\|P\| \to 0} R(f, P, C) + \lim_{\|P\| \to 0} R(g, P, C)$$

$$= \int_a^b f(x) \, dx + \int_a^b g(x) \, dx$$

The second property is proved similarly. ∎

■ **EXAMPLE 3** Calculate $\int_0^3 (2x^2 - 5) \, dx$ using the formula

Eq. (5) was verified in Example 5 of Section 5.1.

$$\int_0^b x^2 \, dx = \frac{b^3}{3}$$

| | 5 |

Solution

$$\int_0^3 (2x^2 - 5) \, dx = 2 \int_0^3 x^2 \, dx + \int_0^3 (-5) \, dx \qquad \text{(linearity)}$$

$$= 2\left(\frac{3^3}{3}\right) - 5(3 - 0) = 3 \qquad \text{[Eqs. (5) and (4)]} ∎$$

So far we have used the notation $\int_a^b f(x) \, dx$ with the understanding that $a < b$. It is convenient to define the definite integral for arbitrary a and b.

According to Eq. (6), the integral changes sign when the limits of integration are reversed. Since we are free to define symbols as we please, why have we chosen to put the minus sign on the right side of Eq. (6)? Because it is only with this definition that the Fundamental Theorem of Calculus holds true.

DEFINITION Reversing the Limits of Integration For $a < b$, we set

$$\int_b^a f(x) \, dx = - \int_a^b f(x) \, dx$$

| | 6 |

For example, by Eq. (5),

$$\int_5^0 x^2\, dx = -\int_0^5 x^2\, dx = -\frac{5^3}{3} = -\frac{125}{3}$$

When $a = b$, the interval $[a, b] = [a, a]$ has length zero and we define the definite integral to be zero:

$$\int_a^a f(x)\, dx = 0$$

EXAMPLE 4 Prove that, for all b (positive or negative),

$$\int_0^b x\, dx = \frac{1}{2}b^2$$

7

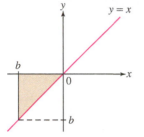

FIGURE 8 Here, $b < 0$ and the signed area is $-\frac{1}{2}b^2$.

Solution If $b > 0$, $\int_0^b x\, dx$ is the area $\frac{1}{2}b^2$ of a triangle of base b and height b. If $b < 0$, $\int_b^0 x\, dx$ is the signed area $-\frac{1}{2}b^2$ of the triangle in Figure 8, and Eq. (7) follows from the rule for reversing limits of integration:

$$\int_0^b x\, dx = -\int_b^0 x\, dx = -\left(-\frac{1}{2}b^2\right) = \frac{1}{2}b^2 \qquad \blacksquare$$

Definite integrals satisfy an important additivity property: If f is integrable and $a \le b \le c$ as in Figure 9, then the integral from a to c is equal to the integral from a to b *plus* the integral from b to c. We state this in the next theorem (a formal proof can be given using Riemann sums).

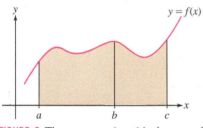

FIGURE 9 The area over $[a, c]$ is the *sum* of the areas over $[a, b]$ and $[b, c]$.

THEOREM 4 **Additivity for Adjacent Intervals** Let $a \le b \le c$, and assume that f is integrable. Then

$$\int_a^c f(x)\, dx = \int_a^b f(x)\, dx + \int_b^c f(x)\, dx$$

This theorem remains true as stated even if the condition $a \le b \le c$ is not satisfied (Exercise 88).

EXAMPLE 5 Calculate $\int_4^7 x^2\, dx$.

Solution Before we can apply the formula $\int_0^b x^2\, dx = b^3/3$ from Example 3, we must use the additivity property for adjacent intervals to write

$$\int_0^4 x^2\, dx + \int_4^7 x^2\, dx = \int_0^7 x^2\, dx$$

Now we can compute our integral as a difference:

$$\int_4^7 x^2\, dx = \int_0^7 x^2\, dx - \int_0^4 x^2\, dx = \left(\frac{1}{3}\right)7^3 - \left(\frac{1}{3}\right)4^3 = 93 \qquad \blacksquare$$

Another basic property of the definite integral is that larger functions have larger integrals (Figure 10).

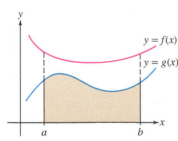

FIGURE 10 The integral of f is larger than the integral of g.

THEOREM 5 **Comparison Theorem** If f and g are integrable and $g(x) \le f(x)$ for x in $[a, b]$, then

$$\int_a^b g(x)\, dx \le \int_a^b f(x)\, dx$$

Proof If $g(x) \le f(x)$, then for any partition and choice of sample points, we have $g(c_i)\Delta x_i \le f(c_i)\Delta x_i$ for all i. Therefore, the Riemann sums satisfy

$$\sum_{i=1}^{N} g(c_i)\Delta x_i \le \sum_{i=1}^{N} f(c_i)\Delta x_i$$

Taking the limit as the norm $\|P\|$ tends to zero, we obtain

$$\int_a^b g(x)\, dx = \lim_{\|P\| \to 0} \sum_{i=1}^{N} g(c_i)\Delta x_i \le \lim_{\|P\| \to 0} \sum_{i=1}^{N} f(c_i)\Delta x_i = \int_a^b f(x)\, dx \qquad \blacksquare$$

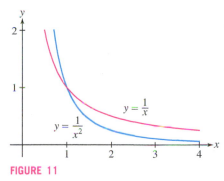

FIGURE 11

■ **EXAMPLE 6** Prove the inequality $\displaystyle\int_1^4 \frac{1}{x^2}\, dx \le \int_1^4 \frac{1}{x}\, dx$.

Solution If $x \ge 1$, then $x^2 \ge x$, and $x^{-2} \le x^{-1}$ (Figure 11). Therefore, the inequality follows from the Comparison Theorem, applied with $g(x) = x^{-2}$ and $f(x) = x^{-1}$. ■

Suppose there are numbers m and M such that $m \le f(x) \le M$ for x in $[a, b]$. We call m and M **lower** and **upper bounds** for $f(x)$ on $[a, b]$. By the Comparison Theorem,

$$\int_a^b m\, dx \le \int_a^b f(x)\, dx \le \int_a^b M\, dx$$

$$\boxed{m(b-a) \le \int_a^b f(x)\, dx \le M(b-a)} \qquad \boxed{8}$$

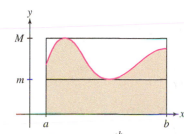

FIGURE 12 The integral $\int_a^b f(x)\, dx$ lies between the areas of the rectangles of heights m and M.

This says simply that the integral of f lies between the areas of two rectangles (Figure 12).

■ **EXAMPLE 7** Prove the inequalities $\displaystyle\frac{3}{4} \le \int_{1/2}^2 \frac{1}{x}\, dx \le 3$.

Solution Because $f(x) = x^{-1}$ is decreasing (Figure 13), its minimum value on $\left[\frac{1}{2}, 2\right]$ is $m = f(2) = \frac{1}{2}$ and its maximum value is $M = f\left(\frac{1}{2}\right) = 2$. By Eq. (8),

$$\underbrace{\frac{1}{2}\left(2 - \frac{1}{2}\right)}_{m(b-a)} = \frac{3}{4} \le \int_{1/2}^2 \frac{1}{x}\, dx \le \underbrace{2\left(2 - \frac{1}{2}\right)}_{M(b-a)} = 3 \qquad \blacksquare$$

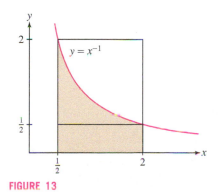

FIGURE 13

5.2 SUMMARY

- A Riemann sum $R(f, P, C)$ for the interval $[a, b]$ is defined by choosing a *partition*

$$P: a = x_0 < x_1 < x_2 < \cdots < x_N = b$$

and *sample points* $C = \{c_i\}$, where $c_i \in [x_{i-1}, x_i]$. Let $\Delta x_i = x_i - x_{i-1}$. Then

$$R(f, P, C) = \sum_{i=1}^{N} f(c_i)\Delta x_i$$

- The maximum of the widths Δx_i is called the norm $\|P\|$ of the partition.
- The *definite integral* is the limit of the Riemann sums (if it exists):

$$\int_a^b f(x)\, dx = \lim_{\|P\| \to 0} R(f, P, C)$$

We say that f is *integrable* over $[a, b]$ if the limit exists.

- Theorem: If f is continuous on $[a, b]$, then f is integrable over $[a, b]$.
- $\displaystyle\int_a^b f(x)\,dx =$ *signed area* of the region between the graph of f and the x-axis.
- Properties of definite integrals:

$$\int_a^b \big(f(x) + g(x)\big)\,dx = \int_a^b f(x)\,dx + \int_a^b g(x)\,dx$$

$$\int_a^b Cf(x)\,dx = C\int_a^b f(x)\,dx \quad \text{for any constant } C$$

$$\int_b^a f(x)\,dx = -\int_a^b f(x)\,dx$$

$$\int_a^a f(x)\,dx = 0$$

$$\int_a^b f(x)\,dx + \int_b^c f(x)\,dx = \int_a^c f(x)\,dx \quad \text{for all } a, b, c$$

- Formulas:

$$\int_a^b C\,dx = C(b - a) \quad (C \text{ any constant})$$

$$\int_0^b x\,dx = \frac{1}{2}b^2$$

$$\int_0^b x^2\,dx = \frac{1}{3}b^3$$

- Comparison Theorem: If $f(x) \le g(x)$ on $[a, b]$, then

$$\int_a^b f(x)\,dx \le \int_a^b g(x)\,dx$$

If $m \le f(x) \le M$ on $[a, b]$, then

$$m(b - a) \le \int_a^b f(x)\,dx \le M(b - a)$$

5.2 EXERCISES

Preliminary Questions

1. What is $\displaystyle\int_3^5 dx$ [the function is $f(x) = 1$]?

2. Let $I = \displaystyle\int_2^7 f(x)\,dx$, where f is continuous. State whether the following are true or false:
(a) I is the area between the graph and the x-axis over $[2, 7]$.
(b) If $f(x) \ge 0$, then I is the area between the graph and the x-axis over $[2, 7]$.

(c) If $f(x) \le 0$, then $-I$ is the area between the graph of f and the x-axis over $[2, 7]$.

3. Explain graphically: $\displaystyle\int_0^\pi \cos x\,dx = 0$.

4. Which is negative, $\displaystyle\int_{-1}^{-5} 8\,dx$ or $\displaystyle\int_{-5}^{-1} 8\,dx$?

Exercises

In Exercises 1–10, draw a graph of the signed area represented by the integral and compute it using geometry.

1. $\displaystyle\int_{-3}^{3} 2x\,dx$

2. $\displaystyle\int_{-2}^{3} (2x + 4)\,dx$

3. $\displaystyle\int_{-2}^{1} (3x + 4)\,dx$

4. $\displaystyle\int_{-2}^{1} 4\,dx$

5. $\displaystyle\int_{6}^{8} (7 - x)\,dx$

6. $\displaystyle\int_{\pi/2}^{3\pi/2} \sin x\,dx$

7. $\displaystyle\int_{0}^{5} \sqrt{25 - x^2}\,dx$

8. $\displaystyle\int_{-2}^{3} |x|\,dx$

9. $\displaystyle\int_{-2}^{2} (2 - |x|)\,dx$

10. $\displaystyle\int_{-2}^{5} (3 + x - 2|x|)\,dx$

11. Calculate $\int_0^{10} (8-x)\,dx$ in two ways:

(a) As the limit $\lim_{N\to\infty} R_N$

(b) By sketching the relevant signed area and using geometry

12. Calculate $\int_{-1}^{4} (4x-8)\,dx$ in two ways:

(a) As the limit $\lim_{N\to\infty} R_N$

(b) By using geometry

In Exercises 13 and 14, refer to Figure 14.

13. Evaluate: (a) $\int_0^2 f(x)\,dx$ (b) $\int_0^6 f(x)\,dx$

14. Evaluate: (a) $\int_1^4 f(x)\,dx$ (b) $\int_1^6 |f(x)|\,dx$

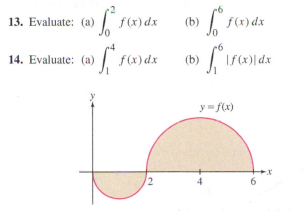

FIGURE 14 The two parts of the graph are semicircles.

In Exercises 15 and 16, refer to Figure 15.

15. Evaluate $\int_0^3 g(t)\,dt$ and $\int_3^5 g(t)\,dt$.

16. Find a, b, and c such that $\int_0^a g(t)\,dt$ and $\int_b^c g(t)\,dt$ are as large as possible.

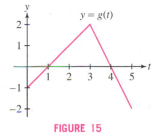

FIGURE 15

17. Describe the partition P and the set of sample points C for the Riemann sum shown in Figure 16. Compute the value of the Riemann sum.

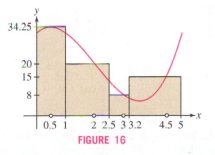

FIGURE 16

18. Compute $R(f, P, C)$ for $f(x) = x^2 + x$ for the partition P and the set of sample points C in Figure 16. [The curve shown is not $f(x) = x^2 + x$.]

In Exercises 19–22, calculate the Riemann sum $R(f, P, C)$ for the given function, partition, and choice of sample points. Also, sketch the graph of f and the rectangles corresponding to $R(f, P, C)$.

19. $f(x) = x$, $P = \{1, 1.2, 1.5, 2\}$, $C = \{1.1, 1.4, 1.9\}$

20. $f(x) = 2x + 3$, $P = \{-4, -1, 1, 4, 8\}$, $C = \{-3, 0, 2, 5\}$

21. $f(x) = x^2 + x$, $P = \{2, 3, 4.5, 5\}$, $C = \{2, 3.5, 5\}$

22. $f(x) = \sin x$, $P = \{0, \frac{\pi}{6}, \frac{\pi}{3}, \frac{\pi}{2}\}$, $C = \{0.4, 0.7, 1.2\}$

In Exercises 23–28, sketch the signed area represented by the integral. Indicate the regions of positive and negative area.

23. $\int_0^5 (4x - x^2)\,dx$

24. $\int_{-\pi/4}^{\pi/4} \tan x\,dx$

25. $\int_\pi^{2\pi} \sin x\,dx$

26. $\int_0^{3\pi} \sin x\,dx$

27. $\int_{1/2}^2 \ln x\,dx$

28. $\int_{-1}^1 \tan^{-1} x\,dx$

In Exercises 29–32, determine the sign of the integral without calculating it. Draw a graph if necessary.

29. $\int_{-2}^1 x^4\,dx$

30. $\int_{-2}^1 x^3\,dx$

31. GU $\int_0^{2\pi} x \sin x\,dx$

32. GU $\int_0^{2\pi} \frac{\sin x}{x}\,dx$

In Exercises 33–42, use properties of the integral and the formulas in the summary to calculate the integrals.

33. $\int_0^4 (6t - 3)\,dt$

34. $\int_{-3}^2 (4x + 7)\,dx$

35. $\int_0^9 x^2\,dx$

36. $\int_2^5 x^2\,dx$

37. $\int_0^1 (u^2 - 2u)\,du$

38. $\int_0^{1/2} (12y^2 + 6y)\,dy$

39. $\int_{-3}^1 (7t^2 + t + 1)\,dt$

40. $\int_{-3}^3 (9x - 4x^2)\,dx$

41. $\int_{-a}^1 (x^2 + x)\,dx$

42. $\int_a^{a^2} x^2\,dx$

In Exercises 43–47, calculate the integral, assuming that

$$\int_0^5 f(x)\,dx = 5, \qquad \int_0^5 g(x)\,dx = 12$$

43. $\int_0^5 (f(x) + g(x))\,dx$

44. $\int_0^5 \left(2f(x) - \frac{1}{3}g(x)\right)\,dx$

45. $\int_5^0 g(x)\,dx$

46. $\int_0^5 (f(x) - x)\,dx$

47. Is it possible to calculate $\int_0^5 g(x)f(x)\,dx$ from the information given?

48. Prove by computing the limit of right-endpoint approximations:

$$\int_0^b x^3 \, dx = \frac{b^4}{4} \qquad \boxed{9}$$

In Exercises 49–54, evaluate the integral using the formulas in the summary and Eq. (9).

49. $\int_0^3 x^3 \, dx$

50. $\int_1^3 x^3 \, dx$

51. $\int_0^3 (x - x^3) \, dx$

52. $\int_0^1 (2x^3 - x + 4) \, dx$

53. $\int_0^1 (12x^3 + 24x^2 - 8x) \, dx$

54. $\int_{-2}^2 (2x^3 - 3x^2) \, dx$

In Exercises 55–58, calculate the integral, assuming that

$$\int_0^1 f(x) \, dx = 1, \qquad \int_0^2 f(x) \, dx = 4, \qquad \int_1^4 f(x) \, dx = 7$$

55. $\int_0^4 f(x) \, dx$

56. $\int_1^2 f(x) \, dx$

57. $\int_4^1 f(x) \, dx$

58. $\int_2^4 f(x) \, dx$

In Exercises 59–62, express each integral as a single integral.

59. $\int_0^3 f(x) \, dx + \int_3^7 f(x) \, dx$

60. $\int_2^9 f(x) \, dx - \int_4^9 f(x) \, dx$

61. $\int_2^9 f(x) \, dx - \int_2^5 f(x) \, dx$

62. $\int_7^3 f(x) \, dx + \int_3^9 f(x) \, dx$

In Exercises 63–66, calculate the integral, assuming that f is integrable and $\int_1^b f(x) \, dx = 1 - b^{-1}$ for all $b > 0$.

63. $\int_1^5 f(x) \, dx$

64. $\int_3^5 f(x) \, dx$

65. $\int_1^6 (3f(x) - 4) \, dx$

66. $\int_{1/2}^1 f(x) \, dx$

67. ✏️ Explain the difference in graphical interpretation between $\int_a^b f(x) \, dx$ and $\int_a^b |f(x)| \, dx$.

68. ✏️ Use the graphical interpretation of the definite integral to explain the inequality

$$\left| \int_a^b f(x) \, dx \right| \le \int_a^b |f(x)| \, dx$$

where f is continuous. Explain also why equality holds if and only if either $f(x) \ge 0$ for all x or $f(x) \le 0$ for all x.

69. ✏️ Let $f(x) = x$. Find an interval $[a, b]$ such that

$$\left| \int_a^b f(x) \, dx \right| = \frac{1}{2} \qquad \text{and} \qquad \int_a^b |f(x)| \, dx = \frac{3}{2}$$

70. ✏️ Evaluate $I = \int_0^{2\pi} \sin^2 x \, dx$ and $J = \int_0^{2\pi} \cos^2 x \, dx$ as follows. First show with a graph that $I = J$. Then prove that $I + J = 2\pi$.

In Exercises 71–74, calculate the integral.

71. $\int_0^6 |3 - x| \, dx$

72. $\int_1^3 |2x - 4| \, dx$

73. $\int_{-1}^1 |x^3| \, dx$

74. $\int_0^2 |x^2 - 1| \, dx$

75. Use the Comparison Theorem to show that

$$\int_0^1 x^5 \, dx \le \int_0^1 x^4 \, dx, \qquad \int_1^2 x^4 \, dx \le \int_1^2 x^5 \, dx$$

76. Prove that $\frac{1}{3} \le \int_4^6 \frac{1}{x} \, dx \le \frac{1}{2}$.

77. Prove that $0.0198 \le \int_{0.2}^{0.3} \sin x \, dx \le 0.0296$. *Hint:* Show that $0.198 \le \sin x \le 0.296$ for x in $[0.2, 0.3]$.

78. Prove that $0.277 \le \int_{\pi/8}^{\pi/4} \cos x \, dx \le 0.363$.

79. Prove that $0 \le \int_{\pi/4}^{\pi/2} \frac{\sin x}{x} \, dx \le \frac{\sqrt{2}}{2}$.

80. Find upper and lower bounds for $\int_0^1 \frac{dx}{\sqrt{5x^3 + 4}}$.

81. ✏️ Suppose that $f(x) \le g(x)$ on $[a, b]$. By the Comparison Theorem, $\int_a^b f(x) \, dx \le \int_a^b g(x) \, dx$. Is it also true that $f'(x) \le g'(x)$ for $x \in [a, b]$? If not, give a counterexample.

82. ✏️ State whether true or false. If false, sketch the graph of a counterexample.

(a) If $f(x) > 0$, then $\int_a^b f(x) \, dx > 0$.

(b) If $\int_a^b f(x) \, dx > 0$, then $f(x) > 0$.

Further Insights and Challenges

83. Explain graphically: If f is an odd function, then $\int_{-a}^a f(x) \, dx = 0$.

84. Compute $\int_{-1}^1 (\sin x)(\sin^2 x + 1) \, dx$.

85. Let k and b be positive. Show, by comparing the right-endpoint approximations, that

$$\int_0^b x^k \, dx = b^{k+1} \int_0^1 x^k \, dx$$

86. Verify for $0 \le b \le 1$ by interpreting in terms of area:

$$\int_0^b \sqrt{1 - x^2}\, dx = \frac{1}{2} b \sqrt{1 - b^2} + \frac{1}{2} \sin^{-1} b$$

87. 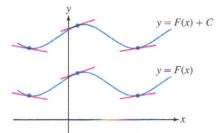 Suppose that f and g are continuous functions such that, for all a,

$$\int_{-a}^a f(x)\, dx = \int_{-a}^a g(x)\, dx$$

Give an *intuitive* argument showing that $f(0) = g(0)$. Explain your idea with a graph.

88. Theorem 4 remains true without the assumption $a \le b \le c$. Verify this for the cases $b < a < c$ and $c < a < b$.

5.3 The Indefinite Integral

In earlier chapters, we have seen how useful it is to be able to find the derivative of a function. But what about the inverse problem? Given the derivative of an unknown function, find the function itself. For example, in physics we may know the velocity $v(t)$ (the derivative) and wish to compute the position $s(t)$ of an object. Since $s'(t) = v(t)$, this amounts to finding a function whose derivative is $v(t)$. A function F whose derivative is f is called an antiderivative of f. Antiderivatives will turn out to be the key to evaluating definite integrals.

> **DEFINITION Antiderivatives**   A function F is an antiderivative of f on an open interval (a, b) if $F'(x) = f(x)$ for all x in (a, b).

Examples:

• $F(x) = -\cos x$ is an antiderivative of $f(x) = \sin x$ because for all values of x,

$$F'(x) = \frac{d}{dx}(-\cos x) = \sin x = f(x)$$

• $F(x) = \frac{1}{3} x^3$ is an antiderivative of $f(x) = x^2$ because for all values of x,

$$F'(x) = \frac{d}{dx}\left(\frac{1}{3} x^3\right) = x^2 = f(x)$$

One critical observation is that antiderivatives are not unique. We are free to add a constant C because the derivative of a constant is zero, and so, if $F'(x) = f(x)$, then $(F(x) + C)' = f(x)$. For example, each of the following is an antiderivative of x^2:

$$\frac{1}{3} x^3, \qquad \frac{1}{3} x^3 + 5, \qquad \frac{1}{3} x^3 - 4$$

Are there any antiderivatives of f other than those obtained by adding a constant to a given antiderivative F? Our next theorem says that the answer is no if f is defined on an open interval (a, b).

> **THEOREM 1 The General Antiderivative**   Let $y = F(x)$ be an antiderivative of $y = f(x)$ on (a, b). Then every other antiderivative on (a, b) is of the form $y = F(x) + C$ for some constant C.

Proof If $y = G(x)$ is a second antiderivative of $y = f(x)$, set $H(x) = G(x) - F(x)$. Then $H'(x) = G'(x) - F'(x) = f(x) - f(x) = 0$. By the Corollary to the Mean Value Theorem in Section 4.3, $H(x)$ must be a constant—say, $H(x) = C$—and therefore $G(x) = F(x) + C$. ∎

> **GRAPHICAL INSIGHT**   The graph of $y = F(x) + C$ is obtained by shifting the graph of $y = F(x)$ vertically by C units. Since vertical shifting moves the tangent lines without changing their slopes, it makes sense that all of the functions $y = F(x) + C$ have the same derivative (Figure 1). Theorem 1 tells us that conversely, if two graphs have parallel tangent lines, then one graph is obtained from the other by a vertical shift.

DF FIGURE 1 The tangent lines to the graphs of $y = F(x)$ and $y = F(x) + C$ are parallel.

We often describe the *general* antiderivative of a function in terms of an arbitrary constant C, as in the following example.

■ **EXAMPLE 1** Find two antiderivatives of $f(x) = \cos x$. Then determine the general antiderivative.

Solution The functions $F(x) = \sin x$ and $G(x) = \sin x + 2$ are both antiderivatives of $f(x) = \cos x$. The general antiderivative is $F(x) = \sin x + C$, where C is any constant.   ■

The process of finding an antiderivative is called **integration**. We will see why in the next section, when we discuss the connection between antiderivatives and areas under curves given by the Fundamental Theorem of Calculus. Anticipating this result, we begin using the integral sign $\int$, the standard notation for antiderivatives.

> **NOTATION** Indefinite Integral   The notation
> $$\int f(x)\,dx = F(x) + C \quad \text{means that} \quad F'(x) = f(x)$$
> We say that $y = F(x) + C$ is the general antiderivative or **indefinite integral** of $y = f(x)$.

The terms "antiderivative" and "indefinite integral" are used interchangeably. In some textbooks, an antiderivative is called a "primitive function."

The expression $f(x)$ appearing in the integral sign is called the **integrand**. The symbol dx is a *differential*. It is part of the integral notation and serves to indicate the independent variable. The constant C is called the *constant of integration*.

Some indefinite integrals can be evaluated by reversing the familiar derivative formulas. For example, we obtain the indefinite integral of $y = x^n$ by reversing the Power Rule for derivatives.

There are no Product, Quotient, or Chain Rules for integrals. However, we will see that the Product Rule for derivatives leads to an important technique called Integration by Parts (Section 7.1) and the Chain Rule leads to the Substitution Method (Section 5.7).

> **THEOREM 2   Power Rule for Integrals**
> $$\int x^n\,dx = \frac{x^{n+1}}{n+1} + C \qquad \text{for } n \neq -1$$

Proof We just need to verify that $F(x) = \dfrac{x^{n+1}}{n+1}$ is an antiderivative of $f(x) = x^n$:

$$F'(x) = \frac{d}{dx}\left(\frac{x^{n+1}}{n+1} + C\right) = \frac{1}{n+1}\left((n+1)x^n\right) = x^n$$ ■

In words, the Power Rule for Integrals says that to integrate a power of x, "add one to the exponent and then divide by the new exponent." Here are some examples:

$$\int x^5\,dx = \frac{1}{6}x^6 + C, \qquad \int x^{-9}\,dx = -\frac{1}{8}x^{-8} + C, \qquad \int x^{3/5}\,dx = \frac{5}{8}x^{8/5} + C$$

The Power Rule is not valid for $n = -1$. In fact, for $n = -1$, we obtain the meaningless result

$$\int x^{-1}\,dx = \frac{x^{n+1}}{n+1} + C = \frac{x^0}{0} + C \qquad \text{(meaningless)}$$

Recall, however, that the derivative of the natural logarithm is $\dfrac{d}{dx}\ln x = \dfrac{1}{x}$. This shows that $F(x) = \ln x$ is an antiderivative of $y = \dfrac{1}{x}$. Thus, for $n = -1$, instead of the Power Rule we have

Notice that in integral notation, we treat dx as a movable variable, and thus we write $\int \frac{1}{x}\,dx$ *as* $\int \frac{dx}{x}$.

$$\int \frac{dx}{x} = \ln x + C$$

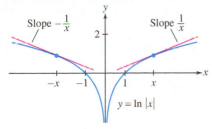

FIGURE 2

This formula is valid for $x > 0$, where $\ln x$ is defined. We would like to have an antiderivative of $y = \frac{1}{x}$ on its full domain, namely on $\{x : x \neq 0\}$. To achieve this end, we extend $F(x)$ to an even function by setting $F(x) = \ln |x|$ (Figure 2). Then $F(x) = F(-x)$, and by the Chain Rule, $F'(x) = -F'(-x)$. For $x < 0$, we obtain

$$\frac{d}{dx} \ln |x| = F'(x) = -F'(-x) = -\frac{1}{-x} = \frac{1}{x}$$

This proves that $\frac{d}{dx} \ln |x| = \frac{1}{x}$ for all $x \neq 0$.

THEOREM 3 Antiderivative of $y = \dfrac{1}{x}$ The function $F(x) = \ln |x|$ is an antiderivative of $y = \dfrac{1}{x}$ in the domain $\{x : x \neq 0\}$; that is,

$$\int \frac{dx}{x} = \ln |x| + C \qquad \boxed{1}$$

The indefinite integral obeys the usual linearity rules that allow us to integrate "term by term." These rules follow from the linearity rules for the derivative (see Exercise 81).

THEOREM 4 Linearity of the Indefinite Integral

- **Sum Rule:** $\displaystyle \int (f(x) + g(x))\,dx = \int f(x)\,dx + \int g(x)\,dx$

- **Multiples Rule:** $\displaystyle \int cf(x)\,dx = c \int f(x)\,dx$

■ **EXAMPLE 2** Evaluate $\int (3x^4 - 5x^{2/3} + x^{-3})\,dx$.

Solution We integrate term by term and use the Power Rule:

$$\int (3x^4 - 5x^{2/3} + x^{-3})\,dx = \int 3x^4\,dx - \int 5x^{2/3}\,dx + \int x^{-3}\,dx \qquad \text{(Sum Rule)}$$

$$= 3\int x^4\,dx - 5\int x^{2/3}\,dx + \int x^{-3}\,dx \qquad \text{(Multiples Rule)}$$

$$= 3\left(\frac{x^5}{5}\right) - 5\left(\frac{x^{5/3}}{5/3}\right) + \frac{x^{-2}}{-2} + C \qquad \text{(Power Rule)}$$

$$= \frac{3}{5}x^5 - 3x^{5/3} - \frac{1}{2}x^{-2} + C$$

When we break up an indefinite integral into a sum of several integrals as in Example 2, it is not necessary to include a separate constant of integration for each integral.

To check the answer, we verify that the derivative is equal to the integrand:

$$\frac{d}{dx}\left(\frac{3}{5}x^5 - 3x^{5/3} - \frac{1}{2}x^{-2} + C\right) = 3x^4 - 5x^{2/3} + x^{-3} \qquad ■$$

■ **EXAMPLE 3** Evaluate $\displaystyle \int \left(\frac{5}{x} - 3x^{-10}\right)dx$.

Solution Apply Eq. (1) and the Power Rule:

$$\int \left(\frac{5}{x} - 3x^{-10}\right)dx = 5\int \frac{dx}{x} - 3\int x^{-10}\,dx$$

$$= 5\ln |x| - 3\left(\frac{x^{-9}}{-9}\right) + C = 5\ln |x| + \frac{1}{3}x^{-9} + C \qquad ■$$

The differentiation formulas for the trigonometric functions give us the following integration formulas. Each formula can be checked by differentiation.

Basic Trigonometric Integrals

$$\int \sin x \, dx = -\cos x + C, \qquad \int \cos x \, dx = \sin x + C$$

$$\int \sec^2 x \, dx = \tan x + C, \qquad \int \csc^2 x \, dx = -\cot x + C$$

$$\int \sec x \tan x \, dx = \sec x + C, \qquad \int \csc x \cot x \, dx = -\csc x + C$$

Similarly, for any constant $k \neq 0$, the formulas

$$\frac{d}{dx} \sin(kx) = k \cos(kx), \qquad \frac{d}{dx} \cos(kx) = -k \sin(kx)$$

translate to the following indefinite integral formulas:

$$\int \cos(kx) \, dx = \frac{1}{k} \sin(kx) + C$$

$$\int \sin(kx) \, dx = -\frac{1}{k} \cos(kx) + C$$

■ **EXAMPLE 4** Evaluate $\int \left(\sin 8t + 20 \cos 9t \right) dt$.

Solution

$$\int \left(\sin 8t + 20 \cos 9t \right) dt = \int \sin 8t \, dt + 20 \int \cos 9t \, dt$$

$$= -\frac{1}{8} \cos 8t + \frac{20}{9} \sin 9t + C \qquad ■$$

Integrals Involving e^x

The formula $(e^x)' = e^x$ says that $f(x) = e^x$ is its own derivative. But this means that $f(x) = e^x$ is also *its own antiderivative*. In other words,

$$\int e^x \, dx = e^x + C$$

More generally, for any constant $k \neq 0$,

$$\int e^{kx} \, dx = \frac{1}{k} e^{kx} + C$$

■ **EXAMPLE 5** Evaluate **(a)** $\int (3e^x - 4) \, dx$ and **(b)** $\int 12e^{7-3x} \, dx$.

Solution

(a) $\int (3e^x - 4) \, dx = 3 \int e^x \, dx - \int 4 \, dx = 3e^x - 4x + C$

(b) $\int 12e^{7-3x} \, dx = 12 \int e^7 e^{-3x} \, dx = 12e^7 \left(\frac{1}{-3} e^{-3x} \right) = -4e^{7-3x} + C \qquad ■$

Initial Conditions

We can think of an antiderivative as a solution to the **differential equation**

$$\frac{dy}{dx} = f(x) \qquad \boxed{2}$$

In general, a differential equation is an equation relating an unknown function and its derivatives. The unknown in Eq. (2) is a function $y = F(x)$ whose derivative is $f(x)$; that is, $y = F(x)$ is an antiderivative of $y = f(x)$.

Eq. (2) has infinitely many solutions (because the antiderivative is not unique), but we can specify a particular solution by imposing an **initial condition**— that is, by requiring that the solution satisfy $y(x_0) = y_0$ for some fixed values x_0 and y_0. A differential equation with an initial condition is called an **initial value problem**.

An initial condition is like the y-intercept of a line, which determines one particular line among all lines with the same slope. The graphs of the antiderivatives of $y = f(x)$ are all parallel (Figure 1), and the initial condition determines one of them. Sometimes, when the variable is not time, the initial condition is called the boundary condition.

■ **EXAMPLE 6** Solve $\dfrac{dy}{dx} = 4x^7$ subject to the initial condition $y(0) = 4$.

Solution First, find the general antiderivative:

$$y(x) = \int 4x^7 \, dx = \frac{1}{2}x^8 + C$$

Then choose C so that the initial condition is satisfied: $y(0) = 0 + C = 4$. This yields $C = 4$, and our solution is $y = \frac{1}{2}x^8 + 4$. ■

■ **EXAMPLE 7** Solve the initial value problem $\dfrac{dy}{dt} = \sin(\pi t), \ y(2) = 2$.

Solution First, find the general antiderivative:

$$y(t) = \int \sin(\pi t) \, dt = -\frac{1}{\pi}\cos(\pi t) + C$$

Then solve for C by evaluating at $t = 2$:

$$y(2) = -\frac{1}{\pi}\cos(2\pi) + C = 2 \quad \Rightarrow \quad C = 2 + \frac{1}{\pi}$$

The solution of the initial value problem is $y(t) = -\frac{1}{\pi}\cos(\pi t) + 2 + \frac{1}{\pi}$. ■

■ **EXAMPLE 8** A car traveling with velocity 24 m/s begins to slow down at time $t = 0$ s with a constant deceleration of $a = -6$ m/s^2. Find **(a)** the velocity $v(t)$ at time t, and **(b)** the distance traveled before the car comes to a halt.

Solution (a) The derivative of velocity is acceleration, so *velocity is the antiderivative of acceleration*:

$$v(t) = \int a \, dt = \int (-6) \, dt = -6t + C$$

The initial condition $v(0) = C = 24$ m/s gives us $v(t) = -6t + 24$.

Relation between position, velocity, and acceleration:

$$s'(t) = v(t), \qquad s(t) = \int v(t)\,dt$$

$$v'(t) = a(t), \qquad v(t) = \int a(t)\,dt$$

(b) Position is the antiderivative of velocity, so the car's position in meters is

$$s(t) = \int v(t)\, dt = \int (-6t + 24)\, dt = -3t^2 + 24t + C_1$$

where C_1 is a constant. We are not told where the car is at $t = 0$, so let us set $s(0) = 0$ for convenience, obtaining $C_1 = 0$. With this choice, $s(t) = -3t^2 + 24t$. This is the distance traveled from time $t = 0$.

The car comes to a halt when its velocity is zero, so we solve

$$v(t) = -6t + 24 = 0 \quad \Rightarrow \quad t = 4 \text{ s}$$

The distance traveled before coming to a halt is $s(4) = -3(4^2) + 24(4) = 48$ m. ■

5.3 SUMMARY

- F is called an *antiderivative* of f if $F'(x) = f(x)$.
- Any two antiderivatives of f on an interval (a, b) differ by a constant.
- The general antiderivative is denoted by the indefinite integral:

$$\int f(x)\, dx = F(x) + C$$

- Integration formulas:

$$\int 0\, dx = C$$

$$\int k\, dx = kx + C \qquad\qquad (k \neq 0)$$

$$\int x^n\, dx = \frac{x^{n+1}}{n+1} + C \qquad\qquad (n \neq -1)$$

$$\int \sin(kx)\, dx = -\frac{1}{k}\cos(kx) + C \qquad\qquad (k \neq 0)$$

$$\int \cos(kx)\, dx = \frac{1}{k}\sin(kx) + C \qquad\qquad (k \neq 0)$$

$$\int e^{kx}\, dx = \frac{1}{k}e^{kx} + C \qquad\qquad (k \neq 0)$$

$$\int \frac{dx}{x} = \ln|x| + C$$

$$\int cf(x)\, dx = c\int f(x)\, dx$$

$$\int f(x) + g(x)\, dx = \int f(x)\, dx + \int g(x)\, dx$$

- To solve an initial value problem $\dfrac{dy}{dx} = f(x)$, $y(x_0) = y_0$, first find the general antiderivative $y = F(x) + C$. Then determine C using the initial condition $F(x_0) + C = y_0$.

5.3 EXERCISES

Preliminary Questions

1. Find an antiderivative of the function $f(x) = 0$.

2. Is there a difference between finding the general antiderivative of a function f and evaluating $\int f(x)\, dx$?

3. Jacques was told that f and g have the same derivative, and he wonders whether $f(x) = g(x)$. Does Jacques have sufficient information to answer his question?

4. Suppose that $F'(x) = f(x)$ and $G'(x) = g(x)$. Which of the following statements are true? Explain.

(a) If $f = g$, then $F = G$.

(b) If F and G differ by a constant, then $f = g$.

(c) If f and g differ by a constant, then $F = G$.

5. Is $y = x$ a solution of the following initial value problem?

$$\frac{dy}{dx} = 1, \qquad y(0) = 1$$

Exercises

In Exercises 1–8, find the general antiderivative of f and check your answer by differentiating.

1. $f(x) = 18x^2$

2. $f(x) = x^{-3/5}$

3. $f(x) = 2x^4 - 24x^2 + 12x^{-1}$

4. $f(x) = 9x + 15x^{-2}$

5. $f(x) = 2\cos x - 9\sin x$

6. $f(x) = 4x^7 - 3\cos x$

7. $f(x) = 12e^x - 5x^{-2}$

8. $f(x) = e^x - 4\sin x$

9. Match functions (a)–(d) with their antiderivatives (i)–(iv).

(a) $f(x) = \sin x$

(b) $f(x) = x\sin(x^2)$

(c) $f(x) = \sin(1 - x)$

(d) $f(x) = x\sin x$

(i) $F(x) = \cos(1 - x)$

(ii) $F(x) = -\cos x$

(iii) $F(x) = -\frac{1}{2}\cos(x^2)$

(iv) $F(x) = \sin x - x\cos x$

In Exercises 10–39, evaluate the indefinite integral.

10. $\int (9x + 2)\, dx$

11. $\int (4 - 18x)\, dx$

12. $\int x^{-3}\, dx$

13. $\int t^{-6/11}\, dt$

14. $\int (5t^3 - t^{-3})\, dt$

15. $\int (18t^5 - 10t^4 - 28t)\, dt$

16. $\int 14s^{9/5}\, ds$

17. $\int (z^{-4/5} - z^{2/3} + z^{5/4})\, dz$

18. $\int \frac{3}{2}\, dx$

19. $\int \frac{1}{\sqrt[3]{x}}\, dx$

20. $\int \frac{dx}{x^{4/3}}$

21. $\int \frac{36\, dt}{t^3}$

22. $\int x(x^2 - 4)\, dx$

23. $\int (t^{1/2} + 1)(t + 1)\, dt$

24. $\int \frac{12 - z}{\sqrt{z}}\, dz$

25. $\int \frac{x^3 + 3x - 4}{x^2}\, dx$

26. $\int \left(\frac{1}{3}\sin x - \frac{1}{4}\cos x \right) dx$

27. $\int 12 \sec x \tan x\, dx$

28. $\int (\theta + \sec^2 \theta)\, d\theta$

29. $\int (\csc t \cot t)\, dt$

30. $\int \sin(7x)\, dx$

31. $\int \sec^2(-3\theta)\, d\theta$

32. $\int (\theta - \cos(-\theta))\, d\theta$

33. $\int 25 \sec^2(3z)\, dz$

34. $\int \sec x \tan x\, dx$

35. $\int \left(\cos(3\theta) - \frac{1}{2}\sec^2\left(\frac{\theta}{4} \right) \right) d\theta$

36. $\int \left(\frac{4}{x} - e^x \right) dx$

37. $\int (3e^{5x})\, dx$

38. $\int e^{3t-4}\, dt$

39. $\int (8x - 4e^{5-2x})\, dx$

40. In Figure 3, is graph (A) or graph (B) the graph of an antiderivative of $y = f(x)$?

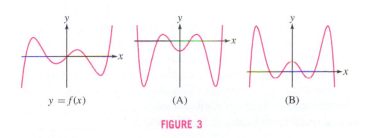

$y = f(x)$ (A)   (B)

FIGURE 3

41. In Figure 4, which of graphs (A), (B), and (C) is *not* the graph of an antiderivative of $y = f(x)$? Explain.

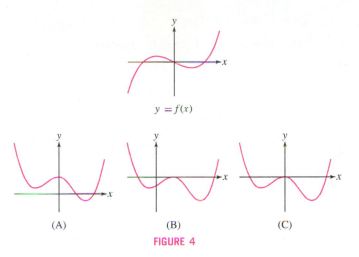

$y = f(x)$

(A)   (B)   (C)

FIGURE 4

42. Show that $F(x) = \frac{1}{3}(x + 13)^3$ is an antiderivative of $f(x) = (x + 13)^2$.

In Exercises 43–46, verify by differentiation.

43. $\int (x + 13)^6\, dx = \frac{1}{7}(x + 13)^7 + C$

44. $\int (x + 13)^{-5}\, dx = -\frac{1}{4}(x + 13)^{-4} + C$

45. $\int (4x + 13)^2\, dx = \frac{1}{12}(4x + 13)^3 + C$

46. $\int (ax + b)^n\, dx = \frac{1}{a(n + 1)}(ax + b)^{n+1} + C$ (for $n \neq -1$)

In Exercises 47–62, solve the initial value problem.

47. $\frac{dy}{dx} = x^3, \ y(0) = 4$

48. $\frac{dy}{dt} = 3 - 2t, \ y(0) = -5$

49. $\frac{dy}{dt} = 2t + 9t^2, \ y(1) = 2$

50. $\frac{dy}{dx} = 8x^3 + 3x^2, \ y(2) = 0$

51. $\frac{dy}{dt} = \sqrt{t}, \ y(1) = 1$

52. $\frac{dz}{dt} = t^{-3/2}, \ z(4) = -1$

53. $\frac{dy}{dx} = (3x + 2)^3, \ y(0) = 1$

54. $\frac{dy}{dt} = (4t + 3)^{-2}, \ y(1) = 0$

55. $\frac{dy}{dx} = \sin x, \ y\left(\frac{\pi}{2} \right) = 1$

56. $\frac{dy}{dz} = \sin 2z, \ y\left(\frac{\pi}{4} \right) = 4$

57. $\frac{dy}{dx} = \cos 5x, \ y(\pi) = 3$

58. $\frac{dy}{dx} = \sec^2 3x, \ y\left(\frac{\pi}{4} \right) = 2$

59. $\frac{dy}{dx} = e^x, \ y(2) = 0$

60. $\frac{dy}{dt} = e^{-t}, \ y(0) = 0$

61. $\frac{dy}{dt} = 9e^{12-3t}, \ y(4) = 7$

62. $\frac{dy}{dt} = t + 2e^{t-9}, \ y(9) = 4$

In Exercises 63–69, first find f' and then find f.

63. $f''(x) = 12x, \ f'(0) = 1, \ f(0) = 2$

64. $f''(x) = x^3 - 2x,$ $f'(1) = 0,$ $f(1) = 2$

65. $f''(x) = x^3 - 2x + 1,$ $f'(0) = 1,$ $f(0) = 0$

66. $f''(x) = x^3 - 2x + 1,$ $f'(1) = 0,$ $f(1) = 4$

67. $f''(t) = t^{-3/2},$ $f'(4) = 1,$ $f(4) = 4$

68. $f''(\theta) = \cos\theta,$ $f'\left(\dfrac{\pi}{2}\right) = 1,$ $f\left(\dfrac{\pi}{2}\right) = 6$

69. $f''(t) = t - \cos t,$ $f'(0) = 2,$ $f(0) = -2$

70. Show that $F(x) = \tan^2 x$ and $G(x) = \sec^2 x$ have the same derivative. What can you conclude about the relation between F and G? Verify this conclusion directly.

71. A particle located at the origin at $t = 1$ s moves along the x-axis with velocity $v(t) = (6t^2 - t)$ m/s. State the differential equation with its initial condition satisfied by the position $s(t)$ of the particle, and find $s(t)$.

72. A particle moves along the x-axis with velocity $v(t) = (6t^2 - t)$ m/s. Find the particle's position $s(t)$, assuming that $s(2) = 4$ m.

73. A water balloon is dropped from a high building. It falls for 5 s before hitting the ground. Determine the velocity it is traveling when it is about to hit the ground, assuming an acceleration due to gravity of -9.8 m/s^2 and no wind resistance.

74. A hammer is dropped and it falls for 2 s before hitting the ground. Determine how far it falls, assuming an acceleration due to gravity of -9.8 m/s^2 and no wind resistance.

75. A mass oscillates at the end of a spring. Let $s(t)$ be the displacement of the mass from the equilibrium position at time t. Assuming that the mass is located at the origin at $t = 0$ and has velocity $v(t) = \sin(\pi t/2)$ m/s, state the differential equation with initial condition satisfied by $s(t)$, and find $s(t)$.

76. Beginning at $t = 0$ s with initial velocity 4 m/s, a particle moves in a straight line with acceleration $a(t) = 3t^{1/2}$ m/s^2. Find the distance traveled after 25 s.

77. A car traveling 25 m/s begins to decelerate at a constant rate of 4 m/s^2. After how many seconds does the car come to a stop and how far will the car have traveled during its deceleration before stopping?

78. At time $t = 1$ s, a particle is traveling at 72 m/s and begins to decelerate at the rate $a(t) = -t^{-1/2}$ until it stops. How far does the particle travel during its deceleration before stopping?

79. A 900-kg rocket is released from a space station. As it burns fuel, the rocket's mass decreases and its velocity increases. Let $v(m)$ be the velocity (in meters per second) as a function of mass m. Find the velocity when $m = 729$ kg if $dv/dm = -50m^{-1/2}$. Assume that $v(900) = 0$ m/s.

80. As water flows through a tube of radius $R = 10$ cm, the velocity v of an individual water particle depends only on its distance r from the center of the tube. The particles at the walls of the tube have zero velocity and $dv/dr = -0.06r$. Determine $v(r)$.

81. Verify the linearity properties of the indefinite integral stated in Theorem 4.

Further Insights and Challenges

82. Find constants c_1 and c_2 such that $F(x) = c_1 x \sin x + c_2 \cos x$ is an antiderivative of $f(x) = x \cos x$.

83. Find constants c_1 and c_2 such that $F(x) = c_1 x e^x + c_2 e^x$ is an antiderivative of $f(x) = x e^x$.

84. Suppose that $F'(x) = f(x)$ and $G'(x) = g(x)$. Is it true that $y = F(x)G(x)$ is an antiderivative of $y = f(x)g(x)$? Confirm or provide a counterexample.

85. Suppose that $F'(x) = f(x)$.

(a) Show that $y = \frac{1}{2}F(2x)$ is an antiderivative of $y = f(2x)$.

(b) Find the general antiderivative of $y = f(kx)$ for $k \neq 0$.

86. Find an antiderivative for $f(x) = |x|$.

87. Using Theorem 1, prove that if $F'(x) = f(x)$, where f is a polynomial of degree $n - 1$, then F is a polynomial of degree n. Then prove that if g is any function such that $g^{(n)}(x) = 0$, then g is a polynomial of degree at most n.

88. Show that $F(x) = \dfrac{x^{n+1} - 1}{n + 1}$ is an antiderivative of $y = x^n$ for $n \neq -1$. Then use L'Hôpital's Rule to prove that

$$\lim_{n \to -1} F(x) = \ln x$$

In this limit, x is fixed and n is the variable. This result shows that, although the Power Rule breaks down for $n = -1$, the antiderivative of $y = x^{-1}$ is a limit of antiderivatives of $y = x^n$ as $n \to -1$.

The FTC was first stated clearly by Isaac Newton in 1666, although other mathematicians, including Newton's teacher Isaac Barrow, had discovered versions of it earlier.

◄··· **REMINDER**

F is called an **antiderivative** of f if $F'(x) = f(x)$. We say also that F is an **indefinite integral** of f, and we use the notation

$$\int f(x)\,dx = F(x) + C$$

5.4 The Fundamental Theorem of Calculus, Part I

Having so far introduced both derivatives and integrals, a very reasonable question is why they appear together in this topic called Calculus. The answer is the Fundamental Theorem of Calculus (FTC), which is one of the most important theorems in all of mathematics. This foundational result reveals an unexpected connection between the two main operations of calculus: differentiation and integration. The theorem has two parts. Although they are closely related, we discuss them in separate sections to emphasize the different ways they are used. The first part of the Fundamental Theorem of Calculus will allow us to compute definite integrals without having to take limits of Riemann sums.

To explain FTC I, recall a result from Example 5 of Section 5.2:

$$\int_4^7 x^2\,dx = \left(\frac{1}{3}\right)7^3 - \left(\frac{1}{3}\right)4^3 = 93$$

Now observe that $F(x) = \frac{1}{3}x^3$ is an antiderivative of $f(x) = x^2$, so we can write

$$\int_4^7 x^2 \, dx = F(7) - F(4)$$

According to FTC I, this is no coincidence; this relation between the definite integral and the antiderivative holds in general.

THEOREM 1 The Fundamental Theorem of Calculus, Part I   Assume that f is continuous on $[a, b]$. If F is an antiderivative of f on $[a, b]$, then

$$\int_a^b f(x) \, dx = F(b) - F(a)$$

$\boxed{1}$

Proof The quantity $F(b) - F(a)$ is the total change in F (also called the "net change") over the interval $[a, b]$. Our task is to relate it to the integral of $F'(x) = f(x)$. There are two main steps.

Step 1. Write total change as a sum of small changes.
    Given any partition P of $[a, b]$:

$$P : x_0 = a < x_1 < x_2 < \cdots < x_N = b$$

we can break up $F(b) - F(a)$ as a sum of changes over the intervals $[x_{i-1}, x_i]$:

$$F(b) - F(a) = \left(\cancel{F(x_1)} - F(a)\right) + \left(\cancel{F(x_2)} - \cancel{F(x_1)}\right) + \cdots + \left(F(b) - \cancel{F(x_{N-1})}\right)$$

On the right-hand side, $F(x_1)$ is canceled by $-F(x_1)$ in the second term, $F(x_2)$ is canceled by $-F(x_2)$, etc. (Figure 1). In summation notation,

$$F(b) - F(a) = \sum_{i=1}^{N} \left(F(x_i) - F(x_{i-1})\right)$$

$\boxed{2}$

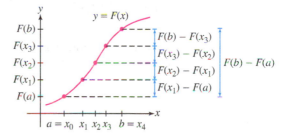

FIGURE 1 Note the cancellation when we write $F(b) - F(a)$ as a sum of small changes $F(x_i) - F(x_{i-1})$.

Step 2. Interpret Eq. (2) as a Riemann sum.
    The Mean Value Theorem tells us that there is a point c_i^* in $[x_{i-1}, x_i]$ such that

$$F(x_i) - F(x_{i-1}) = F'(c_i^*)(x_i - x_{i-1}) = f(c_i^*)(x_i - x_{i-1}) = f(c_i^*)\,\Delta x_i$$

Therefore, Eq. (2) can be written

$$F(b) - F(a) = \sum_{i=1}^{N} f(c_i^*)\,\Delta x_i$$

This sum is the Riemann sum $R(f, P, C^*)$ with sample points $C^* = \{c_i^*\}$.

Now, f is integrable (Theorem 1, Section 5.2), so $R(f, P, C^*)$ approaches $\int_a^b f(x)\,dx$ as the norm $\|P\|$ tends to zero. On the other hand, $R(f, P, C^*)$ is *equal* to $F(b) - F(a)$ with our particular choice C^* of sample points. This proves the desired result:

$$F(b) - F(a) = \lim_{\|P\| \to 0} R(f, P, C^*) = \int_a^b f(x)\,dx$$

■

CONCEPTUAL INSIGHT   **A Tale of Two Graphs** In the proof of FTC I, we used the MVT to write a small change in $F(x)$ in terms of the derivative $F'(x) = f(x)$:

$$F(x_i) - F(x_{i-1}) = f(c_i^*)\Delta x_i$$

But $f(c_i^*)\Delta x_i$ is the area of a thin rectangle that approximates a sliver of area under the graph of f (Figure 2). *This is the essence of the Fundamental Theorem*: the total change $F(b) - F(a)$ is equal to the sum of small changes $F(x_i) - F(x_{i-1})$, which in turn is equal to the sum of the areas of rectangles in a Riemann sum approximation for $f(x)$. We derive the Fundamental Theorem itself by taking the limit as the width of the rectangles tends to zero.

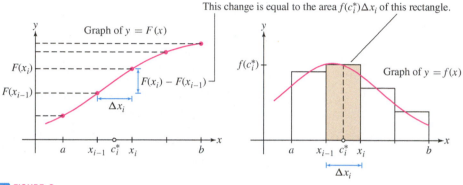

This change is equal to the area $f(c_i^*)\Delta x_i$ of this rectangle.

DF **FIGURE 2**

FTC I tells us that if we can find an antiderivative of f, then we can compute the definite integral easily, without calculating any limits. It is for this reason that we use the integral sign $\int$ for both the definite integral $\int_a^b f(x)\,dx$ and the indefinite integral (antiderivative) $\int f(x)\,dx$.

Notation $F(b) - F(a)$ is denoted $F(x)\Big|_a^b$.

In this notation, the FTC reads

$$\int_a^b f(x)\,dx = F(x)\Big|_a^b$$

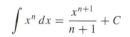

◄┈ **REMINDER** *The Power Rule for Integrals (valid for $n \neq -1$) states*

$$\int x^n\,dx = \frac{x^{n+1}}{n+1} + C$$

■ **EXAMPLE 1** Calculate the area under the graph of $f(x) = x^3$ over $[2, 4]$.

Solution Since $F(x) = \frac{1}{4}x^4$ is an antiderivative of $f(x) = x^3$, FTC I gives us

$$\int_2^4 x^3\,dx = F(4) - F(2) = \frac{1}{4}x^4\Big|_2^4 = \frac{1}{4}4^4 - \frac{1}{4}2^4 = 60 \qquad ■$$

■ **EXAMPLE 2** Find the area under $g(x) = x^{-3/4} + 3x^{5/3}$ over $[1, 3]$.

Solution The function $G(x) = 4x^{1/4} + \frac{9}{8}x^{8/3}$ is an antiderivative of g. The area (Figure 3) is equal to

$$\int_1^3 (x^{-3/4} + 3x^{5/3})\,dx = G(x)\Big|_1^3 = \left(4x^{1/4} + \frac{9}{8}x^{8/3}\right)\Big|_1^3$$

$$= \left(4 \cdot 3^{1/4} + \frac{9}{8} \cdot 3^{8/3}\right) - \left(4 \cdot 1^{1/4} + \frac{9}{8} \cdot 1^{8/3}\right)$$

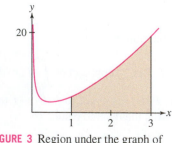

FIGURE 3 Region under the graph of $g(x) = x^{-3/4} + 3x^{5/3}$ over $[1, 3]$.

$$\approx 26.325 - 5.125 = 21.2 \qquad ■$$

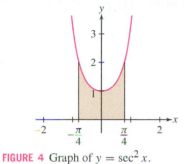

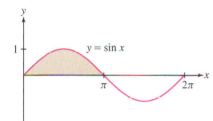

FIGURE 4 Graph of $y = \sec^2 x$.

EXAMPLE 3 Calculate $\displaystyle\int_{-\pi/4}^{\pi/4} \sec^2 x \, dx$ and sketch the corresponding region.

Solution Figure 4 shows the region. Recall that $(\tan x)' = \sec^2 x$. Therefore,

$$\int_{-\pi/4}^{\pi/4} \sec^2 x \, dx = \tan x \Big|_{-\pi/4}^{\pi/4} = \tan\left(\frac{\pi}{4}\right) - \tan\left(-\frac{\pi}{4}\right) = 1 - (-1) = 2 \quad\blacksquare$$

We know that the definite integral is equal to the *signed* area between the graph and the x-axis. Needless to say, the FTC "knows" this also: When you evaluate an integral using the FTC, you obtain the signed area.

EXAMPLE 4 Evaluate (a) $\displaystyle\int_0^{\pi} \sin x \, dx$ and (b) $\displaystyle\int_0^{2\pi} \sin x \, dx$.

Solution

(a) Since $(-\cos x)' = \sin x$, the area of one "hump" (Figure 5) is

$$\int_0^{\pi} \sin x \, dx = -\cos x \Big|_0^{\pi} = -\cos\pi - (-\cos 0) = -(-1) - (-1) = 2$$

(b) We expect the signed area over $[0, 2\pi]$ to be zero since the second hump lies below the x-axis, and indeed,

$$\int_0^{2\pi} \sin x \, dx = -\cos x \Big|_0^{2\pi} = (-\cos(2\pi)) - (-\cos 0)) = -1 - (-1) = 0 \quad\blacksquare$$

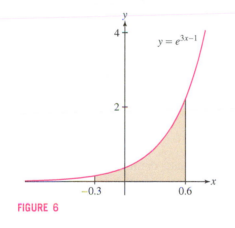

DF **FIGURE 5** The area of one hump is 2. The signed area over $[0, 2\pi]$ is zero.

EXAMPLE 5 **Exponential Function** Evaluate $\displaystyle\int_{-0.3}^{0.6} e^{3x-1} \, dx$.

Solution The function $F(x) = \frac{1}{3}e^{3x-1}$ is an antiderivative of $f(x) = e^{3x-1}$, so the definite integral (the shaded area in Figure 6) is

$$\int_{-0.3}^{0.6} e^{3x-1} \, dx = \frac{1}{3}e^{3x-1} \Big|_{-0.3}^{0.6} = \frac{1}{3}e^{3(0.6)-1} - \frac{1}{3}e^{3(-0.3)-1}$$

$$\approx 0.742 - 0.050 = 0.692 \quad\blacksquare$$

FIGURE 6

Recall (Section 5.3) that $F(x) = \ln|x|$ is an antiderivative of $f(x) = x^{-1}$ in the domain $\{x : x \neq 0\}$. Therefore, the FTC yields the following formula [Figure 7(A)], which is valid if both a and b are positive or both are negative. Note that when $a < 0 < b$, the formula does not hold, since 0 is not in the domain of $\frac{1}{x}$, as in Figure 7(C).

$$\int_a^b \frac{dx}{x} = \ln|b| - \ln|a| = \ln\frac{b}{a} \qquad \boxed{3}$$

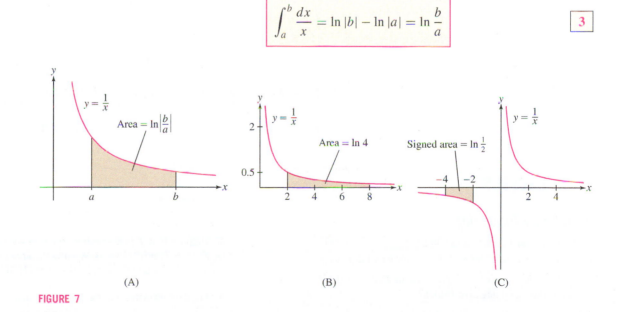

(A)   (B)   (C)

FIGURE 7

■ **EXAMPLE 6** The Logarithm as an Antiderivative Evaluate (a) $\int_2^8 \dfrac{dx}{x}$ and

(b) $\int_{-4}^{-2} \dfrac{dx}{x}$.

Solution By Eq. (3),

(a) $\int_2^8 \dfrac{dx}{x} = \ln\dfrac{8}{2} = \ln 4 \approx 1.39$

(b) $\int_{-4}^{-2} \dfrac{dx}{x} = \ln\left(\dfrac{-2}{-4}\right) = \ln\dfrac{1}{2} \approx -0.69$

The areas represented by these integrals are shown in Figures 7(B) and (C). ■

CONCEPTUAL INSIGHT **Which Antiderivative?** Antiderivatives are unique only to within an additive constant (Section 5.3). Does it matter which antiderivative is used in the FTC? The answer is no. If F and G are both antiderivatives of f, then $F(x) = G(x) + C$ for some constant C, and

$$F(b) - F(a) = \underbrace{(G(b) + C) - (G(a) + C)}_{\text{The constant cancels}} = G(b) - G(a)$$

The two antiderivatives yield the same value for the definite integral:

$$\int_a^b f(x)\,dx = F(b) - F(a) = G(b) - G(a)$$

5.4 SUMMARY

- The Fundamental Theorem of Calculus, Part I, states that

$$\int_a^b f(x)\,dx = F(b) - F(a)$$

where F is an antiderivative of f. FTC I is used to evaluate definite integrals in cases where we can find an antiderivative of the integrand.
- Basic antiderivative formulas for evaluating definite integrals:

$$\int x^n dx = \frac{x^{n+1}}{n+1} + C \qquad \text{for } n \neq -1$$

$$\int e^x\,dx = e^x + C, \qquad\qquad \int \frac{dx}{x} = \ln|x| + C$$

$$\int \sin x\,dx = -\cos x + C, \qquad \int \cos x\,dx = \sin x + C$$

$$\int \sec^2 x\,dx = \tan x + C, \qquad \int \csc^2 x\,dx = -\cot x + C$$

$$\int \sec x \tan x\,dx = \sec x + C, \qquad \int \csc x \cot x\,dx = -\csc x + C$$

5.4 EXERCISES

Preliminary Questions

1. Suppose that $F'(x) = f(x)$ and $F(0) = 3$, $F(2) = 7$.

(a) What is the area under $y = f(x)$ over $[0, 2]$ if $f(x) \geq 0$?

(b) What is the graphical interpretation of $F(2) - F(0)$ if $f(x)$ takes on both positive and negative values?

2. Suppose that f is a *negative* function with antiderivative F such that $F(1) = 7$ and $F(3) = 4$. What is the area (a positive number) between the x-axis and the graph of f over $[1, 3]$?

3. Are the following statements true or false? Explain.

(a) FTC I is valid only for positive functions.
(b) To use FTC I, you have to choose the right antiderivative.
(c) If you cannot find an antiderivative of f, then the definite integral does not exist.

4. Evaluate $\int_{2}^{9} f'(x)\,dx$, where f is differentiable and $f(2) = f(9) = 4$.

Exercises

In Exercises 1–4, sketch the region under the graph of the function and find its area using FTC I.

1. $f(x) = x^2$, $[0, 1]$

2. $f(x) = 2x - x^2$, $[0, 2]$

3. $f(x) = x^{-2}$, $[1, 2]$

4. $f(x) = \cos x$, $[0, \frac{\pi}{2}]$

In Exercises 5–42, evaluate the integral using FTC I.

5. $\int_{3}^{6} x\,dx$

6. $\int_{0}^{9} 2\,dx$

7. $\int_{0}^{1} (4x - 9x^2)\,dx$

8. $\int_{-3}^{2} u^2\,du$

9. $\int_{0}^{2} (12x^5 + 3x^2 - 4x)\,dx$

10. $\int_{-2}^{2} (10x^9 + 3x^5)\,dx$

11. $\int_{3}^{0} (2t^3 - 6t^2)\,dt$

12. $\int_{-1}^{1} (5u^4 + u^2 - u)\,du$

13. $\int_{0}^{4} \sqrt{y}\,dy$

14. $\int_{1}^{8} x^{4/3}\,dx$

15. $\int_{1/16}^{1} t^{1/4}\,dt$

16. $\int_{4}^{1} t^{5/2}\,dt$

17. $\int_{1}^{3} \frac{dt}{t^2}$

18. $\int_{1}^{4} x^{-4}\,dx$

19. $\int_{1/2}^{1} \frac{8}{x^3}\,dx$

20. $\int_{-2}^{-1} \frac{1}{x^3}\,dx$

21. $\int_{1}^{2} (x^2 - x^{-2})\,dx$

22. $\int_{1}^{9} t^{-1/2}\,dt$

23. $\int_{1}^{27} \frac{t+1}{\sqrt{t}}\,dt$

24. $\int_{8/27}^{1} \frac{10t^{4/3} - 8t^{1/3}}{t^2}\,dt$

25. $\int_{\pi/4}^{3\pi/4} \sin\theta\,d\theta$

26. $\int_{2\pi}^{4\pi} \sin x\,dx$

27. $\int_{0}^{\pi/2} \cos\left(\frac{1}{3}\theta\right) d\theta$

28. $\int_{\pi/4}^{5\pi/8} \cos 2x\,dx$

29. $\int_{0}^{\pi/6} \sec^2\left(3t - \frac{\pi}{6}\right) dt$

30. $\int_{0}^{\pi/6} \sec\theta \tan\theta\,d\theta$

31. $\int_{\pi/20}^{\pi/10} \csc 5x \cot 5x\,dx$

32. $\int_{\pi/28}^{\pi/14} \csc^2 7y\,dy$

33. $\int_{0}^{1} e^x\,dx$

34. $\int_{3}^{5} e^{-4x}\,dx$

35. $\int_{0}^{3} e^{1-6t}\,dt$

36. $\int_{2}^{3} e^{4t-3}\,dt$

37. $\int_{2}^{10} \frac{dx}{x}$

38. $\int_{-12}^{-4} \frac{dx}{x}$

39. $\int_{0}^{1} \frac{dt}{t+1}$

40. $\int_{1}^{4} \frac{dt}{5t+4}$

41. $\int_{-2}^{0} (3x - 9e^{3x})\,dx$

42. $\int_{2}^{6} \left(x + \frac{1}{x}\right) dx$

In Exercises 43–48, write the integral as a sum of integrals without absolute values and evaluate.

43. $\int_{-2}^{1} |x|\,dx$

44. $\int_{0}^{5} |3 - x|\,dx$

45. $\int_{-2}^{3} |x^3|\,dx$

46. $\int_{0}^{3} |x^2 - 1|\,dx$

47. $\int_{0}^{\pi} |\cos x|\,dx$

48. $\int_{0}^{5} |x^2 - 4x + 3|\,dx$

In Exercises 49–54, evaluate the integral in terms of the constants.

49. $\int_{1}^{b} x^3\,dx$

50. $\int_{b}^{a} x^4\,dx$

51. $\int_{1}^{b} x^5\,dx$

52. $\int_{-x}^{x} (t^3 + t)\,dt$

53. $\int_{a}^{5a} \frac{dx}{x}$

54. $\int_{b}^{b^2} \frac{dx}{x}$

55. Calculate $\int_{-2}^{3} f(x)\,dx$, where

$$f(x) = \begin{cases} 12 - x^2 & \text{for } x \le 2 \\ x^3 & \text{for } x > 2 \end{cases}$$

56. Calculate $\int_{0}^{2\pi} f(x)\,dx$, where

$$f(x) = \begin{cases} \cos x & \text{for } x \le \pi \\ \cos x - \sin 2x & \text{for } x > \pi \end{cases}$$

57. Use FTC I to show that $\int_{-1}^{1} x^n\,dx = 0$ if n is an odd whole number. Explain graphically.

58. **CAS** Plot the function $f(x) = \sin 3x - x$. Find the positive root of f to three decimal places and use it to find the area under the graph of f in the first quadrant.

59. Calculate $F(4)$ given that $F(1) = 3$ and $F'(x) = x^2$. *Hint:* Express $F(4) - F(1)$ as a definite integral.

60. Calculate $G(16)$, where $dG/dt = t^{-1/2}$ and $G(9) = -5$.

61. 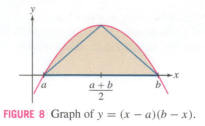 Does $\int_0^1 x^n \, dx$ get larger or smaller as n increases? Explain graphically.

62. Show that the area of the shaded parabolic arch in Figure 8 is equal to four-thirds the area of the triangle shown.

FIGURE 8 Graph of $y = (x - a)(b - x)$.

Further Insights and Challenges

63. Prove a famous result of Archimedes (generalizing Exercise 62): For $r < s$, the area of the shaded region in Figure 9 is equal to four-thirds the area of triangle $\triangle ACE$, where C is the point on the parabola at which the tangent line is parallel to secant line $\overline{AE}$.

(a) Show that C has x-coordinate $(r + s)/2$.

(b) Show that $ABDE$ has area $(s - r)^3/4$ by viewing it as a parallelogram of height $s - r$ and base of length $\overline{CF}$.

(c) Show that $\triangle ACE$ has area $(s - r)^3/8$ by observing that it has the same base and height as the parallelogram.

(d) Compute the shaded area as the area under the graph minus the area of a trapezoid, and prove Archimedes's result.

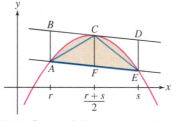

FIGURE 9 Graph of $f(x) = (x - a)(b - x)$.

64. (a) Apply the Comparison Theorem (Theorem 5 in Section 5.2) to the inequality $\sin x \le x$ (valid for $x \ge 0$) to prove that

$$1 - \frac{x^2}{2} \le \cos x \le 1$$

(b) Apply it again to prove that

$$x - \frac{x^3}{6} \le \sin x \le x \quad \text{(for } x \ge 0\text{)}$$

(c) Verify these inequalities for $x = 0.3$.

65. Use the method of Exercise 64 to prove that

$$1 - \frac{x^2}{2} \le \cos x \le 1 - \frac{x^2}{2} + \frac{x^4}{24}$$

$$x - \frac{x^3}{6} \le \sin x \le x - \frac{x^3}{6} + \frac{x^5}{120} \quad \text{(for } x \ge 0\text{)}$$

Verify these inequalities for $x = 0.1$. Why have we specified $x \ge 0$ for $\sin x$ but not for $\cos x$?

66. Calculate the next pair of inequalities for $\sin x$ and $\cos x$ by integrating the results of Exercise 65. Can you guess the general pattern?

67. Use FTC I to prove that if $|f'(x)| \le K$ for $x \in [a, b]$, then $|f(x) - f(a)| \le K|x - a|$ for $x \in [a, b]$.

68. (a) Use Exercise 67 to prove that $|\sin a - \sin b| \le |a - b|$ for all a, b.

(b) Let $f(x) = \sin(x + a) - \sin x$. Use part (a) to show that the graph of f lies between the horizontal lines $y = \pm a$.

(c) GU Plot $y = f(x)$ and the lines $y = \pm a$ to verify (b) for $a = 0.5$ and $a = 0.2$.

5.5 The Fundamental Theorem of Calculus, Part II

Part I of the Fundamental Theorem says that we can use antiderivatives to compute definite integrals. This is a huge advantage over having to take limits of Riemann sums. Part II of the Fundamental Theorem says that the derivative of a certain integral is the integrand. That is to say the derivative cancels out the action of the integral on the original function. However, it is a very particular integral for which this works.

To state Part II correctly, we introduce the **area function** of f with lower limit a:

*A is sometimes called the **cumulative area function**. In the definition of $A(x)$, we use t as the variable of integration to avoid confusion with x, which is the upper limit of integration. In fact, t is a dummy variable and may be replaced by any other variable.*

$$A(x) = \int_a^x f(t) \, dt = \text{signed area from } a \text{ to } x$$

In essence, we turn the definite integral into a function by treating the upper limit x as a variable (Figure 1). Note that $A(a) = 0$ because $A(a) = \int_a^a f(t) \, dt = 0$.

In some cases, we can find an explicit formula for $A(x)$ (Figure 2).

■ **EXAMPLE 1** Find a formula for the area function $A(x) = \int_3^x t^2 \, dt$.

Solution The function $F(t) = \frac{1}{3}t^3$ is an antiderivative for $f(t) = t^2$. By FTC I,

$$A(x) = \int_3^x t^2 \, dt = F(x) - F(3) = \frac{1}{3}x^3 - \frac{1}{3} \cdot 3^3 = \frac{1}{3}x^3 - 9 \qquad ■$$

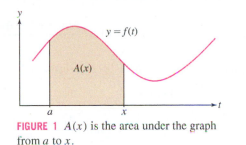

FIGURE 1 $A(x)$ is the area under the graph from a to x.

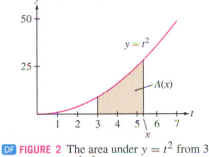

DF FIGURE 2   The area under $y = t^2$ from 3 to x is $A(x) = \frac{1}{3}x^3 - 9$.

Note, in the previous example, that *the derivative of A is f itself:*

$$A'(x) = \frac{d}{dx}\left(\frac{1}{3}x^3 - 9\right) = x^2$$

FTC II states that this relation always holds: The derivative of the area function is equal to the original function.

THEOREM 1   Fundamental Theorem of Calculus, Part II   Assume that f is continuous on an open interval I and let a be a point in I. Then the area function

$$A(x) = \int_a^x f(t)\, dt$$

is an antiderivative of f on I; that is, $A'(x) = f(x)$. Equivalently,

$$\frac{d}{dx}\int_a^x f(t)\, dt = f(x)$$

Furthermore, $A(x)$ satisfies the initial condition $A(a) = 0$.

In this proof,

$$A(x) = \int_a^x f(t)\, dt$$

$$A(x+h) - A(x) = \int_x^{x+h} f(t)\, dt$$

$$A'(x) = \lim_{h\to 0}\frac{A(x+h) - A(x)}{h}$$

Proof   First, we use the additivity property of the definite integral to write the change in A over $[x, x+h]$ as an integral:

$$A(x+h) - A(x) = \int_a^{x+h} f(t)\, dt - \int_a^x f(t)\, dt = \int_x^{x+h} f(t)\, dt$$

In other words, $A(x+h) - A(x)$ is equal to the area of the thin sliver between the graph and the x-axis from x to $x+h$ in Figure 3.

To simplify the rest of the proof, we assume that f is increasing (see Exercise 50 for the general case). Then, if $h > 0$, this thin sliver lies between the two rectangles of heights $f(x)$ and $f(x+h)$ in Figure 4, and we have

$$\underbrace{hf(x)}_{\text{Area of smaller rectangle}} \leq \underbrace{A(x+h) - A(x)}_{\text{Area of sliver}} \leq \underbrace{hf(x+h)}_{\text{Area of larger rectangle}}$$

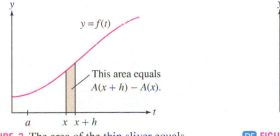

FIGURE 3   The area of the thin sliver equals $A(x+h) - A(x)$.

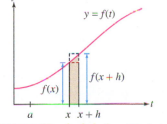

DF FIGURE 4   The shaded sliver lies between the rectangles of heights $f(x)$ and $f(x+h)$.

Now divide by h to squeeze the difference quotient between $f(x)$ and $f(x + h)$:

$$f(x) \leq \frac{A(x + h) - A(x)}{h} \leq f(x + h)$$

We have $\lim\limits_{h \to 0^+} f(x + h) = f(x)$ because f is continuous, and $\lim\limits_{h \to 0^+} f(x) = f(x)$, so the Squeeze Theorem gives us

$$\lim\limits_{h \to 0^+} \frac{A(x + h) - A(x)}{h} = f(x) \qquad \boxed{1}$$

A similar argument shows that for $h < 0$,

$$f(x + h) \leq \frac{A(x + h) - A(x)}{h} \leq f(x)$$

Again, the Squeeze Theorem gives us

$$\lim\limits_{h \to 0^-} \frac{A(x + h) - A(x)}{h} = f(x) \qquad \boxed{2}$$

Equations (1) and (2) show that $A'(x)$ exists and that $A'(x) = f(x)$. ■

In the previous section, we saw that FTC I says that in order to evaluate $\int_a^b f(x)\, dx$, if we can find an antiderivative F of f, we just take $F(b) - F(a)$. FTC II assures us that an antiderivative of f exists.

CONCEPTUAL INSIGHT   Many applications (in the sciences, engineering, and statistics) involve functions for which there is no explicit formula. Often, however, these functions can be expressed as definite integrals (or as infinite series). This enables us to compute their values numerically and create plots using a computer algebra system. Figure 5 shows a computer-generated graph of an antiderivative of $f(x) = \sin(x^2)$, for which there is no explicit formula.

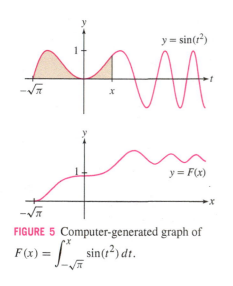

FIGURE 5 Computer-generated graph of $F(x) = \displaystyle\int_{-\sqrt{\pi}}^{x} \sin(t^2)\, dt$.

■ **EXAMPLE 2** Antiderivative as an Integral   Let F be the particular antiderivative of $f(x) = \sin(x^2)$ satisfying $F(-\sqrt{\pi}) = 0$. Express $F(x)$ as an integral.

Solution   According to FTC II, the area function with lower limit $a = -\sqrt{\pi}$ is an antiderivative satisfying $F(-\sqrt{\pi}) = 0$:

$$F(x) = \int_{-\sqrt{\pi}}^{x} \sin(t^2)\, dt \qquad ■$$

■ **EXAMPLE 3** Differentiating an Integral   Find the derivative of

$$A(x) = \int_2^x \sqrt{1 + t^3}\, dt$$

and calculate $A'(2)$, $A'(3)$, and $A(2)$.

Solution   By FTC II, $A'(x) = \sqrt{1 + x^3}$. In particular,

$$A'(2) = \sqrt{1 + 2^3} = 3 \qquad \text{and} \qquad A'(3) = \sqrt{1 + 3^3} = \sqrt{28}$$

On the other hand, $A(2) = \displaystyle\int_2^2 \sqrt{1 + t^3}\, dt = 0$. ■

CONCEPTUAL INSIGHT   The FTC shows that integration and differentiation are *inverse operations*. By FTC II, if you start with a continuous function f and form the integral $\int_a^x f(t)\, dt$, then you get back the original function by differentiating:

$$f(x) \xrightarrow{\text{Integrate}} \int_a^x f(t)\, dt \xrightarrow{\text{Differentiate}} \frac{d}{dx} \int_a^x f(t)\, dt = f(x)$$

On the other hand, by FTC I, if you differentiate first and then integrate, you also recover $f(x)$ [but only up to a constant $f(a)$]:

$$f(x) \xrightarrow{\text{Differentiate}} f'(x) \xrightarrow{\text{Integrate}} \int_a^x f'(t)\, dt = f(x) - f(a)$$

When the upper limit of the integral is a *function* of x rather than x itself, we use FTC II together with the Chain Rule to differentiate the integral.

■ **EXAMPLE 4 The FTC and the Chain Rule**   Find the derivative of

$$G(x) = \int_{-2}^{x^2} \sin t\, dt$$

Solution FTC II does not apply directly because the upper limit is x^2 rather than x. It is necessary to recognize that G is a *composite function* with outer function $A(x) = \int_{-2}^{x} \sin t\, dt$:

$$G(x) = A(x^2) = \int_{-2}^{x^2} \sin t\, dt$$

FTC II tells us that $A'(x) = \sin x$, so by the Chain Rule,

$$G'(x) = A'(x^2) \cdot (x^2)' = \sin(x^2) \cdot (2x) = 2x \sin(x^2)$$

Alternatively, we may set $u = x^2$ and use the Chain Rule as follows:

$$\frac{dG}{dx} = \frac{d}{dx} \int_{-2}^{x^2} \sin t\, dt = \left(\frac{d}{du} \int_{-2}^{u} \sin t\, dt \right) \frac{du}{dx} = (\sin u)2x = 2x \sin(x^2) \quad ■$$

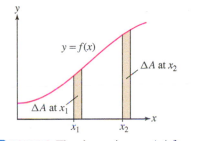

DF FIGURE 6 The change in area ΔA for a given Δx is larger when $f(x)$ is larger.

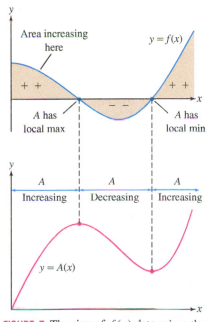

FIGURE 7 The sign of $f(x)$ determines the increasing/decreasing behavior of A.

GRAPHICAL INSIGHT Another Tale of Two Graphs   FTC II tells us that $A'(x) = f(x)$, or in other words, $f(x)$ is the rate of change of $A(x)$. If we did not know this result, we might come to suspect it by comparing the graphs of A and f. Consider the following:

- Figure 6 shows that the increase in area ΔA for a given Δx is larger at x_2 than at x_1 because $f(x_2) > f(x_1)$. So the size of $f(x)$ determines how quickly $A(x)$ changes, as we would expect if $A'(x) = f(x)$.
- Figure 7 shows that the sign of $f(x)$ determines whether A is increasing or decreasing. If $f(x) > 0$, then A is increasing because positive area is added as we move to the right. When $f(x)$ turns negative, A begins to decrease because we start adding negative area.
- A has a local max at points where $f(x)$ changes sign from $+$ to $-$ (the points where the area turns negative), and has a local min when $f(x)$ changes from $-$ to $+$. This agrees with the First Derivative Test.

These observations show that f "behaves" like A', as claimed by FTC II.

5.5 SUMMARY

- The *area function* with lower limit a: $A(x) = \int_a^x f(t)\, dt$. It satisfies $A(a) = 0$.

- FTC II: $A'(x) = f(x)$, or equivalently, $\dfrac{d}{dx} \int_a^x f(t)\, dt = f(x)$.

- **FTC II** shows that every continuous function has an antiderivative—namely, its area function (with any lower limit).

- To differentiate the function $G(x) = \displaystyle\int_a^{g(x)} f(t)\,dt$, write $G(x) = A(g(x))$, where
$A(x) = \displaystyle\int_a^x f(t)\,dt$. Then use the Chain Rule:

$$G'(x) = A'(g(x))g'(x) = f(g(x))g'(x)$$

5.5 EXERCISES

Preliminary Questions

1. Let $G(x) = \displaystyle\int_4^x \sqrt{t^3 + 1}\,dt$.

(a) Is the FTC II needed to calculate $G(4)$?

(b) Is the FTC II needed to calculate $G'(4)$?

2. Which of the following is an antiderivative F of $f(x) = x^2$ satisfying $F(2) = 0$?

(a) $\displaystyle\int_2^x 2t\,dt$ (b) $\displaystyle\int_0^2 t^2\,dt$ (c) $\displaystyle\int_2^x t^2\,dt$

3. Does every continuous function have an antiderivative? Explain.

4. Let $G(x) = \displaystyle\int_4^{x^3} \sin t\,dt$. Which of the following statements are correct?

(a) G is the composite function $\sin(x^3)$.

(b) G is the composite function $A(x^3)$, where

$$A(x) = \int_4^x \sin t\,dt$$

(c) $G(x)$ is too complicated to differentiate.

(d) The Product Rule is used to differentiate G.

(e) The Chain Rule is used to differentiate G.

(f) $G'(x) = 3x^2 \sin(x^3)$.

Exercises

1. Write the area function of $f(x) = 2x + 4$ with lower limit $a = -2$ as an integral and find a formula for it.

2. Find a formula for the area function of $f(x) = 2x + 4$ with lower limit $a = 0$.

3. Let $G(x) = \int_1^x (t^2 - 2)\,dt$. Calculate $G(1)$, $G'(1)$, and $G'(2)$. Then find a formula for $G(x)$.

4. Find $F(0)$, $F'(0)$, and $F'(3)$, where $F(x) = \displaystyle\int_0^x \sqrt{t^2 + t}\,dt$.

5. Find $G(1)$, $G'(0)$, and $G'(\pi/4)$, where $G(x) = \displaystyle\int_1^x \tan t\,dt$.

6. Find $H(-2)$ and $H'(-2)$, where $H(x) = \displaystyle\int_{-2}^x \frac{du}{u^2 + 1}$.

In Exercises 7–16, find formulas for the functions represented by the integrals.

7. $\displaystyle\int_2^x u^4\,du$ **8.** $\displaystyle\int_2^x (12t^2 - 8t)\,dt$

9. $\displaystyle\int_0^x \sin u\,du$ **10.** $\displaystyle\int_{-\pi/4}^x \sec^2 \theta\,d\theta$

11. $\displaystyle\int_4^x e^{3u}\,du$ **12.** $\displaystyle\int_x^0 e^{-t}\,dt$

13. $\displaystyle\int_1^{x^2} t\,dt$ **14.** $\displaystyle\int_{x/2}^{x/4} \sec^2 u\,du$

15. $\displaystyle\int_{3x}^{9x+2} e^{-u}\,du$ **16.** $\displaystyle\int_2^{\sqrt{x}} \frac{dt}{t}$

In Exercises 17–20, express the antiderivative F of f satisfying the given initial condition as an integral.

17. $f(x) = \sqrt{x^3 + 1}$, $F(5) = 0$

18. $f(x) = \dfrac{x + 1}{x^2 + 9}$, $F(7) = 0$ **19.** $f(x) = \sec x$, $F(0) = 0$

20. $f(x) = e^{-x^2}$, $F(-4) = 0$

In Exercises 21–24, calculate the derivative.

21. $\dfrac{d}{dx} \displaystyle\int_0^x (t^5 - 9t^3)\,dt$ **22.** $\dfrac{d}{d\theta} \displaystyle\int_1^\theta \cot u\,du$

23. $\dfrac{d}{dt} \displaystyle\int_{100}^t \sec(5x - 9)\,dx$ **24.** $\dfrac{d}{ds} \displaystyle\int_{-2}^s \tan\left(\frac{1}{1 + u^2}\right) du$

25. Let $A(x) = \displaystyle\int_0^x f(t)\,dt$ for $f(x)$ in Figure 8.

(a) Calculate $A(2)$, $A(3)$, $A'(2)$, and $A'(3)$.

(b) Find formulas for $A(x)$ on $[0, 2]$ and $[2, 4]$, and sketch the graph of A.

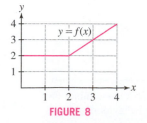

FIGURE 8

26. Make a rough sketch of the graph of $A(x) = \int_0^x g(t)\,dt$ for $y = g(x)$ in Figure 9.

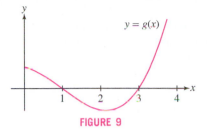

FIGURE 9

27. Verify $\int_0^x |t|\,dt = \frac{1}{2}x|x|$. *Hint:* Consider $x \geq 0$ and $x \leq 0$ separately.

28. Find $G'(1)$, where $G(x) = \int_0^{x^2} \sqrt{t^3 + 3}\,dt$.

In Exercises 29–34, calculate the derivative.

29. $\dfrac{d}{dx} \int_0^{x^2} \dfrac{t\,dt}{t+1}$

30. $\dfrac{d}{dx} \int_1^{1/x} \cos^3 t\,dt$

31. $\dfrac{d}{ds} \int_{-6}^{\cos s} u^4\,du$

32. $\dfrac{d}{dx} \int_{x^2}^{x^4} \sqrt{t}\,dt$

Hint for Exercise 32: $F(x) = A(x^4) - A(x^2)$.

33. $\dfrac{d}{dx} \int_{\sqrt{x}}^{x^2} \tan t\,dt$

34. $\dfrac{d}{du} \int_{-u}^{3u} \sqrt{x^2 + 1}\,dx$

In Exercises 35–38, with $y = f(x)$ as in Figure 10, let

$$A(x) = \int_0^x f(t)\,dt \quad \text{and} \quad B(x) = \int_2^x f(t)\,dt$$

35. Find the min and max of A on $[0, 6]$.

36. Find the min and max of B on $[0, 6]$.

37. Find formulas for $A(x)$ and $B(x)$ valid on $[2, 4]$.

38. Find formulas for $A(x)$ and $B(x)$ valid on $[4, 5]$.

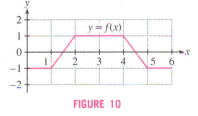

FIGURE 10

39. Let $A(x) = \int_0^x f(t)\,dt$, with $y = f(x)$ as in Figure 11.

(a) Does A have a local maximum at P?

(b) Where does A have a local minimum?

(c) Where does A have a local maximum?

(d) True or false? $A(x) < 0$ for all x in the interval shown.

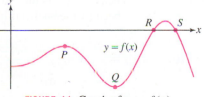

FIGURE 11 Graph of $y = f(x)$.

40. Determine $f(x)$, given that $\int_0^x f(t)\,dt = x^2 + x$.

41. Determine $g(x)$ and all values of c such that

$$\int_c^x g(t)\,dt = x^2 + x - 6$$

42. Find $a \leq b$ such that $\int_a^b (x^2 - 9)\,dx$ has minimal value.

In Exercises 43–44, let $A(x) = \int_a^x f(t)\,dt$.

43. **Area Functions and Concavity** Explain why the following statements are true. Assume f is differentiable.

(a) If c is an inflection point of A, then $f'(c) = 0$.

(b) A is concave up if f is increasing.

(c) A is concave down if f is decreasing.

44. Match the property of A with the corresponding property of the graph of f. Assume f is differentiable.

Area function A

(a) A is decreasing.

(b) A has a local maximum.

(c) A is concave up.

(d) A goes from concave up to concave down.

Graph of f

(i) Lies below the x-axis.

(ii) Crosses the x-axis from positive to negative.

(iii) Has a local maximum.

(iv) f is increasing.

45. Let $A(x) = \int_0^x f(t)\,dt$, with $y = f(x)$ as in Figure 12. Determine:

(a) The intervals on which A is increasing and decreasing

(b) The values x where A has a local min or max

(c) The inflection points of A

(d) The intervals where A is concave up or concave down

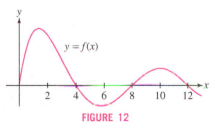

FIGURE 12

46. Let $f(x) = x^2 - 5x - 6$ and $F(x) = \int_0^x f(t)\,dt$.

(a) Find the critical points of F and determine whether they are local minima or local maxima.

(b) Find the points of inflection of F and determine whether the concavity changes from up to down or from down to up.

(c) GU   Plot $y = f(x)$ and $y = F(x)$ on the same set of axes and confirm your answers to (a) and (b).

47. Sketch the graph of an increasing function f such that both $f'(x)$ and $A(x) = \int_0^x f(t)\,dt$ are decreasing.

48. Figure 13 shows the graph of $f(x) = x \sin x$. Let $F(x) = \int_0^x t \sin t \, dt$.

(a) Locate the local max and absolute max of F on $[0, 3\pi]$.
(b) Justify graphically: F has precisely one zero in $[\pi, 2\pi]$.
(c) How many zeros does F have in $[0, 3\pi]$?
(d) Find the inflection points of F on $[0, 3\pi]$. For each one, state whether the concavity changes from up to down or from down to up.

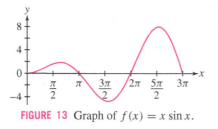

FIGURE 13 Graph of $f(x) = x \sin x$.

49. GU Find the smallest positive critical point of

$$F(x) = \int_0^x \cos(t^{3/2}) \, dt$$

and determine whether it is a local min or max. Then find the smallest positive inflection point of $F(x)$ and use a graph of $y = \cos(x^{3/2})$ to determine whether the concavity changes from up to down or from down to up.

Further Insights and Challenges

50. Proof of FTC II The proof in the text assumes that f is increasing. To prove it for all continuous functions, let $m(h)$ and $M(h)$ denote the *minimum* and *maximum* of f on $[x, x + h]$ (Figure 14). The continuity of f implies that $\lim_{h \to 0} m(h) = \lim_{h \to 0} M(h) = f(x)$. Show that for $h > 0$,

$$hm(h) \le A(x + h) - A(x) \le hM(h)$$

For $h < 0$, the inequalities are reversed. Prove that $A'(x) = f(x)$.

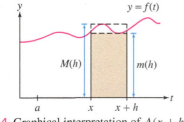

FIGURE 14 Graphical interpretation of $A(x + h) - A(x)$.

51. Proof of FTC I FTC I asserts that $\int_a^b f(t) \, dt = F(b) - F(a)$ if $F'(x) = f(x)$. Use FTC II to give a new proof of FTC I as follows. Set $A(x) = \int_a^x f(t) \, dt$.

(a) Show that $F(x) = A(x) + C$ for some constant.

(b) Show that $F(b) - F(a) = A(b) - A(a) = \int_a^b f(t) \, dt$.

52. Can Every Antiderivative Be Expressed as an Integral? The area function $A(x) = \int_a^x f(t) \, dt$ is an antiderivative of f for every value of a. However, not all antiderivatives are obtained in this way. The general antiderivative of $f(x) = x$ is $F(x) = \frac{1}{2}x^2 + C$. Show that F is an area function if $C \le 0$ but not if $C > 0$.

53. Prove the formula

$$\frac{d}{dx} \int_{u(x)}^{v(x)} f(t) \, dt = f(v(x))v'(x) - f(u(x))u'(x)$$

54. Use the result of Exercise 53 to calculate

$$\frac{d}{dx} \int_{\ln x}^{e^x} \sin t \, dt$$

5.6 Net Change as the Integral of a Rate of Change

So far we have focused on the area interpretation of the integral. In this section, we use the integral to compute net change.

Consider the following problem: Water flows into an empty bucket at a rate of $r(t)$ liters per second. How much water is in the bucket after 4 seconds? If the rate of water flow were *constant*—say, 1.5 liters/second (L/s)—we would have

$$\text{Quantity of water} = \text{flow rate} \times \text{time elapsed} = (1.5)4 = 6 \text{ liters}$$

Suppose, however, that the flow rate $r(t)$ varies as in Figure 1. Then *the quantity of water is equal to the area under the graph of $y = r(t)$*. To prove this, let $s(t)$ be the amount of water in the bucket at time t. Then $s'(t) = r(t)$ because $s'(t)$ is the rate at which the quantity of water is changing. Furthermore, $s(0) = 0$ because the bucket is initially empty. By FTC I,

$$\underbrace{\int_0^4 s'(t) \, dt}_{\substack{\text{Area under the graph} \\ \text{of the flow rate}}} = s(4) - s(0) = \underbrace{s(4)}_{\substack{\text{Water in bucket} \\ \text{at } t = 4}}$$

FIGURE 1 The quantity of water in the bucket is equal to the area under the graph of the flow rate $r(t)$.

More generally, $s(t_2) - s(t_1)$ is the **net change** in $s(t)$ over the interval $[t_1, t_2]$. FTC I yields the following result.

THEOREM 1 Net Change as the Integral of a Rate of Change   The net change in $s(t)$ over an interval $[t_1, t_2]$ is given by the integral

$$\underbrace{\int_{t_1}^{t_2} s'(t)\, dt}_{\text{Integral of the rate of change}} = \underbrace{s(t_2) - s(t_1)}_{\text{Net change over } [t_1, t_2]}$$

In Theorem 1, the variable t does not have to be a time variable.

■ **EXAMPLE 1**   Water leaks from a tank at a rate of $2 + 5t$ liters/hour (L/h), where t is the number of hours after 7 AM. How much water is lost between 9 and 11 AM?

Solution   Let $s(t)$ be the quantity of water in the tank at time t. Then $s'(t) = -(2 + 5t)$, where the minus sign occurs because $s(t)$ is decreasing. Since 9 AM and 11 AM correspond to $t = 2$ and $t = 4$, respectively, the net change in $s(t)$ between 9 and 11 AM is

$$s(4) - s(2) = \int_2^4 s'(t)\, dt = -\int_2^4 (2 + 5t)\, dt$$

$$= -\left(2t + \frac{5}{2}t^2\right)\Big|_2^4 = (-48) - (-14) = -34 \text{ liters}$$

The tank lost 34 liters between 9 and 11 AM.   ■

In the next example, we estimate an integral using numerical data. We shall compute the average of the left- and right-endpoint approximations, because this is usually more accurate than either endpoint approximation alone. (In Section 7.8, this average is called the Trapezoidal Approximation.)

■ **EXAMPLE 2   Traffic Flow**   The number of cars per hour passing an observation point along a highway is called the traffic flow rate $q(t)$ (in cars per hour).

(a) Which quantity is represented by the integral $\int_{t_1}^{t_2} q(t)\, dt$?

(b) The flow rate is recorded at 15-minute intervals between 7:00 and 9:00 AM. Estimate the number of cars using the highway during this 2-hour (h) period.

t	7:00	7:15	7:30	7:45	8:00	8:15	8:30	8:45	9:00
$q(t)$	1044	1297	1478	1844	1451	1378	1155	802	542

Solution

(a) The integral $\int_{t_1}^{t_2} q(t)\, dt$ represents the total number of cars that passed the observation point during the time interval $[t_1, t_2]$.

(b) The data values are spaced at intervals of $\Delta t = 0.25$ h. Thus,

$$L_N = 0.25\left(1044 + 1297 + 1478 + 1844 + 1451 + 1378 + 1155 + 802\right)$$

$$\approx 2612$$

$$R_N = 0.25\left(1297 + 1478 + 1844 + 1451 + 1378 + 1155 + 802 + 542\right)$$

$$\approx 2487$$

In Example 2, L_N is the sum of the values of $q(t)$ at the left endpoints

7:00, 7:15, ..., 8:45

and R_N is the sum of the values of $q(t)$ at the right endpoints

7:15, ..., 8:45, 9:00

We estimate the number of cars that passed the observation point between 7 and 9 AM by taking the average of R_N and L_N:

$$\int_7^9 q(t)\, dt \approx \frac{1}{2}(R_N + L_N) = \frac{1}{2}(2612 + 2487) \approx 2550$$

Approximately 2550 cars used the highway between 7 and 9 AM.   ■

The Integral of Velocity

Let $s(t)$ be the position at time t of an object in linear motion. Then the object's velocity is $v(t) = s'(t)$, and the integral of v is equal to the *net change in position* or *displacement* over a time interval $[t_1, t_2]$:

$$\int_{t_1}^{t_2} v(t)\,dt = \int_{t_1}^{t_2} s'(t)\,dt = \underbrace{s(t_2) - s(t_1)}_{\substack{\text{Displacement or net} \\ \text{change in position}}}$$

We must distinguish between displacement and *distance traveled*. If you travel 10 km and then return to your starting point, your displacement is zero but your distance traveled is 20 km. To compute distance traveled rather than displacement, we integrate the *speed* $|v(t)|$.

> **THEOREM 2   The Integral of Velocity**   For an object in linear motion with velocity $v(t)$, then
>
> $$\text{Displacement during } [t_1, t_2] = \int_{t_1}^{t_2} v(t)\,dt$$
>
> $$\text{Distance traveled during } [t_1, t_2] = \int_{t_1}^{t_2} |v(t)|\,dt$$

■ **EXAMPLE 3**   A particle has velocity $v(t) = t^3 - 10t^2 + 24t$ m/s. Compute:

(a) Displacement over $[0, 6]$ **(b)** Total distance traveled over $[0, 6]$

Indicate the particle's trajectory with a motion diagram.

Solution   First, we compute the indefinite integral:

$$\int v(t)\,dt = \int (t^3 - 10t^2 + 24t)\,dt = \frac{1}{4}t^4 - \frac{10}{3}t^3 + 12t^2 + C$$

(a)   The displacement over the time interval $[0, 6]$ is

$$\int_0^6 v(t)\,dt = \left(\frac{1}{4}t^4 - \frac{10}{3}t^3 + 12t^2 \right)\Big|_0^6 = 36 \text{ m}$$

(b)   The factorization $v(t) = t(t-4)(t-6)$ shows that $v(t)$ changes sign at $t = 4$. It is positive on $[0, 4]$ and negative on $[4, 6]$ as we see in Figure 2. Therefore, the total distance traveled is

$$\int_0^6 |v(t)|\,dt = \int_0^4 v(t)\,dt - \int_4^6 v(t)\,dt$$

We evaluate these two integrals separately:

$$[0, 4]: \quad \int_0^4 v(t)\,dt = \left(\frac{1}{4}t^4 - \frac{10}{3}t^3 + 12t^2 \right)\Big|_0^4 = \frac{128}{3} \text{ m}$$

$$[4, 6]: \quad \int_4^6 v(t)\,dt = \left(\frac{1}{4}t^4 - \frac{10}{3}t^3 + 12t^2 \right)\Big|_4^6 = -\frac{20}{3} \text{ m}$$

The total distance traveled is $\frac{128}{3} + \frac{20}{3} = \frac{148}{3} = 49\frac{1}{3}$ m.

Figure 3 is a motion diagram indicating the particle's trajectory. The particle travels $\frac{128}{3}$ m during the first 4 s and then backtracks $\frac{20}{3}$ m over the next 2 s.   ■

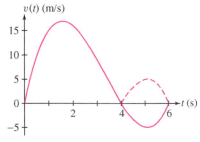

FIGURE 2 Graph of $v(t) = t^3 - 10t^2 + 24t$. Over $[4, 6]$, the dashed curve is the graph of $|v(t)|$.

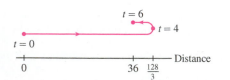

FIGURE 3 Path of the particle along a straight line.

Total Versus Marginal Cost

In Section 3.4, we defined the marginal cost at production level x_0 as the cost

$$C(x_0 + 1) - C(x_0)$$

of producing one additional unit. Since this marginal cost is approximated well by the derivative $C'(x_0)$ for large values of x_0 compared to 1, economists often refer to $C'(x)$ itself as the marginal cost.

Consider the cost $C(x)$ of a manufacturer (the dollar cost of producing x units of a particular product or commodity). The derivative $C'(x)$ is called the **marginal cost**. The cost of increasing production from a units to b units is the net change $C(b) - C(a)$, which is equal to the integral of the marginal cost:

$$\text{Cost of increasing production from } a \text{ units to } b \text{ units} = \int_a^b C'(x)\,dx$$

■ **EXAMPLE 4** The marginal cost of producing x computer chips (in units of 1000) is $C'(x) = 300x^2 - 4000x + 40{,}000$ (dollars per thousand chips).

(a) Find the cost of increasing production from 10,000 to 15,000 chips.

(b) Determine the total cost of producing 15,000 chips, assuming that it costs $30,000 to set up the manufacturing run [i.e., $C(0) = 30{,}000$].

Solution

(a) The cost of increasing production from 10,000 ($x = 10$) to 15,000 ($x = 15$) is

$$C(15) - C(10) = \int_{10}^{15} (300x^2 - 4000x + 40{,}000)\,dx$$

$$= (100x^3 - 2000x^2 + 40{,}000x)\Big|_{10}^{15}$$

$$= 487{,}500 - 300{,}000 = \$187{,}500$$

(b) The cost of increasing production from 0 to 15,000 chips is

$$C(15) - C(0) = \int_0^{15} (300x^2 - 4000x + 40{,}000)\,dx$$

$$= (100x^3 - 2000x^2 + 40{,}000x)\Big|_0^{15} = \$487{,}500$$

The total cost of producing 15,000 chips includes the set-up costs of $30,000:

$$C(15) = C(0) + 487{,}500 = 30{,}000 + 487{,}500 = \$517{,}500 \qquad ■$$

5.6 SUMMARY

- Many applications are based on the following principle: *The net change in a quantity $s(t)$ is equal to the integral of its rate of change:*

$$\underbrace{s(t_2) - s(t_1)}_{\text{Net change over } [t_1, t_2]} = \int_{t_1}^{t_2} s'(t)\,dt$$

- For an object traveling in a straight line at velocity $v(t)$,

$$\text{Displacement during } [t_1, t_2] = \int_{t_1}^{t_2} v(t)\,dt$$

$$\text{Total distance traveled during } [t_1, t_2] = \int_{t_1}^{t_2} |v(t)|\,dt$$

- If $C(x)$ is the cost of producing x units of a commodity, then $C'(x)$ is the marginal cost and

$$\text{Cost of increasing production from } a \text{ units to } b \text{ units} = \int_a^b C'(x)\,dx$$

5.6 EXERCISES

Preliminary Questions

1. A hot metal object is submerged in cold water. The rate at which the object cools (in degrees per minute) is a function $f(t)$ of time. Which quantity is represented by the integral $\int_0^T f(t)\,dt$?

2. A plane travels 560 km from Los Angeles to San Francisco in 1 hour (h). If the plane's velocity at time t is $v(t)$ km/h, what is the value of $\int_0^1 v(t)\,dt$?

3. Which of the following quantities would be naturally represented as derivatives and which as integrals?

(a) Velocity of a train

(b) Rainfall during a 6-month period

(c) Mileage per gallon of an automobile

(d) Increase in the U.S. population from 1990 to 2010

Exercises

1. Water flows into an empty reservoir at a rate of $3000 + 20t$ liters per hour (L/h; t is in hours). What is the quantity of water in the reservoir after 5 h?

2. A population of insects increases at a rate of $200 + 10t + 0.25t^2$ insects per day (t in days). Find the insect population after 3 days, assuming that there are 35 insects at $t = 0$.

3. A survey shows that a mayoral candidate is gaining votes at a rate of $2000t + 1000$ votes per day, where t is the number of days since she announced her candidacy. How many supporters will the candidate have after 60 days, assuming that she had no supporters at $t = 0$?

4. A factory produces bicycles at a rate of $95 + 3t^2 - t$ bicycles per week (t in weeks). How many bicycles were produced from the beginning of week 2 to the end of week 3?

5. Find the displacement of a particle moving in a straight line with velocity $v(t) = 4t - 3$ m/s over the time interval $[2, 5]$.

6. Find the displacement over the time interval $[1, 6]$ of a helicopter whose (vertical) velocity at time t is $v(t) = 0.02t^2 + t$ m/s.

7. A cat falls from a tree (with zero initial velocity) at time $t = 0$. How far does the cat fall between $t = 0.5$ and $t = 1$ s? Use Galileo's formula $v(t) = -9.8t$ m/s.

8. A projectile is released with an initial (vertical) velocity of 100 m/s. Use the formula $v(t) = 100 - 9.8t$ for velocity to determine the distance traveled during the first 15 s.

In Exercises 9–12, a particle moves in a straight line with the given velocity (in meters per second). Find the displacement and distance traveled over the time interval, and draw a motion diagram like Figure 3 (with distance and time labels).

9. $v(t) = 12 - 4t, \quad [0, 5]$

10. $v(t) = 36 - 24t + 3t^2, \quad [0, 10]$

11. $v(t) = t^{-2} - 1, \quad [0.5, 2]$ **12.** $v(t) = \cos t, \quad [0, 3\pi]$

13. Find the net change in velocity over $[1, 4]$ of an object with $a(t) = 8t - t^2$ m/s^2.

14. Show that if acceleration is constant, then the change in velocity is proportional to the length of the time interval.

15. The traffic flow rate past a certain point on a highway is $q(t) = 3000 + 2000t - 300t^2$ (t in hours), where $t = 0$ is 8 AM. How many cars pass by in the time interval from 8 to 10 AM?

16. The marginal cost of producing x tablet computers is $C'(x) = 120 - 0.06x + 0.00001x^2$ What is the cost of producing 3000 units if the set-up cost is \$90,000? If production is set at 3000 units, what is the cost of producing 200 additional units?

17. A small boutique produces wool sweaters at a marginal cost of $40 - 5\lfloor x/5 \rfloor$ for $0 \leq x \leq 20$, where $\lfloor x \rfloor$ is the greatest integer function. Find the cost of producing 20 sweaters. Then compute the average cost of the first 10 sweaters and the last 10 sweaters.

18. The rate (in liters per minute) at which water drains from a tank is recorded at half-minute intervals. Compute the average of the left- and right-endpoint approximations to estimate the total amount of water drained during the first 3 min.

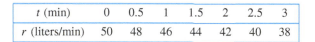

t (min)	0	0.5	1	1.5	2	2.5	3
r (liters/min)	50	48	46	44	42	40	38

19. The velocity of a car is recorded at half-second intervals (in feet per second). Use the average of the left- and right-endpoint approximations to estimate the total distance traveled during the first 4 s.

t	0	0.5	1	1.5	2	2.5	3	3.5	4
$v(t)$	0	12	20	29	38	44	32	35	30

20. To model the effects of a **carbon tax** on CO_2 emissions, policy-makers study the *marginal cost of abatement* $B(x)$, defined as the cost of increasing CO_2 reduction from x to $x + 1$ tons (in units of ten thousand tons—Figure 4). Which quantity is represented by the area under the curve over $[0, 3]$ in Figure 4?

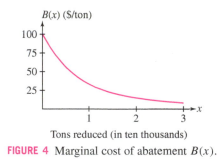

FIGURE 4 Marginal cost of abatement $B(x)$.

21. A megawatt of power is 10^6 W, or 3.6×10^9 joules/hour (J/h). Which quantity is represented by the area under the graph in Figure 5? Estimate the energy (in joules) consumed during the period 4 PM to 8 PM.

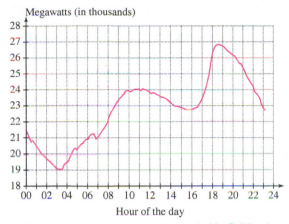

FIGURE 5 Power consumption over 1-day period in California (February 2010).

22. Figure 6 shows the migration rate $M(t)$ of Ireland in the period 1988–1998. This is the rate at which people (in thousands per year) move into or out of the country.

(a) Is the following integral positive or negative? What does this quantity represent?

$$\int_{1988}^{1998} M(t)\,dt$$

(b) Did migration in the period 1988–1998 result in a net influx of people into Ireland or a net outflow of people from Ireland?

(c) During which two years could the Irish prime minister announce, "We've hit an inflection point. We are still losing population, but the trend is now improving."

FIGURE 6 Irish migration rate (in thousands per year).

23. Let $N(d)$ be the number of asteroids of diameter $\le d$ kilometers. Data suggest that the diameters are distributed according to a piecewise power law:

$$N'(d) = \begin{cases} 1.9 \times 10^9 d^{-2.3} & \text{for } d < 70 \\ 2.6 \times 10^{12} d^{-4} & \text{for } d \ge 70 \end{cases}$$

(a) Compute the number of asteroids with a diameter between 0.1 km and 100 km.

(b) Using the approximation $N(d+1) - N(d) \approx N'(d)$, estimate the number of asteroids of diameter 50 km.

24. Heat Capacity The heat capacity $C(T)$ of a substance is the amount of energy (in joules) required to raise the temperature of 1 g by 1°C at temperature T.

(a) Explain why the energy required to raise the temperature from T_1 to T_2 is the area under the graph of $C(T)$ over $[T_1, T_2]$.

(b) How much energy is required to raise the temperature from 50°C to 100°C if $C(T) = 6 + 0.2\sqrt{T}$?

25. Figure 7 shows the rate $R(t)$ of natural gas consumption (in billions of cubic feet per day) in the mid-Atlantic states (New York, New Jersey, Pennsylvania). Express the total quantity of natural gas consumed in

2009 as an integral (with respect to time t in days). Then estimate this quantity, given the following monthly values of $R(t)$:

$$3.18, \quad 2.86, \quad 2.39, \quad 1.49, \quad 1.08, \quad 0.80,$$
$$1.01, \quad 0.89, \quad 0.89, \quad 1.20, \quad 1.64, \quad 2.52$$

Keep in mind that the number of days in a month varies with the month.

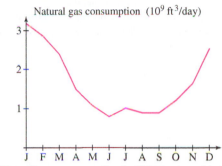

FIGURE 7 Natural gas consumption in 2009 in the mid-Atlantic states.

26. Cardiac output is the rate R of volume of blood pumped by the heart per unit time (in liters per minute). Doctors measure R by injecting A mg of dye into a vein leading into the heart at $t = 0$ and recording the concentration $c(t)$ of dye (in milligrams per liter) pumped out at short regular time intervals (Figure 8).

(a) Explain: The quantity of dye pumped out in a small time interval $[t, t + \Delta t]$ is approximately $Rc(t)\Delta t$.

(b) Show that $A = R \int_0^T c(t)\,dt$, where T is large enough that all of the dye is pumped through the heart but not so large that the dye returns by recirculation.

(c) Assume $A = 5$ mg. Estimate R using the following values of $c(t)$ recorded at 1-second intervals from $t = 0$ to $t = 10$:

$$0, \quad 0.4, \quad 2.8, \quad 6.5, \quad 9.8, \quad 8.9,$$
$$6.1, \quad 4, \quad 2.3, \quad 1.1, \quad 0$$

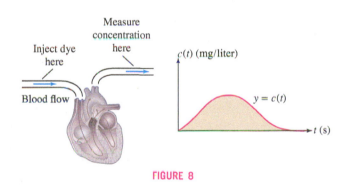

FIGURE 8

Exercises 27 and 28: A study suggests that the extinction rate $r(t)$ of marine animal families during the Phanerozoic Eon can be modeled by the function $r(t) = 3130/(t + 262)$ for $0 \le t \le 544$, where t is time elapsed (in millions of years) since the beginning of the eon 544 million years ago. Thus, $t = 544$ refers to the present time, $t = 540$ is 4 million years ago, and so on.

27. Compute the average of R_N and L_N with $N = 5$ to estimate the total number of families that became extinct in the periods $100 \le t \le 150$ and $350 \le t \le 400$.

28. CAS Estimate the total number of extinct families from $t = 0$ to the present, using M_N with $N = 544$.

Further Insights and Challenges

29. Show that a particle, located at the origin at $t = 1$ and moving along the x-axis with velocity $v(t) = t^{-2}$, will never pass the point $x = 2$.

30. Show that a particle, located at the origin at $t = 1$ and moving along the x-axis with velocity $v(t) = t^{-1/2}$, moves arbitrarily far from the origin after sufficient time has elapsed.

31. In a free market economy, the demand curve is the graph of the function D that represents the demand for a specific product by the consumers in the economy at price q. It is not surprising that the curve is decreasing, as the demand drops as the price goes up. The supply curve is the graph of the function S that represents the supply of the product that the producers are willing to produce as a function of the price q. As the price goes up, the producers are willing to produce more of the item, and therefore, this curve is increasing. The point (p^*, q^*) at which the two curves cross is called the equilibrium point, where the supply and demand balance. Tradition in economics is to make the horizontal axis the quantity q of the item and the vertical axis the price p. We define $p = S(q)$ to correspond to the supply curve and $p = D(q)$ to correspond to the demand curve. In other words, we have inverted the formula for these two functions from giving quantity in terms of price to giving price in terms of quantity. The areas depicted in Figure 9 represent the excess supply and excess demand.

(a) The consumer surplus represents the savings on the part of consumers if they pay price p^* rather than the price greater than p^* that many were willing to pay. Write down a formula for this consumer

surplus. The formula will include a definite integral and it will depend on p^* and q^*.

(b) The producer surplus represents the savings on the part of producers if they sell at price p^* rather than the price less than p^* that some producers were willing to accept. Write down a formula for this producer surplus.

(c) A variety of coffee shops in a town sell mocha latte supreme coffees. If the supply curve is given by $p = \frac{q}{100} + 1$ and demand curve is given by $p = \frac{10}{q/100+1}$, determine the equilibrium point (p^*, q^*) and the consumer surplus and producer surplus when the mocha latte supreme coffees are sold at price p^*.

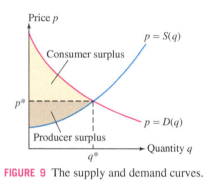

FIGURE 9 The supply and demand curves.

5.7 Substitution Method

Integration (antidifferentiation) is generally more difficult than differentiation. There are no sure-fire methods, and many antiderivatives cannot be expressed in terms of elementary functions. However, there are a few important general techniques. One such technique is the **Substitution Method**, which uses the Chain Rule "in reverse."

Consider the integral $\int 2x \cos(x^2)\,dx$. We can evaluate it if we remember the Chain Rule calculation

$$\frac{d}{dx} \sin(x^2) = 2x \cos(x^2)$$

This tells us that $\sin(x^2)$ is an antiderivative of $2x \cos(x^2)$, and therefore,

$$\int \underbrace{2x}_{\substack{\text{Derivative of}\\\text{inside function}}} \underbrace{\cos(x^2)}_{\substack{\text{Inside}\\\text{function}}} dx = \sin(x^2) + C$$

A similar Chain Rule calculation shows that

$$\int \underbrace{(1 + 3x^2)}_{\substack{\text{Derivative of}\\\text{inside function}}} \underbrace{\cos(x + x^3)}_{\substack{\text{Inside}\\\text{function}}} dx = \sin(x + x^3) + C$$

◄··· *REMINDER A "composite function" has the form $f(g(x))$. For convenience, we call $u = g(x)$ the inside function and $f(u)$ the outside function.*

In both cases, the integrand is the product of a composite function and the derivative of the inside function. The Chain Rule does not help if the derivative of the inside function is missing. For instance, we cannot use the Chain Rule to compute $\int \cos(x + x^3)\,dx$ because the factor $(1 + 3x^2)$ does not appear.

In general, if $F'(u) = f(u)$, then by the Chain Rule,

$$\frac{d}{dx} F(u(x)) = F'(u(x))u'(x) = f(u(x))u'(x)$$

This translates into the following integration formula.

> **THEOREM 1** **The Substitution Method** If $F'(x) = f(x)$, then
>
> $$\int f(u(x))u'(x)\,dx = F(u(x)) + C$$

Substitution Using Differentials

Before proceeding to the examples, we discuss the procedure for carrying out substitution using differentials. Differentials are symbols such as du or dx that occur in the Leibniz notations du/dx and $\int f(x)\,dx$. In our calculations, we shall manipulate them as though they are related by an equation in which the dx "cancels":

$$du = \frac{du}{dx}\,dx$$

Equivalently, du and dx are related by

$$du = u'(x)\,dx \qquad \boxed{1}$$

For example,

$$\text{If } u = x^2, \qquad \text{then} \qquad du = 2x\,dx$$

$$\text{If } u = \cos(x^3), \qquad \text{then} \qquad du = -3x^2\sin(x^3)\,dx$$

Now when the integrand has the form $f(u(x))\,u'(x)$, we can use Eq. (1) to rewrite the entire integral (including the dx term) in terms of u and its differential du:

$$\int \underbrace{f(u(x))}_{f(u)}\ \underbrace{u'(x)\,dx}_{du} = \int f(u)\,du$$

The symbolic calculus of substitution using differentials was invented by Leibniz and is considered one of his most important achievements. It reduces the otherwise complicated process of transforming integrals to a convenient set of rules.

This equation is called the **Change of Variables Formula**. It transforms an integral in the variable x into a (hopefully simpler) integral in the new variable u.

■ **EXAMPLE 1** Evaluate $\displaystyle\int 3x^2 \sin(x^3)\,dx$.

In substitution, the key step is to choose the appropriate inside function u.

Solution The integrand contains the composite function $\sin(x^3)$, so we set $u = x^3$. The differential $du = 3x^2\,dx$ also appears, so we can carry out the substitution:

$$\int 3x^2 \sin(x^3)\,dx = \int \underbrace{\sin(x^3)}_{\sin u}\ \underbrace{3x^2\,dx}_{du} = \int \sin u\,du$$

Now evaluate the integral in the u-variable and replace u by x^3 in the answer:

$$\int 3x^2 \sin(x^3)\,dx = \int \sin u\,du = -\cos u + C = -\cos(x^3) + C$$

Let's check our answer by differentiating:

$$\frac{d}{dx}(-\cos(x^3)) = \sin(x^3)\frac{d}{dx}x^3 = 3x^2 \sin(x^3) \qquad\qquad ■$$

■ **EXAMPLE 2** **Multiplying du by a Constant** Evaluate $\displaystyle\int x(x^2 + 9)^5\,dx$.

Solution We let $u = x^2 + 9$ because the composite $u^5 = (x^2 + 9)^5$ appears in the integrand. The differential $du = 2x\,dx$ does not appear as is, but we can multiply by $\frac{1}{2}$ to obtain

$$\frac{1}{2}du = x\,dx \quad\Rightarrow\quad \frac{1}{2}u^5\,du = x(x^2 + 9)^5\,dx$$

Now we can apply substitution:

$$\int x(x^2+9)^5 \, dx = \int \overbrace{(x^2+9)^5}^{u^5} \overbrace{x\,dx}^{\frac{1}{2}du} = \frac{1}{2}\int u^5 \, du = \frac{1}{12}u^6 + C$$

Finally, we express the answer in terms of x by substituting $u = x^2 + 9$:

$$\int x(x^2+9)^5 \, dx = \frac{1}{12}u^6 + C = \frac{1}{12}(x^2+9)^6 + C \qquad \blacksquare$$

Substitution Method:

(1) Choose u and compute du.

(2) Rewrite the integral in terms of u and du, and evaluate.

(3) Express the final answer in terms of x.

■ **EXAMPLE 3** Evaluate $\displaystyle\int \frac{(x^2+2x)\,dx}{(x^3+3x^2+12)^6}$.

Solution The appearance of $(x^3+3x^2+12)^{-6}$ in the integrand suggests that we try $u = x^3 + 3x^2 + 12$. With this choice,

$$du = (3x^2+6x)\,dx = 3(x^2+2x)\,dx \quad\Rightarrow\quad \frac{1}{3}du = (x^2+2x)\,dx$$

$$\int \frac{(x^2+2x)\,dx}{(x^3+3x^2+12)^6} = \int \overbrace{(x^3+3x^2+12)^{-6}}^{u^{-6}} \overbrace{(x^2+2x)\,dx}^{\frac{1}{3}du}$$

$$= \frac{1}{3}\int u^{-6}\,du = \left(\frac{1}{3}\right)\left(\frac{u^{-5}}{-5}\right) + C$$

$$= -\frac{1}{15}(x^3+3x^2+12)^{-5} + C \qquad \blacksquare$$

CONCEPTUAL INSIGHT An integration method that works for a given function may fail if we change the function even slightly. In the previous example, if we replace 2 by 2.1 and consider instead $\displaystyle\int \frac{(x^2+2.1x)\,dx}{(x^3+3x^2+12)^6}$, the Substitution Method does not work. The problem is that $(x^2+2.1x)\,dx$ is *not* a multiple of $du = (3x^2+6x)\,dx$.

■ **EXAMPLE 4** Evaluate $\displaystyle\int \sin(7\theta+5)\,d\theta$.

Solution Let $u = 7\theta + 5$. Then $du = 7\,d\theta$ and $\frac{1}{7}du = d\theta$. We obtain

$$\int \underbrace{\sin(7\theta+5)}_{\sin u}\,\underbrace{d\theta}_{\frac{1}{7}du} = \frac{1}{7}\int \sin u \, du = -\frac{1}{7}\cos u + C = -\frac{1}{7}\cos(7\theta+5) + C \quad \blacksquare$$

■ **EXAMPLE 5** Evaluate $\displaystyle\int e^{-9t}\,dt$.

Solution Use the substitution $u = -9t$, $du = -9\,dt$. Then $dt = -\frac{1}{9}du$:

$$\int e^{-9t}\,dt = \int e^u\left(-\frac{1}{9}\,du\right) = -\frac{1}{9}\int e^u\,du = -\frac{1}{9}e^u + C = -\frac{1}{9}e^{-9t} + C \quad \blacksquare$$

■ **EXAMPLE 6** **Integral of $\tan\theta$** Evaluate $\displaystyle\int \tan\theta\,d\theta$.

Solution In this case, the idea is to write $\tan\theta\,d\theta = \dfrac{\sin\theta\,d\theta}{\cos\theta}$ and to note that if $u = \cos\theta$, then $du = -\sin\theta\,d\theta$ so $\frac{1}{-1}du = \sin\theta\,d\theta$:

$$\int \tan\theta\,d\theta = \int \frac{\sin\theta\,d\theta}{\cos\theta} = -\int \frac{du}{u} = -\ln|u| + C = -\ln|\cos\theta| + C$$

Now recall that $-\ln u = \ln \frac{1}{u}$. Thus, $-\ln|\cos\theta| = \ln \frac{1}{|\cos\theta|}$, and we obtain

$$\int \tan\theta \, d\theta = \ln\left|\frac{1}{\cos\theta}\right| + C = \ln|\sec\theta| + C \qquad \blacksquare$$

■ **EXAMPLE 7** **Additional Step Necessary** Evaluate $\int x\sqrt{5x+1}\,dx$.

Solution Since $\sqrt{5x+1}$ appears, we are tempted to set $u = 5x+1$. Then

$$du = 5\,dx \quad \Rightarrow \quad \sqrt{5x+1}\,dx = \frac{1}{5}\sqrt{u}\,du$$

Unfortunately, the integrand is not $\sqrt{5x+1}$ but $x\sqrt{5x+1}$. To take care of the extra factor of x, we solve $u = 5x+1$ to obtain $x = \frac{1}{5}(u-1)$. Then

$$x\sqrt{5x+1}\,dx = \left(\frac{1}{5}(u-1)\right)\frac{1}{5}\sqrt{u}\,du = \frac{1}{25}(u^{3/2} - u^{1/2})\,du$$

$$\int x\sqrt{5x+1}\,dx = \frac{1}{25}\int(u^{3/2} - u^{1/2})\,du$$

$$= \frac{1}{25}\left(\frac{2}{5}u^{5/2} - \frac{2}{3}u^{3/2}\right) + C$$

$$= \frac{2}{125}(5x+1)^{5/2} - \frac{2}{75}(5x+1)^{3/2} + C \qquad \blacksquare$$

> *The Substitution Method does not always work, even when the integral looks relatively simple. For example, $\int \sin(x^2)\,dx$ cannot be evaluated explicitly by substitution, or any other method. With experience, you will learn to recognize when substitution is likely to be successful.*

Change of Variables Formula for Definite Integrals

The Change of Variables Formula can be applied to definite integrals provided that the limits of integration are changed, as indicated in the next theorem.

> *The new limits of integration with respect to the u-variable are $u(a)$ and $u(b)$. Think of it this way: As x varies from a to b, the variable $u = u(x)$ varies from $u(a)$ to $u(b)$.*

Change of Variables Formula for Definite Integrals

$$\int_a^b f(u(x))u'(x)\,dx = \int_{u(a)}^{u(b)} f(u)\,du \qquad \boxed{2}$$

Proof If $F(x)$ is an antiderivative of $f(x)$, then $F(u(x))$ is an antiderivative of $f(u(x))u'(x)$. FTC I shows that the two integrals are equal:

$$\int_a^b f(u(x))u'(x)\,dx = F(u(b)) - F(u(a))$$

$$\int_{u(a)}^{u(b)} f(u)\,du = F(u(b)) - F(u(a)) \qquad \blacksquare$$

■ **EXAMPLE 8** Evaluate $\int_0^2 x^2\sqrt{x^3+1}\,dx$.

Solution Use the substitution $u = x^3 + 1$, $du = 3x^2\,dx$:

$$x^2\sqrt{x^3+1}\,dx = \frac{1}{3}\sqrt{u}\,du$$

By Eq. (2), the new limits of integration

$$u(0) = 0^3 + 1 = 1 \qquad \text{and} \qquad u(2) = 2^3 + 1 = 9$$

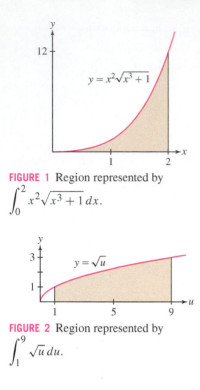

FIGURE 1 Region represented by $\int_0^2 x^2\sqrt{x^3+1}\,dx$.

FIGURE 2 Region represented by $\int_1^9 \sqrt{u}\,du$.

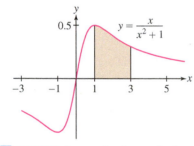

DF **FIGURE 3** Area under the graph of $y = \dfrac{x}{x^2+1}$ over $[1, 3]$.

Thus,

$$\int_0^2 x^2\sqrt{x^3+1}\,dx = \frac{1}{3}\int_1^9 \sqrt{u}\,du = \frac{2}{9}u^{3/2}\Big|_1^9 = \frac{52}{9}$$

This substitution shows that the area in Figure 1 is equal to one-third of the area in Figure 2 (but note that the figures are drawn to different scales). ■

In the previous example, we can avoid changing the limits of integration by evaluating the integral in terms of x:

$$\int x^2\sqrt{x^3+1}\,dx = \frac{1}{3}\int \sqrt{u}\,du = \frac{2}{9}u^{3/2} = \frac{2}{9}(x^3+1)^{3/2}$$

This leads to the same result: $\displaystyle\int_0^2 x^2\sqrt{x^3+1}\,dx = \frac{2}{9}(x^3+1)^{3/2}\Big|_0^2 = \frac{52}{9}$.

■ **EXAMPLE 9** Evaluate $\displaystyle\int_0^{\pi/4} \tan^3\theta \sec^2\theta\,d\theta$.

Solution The substitution $u = \tan\theta$ makes sense because $du = \sec^2\theta\,d\theta$, and therefore, $u^3\,du = \tan^3\theta \sec^2\theta\,d\theta$. The new limits of integration are

$$u(0) = \tan 0 = 0 \qquad \text{and} \qquad u\left(\frac{\pi}{4}\right) = \tan\left(\frac{\pi}{4}\right) = 1$$

Thus,

$$\int_0^{\pi/4} \tan^3\theta \sec^2\theta\,d\theta = \int_0^1 u^3\,du = \frac{u^4}{4}\Big|_0^1 = \frac{1}{4}$$ ■

■ **EXAMPLE 10** Calculate the area under the graph of $y = \dfrac{x}{x^2+1}$ over $[1, 3]$.

Solution The area (Figure 3) is equal to $\displaystyle\int_1^3 \frac{x}{x^2+1}\,dx$. We use the substitution

$$u = x^2+1, \qquad du = 2x\,dx, \qquad \frac{1}{2}\frac{du}{u} = \frac{x\,dx}{x^2+1}$$

The new limits of integration are $u(1) = 1^2 + 1 = 2$ and $u(3) = 3^2 + 1 = 10$, so

$$\int_1^3 \frac{x}{x^2+1}\,dx = \frac{1}{2}\int_2^{10} \frac{du}{u} = \frac{1}{2}\ln|u|\,\Big|_2^{10} = \frac{1}{2}\ln 10 - \frac{1}{2}\ln 2 \approx 0.805$$ ■

5.7 SUMMARY

- Try the Substitution Method when the integrand has the form $f(u(x))\,u'(x)$. If F is an antiderivative of f, then

$$\int f(u(x))\,u'(x)\,dx = F(u(x)) + C$$

- The differential of $u(x)$ is related to dx by $du = u'(x)\,dx$.
- The Substitution Method is expressed by the Change of Variables Formula:

$$\int f(u(x))\,u'(x)\,dx = \int f(u)\,du$$

- Change of Variables Formula for definite integrals:

$$\int_a^b f(u(x))\,u'(x)\,dx = \int_{u(a)}^{u(b)} f(u)\,du$$

5.7 EXERCISES

Preliminary Questions

1. Which of the following integrals is a candidate for the Substitution Method?

(a) $\int 5x^4 \sin(x^5)\, dx$

(b) $\int \sin^5 x \, \cos x \, dx$

(c) $\int x^5 \sin x \, dx$

2. Find an appropriate choice of u for evaluating the following integrals by substitution:

(a) $\int x(x^2 + 9)^4 \, dx$

(b) $\int x^2 \sin(x^3) \, dx$

(c) $\int \sin x \, \cos^2 x \, dx$

3. Which of the following is equal to $\int_0^2 x^2(x^3 + 1)\, dx$ for a suitable substitution?

(a) $\frac{1}{3} \int_0^2 u \, du$

(b) $\int_0^9 u \, du$

(c) $\frac{1}{3} \int_1^9 u \, du$

Exercises

In Exercises 1–6, calculate du.

1. $u = x^3 - x^2$

2. $u = 2x^4 + 8x^{-1}$

3. $u = \cos(x^2)$

4. $u = \tan x$

5. $u = e^{4x+1}$

6. $u = \ln(x^4 + 1)$

In Exercises 7–24, write the integral in terms of u and du. Then evaluate.

7. $\int (x - 7)^3 \, dx, \quad u = x - 7$

8. $\int (x + 25)^{-2} \, dx, \quad u = x + 25$

9. $\int (3t - 4)^5 \, dt, \quad u = 3t - 4$

10. $\int (2x + 7)^{3/2} \, dx, \quad u = 2x + 7$

11. $\int t\sqrt{t^2 + 1} \, dt, \quad u = t^2 + 1$

12. $\int (x^3 + 1) \cos(x^4 + 4x) \, dx, \quad u = x^4 + 4x$

13. $\int \frac{t^3}{(4 - 2t^4)^{11}} \, dt, \quad u = 4 - 2t^4$

14. $\int \sqrt{4x - 1} \, dx, \quad u = 4x - 1$

15. $\int x(x + 1)^9 \, dx, \quad u = x + 1$

16. $\int x\sqrt{4x - 1} \, dx, \quad u = 4x - 1$

17. $\int x^2\sqrt{x + 1} \, dx, \quad u = x + 1$

18. $\int \sin(4\theta - 7) \, d\theta, \quad u = 4\theta - 7$

19. $\int \sin^2 \theta \cos \theta \, d\theta, \quad u = \sin \theta$

20. $\int \sec^2 x \tan x \, dx, \quad u = \tan x$

21. $\int xe^{-x^2} \, dx, \quad u = -x^2$

22. $\int (\sec^2 t)e^{\tan t} \, dt, \quad u = \tan t$

23. $\int \frac{(\ln x)^2 \, dx}{x}, \quad u = \ln x$

24. $\int \frac{(\tan^{-1} x)^2 \, dx}{x^2 + 1}, \quad u = \tan^{-1} x$

In Exercises 25–28, evaluate the integral in the form $a \sin(u(x)) + C$ for an appropriate choice of u(x) and constant a.

25. $\int x^3 \cos(x^4) \, dx$

26. $\int x^2 \cos(x^3 + 1) \, dx$

27. $\int x^{1/2} \cos(x^{3/2}) \, dx$

28. $\int \cos x \cos(\sin x) \, dx$

In Exercises 29–74, evaluate the indefinite integral.

29. $\int (4x + 5)^9 \, dx$

30. $\int \frac{dx}{(x - 9)^5}$

31. $\int \frac{dt}{\sqrt{t + 12}}$

32. $\int (9t + 2)^{2/3} \, dt$

33. $\int \frac{x + 1}{(x^2 + 2x)^3} \, dx$

34. $\int (x + 1)(x^2 + 2x)^{3/4} \, dx$

35. $\int \frac{x}{\sqrt{x^2 + 9}} \, dx$

36. $\int \frac{2x^2 + x}{(4x^3 + 3x^2)^2} \, dx$

37. $\int (3x^2 + 1)(x^3 + x)^2 \, dx$

38. $\int \frac{5x^4 + 2x}{(x^5 + x^2)^3} \, dx$

39. $\int (3x + 8)^{11} \, dx$

40. $\int x(3x + 8)^{11} \, dx$

41. $\int x^2\sqrt{x^3 + 1} \, dx$

42. $\int x^5\sqrt{x^3 + 1} \, dx$

43. $\int \frac{dx}{(x + 5)^3}$

44. $\int \frac{x^2 \, dx}{(x + 5)^3}$

45. $\int z^2(z^3 + 1)^{12} \, dz$

46. $\int (z^5 + 4z^2)(z^3 + 1)^{12} \, dz$

47. $\int (x + 2)(x + 1)^{1/4} \, dx$

48. $\int x^3(x^2 - 1)^{3/2} \, dx$

49. $\int \sin(8 - 3\theta) \, d\theta$

50. $\int \theta \sin(\theta^2) \, d\theta$

51. $\int \dfrac{\cos \sqrt{t}}{\sqrt{t}}\, dt$

52. $\int x^2 \sin(x^3 + 1)\, dx$

53. $\int \tan(4\theta + 9)\, d\theta$

54. $\int \sin^8 \theta \cos \theta\, d\theta$

55. $\int \cot x\, dx$

56. $\int x^{-1/5} \tan\left(x^{4/5}\right) dx$

57. $\int \sec^2(4x + 9)\, dx$

58. $\int \sec^2 x \tan^4 x\, dx$

59. $\int \dfrac{\sec^2(\sqrt{x})\, dx}{\sqrt{x}}$

60. $\int \dfrac{\cos 2x}{(1 + \sin 2x)^2}\, dx$

61. $\int \sin 4x \sqrt{\cos 4x + 1}\, dx$

62. $\int \cos x (3 \sin x - 1)\, dx$

63. $\int \sec \theta \tan \theta (\sec \theta - 1)\, d\theta$

64. $\int \cos t \cos(\sin t)\, dt$

65. $\int e^{14x-7}\, dx$

66. $\int (x + 1) e^{x^2 + 2x}\, dx$

67. $\int \dfrac{e^x\, dx}{(e^x + 1)^4}$

68. $\int (\sec^2 \theta)\, e^{\tan \theta}\, d\theta$

69. $\int \dfrac{e^t\, dt}{e^{2t} + 2e^t + 1}$

70. $\int \dfrac{dx}{x(\ln x)^2}$

71. $\int \dfrac{(\ln x)^4\, dx}{x}$

72. $\int \dfrac{dx}{x \ln x}$

73. $\int \dfrac{\tan(\ln x)}{x}\, dx$

74. $\int (\cot x) \ln(\sin x)\, dx$

75. Evaluate $\int \dfrac{dx}{(1 + \sqrt{x})^3}$ using $u = 1 + \sqrt{x}$. *Hint:* Show that $dx = 2(u - 1)\, du$.

76. Can They Both Be Right? Hannah uses the substitution $u = \tan x$ and Akiva uses $u = \sec x$ to evaluate $\int \tan x \sec^2 x\, dx$. Show that they obtain different answers and explain the apparent contradiction.

77. Evaluate $\int \sin x \cos x\, dx$ using substitution in two different ways: first using $u = \sin x$ and then using $u = \cos x$. Reconcile the two different answers.

78. Some Choices Are Better Than Others Evaluate

$$\int \sin x \, \cos^2 x\, dx$$

in two ways. First use $u = \sin x$ to show that

$$\int \sin x \, \cos^2 x\, dx = \int u\sqrt{1 - u^2}\, du$$

and evaluate the integral on the right by a further substitution. Then show that $u = \cos x$ is a better choice.

79. What are the new limits of integration if we apply the substitution $u = 3x + \pi$ to the integral $\int_0^\pi \sin(3x + \pi)\, dx$?

80. Which of the following is the result of applying the substitution $u = 4x - 9$ to the integral $\int_2^8 (4x - 9)^{20}\, dx$?

(a) $\int_2^8 u^{20}\, du$

(b) $\dfrac{1}{4} \int_2^8 u^{20}\, du$

(c) $4 \int_{-1}^{23} u^{20}\, du$

(d) $\dfrac{1}{4} \int_{-1}^{23} u^{20}\, du$

In Exercises 81–96, use the Change of Variables Formula to evaluate the definite integral.

81. $\int_1^3 (x + 2)^3\, dx$

82. $\int_0^1 (3t - 1)^2\, dt$

83. $\int_0^2 \dfrac{dx}{\sqrt{2x + 5}}$

84. $\int_1^6 \sqrt{x + 3}\, dx$

85. $\int_0^1 \dfrac{x}{(x^2 + 1)^3}\, dx$

86. $\int_{-1}^2 \sqrt{5x + 6}\, dx$

87. $\int_0^4 x\sqrt{x^2 + 9}\, dx$

88. $\int_1^2 \dfrac{4x + 12}{(x^2 + 6x + 1)^2}\, dx$

89. $\int_0^1 (x + 1)(x^2 + 2x)^5\, dx$

90. $\int_{10}^{17} (x - 9)^{-2/3}\, dx$

91. $\int_1^e \dfrac{\ln x}{x}\, dx$

92. $\int_0^{\sqrt{e-1}} \dfrac{x^3}{x^2 + 1}\, dx$

93. $\int_0^1 \theta \tan(\theta^2)\, d\theta$

94. $\int_0^{\pi/6} \sec^2\left(2x - \dfrac{\pi}{6}\right) dx$

95. $\int_0^{\pi/2} \cos^3 x \sin x\, dx$

96. $\int_{\pi/3}^{\pi/2} \cot^2 \dfrac{x}{2} \csc^2 \dfrac{x}{2}\, dx$

97. Evaluate $\int_0^2 r\sqrt{5 - \sqrt{4 - r^2}}\, dr$.

98. Find numbers a and b such that

$$\int_a^b (u^2 + 1)\, du = \int_{-\pi/4}^{\pi/4} \sec^4 \theta\, d\theta$$

and evaluate. *Hint:* Use the identity $\sec^2 \theta = \tan^2 \theta + 1$.

99. Wind engineers have found that wind speed v (in meters per second) at a given location follows a **Rayleigh distribution** of the type

$$W(v) = \dfrac{1}{32} v e^{-v^2/64}$$

This means that at a given moment in time, the probability that v lies between a and b is equal to the shaded area in Figure 4.

(a) Show that the probability that $v \in [0, b]$ is $1 - e^{-b^2/64}$.

(b) Calculate the probability that $v \in [2, 5]$.

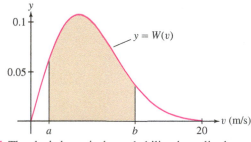

FIGURE 4 The shaded area is the probability that v lies between a and b.

100. Evaluate $\int_0^{\pi/2} \sin^n x \cos x\, dx$ for $n \geq 0$.

In Exercises 101–104, use substitution to evaluate the integral in terms of $f(x)$.

101. $\int f(x)^3 f'(x)\, dx$

102. $\int \dfrac{f'(x)}{f(x)^2}\, dx$

103. $\int \dfrac{f'(x)}{f(x)}\, dx$

104. $\int f'(-x + 7)\, dx$

105. Show that $\displaystyle\int_0^{\pi/6} f(\sin \theta)\, d\theta = \int_0^{1/2} f(u)\, \dfrac{1}{\sqrt{1 - u^2}}\, du$.

Further Insights and Challenges

106. Use the substitution $u = 1 + x^{1/n}$ to show that

$$\int \sqrt{1 + x^{1/n}} \, dx = n \int u^{1/2} (u - 1)^{n-1} \, du$$

Evaluate for $n = 2, 3$.

107. Evaluate $I = \int_0^{\pi/2} \dfrac{d\theta}{1 + \tan^{6000} \theta}$. *Hint:* Use substitution to

show that I is equal to $J = \int_0^{\pi/2} \dfrac{d\theta}{1 + \cot^{6000} \theta}$ and then check that

$I + J = \int_0^{\pi/2} d\theta$.

108. Use substitution to prove that $\int_{-a}^{a} f(x) \, dx = 0$ if f is an odd function.

109. Prove that $\int_a^b \dfrac{1}{x} \, dx = \int_1^{b/a} \dfrac{1}{x} \, dx$ for $a, b > 0$. Then show that the regions under the hyperbola over the intervals $[1, 2]$, $[2, 4]$, $[4, 8]$, ..., all have the same area (Figure 5).

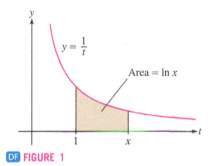

FIGURE 5 The area under $y = \frac{1}{x}$ over $[2^n, 2^{n+1}]$ is the same for all $n = 0, 1, 2, \ldots$.

110. Show that the two regions in Figure 6 have the same area. Then use the identity $\cos^2 u = \frac{1}{2}(1 + \cos 2u)$ to compute the second area.

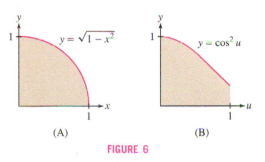

(A)   (B)

FIGURE 6

111. Area of an Ellipse   Prove the formula $A = \pi ab$ for the area of the ellipse with equation (Figure 7)

$$\frac{x^2}{a^2} + \frac{y^2}{b^2} = 1$$

Hint: Use a change of variables to show that A is equal to ab times the area of the unit circle.

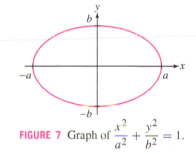

FIGURE 7   Graph of $\dfrac{x^2}{a^2} + \dfrac{y^2}{b^2} = 1$.

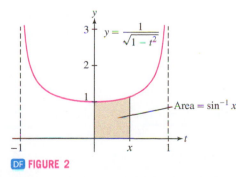

DF **FIGURE 1**

*It is possible (and mathematically, it is more efficient) to take Eq. (1) as the **definition** of $\ln x$ and to define e^x as the corresponding inverse function (see Exercises 78-79).*

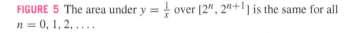

DF **FIGURE 2**

5.8 Further Transcendental Functions

In Section 5.4, we used FTC I to show that when a and b have the same sign,

$$\int_a^b \frac{dx}{x} = \ln \frac{b}{a}$$

We obtain a formula for $\ln x$ as a definite integral by setting $a = 1$ and $b = x$ and keeping in mind that $\ln 1 = 0$:

$$\boxed{\ln x = \int_1^x \frac{dt}{t} \qquad \text{for } x > 0} \qquad \boxed{1}$$

Thus, $\ln x$ is equal to an area under the hyperbola $y = 1/t$ (Figure 1).

In a similar fashion, we can express $\sin^{-1} x$ as a definite integral using the derivative formula from Section 3.8 (Figure 2):

$$\frac{d}{dx} \sin^{-1} x = \frac{1}{\sqrt{1 - x^2}} \quad \Rightarrow \quad \int \frac{dx}{\sqrt{1 - x^2}} = \sin^{-1} x + C$$

Since $\sin^{-1} 0 = 0$, we have

$$\boxed{\sin^{-1} x = \int_0^x \frac{dt}{\sqrt{1 - t^2}} \qquad \text{for } -1 < x < 1}$$

On the other hand, the derivative formulas from Section 3.8 yield integration formulas that are useful for evaluating new types of integrals.

Inverse Trigonometric Functions

$$\frac{d}{dx}\sin^{-1}x = \frac{1}{\sqrt{1-x^2}}, \qquad \int \frac{dx}{\sqrt{1-x^2}} = \sin^{-1}x + C \qquad \boxed{2}$$

$$\frac{d}{dx}\tan^{-1}x = \frac{1}{x^2+1}, \qquad \int \frac{dx}{x^2+1} = \tan^{-1}x + C \qquad \boxed{3}$$

$$\frac{d}{dx}\sec^{-1}x = \frac{1}{|x|\sqrt{x^2-1}}, \qquad \int \frac{dx}{|x|\sqrt{x^2-1}} = \sec^{-1}x + C \qquad \boxed{4}$$

In this list, we omit the integral formulas corresponding to the derivatives of $y = \cos^{-1}x$, $y = \cot^{-1}x$, and $y = \csc^{-1}x$ because the integrals differ only by a minus sign from those already on the list. For example,

$$\frac{d}{dx}\cos^{-1}x = -\frac{1}{\sqrt{1-x^2}}, \qquad \int \frac{dx}{\sqrt{1-x^2}} = -\cos^{-1}x + C$$

■ **EXAMPLE 1** Evaluate $\displaystyle\int_0^1 \frac{dx}{x^2+1}$.

Solution This integral is the area of the region in Figure 3. By Eq. (3),

$$\int_0^1 \frac{dx}{x^2+1} = \tan^{-1}x \Big|_0^1 = \tan^{-1}1 - \tan^{-1}0 = \frac{\pi}{4} - 0 = \frac{\pi}{4} \qquad ■$$

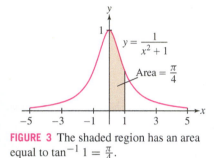

FIGURE 3 The shaded region has an area equal to $\tan^{-1}1 = \frac{\pi}{4}$.

■ **EXAMPLE 2** Using Substitution Evaluate $\displaystyle\int_{1/\sqrt{2}}^1 \frac{dx}{x\sqrt{4x^2-1}}$.

Solution Notice that $\sqrt{4x^2-1}$ can be written as $\sqrt{(2x)^2-1}$, so it makes sense to try the substitution $u = 2x$, $du = 2\,dx$. Then

$$u^2 = 4x^2 \qquad \text{and} \qquad \sqrt{4x^2-1} = \sqrt{u^2-1}$$

The new limits of integration are $u(1/\sqrt{2}) = 2(1/\sqrt{2}) = \sqrt{2}$ and $u(1) = 2$. By Eq. (4),

$$\int_{1/\sqrt{2}}^1 \frac{dx}{x\sqrt{4x^2-1}} = \int_{\sqrt{2}}^2 \frac{\frac{1}{2}\,du}{\frac{1}{2}u\sqrt{u^2-1}} = \int_{\sqrt{2}}^2 \frac{du}{u\sqrt{u^2-1}}$$

$$= \sec^{-1}2 - \sec^{-1}\sqrt{2}$$

$$= \frac{\pi}{3} - \frac{\pi}{4} = \frac{\pi}{12} \qquad ■$$

■ **EXAMPLE 3** Using Substitution Evaluate $\displaystyle\int_0^{3/4} \frac{dx}{\sqrt{9-16x^2}}$.

Solution Let us first rewrite the integrand:

$$\sqrt{9-16x^2} = \sqrt{9\left(1-\frac{16x^2}{9}\right)} = 3\sqrt{1-\left(\frac{4x}{3}\right)^2}$$

Thus, it makes sense to use the substitution $u = \frac{4}{3}x$. Then $du = \frac{4}{3}dx$ and

$$x = \frac{3}{4}u, \qquad dx = \frac{3}{4}du, \qquad \sqrt{9-16x^2} = 3\sqrt{1-u^2}$$

The new limits of integration are $u(0) = 0$ and $u\left(\frac{3}{4}\right) = 1$:

$$\int_0^{3/4} \frac{dx}{\sqrt{9-16x^2}} = \int_0^1 \frac{\frac{3}{4}\,du}{3\sqrt{1-u^2}} = \frac{1}{4}\sin^{-1}x \Big|_0^1 = \frac{1}{4}(\sin^{-1}1 - \sin^{-1}0)$$

$$= \frac{1}{4}\left(\frac{\pi}{2}\right) = \frac{\pi}{8} \qquad ■$$

Integrals Involving $f(x) = b^x$

The exponential function $f(x) = e^x$ is particularly convenient because e^x is both its own derivative and its own antiderivative. For other bases b, we have

$$\frac{d}{dx} b^x = \frac{d}{dx} e^{(\ln b)x} = (\ln b)e^{(\ln b)x} = (\ln b)b^x \quad \Rightarrow \quad \frac{d}{dx}\left(\frac{b^x}{\ln b}\right) = b^x$$

This translates into the integral formula

$$\int b^x \, dx = \frac{b^x}{\ln b} + C$$

| 5 |

■ **EXAMPLE 4** Evaluate $\displaystyle\int_3^5 7^x \, dx$.

Solution Apply Eq. (5) with $b = 7$:

$$\int_3^5 7^x \, dx = \frac{7^x}{\ln 7}\bigg|_3^5 = \frac{7^5 - 7^3}{\ln 7} \approx 8460.8 \qquad ■$$

■ **EXAMPLE 5** Evaluate $\displaystyle\int_0^{\pi/2} (\cos\theta)10^{\sin\theta} \, d\theta$.

Solution Use the substitution $u = \sin\theta$, $du = \cos\theta \, d\theta$. The new limits of integration become $u(0) = 0$ and $u(\pi/2) = 1$:

$$\int_0^{\pi/2} (\cos\theta)10^{\sin\theta} \, d\theta = \int_0^1 10^u \, du = \frac{10^u}{\ln 10}\bigg|_0^1 = \frac{10^1 - 10^0}{\ln 10} \approx 3.91 \qquad ■$$

5.8 SUMMARY

- Integral formula for the natural logarithm:

$$\ln x = \int_1^x \frac{dt}{t}$$

- Integral formulas:

$$\int \frac{dx}{\sqrt{1-x^2}} = \sin^{-1} x + C$$

$$\int \frac{dx}{x^2 + 1} = \tan^{-1} x + C$$

$$\int \frac{dx}{|x|\sqrt{x^2 - 1}} = \sec^{-1} x + C$$

- Integrals of exponential functions ($b > 0$, $b \neq 1$):

$$\int e^x \, dx = e^x + C, \qquad \int b^x \, dx = \frac{b^x}{\ln b} + C$$

5.8 EXERCISES

Preliminary Questions

1. Find b such that $\displaystyle\int_1^b \frac{dx}{x}$ is equal to

(a) $\ln 3$.

(b) 3.

2. Find b such that $\displaystyle\int_0^b \frac{dx}{1+x^2} = \frac{\pi}{3}$.

3. Which integral should be evaluated using substitution?

(a) $\displaystyle\int \frac{9 \, dx}{1+x^2}$

(b) $\displaystyle\int \frac{dx}{1+9x^2}$

4. Which relation between x and u yields $\sqrt{16 + x^2} = 4\sqrt{1 + u^2}$?

Exercises

In Exercises 1–10, evaluate the definite integral.

1. $\int_1^9 \dfrac{dx}{x}$

2. $\int_4^{20} \dfrac{dx}{x}$

3. $\int_1^{e^3} \dfrac{1}{t}\,dt$

4. $\int_{-e^2}^{-e} \dfrac{1}{t}\,dt$

5. $\int_2^{12} \dfrac{dt}{3t+4}$

6. $\int_e^{e^3} \dfrac{dt}{t \ln t}$

7. $\int_1^{\sqrt{3}} \dfrac{dx}{x^2+1}$

8. $\int_2^7 \dfrac{x\,dx}{x^2+1}$

9. $\int_0^{1/2} \dfrac{dx}{\sqrt{1-x^2}}$

10. $\int_{-2}^{-2/\sqrt{3}} \dfrac{dx}{|x|\sqrt{x^2-1}}$

11. Use the substitution $u = x/3$ to prove

$$\int \frac{dx}{9+x^2} = \frac{1}{3}\tan^{-1}\frac{x}{3} + C$$

12. Use the substitution $u = 2x$ to evaluate $\displaystyle\int \frac{dx}{4x^2+1}$.

In Exercises 13–32, calculate the integral.

13. $\int_0^3 \dfrac{dx}{x^2+3}$

14. $\int_0^4 \dfrac{dt}{4t^2+9}$

15. $\int \dfrac{dt}{\sqrt{1-16t^2}}$

16. $\int_{-1/5}^{1/5} \dfrac{dx}{\sqrt{4-25x^2}}$

17. $\int \dfrac{dt}{\sqrt{5-3t^2}}$

18. $\int_{1/(2\sqrt{2})}^{1/2} \dfrac{dx}{x\sqrt{16x^2-1}}$

19. $\int \dfrac{dx}{x\sqrt{12x^2-3}}$

20. $\int \dfrac{x\,dx}{x^4+1}$

21. $\int \dfrac{dx}{x\sqrt{x^4-1}}$

22. $\int_{-1/2}^0 \dfrac{(x+1)\,dx}{\sqrt{1-x^2}}$

23. $\int_{-\ln 2}^0 \dfrac{e^x\,dx}{1+e^{2x}}$

24. $\int \dfrac{\ln(\cos^{-1}x)\,dx}{(\cos^{-1}x)\sqrt{1-x^2}}$

25. $\int \dfrac{\tan^{-1}x\,dx}{1+x^2}$

26. $\int_1^{\sqrt{3}} \dfrac{dx}{(\tan^{-1}x)(1+x^2)}$

27. $\int_0^1 3^x\,dx$

28. $\int_0^1 3^{-x}\,dx$

29. $\int_0^{\log_4(3)} 4^x\,dx$

30. $\int_0^1 t\,5^{t^2}\,dt$

31. $\int 9^x \sin(9^x)\,dx$

32. $\int \dfrac{dx}{\sqrt{5^{2x}-1}}$

In Exercises 33–70, evaluate the integral using the methods covered in the text so far.

33. $\int y e^{y^2}\,dy$

34. $\int \dfrac{dx}{3x+5}$

35. $\int \dfrac{x\,dx}{\sqrt{4x^2+9}}$

36. $\int (x-x^{-2})^2\,dx$

37. $\int 7^{-x}\,dx$

38. $\int e^{9-12t}\,dt$

39. $\int \sec^2\theta \tan^7\theta\,d\theta$

40. $\int \dfrac{\cos(\ln t)\,dt}{t}$

41. $\int \dfrac{t\,dt}{\sqrt{7-t^2}}$

42. $\int 2^x e^{4x}\,dx$

43. $\int \dfrac{(3x+2)\,dx}{x^2+4}$

44. $\int \tan(4x+1)\,dx$

45. $\int \dfrac{dx}{\sqrt{9-4x^2}}$

46. $\int e^t\sqrt{e^t+1}\,dt$

47. $\int (e^{-x} - 4x)\,dx$

48. $\int (7 - e^{10x})\,dx$

49. $\int \dfrac{e^{2x}-e^{4x}}{e^x}\,dx$

50. $\int \dfrac{dx}{x\sqrt{25x^2-1}}$

51. $\int \dfrac{(x+5)\,dx}{\sqrt{4-x^2}}$

52. $\int (t+1)\sqrt{t+1}\,dt$

53. $\int e^x \cos(e^x)\,dx$

54. $\int \dfrac{e^x}{\sqrt{e^x+1}}\,dx$

55. $\int \dfrac{dx}{\sqrt{9-16x^2}}$

56. $\int \dfrac{dx}{(4x-1)\ln(8x-2)}$

57. $\int e^x(e^{2x}+1)^3\,dx$

58. $\int \dfrac{dx}{x(\ln x)^5}$

59. $\int \dfrac{x^2\,dx}{x^3+2}$

60. $\int \dfrac{(3x-1)\,dx}{9-2x+3x^2}$

61. $\int \cot x\,dx$

62. $\int \dfrac{\cos x}{2\sin x+3}\,dx$

63. $\int \dfrac{4\ln x+5}{x}\,dx$

64. $\int (\sec\theta\tan\theta)5^{\sec\theta}\,d\theta$

65. $\int x3^{x^2}\,dx$

66. $\int \dfrac{\ln(\ln x)}{x\ln x}\,dx$

67. $\int \cot x \ln(\sin x)\,dx$

68. $\int \dfrac{t\,dt}{\sqrt{1-t^4}}$

69. $\int t^2\sqrt{t-3}\,dt$

70. $\int \cos x\,5^{-2\sin x}\,dx$

71. Use Figure 4 to prove

$$\int_0^x \sqrt{1-t^2}\,dt = \frac{1}{2}x\sqrt{1-x^2} + \frac{1}{2}\sin^{-1}x$$

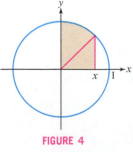

FIGURE 4

72. Use the substitution $u = \tan x$ to evaluate

$$\int \frac{dx}{1 + \sin^2 x}$$

Hint: Show that

$$\frac{dx}{1 + \sin^2 x} = \frac{du}{1 + 2u^2}$$

73. Prove

$$\int \sin^{-1} t \, dt = \sqrt{1 - t^2} + t \sin^{-1} t$$

74. (a) Verify for $r \neq 0$:

$$\int_0^T te^{rt} \, dt = \frac{e^{rT}(rT - 1) + 1}{r^2} \qquad \boxed{6}$$

Hint: For fixed r, let $F(T)$ be the value of the integral on the left. By FTC II, $F'(t) = te^{rt}$ and $F(0) = 0$. Show that the same is true of the function on the right.

(b) Use L'Hôpital's Rule to show that for fixed T, the limit as $r \to 0$ of the right-hand side of Eq. (6) is equal to the value of the integral for $r = 0$.

Further Insights and Challenges

75. Recall that if $f(t) \geq g(t)$ for $t \geq 0$, then for all $x \geq 0$,

$$\int_0^x f(t) \, dt \geq \int_0^x g(t) \, dt \qquad \boxed{7}$$

The inequality $e^t \geq 1$ holds for $t \geq 0$ because $e > 1$. Use (7) to prove that $e^x \geq 1 + x$ for $x \geq 0$. Then prove, by successive integration, the following inequalities (for $x \geq 0$):

$$e^x \geq 1 + x + \frac{1}{2}x^2, \qquad e^x \geq 1 + x + \frac{1}{2}x^2 + \frac{1}{6}x^3$$

76. Generalize Exercise 75; that is, use induction (if you are familiar with this method of proof) to prove that for all $n \geq 0$,

$$e^x \geq 1 + x + \frac{1}{2}x^2 + \frac{1}{6}x^3 + \cdots + \frac{1}{n!}x^n \qquad (x \geq 0)$$

77. Use Exercise 75 to show that $e^x/x^2 \geq x/6$ and conclude that $\lim\limits_{x \to \infty} e^x/x^2 = \infty$. Then use Exercise 76 to prove more generally that $\lim\limits_{x \to \infty} e^x/x^n = \infty$ for all n.

Exercises 78–80 develop an elegant approach to the exponential and logarithm functions. Define a function $G(x)$ for $x > 0$:

$$G(x) = \int_1^x \frac{1}{t} \, dt$$

78. Defining $\ln x$ as an Integral This exercise proceeds as if we didn't know that $G(x) = \ln x$ and shows directly that G has all the basic properties of the logarithm. Prove the following statements:

(a) $\displaystyle\int_a^{ab} \frac{1}{t} \, dt = \int_1^b \frac{1}{t} \, dt$ for all $a, b > 0$. *Hint:* Use the substitution $u = t/a$.

(b) $G(ab) = G(a) + G(b)$. *Hint:* Break up the integral from 1 to ab into two integrals and use (a).

(c) $G(1) = 0$ and $G(a^{-1}) = -G(a)$ for $a > 0$.

(d) $G(a^n) = nG(a)$ for all $a > 0$ and integers n.

(e) $G(a^{1/n}) = \dfrac{1}{n}G(a)$ for all $a > 0$ and integers $n \neq 0$.

(f) $G(a^r) = rG(a)$ for all $a > 0$ and rational numbers r.

(g) G is increasing. *Hint:* Use FTC II.

(h) There exists a number a such that $G(a) > 1$. *Hint:* Show that $G(2) > 0$ and take $a = 2^m$ for $m > 1/G(2)$.

(i) $\lim\limits_{x \to \infty} G(x) = \infty$ and $\lim\limits_{x \to 0+} G(x) = -\infty$.

(j) There exists a unique number E such that $G(E) = 1$.

(k) $G(E^r) = r$ for every rational number r.

79. Defining e^x Use Exercise 78 to prove the following statements:

(a) G has an inverse with domain $\mathbf{R}$ and range $\{x : x > 0\}$. Denote the inverse by F.

(b) $F(x + y) = F(x)F(y)$ for all x, y. *Hint:* It suffices to show that $G(F(x)F(y)) = G(F(x + y))$.

(c) $F(r) = E^r$ for all numbers. In particular, $F(0) = 1$.

(d) $F'(x) = F(x)$. *Hint:* Use the formula for the derivative of an inverse function that appears in Exercise 28 of the Chapter 3 Review Exercises.

This shows that $E = e$ and $F(x) = e^x$ as defined in the text.

80. Defining b^x Let $b > 0$ and let $f(x) = F(xG(b))$ with F as in Exercise 79. Use Exercise 78 (f) to prove that $f(r) = b^r$ for every rational number r. This gives us a way of defining b^x for irrational x, namely $b^x = f(x)$. With this definition, $y = b^x$ is a differentiable function of x (because F is differentiable).

81. The formula $\displaystyle\int x^n \, dx = \frac{x^{n+1}}{n + 1} + C$ is valid for $n \neq -1$. Show that the exceptional case $n = -1$ is a limit of the general case by applying L'Hôpital's Rule to the limit on the left:

$$\lim_{n \to -1} \int_1^x t^n \, dt = \int_1^x t^{-1} \, dt \qquad \text{(for fixed } x > 0\text{)}$$

Note that the integral on the left is equal to $\dfrac{x^{n+1} - 1}{n + 1}$.

82. CAS The integral inside the limit on the left in Exercise 81 is equal to $f_n(x) = \dfrac{x^{n+1} - 1}{n + 1}$ for $x \neq -1$. Investigate the limit graphically by plotting $y = f_n(x)$ for $n = 0, -0.3, -0.6,$ and -0.9 together with $y = \ln x$ on a single plot.

83. (a) Explain why the shaded region in Figure 5 has area $\displaystyle\int_0^{\ln a} e^y \, dy$.

(b) Prove the formula $\displaystyle\int_1^a \ln x \, dx = a \ln a - \int_0^{\ln a} e^y \, dy$.

(c) Conclude that $\displaystyle\int_1^a \ln x \, dx = a \ln a - a + 1$.

(d) Use the result of (a) to find an antiderivative of $y = \ln x$.

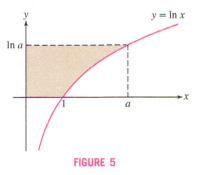

FIGURE 5

5.9 Exponential Growth and Decay

In this section, we explore some applications of the exponential function. Consider a quantity $P(t)$ that depends exponentially on time:

$$P(t) = P_0 e^{kt}$$

The constant k has units of "inverse time"; if t is measured in days, then k has units of $(\text{days})^{-1}$.

If $k > 0$, then $P(t)$ *grows exponentially* and k is called the *growth constant*. Note that P_0 is the initial size (the size at $t = 0$):

$$P(0) = P_0 e^{k \cdot 0} = P_0$$

We can also write $P(t) = P_0 b^t$ with $b = e^k$, because $b^t = (e^k)^t = e^{kt}$.

A quantity that decreases exponentially is said to have *exponential decay*. In this case, we write $P(t) = P_0 e^{-kt}$ with $k > 0$; k is then called the *decay constant*.

Population is a typical example of a quantity that grows exponentially, at least under suitable conditions. To understand why, consider a cell colony with initial population $P_0 = 100$ and assume that each cell divides into two cells after 1 hour. Then population $P(t)$ doubles with each passing hour:

$$P(0) \qquad\qquad = 100 \quad \text{(initial population)}$$
$$P(1) = 2(100) = 200 \quad \text{(population doubles)}$$
$$P(2) = 2(200) = 400 \quad \text{(population doubles again)}$$

After t hours, $P(t) = (100)2^t$.

FIGURE 1 *E. coli* bacteria, found in the human intestine. *(Dr. Gary Gaugler/Science Source)*

Exponential growth cannot continue over long periods of time. A colony starting with one E. coli cell would grow to 5×10^{89} cells after 3 weeks—much more than the estimated number of atoms in the observable universe. In actual cell growth, the exponential phase is followed by a period in which growth slows and may decline.

■ **EXAMPLE 1** In the laboratory, the number of *Escherichia coli* bacteria (Figure 1) grows exponentially with growth constant of $k = 0.41$ $(\text{hours})^{-1}$. Assume that 1000 bacteria are present at time $t = 0$.

(a) Find the formula for the number of bacteria $P(t)$ at time t.

(b) How large is the population after 5 hours (h)?

(c) When will the population reach 10,000?

Solution The growth is exponential, so $P(t) = P_0 e^{kt}$.

(a) The initial size is $P_0 = 1000$ and $k = 0.41$, so $P(t) = 1000 e^{0.41t}$ (t in hours).

(b) After 5 h, $P(5) = 1000 e^{0.41 \cdot 5} = 1000 e^{2.05} \approx 7767.9$. Because the number of bacteria is a whole number, we round off the answer to 7768.

(c) The problem asks for the time t such that $P(t) = 10{,}000$, so we solve

$$1000 e^{0.41t} = 10{,}000 \quad \Rightarrow \quad e^{0.41t} = \frac{10{,}000}{1000} = 10$$

Taking the logarithm of both sides, we obtain $\ln\left(e^{0.41t}\right) = \ln 10$, or

$$0.41t = \ln 10 \quad \Rightarrow \quad t = \frac{\ln 10}{0.41} \approx 5.62 \text{ h}$$

Therefore, $P(t)$ reaches 10,000 after approximately 5 h, 37 min (Figure 2).   ■

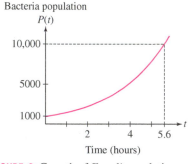

Bacteria population

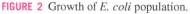

FIGURE 2 Growth of *E. coli* population.

The important role played by exponential functions is best understood in terms of the *differential equation* $y' = ky$. The function $y = P_0 e^{kt}$ satisfies this differential equation, as we can check directly:

$$y' = \frac{d}{dt}\left(P_0 e^{kt}\right) = k P_0 e^{kt} = ky$$

A differential equation is an equation relating a function $y = f(x)$ to its derivative y' (or higher derivatives y', y'', y''', . . .).

Theorem 1 goes further and asserts that the exponential functions are the *only* functions that satisfy this differential equation.

THEOREM 1 If y is a differentiable function satisfying the differential equation

$$y' = ky$$

then $y(t) = P_0 e^{kt}$, where P_0 is the initial value $P_0 = y(0)$.

Proof Compute the derivative of ye^{-kt}. If $y' = ky$, then

$$\frac{d}{dt}\left(ye^{-kt}\right) = y'e^{-kt} - ke^{-kt}y = (ky)e^{-kt} - ke^{-kt}y = 0$$

Because the derivative is zero, $y(t)e^{-kt} = P_0$ for some constant P_0, and $y(t) = P_0 e^{kt}$ as claimed. The initial value is $y(0) = P_0 e^0 = P_0$. ∎

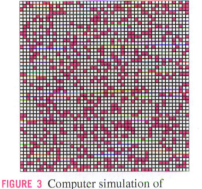

FIGURE 3 Computer simulation of radioactive decay as a random process. The red squares are atoms that have not yet decayed. A fixed fraction of red squares turns white in each unit of time. *(Courtesy of Michael Zingale)*

CONCEPTUAL INSIGHT Theorem 1 tells us that a process obeys an exponential law precisely when *its rate of change is proportional to the amount present*. This helps us understand why certain quantities grow or decay exponentially.

A population grows exponentially because each organism contributes to growth through reproduction, and thus the growth rate is proportional to the population size. However, this is true only under certain conditions. If the organisms interact—say, by competing for food or mates—then the growth rate may not be proportional to population size and we cannot expect exponential growth.

Similarly, experiments show that radioactive substances decay exponentially. This suggests that radioactive decay is a random process in which a fixed fraction of atoms, randomly chosen, decays per unit time (Figure 3). If exponential decay were not observed, we might suspect that the decay was influenced by some interaction between the atoms.

■ **EXAMPLE 2** Find all solutions of $y' = 3y$. Which solution satisfies $y(0) = 9$?

Solution The solutions to $y' = 3y$ are the functions $y(t) = Ce^{3t}$, where C is the initial value $C = y(0)$. The particular solution satisfying $y(0) = 9$ is $y(t) = 9e^{3t}$. ■

■ **EXAMPLE 3** **Modeling Penicillin** Pharmacologists have shown that penicillin leaves a person's bloodstream at a rate proportional to the amount present.

(a) Express this statement as a differential equation.

(b) Find the decay constant if 50 mg of penicillin remain in the bloodstream 7 hours (h) after an initial injection of 450 mg.

(c) Under the hypothesis of (b), at what time were 200 mg of penicillin present?

Solution

(a) Let $A(t)$ be the quantity of penicillin present in the bloodstream at time t. Since the rate at which penicillin leaves the bloodstream is proportional to $A(t)$,

$$A'(t) = -kA(t) \qquad \boxed{1}$$

where $k > 0$ because $A(t)$ is decreasing.

(b) Eq. (1) and the condition $A(0) = 450$ tell us that $A(t) = 450e^{-kt}$. The additional condition $A(7) = 50$ enables us to solve for k:

$$A(7) = 450e^{-7k} = 50 \quad \Rightarrow \quad e^{-7k} = \frac{1}{9} \quad \Rightarrow \quad -7k = \ln\frac{1}{9}$$

Thus, $k = -\frac{1}{7}\ln\frac{1}{9} \approx 0.31$.

(c) To find the time t at which 200 mg were present, we solve

$$A(t) = 450e^{-0.31t} = 200 \quad \Rightarrow \quad e^{-0.31t} = \frac{4}{9}$$

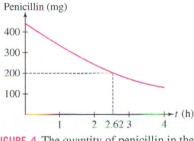

Penicillin (mg)

FIGURE 4 The quantity of penicillin in the bloodstream decays exponentially.

Therefore, $t = -\frac{1}{0.31}\ln\left(\frac{4}{9}\right) \approx 2.62$ h (Figure 4). ■

The constant k has units of time^{-1}, so the doubling time $T = (\ln 2)/k$ has units of time, as we should expect. A similar calculation shows that the tripling time is $(\ln 3)/k$, the quadrupling time is $(\ln 4)/k$, and, in general, the time to n-fold increase is $(\ln n)/k$.

Quantities that grow exponentially possess an important property: There is a doubling time T such that $P(t)$ doubles in size over every time interval of length T. To prove this, let $P(t) = P_0 e^{kt}$ and solve for T in the equation $P(t + T) = 2P(t)$:

$$P_0 e^{k(t+T)} = 2P_0 e^{kt}$$

$$e^{kt} e^{kT} = 2e^{kt}$$

$$e^{kT} = 2$$

We obtain $kT = \ln 2$ or $T = (\ln 2)/k$.

Doubling Time If $P(t) = P_0 e^{kt}$ with $k > 0$, then the doubling time of P is

$$\text{Doubling time} = \frac{\ln 2}{k}$$

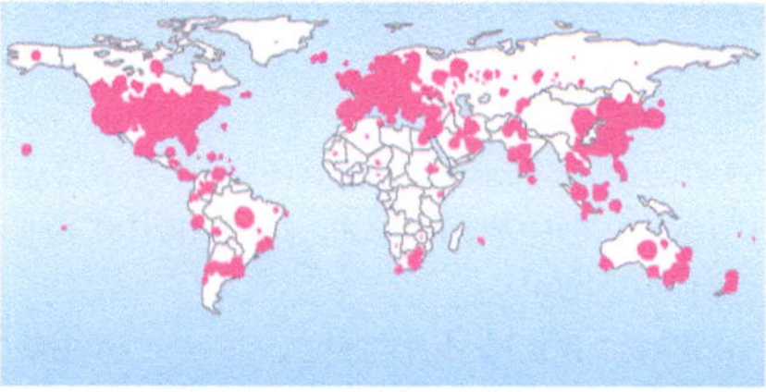

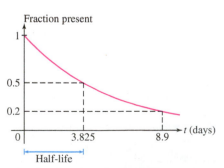

Number of hosts infected with Sapphire: 74855

FIGURE 5 Spread of the Sapphire computer virus 30 minutes after release. The infected hosts spewed billions of copies of the virus into cyberspace, significantly slowing Internet traffic and interfering with businesses, flight schedules, and automated teller machines. (*Copyright © 2010 The Regents of the University of California. All rights reserved. Used with permission.*)

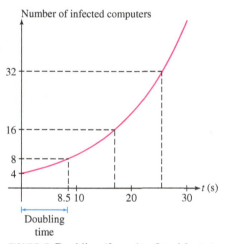

FIGURE 6 Doubling (from 4 to 8 to 16, etc.) occurs at equal time intervals.

■ EXAMPLE 4 Spread of the Sapphire Worm A computer virus nicknamed the *Sapphire Worm* spread throughout the Internet on January 25, 2003 (Figure 5). Studies suggest that during the first few minutes, the population of infected computer hosts increased exponentially with growth constant $k = 0.0815 \text{ s}^{-1}$.

(a) What was the doubling time of the virus?

(b) If the virus began in four computers, how many hosts were infected after 2 minutes? After 3 minutes (min)?

Solution

(a) The doubling time is $(\ln 2)/0.0815 \approx 8.5$ s (Figure 6).

(b) If $P_0 = 4$, the number of infected hosts after t seconds is $P(t) = 4e^{0.0815t}$. After 2 min (120 s), the number of infected hosts is

$$P(120) = 4e^{0.0815(120)} \approx 70,700$$

After 3 min, the number would have been $P(180) = 4e^{0.0815(180)} \approx 9.4$ million. However, it is estimated that a total of around 75,000 hosts were infected, so the exponential phase of the virus could not have lasted much more than 2 min. ■

In the situation of exponential decay $P(t) = P_0 e^{-kt}$, the **half-life** is the time it takes for the quantity to decrease by a factor of $\frac{1}{2}$. The calculation similar to that of doubling time above shows that

$$\text{Half-life} = \frac{\ln 2}{k}$$

■ EXAMPLE 5 The isotope radon-222 decays exponentially with a half-life of 3.825 days. How long will it take for 80% of the isotope to decay?

Solution By the equation for half-life, k equals $\ln 2$ divided by half-life:

$$k = \frac{\ln 2}{3.825} \approx 0.181$$

Therefore, the quantity of radon-222 at time t is $R(t) = R_0 e^{-0.181t}$, where R_0 is the amount present at $t = 0$ (Figure 7). When 80% has decayed, 20% remains, so we solve for t in the equation $R_0 e^{-0.181t} = 0.2R_0$:

$$e^{-0.181t} = 0.2$$

$$-0.181t = \ln(0.2) \quad \Rightarrow \quad t = \frac{\ln(0.2)}{-0.181} \approx 8.9 \text{ days}$$

The quantity of radon-222 decreases by 80% after 8.9 days. ■

FIGURE 7 Fraction of radon-222 present at time t.

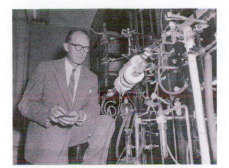

FIGURE 8 American chemist Willard Libby (1908–1980) developed the technique of carbon dating in 1946 to determine the age of fossils and was awarded the Nobel Prize in Chemistry for this work in 1960. Since then the technique has been refined considerably. (© *Bettmann/Corbis*)

Carbon Dating

Carbon dating (Figure 8) relies on the fact that all living organisms contain carbon that enters the food chain through the carbon dioxide absorbed by plants from the atmosphere. Carbon in the atmosphere is made up of nonradioactive C^{12} and a minute amount of radioactive C^{14} that decays into nitrogen. The ratio of C^{14} to C^{12} is approximately $R_{atm} = 10^{-12}$.

The carbon in a living organism has the same ratio R_{atm} because this carbon originates in the atmosphere, but when the organism dies, its carbon is no longer replenished. The C^{14} begins to decay exponentially while the C^{12} remains unchanged. Therefore, the ratio of C^{14} to C^{12} in the organism decreases exponentially. By measuring this ratio, we can determine when the death occurred. The decay constant for C^{14} is $k = 0.000121 \text{ year}^{-1}$, so

$$\text{Ratio of } C^{14} \text{ to } C^{12} \text{ after } t \text{ years} = R_{atm}e^{-0.000121t}$$

■ **EXAMPLE 6** **Cave Paintings** In 1940 a remarkable gallery of prehistoric animal paintings was discovered in the Lascaux cave in Dordogne, France (Figure 9). A charcoal sample from the cave walls had a C^{14}-to-C^{12} ratio equal to 15% of that found in the atmosphere. Approximately how old are the paintings?

Solution The C^{14}-to-C^{12} ratio in the charcoal is now equal to $0.15R_{atm}$, so

$$R_{atm}e^{-0.000121t} = 0.15R_{atm}$$

where t is the age of the paintings. We solve for t:

$$e^{-0.000121t} = 0.15$$

$$-0.000121t = \ln(0.15) \quad \Rightarrow \quad t = -\frac{\ln(0.15)}{0.000121} \approx 15,700$$

The cave paintings are approximately 16,000 years old (Figure 10). ■

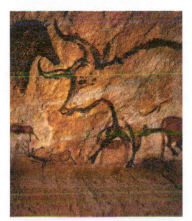

FIGURE 9 Detail of bison and other animals from a replica of the Lascaux cave mural. (© *Gianni Dagli Orti/Corbis*)

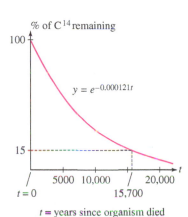

DF FIGURE 10 If only 15% of the C^{14} remains, the object is approximately 16,000 years old.

Compound Interest and Present Value

Exponential functions are used extensively in financial calculations. Two basic applications are compound interest and present value.

When a sum of money P_0, called the **principal**, is deposited into an interest-bearing account, the amount or **balance** in the account at time t depends on two factors: the **interest rate** r and frequency with which interest is **compounded**. Interest paid out once a year at the end of the year is said to be *compounded annually*. The balance increases by the factor $(1 + r)$ after each year, leading to exponential growth:

Convention: Time t is measured in years and interest rates are given as yearly rates, either as a decimal or as a percentage. Thus, $r = 0.05$ corresponds to an interest rate of 5% per year.

	Principal	+	Interest	=	Balance
After 1 year	P_0	+	$r P_0$	=	$P_0(1+r)$
After 2 years	$P_0(1+r)$	+	$r P_0(1+r)$	=	$P_0(1+r)^2$
$\cdots$			$\cdots$		
After t years	$P_0(1+r)^{t-1}$	+	$r P_0(1+r)^{t-1}$	=	$P_0(1+r)^t$

Suppose that interest is paid out quarterly (every 3 months). Then the interest earned after 3 months is $\frac{r}{4} P_0$ dollars and the balance increases by the factor $\left(1 + \frac{r}{4}\right)$. After 1 year (four quarters), the balance increases to $P_0 \left(1 + \frac{r}{4}\right)^4$ and after t years,

$$\text{Balance after } t \text{ years} = P_0 \left(1 + \frac{r}{4}\right)^{4t}$$

For example, if $P_0 = \$100$ and $r = 0.09$, then the balance after 1 year is

$$100 \left(1 + \frac{0.09}{4}\right)^4 = 100(1.0225)^4 \approx 100(1.09308) \approx \$109.31$$

More generally,

Compound Interest If P_0 dollars are deposited into an account earning interest at an annual rate r, compounded M times yearly, then the value of the account after t years is

$$\boxed{P(t) = P_0 \left(1 + \frac{r}{M}\right)^{Mt}}$$

The factor $\left(1 + \frac{r}{M}\right)^M$ is called the **yearly multiplier**.

TABLE 1 **Compound Interest with Principal $P_0 = \$100$ and $r = 0.09$**

	Principal after 1 Year
Annual	$100(1 + 0.09) = \$109$
Quarterly	$100\left(1 + \frac{0.09}{4}\right)^4 \approx \$109.31$
Monthly	$100\left(1 + \frac{0.09}{12}\right)^{12} \approx \$109.38$
Weekly	$100\left(1 + \frac{0.09}{52}\right)^{52} \approx \$109.41$
Daily	$100\left(1 + \frac{0.09}{365}\right)^{365} \approx \$109.42$

Table 1 shows the effect of more frequent compounding. What happens in the limit as M tends to infinity? This question is answered by the next theorem (a proof is given at the end of this section).

THEOREM 2 **Limit Formula for e and e^x**

$$\boxed{e = \lim_{n \to \infty} \left(1 + \frac{1}{n}\right)^n \quad \text{and} \quad e^x = \lim_{n \to \infty} \left(1 + \frac{x}{n}\right)^n \quad \text{for all } x}$$

Figure 11 illustrates the first limit graphically. To compute the limit of the yearly multiplier as $M \to \infty$, we apply the second limit with $x = r$ and $n = M$:

$$\lim_{M \to \infty} \left(1 + \frac{r}{M}\right)^M = e^r \qquad \boxed{2}$$

The multiplier after t years is $(e^r)^t = e^{rt}$. This leads to the following definition.

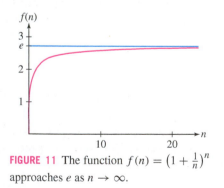

$f(n)$

FIGURE 11 The function $f(n) = \left(1 + \frac{1}{n}\right)^n$ approaches e as $n \to \infty$.

Continuously Compounded Interest If P_0 dollars are deposited into an account earning interest at an annual rate r, compounded continuously, then the value of the account after t years is

$$\boxed{P(t) = P_0 e^{rt}}$$

■ **EXAMPLE 7** A principal of $P_0 = ¥\,100{,}000$ (Japanese yen) is deposited into an account paying 6% interest. Find the balance after 3 years if interest is compounded quarterly and if interest is compounded continuously.

Note: The mathematics of interest rates is the same for all currencies (dollars, euros, pesos, yen, etc.).

Solution After 3 years, the balance is

Quarterly compounding:
$$100,000\left(1+\frac{0.06}{4}\right)^{4(3)} \approx ¥119,562$$

Continuous compounding:
$$100,000e^{(0.06)3} \approx ¥119,722 \quad\blacksquare$$

Present Value

The concept of *present value* (PV) is used in business and finance to compare payments made at different times. Assume that there is an interest rate r (continuously compounded) at which an investor can lend or borrow money. By definition, the PV of P dollars to be received t years in the future is Pe^{-rt}:

> The PV of P dollars received at time t is Pe^{-rt}.

In the financial world, there are many different interest rates (federal funds rate, prime rate, LIBOR, etc.). We simplify the discussion by assuming that there is just one rate.

What is the reasoning behind this definition? When you invest at the rate r for t years, your principal increases by the factor e^{rt}, so if you invest Pe^{-rt} dollars, your principal grows to $(Pe^{-rt})e^{rt} = P$ dollars at time t. The present value Pe^{-rt} is the amount you would have to invest *today* in order to have P dollars at time t.

■ **EXAMPLE 8** Is it better to receive \$2000 today or \$2200 in 2 years? Consider $r = 0.03$ and $r = 0.07$.

Solution We compare \$2000 today with the PV of \$2200 received in 2 years.

- If $r = 0.03$, the PV is $2200e^{-(0.03)2} \approx \$2071.88$. This is more than \$2000, so a payment of \$2200 in 2 years is preferable to a \$2000 payment today.
- If $r = 0.07$, the PV is $2200e^{-(0.07)2} \approx \$1912.59$. This PV is less than \$2000, so it is better to receive \$2000 today if $r = 0.07$. ■

■ **EXAMPLE 9** **Deciding Whether to Invest** Chief Operating Officer Ryan Martinez must decide whether to upgrade his company's computer system. The upgrade costs \$400,000 and will save \$150,000 a year for each of the next 3 years. Is this a good investment if $r = 7\%$?

Solution Ryan must compare today's cost of the upgrade with the PV of the money saved. For simplicity, assume that the annual savings of \$150,000 is received as a lump sum at the end of each year.

If $r = 0.07$, the PV of the savings over 3 years is

$$150,000e^{-(0.07)} + 150,000e^{-(0.07)2} + 150,000e^{-(0.07)3} \approx \$391,850$$

The amount saved is *less* than the cost \$400,000, so the upgrade is not worthwhile. ■

An **income stream** is a sequence of periodic payments that continue over an interval of T years. Consider an investment that produces income at a rate of \$800/year for 5 years. A total of \$4000 is paid out over 5 years, but the PV of the income stream is less. For instance, if $r = 0.06$ and payments are made at the end of the year, then the PV is

$$800e^{-0.06} + 800e^{-(0.06)2} + 800e^{-(0.06)3} + 800e^{-(0.06)4} + 800e^{-(0.06)5} \approx \$3353.12$$

It is more convenient mathematically to assume that payments are made *continuously* at a rate of $R(t)$ dollars per year. We can then calculate PV as an integral. Divide the time interval $[0, T]$ into N subintervals of length $\Delta t = T/N$. If Δt is small, the amount paid out between time t and $t + \Delta t$ is approximately

$$\underbrace{R(t)}_{\text{Rate}} \times \underbrace{\Delta t}_{\text{Time interval}} = R(t)\Delta t$$

The PV of this payment is approximately $e^{-rt} R(t)\Delta t$. Setting $t_i = i\,\Delta t$, we obtain the approximation

$$\text{PV of income stream} \approx \sum_{i=1}^{N} e^{-rt_i} R(t_i)\Delta t$$

This is a Riemann sum whose value approaches $\int_0^T R(t)e^{-rt}\,dt$ as $\Delta t \to 0$.

> **PV of an Income Stream** If the interest rate is r, the present value of an income stream paying out $R(t)$ dollars per year continuously for T years is
>
> $$\text{PV} = \int_0^T R(t)e^{-rt}\,dt$$
> **3**

■ EXAMPLE 10 An investment pays out 800,000 Mexican pesos per year, continuously for 5 years. Find the PV of the investment for $r = 0.04$ and $r = 0.06$.

Solution In this case, $R(t) = 800,000$. If $r = 0.04$, the PV of the income stream is equal (in pesos) to

$$\int_0^5 800,000e^{-0.04t}\,dt = -800,000\frac{e^{-0.04t}}{0.04}\bigg|_0^5 \approx -16,374,615 - (-20,000,000)$$

$$= 3,625,385$$

If $r = 0.06$, the PV is equal (in pesos) to

$$\int_0^5 800,000e^{-0.06t}\,dt = -800,000\frac{e^{-0.06t}}{0.06}\bigg|_0^5 \approx -9,877,576 - (-13,333,333)$$

$$= 3,455,757 \qquad ■$$

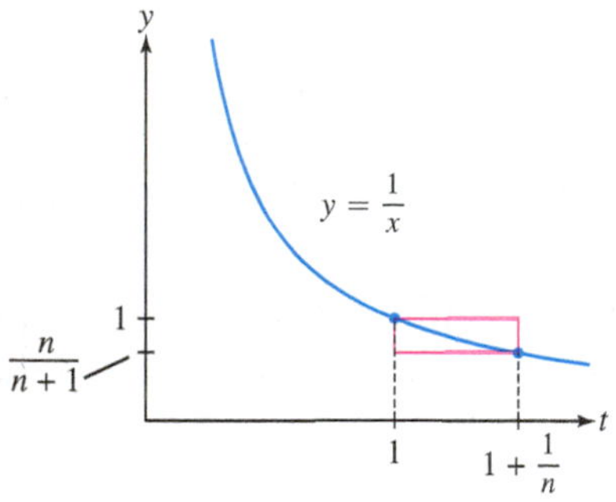

Proof of Theorem 2 Apply the formula $\ln b = \int_1^b t^{-1}\,dt$ with $b = 1 + 1/n$:

$$\ln\left(1 + \frac{1}{n}\right) = \int_1^{1+1/n} \frac{dt}{t}$$

Figure 12 shows that the area represented by this integral lies between the areas of two rectangles of heights $n/(n+1)$ and 1, both of base $1/n$. These rectangles have areas $1/(n+1)$ and $1/n$, so

$$\frac{1}{n+1} \le \ln\left(1 + \frac{1}{n}\right) \le \frac{1}{n}$$
4

Multiply through by n, using the rule $n \ln a = \ln a^n$:

$$\frac{n}{n+1} \le \ln\left(\left(1 + \frac{1}{n}\right)^n\right) \le 1$$

Since $\lim_{n\to\infty} \frac{n}{n+1} = 1$, the middle quantity must approach 1 by the Squeeze Theorem:

$$\lim_{n\to\infty} \ln\left(\left(1 + \frac{1}{n}\right)^n\right) = 1$$

Now we can apply e^x (because it is continuous) to obtain the desired result:

$$e^1 = e^{\lim_{n\to\infty} \ln\left(\left(1+\frac{1}{n}\right)^n\right)} = \lim_{n\to\infty} e^{\ln\left(\left(1+\frac{1}{n}\right)^n\right)} = \lim_{n\to\infty}\left(1 + \frac{1}{n}\right)^n$$

See Exercise 61 for a proof of the more general formula $e^x = \lim_{n\to\infty}\left(1 + \frac{x}{n}\right)^n$. ■

In April 1720 Isaac Newton doubled his money by investing in the South Sea Company, an English company set up to conduct trade with the West Indies and South America. Having gained 7000 pounds, Newton invested a second time, but like many others, he did not realize that the company was built on fraud and manipulation. In what became known as the South Sea Bubble, the stock lost 80% of its value, and the famous scientist suffered a loss of 20,000 pounds.

FIGURE 12

A second method of proof of Theorem 2 is to apply the methods of Section 4.5 to $\lim_{x\to\infty} f(x)^{g(x)}$, which, when we take $f(x) = 1 + \frac{1}{x}$ and $g(x) = x$, is an indeterminate form of type 1^∞.

5.9 SUMMARY

- *Exponential growth* with growth constant $k > 0$: $P(t) = P_0 e^{kt}$.
- *Exponential decay* with decay constant $k > 0$: $P(t) = P_0 e^{-kt}$.
- The solutions of the differential equation $y' = ky$ are the exponential functions $y = Ce^{kt}$, where C is a constant.
- A quantity $P(t)$ grows exponentially if it grows at a rate proportional to its size—that is, if $P'(t) = kP(t)$.
- The *doubling time* for exponential growth and the *half-life* for exponential decay are both equal to $(\ln 2)/k$.
- For use in carbon dating: The decay constant of C^{14} is $k = 0.000121$.
- Interest rate r, compounded M times per year:

$$P(t) = P_0(1 + r/M)^{Mt}$$

- Interest rate r, compounded continuously: $P(t) = P_0 e^{rt}$.
- The *present value* (PV) of P dollars (or other currency), to be paid t years in the future, is Pe^{-rt}.
- Present value of an income stream paying $R(t)$ dollars per year continuously for T years:

$$\text{PV} = \int_0^T R(t)e^{-rt}\, dt$$

5.9 EXERCISES

Preliminary Questions

1. Two quantities increase exponentially with growth constants $k = 1.2$ and $k = 3.4$, respectively. Which quantity doubles more rapidly?

2. A cell population grows exponentially beginning with one cell. Which takes longer: increasing from one to two cells or increasing from 15 million to 20 million cells?

3. Referring to his popular book *A Brief History of Time*, the renowned physicist Stephen Hawking said, "Someone told me that each equation I included in the book would halve its sales." Find a differential equation satisfied by $S(n)$, the number of copies sold if the book has n equations.

4. The PV of N dollars received at time T is (choose the correct answer):

(a) The value at time T of N dollars invested today

(b) The amount you would have to invest today in order to receive N dollars at time T

5. In 1 year, you will be paid \$1. Will the PV increase or decrease if the interest rate goes up?

Exercises

1. A certain population P of bacteria obeys the exponential growth law $P(t) = 2000e^{1.3t}$ (t in hours).

(a) How many bacteria are present initially?

(b) At what time will there be 10,000 bacteria?

2. A quantity P obeys the exponential growth law $P(t) = e^{5t}$ (t in years).

(a) At what time t is $P = 10$?

(b) What is the doubling time for P?

3. Write $f(t) = 5(7)^t$ in the form $f(t) = P_0 e^{kt}$ for some P_0 and k.

4. Write $f(t) = 9e^{1.4t}$ in the form $f(t) = P_0 b^t$ for some P_0 and b.

5. A certain RNA molecule replicates every 3 minutes. Find the differential equation for the number $N(t)$ of molecules present at time t (in minutes). How many molecules will be present after 1 hour if there is one molecule at $t = 0$?

6. A quantity P obeys the exponential growth law $P(t) = Ce^{kt}$ (t in years). Find the formula for $P(t)$, assuming that the doubling time is 7 years and $P(0) = 100$.

7. Find all solutions to the differential equation $y' = -5y$. Which solution satisfies the initial condition $y(0) = 3.4$?

8. Find the solution to $y' = \sqrt{2}y$ satisfying $y(0) = 20$.

9. Find the solution to $y' = 3y$ satisfying $y(2) = 1000$.

10. Find the function $y = f(t)$ that satisfies the differential equation $y' = -0.7y$ and the initial condition $y(0) = 10$.

11. The decay constant of cobalt-60 is 0.13 year^{-1}. Find its half-life.

12. The half-life radium-226 is 1622 years. Find its decay constant.

13. One of the world's smallest flowering plants, *Wolffia globosa* (Figure 13), has a doubling time of approximately 30 hours (h). Find the growth constant k and determine the initial population if the population grew to 1000 after 48 h.

FIGURE 13 The tiny plants are *Wolffia*, with plant bodies smaller than the head of a pin. *(Gerald D. Carr)*

14. A 10-kg quantity of a radioactive isotope decays to 3 kg after 17 years. Find the decay constant of the isotope.

15. The population of a city is $P(t) = 2 \cdot e^{0.06t}$ (in millions), where t is measured in years. Calculate the time it takes for the population to double, to triple, and to increase 7 fold.

16. What is the differential equation satisfied by $P(t)$, the number of infected computer hosts in Example 4? Over which time interval would $P(t)$ increase 100 fold?

17. The decay constant for a certain drug is $k = 0.35 \text{ day}^{-1}$. Calculate the time it takes for the quantity present in the bloodstream to decrease by half, by one-third, and by one-tenth.

18. Light Intensity The intensity of light passing through an absorbing medium decreases exponentially with the distance traveled. Suppose the decay constant for a certain plastic block is $k = 4 \text{ m}^{-1}$. How thick must the block be to reduce the intensity by a factor of one-third?

19. Assuming that population growth is approximately exponential, which of the following two sets of data is most likely to represent the population (in millions) of a city over a 5-year period?

Year	2000	2001	2002	2003	2004
Set I	3.14	3.36	3.60	3.85	4.11
Set II	3.14	3.24	3.54	4.04	4.74

20. The **atmospheric pressure** $P(h)$ (in kilopascals) at a height h meters above sea level satisfies a differential equation $P' = -kP$ for some positive constant k.

(a) Barometric measurements show that $P(0) = 101.3$ and $P(30,900) = 1.013$. What is the decay constant k?

(b) Determine the atmospheric pressure at $h = 500$.

21. Degrees in Physics One study suggests that from 1955 to 1970, the number of bachelor's degrees in physics awarded per year by U.S. universities grew exponentially, with growth constant $k = 0.1$.

(a) If exponential growth continues, how long will it take for the number of degrees awarded per year to increase 14-fold?

(b) If 2500 degrees were awarded in 1955, in which year were 10,000 degrees awarded?

22. The **Beer–Lambert Law** is used in spectroscopy to determine the molar absorptivity α or the concentration c of a compound dissolved in a solution at low concentrations (Figure 14). The law states that the intensity I of light as it passes through the solution satisfies $\ln(I/I_0) = \alpha c x$, where I_0 is the initial intensity and x is the distance traveled by the light. Show that I satisfies a differential equation $dI/dx = -kI$ for some constant k.

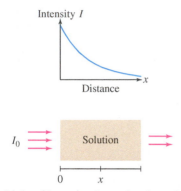

FIGURE 14 Light of intensity I_0 passing through a solution.

23. A sample of sheepskin parchment discovered by archaeologists had a C^{14}-to-C^{12} ratio equal to 40% of that found in the atmosphere. Approximately how old is the parchment?

24. Chauvet Caves In 1994 three French speleologists (geologists specializing in caves) discovered a cave in southern France containing prehistoric cave paintings. A C^{14} analysis carried out by archeologist Helene Valladas showed the paintings to be between 29,700 and 32,400 years old, much older than any previously known human art. Given that the C^{14}-to-C^{12} ratio of the atmosphere is $R = 10^{-12}$, what range of C^{14}-to-C^{12} ratios did Valladas find in the charcoal specimens?

25. A paleontologist discovers remains of animals that appear to have died at the onset of the Holocene ice age, between 10,000 and 12,000 years ago. What range of C^{14}-to-C^{12} ratio would the scientist expect to find in the animal remains?

26. Inversion of Sugar When cane sugar is dissolved in water, it converts to invert sugar over a period of several hours. The percentage $f(t)$ of unconverted cane sugar at time t (in hours) satisfies $f' = -0.2f$. What percentage of cane sugar remains after 5 hours (h)? After 10 h?

27. Continuing with Exercise 26, suppose that 50 g of sugar are dissolved in a container of water. After how many hours will 20 g of invert sugar be present?

28. Two bacteria colonies are cultivated in a laboratory. The first colony has a doubling time of 2 hours (h) and the second a doubling time of 3 h. Initially, the first colony contains 1000 bacteria and the second colony 3000 bacteria. At what time t will the sizes of the colonies be equal?

29. Moore's Law In 1965 Gordon Moore predicted that the number N of transistors on a microchip would increase exponentially.

(a) Does the table of data below confirm Moore's prediction for the period from 1971 to 2000? If so, estimate the growth constant k.

(b) *CAS* Plot the data in the table.

(c) Let $N(t)$ be the number of transistors t years after 1971. Find an approximate formula $N(t) \approx C e^{kt}$, where t is the number of years after 1971.

(d) Estimate the doubling time in Moore's Law for the period from 1971 to 2000.

(e) How many transistors will a chip contain in 2020 if Moore's Law continues to hold?

(f) Can Moore have expected his prediction to hold indefinitely?

Processor	Year	No. Transistors
4004	1971	2250
8008	1972	2500
8080	1974	5000
8086	1978	29,000
286	1982	120,000
386 processor	1985	275,000
486 DX processor	1989	1,180,000
Pentium processor	1993	3,100,000
Pentium II processor	1997	7,500,000
Pentium III processor	1999	24,000,000
Pentium 4 processor	2000	42,000,000
Xeon processor	2008	1,900,000,000

30. Assume that in a certain country, the rate at which jobs are created is proportional to the number of people who already have jobs. If there are 15 million jobs at $t = 0$ and 15.1 million jobs 3 months later, how many jobs will there be after 2 years?

31. The only functions with a *constant* doubling time are the exponential functions $y = P_0 e^{kt}$ with $k > 0$. Show that the doubling time of linear function $f(t) = at + b$ at time t_0 is $t_0 + b/a$ (which increases with t_0). Compute the doubling times of $f(t) = 3t + 12$ at $t_0 = 10$ and $t_0 = 20$.

32. Verify that the half-life of a quantity that decays exponentially with decay constant k is equal to $(\ln 2)/k$.

33. Compute the balance after 10 years if $2000 is deposited in an account paying 9% interest and interest is compounded (a) quarterly, (b) monthly, and (c) continuously.

34. Suppose $500 is deposited into an account paying interest at a rate of 7%, continuously compounded. Find a formula for the value of the account at time t. What is the value of the account after 3 years?

35. A bank pays interest at a rate of 5%. What is the yearly multiplier if interest is compounded

(a) three times a year? **(b)** continuously?

36. How long will it take for $4000 to double in value if it is deposited in an account bearing 7% interest, continuously compounded?

37. How much must one invest today in order to receive $20,000 after 5 years if interest is compounded continuously at the rate $r = 9\%$?

38. An investment increases in value at a continuously compounded rate of 9%. How large must the initial investment be in order to build up a value of $50,000 over a 7-year period?

39. Compute the PV of $5000 received in 3 years if the interest rate is (a) 6% and (b) 11%. What is the PV in these two cases if the sum is instead received in 5 years?

40. Is it better to receive $1000 today or $1300 in 4 years? Consider $r = 0.08$ and $r = 0.03$.

41. Find the interest rate r if the PV of $8000 to be received in 1 year is $7300.

42. A company can earn additional profits of $500,000/year for 5 years by investing $2 million to upgrade its factory. Is the investment worthwhile if the interest rate is 6%? (Assume the savings are received as a lump sum at the end of each year.)

43. A new computer system costing $25,000 will reduce labor costs by $7000/year for 5 years.

(a) Is it a good investment if $r = 8\%$?

(b) How much money will the company actually save?

44. After winning $25 million in the state lottery, Jessica learns that she will receive five yearly payments of $5 million beginning immediately.

(a) What is the PV of Jessica's prize if $r = 6\%$?

(b) How much more would the prize be worth if the entire amount were paid today?

45. Use Eq. (3) to compute the PV of an income stream paying out $R(t) = \$5000$/year continuously for 10 years, assuming $r = 0.05$.

46. Find the PV of an investment that pays out continuously at a rate of $800/year for 5 years, assuming $r = 0.08$.

47. Find the PV of an income stream that pays out continuously at a rate $R(t) = \$5000e^{0.1t}$/year for 7 years, assuming $r = 0.05$.

48. A commercial property generates income at the rate $R(t)$. Suppose that $R(0) = \$70,000$/year and that $R(t)$ increases at a continuously compounded rate of 5%. Find the PV of the income generated in the first 4 years if $r = 6\%$.

49. Show that an investment that pays out R dollars per year continuously for T years has a PV of $R(1 - e^{-rT})/r$.

50. Explain this statement: If T is very large, then the PV of the income stream described in Exercise 49 is approximately R/r.

51. Suppose that $r = 0.06$. Use the result of Exercise 50 to estimate the payout rate R needed to produce an income stream whose PV is $20,000, assuming that the stream continues for a large number of years.

52. Verify by differentiation:

$$\int te^{-rt}\, dt = -\frac{e^{-rt}(1 + rt)}{r^2} + C \qquad \boxed{5}$$

Use Eq. (5) to compute the PV of an investment that pays out income continuously at a rate $R(t) = (5000 + 1000t)$ dollars per year for 5 years, assuming $r = 0.05$.

53. Use Eq. (5) to compute the PV of an investment that pays out income continuously at a rate $R(t) = (5000 + 1000t)e^{0.02t}$ dollars per year for 10 years, assuming $r = 0.08$.

54. **Banker's Rule of 70** If you earn an interest rate of R percent, continuously compounded, your money doubles after approximately $70/R$ years. For example, at $R = 5\%$, your money doubles after 70/5 or 14 years. Use the concept of doubling time to justify the Banker's Rule. (*Note:* Sometimes, the rule $72/R$ is used. It is less accurate but easier to apply because 72 is divisible by more numbers than 70.)

55. **Drug Dosing Interval** Let $y(t)$ be the drug concentration (in micrograms per kilogram) in a patient's body at time t. The initial concentration is $y(0) = L$. Additional doses that increase the concentration by an amount d are administered at regular time intervals of length T. In between doses, $y(t)$ decays exponentially—that is, $y' = -ky$. Find the value of T (in terms of k and d) for which the concentration varies between L and $L - d$ as in Figure 15.

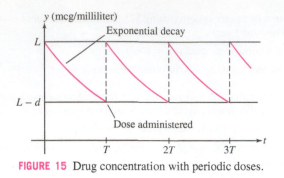

FIGURE 15 Drug concentration with periodic doses.

*Exercises 56 and 57: The **Gompertz differential equation***

$$\frac{dy}{dt} = ky \ln\left(\frac{y}{M}\right) \qquad \boxed{6}$$

(where M and k are constants) was introduced in 1825 by the English mathematician Benjamin Gompertz and is still used today to model aging and mortality.

56. Show that $y = Me^{ae^{kt}}$ satisfies Eq. (6) for any constant a.

57. To model mortality in a population of 200 laboratory rats, a scientist assumes that the number $P(t)$ of rats alive at time t (in months) satisfies

Eq. (6) with $M = 204$ and $k = 0.15$ month^{-1} (Figure 16). Find $P(t)$ [note that $P(0) = 200$] and determine the population after 20 months.

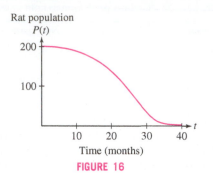

FIGURE 16

58. **Isotopes for Dating** Given that the age of the earth is approximately 4.5 billion years, which of the following would be most suitable for dating extremely old rocks: carbon-14 (half-life 5570 years), lead-210 (half-life 22.26 years), or potassium-49 (half-life 1.3 billion years)? Explain why.

59. Let $P = P(t)$ be a quantity that obeys an exponential growth law with growth constant k. Show that P increases m-fold after an interval of $(\ln m)/k$ years.

Further Insights and Challenges

60. **Average Time of Decay** Physicists use the radioactive decay law $R = R_0 e^{-kt}$ to compute the average or *mean time* M until an atom decays. Let $F(t) = R/R_0 = e^{-kt}$ be the fraction of atoms that have survived to time t without decaying.

(a) Find the inverse function $t(F)$.

(b) By definition of $t(F)$, a fraction $1/N$ of atoms decays in the time interval

$$\left[t\left(\frac{j}{N}\right), t\left(\frac{j-1}{N}\right) \right]$$

Use this to justify the approximation $M \approx \frac{1}{N} \sum_{j=1}^{N} t\left(\frac{j}{N}\right)$. Then argue,

by passing to the limit as $N \to \infty$, that $M = \int_0^1 t(F)\,dF$. Strictly speaking, this is an *improper integral* because $t(0)$ is infinite (it takes an infinite amount of time for all atoms to decay). Therefore, we define M as a limit

$$M = \lim_{c \to 0} \int_c^1 t(F)\,dF$$

(c) Verify the formula $\int \ln x\,dx = x \ln x - x$ by differentiation and use it to show that for $c > 0$,

$$M = \lim_{c \to 0}\left(\frac{1}{k} + \frac{1}{k}(c \ln c - c) \right)$$

(d) Show that $M = 1/k$ by evaluating the limit (use L'Hôpital's Rule to compute $\lim_{c \to 0} c \ln c$).

(e) What is the mean time to decay for radon (with a half-life of 3.825 days)?

61. Modify the proof of the relation $e = \lim_{n \to \infty}\left(1 + \frac{1}{n}\right)^n$ given in the text to prove $e^x = \lim_{n \to \infty}\left(1 + \frac{x}{n}\right)^n$. *Hint:* Express $\ln(1 + xn^{-1})$ as an integral and estimate above and below by rectangles.

62. Prove that, for $n > 0$,

$$\left(1 + \frac{1}{n}\right)^n \le e \le \left(1 + \frac{1}{n}\right)^{n+1}$$

Hint: Take logarithms and use Eq. (4).

63. A bank pays interest at the rate r, compounded M times yearly. The **effective interest rate** r_e is the rate at which interest, if compounded annually, would have to be paid to produce the same yearly return.

(a) Find r_e if $r = 9\%$ compounded monthly.

(b) Show that $r_e = (1 + r/M)^M - 1$ and that $r_e = e^r - 1$ if interest is compounded continuously.

(c) Find r_e if $r = 11\%$ compounded continuously.

(d) Find the rate r that, compounded weekly, would yield an effective rate of 20%.

CHAPTER REVIEW EXERCISES

In Exercises 1–4, refer to the function f whose graph is shown in Figure 1.

1. Estimate L_4 and M_4 on $[0, 4]$.

2. Estimate R_4, L_4, and M_4 on $[1, 3]$.

3. Find an interval $[a, b]$ on which R_4 is larger than $\int_a^b f(x)\,dx$. Do the same for L_4.

4. Justify $\frac{3}{2} \le \int_1^2 f(x)\,dx \le \frac{9}{4}$.

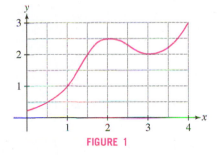

FIGURE 1

In Exercises 5–8, let $f(x) = x^2 + 3x$.

5. Calculate R_6, M_6, and L_6 for f on the interval $[2, 5]$. Sketch the graph of f and the corresponding rectangles for each approximation.

6. Use FTC I to evaluate $A(x) = \displaystyle\int_{-2}^{x} f(t)\,dt$.

7. Find a formula for R_N for f on $[2, 5]$ and compute $\displaystyle\int_{2}^{5} f(x)\,dx$ by taking the limit.

8. Find a formula for L_N for f on $[0, 2]$ and compute $\displaystyle\int_{0}^{2} f(x)\,dx$ by taking the limit.

9. Calculate R_5, M_5, and L_5 for $f(x) = (x^2 + 1)^{-1}$ on the interval $[0, 1]$.

10. Let R_N be the Nth right-endpoint approximation for $f(x) = x^3$ on $[0, 4]$ (Figure 2).

(a) Prove that $R_N = \dfrac{64(N+1)^2}{N^2}$.

(b) Prove that the area of the region within the right-endpoint rectangles above the graph is equal to

$$\frac{64(2N+1)}{N^2}$$

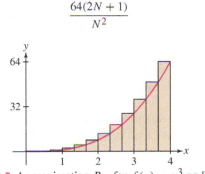

FIGURE 2 Approximation R_N for $f(x) = x^3$ on $[0, 4]$.

11. Which approximation to the area is represented by the shaded rectangles in Figure 3? Compute R_5 and L_5.

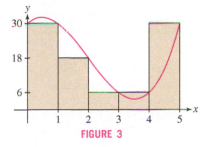

FIGURE 3

12. Calculate any two Riemann sums for $f(x) = x^2$ on the interval $[2, 5]$, but choose partitions with at least five subintervals of unequal widths and intermediate points that are neither endpoints nor midpoints.

In Exercises 13–16, express the limit as an integral (or multiple of an integral) and evaluate.

13. $\displaystyle\lim_{N\to\infty} \frac{\pi}{6N} \sum_{j=1}^{N} \sin\left(\frac{\pi}{3} + \frac{\pi j}{6N}\right)$

14. $\displaystyle\lim_{N\to\infty} \frac{3}{N} \sum_{k=0}^{N-1} \left(10 + \frac{3k}{N}\right)$

15. $\displaystyle\lim_{N\to\infty} \frac{5}{N} \sum_{j=1}^{N} \sqrt{4 + 5j/N}$

16. $\displaystyle\lim_{N\to\infty} \frac{1^k + 2^k + \cdots + N^k}{N^{k+1}} \quad (k > 0)$

In Exercises 17–30, calculate the indefinite integral.

17. $\displaystyle\int \left(4x^3 - 2x^2\right) dx$

18. $\displaystyle\int x^{9/4}\,dx$

19. $\displaystyle\int \sin(\theta - 8)\,d\theta$

20. $\displaystyle\int \cos(5 - 7\theta)\,d\theta$

21. $\displaystyle\int \left(4t^{-3} - 12t^{-4}\right) dt$

22. $\displaystyle\int \left(9t^{-2/3} + 4t^{7/3}\right) dt$

23. $\displaystyle\int \sec^2 x\,dx$

24. $\displaystyle\int \tan 3\theta \sec 3\theta\,d\theta$

25. $\displaystyle\int (y + 2)^4\,dy$

26. $\displaystyle\int \frac{3x^3 - 9}{x^2}\,dx$

27. $\displaystyle\int (e^x - x)\,dx$

28. $\displaystyle\int e^{-4x}\,dx$

29. $\displaystyle\int 4x^{-1}\,dx$

30. $\displaystyle\int \sin(4x - 9)\,dx$

In Exercises 31–36, solve the differential equation with the given initial condition.

31. $\dfrac{dy}{dx} = 4x^3, \quad y(1) = 4$

32. $\dfrac{dy}{dt} = 3t^2 + \cos t, \quad y(0) = 12$

33. $\dfrac{dy}{dx} = x^{-1/2}, \quad y(1) = 1$

34. $\dfrac{dy}{dx} = \sec^2 x, \quad y\left(\frac{\pi}{4}\right) = 2$

35. $\dfrac{dy}{dx} = e^{-x}, \quad y(0) = 3$

36. $\dfrac{dy}{dx} = e^{4x}, \quad y(1) = 1$

37. Find $f(t)$ if $f''(t) = 1 - 2t$, $f(0) = 2$, and $f'(0) = -1$.

38. At time $t = 0$, a driver begins decelerating at a constant rate of -10 m/s^2 and comes to a halt after traveling 500 m. Find the velocity at $t = 0$.

In Exercises 39–42, use the given substitution to evaluate the integral.

39. $\displaystyle\int_{0}^{2} \frac{dt}{4t + 12}, \quad u = 4t + 12$

40. $\displaystyle\int \frac{(x^2 + 1)\,dx}{(x^3 + 3x)^4}, \quad u = x^3 + 3x$

41. $\int_0^{\pi/6} \sin x \cos^4 x \, dx,$ $u = \cos x$

42. $\int \sec^2(2\theta) \tan(2\theta) \, d\theta,$ $u = \tan(2\theta)$

In Exercises 43–92, evaluate the integral.

43. $\int (20x^4 - 9x^3 - 2x) \, dx$

44. $\int_0^2 (12x^3 - 3x^2) \, dx$

45. $\int (2x^2 - 3x)^2 \, dx$

46. $\int_0^1 (x^{7/3} - 2x^{1/4}) \, dx$

47. $\int \frac{x^5 + 3x^4}{x^2} \, dx$

48. $\int_1^3 r^{-4} \, dr$

49. $\int_{-3}^3 |x^2 - 4| \, dx$

50. $\int_{-2}^4 |(x-1)(x-3)| \, dx$

51. $\int_1^3 [t] \, dt$

52. $\int_0^2 (t - [t])^2 \, dt$

53. $\int (10t - 7)^{14} \, dt$

54. $\int_2^3 \sqrt{7y - 5} \, dy$

55. $\int \frac{(2x^3 + 3x) \, dx}{(3x^4 + 9x^2)^5}$

56. $\int_{-3}^{-1} \frac{x \, dx}{(x^2 + 5)^2}$

57. $\int_0^5 15x\sqrt{x + 4} \, dx$

58. $\int t^2 \sqrt{t + 8} \, dt$

59. $\int_0^1 \cos\left(\frac{\pi}{3}(t + 2)\right) dt$

60. $\int_{\pi/2}^\pi \sin\left(\frac{5\theta - \pi}{6}\right) d\theta$

61. $\int t^2 \sec^2(9t^3 + 1) \, dt$

62. $\int \sin^2(3\theta) \cos(3\theta) \, d\theta$

63. $\int \csc^2(9 - 2\theta) \, d\theta$

64. $\int \sin\theta \sqrt{4 - \cos\theta} \, d\theta$

65. $\int_0^{\pi/3} \frac{\sin\theta}{\cos^{2/3}\theta} \, d\theta$

66. $\int \frac{\sec^2 t \, dt}{(\tan t - 1)^2}$

67. $\int e^{9 - 2x} \, dx$

68. $\int_1^3 e^{4x - 3} \, dx$

69. $\int x^2 e^{x^3} \, dx$

70. $\int_0^{\ln 3} e^{x - e^x} \, dx$

71. $\int e^x 10^x \, dx$

72. $\int e^{-2x} \sin(e^{-2x}) \, dx$

73. $\int \frac{e^{-x} \, dx}{(e^{-x} + 2)^3}$

74. $\int \sin\theta \cos\theta e^{\cos^2\theta + 1} \, d\theta$

75. $\int_0^{\pi/6} \tan 2\theta \, d\theta$

76. $\int_{\pi/3}^{2\pi/3} \cot\left(\frac{1}{2}\theta\right) d\theta$

77. $\int \frac{dt}{t(1 + (\ln t)^2)}$

78. $\int \frac{\cos(\ln x) \, dx}{x}$

79. $\int_1^e \frac{\ln x \, dx}{x}$

80. $\int \frac{dx}{x\sqrt{\ln x}}$

81. $\int \frac{dx}{4x^2 + 9}$

82. $\int_0^{0.8} \frac{dx}{\sqrt{1 - x^2}}$

83. $\int_4^{12} \frac{dx}{x\sqrt{x^2 - 1}}$

84. $\int_0^3 \frac{x \, dx}{x^2 + 9}$

85. $\int_0^3 \frac{dx}{x^2 + 9}$

86. $\int \frac{dx}{\sqrt{e^{2x} - 1}}$

87. $\int \frac{x \, dx}{\sqrt{1 - x^4}}$

88. $\int_{-5/\sqrt{2}}^{5/\sqrt{2}} \frac{dx}{\sqrt{25 - x^2}}$

89. $\int_0^4 \frac{dx}{2x^2 + 1}$

90. $\int_5^8 \frac{dx}{x\sqrt{x^2 - 16}}$

91. $\int_0^1 \frac{(\tan^{-1} x)^3 \, dx}{1 + x^2}$

92. $\int \frac{\cos^{-1} t \, dt}{\sqrt{1 - t^2}}$

93. Combine to write as a single integral:

$$\int_0^8 f(x) \, dx + \int_{-2}^0 f(x) \, dx + \int_8^6 f(x) \, dx$$

94. Let $A(x) = \int_0^x f(x) \, dx$, where f is the function shown in Figure 4. Identify the location of the local minima, the local maxima, and points of inflection of A on the interval $[0, E]$, as well as the intervals where A is increasing, decreasing, concave up, or concave down. Where does the absolute maximum of A occur?

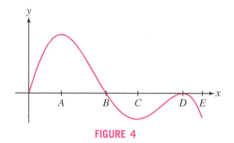

FIGURE 4

95. Find the local minima, the local maxima, and the inflection points of $A(x) = \int_3^x \frac{t \, dt}{t^2 + 1}$.

96. A particle starts at the origin at time $t = 0$ and moves with velocity $v(t)$ as shown in Figure 5.

(a) How many times does the particle return to the origin in the first 12 seconds?

(b) What is the particle's maximum distance from the origin?

(c) What is the particle's maximum distance to the left of the origin?

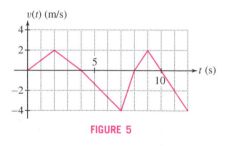

FIGURE 5

97. On a typical day, a city consumes water at the rate of $r(t) = 100 + 72t - 3t^2$ (in thousands of gallons per hour), where t is the number of hours past midnight. What is the daily water consumption? How much water is consumed between 6 PM and midnight?

98. The learning curve in a certain bicycle factory is $L(x) = 12x^{-1/5}$ (in hours per bicycle), which means that it takes a bike mechanic $L(n)$ hours to assemble the nth bicycle. If a mechanic has produced 24 bicycles, how long does it take her or him to produce a subsequent batch of 12?

99. Cost engineers at NASA have the task of projecting the cost P of major space projects. It has been found that the cost C of developing a projection increases with P at the rate $dC/dP \approx 21P^{-0.65}$, where C is in thousands of dollars and P in millions of dollars. What is the cost of developing a projection for a project whose cost turns out to be $P = \$35$ million?

100. An astronomer estimates that in a certain constellation, the number of stars per magnitude m, per degree-squared of sky, is equal to $A(m) = 2.4 \times 10^{-6} m^{7.4}$ (fainter stars have higher magnitudes). Determine the total number of stars of magnitude between 6 and 15 in a 1-degree-squared region of sky.

101. Evaluate $\displaystyle\int_{-8}^{8} \frac{x^{15}\, dx}{3 + \cos^2 x}$, using the properties of odd functions.

102. Evaluate $\displaystyle\int_{0}^{1} f(x)\, dx$, assuming that f is an even continuous function such that

$$\int_{1}^{2} f(x)\, dx = 5, \qquad \int_{-2}^{1} f(x)\, dx = 8$$

103. GU Plot the graph of $f(x) = \sin mx \sin nx$ on $[0, \pi]$ for the pairs $(m, n) = (2, 4), (3, 5)$ and in each case guess the value of $I = \displaystyle\int_{0}^{\pi} f(x)\, dx$. Experiment with a few more values (including two cases with $m = n$) and formulate a conjecture for when I is zero.

104. Show that

$$\int x\, f(x)\, dx = x F(x) - G(x)$$

where $F'(x) = f(x)$ and $G'(x) = F(x)$. Use this to evaluate $\displaystyle\int x \cos x\, dx$.

105. Prove

$$2 \le \int_{1}^{2} 2^x\, dx \le 4 \qquad \text{and} \qquad \frac{1}{9} \le \int_{1}^{2} 3^{-x}\, dx \le \frac{1}{3}$$

106. GU Plot the graph of $f(x) = x^{-2} \sin x$, and show that

$$0.2 \le \int_{1}^{2} f(x)\, dx \le 0.9.$$

107. Find upper and lower bounds for $\displaystyle\int_{0}^{1} f(x)\, dx$, for $y = f(x)$ in Figure 6.

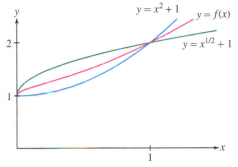

FIGURE 6

In Exercises 108–113, find the derivative.

108. $A'(x)$, where $A(x) = \displaystyle\int_{3}^{x} \sin(t^3)\, dt$

109. $A'(\pi)$, where $A(x) = \displaystyle\int_{2}^{x} \frac{\cos t}{1 + t}\, dt$

110. $\dfrac{d}{dy} \displaystyle\int_{-2}^{y} 3^x\, dx$

111. $G'(x)$, where $G(x) = \displaystyle\int_{-2}^{\sin x} t^3\, dt$

112. $G'(2)$, where $G(x) = \displaystyle\int_{0}^{x^3} \sqrt{t + 1}\, dt$

113. $H'(1)$, where $H(x) = \displaystyle\int_{4x^2}^{9} \frac{1}{t}\, dt$

114. ✎ Explain with a graph: If f is increasing and concave up on $[a, b]$, then L_N is more accurate than R_N. Which is more accurate if f is increasing and concave down?

115. ✎ Explain with a graph: If f is linear on $[a, b]$, then the $\displaystyle\int_{a}^{b} f(x)\, dx = \frac{1}{2}(R_N + L_N)$ for all N.

116. In this exercise, we prove

$$x - \frac{x^2}{2} \le \ln(1 + x) \le x \qquad \text{(for } x > 0\text{)} \qquad \boxed{1}$$

(a) Show that $\ln(1 + x) = \displaystyle\int_{0}^{x} \frac{dt}{1 + t}$ for $x > 0$.

(b) Verify that $1 - t \le \dfrac{1}{1 + t} \le 1$ for all $t > 0$.

(c) Use (b) to prove Eq. (1).

(d) Verify Eq. (1) for $x = 0.5, 0.1$, and 0.01.

117. Let

$$F(x) = x\sqrt{x^2 - 1} - 2 \int_{1}^{x} \sqrt{t^2 - 1}\, dt$$

Prove that $F(x)$ and $y = \cosh^{-1} x$ differ by a constant by showing that they have the same derivative. Then prove they are equal by evaluating both at $x = 1$.

118. ✎ Let f be a positive increasing continuous function on $[a, b]$, where $0 \le a < b$ as in Figure 7. Show that the shaded region has area

$$I = b f(b) - a f(a) - \int_{a}^{b} f(x)\, dx \qquad \boxed{2}$$

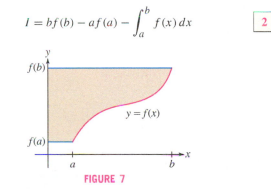

FIGURE 7

119. ✎ How can we interpret the quantity I in Eq. (2) if $a < b \le 0$? Explain with a graph.

120. The isotope thorium-234 has a half-life of 24.5 days.

(a) What is the differential equation satisfied by $y(t)$, the amount of thorium-234 in a sample at time t?

(b) At $t = 0$, a sample contains 2 kg of thorium-234. How much remains after 40 days?

121. The Oldest Snack Food? In Bat Cave, New Mexico, archaeologists found ancient human remains, including cobs of popping corn whose C^{14}-to-C^{12} ratio was approximately 48% of that found in living matter. Estimate the age of the corn cobs.

122. The C^{14}-to-C^{12} ratio of a sample is proportional to the disintegration rate (number of beta particles emitted per minute) that is measured directly with a Geiger counter. The disintegration rate of carbon in a living organism is 15.3 beta particles per minute per gram. Find the age of a sample that emits 9.5 beta particles per minute per gram.

123. What is the interest rate if the PV of $50,000 to be delivered in 3 years is $43,000?

124. An equipment upgrade costing $1 million will save a company $320,000 per year for 4 years. Is this a good investment if the interest rate is $r = 5\%$? What is the largest interest rate that would make the investment worthwhile? Assume that the savings are received as a lump sum at the end of each year.

125. Find the PV of an income stream paying out continuously at a rate of $5000e^{-0.1t}$ dollars per year for 5 years, assuming an interest rate of $r = 4\%$.

126. Calculate the limit:

(a) $\displaystyle\lim_{n\to\infty} \left(1 + \frac{4}{n}\right)^n$

(b) $\displaystyle\lim_{n\to\infty} \left(1 + \frac{1}{n}\right)^{4n}$

(c) $\displaystyle\lim_{n\to\infty} \left(1 + \frac{4}{n}\right)^{3n}$

6 APPLICATIONS OF THE INTEGRAL

In the previous chapter, we used the integral to compute areas under curves and net change. In this chapter, we discuss some of the other quantities that are represented by integrals, including volume, average value, work, total mass, population, and fluid flow.

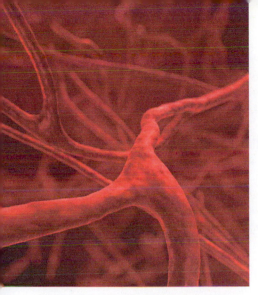

Magnetic Resonance Image (MRI) of veins in a patient's heart. MRI scanners use the mathematics of Fourier transforms, which are based on integrals, to construct two- and three-dimensional images. (*Sergey Panteleev/Getty Images*)

6.1 Area Between Two Curves

Sometimes we are interested in the area between two curves. Figure 1 shows projected electric power generation in the United States through renewable resources (wind, solar, biofuels, etc.) under two scenarios: with and without government stimulus spending. The area of the shaded region between the two graphs represents the additional energy projected to result from stimulus spending.

U. S. Renewable Generating Capacity Forecast Through 2030

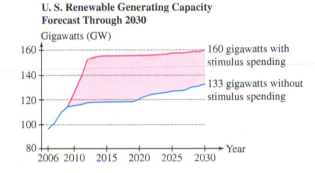

FIGURE 1 The area of the shaded region (which has units of *power × time*, or *energy*) represents the additional energy from renewable generating capacity projected to result from government stimulus spending in 2009–2010. *Source:* Energy Information Agency.

Now suppose that we are given two functions $y = f(x)$ and $y = g(x)$ such that $f(x) \geq g(x)$ for all x in an interval $[a, b]$. We call such a region **vertically simple** since any vertical line that intersects the region does so in a single vertical line segment or point with its lower endpoint on the graph of $y = g(x)$ and upper endpoint on the graph of $y = f(x)$, as appears, for example, in Figure 2. Then the graph of $y = f(x)$ lies above the graph of $y = g(x)$ (Figure 3), and the area between the graphs is equal to the area under the top function minus the area under the bottom function, which is the integral of $f(x) - g(x)$:

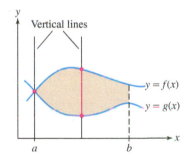

FIGURE 2 A vertically simple region.

$$\text{Area between the graphs} = \int_a^b f(x)\,dx - \int_a^b g(x)\,dx$$

$$= \int_a^b \big(f(x) - g(x)\big)\,dx \qquad \boxed{1}$$

Figure 3 illustrates this formula in the case that both graphs lie above the x-axis. We see that the region between the graphs is obtained by removing the region under $y = g(x)$ from the region under $y = f(x)$.

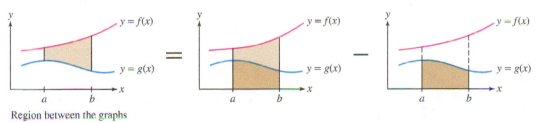

Region between the graphs

FIGURE 3 The area between the graphs is a difference of two areas.

■ **EXAMPLE 1** Find the area of the region between the graphs of the functions

$$f(x) = x^2 - 4x + 10, \qquad g(x) = 4x - x^2, \qquad 1 \le x \le 3$$

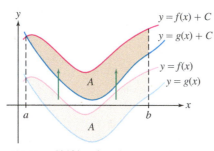

DF FIGURE 4

Solution First, we must determine which graph lies on top. Figure 4 shows that $f(x) \ge g(x)$, as we can verify directly by completing the square:

$$f(x) - g(x) = (x^2 - 4x + 10) - (4x - x^2) = 2x^2 - 8x + 10 = 2(x - 2)^2 + 2 > 0$$

Therefore, the region is vertically simple and by Eq. (1), the area between the graphs is

$$\int_1^3 \left(f(x) - g(x) \right) dx = \int_1^3 \left((x^2 - 4x + 10) - (4x - x^2) \right) dx$$

$$= \int_1^3 (2x^2 - 8x + 10) \, dx = \left(\frac{2}{3}x^3 - 4x^2 + 10x \right) \Big|_1^3 = 12 - \frac{20}{3} = \frac{16}{3} \qquad ■$$

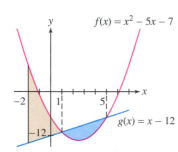

FIGURE 5 Shifting functions up to become positive-valued.

Before continuing with more examples, we note that Eq. (1) remains valid whenever $f(x) \ge g(x)$, even if $f(x)$ and $g(x)$ are not assumed to be positive. As in Figure 5, we can simply shift the two functions up by adding to each a constant C big enough so that both functions are positive over the interval $[a, b]$. This does not change the area between them. Then by our previous result, we have

$$\text{Area between the graphs} = \int_a^b \left((f(x) + C) - (g(x) + C) \right) dx$$

$$= \int_a^b \left(f(x) - g(x) \right) dx$$

Writing $y_{\text{top}} = f(x)$ for the top curve and $y_{\text{bot}} = g(x)$ for the bottom curve, we obtain

$$\boxed{\text{Area between the graphs} = \int_a^b (y_{\text{top}} - y_{\text{bot}}) \, dx = \int_a^b \left(f(x) - g(x) \right) dx} \qquad \boxed{2}$$

■ **EXAMPLE 2** Find the area between the graphs of $f(x) = x^2 - 5x - 7$ and $g(x) = x - 12$ over $[-2, 5]$.

Solution First, we must determine which graph lies on top.

Step 1. Sketch the region (especially, find any points of intersection).
We know that $y = f(x)$ is a parabola with y-intercept -7 and that $y = g(x)$ is a line with y-intercept -12 (Figure 6). To determine where the graphs intersect, we look for values of x where $f(x) = g(x)$, or equivalently, where $f(x) - g(x) = 0$:

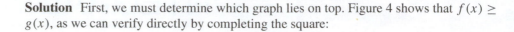

$$f(x) - g(x) = (x^2 - 5x - 7) - (x - 12) = x^2 - 6x + 5 = (x - 1)(x - 5)$$

The graphs intersect where $(x - 1)(x - 5) = 0$, that is, at $x = 1$ and $x = 5$.

FIGURE 6

Step 2. Set up the integrals and evaluate.
We also see that $f(x) - g(x) \le 0$ for $1 \le x < 5$, and thus,

$$f(x) \ge g(x) \text{ on } [-2, 1] \qquad \text{and} \qquad g(x) \ge f(x) \text{ on } [1, 5]$$

In Example 2, we found the intersection points of $y = f(x)$ and $y = g(x)$ algebraically. For more complicated functions, it may be necessary to use a computer algebra system.

This tells us to subdivide our region into two verticaly simple regions over $[-2, 1]$ and $[1, 5]$. Therefore, we write the area as a sum of integrals over the two intervals:

$$\int_{-2}^{5} (y_{\text{top}} - y_{\text{bot}}) \, dx = \int_{-2}^{1} \left(f(x) - g(x) \right) dx + \int_{1}^{5} \left(g(x) - f(x) \right) dx$$

$$= \int_{-2}^{1} \left((x^2 - 5x - 7) - (x - 12) \right) dx + \int_{1}^{5} \left((x - 12) - (x^2 - 5x - 7) \right) dx$$

$$= \int_{-2}^{1} (x^2 - 6x + 5) \, dx + \int_{1}^{5} (-x^2 + 6x - 5) \, dx$$

$$= \left(\frac{1}{3}x^3 - 3x^2 + 5x \right) \Big|_{-2}^{1} + \left(-\frac{1}{3}x^3 + 3x^2 - 5x \right) \Big|_{1}^{5}$$

$$= \left(\frac{7}{3} - \frac{(-74)}{3} \right) + \left(\frac{25}{3} - \frac{(-7)}{3} \right) = \frac{113}{3}$$ ∎

■ **EXAMPLE 3** **Calculating Area by Dividing the Region** Find the area of the region bounded by the graphs of $y = 8/x^2$, $y = 8x$, and $y = x$.

Solution

Step 1. **Sketch the region (especially, find any points of intersection).**

The curve $y = 8/x^2$ cuts off a region in the sector between the two lines $y = 8x$ and $y = x$ (Figure 7). We find the intersection of $y = 8/x^2$ and $y = 8x$ by solving

$$\frac{8}{x^2} = 8x \quad \Rightarrow \quad x^3 = 1 \quad \Rightarrow \quad x = 1$$

and the intersection of $y = 8/x^2$ and $y = x$ by solving

$$\frac{8}{x^2} = x \quad \Rightarrow \quad x^3 = 8 \quad \Rightarrow \quad x = 2$$

Step 2. **Set up the integrals and evaluate.**

Figure 7 shows that $y_{\text{bot}} = x$, but y_{top} changes at $x = 1$ from $y_{\text{top}} = 8x$ to $y_{\text{top}} = 8/x^2$. So the region is not vertically simple. Therefore, we break up the regions into two parts, A and B, each vertically simple, and compute their areas separately.

$$\text{Area of } A = \int_{0}^{1} \left(y_{\text{top}} - y_{\text{bot}} \right) dx = \int_{0}^{1} (8x - x) \, dx = \int_{0}^{1} 7x \, dx = \frac{7}{2}x^2 \Big|_{0}^{1} = \frac{7}{2}$$

$$\text{Area of } B = \int_{1}^{2} \left(y_{\text{top}} - y_{\text{bot}} \right) dx = \int_{1}^{2} \left(\frac{8}{x^2} - x \right) dx = \left(-\frac{8}{x} - \frac{1}{2}x^2 \right) \Big|_{1}^{2} = \frac{5}{2}$$

The total area bounded by the curves is the sum $\frac{7}{2} + \frac{5}{2} = 6$. ∎

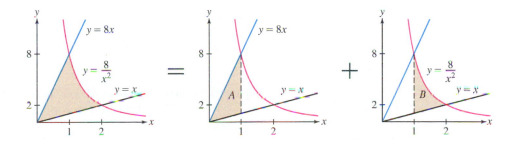

FIGURE 7 Area bounded by $y = 8/x^2$, $y = 8x$, and $y = x$ as a sum of two areas.

Integration Along the *y*-Axis

Suppose we are given x as a function of y, say, $x = g(y)$. What is the meaning of the integral $\int_{c}^{d} g(y) \, dy$? This integral can be interpreted as *signed area*, where regions to the *right* of the y-axis have positive area and regions to the *left* have negative area:

$$\int_{c}^{d} g(y) \, dy = \text{signed area between graph and } y\text{-axis for } c \leq y \leq d$$

In Figure 8(A), the part of the shaded region to the left of the y-axis has negative signed area. The signed area of the entire region is

$$\underbrace{\int_{-6}^{6} (y^2 - 9)\,dy}_{\substack{\text{Area to the right of } y\text{-axis minus} \\ \text{area to the left of } y\text{-axis}}} = \left(\frac{1}{3}y^3 - 9y\right)\Big|_{-6}^{6} = 36$$

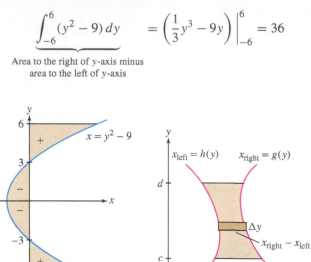

(A) Region between $x = y^2 - 9$ and the y-axis

(B) Region between $x = h(y)$ and $x = g(y)$

FIGURE 8

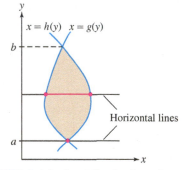

FIGURE 9 A horizontally simple region.

More generally, if $g(y) \geq h(y)$ as in Figure 8(B), then the graph of $x = g(y)$ lies to the right of the graph of $x = h(y)$. In this case, we write $x_{\text{right}} = g(y)$ and $x_{\text{left}} = h(y)$ and we call the region **horizontally simple**, since every horizontal line that intersects the region does so in a single line segment or point such that the left endpoint is on the curve $x = h(y)$ and the right endpoint is on the curve $x = g(y)$, as in Figure 9. The formula for the area corresponding to Eq. (2) is

$$\boxed{\text{Area between the graphs} = \int_{c}^{d} (x_{\text{right}} - x_{\text{left}})\,dy = \int_{c}^{d} \big(g(y) - h(y)\big)\,dy} \qquad \boxed{3}$$

■ **EXAMPLE 4** Calculate the area enclosed by the graphs of $h(y) = y^2 - 1$ and $g(y) = y^2 - \frac{1}{8}y^4 + 1$.

Solution Note that to get a sense of the graph of the second function, we can think of $g(x) = x^2 - \frac{1}{8}x^4 + 1$, and graph it using the techniques we have developed, and then reflect the graph over the line $y = x$ to obtain the graph of $g(y) = y^2 - \frac{1}{8}y^4 + 1$, as in Figure 10.

We find the points where the graphs intersect by solving $g(y) = h(y)$ for y:

$$y^2 - \frac{1}{8}y^4 + 1 = y^2 - 1 \quad \Rightarrow \quad \frac{1}{8}y^4 - 2 = 0 \quad \Rightarrow \quad y = \pm 2$$

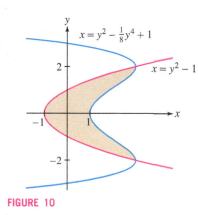

FIGURE 10

Figure 10 shows that the enclosed region stretches from $y = -2$ to $y = 2$. On this interval, $g(y) \geq h(y)$ and therefore the region is horizontally simple. Thus, $x_{\text{right}} = g(y)$, $x_{\text{left}} = h(y)$, and

$$x_{\text{right}} - x_{\text{left}} = \left(y^2 - \frac{1}{8}y^4 + 1\right) - (y^2 - 1) = 2 - \frac{1}{8}y^4$$

It would be more difficult to calculate the area of the region in Figure 10 as an integral with respect to x because the curves are not graphs of functions of x.

The enclosed area is

$$\int_{-2}^{2} (x_{\text{right}} - x_{\text{left}})\,dy = \int_{-2}^{2} \left(2 - \frac{1}{8}y^4\right) dy = \left(2y - \frac{1}{40}y^5\right)\Big|_{-2}^{2}$$

$$= \frac{16}{5} - \left(-\frac{16}{5}\right) = \frac{32}{5} \qquad ■$$

For many regions, we have a choice of whether to find the area by integrating with respect to x or with respect to y. The decision is usually based on how easy it is to obtain the curves as functions of one variable in terms of the other, together with how easy it is to subdivide the region into simple regions and then to integrate the functions involved.

■ **EXAMPLE 5** Find the area of the region that is bounded by the three curves $y = x^2$, $y = (x - 2)^2$, and $y = 0$.

Solution The area appears in Figure 11. Notice immediately that it is not vertically simple, since the top function changes over the interval $[0, 2]$. It is horizontally simple, but to calculate the area using the fact it is a horizontally simple region will take a bit of work. So first, we do it by splitting the region into two vertically simple regions. Then the area is given by

$$\int_0^1 x^2 \, dx + \int_1^2 (x-2)^2 \, dx = \frac{x^3}{3}\Big|_0^1 + \frac{(x-2)^3}{3}\Big|_1^2 = \frac{1}{3} + 0 - \left(-\frac{1}{3}\right) = \frac{2}{3}$$

Now, let's redo the problem, using the fact the region is horizontally simple. We must invert the formulas for the parabolas. The left boundary of the region, which is the right side of the parabola given by $y = x^2$, becomes $x = \sqrt{y}$. To determine the formula for the right boundary of the region, which is the left side of the parabola $y = (x - 2)^2$, we solve for x:

$$x - 2 = \pm\sqrt{y}$$
$$x = 2 \pm \sqrt{y}$$

The equation for the left side of the parabola is given by choosing the minus sign, $x = 2 - \sqrt{y}$.

Then the area is given by

$$\int_0^1 \left((2 - \sqrt{y}) - \sqrt{y}\right) dy = 2y - \frac{4}{3}y^{3/2}\Big|_0^1 = \frac{2}{3} \qquad ■$$

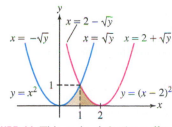

FIGURE 11 This region is horizontally simple but it is easier to cut it into two vertically simple regions.

6.1 SUMMARY

- If $f(x) \geq g(x)$ on $[a, b]$, then the area between the graphs is vertically simple and we have

$$\text{Area between the graphs} = \int_a^b \left(y_{\text{top}} - y_{\text{bot}}\right) dx = \int_a^b \left(f(x) - g(x)\right) dx$$

- To calculate the area between $y = f(x)$ and $y = g(x)$, sketch the region to find y_{top}. If necessary, find points of intersection by solving $f(x) = g(x)$.

- Integral along the y-axis: $\int_c^d g(y) \, dy$ is equal to the signed area between the graph and the y-axis for $c \leq y \leq d$. Area to the right of the y-axis is positive and the area to the left is negative.

- If $g(y) \geq h(y)$ on $[c, d]$, then $x = g(y)$ lies to the right of $x = h(y)$ and the region is horizontally simple.

$$\text{Area between the graphs} = \int_c^d \left(x_{\text{right}} - x_{\text{left}}\right) dy = \int_c^d \left(g(y) - h(y)\right) dy$$

6.1 EXERCISES

Preliminary Questions

1. What is the area interpretation of $\int_a^b \left(f(x) - g(x)\right) dx$ if $f(x) \geq g(x)$?

2. Is $\int_a^b \left(f(x) - g(x)\right) dx$ still equal to the area between the graphs of f and g if $f(x) \geq 0$ but $g(x) \leq 0$?

3. Suppose that $f(x) \geq g(x)$ on $[0, 3]$ and $g(x) \geq f(x)$ on $[3, 5]$. Express the area between the graphs over $[0, 5]$ as a sum of integrals.

4. Suppose that the graph of $x = f(y)$ lies to the left of the y-axis. Is $\int_a^b f(y) \, dy$ positive or negative?

5. Explain what $\int_a^b |f(x) - g(x)| \, dx$ represents.

6. Draw a region that is both vertically simple and horizontally simple.

Exercises

1. Find the area of the region between $y = 3x^2 + 12$ and $y = 4x + 4$ over $[-3, 3]$ (Figure 12).

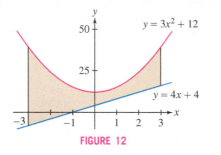

FIGURE 12

2. Find the area of the region between the graphs of $f(x) = 3x + 8$ and $g(x) = x^2 + 2x + 2$ over $[0, 2]$.

3. Find the area of the region enclosed by the graphs of $f(x) = x^2 + 2$ and $g(x) = 2x + 5$ (Figure 13).

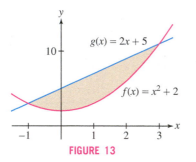

FIGURE 13

4. Find the area of the region enclosed by the graphs of $f(x) = x^3 - 10x$ and $g(x) = 6x$ (Figure 14).

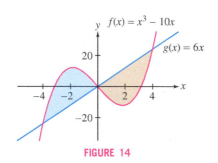

FIGURE 14

In Exercises 5 and 6, sketch the region between $y = \sin x$ and $y = \cos x$ over the interval and find its area.

5. $\left[\dfrac{\pi}{4}, \dfrac{\pi}{2}\right]$

6. $[0, \pi]$

In Exercises 7 and 8, let $f(x) = 20 + x - x^2$ and $g(x) = x^2 - 5x$.

7. Sketch the region enclosed by the graphs of f and g, and compute its area.

8. Sketch the region between the graphs of f and g over $[4, 8]$, and compute its area as a sum of two integrals.

9. Find the area between $y = e^x$ and $y = e^{2x}$ over $[0, 1]$.

10. Find the area of the region bounded by $y = e^x$ and $y = 12 - e^x$ and the y-axis.

11. Sketch the region bounded by the line $y = 2$ and the graph of $y = \sec^2 x$ for $-\dfrac{\pi}{2} < x < \dfrac{\pi}{2}$ and find its area.

12. Sketch the region bounded by

$$y = \sqrt{4 - x^2} \quad \text{and} \quad y = -\sqrt{4 - x^2}$$

for $-2 \leq x \leq 2$. Write down a definite integral that gives its area, but then use geometry to find its area (and thereby determine the integral).

In Exercises 13–16, determine whether or not the region bounded by the curves is vertically simple and/or horizontally simple.

13. $x = y^2, x = 2 - y^2$

14. $y = x^2, x = y^2$

15. $y = x, y = 2x, y = \dfrac{1}{x}$

16. $y = \sin x$ for $0 \leq x \leq \pi$ and $y = 1 - x, y = 0$ (Of the two regions that are bounded by these curves, take the one that does not contain the origin on its boundary.)

In Exercises 17–20, find the area of the shaded region in Figures 15–18.

17.

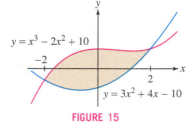

FIGURE 15

18.

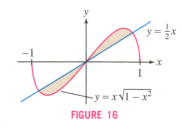

FIGURE 16

19.

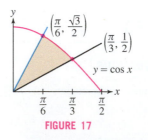

FIGURE 17

20.

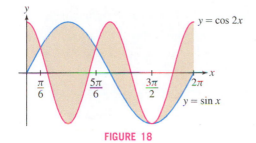

FIGURE 18

In Exercises 21 and 22, find the area between the graphs of $x = \sin y$ and $x = 1 - \cos y$ over the given interval (Figure 19).

21. $0 \le y \le \dfrac{\pi}{2}$

22. $-\dfrac{\pi}{2} \le y \le \dfrac{\pi}{2}$

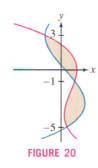

FIGURE 19

23. Find the area of the region lying to the right of $x = y^2 + 4y - 22$ and to the left of $x = 3y + 8$.

24. Find the area of the region lying to the right of $x = y^2 - 5$ and to the left of $x = 3 - y^2$.

25. Figure 20 shows the region enclosed by $x = y^3 - 26y + 10$ and $x = 40 - 6y^2 - y^3$. Match the equations with the curves and compute the area of the region.

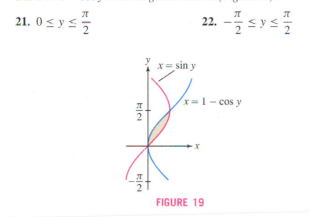

FIGURE 20

26. Figure 21 shows the region enclosed by $y = x^3 - 6x$ and $y = 8 - 3x^2$. Match the equations with the curves and compute the area of the region.

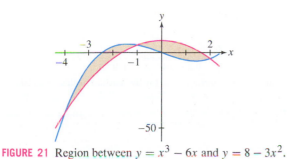

FIGURE 21 Region between $y = x^3 - 6x$ and $y = 8 - 3x^2$.

In Exercises 27 and 28, find the area enclosed by the graphs in two ways: by integrating along the x-axis and by integrating along the y-axis.

27. $x = 9 - y^2$, $x = 5$

28. The *semicubical parabola* $y^2 = x^3$ and the line $x = 1$

In Exercises 29 and 30, find the area of the region using the method (integration along either the x- or the y-axis) that requires you to evaluate just one integral.

29. Region between $y^2 = x + 5$ and $y^2 = 3 - x$

30. Region between $y = x$ and $x + y = 8$ over $[2, 3]$

In Exercises 31–48, sketch the region enclosed by the curves and compute its area as an integral along the x- or y-axis.

31. $y = 4 - x^2$, $y = x^2 - 4$

32. $y = x^2 - 6$, $y = 6 - x^3$, $x = 0$

33. $x + y = 4$, $x - y = 0$, $y + 3x = 4$

34. $y = 8 - 3x$, $y = 6 - x$, $y = 2$

35. $y = 8 - \sqrt{x}$, $y = \sqrt{x}$, $x = 0$

36. $y = \dfrac{x}{x^2 + 1}$, $y = \dfrac{x}{5}$

37. $x = |y|$, $x = 1 - |y|$

38. $y = |x|$, $y = x^2 - 6$

39. $x = y^3 - 18y$, $y + 2x = 0$

40. $y = x\sqrt{x - 2}$, $y = -x\sqrt{x - 2}$, $x = 4$

41. $x = 2y$, $x + 1 = (y - 1)^2$

42. $x + y = 1$, $x^{1/2} + y^{1/2} = 1$

43. $y = \cos x$, $y = \cos 2x$, $x = 0$, $x = \dfrac{2\pi}{3}$

44. $y = \tan x$, $y = -\tan x$, $x = \dfrac{\pi}{4}$

45. $y = \sin x$, $y = \csc^2 x$, $x = \dfrac{\pi}{4}$

46. $x = \sin y$, $x = \dfrac{2}{\pi} y$

47. $y = e^x$, $y = e^{-x}$, $y = 2$

48. $y = \dfrac{\ln x}{x}$, $y = \dfrac{(\ln x)^2}{x}$

49. $\boxed{\text{CAS}}$ Plot

$$y = \dfrac{x}{\sqrt{x^2 + 1}} \quad \text{and} \quad y = (x - 1)^2$$

on the same set of axes. Use a computer algebra system to find the points of intersection numerically and compute the area between the curves.

50. Sketch a region whose area is represented by

$$\int_{-\sqrt{2}/2}^{\sqrt{2}/2} \left(\sqrt{1 - x^2} - |x|\right) dx$$

and evaluate using geometry.

51. Athletes 1 and 2 run along a straight track with velocities $v_1(t)$ and $v_2(t)$ (in meters per second) as shown in Figure 22.

(a) Which of the following is represented by the area of the shaded region over $[0, 10]$?

  i. The distance between athletes 1 and 2 at time $t = 10$ s

  ii. The difference in the distance traveled by the athletes over the time interval $[0, 10]$

(b) Does Figure 22 give us enough information to determine who is ahead at time $t = 10$ s?

(c) If the athletes begin at the same time and place, who is ahead at $t = 10$ s? At $t = 25$ s?

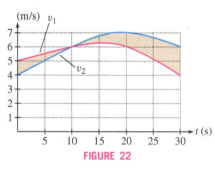

FIGURE 22

52. Express the area (not signed) of the shaded region in Figure 23 as a sum of three integrals involving $f(x)$ and $g(x)$.

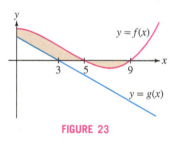

FIGURE 23

53. Find the area enclosed by the curves $y = c - x^2$ and $y = x^2 - c$ as a function of c. Find the value of c for which this area is equal to 1.

54. Set up (but do not evaluate) an integral that expresses the area between the circles $x^2 + y^2 = 2$ and $x^2 + (y - 1)^2 = 1$.

55. Set up (but do not evaluate) an integral that expresses the area between the graphs of $y = (1 + x^2)^{-1}$ and $y = x^2$.

56. ⌂⌂⌂ Find a numerical approximation to the area above $y = 1 - (x/\pi)$ and below $y = \sin x$ (find the points of intersection numerically).

57. ⌂⌂⌂ Find a numerical approximation to the area above $y = |x|$ and below $y = \cos x$.

58. ⌂⌂⌂ Use a computer algebra system to find a numerical approximation to the number c (besides zero) in $\left[0, \frac{\pi}{2}\right]$, where the curves $y = \sin x$ and $y = \tan^2 x$ intersect. Then find the area enclosed by the graphs over $[0, c]$.

59. The back of Jon's guitar (Figure 24) is 19 inches (in.) long. Jon measured the width at 1-in. intervals, beginning and ending $\frac{1}{2}$ in. from the ends, obtaining the results

  6, 9, 10.25, 10.75, 10.75, 10.25, 9.75, 9.5, 10, 11.25,

  12.75, 13.75, 14.25, 14.5, 14.5, 14, 13.25, 11.25, 9

Use the midpoint rule to estimate the area of the back.

FIGURE 24 Back of guitar.

60. Referring to Figure 1 at the beginning of this section, estimate the projected number of additional joules produced in the years 2009–2030 as a result of government stimulus spending in 2009–2010. *Note:* One watt (W) is equal to 1 joule/second (J/s), and 1 gigawatt (GW) is 10^9 watts.

Exercises 61 and 62 use the notation and results of Exercises 49–51 of Section 3.4. For a given country, $F(r)$ is the fraction of total income that goes to the bottom rth fraction of households. The graph of $y = F(r)$ is called the Lorenz curve.

61. Let A be the area between $y = r$ and $y = F(r)$ over the interval $[0, 1]$ (Figure 25). The **Gini index** is the ratio $G = A/B$, where B is the area under $y = r$ over $[0, 1]$.

(a) Show that

$$G = 2 \int_0^1 (r - F(r)) \, dr$$

(b) Calculate G if

$$F(r) = \begin{cases} \frac{1}{3}r & \text{for } 0 \le r \le \frac{1}{2} \\ \frac{5}{3}r - \frac{2}{3} & \text{for } \frac{1}{2} \le r \le 1 \end{cases}$$

(c) The Gini index is a measure of income distribution, with a lower value indicating a more equal distribution. Calculate G if $F(r) = r$ (in this case, all households have the same income by Exercise 51(b) of Section 3.4).

(d) What is G if all of the income goes to one household? *Hint:* In this extreme case, $F(r) = 0$ for $0 \le r < 1$.

62. Calculate the Gini index of the United States in the year 2010 from the Lorenz curve in Figure 25, which consists of segments joining the data points in the following table:

r	0	0.2	0.4	0.6	0.8	1
$F(r)$	0	0.033	0.118	0.264	0.480	1

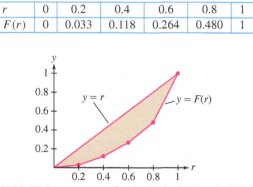

FIGURE 25 Lorenz curve for the United States in 2010.

Further Insights and Challenges

63. Find the line $y = mx$ that divides the area under the curve $y = x(1 - x)$ over $[0, 1]$ into two regions of equal area.

64. *CAS* Let c be the number such that the area under $y = \sin x$ over $[0, \pi]$ is divided in half by the line $y = cx$ (Figure 26). Find an equation for c and solve this equation *numerically* using a computer algebra system.

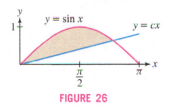

FIGURE 26

65. Explain geometrically (without calculation):

$$\int_0^1 x^n \, dx + \int_0^1 x^{1/n} \, dx = 1 \qquad \text{(for } n > 0\text{)}$$

66. Let f be an increasing function with inverse g. Explain geometrically:

$$\int_0^a f(x) \, dx + \int_{f(0)}^{f(a)} g(x) \, dx = af(a)$$

6.2 Setting Up Integrals: Volume, Density, Average Value

Which quantities are represented by integrals? Roughly speaking, integrals represent quantities that are the "total amount" of something such as area, volume, or total mass. There is a two-step procedure for computing such quantities: (1) Approximate the quantity by a sum of N terms, and (2) pass to the limit as $N \to \infty$ to obtain an integral. We'll use this procedure often in this and other sections.

Volume

The term "solid" or "solid body" refers to a solid three-dimensional object.

Our first example is the **volume** of a solid body. Before proceeding, let's recall that the volume of a *right cylinder* (Figure 1) is Ah, where A is the area of the base and h is the height, measured perpendicular to the base. Here, we use the "right cylinder" in the general sense; the base does not have to be circular, but the sides are perpendicular to the base.

Suppose that the solid body extends from height $y = a$ to $y = b$ along the y-axis as in Figure 2. Let $A(y)$ be the area of the **horizontal cross section** at height y (the intersection of the solid with the horizontal plane at height y).

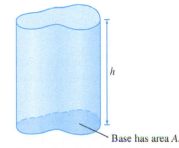

FIGURE 1 The volume of a right cylinder is Ah.

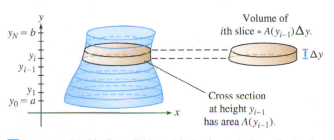
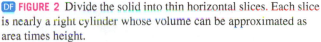

FIGURE 2 Divide the solid into thin horizontal slices. Each slice is nearly a right cylinder whose volume can be approximated as area times height.

To compute the volume V of the body, divide the body into N horizontal slices of thickness $\Delta y = (b - a)/N$. The ith slice extends from y_{i-1} to y_i, where $y_i = a + i\Delta y$. Let V_i be the volume of the slice.

If N is very large, then Δy is very small and the slices are very thin. In this case, the ith slice is nearly a right cylinder of base $A(y_{i-1})$ and height Δy, and therefore, $V_i \approx A(y_{i-1})\Delta y$. Summing up, we obtain

$$V = \sum_{i=1}^{N} V_i \approx \sum_{i=1}^{N} A(y_{i-1})\Delta y$$

The sum on the right is a left-endpoint approximation to the integral $\int_a^b A(y)\,dy$. If we assume that A is a continuous function, then the approximation improves in accuracy and converges to the integral as $N \to \infty$. We conclude that *the volume of the solid is equal to the integral of its cross-sectional area.*

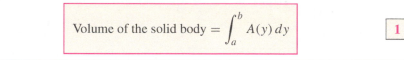

Volume as the Integral of Cross-Sectional Area   Let $A(y)$ be the area of the horizontal cross section at height y of a solid body extending from $y = a$ to $y = b$. Then

$$\text{Volume of the solid body} = \int_a^b A(y)\,dy \qquad \boxed{1}$$

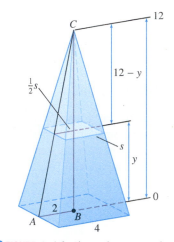

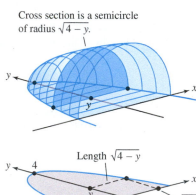

Cross section is a semicircle of radius $\sqrt{4-y}$.

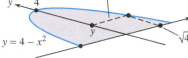

Length $\sqrt{4-y}$

$y = 4 - x^2$

FIGURE 4

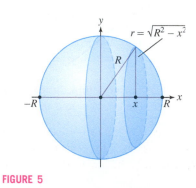

FIGURE 5

■ **EXAMPLE 1   Volume of a Pyramid**   Calculate the volume V of a pyramid of height 12 m whose base is a square of side 4 m.

Solution   To use Eq. (1), we need a formula for the horizontal cross section $A(y)$.

Step 1. **Find a formula for $A(y)$.**
Figure 3 shows that the horizontal cross section at height y is a square. To find the side s of this square, apply the law of similar triangles to $\triangle ABC$ and to the triangle of height $12 - y$ whose base of length $\frac{1}{2}s$ lies on the cross section:

$$\frac{\text{Base}}{\text{Height}} = \frac{2}{12} = \frac{\frac{1}{2}s}{12 - y} \quad \Rightarrow \quad 2(12 - y) = 6s$$

We find that $s = \frac{1}{3}(12 - y)$, and therefore, $A(y) = s^2 = \frac{1}{9}(12 - y)^2$.

Step 2. **Compute V as the integral of $A(y)$.**

$$V = \int_0^{12} A(y)\,dy = \int_0^{12} \frac{1}{9}(12 - y)^2\,dy = -\frac{1}{27}(12 - y)^3 \Big|_0^{12} = 64 \text{ m}^3$$

This agrees with the result obtained using the formula $V = \frac{1}{3}Ah$ for the volume of a pyramid of base A and height h, since $\frac{1}{3}Ah = \frac{1}{3}(4^2)(12) = 64$.   ■

■ **EXAMPLE 2**   Compute the volume V of the solid in Figure 4, whose base is the region between the inverted parabola $y = 4 - x^2$ and the x-axis, and whose vertical cross sections perpendicular to the y-axis are semicircles.

Solution   To find a formula for the area $A(y)$ of the cross section, observe that $y = 4 - x^2$ can be written $x = \pm\sqrt{4 - y}$. We see in Figure 4 that the cross section at y is a semicircle of radius $r = \sqrt{4 - y}$. This semicircle has area $A(y) = \frac{1}{2}\pi r^2 = \frac{\pi}{2}(4 - y)$. Therefore,

$$V = \int_0^4 A(y)\,dy = \frac{\pi}{2}\int_0^4 (4 - y)\,dy = \frac{\pi}{2}\left(4y - \frac{1}{2}y^2\right)\Big|_0^4 = 4\pi \quad ■$$

In the next example, we compute volume using vertical rather than horizontal cross sections. This leads to an integral with respect to x rather than y.

■ **EXAMPLE 3   Volume of a Sphere: Vertical Cross Sections**   Compute the volume of a sphere of radius R.

Solution   As we see in Figure 5, the vertical cross section of the sphere at x is a circle whose radius r satisfies $x^2 + r^2 = R^2$ or $r = \sqrt{R^2 - x^2}$. The area of the cross section is $A(x) = \pi r^2 = \pi(R^2 - x^2)$. Therefore, the sphere has volume

$$\int_{-R}^{R} \pi(R^2 - x^2)\,dx = \pi\left(R^2 x - \frac{x^3}{3}\right)\Big|_{-R}^{R} = 2\left(\pi R^3 - \pi\frac{R^3}{3}\right) = \frac{4}{3}\pi R^3 \quad ■$$

DF **FIGURE 3** A horizontal cross section of the pyramid is a square.

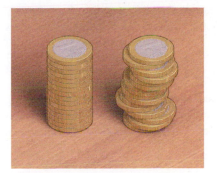

FIGURE 6 The two stacks of coins have equal cross sections, hence equal volumes by Cavalieri's principle.

CONCEPTUAL INSIGHT **Cavalieri's principle** states: Solids with equal cross-sectional areas have equal volume. It is often illustrated convincingly with two stacks of coins (Figure 6). Our formula $V = \int_a^b A(y)\,dy$ includes Cavalieri's principle, because the volumes V are certainly equal if the cross-sectional areas $A(y)$ are equal.

Density

Next, we show that the total mass of an object can be expressed as the integral of its mass density. Consider a rod of length ℓ. The rod's **linear mass density** ρ is defined as the mass per unit length. If ρ is constant, then by definition,

The symbol ρ (lowercase Greek letter rho) is used often to denote density.

$$\text{Total mass} = \text{linear mass density} \times \text{length} = \rho \cdot \ell \qquad \boxed{2}$$

For example, if $\ell = 10$ cm and $\rho = 9$ g/cm, then the total mass is $\rho\ell = 9 \cdot 10 = 90$ g.

Now consider a rod extending along the x-axis from $x = a$ to $x = b$ whose density $y = \rho(x)$ is a continuous function of x, as in Figure 7. To compute the total mass M, we break up the rod into N small segments of length $\Delta x = (b - a)/N$. Then $M = \sum_{i=1}^{N} M_i$, where M_i is the mass of the ith segment.

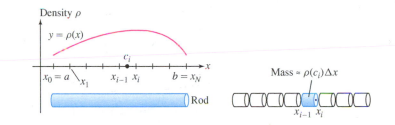

FIGURE 7 The total mass of the rod is equal to the area under the graph of mass density ρ.

We cannot use Eq. (2) because $\rho(x)$ is not constant, but we can argue that if Δx is small, then $\rho(x)$ is *nearly constant* along the ith segment. If the ith segment extends from x_{i-1} to x_i and c_i is any sample point in $[x_{i-1}, x_i]$, then $M_i \approx \rho(c_i)\Delta x$ and

$$\text{Total mass } M = \sum_{i=1}^{N} M_i \approx \sum_{i=1}^{N} \rho(c_i)\Delta x$$

As $N \to \infty$, the accuracy of the approximation improves. However, the sum on the right is a Riemann sum whose value approaches $\int_a^b \rho(x)\,dx$, and thus it makes sense to define *the total mass of a rod as the integral of its linear mass density*:

$$\boxed{\text{Total mass } M = \int_a^b \rho(x)\,dx} \qquad \boxed{3}$$

Note the similarity in the way we use thin slices to compute volume and small pieces to compute total mass.

■ **EXAMPLE 4** **Total Mass** Find the total mass M of a 2-m rod of linear density $\rho(x) = 1 + x(2 - x)$ kg/m, where x is the distance from one end of the rod.

Solution

$$M = \int_0^2 \rho(x)\,dx = \int_0^2 \left(1 + x(2 - x)\right)dx = \left(x + x^2 - \frac{1}{3}x^3\right)\Big|_0^2 = \frac{10}{3}\text{ kg} \qquad ■$$

In general, density is a function $\rho(x, y)$ that depends not just on the distance to the origin but also on the coordinates (x, y). Total mass or population is then computed using double integration, a topic in multivariable calculus.

In some situations, density is a function of distance to the origin. For example, in the study of urban populations, it might be assumed that the population density $\rho(r)$ (in people per square kilometer) depends only on the distance r from the center of a city. Such a density function is called a **radial density function**.

We now derive a formula for the total population P within a radius R of the city center, assuming a radial density $\rho(r)$. First, divide the circle of radius R into N thin rings of equal width $\Delta r = R/N$ as in Figure 8.

Let P_i be the population within the ith ring, so that $P = \sum_{i=1}^{N} P_i$. If the outer radius of the ith ring is r_i, then the circumference is $2\pi r_i$, and if Δr is small, the area of this ring is *approximately* $2\pi r_i \Delta r$ (outer circumference times width). Furthermore, the population density within the thin ring is nearly constant with value $\rho(r_i)$. With these approximations,

$$P_i \approx \underbrace{2\pi r_i \Delta r}_{\text{Area of ring}} \times \underbrace{\rho(r_i)}_{\substack{\text{Population} \\ \text{density}}} = 2\pi r_i \rho(r_i) \Delta r$$

$$P = \sum_{i=1}^{N} P_i \approx 2\pi \sum_{i=1}^{N} r_i \rho(r_i) \Delta r$$

This last sum is a right-endpoint approximation to the integral $2\pi \int_0^R r\rho(r)\,dr$. As N tends to ∞, the approximation improves in accuracy and the sum converges to the integral. Thus, for a population with a radial density function ρ,

> Population P within a radius $R = 2\pi \int_0^R r\rho(r)\,dr$ **4**

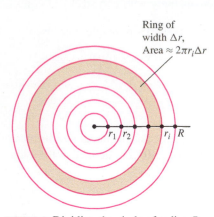

FIGURE 8 Dividing the circle of radius R into N thin rings of width $\Delta r = R/N$.

Remember that for a radial density function, the total population is obtained by integrating $2\pi r\rho(r)$ rather than $\rho(r)$.

■ **EXAMPLE 5** Computing Total Population   The population in a certain city has radial density function $\rho(r) = 15(1 + r^2)^{-1/2}$, where r is the distance from the city center in kilometers and ρ has units of thousands per square kilometer. How many people live in the ring between 10 and 30 km from the city center?

Solution   The population P (in thousands) within the ring is

$$P = 2\pi \int_{10}^{30} r\left(15(1+r^2)^{-1/2}\right) dr = 2\pi(15) \int_{10}^{30} \frac{r}{(1+r^2)^{1/2}}\,dr$$

Now use the substitution $u = 1 + r^2$, $du = 2r\,dr$. The limits of integration become $u(10) = 101$ and $u(30) = 901$:

$$P = 30\pi \int_{101}^{901} u^{-1/2} \left(\frac{1}{2}\right) du = 30\pi u^{1/2} \Big|_{101}^{901} \approx 1881 \text{ thousand}$$

In other words, the population is approximately 1.9 million people.   ■

Flow Rate

When fluid flows through a tube, the **flow rate** Q is the *volume per unit time* of fluid passing through the tube (Figure 9). The flow rate depends on the velocity of the fluid particles. If all particles of the fluid travel with the same velocity v (say, in units of cubic meters per minute), and the tube has radius R, then

$$\underbrace{\text{Flow rate } Q}_{\text{Volume per unit time}} = \text{cross-sectional area} \times \text{velocity} = \pi R^2 v \text{ cm}^3/\text{min}$$

Why is this formula true? Let's fix an observation point P in the tube and ask: Which fluid particles flow past P in a 1-min interval? A particle travels v centimeters each minute, so it flows past P during this minute if it is located not more than v centimeters to the left of P (assuming the fluid flows from left to right). Therefore, the column of fluid flowing past P in a 1-min interval is a cylinder of radius R, length v, and volume $\pi R^2 v$ (Figure 9).

In reality, the fluid particles do not all travel at the same velocity because of friction. However, for a slowly moving fluid, the flow is **laminar**, by which we mean that the

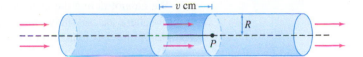

FIGURE 9 The column of fluid flowing past P in 1 unit of time is a cylinder of volume $\pi R^2 v$.

velocity $v(r)$ depends only on the distance r from the center of the tube. The particles at the center of the tube travel most quickly, and the velocity tapers off to zero near the walls of the tube (Figure 10).

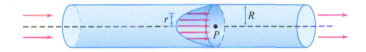

FIGURE 10 Laminar flow: Velocity of fluid increases toward the center of the tube.

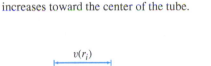

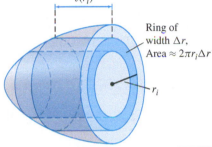

FIGURE 11 In a laminar flow, the fluid particles passing through a thin ring at distance r_i from the center all travel at nearly the same velocity $v(r_i)$.

If the flow is laminar, we can express the flow rate Q as an integral. We divide the circular cross section of the tube into N thin concentric rings of width $\Delta r = R/N$ (Figure 11). The area of the ith ring is approximately $2\pi r_i \Delta r$ and the fluid particles flowing past this ring have velocity that is nearly constant with value $v(r_i)$. Therefore, we can approximate the flow rate Q_i through the ith ring by

$$Q_i \approx \text{cross-sectional area} \times \text{velocity} \approx (2\pi r_i \Delta r) v(r_i)$$

We obtain

$$Q = \sum_{i=1}^{N} Q_i \approx 2\pi \sum_{i=1}^{N} r_i v(r_i) \Delta r$$

The sum on the right is a right-endpoint approximation to the integral $2\pi \int_0^R r v(r)\, dr$. Once again, we let N tend to ∞ to obtain the formula

$$\boxed{\text{Flow rate } Q = 2\pi \int_0^R r v(r)\, dr} \qquad \boxed{5}$$

Note the similarity of this formula and its derivation to that of population with a radial density function.

The French physician Jean Poiseuille (1799–1869) discovered the law of laminar flow that cardiologists use to study blood flow in humans. Poiseuille's Law highlights the danger of cholesterol buildup in blood vessels: The flow through a blood vessel of radius R is proportional to R^4, so if R is reduced by one-half, the flow is reduced by a factor of 16.

■ **EXAMPLE 6** **Laminar Flow** According to **Poiseuille's Law**, the velocity of blood flowing in a blood vessel of radius R centimeters is $v(r) = k(R^2 - r^2)$, where r is the distance from the center of the vessel (in centimeters) and k is a constant. Calculate the flow Q as function of R, assuming that $k = 0.5$ (cm-s)$^{-1}$.

Solution By Eq. (5),

$$Q = 2\pi \int_0^R (0.5) r (R^2 - r^2)\, dr = \pi \left(R^2 \frac{r^2}{2} - \frac{r^4}{4} \right) \Bigg|_0^R = \frac{\pi}{4} R^4 \text{ cm}^3/\text{s}$$

Note that Q is proportional to R^4 (this is true for any value of k). ■

Average Value

As a final example, we discuss the *average value* of a function. Recall that the average of N numbers $a_1, a_2, \ldots, a_N$ is the sum divided by N:

$$\frac{a_1 + a_2 + \cdots + a_N}{N} = \frac{1}{N} \sum_{j=1}^{N} a_j$$

For example, the average of 18, 25, 22, and 31 is $\frac{1}{4}(18 + 25 + 22 + 31) = 24$.

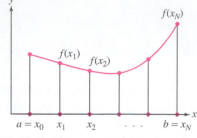

FIGURE 12 The average of the values of $f(x)$ at the points $x_1, x_2, \ldots, x_N$ is equal to $\dfrac{R_N}{b-a}$.

We cannot define the average value of a function f on an interval $[a, b]$ as a sum because there are infinitely many values of x to consider. But recall the formula for the right-endpoint approximation R_N (Figure 12):

$$R_N = \frac{b-a}{N}\big(f(x_1) + f(x_2) + \cdots + f(x_N)\big)$$

where $x_i = a + i\left(\dfrac{b-a}{N}\right)$. We see that R_N divided by $(b-a)$ is equal to the average of the equally spaced function values $f(x_i)$:

$$\frac{1}{b-a} R_N = \underbrace{\frac{f(x_1) + f(x_2) + \cdots + f(x_N)}{N}}_{\text{Average of the function values}}$$

If N is large, it is reasonable to think of this quantity as an *approximation* to the average of $f(x)$ on $[a, b]$. Therefore, we define the average value itself as the limit:

$$\text{Average value} = \lim_{N \to \infty} \frac{1}{b-a} R_N(f) = \frac{1}{b-a} \int_a^b f(x)\,dx$$

DEFINITION Average Value   The **average value** of an integrable function f on $[a, b]$ is the quantity

$$\boxed{\text{Average value} = \frac{1}{b-a} \int_a^b f(x)\,dx} \qquad \boxed{6}$$

The average value of a function is also called the **mean value**.

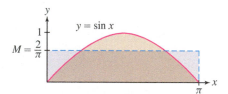

FIGURE 13 The area under the graph is equal to the area of the rectangle whose height is the average value M.

GRAPHICAL INSIGHT   Think of the average value M of a function as the average height of its graph (Figure 13). The region under the graph has the same signed area as the rectangle of height M, because $\displaystyle\int_a^b f(x)\,dx = M(b-a)$.

■ **EXAMPLE 7**   Find the average value of $f(x) = \sin x$ on $[0, \pi]$.

Solution   The average value of $f(x) = \sin x$ on $[0, \pi]$ is

$$\frac{1}{\pi} \int_0^\pi \sin x \, dx = -\frac{1}{\pi} \cos x \Big|_0^\pi = \frac{1}{\pi}\big(-(-1) - (-1)\big) = \frac{2}{\pi} \approx 0.637$$

This answer is reasonable because $\sin x$ varies from 0 to 1 on the interval $[0, \pi]$ and the average 0.637 lies somewhere between the two extremes (Figure 13). ■

■ **EXAMPLE 8   Vertical Jump of a Bushbaby**   The bushbaby (*Galago senegalensis*) is a small primate with remarkable jumping ability (Figure 14). Find the average speed during a jump if the initial vertical velocity is $v_0 = 600$ cm/s. Use Galileo's formula for the height $h(t) = v_0 t - \frac{1}{2} g t^2$ (in centimeters, where $g = 980$ cm/s^2).

Solution   The bushbaby's height is $h(t) = v_0 t - \frac{1}{2} g t^2 = t\big(v_0 - \frac{1}{2} g t\big)$. The height is zero at $t = 0$ and at $t = 2v_0/g = \frac{1200}{980} = \frac{6}{4.9}$ s, when jump ends.

The bushbaby's velocity is $h'(t) = v_0 - gt = 600 - 980t$. The velocity is negative for $t > v_0/g = \frac{6}{9.8}$, so as we see in Figure 15, the integral of speed $|h'(t)|$ is equal to the sum of the areas of two triangles of base $\frac{6}{9.8}$ and height 600:

$$\int_0^{6/4.9} |600 - 980t|\,dt = \frac{1}{2}\left(\frac{6}{9.8}\right)(600) + \frac{1}{2}\left(\frac{6}{9.8}\right)(600) = \frac{3600}{9.8}$$

FIGURE 14 A bushbaby can jump as high as 2 m (its center of mass rises more than five bodylengths). By contrast, NBA star Blake Griffin rises at most 0.6 his body length when executing a slam dunk. (© *Corbis/Alamy*)

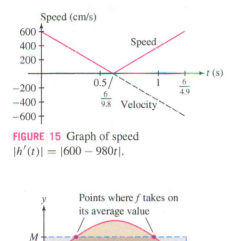

FIGURE 15 Graph of speed $|h'(t)| = |600 - 980t|$.

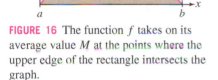

Points where f takes on its average value

FIGURE 16 The function f takes on its average value M at the points where the upper edge of the rectangle intersects the graph.

The average speed $\overline{s}$ is

$$\overline{s} = \frac{1}{\frac{6}{4.9}} \int_0^{6/4.9} |600 - 980t|\, dt = \frac{1}{\frac{6}{4.9}} \left(\frac{3600}{9.8} \right) = 300 \text{ cm/s} \qquad \blacksquare$$

There is an important difference between the average of a list of numbers and the average value of a continuous function. If the average score on an exam is 84, then 84 lies between the highest and lowest scores, but it is possible that no student received a score of 84. By contrast, the Mean Value Theorem (MVT) for Integrals asserts that a continuous function always takes on its average value somewhere in the interval (Figure 16).

For example, the average of $f(x) = \sin x$ on $[0, \pi]$ is $2/\pi$ by Example 7. We have $f(c) = 2/\pi$ for $c = \sin^{-1}(2/\pi) \approx 0.69$. Since 0.69 lies in $[0, \pi]$, $f(x) = \sin x$ indeed takes on its average value at a point in the interval.

THEOREM 1 Mean Value Theorem for Integrals If f is continuous on $[a, b]$, then there exists a value $c \in [a, b]$ such that

$$f(c) = \frac{1}{b - a} \int_a^b f(x)\, dx$$

Proof Let $M = \dfrac{1}{b - a} \displaystyle\int_a^b f(x)\, dx$ be the average value. Because f is continuous, we can apply Theorem 1 of Section 4.2 to conclude that f takes on a minimum value $m_{\min}$ and a maximum value $M_{\max}$ on the closed interval $[a, b]$. Furthermore, by Eq. (8) of Section 5.2,

$$m_{\min}(b - a) \le \int_a^b f(x)\, dx \le M_{\max}(b - a)$$

Dividing by $(b - a)$, we find

$$m_{\min} \le M \le M_{\max}$$

In other words, the average value M lies between $m_{\min}$ and $M_{\max}$. The Intermediate Value Theorem guarantees that $f(x)$ takes on every value between its min and max, so $f(c) = M$ for some c in $[a, b]$. $\qquad \blacksquare$

6.2 SUMMARY

- Formulas:

Volume	$V = \displaystyle\int_a^b A(y)\, dy,$	$A(y) = $ cross-sectional area
Total Mass	$M = \displaystyle\int_a^b \rho(x)\, dx,$	$\rho(x) = $ linear mass density
Total Population	$P = 2\pi \displaystyle\int_0^R r\rho(r)\, dr,$	$\rho(r) = $ radial density
Laminar Flow Rate	$Q = 2\pi \displaystyle\int_0^R rv(r)\, dr,$	$v(r) = $ velocity at radius r
Average value	$M = \dfrac{1}{b - a} \displaystyle\int_a^b f(x)\, dx,$	$f = $ any continuous function

- The MVT for Integrals: If f is continuous on $[a, b]$ with average (or mean) value M, then $f(c) = M$ for some $c \in [a, b]$.

6.2 EXERCISES

Preliminary Questions

1. What is the average value of f on $[0, 4]$ if the area between the graph of f and the x-axis is equal to 12?

2. Find the volume of a solid extending from $y = 2$ to $y = 5$ if every cross section has area $A(y) = 5$.

3. What is the definition of flow rate?

4. Which assumption about fluid velocity did we use to compute the flow rate as an integral?

5. The average value of f on $[1, 4]$ is 5. Find $\int_1^4 f(x)\,dx$.

Exercises

1. Let V be the volume of a pyramid of height 20 whose base is a square of side 8.

(a) Use similar triangles as in Example 1 to find the area of the horizontal cross section at a height y.

(b) Calculate V by integrating the cross-sectional area.

2. Let V be the volume of a right circular cone of height 10 whose base is a circle of radius 4 [Figure 17(A)].

(a) Use similar triangles to find the area of a horizontal cross section at a height y.

(b) Calculate V by integrating the cross-sectional area.

3. Use the method of Exercise 2 to find the formula for the volume of a right circular cone of height h whose base is a circle of radius R [Figure 17(B)].

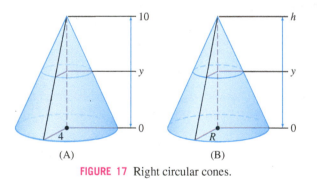

(A)   (B)

FIGURE 17 Right circular cones.

4. Calculate the volume of the ramp in Figure 18 in three ways by integrating the area of the cross sections:

(a) Perpendicular to the x-axis (rectangles)

(b) Perpendicular to the y-axis (triangles)

(c) Perpendicular to the z-axis (rectangles)

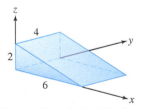

FIGURE 18 Ramp of length 6, width 4, and height 2.

5. Find the volume of liquid needed to fill a sphere of radius R to height h (Figure 19).

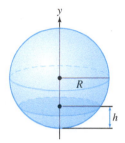

FIGURE 19 Sphere filled with liquid to height h.

6. Find the volume of the wedge in Figure 20(A) by integrating the area of vertical cross sections.

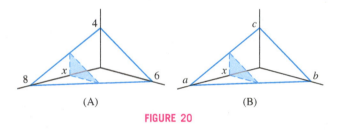

(A)   (B)

FIGURE 20

7. Derive a formula for the volume of the wedge in Figure 20(B) in terms of the constants a, b, and c.

8. Let B be the solid whose base is the unit circle $x^2 + y^2 = 1$ and whose vertical cross sections perpendicular to the x-axis are equilateral triangles. Show that the vertical cross sections have area $A(x) = \sqrt{3}(1 - x^2)$ and compute the volume of B.

In Exercises 9–14, find the volume of the solid with the given base and cross sections.

9. The base is the unit circle $x^2 + y^2 = 1$, and the cross sections perpendicular to the x-axis are triangles whose height and base are equal.

10. The base is the triangle enclosed by $x + y = 1$, the x-axis, and the y-axis. The cross sections perpendicular to the y-axis are semicircles.

11. The base is the semicircle $y = \sqrt{9 - x^2}$, where $-3 \le x \le 3$. The cross sections perpendicular to the x-axis are squares.

12. The base is a square, one of whose sides is the interval $[0, \ell]$ along the x-axis. The cross sections perpendicular to the x-axis are rectangles of height $f(x) = x^2$.

13. The base is the region enclosed by $y = x^2$ and $y = 3$. The cross sections perpendicular to the y-axis are squares.

14. The base is the region enclosed by $y = x^2$ and $y = 3$. The cross sections perpendicular to the y-axis are rectangles of height y^3.

15. Find the volume of the solid whose base is the region $|x| + |y| \le 1$ and whose vertical cross sections perpendicular to the y-axis are semicircles (with diameter along the base).

16. Show that a pyramid of height h whose base is an equilateral triangle of side s has volume $\frac{\sqrt{3}}{12}hs^2$.

17. The area of an ellipse is πab, where a and b are the lengths of the semimajor and semiminor axes (Figure 21). Compute the volume of a cone of height 12 whose base is an ellipse with semimajor axis $a = 6$ and semiminor axis $b = 4$.

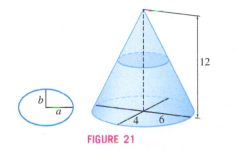

FIGURE 21

18. Find the volume V of a *regular* tetrahedron (Figure 22) whose face is an equilateral triangle of side s. The tetrahedron has height $h = \sqrt{2/3}s$.

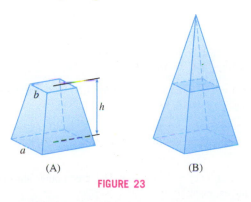

FIGURE 22 Regular tetrahedron.

19. A frustum of a pyramid is a pyramid with its top cut off [Figure 23(A)]. Let V be the volume of a frustum of height h whose base is a square of side a and whose top is a square of side b with $a > b \ge 0$.

(a) Show that if the frustum were continued to a full pyramid, it would have height $ha/(a - b)$ [Figure 23(B)].

(b) Show that the cross section at height x is a square of side $(1/h)(a(h - x) + bx)$.

(c) Show that $V = \frac{1}{3}h(a^2 + ab + b^2)$. A papyrus dating to the year 1850 BCE indicates that Egyptian mathematicians had discovered this formula almost 4000 years ago.

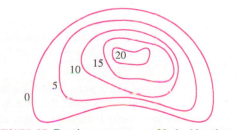

(A)   **(B)**

FIGURE 23

20. A plane inclined at an angle of $45°$ passes through a diameter of the base of a cylinder of radius r. Find the volume of the region within the cylinder and below the plane (Figure 24).

FIGURE 24

21. The solid S in Figure 25 is the intersection of two cylinders of radius r whose axes are perpendicular.

(a) The horizontal cross section of each cylinder at distance y from the central axis is a rectangular strip. Find the strip's width.

(b) Find the area of the horizontal cross section of S at distance y.

(c) Find the volume of S as a function of r.

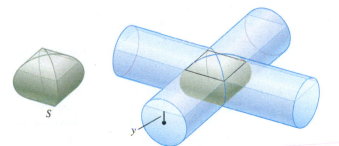

FIGURE 25 Two cylinders intersecting at right angles.

22. Let S be the intersection of two cylinders of radius r whose axes intersect at an angle θ. Find the volume of S as a function of r and θ.

23. Calculate the volume of a cylinder inclined at an angle $\theta = 30°$ with height 10 and base of radius 4 (Figure 26).

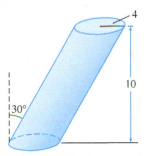

FIGURE 26 Cylinder inclined at an angle $\theta = 30°$.

24. The areas of cross sections of Lake Nogebow at 5-m intervals are given in the table below. Figure 27 shows a contour map of the lake. Estimate the volume V of the lake by taking the average of the right- and left-endpoint approximations to the integral of cross-sectional area.

Depth (m)	0	5	10	15	20
Area (million m²)	2.1	1.5	1.1	0.835	0.217

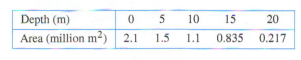

FIGURE 27 Depth contour map of Lake Nogebow.

25. Find the total mass of a 1-m rod whose linear density function is $\rho(x) = 10(x + 1)^{-2}$ kg/m for $0 \le x \le 1$.

26. Find the total mass of a 2-m rod whose linear density function is $\rho(x) = 1 + 0.5\sin(\pi x)$ kg/m for $0 \le x \le 2$.

27. A mineral deposit along a strip of length 6 cm has density $s(x) = 0.01x(6 - x)$ g/cm for $0 \le x \le 6$. Calculate the total mass of the deposit.

28. Charge is distributed along a glass tube of length 10 cm with linear charge density $\rho(x) = x(x^2 + 1)^{-2} \times 10^{-4}$ coulombs per centimeter (C/cm) for $0 \le x \le 10$. Calculate the total charge.

29. Calculate the population within a 10-mile radius of the city center if the radial population density is $\rho(r) = 4(1 + r^2)^{1/3}$ (in thousands per square mile).

30. Odzala National Park in the Republic of the Congo has a high density of gorillas. Suppose that the radial population density is $\rho(r) = 52(1 + r^2)^{-2}$ gorillas per square kilometer, where r is the distance from a grassy clearing with a source of water. Calculate the number of gorillas within a 5-km radius of the clearing.

31. Table 1 lists the population density (in people per square kilometer) as a function of distance r (in kilometers) from the center of a rural town. Estimate the total population within a 1.2-km radius of the center by taking the average of the left- and right-endpoint approximations.

TABLE 1   Population Density

r	$\rho(r)$	r	$\rho(r)$
0.0	125.0	0.8	56.2
0.2	102.3	1.0	46.0
0.4	83.8	1.2	37.6
0.6	68.6		

32. Find the total mass of a circular plate of radius 20 cm whose mass density is the radial function $\rho(r) = 0.03 + 0.01\cos(\pi r^2)$ g/cm².

33. The density of deer in a forest is the radial function $\rho(r) = 150(r^2 + 2)^{-2}$ deer per square kilometer, where r is the distance (in kilometers) to a small meadow. Calculate the number of deer in the region $2 \le r \le 5$ km.

34. Show that a circular plate of radius 2 cm with radial mass density $\rho(r) = \frac{4}{r}$ g/cm² has finite total mass, even though the density becomes infinite at the origin.

35. Find the flow rate through a tube of radius 4 cm, assuming that the velocity of fluid particles at a distance r centimeters from the center is $v(r) = (16 - r^2)$ cm/s.

36. The velocity of fluid particles flowing through a tube of radius 5 cm is $v(r) = (10 - 0.3r - 0.34r^2)$ cm/s, where r centimeters is the distance from the center. What quantity per second of fluid flows through the portion of the tube where $0 \le r \le 2$?

37. A solid rod of radius 1 cm is placed in a pipe of radius 3 cm so that their axes are aligned. Water flows through the pipe and around the rod. Find the flow rate if the velocity of the water is given by the radial function $v(r) = 0.5(r - 1)(3 - r)$ cm/s.

38. Let $v(r)$ be the velocity of blood in an arterial capillary of radius $R = 4 \times 10^{-5}$ m. Use Poiseuille's Law (Example 6) with $k = 10^6$ (m-s)$^{-1}$ to determine the velocity at the center of the capillary and the flow rate (use correct units).

In Exercises 39–48, calculate the average over the given interval.

39. $f(x) = x^3$, $[0, 4]$

40. $f(x) = x^3$, $[-1, 1]$

41. $f(x) = \cos x$, $\left[0, \frac{\pi}{6}\right]$

42. $f(x) = \sec^2 x$, $\left[\frac{\pi}{6}, \frac{\pi}{3}\right]$

43. $f(s) = s^{-2}$, $[2, 5]$

44. $f(x) = \dfrac{\sin(\pi/x)}{x^2}$, $[1, 2]$

45. $f(x) = 2x^3 - 6x^2$, $[-1, 3]$

46. $f(x) = \dfrac{1}{x^2 + 1}$, $[-1, 1]$

47. $f(x) = x^n$ for $n \ge 0$, $[0, 1]$

48. $f(x) = e^{-nx}$, $[-1, 1]$

49. The temperature (in degrees Celsius) at time t (in hours) in an art museum varies according to $T(t) = 20 + 5\cos\left(\frac{\pi}{12}t\right)$. Find the average over the time periods $[0, 24]$ and $[2, 6]$.

50. A steel bar of length 3 m experiences extreme heat at its center, so that the temperature at coordinate x on the bar is given by $T(x) = 40\sin\left(\frac{\pi x}{3}\right) + 50°$C where the bar sits along the interval $[0, 3]$ on the x-axis. Determine the average temperature of the bar.

51. Temperature in the town of Walla Walla during the month of July follows a pattern given by $T(t) = 10\sin\left(\frac{t\pi}{31}\right) + 14\sin\left(\frac{t\pi}{2}\right) + 73°$F. Here, t is measured in days, and there are 31 days in July. Explain why you might see a pattern like this and compute the average temperature during the month of July.

52. The door to the garage is left open and over the next 4 hours (h), the temperature in a house in degrees Celsius is given by $T(t) = 20e^{-t/4}$. Determine the average temperature over those 4 h.

53. A 10-cm copper wire with one end in an ice bath is heated at the other end, so that the temperature at each point x along the wire (in degrees Celsius) is given by $T(x) = 50\cos\frac{\pi x}{20}$. Find the average temperature over the wire.

54. A ball thrown in the air vertically from ground level with initial velocity 18 m/s has height $h(t) = 18t - 9.8t^2$ at time t (in seconds). Find the average height and the average speed over the time interval extending from the ball's release to its return to ground level.

55. Find the average speed over the time interval $[1, 5]$ (time in seconds) of a particle whose position at time t is $s(t) = t^3 - 6t^2$ m.

56. An object with zero initial velocity accelerates at a constant rate of 10 m/s². Find its average velocity during the first 15 s.

57. The acceleration of a particle is $a(t) = 60t - 4t^3$ m/s². Compute the average acceleration and the average speed over the time interval $[2, 6]$, assuming that the particle's initial velocity is zero.

58. What is the average area of the circles whose radii vary from 0 to R?

59. Let M be the average value of $f(x) = x^4$ on $[0, 3]$. Find a value of c in $[0, 3]$ such that $f(c) = M$.

60. Let $f(x) = \sqrt{x}$. Find a value of c in $[4, 9]$ such that $f(c)$ is equal to the average of f on $[4, 9]$.

61. Let M be the average value of $f(x) = x^3$ on $[0, A]$, where $A > 0$. Which theorem guarantees that $f(c) = M$ has a solution c in $[0, A]$? Find c.

62. CAS Let $f(x) = 2\sin x - x$. Use a computer algebra system to plot f and estimate:

(a) the positive root α of f.

(b) the average value M of f on $[0, \alpha]$.

(c) a value $c \in [0, \alpha]$ such that $f(c) = M$.

63. Which of $f(x) = x \sin^2 x$ and $g(x) = x^2 \sin^2 x$ has a larger average value over $[0, 1]$? Over $[1, 2]$?

64. Find the average of $f(x) = ax + b$ over the interval $[-M, M]$, where a, b, and M are arbitrary constants.

65. Sketch the graph of a function f such that $f(x) \geq 0$ on $[0, 1]$ and $f(x) \leq 0$ on $[1, 2]$, whose average on $[0, 2]$ is negative.

66. Give an example of a function (necessarily discontinuous) that does not satisfy the conclusion of the MVT for Integrals.

Further Insights and Challenges

67. An object is tossed into the air vertically from ground level with initial velocity v_0 ft/s at time $t = 0$. Find the average speed of the object over the time interval $[0, T]$, where T is the time the object returns to Earth.

68. Review the MVT stated in Section 4.3 (Theorem 1, p. 226) and show how it can be used, together with the Fundamental Theorem of Calculus, to prove the MVT for Integrals.

6.3 Volumes of Revolution

We use the terms "revolve" and "rotate" interchangeably.

A **solid of revolution** is a solid obtained by rotating a region in the plane about an axis. The sphere and right circular cone are familiar examples of such solids. Each of these is "swept out" as a plane region revolves around an axis (Figure 1).

DF **FIGURE 1** The right circular cone and the sphere are solids of revolution.

This method for computing the volume is referred to as the disk method because the vertical slices of the solid are circular disks.

Suppose that $f(x) \geq 0$ for $a \leq x \leq b$. The solid obtained by rotating the region under the graph about the x-axis has a special feature: All vertical cross sections are circles (Figure 2). In fact, the vertical cross section at location x is a circle of radius $R = f(x)$ and thus

$$\text{Area of the vertical cross section} = \pi R^2 = \pi f(x)^2$$

We know from Section 6.2 that the total volume V is equal to the integral of cross-sectional area. Therefore, $V = \int_a^b \pi f(x)^2 \, dx$.

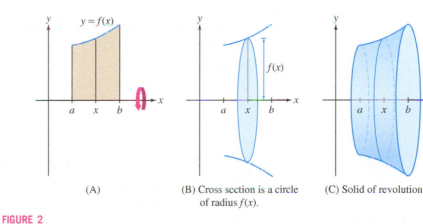

(A)

(B) Cross section is a circle of radius $f(x)$.

(C) Solid of revolution

FIGURE 2

The cross sections of a solid of revolution are circles of radius $R = f(x)$ and area $\pi R^2 = \pi f(x)^2$. The volume, given by Eq. (1), is the integral of cross-sectional area.

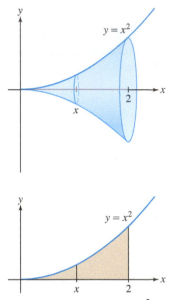

Volume of Revolution: Disk Method   If f is continuous and $f(x) \geq 0$ on $[a, b]$, then the solid obtained by rotating the region under the graph about the x-axis has volume [with $R = f(x)$]

$$V = \pi \int_a^b R^2 \, dx = \pi \int_a^b f(x)^2 \, dx \qquad \boxed{1}$$

■ **EXAMPLE 1**   Calculate the volume V of the solid obtained by rotating the region under $y = x^2$ about the x-axis for $0 \leq x \leq 2$.

Solution   The solid is shown in Figure 3. By Eq. (1) with $f(x) = x^2$, its volume is

$$V = \pi \int_0^2 R^2 \, dx = \pi \int_0^2 (x^2)^2 \, dx = \pi \int_0^2 x^4 \, dx = \pi \left. \frac{x^5}{5} \right|_0^2 = \pi \frac{2^5}{5} = \frac{32}{5} \pi \qquad ■$$

There are some useful variations on the formula for a volume of revolution. First, consider the region *between* two curves $y = f(x)$ and $y = g(x)$, where $f(x) \geq g(x) \geq 0$ as in Figure 5(A). When this region is rotated about the x-axis, segment $\overline{AB}$ sweeps out the **washer** shown in Figure 5(B). The inner and outer radii of this washer (also called an annulus; see Figure 4) are

$$R_{\text{outer}} = f(x), \qquad R_{\text{inner}} = g(x)$$

The washer has area $\pi R_{\text{outer}}^2 - \pi R_{\text{inner}}^2$ or $\pi(f(x)^2 - g(x)^2)$, and the volume of the solid of revolution [Figure 5(C)] is the integral of this cross-sectional area:

$$V = \pi \int_a^b \left(R_{\text{outer}}^2 - R_{\text{inner}}^2 \right) dx = \pi \int_a^b \left(f(x)^2 - g(x)^2 \right) dx \qquad \boxed{2}$$

Keep in mind that the integrand is $(f(x)^2 - g(x)^2)$, not $(f(x) - g(x))^2$.

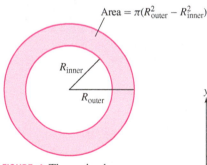

DF FIGURE 3 Region under $y = x^2$ rotated about the x-axis.

Area $= \pi(R_{\text{outer}}^2 - R_{\text{inner}}^2)$

FIGURE 4 The region between two concentric circles is called an annulus, or more informally, a washer.

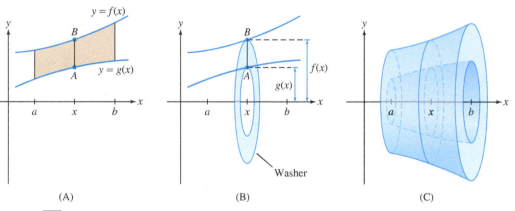

(A)   (B)   (C)

FIGURE 5 $\overline{AB}$ generates a washer when rotated about the x-axis.

CAUTION When using the washer method, make sure you use $(f(x)^2 - g(x)^2)$ in the integrand, not $(f(x) - g(x))^2$.

■ **EXAMPLE 2**   **Region Between Two Curves**   Find the volume V obtained by revolving the region between $y = x^2 + 4$ and $y = 2$ about the x-axis for $1 \leq x \leq 3$.

Solution   The graph of $y = x^2 + 4$ lies above the graph of $y = 2$ (Figure 6). Therefore, $R_{\text{outer}} = x^2 + 4$ and $R_{\text{inner}} = 2$. By Eq. (2),

$$V = \pi \int_1^3 \left(R_{\text{outer}}^2 - R_{\text{inner}}^2 \right) dx = \pi \int_1^3 \left((x^2 + 4)^2 - 2^2 \right) dx$$

$$= \pi \int_1^3 \left(x^4 + 8x^2 + 12 \right) dx = \pi \left(\frac{1}{5}x^5 + \frac{8}{3}x^3 + 12x \right) \Big|_1^3 = \frac{2126}{15} \pi \qquad ■$$

In the next example, we calculate a volume of revolution about a horizontal axis parallel to the x-axis.

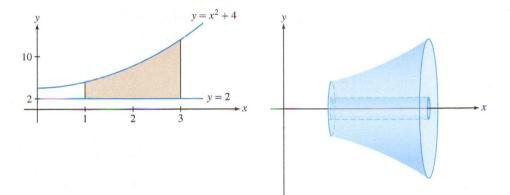

DF FIGURE 6 The area between $y = x^2 + 4$ and $y = 2$ over [1, 3] rotated about the x-axis.

■ **EXAMPLE 3** **Revolving About a Horizontal Axis** Find the volume V of the "wedding band" [Figure 7(C)] obtained by rotating the region between the graphs of $f(x) = x^2 + 2$ and $g(x) = 4 - x^2$ about the horizontal line $y = -3$.

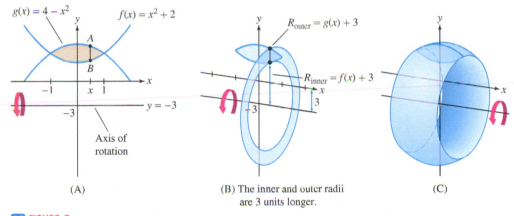

(A)   (B) The inner and outer radii are 3 units longer.   (C)

DF FIGURE 7

When you set up the integral for a volume of revolution, visualize the cross sections. These cross sections are washers (or disks) whose inner and outer radii depend on the axis of rotation.

Solution First, let's find the points of intersection of the two graphs by solving

$$f(x) = g(x) \quad \Rightarrow \quad x^2 + 2 = 4 - x^2 \quad \Rightarrow \quad x^2 = 1 \quad \Rightarrow \quad x = \pm 1$$

Figure 7(A) shows that $g(x) \geq f(x)$ for $-1 \leq x \leq 1$.

If we wanted to revolve about the x-axis, we would use Eq. (2). Since we want to revolve around $y = -3$, we must determine how the radii are affected. Figure 7(B) shows that when we rotate about $y = -3$, $\overline{AB}$ generates a washer whose outer and inner radii are both 3 units longer:

- $R_{\text{outer}} = g(x) - (-3) = (4 - x^2) + 3 = 7 - x^2$
- $R_{\text{inner}} = f(x) - (-3) = (x^2 + 2) + 3 = x^2 + 5$

The volume of revolution is equal to the integral of the area of this washer:

We get R_{outer} by subtracting $y = -3$ from $y = g(x)$ because vertical distance is the difference of the y-coordinates. Similarly, we subtract -3 from $f(x)$ to get R_{inner}.

$$V \text{ (about } y = -3) = \pi \int_{-1}^{1} \left(R_{\text{outer}}^2 - R_{\text{inner}}^2 \right) dx = \pi \int_{-1}^{1} \left((g(x) + 3)^2 - (f(x) + 3)^2 \right) dx$$

$$= \pi \int_{-1}^{1} \left((7 - x^2)^2 - (x^2 + 5)^2 \right) dx$$

$$= \pi \int_{-1}^{1} \left((49 - 14x^2 + x^4) - (x^4 + 10x^2 + 25) \right) dx$$

$$= \pi \int_{-1}^{1} (24 - 24x^2) \, dx = \pi (24x - 8x^3) \Big|_{-1}^{1} = 32\pi \quad ■$$

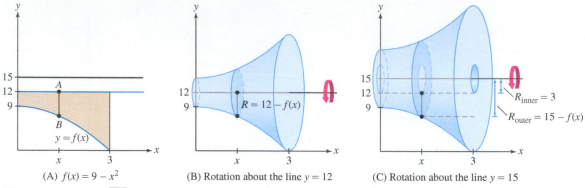

(A) $f(x) = 9 - x^2$ (B) Rotation about the line $y = 12$ (C) Rotation about the line $y = 15$

DF FIGURE 8 Segment $\overline{AB}$ generates a disk when rotated about $y = 12$, but it generates a washer when rotated about $y = 15$.

■ **EXAMPLE 4** Find the volume obtained by rotating the graphs of $f(x) = 9 - x^2$ and $y = 12$ for $0 \le x \le 3$ about:

(a) the line $y = 12$.   **(b)** the line $y = 15$.

Solution To set up the integrals, let's visualize the cross section. Is it a disk or a washer?

In Figure 8, the length of $\overline{AB}$ is $12 - f(x)$ rather than $f(x) - 12$ because the line $y = 12$ lies above the graph of f.

(a) Figure 8(B) shows that $\overline{AB}$ rotated about $y = 12$ generates a *disk* of radius

$$R = \text{length of } \overline{AB} = 12 - f(x) = 12 - (9 - x^2) = 3 + x^2$$

The volume when we rotate about $y = 12$ is

$$V = \pi \int_0^3 R^2 \, dx = \pi \int_0^3 (3 + x^2)^2 \, dx = \pi \int_0^3 (9 + 6x^2 + x^4) \, dx$$

$$= \pi \left(9x + 2x^3 + \frac{1}{5}x^5 \right) \Big|_0^3 = \frac{648}{5} \pi$$

(b) Figure 8(C) shows that $\overline{AB}$ rotated about $y = 15$ generates a *washer*. The outer radius of this washer is the distance from B to the line $y = 15$:

$$R_{\text{outer}} = 15 - f(x) = 15 - (9 - x^2) = 6 + x^2$$

The inner radius is $R_{\text{inner}} = 3$, so the volume of revolution about $y = 15$ is

$$V = \pi \int_0^3 \left(R_{\text{outer}}^2 - R_{\text{inner}}^2 \right) dx = \pi \int_0^3 \left((6 + x^2)^2 - 3^2 \right) dx$$

$$= \pi \int_0^3 (36 + 12x^2 + x^4 - 9) \, dx$$

$$= \pi \left(27x + 4x^3 + \frac{1}{5}x^5 \right) \Big|_0^3 = \frac{1188}{5} \pi \qquad ■$$

We can use the disk and washer methods for solids of revolution about vertical axes, but it is necessary to describe the graph as a function of y—that is, $x = g(y)$.

■ **EXAMPLE 5** **Revolving About a Vertical Axis** Find the volume of the solid obtained by rotating the region under the graph of $f(x) = 9 - x^2$ for $0 \le x \le 3$ about the vertical axis $x = -2$.

Solution Figure 9 shows that $\overline{AB}$ sweeps out a horizontal washer when rotated about the vertical line $x = -2$. We are going to integrate with respect to y, so we need the inner and outer radii of this washer as functions of y. Solving for x in $y = 9 - x^2$, we obtain $x^2 = 9 - y$, or $x = \sqrt{9 - y}$ (since $x \ge 0$). Therefore,

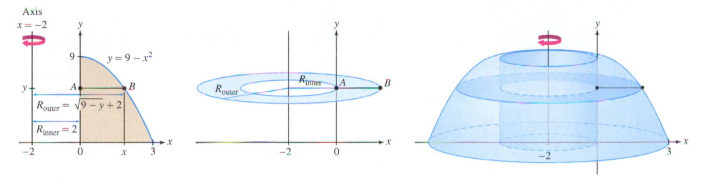

Axis
$x = -2$

DF **FIGURE 9**

$$R_{\text{outer}} = \sqrt{9 - y} + 2, \qquad R_{\text{inner}} = 2$$

$$R_{\text{outer}}^2 - R_{\text{inner}}^2 = \left(\sqrt{9 - y} + 2\right)^2 - 2^2 = (9 - y) + 4\sqrt{9 - y} + 4 - 4$$

$$= 9 - y + 4\sqrt{9 - y}$$

The region extends from $y = 0$ to $y = 9$ along the y-axis, so

$$V = \pi \int_0^9 \left(R_{\text{outer}}^2 - R_{\text{inner}}^2\right) dy = \pi \int_0^9 \left(9 - y + 4\sqrt{9 - y}\right) dy$$

$$= \pi \left(9y - \frac{1}{2}y^2 - \frac{8}{3}(9 - y)^{3/2}\right)\Big|_0^9 = \frac{225}{2}\pi \qquad \blacksquare$$

6.3 SUMMARY

- Disk method: When you rotate the region between two graphs about an axis, the segments *perpendicular* to the axis generate disks or washers. The volume V of the solid of revolution is the integral of the areas of these disks or washers.
- Sketch the graphs to visualize the disks or washers.
- Figure 10(A): Region between $y = f(x)$ and the x-axis, rotated about the x-axis.

    - Vertical cross section: a circle of radius $R = f(x)$ and area $\pi R^2 = \pi f(x)^2$:

$$V = \pi \int_a^b R^2 \, dx = \pi \int_a^b f(x)^2 \, dx$$

- Figure 10(B): Region between $y = f(x)$ and $y = g(x)$, rotated about the x-axis.

    - Vertical cross section: a washer of outer radius $R_{\text{outer}} = f(x)$ and inner radius $R_{\text{inner}} = g(x)$:

$$V = \pi \int_a^b \left(R_{\text{outer}}^2 - R_{\text{inner}}^2\right) dx = \pi \int_a^b \left(f(x)^2 - g(x)^2\right) dx$$

- To rotate about a horizontal line $y = c$, modify the radii appropriately.

    - Figure 10(C): $c \geq f(x) \geq g(x)$:

$$R_{\text{outer}} = c - g(x), \qquad R_{\text{inner}} = c - f(x)$$

    - Figure 10(D): $f(x) \geq g(x) \geq c$:

$$R_{\text{outer}} = f(x) - c, \qquad R_{\text{inner}} = g(x) - c$$

- To rotate about a vertical line $x = c$, express R_{outer} and R_{inner} as functions of y and integrate along the y axis.

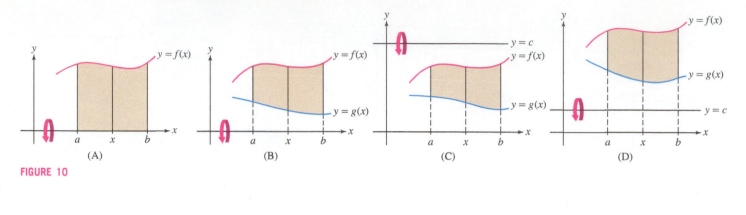

FIGURE 10

6.3 EXERCISES

Preliminary Questions

1. Which of the following is a solid of revolution?

(a) Sphere **(b)** Pyramid **(c)** Cylinder **(d)** Cube

2. True or false? When the region under a single graph is rotated about the x-axis, the cross sections of the solid perpendicular to the x-axis are circular disks.

3. True or false? When the region between two graphs is rotated about the x-axis, the cross sections of the solid perpendicular to the x-axis are circular disks.

4. Which of the following integrals expresses the volume obtained by rotating the area between $y = f(x)$ and $y = g(x)$ over $[a, b]$ around the x-axis? [Assume $f(x) \geq g(x) \geq 0$.]

(a) $\pi \displaystyle\int_a^b \left(f(x) - g(x)\right)^2 dx$

(b) $\pi \displaystyle\int_a^b \left(f(x)^2 - g(x)^2\right) dx$

Exercises

In Exercises 1–4, (a) sketch the solid obtained by revolving the region under the graph of f about the x-axis over the given interval, (b) describe the cross section perpendicular to the x-axis located at x, and (c) calculate the volume of the solid.

1. $f(x) = x + 1$, $[0, 3]$

2. $f(x) = x^2$, $[1, 3]$

3. $f(x) = \sqrt{x + 1}$, $[1, 4]$

4. $f(x) = x^{-1}$, $[1, 4]$

In Exercises 5–12, find the volume of revolution about the x-axis for the given function and interval.

5. $f(x) = 3x - x^2$, $[0, 3]$

6. $f(x) = \dfrac{1}{x^2}$, $[1, 4]$

7. $f(x) = x^{5/3}$, $[1, 8]$

8. $f(x) = 4 - x^2$, $[0, 2]$

9. $f(x) = \dfrac{2}{x + 1}$, $[1, 3]$

10. $f(x) = \sqrt{x^4 + 1}$, $[1, 3]$

11. $f(x) = e^x$, $[0, 1]$

12. $f(x) = \sqrt{\cos x \sin x}$, $\left[0, \frac{\pi}{2}\right]$

In Exercises 13 and 14, R is the shaded region in Figure 11.

13. Which of the integrands (i)–(iv) is used to compute the volume obtained by rotating region R about $y = -2$?

    (i) $(f(x)^2 + 2^2) - (g(x)^2 + 2^2)$

    (ii) $(f(x) + 2)^2 - (g(x) + 2)^2$

    (iii) $(f(x)^2 - 2^2) - (g(x)^2 - 2^2)$

    (iv) $(f(x) - 2)^2 - (g(x) - 2)^2$

14. Which of the integrands (i)–(iv) is used to compute the volume obtained by rotating R about $y = 9$ in Figure 11?

    (i) $(9 + f(x))^2 - (9 + g(x))^2$

    (ii) $(9 + g(x))^2 - (9 + f(x))^2$

    (iii) $(9 - f(x))^2 - (9 - g(x))^2$

    (iv) $(9 - g(x))^2 - (9 - f(x))^2$

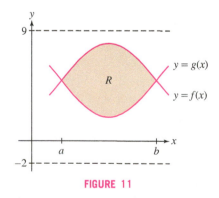

FIGURE 11

In Exercises 15–20, (a) sketch the region enclosed by the curves, (b) describe the cross section perpendicular to the x-axis located at x, and (c) find the volume of the solid obtained by rotating the region about the x-axis.

15. $y = x^2 + 2$, $y = 10 - x^2$

16. $y = x^2$, $y = 2x + 3$

17. $y = 16 - x$, $y = 3x + 12$, $x = -1$

18. $y = \dfrac{1}{x}$, $y = \dfrac{5}{2} - x$

19. $y = \sec x$, $y = 0$, $x = -\dfrac{\pi}{4}$, $x = \dfrac{\pi}{4}$

20. $y = \sec x$, $y = 0$, $x = 0$, $x = \dfrac{\pi}{4}$

In Exercises 21–24, find the volume of the solid obtained by rotating the region enclosed by the graphs about the y-axis over the given interval.

21. $x = \sqrt{y}$, $x = 0$; $1 \le y \le 4$

22. $x = \sqrt{\sin y}$, $x = 0$; $0 \le y \le \pi$

23. $x = y^2$, $x = \sqrt{y}$

24. $x = 4 - y$, $x = 16 - y^2$

25. Rotation of the region in Figure 12 about the y-axis produces a solid with two types of different cross sections. Compute the volume as a sum of two integrals, one for $-12 \le y \le 4$ and one for $4 \le y \le 12$.

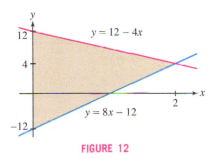

FIGURE 12

26. Let R be the region enclosed by $y = x^2 + 2$, $y = (x - 2)^2$ and the axes $x = 0$ and $y = 0$. Compute the volume V obtained by rotating R about the x-axis. *Hint: Express V as a sum of two integrals.*

In Exercises 27–32, find the volume of the solid obtained by rotating region A in Figure 13 about the given axis.

27. x-axis   **28.** $y = -2$ **29.** $y = 2$

30. y-axis   **31.** $x = -3$ **32.** $x = 2$

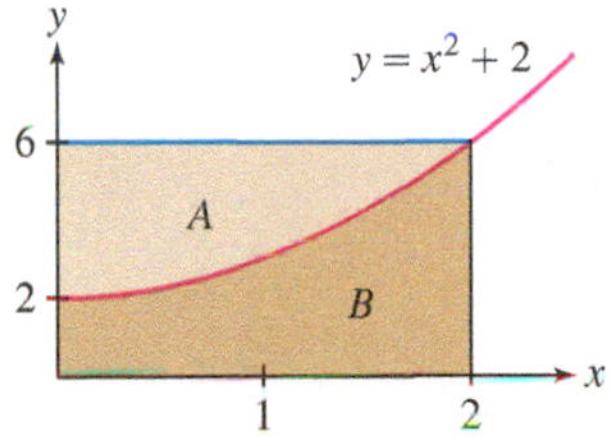

FIGURE 13

In Exercises 33–38, find the volume of the solid obtained by rotating region B in Figure 13 about the given axis.

33. x-axis   **34.** $y = -2$

35. $y = 6$ **36.** y-axis

Hint for Exercise 36: Express the volume as a sum of two integrals along the y-axis or use Exercise 30.

37. $x = 2$ **38.** $x = -3$

In Exercises 39–52, find the volume of the solid obtained by rotating the region enclosed by the graphs about the given axis.

39. $y = x^2$, $y = 12 - x$, $x = 0$, about $y = -2$ $x \ge 0$

40. $y = x^2$, $y = 12 - x$, $x = 0$, about $y = 15$

41. $y = 16 - 2x$, $y = 6$, $x = 0$, about x-axis

42. $y = 32 - 2x$, $y = 2 + 4x$, $x = 0$, about y-axis

43. $y = \sec x$, $y = 1 + \dfrac{3}{\pi}x$, about x-axis

44. $x = 2$, $x = 3$, $y = 16 - x^4$, $y = 0$, about y-axis

45. $y = 2\sqrt{x}$, $y = x$, about $x = -2$

46. $y = 2\sqrt{x}$, $y = x$, about $y = 4$

47. $y = x^3$, $y = x^{1/3}$, for $x \ge 0$, about y-axis

48. $y = x^2$, $y = x^{1/2}$, about $x = -2$

49. $y = \dfrac{9}{x^2}$, $y = 10 - x^2$, $x \ge 0$, about $y = 12$

50. $y = \dfrac{9}{x^2}$, $y = 10 - x^2$, $x \ge 0$, about $x = -1$

51. $y = e^{-x}$, $y = 1 - e^{-x}$, $x = 0$, about $y = 4$

52. $y = \cosh x$, $x = \pm 2$, about x-axis

53. The bowl in Figure 14(A) is 21 cm high, obtained by rotating the curve in Figure 14(B) as indicated. Estimate the volume capacity of the bowl shown by taking the average of right- and left-endpoint approximations to the integral with $N = 7$. The inner radii (in centimeters) starting from the top are 0, 4, 7, 8, 10, 13, 14, 20.

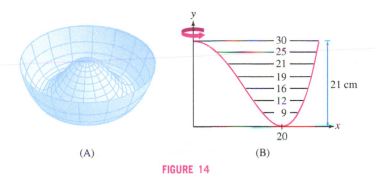

(A)   (B)

FIGURE 14

54. The region between the graphs of f and g over $[0, 1]$ is revolved about the line $y = -3$. Use the midpoint approximation with values from the following table to estimate the volume V of the resulting solid:

x	0.1	0.3	0.5	0.7	0.9
$f(x)$	8	7	6	7	8
$g(x)$	2	3.5	4	3.5	2

55. Find the volume of the cone obtained by rotating the region under the segment joining $(0, h)$ and $(r, 0)$ about the y-axis.

56. The **torus** (doughnut-shaped solid) in Figure 15 is obtained by rotating the circle $(x - a)^2 + y^2 = b^2$ around the y-axis (assume that $a > b$). Show that it has volume $2\pi^2 ab^2$. *Hint: After simplifying it, evaluate the integral by interpreting it as the area of a circle.*

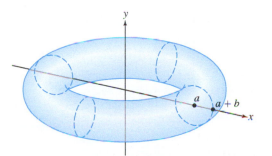

FIGURE 15 Torus obtained by rotating a circle about the y-axis.

57. GU Sketch the hypocycloid $x^{2/3} + y^{2/3} = 1$ and find the volume of the solid obtained by revolving it about the x-axis.

58. The solid generated by rotating the region between the branches of the hyperbola $y^2 - x^2 = 1$ about the x-axis is called a **hyperboloid** (Figure 16). Find the volume of the hyperboloid for $-a \le x \le a$.

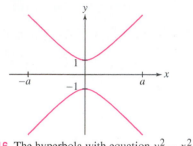

FIGURE 16 The hyperbola with equation $y^2 - x^2 = 1$.

59. A "bead" is formed by removing a cylinder of radius r from the center of a sphere of radius R (Figure 17). Find the volume of the bead with $r = 1$ and $R = 2$.

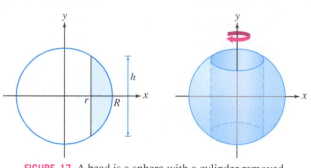

FIGURE 17 A bead is a sphere with a cylinder removed.

Further Insights and Challenges

60. 📖 Find the volume V of the bead (Figure 17) in terms of r and R. Then show that $V = \frac{\pi}{6} h^3$, where h is the height of the bead. This formula has a surprising consequence: Since V can be expressed in terms of h alone, it follows that two beads of height 1 cm, one formed from a sphere the size of an orange and the other from a sphere the size of the earth, would have the same volume! Can you explain intuitively how this is possible?

61. The solid generated by rotating the region inside the ellipse with equation $\left(\frac{x}{a}\right)^2 + \left(\frac{y}{b}\right)^2 = 1$ around the x-axis is called an **ellipsoid**. Show that the ellipsoid has volume $\frac{4}{3}\pi ab^2$. What is the volume if the ellipse is rotated around the y-axis?

62. The curve $y = f(x)$ in Figure 18, called a **tractrix**, has the following property: The tangent line at each point (x, y) on the curve has slope

$$\frac{dy}{dx} = \frac{-y}{\sqrt{1 - y^2}}$$

Let R be the shaded region under the graph of $y = f(x)$ for $0 \le x \le a$ in Figure 18. Compute the volume V of the solid obtained by revolving R around the x-axis in terms of the constant $c = f(a)$. *Hint:* Use the substitution $u = f(x)$ to show that

$$V = \pi \int_c^1 u\sqrt{1 - u^2}\, du$$

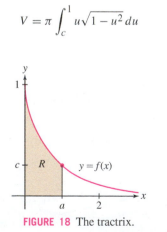

FIGURE 18 The tractrix.

63. Verify the formula

$$\int_{x_1}^{x_2} (x - x_1)(x - x_2)\, dx = \frac{1}{6}(x_1 - x_2)^3 \qquad \boxed{3}$$

Then prove that the solid obtained by rotating the shaded region in Figure 19 about the x-axis has volume $V = \frac{\pi}{6} B H^2$, with B and H as in the figure. *Hint:* Let x_1 and x_2 be the roots of $f(x) = ax + b - (mx + c)^2$, where $x_1 < x_2$. Show that

$$V = \pi \int_{x_1}^{x_2} f(x)\, dx$$

and use Eq. (3).

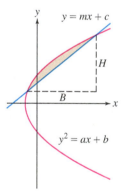

FIGURE 19 The line $y = mx + c$ intersects the parabola $y^2 = ax + b$ at two points above the x-axis.

64. Let R be the region in the unit circle lying above the cut with the line $y = mx + b$ (Figure 20). Assume that the points where the line intersects the circle lie above the x-axis. Use the method of Exercise 63 to show that the solid obtained by rotating R about the x-axis has volume $V = \frac{\pi}{6} h d^2$, with h and d as in the figure.

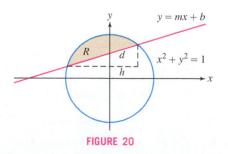

FIGURE 20

6.4 The Method of Cylindrical Shells

In the previous two sections, we computed volumes by integrating cross-sectional area. The **Shell Method**, based on cylindrical shells, is more convenient in some cases.

Consider a cylindrical shell (Figure 1) of height h, with outer radius R and inner radius r. Because the shell is obtained by removing a cylinder of radius r from the wider cylinder of radius R, it has volume

$$\pi R^2 h - \pi r^2 h = \pi h(R^2 - r^2) = \pi h(R + r)(R - r) = \pi h(R + r)\Delta r$$

where $\Delta r = R - r$ is the width of the shell. If the shell is very thin, then R and r are nearly equal and we may replace $(R + r)$ by $2R$ to obtain

$$\text{Volume of shell} \approx 2\pi Rh\Delta r = 2\pi\,(\text{radius}) \times (\text{height of shell}) \times (\text{thickness}) \qquad \boxed{1}$$

This is the surface area of the cylinder times its thickness Δr.

Now, let us rotate the region under $y = f(x)$ from $x = a$ to $x = b$ about the y-axis as in Figure 2. The resulting solid can be divided into thin concentric shells. More precisely, we divide $[a, b]$ into N subintervals of length $\Delta x = (b - a)/N$ with endpoints $x_0, x_1, \ldots, x_N$. When we rotate the thin strip of area above $[x_{i-1}, x_i]$ about the y-axis, we obtain a thin shell whose volume we denote by V_i. The volume of the solid is equal to the sum $V = \sum_{i=1}^{N} V_i$.

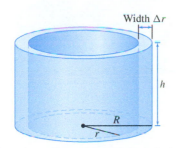

DF FIGURE 1 The volume of the cylindrical shell is approximately $2\pi Rh\Delta r$ where $\Delta r = R - r$.

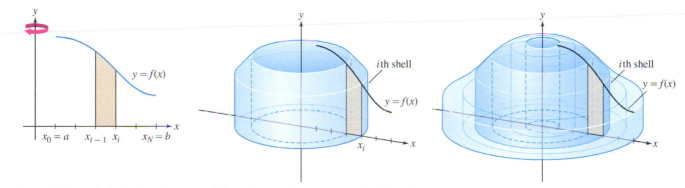

FIGURE 2 The shaded strip, when rotated about the y-axis, generates a "thin shell."

The top rim of the ith thin shell in Figure 2 is curved. However, when Δx is small, we can approximate this thin shell by the cylindrical shell (with flat rim) of height $f(x_i)$. Then, using Eq. (1), we obtain

$$V_i \approx 2\pi x_i f(x_i)\Delta x = 2\pi\,(\text{radius})\,(\text{height of shell})\,(\text{thickness})$$

$$V = \sum_{i=1}^{N} V_i \approx 2\pi \sum_{i=1}^{N} x_i f(x_i)\Delta x$$

The sum on the right is the volume of a cylindrical approximation that converges to V as $N \to \infty$ (Figure 3). This sum is also a right-endpoint approximation that converges to $2\pi \displaystyle\int_a^b x f(x)\,dx$. Thus, we obtain Eq. (2) for the volume of the solid.

Note: In the Shell Method, we integrate with respect to x when the region is rotated about the y-axis.

Volume of Revolution: The Shell Method The solid obtained by rotating the region under $y = f(x)$ over the interval $[a, b]$ *about the y-axis* has volume

$$V = 2\pi \int_a^b (\text{radius})\left(\text{height of shell}\right) dx = 2\pi \int_a^b x f(x)\,dx \qquad \boxed{2}$$

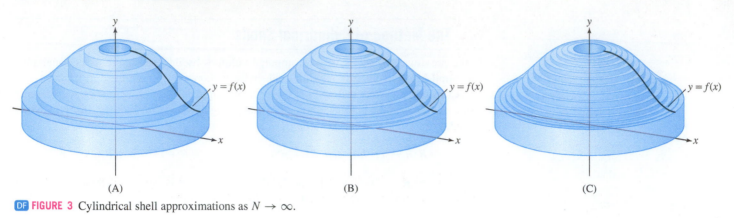

DF FIGURE 3 Cylindrical shell approximations as $N \to \infty$.

■ **EXAMPLE 1** Find the volume V of the solid obtained by rotating the region under the graph of $f(x) = 1 - 2x + 3x^2 - 2x^3$ over $[0, 1]$ about the y-axis.

Solution The solid is shown in Figure 4. By Eq. (2),

$$V = 2\pi \int_0^1 x f(x)\, dx$$

$$= 2\pi \int_0^1 x(1 - 2x + 3x^2 - 2x^3)\, dx$$

$$= 2\pi \left(\frac{1}{2}x^2 - \frac{2}{3}x^3 + \frac{3}{4}x^4 - \frac{2}{5}x^5 \right)\Big|_0^1 = \frac{11}{30}\pi$$ ■

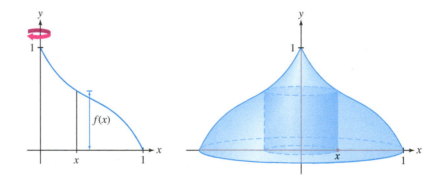

DF FIGURE 4 The graph of
$f(x) = 1 - 2x + 3x^2 - 2x^3$ rotated about
the y-axis.

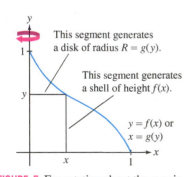

FIGURE 5 For rotation about the y-axis, the
Shell Method uses $y = f(x)$ but the Disk
Method requires the inverse function
$x = g(y)$.

CONCEPTUAL INSIGHT **Shells Versus Disks and Washers** When finding a volume using the Shell Method, the shell height is always *parallel* to the axis of rotation. When finding a volume using the Disk and Washer Method, the disk radius is always *perpendicular* to the axis of rotation. Some volumes can be computed equally well using either the Shell Method or the Disk and Washer Method, but in Example 1, the Shell Method is much easier. To use the Disk Method, we would need to know the radius of the disk generated at height y because we're rotating about the y-axis (Figure 5). This would require finding the inverse $g(y) = f^{-1}(y)$. **In general**: Use the Shell Method if finding the shell height is easier than finding the disk radius. Use the Disk Method when finding the disk radius is easier.

When we rotate the region between the graphs of two functions f and g satisfying $f(x) \geq g(x)$, the vertical segment at location x generates a cylindrical shell of radius x and height $f(x) - g(x)$ (Figure 6). Therefore, the volume is

$$V = 2\pi \int_a^b (\text{radius})\,(\text{height of shell})\, dx = 2\pi \int_a^b x\big(f(x) - g(x)\big)\, dx \qquad \boxed{3}$$

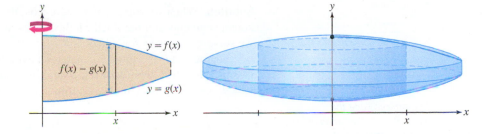

FIGURE 6 The vertical segment at location x generates a shell of radius x and height $f(x) - g(x)$.

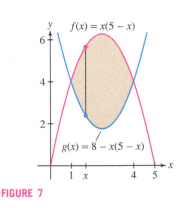

FIGURE 7

The reasoning in Example 3 shows that if we rotate the region under $y = f(x)$ over $[a, b]$ about the vertical line $x = c$, then the volume is

$$V = 2\pi \int_a^b (x - c) f(x) \, dx \quad if \ c \le a$$

$$V = 2\pi \int_a^b (c - x) f(x) \, dx \quad if \ c \ge b$$

■ **EXAMPLE 2** **Region Between Two Curves** Find the volume V obtained by rotating the region enclosed by the graphs of $f(x) = x(5 - x)$ and $g(x) = 8 - x(5 - x)$ about the y-axis.

Solution First, find the points of intersection by solving $x(5 - x) = 8 - x(5 - x)$. We obtain $x^2 - 5x + 4 = (x - 1)(x - 4) = 0$, so the curves intersect at $x = 1, 4$. Sketching the graphs (Figure 7), we see that $f(x) \ge g(x)$ on the interval $[1, 4]$ and

$$\text{Height of shell} = f(x) - g(x) = x(5 - x) - \big(8 - x(5 - x)\big) = 10x - 2x^2 - 8$$

$$V = 2\pi \int_1^4 (\text{radius})(\text{height of shell}) \, dx = 2\pi \int_1^4 x(10x - 2x^2 - 8) \, dx$$

$$= 2\pi \left(\frac{10}{3} x^3 - \frac{1}{2} x^4 - 4x^2 \right) \bigg|_1^4 = 2\pi \left(\frac{64}{3} - \left(-\frac{7}{6} \right) \right) = 45\pi \qquad ■$$

■ **EXAMPLE 3** **Rotating About a Vertical Axis** Use the Shell Method to calculate the volume V obtained by rotating the region under the graph of $f(x) = x^{-1/2}$ over $[1, 4]$ about the axis $x = -3$.

Solution If we were rotating this region about the y-axis (i.e., $x = 0$), we would use Eq. (3). To rotate it around the line $x = -3$, we must take into account that the radius of revolution is now 3 units longer.

Figure 8 shows that the radius of the shell is now $x - (-3) = x + 3$. The height of the shell is still $f(x) = x^{-1/2}$, so

$$V = 2\pi \int_1^4 (\text{radius}) \, (\text{height of shell}) \, dx$$

$$= 2\pi \int_1^4 (x + 3) x^{-1/2} \, dx = 2\pi \left(\frac{2}{3} x^{3/2} + 6x^{1/2} \right) \bigg|_1^4 = \frac{64\pi}{3} \qquad ■$$

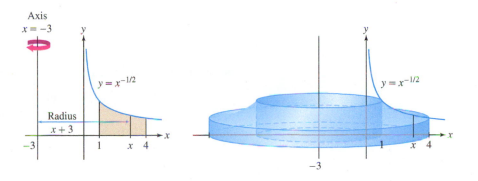

DF FIGURE 8 Rotation about the axis $x = -3$.

The method of cylindrical shells can be applied to rotations about horizontal axes, but in this case, the graph must be described in the form $x = g(y)$.

■ **EXAMPLE 4** **Rotating About the x-Axis** Use the Shell Method to compute the volume V obtained by rotating the region under $y = 9 - x^2$ over $[0, 3]$ about the x-axis.

Solution When we rotate about the x-axis, the cylindrical shells are generated by horizontal segments and the Shell Method gives us an integral with respect to y. Therefore, we solve $y = 9 - x^2$ for x to obtain $x = \sqrt{9 - y}$.

Segment $\overline{AB}$ in Figure 9 generates a cylindrical shell of radius y and height $\sqrt{9 - y}$ (we use the term "height" even though the shell is horizontal). Using the substitution $u = 9 - y$, $du = -dy$ in the resulting integral, we obtain

$$V = 2\pi \int_0^9 (\text{radius})(\text{height of shell})\, dy = 2\pi \int_0^9 y \sqrt{9 - y}\, dy = -2\pi \int_9^0 (9 - u) \sqrt{u}\, du$$

$$= 2\pi \int_0^9 (9u^{1/2} - u^{3/2})\, du = 2\pi \left(6u^{3/2} - \frac{2}{5} u^{5/2} \right) \Bigg|_0^9 = \frac{648}{5} \pi \qquad \blacksquare$$

> ◄·· **REMINDER** *After making the substitution $u = 9 - y$, the limits of integration must be changed. Since $u(0) = 9$ and $u(9) = 0$, we change $\int_0^9$ to $\int_9^0$.*

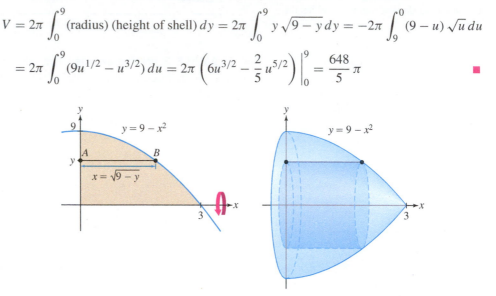

DF **FIGURE 9** Shell generated by a horizontal segment in the region under the graph of $y = 9 - x^2$.

6.4 SUMMARY

- Shell Method: When you rotate the region between two graphs about an axis, the segments *parallel* to the axis generate cylindrical shells [Figure 10(A)]. The volume V of the solid of revolution is the integral of the surface areas of these shells:

$$\text{Surface area of shell} = 2\pi (\text{radius}) (\text{height of shell})$$

- Sketch the graphs to visualize the shells.
- Figure 10(B): Region between $y = f(x)$ (with $f(x) \geq 0$) and the y-axis, rotated about the y-axis:

$$V = 2\pi \int_a^b (\text{radius}) (\text{height of shell})\, dx = 2\pi \int_a^b x f(x)\, dx$$

- Figure 10(C): Region between $y = f(x)$ and $y = g(x)$ (with $f(x) \geq g(x) \geq 0$), rotated about the y-axis:

$$V = 2\pi \int_a^b (\text{radius}) (\text{height of shell})\, dx = 2\pi \int_a^b x(f(x) - g(x))\, dx$$

- Rotation about a vertical axis $x = c$.

   - Figure 10(D): $c \leq a$, radius of shell is $(x - c)$:

$$V = 2\pi \int_a^b (x - c) f(x)\, dx$$

   - Figure 10(E): $c \geq a$, radius of shell is $(c - x)$:

$$V = 2\pi \int_a^b (c - x) f(x)\, dx$$

- Rotation about the x-axis using the Shell Method: Write the graph as $x = g(y)$:

$$V = 2\pi \int_c^d (\text{radius})(\text{height of shell})\, dy = 2\pi \int_c^d y g(y)\, dy$$

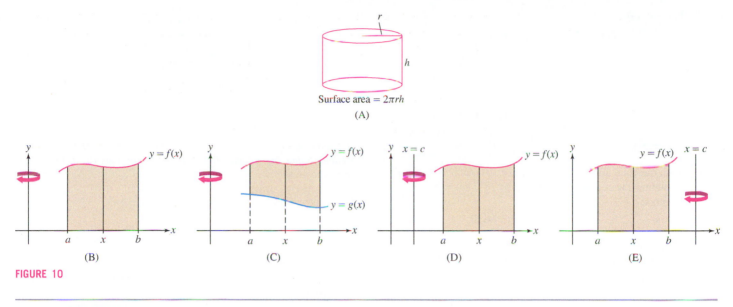

Surface area $= 2\pi rh$

(A)

(B)   (C)   (D)   (E)

FIGURE 10

6.4 EXERCISES

Preliminary Questions

1. Consider the region $\mathcal{R}$ under the graph of the constant function $f(x) = h$ over the interval $[0, r]$. Give the height and the radius of the cylinder generated when $\mathcal{R}$ is rotated about:

(a) the x-axis.   **(b)** the y-axis.

2. Let V be the volume of a solid of revolution about the y-axis.

(a) Does the Shell Method for computing V lead to an integral with respect to x or y?

(b) Does the Disk or Washer Method for computing V lead to an integral with respect to x or y?

3. If we rotate the region under the curve $y = 8$ between $x = 2$ and $x = 3$ about the x-axis, what answer should the Shell Method give us?

Exercises

In Exercises 1–6, sketch the solid obtained by rotating the region underneath the graph of the function over the given interval about the y-axis, and find its volume.

1. $f(x) = x^3$, $[0, 1]$

2. $f(x) = \sqrt{x}$, $[0, 4]$

3. $f(x) = x^{-1}$, $[1, 3]$

4. $f(x) = 4 - x^2$, $[0, 2]$

5. $f(x) = \sqrt{x^2 + 9}$, $[0, 3]$

6. $f(x) = \dfrac{x}{\sqrt{1 + x^3}}$, $[1, 4]$

In Exercises 7–12, use the Shell Method to compute the volume obtained by rotating the region enclosed by the graphs as indicated, about the y-axis.

7. $y = 3x - 2$, $y = 6 - x$, $x = 0$

8. $y = \sqrt{x}$, $y = x^2$

9. $y = x^2$, $y = 8 - x^2$, $x = 0$, for $x \geq 0$

10. $y = 8 - x^3$, $y = 8 - 4x$, for $x \geq 0$

11. $y = (x^2 + 1)^{-2}$, $y = 2 - (x^2 + 1)^{-2}$, $x = 2$

12. $y = 1 - |x - 1|$, $y = 0$

In Exercises 13 and 14, use a graphing utility to find the points of intersection of the curves numerically and then compute the volume of rotation of the enclosed region about the y-axis.

13. GU $y = \frac{1}{2}x^2$, $y = \sin(x^2)$, $x \geq 0$

14. GU $y = e^{-x^2/2}$, $y = x$, $x = 0$

In Exercises 15–20, sketch the solid obtained by rotating the region underneath the graph of f over the interval about the given axis, and calculate its volume using the Shell Method.

15. $f(x) = x^3$, $[0, 1]$, about $x = 2$

16. $f(x) = x^3$, $[0, 1]$, about $x = -2$

17. $f(x) = x^{-4}$, $[-3, -1]$, about $x = 4$

18. $f(x) = \dfrac{1}{\sqrt{x^2 + 1}}$, $[0, 2]$, about $x = 0$

19. $f(x) = a - x$ with $a > 0$, $[0, a]$, about $x = -1$

20. $f(x) = 1 - x^2$, $[-1, 1]$, $x = c$ with $c > 1$

In Exercises 21–26, sketch the enclosed region and use the Shell Method to calculate the volume of rotation about the x-axis.

21. $x = y$, $y = 0$, $x = 1$

22. $x = \frac{1}{4}y + 1$, $x = 3 - \frac{1}{4}y$, $y = 0$

23. $x = y(4 - y)$, $x = 0$

24. $x = y(4 - y)$, $x = (y - 2)^2$

25. $y = 4 - x^2$, $x = 0$, $y = 0$

26. $y = x^{1/3} - 2$, $y = 0$, $x = 27$

27. Determine which of the following is the appropriate integrand needed to determine the volume of the solid obtained by rotating around the vertical axis given by $x = -1$ the area that is between the curves $y = f(x)$ and $y = g(x)$ over the interval $[a, b]$, where $a \geq 0$ and $f(x) \geq g(x)$ over that interval.

(a) $x(f(x) - g(x))$

(b) $(x + 1)(f(x) - g(x))$

(c) $x((f(x) - 1) - (g(x) - 1))$

(d) $(x - 1)(f(x) - g(x))$

(e) $x(f(x + 1) - g(x + 1))$

28. Let $y = f(x)$ be a decreasing function on $[0, b]$, such that $f(b) = 0$. Explain why $2\pi \int_0^b x f(x)\, dx = \pi \int_0^{f(0)} (h(x))^2\, dx$, where h denotes the inverse of f.

29. Use both the Shell and Disk Methods to calculate the volume obtained by rotating the region under the graph of $f(x) = 8 - x^3$ for $0 \le x \le 2$ about:

(a) the x-axis.   (b) the y-axis.

30. Sketch the solid of rotation about the y-axis for the region under the graph of the constant function $f(x) = c$ (where $c > 0$) for $0 \le x \le r$.

(a) Find the volume without using integration.

(b) Use the Shell Method to compute the volume.

31. The graph in Figure 11(A) can be described by both $y = f(x)$ and $x = h(y)$, where h is the inverse of f. Let V be the volume obtained by rotating the region under the graph about the y-axis.

(a) Describe the figures generated by rotating segments $\overline{AB}$ and $\overline{CB}$ about the y-axis.

(b) Set up integrals that compute V by the Shell and Disk Methods.

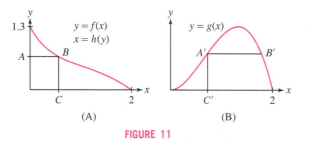

(A)   (B)

FIGURE 11

32. ✎ Let W be the volume of the solid obtained by rotating the region under the graph in Figure 11(B) about the y-axis.

(a) Describe the figures generated by rotating segments $\overline{A'B'}$ and $\overline{A'C'}$ about the y-axis.

(b) Set up an integral that computes W by the Shell Method.

(c) Explain the difficulty in computing W by the Washer Method.

33. Let R be the region under the graph of $y = 9 - x^2$ for $0 \le x \le 2$. Use the Shell Method to compute the volume of rotation of R about the x-axis as a sum of two integrals along the y-axis. *Hint:* The shells generated depend on whether $y \in [0, 5]$ or $y \in [5, 9]$.

34. Let R be the region under the graph of $y = 4x^{-1}$ for $1 \le y \le 4$. Use the Shell Method to compute the volume of rotation of R about the y-axis as a sum of two integrals along the x-axis.

In Exercises 35–40, use the Shell Method to find the volume obtained by rotating region A in Figure 12 about the given axis.

35. y-axis   **36.** $x = -3$

37. $x = 2$ **38.** x-axis

39. $y = -2$ **40.** $y = 6$

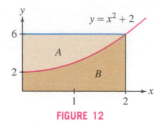

FIGURE 12

In Exercises 41–46, use the most convenient method (Disk or Shell Method) to find the volume obtained by rotating region B in Figure 12 about the given axis.

41. y-axis   **42.** $x = -3$

43. $x = 2$ **44.** x-axis

45. $y = -2$ **46.** $y = 8$

In Exercises 47–54, use the most convenient method (Disk or Shell Method) to find the given volume of rotation.

47. Region between $x = y(5 - y)$ and $x = 0$, rotated about the y-axis

48. Region between $x = y(5 - y)$ and $x = 0$, rotated about the x-axis

49. Region bounded by $y = x^2$ and $x = y^2$, rotated about the y-axis

50. Region bounded by $y = x^2$ and $x = y^2$, rotated about $x = 3$

51. Region in Figure 13, rotated about the x-axis

52. Region in Figure 13, rotated about the y-axis

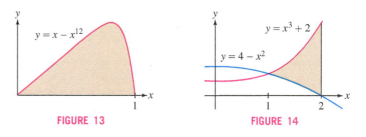

FIGURE 13   **FIGURE 14**

53. Region in Figure 14, rotated about $x = 4$

54. Region in Figure 14, rotated about $y = -2$

In Exercises 55–58, use the Shell Method to find the given volume of rotation.

55. A sphere of radius r

56. The "bead" formed by removing a cylinder of radius r from the center of a sphere of radius R (compare with Exercise 59 in Section 6.3)

57. The torus obtained by rotating the circle $(x - a)^2 + y^2 = b^2$ about the y-axis, where $a > b$ (compare with Exercise 56 in Section 6.3). *Hint:* Evaluate the integral by interpreting part of it as the area of a circle.

58. The "paraboloid" obtained by rotating the region between $y = x^2$ and $y = c$ ($c > 0$) about the y-axis

59. Given a and b, $0 \le a \le b$, find a function f such that the volume obtained by rotating about the x-axis the region R under the graph of $y = f(x)$ over the interval $[a, b]$ equals the volume obtained by rotating that same region R about the y-axis.

Further Insights and Challenges

60. 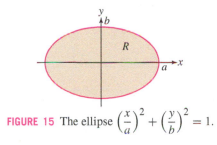   The surface area of a sphere of radius r is $4\pi r^2$. Use this to derive the formula for the volume V of a sphere of radius R in a new way.

(a) Show that the volume of a thin spherical shell of inner radius r and thickness Δr is approximately $4\pi r^2 \Delta r$.

(b) Approximate V by decomposing the sphere of radius R into N thin spherical shells of thickness $\Delta r = R/N$.

(c) Show that the approximation is a Riemann sum that converges to an integral. Evaluate the integral.

61. Show that the solid (an **ellipsoid**) obtained by rotating the region R in Figure 15 about the y-axis has volume $\frac{4}{3}\pi a^2 b$.

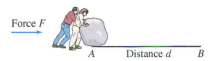

FIGURE 15 The ellipse $\left(\dfrac{x}{a}\right)^2 + \left(\dfrac{y}{b}\right)^2 = 1$.

62. The bell-shaped curve $y = f(x)$ in Figure 16 satisfies $dy/dx = -xy$. Use the Shell Method and the substitution $u = f(x)$ to show that the solid obtained by rotating the region R about the y-axis has volume $V = 2\pi(1 - c)$, where $c = f(a)$. Observe that as $c \to 0$, the region R becomes infinite but the volume V approaches 2π.

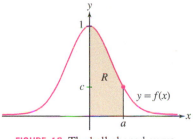

FIGURE 16 The bell-shaped curve.

6.5 Work and Energy

All physical tasks, from running up a hill to turning on a computer, require an expenditure of energy. When a force is applied to an object to move it, the energy expended is called **work**. When a *constant* force F is applied to move the object a distance d in the direction of the force, the work W is defined as "force times distance" (Figure 1):

$$\boxed{W = F \cdot d} \qquad \boxed{1}$$

> For those who want some proof that physicists are human, the proof is in the idiocy of all the different units which they use for measuring energy.
>
> —Richard Feynman,
> *The Character of Physical Law*

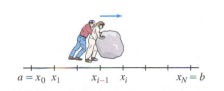

FIGURE 1 The work expended to move the object from A to B is $W = F \cdot d$.

The International System (SI) unit of force is the *newton* (abbreviated N), defined as 1 kg-m/s^2. Energy and work are both measured in units of the *joule* (J), equal to 1 N-m. In the British system, the unit of force is the pound, and both energy and work are measured in foot-pounds (ft-lb). Another unit of energy is the *calorie*. One ft-lb is approximately 1.356 J or 0.324 calories.

To become familiar with the units, let's calculate the work W required to lift a 2-kg stone 3 m above the ground. Gravity pulls down on the stone of mass m with a force equal to $-mg$, where $g = 9.8$ m/s^2. Therefore, lifting the stone requires an upward vertical force $F = mg$, and the work expended is

$$W = \underbrace{(mg)h}_{F \cdot d} = (2 \text{ kg})(9.8 \text{ m/s}^2)(3 \text{ m}) = 58.8 \text{ J}$$

The kilogram is a unit of mass, but the pound is a unit of force. Therefore, the factor g does not appear when work against gravity is computed in the British system. The work required to lift a 2-lb stone 3 ft is

$$W = \underbrace{(2 \text{ lb})(3 \text{ ft})}_{F \cdot d} = 6 \text{ ft-lb}$$

We are interested in the case where the force $F(x)$ varies as the object moves from a to b along the x-axis. Eq. (1) does not apply directly, but we can break up the task into a large number of smaller tasks for which Eq. (1) gives a good approximation. Divide $[a, b]$ into N subintervals of length $\Delta x = (b - a)/N$ as in Figure 2 and let W_i be the work required to move the object from x_{i-1} to x_i. If Δx is small, then the force $F(x)$ is

FIGURE 2 The work to move an object from x_{i-1} to x_i is approximately $F(x_i)\Delta x$.

nearly constant on the interval $[x_{i-1}, x_i]$ with value $F(x_i)$, so $W_i \approx F(x_i)\Delta x$. Summing the contributions, we obtain

$$W = \sum_{i=1}^{N} W_i \approx \underbrace{\sum_{i=1}^{N} F(x_i)\Delta x}_{\text{Right-endpoint approximation}}$$

The sum on the right is a right-endpoint approximation that converges to $\int_a^b F(x)\, dx$. This leads to the following definition.

> **DEFINITION Work** The work performed in moving an object along the x-axis from a to b by applying a force of magnitude $F(x)$ is
>
> $$W = \int_a^b F(x)\, dx$$
>
> **2**

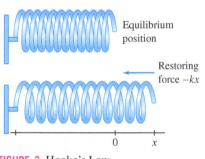

Equilibrium position

Restoring force $-kx$

FIGURE 3 Hooke's Law.

Hooke's Law is named after the English scientist, inventor, and architect Robert Hooke (1635–1703), who made important discoveries in physics, astronomy, chemistry, and biology. He was a pioneer in the use of the microscope to study organisms. Unfortunately, Hooke was involved in several bitter disputes with other scientists, most notably with his contemporary Isaac Newton. Newton was furious when Hooke criticized his work on optics. Later, Hooke told Newton that he believed Kepler's Laws would follow from an inverse square law of gravitation, but Newton refused to acknowledge Hooke's contributions in his masterwork Principia. *Shortly before his death in 1955, Albert Einstein commented on Newton's behavior: "That, alas, is vanity. You find it in so many scientists.... It has always hurt me to think that Galileo did not acknowledge the work of Kepler".*

One typical calculation involves finding the work required to stretch a spring. Assume that the free end of the spring has position $x = 0$ at equilibrium, when no force is acting (Figure 3). According to **Hooke's Law**, when the spring is stretched (or compressed) to position x, it exerts a restoring force of magnitude $-kx$ in the opposite direction, where k is the **spring constant**. If we want to stretch the spring further, we must apply a force $F(x) = kx$ to counteract the force exerted by the spring.

■ **EXAMPLE 1** **Hooke's Law** Assuming a spring constant of $k = 400\,\text{N/m}$, find the work required to:

(a) Stretch the spring 10 cm beyond equilibrium.

(b) Compress the spring 2 cm more when it is already compressed 3 cm.

Solution A force $F(x) = 400x$ N is required to stretch the spring (with x in meters). Note that centimeters must be converted to meters.

(a) The work required to stretch the spring 10 cm (0.1 m) beyond equilibrium is

$$W = \int_0^{0.1} 400x\, dx = 200x^2 \Big|_0^{0.1} = 2\,\text{J}$$

(b) If the spring is at position $x = -3$ cm, then the work W required to compress it further to $x = -5$ cm is

$$W = \int_{-0.03}^{-0.05} 400x\, dx = 200x^2 \Big|_{-0.03}^{-0.05} = 0.5 - 0.18 = 0.32\,\text{J}$$

Observe that we integrate from the starting point $x = -0.03$ to the ending point $x = -0.05$ (even though the lower limit of the integral is larger than the upper limit in this case). ■

In the next two examples, we are not moving a single object through a fixed distance, so we cannot apply Eq. (2). Rather, each thin layer of the object is moved through a different distance. The work performed is computed by "summing" (i.e., *integrating*) the work performed on the thin layers.

■ **EXAMPLE 2** **Building a Concrete Column** Compute the work (against gravity) required to build a concrete column of height 5 m and square base of side 2 m. Assume that concrete has density 1500 kg/m^3.

Solution Think of the column as a stack of n thin layers of width $\Delta y = 5/n$. The work consists of lifting up these layers and placing them on the stack (Figure 4), but the work performed on a given layer depends on how high we lift it. First, let us compute the gravitational force on a thin layer of width Δy:

On the earth's surface, work against gravity is equal to the force mg times the vertical distance through which the object is lifted. No work against gravity is done when an object is moved sideways.

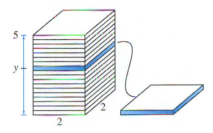

FIGURE 4 Total work is the sum of the work performed on each layer of the column.

In Examples 2 and 3, the work performed on a thin layer is written

$$L(y)\Delta y$$

When we take the sum and let Δy approach zero, we obtain the integral of $L(y)$.

Symbolically, the Δy "becomes" the dy of the integral. Note that

$$L(y) = g \times density \times A(y)$$
$$\times (vertical\ distance\ lifted)$$

where $A(y)$ is the area of the cross section.

Volume of layer = area × width $= 4\Delta y$ m^3

Mass of layer = density × volume = $1500 \cdot 4\Delta y$ kg

Force on layer = g × mass $= 9.8 \cdot 1500 \cdot 4\Delta y = 58,800\,\Delta y$ N

The work performed in lifting this layer to height y is equal to the force times the distance y, which is $(58,800\Delta y)y$. Setting $L(y) = 58,800y$, we have

$$\boxed{\text{Work lifting layer to height } y \approx (58,800\Delta y)y = L(y)\Delta y}$$

This is only an approximation (although a very good one if Δy is small) because the layer has nonzero width and the cement particles at the top have been lifted a little bit higher than those at the bottom. The ith layer is lifted to height y_i, so the total work performed is

$$W \approx \sum_{i=1}^{n} L(y_i)\,\Delta y$$

This sum is a right-endpoint approximation to $\int_0^5 L(y)\,dy$. Letting $n \to \infty$, we obtain

$$W = \int_0^5 L(y)\,dy = \int_0^5 58,800y\,dy = 58,800\frac{y^2}{2}\bigg|_0^5 = 735,000 \text{ J} \qquad \blacksquare$$

■ **EXAMPLE 3 Pumping Water out of a Tank**   A spherical tank of radius 5 m is filled with water. Calculate the work W performed (against gravity) in pumping out the water through a spout of height 1 m at the top. The density of water is 1000 kg/m^3.

Solution The first step, as in the previous example, is to compute the work against gravity performed on a thin layer of water of width Δy. We place the origin of our coordinate system at the center of the sphere because this leads to a simple formula for the radius r of the cross section at height y (Figure 5).

Step 1. Compute work performed on a layer.
Figure 5 shows that the cross section at height y is a circle of radius $r = \sqrt{25 - y^2}$ and area $A(y) = \pi r^2 = \pi(25 - y^2)$. A thin layer has volume $A(y)\Delta y$, and to lift it, we must exert a force against gravity equal to

$$\text{Force on layer} = g \times \overbrace{\text{density} \times A(y)\Delta y}^{\text{mass}} \approx (9.8)1000\pi(25 - y^2)\Delta y$$

The layer has to be lifted a vertical distance $6 - y$, so

$$\boxed{\text{Work on layer} \approx \overbrace{9800\pi(25 - y^2)\Delta y}^{\text{Force against gravity}} \times \overbrace{(6 - y)}^{\text{Vertical distance lifted}} = L(y)\Delta y}$$

where $L(y) = 9800\pi(150 - 25y - 6y^2 + y^3)$.

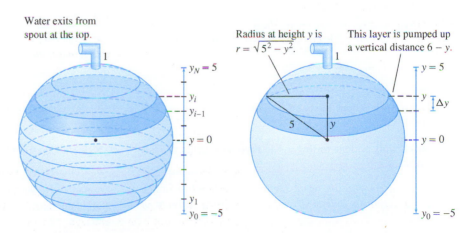

FIGURE 5 The sphere is divided into N thin layers.

Step 2. **Compute total work.**

Now divide the sphere into N layers and let y_i be the height of the ith layer. The work performed on ith layer is approximately $L(y_i)\,\Delta y$, and therefore,

$$W \approx \sum_{i=1}^{N} L(y_i)\,\Delta y$$

This sum approaches the integral of $L(y)$ as $N \to \infty$ (i.e., $\Delta y \to 0$), so

$$W = \int_{-5}^{5} L(y)\,dy = 9800\pi \int_{-5}^{5} (150 - 25y - 6y^2 + y^3)\,dy$$

$$= 9800\pi \left(150y - \frac{25}{2} y^2 - 2y^3 + \frac{1}{4} y^4 \right) \Bigg|_{-5}^{5} = 9{,}800{,}000\pi \text{ J}$$

Note that the integral extends from -5 to 5 because the y-coordinate along the sphere varies from -5 to 5.   ∎

A liter of gasoline has an energy content of approximately 3.4×10^7 joules. Hence, the work required to pump the water out of the spout is equal to the energy content of roughly 0.9 liters (L) of gasoline.

6.5 SUMMARY

- Work performed to move an object:

$$\text{Constant force:} \quad W = F \cdot d, \qquad \text{Variable force:} \quad W = \int_a^b F(x)\,dx$$

- Hooke's Law: A spring stretched x units past equilibrium exerts a restoring force of magnitude $-kx$. A force $F(x) = kx$ is required to stretch the spring further.
- To compute work against gravity by decomposing an object into N thin layers of thickness Δy, express the work performed on a thin layer as $L(y)\Delta y$, where

$$L(y) = g \times \text{density} \times A(y) \times (\text{vertical distance lifted})$$

The total work performed is $W = \int_a^b L(y)\,dy$.

6.5 EXERCISES

Preliminary Questions

1. Why is integration needed to compute the work performed in stretching a spring?

2. Why is integration needed to compute the work performed in pumping water out of a tank but not to compute the work performed in lifting up the tank?

3. Which of the following represents the work required to stretch a spring (with spring constant k) a distance x beyond its equilibrium position: kx, $-kx$, $\frac{1}{2}mk^2$, $\frac{1}{2}kx^2$, or $\frac{1}{2}mx^2$?

4. What does it mean when the integral used to calculate work gives a negative answer?

Exercises

1. How much work is done raising a 4-kg mass to a height of 16 m above ground?

2. How much work is done raising a 4-lb mass to a height of 16 ft above ground?

In Exercises 3–6, compute the work (in joules) required to stretch or compress a spring as indicated, assuming a spring constant of $k = 800$ N/m.

3. Stretching from equilibrium to 12 cm past equilibrium

4. Compressing from equilibrium to 4 cm past equilibrium

5. Stretching from 5 cm to 15 cm past equilibrium

6. Compressing 4 cm more when it is already compressed 5 cm

7. If 5 J of work are needed to stretch a spring 10 cm beyond equilibrium, how much work is required to stretch it 15 cm beyond equilibrium?

8. To create images of samples at the molecular level, atomic force microscopes use silicon micro-cantilevers that obey Hooke's Law $F(x) = -kx$, where x is the distance through which the tip is deflected (Figure 6). Suppose that 10^{-17} J of work are required to deflect the tip a distance 10^{-8} m. Find the deflection if a force of 10^{-9} N is applied to the tip.

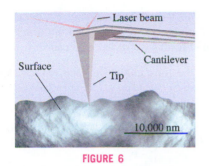

FIGURE 6

9. A spring obeys a force law $F(x) = -kx^{1.1}$ with $k = 100$ N/m$^{1.1}$. Find the work required to stretch the spring 0.3 m past equilibrium.

10. 🕮 Show that the work required to stretch a spring from position a to position b is $\frac{1}{2}k(b^2 - a^2)$, where k is the spring constant. How do you interpret the negative work obtained when $|b| < |a|$?

In Exercises 11–14, use the method of Examples 2 and 3 to calculate the work against gravity required to build the structure out of a lightweight material of density 600 kg/m^3.

11. Box of height 3 m and square base of side 2 m

12. Cylindrical column of height 4 m and radius 0.8 m

13. Right circular cone of height 4 m and base of radius 1.2 m

14. Hemisphere of radius 0.8 m

15. Built around 2600 BCE, the Great Pyramid of Giza in Egypt (Figure 7) is 146 m high and has a square base of side 230 m. Find the work (against gravity) required to build the pyramid if the density of the stone is estimated at 2000 kg/m^3.

FIGURE 7 The Great Pyramid in Giza, Egypt. (© *Elvele Images Ltd/Alamy*)

16. Calculate the work (against gravity) required to build a box of height 3 m and square base of side 2 m out of material of variable density, assuming that the density at height y is $f(y) = 1000 - 100y$ kg/m^3.

In Exercises 17–22, calculate the work (in joules) required to pump all of the water out of a full tank. Distances are in meters, and the density of water is 1000 kg/m^3.

17. Rectangular tank in Figure 8; water exits from a small hole at the top.

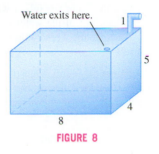

FIGURE 8

18. Rectangular tank in Figure 8; water exits through the spout.

19. Hemisphere in Figure 9; water exits through the spout.

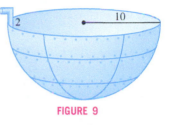

FIGURE 9

20. Conical tank in Figure 10; water exits through the spout.

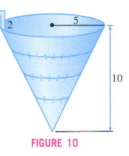

FIGURE 10

21. Horizontal cylinder in Figure 11; water exits from a small hole at the top. *Hint:* Evaluate the integral by interpreting part of it as the area of a circle.

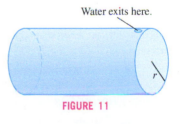

FIGURE 11

22. Trough in Figure 12; water exits by pouring over the sides.

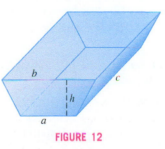

FIGURE 12

23. Find the work W required to empty the tank in Figure 8 through the hole at the top if the tank is half full of water.

24. Assume the tank in Figure 8 is full of water and let W be the work required to pump out half of the water through the hole at the top. Do you expect W to equal the work computed in Exercise 23? Explain and then compute W.

25. Assume the tank in Figure 10 is full. Find the work required to pump out half of the water. *Hint:* First, determine the level H at which the water remaining in the tank is equal to one-half the total capacity of the tank.

26. Assume that the tank in Figure 10 is full.
(a) Calculate the work $F(y)$ required to pump out water until the water level has reached level y.
(b) *CAS*   Plot F.
(c) What is the significance of $F'(y)$ as a rate of change?
(d) *CAS*   If your goal is to pump out all of the water, at which water level y_0 will half of the work be done?

27. Calculate the work required to lift a 10-m chain over the side of a building (Figure 13). Assume that the chain has a density of 8 kg/m. *Hint:* Break up the chain into N segments, estimate the work performed on a segment, and compute the limit as $N \to \infty$ as an integral.

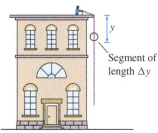

FIGURE 13 The small segment of the chain of length Δy located y meters from the top is lifted through a vertical distance y.

28. How much work is done lifting a 3-m chain over the side of a building if the chain has mass density 4 kg/m?

29. A 6-m chain has mass 18 kg. Find the work required to lift the chain over the side of a building.

30. A 10-m chain with mass density 4 kg/m is initially coiled on the ground. How much work is performed in lifting the chain so that it is fully extended (and one end touches the ground)?

31. How much work is done lifting a 12-m chain that has mass density 3 kg/m (initially coiled on the ground) so that its top end is 10 m above the ground?

32. A 500-kg wrecking ball hangs from a 12-m cable of density 15 kg/m attached to a crane. Calculate the work done if the crane lifts the ball from ground level to 12 m in the air by drawing in the cable.

33. Calculate the work required to lift a 3-m chain over the side of a building if the chain has a variable density of $\rho(x) = x^2 - 3x + 10$ kg/m for $0 \le x \le 3$.

34. A 3-m chain with linear mass density $\rho(x) = 2x(4 - x)$ kg/m lies on the ground. Calculate the work required to lift the chain from its front end so that its bottom is 2 m above ground.

Exercises 35–37: The gravitational force between two objects of mass m and M, separated by a distance r, has magnitude GMm/r^2, where $G = 6.67 \times 10^{-11}$ $m^3kg^{-1}s^{-1}$.

35. Show that if two objects of mass M and m are separated by a distance r_1, then the work required to increase the separation to a distance r_2 is equal to $W = GMm(r_1^{-1} - r_2^{-1})$.

36. Use the result of Exercise 35 to calculate the work required to place a 2000-kg satellite in an orbit 1200 km above the surface of the earth. Assume that the earth is a sphere of radius $R_e = 6.37 \times 10^6$ m and mass $M_e = 5.98 \times 10^{24}$ kg. Treat the satellite as a point mass.

37. Use the result of Exercise 35 to compute the work required to move a 1500-kg satellite from an orbit 1000 to an orbit 1500 km above the surface of the earth.

38. The pressure P and volume V of the gas in a cylinder of length 0.8 m and radius 0.2 m, with a movable piston, are related by $PV^{1.4} = k$, where k is a constant (Figure 14). When the piston is fully extended, the gas pressure is 2000 kilopascals (kPa; 1 kilopascal is 10^3 newtons per square meter).
(a) Calculate k.
(b) The force on the piston is PA, where A is the piston's area. Calculate the force as a function of the length x of the column of gas.
(c) Calculate the work required to compress the gas column from 0.8 m to 0.5 m.

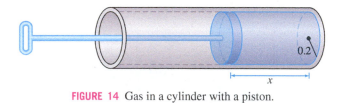

FIGURE 14 Gas in a cylinder with a piston.

Further Insights and Challenges

39. Work-Energy Theorem   An object of mass m moves from x_1 to x_2 during the time interval $[t_1, t_2]$ due to a force $F(x)$ acting in the direction of motion. Let $x(t)$, $v(t)$, and $a(t)$ be the position, velocity, and acceleration at time t. The object's kinetic energy is $KE = \frac{1}{2}mv^2$.
(a) Use the Change of Variables Formula to show that the work performed is equal to

$$W = \int_{x_1}^{x_2} F(x)\,dx = \int_{t_1}^{t_2} F(x(t))v(t)\,dt$$

(b) Use Newton's Second Law, $F(x(t)) = ma(t)$, to show that

$$\frac{d}{dt}\left(\frac{1}{2}mv(t)^2\right) = F(x(t))v(t)$$

(c) Use the FTC to prove the Work-Energy Theorem: The change in kinetic energy during the time interval $[t_1, t_2]$ is equal to the work performed.

40. A model train of mass 0.5 kg is placed at one end of a straight 3-m electric track. Assume that a force $F(x) = (3x - x^2)$ N acts on the train at distance x along the track. Use the Work-Energy Theorem (Exercise 39) to determine the velocity of the train when it reaches the end of the track.

41. With what initial velocity v_0 must we fire a rocket so it attains a maximum height r above the earth? *Hint:* Use the results of Exercises 35 and 39. As the rocket reaches its maximum height, its KE decreases from $\frac{1}{2}mv_0^2$ to zero.

42. With what initial velocity must we fire a rocket so it attains a maximum height of $r = 20$ km above the surface of the earth?

43. Calculate **escape velocity,** the minimum initial velocity of an object to ensure that it will continue traveling into space and never fall back to earth (assuming that no force is applied after takeoff). *Hint:* Take the limit as $r \to \infty$ in Exercise 41.

CHAPTER REVIEW EXERCISES

1. Compute the area of the region in Figure 1(A) enclosed by $y = 2 - x^2$ and $y = -2$.

2. Compute the area of the region in Figure 1(B) enclosed by $y = 2 - x^2$ and $y = x$.

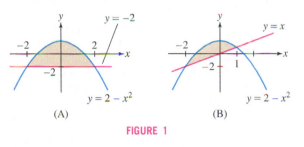

(A)   (B)

FIGURE 1

In Exercises 3–12, find the area of the region enclosed by the graphs of the functions.

3. $y = x^3 - 2x^2 + x$, $y = x^2 - x$

4. $y = x^2 + 2x$, $y = x^2 - 1$, $h(x) = x^2 + x - 2$

5. $x = 4y$, $x = 24 - 8y$, $y = 0$

6. $x = y^2 - 9$, $x = 15 - 2y$

7. $y = 4 - x^2$, $y = 3x$, $y = 4$

8. $\boxed{GU}$ $x = \dfrac{1}{2}y$, $x = y\sqrt{1 - y^2}$, $0 \le y \le 1$

9. $y = \sin x$, $y = \cos x$, $0 \le x \le \dfrac{5\pi}{4}$

10. $f(x) = \sin x$, $g(x) = \sin 2x$, $\dfrac{\pi}{3} \le x \le \pi$

11. $y = e^x$, $y = 1 - x$, $x = 1$

12. $y = \cosh 1 - \cosh x$, $y = \cosh x - \cosh 1$

13. $\boxed{GU}$ Use a graphing utility to locate the points of intersection of $y = e^{-x}$ and $y = 1 - x^2$ and find the area between the two curves (approximately).

14. Figure 2 shows a solid whose horizontal cross section at height y is a circle of radius $(1 + y)^{-2}$ for $0 \le y \le H$. Find the volume of the solid.

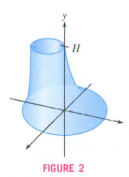

FIGURE 2

15. The base of a solid is the unit circle $x^2 + y^2 = 1$, and its cross sections perpendicular to the x-axis are rectangles of height 4. Find its volume.

16. The base of a solid is the triangle bounded by the axes and the line $2x + 3y = 12$, and its cross sections perpendicular to the y-axis have area $A(y) = (y + 2)$. Find its volume.

17. Find the total mass of a rod of length 1.2 m with linear density $\rho(x) = (1 + 2x + \frac{2}{9}x^3)$ kg/m.

18. Find the flow rate (in the correct units) through a pipe of diameter 6 cm if the velocity of fluid particles at a distance r from the center of the pipe is $v(r) = (3 - r)$ cm/s.

In Exercises 19–24, find the average value of the function over the interval.

19. $f(x) = x^3 - 2x + 2$, $[-1, 2]$ **20.** $f(x) = |x|$, $[-4, 4]$

21. $f(x) = x \cosh(x^2)$, $[0, 1]$ **22.** $f(x) = \dfrac{e^x}{1 + e^{2x}}$, $\left[0, \dfrac{1}{2}\right]$

23. $f(x) = \sqrt{9 - x^2}$, $[0, 3]$ *Hint:* Use geometry to evaluate the integral.

24. $f(x) = x \lfloor x \rfloor$, $[0, 3]$, where $\lfloor x \rfloor$ is the greatest integer function

25. Find $\displaystyle\int_2^5 g(t)\, dt$ if the average value of g on $[2, 5]$ is 9.

26. The average value of R over $[0, x]$ is equal to x for all x. Use the FTC to determine $R(x)$.

27. Use the Washer Method to find the volume obtained by rotating the region in Figure 3 about the x-axis.

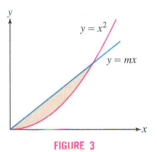

FIGURE 3

28. Use the Shell Method to find the volume obtained by rotating the region in Figure 3 about the x-axis.

In Exercises 29–40, use any method to find the volume of the solid obtained by rotating the region enclosed by the curves about the given axis.

29. $y = x^2 + 2$, $y = x + 4$, x-axis

30. $y = x^2 + 6$, $y = 8x - 1$, y-axis

31. $x = y^2 - 3$, $x = 2y$, axis $y = 4$

32. $y = 2x$, $y = 0$, $x = 8$, axis $x = -3$

33. $y = x^2 - 1$, $y = 2x - 1$, axis $x = -2$

34. $y = x^2 - 1$, $y = 2x - 1$,   axis $y = 4$

35. $y = -x^2 + 4x - 3$, $y = 0$,   axis $y = -1$

36. $y = -x^2 + 4x - 3$, $y = 0$,   axis $x = 4$

37. $x = 4y - y^3$, $x = 0$, $y \geq 0$, x-axis

38. $y^2 = x^{-1}$, $x = 1$, $x = 3$,   axis $y = -3$

39. $y = e^{-x^2/2}$, $y = -e^{-x^2/2}$, $x = 0$, $x = 1$, y-axis

40. $y = \sec x$, $y = \csc x$, $y = 0$, $x = 0$, $x = \dfrac{\pi}{2}$, x-axis

In Exercises 41–44, find the volume obtained by rotating the region about the given axis. The regions refer to the graph of the hyperbola $y^2 - x^2 = 1$ in Figure 4.

41. The shaded region between the upper branch of the hyperbola and the x-axis for $-c \leq x \leq c$, about the x-axis

42. The region between the upper branch of the hyperbola and the x-axis for $0 \leq x \leq c$, about the y-axis

43. The region between the upper branch of the hyperbola and the line $y = x$ for $0 \leq x \leq c$, about the x-axis

44. The region between the upper branch of the hyperbola and $y = 2$, about the y-axis

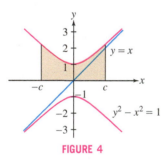

FIGURE 4

45. Let R be the intersection of the circles of radius 1 centered at $(1, 0)$ and $(0, 1)$. Express as an integral (but do not evaluate): **(a)** the area of R and **(b)** the volume of revolution of R about the x-axis.

46. Let R be the intersection of the circles of radius 1 centered at $(0, 0)$ and $(0, 1)$. Express an integral that gives the volume of revolution of R about the x-axis. (Do not evaluate the integral.)

47. Let $a > 0$. Show that the volume obtained when the region between $y = a\sqrt{x - ax^2}$ and the x-axis is rotated about the x-axis is independent of the constant a.

48. If 12 J of work are needed to stretch a spring 20 cm beyond equilibrium, how much work is required to compress it 6 cm beyond equilibrium?

49. A spring whose equilibrium length is 15 cm exerts a force of 50 N when it is stretched to 20 cm. Find the work required to stretch the spring from 22 cm to 24 cm.

50. If 18 ft-lb of work are needed to stretch a spring 1.5 ft beyond equilibrium, how far will the spring stretch if a 12-lb weight is attached to its end?

51. Let W be the work (against the Sun's gravitational force) required to transport an 80-kg person from Earth to Mars when the two planets are aligned with the Sun at their minimal distance of 55.7×10^6 km. Use Newton's Universal Law of Gravity (see Exercises 35–37 in Section 6.5) to express W as an integral and evaluate it. The Sun has mass $M_s = 1.99 \times 10^{30}$ kg, and the distance from the Sun to Earth is 149.6×10^6 km.

In Exercises 52 and 53, water is pumped into a spherical tank of radius 2 m from a source located 1 m below a hole at the bottom (Figure 5). The density of water is 1000 kg/m^3.

52. Calculate the work required to fill the tank.

53. Calculate the work $F(h)$ required to fill the tank to level h meters in the sphere.

54. A tank of mass 20 kg containing 100 kg of water (density 1000 kg/m^3) is raised vertically at a constant speed of 100 m/min for 1 min, during which time it leaks water at a rate of 40 kg/min. Calculate the total work performed in raising the container.

Water source

FIGURE 5

In a Mercator projection of Earth, a point located y radial units from the equator corresponds to a point on the globe with latitude given by the Gudermannian, a function given by $gd(y) =$ $\int_0^y \dfrac{dt}{\cosh t} = \tan^{-1}(\sinh y)$. *(Photodisc/Getty Images)*

7 TECHNIQUES OF INTEGRATION

In Section 5.7, we introduced substitution, one of the most important techniques of integration. In this chapter, we develop a second fundamental technique, Integration by Parts, as well as several techniques for treating particular classes of functions such as trigonometric and rational functions. However, there is no surefire method, and in fact, many important antiderivatives cannot be expressed in elementary terms. Therefore, we discuss numerical integration in the last section. Every definite integral can be approximated numerically to any desired degree of accuracy.

7.1 Integration by Parts

In this section, we derive a formula that often allows us to convert an integral that we cannot evaluate as is into an integral that we can evaluate. The Integration by Parts formula is derived from the Product Rule.

Let u and v be functions of x:

$$\frac{d}{dx}(uv) = u\frac{dv}{dx} + \frac{du}{dx}v$$

According to this formula, uv is an antiderivative of the right-hand side, so

$$uv = \int u\frac{dv}{dx}\,dx + \int v\frac{du}{dx}\,dx$$

Moving the second integral on the right to the other side, we obtain

$$\int u\frac{dv}{dx}\,dx = uv - \int v\frac{du}{dx}\,dx$$

By letting $du = \frac{du}{dx}\,dx$ and $dv = \frac{dv}{dx}\,dx$, we find the following

We can keep track of our choices using the pattern

$$u = \square \qquad dv = \square$$
$$du = \square \qquad v = \square$$

The original integral is the product of the terms on the top line. The resulting expression is the product of the terms on the main diagonal minus the integral of the product of the terms on the bottom line. If we shade the terms that are multiplied together for each term, we have

Integration by Parts Formula

$$\boxed{\int u\,dv = uv - \int v\,du} \qquad \boxed{1}$$

Because the Integration by Parts formula applies to a product, we should consider using it when the integrand is a product of two functions.

■ **EXAMPLE 1** Evaluate $\displaystyle\int x\cos x\,dx$.

Solution The integrand is a product, so we try writing $x\cos x\,dx = u\,dv$ with

$$u = x \qquad\qquad dv = \cos x\,dx$$
$$du = \frac{du}{dx}\,dx = 1\,dx \qquad\qquad v = \sin x$$

In applying Eq. (1), any antiderivative v may be used.

By the Integration by Parts formula,

$$\int \underbrace{x}_{u}\,\underbrace{\cos x\,dx}_{dv} = \underbrace{x\sin x}_{uv} - \int \underbrace{\sin x}_{v}\,\underbrace{dx}_{du} = x\sin x + \cos x + C$$

Let's check the answer by taking the derivative:

$$\frac{d}{dx}(x \sin x + \cos x + C) = x \cos x + \sin x - \sin x = x \cos x \qquad \blacksquare$$

The key step in Integration by Parts is deciding how to write the integral as a product $u\, dv$. Keep in mind that Integration by Parts expresses $\int u\, dv$ in terms of uv and $\int v\, du$. *This is only useful if $v\, du$ is easier to integrate than $u\, dv$.* Here are two guidelines:

- Choose dv so that $v = \displaystyle\int dv$ can be evaluated.
- Choose u so that $\frac{du}{dx}$ is "simpler" than u itself.

■ **EXAMPLE 2** **Good Versus Bad Choices of u and dv** Evaluate $\displaystyle\int xe^x\, dx$.

Solution Based on our guidelines, it makes sense to write $u\, dv = xe^x\, dx$ with

$$\begin{aligned} u &= x & dv &= e^x\, dx \\ du &= 1\, dx & v &= e^x \end{aligned}$$

This is a good choice since $\frac{du}{dx} = 1$ is simpler than u and we can evaluate

$$\int v\, du = \int e^x\, dx = e^x + C$$

Integration by Parts gives us

$$\int xe^x\, dx = uv - \int v\, du = xe^x - \int e^x\, dx = xe^x - e^x + C$$

Let's see what happens if we write $xe^x\, dx = u\, dv$ with

$$\begin{aligned} u &= e^x & dv &= x\, dx \\[6pt] du &= e^x\, dx & v &= \int x\, dx = \frac{1}{2}x^2 \end{aligned}$$

$$\int \underbrace{e^x}_{u}\ \underbrace{x\, dx}_{dv} = \underbrace{\frac{1}{2}x^2 e^x}_{uv} - \int \underbrace{\frac{1}{2}x^2}_{v}\ \underbrace{e^x\, dx}_{du}$$

This is a poor choice of u and dv because the integral on the right is more complicated than our original integral. $\qquad \blacksquare$

■ **EXAMPLE 3** **Integrating by Parts More Than Once** Evaluate $\displaystyle\int x^2 \cos x\, dx$.

In Example 3, it makes sense to take $u = x^2$ because Integration by Parts reduces the integration of $x^2 \cos x$ to the integration of $2x \sin x$, which is easier.

Solution Apply Integration by Parts a first time with $u = x^2$ and $dv = \cos x\, dx$:

$$\int \underbrace{x^2 \cos x\, dx}_{u\, dv} = \underbrace{x^2 \sin x}_{uv} - \int \underbrace{\sin x}_{v}\ \underbrace{2x\, dx}_{du} = x^2 \sin x - 2 \int x \sin x\, dx \qquad \boxed{2}$$

Now apply it again to the integral on the right, this time with $u = x$ and $dv = \sin x\, dx$:

$$\int \underbrace{x \sin x\, dx}_{u\, dv} = \underbrace{-x \cos x}_{uv} - \int \underbrace{(-\cos x)}_{v}\ \underbrace{dx}_{du} = -x \cos x + \sin x + C$$

Using this result in Eq. (2), we obtain

$$\int x^2 \cos x\, dx = x^2 \sin x - 2 \int x \sin x\, dx = x^2 \sin x - 2(-x \cos x + \sin x) + C$$

$$= x^2 \sin x + 2x \cos x - 2 \sin x + C \qquad \blacksquare$$

Integration by Parts applies to *definite integrals*:

$$\int_a^b u\, dv = uv \Big|_a^b - \int_a^b v\, du$$

■ **EXAMPLE 4** Taking $dv = dx$ Evaluate $\displaystyle\int_1^3 \ln x\, dx$.

Surprisingly, the choice $dv = dx$ is effective in some cases. Using it as in Example 4, we find that

$$\int \ln x\, dx = x \ln x - x + C$$

This choice also works for the inverse trigonometric functions (see Exercise 6).

Solution The integrand is not a product, so at first glance, this integral does not look like a candidate for Integration by Parts. However, we can treat $\ln x\, dx$ as a product of $\ln x$ and dx. Then

$$u = \ln x \qquad dv = dx$$
$$du = \frac{1}{x}\, dx \qquad v = x$$

$$\int_1^3 \underbrace{\ln x\, dx}_{u\, dv} = \underbrace{x \ln x}_{uv}\Big|_1^3 - \int_1^3 x\underbrace{\frac{1}{x}\, dx}_{v\, du} = x \ln x \Big|_1^3 - \int_1^3 dx$$

$$= x \ln x - x \Big|_1^3 = (3 \ln 3 - 3) - (1 \ln 1 - 1) = 3 \ln 3 - 2 \qquad ■$$

■ **EXAMPLE 5** Going in a Circle? Evaluate $\displaystyle\int e^x \cos x\, dx$.

Solution There are two reasonable ways of writing $e^x \cos x\, dx$ as $u\, dv$. Let's try setting $u = \cos x$. Then we have

$$u = \cos x \qquad dv = e^x\, dx$$
$$du = -\sin x\, dx \qquad v = e^x$$

In Example 5, the choice $u = e^x$, $dv = \cos x\, dx$ works equally well.

Thus,

$$\int \underbrace{e^x \cos x\, dx}_{u\, dv} = \underbrace{e^x \cos x}_{uv} - \int \underbrace{e^x(-\sin x)\, dx}_{v\, du} \qquad \boxed{3}$$

Now use Integration by Parts on the integral on the right with $u = \sin x$:

$$u = \sin x \qquad dv = e^x\, dx$$
$$du = \cos x\, dx \qquad v = e^x$$

$$\int e^x \sin x\, dx = e^x \sin x - \int e^x \cos x\, dx \qquad \boxed{4}$$

Eq. (4) brings us back to our original integral of $e^x \cos x$, so it looks as if we're going in a circle. But we can substitute Eq. (4) in Eq. (3) and solve for the integral of $e^x \cos x$:

$$\int e^x \cos x\, dx = e^x \cos x + \int e^x \sin x\, dx = e^x \cos x + \left(e^x \sin x - \int e^x \cos x\, dx \right)$$

Now we can add $\int e^x \cos x\, dx$ to both sides. Note that we add a "+ C" to the right side since we no longer have an integral on that side of the equation that will generate the necessary arbitrary constant:

Dividing an arbitrary constant by 2 still leaves an arbitrary constant, so we continue to denote it by C, absorbing the $\frac{1}{2}$ into the C.

$$2 \int e^x \cos x\, dx = e^x \cos x + e^x \sin x + C$$

$$\int e^x \cos x\, dx = \frac{1}{2} e^x (\cos x + \sin x) + C \qquad ■$$

*A reduction formula (also called a **recursive formula**) expresses the integral for a given value of n in terms of a similar integral for a smaller value of n. The desired integral is evaluated by applying the reduction formula repeatedly.*

Integration by Parts can be used to derive **reduction formulas** for integrals that depend on a positive integer n such as $\int x^n e^x \, dx$, $\int \ln^n x \, dx$, or the following example.

■ **EXAMPLE 6** **A Reduction Formula** Derive the reduction formula

$$\boxed{\int \sin^n x \, dx = -\frac{1}{n} \sin^{n-1} x \cos x + \frac{n-1}{n} \int \sin^{n-2} x \, dx} \quad \boxed{5}$$

Then evaluate $\int \sin^3 x \, dx$.

Solution Although we do not know how to integrate $\sin^n x$, we do know how to integrate $\sin x$. So we apply Integration by Parts as follows:

$$u = \sin^{n-1} x \qquad\qquad dv = \sin x \, dx$$

$$du = (n-1)\sin^{n-2} x \cos x \, dx \qquad v = -\cos x$$

Then we have

$$\int \sin^n x \, dx = \underbrace{-\sin^{n-1} x \cos x}_{uv} - \int \underbrace{(-\cos x)(n-1)\sin^{n-2} x \cos x \, dx}_{v\,du}$$

$$= -\sin^{n-1} x \cos x + (n-1)\int \sin^{n-2} x \cos^2 x \, dx$$

Using the fact $\cos^2 x = 1 - \sin^2 x$, we obtain

$$\int \sin^n x \, dx = -\sin^{n-1} x \cos x + (n-1)\int \sin^{n-2} x \, dx - (n-1)\int \sin^n x \, dx$$

Adding $(n-1)\int \sin^n x \, dx$ to both sides, we have

$$n\int \sin^n x \, dx = -\sin^{n-1} x \cos x + (n-1)\int \sin^{n-2} x \, dx$$

$$\int \sin^n x \, dx = -\frac{1}{n}\sin^{n-1} x \cos x + \frac{n-1}{n}\int \sin^{n-2} x \, dx$$

Applying this formula in the case $n = 3$, we obtain

$$\int \sin^3 x \, dx = -\frac{1}{3}\sin^2 x \cos x + \frac{2}{3}\int \sin x \, dx = -\frac{1}{3}\sin^2 x \cos x - \frac{2}{3}\cos x + C$$

■

In Exercise 58, you will be asked to derive the following reduction formula.

$$\boxed{\int \cos^n x \, dx = \frac{1}{n}\cos^{n-1} x \sin x + \frac{n-1}{n}\int \cos^{n-2} x \, dx} \quad \boxed{6}$$

7.1 SUMMARY

- Integration by Parts formula: $\int u \, dv = uv - \int v \, du$.
- The key step is deciding how to write the integrand as a product $u \, dv$. Keep in mind that Integration by Parts is useful when $v \, du$ is easier (or, at least, not more difficult) to integrate than $u \, dv$. Here are some guidelines:
  - Choose u so that $\frac{du}{dx}$ is simpler than u itself.
  - Choose dv so that $v = \int dv$ can be evaluated.
  - Sometimes, $dv = dx$ is a good choice.
  - Good choices for u include x^n, $\ln x$, and inverse trig functions.

7.1 EXERCISES

Preliminary Questions

1. Which derivative rule is used to derive the Integration by Parts formula?

2. For each of the following integrals, state whether substitution or Integration by Parts should be used:

$$\int x\cos(x^2)\,dx, \qquad \int x\cos x\,dx, \qquad \int x^2 e^x\,dx, \qquad \int xe^{x^2}\,dx$$

3. Why is $u=\cos x, dv = x\,dx$ a poor choice for evaluating
$$\int x\cos x\,dx?$$

Exercises

In Exercises 1–6, evaluate the integral using the Integration by Parts formula with the given choice of u and dv.

1. $\int x\sin x\,dx;\quad u=x, dv = \sin x\,dx$

2. $\int xe^{2x}\,dx;\quad u=x, dv = e^{2x}\,dx$

3. $\int (2x+9)e^x\,dx;\quad u=2x+9, dv = e^x\,dx$

4. $\int x\cos 4x\,dx;\quad u=x, dv = \cos 4x\,dx$

5. $\int x^3\ln x\,dx;\quad u=\ln x, dv = x^3\,dx$

6. $\int \tan^{-1}x\,dx;\quad u=\tan^{-1}x, dv = dx$

In Exercises 7–34, evaluate using Integration by Parts.

7. $\int (4x-3)e^{-x}\,dx$

8. $\int (2x+1)e^x\,dx$

9. $\int x\,e^{5x+2}\,dx$

10. $\int x^2 e^x\,dx$

11. $\int x\cos 2x\,dx$

12. $\int x\sin(3-x)\,dx$

13. $\int x^2\sin x\,dx$

14. $\int x^2\cos 3x\,dx$

15. $\int e^{-x}\sin x\,dx$

16. $\int e^x\sin 2x\,dx$

17. $\int e^{-5x}\sin x\,dx$

18. $\int e^{3x}\cos 4x\,dx$

19. $\int x\ln x\,dx$

20. $\int \frac{\ln x}{x^2}\,dx$

21. $\int x^2\ln x\,dx$

22. $\int x^{-5}\ln x\,dx$

23. $\int (\ln x)^2\,dx$

24. $\int x(\ln x)^2\,dx$

25. $\int \cos^{-1}x\,dx$

26. $\int \sin^{-1}x\,dx$

27. $\int \sec^{-1}x\,dx$

28. $\int x5^x\,dx$

29. $\int 3^x\cos x\,dx$

30. $\int x\sinh x\,dx$

31. $\int x^2\cosh x\,dx$

32. $\int \cos x\cosh x\,dx$

33. $\int \tanh^{-1}4x\,dx$

34. $\int \sinh^{-1}x\,dx$

In Exercises 35–36, evaluate using substitution and then Integration by Parts.

35. $\int e^{\sqrt{x}}\,dx$ *Hint:* Let $u=x^{1/2}$. **36.** $\int x^3 e^{x^2}\,dx$

In Exercises 37–46, evaluate using Integration by Parts, substitution, or both if necessary.

37. $\int x\cos 4x\,dx$

38. $\int \frac{\ln(\ln x)\,dx}{x}$

39. $\int \frac{x\,dx}{\sqrt{x+1}}$

40. $\int x^2(x^3+9)^{15}\,dx$

41. $\int \cos x\ln(\sin x)\,dx$

42. $\int \sin\sqrt{x}\,dx$

43. $\int \sqrt{x}e^{\sqrt{x}}\,dx$

44. $\int \frac{\tan\sqrt{x}\,dx}{\sqrt{x}}$

45. $\int \frac{\ln(\ln x)\ln x\,dx}{x}$

46. $\int \sin(\ln x)\,dx$

In Exercises 47–56, compute the definite integral.

47. $\int_0^3 xe^{4x}\,dx$

48. $\int_0^{\pi/4} x\sin 2x\,dx$

49. $\int_1^2 x\ln x\,dx$

50. $\int_1^e \frac{\ln x\,dx}{x^2}$

51. $\int_0^1 xe^{-x}\,dx$

52. $\int_0^1 \frac{x^3}{\sqrt{9+x^2}}\,dx$

53. $\int_0^1 x3^x\,dx$

54. $\int_0^1 x\cos(\pi x)\,dx$

55. $\int_0^\pi e^x\sin x\,dx$

56. $\int_0^1 \tan^{-1}x\,dx$

57. Use Eq. (5) to find $\int \sin^5 x\,dx$.

58. Derive the reduction formula

$$\int \cos^n x\,dx = \frac{1}{n}\cos^{n-1}x\sin x + \frac{n-1}{n}\int \cos^{n-2}x\,dx$$

that appears in Eq. (6).

59. Use the reduction formula from the problem above [Eq. (6)] to find
$$\int \cos^3 x\,dx.$$

60. Derive the reduction formula

$$\int x^n e^x\,dx = x^n e^x - n\int x^{n-1}e^x\,dx.$$

61. Use the reduction formula from Exercise 60 to find $\int x^3 e^x \, dx$.

62. Use substitution and the reduction formula from Exercise 60 to evaluate $\int x^4 e^{7x} \, dx$.

63. Find a reduction formula for $\int x^n e^{-x} \, dx$ similar to the formula appearing in Exercise 60.

64. Evaluate $\int x^n \ln x \, dx$ for $n \neq -1$. Which method should be used to evaluate $\int x^{-1} \ln x \, dx$?

65. Find the volume of the solid of revolution that results when the region under the graph of $f(x) = x\sqrt{\sin x}$ for $0 \leq x \leq \pi$ is revolved around the x-axis.

66. Find the volume of the solid of revolution that results when the region under the graph of $f(x) = \ln x$ for $1 \leq x \leq e$ is revolved around the x-axis.

In Exercises 67–74, indicate a good method for evaluating the integral (but do not evaluate). Your choices are algebraic manipulation, substitution (specify u and du), and Integration by Parts (specify u and dv). If it appears that the techniques you have learned thus far are not sufficient, state this.

67. $\int \sqrt{x} \ln x \, dx$

68. $\int \dfrac{x^2 - \sqrt{x}}{2x} \, dx$

69. $\int \dfrac{x^3 \, dx}{\sqrt{4 - x^2}}$

70. $\int \dfrac{dx}{\sqrt{4 - x^2}}$

71. $\int \dfrac{x + 2}{x^2 + 4x + 3} \, dx$

72. $\int \dfrac{dx}{(x + 2)(x^2 + 4x + 3)}$

73. $\int x \sin(3x + 4) \, dx$

74. $\int x \cos(9x^2) \, dx$

75. Evaluate $\int (\sin^{-1} x)^2 \, dx$. *Hint:* Use Integration by Parts first and then substitution.

76. Evaluate $\int \dfrac{(\ln x)^2 \, dx}{x^2}$. *Hint:* Use substitution first and then Integration by Parts.

77. Evaluate $\int x^7 \cos(x^4) \, dx$.

78. Find $f(x)$, assuming that
$$\int f(x) e^x \, dx = f(x) e^x - \int x^{-1} e^x \, dx$$

79. Find the volume of the solid obtained by revolving the region under $y = e^x$ for $0 \leq x \leq 2$ about the y-axis.

80. Find the area enclosed by $y = \ln x$ and $y = (\ln x)^2$.

81. Recall that the *present value* (PV) of an investment that pays out income continuously at a rate $R(t)$ for T years is $\int_0^T R(t) e^{-rt} \, dt$, where r is the interest rate. Find the PV if $R(t) = 5000 + 100t$ \$/year, $r = 0.05$, and $T = 10$ years.

82. Derive the reduction formula
$$\int (\ln x)^k \, dx = x(\ln x)^k - k \int (\ln x)^{k-1} \, dx \qquad \boxed{7}$$

83. Use Eq. (7) to calculate $\int (\ln x)^k \, dx$ for $k = 2, 3$.

84. Derive the reduction formulas
$$\int x^n \cos x \, dx = x^n \sin x - n \int x^{n-1} \sin x \, dx$$
$$\int x^n \sin x \, dx = -x^n \cos x + n \int x^{n-1} \cos x \, dx$$

85. Prove that $\int x b^x \, dx = b^x \left(\dfrac{x}{\ln b} - \dfrac{1}{\ln^2 b} \right) + C$.

86. Define $P_n(x)$ by
$$\int x^n e^x \, dx = P_n(x) e^x + C$$
Use the reduction formula in Problem 60 to prove that $P_n(x) = x^n - n P_{n-1}(x)$. Use this recursion relation to find $P_n(x)$ for $n = 1, 2, 3, 4$. Note that $P_0(x) = 1$.

Further Insights and Challenges

87. The Integration by Parts formula can be written
$$\int u v \, dx = u V - \int V \, du \qquad \boxed{8}$$
where $V(x)$ satisfies $V'(x) = v(x)$.

(a) Show directly that the right-hand side of Eq. (8) does not change if $V(x)$ is replaced by $V(x) + C$, where C is a constant.

(b) Use $u = \tan^{-1} x$ and $v = x$ in Eq. (8) to calculate $\int x \tan^{-1} x \, dx$, but carry out the calculation twice: first with $V(x) = \frac{1}{2}x^2$ and then with $V(x) = \frac{1}{2}x^2 + \frac{1}{2}$. Which choice of $V(x)$ results in a simpler calculation?

88. Prove in two ways that
$$\int_0^a f(x) \, dx = af(a) - \int_0^a x f'(x) \, dx \qquad \boxed{9}$$

First use Integration by Parts. Then assume f is increasing. Use the substitution $u = f(x)$ to prove that $\int_0^a x f'(x) \, dx$ is equal to the area

of the shaded region in Figure 1 and derive Eq. (9) a second time.

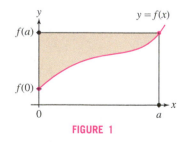

FIGURE 1

89. Assume that $f(0) = f(1) = 0$ and that f'' exists. Prove
$$\int_0^1 f''(x) f(x) \, dx = - \int_0^1 f'(x)^2 \, dx \qquad \boxed{10}$$

Use this to prove that if $f(0) = f(1) = 0$ and $f''(x) = \lambda f(x)$ for some constant λ, then $\lambda < 0$. Can you think of a function satisfying these conditions for some λ?

90. Set $I(a, b) = \int_0^1 x^a (1 - x)^b \, dx$, where a, b are whole numbers.

(a) Use substitution to show that $I(a, b) = I(b, a)$.

(b) Show that $I(a, 0) = I(0, a) = \dfrac{1}{a + 1}$.

(c) Prove that for $a \geq 1$ and $b \geq 0$,

$$I(a, b) = \frac{a}{b + 1} I(a - 1, b + 1)$$

(d) Use (b) and (c) to calculate $I(1, 1)$ and $I(3, 2)$.

(e) Show that $I(a, b) = \dfrac{a! \, b!}{(a + b + 1)!}$.

91. Let $I_n = \int x^n \cos(x^2) \, dx$ and $J_n = \int x^n \sin(x^2) \, dx$.

(a) Find a reduction formula that expresses I_n in terms of J_{n-2}. *Hint:* Write $x^n \cos(x^2)$ as $x^{n-1}(x \cos(x^2))$.

(b) Use the result of (a) to show that I_n can be evaluated explicitly if n is odd.

(c) Evaluate I_3.

7.2 Trigonometric Integrals

Many trigonometric functions can be integrated by combining substitution and Integration by Parts with the appropriate trigonometric identities. In this section, we consider techniques for integrating various types of trigonometric functions. We begin with integrals of the form

$$\int \sin^m x \cos^n x \, dx$$

where m, n are whole numbers. The easier case is when at least one of m, n is *odd*.

■ **EXAMPLE 1** Odd Power of $\sin x$ Evaluate $\int \sin^3 x \, dx$.

Solution We did this integral in Example 6 of the last section. However, we will use a different method that is more broadly applicable. Because $\sin^3 x$ is an odd power, we split off one power of $\sin x$ and use the identity $\sin^2 x = 1 - \cos^2 x$ to convert the rest of the integrand into an expression in $\cos x$:

$$\sin^3 x = (\sin^2 x)(\sin x) = (1 - \cos^2 x) \sin x$$

We then use the substitution $u = \cos x$, $du = -\sin x \, dx$:

$$\int \sin^3 x \, dx = \int (1 - \cos^2 x) \sin x \, dx = -\int (1 - u^2) \, du$$

$$= \frac{u^3}{3} - u + C = \frac{\cos^3 x}{3} - \cos x + C \qquad ■$$

The strategy of the previous example works when $\sin^m x$ appears with m odd, no matter what power of $\cos x$ is present. Similarly, if n is odd, we write $\cos^n x$ as a power of $(1 - \sin^2 x)$ times $\cos x$.

■ **EXAMPLE 2** Odd Power of $\cos x$ Evaluate $\int \sin^4 x \cos^5 x \, dx$.

Solution We take advantage of the fact that $\cos^5 x$ is an odd power to write

$$\sin^4 x \cos^5 x = \sin^4 x \, \cos^4 x (\cos x) = \sin^4 x (1 - \sin^2 x)^2 (\cos x)$$

$$= (\sin^4 x - 2 \sin^6 x + \sin^8 x) \cos x$$

This allows us to use the substitution $u = \sin x$, $du = \cos x \, dx$:

$$\int \sin^4 x \cos^5 x \, dx = \int (\sin^4 x - 2 \sin^6 x + \sin^8 x) \cos x \, dx$$

$$= \int (u^4 - 2u^6 + u^8) \, du$$

$$= \frac{u^5}{5} - \frac{2u^7}{7} + \frac{u^9}{9} + C = \frac{\sin^5 x}{5} - \frac{2 \sin^7 x}{7} + \frac{\sin^9 x}{9} + C \qquad ■$$

We will need a different strategy when neither $\sin x$ nor $\cos x$ appears with an odd power.

■ **EXAMPLE 3** Evaluate $\int \sin^2 x\, dx$.

Solution We could apply the reduction formula Eq. (5) from the last section. However, instead, we apply a method that does not rely on knowing that formula. We utilize the trigonometric identity called the double angle formula $\sin^2 x = \frac{1}{2}(1 - \cos 2x)$. Then

$$\int \sin^2 x\, dx = \int \frac{1}{2}(1 - \cos 2x)\, dx = \frac{x}{2} - \frac{\sin 2x}{4} + C \quad ■$$

Using the trigonometric identities in the margin, we can also integrate $\cos^2 x$, obtaining the following:

$$\boxed{\int \sin^2 x\, dx = \frac{x}{2} - \frac{\sin 2x}{4} + C = \frac{x}{2} - \frac{1}{2}\sin x \cos x + C} \qquad \boxed{1}$$

$$\boxed{\int \cos^2 x\, dx = \frac{x}{2} + \frac{\sin 2x}{4} + C = \frac{x}{2} + \frac{1}{2}\sin x \cos x + C} \qquad \boxed{2}$$

← **REMINDER** *Useful Identities:*

$$\sin^2 x = \frac{1}{2}(1 - \cos 2x)$$

$$\cos^2 x = \frac{1}{2}(1 + \cos 2x)$$

$$\sin 2x = 2\sin x \cos x$$

$$\cos 2x = \cos^2 x - \sin^2 x$$

■ **EXAMPLE 4** Evaluate $\int \sin^4 x\, dx$.

Solution Again, we could use the reduction formula that appears as Eq. (5) in the previous section, but rather than rely on that formula, we proceed as in the previous example. Using the double angle formula $\sin^2 x = \frac{1}{2}(1 - \cos 2x)$, we obtain

$$\int \sin^4 x\, dx = \int (\sin^2 x)^2\, dx = \int \left(\frac{1}{2}(1 - \cos 2x)\right)^2 dx$$

$$= \frac{1}{4}\int (1 - 2\cos 2x + \cos^2 2x)\, dx$$

Applying the double angle formula to $\cos^2 2x$, we get

$$\int \sin^4 x\, dx = \frac{1}{4}\int (1 - 2\cos 2x + \cos^2 2x)\, dx$$

$$= \frac{1}{4}\int \left(1 - 2\cos 2x + \frac{1 + \cos 4x}{2}\right) dx$$

$$= \frac{1}{4}\int \left(\frac{3}{2} - 2\cos 2x + \frac{\cos 4x}{2}\right) dx$$

Integrating $\sin^m x \cos^n x$

Case 1: *m* odd

Split off one power of $\sin x$, *and use* $\sin^2 x = 1 - \cos^2 x$ *to express the remaining powers of* $\sin x$ *in terms of* $\cos x$. *Then substitute* $u = \cos x, du = -\sin x\, dx$.

Case 2: *n* odd

Split off one power of $\cos x$, *and use* $\cos^2 x = 1 - \sin^2 x$ *to express the remaining powers of* $\cos x$ *in terms of* $\sin x$. *Then substitute* $u = \sin x, du = \cos x\, dx$.

Case 3: *m, n* both even

Either use the double angle formulas repeatedly (see Exercises 67–70), or convert the integrand into an expression entirely in terms of $\sin x$ *or* $\cos x$ *and then apply the reduction formulas [Eqs. (5) or (6)] from the previous section.*

By simple u-substitutions, this yields

$$\int \sin^4 x\, dx = \frac{1}{4}\left(\frac{3x}{2} - \sin 2x + \frac{\sin 4x}{8}\right) + C = \frac{3x}{8} - \frac{\sin 2x}{4} + \frac{\sin 4x}{32} + C \quad ■$$

Trigonometric integrals can be expressed in many different ways because trigonometric functions satisfy a large number of identities. For example, we can determine the integral in the previous example by applying the reduction formula in Eq. (5) from the previous section twice to obtain

$$\int \sin^4 x\, dx = -\frac{1}{4}\sin^3 x \cos x - \frac{3}{8}\sin x \cos x + \frac{3}{8}x + C$$

You can check that this agrees with the result in Example 4 (see Exercise 63).

More work is required to integrate $\sin^m x \cos^n x$ when both m and n are even.

■ **EXAMPLE 5** **Even Powers of $\sin x$ and $\cos x$** Evaluate $\int \sin^2 x \cos^4 x\, dx$.

Solution Here, $m = 2$ and $n = 4$. Since $m < n$, we replace $\sin^2 x$ by $1 - \cos^2 x$:

$$\int \sin^2 x \cos^4 x\, dx = \int (1 - \cos^2 x)\cos^4 x\, dx = \int \cos^4 x\, dx - \int \cos^6 x\, dx$$

The reduction formula for $n = 6$ gives

$$\int \cos^6 x \, dx = \frac{1}{6} \cos^5 x \sin x + \frac{5}{6} \int \cos^4 x \, dx$$

Using this result in the right-hand side of the first equation in this solution, we obtain

$$\int \sin^2 x \cos^4 x \, dx = \int \cos^4 x \, dx - \left(\frac{1}{6} \cos^5 x \sin x + \frac{5}{6} \int \cos^4 x \, dx \right)$$

$$= -\frac{1}{6} \cos^5 x \sin x + \frac{1}{6} \int \cos^4 x \, dx$$

Next, we evaluate $\int \cos^4 x \, dx$ using the reduction formulas for $n = 4$ and $n = 2$:

$$\int \cos^4 x \, dx = \frac{1}{4} \cos^3 x \sin x + \frac{3}{4} \int \cos^2 x \, dx$$

$$= \frac{1}{4} \cos^3 x \sin x + \frac{3}{4} \left(\frac{1}{2} \cos x \sin x + \frac{1}{2} x \right) + C$$

$$= \frac{1}{4} \cos^3 x \sin x + \frac{3}{8} \cos x \sin x + \frac{3}{8} x + C$$

As we have noted, trigonometric integrals can be expressed in more than one way. According to Mathematica,

$$\int \sin^2 x \cos^4 x \, dx$$

$$= \tfrac{1}{16} x + \tfrac{1}{64} \sin 2x - \tfrac{1}{64} \sin 4x - \tfrac{1}{192} \sin 6x$$

Trigonometric identities show that this agrees with the solution for Example 5.

Altogether,

$$\int \sin^2 x \cos^4 x \, dx = -\frac{1}{6} \cos^5 x \sin x + \frac{1}{6} \left(\frac{1}{4} \cos^3 x \sin x + \frac{3}{8} \cos x \sin x + \frac{3}{8} x \right) + C$$

$$= -\frac{1}{6} \cos^5 x \sin x + \frac{1}{24} \cos^3 x \sin x + \frac{1}{16} \cos x \sin x + \frac{1}{16} x + C \quad \blacksquare$$

We now consider integrals of the remaining trigonometric functions.

■ **EXAMPLE 6** Integral of the Tangent and Secant Derive the formulas

$$\int \tan x \, dx = \ln | \sec x | + C \qquad \text{and} \qquad \int \sec x \, dx = \ln \left| \sec x + \tan x \right| + C$$

Solution To integrate $\tan x$, use the substitution $u = \cos x$, $du = -\sin x \, dx$:

$$\int \tan x \, dx = \int \frac{\sin x}{\cos x} \, dx = -\int \frac{du}{u} = -\ln |u| + C = -\ln |\cos x| + C$$

$$= \ln \frac{1}{|\cos x|} + C = \ln |\sec x| + C$$

The integral $\int \sec x \, dx$ was first computed numerically in the 1590s by the English mathematician Edward Wright, decades before the invention of calculus. Although he did not invent the concept of an integral, Wright realized that the sums that approximate the integral hold the key to understanding the Mercator map projection, of great importance in sea navigation because it enabled sailors to reach their destinations along lines of fixed compass direction. The formula for the integral was first proved by James Gregory in 1668.

To integrate $\sec x$, we employ a clever and highly non-obvious trick: Multiply the integrand by

$$1 = \frac{\sec x + \tan x}{\sec x + \tan x}$$

Then

$$\int \sec x \, dx = \int \sec x \left(\frac{\sec x + \tan x}{\sec x + \tan x} \right) dx = \int \frac{\sec^2 x + \sec x \tan x}{\sec x + \tan x} \, dx$$

Noting that the numerator is the derivative of the denominator, we let the denominator be u:

$$u = \sec x + \tan x \qquad \text{and} \qquad du = (\sec x \tan x + \sec^2 x) \, dx$$

Then our integral becomes

$$\int \frac{du}{u} = \ln |u| + C = \ln | \sec x + \tan x | + C \qquad \blacksquare$$

The table of integrals at the end of this section (page 383) contains a list of additional trigonometric integrals and reduction formulas.

■ **EXAMPLE 7 Using a Table of Integrals** Evaluate $\displaystyle\int_0^{\pi/4} \tan^3 x \, dx$.

Solution We use reduction formula (10) in the table with $k = 3$.

$$\int_0^{\pi/4} \tan^3 x \, dx = \frac{\tan^2 x}{2}\Big|_0^{\pi/4} - \int_0^{\pi/4} \tan x \, dx = \left(\frac{1}{2}\tan^2 x - \ln|\sec x|\right)\Big|_0^{\pi/4}$$

$$= \left(\frac{1}{2}\tan^2\frac{\pi}{4} - \ln\left|\sec\frac{\pi}{4}\right|\right) - \left(\frac{1}{2}\tan^2 0 - \ln|\sec 0|\right)$$

$$= \left(\frac{1}{2}(1)^2 - \ln\sqrt{2}\right) - \left(\frac{1}{2}0^2 - \ln|1|\right) = \frac{1}{2} - \ln\sqrt{2}$$ ■

In the margin, we describe a method for integrating $\tan^m x \sec^n x$.

■ **EXAMPLE 8** Evaluate $\displaystyle\int \tan^3 x \sec^5 x \, dx$.

Solution We note that we have a copy of $\sec x \tan x$ in the integrand, and the remaining powers of $\tan x$ are even. So we separate out one copy of $\sec x \tan x$ and convert the rest of the integrand into powers of $\sec x$. Then since the derivative of $\sec x$ is $\sec x \tan x$, we are set up for a u-substitution.

The first step is to use the identity $\tan^2 x = \sec^2 x - 1$:

$$\int \tan^3 x \sec^5 x \, dx = \int (\sec^2 x - 1)(\sec^4 x)(\sec x \tan x) \, dx$$

$$= \int (\sec^6 x - \sec^4 x) \sec x \tan x \, dx$$

Letting $u = \sec x$, so $du = \sec x \tan x \, dx$, we have

$$\int \tan^3 x \sec^5 x \, dx = \int (u^6 - u^4) \, du = \frac{u^7}{7} - \frac{u^5}{5} + C$$

$$= \frac{\sec^7 x}{7} - \frac{\sec^5 x}{5} + C$$ ■

Note that the above method works whenever we integrate $\int \tan^m x \sec^n x \, dx$, where m is odd and $n > 0$.

■ **EXAMPLE 9** Evaluate $\displaystyle\int \tan^2 x \sec^4 x \, dx$.

Solution In this case, we separate $\sec^2 x$ and convert the rest of the integrand into powers of $\tan x$, since the derivative of $\tan x$ is $\sec^2 x$. Using the fact $\sec^2 x = \tan^2 x + 1$ yields

$$\int \tan^2 x \sec^4 x \, dx = \int (\tan^2 x)(\tan^2 x + 1)(\sec^2 x) \, dx = \int (\tan^4 x + \tan^2 x) \sec^2 x \, dx$$

Setting $u = \tan x$ and therefore $du = \sec^2 x \, dx$, we have

$$\int \tan^2 x \sec^4 x \, dx = \int (u^4 + u^2) \, du = \frac{u^5}{5} + \frac{u^3}{3} + C = \frac{\tan^5 x}{5} + \frac{\tan^3 x}{3} + C$$ ■

The above method works to integrate $\int \tan^m x \sec^n x \, dx$ whenever $n > 0$ is even. The last case to deal with is when m is even and n is odd. Then we can convert the integrand to be entirely in terms of powers of $\sec x$ and apply reduction formulas.

Formulas (17)–(19) in the table describe the integrals of the products $\sin mx \sin nx$, $\cos mx \cos nx$, and $\sin mx \cos nx$. These integrals appear in the theory of Fourier Series, which is a fundamental technique used extensively in engineering and physics.

Integrating $\tan^m x \sec^n x$

Case 1: m odd and $n \geq 1$

Separate out the factor $\sec x \tan x$ and use the identity $\tan^2 x = \sec^2 x - 1$ to express the rest of the integrand in terms of $\sec x$. Then use the substitution $u = \sec x$, $du = \sec x \tan x \, dx$ to obtain an integral involving only powers of u.

Case 2: n even

Separate out a factor of $\sec^2 x$ and use the identity $\sec^2 x = 1 + \tan^2 x$ to express the rest of the integrand in terms of $\tan x$. Then substitute $u = \tan x$, $du = \sec^2 x \, dx$ to obtain an integral involving only powers of u.

Case 3: m even and n odd

Use the identity $\tan^2 x = \sec^2 x - 1$ to obtain an integral involving only powers of $\sec x$ and use the reduction formula (14).

■ **EXAMPLE 10** Integral of sin mx cos nx Evaluate $\int_0^\pi \sin 4x \cos 3x \, dx$.

Solution Apply formula (18), with $m = 4$ and $n = 3$:

$$\int_0^\pi \sin 4x \cos 3x \, dx = \left(-\frac{\cos(4-3)x}{2(4-3)} - \frac{\cos(4+3)x}{2(4+3)} \right) \Bigg|_0^\pi$$

$$= \left(-\frac{\cos x}{2} - \frac{\cos 7x}{14} \right) \Bigg|_0^\pi$$

$$= \left(\frac{1}{2} + \frac{1}{14} \right) - \left(-\frac{1}{2} - \frac{1}{14} \right) = \frac{8}{7} \qquad ■$$

TABLE OF TRIGONOMETRIC INTEGRALS

$$\int \sin^2 x \, dx = \frac{x}{2} - \frac{\sin 2x}{4} + C = \frac{x}{2} - \frac{1}{2} \sin x \cos x + C \qquad \boxed{3}$$

$$\int \cos^2 x \, dx = \frac{x}{2} + \frac{\sin 2x}{4} + C = \frac{x}{2} + \frac{1}{2} \sin x \cos x + C \qquad \boxed{4}$$

$$\int \sin^n x \, dx = -\frac{\sin^{n-1} x \cos x}{n} + \frac{n-1}{n} \int \sin^{n-2} x \, dx \qquad \boxed{5}$$

$$\int \cos^n x \, dx = \frac{\cos^{n-1} x \sin x}{n} + \frac{n-1}{n} \int \cos^{n-2} x \, dx \qquad \boxed{6}$$

$$\int \sin^m x \cos^n x \, dx = \frac{\sin^{m+1} x \cos^{n-1} x}{m+n} + \frac{n-1}{m+n} \int \sin^m x \cos^{n-2} x \, dx \qquad \boxed{7}$$

$$\int \sin^m x \cos^n x \, dx = -\frac{\sin^{m-1} x \cos^{n+1} x}{m+n} + \frac{m-1}{m+n} \int \sin^{m-2} x \cos^n x \, dx \qquad \boxed{8}$$

$$\int \tan x \, dx = \ln|\sec x| + C = -\ln|\cos x| + C \qquad \boxed{9}$$

$$\int \tan^m x \, dx = \frac{\tan^{m-1} x}{m-1} - \int \tan^{m-2} x \, dx \qquad \boxed{10}$$

$$\int \cot x \, dx = -\ln|\csc x| + C = \ln|\sin x| + C \qquad \boxed{11}$$

$$\int \cot^m x \, dx = -\frac{\cot^{m-1} x}{m-1} - \int \cot^{m-2} x \, dx \qquad \boxed{12}$$

$$\int \sec x \, dx = \ln|\sec x + \tan x| + C \qquad \boxed{13}$$

$$\int \sec^m x \, dx = \frac{\tan x \sec^{m-2} x}{m-1} + \frac{m-2}{m-1} \int \sec^{m-2} x \, dx \qquad \boxed{14}$$

$$\int \csc x \, dx = \ln|\csc x - \cot x| + C \qquad \boxed{15}$$

$$\int \csc^m x \, dx = -\frac{\cot x \csc^{m-2} x}{m-1} + \frac{m-2}{m-1} \int \csc^{m-2} x \, dx \qquad \boxed{16}$$

$$\int \sin mx \sin nx \, dx = \frac{\sin(m-n)x}{2(m-n)} - \frac{\sin(m+n)x}{2(m+n)} + C \quad (m \neq \pm n) \qquad \boxed{17}$$

$$\int \sin mx \cos nx \, dx = -\frac{\cos(m-n)x}{2(m-n)} - \frac{\cos(m+n)x}{2(m+n)} + C \quad (m \neq \pm n) \qquad \boxed{18}$$

$$\int \cos mx \cos nx \, dx = \frac{\sin(m-n)x}{2(m-n)} + \frac{\sin(m+n)x}{2(m+n)} + C \quad (m \neq \pm n) \qquad \boxed{19}$$

Although we include this table of trigonometric integrals, it makes more sense to understand how to obtain a given integral than to simply rely on the table.

7.2 SUMMARY

- To integrate an odd power of $\sin x$ times $\cos^n x$, write

$$\int \sin^{2k+1} x \cos^n x \, dx = \int (1 - \cos^2 x)^k \cos^n x \sin x \, dx$$

Then use the substitution $u = \cos x$, $du = -\sin x \, dx$.
- To integrate an odd power of $\cos x$ times $\sin^m x$, write

$$\int \sin^m x \cos^{2k+1} x \, dx = \int (\sin^m x)(1 - \sin^2 x)^k \cos x \, dx$$

Then use the substitution $u = \sin x$, $du = \cos x \, dx$.
- If both $\sin x$ and $\cos x$ occur to an even power, write

$$\int \sin^m x \cos^n x \, dx = \int (1 - \cos^2 x)^{m/2} \cos^n x \, dx \quad \text{(if } m \leq n)$$

$$\int \sin^m x \cos^n x \, dx = \int (\sin^m x)(1 - \sin^2 x)^{n/2} \, dx \quad \text{(if } m \geq n)$$

Expand the right-hand side to obtain a sum of powers of $\cos x$ or powers of $\sin x$. Then use the reduction formulas

$$\int \sin^n x \, dx = -\frac{1}{n} \sin^{n-1} x \cos x + \frac{n-1}{n} \int \sin^{n-2} x \, dx$$

$$\int \cos^n x \, dx = \frac{1}{n} \cos^{n-1} x \sin x + \frac{n-1}{n} \int \cos^{n-2} x \, dx$$

One can also apply the double angle trigonometric identities to obtain an integrand with smaller powers of sine and cosine in the variable $2x$.
- The integral $\int \tan^m x \sec^n x \, dx$ can be evaluated as in the marginal note on page 382.

7.2 EXERCISES

Preliminary Questions

1. Describe the technique used to evaluate $\int \sin^5 x \, dx$.

2. Describe a way of evaluating $\int \sin^6 x \, dx$.

3. Are reduction formulas needed to evaluate $\int \sin^7 x \cos^2 x \, dx$? Why or why not?

4. Describe a way of evaluating $\int \sin^6 x \cos^2 x \, dx$.

5. Which integral requires more work to evaluate?

$$\int \sin^{798} x \cos x \, dx \quad \text{or} \quad \int \sin^4 x \cos^4 x \, dx$$

Explain your answer.

Exercises

In Exercises 1–6, use the method for odd powers to evaluate the integral.

1. $\int \cos^3 x \, dx$

2. $\int \sin^5 x \, dx$

3. $\int \sin^3 \theta \cos^2 \theta \, d\theta$

4. $\int \sin^5 x \cos x \, dx$

5. $\int \sin^3 t \cos^3 t \, dt$

6. $\int \sin^2 x \cos^5 x \, dx$

7. Find the area of the shaded region in Figure 1.

8. Use the identity $\sin^2 x = 1 - \cos^2 x$ to write $\int \sin^2 x \cos^2 x \, dx$ as a sum of two integrals, and then evaluate using the reduction formula.

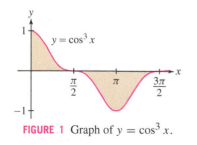

FIGURE 1 Graph of $y = \cos^3 x$.

In Exercises 9–12, evaluate the integral using methods employed in Examples 4 and 5.

9. $\int \cos^4 y \, dy$

10. $\int \cos^2 \theta \sin^2 \theta \, d\theta$

11. $\int \sin^4 x \cos^2 x \, dx$

12. $\int \sin^2 x \cos^6 x \, dx$

In Exercises 13 and 14, evaluate using Eq. (7).

13. $\int \sin^3 x \cos^2 x \, dx$

14. $\int \sin^2 x \cos^4 x \, dx$

In Exercises 15–18, evaluate the integral using the method described on page 382 and the reduction formulas on page 383 as necessary.

15. $\int \tan^3 x \sec x \, dx$

16. $\int \tan^2 x \sec x \, dx$

17. $\int \tan^2 x \sec^4 x \, dx$

18. $\int \tan^8 x \sec^2 x \, dx$

In Exercises 19–22, evaluate using methods similar to those that apply to integral $\tan^m x \sec^n x$.

19. $\int \cot^3 x \, dx$

20. $\int \sec^3 x \, dx$

21. $\int \cot^5 x \csc^2 x \, dx$

22. $\int \cot^4 x \csc x \, dx$

In Exercises 23–48, evaluate the integral.

23. $\int \cos^5 x \sin x \, dx$

24. $\int \cos^3(2-x) \sin(2-x) \, dx$

25. $\int \cos^4(3x+2) \, dx$

26. $\int \cos^7 3x \, dx$

27. $\int \cos^3(\pi\theta) \sin^4(\pi\theta) \, d\theta$

28. $\int \cos^{498} y \sin^3 y \, dy$

29. $\int \sin^4(3x) \, dx$

30. $\int \sin^2 x \cos^6 x \, dx$

31. $\int \frac{\cos^5 x}{\sin^3 x} \, dx$

32. $\int \frac{\sin^7 x}{\cos^4 x} \, dx$

33. $\int \csc^2(3-2x) \, dx$

34. $\int \csc^3 x \, dx$

35. $\int \tan x \sec^2 x \, dx$

36. $\int \tan^3 \theta \sec^3 \theta \, d\theta$

37. $\int \tan^5 x \sec^4 x \, dx$

38. $\int \tan^4 x \sec x \, dx$

39. $\int \tan^6 x \sec^4 x \, dx$

40. $\int \tan^2 x \sec^3 x \, dx$

41. $\int \cot^5 x \csc^5 x \, dx$

42. $\int \cot^2 x \csc^4 x \, dx$

43. $\int \sin 2x \cos 2x \, dx$

44. $\int \cos 4x \cos 6x \, dx$

45. $\int t \cos^3(t^2) \, dt$

46. $\int \frac{\tan^3(\ln t)}{t} \, dt$

47. $\int \cos^2(\sin t) \cos t \, dt$

48. $\int e^x \tan^2(e^x) \, dx$

In Exercises 49–62, evaluate the definite integral.

49. $\int_0^{2\pi} \sin^2 x \, dx$

50. $\int_0^{\pi/2} \cos^3 x \, dx$

51. $\int_0^{\pi/2} \sin^5 x \, dx$

52. $\int_0^{\pi/2} \sin^2 x \cos^3 x \, dx$

53. $\int_0^{\pi/4} \frac{dx}{\cos x}$

54. $\int_{\pi/4}^{\pi/2} \frac{dx}{\sin x}$

55. $\int_0^{\pi/3} \tan x \, dx$

56. $\int_0^{\pi/4} \tan^5 x \, dx$

57. $\int_{-\pi/4}^{\pi/4} \sec^4 x \, dx$

58. $\int_{\pi/4}^{3\pi/4} \cot^4 x \csc^2 x \, dx$

59. $\int_0^{\pi} \sin 3x \cos 4x \, dx$

60. $\int_0^{\pi} \sin x \sin 3x \, dx$

61. $\int_0^{\pi/6} \sin 2x \cos 4x \, dx$

62. $\int_0^{\pi/4} \sin 7x \cos 2x \, dx$

63. Use the identities for $\sin 2x$ and $\cos 2x$ on page 380 to verify that the following formulas are equivalent:

$$\int \sin^4 x \, dx = \frac{1}{32}(12x - 8\sin 2x + \sin 4x) + C$$

$$\int \sin^4 x \, dx = -\frac{1}{4}\sin^3 x \cos x - \frac{3}{8}\sin x \cos x + \frac{3}{8}x + C$$

64. Evaluate $\int \sin^2 x \cos^3 x \, dx$ using the method described in the text and verify that your result is equivalent to the following result produced by a computer algebra system:

$$\int \sin^2 x \cos^3 x \, dx = \frac{1}{30}(7 + 3\cos 2x) \sin^3 x + C$$

65. Find the volume of the solid obtained by revolving $y = \sin x$ for $0 \le x \le \pi$ about the x-axis.

66. Use Integration by Parts to prove Eqs. (1) and (2).

In Exercises 67–70, use the following alternative method for evaluating the integral $J = \int \sin^m x \cos^n x \, dx$ when m and n are both even. Use the identities

$$\sin^2 x = \frac{1}{2}(1 - \cos 2x), \qquad \cos^2 x = \frac{1}{2}(1 + \cos 2x)$$

to write $J = \frac{1}{4} \int (1 - \cos 2x)^{m/2}(1 + \cos 2x)^{n/2} \, dx$, and expand the right-hand side as a sum of integrals involving smaller powers of sine and cosine in the variable $2x$.

67. $\int \sin^2 x \cos^2 x \, dx$

68. $\int \cos^4 x \, dx$

69. $\int \sin^4 x \cos^2 x \, dx$

70. $\int \sin^6 x \, dx$

71. Prove the reduction formula

$$\int \tan^k x \, dx = \frac{\tan^{k-1} x}{k-1} - \int \tan^{k-2} x \, dx$$

Hint: $\tan^k x = (\sec^2 x - 1) \tan^{k-2} x$.

72. Use the substitution $u = \csc x - \cot x$ to evaluate $\int \csc x \, dx$ (see Example 6).

73. Let $I_m = \int_0^{\pi/2} \sin^m x \, dx$.

(a) Show that $I_0 = \frac{\pi}{2}$ and $I_1 = 1$.

(b) Prove that, for $m \ge 2$,

$$I_m = \frac{m-1}{m} I_{m-2}$$

(c) Use (a) and (b) to compute I_m for $m = 2, 3, 4, 5$.

74. Evaluate $\int_0^{\pi} \sin^2 mx \, dx$ for m an arbitrary integer.

75. Evaluate $\int \sin x \ln(\sin x) \, dx$. *Hint:* Use Integration by Parts as a first step.

76. Total Energy   A 100-watt (W) light bulb has resistance $R = 144$ ohms (Ω) when attached to household current, where the voltage varies as $V = V_0 \sin(2\pi f t)$ ($V_0 = 110$ V, $f = 60$ Hz). The energy (in joules) expended by the bulb over a period of T seconds is

$$U = \int_0^T P(t)\,dt$$

where $P = V^2/R$ (J/s) is the power. Compute U if the bulb remains on for 5 h.

77. Let m, n be integers with $m \neq \pm n$. Use Eqs. (17)–(19) to prove the so-called **orthogonality relations** that play a basic role in the theory of Fourier Series (Figure 2):

$$\int_0^\pi \sin mx \sin nx\,dx = 0$$

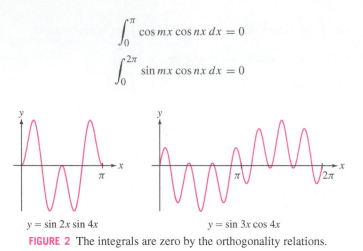

$$\int_0^\pi \cos mx \cos nx\,dx = 0$$

$$\int_0^{2\pi} \sin mx \cos nx\,dx = 0$$

$y = \sin 2x \sin 4x$ $y = \sin 3x \cos 4x$

FIGURE 2 The integrals are zero by the orthogonality relations.

Further Insights and Challenges

78. Use the trigonometric identity

$$\sin mx \cos nx = \frac{1}{2}\big(\sin(m-n)x + \sin(m+n)x\big)$$

to prove Eq. (18) in the table of integrals on page 383.

79. Use Integration by Parts to prove that (for $m \neq 1$)

$$\int \sec^m x\,dx = \frac{\tan x \sec^{m-2} x}{m-1} + \frac{m-2}{m-1}\int \sec^{m-2} x\,dx$$

80. Set $I_m = \int_0^{\pi/2} \sin^m x\,dx$. Use Exercise 73 to prove that

$$I_{2m} = \frac{2m-1}{2m}\,\frac{2m-3}{2m-2}\cdots\frac{1}{2}\cdot\frac{\pi}{2}$$

$$I_{2m+1} = \frac{2m}{2m+1}\,\frac{2m-2}{2m-1}\cdots\frac{2}{3}$$

Conclude that

$$\frac{\pi}{2} = \frac{2\cdot 2}{1\cdot 3}\cdot\frac{4\cdot 4}{3\cdot 5}\cdots\frac{2m\cdot 2m}{(2m-1)(2m+1)}\,\frac{I_{2m}}{I_{2m+1}}$$

81. This is a continuation of Exercise 80.

(a) Prove that $I_{2m+1} \leq I_{2m} \leq I_{2m-1}$. *Hint:*

$$\sin^{2m+1} x \leq \sin^{2m} x \leq \sin^{2m-1} x \quad \text{for} \quad 0 \leq x \leq \tfrac{\pi}{2}$$

(b) Show that

$$\frac{I_{2m-1}}{I_{2m+1}} = 1 + \frac{1}{2m}.$$

(c) Show that $1 \leq \dfrac{I_{2m}}{I_{2m+1}} \leq 1 + \dfrac{1}{2m}$.

(d) Prove that $\displaystyle\lim_{m\to\infty} \frac{I_{2m}}{I_{2m+1}} = 1$.

(e) Finally, deduce the infinite product for $\frac{\pi}{2}$ discovered by English mathematician John Wallis (1616–1703):

$$\frac{\pi}{2} = \lim_{m\to\infty} \frac{2}{1}\cdot\frac{2}{3}\cdot\frac{4}{3}\cdot\frac{4}{5}\cdots\frac{2m\cdot 2m}{(2m-1)(2m+1)}$$

7.3 Trigonometric Substitution

Our next goal is to integrate functions involving one of the square root expressions:

$$\sqrt{a^2 - x^2},\qquad \sqrt{x^2 + a^2},\qquad \sqrt{x^2 - a^2}$$

More generally, we will see that we can also integrate functions involving $\sqrt{ax^2 + bx + c}$. In each case, a substitution transforms the integral into a trigonometric integral. For example, if we let $x = a\sin\theta$ in the first case, we can use the fact $\sin^2\theta + \cos^2\theta = 1$ to obtain

$$\sqrt{a^2 - x^2} = \sqrt{a^2 - a^2\sin^2\theta} = a\sqrt{1 - \sin^2\theta} = a\sqrt{\cos^2\theta}$$

$$= a\cos\theta \quad \text{for} \quad -\frac{\pi}{2} \leq \theta \leq \frac{\pi}{2}$$

The other root functions listed above can also be integrated by applying identities involving the other trigonometric functions.

■ **EXAMPLE 1** Evaluate $\displaystyle\int \frac{1}{\sqrt{1-x^2}}\,dx$.

Solution

Step 1. Substitute to eliminate the square root.

The integrand is defined for $-1 < x < 1$, so we may set $x = \sin\theta$, where $-\frac{\pi}{2} < \theta < \frac{\pi}{2}$. Because $\cos\theta > 0$ for such θ, we obtain the positive square root

$$\sqrt{1-x^2} = \sqrt{1-\sin^2\theta} = \sqrt{\cos^2\theta} = \cos\theta \qquad \boxed{1}$$

Step 2. Evaluate the trigonometric integral.

Since $x = \sin\theta$, we have $dx = \cos\theta\,d\theta$, and $\dfrac{1}{\sqrt{1-x^2}}\,dx = \dfrac{1}{\cos\theta}(\cos\theta\,d\theta) = d\theta$. Thus,

$$\int \frac{1}{\sqrt{1-x^2}}\,dx = \int d\theta = \theta + C$$

Step 3. Convert back to the original variable.

Since $x = \sin\theta$ for $-\frac{\pi}{2} < \theta < \frac{\pi}{2}$, the inverse of $\sin\theta$ is defined, and $\theta = \sin^{-1}x$. Therefore,

$$\int \frac{1}{\sqrt{1-x^2}}\,dx = \sin^{-1}x + C \qquad ■$$

■ **EXAMPLE 2** Evaluate $\displaystyle\int \sqrt{1-x^2}\,dx$.

Solution

Step 1. Substitute to eliminate the square root.

The integrand is defined for $-1 \le x \le 1$, so, as in the previous example, we set $x = \sin\theta$, where $-\frac{\pi}{2} \le \theta \le \frac{\pi}{2}$. Because $\cos\theta \ge 0$ for such θ, we again obtain the positive square root

$$\sqrt{1-x^2} = \sqrt{1-\sin^2\theta} = \sqrt{\cos^2\theta} = \cos\theta \qquad \boxed{2}$$

Step 2. Evaluate the trigonometric integral.

Since $x = \sin\theta$, we have $dx = \cos\theta\,d\theta$, and $\sqrt{1-x^2}\,dx = \cos\theta(\cos\theta\,d\theta)$. Thus,

$$\int \sqrt{1-x^2}\,dx = \int \cos^2\theta\,d\theta = \frac{1}{2}\theta + \frac{1}{2}\sin\theta\cos\theta + C$$

Step 3. Convert back to the original variable.

It remains to express the answer in terms of x (see Figure 1):

$$x = \sin\theta, \qquad \theta = \sin^{-1}x, \qquad \sqrt{1-x^2} = \cos\theta$$

$$\int \sqrt{1-x^2}\,dx = \frac{1}{2}\theta + \frac{1}{2}\sin\theta\cos\theta + C = \frac{1}{2}\sin^{-1}x + \frac{1}{2}x\sqrt{1-x^2} + C \qquad ■$$

$$\int \cos^2\theta\,d\theta = \frac{1}{2}\theta + \frac{1}{2}\sin\theta\cos\theta + C$$

FIGURE 1 Right triangle with $x = \sin\theta$, from which $\cos\theta = \sqrt{1-x^2}$.

Note: If $x = a\sin\theta$ and $a > 0$, then

$$a^2 - x^2 = a^2(1-\sin^2\theta) = a^2\cos^2\theta$$

For $-\frac{\pi}{2} \le \theta \le \frac{\pi}{2}$, $\cos\theta \ge 0$ and thus

$$\sqrt{a^2 - x^2} = a\cos\theta$$

Integrals Involving $\sqrt{a^2 - x^2}$ If $\sqrt{a^2 - x^2}$ occurs in an integral where $a > 0$, try the substitution

$$x = a\sin\theta, \qquad dx = a\cos\theta\,d\theta, \qquad \sqrt{a^2 - x^2} = a\cos\theta$$

The next example shows that trigonometric substitution can be used with integrands involving $(a^2 - x^2)^{n/2}$, where n is any integer.

■ **EXAMPLE 3** Integrand Involving $(a^2 - x^2)^{3/2}$ Evaluate $\displaystyle\int \frac{x^2}{(4 - x^2)^{3/2}}\, dx$.

Solution

***Step 1.* Substitute to eliminate the square root.**

In this case, $a = 2$ since $\sqrt{4 - x^2} = \sqrt{2^2 - x^2}$. Therefore, we use

$$x = 2\sin\theta, \qquad dx = 2\cos\theta\, d\theta, \qquad \sqrt{4 - x^2} = 2\cos\theta$$

$$\int \frac{x^2}{(4 - x^2)^{3/2}}\, dx = \int \frac{4\sin^2\theta}{2^3\cos^3\theta}\, 2\cos\theta\, d\theta = \int \frac{\sin^2\theta}{\cos^2\theta}\, d\theta = \int \tan^2\theta\, d\theta$$

***Step 2.* Evaluate the trigonometric integral.**

Use the trigonometric identity $\tan^2 x = \sec^2 x - 1$:

$$\int \tan^2\theta\, d\theta = \int (\sec^2\theta - 1)\, d\theta = \tan\theta - \theta + C$$

***Step 3.* Convert back to the original variable.**

We must write $\tan\theta$ and θ in terms of x. By definition, $x = 2\sin\theta$, so

$$\sin\theta = \frac{x}{2}, \qquad \theta = \sin^{-1}\frac{x}{2}$$

To express $\tan\theta$ in terms of x, we use the right triangle in Figure 2. The angle θ satisfies $\sin\theta = \dfrac{x}{2}$ and

$$\tan\theta = \frac{\text{opposite}}{\text{adjacent}} = \frac{x}{\sqrt{4 - x^2}}$$

Thus, we have

$$\int \frac{x^2}{(4 - x^2)^{3/2}}\, dx = \tan\theta - \theta + C = \frac{x}{\sqrt{4 - x^2}} - \sin^{-1}\frac{x}{2} + C \qquad ■$$

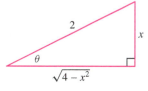

FIGURE 2 Right triangle with $\sin\theta = \frac{x}{2}$.

When the integrand involves $\sqrt{x^2 + a^2}$, try the substitution $x = a\tan\theta$. Then

$$x^2 + a^2 = a^2\tan^2\theta + a^2 = a^2(1 + \tan^2\theta) = a^2\sec^2\theta$$

and thus $\sqrt{x^2 + a^2} = a\sec\theta$.

In the substitution $x = a\tan\theta$, we choose $-\frac{\pi}{2} < \theta < \frac{\pi}{2}$. Therefore, $a\sec\theta$ is the positive square root $\sqrt{x^2 + a^2}$.

> **Integrals Involving $\sqrt{x^2 + a^2}$** If $\sqrt{x^2 + a^2}$ occurs in an integral where $a > 0$, try the substitution
>
> $$x = a\tan\theta, \qquad dx = a\sec^2\theta\, d\theta, \qquad \sqrt{x^2 + a^2} = a\sec\theta$$

■ **EXAMPLE 4** Evaluate $\displaystyle\int \frac{1}{\sqrt{x^2 + 9}}\, dx$.

Solution We have the form $\sqrt{x^2 + a^2}$ with $a = 3$.

***Step 1.* Substitute to eliminate the square root.**

$$x = 3\tan\theta, \qquad dx = 3\sec^2\theta\, d\theta, \qquad \sqrt{x^2 + 9} = 3\sec\theta$$

$$\int \frac{1}{\sqrt{x^2 + 9}}\, dx = \int \left(\frac{1}{3\sec\theta}\right) 3\sec^2\theta\, d\theta = \int \sec\theta\, d\theta$$

***Step 2.* Evaluate the trigonometric integral.**

$$\int \frac{1}{\sqrt{x^2 + 9}}\, dx = \int \sec\theta\, d\theta = \ln|\sec\theta + \tan\theta| + C$$

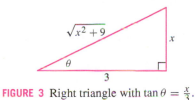

FIGURE 3 Right triangle with $\tan\theta = \frac{x}{3}$.

Step 3. **Convert back to the original variable.**

Since $x = 3\tan\theta$, we use the right triangle in Figure 3.

$$\tan\theta = \frac{\text{opposite}}{\text{adjacent}} = \frac{x}{3}, \qquad \sec\theta = \frac{\text{hypotenuse}}{\text{adjacent}} = \frac{\sqrt{x^2+9}}{3}$$

$$\int \frac{1}{\sqrt{x^2+9}}\,dx = \ln\left|\frac{\sqrt{x^2+9}}{3} + \frac{x}{3}\right| + C$$

Notice that we can rewrite this answer as follows, absorbing the constant term into the arbitrary constant:

$$\ln\left|\frac{\sqrt{x^2+9}}{3} + \frac{x}{3}\right| + C = \ln\left|\sqrt{x^2+9} + x\right| - \ln 3 + C = \ln\left|\sqrt{x^2+9} + x\right| + C \quad\blacksquare$$

Our last trigonometric substitution $x = a\sec\theta$ transforms $\sqrt{x^2-a^2}$ into $a\tan\theta$ because

$$x^2 - a^2 = a^2\sec^2\theta - a^2 = a^2(\sec^2\theta - 1) = a^2\tan^2\theta$$

> In the substitution $x = a\sec\theta$, we choose $0 \le \theta < \frac{\pi}{2}$ if $x \ge a$ and $\pi \le \theta < \frac{3\pi}{2}$ if $x \le -a$. With these choices, $a\tan\theta$ is the positive square root $\sqrt{x^2-a^2}$. Note that either $x \ge a$ or $x \le -a$ for the square root to be defined.

Integrals Involving $\sqrt{x^2 - a^2}$ If $\sqrt{x^2 - a^2}$ occurs in an integral where $a > 0$, try the substitution

$$x = a\sec\theta, \qquad dx = a\sec\theta\tan\theta\,d\theta, \qquad \sqrt{x^2 - a^2} = a\tan\theta$$

■ **EXAMPLE 5** Evaluate $\displaystyle\int \frac{dx}{x^2\sqrt{4x^2 - 36}}$.

Solution First, factor out the 4 to obtain $x^2\sqrt{4x^2 - 36} = 2x^2\sqrt{x^2 - 9}$. In this case, make the substitution

$$x = 3\sec\theta, \qquad dx = 3\sec\theta\tan\theta\,d\theta, \qquad \sqrt{x^2 - 9} = 3\tan\theta$$

$$\int \frac{dx}{x^2\sqrt{4x^2 - 36}} = \frac{1}{2}\int \frac{3\sec\theta\tan\theta\,d\theta}{(9\sec^2\theta)(3\tan\theta)} = \frac{1}{18}\int \cos\theta\,d\theta = \frac{1}{18}\sin\theta + C$$

Since $x = 3\sec\theta$, we use the right triangle in Figure 4:

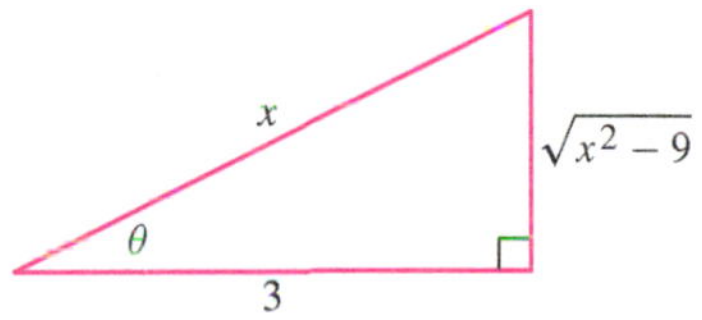

$$\sec\theta = \frac{\text{hypotenuse}}{\text{adjacent}} = \frac{x}{3}, \qquad \sin\theta = \frac{\text{opposite}}{\text{hypotenuse}} = \frac{\sqrt{x^2-9}}{x}$$

Therefore,

$$\int \frac{dx}{x^2\sqrt{4x^2 - 36}} = \frac{1}{18}\sin\theta + C = \frac{\sqrt{x^2-9}}{18x} + C \quad\blacksquare$$

FIGURE 4 Right triangle with $\sec\theta = \frac{x}{3}$.

Square Roots of General Quadratic Functions

So far we have dealt with the expressions $\sqrt{x^2 \pm a^2}$ and $\sqrt{a^2 - x^2}$. By completing the square (Section 1.2), we can treat the more general form $\sqrt{ax^2 + bx + c}$.

■ **EXAMPLE 6** **Completing the Square** Evaluate $\displaystyle\int \frac{dx}{(x^2 - 6x + 11)^2}$.

Solution

Step 1. **Complete the square.**

$$x^2 - 6x + 11 = (x^2 - 6x + 9) + 2 = \underbrace{(x - 3)^2}_{u^2} + 2$$

Step 2. **Use substitution.**

Let $u = x - 3$, $du = dx$:

$$\int \frac{dx}{(x^2 - 6x + 11)^2} = \int \frac{du}{(u^2 + 2)^2}$$

3

◀··· **REMINDER**

$$\int \cos^2 \theta \, d\theta = \frac{\theta}{2} + \frac{\sin \theta \cos \theta}{2} + C$$

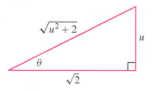

FIGURE 5 Right triangle with $\tan \theta = \frac{u}{\sqrt{2}}$.

Step 3. Trigonometric substitution.

Evaluate the u-integral using trigonometric substitution:

$$u = \sqrt{2} \tan \theta, \qquad u^2 + 2 = 2 \sec^2 \theta, \qquad du = \sqrt{2} \sec^2 \theta \, d\theta$$

$$\int \frac{du}{(u^2 + 2)^2} = \int \frac{\sqrt{2} \sec^2 \theta \, d\theta}{4 \sec^4 \theta} = \frac{1}{2\sqrt{2}} \int \cos^2 \theta \, d\theta$$

$$= \frac{1}{2\sqrt{2}} \left(\frac{\theta}{2} + \frac{\sin \theta \cos \theta}{2} \right) + C \qquad \boxed{4}$$

Since $\theta = \tan^{-1} \dfrac{u}{\sqrt{2}}$, we use the right triangle in Figure 5 to obtain

$$\sin \theta \cos \theta = \left(\frac{\text{opposite}}{\text{hypotenuse}} \right) \left(\frac{\text{adjacent}}{\text{hypotenuse}} \right) = \frac{u}{\sqrt{u^2 + 2}} \cdot \frac{\sqrt{2}}{\sqrt{u^2 + 2}} = \frac{\sqrt{2}u}{u^2 + 2}$$

Thus, Eq. (4) becomes

$$\int \frac{du}{(u^2 + 2)^2} = \frac{1}{4\sqrt{2}} \left(\tan^{-1} \frac{u}{\sqrt{2}} + \frac{\sqrt{2}u}{u^2 + 2} \right) + C$$

$$= \frac{1}{4\sqrt{2}} \tan^{-1} \frac{u}{\sqrt{2}} + \frac{u}{4(u^2 + 2)} + C \qquad \boxed{5}$$

Step 4. Convert to the original variable.

Since $u = x - 3$ and $u^2 + 2 = x^2 - 6x + 11$, Eq. (5) becomes

$$\int \frac{du}{(u^2 + 2)^2} = \frac{1}{4\sqrt{2}} \tan^{-1} \frac{x - 3}{\sqrt{2}} + \frac{x - 3}{4(x^2 - 6x + 11)} + C$$

This is our final answer by Eq. (3):

$$\int \frac{dx}{(x^2 - 6x + 11)^2} = \frac{1}{4\sqrt{2}} \tan^{-1} \frac{x - 3}{\sqrt{2}} + \frac{x - 3}{4(x^2 - 6x + 11)} + C \qquad ■$$

7.3 SUMMARY

• Trigonometric substitution:

Square root form in integrand	Trigonometric substitution
$\sqrt{a^2 - x^2}$	$x = a \sin \theta, \quad dx = a \cos \theta \, d\theta, \quad \sqrt{a^2 - x^2} = a \cos \theta$
$\sqrt{x^2 + a^2}$	$x = a \tan \theta, \quad dx = a \sec^2 \theta \, d\theta, \quad \sqrt{x^2 + a^2} = a \sec \theta$
$\sqrt{x^2 - a^2}$	$x = a \sec \theta, \quad dx = a \sec \theta \tan \theta \, d\theta, \quad \sqrt{x^2 - a^2} = a \tan \theta$

Step 1. Substitute to eliminate the square root.

Step 2. Evaluate the trigonometric integral.

Step 3. Convert back to the original variable.

• The three trigonometric substitutions correspond to three right triangles (Figure 6) that we use to express the trigonometric functions of θ in terms of x.

• Integrands involving $\sqrt{x^2 + bx + c}$ are treated by completing the square (see Example 6).

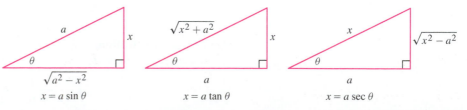

FIGURE 6 Right triangles used in trigonometric substitution.

7.3 EXERCISES

Preliminary Questions

1. State the trigonometric substitution appropriate to the given integral:

(a) $\displaystyle\int \sqrt{9-x^2}\,dx$

(b) $\displaystyle\int x^2(x^2-16)^{3/2}\,dx$

(c) $\displaystyle\int x^2(x^2+16)^{3/2}\,dx$

(d) $\displaystyle\int (x^2-5)^{-2}\,dx$

2. Is trigonometric substitution needed to evaluate $\displaystyle\int x\sqrt{9-x^2}\,dx$?

3. Express $\sin 2\theta$ in terms of $x=\sin\theta$.

4. Draw a triangle that would be used together with the substitution $x=3\sec\theta$.

Exercises

In Exercises 1–4, evaluate the integral by following the steps given.

1. $\displaystyle I=\int \frac{dx}{\sqrt{9-x^2}}$

(a) Show that the substitution $x=3\sin\theta$ transforms I into $\displaystyle\int d\theta$, and evaluate I in terms of θ.

(b) Evaluate I in terms of x.

2. $\displaystyle I=\int \frac{dx}{x^2\sqrt{x^2-2}}$

(a) Show that the substitution $x=\sqrt{2}\sec\theta$ transforms the integral I into $\dfrac{1}{2}\displaystyle\int \cos\theta\,d\theta$, and evaluate I in terms of θ.

(b) Use a right triangle to show that with the above substitution, $\sin\theta=\sqrt{x^2-2}/x$.

(c) Evaluate I in terms of x.

3. $\displaystyle I=\int \frac{dx}{\sqrt{4x^2+9}}$

(a) Show that the substitution $x=\frac{3}{2}\tan\theta$ transforms I into $\dfrac{1}{2}\displaystyle\int \sec\theta\,d\theta$.

(b) Evaluate I in terms of θ (refer to the table of integrals on page 383 in Section 7.2 if necessary).

(c) Express I in terms of x.

4. $\displaystyle I=\int \frac{dx}{(x^2+4)^2}$

(a) Show that the substitution $x=2\tan\theta$ transforms the integral I into $\dfrac{1}{8}\displaystyle\int \cos^2\theta\,d\theta$.

(b) Use the formula $\displaystyle\int \cos^2\theta\,d\theta=\frac{1}{2}\theta+\frac{1}{2}\sin\theta\cos\theta$ to evaluate I in terms of θ.

(c) Show that $\sin\theta=\dfrac{x}{\sqrt{x^2+4}}$ and $\cos\theta=\dfrac{2}{\sqrt{x^2+4}}$.

(d) Express I in terms of x.

In Exercises 5–10, use the indicated substitution to evaluate the integral.

5. $\displaystyle\int \sqrt{16-5x^2}\,dx,\quad x=\frac{4}{\sqrt{5}}\sin\theta$

6. $\displaystyle\int_0^{1/2} \frac{x^2}{\sqrt{1-x^2}}\,dx,\quad x=\sin\theta$

7. $\displaystyle\int \frac{dx}{x\sqrt{x^2-9}},\quad x=3\sec\theta$

8. $\displaystyle\int_{1/2}^1 \frac{dx}{x^2\sqrt{x^2+4}},\quad x=2\tan\theta$

9. $\displaystyle\int \frac{dx}{(x^2-4)^{3/2}},\quad x=2\sec\theta$

10. $\displaystyle\int_0^1 \frac{dx}{(4+4x^2)^2},\quad x=\tan\theta$

11. Evaluate $\displaystyle\int \frac{x\,dx}{\sqrt{x^2-4}}$ in two ways: using the direct substitution $u=x^2-4$ and by trigonometric substitution.

12. Is the substitution $u=x^2-4$ effective for evaluating the integral $\displaystyle\int \frac{x^2\,dx}{\sqrt{x^2-4}}$? If not, evaluate using trigonometric substitution.

13. Evaluate using the substitution $u=1-x^2$ or trigonometric substitution.

(a) $\displaystyle\int \frac{x}{\sqrt{1-x^2}}\,dx$

(b) $\displaystyle\int x^2\sqrt{1-x^2}\,dx$

(c) $\displaystyle\int x^3\sqrt{1-x^2}\,dx$

(d) $\displaystyle\int \frac{x^4}{\sqrt{1-x^2}}\,dx$

14. Evaluate:

(a) $\displaystyle\int_0^1 \frac{dt}{(t^2+1)^{3/2}}$

(b) $\displaystyle\int_0^1 \frac{t\,dt}{(t^2+1)^{3/2}}$

In Exercises 15–32, evaluate using trigonometric substitution. Refer to the table of trigonometric integrals as necessary.

15. $\displaystyle\int \frac{x^2\,dx}{\sqrt{9-x^2}}$

16. $\displaystyle\int \frac{dt}{(16-t^2)^{3/2}}$

17. $\displaystyle\int \frac{dx}{x\sqrt{x^2+16}}$

18. $\displaystyle\int \sqrt{12+4t^2}\,dt$

19. $\displaystyle\int \frac{dx}{\sqrt{x^2-9}}$

20. $\displaystyle\int \frac{dt}{t^2\sqrt{t^2-25}}$

21. $\displaystyle\int \frac{dy}{y^2\sqrt{5-y^2}}$

22. $\displaystyle\int x^3\sqrt{9-x^2}\,dx$

23. $\displaystyle\int \frac{dx}{\sqrt{25x^2+2}}$

24. $\displaystyle\int \frac{dt}{(9t^2+4)^2}$

25. $\displaystyle\int \frac{dz}{z^3\sqrt{z^2-4}}$

26. $\displaystyle\int \frac{dy}{\sqrt{y^2-9}}$

27. $\displaystyle\int \frac{x^2\,dx}{(6x^2-49)^{1/2}}$

28. $\displaystyle\int \frac{dx}{(x^2-4)^2}$

29. $\displaystyle\int_0^1 \frac{dt}{(t^2+9)^2}$

30. $\displaystyle\int_0^1 \frac{dx}{(x^2+1)^3}$

31. $\displaystyle\int \frac{x^2\,dx}{(x^2-1)^{3/2}}$

32. $\displaystyle\int \frac{x^2\,dx}{(x^2+1)^{3/2}}$

33. Prove for $a>0$:

$$\int \frac{dx}{x^2+a}=\frac{1}{\sqrt{a}}\tan^{-1}\frac{x}{\sqrt{a}}+C$$

34. Prove for $a > 0$:

$$\int \frac{dx}{(x^2 + a)^2} = \frac{1}{2a}\left(\frac{x}{x^2+a} + \frac{1}{\sqrt{a}}\tan^{-1}\frac{x}{\sqrt{a}}\right) + C$$

35. Let $I = \displaystyle\int \frac{dx}{\sqrt{x^2 - 4x + 8}}$.

(a) Complete the square to show that $x^2 - 4x + 8 = (x - 2)^2 + 4$.

(b) Use the substitution $u = x - 2$ to show that $I = \displaystyle\int \frac{du}{\sqrt{u^2 + 2^2}}$. Evaluate the u-integral.

(c) Show that $I = \ln\left|\sqrt{(x-2)^2 + 4} + x - 2\right| + C$.

36. Evaluate $\displaystyle\int \frac{dx}{\sqrt{12x - x^2}}$. First complete the square to write $12x - x^2 = 36 - (x - 6)^2$.

In Exercises 37–42, evaluate the integral by completing the square and using trigonometric substitution.

37. $\displaystyle\int \frac{dx}{\sqrt{x^2 + 4x + 13}}$

38. $\displaystyle\int \frac{dx}{\sqrt{2 + x - x^2}}$

39. $\displaystyle\int \frac{dx}{\sqrt{x + 6x^2}}$

40. $\displaystyle\int \sqrt{x^2 - 4x + 7}\,dx$

41. $\displaystyle\int \sqrt{x^2 - 4x + 3}\,dx$

42. $\displaystyle\int \frac{dx}{(x^2 + 6x + 6)^2}$

In Exercises 43–46, evaluate using Integration by Parts as a first step.

43. $\displaystyle\int \sec^{-1} x\,dx$

44. $\displaystyle\int \frac{\sin^{-1} x}{x^2}\,dx$

45. $\displaystyle\int \ln(x^2 + 1)\,dx$

46. $\displaystyle\int x^2 \ln(x^2 + 1)\,dx$

47. Find the average height of a point on the semicircle $y = \sqrt{1 - x^2}$ for $-1 \le x \le 1$.

48. Find the volume of the solid obtained by revolving the graph of $y = x\sqrt{1 - x^2}$ over $[0, 1]$ about the y-axis.

49. Find the volume of the solid obtained by revolving the region between the graph of $y^2 - x^2 = 1$ and the line $y = 2$ about the line $y = 2$.

50. Find the volume of revolution for the region in Exercise 49, but revolve around $y = 3$.

51. Compute $\displaystyle\int \frac{dx}{x^2 - 1}$ in two ways and verify that the answers agree: first via trigonometric substitution and then using the identity

$$\frac{1}{x^2 - 1} = \frac{1}{2}\left(\frac{1}{x - 1} - \frac{1}{x + 1}\right)$$

52. 𝖢𝖠𝖲   You want to divide an 18-in. pizza equally among three friends using vertical slices at $\pm x$ as in Figure 7. Find an equation satisfied by x and find the approximate value of x using a computer algebra system.

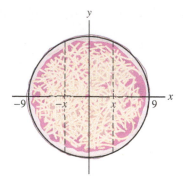

FIGURE 7 Dividing a pizza into three equal parts.

53. A charged wire creates an electric field at a point P located at a distance D from the wire (Figure 8). The component $E_\perp$ of the field perpendicular to the wire (in newtons per coulomb) is

$$E_\perp = \int_{x_1}^{x_2} \frac{k\lambda D}{(x^2 + D^2)^{3/2}}\,dx$$

where λ is the charge density (coulombs per meter), $k = 8.99 \times 10^9$ N·m^2/C^2 (Coulomb constant), and x_1, x_2 are as in the figure. Suppose that $\lambda = 6 \times 10^{-4}$ C/m, and $D = 3$ m. Find $E_\perp$ if (a) $x_1 = 0$ and $x_2 = 30$ m, and (b) $x_1 = -15$ m and $x_2 = 15$ m.

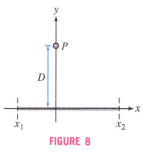

FIGURE 8

Further Insights and Challenges

54. Let $J_n = \displaystyle\int \frac{dx}{(x^2 + 1)^n}$. Use Integration by Parts to prove

$$J_{n+1} = \left(1 - \frac{1}{2n}\right) J_n + \left(\frac{1}{2n}\right)\frac{x}{(x^2 + 1)^n}$$

Then use this recursion relation to calculate J_2 and J_3.

55. Prove the formula

$$\int \sqrt{1 - x^2}\,dx = \frac{1}{2}\sin^{-1} x + \frac{1}{2}x\sqrt{1 - x^2} + C$$

using geometry by interpreting the integral as the area of part of the unit circle.

7.4 Integrals Involving Hyperbolic and Inverse Hyperbolic Functions

In Section 1.6, we noted the similarities between hyperbolic and trigonometric functions. We also saw in Section 3.9 that the formulas for their derivatives resemble each other, differing in at most a sign. The derivative formulas for the hyperbolic functions are equivalent to the following integral formulas.

◄·· REMINDER

$$\sinh x = \frac{e^x - e^{-x}}{2} \qquad \cosh x = \frac{e^x + e^{-x}}{2}$$

$$\frac{d}{dx}\sinh x = \cosh x \qquad \frac{d}{dx}\cosh x = \sinh x$$

$$\frac{d}{dx}\tanh x = \operatorname{sech}^2 x$$

$$\frac{d}{dx}\coth x = -\operatorname{csch}^2 x$$

$$\frac{d}{dx}\operatorname{sech} x = -\operatorname{sech} x \tanh x$$

$$\frac{d}{dx}\operatorname{csch} x = -\operatorname{csch} x \coth x$$

Hyperbolic Identities:

$$\cosh^2 x - \sinh^2 x = 1$$

$$\cosh^2 x = 1 + \sinh^2 x$$

$$\cosh^2 x = \tfrac{1}{2}(\cosh 2x + 1)$$

$$\sinh^2 x = \tfrac{1}{2}(\cosh 2x - 1)$$

$$\sinh 2x = 2\sinh x \cosh x$$

$$\cosh 2x = \cosh^2 x + \sinh^2 x$$

Hyperbolic Integral Formulas

$$\int \sinh x \, dx = \cosh x + C, \qquad \int \cosh x \, dx = \sinh x + C$$

$$\int \operatorname{sech}^2 x \, dx = \tanh x + C, \qquad \int \operatorname{csch}^2 x \, dx = -\coth x + C$$

$$\int \operatorname{sech} x \tanh x \, dx = -\operatorname{sech} x + C, \qquad \int \operatorname{csch} x \coth x \, dx = -\operatorname{csch} x + C$$

■ **EXAMPLE 1**   Calculate $\int x \cosh(x^2)\, dx$.

Solution   The substitution $u = x^2$, $du = 2x\, dx$ yields

$$\int x \cosh(x^2)\, dx = \frac{1}{2} \int \cosh u \, du = \frac{1}{2}\sinh u + C = \frac{1}{2}\sinh(x^2) + C \qquad ■$$

The techniques for computing trigonometric integrals discussed in Section 7.2 apply with little change to hyperbolic integrals. In place of trigonometric identities, we use the corresponding hyperbolic identities (see margin).

■ **EXAMPLE 2** **Powers of $\sinh x$ and $\cosh x$**   Calculate: **(a)** $\int \sinh^4 x \cosh^5 x \, dx$

and   **(b)** $\int \cosh^2 x \, dx$.

Solution

(a)   Since $\cosh x$ appears to an odd power, use $\cosh^2 x = 1 + \sinh^2 x$ to write

$$\cosh^5 x = \cosh^4 x \cdot \cosh x = (\sinh^2 x + 1)^2 \cosh x$$

Then use the substitution $u = \sinh x$, $du = \cosh x \, dx$:

$$\int \sinh^4 x \cosh^5 x \, dx = \int \underbrace{\sinh^4 x}_{u^4} \underbrace{(\sinh^2 x + 1)^2}_{(u^2+1)^2} \underbrace{\cosh x \, dx}_{du}$$

$$= \int u^4 (u^2 + 1)^2 \, du = \int (u^8 + 2u^6 + u^4)\, du$$

$$= \frac{u^9}{9} + \frac{2u^7}{7} + \frac{u^5}{5} + C = \frac{\sinh^9 x}{9} + \frac{2\sinh^7 x}{7} + \frac{\sinh^5 x}{5} + C$$

(b)   Use the identity $\cosh^2 x = \frac{1}{2}(\cosh 2x + 1)$:

$$\int \cosh^2 x \, dx = \frac{1}{2} \int (\cosh 2x + 1)\, dx = \frac{1}{2}\left(\frac{\sinh 2x}{2} + x\right) + C$$

$$= \frac{1}{4}\sinh 2x + \frac{1}{2}x + C \qquad ■$$

Hyperbolic substitution may be used as an alternative to trigonometric substitution to integrate functions involving the following square root expressions:

In trigonometric substitution, we treat $\sqrt{x^2 + a^2}$ using the substitution $x = a\tan\theta$ and $\sqrt{x^2 - a^2}$ using $x = a\sec\theta$. Identities can be used to show that the results coincide with those obtained from hyperbolic substitution (see Exercises 31–35).

Square root form	Hyperbolic substitution
$\sqrt{x^2 + a^2}$	$x = a\sinh u,\ dx = a\cosh u,\ \sqrt{x^2 + a^2} = a\cosh u$
$\sqrt{x^2 - a^2}$	$x = a\cosh u,\ dx = a\sinh u,\ \sqrt{x^2 - a^2} = a\sinh u$

■ **EXAMPLE 3** **Hyperbolic Substitution** Calculate $\int \sqrt{x^2 + 16}\,dx$.

Solution

Step 1. Substitute to eliminate the square root.
Use the hyperbolic substitution $x = 4\sinh u$, $dx = 4\cosh u\,du$. Then

$$x^2 + 16 = 16(\sinh^2 u + 1) = (4\cosh u)^2$$

Furthermore, $4\cosh u > 0$, so $\sqrt{x^2 + 16} = 4\cosh u$, and thus,

$$\int \sqrt{x^2 + 16}\,dx = \int (4\cosh u)\,4\cosh u\,du = 16\int \cosh^2 u\,du$$

Step 2. Evaluate the hyperbolic integral.
We evaluated the integral of $\cosh^2 u$ in Example 2(b):

$$\int \sqrt{x^2 + 16}\,dx = 16\int \cosh^2 u\,du = 16\left(\frac{1}{4}\sinh 2u + \frac{1}{2}u + C\right)$$

$$= 4\sinh 2u + 8u + C \qquad \boxed{1}$$

Step 3. Convert back to the original variable.
To write the answer in terms of the original variable x, we note that

$$\sinh u = \frac{x}{4}, \qquad u = \sinh^{-1}\frac{x}{4}$$

Use the identities recalled in the margin to write

$$4\sinh 2u = 4(2\sinh u\cosh u) = 8\sinh u\sqrt{\sinh^2 u + 1}$$

<-·· *REMINDER*

$$\sinh 2u = 2\sinh u\cosh u$$

$$\cosh u = \sqrt{\sinh^2 u + 1}$$

$$= 8\left(\frac{x}{4}\right)\sqrt{\left(\frac{x}{4}\right)^2 + 1} = 2x\sqrt{\frac{x^2}{16} + 1} = \frac{1}{2}x\sqrt{x^2 + 16}$$

Then Eq. (1) becomes

$$\int \sqrt{x^2 + 16}\,dx = 4\sinh 2u + 8u + C = \frac{1}{2}x\sqrt{x^2 + 16} + 8\sinh^{-1}\frac{x}{4} + C \qquad ■$$

In a similar manner, we could derive each of the following integral formulas. Alternatively, we can realize them from the corresponding derivative formulas for the inverse hyperbolic functions recorded in Section 3.9. Each formula is valid on the domain where the integrand and inverse hyperbolic function are defined.

THEOREM 1 **Integrals Involving Inverse Hyperbolic Functions**

$$\int \frac{dx}{\sqrt{x^2 + 1}} = \sinh^{-1} x + C$$

$$\int \frac{dx}{\sqrt{x^2 - 1}} = \cosh^{-1} x + C \qquad \text{(for } x > 1)$$

$$\int \frac{dx}{1 - x^2} = \tanh^{-1} x + C \qquad \text{(for } |x| < 1)$$

$$\int \frac{dx}{1 - x^2} = \coth^{-1} x + C \qquad \text{(for } |x| > 1)$$

$$\int \frac{dx}{x\sqrt{1 - x^2}} = -\text{sech}^{-1} x + C \qquad \text{(for } 0 < x < 1)$$

$$\int \frac{dx}{|x|\sqrt{1 + x^2}} = -\text{csch}^{-1} x + C \qquad \text{(for } x \neq 0)$$

If your calculator does not provide values of inverse hyperbolic functions, you can use an online resource such as http://wolframalpha.com.

■ **EXAMPLE 4** Evaluate: **(a)** $\displaystyle\int_2^4 \frac{dx}{\sqrt{x^2-1}}$ and **(b)** $\displaystyle\int_{0.2}^{0.6} \frac{x\,dx}{1-x^4}$.

Solution

(a) By Theorem 1,

$$\int_2^4 \frac{dx}{\sqrt{x^2-1}} = \cosh^{-1} x \Big|_2^4 = \cosh^{-1} 4 - \cosh^{-1} 2 \approx 0.75$$

(b) First use the substitution $u = x^2$, $du = 2x\,du$. The new limits of integration become $u = (0.2)^2 = 0.04$ and $u = (0.6)^2 = 0.36$, so

$$\int_{0.2}^{0.6} \frac{x\,dx}{1-x^4} = \int_{0.04}^{0.36} \frac{\frac{1}{2}du}{1-u^2} = \frac{1}{2}\int_{0.04}^{0.36} \frac{du}{1-u^2}$$

By Theorem 1, both $\tanh^{-1} u$ and $\coth^{-1} u$ are antiderivatives of $f(u) = (1-u^2)^{-1}$. We use $\tanh^{-1} u$ because the interval of integration $[0.04, 0.36]$ is contained in the domain $(-1, 1)$ of $f(u) = \tanh^{-1} u$. If the limits of integration were contained in $(1, \infty)$ or $(-\infty, -1)$, we would use $\coth^{-1} u$. The result is

$$\frac{1}{2}\int_{0.04}^{0.36} \frac{du}{1-u^2} = \frac{1}{2}\Big(\tanh^{-1}(0.36) - \tanh^{-1}(0.04)\Big) \approx 0.1684 \quad ■$$

Excursion: A Leap of Imagination

The terms "hyperbolic sine" and "hyperbolic cosine" suggest a connection between the hyperbolic and trigonometric functions. This excursion explores the source of this connection, which leads us to **complex numbers** and a famous formula of Euler (Figure 1).

Recall that $y = e^t$ satisfies the differential equation $y' = y$. In fact, we know that *every* solution is of the form $y = Ce^t$ for some constant C. Observe that both $y = e^t$ and $y = e^{-t}$ satisfy the **second-order differential equation**

This differential equation is called "second-order" because it involves the second derivative y''.

$$\boxed{y'' = y} \qquad \boxed{2}$$

Indeed, $(e^t)'' = e^t$ and $(e^{-t})'' = (-e^{-t})' = e^{-t}$. Furthermore, *every* solution of Eq. (2) has the form $y = Ae^t + Be^{-t}$ for some constants A and B (Exercise 44).

Now let's see what happens when we change Eq. (2) by a minus sign:

$$\boxed{y'' = -y} \qquad \boxed{3}$$

In this case, $y = \sin t$ and $y = \cos t$ are solutions because

$$(\sin t)'' = (\cos t)' = -\sin t, \qquad (\cos t)'' = (-\sin t)' = -\cos t$$

And as before, every solution of Eq. (3) has the form

$$\boxed{y = A\cos t + B\sin t}$$

This might seem to be the end of the story. However, we can also write down solutions of Eq. (3) using the exponential functions $y = e^{it}$ and $y = e^{-it}$. Here,

$$\boxed{i = \sqrt{-1}}$$

FIGURE 1 Leonhard Euler (1707–1783). Euler (pronounced "oi-ler") ranks among the greatest mathematicians of all time. His work (printed in more than 70 volumes) contains fundamental contributions to almost every aspect of the mathematics and physics of his time. The French mathematician Pierre Simon de Laplace once declared: "Read Euler, he is our master in everything." *(The Granger Collection, NYC. All rights reserved.)*

is an *imaginary* complex number satisfying $i^2 = -1$. Since i is not a real number, e^{it} is not defined without further explanation. But let's assume that e^{it} *can* be defined and that the usual rules of calculus apply:

$$(e^{it})' = ie^{it}$$

$$(e^{it})'' = (ie^{it})' = i^2 e^{it} = -e^{it}$$

This shows that $y = e^{it}$ is a solution of $y'' = -y$, so there must exist constants A and B such that

$$e^{it} = A \cos t + B \sin t \qquad \boxed{4}$$

The constants are determined by initial conditions. First, set $t = 0$ in Eq. (4):

$$1 = e^{i0} = A \cos 0 + B \sin 0 = A$$

Then take the derivative of Eq. (4) and set $t = 0$:

$$i e^{it} = \frac{d}{dt} e^{it} = A \cos' t + B \sin' t = -A \sin t + B \cos t$$

$$i = i e^{i0} = -A \sin 0 + B \cos 0 = B$$

Thus, $A = 1$ and $B = i$, and Eq. (4) yields **Euler's Formula**:

$$\boxed{e^{it} = \cos t + i \sin t}$$

Euler proved his formula using power series, which may be used to define e^{it} in a precise fashion. At $t = \pi$, Euler's Formula yields

$$\boxed{e^{i\pi} = -1}$$

Here, we have a simple but surprising relation among the four important numbers e, i, π, and -1.

Euler's Formula also reveals the source of the analogy between hyperbolic and trigonometric functions. Let us calculate the hyperbolic cosine at $x = it$:

$$\cosh(it) = \frac{e^{it} + e^{-it}}{2} = \frac{\cos t + i \sin t}{2} + \frac{\cos(-t) + i \sin(-t)}{2} = \cos t$$

A similar calculation shows that $\sinh(it) = i \sin t$. In other words, the hyperbolic and trigonometric functions are not merely analogous—once we introduce complex numbers, we see that they are very nearly the same functions.

7.4 SUMMARY

- Integrals of hyperbolic functions:

$$\int \sinh x \, dx = \cosh x + C, \qquad \int \cosh x \, dx = \sinh x + C$$

$$\int \text{sech}^2 x \, dx = \tanh x + C, \qquad \int \text{csch}^2 x \, dx = -\coth x + C$$

$$\int \text{sech} \, x \tanh x \, dx = -\text{sech} \, x + C, \qquad \int \text{csch} \, x \coth x \, dx = -\text{csch} \, x + C$$

- Integrals involving inverse hyperbolic functions:

$$\int \frac{dx}{\sqrt{x^2 + 1}} = \sinh^{-1} x + C$$

$$\int \frac{dx}{\sqrt{x^2 - 1}} = \cosh^{-1} x + C \quad (\text{for } x > 1)$$

$$\int \frac{dx}{1-x^2} = \tanh^{-1} x + C \qquad \text{(for } |x| < 1)$$

$$\int \frac{dx}{1-x^2} = \coth^{-1} x + C \qquad \text{(for } |x| > 1)$$

$$\int \frac{dx}{x\sqrt{1-x^2}} = -\operatorname{sech}^{-1} x + C \qquad \text{(for } 0 < x < 1)$$

$$\int \frac{dx}{|x|\sqrt{1+x^2}} = -\operatorname{csch}^{-1} x + C \qquad \text{(for } x \neq 0)$$

7.4 EXERCISES

Preliminary Questions

1. Which hyperbolic substitution can be used to evaluate the following integrals?

(a) $\displaystyle\int \frac{dx}{\sqrt{x^2+1}}$ **(b)** $\displaystyle\int \frac{dx}{\sqrt{x^2+9}}$ **(c)** $\displaystyle\int \frac{dx}{\sqrt{9x^2+1}}$

2. Which two of the hyperbolic integration formulas differ from their trigonometric counterparts by a minus sign?

3. Which antiderivative of $y = (1-x^2)^{-1}$ should we use to evaluate the integral $\displaystyle\int_3^5 (1-x^2)^{-1}\,dx$?

Exercises

In Exercises 1–16, calculate the integral.

1. $\displaystyle\int \cosh(3x)\,dx$

2. $\displaystyle\int \sinh(x+1)\,dx$

3. $\displaystyle\int x \sinh(x^2+1)\,dx$

4. $\displaystyle\int \sinh^2 x \cosh x\,dx$

5. $\displaystyle\int \operatorname{sech}^2(1-2x)\,dx$

6. $\displaystyle\int \tanh(3x) \operatorname{sech}(3x)\,dx$

7. $\displaystyle\int \tanh x \operatorname{sech}^2 x\,dx$

8. $\displaystyle\int \frac{\cosh x}{3 \sinh x + 4}\,dx$

9. $\displaystyle\int \tanh x\,dx$

10. $\displaystyle\int x \operatorname{csch}(x^2) \coth(x^2)\,dx$

11. $\displaystyle\int \frac{\cosh x}{\sinh x}\,dx$

12. $\displaystyle\int \frac{\cosh x}{\sinh^2 x}\,dx$

13. $\displaystyle\int \sinh^2(4x-9)\,dx$

14. $\displaystyle\int \sinh^3 x \cosh^6 x\,dx$

15. $\displaystyle\int \sinh^2 x \cosh^2 x\,dx$

16. $\displaystyle\int \tanh^3 x\,dx$

In Exercises 17–30, calculate the integral in terms of the inverse hyperbolic functions.

17. $\displaystyle\int \frac{dx}{\sqrt{x^2-1}}$

18. $\displaystyle\int \frac{dx}{\sqrt{9x^2-4}}$

19. $\displaystyle\int \frac{dx}{\sqrt{16+25x^2}}$

20. $\displaystyle\int \frac{dx}{\sqrt{1+3x^2}}$

21. $\displaystyle\int \sqrt{x^2-1}\,dx$

22. $\displaystyle\int \frac{x^2\,dx}{\sqrt{x^2+1}}$

23. $\displaystyle\int_{-1/2}^{1/2} \frac{dx}{1-x^2}$

24. $\displaystyle\int_4^5 \frac{dx}{1-x^2}$

25. $\displaystyle\int_0^1 \frac{dx}{\sqrt{1+x^2}}$

26. $\displaystyle\int_2^{10} \frac{dx}{4x^2-1}$

27. $\displaystyle\int_{-3}^{-1} \frac{dx}{x\sqrt{x^2+16}}$

28. $\displaystyle\int_{0.2}^{0.8} \frac{dx}{x\sqrt{1-x^2}}$

29. $\displaystyle\int \frac{\sqrt{x^2-1}\,dx}{x^2}$

30. $\displaystyle\int_1^9 \frac{dx}{x\sqrt{x^4+1}}$

31. Verify the formulas

$$\sinh^{-1} x = \ln|x + \sqrt{x^2+1}|$$

$$\cosh^{-1} x = \ln|x + \sqrt{x^2-1}| \qquad \text{(for } x \geq 1)$$

32. Verify that $\tanh^{-1} x = \dfrac{1}{2} \ln\left|\dfrac{1+x}{1-x}\right|$ for $|x| < 1$.

33. Evaluate $\displaystyle\int \sqrt{x^2+16}\,dx$ using trigonometric substitution. Then use Exercise 31 to verify that your answer agrees with the answer in Example 3.

34. Evaluate $\displaystyle\int \sqrt{x^2-9}\,dx$ in two ways: using trigonometric substitution and using hyperbolic substitution. Then use Exercise 31 to verify that the two answers agree.

35. Prove the reduction formula for $n \geq 2$:

$$\int \cosh^n x\,dx = \frac{1}{n} \cosh^{n-1} x \sinh x + \frac{n-1}{n} \int \cosh^{n-2} x\,dx \qquad \boxed{5}$$

36. Use Eq. (5) to evaluate $\displaystyle\int \cosh^4 x\,dx$.

In Exercises 37–40, evaluate the integral.

37. $\displaystyle\int \frac{\tanh^{-1} x\,dx}{x^2-1}$

38. $\displaystyle\int \sinh^{-1} x\,dx$

39. $\displaystyle\int \tanh^{-1} x\,dx$

40. $\displaystyle\int x \tanh^{-1} x\,dx$

Further Insights and Challenges

41. Show that if $u = \tanh(x/2)$, then

$$\cosh x = \frac{1+u^2}{1-u^2}, \qquad \sinh x = \frac{2u}{1-u^2}, \qquad dx = \frac{2\,du}{1-u^2}$$

Hint: For the first relation, use the identities

$$\sinh^2\left(\frac{x}{2}\right) = \frac{1}{2}(\cosh x - 1), \qquad \cosh^2\left(\frac{x}{2}\right) = \frac{1}{2}(\cosh x + 1)$$

In Exercises 42 and 43, evaluate using the substitution of Exercise 41.

42. $\displaystyle\int \operatorname{sech} x\, dx$

43. $\displaystyle\int \frac{dx}{1 + \cosh x}$

44. Suppose that $y = f(x)$ satisfies $y'' = y$. Prove:
(a) $f(x)^2 - (f'(x))^2$ is constant.
(b) If $f(0) = f'(0) = 0$, then f is the zero function.
(c) $f(x) = f(0)\cosh x + f'(0)\sinh x$.
Hint: Refer to Theorem 1 in Section 5.9.

*Exercises 45–48 refer to the function $gd(y) = \tan^{-1}(\sinh y)$, called the **Gudermannian**. In a map of the earth constructed by Mercator projection, points located y radial units from the equator correspond to points on the globe of latitude $gd(y)$.*

45. Prove that $\dfrac{d}{dy} gd(y) = \operatorname{sech} y$.

46. Let $f(y) = 2\tan^{-1}(e^y) - \pi/2$. Prove that $gd(y) = f(y)$. *Hint:* Show that $gd'(y) = f'(y)$ and $f(0) = gd(0)$.

47. Let $t(y) = \sinh^{-1}(\tan y)$ Show that $t(y)$ is the inverse of $gd(y)$ for $0 \le y < \pi/2$.

48. Verify that $t(y)$ in Exercise 47 satisfies $t'(y) = \sec y$, and find a value of a such that

$$t(y) = \int_a^y \frac{dt}{\cos t}$$

49. The relations $\cosh(it) = \cos t$ and $\sinh(it) = i\sin t$ were discussed in the Excursion. Use these relations to show that the identity $\cos^2 t + \sin^2 t = 1$ results from setting $x = it$ in the identity $\cosh^2 x - \sinh^2 x = 1$.

7.5 The Method of Partial Fractions

The Method of Partial Fractions is used to integrate rational functions:

$$f(x) = \frac{P(x)}{Q(x)}$$

where P and Q are polynomials. The idea is to write f as a sum of simpler rational functions that can be integrated directly. For example, in the simplest case we use the identity

$$\frac{1}{x^2 - 1} = \frac{\frac{1}{2}}{x - 1} - \frac{\frac{1}{2}}{x + 1}$$

to evaluate the integral

$$\int \frac{dx}{x^2 - 1} = \frac{1}{2}\int \frac{dx}{x-1} - \frac{1}{2}\int \frac{dx}{x+1} = \frac{1}{2}\ln|x-1| - \frac{1}{2}\ln|x+1|$$

> It is a fact from algebra (known as the "Fundamental Theorem of Algebra") that every polynomial Q with real coefficients can be written as a product of linear and quadratic factors with real coefficients. However, it is not always possible to find these factors explicitly.

A rational function P/Q is called **proper** if the degree of P [denoted $\deg(P)$] is *less than* the degree of Q. For example,

$$\underbrace{\frac{x^2 - 3x + 7}{x^4 - 16}}_{\text{Proper}}, \qquad \underbrace{\frac{2x^2 + 7}{x - 5}, \qquad \frac{x - 2}{x - 5}}_{\text{Not proper}}$$

Suppose first that P/Q is proper and that the denominator $Q(x)$ factors as a product of *distinct linear factors*. In other words,

> Each distinct linear factor $(x - a)$ in the denominator contributes a term
> $$\frac{A}{x - a}$$
> to the partial fraction decomposition.

$$\frac{P(x)}{Q(x)} = \frac{P(x)}{(x - a_1)(x - a_2)\cdots(x - a_n)}$$

where the roots $a_1, a_2, \ldots, a_n$ are all distinct and $\deg(P) < n$. Then there is a **partial fraction decomposition**:

$$\frac{P(x)}{Q(x)} = \frac{A_1}{(x - a_1)} + \frac{A_2}{(x - a_2)} + \cdots + \frac{A_n}{(x - a_n)}$$

for suitable constants $A_1, \ldots, A_n$. For example,

$$\frac{5x^2 + x - 28}{(x + 1)(x - 2)(x - 3)} = -\frac{2}{x + 1} + \frac{2}{x - 2} + \frac{5}{x - 3}$$

Once we have found the partial fraction decomposition, we can integrate the individual terms.

■ **EXAMPLE 1** **Finding the Constants** Evaluate $\int \dfrac{dx}{x^2 - 7x + 10}$.

Solution The denominator factors as $x^2 - 7x + 10 = (x - 2)(x - 5)$, so we look for a partial fraction decomposition:

$$\frac{1}{(x - 2)(x - 5)} = \frac{A}{x - 2} + \frac{B}{x - 5}$$

To find A and B, first multiply by $(x - 2)(x - 5)$ to clear denominators:

$$1 = (x - 2)(x - 5)\left(\frac{A}{x - 2} + \frac{B}{x - 5}\right)$$

$$1 = A(x - 5) + B(x - 2) \qquad \boxed{1}$$

A second method for determining A and B, once we have the equation $1 = A(x - 5) + B(x - 2)$ is to rewrite it as $1 = (A + B)x + (-5A - 2B)$. Since there is no x on the left side of the equation, it must be that $A + B = 0$. Then what remains is $1 = -5A - 2B$. Solving these two equations for A and B yields $A = -\frac{1}{3}$ and $B = \frac{1}{3}$.

This equation holds for all values of x (including $x = 2$ and $x = 5$, by continuity). We determine A by setting $x = 2$ (this makes the second term disappear):

$$1 = A(2 - 5) + \underbrace{B(2 - 2)}_{\text{This is zero}} = -3A \quad \Rightarrow \quad \boxed{A = -\frac{1}{3}}$$

Similarly, to calculate B, set $x = 5$ in Eq. (1):

$$1 = A(5 - 5) + B(5 - 2) = 3B \quad \Rightarrow \quad \boxed{B = \frac{1}{3}}$$

The resulting partial fraction decomposition is

$$\boxed{\frac{1}{(x - 2)(x - 5)} = \frac{-\frac{1}{3}}{x - 2} + \frac{\frac{1}{3}}{x - 5}}$$

The integration can now be carried out:

$$\int \frac{dx}{(x - 2)(x - 5)} = -\frac{1}{3}\int \frac{dx}{x - 2} + \frac{1}{3}\int \frac{dx}{x - 5}$$

$$= -\frac{1}{3}\ln|x - 2| + \frac{1}{3}\ln|x - 5| + C \qquad ■$$

■ **EXAMPLE 2** Evaluate $\int \dfrac{x^2 + 2}{(x - 1)(2x - 8)(x + 2)}\, dx$.

Solution

Step 1. **Find the partial fraction decomposition.**

The decomposition has the form

In Eq. (2), the linear factor $2x - 8$ does not have the form $(x - a)$ used previously, but the partial fraction decomposition can be carried out in the same way.

$$\frac{x^2 + 2}{(x - 1)(2x - 8)(x + 2)} = \frac{A}{x - 1} + \frac{B}{2x - 8} + \frac{C}{x + 2} \qquad \boxed{2}$$

As before, multiply by $(x - 1)(2x - 8)(x + 2)$ to clear denominators:

$$x^2 + 2 = A(2x - 8)(x + 2) + B(x - 1)(x + 2) + C(x - 1)(2x - 8) \qquad \boxed{3}$$

Since A goes with the factor $(x - 1)$, we set $x = 1$ in Eq. (3) to compute A:

$$1^2 + 2 = A(2 - 8)(1 + 2) + \overbrace{B(1 - 1)(1 + 2) + C(1 - 1)(2 - 8)}^{\text{Zero}}$$

$$3 = -18A \quad \Rightarrow \quad \boxed{A = -\frac{1}{6}}$$

Similarly, 4 is the root of $2x - 8$, so we compute B by setting $x = 4$ in Eq. (3):

$$4^2 + 2 = A(8 - 8)(4 + 2) + B(4 - 1)(4 + 2) + C(4 - 1)(8 - 8)$$

$$18 = 18B \quad \Rightarrow \quad \boxed{B = 1}$$

Finally, C is determined by setting $x = -2$ in Eq. (3):

$$(-2)^2 + 2 = A(-4 - 8)(-2 + 2) + B(-2 - 1)(-2 + 2) + C(-2 - 1)(-4 - 8)$$

$$6 = 36C \quad \Rightarrow \quad \boxed{C = \frac{1}{6}}$$

The result is

$$\boxed{\frac{x^2 + 2}{(x - 1)(2x - 8)(x + 2)} = -\frac{\frac{1}{6}}{x - 1} + \frac{1}{2x - 8} + \frac{\frac{1}{6}}{x + 2}}$$

Step 2. Carry out the integration.

$$\int \frac{x^2 + 2}{(x - 1)(2x - 8)(x + 2)} \, dx = -\frac{1}{6} \int \frac{dx}{x - 1} + \int \frac{dx}{2x - 8} + \frac{1}{6} \int \frac{dx}{x + 2}$$

$$= -\frac{1}{6} \ln|x - 1| + \frac{1}{2} \ln|2x - 8| + \frac{1}{6} \ln|x + 2| + C \quad \blacksquare$$

Now suppose that the denominator has repeated linear factors:

$$\frac{P(x)}{Q(x)} = \frac{P(x)}{(x - a_1)^{M_1}(x - a_2)^{M_2} \cdots (x - a_n)^{M_n}}$$

Each factor $(x - a_i)^{M_i}$ contributes the following sum of terms to the partial fraction decomposition:

$$\frac{B_1}{(x - a_i)} + \frac{B_2}{(x - a_i)^2} + \cdots + \frac{B_{M_i}}{(x - a_i)^{M_i}}$$

■ **EXAMPLE 3** **Repeated Linear Factors** Evaluate $\displaystyle\int \frac{3x - 9}{(x - 1)(x + 2)^2} \, dx$.

Solution We are looking for a partial fraction decomposition of the form

$$\frac{3x - 9}{(x - 1)(x + 2)^2} = \frac{A}{x - 1} + \frac{B}{x + 2} + \frac{C}{(x + 2)^2}$$

Let's clear denominators to obtain

$$3x - 9 = A(x + 2)^2 + B(x - 1)(x + 2) + C(x - 1) \qquad \boxed{4}$$

We compute A and C by substituting in Eq. (4) in the usual way:

• Set $x = 1$: This gives $-6 = 9A$, or $\boxed{A = -\dfrac{2}{3}}$.

• Set $x = -2$: This gives $-15 = -3C$, or $\boxed{C = 5}$.

With these constants, Eq. (4) becomes

$$3x - 9 = -\frac{2}{3}(x + 2)^2 + B(x - 1)(x + 2) + 5(x - 1) \qquad \boxed{5}$$

We cannot determine B in the same way as A and C. Here are two ways to proceed.

• **First method (substitution):** There is no use substituting $x = 1$ or $x = -2$ in Eq. (5) because the term involving B drops out. But we are free to plug in any other value of x. Let's try $x = 2$ in Eq. (5):

$$3(2) - 9 = -\frac{2}{3}(2 + 2)^2 + B(2 - 1)(2 + 2) + 5(2 - 1)$$

$$-3 = -\frac{32}{3} + 4B + 5$$

$$B = \frac{1}{4}\left(-8 + \frac{32}{3}\right) = \frac{2}{3}$$

• **Second method (undetermined coefficients):** Expand the terms in Eq. (5):

$$3x - 9 = -\frac{2}{3}(x^2 + 4x + 4) + B(x^2 + x - 2) + 5(x - 1)$$

$$3x - 9 = \left(-\frac{2}{3} + B\right)x^2 + \left(-\frac{8}{3} + B + 5\right)x + \left(-\frac{8}{3} - 2B - 5\right)$$

The coefficients of the powers of x on each side of the equation must be equal. Since x^2 does not occur on the left-hand side, $0 = -\frac{2}{3} + B$, or $B = \frac{2}{3}$.

Either way, we have shown that

$$\frac{3x - 9}{(x - 1)(x + 2)^2} = -\frac{\frac{2}{3}}{x - 1} + \frac{\frac{2}{3}}{x + 2} + \frac{5}{(x + 2)^2}$$

$$\int \frac{3x - 9}{(x - 1)(x + 2)^2}\,dx = -\frac{2}{3}\int \frac{dx}{x - 1} + \frac{2}{3}\int \frac{dx}{x + 2} + 5\int \frac{dx}{(x + 2)^2}$$

$$= -\frac{2}{3}\ln|x - 1| + \frac{2}{3}\ln|x + 2| - \frac{5}{x + 2} + C \qquad \blacksquare$$

If P/Q is not proper—that is, if $\deg(P) \geq \deg(Q)$—we use long division to write

$$\frac{P(x)}{Q(x)} = g(x) + \frac{R(x)}{Q(x)}$$

where g is a polynomial and R/Q is proper. We may then integrate $P(x)/Q(x)$ using the partial fraction decomposition of $R(x)/Q(x)$.

■ **EXAMPLE 4** **Long Division Necessary** Evaluate $\displaystyle\int \frac{x^3 + 1}{x^2 - 4}\,dx$.

Solution Using long division, we write

$$\frac{x^3 + 1}{x^2 - 4} = x + \frac{4x + 1}{x^2 - 4} = x + \frac{4x + 1}{(x - 2)(x + 2)}$$

It is not difficult to show that the second term has a partial fraction decomposition:

$$\frac{4x + 1}{(x - 2)(x + 2)} = \frac{\frac{9}{4}}{x - 2} + \frac{\frac{7}{4}}{x + 2}$$

Therefore,

$$\int \frac{(x^3 + 1)\,dx}{x^2 - 4} = \int x\,dx + \frac{9}{4}\int \frac{dx}{x - 2} + \frac{7}{4}\int \frac{dx}{x + 2}$$

$$= \frac{1}{2}x^2 + \frac{9}{4}\ln|x - 2| + \frac{7}{4}\ln|x + 2| + C \qquad \blacksquare$$

Long division:

$$\begin{array}{r} x \\ x^2 - 4 \overline{\smash{\big)}\, x^3 + 1} \\ \underline{x^3 - 4x} \\ 4x + 1 \end{array}$$

The quotient $\dfrac{x^3 + 1}{x^2 - 4}$ is equal to x with remainder $4x + 1$.

Quadratic Factors

A quadratic polynomial $ax^2 + bx + c$ is called **irreducible** if it cannot be written as a product of two linear factors (without using complex numbers). A power of an irreducible quadratic factor $(ax^2 + bx + c)^M$ contributes a sum of the following type to a partial fraction decomposition:

$$\frac{A_1 x + B_1}{ax^2 + bx + c} + \frac{A_2 x + B_2}{(ax^2 + bx + c)^2} + \cdots + \frac{A_M x + B_M}{(ax^2 + bx + c)^M}$$

For example,

$$\frac{4 - 12x}{(x + 1)(x^2 + x + 4)^2} = \frac{1}{x + 1} - \frac{x}{x^2 + x + 4} - \frac{4x + 12}{(x^2 + x + 4)^2}$$

You may need to use trigonometric substitution to integrate these terms. In particular, the following result may be useful (see Exercise 33 in Section 7.3):

⟵·· *REMINDER If $b > 0$, then $x^2 + b$ is irreducible, but $x^2 - b$ is reducible because*

$$x^2 - b = (x + \sqrt{b})(x - \sqrt{b})$$

$$\int \frac{dx}{x^2 + a} = \frac{1}{\sqrt{a}} \tan^{-1}\left(\frac{x}{\sqrt{a}}\right) + C \quad \text{(for } a > 0\text{)} \qquad \boxed{6}$$

■ **EXAMPLE 5** **Irreducible Versus Reducible Quadratic Factors** Evaluate:

(a) $\displaystyle\int \frac{18}{(x + 3)(x^2 + 9)}\, dx$

(b) $\displaystyle\int \frac{18}{(x + 3)(x^2 - 9)}\, dx$

Solution

(a) The quadratic factor $x^2 + 9$ is irreducible, so the partial fraction decomposition has the form

$$\frac{18}{(x + 3)(x^2 + 9)} = \frac{A}{x + 3} + \frac{Bx + C}{x^2 + 9}$$

Clear the denominators to obtain

$$18 = A(x^2 + 9) + (Bx + C)(x + 3) \qquad \boxed{7}$$

To find A, set $x = -3$:

$$18 = A\big((-3)^2 + 9\big) + 0 \quad \Rightarrow \quad \boxed{A = 1}$$

Then substitute $A = 1$ in Eq. (7) to obtain

$$18 = (x^2 + 9) + (Bx + C)(x + 3) = (B + 1)x^2 + (C + 3B)x + (9 + 3C)$$

Equating coefficients, we get $B + 1 = 0$ and $9 + 3C = 18$. Hence,

$$\boxed{B = -1, \qquad C = 3}$$

In the second equality, we use

$$\int \frac{x\, dx}{x^2 + 9} = \frac{1}{2}\int \frac{du}{u} = \frac{1}{2}\ln(x^2 + 9) + C$$

and Eq. (6):

$$\int \frac{dx}{x^2 + 9} = \frac{1}{3}\tan^{-1}\frac{x}{3} + C$$

$$\int \frac{18\, dx}{(x + 3)(x^2 + 9)} = \int \frac{dx}{x + 3} + \int \frac{(-x + 3)\, dx}{x^2 + 9}$$

$$= \int \frac{dx}{x + 3} - \int \frac{x\, dx}{x^2 + 9} + \int \frac{3\, dx}{x^2 + 9}$$

$$= \ln|x + 3| - \frac{1}{2}\ln(x^2 + 9) + \tan^{-1}\frac{x}{3} + C$$

The last line comes from applying the formulas in the margin.

(b) The polynomial $x^2 - 9$ is not irreducible because $x^2 - 9 = (x-3)(x+3)$. Therefore, the partial fraction decomposition has the form

$$\frac{18}{(x+3)(x^2-9)} = \frac{18}{(x+3)^2(x-3)} = \frac{A}{x-3} + \frac{B}{x+3} + \frac{C}{(x+3)^2}$$

Clear the denominators:

$$18 = A(x+3)^2 + B(x+3)(x-3) + C(x-3)$$

For $x = 3$, this yields $18 = (6^2)A$, and for $x = -3$, this yields $18 = -6C$. Therefore,

$$A = \frac{1}{2}, \quad C = -3 \quad \Rightarrow \quad 18 = \frac{1}{2}(x+3)^2 + B(x+3)(x-3) - 3(x-3)$$

To solve for B, we can plug in any value of x other than ± 3. The choice $x = 2$ yields $18 = \frac{1}{2}(25) - 5B + 3$, or $B = -\frac{1}{2}$, and

$$\int \frac{18}{(x+3)(x^2-9)} \, dx = \frac{1}{2} \int \frac{dx}{x-3} - \frac{1}{2} \int \frac{dx}{x+3} - 3 \int \frac{dx}{(x+3)^2}$$

$$= \frac{1}{2} \ln|x-3| - \frac{1}{2} \ln|x+3| + 3(x+3)^{-1} + C \qquad \blacksquare$$

■ **EXAMPLE 6** **Repeated Quadratic Factor** Evaluate $\displaystyle\int \frac{4-x}{x(x^2+2)^2} \, dx$.

Solution The partial fraction decomposition has the form

$$\frac{4-x}{x(x^2+2)^2} = \frac{A}{x} + \frac{Bx+C}{x^2+2} + \frac{Dx+E}{(x^2+2)^2}$$

Clear the denominators by multiplying through by $x(x^2+2)^2$:

$$4 - x = A(x^2+2)^2 + (Bx+C)\big(x(x^2+2)\big) + (Dx+E)x \qquad \boxed{8}$$

We compute A directly by setting $x = 0$. Then Eq. (8) reduces to $4 = 4A$, or $A = 1$. We find the remaining coefficients by the method of undetermined coefficients. Set $A = 1$ in Eq. (8) and expand:

$$4 - x = (x^4 + 4x^2 + 4) + (Bx^4 + 2Bx^2 + Cx^3 + 2Cx) + (Dx^2 + Ex)$$

$$= (1+B)x^4 + Cx^3 + (4+2B+D)x^2 + (2C+E)x + 4$$

Now equate the coefficients on the two sides of the equation:

$$1 + B = 0 \qquad \text{(Coefficient of } x^4\text{)}$$

$$C = 0 \qquad \text{(Coefficient of } x^3\text{)}$$

$$4 + 2B + D = 0 \qquad \text{(Coefficient of } x^2\text{)}$$

$$2C + E = -1 \qquad \text{(Coefficient of } x\text{)}$$

$$2C + 4 = 4 \qquad \text{(Constant term)}$$

These equations yield $B = -1$, $C = 0$, $D = -2$, and $E = -1$. Thus,

$$\int \frac{(4-x)\,dx}{x(x^2+2)^2} = \int \frac{dx}{x} - \int \frac{x\,dx}{x^2+2} - \int \frac{(2x+1)\,dx}{(x^2+2)^2}$$

$$= \ln|x| - \frac{1}{2}\ln(x^2+2) - \int \frac{(2x+1)dx}{(x^2+2)^2}$$

The middle integral was evaluated using the substitution $u = x^2 + 2$, $du = 2x\,dx$. The third integral breaks up as a sum:

$$\int \frac{(2x + 1)\,dx}{(x^2 + 2)^2} = \overbrace{\int \frac{2x\,dx}{(x^2 + 2)^2}}^{\text{Use substitution } u = x^2 + 2} + \int \frac{dx}{(x^2 + 2)^2}$$

$$= -(x^2 + 2)^{-1} + \int \frac{dx}{(x^2 + 2)^2} \qquad \boxed{9}$$

To evaluate the integral in Eq. (9), we use the trigonometric substitution

$$x = \sqrt{2} \tan\theta, \qquad dx = \sqrt{2} \sec^2\theta\,d\theta, \qquad x^2 + 2 = 2\tan^2\theta + 2 = 2\sec^2\theta$$

Referring to Figure 1, we obtain

$$\int \frac{dx}{(x^2 + 2)^2} = \int \frac{\sqrt{2}\sec^2\theta\,d\theta}{(2\tan^2\theta + 2)^2} = \int \frac{\sqrt{2}\sec^2\theta\,d\theta}{4\sec^4\theta}$$

$$= \frac{\sqrt{2}}{4} \int \cos^2\theta\,d\theta = \frac{\sqrt{2}}{4}\left(\frac{1}{2}\theta + \frac{1}{2}\sin\theta\cos\theta\right) + C$$

$$= \frac{\sqrt{2}}{8}\tan^{-1}\frac{x}{\sqrt{2}} + \frac{\sqrt{2}}{8}\frac{x}{\sqrt{x^2 + 2}}\frac{\sqrt{2}}{\sqrt{x^2 + 2}} + C$$

$$= \frac{1}{4\sqrt{2}}\tan^{-1}\frac{x}{\sqrt{2}} + \frac{1}{4}\frac{x}{x^2 + 2} + C$$

Collecting all the terms, we have

$$\int \frac{4 - x}{x(x^2 + 2)^2}\,dx = \ln|x| - \frac{1}{2}\ln(x^2 + 2) + \frac{1 - \frac{1}{4}x}{x^2 + 2} - \frac{1}{4\sqrt{2}}\tan^{-1}\frac{x}{\sqrt{2}} + C \qquad \blacksquare$$

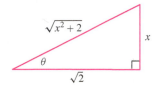

FIGURE 1 Right triangle with $\tan\theta = \frac{x}{\sqrt{2}}$.

> **CONCEPTUAL INSIGHT**   The examples in this section illustrate a general fact: The integral of a rational function can be expressed as a sum of rational functions, arctangents of linear or quadratic polynomials, and logarithms of linear or quadratic polynomials. Other types of functions, such as exponential and trigonometric functions, do not appear.

Using a Computer Algebra System

Finding partial fraction decompositions often requires laborious computation. Fortunately, most computer algebra systems have a command that produces partial fraction decompositions (with names such as "Apart," "parfrac," or simply "partial fractions"). For example in Mathematica, the command

```
Apart[(x^2 − 2)/((x + 2)(x^2 + 4)^3)]
```

produces the partial fraction decomposition

$$\frac{x^2 - 2}{(x + 2)(x^2 + 4)^3} = \frac{1}{256(2 + x)} + \frac{3(x - 2)}{4(4 + x^2)^3} + \frac{2 - x}{32(4 + x^2)^2} + \frac{2 - x}{256(4 + x^2)}$$

However, a computer algebra system cannot produce a partial fraction decomposition in cases where $Q(x)$ cannot be factored explicitly.

The polynomial $x^5 + 2x + 2$ cannot be factored explicitly, so the command

```
Apart[1/(x^5 + 2x + 2)]
```

returns the useless response

$$\frac{1}{x^5 + 2x + 2}$$

7.5 SUMMARY

Method of Partial Fractions: Assume first that P/Q is a *proper* rational function [i.e., $\deg(P) < \deg(Q)$] and that $Q(x)$ can be factored explicitly as a product of linear and irreducible quadratic terms.

- If $Q(x) = (x - a_1)(x - a_2) \cdots (x - a_n)$, where the roots a_j are distinct, then

$$\frac{P(x)}{(x - a_1)(x - a_2) \cdots (x - a_n)} = \frac{A_1}{x - a_1} + \frac{A_2}{x - a_2} + \cdots + \frac{A_n}{x - a_n}$$

  To calculate the constants, clear denominators and substitute, in turn, the values $x = a_1$, $a_2, \ldots, a_n$.

- If $Q(x)$ is equal to a product of powers of linear factors $(x - a)^M$ and irreducible quadratic factors $(x^2 + b)^N$ with $b > 0$, then the partial fraction decomposition of $P(x)/Q(x)$ is a sum of terms of the following type:

$$(x - a)^M \quad \text{contributes} \quad \frac{A_1}{x - a} + \frac{A_2}{(x - a)^2} + \cdots + \frac{A_M}{(x - a)^M}$$

$$(x^2 + b)^N \quad \text{contributes} \quad \frac{A_1 x + B_1}{x^2 + b} + \frac{A_2 x + B_2}{(x^2 + b)^2} + \cdots + \frac{A_N x + B_N}{(x^2 + b)^N}$$

  Substitution and trigonometric substitution may be needed to integrate the terms corresponding to $(x^2 + b)^N$ (see Example 6).

- If P/Q is improper, use long division (see Example 4).

7.5 EXERCISES

Preliminary Questions

1. Suppose that $\int f(x)\,dx = \ln x + \sqrt{x + 1} + C$. Can f be a rational function? Explain.

2. Which of the following are *proper* rational functions?

(a) $\dfrac{x}{x - 3}$

(b) $\dfrac{4}{9 - x}$

(c) $\dfrac{x^2 + 12}{(x + 2)(x + 1)(x - 3)}$

(d) $\dfrac{4x^3 - 7x}{(x - 3)(2x + 5)(9 - x)}$

3. Which of the following quadratic polynomials are irreducible? To check, complete the square if necessary.

(a) $x^2 + 5$

(b) $x^2 - 5$

(c) $x^2 + 4x + 6$

(d) $x^2 + 4x + 2$

4. Let P/Q be a proper rational function where $Q(x)$ factors as a product of distinct linear factors $(x - a_i)$. Then

$$\int \frac{P(x)\,dx}{Q(x)}$$

(choose the correct answer):

(a) is a sum of logarithmic terms $A_i \ln(x - a_i)$ for some constants A_i.

(b) may contain a term involving the arctangent.

Exercises

1. Match the rational functions (a)–(d) with the corresponding partial fraction decompositions (i)–(iv).

(a) $\dfrac{x^2 + 4x + 12}{(x + 2)(x^2 + 4)}$

(b) $\dfrac{2x^2 + 8x + 24}{(x + 2)^2(x^2 + 4)}$

(c) $\dfrac{x^2 - 4x + 8}{(x - 1)^2(x - 2)^2}$

(d) $\dfrac{x^4 - 4x + 8}{(x + 2)(x^2 + 4)}$

(i) $x - 2 + \dfrac{4}{x + 2} - \dfrac{4x - 4}{x^2 + 4}$

(ii) $\dfrac{-8}{x - 2} + \dfrac{4}{(x - 2)^2} + \dfrac{8}{x - 1} + \dfrac{5}{(x - 1)^2}$

(iii) $\dfrac{1}{x + 2} + \dfrac{2}{(x + 2)^2} + \dfrac{-x + 2}{x^2 + 4}$

(iv) $\dfrac{1}{x + 2} + \dfrac{4}{x^2 + 4}$

2. Determine the constants A, B:

$$\frac{2x - 3}{(x - 3)(x - 4)} = \frac{A}{x - 3} + \frac{B}{x - 4}$$

3. Clear the denominators in the following partial fraction decomposition and determine the constant B (substitute a value of x or use the method of undetermined coefficients).

$$\frac{3x^2 + 11x + 12}{(x + 1)(x + 3)^2} = \frac{1}{x + 1} - \frac{B}{x + 3} - \frac{3}{(x + 3)^2}$$

4. Find the constants in the partial fraction decomposition

$$\frac{2x + 4}{(x - 2)(x^2 + 4)} = \frac{A}{x - 2} + \frac{Bx + C}{x^2 + 4}$$

In Exercises 5–8, evaluate using long division first to write $f(x)$ as the sum of a polynomial and a proper rational function.

5. $\displaystyle\int \frac{x\,dx}{3x - 4}$

6. $\displaystyle\int \frac{(x^2 + 2)\,dx}{x + 3}$

7. $\displaystyle\int \frac{(x^3 + 2x^2 + 1)\,dx}{x + 2}$

8. $\displaystyle\int \frac{(x^3 + 1)\,dx}{x^2 + 1}$

In Exercises 9–46, evaluate the integral.

9. $\displaystyle\int \frac{dx}{(x - 2)(x - 4)}$

10. $\displaystyle\int \frac{(x + 3)\,dx}{x + 4}$

11. $\displaystyle\int \frac{dx}{x(2x + 1)}$

12. $\displaystyle\int \frac{(2x - 1)\,dx}{x^2 - 5x + 6}$

13. $\displaystyle\int \frac{x^2\,dx}{x^2 + 9}$

14. $\displaystyle\int \frac{dx}{(x - 2)(x - 3)(x + 2)}$

15. $\displaystyle\int \frac{(x^2 + 3x - 44)\,dx}{(x + 3)(x + 5)(3x - 2)}$

16. $\displaystyle\int \frac{3\,dx}{(x + 1)(x^2 + x)}$

17. $\displaystyle\int \frac{(x^2 + 11x)\,dx}{(x-1)(x+1)^2}$

18. $\displaystyle\int \frac{(4x^2 - 21x)\,dx}{(x-3)^2(2x+3)}$

19. $\displaystyle\int \frac{dx}{(x-1)^2(x-2)^2}$

20. $\displaystyle\int \frac{(x^2 - 8x)\,dx}{(x+1)(x+4)^3}$

21. $\displaystyle\int \frac{8\,dx}{x(x+2)^3}$

22. $\displaystyle\int \frac{x^2\,dx}{x^2 + 3}$

23. $\displaystyle\int \frac{dx}{2x^2 - 3}$

24. $\displaystyle\int \frac{dx}{(x-4)^2(x-1)}$

25. $\displaystyle\int \frac{dx}{x^3 + x^2 - x - 1}$

26. $\displaystyle\int \frac{dx}{x^3 - 3x^2 + 4}$

27. $\displaystyle\int \frac{4x^2 - 20}{(2x+5)^3}\,dx$

28. $\displaystyle\int \frac{3x+6}{x^2(x-1)(x-3)}\,dx$

29. $\displaystyle\int \frac{dx}{x(x-1)^3}$

30. $\displaystyle\int \frac{(3x^2 - 2)\,dx}{x - 4}$

31. $\displaystyle\int \frac{(x^2 - x + 1)\,dx}{x^2 + x}$

32. $\displaystyle\int \frac{dx}{x(x^2 + 1)}$

33. $\displaystyle\int \frac{(3x^2 - 4x + 5)\,dx}{(x-1)(x^2+1)}$

34. $\displaystyle\int \frac{x^2}{(x+1)(x^2+1)}\,dx$

35. $\displaystyle\int \frac{dx}{x(x^2 + 25)}$

36. $\displaystyle\int \frac{dx}{x^2(x^2 + 25)}$

37. $\displaystyle\int \frac{(6x^2 + 2)\,dx}{x^2 + 2x - 3}$

38. $\displaystyle\int \frac{6x^2 + 7x - 6}{(x^2 - 4)(x + 2)}\,dx$

39. $\displaystyle\int \frac{10\,dx}{(x-1)^2(x^2+9)}$

40. $\displaystyle\int \frac{10\,dx}{(x+1)(x^2+9)^2}$

41. $\displaystyle\int \frac{dx}{x(x^2 + 8)^2}$

42. $\displaystyle\int \frac{100x\,dx}{(x-3)(x^2+1)^2}$

43. $\displaystyle\int \frac{dx}{(x+2)(x^2+4x+10)}$

44. $\displaystyle\int \frac{9\,dx}{(x+1)(x^2-2x+6)}$

45. $\displaystyle\int \frac{25\,dx}{x(x^2+2x+5)^2}$

46. $\displaystyle\int \frac{(x^2+3)\,dx}{(x^2+2x+3)^2}$

In Exercises 47–50, evaluate by using first substitution and then partial fractions if necessary.

47. $\displaystyle\int \frac{x\,dx}{x^4 + 1}$

48. $\displaystyle\int \frac{x\,dx}{(x+2)^4}$

49. $\displaystyle\int \frac{e^x\,dx}{e^{2x} - e^x}$

50. $\displaystyle\int \frac{\sec^2\theta\,d\theta}{\tan^2\theta - 1}$

51. Evaluate $\displaystyle\int \frac{\sqrt{x}\,dx}{x - 1}$. *Hint:* Use the substitution $u = \sqrt{x}$ (sometimes called a **rationalizing substitution**).

52. Evaluate $\displaystyle\int \frac{dx}{x^{1/2} - x^{1/3}}$. *Hint:* Use the substitution $u = x^{1/6}$.

53. Evaluate $\displaystyle\int \frac{dx}{x^{5/4} - 4x^{3/4}}$.

54. Evaluate $\displaystyle\int \frac{dx}{x^{4/3} + x - 2x^{2/3}}$.

55. Evaluate $\displaystyle\int \frac{dx}{x^2 - 1}$ in two ways: using partial fractions and using trigonometric substitution. Verify that the two answers agree.

56. $\boxed{\text{GU}}$ Graph the equation $(x - 40)y^2 = 10x(x - 30)$ and find the volume of the solid obtained by revolving the region between the graph and the x-axis for $0 \le x \le 30$ around the x-axis.

57. Show that the substitution $\theta = 2\tan^{-1} t$ (Figure 2) yields the formulas

$$\cos\theta = \frac{1 - t^2}{1 + t^2}, \qquad \sin\theta = \frac{2t}{1 + t^2}, \qquad d\theta = \frac{2\,dt}{1 + t^2} \qquad \boxed{10}$$

This substitution transforms the integral of any rational function of $\cos\theta$ and $\sin\theta$ into an integral of a rational function of t (which can then be evaluated using partial fractions). Use it to evaluate $\displaystyle\int \frac{d\theta}{\cos\theta + \frac{3}{4}\sin\theta}$.

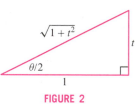

FIGURE 2

58. Use the substitution of Exercise 57 to evaluate $\displaystyle\int \frac{d\theta}{\cos\theta + \sin\theta}$.

Further Insights and Challenges

59. Prove the general formula

$$\int \frac{dx}{(x-a)(x-b)} = \frac{1}{a - b} \ln\left|\frac{x - a}{x - b}\right| + C$$

where a, b are constants such that $a \ne b$.

60. The method of partial fractions shows that

$$\int \frac{dx}{x^2 - 1} = \frac{1}{2}\ln|x - 1| - \frac{1}{2}\ln|x + 1| + C$$

The computer algebra system Mathematica evaluates this integral as $-\tanh^{-1} x$, where $\tanh^{-1} x$ is the inverse hyperbolic tangent function. Can you reconcile the two answers?

61. Suppose that $Q(x) = (x - a)(x - b)$, where $a \ne b$, and let P/Q be a proper rational function so that

$$\frac{P(x)}{Q(x)} = \frac{A}{(x - a)} + \frac{B}{(x - b)}$$

(a) Show that $A = \dfrac{P(a)}{Q'(a)}$ and $B = \dfrac{P(b)}{Q'(b)}$.

(b) Use this result to find the partial fraction decomposition for $P(x) = 3x - 2$ and $Q(x) = x^2 - 4x - 12$.

62. Suppose that $Q(x) = (x - a_1)(x - a_2)\cdots(x - a_n)$, where the roots a_j are all distinct. Let P/Q be a proper rational function so that

$$\frac{P(x)}{Q(x)} = \frac{A_1}{(x - a_1)} + \frac{A_2}{(x - a_2)} + \cdots + \frac{A_n}{(x - a_n)}$$

(a) Show that $A_j = \dfrac{P(a_j)}{Q'(a_j)}$ for $j = 1, \ldots, n$.

(b) Use this result to find the partial fraction decomposition for $P(x) = 2x^2 - 1$, $Q(x) = x^3 - 4x^2 + x + 6 = (x + 1)(x - 2)(x - 3)$.

7.6 Strategies for Integration

In Chapter 5, and in the preceding sections of this chapter, we have seen a variety of techniques for evaluating various integrals. However, in the real world, when confronted with a given integral, it will not appear in a particular section devoted to a particular technique of integration. Hence, it is important to be able to recognize which technique of integration is likely to apply. This section is devoted to that topic. In addition to considering how to recognize what technique to apply, we also discuss how tables of integrals and how computer algebra systems can be utilized to find an integral.

Often, an integral that comes from a particular application of calculus is a definite integral, with limits of integration. In the worst-case scenario, when none of the techniques of integration yield the answer, numerical approximation can be used to find a numerical estimate of the value, which can be made accurate enough for the particular application. Section 7.9 will discuss those techniques.

In general, there are no hard and fast rules for evaluating a given integral. But there are various heuristics that help us to determine which techniques are likely to apply.

Given an integral $\int f(x)\,dx$, here are steps that one might apply to try to compute it.

1. Do any algebraic simplification possible. Cancel terms in fractions when possible. For instance:

(a) $\int \dfrac{x^3-1}{x-1}\,dx = \int \dfrac{(x-1)(x^2+x+1)}{x-1}\,dx = \int (x^2+x+1)\,dx$
$$= \dfrac{x^3}{3} + \dfrac{x^2}{2} + x + C$$

(b) $\int \dfrac{x-x^3}{\sqrt{x}}\,dx = \int (x^{1/2} - x^{5/2})\,dx$, which can be easily integrated.

(c) $\int \dfrac{1}{\sin^2 x}\,dx = \int \csc^2 x\,dx$, which we have seen to be equal to $-\cot x + C$.

(d) $\int e^x \sinh x\,dx$. It is tempting to try to apply Integration by Parts to this integral, but it turns out not to work. Instead, we replace $\sinh x$ with its expression in terms of exponential functions to obtain

$$\int e^x \sinh x\,dx = \int e^x \left(\dfrac{e^x - e^{-x}}{2} \right) dx = \dfrac{1}{2} \int \left(e^{2x} - 1 \right) dx = \dfrac{e^{2x}}{4} - \dfrac{x}{2} + C$$

2. Look for a potential substitution.

Any time we recognize that our integral is of the form $\int f(g(x))g'(x)\,dx$, we can use substitution. The key to successfully substituting is that if we want to replace $g(x)$ by u, we need a $g'(x)$ to exist that we can pair with dx to obtain the necessary du.

So, for example, $\int x^2 \sin(x^3)\,dx$ is set up for substitution since if we let $u = x^3$, then $du = 3x^2\,dx$, and we have the requisite x^2 to make this work. The constant 3 that we also need is not a problem. We can always multiply by $1/3$ to compensate for the 3 we need.

In the case of $\int e^{\sin x} \cos x\,dx$, we can use $u = \sin x$ and $du = \cos x\,dx$. In the case of $\int e^x \sqrt{e^x + 1}\,dx$, we can let $u = e^x + 1$ and then $du = e^x\,dx$, which we have.

Sometimes, a less obvious substitution does the trick.

■ **EXAMPLE 1** Find $\int x^3 \sqrt{1+x^2}\,dx$.

Solution Our first inclination is to try the substitution $u = 1+x^2$ since that quantity is inside the square root. But then $du = 2x\,dx$, and we have an extra factor of x^2 left over. Instead of turning to another method of integration, though, we can note that since $u = 1+x^2$, $x^2 = u - 1$. Thus, we have

$$\int x^3 \sqrt{1+x^2}\, dx = \int x^2 \sqrt{1+x^2}\, x\, dx = \int (u-1)\sqrt{u}\, \frac{du}{2}$$

$$= \frac{1}{2}\int \left(u^{3/2} - u^{1/2}\right) du = \frac{u^{5/2}}{5} - \frac{u^{3/2}}{3} + C$$

$$= \frac{(1+x^2)^{5/2}}{5} - \frac{(1+x^2)^{3/2}}{3} + C \qquad\blacksquare$$

This integral could also be evaluated using Integration by Parts, but that would be more work.

■ **EXAMPLE 2** Evaluate $\displaystyle\int \frac{dx}{\sqrt{\sqrt{x}+1}}$.

Solution Our first inclination is to try the substitution $u = \sqrt{x}$. However, then $du = \frac{dx}{2\sqrt{x}}$, and we cannot perform the substitution. Instead, let's try the substitution $u = \sqrt{x}+1$. Before we find du, we first solve for x to obtain $x = (u-1)^2$. Then $dx = 2(u-1)\, du$. Substituting, we obtain

$$\int \frac{dx}{\sqrt{\sqrt{x}+1}} = \int \frac{2(u-1)\, du}{\sqrt{u}} = 2\int u^{1/2} - u^{-1/2}\, du$$

$$= 2\left(\frac{2}{3}u^{3/2} - 2u^{1/2}\right) + C = \frac{4}{3}(\sqrt{x}+1)^{3/2} - 4(\sqrt{x}+1)^{1/2} + C \qquad\blacksquare$$

3. Consider Integration by Parts. As we have seen, if there is a product in the integrand, we can split the integral into two pieces u and dv and then apply Integration by Parts to obtain

$$\int u\, dv = uv - \int v\, du$$

For example, look at $\displaystyle\int xe^x\, dx$. In this case, it is obvious that Integration by Parts applies. If the x were not present, we could easily integrate e^x. So choosing $u = x$ and $dv = e^x\, dx$ will yield $\displaystyle\int xe^x\, dx = xe^x - \int e^x\, dx$. The remaining integral is now one we can compute.

In particular, we are looking for integrands that are products such that differentiating one term in the product and integrating the other yields an integral that is easier to evaluate. Here are some cases to look for:

(a) $\displaystyle\int x^n f(x)\, dx$. Assuming $f(x)$ can be repeatedly integrated, we may use repeated applications of Integration by Parts, always choosing u equal to the remaining power of x, until we eliminate the powers of x, leaving something we can integrate. Candidates for $f(x)$ include $\sin x, \cos x, e^x, a^x, \sinh x, \cosh x, \sec^2 x$, and $\csc^2 x$, among others. Note that the same technique applies when we replace x^n in the integrand with a more complicated polynomial. Repeated application of Integration by Parts can be used to eliminate the polynomial.

(b) $\displaystyle\int e^x \sin x\, dx$ Two applications of Integration by Parts, choosing $u = e^x$ both times, yields

$$\int e^x \sin x\, dx = -e^x \cos x + e^x \sin x - \int e^x \sin x\, dx$$

Instead of obtaining a simpler integral on the right, we obtained the exact same integral we started with on the left. Adding it to both sides, we get

$$2\int e^x \sin x\, dx = -e^x \cos x + e^x \sin x + C$$

$$\int e^x \sin x\, dx = \frac{1}{2}(-e^x \cos x + e^x \sin x) + C$$

(c) Although we usually apply it to a product, Integration by Parts also sometimes works on single functions such as $\int \ln x \, dx$, $\int \tan^{-1} x \, dx$, and $\int \sin^{-1} x \, dx$. Here, we pick u to be the function in the integrand and set $dv = dx$.

4. Consider the form of the integral. Does it fall into one of the categories for which we have a technique?

(a) Is it a trigonometric integral? If it is of the form $\int \sin^n x \cos^m x \, dx$, then if n is odd, use the identity $\sin^2 x = 1 - \cos^2 x$ to remove all but one copy of $\sin x$. Then use the substitution $u = \cos x$ and $du = -\sin x \, dx$. If m is odd, use the identity $\cos^2 x = 1 - \sin^2 x$ to remove all but one copy of $\cos x$. Then use the substitution $u = \sin x$ so $du = \cos x \, dx$. If m and n are both even, then use trigonometric identities to decrease the powers of the trigonometric functions. If it is of the form $\int \tan^n x \sec^m x \, dx$, trigonometric identities can be similarly applied.

> These trigonometric identities can be used to reduce the powers of the sin and cos functions: $\sin^2 x = \dfrac{1 - \cos(2x)}{2}$ and $\cos^2 x = \dfrac{1 + \cos(2x)}{2}$

(b) Does the integrand contain $\sqrt{a^2 - x^2}$, $\sqrt{x^2 - a^2}$, or $\sqrt{x^2 + a^2}$?

If so, then try a trigonometric substitution. Use the fact that if $x = a \sin \theta$, then $\sqrt{a^2 - x^2} = a \cos \theta$. If $x = a \sec \theta$, then $\sqrt{x^2 - a^2} = a \tan \theta$. And if $x = a \tan \theta$, then $\sqrt{x^2 + a^2} = a \sec \theta$. Note that a similar technique works for other half-integer powers besides just the square root.

■ **EXAMPLE 3** Evaluate $\int (x^2 + 16)^{3/2} \, dx$.

Solution Let $x = 4 \tan \theta$. Then $(x^2 + 16)^{3/2} = (16 \tan^2 \theta + 16)^{3/2} = (16 \sec^2 \theta)^{3/2} = 64 \sec^3 \theta$ and $dx = 4 \sec^2 \theta \, d\theta$. Hence, we have

$$\int (x^2 + 16)^{3/2} \, dx = \int 64 \sec^3 \theta \, (4 \sec^2 \theta) \, d\theta = 256 \int \sec^5 \theta \, d\theta$$

Now, we can use reduction formula (15) from Section 7.2 to obtain

$$256 \int \sec^5 \theta \, d\theta = 256 \left(\frac{\tan \theta \sec^3 \theta}{4} + \frac{3}{4} \int \sec^3 \theta \, d\theta \right)$$

$$= 256 \left(\frac{\tan \theta \sec^3 \theta}{4} + \frac{3}{4} \left(\frac{\tan \theta \sec \theta}{2} + \frac{1}{2} \int \sec \theta \, d\theta \right) \right)$$

$$= 256 \left(\frac{\tan \theta \sec^3 \theta}{4} + \frac{3}{4} \left(\frac{\tan \theta \sec \theta}{2} + \frac{1}{2} \ln |\sec \theta + \tan \theta| \right) \right) + C$$

$$= 64 \tan \theta \sec^3 \theta + 96 \tan \theta \sec \theta + 96 \ln |\sec \theta + \tan \theta| + C$$

Since $x = 4 \tan \theta$, we know $\tan \theta = \frac{x}{4}$, and by the triangle in Figure 1, $\sec \theta = \frac{\sqrt{x^2+16}}{4}$. Thus, we have

$$\int (x^2 + 16)^{3/2} \, dx$$

$$= 64 \frac{x}{4} (x^2 + 16)^{3/2} + 96 \frac{x}{4} (x^2 + 16)^{1/2} + 96 \ln \left| (x^2 + 16)^{1/2} + \frac{x}{4} \right| + C$$

$$= 16x(x^2 + 16)^{3/2} + 24x(x^2 + 16)^{1/2} + 96 \ln \left| (x^2 + 16)^{1/2} + \frac{x}{4} \right| + C \qquad ■$$

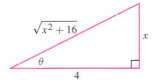

FIGURE 1 Since $x/4 = \tan \theta$, we can complete the triangle using the Pythagorean Theorem to find $\sec \theta = \dfrac{\sqrt{x^2 + 16}}{4}$.

If the integral contains an expression of the form $\sqrt{ax^2 + bx + c}$, we can complete the square inside the square root in order to obtain one of the cases discussed above.

(c) Is it a rational function $\dfrac{P}{Q}$ for polynomials P and Q? If so, we can often apply the method of partial fractions. First, if P has a degree that is at least as large as the degree of Q, we divide $P(x)$ by $Q(x)$ to obtain a polynomial (which is easily integrated) together with a remainder term $\dfrac{R(x)}{Q(x)}$, to which the method of partial fractions can be applied.

■ **EXAMPLE 4** Find $\displaystyle\int \frac{x^2 + 2x + 10}{x^2 + x - 6}\,dx$.

Solution Noting that both the numerator and the denominator have degree 2, we divide to obtain

$$\frac{x^2 + 2x + 10}{x^2 + x - 6} = 1 + \frac{x + 16}{x^2 + x - 6}$$

Then we can write

$$\frac{x + 16}{x^2 + x - 6} = \frac{x + 16}{(x + 3)(x - 2)} = \frac{A}{x + 3} + \frac{B}{x - 2}$$

Clearing the denominator, this yields

$$x + 16 = A(x - 2) + B(x + 3)$$

When $x = 2$, this equation becomes

$$18 = 5B \quad \Rightarrow \quad B = \frac{18}{5}$$

When $x = -3$, this equation becomes

$$13 = -5A \quad \Rightarrow \quad A = -\frac{13}{5}$$

Thus,

$$\int \frac{x^2 + 2x + 10}{x^2 + x - 6}\,dx = \int 1 + \frac{-13}{5(x + 3)} + \frac{18}{5(x - 2)}\,dx$$

$$= x - \frac{13}{5}\ln|x + 3| + \frac{18}{5}\ln|x - 2| + C \qquad ■$$

Using an Integral Table

The final three pages of this book contain a list of forms of many common integrals. Given a particular integral that we may want to evaluate, if we can get it into the form of one of the integrals in the table, then we can apply the given formula to evaluate the integral. Let's consider a few examples.

■ **EXAMPLE 5** Evaluate $\displaystyle\int \frac{\sqrt{9 - 4x^2}}{x}\,dx$.

Solution Looking down the list of integrals given at the end of the book, the formula with the integral that most resembles this one is #69:

$$\int \frac{\sqrt{a^2 - u^2}}{u}\,du = \sqrt{a^2 - u^2} - a\ln\left|\frac{a + \sqrt{a^2 - u^2}}{u}\right| + C$$

However, to get our integral into that form, we will need to manipulate it a bit. We begin with the substitution $u = 2x$. Then $du = 2dx$ and $x = \frac{u}{2}$, so we have

$$\int \frac{\sqrt{9 - 4x^2}}{x}\,dx = \int \frac{\sqrt{9 - u^2}}{u/2}\frac{du}{2} = \int \frac{\sqrt{9 - u^2}}{u}\,du$$

This is in the correct form, with $a = 3$; hence, we find

$$\int \frac{\sqrt{9 - 4x^2}}{x}\,dx = \int \frac{\sqrt{9 - u^2}}{u}\,du = \sqrt{9 - u^2} - 3\ln\left|\frac{3 + \sqrt{9 - u^2}}{u}\right| + C$$

$$= \sqrt{9 - 4x^2} - 3\ln\left|\frac{3 + \sqrt{9 - 4x^2}}{2x}\right| + C$$

Notice that this integral could also be evaluated by applying a trigonometric substitution followed by a second substitution. ■

■ **EXAMPLE 6** Evaluate $\int x \sin^2 x \cos x \, dx$.

Solution This integral does not appear in the table of integrals at the back of the book, but if the x were not in the integral, we could compute it relatively easily. So let's try applying Integration by Parts with $u = x$ and $dv = \sin^2 x \cos x \, dx$.

Then $du = dx$ and $v = \dfrac{\sin^3 x}{3}$. Thus, we have

$$\int x \sin^2 x \cos x \, dx = x \frac{\sin^3 x}{3} - \int \frac{\sin^3 x}{3} \, dx$$

This last integral matches the integral in the integral table appearing in this formula:

#38. $\displaystyle \int \sin^3 u \, dx = -\frac{1}{3}(2 + \sin^2 u) \cos u + C$

Note that we could also find $\int \sin^3 u \, dx$ by replacing $\sin^2 u$ by $1 - \cos^2 x$, and integrating the result using u-substitution.

Hence, we have

$$\int x \sin^2 x \cos x \, dx = \frac{1}{3} x \sin^3 x + \frac{1}{9}(2 + \sin^2 x) \cos x + C$$ ■

■ **EXAMPLE 7** Evaluate $\int x \sqrt{x^2 - 4x - 5} \, dx$.

Solution None of the integrals in the back of the book are going to match this. However, if we complete the square on the quantity inside the square root, we can make the integral look more similar to the integrals in the table:

$$x^2 - 4x - 5 = (x - 2)^2 - 9$$

Then we can do a u substitution with $u = x - 2$. Thus, $du = dx$ and $x = u + 2$:

$$\int x \sqrt{x^2 - 4x - 5} \, dx = \int x \sqrt{(x - 2)^2 - 9} \, dx = \int (u + 2)\sqrt{u^2 - 9} \, du$$

$$= \int u \sqrt{u^2 - 9} \, du + 2 \int \sqrt{u^2 - 9} \, du$$

The first of these integrals can be computed using the substitution $w = u^2 - 9$, and $dw = 2u \, du$, so we have

$$\int u \sqrt{u^2 - 9} \, du = \int \sqrt{w} \, \frac{dw}{2} = \frac{1}{3} w^{3/2} + C = \frac{1}{3}(u^2 - 9)^{3/2} + C$$

The second integral appears in this formula from the back of the book:

#76. $\displaystyle \int \sqrt{u^2 - a^2} \, du = \frac{u}{2}\sqrt{u^2 - a^2} - \frac{a^2}{2} \ln \left| u + \sqrt{u^2 - a^2} \right| + C$

with $a = 3$. Hence, we obtain

$$\int x \sqrt{x^2 - 4x - 5} \, dx$$

$$= \int u \sqrt{u^2 - 9} \, du + 2 \int \sqrt{u^2 - 9} \, du$$

$$= \frac{1}{3}(u^2 - 9)^{3/2} + u\sqrt{u^2 - 9} - 9 \ln \left| u + \sqrt{u^2 - 9} \right| + C$$

$$= \frac{1}{3}((x - 2)^2 - 9)^{3/2} + (x - 2)\sqrt{(x - 2)^2 - 9} - 9 \ln \left| x - 2 + \sqrt{(x - 2)^2 - 9} \right| + C$$

$$= \frac{1}{3}(x^2 - 4x + 5)^{3/2} + (x - 2)\sqrt{x^2 - 4x + 5} - 9 \ln \left| x - 2 + \sqrt{x^2 - 4x + 5} \right| + C$$ ■

Using a Computer Algebra System

In order to find these last integrals using the tables at the back of the book, we needed to do some algebra and/or apply some integration techniques to get them to match the pattern for one of the formulas at the back of the book. Computers are particularly adept at attempting various rearrangements and integration techniques and then matching the resulting integrals. For this reason, computer algebra systems such as Mathematica, Maple, Sage, and WolframAlpha are very good at evaluating integrals. However, keep in mind that the output that they generate may not be the form that is the most convenient to use.

For example, if we want to find $\int x(x-3)^{15}\,dx$, we do a substitution for $u = x - 3$ (and therefore $x = u + 3$) and find

$$\int x(x-3)^{15}\,dx = \int (u+3)u^{15}\,du = \int u^{16} + 3u^{15}\,du \quad = \frac{1}{17}u^{17} + \frac{3}{16}u^{16} + C$$

$$= \frac{1}{17}(x-3)^{17} + \frac{3}{16}(x-3)^{16} + C$$

On the other hand, the computer algebra system generates the following answer:

$$\int x(x-3)^{15}\,dx = \frac{x^{17}}{17} - \frac{45x^{16}}{16} + 63x^{15} - \frac{1755x^{14}}{2} + 8505x^{13} - \frac{243,243x^{12}}{4}$$

$$+ 331,695x^{11} - \frac{2,814,669x^{10}}{2} + 4,691,115x^9 - \frac{98,513,415x^8}{8}$$

$$+ 25,332,021x^7 - \frac{80,601,885x^6}{2} + 48361,131x^5 - \frac{167,403,915x^4}{4}$$

$$+ 23,914,845x^3 - \frac{14,348,907x^2}{2} + C$$

As you can see, the answer that we obtained by substitution may be more useful in most situations.

7.6 SUMMARY

Strategy for integration:

- Do any algebraic simplification possible.
- Consider possible substitutions, $u = g(x)$, keeping in mind that we will need the presence of $g'(x)$ since $du = g'(x)\,dx$.
- Consider Integration by Parts, choosing u to be a function that can be differentiated, dv to be something that can be integrated, and such that $\int v\,du$ is an easier integral than the original. Keep in mind that Integration by Parts can even work when the integrand is not an obvious product.
- Consider if the integral is a trigonometric integral of the form $\int \sin^n x \cos^m x\,dx$ or $\int \tan^m x \sec^m x\,dx$. Then trigonometric identities can be applied.
- Does the integral contain $\sqrt{a^2 - x^2}$, $\sqrt{x^2 - a^2}$ or $\sqrt{x^2 + a^2}$? If so, then try a trigonometric substitution.
- Is the integrand a rational function $\dfrac{P}{Q}$ for polynomials P and Q? If so, we can often apply the method of partial fractions.
- Determine whether the integral, after suitable manipulation, appears in the integral table at the back of the book.
- Consider using a computer algebra system to determine the integral.
- If the integral is a definite integral, and none of the above methods work, use a computer algebra system to approximate the integral numerically.

7.6 EXERCISES

Preliminary Questions

For each of the following, state what method applies and how one applies it, but do not evaluate the integral:

1. $\int x \sin x \, dx$

2. $\int \sqrt{1 + x^2} \, dx$

3. $\int \frac{1 + x^2}{1 - x^2} \, dx$

4. $\int \cos^2 x \sin x \, dx$

5. $\int x \ln x \, dx$

6. $\int \sqrt{1 - x^2} \, dx$

7. $\int \sin^3 x \cos^2 x \, dx$

For each of the following, find the formula in the integral table at the back of the book that can be applied to find the integral:

8. $\int \frac{3x^2 \, dx}{5x + 2}$

9. $\int \frac{\sqrt{25 + 16x^2}}{x^2} \, dx$

10. $\int \sec^3(4x) \, dx$

11. $\int \frac{x^2}{\sqrt{x^2 + 2x + 5}} \, dx$

Exercises

In Exercises 1–10, indicate a good method for evaluating the integral (but do not evaluate). Your choices are: substitution (specify u and du), Integration by Parts (specify u and dv), a trigonometric method, or trigonometric substitution (specify). If it appears that these techniques are not sufficient, state this.

1. $\int \frac{x \, dx}{\sqrt{12 - 6x - x^2}}$

2. $\int \sqrt{4x^2 - 1} \, dx$

3. $\int \sin^3 x \cos^3 x \, dx$

4. $\int x \sec^2 x \, dx$

5. $\int \frac{dx}{\sqrt{9 - x^2}}$

6. $\int \sqrt{1 - x^3} \, dx$

7. $\int \sin^{3/2} x \, dx$

8. $\int x^2 \sqrt{x + 1} \, dx$

9. $\int \frac{dx}{(x + 1)(x + 2)^3}$

10. $\int \frac{dx}{(x + 12)^4}$

In Exercises 11–57, evaluate the integral using the appropriate method or combination of methods covered thus far in the text. You may use the integral tables at the end of the book, but do not use a computer algebra system.

11. $\int \frac{dx}{x^2 \sqrt{4 - x^2}}$

12. $\int \frac{dx}{x(x - 1)^2}$

13. $\int \cos^2 4x \, dx$

14. $\int x \csc x \cot x \, dx$

15. $\int \frac{dx}{(x^2 + 9)^2}$

16. $\int \theta \sec^{-1} \theta \, d\theta$

17. $\int \tan^5 x \sec x \, dx$

18. $\int \frac{(3x^2 - 1) \, dx}{x(x^2 - 1)}$

19. $\int \ln(x^4 - 1) \, dx$

20. $\int \frac{x \, dx}{(x^2 - 1)^{3/2}}$

21. $\int \frac{x^2 \, dx}{(x^2 - 1)^{3/2}}$

22. $\int \sqrt{(x^2 - 6x + 5)} \, dx$

23. $\int \frac{(x + 1) \, dx}{(x^2 + 4x + 8)^2}$

24. $\int \frac{\sqrt{x} \, dx}{x^3 + 1}$

25. $\int \frac{x^{1/2} \, dx}{x^{1/3} + 1}$

26. $\int \frac{dx}{\sqrt{16 + x^2}}$

27. $\int \frac{dt}{3 + e^{-t}}$

28. $\int \frac{dt}{(1 + 4t^2)^{3/2}}$

29. $\int x^2 \ln x \, dx$

30. $\int \sec^5 y \tan y \, dy$

31. $\int \frac{dx}{x^2 + 2x + 5}$

32. $\int \frac{x^4 + 1}{x^2 + 1} \, dx$

33. $\int \sqrt{x^4 + x^7} \, dx$

34. $\int \sqrt{x^2 + 6x} \, dx$

35. $\int \frac{dx}{1 + e^x}$

36. $\int \frac{x^5}{x^3 - 1} \, dx$

37. $\int \frac{x}{\sqrt{x} - 1} \, dx$

38. $\int \frac{x}{\sqrt{x + 2}} \, dx$

39. $\int \frac{x^2}{\sqrt{x + 1}} \, dx$

40. $\int \sqrt{x^2 - 16} \, dx$

41. $\int (\sin x + \cos 2x)^2 \, dx$

42. $\int \sqrt{1 + \sqrt{x}} \, dx$

43. $\int \sin^2 x \tan x \, dx$

44. $\int x \ln(x + 12) \, dx$

45. $\int \ln(x^2 + 9) \, dx$

46. $\int \frac{dx}{x(x^2 - 6x - 7)}$

47. $\int \sin^5 x \cos^2 x \, dx$

48. $\int e^x \sqrt{e^{2x} - 1} \, dx$

49. $\int \cos^7 x \, dx$

50. $\int \frac{x^{11}}{x^4 - 1} \, dx$

51. $\int \frac{x^5}{x^4 - 1} \, dx$

52. $\int \tan x \sec^{5/4} x \, dx$

53. $\int (3 \sec x - \cos x)^2 \, dx$

54. $\int x^3 \ln x \, dx$

55. $\int x^3 \ln^2 x \, dx$

56. $\int \sqrt{e^x + 1} \, dx$

57. $\int \frac{dx}{\sqrt{x^2 - 36}}$

58. Attempt to use a computer algebra system to find $\int (\ln x + 1)\sqrt{(x \ln x)^2 + 1} \, dx$. (This will most likely fail.) Now, compute it yourself.

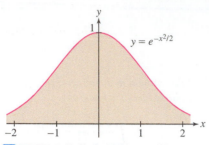

DF **FIGURE 1** Bell-shaped curve. The region extends infinitely far in both directions, but the total area is finite.

*The great British mathematician G. H. Hardy (1877–1947) observed that in calculus, we learn to ask not "What is it?" but rather "How shall we **define** it?" We saw that tangent lines and areas under curves have no clear meaning until we define them precisely using limits. Here again, the key question is "How shall we define the area of an unbounded region?"*

7.7 Improper Integrals

The integrals we have studied so far represent signed areas of bounded regions. However, areas of unbounded regions (Figure 1) also arise in applications and are represented by **improper integrals**.

There are two ways an integral can be improper: (1) The interval of integration may be infinite, or (2) the integrand may tend to infinity. We deal first with improper integrals over infinite intervals. One or both endpoints may be infinite:

$$\int_{-\infty}^{a} f(x)\,dx, \qquad \int_{a}^{\infty} f(x)\,dx, \qquad \int_{-\infty}^{\infty} f(x)\,dx$$

How can an unbounded region have finite area? To answer this question, we must specify what we mean by the area of an unbounded region. Consider the area [Figure 2(A)] under the graph of $f(x) = e^{-x}$ over the finite interval $[0, R]$:

$$\int_{0}^{R} e^{-x}\,dx = -e^{-x}\Big|_{0}^{R} = -e^{-R} + e^{0} = 1 - e^{-R}$$

As $R \to \infty$, this area approaches a finite value [Figure 2(B)]:

$$\int_{0}^{\infty} e^{-x}\,dx = \lim_{R\to\infty} \int_{0}^{R} e^{-x}\,dx = \lim_{R\to\infty}\left(1 - e^{-R}\right) = 1 \qquad \boxed{1}$$

It seems reasonable to take this limit as the *definition* of the area under the graph over the infinite interval $[0, \infty)$. Thus, the unbounded region in Figure 2(C) has area 1.

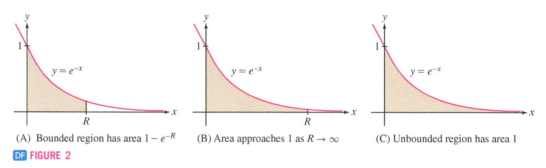

(A) Bounded region has area $1 - e^{-R}$ (B) Area approaches 1 as $R \to \infty$ (C) Unbounded region has area 1

DF **FIGURE 2**

⬅··· **REMINDER** *When we say, "The limit does not exist," this includes the possibilities that the function is $+\infty$, $-\infty$, or that the function does not approach any single real number*

> **DEFINITION Improper Integral** Fix a number a and assume that f is integrable over $[a, b]$ for all $b > a$. The *improper integral* of f over $[a, \infty)$ is defined as the following limit (if it exists):
>
> $$\int_{a}^{\infty} f(x)\,dx = \lim_{R\to\infty} \int_{a}^{R} f(x)\,dx$$
>
> We say that the improper integral *converges* if the limit exists (and is finite) and that it *diverges* if the limit does not exist.

Similarly, we define

$$\int_{-\infty}^{a} f(x)\,dx = \lim_{R\to-\infty} \int_{R}^{a} f(x)\,dx$$

■ **EXAMPLE 1** Show that $\displaystyle\int_{2}^{\infty} \frac{dx}{x^3}$ converges and compute its value.

Solution

$$\int_{2}^{\infty} \frac{dx}{x^3} = \lim_{R\to\infty} \int_{2}^{R} \frac{dx}{x^3} = \lim_{R\to\infty} -\frac{1}{2}x^{-2}\Big|_{2}^{R} = \lim_{R\to\infty} -\frac{1}{2}\left(R^{-2}\right) + \frac{1}{2}\left(2^{-2}\right)$$

$$= \lim_{R\to\infty}\left(\frac{1}{8} - \frac{1}{2R^2}\right) = \frac{1}{8}$$

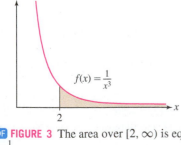

DF **FIGURE 3** The area over $[2, \infty)$ is equal to $\frac{1}{8}$.

We conclude that the unbounded shaded region in Figure 3 has area $\frac{1}{8}$. ■

■ **EXAMPLE 2** Determine whether $\displaystyle\int_{-\infty}^{-1} \frac{dx}{x}$ converges.

Solution Since the lower limit of the integral is $-\infty$, we take $R < -1$. Then we compute the limit as $R \to -\infty$:

$$\int_{-\infty}^{-1} \frac{dx}{x} = \lim_{R \to -\infty} \int_{R}^{-1} \frac{dx}{x} = \lim_{R \to -\infty} \ln|x| \Big|_{R}^{-1}$$

$$= \lim_{R \to -\infty} \ln|-1| - \ln|R| = \lim_{R \to -\infty} (-\ln|R|) = -\lim_{R \to -\infty} \ln|R| = -\infty$$

The limit is infinite, so the improper integral diverges. We conclude that the area of the unbounded region in Figure 4 is infinite. ■

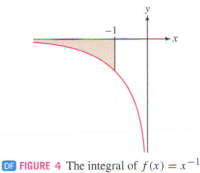

DF **FIGURE 4** The integral of $f(x) = x^{-1}$ over $(-\infty, -1]$ is infinite.

A doubly infinite improper integral is defined as a sum (provided that both integrals on the right converge):

$$\int_{-\infty}^{\infty} f(x)\,dx = \int_{-\infty}^{0} f(x)\,dx + \int_{0}^{\infty} f(x)\,dx \qquad \boxed{2}$$

We could have used some number other than 0 as a choice of where to split the integral, if it were more convenient to do so.

■ **EXAMPLE 3** Determine if $\displaystyle\int_{-\infty}^{\infty} \frac{1}{1+x^2}\,dx$ converges and, if so, compute its value.

Solution

$$\int_{-\infty}^{\infty} \frac{1}{1+x^2}\,dx = \int_{-\infty}^{0} \frac{1}{1+x^2}\,dx + \int_{0}^{\infty} \frac{1}{1+x^2}\,dx$$

assuming both of these integrals converge. For the second of these:

$$\int_{0}^{\infty} \frac{1}{1+x^2}\,dx = \lim_{R \to \infty} \int_{0}^{R} \frac{1}{1+x^2}\,dx = \lim_{R \to \infty} \tan^{-1} x \Big|_{0}^{R} = \lim_{R \to \infty} \tan^{-1} R - 0 = \frac{\pi}{2}$$

Similarly,

$$\int_{-\infty}^{0} \frac{1}{1+x^2}\,dx = \lim_{R \to -\infty} \int_{R}^{0} \frac{1}{1+x^2}\,dx = \lim_{R \to -\infty} \tan^{-1} x \Big|_{R}^{0} = \lim_{R \to -\infty} 0 - \tan^{-1} R = \frac{\pi}{2}$$

Thus, since both integrals converge,

$$\int_{-\infty}^{\infty} \frac{1}{1+x^2}\,dx = \int_{-\infty}^{0} \frac{1}{1+x^2}\,dx + \int_{0}^{\infty} \frac{1}{1+x^2}\,dx = \frac{\pi}{2} + \frac{\pi}{2} = \pi \qquad ■$$

CONCEPTUAL INSIGHT If you compare the unbounded shaded regions in Figures 3 and 4, you may wonder why one has finite area and the other has infinite area. Convergence of an improper integral depends on how rapidly $f(x)$ tends to zero as $x \to \infty$ (or $x \to -\infty$). Our calculations show that x^{-2} tends to zero quickly enough for convergence, whereas x^{-1} does not.

An improper integral of a power function $f(x) = x^{-p}$ is called a *p-integral*. Note that $f(x) = x^{-p}$ decreases more rapidly as p gets larger. Interestingly, our next theorem shows that the exponent $p = -1$ is the dividing line between convergence and divergence.

p-integrals are particularly important because they are often used to determine the convergence or divergence of more complicated improper integrals by means of the Comparison Test (see Example 10).

THEOREM 1 The *p*-Integral over $[a, \infty)$ For $a > 0$,

$$\int_a^\infty \frac{dx}{x^p} = \begin{cases} \dfrac{a^{1-p}}{p-1} & \text{if } p > 1 \\ \text{diverges} & \text{if } p \le 1 \end{cases}$$

Proof   Denote the *p*-integral by J. Then

$$J = \lim_{R \to \infty} \int_a^R x^{-p}\, dx = \lim_{R \to \infty} \frac{x^{1-p}}{1-p} \bigg|_a^R = \lim_{R \to \infty} \left(\frac{R^{1-p}}{1-p} - \frac{a^{1-p}}{1-p} \right)$$

If $p > 1$, then $1 - p < 0$ and R^{1-p} tends to zero as $R \to \infty$. In this case, $J = \dfrac{a^{1-p}}{p-1}$. If $p < 1$, then $1 - p > 0$ and R^{1-p} tends to ∞. In this case, J diverges. If $p = 1$, then J diverges because $\displaystyle\lim_{R \to \infty} \int_a^R x^{-1}\, dx = \lim_{R \to \infty} (\ln R - \ln a) = \infty$. ∎

Sometimes it is necessary to use L'Hôpital's Rule to determine the limits that arise in improper integrals.

■ **EXAMPLE 4** Using L'Hôpital's Rule   Calculate $\displaystyle\int_0^\infty x e^{-x}\, dx$.

Solution   First, use Integration by Parts with $u = x$ and $dv = e^{-x}\, dx$:

$$\int x e^{-x}\, dx = -x e^{-x} + \int e^{-x}\, dx = -x e^{-x} - e^{-x} = -(x+1) e^{-x} + C$$

$$\int_0^R x e^{-x}\, dx = -(x+1) e^{-x} \bigg|_0^R = -(R+1) e^{-R} + 1 = 1 - \frac{R+1}{e^R}$$

Then compute the improper integral as a limit using L'Hôpital's Rule:

$$\int_0^\infty x e^{-x}\, dx = 1 - \lim_{R \to \infty} \frac{R+1}{e^R} = 1 - \underbrace{\lim_{R \to \infty} \frac{1}{e^R}}_{\text{L'Hôpital's Rule}} = 1 - 0 = 1 \qquad ■$$

Improper integrals arise in applications when it makes sense to treat certain large quantities as if they were infinite. For example, an object launched with escape velocity never falls back to the earth but, rather, travels "infinitely far" into space.

■ **EXAMPLE 5** Escape Velocity   The earth exerts a gravitational force of magnitude $F(r) = GM_e m / r^2$ on an object of mass m at distance r from the center of the earth.

(a) Find the work required to move the object infinitely far from the earth.

(b) Calculate the escape velocity v_{esc} on the earth's surface.

In physics, we speak of moving an object "infinitely far away." In practice, this means "very far away," but it is more convenient to work with an improper integral.

Solution   This amounts to computing a *p*-integral with $p = 2$. Recall that work is the integral of force as a function of distance (Section 6.5).

(a) The work required to move an object from the earth's surface ($r = r_e$) to a distance R from the center is

$$\int_{r_e}^R \frac{GM_e m}{r^2}\, dr = -\frac{GM_e m}{r} \bigg|_{r_e}^R = GM_e m \left(\frac{1}{r_e} - \frac{1}{R} \right) \text{ joules}$$

←·· **REMINDER** *The mass of the earth is*

$$M_e \approx 5.98 \cdot 10^{24} \text{ kg}$$

The radius of the earth is

$$r_e \approx 6.37 \cdot 10^6 \text{ m}$$

The universal gravitational constant is

$$G \approx 6.67 \cdot 10^{-11} \text{ N-m}^2/\text{kg}^2$$

A newton is 1 kg-m/s^2 and a joule is 1 N-m.

The work moving the object "infinitely far away" is the improper integral

$$GM_e m \int_{r_e}^\infty \frac{dr}{r^2} = \lim_{R \to \infty} GM_e m \left(\frac{1}{r_e} - \frac{1}{R} \right) = \frac{GM_e m}{r_e} \text{ joules}$$

(b) By the principle of Conservation of Energy, an object launched with velocity v_0 will escape the earth's gravitational field if its kinetic energy $\frac{1}{2} m v_0^2$ is at least as large as the work required to move the object to infinity—that is, if

$$\frac{1}{2}mv_0^2 \geq \frac{GM_e m}{r_e} \quad \Rightarrow \quad v_0 \geq \left(\frac{2GM_e}{r_e}\right)^{1/2}$$

Escape velocity in miles per hour is approximately 25,000 mph.

Using the values recalled in the marginal note, we find that $v_0 \geq 11{,}200$ m/s. The minimal velocity is the escape velocity $v_{esc} = 11{,}200$ m/s. ∎

EXAMPLE 6 Perpetual Annuity An investment pays a dividend continuously at a rate of \$6000/year. Compute the present value of the income stream if the interest rate is 4% and the dividends continue forever.

In practice, the word "forever" means "a long but unspecified length of time." For example, if the investment pays out dividends for 100 years, then its present value is

$$\int_0^{100} 6000 e^{-0.04t}\, dt \approx \$147{,}253$$

The improper integral (\$150,000) gives a useful and convenient approximation to this value.

Solution Recall from Section 5.9 that the present value (PV) after T years at interest rate $r = 0.04$ is $\displaystyle \int_0^T 6000 e^{-0.04t}\, dt$. Over an infinite time interval,

$$\text{PV} = \int_0^\infty 6000 e^{-0.04t}\, dt = \lim_{T\to\infty} \left.\frac{6000 e^{-0.04t}}{-0.04}\right|_0^T = \frac{6000}{0.04} = \$150{,}000$$

Although an infinite number of dollars are paid out during the infinite time interval, their total present value is finite. ∎

Unbounded Functions

An integral over a finite interval $[a, b]$ is improper if the integrand is unbounded. In this case, the region in question is unbounded in the vertical direction. For example, $\displaystyle \int_0^9 \frac{dx}{\sqrt{x}}$ is improper because the integrand $f(x) = x^{-1/2}$ tends to ∞ as $x \to 0+$ (Figure 5). Improper integrals of this type are defined as one-sided limits.

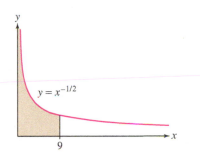

FIGURE 5 The shaded region has area 6 by Example 7(a).

DEFINITION Unbounded Integrands If f is continuous on $[a, b)$ and $\displaystyle \lim_{x\to b^-} f(x) = \pm\infty$, we define

$$\int_a^b f(x)\, dx = \lim_{R\to b^-} \int_a^R f(x)\, dx$$

Similarly, if f is continuous on $(a, b]$ and $\displaystyle \lim_{x\to a^+} f(x) = \pm\infty$,

$$\int_a^b f(x)\, dx = \lim_{R\to a^+} \int_R^b f(x)\, dx$$

In both cases, we say that the improper integral converges if the limit exists and that it diverges otherwise.

Note that if there is a single point c in the interval $[a, b]$ such that $\displaystyle \lim_{x\to c^-} f(x) = \pm\infty$, or $\displaystyle \lim_{x\to c^+} f(x) = \pm\infty$, and if $\displaystyle \int_a^c f(x)\, dx$ and $\displaystyle \int_c^b f(x)\, dx$ both converge, then we define

$$\int_a^b f(x)\, dx = \int_a^c f(x)\, dx + \int_c^b f(x)\, dx.$$

EXAMPLE 7 Calculate: **(a)** $\displaystyle \int_0^9 \frac{dx}{\sqrt{x}}$ and **(b)** $\displaystyle \int_0^{1/2} \frac{dx}{x}$.

Solution Both integrals are improper because the integrands have infinite discontinuities at $x = 0$. The first integral converges:

$$\int_0^9 \frac{dx}{\sqrt{x}} = \lim_{R\to 0^+} \int_R^9 x^{-1/2}\, dx = \lim_{R\to 0^+} \left. 2x^{1/2}\right|_R^9$$

$$= \lim_{R\to 0^+} (6 - 2R^{1/2}) = 6$$

The second integral diverges:

$$\int_0^{1/2} \frac{dx}{x} = \lim_{R \to 0^+} \int_R^{1/2} \frac{dx}{x} = \lim_{R \to 0^+} \left(\ln \frac{1}{2} - \ln R \right)$$

$$= \ln \frac{1}{2} - \lim_{R \to 0^+} \ln R = \infty$$ ■

■ **EXAMPLE 8** Calculate $\displaystyle\int_0^2 \frac{dx}{(x-1)^{\frac{2}{3}}}$.

Solution This integral is improper with an infinite discontinuity at $x = 1$ (Figure 6). Therefore, we write

$$\int_0^2 \frac{dx}{(x-1)^{\frac{2}{3}}} = \int_0^1 \frac{dx}{(x-1)^{\frac{2}{3}}} + \int_1^2 \frac{dx}{(x-1)^{\frac{2}{3}}}$$

We consider each integral individually:

$$\int_0^1 \frac{dx}{(x-1)^{\frac{2}{3}}} = \lim_{R \to 1^-} \int_0^R \frac{dx}{(x-1)^{\frac{2}{3}}} = \lim_{R \to 1^-} 3(x-1)^{\frac{1}{3}} \Big|_0^R$$

$$= \lim_{R \to 1^-} 3(R-1)^{\frac{1}{3}} - 3(-1)^{\frac{1}{3}} = 3$$

$$\int_1^2 \frac{dx}{(x-1)^{\frac{2}{3}}} = \lim_{R \to 1^+} \int_R^2 \frac{dx}{(x-1)^{\frac{2}{3}}} = \lim_{R \to 1^+} 3(x-1)^{\frac{1}{3}} \Big|_R^2$$

$$= \lim_{R \to 1^+} 3(1)^{\frac{1}{3}} - 3(R-1)^{\frac{1}{3}} = 3$$

Therefore, we obtain

$$\int_0^2 \frac{dx}{(x-1)^{\frac{2}{3}}} = \int_0^1 \frac{dx}{(x-1)^{\frac{2}{3}}} + \int_1^2 \frac{dx}{(x-1)^{\frac{2}{3}}} = 3 + 3 = 6$$ ■

The proof of the next theorem is similar to the proof of Theorem 1 (see Exercise 52).

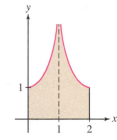

FIGURE 6 The unbounded shaded region has area 6.

Theorem 2 is valid for all exponents p. However, the integral is not improper if $p < 0$.

> **THEOREM 2 The p-Integral over $[0, a]$**   For $a > 0$,
>
> $$\int_0^a \frac{dx}{x^p} = \begin{cases} \dfrac{a^{1-p}}{1-p} & \text{if } p < 1 \\[2ex] \text{diverges} & \text{if } p \geq 1 \end{cases}$$

GRAPHICAL INSIGHT   The p-integrals $\displaystyle\int_a^\infty x^{-p}\, dx$ and $\displaystyle\int_0^a x^{-p}\, dx$ have opposite behavior for $p \neq 1$. The first converges only for $p > 1$, and the second converges only for $p < 1$ (both diverge for $p = 1$). This is reflected in the graphs of $y = x^{-p}$ and $y = x^{-q}$, which switch places at $x = 1$ (Figure 7). We see that a large value of p helps $\displaystyle\int_a^\infty x^{-p}\, dx$ to converge but causes $\displaystyle\int_0^a x^{-p}\, dx$ to diverge.

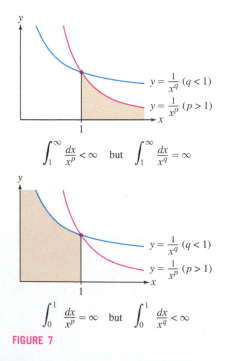

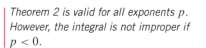

$$\int_1^\infty \frac{dx}{x^p} < \infty \quad \text{but} \quad \int_1^\infty \frac{dx}{x^q} = \infty$$

$y = \dfrac{1}{x^q} \ (q < 1)$

$y = \dfrac{1}{x^p} \ (p > 1)$

$$\int_0^1 \frac{dx}{x^p} = \infty \quad \text{but} \quad \int_0^1 \frac{dx}{x^q} < \infty$$

FIGURE 7

In Section 8.1, we will compute the length of a curve as an integral. It turns out that the improper integral in our next example represents the length of one-quarter of a unit circle. Thus, we can expect its value to be $\frac{1}{4}(2\pi) = \pi/2$.

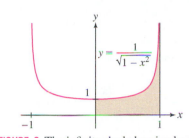

■ **EXAMPLE 9** Evaluate $\int_0^1 \dfrac{dx}{\sqrt{1-x^2}}$.

Solution This integral is improper with an infinite discontinuity at $x = 1$ (Figure 8). Using the formula $\int dx/\sqrt{1-x^2} = \sin^{-1} x + C$, we find

$$\int_0^1 \frac{dx}{\sqrt{1-x^2}} = \lim_{R \to 1-} \int_0^R \frac{dx}{\sqrt{1-x^2}}$$

$$= \lim_{R \to 1-} (\sin^{-1} R - \sin^{-1} 0)$$

$$= \sin^{-1} 1 - \sin^{-1} 0 = \frac{\pi}{2} - 0 = \frac{\pi}{2} \qquad ■$$

FIGURE 8 The infinite shaded region has area $\frac{\pi}{2}$.

Comparing Integrals

Sometimes we are interested in determining whether an improper integral converges, even if we cannot find its exact value. For instance, the integral

$$\int_1^\infty \frac{e^{-x}}{x}\, dx$$

cannot be evaluated explicitly. However, if $x \geq 1$, then

$$0 \leq \frac{1}{x} \leq 1 \quad \Rightarrow \quad 0 \leq \frac{e^{-x}}{x} \leq e^{-x}$$

In other words, the graph of $y = e^{-x}/x$ lies *underneath* the graph of $y = e^{-x}$ for $x \geq 1$ (Figure 9). Therefore,

$$0 \quad \leq \quad \int_1^\infty \frac{e^{-x}}{x}\, dx \quad \leq \quad \underbrace{\int_1^\infty e^{-x}\, dx = e^{-1}}_{\text{Converges by direct computation}}$$

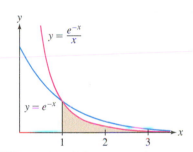

FIGURE 9 There is less area under $y = e^{-x}/x$ than $y = e^{-x}$ over the interval $[1, \infty)$.

Since the larger integral converges, we can expect that the smaller integral also converges (and that its value is some positive number less than e^{-1}). This type of conclusion is stated in the next theorem. A proof is provided in a supplement on the text's web site.

> **THEOREM 3 Comparison Test for Improper Integrals**
> Assume that f and g are continuous functions such that $f(x) \geq g(x) \geq 0$ for $x \geq a$:
>
> • If $\displaystyle\int_a^\infty f(x)\, dx$ converges, then $\displaystyle\int_a^\infty g(x)\, dx$ also converges.
>
> • If $\displaystyle\int_a^\infty g(x)\, dx$ diverges, then $\displaystyle\int_a^\infty f(x)\, dx$ also diverges.
>
> The Comparison Test is also valid for improper integrals with infinite discontinuities at the endpoints.

■ **EXAMPLE 10** Show that $\displaystyle\int_1^\infty \dfrac{dx}{\sqrt{x^3 + 1}}$ converges.

Solution We cannot evaluate this integral, but we can use the Comparison Test. To show convergence, we must compare the integrand $(x^3 + 1)^{-1/2}$ with a *larger* function whose integral we can compute.

It makes sense to compare with $x^{-3/2}$ because $\sqrt{x^3} \leq \sqrt{x^3 + 1}$. Therefore,

$$\frac{1}{\sqrt{x^3 + 1}} \leq \frac{1}{\sqrt{x^3}} = x^{-3/2}$$

The integral of the larger function converges, so the integral of the smaller function also converges:

$$\underbrace{\int_1^\infty \frac{dx}{x^{3/2}}}_{p\text{-integral with } p > 1} \quad \text{converges} \quad \Rightarrow \quad \underbrace{\int_1^\infty \frac{dx}{\sqrt{x^3+1}}}_{\text{Integral of smaller function}} \quad \text{converges} \quad \blacksquare$$

What the Comparison Test says (for nonnegative functions):

- If the integral of the larger function converges, then the integral of the smaller function also converges.
- If the integral of the smaller function diverges, then the integral of the larger function also diverges.

■ **EXAMPLE 11   Choosing the Right Comparison**   Does $\displaystyle\int_1^\infty \frac{dx}{\sqrt{x}+e^{3x}}$ converge?

Solution   Since $\sqrt{x} \geq 0$, we have $\sqrt{x} + e^{3x} \geq e^{3x}$ and therefore

$$\frac{1}{\sqrt{x}+e^{3x}} \leq \frac{1}{e^{3x}}$$

Furthermore,

$$\int_1^\infty \frac{dx}{e^{3x}} = \lim_{R\to\infty} -\frac{1}{3}e^{-3x}\Big|_1^R = \lim_{R\to\infty} \frac{1}{3}\left(e^{-3} - e^{-3R}\right) = \frac{1}{3}e^{-3} \quad \text{(converges)}$$

Our integral converges by the Comparison Test:

$$\underbrace{\int_1^\infty \frac{dx}{e^{3x}}}_{\text{Integral of larger function}} \quad \text{converges} \quad \Rightarrow \quad \underbrace{\int_1^\infty \frac{dx}{\sqrt{x}+e^{3x}}}_{\text{Integral of smaller function}} \quad \text{also converges}$$

Had we not been thinking, we might have tried to use the inequality

$$\frac{1}{\sqrt{x}+e^{3x}} \leq \frac{1}{\sqrt{x}}$$

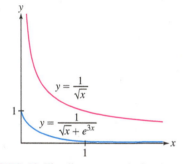

FIGURE 10 The divergence of a larger integral says nothing about the smaller integral.

However, $\displaystyle\int_1^\infty \frac{dx}{\sqrt{x}}$ diverges (p-integral with $p < 1$), and this says nothing about our smaller integral (Figure 10).   ■

■ **EXAMPLE 12   Endpoint Discontinuity**   Does $J = \displaystyle\int_0^{0.5} \frac{dx}{x^8 + x^2}$ converge?

Solution   This integrand has a discontinuity at $x = 0$, since $\displaystyle\lim_{x\to 0^+} \frac{1}{x^8 + x^2} = +\infty$. We might try the comparison

$$x^8 + x^2 > x^2 \quad \Rightarrow \quad \frac{1}{x^8 + x^2} < \frac{1}{x^2}$$

However, the p-integral $\displaystyle\int_0^{0.5} \frac{dx}{x^2}$ diverges, so this says nothing about our integral J, which is smaller. But notice that if $0 < x < 0.5$, then $x^8 < x^2$, and therefore

$$x^8 + x^2 < 2x^2 \quad \Rightarrow \quad \frac{1}{x^8 + x^2} > \frac{1}{2x^2}$$

Since $\displaystyle\int_0^{0.5} \frac{dx}{2x^2}$ diverges, the larger integral J also diverges.   ■

7.7 SUMMARY

- An *improper integral* is defined as the limit of ordinary definite integrals:

$$\int_a^\infty f(x)\, dx = \lim_{R\to\infty} \int_a^R f(x)\, dx$$

The improper integral *converges* if this limit exists, and it *diverges* otherwise.

- If f is continuous on $[a, b)$ and $\lim\limits_{x \to b^-} f(x) = \pm\infty$, then

$$\int_a^b f(x)\, dx = \lim_{R \to b^-} \int_a^R f(x)\, dx$$

- If f is continuous on $[a, b]$ and $\lim\limits_{x \to c^-} f(x) = \pm\infty$ or $\lim\limits_{x \to c^+} f(x) = \pm\infty$, where $a < c < b$ and if $\int_a^c f(x)\, dx$ and $\int_c^b f(x)\, dx$ converge, then

$$\int_a^b f(x)\, dx = \int_a^c f(x)\, dx + \int_c^b f(x)\, dx$$

- An improper integral of x^{-p} is called a p-integral. For $a > 0$,

$$p > 1: \quad \int_a^\infty \frac{dx}{x^p} \quad \text{converges} \quad \text{and} \quad \int_0^a \frac{dx}{x^p} \quad \text{diverges}$$

$$p < 1: \quad \int_a^\infty \frac{dx}{x^p} \quad \text{diverges} \quad \text{and} \quad \int_0^a \frac{dx}{x^p} \quad \text{converges}$$

$$p = 1: \quad \int_a^\infty \frac{dx}{x} \quad \text{and} \quad \int_0^a \frac{dx}{x} \quad \text{both diverge}$$

- The Comparison Test: Assume that f and g are continuous functions such that $f(x) \geq g(x) \geq 0$ for $x \geq a$. Then

$$\text{If} \quad \int_a^\infty f(x)\, dx \text{ converges,} \quad \text{then} \quad \int_a^\infty g(x)\, dx \text{ converges.}$$

$$\text{If} \quad \int_a^\infty g(x)\, dx \text{ diverges,} \quad \text{then} \quad \int_a^\infty f(x)\, dx \text{ diverges.}$$

- Remember that the Comparison Test provides no information if the larger integral $\int_a^\infty f(x)\, dx$ diverges or the smaller integral $\int_a^\infty g(x)\, dx$ converges.
- The Comparison Test is also valid for improper integrals with infinite discontinuities at endpoints.

7.7 EXERCISES

Preliminary Questions

1. State whether each of the following integrals converges or diverges:

(a) $\int_1^\infty x^{-3}\, dx$

(b) $\int_0^1 x^{-3}\, dx$

(c) $\int_1^\infty x^{-2/3}\, dx$

(d) $\int_0^1 x^{-2/3}\, dx$

2. Is $\int_0^{\pi/2} \cot x\, dx$ an improper integral? Explain.

3. Find a value of $b > 0$ that makes $\int_0^b \frac{1}{x^2 - 4}\, dx$ an improper integral.

4. Which comparison would show that $\int_0^\infty \frac{dx}{x + e^x}$ converges?

5. Explain why it is not possible to draw any conclusions about the convergence of $\int_1^\infty \frac{e^{-x}}{x}\, dx$ by comparing with the integral $\int_1^\infty \frac{dx}{x}$.

Exercises

1. Which of the following integrals is improper? Explain your answer, but do not evaluate the integral.

(a) $\int_0^2 \frac{dx}{x^{1/3}}$

(b) $\int_1^\infty \frac{dx}{x^{0.2}}$

(c) $\int_{-1}^\infty e^{-x}\,dx$

(d) $\int_0^1 e^{-x}\,dx$

(e) $\int_0^\pi \sec x\,dx$

(f) $\int_0^\infty \sin x\,dx$

(g) $\int_0^1 \sin x\,dx$

(h) $\int_0^1 \frac{dx}{\sqrt{3-x^2}}$

(i) $\int_1^\infty \ln x\,dx$

(j) $\int_0^3 \ln x\,dx$

2. Let $f(x) = x^{-4/3}$.

(a) Evaluate $\int_1^R f(x)\,dx$.

(b) Evaluate $\int_1^\infty f(x)\,dx$ by computing the limit

$$\lim_{R\to\infty} \int_1^R f(x)\,dx$$

3. Prove that $\int_1^\infty x^{-2/3}\,dx$ diverges by showing that

$$\lim_{R\to\infty} \int_1^R x^{-2/3}\,dx = \infty$$

4. Determine whether $\int_0^3 \frac{dx}{(3-x)^{3/2}}$ converges by computing

$$\lim_{R\to 3-} \int_0^R \frac{dx}{(3-x)^{3/2}}$$

In Exercises 5–40, determine whether the improper integral converges and, if so, evaluate it.

5. $\int_1^\infty \frac{dx}{x^{19/20}}$

6. $\int_1^\infty \frac{dx}{x^{20/19}}$

7. $\int_{-\infty}^4 e^{0.0001t}\,dt$

8. $\int_{20}^\infty \frac{dt}{t}$

9. $\int_0^5 \frac{dx}{x^{20/19}}$

10. $\int_0^5 \frac{dx}{x^{19/20}}$

11. $\int_0^4 \frac{dx}{\sqrt{4-x}}$

12. $\int_5^6 \frac{dx}{(x-5)^{3/2}}$

13. $\int_2^\infty x^{-3}\,dx$

14. $\int_0^\infty \frac{dx}{(x+1)^3}$

15. $\int_{-3}^\infty \frac{dx}{(x+4)^{3/2}}$

16. $\int_2^\infty e^{-2x}\,dx$

17. $\int_{-1}^1 \frac{dx}{x^{0.2}}$

18. $\int_2^\infty x^{-1/3}\,dx$

19. $\int_4^\infty e^{-3x}\,dx$

20. $\int_4^\infty e^{3x}\,dx$

21. $\int_{-\infty}^0 e^{3x}\,dx$

22. $\int_1^2 \frac{dx}{(x-1)^2}$

23. $\int_1^3 \frac{dx}{\sqrt{3-x}}$

24. $\int_{-4}^0 \frac{dx}{(x+2)^{1/3}}$

25. $\int_0^\infty \frac{dx}{1+x}$

26. $\int_{-\infty}^0 xe^{-x^2}\,dx$

27. $\int_0^\infty \frac{x\,dx}{(1+x^2)^2}$

28. $\int_3^6 \frac{x\,dx}{\sqrt{x-3}}$

29. $\int_0^\infty e^{-x}\cos x\,dx$

30. $\int_1^\infty xe^{-2x}\,dx$

31. $\int_0^3 \frac{dx}{\sqrt{9-x^2}}$

32. $\int_0^1 \frac{e^{\sqrt{x}}\,dx}{\sqrt{x}}$

33. $\int_1^\infty \frac{e^{\sqrt{x}}\,dx}{\sqrt{x}}$

34. $\int_0^\pi \sec\theta\,d\theta$

35. $\int_0^\infty \sin x\,dx$

36. $\int_0^{\pi/2} \tan x\,dx$

37. $\int_0^1 \ln x\,dx$

38. $\int_1^2 \frac{dx}{x\ln x}$

39. $\int_0^1 \frac{\ln x}{x^2}\,dx$

40. $\int_1^\infty \frac{\ln x}{x^2}\,dx$

41. Let $I = \int_4^\infty \frac{dx}{(x-2)(x-3)}$.

(a) Show that for $R > 4$,

$$\int_4^R \frac{dx}{(x-2)(x-3)} = \ln\left|\frac{R-3}{R-2}\right| - \ln\frac{1}{2}$$

(b) Then show that $I = \ln 2$.

42. Evaluate the integral $I = \int_1^\infty \frac{dx}{x(2x+5)}$.

43. Evaluate $I = \int_0^1 \frac{dx}{x(2x+5)}$ or state that it diverges.

44. Evaluate $I = \int_2^\infty \frac{dx}{(x+3)(x+1)^2}$ or state that it diverges.

In Exercises 45–48, determine whether the doubly infinite improper integral converges and, if so, evaluate it. Use definition (2).

45. $\int_{-\infty}^\infty \frac{x\,dx}{1+x^2}$

46. $\int_{-\infty}^\infty e^{-|x|}\,dx$

47. $\int_{-\infty}^\infty xe^{-x^2}\,dx$

48. $\int_{-\infty}^\infty \frac{dx}{(x^2+1)^{3/2}}$

49. Determine whether $J = \int_{-1}^1 \frac{dx}{x^{1/3}}$ converges and, if so, to what.

50. Consider the integral $\int_{-\infty}^\infty x\,dx$.

(a) Show that it diverges.

(b) Show that $\lim_{R\to\infty} \int_{-R}^R x\,dx$ converges, thereby demonstrating that the definition of $\int_{-\infty}^\infty f(x)\,dx$ needs to be adhered to carefully.

51. For which values of a does $\int_0^\infty e^{ax}\,dx$ converge?

52. Show that $\int_0^1 \frac{dx}{x^p}$ converges if $p < 1$ and diverges if $p \geq 1$.

53. Sketch the region under the graph of $f(x) = \dfrac{1}{1+x^2}$ for $-\infty < x < \infty$, and show that its area is π.

54. Show that $\dfrac{1}{\sqrt{x^4+1}} \le \dfrac{1}{x^2}$ for all x, and use this to prove that

$$\int_1^\infty \frac{dx}{\sqrt{x^4+1}} \text{ converges.}$$

55. Show that $\displaystyle\int_1^\infty \frac{dx}{x^3+4}$ converges by comparing with

$$\int_1^\infty x^{-3}\,dx.$$

56. Show that $\displaystyle\int_2^\infty \frac{dx}{x^3-4}$ converges by comparing with

$$\int_2^\infty 2x^{-3}\,dx.$$

57. ▱ Show that $0 \le e^{-x^2} \le e^{-x}$ for $x \ge 1$ (Figure 11). Use the Comparison Test to show that $\int_0^\infty e^{-x^2}\,dx$ converges. *Hint:* It suffices (why?) to make the comparison for $x \ge 1$ because

$$\int_0^\infty e^{-x^2}\,dx = \int_0^1 e^{-x^2}\,dx + \int_1^\infty e^{-x^2}\,dx$$

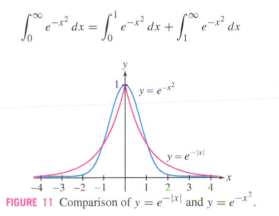

FIGURE 11 Comparison of $y = e^{-|x|}$ and $y = e^{-x^2}$.

58. Prove that $\displaystyle\int_{-\infty}^\infty e^{-x^2}\,dx$ converges by comparing with

$$\int_{-\infty}^\infty e^{-|x|}\,dx \text{ (Figure 11).}$$

59. Show that $\displaystyle\int_1^\infty \frac{1-\sin x}{x^2}\,dx$ converges.

60. Let $a > 0$. Recall that $\displaystyle\lim_{x\to\infty} \frac{x^a}{\ln x} = \infty$ (by Exercise 66 in Section 4.5).
(a) Show that $x^a > 2\ln x$ for all x sufficiently large.
(b) Show that $e^{-x^a} < x^{-2}$ for all x sufficiently large.
(c) Show that $\displaystyle\int_1^\infty e^{-x^a}\,dx$ converges.

In Exercises 61–75, use the Comparison Test to determine whether or not the integral converges.

61. $\displaystyle\int_1^\infty \frac{1}{\sqrt{x^5+2}}\,dx$

62. $\displaystyle\int_1^\infty \frac{dx}{(x^3+2x+4)^{1/2}}$

63. $\displaystyle\int_3^\infty \frac{dx}{\sqrt{x}-1}$

64. $\displaystyle\int_0^5 \frac{dx}{x^{1/3}+x^3}$

65. $\displaystyle\int_1^\infty e^{-(x+x^{-1})}\,dx$

66. $\displaystyle\int_0^1 \frac{|\sin x|}{\sqrt{x}}\,dx$

67. $\displaystyle\int_0^1 \frac{e^x}{x^2}\,dx$

68. $\displaystyle\int_1^\infty \frac{1}{x^4+e^x}\,dx$

69. $\displaystyle\int_0^1 \frac{1}{x^4+\sqrt{x}}\,dx$

70. $\displaystyle\int_1^\infty \frac{\ln x}{\sinh x}\,dx$

71. $\displaystyle\int_5^\infty \frac{1}{x^2\ln x}\,dx$

72. $\displaystyle\int_1^\infty \frac{dx}{\sqrt{x^{1/3}+x^3}}$

73. $\displaystyle\int_0^1 \frac{dx}{(8x^2+x^4)^{1/3}}$

74. $\displaystyle\int_1^\infty \frac{dx}{(x+x^2)^{1/3}}$

75. $\displaystyle\int_0^1 \frac{dx}{xe^x+x^2}$

Hint for Exercise 74: Show that for $x \ge 1$,

$$\frac{1}{(x+x^2)^{1/3}} \ge \frac{1}{2^{1/3}x^{2/3}}$$

Hint for Exercise 75: Show that for $0 \le x \le 1$,

$$\frac{1}{xe^x+x^2} \ge \frac{1}{(e+1)x}$$

76. Use the Comparison Test to determine for what values of p this integral converges: $\displaystyle\int_5^\infty \frac{1}{x^p\ln x}\,dx$.

77. Define $J = \displaystyle\int_0^\infty \frac{dx}{x^{1/2}(x+1)}$ as the sum of the two improper integrals

$$\int_0^1 \frac{dx}{x^{1/2}(x+1)} + \int_1^\infty \frac{dx}{x^{1/2}(x+1)}$$

Use the Comparison Test to show that J converges.

78. Determine whether $J = \displaystyle\int_0^\infty \frac{dx}{x^{3/2}(x+1)}$ (defined as in Exercise 77) converges.

79. An investment pays a dividend of \$250/year continuously forever. If the interest rate is 7%, what is the present value of the entire income stream generated by the investment?

80. An investment is expected to earn profits at a rate of $10{,}000e^{0.01t}$ dollars per year forever. Find the present value of the income stream if the interest rate is 4%.

81. Compute the present value of an investment that generates income at a rate of $5000te^{0.01t}$ dollars per year forever, assuming an interest rate of 6%.

82. Find the volume of the solid obtained by rotating the region below the graph of $y = e^{-x}$ about the x-axis for $0 \le x < \infty$.

83. When a capacitor of capacitance C is charged by a source of voltage V, the power expended at time t is

$$P(t) = \frac{V^2}{R}(e^{-t/RC} - e^{-2t/RC})$$

where R is the resistance in the circuit. The total energy stored in the capacitor is

$$W = \int_0^\infty P(t)\,dt$$

Show that $W = \frac{1}{2}CV^2$.

84. The solid S obtained by rotating the region below the graph of $y = x^{-1}$ about the x-axis for $1 \le x < \infty$ is called **Gabriel's Horn** (Figure 12).

(a) Use the Disk Method (Section 6.3) to compute the volume of S. Note that the volume is finite even though S is an infinite region.

(b) It can be shown that the surface area of S is

$$A = 2\pi \int_1^\infty x^{-1}\sqrt{1 + x^{-4}}\, dx$$

Show that A is infinite. If S were a container, you could fill its interior with a finite amount of paint, but you could not paint its surface with a finite amount of paint.

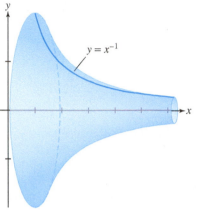

y = x⁻¹

FIGURE 12

85. Compute the volume of the solid obtained by rotating the region below the graph of $y = e^{-|x|/2}$ about the x-axis for $-\infty < x < \infty$.

86. For which integers p does $\displaystyle\int_0^{1/2} \frac{dx}{x(\ln x)^p}$ converge?

87. Conservation of Energy can be used to show that when a mass m oscillates at the end of a spring with spring constant k, the period of oscillation is

$$T = 4\sqrt{m} \int_0^{\sqrt{2E/k}} \frac{dx}{\sqrt{2E - kx^2}}$$

where E is the total energy of the mass. Show that this is an improper integral with value $T = 2\pi\sqrt{m/k}$.

*In Exercises 88–91, the **Laplace transform** of a function f is the function $\mathcal{L}f(s)$ of the variable s defined by the improper integral (if it converges):*

$$\mathcal{L}f(s) = \int_0^\infty f(x)e^{-sx}\, dx$$

Laplace transforms are widely used in physics and engineering.

88. Show that if $f(x) = C$, where C is a constant, then $\mathcal{L}f(s) = C/s$ for $s > 0$.

89. Show that if $f(x) = \sin \alpha x$, then $\mathcal{L}f(s) = \dfrac{\alpha}{s^2 + \alpha^2}$.

90. Compute $\mathcal{L}f(s)$, where $f(x) = e^{\alpha x}$ and $s > \alpha$.

91. Compute $\mathcal{L}f(s)$, where $f(x) = \cos \alpha x$ and $s > 0$.

92. When a radioactive substance decays, the fraction of atoms present at time t is $f(t) = e^{-kt}$, where $k > 0$ is the decay constant. It can be shown that the *average* life of an atom (until it decays) is $A = -\int_0^\infty t f'(t)\, dt$. Use Integration by Parts to show that $A = \int_0^\infty f(t)\, dt$ and compute A. What is the average decay time of radon-222, whose half-life is 3.825 days?

93. Let $J_n = \displaystyle\int_0^\infty x^n e^{-\alpha x}\, dx$, where $n \ge 1$ is an integer and $\alpha > 0$. Prove that

$$J_n = \frac{n}{\alpha} J_{n-1}$$

and $J_0 = 1/\alpha$. Use this to compute J_4. Show that $J_n = n!/\alpha^{n+1}$.

94. Let $a > 0$ and $n > 1$. Define $f(x) = \dfrac{x^n}{e^{ax} - 1}$ for $x \ne 0$ and $f(0) = 0$.

(a) Use L'Hôpital's Rule to show that f is continuous at $x = 0$.

(b) Show that $\int_0^\infty f(x)\, dx$ converges. *Hint:* Show that $f(x) \le 2x^n e^{-ax}$ if x is large enough. Then use the Comparison Test and Exercise 93.

95. According to **Planck's Radiation Law**, the amount of electromagnetic energy with frequency between ν and $\nu + \Delta\nu$ that is radiated by a so-called black body at temperature T is proportional to $F(\nu)\,\Delta\nu$, where

$$F(\nu) = \left(\frac{8\pi h}{c^3}\right) \frac{\nu^3}{e^{h\nu/kT} - 1}$$

where c, h, k are physical constants. Use Exercise 94 to show that the total radiated energy

$$E = \int_0^\infty F(\nu)\, d\nu$$

is finite. To derive his law, Planck introduced the quantum hypothesis in 1900, which marked the birth of quantum mechanics.

Further Insights and Challenges

96. Let $I = \displaystyle\int_0^1 x^p \ln x\, dx$.

(a) Show that I diverges for $p = -1$.

(b) Show that if $p \ne -1$, then

$$\int x^p \ln x\, dx = \frac{x^{p+1}}{p+1}\left(\ln x - \frac{1}{p+1}\right) + C$$

(c) Use L'Hôpital's Rule to show that I converges if $p > -1$ and diverges if $p < -1$.

97. Let

$$F(x) = \int_2^x \frac{dt}{\ln t} \qquad \text{and} \qquad G(x) = \frac{x}{\ln x}$$

Verify that L'Hôpital's Rule applies to the limit $L = \displaystyle\lim_{x\to\infty} \frac{F(x)}{G(x)}$ and evaluate L.

*In Exercises 98–100, an improper integral $I = \int_a^\infty f(x)\, dx$ is called **absolutely convergent** if $\int_a^\infty |f(x)|\, dx$ converges. It can be shown that if I is absolutely convergent, then it is convergent.*

98. Show that $\displaystyle\int_1^\infty \frac{\sin x}{x^2}\, dx$ is absolutely convergent.

99. Show that $\int_1^\infty e^{-x^2}\cos x\, dx$ is absolutely convergent.

100. Let $f(x) = \sin x / x$ and $I = \int_0^\infty f(x)\, dx$. We define $f(0) = 1$. Then f is continuous and I is not improper at $x = 0$.

(a) Show that

$$\int_1^R \frac{\sin x}{x}\, dx = -\left.\frac{\cos x}{x}\right|_1^R - \int_1^R \frac{\cos x}{x^2}\, dx$$

(b) Show that $\int_1^\infty (\cos x / x^2)\, dx$ converges. Conclude that the limit as $R \to \infty$ of the integral in (a) exists and is finite.

(c) Show that I converges.

It is known that $I = \frac{\pi}{2}$. However, I is *not* absolutely convergent. The convergence depends on cancellation, as shown in Figure 13.

101. The **gamma function**, which plays an important role in advanced applications, is defined for $n \geq 1$ by

$$\Gamma(n) = \int_0^\infty t^{n-1} e^{-t}\, dt$$

(a) Show that the integral defining $\Gamma(n)$ converges for $n \geq 1$ (it actually converges for all $n > 0$). *Hint:* Show that $t^{n-1} e^{-t} < t^{-2}$ for t sufficiently large.

(b) Show that $\Gamma(n + 1) = n\Gamma(n)$ using Integration by Parts.

(c) Show that $\Gamma(n + 1) = n!$ if $n \geq 1$ is an integer. *Hint:* Use (b) repeatedly. Thus, $\Gamma(n)$ provides a way of defining n-factorial when n is not an integer.

102. Use the results of Exercise 101 to show that the Laplace transform (see Exercises 88–91 above) of x^n is $\dfrac{n!}{s^{n+1}}$.

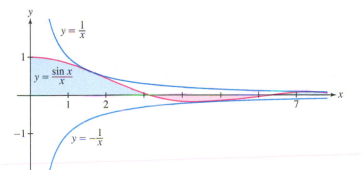

FIGURE 13 Convergence of $\int_1^\infty (\sin x / x)\, dx$ is due to the cancellation arising from the periodic change of sign.

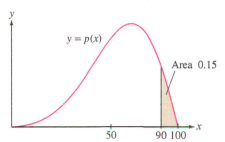

DF **FIGURE 1** Probability density function for scores on an exam. The shaded region has area 0.15, so there is a 15% chance that a randomly chosen exam has a score above 90.

7.8 Probability and Integration

What is the probability that a customer will arrive at a fast-food restaurant in the next 45 seconds? Or, of scoring above 90% on a standardized test? Probabilities such as these are given by a number between 0 and 1, where 0 means there is no probability the event will occur and 1 means that the event is sure to happen. These probabilities are best described as areas under the graph of a function $y = p(x)$ called a **probability density function** (Figure 1). The methods of integration developed in this chapter are used extensively in the study of such functions.

In probability theory, the quantity X that we are trying to predict (time to arrival, exam score, etc.) is called a **random variable**. The probability that X lies in a given range $[a, b]$ is denoted

$$P(a \leq X \leq b)$$

For example, the probability of a customer arriving within the next 30 to 45 seconds (s) is denoted $P(30 \leq X \leq 45)$.

We say that p is a **probability density function** for X if it is a continuous function such that

$$P(a \leq X \leq b) = \int_a^b p(x)\, dx$$

The probability density function p must also satisfy two conditions. First, it must satisfy $p(x) \geq 0$ for all x, because a probability cannot be negative. Second,

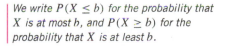

We write $P(X \leq b)$ for the probability that X is at most b, and $P(X \geq b)$ for the probability that X is at least b.

$$\int_{-\infty}^\infty p(x)\, dx = 1 \qquad \boxed{1}$$

The integral represents $P(-\infty < X < \infty)$. It must equal 1 because it is certain (the probability is 1) that the value of X lies between $-\infty$ and ∞. This also ensures that $P(a \leq X \leq b)$ is a number in the interval $[0, 1]$, even when we allow $a = -\infty$ and/or $b = \infty$.

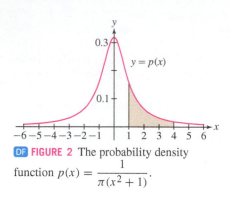

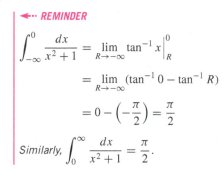

DF **FIGURE 2** The probability density
function $p(x) = \dfrac{1}{\pi(x^2 + 1)}$.

←·· REMINDER

$$\int_{-\infty}^{0} \frac{dx}{x^2 + 1} = \lim_{R \to -\infty} \tan^{-1} x \Big|_{R}^{0}$$

$$= \lim_{R \to -\infty} (\tan^{-1} 0 - \tan^{-1} R)$$

$$= 0 - \left(-\frac{\pi}{2}\right) = \frac{\pi}{2}$$

Similarly, $\displaystyle\int_{0}^{\infty} \frac{dx}{x^2 + 1} = \frac{\pi}{2}.$

■ **EXAMPLE 1** Find a constant C for which $p(x) = \dfrac{C}{x^2 + 1}$ is a probability density function. Then compute $P(1 \le X \le 4)$.

Solution We must choose C so that Eq. (1) is satisfied. The improper integral is a sum of two integrals (see Example 3 of Section 7.7).

$$\int_{-\infty}^{\infty} p(x)\, dx = C \int_{-\infty}^{0} \frac{dx}{x^2 + 1} + C \int_{0}^{\infty} \frac{dx}{x^2 + 1} = C\frac{\pi}{2} + C\frac{\pi}{2} = C\pi$$

Therefore, Eq. (1) is satisfied if $C\pi = 1$ or $C = \pi^{-1}$. We have

$$P(1 < X < 4) = \int_{1}^{4} p(x)\, dx = \int_{1}^{4} \frac{\pi^{-1}\, dx}{x^2 + 1} = \pi^{-1}(\tan^{-1} 4 - \tan^{-1} 1) \approx 0.17$$

Therefore, X lies between 1 and 4 with probability 0.17, or a 17% chance (Figure 2). ■

CONCEPTUAL INSIGHT If X is a random variable with probability density function p, then the probability of X taking on any specific value a is *zero* because $\displaystyle\int_{a}^{a} p(x)\, dx = 0$. So what is the meaning of $p(a)$? We must think of it this way: The probability that X lies in a *small interval* $[a, a + \Delta x]$ is approximately $p(a)\Delta x$:

$$P(a \le X \le a + \Delta x) = \int_{a}^{a+\Delta x} p(x)\, dx \approx p(a)\Delta x$$

A probability density is similar to a linear mass density $\rho(x)$. The mass of a small segment $[a, a + \Delta x]$ is approximately $\rho(a)\Delta x$, but the mass of any particular point $x = a$ is zero.

The *mean* or *average value* of a random variable is the quantity

$$\boxed{\mu = \mu(X) = \int_{-\infty}^{\infty} x p(x)\, dx}$$

2

if this integral converges.

The symbol μ is a lowercase Greek letter mu. If $p(x)$ is defined on $[0, \infty)$ instead of $(-\infty, \infty)$, or on some other interval, then μ is computed by integrating over that interval. Similarly, in Eq. (1) we integrate over the interval on which $p(x)$ is defined.

In the next example, we consider the **exponential probability density** with parameter $r > 0$, defined on $[0, \infty)$ by

$$\boxed{p(t) = \frac{1}{r}e^{-t/r}}$$

This density function is often used to model "waiting times" between events that occur randomly. Exercise 10 asks you to verify that $p(t)$ satisfies Eq. (1).

■ **EXAMPLE 2** **Mean of an Exponential Density** Let $r > 0$. Calculate the mean of the exponential probability density $p(t) = \frac{1}{r}e^{-t/r}$ on $[0, \infty)$.

Solution The mean is the integral of $tp(t)$ over $[0, \infty)$. Using Integration by Parts with $u = t/r$ and $dv = e^{-t/r}\, dt$, we have $du = dt/r$, $v = -re^{-t/r}$, and

$$\int tp(t)\, dt = \int \left(\frac{t}{r}e^{-t/r}\right) dt = -te^{-t/r} + \int e^{-t/r}\, dt = -(r + t)e^{-t/r}$$

Thus (since $re^{-R/r}$ and $Re^{-R/r}$ both tend to zero as $R \to \infty$ in the last step),

$$\mu = \int_{0}^{\infty} tp(t)\, dt = \int_{0}^{\infty} t\left(\frac{1}{r}e^{-t/r}\right) dt = \lim_{R \to \infty} -(r + t)e^{-t/r}\Big|_{0}^{R}$$

$$= \lim_{R \to \infty}\left(r - (r + R)\, e^{-R/r}\right) = r$$

■

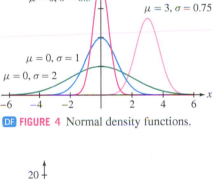

DF FIGURE 3 Customer arrivals have an exponential distribution.

■ **EXAMPLE 3** Waiting Time The waiting time T between customer arrivals in a drive-through fast-food restaurant is a random variable with exponential probability density. If the average waiting time is 60 seconds (s), what is the probability that a customer will arrive within 30 to 45 s after another customer?

Solution If the average waiting time is 60 s, then $r = 60$ and $p(t) = \frac{1}{60}e^{-t/60}$ because the mean of $\frac{1}{r}e^{-t/r}$ is r by the previous example. Therefore, the probability of waiting between 30 and 45 s for the next customer is

$$P(30 \le T \le 45) = \int_{30}^{45} \frac{1}{60}e^{-t/60} = -e^{-t/60}\Big|_{30}^{45} = -e^{-3/4} + e^{-1/2} \approx 0.134$$

This probability is the area of the shaded region in Figure 3. ■

The **normal density functions**, whose graphs are the familiar bell-shaped curves, appear in a surprisingly wide range of applications. The **standard normal** density is defined by

$$\boxed{p(x) = \frac{1}{\sqrt{2\pi}}e^{-x^2/2}} \quad \boxed{3}$$

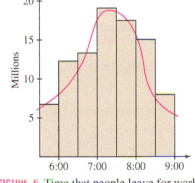

DF FIGURE 4 Normal density functions.

We can prove that $p(x)$ satisfies Eq. (1) using multivariable calculus.

More generally, we define the normal density function with mean μ and standard deviation σ:

$$\boxed{p(x) = \frac{1}{\sigma\sqrt{2\pi}}e^{-(x-\mu)^2/(2\sigma^2)}}$$

The standard deviation σ measures the spread; for larger values of σ, the graph is more spread out about the mean μ (Figure 4). The standard normal density in Eq. (3) has mean $\mu = 0$ and $\sigma = 1$. A random variable with a normal density function is said to have a **normal** or **Gaussian distribution**. Examples of data that fall in a normal distribution include sale prices for houses in Denver, heights of female children of age 11 in Chicago, and blood pressure readings for adults in Akron, Ohio. The normal distribution is ubiquitous in everyday life.

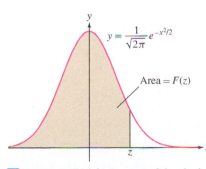

FIGURE 5 Time that people leave for work.

■ **EXAMPLE 4** In Figure 5, we see data from the 2012 American Community Survey on the time of day that workers in the United States leave for work. Note the bell-shaped curve generated by the data.

One difficulty with normal density functions is that they do not have elementary antiderivatives. As a result, we cannot evaluate the probabilities

$$P(a \le X \le b) = \frac{1}{\sigma\sqrt{2\pi}}\int_a^b e^{-(x-\mu)^2/(2\sigma^2)}\,dx$$

explicitly. However, the next theorem shows that these probabilities can all be expressed in terms of a single function called the **standard normal cumulative distribution function**:

$$\boxed{F(z) = \frac{1}{\sqrt{2\pi}}\int_{-\infty}^{z} e^{-x^2/2}\,dx}$$

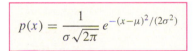

DF FIGURE 6 $F(z)$ is the area of the shaded region.

Observe that $F(z)$ is equal to the area under the graph over $(-\infty, z]$ in Figure 6. Numerical values of $F(z)$ are widely available on scientific calculators, on computer algebra systems, and online (search "standard cumulative normal distribution").

THEOREM 1   If X has a normal distribution with mean μ and standard deviation σ, then for all $a \le b$,

$$P(X \le b) = F\left(\frac{b-\mu}{\sigma}\right) \qquad \boxed{4}$$

$$P(a \le X \le b) = F\left(\frac{b-\mu}{\sigma}\right) - F\left(\frac{a-\mu}{\sigma}\right) \qquad \boxed{5}$$

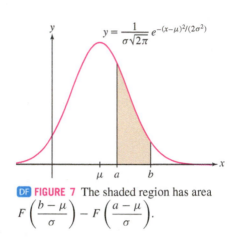

FIGURE 7 The shaded region has area $F\left(\dfrac{b-\mu}{\sigma}\right) - F\left(\dfrac{a-\mu}{\sigma}\right)$.

Proof   We use two changes of variables, first $u = x - \mu$ and then $t = u/\sigma$:

$$P(X \le b) = \frac{1}{\sigma\sqrt{2\pi}} \int_{-\infty}^{b} e^{-(x-\mu)^2/(2\sigma^2)}\,dx = \frac{1}{\sigma\sqrt{2\pi}} \int_{-\infty}^{b-\mu} e^{-u^2/(2\sigma^2)}\,du$$

$$= \frac{1}{\sqrt{2\pi}} \int_{-\infty}^{(b-\mu)/\sigma} e^{-t^2/2}\,dt = F\left(\frac{b-\mu}{\sigma}\right)$$

This proves Eq. (4). Eq. (5) follows because $P(a \le X \le b)$ is the area under the graph between a and b, and this is equal to the area to the left of b minus the area to the left of a (Figure 7).   ■

■ **EXAMPLE 5**   Assume that the scores X on a standardized test are normally distributed with mean $\mu = 500$ and standard deviation $\sigma = 100$. Find the probability that a test chosen at random has score

(a) at most 600.

(b) between 450 and 650.

Solution   We use a computer algebra system to evaluate $F(z)$ numerically.

(a) Apply Eq. (4) with $\mu = 500$ and $\sigma = 100$:

$$P(x \le 600) = F\left(\frac{600 - 500}{100}\right) = F(1) \approx 0.84$$

Thus, a randomly chosen score is 600 or less with a probability of 0.84, or 84%.

(b) Applying Eq. (5), we find that a randomly chosen score lies between 450 and 650 with a probability of 62.5%:

$$P(450 \le x \le 650) = F(1.5) - F(-0.5) \approx 0.933 - 0.308 = 0.625$$ ■

CONCEPTUAL INSIGHT   Why have we defined the mean of a continuous random variable X as the integral $\mu = \displaystyle\int_{-\infty}^{\infty} x p(x)\,dx$? Suppose first we are given N numbers $a_1, a_2, \ldots, a_N$, and for each value x, let $N(x)$ be the number of times x occurs among the a_j. Then a randomly chosen a_j has value x with probability $p(x) = N(x)/N$. For example, given the numbers 4, 4, 5, 5, 5, 8, we have $N = 6$ and $N(5) = 3$. The probability of choosing a 5 is $p(5) = N(5)/N = \frac{3}{6} = \frac{1}{2}$. Now observe that we can write the mean (average value) of the a_j in terms of the probabilities $p(x)$:

$$\frac{a_1 + a_2 + \cdots + a_N}{N} = \frac{1}{N} \sum_x N(x)x = \sum_x x p(x)$$

For example,

$$\frac{4 + 4 + 5 + 5 + 5 + 8}{6} = \frac{1}{6}(2 \cdot 4 + 3 \cdot 5 + 1 \cdot 8) = 4p(4) + 5p(5) + 8p(8)$$

In defining the mean of a continuous random variable X, we replace the sum $\sum_x x p(x)$ with the integral $\mu = \displaystyle\int_{-\infty}^{\infty} x p(x)\,dx$. This makes sense because the integral is the limit of sums $\sum x_i p(x_i)\Delta x$, and as we have seen, $p(x_i)\Delta x$ is the approximate probability that X lies in $[x_i, x_i + \Delta x]$.

7.8 SUMMARY

- If X is a continuous random variable with probability density function p, then

$$P(a \le X \le b) = \int_a^b p(x)\,dx$$

- Probability densities satisfy two conditions: $p(x) \ge 0$ and $\int_{-\infty}^{\infty} p(x)\,dx = 1$.
- Mean (or average) value of X:

$$\mu = \int_{-\infty}^{\infty} xp(x)\,dx$$

- Exponential density function of mean r:

$$p(x) = \frac{1}{r}e^{-x/r}$$

- Normal density of mean μ and standard deviation σ:

$$p(x) = \frac{1}{\sigma\sqrt{2\pi}}e^{-(x-\mu)^2/(2\sigma^2)}$$

- Standard cumulative normal distribution function:

$$F(z) = \frac{1}{\sqrt{2\pi}}\int_{-\infty}^{z} e^{-t^2/2}\,dt$$

- If X has a normal distribution of mean μ and standard deviation σ, then

$$P(X \le b) = F\left(\frac{b-\mu}{\sigma}\right)$$

$$P(a \le X \le b) = F\left(\frac{b-\mu}{\sigma}\right) - F\left(\frac{a-\mu}{\sigma}\right)$$

7.8 EXERCISES

Preliminary Questions

1. The function $p(x) = \cos x$ satisfies $\int_{-\pi/2}^{\pi} p(x)\,dx = 1$. Is p a probability density function on $[-\pi/2, \pi]$?

2. Estimate $P(2 \le X \le 2.1)$ assuming that the probability density function of X satisfies $p(2) = 0.2$.

3. Which exponential probability density has mean $\mu = \frac{1}{4}$?

Exercises

In Exercises 1–6, find a constant C such that p is a probability density function on the given interval, and compute the probability indicated.

1. $p(x) = \dfrac{C}{(x+1)^3}$ on $[0, \infty)$; $P(0 \le X \le 1)$

2. $p(x) = Cx(4-x)$ on $[0, 4]$; $P(3 \le X \le 4)$

3. $p(x) = \dfrac{C}{\sqrt{1-x^2}}$ on $(-1, 1)$; $P\left(-\frac{1}{2} \le X \le \frac{1}{2}\right)$

4. $p(x) = \dfrac{Ce^{-x}}{1+e^{-2x}}$ on $(-\infty, \infty)$; $P(X \le -4)$

5. $p(x) = C\sqrt{1-x^2}$ on $(-1, 1)$; $P\left(-\frac{1}{2} \le X \le 1\right)$

6. $p(x) = Ce^{-x}e^{-e^{-x}}$ on $(-\infty, \infty)$; $P(-4 \le X \le 4)$

This function, called the **Gumbel density**, is used to model extreme events such as floods and earthquakes.

7. Verify that $p(x) = 3x^{-4}$ is a probability density function on $[1, \infty)$ and calculate its mean value.

8. Show that the density function $p(x) = \dfrac{2}{\pi(x^2+1)}$ on $[0, \infty)$ has infinite mean.

9. Verify that $p(t) = \frac{1}{50}e^{-t/50}$ satisfies the condition

$$\int_0^{\infty} p(t)\,dt = 1$$

10. Verify that for all $r > 0$, the exponential density function $p(t) = \frac{1}{r}e^{-t/r}$ satisfies the condition

$$\int_0^\infty p(t)\, dt = 1$$

11. The life X (in hours) of a battery in constant use is a random variable with exponential density. What is the probability that the battery will last more than 12 hours (h) if the average life is 8 h?

12. The time between incoming phone calls at a call center is a random variable with exponential density. There is a 50% probability of waiting 20 seconds or more between calls. What is the average time between calls?

13. The distance r between the electron and the nucleus in a hydrogen atom (in its lowest energy state) is a random variable with probability density $p(r) = 4a_0^{-3}r^2 e^{-2r/a_0}$ for $r \geq 0$, where a_0 is the Bohr radius (Figure 8). Calculate the probability P that the electron is within one Bohr radius of the nucleus. The value of a_0 is approximately 5.29×10^{-11} m, but this value is not needed to compute P.

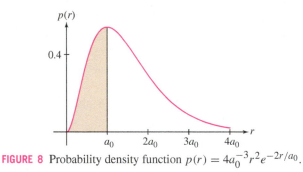

FIGURE 8 Probability density function $p(r) = 4a_0^{-3}r^2 e^{-2r/a_0}$.

14. Show that the distance r between the electron and the nucleus in Exercise 13 has mean $\mu = 3a_0/2$.

In Exercises 15–21, $F(z)$ denotes the cumulative normal distribution function. Refer to a calculator, computer algebra system, or online resource to obtain values of $F(z)$.

15. Express the area of region A in Figure 9 in terms of $F(z)$ and compute its value.

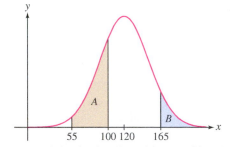

FIGURE 9 Normal density function with $\mu = 120$ and $\sigma = 30$.

16. Show that the area of region B in Figure 9 is equal to $1 - F(1.5)$ and compute its value. Verify numerically that this area is also equal to $F(-1.5)$ and explain why graphically.

17. Assume X has a standard normal distribution ($\mu = 0, \sigma = 1$). Find:

(a) $P(X \leq 1.2)$ **(b)** $P(X \geq -0.4)$

18. Evaluate numerically $\dfrac{1}{3\sqrt{2\pi}} \displaystyle\int_{14.5}^\infty e^{-(z-10)^2/18}\, dz$.

19. Use a graph to show that $F(-z) = 1 - F(z)$ for all z. Then show that if $p(x)$ is a normal density function with mean μ and standard deviation σ, then for all $r \geq 0$,

$$P(\mu - r\sigma \leq X \leq \mu + r\sigma) = 2F(r) - 1$$

20. The average September rainfall in Erie, Pennsylvania, is a random variable X with mean $\mu = 102$ mm. Assume that the amount of rainfall is normally distributed with standard deviation $\sigma = 48$.

(a) Express $P(128 \leq X \leq 150)$ in terms of $F(z)$ and compute its value numerically.

(b) Let P be the probability that September rainfall will be at least 120 mm. Express P as an integral of an appropriate density function and compute its value numerically.

21. A bottling company produces bottles of fruit juice that are filled, on average, with 32 ounces (oz) of juice. Due to random fluctuations in the machinery, the actual volume of juice is normally distributed with a standard deviation of 0.4 ounce. Let P be the probability of a bottle having less than 31 ounces. Express P as an integral of an appropriate density function and compute its value numerically.

22. According to **Maxwell's Distribution Law**, in a gas of molecular mass m, the speed v of a molecule in a gas at temperature T (kelvins) is a random variable with density

$$p(v) = 4\pi \left(\frac{m}{2\pi kT}\right)^{3/2} v^2 e^{-mv^2/(2kT)} \quad (v \geq 0)$$

where k is Boltzmann's constant. Show that the average molecular speed is equal to $(8kT/\pi m)^{1/2}$. The average speed of oxygen molecules at room temperature is around 450 m/s.

23. Define the median of a probability distribution to be that value a such that $\displaystyle\int_a^\infty p(x)\, dx = \int_{-\infty}^a p(x)\, dx = \frac{1}{2}$. Show that if a probability function is symmetric about the line $x = m$, then m is both the mean and the median.

24. Define the quartiles of a probability function to be those values a_1, a_2 and a_3 such that $P(-\infty < x \leq a_1) = P(a_1 \leq x \leq a_2) = P(a_2 \leq x \leq a_3) = P(a_3 \leq x < \infty) = \frac{1}{4}$. Find the quartile values for the probability function $p(x) = \frac{1}{1+x^2}$.

*In Exercises 25–28, calculate μ and σ, where σ is the **standard deviation**, defined by*

$$\sigma^2 = \int_{-\infty}^\infty (x - \mu)^2\, p(x)\, dx$$

The smaller the value of σ, the more tightly clustered are the values of the random variable X about the mean μ. (The limits of integration need not be $\pm\infty$ if p is defined over a smaller domain.)

25. $p(x) = \dfrac{5}{2x^{7/2}}$ on $[1, \infty)$

26. $p(x) = \dfrac{1}{\pi\sqrt{1 - x^2}}$ on $(-1, 1)$

27. $p(x) = \dfrac{1}{3}e^{-x/3}$ on $[0, \infty)$

28. $p(x) = \dfrac{1}{r}e^{-x/r}$ on $[0, \infty)$, where $r > 0$

Further Insights and Challenges

29. The time to decay of an atom in a radioactive substance is a random variable X. The law of radioactive decay states that if N atoms are present at time $t = 0$, then $N f(t)$ atoms will be present at time t, where $f(t) = e^{-kt}$ ($k > 0$ is the decay constant). Explain the following statements:

(a) The fraction of atoms that decay in a small time interval $[t, t + \Delta t]$ is approximately $-f'(t)\Delta t$.

(b) The probability density function of X is $y = -f'(t)$.

(c) The average time to decay is $1/k$.

30. The half-life of radon-222 is 3.825 days. Use Exercise 29 to compute:

(a) the average time to decay of a radon-222 atom.

(b) the probability that a given atom will decay in the next 24 hours.

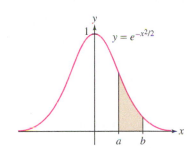

FIGURE 1 Areas under the bell-shaped curve are computed using numerical integration.

7.9 Numerical Integration

Numerical integration is the process of approximating a definite integral using well-chosen sums of function values. It is needed when we cannot find an antiderivative explicitly, as in the case of the Gaussian function $f(x) = e^{-x^2/2}$ (Figure 1).

In Section 5.1, we saw that we can approximate a definite integral by splitting the interval of integration $[a, b]$ into N subintervals, each of size Δx. Then we take the value of the function at each left-hand endpoint, multiply that by the width of the interval Δx, and sum over the intervals. This approximation is known as the left-endpoint approximation. Similarly, we saw a right-hand approximation. The third method that we introduced in that section, which we reconsider here, uses the midpoints of the intervals, and usually gives a better approximation.

The Midpoint Rule

To approximate the definite integral $\displaystyle\int_a^b f(x)\,dx$, we fix a whole number N and divide $[a, b]$ into N subintervals of length $\Delta x = (b - a)/N$. The endpoints of the subintervals (Figure 4) are

$$x_0 = a, \qquad x_1 = a + \Delta x, \qquad x_2 = a + 2\Delta x, \qquad \ldots, \qquad x_N = b$$

We shall denote the values of $f(x)$ at these endpoints by y_j:

$$y_j = f(x_j) = f(a + j\Delta x)$$

In particular, $y_0 = f(a)$ and $y_N = f(b)$.

The midpoint approximation M_N, is the sum of the areas of the rectangles of height $f(c_j)$ and base Δx, where c_j is the midpoint of the interval $[x_{j-1}, x_j]$ [Figure 2(A)].

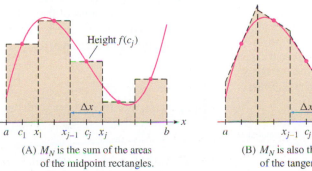

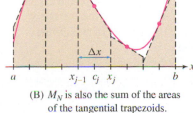

(A) M_N is the sum of the areas of the midpoint rectangles.

(B) M_N is also the sum of the areas of the tangential trapezoids.

FIGURE 2 Two interpretations of M_N.

Midpoint Rule The Nth midpoint approximation to $\int_a^b f(x)\,dx$ is

$$M_N = \Delta x \big(f(c_1) + f(c_2) + \cdots + f(c_N) \big)$$

where $\Delta x = \dfrac{b-a}{N}$ and $c_j = a + \big(j - \tfrac{1}{2}\big)\Delta x$ is the midpoint of $[x_{j-1}, x_j]$.

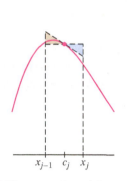

FIGURE 3 The rectangle and the trapezoid have the same area.

GRAPHICAL INSIGHT M_N has a second interpretation as the sum of the areas of tangential trapezoids—that is, trapezoids whose top edges are tangent to the graph of f at the midpoints c_j [Figure 2(B)]. The trapezoids have the same area as the rectangles because the top edge of the trapezoid passes through the midpoint of the top edge of the rectangle, as shown in Figure 3.

The Trapezoid Rule

The **Trapezoidal Rule** T_N approximates $\int_a^b f(x)\,dx$ by the area of the trapezoids obtained by joining the points (x_0, y_0), (x_1, y_1), ..., (x_N, y_N) with line segments, as in Figure 4. The area of the jth trapezoid is $\frac{1}{2}\Delta x(y_{j-1} + y_j)$, and therefore,

$$T_N = \frac{1}{2}\Delta x(y_0 + y_1) + \frac{1}{2}\Delta x(y_1 + y_2) + \cdots + \frac{1}{2}\Delta x(y_{N-1} + y_N)$$

$$= \frac{1}{2}\Delta x\Big((y_0 + y_1) + (y_1 + y_2) + \cdots + (y_{N-1} + y_N)\Big)$$

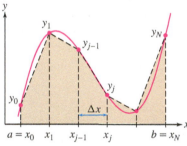

DF FIGURE 4 T_N approximates the area under the graph by trapezoids.

Note that each value y_j occurs twice except for y_0 and y_N, so we obtain

$$T_N = \frac{1}{2}\Delta x\Big(y_0 + 2y_1 + 2y_2 + \cdots + 2y_{N-1} + y_N\Big)$$

Trapezoidal Rule The Nth trapezoidal approximation to $\int_a^b f(x)\,dx$ is

$$T_N = \frac{1}{2}\Delta x\big(y_0 + 2y_1 + \cdots + 2y_{N-1} + y_N\big)$$

where $\Delta x = \dfrac{b-a}{N}$ and $y_j = f(x_j)$.

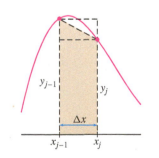

DF FIGURE 5 The shaded trapezoid has area $\frac{1}{2}\Delta x(y_{j-1} + y_j)$. This is the average of the areas of the left- and right-endpoint rectangles.

CONCEPTUAL INSIGHT We see in Figure 5 that the area of the jth trapezoid is equal to the average of the areas of the endpoint rectangles with heights y_{j-1} and y_j. It follows that T_N is equal to the average of the right- and left-endpoint approximations R_N and L_N introduced in Section 5.1:

$$T_N = \frac{1}{2}(R_N + L_N)$$

In general, this average is a better approximation than either R_N alone or L_N alone.

■ **EXAMPLE 1** *CAS*   Calculate T_8 for the integral $\int_1^3 \sin(x^2)\,dx$. Then use a computer algebra system to calculate T_N for $N = 50, 100, 500, 1000,$ and $10,000$.

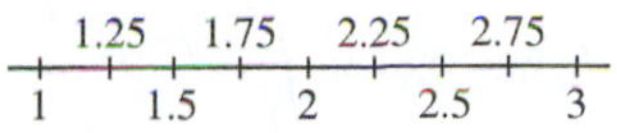

1.25 1.75 2.25 2.75

‖———‖———‖———‖———‖
1   1.5   2   2.5   3

FIGURE 6 Division of $[1, 3]$ into $N = 8$ subintervals.

Solution   Divide $[1, 3]$ into $N = 8$ subintervals of length $\Delta x = \frac{3-1}{8} = \frac{1}{4}$. Then sum the function values at the endpoints (Figure 6) with the appropriate coefficients:

$$T_8 = \frac{1}{2}\left(\frac{1}{4}\right)\left[\sin(1^2) + 2\sin(1.25^2) + 2\sin(1.5^2) + 2\sin(1.75^2)\right.$$

$$\left. + 2\sin(2^2) + 2\sin(2.25^2) + 2\sin(2.5^2) + 2\sin(2.75^2) + \sin(3^2)\right]$$

$$\approx 0.4281$$

In general, $\Delta x = (3 - 1)/N = 2/N$ and $x_j = 1 + 2j/N$. In summation notation,

$$T_N = \frac{1}{2}\left(\frac{2}{N}\right)\left[\sin(1^2) + 2\underbrace{\sum_{j=1}^{N-1} \sin\left(\left(1 + \frac{2j}{N}\right)^2\right)}_{\text{Sum of terms with coefficient 2}} + \sin(3^2)\right]$$

We evaluate the inner sum on a CAS, using a command such as

`Sum[Sin[(1 + 2j/N)^2], {j, 1, N − 1}]`

The results in Table 1 suggest that $\int_1^3 \sin(x^2)\,dx$ is approximately 0.4633. ∎

TABLE 1

N	T_N
50	0.4624205
100	0.4630759
500	**0.4632855**
1000	**0.4632920**
10,000	**0.4632942**

Error Bounds

In applications, it is important to know the accuracy of a numerical approximation. We define the error in M_N and T_N by

$$\text{Error}(M_N) = \left|\int_a^b f(x)\,dx - M_N\right|, \qquad \text{Error}(T_N) = \left|\int_a^b f(x)\,dx - T_N\right|$$

According to the next theorem, the magnitudes of these errors are related to the size of the *second* derivative $f''(x)$. A proof of Theorem 1 is provided in a supplement on the text's web site.

In the error bound, you can let K_2 be the maximum of $|f''(x)|$ on $[a, b]$, but if it is inconvenient to find this maximum exactly, take K_2 to be any number that is definitely larger than the maximum.

THEOREM 1   Error Bound for M_N and T_N   Assume f'' exists and is continuous. Let K_2 be a number such that $|f''(x)| \le K_2$ for all x in $[a, b]$. Then

$$\text{Error}(M_N) \le \frac{K_2(b-a)^3}{24N^2}, \qquad \text{Error}(T_N) \le \frac{K_2(b-a)^3}{12N^2}$$

GRAPHICAL INSIGHT   Note that the error bound for M_N is one-half of the error bound for T_N, suggesting that M_N is generally more accurate than T_N. Why do both error bounds depend on $f''(x)$? The second derivative measures concavity, so if $|f''(x)|$ is large, then the graph of f bends a lot and trapezoids do a poor job of approximating the region under the graph. Thus the errors in both T_N and M_N (which uses tangential trapezoids) are likely to be large (Figure 7).

FIGURE 7 M_N and T_N are more accurate when $|f''(x)|$ is small.

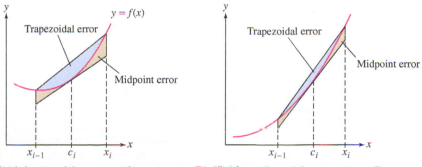

(A) $f''(x)$ is larger and the errors are larger.

(B) $f''(x)$ is smaller and the errors are smaller.

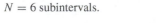

Midpoints 1.25 1.75 2.25 2.75 3.25 3.75

Endpoints 1 1.5 2 2.5 3 3.5 4

FIGURE 8 Interval [1, 4] divided into $N = 6$ subintervals.

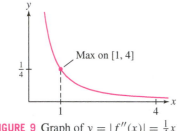

FIGURE 9 Graph of $y = |f''(x)| = \frac{1}{4}x^{-3/2}$ for $f(x) = \sqrt{x}$.

In Example 2, the error in T_6 is approximately twice as large as the error in M_6. In practice, this is often the case.

■ **EXAMPLE 2 Checking the Error Bound** Calculate M_6 and T_6 for $\int_1^4 \sqrt{x}\,dx$.

(a) Calculate the error bounds.

(b) Calculate the integral exactly and verify that the error bounds are satisfied.

Solution Divide [1, 4] into six subintervals of width $\Delta x = \frac{4-1}{6} = \frac{1}{2}$. Using the endpoints and midpoints shown in Figure 8, we obtain

$$M_6 = \frac{1}{2}\left(\sqrt{1.25} + \sqrt{1.75} + \sqrt{2.25} + \sqrt{2.75} + \sqrt{3.25} + \sqrt{3.75}\right) \approx 4.669245$$

$$T_6 = \frac{1}{2}\left(\frac{1}{2}\right)\left(\sqrt{1} + 2\sqrt{1.5} + 2\sqrt{2} + 2\sqrt{2.5} + 2\sqrt{3} + 2\sqrt{3.5} + \sqrt{4}\right) \approx 4.661488$$

(a) Let $f(x) = \sqrt{x}$. We must find a number K_2 such that $|f''(x)| \le K_2$ for $1 \le x \le 4$. We have $f''(x) = -\frac{1}{4}x^{-3/2}$. The absolute value $|f''(x)| = \frac{1}{4}x^{-3/2}$ is decreasing on [1, 4], so its maximum occurs at $x = 1$ (Figure 9). Thus, we may take $K_2 = |f''(1)| = \frac{1}{4}$. By Theorem 1,

$$\text{Error}(M_6) \le \frac{K_2(b-a)^3}{24N^2} = \frac{\frac{1}{4}(4-1)^3}{24(6)^2} = \frac{1}{128} \approx 0.0078$$

$$\text{Error}(T_6) \le \frac{K_2(b-a)^3}{12N^2} = \frac{\frac{1}{4}(4-1)^3}{12(6)^2} = \frac{1}{64} \approx 0.0156$$

(b) The exact value is $\int_1^4 \sqrt{x}\,dx = \frac{2}{3}x^{3/2}\Big|_1^4 = \frac{14}{3}$, so the actual errors are

$$\text{Error}(M_6) \approx \left|\frac{14}{3} - 4.669245\right| \approx 0.00258 \quad \text{(less than error bound 0.0078)}$$

$$\text{Error}(T_6) \approx \left|\frac{14}{3} - 4.661488\right| \approx 0.00518 \quad \text{(less than error bound 0.0156)}$$

The actual errors are less than the error bound, so Theorem 1 is verified. ■

The error bound can be used to determine values of N that provide a given accuracy.

■ **EXAMPLE 3 Obtaining the Desired Accuracy** Find N such that T_N approximates $\int_0^3 e^{-x^2}\,dx$ with an error of at most 10^{-4}.

A quick way to find a value for K_2 is to plot f'' using a graphing utility and find a bound for $|f''(x)|$ visually, as we do in Example 3.

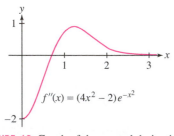

FIGURE 10 Graph of the second derivative of $f(x) = e^{-x^2}$.

Solution Let $f(x) = e^{-x^2}$. To apply the error bound, we must find a number K_2 such that $|f''(x)| \le K_2$ for all $x \in [0, 3]$. We have $f'(x) = -2xe^{-x^2}$ and

$$f''(x) = (4x^2 - 2)e^{-x^2}$$

A graphing utility was used to plot f'' (Figure 10). The graph shows that the maximum value of $|f''(x)|$ on [0, 3] is $|f''(0)| = |-2| = 2$, so we take $K_2 = 2$ in the error bound:

$$\text{Error}(T_N) \le \frac{K_2(b-a)^3}{12N^2} = \frac{2(3-0)^3}{12N^2} = \frac{9}{2N^2}$$

The error is at most 10^{-4} if

$$\frac{9}{2N^2} \le 10^{-4} \quad \Rightarrow \quad N^2 \ge \frac{9 \times 10^4}{2} \quad \Rightarrow \quad N \ge \frac{300}{\sqrt{2}} \approx 212.1$$

We conclude that T_{213} has an error of at most 10^{-4}. We can confirm this using a computer algebra system. A CAS shows that $T_{213} \approx 0.886207$, whereas the value of the integral to nine places is 0.886207348. Thus, the error is less than 10^{-6}. ■

Simpson's Rule

As we have seen, the Midpoint Rule uses trapezoids that are tangent to the curve to approximate the area under the curve. The Trapezoid Rule uses trapezoids with vertices on the curve to approximate the area. In both cases, the top edge of each trapezoid is a straight line segment. One must wonder whether we could do better using some other curve at the top of each region. In **Simpson's Rule**, we replace the line segments with parabolas, allowing us to obtain an approximation that is usually substantially more accurate.

To begin, we again subdivide $[a, b]$ into N subintervals, each of length $\Delta x = \frac{b-a}{N}$. However, we require N to be even. Then we pair up the resulting intervals, $[x_0, x_1]$ with $[x_1, x_2]$, $[x_2, x_3]$ with $[x_3, x_4]$, and so on. For each pair of intervals, we find a parabola that passes through the three points on the curve above the endpoints of the two intervals, as in Figure 11. Then we take the sum of the areas that are under the parabolas and over the corresponding intervals to approximate the area under the curve.

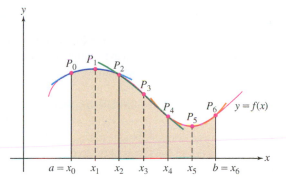

DF **FIGURE 11** Approximating the area under the curve using parabolas.

In order to determine the resulting formula, we begin with the case of a pair of intervals $[-\Delta x, 0]$ and $[0, \Delta x]$ centered around the origin. We assume that the corresponding three points on the curve are $P_0(-\Delta x, y_0)$, $P_1(0, y_1)$, and $P_2(\Delta x, y_2)$. See Figure 12.

The general equation for a parabola is $y = Cx^2 + Dx + E$ for constants C, D, and E. The area that is under the parabola and above the two intervals $[-\Delta x, 0]$ and $[0, \Delta x]$ is obtained by integrating:

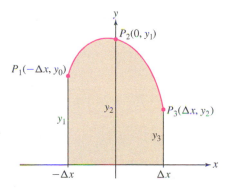

FIGURE 12 Finding the area under the parabola over the interval $[-\Delta x, \Delta x]$.

$$\text{Area} = \int_{-\Delta x}^{\Delta x} Cx^2 + Dx + E \, dx = \frac{Cx^3}{3} + \frac{Dx^2}{2} + Ex \bigg|_{-\Delta x}^{\Delta x}$$

$$= 2\left(\frac{C(\Delta x)^3}{3} + E\Delta x\right) = \frac{\Delta x}{3}(2C(\Delta x)^2 + 6E)$$

Because the parabola must pass through the three points P_0, P_1, and P_2, we know the coordinates of each must satisfy the equation of the parabola. Hence, we obtain the three equations:

$$y_0 = C(\Delta x)^2 - D\Delta x + E$$

$$y_1 = E$$

$$y_2 = C(\Delta x)^2 + D\Delta x + E$$

Adding the first and third equation yields

$$y_0 + y_2 = 2C(\Delta x)^2 + 2E$$

Adding four copies of the second equation gives the expression that appears in the area of the integral:

$$y_0 + 4y_1 + y_2 = 2C(\Delta x)^2 + 6E$$

Thus, the area under the parabola is given by $\frac{\Delta x}{3}(y_0 + 4y_1 + y_2)$.

This area only depends on the y-coordinates of the three points, so we obtain a similar expression if we site the parabola over any of the subsequent pairs of adjacent subintervals. Therefore, we can approximate the area under the curve by

$$\frac{\Delta x}{3}(y_0 + 4y_1 + y_2) + \frac{\Delta x}{3}(y_2 + 4y_3 + y_4) + \cdots + \frac{\Delta x}{3}(y_{N-2} + 4y_{N-1} + y_N)$$

Simplifying, we obtain the following:

> **Simpson's Rule** For N even, the Nth approximation to $\displaystyle\int_a^b f(x)\,dx$ by Simpson's Rule is
>
> $$S_N = \frac{1}{3}\Delta x\big[y_0 + 4y_1 + 2y_2 + \cdots + 4y_{N-3} + 2y_{N-2} + 4y_{N-1} + y_N\big] \qquad \boxed{1}$$
>
> where $\Delta x = \dfrac{b-a}{N}$ and $y_j = f(x_j)$.

Although we derived this rule in the case $f(x) \geq 0$ in order to make it easier to picture as an area, Simpson's Rule holds even when we are integrating a continuous function that is sometimes negative.

CONCEPTUAL INSIGHT Comparing Simpson's Rule to the Midpoint Rule and the Trapezoid Rule, we see that Simpson's Rule is a linear combination of the other two rules. That is to say, Simpson's Rule is given by $S_N = \frac{2}{3}M_{N/2} + \frac{1}{3}T_{N/2}$. When a function is always concave up or always concave down, the value of the actual integral is sandwiched between $M_{N/2}$ and $T_{N/2}$. So a linear combination of the two should do better in this and many other cases. That $M_{N/2}$ is more heavily weighted in the linear combination is advantageous, as we have seen its error bound is half that of $T_{N/2}$.

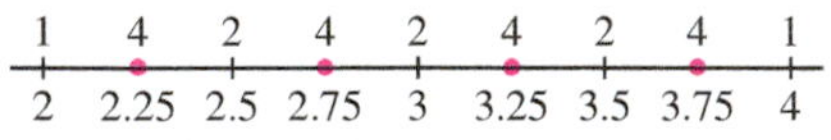

$$
\begin{array}{ccccccccc}
1 & 4 & 2 & 4 & 2 & 4 & 2 & 4 & 1 \\
\hline
2 & 2.25 & 2.5 & 2.75 & 3 & 3.25 & 3.5 & 3.75 & 4
\end{array}
$$

FIGURE 13 Coefficients for S_8 on $[2, 4]$ shown above the corresponding endpoint.

The accuracy of Simpson's Rule is impressive. Using a computer algebra system, we find that the approximation in Example 4 has an error of less than 3×10^{-6}.

■ **EXAMPLE 4** Use Simpson's Rule with $N = 8$ to approximate $\displaystyle\int_2^4 \sqrt{1 + x^3}\,dx$.

Solution We have $\Delta x = \frac{4-2}{8} = \frac{1}{4}$. Figure 13 shows the endpoints and coefficients needed to compute S_8 using Eq. (1):

$$\frac{1}{3}\left(\frac{1}{4}\right)\big[\sqrt{1 + 2^3} + 4\sqrt{1 + 2.25^3} + 2\sqrt{1 + 2.5^3} + 4\sqrt{1 + 2.75^3} + 2\sqrt{1 + 3^3}$$

$$+ 4\sqrt{1 + 3.25^3} + 2\sqrt{1 + 3.5^3} + 4\sqrt{1 + 3.75^3} + \sqrt{1 + 4^3}\big]$$

$$\approx \frac{1}{12}\big[3 + 4(3.52003) + 2(4.07738) + 4(4.66871) + 2(5.2915)$$

$$+ 4(5.94375) + 2(6.62382) + 4(7.33037) + 8.06226\big] \approx 10.74159 \qquad ■$$

■ **EXAMPLE 5** **Estimating Integrals from Numerical Data** The velocity (in kilometers per hour) of a Piper Cub aircraft traveling due west is recorded every minute during the first 10 minutes after takeoff. Use Simpson's Rule to estimate the distance traveled.

t (min)	0	1	2	3	4	5	6	7	8	9	10
$v(t)$ (km/h)	0	80	100	128	144	160	152	136	128	120	136

Solution The distance traveled is the integral of velocity. We convert from minutes to hours because velocity is given in kilometers per hour, and thus we apply Simpson's Rule, where the number of intervals is $N = 10$ and each interval has length $\Delta t = \frac{1}{60}$ h:

$$S_{10} = \left(\frac{1}{3}\right)\left(\frac{1}{60}\right)\big(0 + 4(80) + 2(100) + 4(128) + 2(144) + 4(160)$$

$$+ 2(152) + 4(136) + 2(128) + 4(120) + 136\big) \approx 21.2 \text{ km}$$

The distance traveled is approximately 21.2 km (Figure 14).   ■

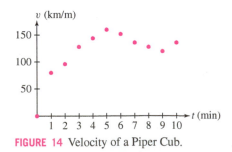

FIGURE 14 Velocity of a Piper Cub.

We now state (without proof) the error bound for Simpson's Rule. Set

$$\text{Error}(S_N) = \left| \int_a^b f(x) - S_N(f) \, dx \right|$$

The error involves the fourth derivative, which we assume exists and is continuous.

Although Simpson's Rule provides good approximations, more sophisticated techniques are implemented in computer algebra systems. These techniques are studied in the area of mathematics called numerical analysis.

> **THEOREM 2** **Error Bound for S_N**   Let K_4 be a number such that $|f^{(4)}(x)| \le K_4$ for all $x \in [a, b]$. Then
>
> $$\text{Error}(S_N) \le \frac{K_4(b-a)^5}{180N^4}$$

■ **EXAMPLE 6**   Calculate S_8 for $\displaystyle\int_1^3 \frac{1}{x} \, dx$.

(a) Find a bound for the error in S_8.

(b) Find N such that S_N has an error of at most 10^{-6}.

Solution The width is $\Delta x = \frac{3-1}{8} = \frac{1}{4}$ and the endpoints in the partition of $[1, 3]$ are $1, 1.25, 1.5, \ldots, 2.75, 3$. Using Eq. (1) with $f(x) = x^{-1}$, we obtain

$$S_8 = \frac{1}{3}\left(\frac{1}{4}\right)\left[\frac{1}{1} + \frac{4}{1.25} + \frac{2}{1.5} + \frac{4}{1.75} + \frac{2}{2} + \frac{4}{2.25} + \frac{2}{2.5} + \frac{4}{2.75} + \frac{1}{3}\right]$$

$$\approx 1.09873$$

(a) The fourth derivative $f^{(4)}(x) = 24x^{-5}$ is decreasing, so the max of $|f^{(4)}(x)|$ on $[1, 3]$ is $|f^{(4)}(1)| = 24$. Therefore, we use the error bound with $K_4 = 24$:

$$\text{Error}(S_N) \le \frac{K_4(b-a)^5}{180N^4} = \frac{24(3-1)^5}{180N^4} = \frac{64}{15N^4}$$

$$\text{Error}(S_8) \le \frac{K_4(b-a)^5}{180(8)^4} = \frac{24(3-1)^5}{180(8^4)} \approx 0.001$$

Using a CAS, we find that

$$S_{46} \approx 1.09861241$$

$$\int_1^3 \frac{1}{x} \, dx = \ln 3 \approx 1.09861229$$

The error is indeed less than 10^{-6}.

(b) The error will be at most 10^{-6} if N satisfies

$$\text{Error}(S_N) = \frac{64}{15N^4} \le 10^{-6}$$

In other words,

$$N^4 \ge 10^6\left(\frac{64}{15}\right) \qquad \text{or} \qquad N \ge \left(\frac{10^6 \cdot 64}{15}\right)^{1/4} \approx 45.45$$

Thus, we may take $N = 46$ (see the marginal comment). ■

7.9 SUMMARY

• We consider three numerical approximations to $\displaystyle\int_a^b f(x)\,dx$: the *Midpoint Rule M_N*, the *Trapezoidal Rule T_N*, and *Simpson's Rule S_N* (for N even).

$$M_N = \Delta x\big(f(c_1) + f(c_2) + \cdots + f(c_N)\big) \qquad \left(c_j = a + \left(j - \frac{1}{2}\right)\Delta x\right)$$

$$T_N = \frac{1}{2}\Delta x\big(y_0 + 2y_1 + 2y_2 + \cdots + 2y_{N-1} + y_N\big)$$

$$S_N = \frac{1}{3}\Delta x\big[y_0 + 4y_1 + 2y_2 + \cdots + 4y_{N-3} + 2y_{N-2} + 4y_{N-1} + y_N\big]$$

where $\Delta x = (b-a)/N$ and $y_j = f(a + j\,\Delta x)$.

- M_N has two geometric interpretations; it may be interpreted either as the sum of the areas of the midpoint rectangles or as the sum of the areas of the tangential trapezoids.
- T_N is equal to the sum of the areas of the trapezoids obtained by connecting the points $(x_0, y_0), (x_1, y_1), \ldots, (x_N, y_N)$ with line segments.
- S_N is equal to $\frac{1}{3} T_{N/2} + \frac{2}{3} M_{N/2}$.
- Error bounds:

$$\text{Error}(M_N) \le \frac{K_2(b-a)^3}{24N^2}, \quad \text{Error}(T_N) \le \frac{K_2(b-a)^3}{12N^2}, \quad \text{Error}(S_N) \le \frac{K_4(b-a)^5}{180N^4}$$

where K_2 is any number such that $|f''(x)| \le K_2$ for all $x \in [a, b]$ and K_4 is any number such that $|f^{(4)}(x)| \le K_4$ for all $x \in [a, b]$.

7.9 EXERCISES

Preliminary Questions

1. What are T_1 and T_2 for a function on $[0, 2]$ such that $f(0) = 3$, $f(1) = 4$, and $f(2) = 3$?

2. For which graph in Figure 15 will T_N overestimate the integral? What about M_N?

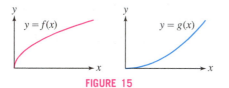

FIGURE 15

3. How large is the error when the Trapezoidal Rule is applied to a linear function? Explain graphically.

4. What is the maximum possible error if T_4 is used to approximate

$$\int_0^3 f(x)\, dx$$

where $|f''(x)| \le 2$ for all x.

5. What are the two graphical interpretations of the Midpoint Rule?

Exercises

In Exercises 1–12, calculate M_N and T_N for the value of N indicated.

1. $\int_0^2 x^2\, dx, \quad N = 4$

2. $\int_0^4 \sqrt{x}\, dx, \quad N = 4$

3. $\int_1^4 x^3\, dx, \quad N = 6$

4. $\int_1^2 \sqrt{x^4 + 1}\, dx, \quad N = 5$

5. $\int_1^4 \frac{dx}{x}, \quad N = 6$

6. $\int_{-2}^{-1} \frac{dx}{x}, \quad N = 5$

7. $\int_0^{\pi/2} \sqrt{\sin x}\, dx, \quad N = 6$

8. $\int_0^{\pi/4} \sec x\, dx, \quad N = 6$

9. $\int_1^2 \ln x\, dx, \quad N = 5$

10. $\int_2^3 \frac{dx}{\ln x}, \quad N = 5$

11. $\int_0^1 e^{-x^2}\, dx, \quad N = 5$

12. $\int_{-2}^1 e^{x^2}\, dx, \quad N = 6$

In Exercises 13–22, calculate S_N given by Simpson's Rule for the value of N indicated.

13. $\int_0^4 \sqrt{x}\, dx, \quad N = 4$

14. $\int_3^5 (9 - x^2)\, dx, \quad N = 4$

15. $\int_0^3 \frac{dx}{x^4 + 1}, \quad N = 6$

16. $\int_0^1 \cos(x^2)\, dx, \quad N = 6$

17. $\int_0^1 e^{-x^2}\, dx, \quad N = 4$

18. $\int_1^2 e^{-x}\, dx, \quad N = 6$

19. $\int_1^4 \ln x\, dx, \quad N = 8$

20. $\int_2^4 \sqrt{x^4 + 1}\, dx, \quad N = 8$

21. $\int_0^{\pi/4} \tan \theta\, d\theta, \quad N = 10$

22. $\int_0^2 (x^2 + 1)^{-1/3}\, dx, \quad N = 10$

In Exercises 23–26, calculate the approximation to the volume of the solid obtained by rotating the graph around the given axis.

23. $y = \cos x$; $[0, \frac{\pi}{2}]$; x-axis; M_8

24. $y = \cos x$; $[0, \frac{\pi}{2}]$; y-axis; S_8

25. $y = e^{-x^2}$; $[0, 1]$; x-axis; T_8

26. $y = e^{-x^2}$; $[0, 1]$; y-axis; S_8

27. An airplane's velocity is recorded at 5-minute (min) intervals during a 1-hour (h) period with the following results, in miles per hour:

550,   575,   600,   580,   610,   640,   625,
595,   590,   620,   640,   640,   630

Use Simpson's Rule to estimate the distance traveled during the hour.

28. Use Simpson's Rule to determine the average temperature in a museum over a 3-h period if the temperatures (in degrees Celsius), recorded at 15-min intervals, are

21,   21.3,   21.5,   21.8,   21.6,   21.2,   20.8,
20.6,   20.9,   21.2,   21.1,   21.3,   21.2

29. **Tsunami Arrival Times** Scientists estimate the arrival times of tsunamis (seismic ocean waves) based on the point of origin P and ocean depths. The speed s of a tsunami in miles per hour is approximately $s = \sqrt{15d}$, where d is the ocean depth in feet.

(a) Let $f(x)$ be the ocean depth x miles from P (in the direction of the coast). Argue using Riemann sums that the time T required for the tsunami to travel M miles toward the coast is

$$T = \int_0^M \frac{dx}{\sqrt{15 f(x)}}$$

(b) Use Simpson's Rule to estimate T if $M = 1000$ and the ocean depths (in feet), measured at 100-mile intervals starting from P, are

13,000,	11,500,	10,500,	9000,	8500,
7000,	6000,	4400,	3800,	3200, 2000

30. Use S_8 to estimate $\int_0^{\pi/2} \frac{\sin x}{x}\, dx$, taking the value of $\frac{\sin x}{x}$ at $x = 0$ to be 1.

31. Calculate T_6 for the integral $I = \int_0^2 x^3\, dx$.

(a) Is T_6 too large or too small? Explain graphically.

(b) Show that $K_2 = |f''(2)|$ may be used in the error bound and find a bound for the error.

(c) Evaluate I and check that the actual error is less than the bound computed in (b).

32. Calculate M_4 for the integral $I = \int_0^1 x \sin(x^2)\, dx$.

(a) [GU] Use a plot of f'' to show that $K_2 = 3.2$ may be used in the error bound and find a bound for the error.

(b) [CAS] Evaluate I numerically and check that the actual error is less than the bound computed in (a).

In Exercises 33–36, state whether T_N or M_N underestimates or overestimates the integral and find a bound for the error (but do not calculate T_N or M_N).

33. $\int_1^4 \frac{1}{x}\, dx, \quad T_{10}$

34. $\int_0^2 e^{-x/4}\, dx, \quad T_{20}$

35. $\int_1^4 \ln x\, dx, \quad M_{10}$

36. $\int_0^{\pi/4} \cos x, \quad M_{20}$

[CAS] *In Exercises 37–40, use the error bound to find a value of N for which $Error(T_N) \le 10^{-6}$. If you have a computer algebra system, calculate the corresponding approximation and confirm that the error satisfies the required bound.*

37. $\int_0^1 x^4\, dx$

38. $\int_0^3 (5x^4 - x^5)\, dx$

39. $\int_2^5 \frac{1}{x}\, dx$

40. $\int_0^3 e^{-x}\, dx$

41. Compute the error bound for the approximations T_{10} and M_{10} to $\int_0^3 (x^3 + 1)^{-1/2}\, dx$, using Figure 16 to determine a value of K_2. Then find a value of N such that the error in M_N is at most 10^{-6}.

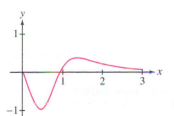

FIGURE 16 Graph of f'', where $f(x) = (x^3 + 1)^{-1/2}$.

42. (a) Compute S_6 for the integral $I = \int_0^1 e^{-2x}\, dx$.

(b) Show that $K_4 = 16$ may be used in the error bound and compute the error bound.

(c) Evaluate I and check that the actual error is less than the bound for the error computed in (b).

43. Calculate S_8 for $\int_1^5 \ln x\, dx$ and calculate the error bound. Then find a value of N such that S_N has an error of at most 10^{-6}.

44. Find a bound for the error in the approximation S_{10} to $\int_0^3 e^{-x^2}\, dx$ (use Figure 17 to determine a value of K_4). Then find a value of N such that S_N has an error of at most 10^{-6}.

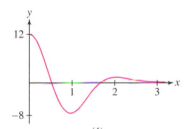

FIGURE 17 Graph of $f^{(4)}$, where $f(x) = e^{-x^2}$.

45. [CAS] Use a computer algebra system to compute and graph $f^{(4)}$ for $f(x) = \sqrt{1 + x^4}$, and find a bound for the error in the approximation S_{40} to $\int_0^5 f(x)\, dx$.

46. [CAS] Use a computer algebra system to compute and graph $f^{(4)}$ for $f(x) = \tan x - \sec x$, and find a bound for the error in the approximation S_{40} to $\int_0^{\pi/4} f(x)\, dx$.

In Exercises 47–50, use the error bound to find a value of N for which $Error(S_N) \le 10^{-9}$.

47. $\int_1^6 x^{4/3}\, dx$

48. $\int_0^4 x e^x\, dx$

49. $\int_0^1 e^{x^2}\, dx$

50. $\int_1^4 \sin(\ln x)\, dx$

51. [CAS] Show that $\int_0^1 \frac{dx}{1 + x^2} = \frac{\pi}{4}$ [use Eq. (3) in Section 5.8].

(a) Use a computer algebra system to graph the function $f^{(4)}$ for $f(x) = (1 + x^2)^{-1}$ and find its maximum on $[0, 1]$.

(b) Find a value of N such that S_N approximates the integral with an error of at most 10^{-6}. Calculate the corresponding approximation and confirm that you have computed $\frac{\pi}{4}$ to at least four places.

52. Let $J = \int_0^\infty e^{-x^2}\, dx$ and $J_N = \int_0^N e^{-x^2}\, dx$. Although e^{-x^2} has no elementary antiderivative, it is known that $J = \sqrt{\pi}/2$. Let T_N be the Nth trapezoidal approximation to J_N. Calculate T_4 and show that T_4 approximates J to three decimal places.

53. Let $f(x) = \sin(x^2)$ and $I = \int_0^1 f(x)\, dx$.

(a) Check that $f''(x) = 2\cos(x^2) - 4x^2 \sin(x^2)$. Then show that $|f''(x)| \le 6$ for $x \in [0, 1]$. *Hint:* Note that $|2\cos(x^2)| \le 2$ and $|4x^2 \sin(x^2)| \le 4$ for $x \in [0, 1]$.

(b) Show that $Error(M_N)$ is at most $\dfrac{1}{4N^2}$.

(c) Find an N such that $|I - M_N| \le 10^{-3}$.

54. _CAS_ 📖 The error bound for M_N is proportional to $1/N^2$, so the error bound decreases by $\frac{1}{4}$ if N is increased to $2N$. Compute the actual error in M_N for $\int_0^\pi \sin x\, dx$ for $N = 4, 8, 16, 32$, and 64. Does the actual error seem to decrease by $\frac{1}{4}$ as N is doubled?

55. _CAS_ 📖 Observe that the error bound for T_N (which has 12 in the denominator) is twice as large as the error bound for M_N (which has 24 in the denominator). Compute the actual error in T_N for $\int_0^\pi \sin x\, dx$ for $N = 4, 8, 16, 32$, and 64 and compare it with the calculations of Exercise 54. Does the actual error in T_N seem to be roughly twice as large as the error in M_N in this case?

56. _CAS_ 📖 Explain why the error bound for S_N decreases by $\frac{1}{16}$ if N is increased to $2N$. Compute the actual error in S_N for $\int_0^\pi \sin x\, dx$ for $N = 4, 8, 16, 32$, and 64. Does the actual error seem to decrease by $\frac{1}{16}$ as N is doubled?

57. Verify that S_2 yields the exact value of $\int_0^1 (x - x^3)\, dx$.

58. Verify that S_2 yields the exact value of $\int_a^b (x - x^3)\, dx$ for all $a < b$.

Further Insights and Challenges

59. Show that if $f(x) = rx + s$ is a linear function (r, s constants), then $T_N = \int_a^b f(x)\, dx$ for all N and all endpoints a, b.

60. Show that if $f(x) = px^2 + qx + r$ is a quadratic polynomial, then $S_2 = \int_a^b f(x)\, dx$. In other words, show that

$$\int_a^b f(x)\, dx = \frac{b - a}{6}(y_0 + 4y_1 + y_2)$$

where $y_0 = f(a)$, $y_1 = f\left(\frac{a + b}{2}\right)$, and $y_2 = f(b)$. _Hint:_ Show this first for $f(x) = 1, x, x^2$ and use linearity.

61. For N even, divide $[a, b]$ into N subintervals of width $\Delta x = \frac{b - a}{N}$. Set $x_j = a + j\,\Delta x$, $y_j = f(x_j)$, and

$$S_2^{2j} = \frac{b - a}{3N}(y_{2j} + 4y_{2j+1} + y_{2j+2})$$

(a) Show that S_N is the sum of the approximations on the intervals $[x_{2j}, x_{2j+2}]$—that is, $S_N = S_2^0 + S_2^2 + \cdots + S_2^{N-2}$.

(b) By Exercise 60, $S_2^{2j} = \int_{x_{2j}}^{x_{2j+2}} f(x)\, dx$ if f is a quadratic polynomial. Use (a) to show that S_N is exact _for all N if f is a quadratic polynomial._

62. Show that S_2 also gives the exact value for $\int_a^b x^3\, dx$ and conclude, as in Exercise 61, that S_N is exact for all cubic polynomials. Show by counterexample that S_2 is not exact for integrals of x^4.

63. Use the error bound for S_N to obtain another proof that Simpson's Rule is exact for all cubic polynomials.

64. 📖 **Sometimes Simpson's Rule Performs Poorly** Calculate M_{10} and S_{10} for the integral $\int_0^1 \sqrt{1 - x^2}\, dx$, whose value we know to be $\frac{\pi}{4}$ (one-quarter of the area of the unit circle).

(a) We usually expect S_N to be more accurate than M_N. Which of M_{10} and S_{10} is more accurate in this case?

(b) How do you explain the result of part (a)? _Hint:_ The error bounds are not valid because $|f''(x)|$ and $|f^{(4)}(x)|$ tend to ∞ as $x \to 1$, but $|f^{(4)}(x)|$ goes to infinity more quickly.

CHAPTER REVIEW EXERCISES

1. Match the integrals (a)–(e) with their antiderivatives (i)–(v) on the basis of the general form (do not evaluate the integrals).

(a) $\displaystyle\int \frac{x\, dx}{x^2 - 4}$

(b) $\displaystyle\int \frac{(2x + 9)\, dx}{x^2 + 4}$

(c) $\displaystyle\int \sin^3 x \cos^2 x\, dx$

(d) $\displaystyle\int \frac{dx}{x\sqrt{16x^2 - 1}}$

(e) $\displaystyle\int \frac{16\, dx}{x(x - 4)^2}$

(i) $\sec^{-1} 4x + C$

(ii) $\log|x| - \log|x - 4| - \dfrac{4}{x - 4} + C$

(iii) $\dfrac{1}{30}(3\cos^5 x - 3\cos^3 x \sin^2 x - 7\cos^3 x) + C$

(iv) $\dfrac{9}{2}\tan^{-1}\dfrac{x}{2} + \ln(x^2 + 4) + C$ **(v)** $\sqrt{x^2 - 4} + C$

2. Evaluate $\displaystyle\int \frac{x\, dx}{x + 2}$ in two ways: using substitution and using the Method of Partial Fractions.

In Exercises 3–12, evaluate using the suggested method.

3. $\displaystyle\int \cos^3\theta \sin^8\theta\, d\theta$ [write $\cos^3\theta$ as $\cos\theta(1 - \sin^2\theta)$]

4. $\displaystyle\int xe^{-12x}\, dx$ (Integration by Parts)

5. $\displaystyle\int \sec^3\theta \tan^4\theta\, d\theta$ (trigonometric identity, reduction formula)

6. $\displaystyle\int \frac{4x + 4}{(x - 5)(x + 3)}\, dx$ (partial fractions)

7. $\displaystyle\int \frac{dx}{x(x^2 - 1)^{3/2}}$ (trigonometric substitution)

8. $\displaystyle\int (1 + x^2)^{-3/2}dx$ (trigonometric substitution)

9. $\displaystyle\int \frac{dx}{x^{3/2} + x^{1/2}}$ (substitution)

10. $\displaystyle\int \frac{dx}{x + x^{-1}}$ (rewrite integrand)

11. $\displaystyle\int x^{-2}\tan^{-1}x\,dx$ (Integration by Parts)

12. $\displaystyle\int \frac{dx}{x^2+4x-5}$ (complete the square, substitution, partial fractions)

In Exercises 13–64, evaluate using the appropriate method or combination of methods.

13. $\displaystyle\int_0^1 x^2 e^{4x}\,dx$

14. $\displaystyle\int \frac{x^2}{\sqrt{9-x^2}}\,dx$

15. $\displaystyle\int \cos^9 6\theta \sin^3 6\theta\,d\theta$

16. $\displaystyle\int \sec^2\theta \tan^4\theta\,d\theta$

17. $\displaystyle\int \frac{(6x+4)\,dx}{x^2-1}$

18. $\displaystyle\int_4^9 \frac{dt}{(t^2-1)^2}$

19. $\displaystyle\int \frac{d\theta}{\cos^4\theta}$

20. $\displaystyle\int \sin 2\theta \sin^2\theta\,d\theta$

21. $\displaystyle\int_0^1 \ln(4-2x)\,dx$

22. $\displaystyle\int (\ln(x+1))^2\,dx$

23. $\displaystyle\int \sin^5\theta\,d\theta$

24. $\displaystyle\int \cos^4(9x-2)\,dx$

25. $\displaystyle\int_0^{\pi/4} \sin 3x \cos 5x\,dx$

26. $\displaystyle\int \sin 2x \sec^2 x\,dx$

27. $\displaystyle\int \sqrt{\tan x}\,\sec^2 x\,dx$

28. $\displaystyle\int (\sec x + \tan x)^2\,dx$

29. $\displaystyle\int \sin^5\theta \cos^3\theta\,d\theta$

30. $\displaystyle\int \cot^3 x \csc x\,dx$

31. $\displaystyle\int \cot^2 x \csc^2 x\,dx$

32. $\displaystyle\int_{\pi/2}^{\pi} \cot^2\frac{\theta}{2}\,d\theta$

33. $\displaystyle\int_{\pi/4}^{\pi/2} \cot^2 x \csc^3 x\,dx$

34. $\displaystyle\int_4^6 \frac{dt}{(t-3)(t+4)}$

35. $\displaystyle\int \frac{dt}{(t-3)^2(t+4)}$

36. $\displaystyle\int \sqrt{x^2+9}\,dx$

37. $\displaystyle\int \frac{dx}{x\sqrt{x^2-4}}$

38. $\displaystyle\int_8^{27} \frac{dx}{x+x^{2/3}}$

39. $\displaystyle\int \frac{dx}{x^{3/2}+ax^{1/2}}$

40. $\displaystyle\int \frac{dx}{(x-b)^2+4}$

41. $\displaystyle\int \frac{(x^2-x)\,dx}{(x+2)^3}$

42. $\displaystyle\int \frac{(7x^2+x)\,dx}{(x-2)(2x+1)(x+1)}$

43. $\displaystyle\int \frac{16\,dx}{(x-2)^2(x^2+4)}$

44. $\displaystyle\int \frac{dx}{(x^2+25)^2}$

45. $\displaystyle\int \frac{dx}{x^2+8x+25}$

46. $\displaystyle\int \frac{dx}{x^2+8x+4}$

47. $\displaystyle\int \frac{x-2}{x^3-2x^2-x+2}\,dx$

48. $\displaystyle\int_0^1 t^2\sqrt{1-t^2}\,dt$

49. $\displaystyle\int \frac{dx}{x^4\sqrt{x^2+4}}$

50. $\displaystyle\int \frac{dx}{(x^2+5)^{3/2}}$

51. $\displaystyle\int (x+1)e^{4-3x}\,dx$

52. $\displaystyle\int x^{-2}\tan^{-1}x\,dx$

53. $\displaystyle\int x^3\cos(x^2)\,dx$

54. $\displaystyle\int x^2(\ln x)^2\,dx$

55. $\displaystyle\int x\tanh^{-1}x\,dx$

56. $\displaystyle\int \frac{\tan^{-1}t\,dt}{1+t^2}$

57. $\displaystyle\int \ln(x^2+9)\,dx$

58. $\displaystyle\int (\sin x)(\cosh x)\,dx$

59. $\displaystyle\int_0^1 \cosh 2t\,dt$

60. $\displaystyle\int \sinh^3 x \cosh x\,dx$

61. $\displaystyle\int \coth^2(1-4t)\,dt$

62. $\displaystyle\int_{-0.3}^{0.3} \frac{dx}{1-x^2}$

63. $\displaystyle\int_0^{3\sqrt{3}/2} \frac{dx}{\sqrt{9-x^2}}$

64. $\displaystyle\int \frac{\sqrt{x^2+1}\,dx}{x^2}$

65. Use the substitution $u=\tanh t$ to evaluate $\displaystyle\int \frac{dt}{\cosh^2 t + \sinh^2 t}$.

66. Find the volume obtained by rotating the region enclosed by $y=\ln x$ and $y=(\ln x)^2$ about the y-axis.

67. Let $\displaystyle I_n = \int \frac{x^n\,dx}{x^2+1}$.

(a) Prove that $\displaystyle I_n = \frac{x^{n-1}}{n-1} - I_{n-2}$.

(b) Use (a) to calculate I_n for $0 \le n \le 5$.

(c) Show that, in general,

$$I_{2n+1} = \frac{x^{2n}}{2n} - \frac{x^{2n-2}}{2n-2} + \cdots$$
$$+ (-1)^{n-1}\frac{x^2}{2} + (-1)^n\frac{1}{2}\ln(x^2+1) + C$$
$$I_{2n} = \frac{x^{2n-1}}{2n-1} - \frac{x^{2n-3}}{2n-3} + \cdots$$
$$+ (-1)^{n-1}x + (-1)^n\tan^{-1}x + C$$

68. Let $\displaystyle J_n = \int x^n e^{-x^2/2}\,dx$.

(a) Show that $J_1 = -e^{-x^2/2}$.

(b) Prove that $J_n = -x^{n-1}e^{-x^2/2} + (n-1)J_{n-2}$.

(c) Use (a) and (b) to compute J_3 and J_5.

69. Compute $p(X \le 1)$, where X is a continuous random variable with probability density $\displaystyle p(x) = \frac{1}{\pi(x^2+1)}$.

70. Show that $p(x) = \frac{1}{4}e^{-x/2} + \frac{1}{6}e^{-x/3}$ is a probability density over the domain $[0,\infty)$ and find its mean.

71. Find a constant C such that $p(x) = Cx^3 e^{-x^2}$ is a probability density over the domain $[0,\infty)$ and compute $P(0 \le X \le 1)$.

72. The interval between patient arrivals in an emergency room is a random variable with exponential density function $p(t) = 0.125e^{-0.125t}$ (t in minutes). What is the average time between patient arrivals? What is the probability of two patients arriving within 3 min of each other?

73. Calculate the following probabilities, assuming that X is normally distributed with mean $\mu = 40$ and $\sigma = 5$.

(a) $P(X \ge 45)$

(b) $P(0 \le X \le 40)$

74. According to kinetic theory, the molecules of ordinary matter are in constant random motion. The energy E of a molecule is a random variable with density function $p(E) = \frac{1}{kT} e^{-E/(kT)}$, where T is the temperature (in kelvins) and k is Boltzmann's constant. Compute the mean kinetic energy $\overline{E}$ in terms of k and T.

In Exercises 75–84, determine whether the improper integral converges and, if so, evaluate it.

75. $\displaystyle\int_0^\infty \frac{dx}{(x+2)^2}$

76. $\displaystyle\int_4^\infty \frac{dx}{x^{2/3}}$

77. $\displaystyle\int_0^4 \frac{dx}{x^{2/3}}$

78. $\displaystyle\int_9^\infty \frac{dx}{x^{12/5}}$

79. $\displaystyle\int_{-\infty}^0 \frac{dx}{x^2+1}$

80. $\displaystyle\int_{-\infty}^9 e^{4x}\,dx$

81. $\displaystyle\int_0^{\pi/2} \cot\theta\,d\theta$

82. $\displaystyle\int_1^\infty \frac{dx}{(x+2)(2x+3)}$

83. $\displaystyle\int_0^\infty (5+x)^{-1/3}\,dx$

84. $\displaystyle\int_2^5 (5-x)^{-1/3}\,dx$

In Exercises 85–90, use the Comparison Test to determine whether the improper integral converges or diverges.

85. $\displaystyle\int_8^\infty \frac{dx}{x^2-4}$

86. $\displaystyle\int_8^\infty (\sin^2 x)e^{-x}\,dx$

87. $\displaystyle\int_3^\infty \frac{dx}{x^4+\cos^2 x}$

88. $\displaystyle\int_1^\infty \frac{dx}{x^{1/3}+x^{2/3}}$

89. $\displaystyle\int_0^1 \frac{dx}{x^{1/3}+x^{2/3}}$

90. $\displaystyle\int_0^\infty e^{-x^3}\,dx$

91. Calculate the volume of the infinite solid obtained by rotating the region under $y = (x^2+1)^{-2}$ for $0 \le x < \infty$ about the y-axis.

92. Let R be the region under the graph of $y = (x+1)^{-1}$ for $0 \le x < \infty$. Which of the following quantities is finite?
(a) The area of R
(b) The volume of the solid obtained by rotating R about the x-axis
(c) The volume of the solid obtained by rotating R about the y-axis

93. Show that $\displaystyle\int_0^\infty x^n e^{-x^2}\,dx$ converges for all $n > 0$. *Hint:* First observe that $x^n e^{-x^2} < x^n e^{-x}$ for $x > 1$. Then show that $x^n e^{-x} < x^{-2}$ for x sufficiently large.

94. Compute the Laplace transform $Lf(s)$ of the function $f(x) = x$ for $s > 0$. See Exercises 88–91 in Section 7.7 for the definition of $Lf(s)$.

95. Compute the Laplace transform $Lf(s)$ of the function $f(x) = x^2 e^{\alpha x}$ for $s > \alpha$.

96. Estimate $\displaystyle\int_2^5 f(x)\,dx$ by computing T_2, M_3, T_6, and S_6 for a function f taking on the values in the following table:

x	2	2.5	3	3.5	4	4.5	5
$f(x)$	$\frac{1}{2}$	2	1	0	$-\frac{3}{2}$	-4	-2

97. State whether the approximation M_N or T_N is larger or smaller than the integral.
(a) $\displaystyle\int_0^\pi \sin x\,dx$
(b) $\displaystyle\int_\pi^{2\pi} \sin x\,dx$
(c) $\displaystyle\int_1^8 \frac{dx}{x^2}$
(d) $\displaystyle\int_2^5 \ln x\,dx$

98. The rainfall rate (in inches per hour) was measured hourly during a 10-h thunderstorm with the following results:

$$0, \quad 0.41, \quad 0.49, \quad 0.32, \quad 0.3, \quad 0.23,$$
$$0.09, \quad 0.08, \quad 0.05, \quad 0.11, \quad 0.12$$

Use Simpson's Rule to estimate the total rainfall during the 10-h period.

In Exercises 99–104, compute the given approximation to the integral.

99. $\displaystyle\int_0^1 e^{-x^2}\,dx, \quad M_5$

100. $\displaystyle\int_2^4 \sqrt{6t^3+1}\,dt, \quad T_3$

101. $\displaystyle\int_{\pi/4}^{\pi/2} \sqrt{\sin\theta}\,d\theta, \quad M_4$

102. $\displaystyle\int_1^4 \frac{dx}{x^3+1}, \quad T_6$

103. $\displaystyle\int_0^1 e^{-x^2}\,dx, \quad S_4$

104. $\displaystyle\int_5^9 \cos(x^2)\,dx, \quad S_8$

105. The following table gives the area $A(h)$ of a horizontal cross section of a pond at depth h. Use the Trapezoidal Rule to estimate the volume V of the pond (Figure 1).

h (ft)	$A(h)$ (acres)	h (ft)	$A(h)$ (acres)
0	2.8	10	0.8
2	2.4	12	0.6
4	1.8	14	0.2
6	1.5	16	0.1
8	1.2	18	0

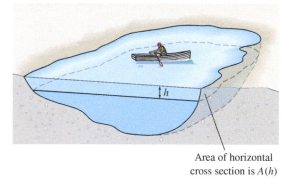

Area of horizontal cross section is $A(h)$

FIGURE 1

106. Suppose that the second derivative of the function A in Exercise 105 satisfies $|A''(h)| \le 1.5$. Use the error bound to find the maximum possible error in your estimate of the volume V of the pond.

107. Find a bound for the error $\left| M_{16} - \displaystyle\int_1^3 x^3\,dx \right|$.

108. GU Let $f(x) = \sin(x^3)$. Find a bound for the error

$$\left| T_{24} - \int_0^{\pi/2} f(x)\,dx \right|$$

Hint: Find a bound K_2 for $|f''(x)|$ by plotting f'' with a graphing utility.

109. Find a value of N such that

$$\left| M_N - \int_0^{\pi/4} \tan x\,dx \right| \le 10^{-4}$$

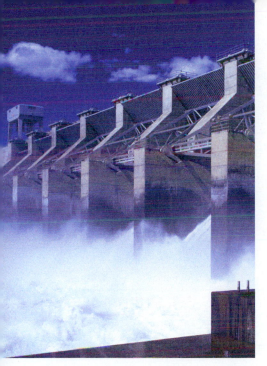

Engineers utilize integrals when designing dams in order to ensure that the dam walls can withstand the force imparted by the water. (*Earl Roberge/Science Source*)

8 FURTHER APPLICATIONS OF THE INTEGRAL AND TAYLOR POLYNOMIALS

The first three sections of this chapter develop some additional uses of integration, including two important physical applications. The last section introduces Taylor polynomials, the higher order generalizations of the linear approximation. Taylor polynomials illustrate beautifully the power of calculus to yield valuable insight into functions.

8.1 Arc Length and Surface Area

We have seen that integrals are used to compute "total amounts" (such as distance traveled, total mass, total cost, etc.). Another such quantity is the length of a curve (also called **arc length**). We derive a formula for arc length using our standard procedure: approximation followed by passage to a limit. In this case, we approximate the curve by a polygonal path made up of straight line segments connecting points on the curve. It is easy to find the length of a collection of straight line segments. Then we can take the limit of the sum of their lengths as the number of line segments grows.

To make this precise, consider the graph of $y = f(x)$ over an interval $[a, b]$. Choose a partition P of $[a, b]$ into N subintervals with endpoints

$$P : a = x_0 < x_1 < \cdots < x_N = b$$

and let $P_i = (x_i, f(x_i))$ be the point on the graph above x_i. Now join these points by line segments $L_i = \overline{P_{i-1}P_i}$. The resulting curve L is called a **polygonal approximation** (Figure 1). The length of L, which we denote $|L|$, is the sum of the lengths $|L_i|$ of the segments:

$$|L| = |L_1| + |L_2| + \cdots + |L_N| = \sum_{i=1}^{N} |L_i|$$

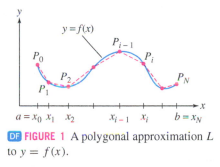

DF **FIGURE 1** A polygonal approximation L to $y = f(x)$.

The letter s is commonly used to denote arc length.

As may be expected, the polygonal approximations L approximate the curve more and more closely as the length of the longest subinterval, which we defined as the norm of the partition $\|P\|$ in Section 5.2, decreases (Figure 2). Based on this idea, we define the arc length s of the graph to be the limit of the lengths $|L|$ as the norm $\|P\|$ of the partition tends to zero:

$$\text{Arc length } s = \lim_{\|P\| \to 0} \sum_{i=1}^{N} |L_i|$$

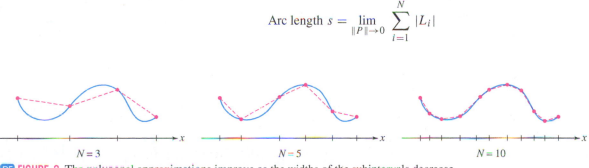

$N = 3$ $N = 5$ $N = 10$

DF **FIGURE 2** The polygonal approximations improve as the widths of the subintervals decrease.

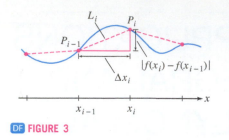

DF FIGURE 3

To compute the arc length s, we express the limit of the polygonal approximations as an integral. Figure 3 shows that the segment L_i is the hypotenuse of a right triangle of base $\Delta x_i = x_i - x_{i-1}$ and height $|f(x_i) - f(x_{i-1})|$. By the Pythagorean Theorem,

$$|L_i| = \sqrt{\Delta x_i^2 + (f(x_i) - f(x_{i-1}))^2}$$

We shall assume that f' exists and is continuous. Then, by the Mean Value Theorem, there is a value c_i in $[x_{i-1}, x_i]$ such that

$$f(x_i) - f(x_{i-1}) = f'(c_i)(x_i - x_{i-1}) = f'(c_i)\Delta x_i$$

and therefore,

$$|L_i| = \sqrt{(\Delta x_i)^2 + (f'(c_i)\Delta x_i)^2} = \sqrt{(\Delta x_i)^2(1 + [f'(c_i)]^2)} = \sqrt{1 + [f'(c_i)]^2}\, \Delta x_i$$

We find that the length $|L|$ is a Riemann sum for $\sqrt{1 + [f'(x)]^2}$:

REMINDER *A Riemann sum for the integral* $\int_a^b g(x)\,dx$ *is a sum*

$$\sum_{i=1}^{N} g(c_i)\Delta x_i$$

where $x_0, x_1, \ldots, x_N$ *is a partition of* $[a, b]$, $\Delta x_i = x_i - x_{i-1}$, *and* c_i *is any number in* $[x_{i-1}, x_i]$.

$$|L| = |L_1| + |L_2| + \cdots + |L_N| = \sum_{i=1}^{N} \sqrt{1 + [f'(c_i)]^2}\, \Delta x_i$$

This function is continuous, and hence integrable, so the Riemann sums approach

$$\int_a^b \sqrt{1 + [f'(x)]^2}\, dx$$

as the norm (maximum of the widths Δx_i) of the partition tends to zero.

THEOREM 1 Formula for Arc Length Assume that f' exists and is continuous on the interval $[a, b]$. Then the arc length s of $y = f(x)$ over $[a, b]$ is equal to

$$s = \int_a^b \sqrt{1 + [f'(x)]^2}\, dx \qquad \boxed{1}$$

In Exercises 22–24, we verify that Eq. (1) correctly gives the lengths of line segments and circles.

■ **EXAMPLE 1** Find the arc length s of the graph of $f(x) = \frac{1}{12}x^3 + x^{-1}$ over the interval $[1, 3]$ (Figure 4).

Solution First, let's calculate $1 + f'(x)^2$. Since $f'(x) = \frac{1}{4}x^2 - x^{-2}$,

$$1 + f'(x)^2 = 1 + \left(\frac{1}{4}x^2 - x^{-2}\right)^2 = 1 + \left(\frac{1}{16}x^4 - \frac{1}{2} + x^{-4}\right)$$

$$= \frac{1}{16}x^4 + \frac{1}{2} + x^{-4} = \left(\frac{1}{4}x^2 + x^{-2}\right)^2$$

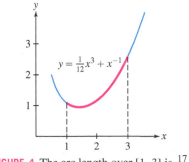

FIGURE 4 The arc length over $[1, 3]$ is $\frac{17}{6}$.

Fortunately, $1 + f'(x)^2$ is a square, so we can easily compute the arc length:

$$s = \int_1^3 \sqrt{1 + f'(x)^2}\, dx = \int_1^3 \left(\frac{1}{4}x^2 + x^{-2}\right) dx = \left(\frac{1}{12}x^3 - x^{-1}\right)\Big|_1^3$$

$$= \left(\frac{9}{4} - \frac{1}{3}\right) - \left(\frac{1}{12} - 1\right) = \frac{17}{6} \qquad ■$$

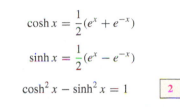

FIGURE 5

$$\cosh x = \frac{1}{2}(e^x + e^{-x})$$

$$\sinh x = \frac{1}{2}(e^x - e^{-x})$$

$$\cosh^2 x - \sinh^2 x = 1 \qquad \boxed{2}$$

■ **EXAMPLE 2** **Arc Length as a Function of the Upper Limit** Find the arc length $s(a)$ of $y = \cosh x$ over $[0, a]$ (Figure 5). Then find the arc length over $[0, 2]$.

Solution Recall that $y' = (\cosh x)' = \sinh x$. By Eq. (2) in the margin,

$$1 + (y')^2 = 1 + \sinh^2 x = \cosh^2 x$$

Because $\cosh x > 0$, we have $\sqrt{1 + (y')^2} = \cosh x$ and

$$s(a) = \int_0^a \sqrt{1 + (y')^2}\, dx = \int_0^a \cosh x\, dx = \sinh x \Big|_0^a = \sinh a$$

The arc length over $[0, 2]$ is $s(2) = \sinh 2 \approx 3.63$. ■

In Examples 1 and 2, the quantity $1 + f'(x)^2$ turned out to be a perfect square, and we were able to compute s exactly. Usually, $\sqrt{1 + f'(x)^2}$ does not have an elementary antiderivative and there is no explicit formula for the arc length. However, we can always approximate arc length using numerical integration.

■ **EXAMPLE 3** **No Exact Formula for Arc Length** ⌐⌐⌐ Approximate the length s of $y = \sin x$ over $[0, \pi]$ using Simpson's Rule S_N with $N = 6$.

Solution We have $y' = \cos x$ and $\sqrt{1 + (y')^2} = \sqrt{1 + \cos^2 x}$. The arc length is

$$s = \int_0^\pi \sqrt{1 + \cos^2 x}\, dx$$

This integral cannot be evaluated explicitly, so we approximate it by applying Simpson's Rule (Section 8.9) to the integrand $g(x) = \sqrt{1 + \cos^2 x}$. Divide $[0, \pi]$ into $N = 6$ subintervals of width $\Delta x = \pi/6$. Then

$$S_6 = \frac{\Delta x}{3}\left(g(0) + 4g\left(\frac{\pi}{6}\right) + 2g\left(\frac{2\pi}{6}\right) + 4g\left(\frac{3\pi}{6}\right) + 2g\left(\frac{4\pi}{6}\right) + 4g\left(\frac{5\pi}{6}\right) + g(\pi)\right)$$

$$\approx \frac{\pi}{18}(1.4142 + 5.2915 + 2.2361 + 4 + 2.2361 + 5.2915 + 1.4142) \approx 3.82$$

Thus, $s \approx 3.82$ (Figure 6). A computer algebra system yields the more accurate approximation $s \approx 3.820198$. ■

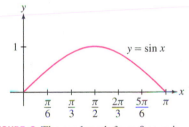

FIGURE 6 The arc length from 0 to π is approximately 3.82.

The surface area S of a surface of revolution (Figure 7) can be computed by an integral that is similar to the arc length integral. Suppose that $f(x) \geq 0$, so the graph lies above the x-axis. We can approximate the surface by rotating a polygonal approximation to $y = f(x)$ about the x-axis. The result is a surface built out of truncated cones (Figure 8).

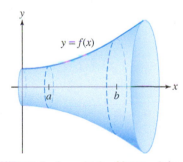

FIGURE 7 Surface obtained by revolving $y = f(x)$ about the x-axis.

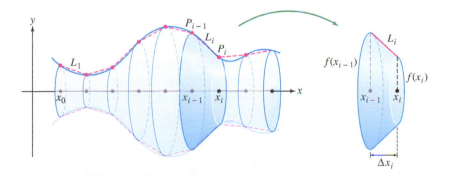

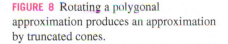

FIGURE 8 Rotating a polygonal approximation produces an approximation by truncated cones.

To determine the surface area of a truncated cone, we first consider a circular cone of radius r and slant length l, as in Figure 9. Cutting it open along the green line, we can then flatten it out in the plane, obtaining the fraction of the circle of radius l shown. Since this sector is a fraction of the entire circle determined by the fraction of the entire circumference to which it corresponds, namely $\dfrac{2\pi r}{2\pi l}$, and since the entire circle has an area of πl^2, the surface area of the cone must be $\dfrac{2\pi r}{2\pi l}\pi l^2 = \pi r l$.

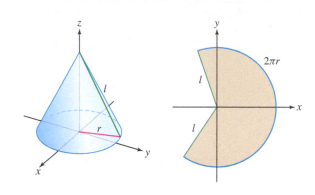

FIGURE 9 The cone has surface area $\pi r l$.

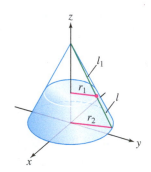

FIGURE 10 The truncated cone has surface area $\pi(r_1 + r_2)l$.

To find the surface area of a truncated cone, we consider Figure 10. The surface area is obtained by subtracting the area of the top cone from the area of the entire bigger cone, yielding $A = \pi r_2(l + l_1) - \pi r_1 l_1 = \pi r_2 l + \pi l_1(r_2 - r_1)$.

By similar triangles,

$$\frac{l_1}{r_1} = \frac{l + l_1}{r_2}$$

This yields $l_1 r_2 = r_1(l + l_1)$ and therefore $l_1(r_2 - r_1) = r_1 l$. So the surface area of the truncated cone is given by

$$A = \pi r_2 l + \pi l_1(r_2 - r_1) = \pi r_2 l + \pi r_1 l = \pi(r_1 + r_2)l$$

In words, the surface area of a truncated cone is equal to π times the sum of the two radii times the length of the slanted side.

Using the notation from the derivation of the arc length formula above, we therefore find that the surface area of the truncated cone corresponding to the subinterval $[x_{i-1}, x_i]$ is

$$\pi \underbrace{\left(f(x_{i-1}) + f(x_i)\right)}_{\text{Sum of radii}} \underbrace{|\overline{P_{i-1}P_i}|}_{\text{Slant length}} = 2\pi \left(\frac{f(x_{i-1}) + f(x_i)}{2}\right)\sqrt{1 + f'(c_i)^2}\,\Delta x_i$$

The surface area S is equal to the limit of the sums of the surface areas of the truncated cones as $N \to \infty$. We can show that the limit is not affected if we replace x_{i-1} and x_i by c_i. Therefore,

$$S = 2\pi \lim_{N \to \infty} \sum_{i=1}^{N} f(c_i)\sqrt{1 + f'(c_i)^2}\,\Delta x_i$$

This is a limit of Riemann sums that converges to the integral in Eq. (3) below.

Area of a Surface of Revolution Assume that $f(x) \geq 0$ and that f' exists and is continuous on the interval $[a, b]$. The surface area S of the surface obtained by rotating the graph of f about the x-axis for $a \leq x \leq b$ is equal to

$$S = 2\pi \int_a^b f(x)\sqrt{1 + f'(x)^2}\,dx \qquad \boxed{3}$$

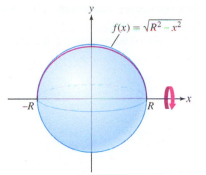

FIGURE 11 A sphere is obtained by revolving the red semicircle about the x-axis.

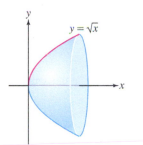

FIGURE 12 A paraboloid results when the top half of the parabola is revolved about the x-axis.

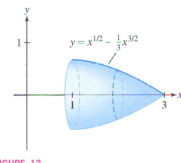

FIGURE 13

■ **EXAMPLE 4** Calculate the surface area of a sphere of radius R.

Solution The graph of $f(x) = \sqrt{R^2 - x^2}$ is a semicircle of radius R (Figure 11). We obtain a sphere by rotating it about the x-axis. We have

$$f'(x) = -\frac{x}{\sqrt{R^2 - x^2}}, \qquad 1 + f'(x)^2 = 1 + \frac{x^2}{R^2 - x^2} = \frac{R^2}{R^2 - x^2}$$

The surface area integral gives us the usual formula for the surface area of a sphere:

$$S = 2\pi \int_{-R}^{R} f(x)\sqrt{1 + f'(x)^2}\, dx = 2\pi \int_{-R}^{R} \sqrt{R^2 - x^2}\, \frac{R}{\sqrt{R^2 - x^2}}\, dx$$

$$= 2\pi R \int_{-R}^{R} dx = 2\pi R(2R) = 4\pi R^2 \qquad ■$$

■ **EXAMPLE 5** Find the surface area of the surface, called a paraboloid, that is obtained by rotating the graph of $f(x) = \sqrt{x}$ about the x-axis for $0 \le x \le 1$.

Solution The graph of $f(x) = \sqrt{x}$ is the top half of a parabola opening along the x-axis, which becomes a paraboloid when rotated about the x-axis (Figure 12). Then $f'(x) = \frac{1}{2\sqrt{x}}$ and hence we obtain

$$S = 2\pi \int_0^1 f(x)\sqrt{1 + f'(x)^2}\, dx = 2\pi \int_0^1 \sqrt{x}\sqrt{1 + \left(\frac{1}{2\sqrt{x}}\right)^2}\, dx$$

$$= 2\pi \int_0^1 \frac{\sqrt{x}}{2\sqrt{x}}\sqrt{4x + 1}\, dx$$

$$= \pi \int_0^1 \sqrt{4x + 1}\, dx$$

$$= \frac{\pi}{6}(4x + 1)^{\frac{3}{2}}\Big|_0^1 = \frac{\pi}{6}(5^{\frac{3}{2}} - 1) \approx 5.3304 \qquad ■$$

■ **EXAMPLE 6** Find the surface area S of the surface obtained by rotating the graph of $y = x^{1/2} - \frac{1}{3}x^{3/2}$ about the x-axis for $1 \le x \le 3$.

Solution Let $f(x) = x^{1/2} - \frac{1}{3}x^{3/2}$. Then $f'(x) = \frac{1}{2}(x^{-1/2} - x^{1/2})$ and

$$1 + f'(x)^2 = 1 + \left(\frac{x^{-1/2} - x^{1/2}}{2}\right)^2 = 1 + \frac{x^{-1} - 2 + x}{4}$$

$$= \frac{x^{-1} + 2 + x}{4} = \left(\frac{x^{1/2} + x^{-1/2}}{2}\right)^2$$

The surface area (Figure 13) is equal to

$$S = 2\pi \int_1^3 f(x)\sqrt{1 + f'(x)^2}\, dx = 2\pi \int_1^3 \left(x^{1/2} - \frac{1}{3}x^{3/2}\right)\left(\frac{x^{1/2} + x^{-1/2}}{2}\right) dx$$

$$= \pi \int_1^3 \left(1 + \frac{2}{3}x - \frac{1}{3}x^2\right) dx = \pi \left(x + \frac{1}{3}x^2 - \frac{1}{9}x^3\right)\Big|_1^3 = \frac{16\pi}{9} \qquad ■$$

Note that we can also find the area of a surface of revolution obtained by revolving a curve around the y-axis. We just reverse the roles of x and y. However, we do need to have the curve in the form $x = f(y)$ for the resulting formula to apply.

8.1 SUMMARY

• The *arc length* of $y = f(x)$ over the interval $[a, b]$ is

$$s = \int_a^b \sqrt{1 + f'(x)^2}\, dx$$

- Use numerical integration to approximate arc length when the arc length integral cannot be evaluated explicitly.
- Assume that $f(x) \geq 0$. The *surface area* of the surface obtained by rotating the graph of f about the x-axis for $a \leq x \leq b$ is

$$\text{Surface area} = 2\pi \int_a^b f(x)\sqrt{1 + f'(x)^2}\, dx$$

8.1 EXERCISES

Preliminary Questions

1. Which integral represents the length of the curve $y = \cos x$ between 0 and $\frac{\pi}{4}$?

$$\int_0^{\frac{\pi}{4}} \sqrt{1 + \cos^2 x}\, dx, \qquad \int_0^{\frac{\pi}{4}} \sqrt{1 + \sin^2 x}\, dx$$

2. By rotating the line $y = r$ about the x-axis, for x in the interval $[0, h]$, and applying the surface area formula, obtain the well-known fact that the surface area of a cylinder of radius r and length h is given by $2\pi r h$.

3. If $0 \leq f(x) \leq g(x)$ for x in the interval $[a, b]$, can the surface obtained by rotating the graph of $y = g(x)$ around the x-axis over the interval have less surface area than the surface obtained by rotating the graph of $y = f(x)$ around the x-axis over the same interval?

4. Use the formula for arc length to show that for any constant C, the graphs $y = f(x)$ and $y = f(x) + C$ have the same length over every interval $[a, b]$. Explain geometrically.

5. Use the formula for arc length to show that the length of a graph over $[1, 4]$ cannot be less than 3.

Exercises

1. Express the arc length of the curve $y = x^4$ between $x = 2$ and $x = 6$ as an integral (but do not evaluate).

2. Express the arc length of the curve $y = \tan x$ for $0 \leq x \leq \frac{\pi}{4}$ as an integral (but do not evaluate).

3. Find the arc length of $y = \frac{1}{12}x^3 + x^{-1}$ for $1 \leq x \leq 2$. *Hint:* Show that $1 + (y')^2 = \left(\frac{1}{4}x^2 + x^{-2}\right)^2$.

4. Find the arc length of $y = \left(\frac{x}{2}\right)^4 + \frac{1}{2x^2}$ over $[1, 4]$. *Hint:* Show that $1 + (y')^2$ is a perfect square.

In Exercises 5–10, calculate the arc length over the given interval.

5. $y = 3x + 1$, $[0, 3]$

6. $y = 9 - 3x$, $[1, 3]$

7. $y = x^{3/2}$, $[1, 2]$

8. $y = \frac{1}{3}x^{3/2} - x^{1/2}$, $[2, 8]$

9. $y = \frac{1}{4}x^2 - \frac{1}{2}\ln x$, $[1, 2e]$

10. $y = \ln(\cos x)$, $\left[0, \frac{\pi}{4}\right]$

In Exercises 11–16, approximate the arc length of the curve over the interval using the Trapezoidal Rule T_N, the Midpoint Rule M_N, or Simpson's Rule S_N as indicated.

11. $y = \frac{1}{4}x^4$, $[1, 2]$, T_5

12. $y = \sin x$, $\left[0, \frac{\pi}{2}\right]$, M_8

13. $y = x^{-1}$, $[1, 2]$, S_8

14. $y = e^{-x^2}$, $[0, 2]$, S_8

15. $y = \ln x$, $[1, 3]$, M_6

16. $y = \cos x$, $[0, 2]$, T_8

17. Calculate the length of the astroid $x^{2/3} + y^{2/3} = 1$ (Figure 14).

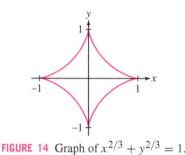

FIGURE 14 Graph of $x^{2/3} + y^{2/3} = 1$.

18. Show that the arc length of the astroid $x^{2/3} + y^{2/3} = a^{2/3}$ (for $a > 0$) is proportional to a.

19. Find the length of the arc of the curve $x^2 = (y - 2)^3$ from $P(1, 3)$ to $Q(8, 6)$.

20. Find the arc length of the curve shown in Figure 15.

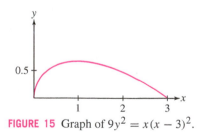

FIGURE 15 Graph of $9y^2 = x(x - 3)^2$.

21. Find the value of a such that the arc length of the *catenary* $y = \cosh x$ for $-a \leq x \leq a$ equals 10.

22. Calculate the arc length of the graph of $f(x) = mx + r$ over $[a, b]$ in two ways: using the Pythagorean theorem (Figure 16) and using the arc length integral.

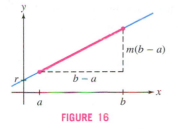

FIGURE 16

23. Show that the circumference of the unit circle is equal to

$$2 \int_{-1}^{1} \frac{dx}{\sqrt{1-x^2}} \quad \text{(an improper integral)}$$

Evaluate, thus verifying that the circumference is 2π.

24. Generalize the result of Exercise 23 to show that the circumference of the circle of radius r is $2\pi r$.

25. Calculate the arc length of $y = x^2$ over $[0, a]$. *Hint:* Use trigonometric substitution. Evaluate for $a = 1$.

26. Express the arc length of $g(x) = \sqrt{x}$ over $[0, 1]$ as a definite integral. Then use the substitution $u = \sqrt{x}$ to show that this arc length is equal to the arc length of $y = x^2$ over $[0, 1]$ (but do not evaluate the integrals). Explain this result graphically.

27. Find the arc length of $y = e^x$ over $[0, a]$. *Hint:* Try the substitution $u = \sqrt{1 + e^{2x}}$ followed by partial fractions.

28. Show that the arc length of $y = \ln(f(x))$ for $a \le x \le b$ is

$$\int_a^b \frac{\sqrt{f(x)^2 + f'(x)^2}}{f(x)} \, dx \qquad \boxed{4}$$

29. Use Eq. (4) to compute the arc length of $y = \ln(\sin x)$ for $\frac{\pi}{4} \le x \le \frac{\pi}{2}$.

30. Use Eq. (4) to compute the arc length of $y = \ln\left(\dfrac{e^x + 1}{e^x - 1}\right)$ over $[1, 3]$.

31. Show that if $0 \le f'(x) \le 1$ for all x, then the arc length of $y = f(x)$ over $[a, b]$ is at most $\sqrt{2}(b - a)$. Show that for $f(x) = x$, the arc length equals $\sqrt{2}(b - a)$.

32. Use the Comparison Theorem (Section 5.2) to prove that the arc length of $y = x^{4/3}$ over $[1, 2]$ is not less than $\frac{5}{3}$.

33. Approximate the arc length of one-quarter of the unit circle (which we know is $\frac{\pi}{2}$) by computing the length of the polygonal approximation with $N = 4$ segments (Figure 17).

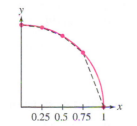

FIGURE 17 One-quarter of the unit circle.

34. ⌐⊏⊐⌐ A merchant intends to produce specialty carpets in the shape of the region in Figure 18, bounded by the axes and graph of $y = 1 - x^n$ (units in yards). Assume that material costs $50/yd^2$ and

that it costs $50L$ dollars to cut the carpet, where L is the length of the curved side of the carpet. The carpet can be sold for $150A$ dollars, where A is the carpet's area. Using numerical integration with a computer algebra system, find the whole number n for which the merchant's profits are maximal.

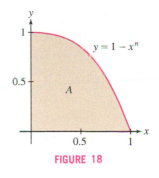

FIGURE 18

In Exercises 35–42, compute the surface area of revolution about the x-axis over the interval.

35. $y = x$, $[0, 4]$

36. $y = 4x + 3$, $[0, 1]$

37. $y = x^3$, $[0, 2]$

38. $y = x^2$, $[0, 4]$

39. $y = (4 - x^{2/3})^{3/2}$, $[0, 8]$

40. $y = e^{-x}$, $[0, 1]$

41. $y = \frac{1}{4}x^2 - \frac{1}{2}\ln x$, $[1, e]$

42. $y = \sin x$, $[0, \pi]$

⌐⊏⊐⌐ *In Exercises 43–46, use a computer algebra system to find the approximate surface area of the solid generated by rotating the curve about the x-axis.*

43. $y = x^{-1}$, $[1, 3]$

44. $y = x^4$, $[0, 1]$

45. $y = e^{-x^2/2}$, $[0, 2]$

46. $y = \tan x$, $\left[0, \frac{\pi}{4}\right]$

47. Find the area of the surface obtained by rotating $y = \cosh x$ over $[-\ln 2, \ln 2]$ around the x-axis.

48. Show that a spherical cap of height h and radius R (Figure 19) has surface area $2\pi Rh$.

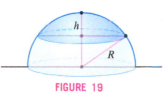

FIGURE 19

49. Find the surface area of the torus obtained by rotating the circle $x^2 + (y - b)^2 = r^2$ about the x-axis (Figure 20).

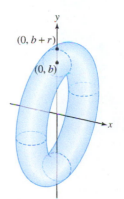

FIGURE 20 Torus obtained by rotating a circle about the x-axis.

50. By rotating the line $y = mx$ about the x-axis for $0 \le a \le x \le b$, we obtain a truncated cone. In this section, we used geometry to show that the surface area of such a truncated cone is $\pi(r_1 + r_2)l$ where r_1 and r_2 are the radii of the boundary circles of the cone and l is the slant length of the cone.

(a) Show that the slant height of the cone is given by $l = \sqrt{1 + m^2}(b - a)$ and that the surface area then becomes $\pi(r_1 + r_2)l = \pi m \sqrt{1 + m^2}(b^2 - a^2)$.

(b) Apply Equation 3 to the surface of revolution to obtain the same result for the surface area.

Further Insights and Challenges

51. Find the surface area of the ellipsoid obtained by rotating the ellipse $\left(\dfrac{x}{a}\right)^2 + \left(\dfrac{y}{b}\right)^2 = 1$ about the x-axis.

52. Show that if the arc length of $y = f(x)$ over $[0, a]$ is proportional to a, then $y = f(x)$ must be a linear function.

53. *CAS* Let L be the arc length of the upper half of the ellipse with equation

$$y = \frac{b}{a}\sqrt{a^2 - x^2}$$

(Figure 21) and let $\eta = \sqrt{1 - (b^2/a^2)}$. Use substitution to show that

$$L = a \int_{-\pi/2}^{\pi/2} \sqrt{1 - \eta^2 \sin^2 \theta}\, d\theta$$

Use a computer algebra system to approximate L for $a = 2, b = 1$.

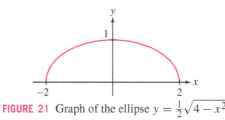

FIGURE 21 Graph of the ellipse $y = \frac{1}{2}\sqrt{4 - x^2}$.

54. Prove that the portion of a sphere of radius R seen by an observer located at a distance d above the North Pole has area $A = 2\pi d R^2/(d + R)$. *Hint:* According to Exercise 48, the cap has surface area $2\pi Rh$. Show that $h = dR/(d + R)$ by applying the Pythagorean Theorem to the three right triangles in Figure 22.

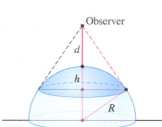

FIGURE 22 Spherical cap observed from a distance d above the North Pole.

55. Suppose that the observer in Exercise 54 moves off to infinity—that is, $d \to \infty$. What do you expect the limiting value of the observed area to be? Check your guess by calculating the limit using the formula for the area in the previous exercise.

56. Let M be the total mass of a metal rod in the shape of the curve $y = f(x)$ over $[a, b]$ whose mass density $\rho(x)$ varies as a function of x. Use Riemann sums to justify the formula

$$M = \int_a^b \rho(x)\sqrt{1 + f'(x)^2}\, dx$$

57. Let f be an increasing function on $[a, b]$ and let g be its inverse. Argue on the basis of arc length that the following equality holds:

$$\int_a^b \sqrt{1 + f'(x)^2}\, dx = \int_{f(a)}^{f(b)} \sqrt{1 + g'(y)^2}\, dy \qquad \boxed{5}$$

Then use the substitution $u = f(x)$ to prove Eq. (5).

8.2 Fluid Pressure and Force

Fluid force is the force on an object submerged in a fluid. Divers feel this force as they descend below the water surface (Figure 1). Our calculation of fluid force is based on two laws that determine the pressure exerted by a fluid:

- Fluid pressure p is proportional to depth.
- Fluid pressure does not act in a specific direction. Rather, a fluid exerts pressure on each side of an object in the perpendicular direction (Figure 2).

This second fact, known as Pascal's principle, points to an important difference between fluid pressure and the pressure exerted by one solid object on another.

FIGURE 1 Since water pressure is proportional to depth, divers breathe compressed air to equalize the pressure and avoid lung injury. (*Paul Soudors/WorldFoto/Aurora Photos*)

Fluid Pressure The pressure p at depth h in a fluid of mass density ρ is

$$\boxed{p = \rho g h} \qquad \boxed{1}$$

The pressure acts at each point on an object in the direction perpendicular to the object's surface at that point.

Pressure, by definition, is force per unit area.

- The SI unit of pressure is the pascal (Pa) ($1 \text{ Pa} = 1 \text{ N/m}^2 = 1 \text{ kg/ms}^2$).
- Mass density (mass per unit volume) is denoted ρ (Greek rho).
- The factor ρg is the density by weight, where $g = 9.8 \text{ m/s}^2$ is the acceleration due to gravity.

Our first example does not require integration because the pressure p is constant. In this case, the total force acting on a surface of area A is

$$\boxed{\text{Force} = \text{pressure} \times \text{area} = pA}$$

■ **EXAMPLE 1** Calculate the fluid force on the top and bottom of a box of dimensions $2 \times 2 \times 5$ m, submerged in a pool of water with its top 3 m below the water surface (Figure 2). The density of water is $\rho = 10^3 \text{ kg/m}^3$.

Solution The top of the box is located at depth $h = 3$ m, so by Eq. (1) with $g = 9.8 \text{ m/s}^2$,

$$\text{Pressure on top } p = \rho g h = 10^3(9.8)(3) = 29{,}400 \text{ Pa}$$

The top has area $A = 4 \text{ m}^2$ and the pressure is constant, so

$$\text{Downward force on top} = pA = 29{,}400 \times 4 = 117{,}600 \text{ N}$$

The bottom of the box is at depth $h = 8$ m, so the total force on the bottom is

$$\text{Upward force on bottom} = pA = \rho g A = 10^3(9.8)(8) \times 4 = 313{,}600 \text{ N} \qquad ■$$

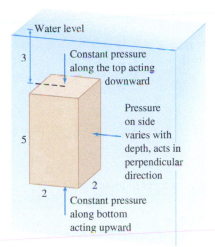

FIGURE 2 Fluid pressure acts on each side in the perpendicular direction.

In the next example, the pressure varies with depth, and it is necessary to calculate the force as an integral.

■ **EXAMPLE 2** **Calculating Force Using Integration** Calculate the fluid force F on the side of the box in Example 1.

Solution Since the pressure varies with depth, we divide the side of the box into N thin horizontal strips (Figure 3). Let F_j be the force on the jth strip. The total force F is equal to the sum of the forces on the strips:

$$F = F_1 + F_2 + \cdots + F_N$$

Step 1. Approximate the force on a strip.

We'll use the variable y to denote depth, where $y = 0$ at the water level and y is positive in the downward direction. Thus, a larger value of y denotes greater depth. Each strip is a rectangle of height $\Delta y = 5/N$ and length 2, so the area of a strip is $2\Delta y$. The bottom edge of the jth strip has depth $y_j = 3 + j\Delta y$.

If Δy is small, the pressure on the jth strip is nearly constant with value $\rho g y_j$ (because all points on the strip lie at nearly the same depth y_j), so we can approximate the force on the jth strip:

$$F_j \approx \underbrace{\rho g y_j}_{\text{Pressure}} \times \underbrace{(2\Delta y)}_{\text{Area}} = (\rho g) 2 y_j \Delta y$$

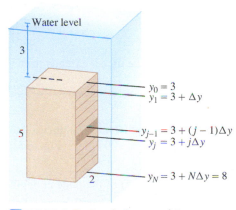

DF **FIGURE 3** Each strip has area $2\Delta y$.

Step 2. Approximate total force as a Riemann sum.

$$F = F_1 + F_2 + \cdots + F_N \approx \rho g \sum_{j=1}^{N} 2 y_j \Delta y$$

The sum on the right is a Riemann sum that converges to the integral $\rho g \displaystyle\int_3^8 2y \, dy$.

The interval of integration is $[3, 8]$ because the box extends from $y = 3$ to $y = 8$ (the Riemann sum has been set up with $y_0 = 3$ and $y_N = 8$).

Step 3. Evaluate total force as an integral.

As Δy tends to zero, the Riemann sum approaches the integral, and we obtain

$$F = \rho g \int_3^8 2y \, dy = (\rho g) y^2 \Big|_3^8 = (10^3)(9.8)(8^2 - 3^2) = 539{,}000 \text{ N} \qquad \blacksquare$$

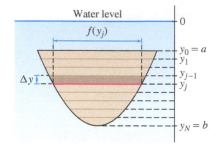

DF FIGURE 4 The area of the shaded strip is approximately $f(y_j) \, \Delta y$.

Now we'll add another complication: allowing the widths of the horizontal strips to vary with depth (Figure 4). Denote the width at depth y by $f(y)$:

$$f(y) = \text{width of the side at depth } y$$

As before, assume that the object extends from $y = a$ to $y = b$. Divide the flat side of the object into N horizontal strips of thickness $\Delta y = (b - a)/N$. If Δy is small, the jth strip is nearly rectangular of area $f(y) \Delta y$. Since the strip lies at depth $y_j = a + j \Delta y$, the force F_j on the jth strip can be approximated:

$$F_j \approx \underbrace{\rho g y_j}_{\text{Pressure}} \times \underbrace{f(y_j) \Delta y}_{\text{Area}} = (\rho g) y_j f(y_j) \Delta y$$

The force F is approximated by a Riemann sum that converges to an integral:

$$F = F_1 + \cdots + F_N \approx \rho g \sum_{j=1}^{N} y_j f(y_j) \Delta y \quad \Rightarrow \quad F = \rho g \int_a^b y f(y) \, dy$$

THEOREM 1 Fluid Force on a Flat Surface Submerged Vertically   The fluid force F on a flat side of an object submerged vertically in a fluid is

$$F = \rho g \int_a^b y f(y) \, dy \qquad \boxed{2}$$

where $f(y)$ is the horizontal width of the side at depth y, and the object extends from depth $y = a$ to depth $y = b$.

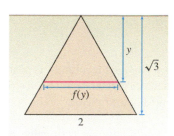

FIGURE 5 Triangular plate submerged in a tank of oil.

■ **EXAMPLE 3**   Calculate the fluid force F on one side of an equilateral triangular plate of side 2 m submerged vertically in a tank of oil of mass density $\rho = 900 \text{ kg/m}^3$ (Figure 5).

Solution   To use Eq. (2), we need to find the horizontal width $f(y)$ of the plate at depth y. An equilateral triangle of side $s = 2$ has height $\sqrt{3}s/2 = \sqrt{3}$. By similar triangles, $y/f(y) = \sqrt{3}/2$ and thus $f(y) = 2y/\sqrt{3}$. By Eq. (2),

$$F = \rho g \int_0^{\sqrt{3}} y \, f(y) \, dy = (900)(9.8) \int_0^{\sqrt{3}} \frac{2}{\sqrt{3}} y^2 \, dy = \left(\frac{17{,}640}{\sqrt{3}} \right) \frac{y^3}{3} \Big|_0^{\sqrt{3}} = 17{,}640 \text{ N}$$

$\blacksquare$

The next example shows how to modify the force calculation when the side of the submerged object is inclined at an angle.

Hoover Dam, with Mike O'Callaghan–Pat Tillman Memorial Bridge. (*Andrew Zarivny/ Shutterstock*)

■ **EXAMPLE 4   Force on an Inclined Surface**   The side of a dam is inclined at an angle of 45°. The dam has height 700 ft and width 1500 ft as in Figure 6. Calculate the force F on the dam if the reservoir is filled to the top of the dam. Water has weight density $w = 62.4 \text{ lb/ft}^3$.

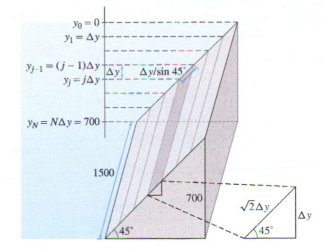

FIGURE 6

Solution The vertical height of the dam is 700 ft, so we divide the vertical axis from 0 to 700 into N subintervals of length $\Delta y = 700/N$. This divides the face of the dam into N strips as in Figure 6. By trigonometry, each strip has a width equal to $\Delta y / \sin(45°) = \sqrt{2}\Delta y$. Therefore,

$$\text{Area of each strip} = \text{length} \times \text{width} = 1500(\sqrt{2}\,\Delta y)$$

As usual, we approximate the force F_j on the jth strip. The term ρg is equal to weight per unit volume, so we use $w = 62.4$ lb/ft^3 in place of ρg:

$$F_j \approx \overbrace{wy_j}^{\text{Pressure}} \times \overbrace{1500\sqrt{2}\Delta y}^{\text{Area of strip}} = wy_j \times 1500\sqrt{2}\,\Delta y \ \text{lb}$$

$$F = \sum_{j=1}^{N} F_j \approx \sum_{j=1}^{N} wy_j\left(1500\sqrt{2}\,\Delta y\right) = 1500\sqrt{2}\,w \sum_{j=1}^{N} y_j \Delta y$$

This is a Riemann sum for the integral $1500\sqrt{2}w \displaystyle\int_0^{700} y\,dy$. Therefore,

$$F = 1500\sqrt{2}w \int_0^{700} y\,dy = 1500\sqrt{2}(62.4)\frac{700^2}{2} \approx 3.24 \times 10^{10} \ \text{lb} \qquad \blacksquare$$

8.2 SUMMARY

- If pressure is constant, then force = pressure × area.
- The fluid pressure at depth h is equal to $\rho g h$, where ρ is the fluid density (mass per unit volume) and $g = 9.8$ m/s^2 is the acceleration due to gravity. Fluid pressure acts on a surface in the direction perpendicular to the surface. Water has mass density 1000 kg/m^3.
- If an object is submerged vertically in a fluid and extends from depth $y = a$ to $y = b$, then the total fluid force on a side of the object is

$$F = \rho g \int_a^b yf(y)\,dy$$

where $f(y)$ is the horizontal width of the side at depth y.
- If fluid density is given as *weight* per unit volume, we use w in place of ρg. Water has weight density 62.4 lb/ft^3.

8.2 EXERCISES

Preliminary Questions

1. How is pressure defined?

2. Fluid pressure is proportional to depth. What is the factor of proportionality?

3. When fluid force acts on the side of a submerged object, in which direction does it act?

4. Why is fluid pressure on a surface calculated using thin horizontal strips rather than thin vertical strips?

5. If a thin plate is submerged horizontally, then the fluid force on one side of the plate is equal to pressure times area. Is this true if the plate is submerged vertically?

Exercises

1. A box of height 6 m and square base of side 3 m is submerged in a pool of water. The top of the box is 2 m below the surface of the water.

(a) Calculate the fluid force on the top and bottom of the box.

(b) Write a Riemann sum that approximates the fluid force on a side of the box by dividing the side into N horizontal strips of thickness $\Delta y = 6/N$.

(c) To which integral does the Riemann sum converge?

(d) Compute the fluid force on a side of the box.

2. A square plate that is 2 meters by 2 meters is submerged in water so that its top edge is level with the surface of the water. Calculate the fuid force on one side of it.

3. If a rectangular plate that is 1 meter by 2 meters is dipped into a pool of water so that initially its top edge of length 1 is even with the surface of the water, and then it is lowered so that its top edge is at a depth of 1 m, calculate the increase in fluid force on one side of it.

4. A plate in the shape of an isosceles triangle with base 1 m and height 2 m is submerged vertically in a tank of water so that its vertex touches the surface of the water (Figure 7).

(a) Show that the width of the triangle at depth y is $f(y) = \frac{1}{2}y$.

(b) Consider a thin strip of thickness Δy at depth y. Explain why the fluid force on a side of this strip is approximately equal to $\rho g \frac{1}{2} y^2 \Delta y$.

(c) Write an approximation for the total fluid force F on a side of the plate as a Riemann sum and indicate the integral to which it converges.

(d) Calculate F.

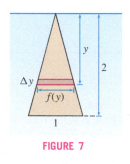

FIGURE 7

5. Repeat Exercise 4, but assume that the top of the triangle is located 3 m below the surface of the water.

6. The plate R in Figure 8, bounded by the parabola $y = x^2$ and $y = 1$, is submerged vertically in water (distance in meters).

(a) Show that the width of R at height y is $f(y) = 2\sqrt{y}$ and the fluid force on a side of a horizontal strip of thickness Δy at height y is approximately $(\rho g) 2 y^{1/2} (1 - y) \Delta y$.

(b) Write a Riemann sum that approximates the fluid force F on a side of R and use it to explain why

$$F = \rho g \int_0^1 2 y^{1/2} (1 - y) \, dy$$

(c) Calculate F.

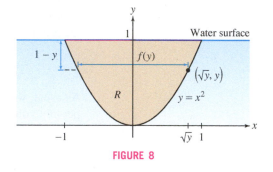

FIGURE 8

7. Let F be the fluid force on a side of a semicircular plate of radius r meters, submerged vertically in water so that its diameter is level with the water's surface (Figure 9).

(a) Show that the width of the plate at depth y is $2\sqrt{r^2 - y^2}$.

(b) Calculate F as a function of r using Eq. (2).

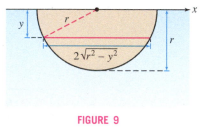

FIGURE 9

8. Calculate the force on one side of a circular plate with radius 2 m, submerged vertically in a tank of water so that the top of the circle is tangent to the water surface.

9. A semicircular plate of radius r meters, oriented as in Figure 9, is submerged in water so that its diameter is located at a depth of m meters. Calculate the fluid force on one side of the plate in terms of m and r.

10. A plate extending from depth $y = 2$ m to $y = 5$ m is submerged in a fluid of density $\rho = 850$ kg/m^3. The horizontal width of the plate at depth y is $f(y) = 2(1 + y^2)^{-1}$. Calculate the fluid force on one side of the plate.

11. Figure 10 shows the wall of a dam on a water reservoir. Use the Trapezoidal Rule and the width and depth measurements in the figure to estimate the fluid force on the wall.

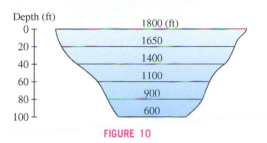

FIGURE 10

12. Calculate the fluid force on a side of the plate in Figure 11(A), submerged in water.

13. Calculate the fluid force on a side of the plate in Figure 11(B), submerged in a fluid of mass density $\rho = 800$ kg/m^3.

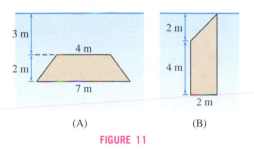

(A)   (B)

FIGURE 11

14. Find the fluid force on the side of the plate in Figure 12, submerged in a fluid of density $\rho = 1200$ kg/m^3. The top of the plate is level with the fluid surface. The edges of the plate are the curves $y = x^{1/3}$ and $y = -x^{1/3}$.

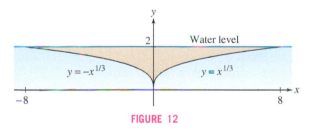

FIGURE 12

15. Let R be the plate in the shape of the region under $y = \sin x$ for $0 \le x \le \frac{\pi}{2}$ in Figure 13(A). If R is rotated counterclockwise by 90° and then submerged in a fluid of density 1100 kg/m^3 with its top edge level with the surface of the fluid as in (B), find the fluid force on a side of R.

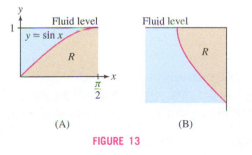

(A)   (B)

FIGURE 13

16. In the notation of Exercise 15, calculate the fluid force on a side of the plate R if it is oriented as in Figure 13(A). You may need to use Integration by Parts and trigonometric substitution.

17. Calculate the fluid force on one side of a plate in the shape of region A shown Figure 14. The water surface is at $y = 1$, and the fluid has density $\rho = 900$ kg/m^3.

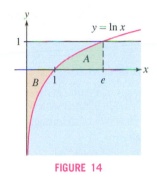

FIGURE 14

18. Calculate the fluid force on one side of the "infinite" plate B in Figure 14, assuming the fluid has density $\rho = 900$ kg/m^3.

19. Figure 15(A) shows a ramp inclined at 30° leading into a swimming pool. Calculate the fluid force on the ramp.

20. Calculate the fluid force on one side of the plate (an isosceles triangle) shown in Figure 15(B).

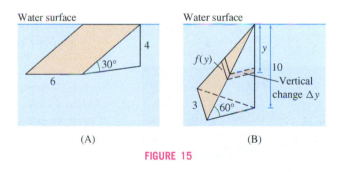

(A)   (B)

FIGURE 15

21. The massive Three Gorges Dam on China's Yangtze River has height 185 m (Figure 16). Calculate the force on the dam, assuming that the dam is a trapezoid of base 2000 m and upper edge 3000 m, inclined at an angle of 55° to the horizontal (Figure 17).

FIGURE 16 Three Gorges Dam on the Yangtze River. (*Reuters/STR/Landov*)

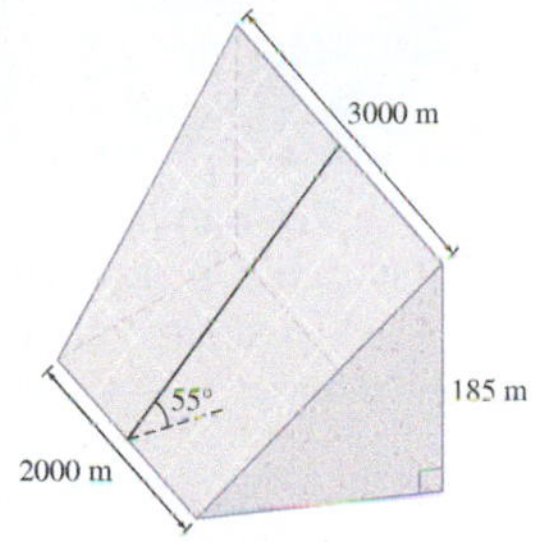

FIGURE 17

22. A square plate of side 3 m is submerged in water at an incline of 30° with the horizontal. Calculate the fluid force on one side of the plate if the top edge of the plate lies at a depth of 6 m.

23. The trough in Figure 18 is filled with corn syrup, whose weight density is 90 lb/ft^3. Calculate the force on the front side of the trough.

24. Calculate the fluid pressure on one of the slanted sides of the trough in Figure 18 when it is filled with corn syrup as in Exercise 23.

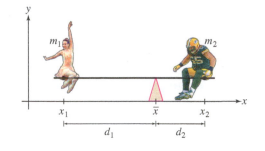

FIGURE 18

Further Insights and Challenges

25. The end of the trough in Figure 19 is an equilateral triangle of side 3. Assume that the trough is filled with water to height H. Calculate the fluid force on each side of the trough as a function of H and the length l of the trough.

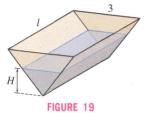

FIGURE 19

26. A rectangular plate of side ℓ is submerged vertically in a fluid of density w, with its top edge at depth h. Show that if the depth is increased by an amount Δh, then the force on a side of the plate increases by $wA\Delta h$, where A is the area of the plate.

27. Prove that the force on the side of a rectangular plate of area A submerged vertically in a fluid is equal to $p_0 A$, where p_0 is the fluid pressure at the center point of the rectangle.

28. If the density of a fluid varies with depth, then the pressure at depth y is $p(y)$ (which need not equal wy as in the case of constant density). Use Riemann sums to argue that the total force F on the flat side of a submerged object submerged vertically is
$$F = \int_a^b f(y)p(y)\,dy,$$ where $f(y)$ is the width of the side at depth y.

8.3 Center of Mass

Every object has a balance point called the *center of mass* (Figure 1). When a rigid object such as a hammer is tossed in the air, it may rotate in a complicated fashion, but its center of mass follows the same simple parabolic trajectory as a stone tossed in the air. In this section we use integration to compute the center of mass of a thin plate (also called a **lamina**) of constant mass density δ.

Consider a seesaw with a linebacker for the Green Bay Packers on one end and a ballerina from the New York City Ballet on the other end, as in Figure 2. Clearly, the balance point $\bar{x}$ must be closer to the linebacker than the ballerina.

FIGURE 1 This acrobat with Cirque du Soleil must distribute his weight so that his arm provides support directly below his center of mass. *(Mark Dadswell/Getty Images)*

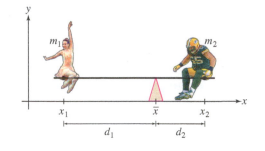

DF **FIGURE 2**

As Archimedes realized over 2000 years ago, the mass of each object times its distance to the balance point must be equal. In this case, $m_1 d_1 = m_2 d_2$. If we use the coordinates given on the number line, this becomes

$$m_1(\bar{x} - x_1) = m_2(x_2 - \bar{x})$$

We solve this for $\bar{x}$:

$$m_1\bar{x} - m_1x_1 = m_2x_2 - m_2\bar{x}$$

$$m_1\bar{x} + m_2\bar{x} = m_1x_1 + m_2x_2$$

$$(m_1 + m_2)\bar{x} = m_1x_1 + m_2x_2$$

$$\bar{x} = \frac{m_1x_1 + m_2x_2}{m_1 + m_2}$$

In other words, the balance point, called the **center of mass (COM)**, occurs at the coordinate given by taking the sum of the product of each mass times its distance (which is x) from the y-axis, and then dividing this sum by the total mass. These quantities m_1x_1 and m_2x_2 are called **moments**. In this case, the center of mass occurs on the x-axis and the distances are measured to the y-axis.

More generally, the moment of a single particle of mass m with respect to a line L is the product of the particle's mass m and its directed distance (positive or negative) to the line:

Moment with respect to line $L = m \times$ directed distance to L

The particular moments with respect to the x- and y-axes are denoted M_x and M_y. For a particle located at the point (x, y) (Figure 3),

$$M_x = my \qquad \text{(mass times directed distance to } x\text{-axis)}$$

$$M_y = mx \qquad \text{(mass times directed distance to } y\text{-axis)}$$

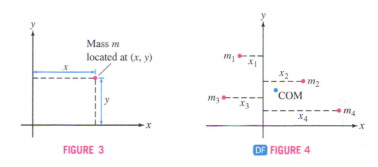

FIGURE 3

DF **FIGURE 4**

Moments are additive: the moment of a system of n particles with coordinates (x_j, y_j) and mass m_j (Figure 4) is the sum

$$M_x = m_1y_1 + m_2y_2 + \cdots + m_ny_n$$

$$M_y = m_1x_1 + m_2x_2 + \cdots + m_nx_n$$

> **CAUTION** The notation is potentially confusing: M_x is defined in terms of the distance to the x-axis (given by y-coordinates), and M_y is defined in terms of the distance to the y-axis (given by x-coordinates).

The COM is the point $P = (\bar{x}, \bar{y})$ with coordinates

$$\boxed{\bar{x} = \frac{M_y}{M}, \qquad \bar{y} = \frac{M_x}{M}}$$

where $M = m_1 + m_2 + \cdots + m_n$ is the total mass of the system.

■ **EXAMPLE 1** Find the COM of the system of three particles in Figure 5, having masses 2, 4, and 8 at locations $(0, 2)$, $(3, 1)$, and $(6, 4)$.

Solution The total mass is $M = 2 + 4 + 8 = 14$ and the moments are

$$M_x = m_1y_1 + m_2y_2 + m_3y_3 = 2\cdot2 + 4\cdot1 + 8\cdot4 = 40$$

$$M_y = m_1x_1 + m_2x_2 + m_3x_3 = 2\cdot0 + 4\cdot3 + 8\cdot6 = 60$$

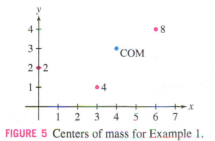

FIGURE 5 Centers of mass for Example 1.

Therefore, $\bar{x} = \frac{60}{14} = \frac{30}{7}$ and $\bar{y} = \frac{40}{14} = \frac{20}{7}$. The COM is $\left(\frac{30}{7}, \frac{20}{7}\right)$. ■

Laminas (Thin Plates)

In this section, we restrict our attention to thin plates of constant mass density (also called "uniform density"). COM computations when mass density is not constant require multiple integration and are covered in Section 15.5.

Now consider a lamina (thin plate) of constant mass density δ occupying the region under the graph of f over an interval $[a, b]$, where f is continuous and $f(x) \geq 0$ (Figure 6). In our calculations, we will use the principle of *additivity of moments* mentioned above for point masses:

If a region is decomposed into smaller, nonoverlapping regions, then the moment of the region is the sum of the moments of the smaller regions.

To compute the moment with respect to the y-axis, M_y, we begin, as usual, by dividing $[a, b]$ into N subintervals of width $\Delta x = (b - a)/N$ and endpoints $x_j = a + j\Delta x$. This divides the lamina into N vertical strips (Figure 7). If Δx is small, the jth strip is nearly rectangular of area $f(x_j)\Delta x$ and mass $\delta f(x_j)\Delta x$. Since all points in the strip lie at approximately the same distance x_j from the y-axis, the moment $M_{y,j}$ of the jth strip is approximately

$$M_{y,j} \approx (\text{mass}) \times (\text{directed distance to } y\text{-axis}) = (\delta f(x_j)\Delta x)x_j$$

By additivity of moments,

$$M_y = \sum_{j=1}^{N} M_{y,j} \approx \delta \sum_{j=1}^{N} x_j f(x_j)\Delta x$$

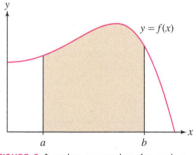

FIGURE 6 Lamina occupying the region under the graph of f over $[a, b]$.

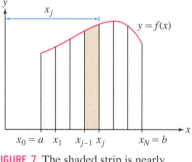

FIGURE 7 The shaded strip is nearly rectangular of area $f(x_j)\Delta x$.

This is a Riemann sum whose value approaches $\delta \int_a^b x f(x)\, dx$ as $N \to \infty$, and thus,

$$\boxed{M_y = \delta \int_a^b x f(x)\, dx}$$

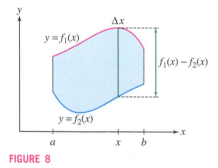

FIGURE 8

More generally, if the lamina occupies the region *between* the graphs of two functions f_1 and f_2 over $[a, b]$, where $f_1(x) \geq f_2(x)$, then

$$\boxed{M_y = \delta \int_a^b x(\text{length of vertical cut})\, dx = \delta \int_a^b x\big(f_1(x) - f_2(x)\big)\, dx} \qquad \boxed{1}$$

Think of the lamina as made up of vertical strips of length $f_1(x) - f_2(x)$ at distance x from the y-axis (Figure 8).

We can compute the x-moment by dividing the lamina into *horizontal* strips, but this requires us to describe the lamina as a region between two curves $x = g_1(y)$ and

$x = g_2(y)$ with $g_1(y) \geq g_2(y)$ over an interval $[c, d]$ along the y-axis (Figure 9):

$$M_x = \delta \int_c^d y(\text{length of horizontal cut}) \, dy = \delta \int_c^d y(g_1(y) - g_2(y)) \, dy \qquad \boxed{2}$$

The total mass of the lamina is $M = \delta A$, where A is the area of the lamina:

$$M = \delta A = \delta \int_a^b (f_1(x) - f_2(x)) \, dx \qquad \text{or} \qquad M = \delta \int_c^d (g_1(y) - g_2(y)) \, dy$$

The center-of-mass coordinates are the moments divided by the total mass:

$$\bar{x} = \frac{M_y}{M}, \qquad \bar{y} = \frac{M_x}{M}$$

The lamina will balance at the point $(\bar{x}, \bar{y})$ as in Figure 10.

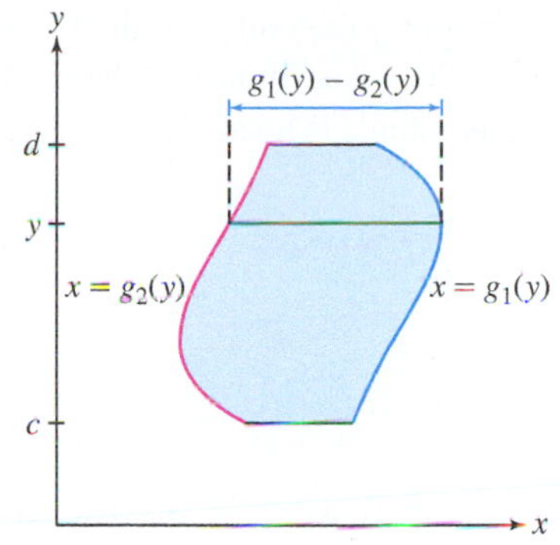

■ **EXAMPLE 2** Find the moments and COM of the lamina of uniform density δ occupying the region underneath the graph of $f(x) = x^2$ and above the x-axis for $0 \leq x \leq 2$.

Solution First, compute M_y using Eq. (1):

$$M_y = \delta \int_0^2 x f(x) \, dx = \delta \int_0^2 x(x^2) \, dx = \delta \frac{x^4}{4} \Big|_0^2 = 4\delta$$

Then compute M_x using Eq. (2), describing the lamina as the region between $x = \sqrt{y}$ and $x = 2$ over the interval $[0, 4]$ along the y-axis (Figure 11). By Eq. (2),

$$M_x = \delta \int_0^4 y(g_1(y) - g_2(y)) \, dy = \delta \int_0^4 y(2 - \sqrt{y}) \, dy$$

$$= \delta \left(y^2 - \frac{2}{5} y^{5/2} \right) \Big|_0^4 = \delta \left(16 - \frac{2}{5} \cdot 32 \right) = \frac{16}{5} \delta$$

The plate has area $A = \int_0^2 x^2 \, dx = \frac{8}{3}$ and total mass $M = \frac{8}{3}\delta$. Therefore,

$$\bar{x} = \frac{M_y}{M} = \frac{4\delta}{\frac{8}{3}\delta} = \frac{3}{2}, \qquad \bar{y} = \frac{M_x}{M} = \frac{\frac{16}{5}\delta}{\frac{8}{3}\delta} = \frac{6}{5}$$

See Figure 12. ■

CONCEPTUAL INSIGHT The COM of a lamina of constant mass density δ is also called the **centroid**. The centroid depends on the shape of the lamina, but not on its mass density because the factor δ cancels in the ratios M_x/M and M_y/M. In particular, *in calculating the centroid, we can take $\delta = 1$.* When mass density is not constant, the COM depends on both shape and mass density. In this case, the COM is computed using multiple integration (Section 16.5).

A drawback of Eq. (2) for M_x is that it requires integration along the y-axis, which may cause difficulties if the lamina is defined in terms of functions of x. Fortunately, there is a second formula for M_x as an integral along the x-axis. As before, divide the region into

---

FIGURE 9

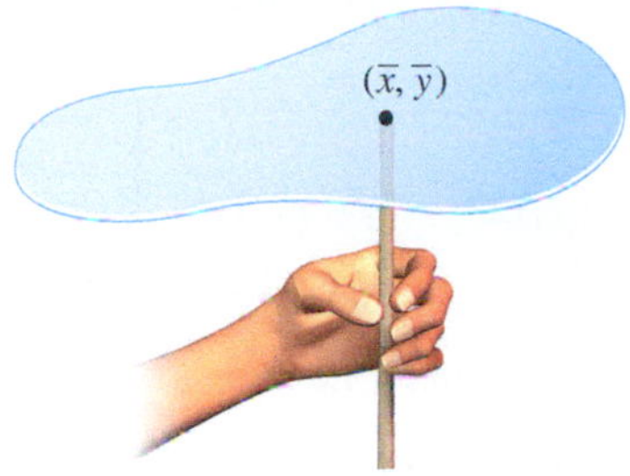

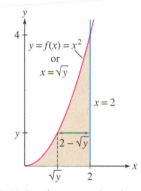

DF **FIGURE 10** A lamina balances at its center of mass $(\bar{x}, \bar{y})$.

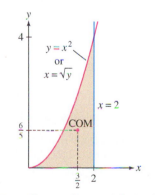

FIGURE 11 Lamina occupying the region under the graph of $f(x) = x^2$ over $[0, 2]$.

FIGURE 12 Center of mass of the lamina.

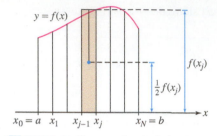

DF **FIGURE 13** Because the shaded strip is nearly rectangular, its COM has an approximate height of $\frac{1}{2}f(x_j)$.

N thin vertical strips of width Δx (see Figure 13). Let $M_{x,j}$ be the moment with respect to the x-axis of the jth strip and let m_j be its mass. We can use the following trick to approximate $M_{x,j}$. The strip is nearly rectangular with height $f(x_j)$ and width Δx, so $m_j \approx \delta f(x_j)\Delta x$. Furthermore, $M_{x,j} = \overline{y_j} m_j$, where $\overline{y_j}$ is the y-coordinate of the COM of the strip. However, $\overline{y_j} \approx \frac{1}{2}f(x_j)$ because the COM of a rectangle is located at its center. Thus,

$$M_{x,j} = m_j \overline{y_j} \approx \delta f(x_j)\Delta x \cdot \frac{1}{2}f(x_j) = \frac{1}{2}\delta f(x_j)^2 \Delta x$$

$$M_x = \sum_{j=1}^{N} M_{x,j} \approx \frac{1}{2}\delta \sum_{j=1}^{N} f(x_j)^2 \Delta x$$

This is a Riemann sum whose value approaches $\frac{1}{2}\delta \int_a^b f(x)^2\,dx$ as $N \to \infty$. The case of a region *between* the graphs of functions f_1 and f_2 where $f_1(x) \geq f_2(x) \geq 0$ is the difference of the moments corresponding to $f_1(x)$ and $f_2(x)$, by the principle of additivity of moments, so we obtain the alternative formulas

$$\boxed{M_x = \frac{1}{2}\delta \int_a^b f(x)^2\,dx \qquad \text{or} \qquad M_x = \frac{1}{2}\delta \int_a^b \left(f_1(x)^2 - f_2(x)^2\right)dx} \qquad \boxed{3}$$

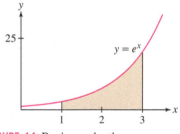

FIGURE 14 Region under the curve $y = e^x$ between $x = 1$ and $x = 3$.

■ **EXAMPLE 3** Find the centroid of the shaded region in Figure 14.

Solution The centroid does not depend on δ, so we may set $\delta = 1$ and apply Eqs. (1) and (3) with $f(x) = e^x$:

$$M_x = \frac{1}{2}\int_1^3 f(x)^2\,dx = \frac{1}{2}\int_1^3 e^{2x}\,dx = \frac{1}{4}e^{2x}\Big|_1^3 = \frac{e^6 - e^2}{4}$$

Using Integration by Parts, we get

$$M_y = \int_1^3 x f(x)\,dx = \int_1^3 x e^x\,dx = (x-1)e^x\Big|_1^3 = 2e^3$$

The total mass is $M = \int_1^3 e^x\,dx = (e^3 - e)$. The centroid has coordinates

$$\bar{x} = \frac{M_y}{M} = \frac{2e^3}{e^3 - e} \approx 2.313, \qquad \bar{y} = \frac{M_x}{M} = \frac{e^6 - e^2}{4(e^3 - e)} \approx 5.701 \qquad ■$$

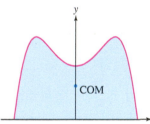

FIGURE 15 The COM of a symmetric plate lies on the axis of symmetry.

The symmetry properties of an object give information about its centroid (Figure 15). For instance, the centroid of a square or circular plate is located at its center. Here is a precise formulation (see Exercise 45).

←⋯ **REMINDER** *A region is symmetric with respect to a line if reflection across the line sends each point of the region to another point of the region.*

THEOREM 1   Symmetry Principle   If a lamina is symmetric with respect to a line, then its centroid lies on that line.

■ **EXAMPLE 4   Using Symmetry**   Find the centroid of a circular half-disk of radius 3.

Solution Symmetry cuts our work in half. The half-disk is symmetric with respect to the y-axis, so the centroid lies on the y-axis, and hence $\bar{x} = 0$. It remains to calculate M_x and $\bar{y}$. The boundary of the half-disk is the graph of $f(x) = \sqrt{9 - x^2}$ (Figure 16). By Eq. (3) with $\delta = 1$,

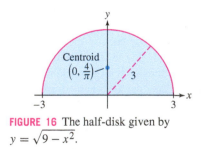

FIGURE 16 The half-disk given by $y = \sqrt{9 - x^2}$.

$$M_x = \frac{1}{2}\int_{-3}^3 f(x)^2\,dx = \frac{1}{2}\int_{-3}^3 (9 - x^2)\,dx = \frac{1}{2}\left(9x - \frac{1}{3}x^3\right)\Big|_{-3}^3 = 9 - (-9) = 18$$

The half-disk has area (and mass) equal to $A = \frac{1}{2}\pi(3^2) = 9\pi/2$, so

$$\bar{y} = \frac{M_x}{M} = \frac{18}{9\pi/2} = \frac{4}{\pi} \approx 1.27 \qquad ■$$

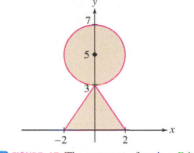

FIGURE 17 The moment of region R is the sum of the moments of the triangle and circle.

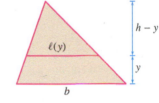

FIGURE 18 By similar triangles,
$\dfrac{\ell(y)}{h-y} = \dfrac{b}{h}$.

■ **EXAMPLE 5** Using Additivity and Symmetry Find the centroid of the region R in Figure 17.

Solution We set $\delta = 1$ because we are computing a centroid. The region R is symmetric with respect to the y-axis, so we know in advance that $\bar{x} = 0$. To find $\bar{y}$, we compute the moment M_x.

Step 1. Use additivity of moments.

Let M_x^{triangle} and M_x^{circle} be the x-moments of the triangle and the circle. Then

$$M_x = M_x^{\text{triangle}} + M_x^{\text{circle}}$$

Step 2. Moment of the circle.

To save work, we use the fact that the centroid of the circle is located at the center $(0, 5)$ by symmetry. Thus, $\bar{y}^{\text{circle}} = 5$ and we can solve for the moment:

$$\bar{y}^{\text{circle}} = \frac{M_x^{\text{circle}}}{M^{\text{circle}}} = \frac{M_x^{\text{circle}}}{4\pi} = 5 \quad \Rightarrow \quad M_x^{\text{circle}} = 20\pi$$

Here, the mass of the circle is its area $M^{\text{circle}} = \pi(2^2) = 4\pi$ (since $\delta = 1$).

Step 3. Moment of a triangle.

Let's compute M_x^{triangle} for an arbitrary triangle of height h and base b (Figure 18). Let $\ell(y)$ be the width of the triangle at height y. By similar triangles,

$$\frac{\ell(y)}{h-y} = \frac{b}{h} \quad \Rightarrow \quad \ell(y) = b - \frac{b}{h}y$$

By Eq. (2),

$$M_x^{\text{triangle}} = \int_0^h y\ell(y)\,dy = \int_0^h y\left(b - \frac{b}{h}y\right)dy = \left(\frac{by^2}{2} - \frac{by^3}{3h}\right)\Bigg|_0^h = \frac{bh^2}{6}$$

In our case, $b = 4$, $h = 3$, and $M_x^{\text{triangle}} = \dfrac{4 \cdot 3^2}{6} = 6$.

Step 4. Computation of $\bar{y}$.

$$M_x = M_x^{\text{triangle}} + M_x^{\text{circle}} = 6 + 20\pi$$

The triangle has mass $\frac{1}{2} \cdot 4 \cdot 3 = 6$, and the circle has mass 4π, so R has mass $M = 6 + 4\pi$ and

$$\bar{y} = \frac{M_x}{M} = \frac{6 + 20\pi}{6 + 4\pi} \approx 3.71 \qquad ■$$

We end this section with the Theorem of Pappus, attributed to Pappus of Alexandria, a mathematician of the fourth century BCE.

> **THEOREM 2 Theorem of Pappus** Let R be a region in the plane of area A that we rotate about an axis that is disjoint from R. Then the volume of the resulting solid is the product of A with the distance traveled by the centroid of R.

Proof Since we assume a uniform density of 1, the area A of the region is equal to the mass M. We will only prove the theorem in the special case that we have a region bounded by $y = f_1(x)$ and $y = f_2(x)$ for $a \le x \le b$ and $f_1(x) \ge f_2(x) > 0$ and we are rotating about the x-axis. In this case, we know that the volume is given by

$$V = \pi \int_a^b (f_1(x)^2 - f_2(x)^2)\,dx = 2\pi\left(\frac{1}{2}\int_a^b (f_1(x)^2 - f_2(x)^2)\,dx\right)$$

$$= 2\pi M_x = A \cdot 2\pi \frac{M_x}{A} = A \cdot 2\pi\bar{x} \qquad ■$$

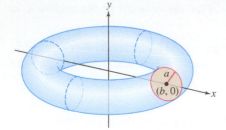

FIGURE 19 Rotating a disk about the y-axis to obtain a solid torus.

■ EXAMPLE 6 Find the formula for the volume of the solid torus obtained by rotating the disk of radius a centered at $(b, 0)$ about the y-axis, where $a < b$, as in Figure 19.

Solution The centroid of the disk occurs at its center. So, $(\bar{x}, \bar{y}) = (b, 0)$. The Theorem of Pappus then says that

$$V = A \cdot 2\pi \bar{x} = \pi a^2 2\pi b = 2\pi^2 a^2 b \qquad ■$$

FIGURE 20 Archimedes's Law of the Lever:
$$m_1 L_1 = m_2 L_2$$

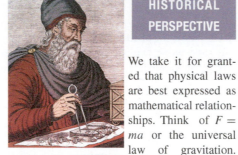
Archimedes (287–212 BCE) (*Roger Viollet Collection/ Getty Images*)

HISTORICAL PERSPECTIVE

We take it for granted that physical laws are best expressed as mathematical relationships. Think of $F = ma$ or the universal law of gravitation. However, the fundamental insight that mathematics could be used to formulate laws of nature (and not just for counting or measuring) developed gradually, beginning with the philosophers of ancient Greece and culminating some 2000 years later in the discoveries of Galileo and Newton. Archimedes was one of the first scientists (perhaps *the* first) to formulate a precise physical law. Concerning the principle of the lever, Archimedes wrote, "Commensurable magnitudes balance at distances reciprocally proportional to their weight." In other words, if weights of mass m_1 and m_2 are placed on a weightless lever at distances L_1 and L_2 from the fulcrum P (Figure 20), then the lever will balance if $m_1/m_2 = L_2/L_1$, or

$$\boxed{m_1 L_1 = m_2 L_2}$$

In our terminology, what Archimedes had discovered was the center of mass P of the system of weights (see Exercises 43 and 44).

8.3 SUMMARY

- The *moments* of a system of particles of mass m_j located at (x_j, y_j) are

$$M_x = m_1 y_1 + \cdots + m_n y_n, \qquad M_y = m_1 x_1 + \cdots + m_n x_n$$

The *center of mass* (COM) has coordinates

$$\bar{x} = \frac{M_y}{M} \qquad \text{and} \qquad \bar{y} = \frac{M_x}{M}$$

where $M = m_1 + \cdots + m_n$.

- Lamina (thin plate) of constant mass density δ [region under the graph of f where $f(x) \geq 0$, or *between* the graphs of f_1 and f_2 where $f_1(x) \geq f_2(x)$]:

$$M_y = \delta \int_a^b x f(x)\, dx \qquad \text{or} \qquad M_y = \delta \int_a^b x\big(f_1(x) - f_2(x)\big)\, dx$$

- There are two ways to compute the x-moment M_x. If the lamina occupies the region between the graph of $x = g(y)$ and the y-axis where $g(y) \geq 0$, or *between* the graphs of $x = g_1(y)$ and $x = g_2(y)$ where $g_1(y) \geq g_2(y)$, then

$$M_x = \delta \int_c^d y g(y)\, dy \qquad \text{or} \qquad M_x = \delta \int_c^d y\big(g_1(y) - g_2(y)\big)\, dy$$

- Alternative (often more convenient) formula for M_x:

$$M_x = \frac{1}{2}\delta \int_a^b f(x)^2\, dx \qquad \text{or} \qquad M_x = \frac{1}{2}\delta \int_a^b \big(f_1(x)^2 - f_2(x)^2\big)\, dx$$

- The total mass of the lamina is $M = \delta \int_a^b \big(f_1(x) - f_2(x)\big)\, dx$. The coordinates of the center of mass (also called the *centroid*) are

$$\bar{x} = \frac{M_y}{M}, \qquad \bar{y} = \frac{M_x}{M}$$

- Additivity: If a region is decomposed into smaller nonoverlapping regions, then the moment of the region is the sum of the moments of the smaller regions.
- Symmetry Principle: If a lamina of constant mass density is symmetric with respect to a given line, then the center of mass (centroid) lies on that line.
- The Theorem of Pappus: If a lamina is rotated about a disjoint axis, then the volume of the resulting solid of revolution is the area of the lamina times the distance traveled by the centroid.

8.3 EXERCISES

Preliminary Questions

1. What are the x- and y-moments of a lamina whose center of mass is located at the origin?

2. A thin plate has mass 3. What is the x-moment of the plate if its center of mass has coordinates $(2, 7)$?

3. The center of mass of a lamina of total mass 5 has coordinates $(2, 1)$. What are the lamina's x- and y-moments?

4. Explain how the Symmetry Principle is used to conclude that the centroid of a rectangle is the center of the rectangle.

5. Give an example of a plate such that its center of mass does not occur at any point on the plate.

6. Draw a plate such that its center of mass occurs on its boundary. (You do not need to verify this fact. It should just be believable from the drawing.)

Exercises

1. Four particles are located at points $(1, 1)$, $(1, 2)$, $(4, 0)$, $(3, 1)$.

(a) Find the moments M_x and M_y and the center of mass of the system, assuming that the particles have equal mass m.

(b) Find the center of mass of the system, assuming the particles have masses 3, 2, 5, and 7, respectively.

2. Find the center of mass for the system of particles of masses 4, 2, 5, 1 located at $(1, 2)$, $(-3, 2)$, $(2, -1)$, $(4, 0)$.

3. Point masses of equal size are placed at the vertices of the triangle with coordinates $(a, 0)$, $(b, 0)$, and $(0, c)$. Show that the center of mass of the system of masses has coordinates $\left(\frac{1}{3}(a + b), \frac{1}{3}c\right)$.

4. Point masses of mass m_1, m_2, and m_3 are placed at the points $(-1, 0)$, $(3, 0)$, and $(0, 4)$.

(a) Suppose that $m_1 = 6$. Find m_2 such that the center of mass lies on the y-axis.

(b) Suppose that $m_1 = 6$ and $m_2 = 4$. Find the value of m_3 such that $\bar{y} = 2$.

5. Sketch the lamina S of constant density $\rho = 3$ g/cm^2 occupying the region beneath the graph of $y = x^2$ for $0 \le x \le 3$.

(a) Use Eqs. (1) and (2) to compute M_x and M_y.

(b) Find the area and the center of mass of S.

6. Use Eqs. (1) and (3) to find the moments and center of mass of the lamina S of constant density $\rho = 2$ g/cm^2 occupying the region between $y = x^2$ and $y = 9x$ over $[0, 3]$. Sketch S, indicating the location of the center of mass.

7. Find the moments and center of mass of the lamina of uniform density ρ occupying the region underneath $y = x^3$ for $0 \le x \le 2$.

8. Calculate M_x (assuming $\rho = 1$) for the region underneath the graph of $y = 1 - x^2$ for $0 \le x \le 1$ in two ways, first using Eq. (2) and then using Eq. (3).

9. Let T be the triangular lamina in Figure 21.

(a) Show that the horizontal cut at height y has length $4 - \frac{2}{3}y$ and use Eq. (2) to compute M_x (with $\rho = 1$).

(b) Use the Symmetry Principle to show that $M_y = 0$ and find the center of mass.

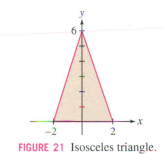

FIGURE 21 Isosceles triangle.

In Exercises 10–17, find the centroid of the region lying underneath the graph of the function over the given interval.

10. $f(x) = 6 - 2x$, $[0, 3]$

11. $f(x) = \sqrt{x}$, $[1, 4]$

12. $f(x) = x^3$, $[0, 1]$

13. $f(x) = 9 - x^2$, $[0, 3]$

14. $f(x) = (1 + x^2)^{-1/2}$, $[0, 3]$

15. $f(x) = e^{-x}$, $[0, 4]$

16. $f(x) = \ln x$, $[1, 2]$

17. $f(x) = \sin x$, $[0, \pi]$

18. Calculate the moments and center of mass of the lamina occupying the region between the curves $y = x$ and $y = x^2$ for $0 \le x \le 1$.

19. Sketch the region between $y = x + 4$ and $y = 2 - x$ for $0 \le x \le 2$. Using symmetry, explain why the centroid of the region lies on the line $y = 3$. Verify this by computing the moments and the centroid.

In Exercises 20–25, find the centroid of the region lying between the graphs of the functions over the given interval.

20. $y = x$, $y = \sqrt{x}$, $[0, 1]$

21. $y = x^2$, $y = \sqrt{x}$, $[0, 1]$

22. $y = x^{-1}$, $y = 2 - x$, $[1, 2]$

23. $y = e^x$, $y = 1$, $[0, 1]$

24. $y = \ln x$, $y = x - 1$, $[1, 3]$

25. $y = \sin x$, $y = \cos x$, $[0, \pi/4]$

26. Sketch the region enclosed by $y = x + 1$ and $y = (x - 1)^2$ and find its centroid.

27. Sketch the region enclosed by $y = 0$, $y = (x + 1)^3$, and $y = (1 - x)^3$, and find its centroid.

In Exercises 28–32, find the centroid of the region.

28. Top half of the ellipse $\left(\frac{x}{2}\right)^2 + \left(\frac{y}{4}\right)^2 = 1$

29. Top half of the ellipse $\left(\frac{x}{a}\right)^2 + \left(\frac{y}{b}\right)^2 = 1$ for arbitrary $a, b > 0$

30. Semicircle of radius r with center at the origin

31. Quarter of the unit circle lying in the first quadrant

32. Region between $y = x(a - x)$ and the x-axis for $a > 0$

33. Find the centroid of the shaded region of the semicircle of radius r in Figure 22. What is the centroid when $r = 1$ and $h = \frac{1}{2}$? *Hint:* Use geometry rather than integration to show that the *area* of the region is $r^2 \sin^{-1}(\sqrt{1 - h^2/r^2}) - h\sqrt{r^2 - h^2}$.

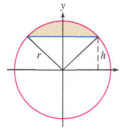

FIGURE 22

34. Sketch the region between $y = x^n$ and $y = x^m$ for $0 \le x \le 1$, where $m > n \ge 0$, and find the COM of the region. Find a pair (n, m) such that the COM lies outside the region.

35. Find the formula for the volume of a right circular cone of height H and radius R using the Theorem of Pappus as applied to the triangle bounded by the x-axis, the y-axis, and the line $y = \frac{-H}{R}x + H$, rotated about the y-axis.

36. Use the Theorem of Pappus to find the centroid of the half-disk bounded by $y = \sqrt{R^2 - x^2}$ and the x-axis.

In Exercises 37–39, use the additivity of moments to find the COM of the region.

37. Isosceles triangle of height 2 on top of a rectangle of base 4 and height 3 (Figure 23)

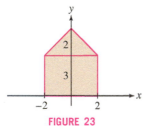

FIGURE 23

38. An ice cream cone consisting of a semicircle on top of an equilateral triangle of side 6 (Figure 24)

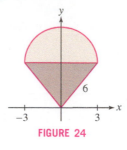

FIGURE 24

39. Three-quarters of the unit circle (remove the part in the fourth quadrant)

40. Let S be the lamina of mass density $\rho = 1$ obtained by removing a circle of radius r from the circle of radius $2r$ shown in Figure 25. Let M_x^S and M_y^S denote the moments of S. Similarly, let M_y^{big} and M_y^{small} be the y-moments of the larger and smaller circles.

(a) Use the Symmetry Principle to show that $M_x^S = 0$.

(b) Show that $M_y^S = M_y^{\text{big}} - M_y^{\text{small}}$ using the additivity of moments.

(c) Find M_y^{big} and M_y^{small} using the fact that the COM of a circle is its center. Then compute M_y^S using (b).

(d) Determine the COM of S.

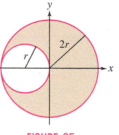

FIGURE 25

41. Find the COM of the laminas in Figure 26 obtained by removing squares of side 2 from a square of side 8.

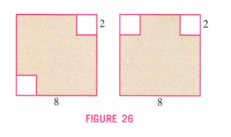

FIGURE 26

Further Insights and Challenges

42. A **median** of a triangle is a segment joining a vertex to the midpoint of the opposite side. Show that the centroid of a triangle lies on each of its medians, at a distance two-thirds down from the vertex. Then use this fact to prove that the three medians intersect at a single point. *Hint:* Simplify the calculation by assuming that one vertex lies at the origin and another on the x-axis.

43. Let P be the COM of a system of two weights with masses m_1 and m_2 separated by a distance d. Prove Archimedes's Law of the (weightless) Lever: P is the point on a line between the two weights such that $m_1 L_1 = m_2 L_2$, where L_j is the distance from mass j to P.

44. Find the COM of a system of two weights of masses m_1 and m_2 connected by a lever of length d whose mass density ρ is uniform. *Hint:* The moment of the system is the sum of the moments of the weights and the lever.

45. ✎ **Symmetry Principle** Let $\mathcal{R}$ be the region under the graph of $y = f(x)$ over the interval $[-a, a]$, where $f(x) \ge 0$. Assume that $\mathcal{R}$ is symmetric with respect to the y-axis.

(a) Explain why $y = f(x)$ is even—that is, why $f(x) = f(-x)$.

(b) Show that $y = xf(x)$ is an *odd* function.

(c) Use (b) to prove that $M_y = 0$.

(d) Prove that the COM of $\mathcal{R}$ lies on the y-axis (a similar argument applies to symmetry with respect to the x-axis).

46. Prove directly that Eqs. (2) and (3) are equivalent in the following situation. Let f be a positive decreasing function on $[0, b]$ such that $f(b) = 0$. Set $d = f(0)$ and $g(y) = f^{-1}(y)$. Show that

$$\frac{1}{2} \int_0^b f(x)^2 \, dx = \int_0^d yg(y) \, dy$$

Hint: First apply the substitution $y = f(x)$ to the integral on the left and observe that $dx = g'(y) \, dy$. Then apply Integration by Parts.

47. Let R be a lamina of uniform density submerged in a fluid of density w (Figure 27). Prove the following law: The fluid force on one side of R is equal to the area of R times the fluid pressure on the centroid. *Hint:* Let $g(y)$ be the horizontal width of R at depth y. Express both the fluid pressure [Eq. (2) in Section 8.2] and y-coordinate of the centroid in terms of $g(y)$.

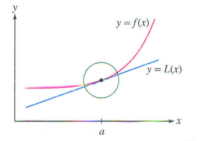

FIGURE 27

8.4 Taylor Polynomials

Many functions are difficult to work with. For instance, $\sin(x^2)$ cannot be integrated using elementary functions. Nor can e^{-x^2}. In fact, even simple functions like $\sin x$, $\cos x$, e^x and $\ln x$ can only be evaluated exactly at relatively few values of x and otherwise they must be numerically approximated. On the other hand, polynomials such as $f(x) = 3x^4 - 7x^3 + 2x - 4$ can be easily differentiated and integrated. They can be evaluated at any value of x using just multiplication and addition. In this section, we introduce a method for approximating a function f by a polynomial, that behaves similarly to the way f behaves.

We have done this before. In Section 4.1, we used the linearization $L(x) = f(a) + f'(a)(x - a)$ to approximate $f(x)$ near a point $x = a$:

$$f(x) \approx f(a) + f'(a)(x - a)$$

We refer to $L(x)$ as a "first-order" approximation to $f(x)$ at $x = a$ because $f(x)$ and $L(x)$ have the same value and the same first derivative at $x = a$ (Figure 1):

$$L(a) = f(a), \qquad L'(a) = f'(a)$$

A first-order approximation is useful only in a small interval around $x = a$. In this section, we learn how to achieve greater accuracy over larger intervals using higher order approximations (Figure 2). These higher order approximations will simply be polynomials with higher powers. Ultimately, we will see that the errors in utilizing these higher-order approximations are determined by integrals.

English mathematician Brook Taylor (1685–1731) made important contributions to calculus and physics, as well as to the theory of linear perspective used in drawing. *(The Granger Collection, NYC. All rights reserved.)*

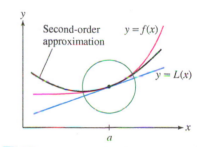

FIGURE 1 The linear approximation $L(x)$ is a first-order approximation to f.

DF FIGURE 2 A second-order approximation is more accurate over a larger interval.

In what follows, assume that f is defined on an open interval I and that all higher derivatives $f^{(k)}$ exist on I. Let $a \in I$. We say that two functions f and g **agree to order n** at $x = a$ if their derivatives up to order n at $x = a$ are equal:

$$f(a) = g(a), \quad f'(a) = g'(a), \quad f''(a) = g''(a), \quad \ldots, \quad f^{(n)}(a) = g^{(n)}(a)$$

We also say that g "approximates f to order n" at $x = a$.

Define the nth **Taylor polynomial T_n of f centered at $x = a$** as follows:

$$T_n(x) = f(a) + \frac{f'(a)}{1!}(x - a) + \frac{f''(a)}{2!}(x - a)^2 + \cdots + \frac{f^{(n)}(a)}{n!}(x - a)^n$$

The first few Taylor polynomials are

$$T_0(x) = f(a)$$

$$T_1(x) = f(a) + f'(a)(x - a)$$

$$T_2(x) = f(a) + f'(a)(x - a) + \frac{1}{2}f''(a)(x - a)^2$$

$$T_3(x) = f(a) + f'(a)(x - a) + \frac{1}{2}f''(a)(x - a)^2 + \frac{1}{6}f'''(a)(x - a)^3$$

Note that T_1 is the linearization of f at a. Note also that T_n is obtained from T_{n-1} by adding on a term of degree n:

$$T_n(x) = T_{n-1}(x) + \frac{f^{(n)}(a)}{n!}(x - a)^n$$

The next theorem justifies our definition of T_n.

<-- **REMINDER** k-factorial *is the number*
$k! = k(k - 1)(k - 2) \cdots (2)(1)$. *Thus,*

$$1! = 1, \quad 2! = (2)1 = 2$$

$$3! = (3)(2)1 = 6$$

By convention, we define $0! = 1$.

THEOREM 1   The polynomial T_n centered at a agrees with f to order n at $x = a$, and it is the only polynomial of degree at most n with this property.

The verification of Theorem 1 is left to the exercises (Exercises 70–71), but we'll illustrate the idea by checking that T_2 agrees with f to order $n = 2$:

$$T_2(x) = f(a) + f'(a)(x - a) + \frac{1}{2}f''(a)(x - a)^2, \quad T_2(a) = f(a)$$

$$T_2'(x) = f'(a) + f''(a)(x - a), \qquad\qquad\qquad T_2'(a) = f'(a)$$

$$T_2''(x) = f''(a), \qquad\qquad\qquad\qquad\qquad\qquad T_2''(a) = f''(a)$$

This shows that the value and the derivatives of order up to $n = 2$ at $x = a$ are equal. Before proceeding to the examples, we write T_n in summation notation:

$$T_n(x) = \sum_{j=0}^{n} \frac{f^{(j)}(a)}{j!}(x - a)^j$$

By convention, we regard f as the *zeroth* derivative, and thus $f^{(0)}$ is f itself. When $a = 0$, T_n is also called the nth **Maclaurin polynomial**.

■ **EXAMPLE 1**   Maclaurin Polynomials for $f(x) = e^x$ Plot the third and fourth Maclaurin polynomials for $f(x) = e^x$. Compare with the linear approximation.

Solution   All higher derivatives coincide with f itself: $f^{(k)}(x) = e^x$. Therefore,

$$f(0) = f'(0) = f''(0) = f'''(0) = f^{(4)}(0) = e^0 = 1$$

The third Maclaurin polynomial (the case $a = 0$) is

$$T_3(x) = f(0) + f'(0)x + \frac{1}{2}f''(0)x^2 + \frac{1}{3!}f'''(0)x^3 = 1 + x + \frac{1}{2}x^2 + \frac{1}{6}x^3$$

We obtain $T_4(x)$ by adding the term of degree 4 to $T_3(x)$:

$$T_4(x) = T_3(x) + \frac{1}{4!}f^{(4)}(0)x^4 = 1 + x + \frac{1}{2}x^2 + \frac{1}{6}x^3 + \frac{1}{24}x^4$$

Figure 3 shows that T_3 and T_4 approximate $f(x) = e^x$ much more closely than the linear approximation T_1 on an interval around $a = 0$. Higher-degree Maclaurin polynomials would provide even better approximations on larger intervals. ∎

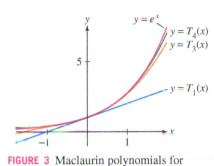

FIGURE 3 Maclaurin polynomials for $f(x) = e^x$.

■ **EXAMPLE 2** **Computing Taylor Polynomials** Compute the Taylor polynomial T_4 centered at $a = 3$ for $f(x) = \sqrt{x + 1}$.

Solution First evaluate the derivatives up to degree 4 at $a = 3$:

$$f(x) = (x + 1)^{1/2}, \qquad\qquad f(3) = 2$$

$$f'(x) = \frac{1}{2}(x + 1)^{-1/2}, \qquad f'(3) = \frac{1}{4}$$

$$f''(x) = -\frac{1}{4}(x + 1)^{-3/2}, \qquad f''(3) = -\frac{1}{32}$$

$$f'''(x) = \frac{3}{8}(x + 1)^{-5/2}, \qquad f'''(3) = \frac{3}{256}$$

$$f^{(4)}(x) = -\frac{15}{16}(x + 1)^{-7/2}, \qquad f^{(4)}(3) = -\frac{15}{2048}$$

Then compute the coefficients $\dfrac{f^{(j)}(3)}{j!}$:

> The first term $f(a)$ in the Taylor polynomial T_n is called the constant term.

Constant term $= f(3) = 2$

Coefficient of $(x - 3)$ $= f'(3) = \dfrac{1}{4}$

Coefficient of $(x - 3)^2 = \dfrac{f''(3)}{2!} = -\dfrac{1}{32} \cdot \dfrac{1}{2!} = -\dfrac{1}{64}$

Coefficient of $(x - 3)^3 = \dfrac{f'''(3)}{3!} = \dfrac{3}{256} \cdot \dfrac{1}{3!} = \dfrac{1}{512}$

Coefficient of $(x - 3)^4 = \dfrac{f^{(4)}(3)}{4!} = -\dfrac{15}{2048} \cdot \dfrac{1}{4!} = -\dfrac{5}{16,384}$

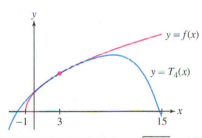

FIGURE 4 Graph of $f(x) = \sqrt{x + 1}$ and T_4 centered at $x = 3$.

The Taylor polynomial T_4 centered at $a = 3$ is (see Figure 4):

$$T_4(x) = 2 + \frac{1}{4}(x - 3) - \frac{1}{64}(x - 3)^2 + \frac{1}{512}(x - 3)^3 - \frac{5}{16,384}(x - 3)^4 \qquad ■$$

■ **EXAMPLE 3** **Finding a General Formula for T_n** Find the Taylor polynomials T_n of $f(x) = \ln x$ centered at $a = 1$.

Solution For $f(x) = \ln x$, the constant term of T_n at $a = 1$ is zero because $f(1) = \ln 1 = 0$. Next, we compute the derivatives:

> After computing several derivatives of $f(x) = \ln x$, we begin to discern the pattern. For many functions of interest, however, the derivatives follow no simple pattern and there is no convenient formula for the general Taylor polynomial.

$$f'(x) = x^{-1}, \qquad f''(x) = -x^{-2}, \qquad f'''(x) = 2x^{-3}, \qquad f^{(4)}(x) = -3 \cdot 2x^{-4}$$

Similarly, $f^{(5)}(x) = 4 \cdot 3 \cdot 2x^{-5}$. The general pattern is that $f^{(k)}(x)$ is a multiple of x^{-k}, with a coefficient $\pm(k-1)!$ that alternates in sign:

$$f^{(k)}(x) = (-1)^{k-1}(k-1)!\,x^{-k} \qquad \boxed{1}$$

The coefficient of $(x-1)^k$ in T_n is

$$\frac{f^{(k)}(1)}{k!} = \frac{(-1)^{k-1}(k-1)!}{k!} = \frac{(-1)^{k-1}}{k} \qquad (\text{for } k \geq 1)$$

Thus, the coefficients for $k \geq 1$ form a sequence $1, -\frac{1}{2}, \frac{1}{3}, -\frac{1}{4}, \ldots$, and

$$T_n(x) = (x-1) - \frac{1}{2}(x-1)^2 + \frac{1}{3}(x-1)^3 - \cdots + (-1)^{n-1}\frac{1}{n}(x-1)^n \qquad \blacksquare$$

■ **EXAMPLE 4** **Cosine** Find the Maclaurin polynomials of $f(x) = \cos x$.

Solution The derivatives form a repeating pattern of period 4:

$$f(x) = \cos x, \qquad f'(x) = -\sin x, \qquad f''(x) = -\cos x, \qquad f'''(x) = \sin x,$$

$$f^{(4)}(x) = \cos x, \qquad f^{(5)}(x) = -\sin x, \qquad \cdots$$

In general, $f^{(j+4)}(x) = f^{(j)}(x)$. The derivatives at $x = 0$ also form a pattern:

$f(0)$	$f'(0)$	$f''(0)$	$f'''(0)$	$f^{(4)}(0)$	$f^{(5)}(0)$	$f^{(6)}(0)$	$f^{(7)}(0)$	$\cdots$
1	0	-1	0	1	0	-1	0	$\cdots$

Therefore, the coefficients of the odd powers x^{2k+1} are zero, and the coefficients of the even powers x^{2k} alternate in sign with value $(-1)^k/(2k)!$:

$$T_0(x) = T_1(x) = 1, \qquad T_2(x) = T_3(x) = 1 - \frac{1}{2!}x^2$$

$$T_4(x) = T_5(x) = 1 - \frac{x^2}{2} + \frac{x^4}{4!}$$

$$T_{2n}(x) = T_{2n+1}(x) = 1 - \frac{1}{2}x^2 + \frac{1}{4!}x^4 - \frac{1}{6!}x^6 + \cdots + (-1)^n\frac{1}{(2n)!}x^{2n}$$

Figure 5 shows that as n increases, T_n approximates $f(x) = \cos x$ well over larger and larger intervals, but outside this interval, the approximation fails. ■

Taylor polynomials for $\ln x$ at $a = 1$:

$$T_1(x) = (x-1)$$

$$T_2(x) = (x-1) - \frac{1}{2}(x-1)^2$$

$$T_3(x) = (x-1) - \frac{1}{2}(x-1)^2 + \frac{1}{3}(x-1)^3$$

Scottish mathematician Colin Maclaurin (1698–1746) was a professor in Edinburgh. Newton was so impressed by his work that he once offered to pay part of Maclaurin's salary. (*The Granger Collection, NYC. All rights reserved.*)

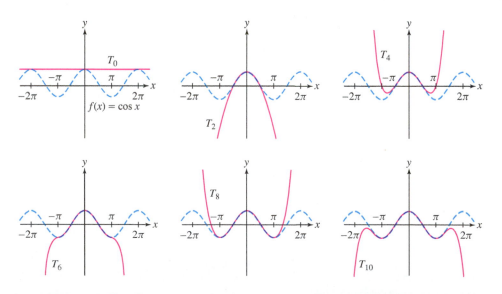

FIGURE 5 Maclaurin polynomials for $f(x) = \cos x$. The graph of f is shown as a dashed curve.

■ **EXAMPLE 5** **How Far Is the Horizon?** Valerie is at the beach, looking out over the ocean (Figure 6). How far can she see? Use Maclaurin polynomials to estimate the distance d, assuming that Valerie's eye level is $h = 1.7$ meters (m) above ground. What if she looks out from a window where her eye level is 20 m?

FIGURE 6 View from the beach. *(Alexandr Ozerov/iStockphoto)*

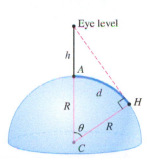

FIGURE 7 Valerie can see a distance $d = R\theta$, the length of arc AH.

Solution Let R be the radius of the earth. Figure 7 shows that Valerie can see a distance $d = R\theta$, the length of the circular arc AH in Figure 7. We have

$$\cos\theta = \frac{R}{R+h}$$

Our key observation is that θ is close to zero (both θ and h are much smaller than shown in the figure), so we lose very little accuracy if we replace $\cos\theta$ by its second Maclaurin polynomial $T_2(\theta) = 1 - \frac{1}{2}\theta^2$, as computed in Example 4:

$$1 - \frac{1}{2}\theta^2 \approx \frac{R}{R+h} \quad \Rightarrow \quad \theta^2 \approx 2 - \frac{2R}{R+h} \quad \Rightarrow \quad \theta \approx \sqrt{\frac{2h}{R+h}}$$

This calculation ignores the bending of light (called refraction) as it passes through the atmosphere. Refraction typically increases d by around 10%, although the actual effect is complex and varies with atmospheric temperature.

Furthermore, h is very small relative to R, so we may replace $R + h$ by R to obtain

$$d = R\theta \approx R\sqrt{\frac{2h}{R}} = \sqrt{2Rh}$$

The earth's radius is approximately $R \approx 6.37 \times 10^6$ m, so

$$d = \sqrt{2Rh} \approx \sqrt{2(6.37 \times 10^6)h} \approx 3569\sqrt{h} \text{ m}$$

In particular, we see that d is proportional to $\sqrt{h}$.

If Valerie's eye level is $h = 1.7$ m, then $d \approx 3569\sqrt{1.7} \approx 4653$ m, or roughly 4.7 km. If $h = 20$ m, then $d \approx 3569\sqrt{20} \approx 15961$ m, or nearly 16 km. ∎

The Error Bound

To use Taylor polynomials effectively, we need a way to estimate the size of the error. This is provided by the next theorem, which shows that the size of this error depends on the size of the $(n+1)$st derivative.

A proof of Theorem 2 is presented at the end of this section.

THEOREM 2 Error Bound Assume that $f^{(n+1)}$ exists and is continuous. Let K be a number such that $|f^{(n+1)}(u)| \leq K$ for all u between a and x. Then

$$|f(x) - T_n(x)| \leq K\frac{|x-a|^{n+1}}{(n+1)!}$$

where T_n is the nth Taylor polynomial centered at $x = a$.

■ **EXAMPLE 6 Using the Error Bound** Apply the error bound to

$$|\ln 1.2 - T_3(1.2)|$$

where T_3 is the third Taylor polynomial for $f(x) = \ln x$ at $a = 1$. Check your result with a calculator.

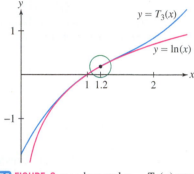

DF **FIGURE 8** $y = \ln x$ and $y = T_3(x)$ are indistinguishable near $x = 1.2$.

Solution

Step 1. Find a value of K.

To use the error bound with $n = 3$, we must find a value of K such that $|f^{(4)}(u)| \le K$ for all u between $a = 1$ and $x = 1.2$. As we computed in Example 3, $f^{(4)}(x) = -6x^{-4}$. The absolute value $|f^{(4)}(x)|$ is decreasing for $x > 0$, so its maximum value on $[1, 1.2]$ is $|f^{(4)}(1)| = 6$. Therefore, we may take $K = 6$.

Step 2. Apply the error bound.

$$|\ln 1.2 - T_3(1.2)| \le K\frac{|x - a|^{n+1}}{(n+1)!} = 6\frac{|1.2 - 1|^4}{4!} \approx 0.0004$$

Step 3. Check the result.

Recall from Example 3 that

$$T_3(x) = (x - 1) - \frac{1}{2}(x - 1)^2 + \frac{1}{3}(x - 1)^3$$

The following values from a calculator confirm that the error is at most 0.0004:

$$|\ln 1.2 - T_3(1.2)| \approx |0.182667 - 0.182322| \approx 0.00035 < 0.0004$$

Observe in Figure 8 that $y = \ln x$ and $y = T_3(x)$ are indistinguishable near $x = 1.2$. ■

■ **EXAMPLE 7**   **Approximating with a Given Accuracy**   Let T_n be the nth Maclaurin polynomial for $f(x) = \cos x$. Find a value of n such that

$$|\cos 0.2 - T_n(0.2)| < 10^{-5}$$

Solution

Step 1. Find a value of K.

Since $|f^{(n)}(x)|$ is $|\cos x|$ or $|\sin x|$, depending on whether n is even or odd, we have $|f^{(n)}(u)| \le 1$ for all u. Thus, we may apply the error bound with $K = 1$.

Step 2. Find a value of n.

The error bound gives us

$$|\cos 0.2 - T_n(0.2)| \le K\frac{|0.2 - 0|^{n+1}}{(n+1)!} = \frac{|0.2|^{n+1}}{(n+1)!}$$

To make the error less than 10^{-5}, we must choose n so that

$$\frac{|0.2|^{n+1}}{(n+1)!} < 10^{-5}$$

It's not possible to solve this inequality for n, but we can find a suitable n by checking several values:

n	2	3	4
$\dfrac{|0.2|^{n+1}}{(n+1)!}$	$\dfrac{0.2^3}{3!} \approx 0.0013$	$\dfrac{0.2^4}{4!} \approx 6.67 \times 10^{-5}$	$\dfrac{0.2^5}{5!} \approx 2.67 \times 10^{-6} < 10^{-5}$

We see that the error is less than 10^{-5} for $n = 4$. ■

The rest of this section is devoted to a proof of the error bound (Theorem 2). Define the ***nth remainder***:

$$\boxed{R_n(x) = f(x) - T_n(x)}$$

The error in $T_n(x)$ is the absolute value $|R_n(x)|$. As a first step in proving the error bound, we show that $R_n(x)$ can be represented as an integral.

Taylor's Theorem   Assume that $f^{(n+1)}$ exists and is continuous. Then

$$\boxed{R_n(x) = \frac{1}{n!}\int_a^x (x - u)^n f^{(n+1)}(u)\, du}$$

2

To use the error bound, it is not necessary to find the smallest possible value of K. In this example, we take $K = 1$. This works for all n, but for odd n we could have used the smaller value $K = \sin 0.2 \approx 0.2$.

Proof Set

$$I_n(x) = \frac{1}{n!} \int_a^x (x - u)^n f^{(n+1)}(u) \, du$$

Our goal is to show that $R_n(x) = I_n(x)$. For $n = 0$, $R_0(x) = f(x) - f(a)$ and the desired result is just a restatement of the Fundamental Theorem of Calculus:

$$I_0(x) = \int_a^x f'(u) \, du = f(x) - f(a) = R_0(x)$$

Exercise 64 reviews this proof for the special case $n = 2$.

To prove the formula for $n > 0$, we apply Integration by Parts to $I_n(x)$ with

$$h(u) = \frac{1}{n!}(x - u)^n, \qquad g(u) = f^{(n)}(u)$$

Then $g'(u) = f^{(n+1)}(u)$, and so

$$I_n(x) = \int_a^x h(u) \, g'(u) \, du = h(u)g(u) \Big|_a^x - \int_a^x h'(u)g(u) \, du$$

$$= \frac{1}{n!}(x - u)^n f^{(n)}(u) \Big|_a^x - \frac{1}{n!} \int_a^x (-n)(x - u)^{n-1} f^{(n)}(u) \, du$$

$$= -\frac{1}{n!}(x - a)^n f^{(n)}(a) + I_{n-1}(x)$$

This can be rewritten as

$$I_{n-1}(x) = \frac{f^{(n)}(a)}{n!}(x - a)^n + I_n(x)$$

Now apply this relation n times, noting that $I_0(x) = f(x) - f(a)$:

$$f(x) = f(a) + I_0(x)$$

$$= f(a) + \frac{f'(a)}{1!}(x - a) + I_1(x)$$

$$= f(a) + \frac{f'(a)}{1!}(x - a) + \frac{f''(a)}{2!}(x - a)^2 + I_2(x)$$

$$\vdots$$

$$= f(a) + \frac{f'(a)}{1!}(x - a) + \cdots + \frac{f^{(n)}(a)}{n!}(x - a)^n + I_n(x)$$

This shows that $f(x) = T_n(x) + I_n(x)$ and hence $I_n(x) = R_n(x)$, as desired. ∎

Proof Now we can prove Theorem 2. Assume first that $x \geq a$. Then

In Eq. (3), we use the inequality

$$\left| \int_a^b f(x) \, dx \right| \leq \int_a^b |f(x)| \, dx$$

which is valid for all integrable functions.

$$|f(x) - T_n(x)| = |R_n(x)| = \left| \frac{1}{n!} \int_a^x (x - u)^n f^{(n+1)}(u) \, du \right|$$

$$\leq \frac{1}{n!} \int_a^x \left| (x - u)^n f^{(n+1)}(u) \right| du \qquad \boxed{3}$$

$$\leq \frac{K}{n!} \int_a^x |x - u|^n \, du \qquad \boxed{4}$$

$$= \frac{K}{n!} \frac{-(x - u)^{n+1}}{n + 1} \Big|_{u=a}^x = K \frac{|x - a|^{n+1}}{(n + 1)!}$$

Note that the absolute value is not needed in Eq. (4) because $x - u \geq 0$ for $a \leq u \leq x$. If $x \leq a$, we must interchange the upper and lower limits of the integral in Eqs. (3) and (4). ∎

8.4 SUMMARY

- The nth *Taylor polynomial* centered at $x = a$ for the function f is

$$T_n(x) = f(a) + \frac{f'(a)}{1!}(x-a)^1 + \frac{f''(a)}{2!}(x-a)^2 + \cdots + \frac{f^{(n)}(a)}{n!}(x-a)^n$$

When $a = 0$, T_n is also called the nth *Maclaurin polynomial*.
- If $f^{(n+1)}$ exists and is continuous, then we have the *error bound*

$$|T_n(x) - f(x)| \le K\,\frac{|x-a|^{n+1}}{(n+1)!}$$

where K is a number such that $|f^{(n+1)}(u)| \le K$ for all u between a and x.
- For reference, we include a table of standard Maclaurin and Taylor polynomials.

$f(x)$	a	Maclaurin or Taylor Polynomial
e^x	0	$T_n(x) = 1 + x + \dfrac{x^2}{2!} + \dfrac{x^3}{3!} + \cdots + \dfrac{x^n}{n!}$
$\sin x$	0	$T_{2n+1}(x) = T_{2n+2}(x) = x - \dfrac{x^3}{3!} + \cdots + (-1)^n \dfrac{x^{2n+1}}{(2n+1)!}$
$\cos x$	0	$T_{2n}(x) = T_{2n+1}(x) = 1 - \dfrac{x^2}{2!} + \dfrac{x^4}{4!} - \cdots + (-1)^n \dfrac{x^{2n}}{(2n)!}$
$\ln x$	1	$T_n(x) = (x-1) - \dfrac{1}{2}(x-1)^2 + \cdots + \dfrac{(-1)^{n-1}}{n}(x-1)^n$
$\dfrac{1}{1-x}$	0	$T_n(x) = 1 + x + x^2 + \cdots + x^n$

8.4 EXERCISES

Preliminary Questions

1. What is T_3 centered at $a = 3$ for a function f such that $f(3) = 9$, $f'(3) = 8$, $f''(3) = 4$, and $f'''(3) = 12$?

2. The dashed graphs in Figure 9 are Taylor polynomials for a function f. Which of the two is a Maclaurin polynomial?

3. For which value of x does the Maclaurin polynomial T_n satisfy $T_n(x) = f(x)$, no matter what f is?

4. Let T_n be the Maclaurin polynomial of a function f satisfying $|f^{(4)}(x)| \le 1$ for all x. Which of the following statements follow from the error bound?

(a) $|T_4(2) - f(2)| \le \frac{2}{3}$

(b) $|T_3(2) - f(2)| \le \frac{2}{3}$

(c) $|T_3(2) - f(2)| \le \frac{1}{3}$

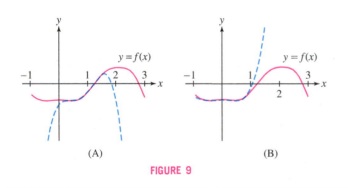

(A)   (B)

FIGURE 9

Exercises

In Exercises 1–14, calculate the Taylor polynomials T_2 and T_3 centered at $x = a$ for the given function and value of a.

1. $f(x) = \sin x, \quad a = 0$

2. $f(x) = \sin x, \quad a = \dfrac{\pi}{2}$

3. $f(x) = \dfrac{1}{1+x}, \quad a = 2$

4. $f(x) = \dfrac{1}{1+x^2}, \quad a = -1$

5. $f(x) = x^4 - 2x, \quad a = 3$

6. $f(x) = \dfrac{x^2+1}{x+1}, \quad a = -2$

7. $f(x) = \tan x, \quad a = 0$

8. $f(x) = \tan x, \quad a = \dfrac{\pi}{4}$

9. $f(x) = e^{-x} + e^{-2x}, \quad a = 0$

10. $f(x) = e^{2x}, \quad a = \ln 2$

11. $f(x) = x^2 e^{-x}, \quad a = 1$

12. $f(x) = \cosh 2x, \quad a = 0$

13. $f(x) = \dfrac{\ln x}{x}, \quad a = 1$

14. $f(x) = \ln(x+1), \quad a = 0$

15. Show that the nth Maclaurin polynomial for $f(x) = e^x$ is

$$T_n(x) = 1 + \frac{x}{1!} + \frac{x^2}{2!} + \cdots + \frac{x^n}{n!}$$

16. Show that the nth Taylor polynomial for $f(x) = \dfrac{1}{x+1}$ at $a = 1$ is

$$T_n(x) = \frac{1}{2} - \frac{(x-1)}{4} + \frac{(x-1)^2}{8} + \cdots + (-1)^n \frac{(x-1)^n}{2^{n+1}}$$

17. Show that the Maclaurin polynomials for $f(x) = \sin x$ are

$$T_{2n+1}(x) = T_{2n+2}(x) = x - \frac{x^3}{3!} + \frac{x^5}{5!} - \cdots + (-1)^n \frac{x^{2n+1}}{(2n+1)!}$$

18. Show that the Maclaurin polynomials for $f(x) = \ln(1+x)$ are

$$T_n(x) = x - \frac{x^2}{2} + \frac{x^3}{3} + \cdots + (-1)^{n-1} \frac{x^n}{n}$$

In Exercises 19–24, find T_n centered at $x = a$ for all n.

19. $f(x) = \dfrac{1}{1+x}$, $a = 0$

20. $f(x) = \dfrac{1}{x-1}$, $a = 4$

21. $f(x) = e^x$, $a = 1$

22. $f(x) = x^{-2}$, $a = 2$

23. $f(x) = \cos x$, $a = \dfrac{\pi}{4}$

24. $f(\theta) = \sin 3\theta$, $a = 0$

In Exercises 25–28, find T_2 and use a calculator to compute the error $|f(x) - T_2(x)|$ for the given values of a and x.

25. $y = e^x$, $a = 0$, $x = -0.5$

26. $y = \cos x$, $a = 0$, $x = \dfrac{\pi}{12}$

27. $y = x^{-2/3}$, $a = 1$, $x = 1.2$

28. $y = e^{\sin x}$, $a = \dfrac{\pi}{2}$, $x = 1.5$

29. $\boxed{\text{GU}}$ Compute T_3 for $f(x) = \sqrt{x}$ centered at $a = 1$. Then use a plot of the error $|f(x) - T_3(x)|$ to find a value $c > 1$ such that the error on the interval $[1, c]$ is at most 0.25.

30. *CAS* Plot $f(x) = 1/(1+x)$ together with the Taylor polynomials T_n at $a = 1$ for $1 \le n \le 4$ on the interval $[-2, 8]$ (be sure to limit the upper plot range).

(a) Over which interval does T_4 appear to approximate f closely?

(b) What happens for $x < -1$?

(c) Use your computer algebra system to produce and plot T_{30} together with f on $[-2, 8]$. Over which interval does T_{30} appear to give a close approximation?

31. Let T_3 be the Maclaurin polynomial of $f(x) = e^x$. Use the error bound to find the maximum possible value of $|f(1.1) - T_3(1.1)|$. Show that we can take $K = e^{1.1}$.

32. Let T_2 be the Taylor polynomial of $f(x) = \sqrt{x}$ at $a = 4$. Apply the error bound to find the maximum possible value of the error $|f(3.9) - T_2(3.9)|$.

In Exercises 33–36, compute the Taylor polynomial indicated and use the error bound to find the maximum possible size of the error. Verify your result with a calculator.

33. $f(x) = \cos x$, $a = 0$; $|\cos 0.25 - T_5(0.25)|$

34. $f(x) = x^{11/2}$, $a = 1$; $|f(1.2) - T_4(1.2)|$

35. $f(x) = x^{-1/2}$, $a = 4$; $|f(4.3) - T_3(4.3)|$

36. $f(x) = \sqrt{1+x}$, $a = 8$; $|\sqrt{9.02} - T_3(8.02)|$

37. Calculate the Maclaurin polynomial T_3 for $f(x) = \tan^{-1} x$. Compute $T_3(\frac{1}{2})$ and use the error bound to find a bound for the error $|\tan^{-1}\frac{1}{2} - T_3(\frac{1}{2})|$. Refer to the graph in Figure 10 to find an acceptable value of K. Verify your result by computing $|\tan^{-1}\frac{1}{2} - T_3(\frac{1}{2})|$ using a calculator.

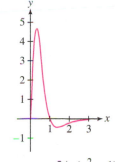

FIGURE 10 Graph of $f^{(4)}(x) = \dfrac{-24x(x^2 - 1)}{(x^2 + 1)^4}$, where $f(x) = \tan^{-1} x$.

38. Let $f(x) = \ln(x^3 - x + 1)$. The third Taylor polynomial at $a = 1$ is

$$T_3(x) = 2(x - 1) + (x - 1)^2 - \frac{7}{3}(x - 1)^3$$

Find the maximum possible value of $|f(1.1) - T_3(1.1)|$, using the graph in Figure 11 to find an acceptable value of K. Verify your result by computing $|f(1.1) - T_3(1.1)|$ using a calculator.

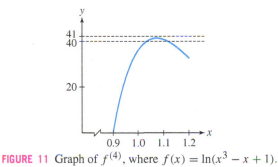

FIGURE 11 Graph of $f^{(4)}$, where $f(x) = \ln(x^3 - x + 1)$.

39. $\boxed{\text{GU}}$ Let T_2 be the Taylor polynomial at $a = 0.5$ for $f(x) = \cos(x^2)$. Use the error bound to find the maximum possible value of $|f(0.6) - T_2(0.6)|$. Plot $f^{(3)}$ to find an acceptable value of K.

40. $\boxed{\text{GU}}$ Calculate the Maclaurin polynomial T_2 for $f(x) = \operatorname{sech} x$ and use the error bound to find the maximum possible value of $|f(\frac{1}{2}) - T_2(\frac{1}{2})|$. Plot f''' to find an acceptable value of K.

In Exercises 41–44, use the error bound to find a value of n for which the given inequality is satisfied. Then verify your result using a calculator.

41. $|\cos 0.1 - T_n(0.1)| \le 10^{-7}$, $a = 0$

42. $|\ln 1.3 - T_n(1.3)| \le 10^{-4}$, $a = 1$

43. $|\sqrt{1.3} - T_n(1.3)| \le 10^{-6}$, $a = 1$

44. $|e^{-0.1} - T_n(-0.1)| \le 10^{-6}$, $a = 0$

45. Let $f(x) = e^{-x}$ and $T_3(x) = 1 - x + \dfrac{x^2}{2} - \dfrac{x^3}{6}$. Use the error bound to show that for all $x \geq 0$,

$$|f(x) - T_3(x)| \leq \frac{x^4}{24}$$

If you have a GU, illustrate this inequality by plotting $y = f(x) - T_3(x)$ and $y = x^4/24$ together over $[0, 1]$.

46. Use the error bound with $n = 4$ to show that

$$\left| \sin x - \left(x - \frac{x^3}{6} \right) \right| \leq \frac{|x|^5}{120} \quad \text{(for all } x\text{)}$$

47. Let T_n be the Taylor polynomial for $f(x) = \ln x$ at $a = 1$, and let $c > 1$. Show that

$$|\ln c - T_n(c)| \leq \frac{|c - 1|^{n+1}}{n + 1}$$

Then find a value of n such that $|\ln 1.5 - T_n(1.5)| \leq 10^{-2}$.

48. Let $n \geq 1$. Show that if $|x|$ is small, then

$$(x + 1)^{1/n} \approx 1 + \frac{x}{n} + \frac{1 - n}{2n^2} x^2$$

Use this approximation with $n = 6$ to estimate $1.5^{1/6}$.

49. Verify that the third Maclaurin polynomial for $f(x) = e^x \sin x$ is equal to the product of the third Maclaurin polynomials of $f(x) = e^x$ and $f(x) = \sin x$ (after discarding terms of degree greater than 3 in the product).

50. Find the fourth Maclaurin polynomial for $f(x) = \sin x \cos x$ by multiplying the fourth Maclaurin polynomials for $f(x) = \sin x$ and $f(x) = \cos x$.

51. Find the Maclaurin polynomials T_n for $f(x) = \cos(x^2)$. You may use the fact that $T_n(x)$ is equal to the sum of the terms up to degree n obtained by substituting x^2 for x in the nth Maclaurin polynomial of $\cos x$.

52. Find the Maclaurin polynomials of $1/(1 + x^2)$ by substituting $-x^2$ for x in the Maclaurin polynomials of $1/(1 - x)$.

53. Let $f(x) = 3x^3 + 2x^2 - x - 4$. Calculate T_j for $j = 1, 2, 3, 4, 5$ at both $a = 0$ and $a = 1$. Show that $T_3(x) = f(x)$ in both cases.

54. Let T_n be the nth Taylor polynomial at $x = a$ for a polynomial f of degree n. Based on the result of Exercise 53, guess the value of $|f(x) - T_n(x)|$. Prove that your guess is correct using the error bound.

55. Let $s(t)$ be the distance of a truck to an intersection. At time $t = 0$, the truck is 60 m from the intersection, travels away from it with a velocity of 24 m/s, and begins to slow down with an acceleration of $a = -3$ m/s^2. Determine the second Maclaurin polynomial of s, and use it to estimate the truck's distance from the intersection after 4 s.

56. A bank owns a portfolio of bonds whose value $P(r)$ depends on the interest rate r (measured in percent; e.g., $r = 5$ means a 5% interest rate). The bank's quantitative analyst determines that

$$P(5) = 100,000, \quad \left. \frac{dP}{dr} \right|_{r=5} = -40,000, \quad \left. \frac{d^2 P}{dr^2} \right|_{r=5} = 50,000$$

In finance, this second derivative is called **bond convexity**. Find the second Taylor polynomial of $P(r)$ centered at $r = 5$ and use it to estimate the value of the portfolio if the interest rate moves to $r = 5.5\%$.

57. A narrow, negatively charged ring of radius R exerts a force on a positively charged particle P located at distance x above the center of the ring of magnitude

$$F(x) = -\frac{kx}{(x^2 + R^2)^{3/2}}$$

where $k > 0$ is a constant (Figure 12).

(a) Compute the third-degree Maclaurin polynomial for F.

(b) Show that $F \approx -(k/R^3)x$ to second order. This shows that when x is small, $F(x)$ behaves like a restoring force similar to the force exerted by a spring.

(c) Show that $F(x) \approx -k/x^2$ when x is large by showing that

$$\lim_{x \to \infty} \frac{F(x)}{-k/x^2} = 1$$

Thus, $F(x)$ behaves like an inverse square law, and the charged ring looks like a point charge from far away.

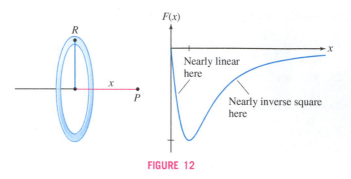

FIGURE 12

58. A light wave of wavelength λ travels from A to B by passing through an aperture (circular region) located in a plane that is perpendicular to $\overline{AB}$ (see Figure 13 for the notation). Let $f(r) = d' + h'$; that is, $f(r)$ is the distance $AC + CB$ as a function of r.

(a) Show that $f(r) = \sqrt{d^2 + r^2} + \sqrt{h^2 + r^2}$, and use the Maclaurin polynomial of order 2 to show that

$$f(r) \approx d + h + \frac{1}{2}\left(\frac{1}{d} + \frac{1}{h} \right) r^2$$

(b) The **Fresnel zones**, used to determine the optical disturbance at B, are the concentric bands bounded by the circles of radius R_n such that $f(R_n) = d + h + n\lambda/2$. Show that $R_n \approx \sqrt{n\lambda L}$, where $L = (d^{-1} + h^{-1})^{-1}$.

(c) Estimate the radii R_1 and R_{100} for blue light ($\lambda = 475 \times 10^{-7}$ cm) if $d = h = 100$ cm.

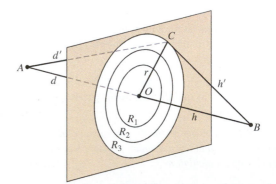

FIGURE 13 The Fresnel zones are the regions between the circles of radius R_n.

59. Referring to Figure 14, let a be the length of the chord $\overline{AC}$ of angle θ of the unit circle. Derive the following approximation for the excess of the arc over the chord:

$$\theta - a \approx \frac{\theta^3}{24}$$

Hint: Show that $\theta - a = \theta - 2\sin(\theta/2)$ and use the third Maclaurin polynomial as an approximation.

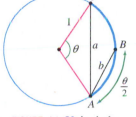

FIGURE 14 Unit circle.

60. To estimate the length θ of a circular arc of the unit circle, the seventeenth-century Dutch scientist Christian Huygens used the approximation $\theta \approx (8b - a)/3$, where a is the length of the chord $\overline{AC}$ of angle θ and b is the length of the chord $\overline{AB}$ of angle $\theta/2$ (Figure 14).

(a) Prove that $a = 2\sin(\theta/2)$ and $b = 2\sin(\theta/4)$, and show that the Huygens approximation amounts to the approximation

$$\theta \approx \frac{16}{3}\sin\frac{\theta}{4} - \frac{2}{3}\sin\frac{\theta}{2}$$

(b) Compute the fifth Maclaurin polynomial of the function on the right.

(c) Use the error bound to show that the error in the Huygens approximation is less than $0.00022|\theta|^5$.

Further Insights and Challenges

61. Show that the nth Maclaurin polynomial of $f(x) = \arcsin x$ for n odd is

$$T_n(x) = x + \frac{1}{2}\frac{x^3}{3} + \frac{1\cdot 3}{2\cdot 4}\frac{x^5}{5} + \cdots + \frac{1\cdot 3\cdot 5\cdots(n-2)}{2\cdot 4\cdot 6\cdots(n-1)}\frac{x^n}{n}$$

62. Let $x \geq 0$ and assume that $f^{(n+1)}(t) \geq 0$ for $0 \leq t \leq x$. Use Taylor's Theorem to show that the nth Maclaurin polynomial T_n satisfies

$$T_n(x) \leq f(x), \quad \text{for all } x \geq 0$$

63. Use Exercise 62 to show that for $x \geq 0$ and all n,

$$e^x \geq 1 + x + \frac{x^2}{2!} + \cdots + \frac{x^n}{n!}$$

Sketch the graphs of $y = e^x$, $y = T_1(x)$, and $y = T_2(x)$ on the same coordinate axes. Does this inequality remain true for $x < 0$?

64. This exercise is intended to reinforce the proof of Taylor's Theorem.

(a) Show that $f(x) = T_0(x) + \int_a^x f'(u)\, du$.

(b) Use Integration by Parts to prove the formula

$$\int_a^x (x - u)f^{(2)}(u)\, du = -f'(a)(x - a) + \int_a^x f'(u)\, du$$

(c) Prove the case $n = 2$ of Taylor's Theorem:

$$f(x) = T_1(x) + \int_a^x (x - u)f^{(2)}(u)\, du$$

In Exercises 65–69, we estimate integrals using Taylor polynomials. Exercise 66 is used to estimate the error.

65. Find the fourth Maclaurin polynomial T_4 for $f(x) = e^{-x^2}$, and calculate $I = \int_0^{1/2} T_4(x)\, dx$ as an estimate for $\int_0^{1/2} e^{-x^2}\, dx$. A CAS yields the value $I \approx 0.461281$. How large is the error in your approximation? *Hint:* T_4 is obtained by substituting $-x^2$ in the second Maclaurin polynomial for e^x.

66. Approximating Integrals Let $L > 0$. Show that if two functions f and g satisfy $|f(x) - g(x)| < L$ for all $x \in [a, b]$, then

$$\left| \int_a^b f(x)\, dx - \int_a^b g(x)\, dx \right| dx < L(b - a)$$

67. Let T_4 be the fourth Maclaurin polynomial for $f(x) = \cos x$.

(a) Show that

$$|\cos x - T_4(x)| \leq \frac{\left(\frac{1}{2}\right)^6}{6!} \quad \text{for all } x \in \left[0, \frac{1}{2}\right]$$

Hint: $T_4(x) = T_5(x)$.

(b) Evaluate $\int_0^{1/2} T_4(x)\, dx$ as an approximation to $\int_0^{1/2} \cos x\, dx$. Use Exercise 66 to find a bound for the size of the error.

68. Let $Q(x) = 1 - x^2/6$. Use the error bound for $f(x) = \sin x$ to show that

$$\left| \frac{\sin x}{x} - Q(x) \right| \leq \frac{|x|^4}{5!}$$

Then calculate $\int_0^1 Q(x)\, dx$ as an approximation to $\int_0^1 (\sin x/x)\, dx$ and find a bound for the error.

69. (a) Compute the sixth Maclaurin polynomial T_6 for $f(x) = \sin(x^2)$ by substituting x^2 in $P(x) = x - x^3/6$, the third Maclaurin polynomial for $f(x) = \sin x$.

(b) Show that $|\sin(x^2) - T_6(x)| \leq \frac{|x|^{10}}{5!}$.

Hint: Substitute x^2 for x in the error bound for $|\sin x - P(x)|$, noting that P is also the fourth Maclaurin polynomial for $f(x) = \sin x$.

(c) Use T_6 to approximate $\int_0^{1/2} \sin(x^2)\, dx$ and find a bound for the error.

70. Prove by induction that for all k,

$$\frac{d^j}{dx^j}\left(\frac{(x - a)^k}{k!}\right) = \frac{k(k - 1)\cdots(k - j + 1)(x - a)^{k-j}}{k!}$$

$$\frac{d^j}{dx^j}\left(\frac{(x - a)^k}{k!}\right)\bigg|_{x=a} = \begin{cases} 1 & \text{for } k = j \\ 0 & \text{for } k \neq j \end{cases}$$

Use this to prove that T_n agrees with f at $x = a$ to order n.

71. Let a be any number and let

$$P(x) = a_n x^n + a_{n-1} x^{n-1} + \cdots + a_1 x + a_0$$

be a polynomial of degree n or less.

(a) Show that if $P^{(j)}(a) = 0$ for $j = 0, 1, \ldots, n$, then $P(x) = 0$, that is, $a_j = 0$ for all j. *Hint:* Use induction, noting that if the statement is true for degree $n - 1$, then $P'(x) = 0$.

(b) Prove that T_n is the only polynomial of degree n or less that agrees with f at $x = a$ to order n. *Hint:* If Q is another such polynomial, apply (a) to $P(x) = T_n(x) - Q(x)$.

CHAPTER REVIEW EXERCISES

In Exercises 1–4, calculate the arc length over the given interval.

1. $y = \dfrac{x^5}{10} + \dfrac{x^{-3}}{6}$, $[1, 2]$

2. $y = e^{x/2} + e^{-x/2}$, $[0, 2]$

3. $y = 4x - 2$, $[-2, 2]$

4. $y = x^{2/3}$, $[1, 8]$

5. Show that the arc length of $y = 2\sqrt{x}$ over $[0, a]$ is equal to $\sqrt{a(a+1)} + \ln(\sqrt{a} + \sqrt{a+1})$. *Hint:* Apply the substitution $x = \tan^2 \theta$ to the arc length integral.

6. 𝖢𝖠𝖲 Compute the trapezoidal approximation T_5 to the arc length s of $y = \tan x$ over $\left[0, \frac{\pi}{4}\right]$.

In Exercises 7–10, calculate the surface area of the solid obtained by rotating the curve over the given interval about the x-axis.

7. $y = x + 1$, $[0, 4]$

8. $y = \dfrac{2}{3}x^{3/4} - \dfrac{2}{5}x^{5/4}$, $[0, 1]$

9. $y = \dfrac{2}{3}x^{3/2} - \dfrac{1}{2}x^{1/2}$, $[1, 2]$ **10.** $y = \dfrac{1}{2}x^2$, $[0, 2]$

11. Compute the total surface area of the coin obtained by rotating the region in Figure 1 about the x-axis. The top and bottom parts of the region are semicircles with a radius of 1 mm.

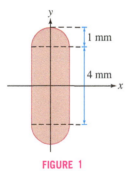

FIGURE 1

12. Calculate the fluid force on the side of a right triangle of height 3 m and base 2 m submerged in water vertically, with its upper vertex at the surface of the water.

13. Calculate the fluid force on the side of a right triangle of height 3 m and base 2 m submerged in water vertically, with its upper vertex located at a depth of 4 m.

14. A plate in the shape of the shaded region in Figure 2 is submerged in water. Calculate the fluid force on a side of the plate if the water surface is $y = 1$.

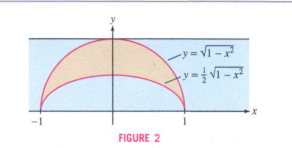

FIGURE 2

15. Figure 3 shows an object whose face is an equilateral triangle with 5-m sides. The object is 2 m thick and is submerged in water with its vertex 3 m below the water surface. Calculate the fluid force on both a triangular face and a slanted rectangular edge of the object.

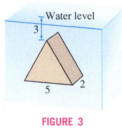

FIGURE 3

16. The end of a horizontal oil tank is an ellipse (Figure 4) with equation $(x/4)^2 + (y/3)^2 = 1$ (length in meters). Assume that the tank is filled with oil of density 900 kg/m³.

(a) Calculate the total force F on the end of the tank when the tank is full.

(b) 📖 Would you expect the total force on the lower half of the tank to be greater than, less than, or equal to $\frac{1}{2}F$? Explain. Then compute the force on the lower half exactly and confirm (or refute) your expectation.

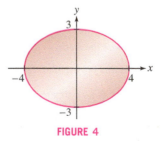

FIGURE 4

17. Calculate the moments and COM of the lamina occupying the region under $y = x(4 - x)$ for $0 \le x \le 4$, assuming a density of $\rho = 1200$ kg/m³.

18. Sketch the region between $y = 4(x + 1)^{-1}$ and $y = 1$ for $0 \le x \le 3$, and find its centroid.

19. Find the centroid of the region between the semicircle $y = \sqrt{1 - x^2}$ and the top half of the ellipse $y = \frac{1}{2}\sqrt{1 - x^2}$ (Figure 2).

20. Find the centroid of the shaded region in Figure 5 bounded on the left by $x = 2y^2 - 2$ and on the right by a semicircle of radius 1. *Hint:* Use symmetry and additivity of moments.

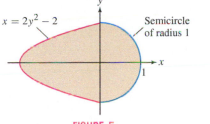

FIGURE 5

21. Use the Theorem of Pappus to find the volume of the solid of revolution obtained by rotating the region in the first quadrant bounded by $y = x^2$ and $y = \sqrt{x}$ about the y-axis.

22. Use the Theorem of Pappus to find a formula for the volume of the solid obtained by rotating the triangle with vertices $(1, 0)$, $(3, 0)$, and $(2, 2)$ about the y-axis.

In Exercises 23–28, find the Taylor polynomial at $x = a$ for the given function.

23. $f(x) = x^3$, T_3, $a = 1$

24. $f(x) = 3(x + 2)^3 - 5(x + 2)$, T_3, $a = -2$

25. $f(x) = x \ln(x)$, T_4, $a = 1$

26. $f(x) = (3x + 2)^{1/3}$, T_3, $a = 2$

27. $f(x) = xe^{-x^2}$, T_4, $a = 0$

28. $f(x) = \ln(\cos x)$, T_3, $a = 0$

29. Find the nth Maclaurin polynomial for $f(x) = e^{3x}$.

30. Use the fifth Maclaurin polynomial of $f(x) = e^x$ to approximate $\sqrt{e}$. Use a calculator to determine the error.

31. Use the third Taylor polynomial of $f(x) = \tan^{-1} x$ at $a = 1$ to approximate $f(1.1)$. Use a calculator to determine the error.

32. Let T_4 be the Taylor polynomial for $f(x) = \sqrt{x}$ at $a = 16$. Use the error bound to find the maximum possible size of $|f(17) - T_4(17)|$.

33. Find n such that $|e - T_n(1)| < 10^{-8}$, where T_n is the nth Maclaurin polynomial for $f(x) = e^x$.

34. Let T_4 be the Taylor polynomial for $f(x) = x \ln x$ at $a = 1$ computed in Exercise 25. Use the error bound to find a bound for $|f(1.2) - T_4(1.2)|$.

35. Verify that $T_n(x) = 1 + x + x^2 + \cdots + x^n$ is the nth Maclaurin polynomial of $f(x) = 1/(1 - x)$. Show using substitution that the nth Maclaurin polynomial for $f(x) = 1/(1 - x/4)$ is

$$T_n(x) = 1 + \frac{1}{4}x + \frac{1}{4^2}x^2 + \cdots + \frac{1}{4^n}x^n$$

What is the nth Maclaurin polynomial for $g(x) = \dfrac{1}{1 + x}$?

36. Let $f(x) = \dfrac{5}{4 + 3x - x^2}$ and let a_k be the coefficient of x^k in the Maclaurin polynomial T_n for $k \leq n$.

(a) Show that $f(x) = \left(\dfrac{1/4}{1 - x/4} + \dfrac{1}{1 + x} \right)$.

(b) Use Exercise 35 to show that $a_k = \dfrac{1}{4^{k+1}} + (-1)^k$.

(c) Compute T_3.

37. Let T_n be the nth Maclaurin polynomial for the function $f(x) = \sin x + \sinh x$.

(a) Show that $T_5(x) = T_6(x) = T_7(x) = T_8(x)$.

(b) Show that $|f^n(x)| \leq 1 + \cosh x$ for all n. *Hint:* Note that $|\sinh x| \leq |\cosh x|$ for all x.

(c) Show that $|T_8(x) - f(x)| \leq \dfrac{2.6}{9!}|x|^9$ for $-1 \leq x \leq 1$.

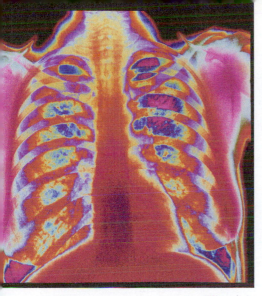

In X-ray computerized tomography, differential equations are used to fill in missing data and provide detailed images. (*M. Kulyk/Science Source*)

9 INTRODUCTION TO DIFFERENTIAL EQUATIONS

Will this airplane fly?…How can we create an image of the interior of the human body using very weak X-rays?…What is a design of a bicycle frame that combines low weight with rigidity?…How much would the mean temperature of the earth increase if the amount of carbon dioxide in the atmosphere increased by 20 percent?

—An overview of applications of differential equations in K. Eriksson, D. Estep, P. Hansbo, and C. Johnson, *Computational Differential Equations*, Cambridge University Press, New York, 1996

Differential equations are among the most powerful tools we have for analyzing the world mathematically. They are used to formulate the fundamental laws of nature (from Newton's laws to Maxwell's equations and the laws of quantum mechanics) and to model the most diverse physical phenomena. The quotation above lists just a few of the myriad applications. This chapter provides an introduction to some elementary techniques and applications of this important subject.

9.1 Solving Differential Equations

A differential equation is an equation that involves an unknown function $y = y(x)$ and its first or higher derivatives. A **solution** is a function $y = f(x)$ satisfying the given equation. As we have seen in previous chapters, solutions usually depend on one or more arbitrary constants (denoted A, B, C, and K in the following examples):

Differential equation	General solution
$y' = -2y$	$y = Ke^{-2x}$
$\dfrac{dy}{dt} = t$	$y = \dfrac{1}{2}t^2 + C$
$y'' + y = 0$	$y = A\sin x + B\cos x$

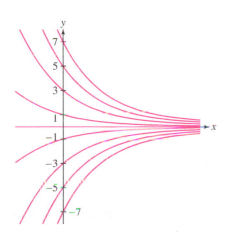

FIGURE 1 Family of solutions of $y' = -2y$.

Notice that for any of these equations, we can easily check that the given function is a solution. For example, if $y = Ke^{-2x}$, then $y' = -2Ke^{-2x} = -2y$ as we wanted to show. An expression such as $y = Ke^{-2x}$ is called a **general solution**. For each value of K, we obtain a **particular solution**. The graphs of the solutions as K varies form a family of curves in the xy-plane (Figure 1).

The first step in any study of differential equations is to classify the equations according to various properties. The most important attributes of a differential equation are its order and whether or not it is linear.

The **order** of a differential equation is the order of the highest derivative appearing in the equation. The general solution of an equation of order n usually involves n arbitrary constants. For example,

$$y'' + y = 0$$

has order 2 and its general solution has two arbitrary constants A and B as listed above.

A differential equation is called **linear** if it can be written in the form

$$a_n(x)y^{(n)} + a_{n-1}(x)y^{(n-1)} + \cdots + a_1(x)y' + a_0(x)y = b(x)$$

The coefficients $a_j(x)$ and $b(x)$ can be arbitrary functions of x, but a linear equation cannot have terms such as y^3, yy', or $\sin y$.

Differential equation	Order	Linear or nonlinear
$x^2 y' + e^x y = 4$	First-order	Linear
$x(y')^2 = y + x$	First-order	Nonlinear [because $(y')^2$ appears]
$y'' = (\sin x)y'$	Second-order	Linear
$y''' = x(\sin y)$	Third-order	Nonlinear (because $\sin y$ appears)

In this chapter, we restrict our attention to first-order equations.

Separation of Variables

We are familiar with the simplest type of differential equation, namely $y' = f(x)$. A solution is simply an antiderivative of f, so we can write the general solution as

$$y = \int f(x)\,dx$$

A more general class of first-order differential equations that can be solved directly by integration are the **separable equations**, which have the form

$$\frac{dy}{dx} = f(x)g(y) \qquad \boxed{1}$$

For example,

- $\dfrac{dy}{dx} = y \sin x$ is separable.

- $\dfrac{dy}{dx} = x + y$ is not separable because $x + y$ is not a *product* $f(x)g(y)$.

In Separation of Variables, we manipulate dx and dy symbolically, just as in the Substitution Rule.

Separable equations are solved using the method of **Separation of Variables**: Move the terms involving y and dy to the left and those involving x and dx to the right. Then integrate both sides:

$$\frac{dy}{dx} = f(x)g(y) \qquad \text{(separable equation)}$$

$$\frac{dy}{g(y)} = f(x)\,dx \qquad \text{(separate the variables)}$$

$$\int \frac{dy}{g(y)} = \int f(x)\,dx \qquad \text{(integrate)}$$

If these integrals can be evaluated, we can try to solve for y as a function of x. One can check that the solution we have found is, in fact, a solution to the original differential equation. Therefore, the symbolic manipulation that we applied does generate a valid solution.

■ **EXAMPLE 1** Show that $y\dfrac{dy}{dx} - x = 0$ is separable but not linear. Then find the general solution and plot the family of solutions.

Solution This differential equation is nonlinear because it contains the term yy'. To show that it is separable, rewrite the equation:

$$y\frac{dy}{dx} - x = 0 \quad \Rightarrow \quad \frac{dy}{dx} = \frac{x}{y} \qquad \text{(separable equation)}$$

Now use Separation of Variables:

$$y \, dy = x \, dx \qquad \text{(separate the variables)}$$

$$\int y \, dy = \int x \, dx \qquad \text{(integrate)}$$

$$\frac{1}{2} y^2 = \frac{1}{2} x^2 + C \qquad \boxed{2}$$

$$y = \pm\sqrt{x^2 + 2C} \quad \text{(solve for } y\text{)}$$

Note that one constant of integration is sufficient in Eq. (2). An additional constant for the integral on the left is not needed as we could subtract it from both sides of the equation and absorb it into the constant on the right.

Since C is arbitrary, we may replace $2C$ by K to obtain (Figure 2)

$$y = \pm\sqrt{x^2 + K}$$

Each choice of sign and value of K yields a solution.

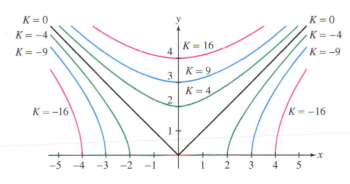

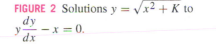

FIGURE 2 Solutions $y = \sqrt{x^2 + K}$ to $y\dfrac{dy}{dx} - x = 0$.

It is a good idea to verify that solutions you have found satisfy the differential equation. In our case, for the positive square root (the negative square root is similar), we have

$$\frac{dy}{dx} = \frac{d}{dx}\sqrt{x^2 + K} = \frac{x}{\sqrt{x^2 + K}}$$

$$y\frac{dy}{dx} = \sqrt{x^2 + K}\left(\frac{x}{\sqrt{x^2 + K}}\right) = x \quad \Rightarrow \quad y\frac{dy}{dx} - x = 0$$

This verifies that $y = \sqrt{x^2 + K}$ is a solution. ∎

Most differential equations arising in applications have an existence and uniqueness property: There exists one and only one solution satisfying a given initial condition. General existence and uniqueness theorems are discussed in textbooks on differential equations.

Although it is useful to find general solutions, in applications we are usually interested in the solution that describes a particular physical situation. The general solution to a first-order equation generally depends on one arbitrary constant, so we can pick out a particular solution $y(x)$ by specifying the value $y(x_0)$ for some fixed x_0 (Figure 3). This specification is called an **initial condition**. A differential equation together with an initial condition is called an **Initial Value Problem**.

■ **EXAMPLE 2** **Initial Value Problem** Solve the Initial Value Problem

$$y' = -ty, \qquad y(0) = 3$$

Solution Use Separation of Variables to find the general solution (assuming for now that $y \neq 0$):

$$\frac{dy}{dt} = -ty \quad \Rightarrow \quad \frac{dy}{y} = -t \, dt$$

$$\int \frac{dy}{y} = -\int t \, dt$$

$$\ln|y| = -\frac{1}{2}t^2 + C$$

$$|y| = e^{-t^2/2 + C} = e^C e^{-t^2/2}$$

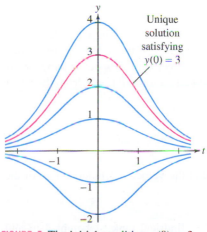

FIGURE 3 The initial condition $y(0) = 3$ determines one curve in the family of solutions to $y' = -ty$.

Thus, $y = \pm e^C e^{-t^2/2}$. Since C is arbitrary, e^C represents an arbitrary *positive number*, and $\pm e^C$ is an arbitrary nonzero number. We replace $\pm e^C$ by K and write the general solution as

$$y = Ke^{-t^2/2} \qquad \boxed{3}$$

Now use the initial condition $y(0) = Ke^{-0^2/2} = 3$. Thus, $K = 3$ and $y = 3e^{-t^2/2}$ is the solution to the Initial Value Problem (Figure 3).   ∎

In the context of differential equations, the term "modeling" means finding a differential equation that describes a given physical situation. As an example, consider water leaking through a hole at the bottom of a tank (Figure 4). The problem is to find the water level $y(t)$ at time t. We solve it by showing that $y(t)$ satisfies a differential equation.

The key observation is that the water lost during the interval from t to $t + \Delta t$ can be computed in two ways. Let

$$v(y) = \text{velocity of the water flowing through the hole}$$
$$\text{when the tank is filled to height } y$$

$$B = \text{area of the hole}$$

$$A(y) = \text{area of horizontal cross section of the tank at height } y$$

First, we observe that the water exiting through the hole during a time interval Δt forms a cylinder of base B and height $v(y)\Delta t$ [because the water travels a distance $v(y)\Delta t$—see Figure 4]. The volume of this cylinder is approximately $Bv(y)\Delta t$ [approximately but not exactly, because $v(y)$ may not be constant]. Thus,

$$\text{Volume of water lost between } t \text{ and } t + \Delta t \approx Bv(y)\Delta t$$

Second, we note that if the water level drops by an amount Δy during the interval Δt, then the volume of water lost is approximately $A(y)\Delta y$ (Figure 4). Therefore,

$$\text{Volume of water lost between } t \text{ and } t + \Delta t \approx A(y)\,\Delta y$$

This is also an approximation because the cross-sectional area may not be constant. Comparing the two results, we obtain $A(y)\Delta y \approx Bv(y)\Delta t$, or

$$\frac{\Delta y}{\Delta t} \approx \frac{Bv(y)}{A(y)}$$

Now take the limit as $\Delta t \to 0$ to obtain the differential equation

$$\boxed{\frac{dy}{dt} = \frac{Bv(y)}{A(y)}} \qquad \boxed{4}$$

To use Eq. (4), we need to know the velocity of the water leaving the hole. This is given by **Torricelli's Law**, which was discovered in a slightly different form by the Italian scientist Evangelista Torricelli in 1643. It states that the velocity of water leaving a hole in a tank is equal to the velocity a drop of water would attain if it fell freely from the water level y to the hole (as in the marginal note):

$$\boxed{v(y) = -\sqrt{2gy} = -\sqrt{2(9.8)y} \approx -4.43\sqrt{y} \text{ m/s}} \qquad \boxed{5}$$

Here, we used the fact $g = 9.8$ m/s^2. Notice that the velocity is independent of the shape of the tank.

■ **EXAMPLE 3** **Application of Torricelli's Law** A cylindrical tank of height 4 m and radius 1 m is filled with water. Water drains through a square hole of side 2 cm in the bottom. Determine the water level $y(t)$ at time t (seconds). How long does it take for the tank to go from full to empty?

If we set $K = 0$ in Eq. (3), we obtain the solution $y = 0$. The Separation of Variables procedure did not directly yield this solution because we divided by y (and thus assumed that $y \neq 0$).

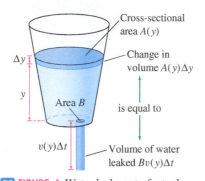

DF **FIGURE 4** Water leaks out of a tank through a hole of area B at the bottom.

Like most if not all mathematical models, our model of water draining from a tank is at best an approximation. The differential equation (4) does not take into account viscosity (resistance of a fluid to flow). This can be remedied by using the differential equation

$$\frac{dy}{dt} = k\frac{Bv(y)}{A(y)}$$

where $k < 1$ is a viscosity constant. Furthermore, Torricelli's Law is valid only when the hole size B is small relative to the cross-sectional areas $A(y)$.

In Eq. (4) of Section 3.4, Galileo's Formula said that an object near the surface of the earth falling under the influence of gravity for a time t falls a distance $y(t) = \frac{1}{2}gt^2$ and reaches a velocity of $v(t) = -gt$. Eliminating the variable t, we find that $v(y) = -\sqrt{2gy}$.

Solution We use units of centimeters.

Step 1. **Write down and solve the differential equation.**

The horizontal cross section of the cylinder is a circle of radius $r = 100$ cm and area $A(y) = \pi r^2 = 10,000\pi$ cm^2 (Figure 5). The hole is a square of side 2 cm and area $B = 4$ cm^2. Since we are using centimeters, we take $g = 980$ cm/s^2 in Eq. (5), which gives us $v(y) = -\sqrt{2(980)y} \approx -44.3\sqrt{y}$ cm/s. Eq. (4) becomes

$$\frac{dy}{dt} = \frac{Bv(y)}{A(y)} = -\frac{4(44.3\sqrt{y})}{10,000\pi} \approx -0.0056\sqrt{y} \qquad \boxed{6}$$

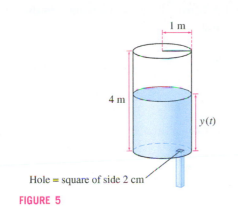

1 m

4 m

$y(t)$

Hole = square of side 2 cm

FIGURE 5

Solve using Separation of Variables:

$$\int \frac{dy}{\sqrt{y}} = -0.0056 \int dt$$

$$2y^{1/2} = -0.0056t + C \qquad \boxed{7}$$

$$y = \left(-0.0028t + \frac{1}{2}C\right)^2$$

Since C is arbitrary, we may replace $\frac{1}{2}C$ by K and write

$$y = (K - 0.0028t)^2$$

Step 2. **Use the initial condition.**

The tank is full at $t = 0$, so we have the initial condition $y(0) = 400$ cm. Thus,

$$y(0) = K^2 = 400 \quad \Rightarrow \quad K = \pm 20$$

Which sign is correct? You might think that both sign choices are possible, but notice that the water level y is a decreasing function of t, and the function $y = (K - 0.0028t)^2$ decreases to 0 only if K is positive. Alternatively, we can see directly from Eq. (7) that $K > 0$, because $2y^{1/2}$ is *nonnegative*. Thus,

$$y(t) = (20 - 0.0028t)^2$$

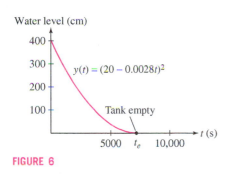

Water level (cm)

400

300

$y(t) = (20 - 0.0028t)^2$

200

100

Tank empty

5000 t_e 10,000 t (s)

FIGURE 6

To determine the time t_e that it takes to empty the tank, we solve

$$y(t_e) = (20 - 0.0028t_e)^2 = 0 \quad \Rightarrow \quad t_e \approx 7142 \text{ s}$$

So, the tank is empty after 7142 s, or nearly 2 hours (Figure 6). ■

CONCEPTUAL INSIGHT The previous example highlights the need to analyze solutions to differential equations rather than relying on algebra alone. The algebra suggested two possible values for K, but further analysis showed that $K = -20$ does not yield a solution for $t \geq 0$. Note also that the function

$$y(t) = (20 - 0.0028t)^2$$

is a solution only for $t \leq t_e$—that is, until the tank is empty. This function cannot satisfy Eq. (6) for $t > t_e$ because its derivative is positive for $t > t_e$ (Figure 6), but solutions of Eq. (6) have nonpositive derivatives.

9.1 SUMMARY

- A differential equation has order n if $y^{(n)}$ is the highest order derivative appearing in the equation.
- A differential equation is *linear* if it can be written as

$$a_n(x)y^{(n)} + a_{n-1}(x)y^{(n-1)} + \cdots + a_1(x)y' + a_0(x)y = b(x)$$

- Separable first-order equation: $\dfrac{dy}{dx} = f(x)g(y)$
- Separation of Variables: (for a separable equation): Move all terms involving y to the left and all terms involving x to the right and integrate

$$\frac{dy}{g(y)} = f(x)\,dx$$

$$\int \frac{dy}{g(y)} = \int f(x)\,dx$$

- Differential equation for water leaking through a hole of area B in a tank of cross-sectional areas $A(y)$:

$$\boxed{\frac{dy}{dt} = \frac{Bv(y)}{A(y)}}$$

Torricelli's Law: $v(y) = -\sqrt{2gy}$, where $g = 9.8$ m/s^2.

9.1 EXERCISES

Preliminary Questions

1. Determine the order of the following differential equations:

(a) $x^5 y' = 1$

(b) $(y')^3 + x = 1$

(c) $y''' + x^4 y' = 2$

(d) $\sin(y'') + x = y$

2. Is $y'' = \sin x$ a linear differential equation?

3. Give an example of a nonlinear differential equation of the form $y' = f(y)$.

4. Can a nonlinear differential equation be separable? If so, give an example.

5. Give an example of a linear, nonseparable differential equation.

Exercises

1. Which of the following differential equations are first-order?

(a) $y' = x^2$

(b) $y'' = y^2$

(c) $(y')^3 + yy' = \sin x$

(d) $x^2 y' - e^x y = \sin y$

(e) $y'' + 3y' = \dfrac{y}{x}$

(f) $yy' + x + y = 0$

2. Which of the equations in Exercise 1 are linear?

In Exercises 3–8, verify that the given function is a solution of the differential equation.

3. $y' - 8x = 0, \quad y = 4x^2$

4. $yy' + 4x = 0, \quad y = \sqrt{12 - 4x^2}$

5. $y' + 4xy = 0, \quad y = 25e^{-2x^2}$

6. $(x^2 - 1)y' + xy = 0, \quad y = 4(x^2 - 1)^{-1/2}$

7. $y'' - 2xy' + 8y = 0, \quad y = 4x^4 - 12x^2 + 3$

8. $y'' - 2y' + 5y = 0, \quad y = e^x \sin 2x$

9. Which of the following equations are separable? Write those that are separable in the form $y' = f(x)g(y)$ (but do not solve).

(a) $xy' - 9y^2 = 0$

(b) $\sqrt{4 - x^2}\,y' = e^{3y} \sin x$

(c) $y' = x^2 + y^2$

(d) $y' = 9 - y^2$

10. The following differential equations appear similar but have very different solutions:

$$\frac{dy}{dx} = x, \qquad \frac{dy}{dx} = y$$

Solve both subject to the initial condition $y(1) = 2$.

11. The following differential equations appear similar but have very different solutions:

$$\frac{dy}{dx} = x^2, \qquad \frac{dy}{dx} = y^2$$

Solve both subject to the initial condition $y(0) = 1$.

12. Consider the differential equation $y^3 y' - 9x^2 = 0$.

(a) Write it as $y^3\,dy = 9x^2\,dx$.

(b) Integrate both sides to obtain $\frac{1}{4}y^4 = 3x^3 + C$.

(c) Verify that $y = (12x^3 + C)^{1/4}$ is the general solution.

(d) Find the particular solution satisfying $y(1) = 2$.

13. Verify that $x^2 y' + e^{-y} = 0$ is separable.

(a) Write it as $e^y\,dy = -x^{-2}\,dx$.

(b) Integrate both sides to obtain $e^y = x^{-1} + C$.

(c) Verify that $y = \ln(x^{-1} + C)$ is the general solution.

(d) Find the particular solution satisfying $y(2) = 4$.

In Exercises 14–30, use Separation of Variables to find the general solution.

14. $y' + 4xy^2 = 0$

15. $y' + x^2 y = 0$

16. $y' - e^{x+y} = 0$

17. $\dfrac{dy}{dt} - 20t^4 e^{-y} = 0$

18. $t^3 y' + 4y^2 = 0$

19. $2y' + 5y = 4$

20. $\dfrac{dy}{dt} = 8\sqrt{y}$

21. $\sqrt{1 - x^2}\,y' = xy$

22. $y' = y^2(1 - x^2)$

23. $yy' = x$

24. $(\ln y)y' - ty = 0$

25. $\dfrac{dx}{dt} = (t + 1)(x^2 + 1)$

26. $(1 + x^2)y' = x^3 y$

27. $y' = x \sec y$

28. $\dfrac{dy}{d\theta} = \tan y$

29. $\dfrac{dy}{dt} = y \tan t$

30. $\dfrac{dx}{dt} = t \tan x$

In Exercises 31–44, solve the Initial Value Problem.

31. $y' + 2y = 0$, $y(\ln 5) = 3$

32. $y' - 3y + 12 = 0$, $y(2) = 1$

33. $yy' = xe^{-y^2}$, $y(0) = -2$ **34.** $y^2 \dfrac{dy}{dx} = x^{-3}$, $y(1) = 0$

35. $y' = (x - 1)(y - 2)$, $y(2) = 4$

36. $y' = (x - 1)(y - 2)$, $y(2) = 2$

37. $y' = x(y^2 + 1)$, $y(0) = 0$

38. $(1 - t)\dfrac{dy}{dt} - y = 0$, $y(2) = -4$

39. $\dfrac{dy}{dt} = ye^{-t}$, $y(0) = 1$ **40.** $\dfrac{dy}{dt} = te^{-y}$, $y(1) = 0$

41. $t^2 \dfrac{dy}{dt} - t = 1 + y + ty$, $y(1) = 0$

42. $\sqrt{1 - x^2}\, y' = y^2 + 1$, $y(0) = 0$

43. $y' = \tan y$, $y(\ln 2) = \dfrac{\pi}{2}$ **44.** $y' = y^2 \sin x$, $y(\pi) = 2$

45. Find all values of a such that $y = x^a$ is a solution of

$$y'' - 12x^{-2}y = 0$$

46. Find all values of a such that $y = e^{ax}$ is a solution of

$$y'' + 4y' - 12y = 0$$

In Exercises 47 and 48, let $y(t)$ be a solution of $(\cos y + 1)\dfrac{dy}{dt} = 2t$ such that $y(2) = 0$.

47. Show that $\sin y + y = t^2 + C$. We cannot solve for y as a function of t, but assuming that $y(2) = 0$, find the values of t at which $y(t) = \pi$.

48. Assuming that $y(6) = \pi/3$, find an equation of the tangent line to the graph of $y(t)$ at $(6, \pi/3)$.

In Exercises 49–54, use Eq. (4) and Torricelli's Law [Eq. (5)].

49. Water leaks through a hole of area $B = 0.002$ m^2 at the bottom of a cylindrical tank that is filled with water and has height 3 m and a base of area 10 m^2. How long does it take (a) for half of the water to leak out and (b) for the tank to empty?

50. At $t = 0$, a conical tank of height 300 cm and top radius 100 cm [Figure 7(A)] is filled with water. Water leaks through a hole in the bottom of area $B = 3$ cm^2. Let $y(t)$ be the water level at time t.

(a) Show that the tank's cross-sectional area at height y is $A(y) = \frac{\pi}{9}y^2$.

(b) Find and solve the differential equation satisfied by $y(t)$.

(c) How long does it take for the tank to empty?

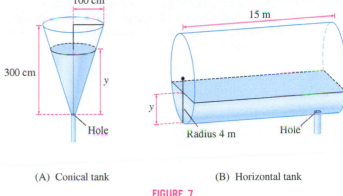

(A) Conical tank   (B) Horizontal tank

FIGURE 7

51. The tank in Figure 7(B) is a cylinder of radius 4 m and height 15 m. Assume that the tank is half-filled with water and that water leaks through a hole in the bottom of area $B = 0.001$ m^2. Determine the water level $y(t)$ and the time t_e when the tank is empty.

52. A tank has the shape of the parabola $y = x^2$, revolved around the y-axis. Water leaks from a hole of area $B = 0.0005$ m^2 at the bottom of the tank. Let $y(t)$ be the water level at time t. How long does it take for the tank to empty if it is initially filled to height $y_0 = 1$ m?

53. A tank has the shape of the parabola $y = ax^2$ (where a is a constant) revolved around the y-axis. Water drains from a hole of area B m^2 at the bottom of the tank.

(a) Show that the water level at time t is

$$y(t) = \left(y_0^{3/2} - \frac{3aB\sqrt{2g}}{2\pi}t\right)^{2/3}$$

where y_0 is the water level at time $t = 0$.

(b) Show that if the total volume of water in the tank has volume V at time $t = 0$, then $y_0 = \sqrt{2aV/\pi}$. *Hint:* Compute the volume of the tank as a volume of rotation.

(c) Show that the tank is empty at time

$$t_e = \left(\frac{2}{3B\sqrt{g}}\right)\left(\frac{2\pi V^3}{a}\right)^{1/4}$$

We see that for fixed initial water volume V, the time t_e is proportional to $a^{-1/4}$. A large value of a corresponds to a tall thin tank. Such a tank drains more quickly than a short wide tank of the same initial volume.

54. A cylindrical tank filled with water has height h and a base of area A. Water leaks through a hole in the bottom of area B.

(a) Show that the time required for the tank to empty is proportional to $A\sqrt{h}/B$.

(b) Show that the emptying time is proportional to $Vh^{-1/2}$, where V is the volume of the tank.

(c) Two tanks have the same volume and a hole of the same size, but they have different heights and bases. Which tank empties first: the taller or the shorter tank?

55. Figure 8 shows a circuit consisting of a resistor of R ohms, a capacitor of C farads, and a battery of voltage V. When the circuit is completed, the amount of charge $q(t)$ (in coulombs) on the plates of the capacitor varies according to the differential equation (t in seconds)

$$R\frac{dq}{dt} + \frac{1}{C}q = V$$

where R, C, and V are constants.

(a) Solve for $q(t)$, assuming that $q(0) = 0$.

(b) Sketch the graph of q.

(c) Show that $\lim_{t\to\infty} q(t) = CV$.

(d) Show that the capacitor charges to approximately 63% of its final value CV after a time period of length $\tau = RC$ (τ is called the time constant of the capacitor).

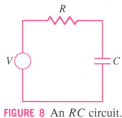

FIGURE 8 An RC circuit.

56. Assume in the circuit of Figure 8 that $R = 200$ ohms, $C = 0.02$ farad, and $V = 12$ volts. How many seconds does it take for the charge on the capacitor plates to reach half of its limiting value?

57. 📖   According to one hypothesis, the growth rate dV/dt of a cell's volume V is proportional to its surface area A. Since V has cubic units such as cm^3 and A has square units such as cm^2, we may assume roughly that $A \propto V^{2/3}$, and hence $dV/dt = kV^{2/3}$ for some constant k. If this hypothesis is correct, which dependence of volume on time would we expect to see (again, roughly speaking) in the laboratory?

(a) Linear   **(b)** Quadratic   **(c)** Cubic

58. We might also guess that the volume V of a melting snowball decreases at a rate proportional to its surface area. Argue as in Exercise 57 to find a differential equation satisfied by V. Suppose the snowball has volume 1000 cm^3 and that it loses half of its volume after 5 minutes. According to this model, when will the snowball disappear?

59. In general, $(fg)'$ is not equal to $f'g'$, but let $f(x) = e^{3x}$ and find a function g such that $(fg)' = f'g'$. Do the same for $f(x) = x$.

60. A boy standing at point B on a dock holds a rope of length ℓ attached to a boat at point A [Figure 9(A)]. As the boy walks along the dock, holding the rope taut, the boat moves along a curve called a **tractrix** (from the Latin *tractus*, meaning "pulled"). The segment from a point P on the curve to the x-axis along the tangent line has constant length ℓ. Let $y = f(x)$ be the equation of the tractrix.

(a) Show that $y^2 + (y/y')^2 = \ell^2$ and conclude $y' = -\dfrac{y}{\sqrt{\ell^2 - y^2}}$. Why must we choose the negative square root?

(b) Prove that the tractrix is the graph of

$$x = \ell \ln\left(\frac{\ell + \sqrt{\ell^2 - y^2}}{y}\right) - \sqrt{\ell^2 - y^2}$$

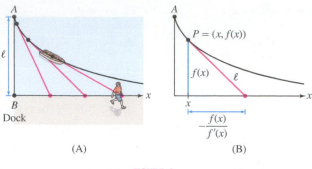

(A)   (B)

FIGURE 9

61. Show that the differential equations $y' = 3y/x$ and $y' = -x/3y$ define **orthogonal families** of curves; that is, the graphs of solutions to the first equation intersect the graphs of the solutions to the second equation in right angles (Figure 10). Find these curves explicitly.

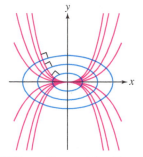

FIGURE 10 Two orthogonal families of curves.

62. Find the family of curves satisfying $y' = x/y$ and sketch several members of the family. Then find the differential equation for the orthogonal family (see Exercise 61), find its general solution, and add some members of this orthogonal family to your plot.

63. A 50-kg model rocket lifts off by expelling fuel downward at a rate of $k = 4.75$ kg/s for 10 s. The fuel leaves the end of the rocket with an exhaust velocity of $b = -100$ m/s. Let $m(t)$ be the mass of the rocket at time t. From the law of conservation of momentum, we find the following differential equation for the rocket's velocity $v(t)$ (in meters per second):

$$m(t)v'(t) = -9.8m(t) + b\frac{dm}{dt}$$

(a) Show that $m(t) = 50 - 4.75t$ kg.

(b) Solve for $v(t)$ and compute the rocket's velocity at rocket burnout (after 10 s).

64. Let $v(t)$ be the velocity of an object of mass m in free-fall near the earth's surface. If we assume that air resistance is proportional to v^2, then v satisfies the differential equation $m\dfrac{dv}{dt} = -g + kv^2$ for some constant $k > 0$.

(a) Set $\alpha = (g/k)^{1/2}$ and rewrite the differential equation as

$$\frac{dv}{dt} = -\frac{k}{m}(\alpha^2 - v^2)$$

Then solve using Separation of Variables with initial condition $v(0) = 0$.

(b) Show that the terminal velocity $\lim_{t\to\infty} v(t)$ is equal to $-\alpha$.

65. If a bucket of water spins about a vertical axis with constant angular velocity ω (in radians per second), the water climbs up the side of the bucket until it reaches an equilibrium position (Figure 11). Two forces act on a particle located at a distance x from the vertical axis: the gravitational force $-mg$ acting downward and the force of the bucket on the particle (transmitted indirectly through the liquid) in the direction perpendicular to the surface of the water. These two forces must combine to supply a centripetal force $m\omega^2 x$, and this occurs if the diagonal of the rectangle in Figure 11 is normal to the water's surface (i.e., perpendicular to the tangent line). Prove that if $y = f(x)$ is the equation of the curve obtained by taking a vertical cross section through the axis, then $-1/y' = -g/(\omega^2 x)$. Show that $y = f(x)$ is a parabola.

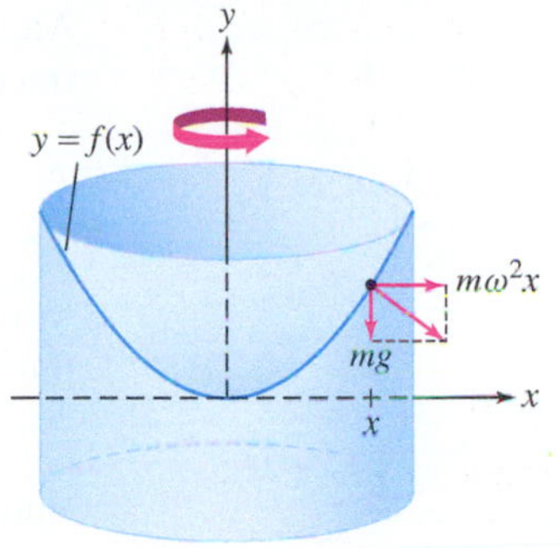

FIGURE 11

Further Insights and Challenges

66. In Section 6.2, we computed the volume V of a solid as the integral of cross-sectional area. Explain this formula in terms of differential equations. Let $V(y)$ be the volume of the solid up to height y, and let $A(y)$ be the cross-sectional area at height y as in Figure 12.

(a) Explain the following approximation for small Δy:

$$V(y + \Delta y) - V(y) \approx A(y)\,\Delta y \qquad \boxed{8}$$

(b) Use Eq. (8) to justify the differential equation $dV/dy = A(y)$. Then derive the formula

$$V = \int_a^b A(y)\,dy$$

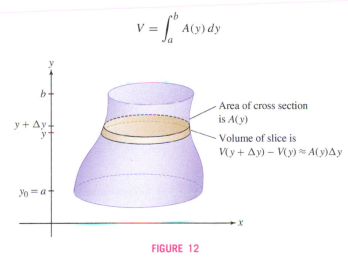

FIGURE 12

67. A basic theorem states that a *linear* differential equation of order n has a general solution that depends on n arbitrary constants. The following examples show that, in general, the theorem does not hold for nonlinear differential equations.

(a) Show that $(y')^2 + y^2 = 0$ is a first-order equation with only one solution $y = 0$.

(b) Show that $(y')^2 + y^2 + 1 = 0$ is a first-order equation with no solutions.

68. Show that $y = Ce^{rx}$ is a solution of $y'' + ay' + by = 0$ if and only if r is a root of $P(r) = r^2 + ar + b$. Then verify directly that $y = C_1 e^{3x} + C_2 e^{-x}$ is a solution of $y'' - 2y' - 3y = 0$ for any constants C_1, C_2.

69. A spherical tank of radius R is half-filled with water. Suppose that water leaks through a hole in the bottom of area B. Let $y(t)$ be the water level at time t (seconds).

(a) Show that $\dfrac{dy}{dt} = \dfrac{\sqrt{2g}\,B\sqrt{y}}{\pi(2Ry - y^2)}$.

(b) Show that for some constant C,

$$\frac{2\pi}{15B\sqrt{2g}}\left(10Ry^{3/2} - 3y^{5/2}\right) = C - t$$

(c) Use the initial condition $y(0) = R$ to compute C, and show that $C = t_e$, the time at which the tank is empty.

(d) Show that t_e is proportional to $R^{5/2}$ and inversely proportional to B.

9.2 Models Involving $y' = k(y - b)$

In Theorem 1 of Section 5.9, we saw a quantity grow or decay exponentially if its *rate of change* is proportional to the amount present. This characteristic property is expressed by the differential equation $y' = ky$. We now study the closely related differential equation

$$\frac{dy}{dt} = k(y - b) \qquad \boxed{1}$$

*Every first-order, linear differential equation with **constant coefficients** can be written in the form of Eq. (1). This equation is used to model a variety of phenomena, such as the cooling process, free-fall with air resistance, and current in a circuit.*

where k and b are constants. This differential equation describes a quantity y whose *rate of change is proportional to the difference* $y - b$. We can use Separation of Variables to show that the general solution is

$$y(t) = b + Ce^{kt} \qquad \boxed{2}$$

Alternatively, we may observe that $(y - b)' = y'$ since b is a constant, so Eq. (1) may be rewritten

$$\frac{d}{dt}(y - b) = k(y - b)$$

In other words, $y - b$ satisfies the differential equation of an exponential function and thus $y - b = Ce^{kt}$, or $y = b + Ce^{kt}$, as claimed.

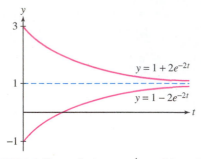

$y = 1 + 2e^{-2t}$

$y = 1 - 2e^{-2t}$

FIGURE 1 Two solutions to $y' = -2(y - 1)$ corresponding to $C = 2$ and $C = -2$.

GRAPHICAL INSIGHT   The behavior of the solution $y(t)$ as $t \to \infty$ depends on whether C and k are positive or negative:

- When $k > 0$, e^{kt} tends to ∞ and, therefore, $y(t)$ tends to ∞ if $C > 0$ and $y(t)$ tends to $-\infty$ if $C < 0$.
- When $k < 0$, we usually rewrite the differential equation as $y' = -k(y - b)$ with $k > 0$. In this case, $y(t) = b + Ce^{-kt}$ and $y(t)$ approaches the horizontal asymptote $y = b$ since Ce^{-kt} tends to zero as $t \to \infty$ (Figure 1). Note that $y(t)$ approaches the asymptote from above or below, depending on whether $C > 0$ or $C < 0$.

We now consider some applications of Eq. (1), beginning with Newton's Law of Cooling. Let $y(t)$ be the temperature of a hot object that is cooling off in an environment where the ambient temperature is T_0. Newton assumed that the *rate of cooling* is proportional to the temperature difference $y - T_0$. We express this hypothesis in a precise way by the differential equation

Newton's Law of Cooling implies that the object cools quickly when it is much hotter than its surroundings (when $y - T_0$ is large). The rate of cooling slows as y approaches T_0. When the object's initial temperature is less than T_0, y' is positive and Newton's Law models warming.

$$\boxed{y' = -k(y - T_0) \qquad (T_0 = \text{ambient temperature})}$$

⟵·· REMINDER *The differential equation*

$$\frac{dy}{dt} = k(y - b)$$

has the general solution

$$y = b + Ce^{kt}$$

The constant k, in units of $(\text{time})^{-1}$, is called the **cooling constant** and depends on the physical properties of the object.

■ EXAMPLE 1 **Newton's Law of Cooling** A hot metal bar with cooling constant $k = 2.1 \text{ min}^{-1}$ is submerged in a large tank of water held at temperature $T_0 = 10°C$. Let $y(t)$ be the bar's temperature at time t (in minutes).

(a) Find the differential equation satisfied by $y(t)$ and find its general solution.

(b) What is the bar's temperature after 1 min if its initial temperature was $180°C$?

(c) What was the bar's initial temperature if it cooled to $80°C$ in 30 seconds?

Solution

(a) Since $k = 2.1 \text{ min}^{-1}$, $y(t)$ (with t in minutes) satisfies

$$y' = -2.1(y - 10)$$

By Eq. (2), the general solution is $y(t) = 10 + Ce^{-2.1t}$ for some constant C.

(b) If the initial temperature was $180°C$, then $y(0) = 10 + C = 180$. Thus, $C = 170$ and $y(t) = 10 + 170e^{-2.1t}$ (Figure 2). After 1 min,

$$y(1) = 10 + 170e^{-2.1(1)} \approx 30.8°C$$

(c) If the temperature after 30 s is $80°C$, then $y(0.5) = 80$, and we have

$$10 + Ce^{-2.1(0.5)} = 80 \quad \Rightarrow \quad Ce^{-1.05} = 70 \quad \Rightarrow \quad C = 70e^{1.05} \approx 200$$

It follows that $y(t) = 10 + 200e^{-2.1t}$ and the initial temperature was

$$y(0) = 10 + 200e^{-2.1(0)} = 10 + 200 = 210°C \qquad ■$$

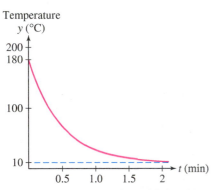

Temperature y (°C)

FIGURE 2 Temperature of metal bar as it cools when the ambient temperature is $10°C$.

The effect of air resistance depends on the physical situation. A high-speed bullet is affected differently than a skydiver. Our model is fairly realistic for a large object such as a skydiver falling from high altitudes.

The differential equation $y' = k(y - b)$ is also used to model free-fall when air resistance is taken into account. Assume that the force due to air resistance is proportional to the velocity v and acts opposite to the direction of the fall. We write this force as $-kv$, where $k > 0$. We take the upward direction to be positive, so $v < 0$ for a falling object and $-kv$ is an upward acting force.

The force due to gravity on a falling object of mass m is $-mg$, where g is the acceleration due to gravity, so the total force is $F = -mg - kv$. By Newton's Law,

$$F = ma = mv' \qquad (a = v' \text{ is the acceleration})$$

Thus, $mv' = -mg - kv$, which can be written as

In this model of free-fall, k has units of mass per time, such as kilograms per second.

$$v' = -\frac{k}{m}\left(v + \frac{mg}{k}\right) \qquad \boxed{3}$$

This equation has the form $v' = -k(v - b)$ with k replaced by k/m and $b = -mg/k$. By Eq. (2), the general solution is

$$v(t) = -\frac{mg}{k} + Ce^{-(k/m)t} \qquad \boxed{4}$$

Since $Ce^{-(k/m)t}$ tends to zero as $t \to \infty$, $v(t)$ tends to a limiting terminal velocity:

$$\text{Terminal velocity} = \lim_{t\to\infty} v(t) = -\frac{mg}{k} \qquad \boxed{5}$$

Without air resistance, the velocity would increase indefinitely until a sudden collision with the ground occurs.

Skydiver in free-fall. (*Hector Mandel/iStockphoto*)

■ **EXAMPLE 2** An 80-kg skydiver steps out of an airplane.

(a) What is her terminal velocity if $k = 8$ kg/s?

(b) What is her velocity after 30 s?

Solution

(a) By Eq. (5), with $k = 8$ kg/s and $g = 9.8$ m/s^2, the terminal velocity is

$$-\frac{mg}{k} = -\frac{(80)9.8}{8} = -98 \text{ m/s}$$

(b) With t in seconds, we have, by Eq. (4),

$$v(t) = -98 + Ce^{-(k/m)t} = -98 + Ce^{-(8/80)t} = -98 + Ce^{-0.1t}$$

We assume that the skydiver leaves the airplane with no initial vertical velocity, so $v(0) = -98 + C = 0$, and $C = 98$. Thus, we have $v(t) = -98(1 - e^{-0.1t})$ (Figure 3). The skydiver's velocity after 30 s is

$$v(30) = -98(1 - e^{-0.1(30)}) \approx -93.1 \text{ m/s} \qquad ■$$

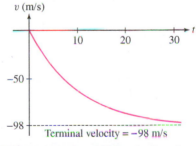
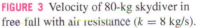

FIGURE 3 Velocity of 80-kg skydiver in free-fall with air resistance ($k = 8$ kg/s).

An **annuity** is an investment in which an amount of money P_0, called the principal, is placed in an account that earns interest (that is compounded at regular intervals) at a rate r, and money is withdrawn at regular intervals. If the compounding and withdrawals occur over short enough intervals, we can assume continuous compounding at rate r and continuous withdrawals at a rate of N dollars per year. This allows us to model the annuity by a differential equation. Let $P(t)$ be the balance in the annuity after t years. Then

Notice in Eq. (6) that $P'(t)$ is determined by the growth rate r and the withdrawal rate N. If no withdrawals occurred, $P(t)$ would grow with compound interest and would satisfy $P'(t) = rP(t)$.

$$\underbrace{P'(t)}_{\substack{\text{Rate of} \\ \text{change}}} = \underbrace{rP(t)}_{\substack{\text{Growth due} \\ \text{to interest}}} - \underbrace{N}_{\substack{\text{Withdrawal} \\ \text{rate}}} = r\left(P(t) - \frac{N}{r}\right) \qquad \boxed{6}$$

This equation has the form $y' = k(y - b)$ with $k = r$ and $b = N/r$, so by Eq. (2), the general solution is

$$P(t) = \frac{N}{r} + Ce^{rt} \qquad \boxed{7}$$

Since $P(0) = P_0$, we know $P_0 = \dfrac{N}{r} + C$ and, therefore, C is given by $C = P_0 - \dfrac{N}{r}$. Because e^{rt} tends to infinity as $t \to \infty$, the balance $P(t)$ tends to ∞ if $C > 0$. If $C < 0$, then $P(t)$ tends to $-\infty$ (i.e., the annuity eventually runs out of money). If $C = 0$, then $P(t)$ remains constant with value N/r.

■ **EXAMPLE 3 Does an Annuity Pay Out Forever?** An annuity earns interest at the rate $r = 0.07$ (7% per year), and withdrawals are made continuously at a rate of $N = \$500$/year.

(a) When will the annuity run out of money if the initial deposit is $P(0) = \$5000$?

(b) Show that the balance increases indefinitely if $P(0) = \$9000$.

Solution We have $N/r = \frac{500}{0.07} \approx 7143$, so $P(t) = 7143 + Ce^{0.07t}$ by Eq. (7).

(a) If $P(0) = 5000 = 7143 + Ce^0$, then $C = -2143$ and

$$P(t) = 7143 - 2143e^{0.07t}$$

The account runs out of money when $P(t) = 7143 - 2143e^{0.07t} = 0$, or

$$e^{0.07t} = \frac{7143}{2143} \quad \Rightarrow \quad 0.07t = \ln\left(\frac{7143}{2143}\right) \approx 1.2$$

The annuity money runs out at time $t = \frac{1.2}{0.07} \approx 17$ years.

(b) If $P(0) = 9000 = 7143 + Ce^0$, then $C = 1857$ and

$$P(t) = 7143 + 1857e^{0.07t}$$

Since the coefficient $C = 1857$ is positive, the account never runs out of money. In fact, $P(t)$ increases indefinitely as $t \to \infty$. Figure 4 illustrates the two cases.   ■

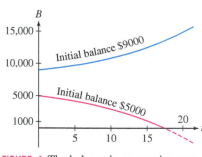

FIGURE 4 The balance in an annuity may increase indefinitely or decrease to zero (eventually becoming negative), depending on the size of initial deposit P_0, the interest rate, and the rate of withdrawal.

9.2 SUMMARY

- The general solution of $y' = k(y - b)$ is $y = b + Ce^{kt}$, where C is a constant.
- The following table describes the solutions to $y' = k(y - b)$ (see Figure 5):

Equation ($k > 0$)	Solution	Behavior as $t \to \infty$
$y' = k(y - b)$	$y(t) = b + Ce^{kt}$	$\displaystyle\lim_{t\to\infty} y(t) = \begin{cases} \infty & \text{if } C > 0 \\ -\infty & \text{if } C < 0 \end{cases}$
$y' = -k(y - b)$	$y(t) = b + Ce^{-kt}$	$\displaystyle\lim_{t\to\infty} y(t) = b$

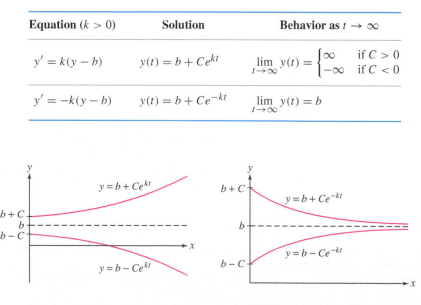

Solutions to $y' = k(y - b)$ with $k, C > 0$ Solutions to $y' = -k(y - b)$ with $k, C > 0$

FIGURE 5

- Three applications:

  - Newton's Law of Cooling: $y' = -k(y - T_0)$, $y(t)$ = temperature of the object, T_0 = ambient temperature, k = cooling constant
  - Free-fall with air resistance: $v' = -\dfrac{k}{m}\left(v + \dfrac{mg}{k}\right)$, $v(t)$ = velocity, m = mass, k = air resistance constant, g = acceleration due to gravity
  - Continuous annuity: $P' = r\left(P - \dfrac{N}{r}\right)$, $P(t)$ = balance in the annuity, r = interest rate, N = withdrawal rate

9.2 EXERCISES

Preliminary Questions

1. Write down a solution to $y' = 4(y - 5)$ that tends to $-\infty$ as $t \to \infty$.

2. Does $y' = -4(y - 5)$ have a solution that tends to ∞ as $t \to \infty$?

3. True or false? If $k > 0$, then all solutions of $y' = -k(y - b)$ approach the same limit as $t \to \infty$.

4. As an object cools, its rate of cooling slows. Explain how this follows from Newton's Law of Cooling.

Exercises

1. Find the general solution of $y' = 2(y - 10)$. Then find the two solutions satisfying $y(0) = 25$ and $y(0) = 5$, and sketch their graphs.

2. Verify directly that $y = 12 + Ce^{-3t}$ satisfies $y' = -3(y - 12)$ for all C. Then find the two solutions satisfying $y(0) = 20$ and $y(0) = 0$, and sketch their graphs.

3. Solve $y' = 4y + 24$ subject to $y(0) = 5$.

4. Solve $y' + 6y = 12$ subject to $y(2) = 10$.

In Exercises 5–12, use Newton's Law of Cooling.

5. A hot anvil with cooling constant $k = 0.02 \text{ s}^{-1}$ is submerged in a large pool of water whose temperature is $10°C$. Let $y(t)$ be the anvil's temperature t seconds later.

(a) What is the differential equation satisfied by $y(t)$?

(b) Find a formula for $y(t)$, assuming the object's initial temperature is $100°C$.

(c) How long does it take the object to cool down to $20°$?

6. Frank's automobile engine runs at $100°C$. On a day when the outside temperature is $21°C$, he turns off the ignition and notes that 5 minutes later, the engine has cooled to $70°C$.

(a) Determine the engine's cooling constant k.

(b) What is the formula for $y(t)$?

(c) When will the engine cool to $40°C$?

7. At 10:30 AM, detectives discover a dead body in a room and measure its temperature at $26°C$. One hour later, the body's temperature had dropped to $24.8°C$. Determine the time of death (when the body temperature was a normal $37°C$), assuming that the temperature in the room was held constant at $20°C$.

8. A cup of coffee with cooling constant $k = 0.09 \text{ min}^{-1}$ is placed in a room at temperature $20°C$.

(a) How fast is the coffee cooling (in degrees per minute) when its temperature is $T = 80°C$?

(b) Use the Linear Approximation to estimate the change in temperature over the next 6 s when $T = 80°C$.

(c) If the coffee is served at $90°C$, how long will it take to reach an optimal drinking temperature of $65°C$?

9. A cold metal bar at $-30°C$ is submerged in a pool maintained at a temperature of $40°C$. Half a minute later, the temperature of the bar is $20°C$. How long will it take for the bar to attain a temperature of $30°C$?

10. When a hot object is placed in a water bath whose temperature is $25°C$, it cools from 100 to $50°C$ in 150 s. In another bath, the same cooling occurs in 120 s. Find the temperature of the second bath.

11. ⎡GU⎤   Objects A and B are placed in a warm bath at temperature $T_0 = 40°C$. Object A has initial temperature $-20°C$ and cooling constant $k = 0.004 \text{ s}^{-1}$. Object B has initial temperature $0°C$ and cooling constant $k = 0.002 \text{ s}^{-1}$. Plot the temperatures of A and B for $0 \le t \le 1000$. After how many seconds will the objects have the same temperature?

12. In Newton's Law of Cooling, the constant $\tau = 1/k$ is called the "characteristic time." Show that τ is the time required for the temperature difference $(y - T_0)$ to decrease by the factor $e^{-1} \approx 0.37$. For example, if $y(0) = 100°C$ and $T_0 = 0°C$, then the object cools to $100/e \approx 37°C$ in time τ, to $100/e^2 \approx 13.5°C$ in time 2τ, and so on.

In Exercises 13–16, use Eq. (3) as a model for free-fall with air resistance.

13. A 60-kg skydiver jumps out of an airplane. What is her terminal velocity, in meters per second, assuming that $k = 10 \text{ kg/s}$ for free-fall (no parachute)?

14. Find the terminal velocity of a skydiver of weight $w = 192$ pounds (lb) if $k = 1.2 \text{ lb-s/ft}$. How long does it take him to reach half of his terminal velocity if his initial velocity is zero? Mass and weight are related by $w = mg$, and Eq. (3) becomes $v' = -(kg/w)(v + w/k)$ with $g = 32 \text{ ft/s}^2$.

15. An 80-kg skydiver jumps out of an airplane (with zero initial velocity). Assume that $k = 12 \text{ kg/s}$ with a closed parachute and $k = 70 \text{ kg/s}$ with an open parachute. What is the skydiver's velocity at $t = 25$ s if the parachute opens after 20 s of free-fall?

16. ⎡◿⎤   Does a heavier or a lighter skydiver reach terminal velocity more quickly?

In Exercises 17(a)–(f), use Formulas 6 and 7 that describe the balance paid by a continuous annuity.

17. (a) A continuous annuity with withdrawal rate $N = \$5000$/year and interest rate $r = 5\%$ is funded by an initial deposit of $P_0 = \$50,000$.

  i. What is the balance in the annuity after 10 years?

  ii. When will the annuity run out of funds?

(b) Show that a continuous annuity with withdrawal rate $N = \$5000$/year and interest rate $r = 8\%$, funded by an initial deposit of $P_0 = \$75,000$, never runs out of money.

(c) Find the minimum initial deposit P_0 that will allow an annuity to pay out $\$6000$/year indefinitely if it earns interest at a rate of 5%.

(d) Find the minimum initial deposit P_0 necessary to fund an annuity for 20 years if withdrawals are made at a rate of $\$10,000$/year and interest is earned at a rate of 7%.

(e) An initial deposit of 100,000 euros is placed in an annuity with a French bank. What is the minimum interest rate the annuity must earn to allow withdrawals at a rate of 8000 euros/year to continue indefinitely?

(f) Show that a continuous annuity never runs out of money if the initial balance is greater than or equal to N/r, where N is the withdrawal rate and r the interest rate.

18. 📖   Sam borrows $\$10,000$ from a bank at an interest rate of 9% and pays back the loan continuously at a rate of N dollars per year. Let $P(t)$ denote the amount still owed at time t.

(a) Explain why $P(t)$ satisfies the differential equation

$$y' = 0.09y - N$$

(b) How long will it take Sam to pay back the loan if $N = \$1200$?

(c) Will the loan ever be paid back if $N = \$800$?

19. April borrows $\$18,000$ at an interest rate of 5% to purchase a new automobile. At what rate (in dollars per year) must she pay back the loan, if the loan must be paid off in 5 years? *Hint:* Set up the differential equation as in Exercise 18.

20. Let $N(t)$ be the fraction of the population who have heard a given piece of news t hours after its initial release. According to one model, the rate $N'(t)$ at which the news spreads is equal to k times the fraction of the population that has not yet heard the news, for some constant $k > 0$.

(a) Determine the differential equation satisfied by $N(t)$.

(b) Find the solution of this differential equation with the initial condition $N(0) = 0$ in terms of k.

(c) Suppose that half of the population is aware of an earthquake 8 hours (h) after it occurs. Use the model to calculate k and estimate the percentage that will know about the earthquake 12 h after it occurs.

21. Current in a Circuit   When the circuit in Figure 6 (which consists of a battery of V volts, a resistor of R ohms, and an inductor of L henries) is connected, the current $I(t)$ flowing in the circuit satisfies

$$L\frac{dI}{dt} + RI = V$$

with the initial condition $I(0) = 0$.

(a) Find a formula for $I(t)$ in terms of L, V, and R.

(b) Show that $\lim\limits_{t \to \infty} I(t) = V/R$.

(c) Show that $I(t)$ reaches approximately 63% of its maximum value at the "characteristic time" $\tau = L/R$.

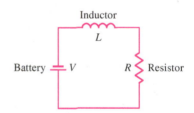

FIGURE 6 Current flow approaches the level $I_{\max} = V/R$.

Further Insights and Challenges

22. Show that the cooling constant of an object can be determined from two temperature readings $y(t_1)$ and $y(t_2)$ at times $t_1 \neq t_2$ by the formula

$$k = \frac{1}{t_1 - t_2} \ln\left(\frac{y(t_2) - T_0}{y(t_1) - T_0}\right)$$

23. Show that by Newton's Law of Cooling, the time required to cool an object from temperature A to temperature B is

$$t = \frac{1}{k}\ln\left(\frac{A - T_0}{B - T_0}\right)$$

where T_0 is the ambient temperature.

24. Air Resistance   A projectile of mass $m = 1$ travels straight up from ground level with initial velocity v_0. Suppose that the velocity v satisfies $v' = -g - kv$.

(a) Find a formula for $v(t)$.

(b) Show that the projectile's height $h(t)$ is given by

$$h(t) = C(1 - e^{-kt}) - \frac{g}{k}t$$

where $C = k^{-2}(g + kv_0)$.

(c) Show that the projectile reaches its maximum height at time $t_{\max} = k^{-1}\ln(1 + kv_0/g)$.

(d) In the absence of air resistance, the maximum height is reached at time $t = v_0/g$. In view of this, explain why we should expect that

$$\lim_{k \to 0}\frac{\ln\left(1 + \frac{kv_0}{g}\right)}{k} = \frac{v_0}{g} \qquad \boxed{8}$$

(e) Verify Eq. (8). *Hint:* Use Theorem 2 in Section 5.9 to show that

$$\lim_{k \to 0}\left(1 + \frac{kv_0}{g}\right)^{1/k} = e^{v_0/g} \text{ or use L'Hôpital's Rule.}$$

9.3 Graphical and Numerical Methods

In the previous two sections, we focused on finding solutions to differential equations. However, most differential equations cannot be solved explicitly. Fortunately, there are techniques for analyzing the solutions that do not rely on explicit formulas. In this section, we discuss the method of slope fields, which provides us with a good visual understanding of first-order equations. We also discuss Euler's Method for finding numerical approximations to solutions.

We use t as the independent variable. A first-order differential equation can then be written in the form

$$\frac{dy}{dt} = F(t, y)$$

1

where $F(t, y)$ is a function of t and y. In other words, the slope of the tangent line to the graph of $y = y(t)$ at a point (t, y) is given by $F(t, y)$.

It is useful to think of Eq. (1) as a set of instructions that "tells a solution" which direction to go in. Thus, a solution passing through a point (t, y) is "instructed" to continue in the direction of slope $F(t, y)$. To visualize this set of instructions, we draw a **slope field**, which is an array of small segments of slope $F(t, y)$ at points (t, y) lying on a rectangular grid in the plane.

To illustrate, let's return to the differential equation:

$$\frac{dy}{dt} = -ty$$

In this case, $F(t, y) = -ty$. According to Example 2 of Section 9.1, the general solution is $y = Ke^{-t^2/2}$. Figure 1(A) shows segments of slope $-ty$ at points (t, y) along the graph of a particular solution $y(t)$. This particular solution passes through $(-1, 3)$, and according to the differential equation, $y'(-1) = -ty = -(-1)3 = 3$. Thus, the segment located at the point $(-1, 3)$ has slope 3. Using the table of values at the left, we can determine the slopes at the various points on the curve. The graph of the solution is tangent to each segment [Figure 1(B)].

t	-2	-1	0	1	2
y	0.5	3	5	3	0.5
$\dfrac{dy}{dt}$	1	3	0	-3	-1

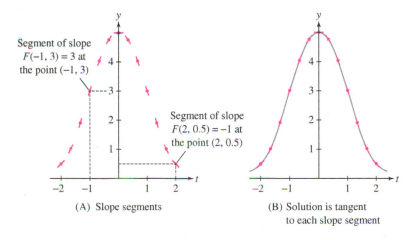

(A) Slope segments

(B) Solution is tangent to each slope segment

DF **FIGURE 1** The solution of $\dfrac{dy}{dt} = -ty$ satisfying $y(-1) = 3$.

Ignoring the fact that we already know the general solution, we will use the slope field to see how we expect solutions to appear. To sketch the slope field for $\dfrac{dy}{dt} = -ty$, we make a table of values of $\dfrac{dy}{dt}$ for certain values of y and t. Then we can draw small segments of slope $-ty$ at some array of points (t, y) in the plane, as in Figure 2(A). *The slope field allows us to visualize all of the solutions at a glance.* Starting at any point, we can sketch a solution by drawing a curve that runs tangent to the slope segments at each point [Figure 2(B)]. The graph of a solution is also called an **integral curve**.

■ **EXAMPLE 1** **Using Isoclines** Draw the slope field for

$$\frac{dy}{dt} = y - t$$

and sketch the integral curves satisfying the initial conditions

(a) $y(0) = 1$ and **(b)** $y(1) = -2$.

To imagine yourself subject to a differential equation, start somewhere. There you are tugged in some direction, so you move that way.... As you move, the tugging forces change, pulling you in a new direction; for your motion to solve the differential equation you must keep drifting with and responding to the ambient forces.

——From the introduction to *Differential Equations*, J. H. Hubbard and Beverly West, Springer-Verlag, New York, 1991

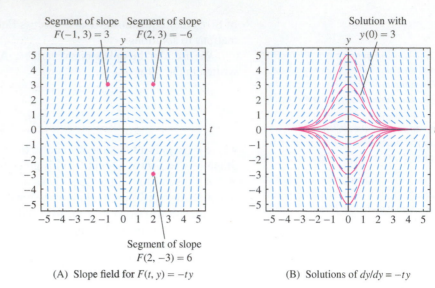

Segment of slope $F(-1, 3) = 3$ Segment of slope $F(2, 3) = -6$

Solution with $y(0) = 3$

Segment of slope $F(2, -3) = 6$

(A) Slope field for $F(t, y) = -ty$

(B) Solutions of $dy/dy = -ty$

DF FIGURE 2 Slope field for $F(t, y) = -ty$.

Solution A good way to sketch the slope field of $\dfrac{dy}{dt} = F(t, y)$ is to choose several values c and identify the curve $F(t, y) = c$, called the **isocline** of slope c. The isocline is the curve consisting of all points where the slope field has slope c.

In our case, $F(t, y) = y - t$, so the isocline of fixed slope c has equation $y - t = c$, or $y = t + c$, which is a line. Consider the following values:

- $c = 0$: This isocline is $y - t = 0$, or $y = t$. We draw segments of slope $c = 0$ at points along the line $y = t$, as in Figure 3(A).

- $c = 1$: This isocline is $y - t = 1$, or $y = t + 1$. We draw segments of slope 1 at points along $y = t + 1$, as in Figure 3(B).

- $c = 2$: This isocline is $y - t = 2$, or $y = t + 2$. We draw segments of slope 2 at points along $y = t + 2$, as in Figure 3(C).

- $c = -1$: This isocline is $y - t = -1$, or $y = t - 1$ [Figure 3(C)].

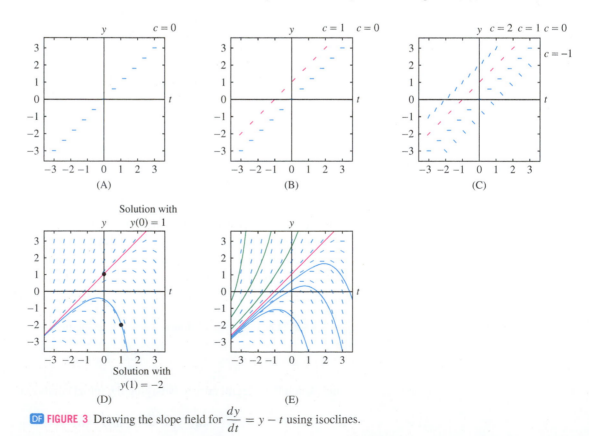

DF FIGURE 3 Drawing the slope field for $\dfrac{dy}{dt} = y - t$ using isoclines.

A more detailed slope field is shown in Figure 3(D). To sketch the solution satisfying $y(0) = 1$, begin at the point $(t_0, y_0) = (0, 1)$ and draw the integral curve that follows the directions indicated by the slope field. Similarly, the graph of the solution satisfying $y(1) = -2$ is the integral curve obtained by starting at $(t_0, y_0) = (1, -2)$ and moving along the slope field. Figure 3(E) shows several other solutions (integral curves). ∎

GRAPHICAL INSIGHT   Slope fields often let us see the *asymptotic* behavior of solutions (as $t \to \infty$) at a glance. Figure 3(E) suggests that the asymptotic behavior depends on the initial value (the y-intercept): If $y(0) > 1$, then $y(t)$ tends to ∞, and if $y(0) < 1$, then $y(t)$ tends to $-\infty$. We can check this using the general solution $y(t) = 1 + t + Ce^t$, where $y(0) = 1 + C$. If $y(0) > 1$, then $C > 0$ and $y(t)$ tends to ∞, but if $y(0) < 1$, then $C < 0$ and $y(t)$ tends to $-\infty$. The solution $y = 1 + t$ with initial condition $y(0) = 1$ is the straight line shown in Figure 3(D).

■ **EXAMPLE 2** **Newton's Law of Cooling Revisited**   The temperature $y(t)$ (in degrees Celsius) of an object placed in a refrigerator satisfies $\dfrac{dy}{dt} = -0.5(y - 4)$ (t in minutes). Draw the slope field and describe the behavior of the solutions.

Solution   The function $F(t, y) = -0.5(y - 4)$ depends only on y, so slopes of the segments in the slope field do not vary in the t-direction. The slope $F(t, y)$ is positive for $y < 4$ and negative for $y > 4$. More precisely, the slope at height y is $-0.5(y - 4) = -0.5y + 2$, so the segments grow steeper with positive slope as $y \to -\infty$, and they grow steeper with negative slope as $y \to \infty$ (Figure 4).

The slope field shows that if the initial temperature satisfies $y_0 > 4$, then $y(t)$ decreases to $y = 4$ as $t \to \infty$. In other words, the object cools down to 4°C when placed in the refrigerator. If $y_0 < 4$, then $y(t)$ increases to $y = 4$ as $t \to \infty$. The object warms up when placed in the refrigerator. If $y_0 = 4$, then y remains at 4°C for all time t. ∎

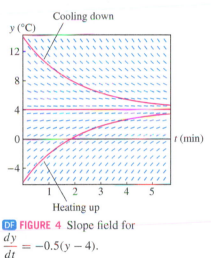

Cooling down

Heating up

 FIGURE 4 Slope field for $\dfrac{dy}{dt} = -0.5(y - 4)$.

CONCEPTUAL INSIGHT   Most first-order equations arising in applications have a uniqueness property: There is precisely one solution $y(t)$ satisfying a given initial condition $y(t_0) = y_0$. Graphically, this means that precisely one integral curve (solution) passes through the point (t_0, y_0). Thus, when uniqueness holds, distinct integral curves never cross or overlap. Figure 5 shows the slope field of $\dfrac{dy}{dt} = -\sqrt{|y|}$, where uniqueness fails. We can prove that once an integral curve touches the t-axis, it either remains on the t-axis or continues along the t-axis for a period of time before moving below the t-axis. Therefore, infinitely many integral curves pass through each point on the t-axis. However, the slope field does not show this clearly. This highlights again the need to analyze solutions rather than rely on visual impressions alone.

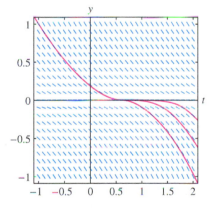

FIGURE 5 Overlapping integral curves for $\frac{dy}{dt} = -\sqrt{|y|}$ (uniqueness fails for this differential equation).

Euler's Method is the simplest method for solving Initial Value Problems numerically, but it is not very efficient. Computer systems use more sophisticated schemes, making it possible to plot and analyze solutions to the complex systems of differential equations arising in areas such as weather prediction, aerodynamic modeling, and economic forecasting.

Euler's Method

Euler's Method produces numerical approximations to the solution of a first-order Initial Value Problem:

$$\frac{dy}{dt} = F(t, y), \qquad y(t_0) = y_0 \qquad \boxed{2}$$

We begin by choosing a small number h, called the **time step**, and consider the sequence of times starting at the initial value t_0 and spaced at intervals of size h:

$$t_0, \qquad t_1 = t_0 + h, \qquad t_2 = t_0 + 2h, \qquad t_3 = t_0 + 3h, \qquad \dots$$

In general, $t_k = t_0 + kh$ for $k = 0, 1, 2, \dots$. Euler's Method consists of computing a sequence of values $y_1, y_2, y_3, \dots, y_n$ successively using the formula

$$\boxed{y_k = y_{k-1} + h\,F(t_{k-1}, y_{k-1})} \qquad \boxed{3}$$

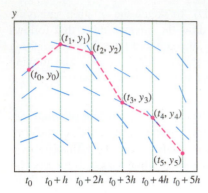

FIGURE 6 In Euler's Method, we move from one point to the next by traveling along the line indicated by the slope field.

Starting with the initial value $y_0 = y(t_0)$, we compute $y_1 = y_0 + hF(t_0, y_0)$, etc. The value y_k is the Euler approximation to $y(t_k)$. We connect the points $P_k = (t_k, y_k)$ by segments to obtain an approximation to the graph of $y(t)$ (Figure 6).

GRAPHICAL INSIGHT   The values y_k are defined so that the segment joining P_{k-1} to P_k has slope

$$\frac{y_k - y_{k-1}}{t_k - t_{k-1}} = \frac{(y_{k-1} + hF(t_{k-1}, y_{k-1})) - y_{k-1}}{h} = F(t_{k-1}, y_{k-1})$$

Thus, in Euler's method, we move from P_{k-1} to P_k by traveling in the direction specified by the slope field at P_{k-1} for a time interval of length h (Figure 6).

■ **EXAMPLE 3**   Use Euler's Method with time step $h = 0.2$ and $n = 4$ steps to approximate the solution of $\dfrac{dy}{dt} = y - t^2$, $y(0) = 3$.

Solution   Our initial value at $t_0 = 0$ is $y_0 = 3$. Since $h = 0.2$, the time values are $t_1 = 0.2$, $t_2 = 0.4$, $t_3 = 0.6$, and $t_4 = 0.8$. We use Eq. (3) with $F(t, y) = y - t^2$ to calculate

$$y_1 = y_0 + hF(t_0, y_0) = 3 + 0.2(3 - (0)^2) = 3.6$$

$$y_2 = y_1 + hF(t_1, y_1) = 3.6 + 0.2(3.6 - (0.2)^2) \approx 4.3$$

$$y_3 = y_2 + hF(t_2, y_2) = 4.3 + 0.2(4.3 - (0.4)^2) \approx 5.14$$

$$y_4 = y_3 + hF(t_3, y_3) = 5.14 + 0.2(5.14 - (0.6)^2) \approx 6.1$$

Figure 7(A) shows the exact solution $y(t) = 2 + 2t + t^2 + e^t$ together with a plot of the points (t_k, y_k) for $k = 0, 1, 2, 3, 4$ connected by line segments.   ■

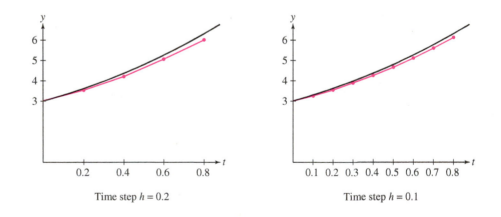

DF **FIGURE 7** Euler's Method applied to $\dfrac{dy}{dt} = y - t^2$, $y(0) = 3$.

Time step $h = 0.2$ Time step $h = 0.1$

CONCEPTUAL INSIGHT   Figure 7(B) shows that the time step $h = 0.1$ gives a better approximation than $h = 0.2$. In general, the smaller the time step, the better the approximation. In fact, if we start at a point $(a, y(a))$ and use Euler's Method to approximate $(b, y(b))$ using N steps with $h = (b - a)/N$, then the error is roughly proportional to $1/N$ [provided that $F(t, y)$ is a well-behaved function]. This is similar to the error size in the Nth left- and right-endpoint approximations to an integral. What this means, however, is that Euler's Method is quite inefficient; to cut the error in half, it is necessary to double the number of steps, and to achieve n-digit accuracy requires roughly 10^n steps. Fortunately, there are several methods that improve on Euler's Method in much the same way as the Midpoint Rule and Simpson's Rule improve on the endpoint approximations (see Exercises 24–29).

■ **EXAMPLE 4** *CAS* Let $y(t)$ be the solution of $\dfrac{dy}{dt} = \sin t \cos y$, $y(0) = 0$.

(a) Use Euler's Method with time step $h = 0.1$ to approximate $y(0.5)$.

(b) Use a computer algebra system to implement Euler's Method with time steps $h = 0.01, 0.001$, and 0.0001 to approximate $y(0.5)$.

Solution

Euler's Method:

$$y_k = y_{k-1} + hF(t_{k-1}, y_{k-1})$$

(a) When $h = 0.1$, y_k is an approximation to $y(0 + k(0.1)) = y(0.1k)$, so y_5 is an approximation to $y(0.5)$. It is convenient to organize calculations in the following table. Note that the value y_{k+1} computed in the last column of each line is used in the next line to continue the process.

t_k	y_k	$F(t_k, y_k) = \sin t_k \cos y_k$	$y_{k+1} = y_k + hF(t_k, y_k)$
$t_0 = 0$	$y_0 = 0$	$(\sin 0)\cos 0 = 0$	$y_1 = 0 + 0.1(0) = 0$
$t_1 = 0.1$	$y_1 = 0$	$(\sin 0.1)\cos 0 \approx 0.1$	$y_2 \approx 0 + 0.1(0.1) = 0.01$
$t_2 = 0.2$	$y_2 \approx 0.01$	$(\sin 0.2)\cos(0.01) \approx 0.2$	$y_3 \approx 0.01 + 0.1(0.2) = 0.03$
$t_3 = 0.3$	$y_3 \approx 0.03$	$(\sin 0.3)\cos(0.03) \approx 0.3$	$y_4 \approx 0.03 + 0.1(0.3) = 0.06$
$t_4 = 0.4$	$y_4 \approx 0.06$	$(\sin 0.4)\cos(0.06) \approx 0.4$	$y_5 \approx 0.06 + 0.1(0.4) = 0.10$

Thus, Euler's Method yields the approximation $y(0.5) \approx y_5 \approx 0.1$.

(b) When the number of steps is large, the calculations are too lengthy to do by hand, but they are easily carried out using a CAS. Note that for $h = 0.01$, the kth value y_k is an approximation to $y(0 + k(0.01)) = y(0.01k)$, and y_{50} gives an approximation to $y(0.5)$. Similarly, when $h = 0.001$, y_{500} is an approximation to $y(0.5)$, and when $h = 0.0001$, $y_{5,000}$ is an approximation to $y(0.5)$. Here are the results obtained using a CAS:

A typical CAS command to implement Euler's Method with time step $h = 0.01$ reads as follows:

```
>> For[n = 0; y = 0, n < 50, n++,
>> y = y + (.01) * (Sin[.01 * n] * Cos[y])]
>> y
>> 0.119746
```

The command `For[...]` *updates the variable y successively through the values $y_1, y_2, \ldots, y_{50}$ according to Euler's Method.*

Time step $h = 0.01$	$y_{50} \approx 0.1197$
Time step $h = 0.001$	$y_{500} \approx 0.1219$
Time step $h = 0.0001$	$y_{5000} \approx 0.1221$

The values appear to converge and we may assume that $y(0.5) \approx 0.12$. However, we see here that Euler's Method converges quite slowly. ■

9.3 SUMMARY

- The *slope field* for a first-order differential equation $\dfrac{dy}{dt} = F(t, y)$ is obtained by drawing small segments of slope $F(t, y)$ at points (t, y) lying on a rectangular grid in the plane.
- The graph of a solution (also called an *integral curve* of the differential equation) satisfying $y(t_0) = y_0$ is a curve through (t_0, y_0) that runs tangent to the segments of the slope field at each point.
- Euler's Method: To approximate a solution to $\dfrac{dy}{dt} = F(t, y)$ with initial condition $y(t_0) = y_0$, fix a time step h and set $t_k = t_0 + kh$. Define $y_1, y_2, \ldots$ successively by the formula

$$\boxed{y_k = y_{k-1} + hF(t_{k-1}, y_{k-1})}$$ **4**

The values $y_0, y_1, y_2, \ldots$ are approximations to the values $y(t_0), y(t_1), y(t_2), \ldots$.

9.3 EXERCISES

Preliminary Questions

1. What is the slope of the segment in the slope field for $\dfrac{dy}{dt} = ty + 1$ at the point $(2, 3)$?

2. What is the equation of the isocline of slope $c = 1$ for $\dfrac{dy}{dt} = y^2 - t$?

3. For which of the following differential equations are the slopes at points on a vertical line $t = C$ all equal?

(a) $\dfrac{dy}{dt} = \ln y$

(b) $\dfrac{dy}{dt} = \ln t$

4. Let $y(t)$ be the solution to $\dfrac{dy}{dt} = F(t, y)$ with $y(1) = 3$. How many iterations of Euler's Method are required to approximate $y(3)$ if the time step is $h = 0.1$?

Exercises

1. Figure 8 shows the slope field for $\dfrac{dy}{dt} = \sin y \sin t$. Sketch the graphs of the solutions with initial conditions $y(0) = 1$ and $y(0) = -1$. Show that $y(t) = 0$ is a solution and add its graph to the plot.

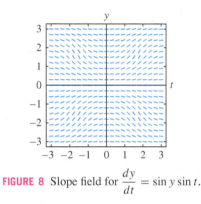

FIGURE 8 Slope field for $\dfrac{dy}{dt} = \sin y \sin t$.

2. Figure 9 shows the slope field for $\dfrac{dy}{dt} = t^2 - y^2$. Sketch the integral curve passing through the point $(0, -1)$, the curve through $(0, 0)$, and the curve through $(0, 2)$. Is $y(t) = 0$ a solution?

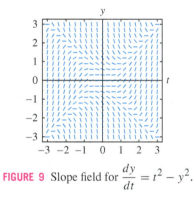

FIGURE 9 Slope field for $\dfrac{dy}{dt} = t^2 - y^2$.

3. Show that $f(t) = \frac{1}{2}\left(t - \frac{1}{2}\right)$ is a solution to $\dfrac{dy}{dt} = t - 2y$. Sketch the four solutions with $y(0) = \pm 0.5, \pm 1$ on the slope field in Figure 10. The slope field suggests that every solution approaches $f(t)$ as $t \to \infty$. Confirm this by showing that $y = f(t) + Ce^{-2t}$ is the general solution.

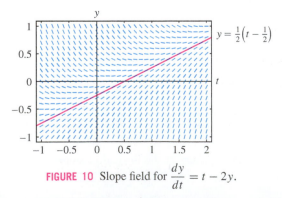

FIGURE 10 Slope field for $\dfrac{dy}{dt} = t - 2y$.

4. One of the slope fields in Figures 11(A) and (B) is the slope field for $\dfrac{dy}{dt} = t^2$. The other is for $\dfrac{dy}{dt} = y^2$. Identify which is which. In each case, sketch the solutions with initial conditions $y(0) = 1$, $y(0) = 0$, and $y(0) = -1$.

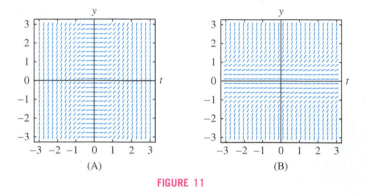

FIGURE 11

5. Consider the differential equation $\dfrac{dy}{dt} = t - y$.

(a) Sketch the slope field of the differential equation $\dfrac{dy}{dt} = t - y$ in the range $-1 \le t \le 3, -1 \le y \le 3$. As an aid, observe that the isocline of slope c is the line $t - y = c$, so the segments have slope c at points on the line $y = t - c$.

(b) Show that $y = t - 1 + Ce^{-t}$ is a solution for all C. Since $\lim\limits_{t\to\infty} e^{-t} = 0$, these solutions approach the particular solution $y = t - 1$ as $t \to \infty$. Explain how this behavior is reflected in your slope field.

6. Show that the isoclines of $\dfrac{dy}{dt} = 1/y$ are horizontal lines. Sketch the slope field for $-2 \le t \le 2, -2 \le y \le 2$ and plot the solutions with initial conditions $y(0) = 0$ and $y(0) = 1$.

7. Sketch the slope field for $\dfrac{dy}{dt} = y + t$ for $-2 \le t \le 2, -2 \le y \le 2$.

8. Sketch the slope field for $\dfrac{dy}{dt} = \dfrac{t}{y}$ for $-2 \le t \le 2, -2 \le y \le 2$.

9. Show that the isoclines of $\dfrac{dy}{dt} = t$ are vertical lines. Sketch the slope field for $-2 \le t \le 2, -2 \le y \le 2$ and plot the integral curves passing through $(0, -1)$ and $(0, 1)$.

10. Sketch the slope field of $\dfrac{dy}{dt} = ty$ for $-2 \le t \le 2, -2 \le y \le 2$. Based on the sketch, determine $\lim\limits_{t\to\infty} y(t)$, where $y(t)$ is a solution with $y(0) > 0$. What is $\lim\limits_{t\to\infty} y(t)$ if $y(0) < 0$?

11. Match each differential equation with its slope field in Figures 12(A)–(F).

(i) $\dfrac{dy}{dt} = -1$

(ii) $\dfrac{dy}{dt} = \dfrac{y}{t}$

(iii) $\dfrac{dy}{dt} = t^2 y$

(iv) $\dfrac{dy}{dt} = ty^2$

(v) $\dfrac{dy}{dt} = t^2 + y^2$

(vi) $\dfrac{dy}{dt} = t$

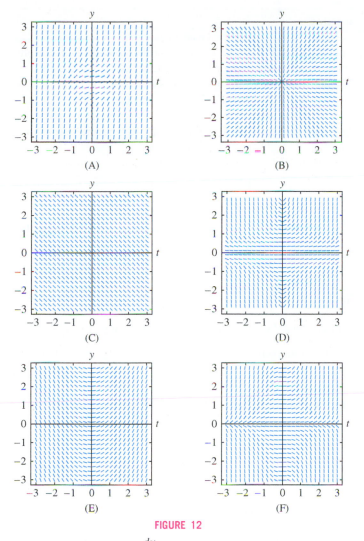

FIGURE 12

12. Sketch the solution of $\dfrac{dy}{dt} = ty^2$ satisfying $y(0) = 1$ in the appropriate slope field of Figure 12(A)–(F). Then show, using Separation of Variables, that if $y(t)$ is a solution such that $y(0) > 0$, then $y(t)$ tends to infinity as $t \to \sqrt{2/y(0)}$.

13. **(a)** Sketch the slope field of $\dfrac{dy}{dt} = t/y$ in the region $-2 \le t \le 2$, $-2 \le y \le 2$.

(b) Check that $y = \pm\sqrt{t^2 + C}$ is the general solution.

(c) Sketch the solutions on the slope field with initial conditions $y(0) = 1$ and $y(0) = -1$.

14. Sketch the slope field of $\dfrac{dy}{dt} = t^2 - y$ in the region $-3 \le t \le 3$, $-3 \le y \le 3$ and sketch the solutions satisfying $y(1) = 0$, $y(1) = 1$, and $y(1) = -1$.

15. Let $F(t, y) = t^2 - y$ and let $y(t)$ be the solution of $\dfrac{dy}{dt} = F(t, y)$ satisfying $y(2) = 3$. Let $h = 0.1$ be the time step in Euler's Method, and set $y_0 = y(2) = 3$.

(a) Calculate $y_1 = y_0 + h F(2, 3)$.

(b) Calculate $y_2 = y_1 + h F(2.1, y_1)$.

(c) Calculate $y_3 = y_2 + h F(2.2, y_2)$ and continue computing y_4, y_5, and y_6.

(d) Find approximations to $y(2.2)$ and $y(2.5)$.

16. Let $y(t)$ be the solution to $\dfrac{dy}{dt} = te^{-y}$ satisfying $y(0) = 0$.

(a) Use Euler's Method with time step $h = 0.1$ to approximate $y(0.1), y(0.2), \ldots, y(0.5)$.

(b) Use Separation of Variables to find $y(t)$ exactly.

(c) Compute the errors in the approximations to $y(0.1)$ and $y(0.5)$.

In Exercises 17–22, use Euler's Method to approximate the given value of $y(t)$ with the time step h indicated.

17. $y(0.5);\quad \dfrac{dy}{dt} = y + t,\quad y(0) = 1,\quad h = 0.1$

18. $y(0.7);\quad \dfrac{dy}{dt} = 2y,\quad y(0) = 3,\quad h = 0.1$

19. $y(3.3);\quad \dfrac{dy}{dt} = t^2 - y,\quad y(3) = 1,\quad h = 0.05$

20. $y(3);\quad \dfrac{dy}{dt} = \sqrt{t + y},\quad y(2.7) = 5,\quad h = 0.05$

21. $y(2);\quad \dfrac{dy}{dt} = t \sin y,\quad y(1) = 2,\quad h = 0.2$

22. $y(5.2);\quad \dfrac{dy}{dt} = t - \sec y,\quad y(4) = -2,\quad h = 0.2$

Further Insights and Challenges

23. If f is continuous on $[a, b]$, then the solution to $\dfrac{dy}{dt} = f(t)$ with initial condition $y(a) = 0$ is $y(t) = \displaystyle\int_a^t f(u)\,du$. Show that Euler's Method with time step $h = (b - a)/N$ for N steps yields the Nth left-endpoint approximation to $y(b) = \displaystyle\int_a^b f(u)\,du$.

*Exercises 24–29: **Euler's Midpoint Method** is a variation on Euler's Method that is significantly more accurate in general. For time step h and initial value $y_0 = y(t_0)$, the values y_k are defined successively by*

$$y_k = y_{k-1} + h m_{k-1}$$

where $m_{k-1} = F\left(t_{k-1} + \dfrac{h}{2},\ y_{k-1} + \dfrac{h}{2} F(t_{k-1}, y_{k-1})\right)$.

24. Apply both Euler's Method and the Euler Midpoint Method with $h = 0.1$ to estimate $y(1.5)$, where $y(t)$ satisfies $\dfrac{dy}{dt} = y$ with $y(0) = 1$. Find $y(t)$ exactly and compute the errors in these two approximations.

In Exercises 25–28, use Euler's Midpoint Method with the time step indicated to approximate the given value of $y(t)$.

25. $y(0.5);\quad \dfrac{dy}{dt} = y + t,\quad y(0) = 1,\quad h = 0.1$

26. $y(2);\quad \dfrac{dy}{dt} = t^2 - y,\quad y(1) = 3,\quad h = 0.2$

27. $y(0.25);\quad \dfrac{dy}{dt} = \cos(y + t),\quad y(0) = 1,\quad h = 0.05$

28. $y(2.3);\quad \dfrac{dy}{dt} = y + t^2,\quad y(2) = 1,\quad h = 0.05$

29. Assume that f is continuous on $[a, b]$. Show that Euler's Midpoint Method applied to $\dfrac{dy}{dt} = f(t)$ with initial condition $y(a) = 0$ and time step $h = (b - a)/N$ for N steps yields the Nth midpoint approximation to

$$y(b) = \int_a^b f(u)\,du$$

9.4 The Logistic Equation

The simplest model of population growth is $dy/dt = ky$, according to which populations grow exponentially. This may be true over short periods of time, but it is clear that no population can increase without limit due to finite resources (such as food and space). Therefore, population biologists use a variety of other differential equations that take into account environmental limitations to growth such as food scarcity and competition between species. One widely used model is based on the **logistic differential equation**:

$$\frac{dy}{dt} = ky\left(1 - \frac{y}{A}\right)$$

1

Here, $k > 0$ is the growth constant, and $A > 0$ is a constant called the **carrying capacity**. Figure 1 shows a typical S-shaped solution of Eq. (1).

CONCEPTUAL INSIGHT The logistic equation $\dfrac{dy}{dt} = ky(1 - y/A)$ differs from the exponential differential equation $\dfrac{dy}{dt} = ky$ only by the additional factor $(1 - y/A)$. As long as y is small relative to A, this factor is close to 1 and can be ignored, yielding $\dfrac{dy}{dt} \approx ky$. Thus, $y(t)$ grows nearly exponentially when the population is small (Figure 1). As $y(t)$ approaches A, the factor $(1 - y/A)$ tends to zero. This causes $\dfrac{dy}{dt}$ to decrease and prevents $y(t)$ from exceeding the carrying capacity A.

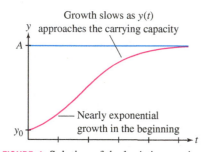

Growth slows as $y(t)$ approaches the carrying capacity

Nearly exponential growth in the beginning

FIGURE 1 Solution of the logistic equation.

The slope field in Figure 2 shows clearly that there are three families of solutions, depending on the initial value $y_0 = y(0)$.

- If $y_0 > A$, then $y(t)$ is decreasing and approaches A as $t \to \infty$.
- If $0 < y_0 < A$, then $y(t)$ is increasing and approaches A as $t \to \infty$.
- If $y_0 < 0$, then $y(t)$ is decreasing and $\lim\limits_{t \to t_b^-} y(t) = -\infty$ for some time t_b.

Equation (1) also has two constant solutions: $y = 0$ and $y = A$. They correspond to the roots of $ky(1 - y/A) = 0$, and they satisfy Eq. (1) because $\dfrac{dy}{dt} = 0$ when y is a constant. Constant solutions are called **equilibrium** or **steady-state** solutions. The equilibrium solution $y = A$ is a **stable equilibrium** because every solution with initial value y_0 close to A approaches the equilibrium $y = A$ as $t \to \infty$. By contrast, $y = 0$ is an **unstable equilibrium** because every nonequilibrium solution with initial value y_0 near $y = 0$ either increases to A or decreases to $-\infty$. These nonequilibrium solutions deviate away from the unstable equilibrium solution as time moves forward.

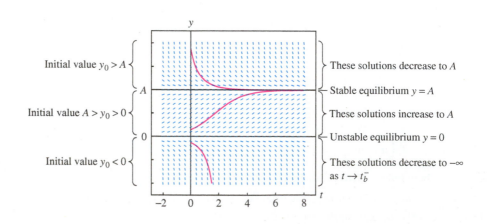

Initial value $y_0 > A$ — These solutions decrease to A

Stable equilibrium $y = A$

Initial value $A > y_0 > 0$ — These solutions increase to A

Unstable equilibrium $y = 0$

Initial value $y_0 < 0$ — These solutions decrease to $-\infty$ as $t \to t_b^-$

FIGURE 2 Slope field for $\dfrac{dy}{dt} = ky\left(1 - \dfrac{y}{A}\right)$.

Having described the solutions qualitatively, let us now find the nonequilibrium solutions explicitly using Separation of Variables. Assuming that $y \neq 0$ and $y \neq A$, we have

$$\frac{dy}{dt} = ky\left(1 - \frac{y}{A}\right)$$

$$\frac{dy}{y(1 - y/A)} = k\,dt$$

$$\int \left(\frac{1}{y} - \frac{1}{y - A}\right) dy = \int k\,dt \qquad \boxed{2}$$

$$\ln|y| - \ln|y - A| = kt + C$$

$$\left|\frac{y}{y - A}\right| = e^{kt+C} \quad \Rightarrow \quad \frac{y}{y - A} = \pm e^C e^{kt}$$

Since $\pm e^C$ takes on arbitrary nonzero values, we replace $\pm e^C$ with B (nonzero):

$$\frac{y}{y - A} = Be^{kt} \qquad \boxed{3}$$

For $t = 0$, this gives a useful relation between B and the initial value $y_0 = y(0)$:

$$\frac{y_0}{y_0 - A} = B \qquad \boxed{4}$$

To solve for y, multiply each side of Eq. (3) by $(y - A)$:

$$y = (y - A)Be^{kt}$$

$$y(1 - Be^{kt}) = -ABe^{kt}$$

$$y = \frac{ABe^{kt}}{Be^{kt} - 1}$$

As $B \neq 0$, we may divide by Be^{kt} to obtain the general nonequilibrium solution:

$$\boxed{\frac{dy}{dt} = ky\left(1 - \frac{y}{A}\right), \qquad y = \frac{A}{1 - e^{-kt}/B}} \qquad \boxed{5}$$

Although we will use the solution we just derived in the next two examples, it is generally easier to rederive the solution in a particular case by Separation of Variables than it is to memorize the solution.

■ **EXAMPLE 1** Solve $\dfrac{dy}{dt} = 0.3y(4 - y)$ with initial condition $y(0) = 1$.

Solution To apply Eq. (5), we must rewrite the equation in the form

$$\frac{dy}{dt} = 1.2y\left(1 - \frac{y}{4}\right)$$

Thus, $k = 1.2$ and $A = 4$, and the general solution is

$$y = \frac{4}{1 - e^{-1.2t}/B}$$

There are two ways to find B. One way is to solve $y(0) = 1$ for B directly. An easier way is to use Eq. (4):

$$B = \frac{y_0}{y_0 - A} = \frac{1}{1 - 4} = -\frac{1}{3}$$

We find that the particular solution is $y = \dfrac{4}{1 + 3e^{-1.2t}}$ (Figure 3). ■

In Eq. (2), we use the the partial fraction decomposition

$$\frac{1}{y(1 - y/A)} = \frac{1}{y} - \frac{1}{y - A}$$

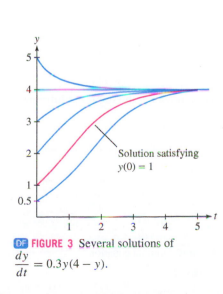

DF **FIGURE 3** Several solutions of $\dfrac{dy}{dt} = 0.3y(4 - y)$.

Solution satisfying $y(0) = 1$

FIGURE 4

The logistic equation may be too simple to describe a real deer population accurately, but it serves as a starting point for more sophisticated models used by ecologists, population biologists, and forestry professionals. (Mr_Jamsey/iStockphoto)

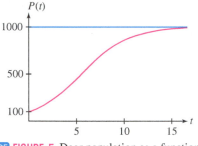

DF **FIGURE 5** Deer population as a function of t (in years).

■ **EXAMPLE 2** **Deer Population** A deer population (Figure 4) grows logistically with growth constant $k = 0.4$ year^{-1} in a forest with a carrying capacity of 1000 deer.

(a) Find the deer population $P(t)$ if the initial population is $P_0 = 100$.

(b) How long does it take for the deer population to reach 500?

Solution The time unit is the year because the unit of k is year^{-1}.

(a) Since $k = 0.4$ and $A = 1000$, $P(t)$ satisfies the differential equation

$$\frac{dP}{dt} = 0.4P\left(1 - \frac{P}{1000}\right)$$

The general solution is given by Eq. (5):

$$P(t) = \frac{1000}{1 - e^{-0.4t}/B} \qquad \boxed{6}$$

Using Eq. (4) to compute B, we find (Figure 5)

$$B = \frac{P_0}{P_0 - A} = \frac{100}{100 - 1000} = -\frac{1}{9} \quad\Rightarrow\quad P(t) = \frac{1000}{1 + 9e^{-0.4t}}$$

(b) To find the time t when $P(t) = 500$, we could solve the equation

$$P(t) = \frac{1000}{1 + 9e^{-0.4t}} = 500$$

But it is easier to use Eq. (3):

$$\frac{P}{P - A} = Be^{kt}$$

$$\frac{P}{P - 1000} = -\frac{1}{9}e^{0.4t}$$

Set $P = 500$ and solve for t:

$$-\frac{1}{9}e^{0.4t} = \frac{500}{500 - 1000} = -1 \quad\Rightarrow\quad e^{0.4t} = 9 \quad\Rightarrow\quad 0.4t = \ln 9$$

This gives $t = (\ln 9)/0.4 \approx 5.5$ years. ■

9.4 SUMMARY

- The *logistic equation* and its general nonequilibrium solution ($k > 0$ and $A > 0$):

$$\frac{dy}{dt} = ky\left(1 - \frac{y}{A}\right), \qquad y = \frac{A}{1 - e^{-kt}/B}, \qquad \text{or equivalently,} \qquad \frac{y}{y - A} = Be^{kt}$$

- Two equilibrium (constant) solutions:
  - $y = 0$ is an unstable equilibrium.
  - $y = A$ is a stable equilibrium.

- If the initial value $y_0 = y(0)$ satisfies $y_0 > 0$, then $y(t)$ approaches the stable equilibrium $y = A$; that is, $\lim_{t \to \infty} y(t) = A$.

9.4 EXERCISES

Preliminary Questions

1. Which of the following differential equations is a logistic differential equation?

(a) $\dfrac{dy}{dt} = 2y(1 - y^2)$

(b) $\dfrac{dy}{dt} = 2y\left(1 - \dfrac{y}{3}\right)$

(c) $\dfrac{dy}{dt} = 2y\left(1 - \dfrac{t}{4}\right)$

(d) $\dfrac{dy}{dt} = 2y(1 - 3y)$

2. Is the logistic equation a linear differential equation?

3. Is the logistic equation separable?

Exercises

1. Find the general solution of the logistic equation

$$\frac{dy}{dt} = 3y\left(1 - \frac{y}{5}\right)$$

Then find the particular solution satisfying $y(0) = 2$.

2. Find the solution of $\dfrac{dy}{dt} = 2y(3 - y)$, $y(0) = 10$.

3. Let $y(t)$ be a solution of $\dfrac{dy}{dt} = 0.5y(1 - 0.5y)$ such that $y(0) = 4$. Determine $\lim\limits_{t \to \infty} y(t)$ without finding $y(t)$ explicitly.

4. Let $y(t)$ be a solution of $\dfrac{dy}{dt} = 5y(1 - y/5)$. State whether y is increasing, decreasing, or constant in the following cases:

(a) $y(0) = 2$ **(b)** $y(0) = 5$ **(c)** $y(0) = 8$

5. A population of squirrels lives in a forest with a carrying capacity of 2000. Assume logistic growth with growth constant $k = 0.6 \text{ yr}^{-1}$.

(a) Find a formula for the squirrel population $P(t)$, assuming an initial population of 500 squirrels.

(b) How long will it take for the squirrel population to double?

6. The population $P(t)$ of mosquito larvae growing in a tree hole increases according to the logistic equation with growth constant $k = 0.3 \text{ day}^{-1}$ and carrying capacity $A = 500$.

(a) Find a formula for the larvae population $P(t)$, assuming an initial population of $P_0 = 50$ larvae.

(b) After how many days will the larvae population reach 200?

7. Sunset Lake is stocked with 2000 rainbow trout, and after 1 year the population has grown to 4500. Assuming logistic growth with a carrying capacity of 20,000, find the growth constant k (specify the units) and determine when the population will increase to 10,000.

8. Spread of a Rumor A rumor spreads through a small town. Let $y(t)$ be the fraction of the population that has heard the rumor at time t and assume that the rate at which the rumor spreads is proportional to the product of the fraction y of the population that has heard the rumor and the fraction $1 - y$ that has not yet heard the rumor.

(a) Write down the differential equation satisfied by y in terms of a proportionality factor k.

(b) Find k (in units of day^{-1}), assuming that 10% of the population knows the rumor at $t = 0$ and 40% knows it at $t = 2$ days.

(c) Using the assumptions of part (b), determine when 75% of the population will know the rumor.

9. A rumor spreads through a school with 1000 students. At 8 AM, 80 students have heard the rumor, and by noon, half the school has heard it. Using the logistic model of Exercise 8, determine when 90% of the students will have heard the rumor.

10. GU A simpler model for the spread of a rumor assumes that the rate at which the rumor spreads is proportional (with factor k) to the fraction of the population that has not yet heard the rumor.

(a) Compute the solutions to this model and the model of Exercise 8 with the values $k = 0.9$ and $y_0 = 0.1$.

(b) Graph the two solutions on the same axis.

(c) Which model seems more realistic? Why?

11. Let $k = 1$ and $A = 1$ in the logistic equation.

(a) Find the solutions satisfying $y_1(0) = 10$ and $y_2(0) = -1$.

(b) Find the time t when $y_1(t) = 5$.

(c) When does $y_2(t)$ become infinite?

12. A tissue culture grows until it has a maximum area of M square centimeters. The area $A(t)$ of the culture at time t may be modeled by the differential equation

$$\frac{dA}{dt} = k\sqrt{A}\left(1 - \frac{A}{M}\right) \qquad \boxed{7}$$

where k is a growth constant.

(a) Show that if we set $A = u^2$, then

$$\frac{du}{dt} = \frac{1}{2}k\left(1 - \frac{u^2}{M}\right)$$

Then find the general solution using Separation of Variables.

(b) Show that the general solution to Eq. (7) is

$$A(t) = M\left(\frac{Ce^{(k/\sqrt{M})t} - 1}{Ce^{(k/\sqrt{M})t} + 1}\right)^2$$

13. GU In the model of Exercise 12, let $A(t)$ be the area at time t (hours) of a growing tissue culture with initial size $A(0) = 1 \text{ cm}^2$, assuming that the maximum area is $M = 16 \text{ cm}^2$ and the growth constant is $k = 0.1$.

(a) Find a formula for $A(t)$. *Note:* The initial condition is satisfied for two values of the constant C. Choose the value of C for which $A(t)$ is increasing.

(b) Determine the area of the culture at $t = 10$ hours.

(c) GU Graph the solution using a graphing utility.

14. Show that if a tissue culture grows according to Eq. (7), then the growth rate reaches a maximum when $A = M/3$.

15. In 1751 Benjamin Franklin predicted that the U.S. population $P(t)$ would increase with growth constant $k = 0.028 \text{ year}^{-1}$. According to the census, the U.S. population was 5 million in 1800 and 76 million in 1900. Assuming logistic growth with $k = 0.028$, find the predicted carrying capacity for the U.S. population. *Hint:* Use Eqs. (3) and (4) to show that

$$\frac{P(t)}{P(t) - A} = \frac{P_0}{P_0 - A}e^{kt}$$

16. **Reverse Logistic Equation** Consider the following logistic equation (with $k, B > 0$):

$$\frac{dP}{dt} = -kP\left(1 - \frac{P}{B}\right) \qquad \boxed{8}$$

(a) Sketch the slope field of this equation.

(b) The general solution is $P(t) = B/(1 - e^{kt}/C)$, where C is a nonzero constant. Show that $P(0) > B$ if $C > 1$ and $0 < P(0) < B$ if $C < 0$.

(c) Show that Eq. (8) models an "extinction–explosion" population. That is, $P(t)$ tends to zero if the initial population satisfies $0 < P(0) < B$, and it tends to ∞ after a finite amount of time if $P(0) > B$.

(d) Show that $P = 0$ is a stable equilibrium and $P = B$ an unstable equilibrium.

Further Insights and Challenges

In Exercises 17 and 18, let $y(t)$ be a solution of the logistic equation

$$\frac{dy}{dt} = ky\left(1 - \frac{y}{A}\right) \qquad \boxed{9}$$

where $A > 0$ and $k > 0$.

17. **(a)** Differentiate Eq. (9) with respect to t and use the Chain Rule to show that

$$\frac{d^2y}{dt^2} = k^2 y\left(1 - \frac{y}{A}\right)\left(1 - \frac{2y}{A}\right)$$

(b) Show that the graph of the function y is concave up if $0 < y < A/2$ and concave down if $A/2 < y < A$.

(c) Show that if $0 < y(0) < A/2$, then y has a point of inflection at $y = A/2$ (Figure 6).

(d) Assume that $0 < y(0) < A/2$. Find the time t when $y(t)$ reaches the inflection point.

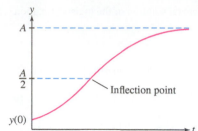

FIGURE 6 An inflection point occurs at $y = A/2$ in the logistic curve.

18. Let $y = \dfrac{A}{1 - e^{-kt}/B}$ be the general nonequilibrium solution to Eq. (9). If $y(t)$ has a vertical asymptote at $t = t_b$, that is, if $\lim\limits_{t \to t_b-} y(t) = \pm\infty$, we say that the solution "blows up" at $t = t_b$.

(a) Show that if $0 < y(0) < A$, then y does not blow up at any time t_b.

(b) Show that if $y(0) > A$, then y blows up at a time t_b, which is negative (and hence does not correspond to a real time).

(c) Show that y blows up at some positive time t_b if and only if $y(0) < 0$ (and hence does not correspond to a real population).

9.5 First-Order Linear Equations

First-Order Differential Equations

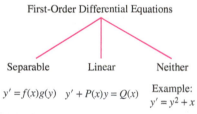

Separable   Linear   Neither

$y' = f(x)g(y)$ $y' + P(x)y = Q(x)$ Example:
$y' = y^2 + x$

FIGURE 1

This section introduces the method of "integrating factors" for solving first-order linear equations. Although we already have a method (Separation of Variables) for solving separable equations, this new method applies to all linear equations, whether separable or not (Figure 1).

A first-order linear equation is one that can be put in the following form:

$$\boxed{y' + P(x)y = Q(x)} \qquad \boxed{1}$$

Note that in this section, x is used as an independent variable (but t is used in Example 3 below). To solve Eq. (1), we shall multiply through by a function $\alpha(x)$, called an **integrating factor**, that turns the left-hand side into the derivative of $\alpha(x)y$:

$$\alpha(x)\big(y' + P(x)y\big) = \big(\alpha(x)y\big)' \qquad \boxed{2}$$

Suppose we can find $\alpha(x)$ satisfying Eq. (2) that is nonzero. Then Eq. (1) yields

$$\alpha(x)\big(y' + P(x)y\big) = \alpha(x)Q(x)$$

$$\big(\alpha(x)y\big)' = \alpha(x)Q(x)$$

We can solve this equation by integration:

$$\alpha(x)y = \int \alpha(x)Q(x)\,dx \qquad \text{or} \qquad y = \frac{1}{\alpha(x)}\left(\int \alpha(x)Q(x)\,dx\right)$$

To find $\alpha(x)$, expand Eq. (2), using the Product Rule on the right-hand side:

$$\alpha(x)y' + \alpha(x)P(x)y = \alpha(x)y' + \alpha'(x)y \quad \Rightarrow \quad \alpha(x)P(x)y = \alpha'(x)y$$

Dividing by y, we obtain

$$\boxed{\frac{d\alpha}{dx} = \alpha(x)P(x)} \qquad \boxed{3}$$

We solve this equation using Separation of Variables:

$$\frac{d\alpha}{\alpha} = P(x)\,dx \quad \Rightarrow \quad \int \frac{d\alpha}{\alpha} = \int P(x)\,dx$$

Therefore, $\ln|\alpha(x)| = \int P(x)\,dx$, and by exponentiation, $\alpha(x) = \pm e^{\int P(x)\,dx}$. Since we need just one solution of Eq. (3), we choose the positive sign in the expression for $\alpha(x)$.

In the formula for the integrating factor $\alpha(x)$, the integral $\int P(x)\,dx$ denotes any antiderivative of P.

THEOREM 1 The general solution of $y' + P(x)y = Q(x)$ is

$$y = \frac{1}{\alpha(x)} \left(\int \alpha(x)Q(x)\,dx \right) \qquad \boxed{4}$$

where $\alpha(x)$ is an integrating factor:

$$\alpha(x) = e^{\int P(x)\,dx} \qquad \boxed{5}$$

Rather than memorizing this theorem, one need only remember that the integrating factor is $\alpha(x) = e^{\int P(x)\,dx}$. Multiplying both sides of the differential equation $y' + P(x)y = Q(x)$ by $\alpha(x)$, so that the left side becomes the derivative of a product, and then integrating and solving for y yields the solution.

■ **EXAMPLE 1** Solve $xy' - 3y = x^2$, $y(1) = 2$.

Solution First divide by x to put the equation in the form $y' + P(x)y = Q(x)$:

$$y' - \frac{3}{x}\,y = x$$

Thus, $P(x) = -3x^{-1}$ and $Q(x) = x$.

Step 1. **Find an integrating factor.**
In our case, $P(x) = -3x^{-1}$, and by Eq. (5),

$$\alpha(x) = e^{\int P(x)\,dx} = e^{\int(-3/x)\,dx} = e^{-3\ln x} = e^{\ln(x^{-3})} = x^{-3}$$

Step 2. **Multiply the equation by the integrating factor.**

$$x^{-3}\left(y' - \frac{3}{x}\,y\right) = x^{-3}(x)$$

$$(x^{-3}y)' = x^{-2}$$

Step 3. **Integrate on both sides.**

$$x^{-3}y = -x^{-1} + C$$

CAUTION *We have to include the constant of integration C, but note that in the general solution, C does not appear added on by itself. The general solution is $y = -x^2 + Cx^3$. It is not correct to write $-x^2 + C$ or $-x^2 + Cx^3 + D$.*

Step 4. **Solve for y.**

$$y = x^3(-x^{-1} + C)$$

$$= -x^2 + Cx^3$$

Step 5. **Solve the Initial Value Problem.**
Now solve for C using the initial condition $y(1) = 2$:

$$y(1) = -1^2 + C \cdot 1^3 = 2 \qquad \text{or} \qquad C = 3$$

Therefore, the solution of the Initial Value Problem is $y = -x^2 + 3x^3$.

Finally, let's check that $y = -x^2 + 3x^3$ satisfies our equation $xy' - 3y = x^2$:

$$xy' - 3y = x(-2x + 9x^2) - 3(-x^2 + 3x^3)$$

$$= (-2x^2 + 9x^3) + (3x^2 - 9x^3) = x^2 \qquad ■$$

■ **EXAMPLE 2** Solve the Initial Value Problem: $y' + (1 - x^{-1})y = x^2$, $y(1) = 2$.

Solution This equation has the form $y' + P(x)y = Q(x)$ with $P(x) = (1 - x^{-1})$. By Eq. (5), an integrating factor is

$$\alpha(x) = e^{\int (1 - x^{-1})\, dx} = e^{x - \ln x} = e^x e^{\ln x^{-1}} = x^{-1} e^x$$

By either multiplying by the integration factor and then integrating both sides of the resulting equation or by applying Eq. (4) with $Q(x) = x^2$, we obtain the general solution:

$$y = \alpha(x)^{-1} \left(\int \alpha(x) Q(x)\, dx \right) = xe^{-x} \left(\int (x^{-1} e^x) x^2\, dx \right)$$

$$= xe^{-x} \left(\int xe^x\, dx \right)$$

Integration by Parts shows that $\int xe^x\, dx = (x - 1)e^x + C$, so we obtain

$$y = xe^{-x}\big((x - 1)e^x + C\big) = x(x - 1) + Cxe^{-x}$$

The initial condition $y(1) = 2$ gives

$$y(1) = 1(1 - 1) + Ce^{-1} = Ce^{-1} = 2 \quad \Rightarrow \quad C = 2e$$

The desired particular solution is

$$y = x(x - 1) + (2e)xe^{-x} = x(x - 1) + 2xe^{1-x} \qquad \blacksquare$$

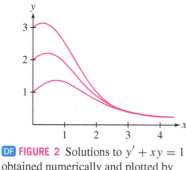

CONCEPTUAL INSIGHT We have expressed the general solution of a first-order linear differential equation in terms of the integrals in Eqs. (4) and (5). Keep in mind, however, that it is not always possible to evaluate these integrals explicitly. For example, the general solution of $y' + xy = 1$ is

$$y = e^{-x^2/2} \left(\int e^{x^2/2}\, dx + C \right)$$

The integral $\int e^{x^2/2}\, dx$ cannot be evaluated in elementary terms. However, we can approximate the integral numerically and plot the solutions by computer (Figure 2).

 FIGURE 2 Solutions to $y' + xy = 1$ obtained numerically and plotted by computer.

In the next example, we use a differential equation to model a "mixing problem," which has applications in biology, chemistry, and medicine.

■ **EXAMPLE 3** **A Mixing Problem** A tank contains 600 liters (L) of water with a sucrose concentration of 0.2 kg/L. We begin adding water with a sucrose concentration of 0.1 kg/L at a rate of $R_{in} = 40$ L/min (Figure 3). The water mixes instantaneously and exits the bottom of the tank at a rate of $R_{out} = 20$ L/min. Let $y(t)$ be the quantity of sucrose in the tank at time t (in minutes). Set up a differential equation for $y(t)$ and solve for $y(t)$.

Solution

Step 1. **Set up the differential equation.**
The derivative dy/dt is the difference of two rates of change, namely the rate at which sucrose enters the tank and the rate at which it leaves:

$$\frac{dy}{dt} = \text{sucrose rate in} - \text{sucrose rate out} \qquad \boxed{6}$$

The rate at which sucrose enters the tank is

$$\text{Sucrose rate in} = \underbrace{(0.1\ \text{kg/L})(40\ \text{L/min})}_{\text{Concentration times water rate in}} = 4\ \text{kg/min}$$

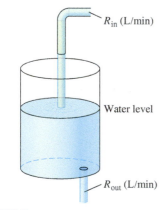

FIGURE 3

Next, we compute the sucrose concentration in the tank at time t. Water flows in at 40 L/min and out at 20 L/min, so there is a net inflow of 20 L/min. The tank has 600 L at time $t = 0$, so it has $600 + 20t$ liters at time t, and

$$\text{Concentration at time } t = \frac{\text{kilograms of sucrose in tank}}{\text{liters of water in tank}} = \frac{y(t)}{600 + 20t} \text{ kg/L}$$

The rate at which sucrose leaves the tank is the product of the concentration and the rate at which water flows out:

$$\text{Sucrose rate out} = \underbrace{\left(\frac{y}{600 + 20t} \frac{\text{kg}}{\text{L}}\right)\left(20 \frac{\text{L}}{\text{min}}\right)}_{\text{Concentration times water rate out}} = \frac{20y}{600 + 20t} = \frac{y}{t + 30} \text{ kg/min}$$

Now Eq. (6) gives us the differential equation

$$\frac{dy}{dt} = 4 - \frac{y}{t + 30} \qquad \boxed{7}$$

Step 2. Find the general solution.

We write Eq. (7) in standard form:

$$\frac{dy}{dt} + \underbrace{\frac{1}{t + 30}}_{P(t)} y = \underbrace{4}_{Q(t)} \qquad \boxed{8}$$

An integrating factor is

$$\alpha(t) = e^{\int P(t)\, dt} = e^{\int dt/(t+30)} = e^{\ln(t+30)} = t + 30$$

The general solution is

$$y(t) = \alpha(t)^{-1}\left(\int \alpha(t) Q(t)\, dt + C\right)$$

$$= \frac{1}{t + 30}\left(\int (t + 30)(4)\, dt + C\right)$$

$$= \frac{1}{t + 30}\left(2(t + 30)^2 + C\right) = 2t + 60 + \frac{C}{t + 30}$$

Step 3. Solve the Initial Value Problem.

At $t = 0$, the tank contains 600 L of water with a sucrose concentration of 0.2 kg/L. Thus, the total sucrose at $t = 0$ is $y(0) = (600)(0.2) = 120$ kg, and

$$y(0) = 2(0) + 60 + \frac{C}{0 + 30} = 60 + \frac{C}{30} = 120 \quad \Rightarrow \quad C = 1800$$

We obtain the following formula (t in minutes), which is valid until the tank overflows:

$$y(t) = 2t + 60 + \frac{1800}{t + 30} \text{ kg sucrose} \qquad ■$$

Summary:

Sucrose rate in = 4 kg/min

Sucrose rate out = $\dfrac{y}{t + 30}$ kg/min

$\dfrac{dy}{dt} = 4 - \dfrac{y}{t + 30}$

$\alpha(t) = t + 30$

$y(t) = 2t + 60 + \dfrac{C}{t + 30}$

9.5 SUMMARY

• A *first-order linear differential equation* can always be written in the form

$$y' + P(x)y = Q(x)$$

• The general solution is

$$y = \alpha(x)^{-1}\left(\int \alpha(x) Q(x)\, dx + C\right)$$

where $\alpha(x)$ is an *integrating factor*: $\alpha(x) = e^{\int P(x)\, dx}$.

9.5 EXERCISES

Preliminary Questions

1. Which of the following are first-order linear equations?

(a) $y' + x^2 y = 1$

(b) $y' + xy^2 = 1$

(c) $x^5 y' + y = e^x$

(d) $x^5 y' + y = e^y$

2. If $\alpha(x)$ is an integrating factor for $y' + A(x)y = B(x)$, then $\alpha'(x)$ is equal to (choose the correct answer):

(a) $B(x)$

(b) $\alpha(x)A(x)$

(c) $\alpha(x)A'(x)$

(d) $\alpha(x)B(x)$

3. For what function P is the integrating factor $\alpha(x)$ equal to x?

4. For what function P is the integrating factor $\alpha(x)$ equal to e^x?

Exercises

1. Consider $y' + x^{-1}y = x^3$.

(a) Verify that $\alpha(x) = x$ is an integrating factor.

(b) Show that when multiplied by $\alpha(x)$, the differential equation can be written $(xy)' = x^4$.

(c) Conclude that xy is an antiderivative of x^4 and use this information to find the general solution.

(d) Find the particular solution satisfying $y(1) = 0$.

2. Consider $\dfrac{dy}{dt} + 2y = e^{-3t}$.

(a) Verify that $\alpha(t) = e^{2t}$ is an integrating factor.

(b) Use Eq. (4) to find the general solution.

(c) Find the particular solution with initial condition $y(0) = 1$.

3. Let $\alpha(x) = e^{x^2}$. Verify the identity

$$(\alpha(x)y)' = \alpha(x)(y' + 2xy)$$

and explain how it is used to find the general solution of

$$y' + 2xy = x$$

4. Find the solution of $y' - y = e^{2x}$, $y(0) = 1$.

In Exercises 5–18, find the general solution of the first-order linear differential equation.

5. $xy' + y = x$

6. $xy' - y = x^2 - x$

7. $3xy' - y = x^{-1}$

8. $y' + xy = x$

9. $y' + 3x^{-1}y = x + x^{-1}$

10. $y' + x^{-1}y = \cos(x^2)$

11. $xy' = y - x$

12. $xy' = x^{-2} - \dfrac{3y}{x}$

13. $y' + y = e^x$

14. $y' + (\sec x)y = \cos x$

15. $y' + (\tan x)y = \cos x$

16. $e^{2x}y' = 1 - e^x y$

17. $y' - (\ln x)y = x^x$

18. $y' + y = \cos x$

In Exercises 19–26, solve the Initial Value Problem.

19. $y' + 3y = e^{2x}$, $y(0) = -1$

20. $xy' + y = e^x$, $y(1) = 3$

21. $y' + \dfrac{1}{x+1}y = x^{-2}$, $y(1) = 2$

22. $y' + y = \sin x$, $y(0) = 1$

23. $(\sin x)y' = (\cos x)y + 1$, $y\left(\dfrac{\pi}{4}\right) = 0$

24. $y' + (\sec t)y = \sec t$, $y\left(\dfrac{\pi}{4}\right) = 1$

25. $y' + (\tanh x)y = 1$, $y(0) = 3$

26. $y' + \dfrac{x}{1+x^2}y = \dfrac{1}{(1+x^2)^{3/2}}$, $y(1) = 0$

27. Find the general solution of $y' + ny = e^{mx}$ for all m, n. *Note:* The case $m = -n$ must be treated separately.

28. Find the general solution of $y' + ny = \cos x$ for all n.

In Exercises 29–32, a 1000-liter (L) tank contains 500 L of water with a salt concentration of 10 g/L. Water with a salt concentration of 50 g/L flows into the tank at a rate of $R_{in} = 80$ L/minutes (min). The fluid mixes instantaneously and is pumped out at a specified rate R_{out}. Let $y(t)$ denote the quantity of salt in the tank at time t.

29. Assume that $R_{out} = 40$ L/min.

(a) Set up and solve the differential equation for $y(t)$.

(b) What is the salt concentration when the tank overflows?

30. Find the salt concentration when the tank overflows, assuming that $R_{out} = 60$ L/min.

31. Find the limiting salt concentration as $t \to \infty$, assuming that $R_{out} = 80$ L/min.

32. Assuming that $R_{out} = 120$ L/min, find $y(t)$. Then calculate the tank volume and the salt concentration at $t = 10$ min.

33. Water flows into a tank at the variable rate of $R_{in} = 20/(1 + t)$ gal/min and out at the constant rate $R_{out} = 5$ gal/min. Let $V(t)$ be the volume of water in the tank at time t.

(a) Set up a differential equation for $V(t)$ and solve it with the initial condition $V(0) = 100$.

(b) Find the maximum value of V.

(c) *CAS* Plot $V(t)$ and estimate the time t when the tank is empty.

34. A stream feeds into a lake at a rate of 1000 m³/day. The stream is polluted with a toxin whose concentration is 5 g/m³. Assume that the lake has volume 10^6 m³ and that water flows out of the lake at the same rate of 1000 m³/day.

(a) Set up a differential equation for the concentration $c(t)$ of toxin in the lake and solve for $c(t)$, assuming that $c(0) = 0$. *Hint:* Find the differential equation for the quantity of toxin $y(t)$, and observe that $c(t) = y(t)/10^6$.

(b) What is the limiting concentration for large t?

In Exercises 35–38, consider a series circuit (Figure 4) consisting of a resistor of R ohms, an inductor of L henries, and a variable voltage source of $V(t)$ volts (time t in seconds). The current through the circuit $I(t)$ (in amperes) satisfies the differential equation

$$\frac{dI}{dt} + \frac{R}{L}I = \frac{1}{L}V(t)$$

9

35. Solve Eq. (9) with initial condition $I(0) = 0$, assuming that $R = 100$ ohms (Ω), $L = 5$ henries (H), and $V(t)$ is constant with $V(t) = 10$ volts (V).

36. Assume that $R = 110$ ohms, $L = 10$ henries, and $V(t) = e^{-t}$ volts.

(a) Solve Eq. (9) with initial condition $I(0) = 0$.

(b) Calculate t_m and $I(t_m)$, where t_m is the time at which $I(t)$ has a maximum value.

(c) GU Use a computer algebra system to sketch the graph of the solution for $0 \le t \le 3$.

37. Assume that $V(t) = V$ is constant and $I(0) = 0$.

(a) Solve for $I(t)$.

(b) Show that $\lim_{t \to \infty} I(t) = V/R$ and that $I(t)$ reaches approximately 63% of its limiting value after L/R seconds.

(c) How long does it take for $I(t)$ to reach 90% of its limiting value if $R = 500$ ohms, $L = 4$ henries, and $V = 20$ volts?

38. Solve for $I(t)$, assuming that $R = 500$ ohms, $L = 4$ henries, and $V = 20\cos(80)$ volts.

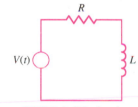

FIGURE 4 RL circuit.

39. 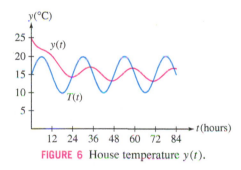 Tank 1 in Figure 5 is filled with V_1 liters of water containing blue dye at an initial concentration of c_0 g/L. Water flows into the tank at a rate of R L/min, is mixed instantaneously with the dye solution, and flows out through the bottom at the same rate R. Let $c_1(t)$ be the dye concentration in the tank at time t.

(a) Explain why c_1 satisfies the differential equation $\dfrac{dc_1}{dt} = -\dfrac{R}{V_1} c_1$.

(b) Solve for $c_1(t)$ with $V_1 = 300$ L, $R = 50$, and $c_0 = 10$ g/L.

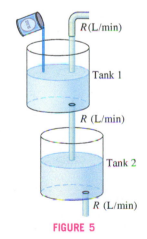

FIGURE 5

40. Continuing with the previous exercise, let Tank 2 be another tank filled with V_2 liters of water. Assume that the dye solution from Tank 1 empties into Tank 2 as in Figure 5, mixes instantaneously, and leaves Tank 2 at the same rate R. Let $c_2(t)$ be the dye concentration in Tank 2 at time t.

(a) Explain why c_2 satisfies the differential equation

$$\frac{dc_2}{dt} = \frac{R}{V_2}(c_1 - c_2)$$

(b) Use the solution to Exercise 39 to solve for $c_2(t)$ if $V_1 = 300$, $V_2 = 200$, $R = 50$, and $c_0 = 10$.

(c) Find the maximum concentration in Tank 2.

(d) GU Plot the solution.

41. Let a, b, r be constants. Show that

$$y = Ce^{-kt} + a + bk\left(\frac{k\sin rt - r\cos rt}{k^2 + r^2}\right)$$

is a general solution of

$$\frac{dy}{dt} = -k\left(y - a - b\sin rt\right)$$

42. Assume that the outside temperature varies as

$$T(t) = 15 + 5\sin(\pi t/12)$$

where $t = 0$ is 12 noon. A house is heated to $25°C$ at $t = 0$ and after that, its temperature $y(t)$ varies according to Newton's Law of Cooling (Figure 6):

$$\frac{dy}{dt} = -0.1\left(y(t) - T(t)\right)$$

Use Exercise 41 to solve for $y(t)$.

FIGURE 6 House temperature $y(t)$.

Further Insights and Challenges

43. Let $\alpha(x)$ be an integrating factor for $y' + P(x)y = Q(x)$. The differential equation $y' + P(x)y = 0$ is called the associated **homogeneous equation**.

(a) Show that $y = 1/\alpha(x)$ is a solution of the associated homogeneous equation.

(b) Show that if $y = f(x)$ is a particular solution of $y' + P(x)y = Q(x)$, then $f(x) + C/\alpha(x)$ is also a solution for any constant C.

44. Use the Fundamental Theorem of Calculus and the Product Rule to verify directly that for any x_0, the function

$$f(x) = \alpha(x)^{-1}\int_{x_0}^{x} \alpha(t)Q(t)\,dt$$

is a solution of the Initial Value Problem

$$y' + P(x)y = Q(x), \qquad y(x_0) = 0$$

where $\alpha(x)$ is an integrating factor [a solution to Eq. (3)].

45. Transient Currents Suppose the circuit described by Eq. (9) is driven by a sinusoidal voltage source $V(t) = V \sin \omega t$ (where V and ω are constant).

(a) Show that

$$I(t) = \frac{V}{R^2 + L^2\omega^2}(R \sin \omega t - L\omega \cos \omega t) + Ce^{-(R/L)t}$$

(b) Let $Z = \sqrt{R^2 + L^2\omega^2}$. Choose θ so that $Z \cos \theta = R$ and $Z \sin \theta = L\omega$. Use the addition formula for the sine function to show that

$$I(t) = \frac{V}{Z} \sin(\omega t - \theta) + Ce^{-(R/L)t}$$

This shows that the current in the circuit varies sinusoidally apart from a DC term (called the **transient current** in electronics) that decreases exponentially.

CHAPTER REVIEW EXERCISES

1. Which of the following differential equations are linear? Determine the order of each equation.

(a) $y' = y^5 - 3x^4 y$

(b) $y' = x^5 - 3x^4 y$

(c) $y = y''' - 3x\sqrt{y}$

(d) $\sin x \cdot y'' = y - 1$

2. Find a value of c such that $y = x - 2 + e^{cx}$ is a solution of $2y' + y = x$.

In Exercises 3–6, solve using Separation of Variables.

3. $\dfrac{dy}{dt} = t^2 y^{-3}$

4. $xyy' = 1 - x^2$

5. $x\dfrac{dy}{dx} - y = 1$

6. $y' = \dfrac{xy^2}{x^2 + 1}$

In Exercises 7–10, solve the Initial Value Problem using Separation of Variables.

7. $y' = \cos^2 x$, $y(0) = \dfrac{\pi}{4}$

8. $y' = \cos^2 y$, $y(0) = \dfrac{\pi}{4}$

9. $y' = xy^2$, $y(1) = 2$

10. $xyy' = 1$, $y(3) = 2$

11. Figure 1 shows the slope field for $\dfrac{dy}{dt} = \sin y + ty$. Sketch the graphs of the solutions with the initial conditions $y(0) = 1$, $y(0) = 0$, and $y(0) = -1$.

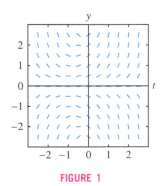

FIGURE 1

12. Sketch the slope field for $\dfrac{dy}{dt} = t^2 y$ for $-2 \le t \le 2, -2 \le y \le 2$.

13. Sketch the slope field for $\dfrac{dy}{dt} = y \sin t$ for $-2\pi \le t \le 2\pi, -2 \le y \le 2$.

14. Which of the equations (i)–(iii) corresponds to the slope field in Figure 2?

(i) $\dfrac{dy}{dt} = 1 - y^2$ **(ii)** $\dfrac{dy}{dt} = 1 + y^2$ **(iii)** $\dfrac{dy}{dt} = y^2$

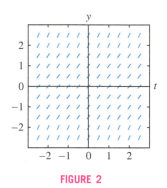

FIGURE 2

15. Let $y(t)$ be the solution to the differential equation with the slope field as shown in Figure 2, satisfying $y(0) = 0$. Sketch the graph of $y(t)$. Then use your answer to Exercise 14 to solve for $y(t)$.

16. Let $y(t)$ be the solution of $4\dfrac{dy}{dt} = y^2 + t$ satisfying $y(2) = 1$. Carry out Euler's Method with time step $h = 0.05$ for $n = 6$ steps.

17. Let $y(t)$ be the solution of $(x^3 + 1)\dfrac{dy}{dt} = y$ satisfying $y(0) = 1$. Compute approximations to $y(0.1)$, $y(0.2)$, and $y(0.3)$ using Euler's Method with time step $h = 0.1$.

In Exercises 18–21, solve using the method of integrating factors.

18. $\dfrac{dy}{dt} = y + t^2$, $y(0) = 4$

19. $\dfrac{dy}{dx} = \dfrac{y}{x} + x$, $y(1) = 3$

20. $\dfrac{dy}{dt} = y - 3t$, $y(-1) = 2$

21. $y' + 2y = 1 + e^{-x}$, $y(0) = -4$

In Exercises 22–29, solve using the appropriate method.

22. $x^2 y' = x^2 + 1$, $y(1) = 10$

23. $y' + (\tan x)y = \cos^2 x$, $y(\pi) = 2$

24. $xy' = 2y + x - 1$, $y(\tfrac{3}{2}) = 9$

25. $(y - 1)y' = t$, $y(1) = -3$

26. $(\sqrt{y} + 1)y' = yte^{t^2}$, $y(0) = 1$

27. $\dfrac{dw}{dx} = k\dfrac{1 + w^2}{x}$, $w(1) = 1$

28. $y' + \dfrac{3y - 1}{t} = t + 2$

29. $y' + \dfrac{y}{x} = \sin x$

30. Find the solutions to $y' = 4(y - 12)$ satisfying $y(0) = 20$ and $y(0) = 0$, and sketch their graphs.

31. Find the solutions to $y' = -2y + 8$ satisfying $y(0) = 3$ and $y(0) = 4$, and sketch their graphs.

32. Show that $y = \sin^{-1} x$ satisfies the differential equation $y' = \sec y$ with initial condition $y(0) = 0$.

33. What is the limit $\lim_{t \to \infty} y(t)$ if $y(t)$ is a solution of each of the following?

(a) $\dfrac{dy}{dt} = -4(y - 12)$ **(b)** $\dfrac{dy}{dt} = 4(y - 12)$

(c) $\dfrac{dy}{dt} = -4y - 12$

In Exercises 34–37, let $P(t)$ denote the balance at time t (years) of an annuity that earns 5% interest continuously compounded and pays out $20,000/year continuously.

34. Find the differential equation satisfied by $P(t)$.

35. Determine $P(5)$ if $P(0) = \$200{,}000$.

36. When does the annuity run out of money if $P(0) = \$300{,}000$?

37. What is the minimum initial balance that will allow the annuity to make payments indefinitely?

38. State whether the differential equation can be solved using Separation of Variables, the method of integrating factors, both, or neither.

(a) $y' = y + x^2$ **(b)** $xy' = y + 1$
(c) $y' = y^2 + x^2$ **(d)** $xy' = y^2$

39. Let A and B be constants. Prove that if $A > 0$, then all solutions of $\dfrac{dy}{dt} + Ay = B$ approach the same limit as $t \to \infty$.

40. At time $t = 0$, a tank of height 5 m in the shape of an inverted pyramid whose cross section at the top is a square of side 2 m is filled with water. Water flows through a hole at the bottom of area 0.002 m^2. Use Torricelli's Law to determine the time required for the tank to empty.

41. The trough in Figure 3 (dimensions in centimeters) is filled with water. At time $t = 0$ (in seconds), water begins leaking through a hole at the bottom of area 4 cm^2. Let $y(t)$ be the water height at time t. Find a differential equation for $y(t)$ and solve it to determine when the water level decreases to 60 cm.

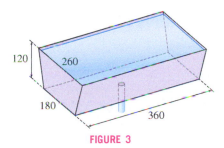

FIGURE 3

42. Find the solution of the logistic equation $\dfrac{dy}{dt} = 0.4y(4 - y)$ satisfying $y(0) = 8$.

43. Let $y(t)$ be the solution of $\dfrac{dy}{dt} = 0.3y(2 - y)$ with $y(0) = 1$. Determine $\lim_{t \to \infty} y(t)$ without solving for y explicitly.

44. Suppose that $y' = ky(1 - y/8)$ has a solution satisfying $y(0) = 12$ and $y(10) = 24$. Find k.

45. A lake has a carrying capacity of 1000 fish. Assume that the fish population grows logistically with growth constant $k = 0.2$ day^{-1}. How many days will it take for the population to reach 900 fish if the initial population is 20 fish?

46. A rabbit population on an island increases exponentially with growth rate $k = 0.12$ months^{-1}. When the population reaches 300 rabbits (say, at time $t = 0$), wolves begin eating the rabbits at a rate of r rabbits per month.

(a) Find a differential equation satisfied by the rabbit population $P(t)$.

(b) How large can r be without the rabbit population becoming extinct?

47. Show that $y = \sin(\tan^{-1} x + C)$ is the general solution of $y' = \sqrt{1 - y^2}/(1 + x^2)$. Then use the addition formula for the sine function to show that the general solution may be written

$$y = \frac{(\cos C)x + \sin C}{\sqrt{1 + x^2}}$$

48. A tank is filled with 300 liters (L) of contaminated water containing 3 kg of toxin. Pure water is pumped in at a rate of 40 L/min, mixes instantaneously, and is then pumped out at the same rate. Let $y(t)$ be the quantity of toxin present in the tank at time t.

(a) Find a differential equation satisfied by $y(t)$.

(b) Solve for $y(t)$.

(c) Find the time at which there is 0.01 kg of toxin present.

49. At $t = 0$, a tank of volume 300 L is filled with 100 L of water containing salt at a concentration of 8 g/L. Fresh water flows in at a rate of 40 L/min, mixes instantaneously, and exits at the same rate. Let $c_1(t)$ be the salt concentration at time t.

(a) Find a differential equation satisfied by $c_1(t)$ *Hint:* Find the differential equation for the quantity of salt $y(t)$, and observe that $c_1(t) = y(t)/100$.

(b) Find the salt concentration $c_1(t)$ in the tank as a function of time.

50. The outflow of the tank in Exercise 49 is directed into a second tank containing V liters of fresh water where it mixes instantaneously and exits at the same rate of 40 L/min. Determine the salt concentration $c_2(t)$ in the second tank as a function of time in the following two cases:

(a) $V = 200$ **(b)** $V = 300$

In each case, determine the maximum concentration.

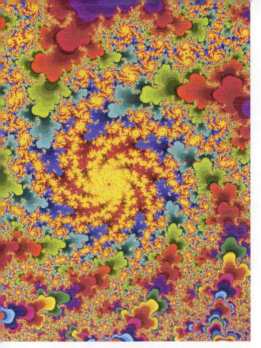

Infinite series allow us to make sense of a process that iterates infinitely, as occurs with fractals, generating pictures such as this one.

(Gregory Sams/Science Source)

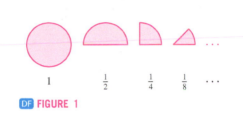

| 1 | $\frac{1}{2}$ | $\frac{1}{4}$ | $\frac{1}{8}$ | $\cdots$ |

DF **FIGURE 1**

| The sequence b_n is the Balmer series of absorption wavelengths of the hydrogen atom in nanometers. It plays a key role in spectroscopy.

10 INFINITE SERIES

The theory of infinite series is a third branch of calculus, in addition to differential and integral calculus. Infinite series yield a new perspective on functions and on many interesting numbers. Two examples are the infinite series for the exponential function

$$e^x = 1 + x + \frac{x^2}{2!} + \frac{x^3}{3!} + \frac{x^4}{4!} + \cdots$$

and the Gregory–Leibniz series (see Exercise 63 in Section 10.2)

$$\frac{\pi}{4} = 1 - \frac{1}{3} + \frac{1}{5} - \frac{1}{7} + \frac{1}{9} - \cdots$$

The first shows that e^x can be expressed as an "infinite polynomial," and the second reveals that π is related to the reciprocals of the odd integers in an unexpected way. To make sense of infinite series, we need to define precisely what it means to add up infinitely many terms. Limits play a key role here, just as they do in differential and integral calculus.

10.1 Sequences

Sequences of numbers appear in diverse situations. If you divide a cake in half, and then divide the remaining half in half, and continue dividing in half indefinitely (Figure 1), then the fraction of cake remaining at each step forms the sequence

$$1, \quad \frac{1}{2}, \quad \frac{1}{4}, \quad \frac{1}{8}, \quad \cdots$$

This is the sequence of values of the function $f(n) = \frac{1}{2^n}$ for $n = 0, 1, 2, \ldots$.

DEFINITION Sequence A **sequence** $\{a_n\}$ is an ordered collection of numbers defined by a function f on a set of sequential integers. The values $a_n = f(n)$ are called the **terms** of the sequence, and n is called the **index**. Informally, we think of a sequence $\{a_n\}$ as a list of terms:

$$a_1, \quad a_2, \quad a_3, \quad a_4, \quad \cdots$$

The sequence does not have to start at $n = 1$. It can start at $n = 0$, $n = 2$, or any other integer.

General term	Domain	Sequence
$a_n = 1 - \dfrac{1}{n}$	$n \geq 1$	$0, \ \dfrac{1}{2}, \ \dfrac{2}{3}, \ \dfrac{3}{4}, \ \dfrac{4}{5}, \ \ldots$
$a_n = (-1)^n n$	$n \geq 0$	$0, \ -1, \ 2, \ -3, \ 4, \ \ldots$
$b_n = \dfrac{364.5n^2}{n^2 - 4}$	$n \geq 3$	$656.1, \ 486, \ 433.9, \ 410.1, \ 396.9, \ \ldots$

Not all sequences are generated by a formula. For instance, here is a sequence:

$$3, \ 1, \ 4, \ 1, \ 5, \ 9, \ 2, \ 6, \ \ldots$$

This sequence is simply the digits of π, and there is no formula for the nth digit of π. When a_n is given by a formula, we refer to a_n as the **general term**.

The sequence in the next example is defined *recursively*. For such a sequence, the first one or more terms may be given, and then the nth term is computed in terms of the preceding terms using some formula.

■ **EXAMPLE 1** **The Fibonacci Sequence** We define the sequence by taking $F_1 = 1$, $F_2 = 1$ and $F_n = F_{n-1} + F_{n-2}$ for all integers $n > 2$. In other words, each subsequent term is obtained by adding together the two preceding terms. From this, every term can easily be determined. Calculating out terms, we find the sequence to be

$$1, \ 1, \ 2, \ 3, \ 5, \ 8, \ 13, \ 21, \ 34 \ldots$$

This particular sequence appears in a surprisingly wide variety of situations, particularly in nature. For instance, the number of spiral arms in a sunflower almost always turns out to be a number from the Fibonacci sequence, as in Figure 2.

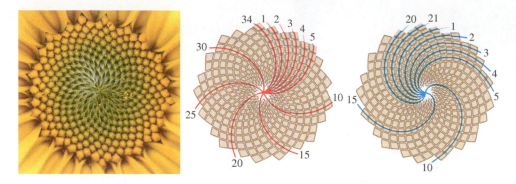

FIGURE 2 The number of spiral arms in a sunflower when counted at different angles are Fibonacci numbers. In the first case, we get 34, and in the second, we get 21. (*Eiji Ueda/Shutterstock*)

If we take the limit of the $n + 1$st term divided by the nth term, we obtain $\phi = \frac{1+\sqrt{5}}{2} \approx$ 1.618, which is known as the Golden Mean. According to the ancient Greeks, this quantity gives the ideal ratio of the sides of a rectangle. The face of the Parthenon was designed to reflect this ratio. ■

■ **EXAMPLE 2** **Recursive Sequence** Compute a_2, a_3, a_4 for the sequence defined recursively by

$$a_1 = 1, \qquad a_n = \frac{1}{2}\left(a_{n-1} + \frac{2}{a_{n-1}}\right)$$

Solution

$$a_2 = \frac{1}{2}\left(a_1 + \frac{2}{a_1}\right) = \frac{1}{2}\left(1 + \frac{2}{1}\right) = \frac{3}{2} = 1.5$$

$$a_3 = \frac{1}{2}\left(a_2 + \frac{2}{a_2}\right) = \frac{1}{2}\left(\frac{3}{2} + \frac{2}{3/2}\right) = \frac{17}{12} \approx 1.4167$$

$$a_4 = \frac{1}{2}\left(a_3 + \frac{2}{a_3}\right) = \frac{1}{2}\left(\frac{17}{12} + \frac{2}{17/12}\right) = \frac{577}{408} \approx 1.414216 \qquad ■$$

You may recognize the sequence in Example 2 as the sequence of approximations to $\sqrt{2} \approx 1.4142136$ produced by Newton's Method with a starting value $a_1 = 1$ (see Section 4.8). As n tends to infinity, a_n approaches $\sqrt{2}$.

Our main goal is to study convergence of sequences. A sequence $\{a_n\}$ converges to a limit L if $|a_n - L|$ becomes arbitrarily small when n is sufficiently large. Here is the formal definition.

DEFINITION **Limit of a Sequence** We say that $\{a_n\}$ **converges to a limit** L, and we write

$$\lim_{n\to\infty} a_n = L \qquad \text{or} \qquad a_n \to L$$

if, for every $\epsilon > 0$, there is a number M such that $|a_n - L| < \epsilon$ for all $n > M$.

• If no limit exists, we say that $\{a_n\}$ **diverges**.
• If the terms increase without bound, we say that $\{a_n\}$ **diverges to infinity**.

If $\{a_n\}$ converges, then its limit L is unique. A good way to visualize the limit is to plot the points $(1, a_1), (2, a_2), (3, a_3), \ldots$, as in Figure 3. The sequence converges to L if, for every $\epsilon > 0$, the plotted points eventually remain within an ϵ-band around the horizontal line $y = L$. Figure 4 shows the plot of a sequence converging to $L = 1$. On the other hand, we can show that the sequence $a_n = \cos n$ in Figure 5 has no limit.

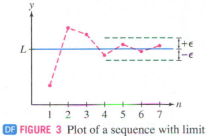

DF FIGURE 3 Plot of a sequence with limit L. For any ϵ, the dots eventually remain within an ϵ-band around L.

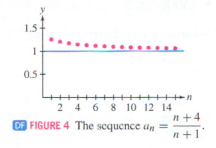

DF FIGURE 4 The sequence $a_n = \dfrac{n+4}{n+1}$.

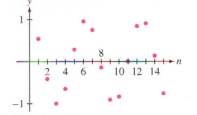

DF FIGURE 5 The sequence $a_n = \cos n$ has no limit.

■ **EXAMPLE 3** **Proving Convergence** Let $a_n = \dfrac{n+4}{n+1}$. Prove formally $\lim\limits_{n\to\infty} a_n = 1$.

Solution The definition requires us to find, for every $\epsilon > 0$, a number M such that

$$|a_n - 1| < \epsilon \qquad \text{for all } n > M \qquad \boxed{1}$$

We have

$$|a_n - 1| = \left| \frac{n+4}{n+1} - 1 \right| = \frac{3}{n+1}$$

Therefore, $|a_n - 1| < \epsilon$ if

$$\frac{3}{n+1} < \epsilon \qquad \text{or} \qquad n > \frac{3}{\epsilon} - 1$$

In other words, Eq. (1) is valid with $M = \frac{3}{\epsilon} - 1$. This proves that $\lim\limits_{n\to\infty} a_n = 1$. ■

Note the following two facts about sequences:

- The limit does not change if we change or drop finitely many terms of the sequence.
- If C is a constant and $a_n = C$ for all n greater than some fixed value N, then $\lim\limits_{n\to\infty} a_n = C$.

Many of the sequences we consider are defined by functions; that is, $a_n = f(n)$ for some function f. For example,

$$a_n = \frac{n-1}{n} \qquad \text{is defined by} \qquad f(x) = \frac{x-1}{x}$$

We will often use the fact that if $f(x)$ approaches a limit L as $x \to \infty$, then the sequence $a_n = f(n)$ approaches the same limit L (Figure 6). Indeed, for all $\epsilon > 0$, we can find a positive real number M so that $|f(x) - L| < \epsilon$ for all $x > M$. It follows automatically that $|f(n) - L| < \epsilon$ for all integers $n > M$.

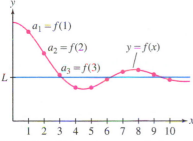

FIGURE 6 If $f(x)$ converges to L, then the sequence $a_n = f(n)$ also converges to L.

> **THEOREM 1** **Sequence Defined by a Function** If $\lim\limits_{x\to\infty} f(x)$ exists, then the sequence $a_n = f(n)$ converges to the same limit:
>
> $$\lim_{n\to\infty} a_n = \lim_{x\to\infty} f(x)$$

■ **EXAMPLE 4** Find the limit of the sequence

$$\frac{2^2 - 2}{2^2}, \quad \frac{3^2 - 2}{3^2}, \quad \frac{4^2 - 2}{4^2}, \quad \frac{5^2 - 2}{5^2}, \quad \cdots$$

Solution This is the sequence with general term

$$a_n = \frac{n^2 - 2}{n^2} = 1 - \frac{2}{n^2}$$

Therefore, we apply Theorem 1 with $f(x) = 1 - \frac{2}{x^2}$:

$$\lim_{n\to\infty} a_n = \lim_{x\to\infty} \left(1 - \frac{2}{x^2} \right) = 1 - \lim_{x\to\infty} \frac{2}{x^2} = 1 - 0 = 1 \qquad ■$$

■ **EXAMPLE 5** Calculate $\lim_{n\to\infty} \dfrac{n+\ln n}{n^2}$.

Solution Apply Theorem 1, using L'Hôpital's Rule in the second step:

$$\lim_{n\to\infty} \frac{n+\ln n}{n^2} = \lim_{x\to\infty} \frac{x+\ln x}{x^2} = \lim_{x\to\infty} \frac{1+(1/x)}{2x} = 0$$ ■

The limit of the Balmer wavelengths b_n in the next example plays a role in physics and chemistry because it determines the ionization energy of the hydrogen atom. Table 1 suggests that b_n approaches 364.5 nanometers (nm). Figure 7 gives the graph, and in Figure 8, the wavelengths are shown "crowding in" toward their limiting value.

TABLE 1
Balmer Wavelengths (nm)

n	b_n
3	656.1
4	486
5	433.9
6	410.1
7	396.9
10	379.7
20	368.2
40	365.4
60	364.9
80	364.7
100	364.6

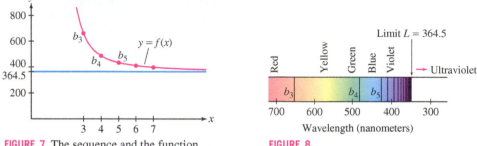

FIGURE 7 The sequence and the function approach the same limit.

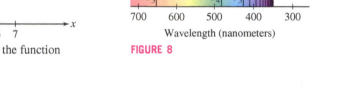

FIGURE 8

■ **EXAMPLE 6** **Balmer Wavelengths** Calculate the limit of the Balmer wavelengths $b_n = \dfrac{364.5n^2}{n^2-4}$ in nanometers, where $n \geq 3$.

Solution Apply Theorem 1 with $f(x) = \dfrac{364.5x^2}{x^2-4}$:

$$\lim_{n\to\infty} b_n = \lim_{x\to\infty} \frac{364.5x^2}{x^2-4} = \lim_{x\to\infty} \frac{364.5x^2\frac{1}{x^2}}{(x^2-4)\frac{1}{x^2}}$$

$$= \lim_{x\to\infty} \frac{364.5}{1-4/x^2} = \frac{364.5}{\lim_{x\to\infty}(1-4/x^2)} = 364.5 \text{ nm}$$ ■

A **geometric sequence** is a sequence $a_n = cr^n$, where c and r are nonzero constants. Each term is r times the previous term; that is, $a_n/a_{n-1} = r$. The number r is called the **common ratio**. For instance, if $r = 3$ and $c = 2$, we obtain the sequence (starting at $n = 0$)

$$2, \quad 2\cdot3, \quad 2\cdot3^2, \quad 2\cdot3^3, \quad 2\cdot3^4, \quad 2\cdot3^5, \quad \ldots$$

In the next example, we determine when a geometric series converges. Recall that $\{a_n\}$ **diverges to** ∞ if the terms a_n increase beyond all bounds (Figure 9); that is,

$$\lim_{n\to\infty} a_n = \infty \quad \text{if, for every number } N, a_n > N \text{ for all sufficiently large } n$$

We define $\lim_{n\to\infty} a_n = -\infty$ similarly.

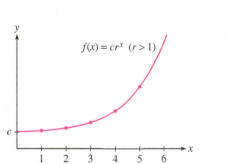

FIGURE 9 If $r > 1$, the geometric sequence $a_n = r^n$ diverges to ∞.

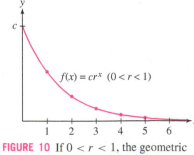

FIGURE 10 If $0 < r < 1$, the geometric sequence $a_n = r^n$ converges to 0.

■ **EXAMPLE 7** **Geometric Sequences with $r \geq 0$** Prove that for $r \geq 0$ and $c > 0$,

$$\lim_{n\to\infty} cr^n = \begin{cases} 0 & \text{if} & 0 \leq r < 1 \\ c & \text{if} & r = 1 \\ \infty & \text{if} & r > 1 \end{cases}$$

Solution Set $f(r) = cr^x$. If $0 \leq r < 1$, then (Figure 10)

$$\lim_{n\to\infty} cr^n = \lim_{x\to\infty} f(x) = c \lim_{x\to\infty} r^x = 0$$

If $r > 1$, then both $f(x)$ and the sequence $\{cr^n\}$ diverge to ∞ (because $c > 0$) (Figure 9). If $r = 1$, then $cr^n = c$ for all n, and the limit is c.   ■

This last example will prove extremely useful when we consider geometric series in Section 10.2.

The limit laws we have used for functions also apply to sequences and are proved in a similar fashion.

THEOREM 2 Limit Laws for Sequences Assume that $\{a_n\}$ and $\{b_n\}$ are convergent sequences with

$$\lim_{n \to \infty} a_n = L, \qquad \lim_{n \to \infty} b_n = M$$

Then:

(i) $\displaystyle\lim_{n \to \infty} (a_n \pm b_n) = \lim_{n \to \infty} a_n \pm \lim_{n \to \infty} b_n = L \pm M$

(ii) $\displaystyle\lim_{n \to \infty} a_n b_n = \left(\lim_{n \to \infty} a_n \right) \left(\lim_{n \to \infty} b_n \right) = LM$

(iii) $\displaystyle\lim_{n \to \infty} \frac{a_n}{b_n} = \frac{\lim\limits_{n \to \infty} a_n}{\lim\limits_{n \to \infty} b_n} = \frac{L}{M}$ if $M \neq 0$

(iv) $\displaystyle\lim_{n \to \infty} c a_n = c \lim_{n \to \infty} a_n = cL$ for any constant c

---

◄·· **REMINDER** $n!$ (*n-factorial*) *is the number*

$$n! = n(n-1)(n-2) \cdots 2 \cdot 1$$

For example, $4! = 4 \cdot 3 \cdot 2 \cdot 1 = 24$. *By definition,* $0! = 1$.

THEOREM 3 Squeeze Theorem for Sequences Let $\{a_n\}$, $\{b_n\}$, $\{c_n\}$ be sequences such that for some number M,

$$b_n \leq a_n \leq c_n \quad \text{for } n > M \qquad \text{and} \qquad \lim_{n \to \infty} b_n = \lim_{n \to \infty} c_n = L$$

Then $\displaystyle\lim_{n \to \infty} a_n = L$.

■ **EXAMPLE 8** Show that if $\displaystyle\lim_{n \to \infty} |a_n| = 0$, then $\displaystyle\lim_{n \to \infty} a_n = 0$.

Solution We have

$$-|a_n| \leq a_n \leq |a_n|$$

By hypothesis, $\displaystyle\lim_{n \to \infty} |a_n| = 0$, and thus also $\displaystyle\lim_{n \to \infty} -|a_n| = -\lim_{n \to \infty} |a_n| = 0$. Therefore, we can apply the Squeeze Theorem to conclude that $\displaystyle\lim_{n \to \infty} a_n = 0$. ■

■ **EXAMPLE 9** Geometric Sequences with $r < 0$ Prove that for $c \neq 0$,

$$\lim_{n \to \infty} cr^n = \begin{cases} 0 & \text{if} \quad -1 < r < 0 \\ \text{diverges} & \text{if} \quad r \leq -1 \end{cases}$$

Solution If $-1 < r < 0$, then $0 < |r| < 1$ and $\displaystyle\lim_{n \to \infty} |cr^n| = 0$ by Example 7. Thus, $\displaystyle\lim_{n \to \infty} cr^n = 0$ by Example 8. If $r = -1$, then the sequence $cr^n = (-1)^n c$ alternates in sign and does not approach a limit. The sequence also diverges if $r < -1$ because cr^n alternates in sign and $|cr^n|$ grows arbitrarily large. ■

As another application of the Squeeze Theorem, consider the sequence

$$a_n = \frac{5^n}{n!}$$

Both the numerator and the denominator grow without bound, so it is not clear in advance whether $\{a_n\}$ converges. Figure 11 and Table 2 suggest that a_n increases initially and then tends to zero. In the next example, we verify that $a_n = R^n/n!$ converges to zero for all R. This fact is used in the discussion of Taylor series in Section 10.7.

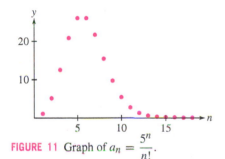

FIGURE 11 Graph of $a_n = \dfrac{5^n}{n!}$.

TABLE 2

n	$a_n = \dfrac{5^n}{n!}$
1	5
2	12.5
3	20.83
4	26.04
10	2.69
15	0.023
20	0.000039
50	2.92×10^{-30}

■ **EXAMPLE 10** Prove that $\lim\limits_{n\to\infty} \dfrac{R^n}{n!} = 0$ for all R.

Solution Assume first that $R > 0$ and let M be the nonnegative integer such that

$$M \le R < M + 1$$

For $n > M$, we write $R^n/n!$ as a product of n factors:

$$\frac{R^n}{n!} = \underbrace{\left(\frac{R}{1}\frac{R}{2}\cdots\frac{R}{M}\right)}_{\text{Call this constant } C}\underbrace{\left(\frac{R}{M+1}\right)\left(\frac{R}{M+2}\right)\cdots\left(\frac{R}{n}\right)}_{\text{Each factor is less than 1}} \le C\left(\frac{R}{n}\right) \qquad \boxed{2}$$

The first M factors are greater than or equal to 1 and the last $n - M$ factors are less than 1. If we lump together the first M factors and call the product C, and drop all the remaining factors except the last factor R/n, we see that

$$0 \le \frac{R^n}{n!} \le \frac{CR}{n}$$

Since $CR/n \to 0$, the Squeeze Theorem gives us $\lim\limits_{n\to\infty} R^n/n! = 0$ as claimed. If $R < 0$, the limit is also zero by Example 8 because $\left|R^n/n!\right|$ tends to zero. ■

Given a sequence $\{a_n\}$ and a function f, we can form the new sequence $f(a_n)$. It is useful to know that if f is continuous and $a_n \to L$, then $f(a_n) \to f(L)$. A proof is given in Appendix D.

THEOREM 4 If f is continuous and $\lim\limits_{n\to\infty} a_n = L$, then

$$\lim_{n\to\infty} f(a_n) = f\left(\lim_{n\to\infty} a_n\right) = f(L)$$

In other words, we may "bring a limit of a sequence inside a continuous function."

■ **EXAMPLE 11** Apply Theorem 4 to the sequence $a_n = \dfrac{3n}{n+1}$ and to the functions

(a) $f(x) = e^x$ and (b) $g(x) = x^2$.

Solution Observe first that

$$L = \lim_{n\to\infty} a_n = \lim_{n\to\infty} \frac{3n}{n+1} = \lim_{n\to\infty} \frac{3}{1 + n^{-1}} = 3$$

(a) With $f(x) = e^x$, we have $f(a_n) = e^{a_n} = e^{\frac{3n}{n+1}}$. According to Theorem 4,

$$\lim_{n\to\infty} f(a_n) = f\left(\lim_{n\to\infty} a_n\right) = e^{\lim\limits_{n\to\infty} \frac{3n}{n+1}} = e^3$$

(b) With $g(x) = x^2$, we have $g(a_n) = a_n^2$. According to Theorem 4,

$$\lim_{n\to\infty} g(a_n) = g\left(\lim_{n\to\infty} a_n\right) = \left(\lim_{n\to\infty} \frac{3n}{n+1}\right)^2 = 3^2 = 9 \qquad ■$$

Of great importance for understanding convergence are the concepts of a bounded sequence and a monotonic sequence.

DEFINITION Bounded Sequences A sequence $\{a_n\}$ is:

- **Bounded from above** if there is a number M such that $a_n \le M$ for all n. The number M is called an *upper bound*.
- **Bounded from below** if there is a number m such that $a_n \ge m$ for all n. The number m is called a *lower bound*.

The sequence $\{a_n\}$ is called **bounded** if it is bounded from above and below. A sequence that is not bounded is called an **unbounded sequence**.

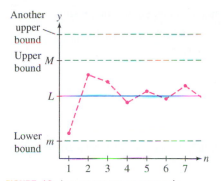

FIGURE 12 A convergent sequence is bounded.

Thus, for instance, the sequence given by $a_n = 3 - \frac{1}{n}$ is clearly bounded above by 3. It is also bounded below by 0, since all the terms are positive. Hence, this sequence is bounded.

Upper and lower bounds are not unique. If M is an upper bound, then any larger number is also an upper bound, and if m is a lower bound, then any smaller number is also a lower bound (Figure 12).

As we might expect, a convergent sequence $\{a_n\}$ is necessarily bounded because the terms a_n get closer and closer to the limit. This fact is recorded in the next theorem.

THEOREM 5 Convergent Sequences Are Bounded If $\{a_n\}$ converges, then $\{a_n\}$ is bounded.

Proof Let $L = \lim\limits_{n\to\infty} a_n$. Then there exists $N > 0$ such that $|a_n - L| < 1$ for $n > N$. In other words,

$$L - 1 < a_n < L + 1 \qquad \text{for } n > N$$

If M is any number larger than $L + 1$ and also larger than the numbers $a_1, a_2, \ldots, a_N$, then $a_n < M$ for all n. Thus, M is an upper bound. Similarly, any number m smaller than $L - 1$ and also smaller than the numbers $a_1, a_2, \ldots, a_N$ is a lower bound. ∎

There are two ways that a sequence $\{a_n\}$ can diverge. One way is by being unbounded. For example, the unbounded sequence $a_n = n$ diverges:

$$1, \quad 2, \quad 3, \quad 4, \quad 5, \quad 6, \quad \ldots$$

However, a sequence can diverge even if it is bounded. This is the case with $a_n = (-1)^{n+1}$, whose terms a_n bounce back and forth but never settle down to approach a limit:

$$1, \quad -1, \quad 1, \quad -1, \quad 1, \quad -1, \quad \ldots$$

There is no surefire method for determining whether a sequence $\{a_n\}$ converges, unless the sequence happens to be both bounded and **monotonic**. By definition, $\{a_n\}$ is monotonic if it is either increasing or decreasing:

- $\{a_n\}$ is *increasing* if $a_n < a_{n+1}$ for all n.
- $\{a_n\}$ is *decreasing* if $a_n > a_{n+1}$ for all n.

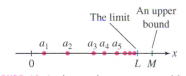

FIGURE 13 An increasing sequence with upper bound M approaches a limit L.

Intuitively, if $\{a_n\}$ is increasing and bounded above by M, then the terms must bunch up near some limiting value L that is not greater than M (Figure 13). See Appendix B for a proof of the next theorem.

THEOREM 6 Bounded Monotonic Sequences Converge

- If $\{a_n\}$ is increasing and $a_n \leq M$, then $\{a_n\}$ converges and $\lim\limits_{n\to\infty} a_n \leq M$.
- If $\{a_n\}$ is decreasing and $a_n \geq m$, then $\{a_n\}$ converges and $\lim\limits_{n\to\infty} a_n \geq m$.

■ **EXAMPLE 12** Verify that $a_n = \sqrt{n+1} - \sqrt{n}$ is decreasing and bounded below. Does $\lim\limits_{n\to\infty} a_n$ exist?

Solution The function $f(x) = \sqrt{x+1} - \sqrt{x}$ is decreasing because its derivative is negative:

$$f'(x) = \frac{1}{2\sqrt{x+1}} - \frac{1}{2\sqrt{x}} < 0 \qquad \text{for } x > 0$$

It follows that $a_n = f(n)$ is decreasing (see Table 3). Furthermore, $a_n > 0$ for all n, so the sequence has lower bound $m = 0$. Theorem 6 guarantees that $L = \lim\limits_{n\to\infty} a_n$ exists and $L \geq 0$. In fact, we can show that $L = 0$ by noting that $f(x)$ can be rewritten as

$$f(x) = \frac{1}{\sqrt{x+1} + \sqrt{x}}. \text{ Hence, } \lim\limits_{x\to\infty} f(x) = 0.$$ ■

DF TABLE 3

$a_n = \sqrt{n+1} - \sqrt{n}$
$a_1 \approx 0.4142$
$a_2 \approx 0.3178$
$a_3 \approx 0.2679$
$a_4 \approx 0.2361$
$a_5 \approx 0.2134$
$a_6 \approx 0.1963$
$a_7 \approx 0.1827$
$a_8 \approx 0.1716$

■ **EXAMPLE 13** Show that the following sequence is bounded and increasing:

$$a_1 = \sqrt{2}, \quad a_2 = \sqrt{2\sqrt{2}}, \quad a_3 = \sqrt{2\sqrt{2\sqrt{2}}}, \quad \dots$$

Then prove that $L = \lim_{n\to\infty} a_n$ exists and compute its value.

Solution

Step 1. **Show that $\{a_n\}$ is bounded above.**

We claim that $M = 2$ is an upper bound. We certainly have $a_1 < 2$ because $a_1 = \sqrt{2} \approx 1.414$. On the other hand,

$$\text{if} \quad a_n < 2, \quad \text{then} \quad a_{n+1} < 2 \qquad \boxed{3}$$

is true because $a_{n+1} = \sqrt{2a_n} < \sqrt{2 \cdot 2} = 2$. Now, since $a_1 < 2$, we can apply (3) to conclude that $a_2 < 2$. Similarly, $a_2 < 2$ implies $a_3 < 2$, etc. for all n. (Formally speaking, this is a proof by induction.)

Step 2. **Show that $\{a_n\}$ is increasing.**

Since a_n is positive and $a_n < 2$, we have

$$a_{n+1} = \sqrt{2a_n} > \sqrt{a_n \cdot a_n} = a_n$$

This shows that $\{a_n\}$ is increasing. Since the sequence is bounded above and increasing, we conclude that the limit L exists.

Now that we know the limit L exists, we can find its value as follows. The idea is that L "contains a copy" of itself under the square root sign:

$$L = \sqrt{2\sqrt{2\sqrt{2\sqrt{2\cdots}}}} = \sqrt{2\left(\sqrt{2\sqrt{2\sqrt{2\cdots}}}\right)} = \sqrt{2L}$$

Thus, $L^2 = 2L$, which implies that $L = 2$ or $L = 0$. We eliminate $L = 0$ because the terms a_n are positive and increasing (as shown below), so we must have $L = 2$ (see Table 4).

This argument is phrased more formally by noting that the sequence is defined recursively by

$$a_1 = \sqrt{2}, \qquad a_{n+1} = \sqrt{2a_n}$$

If a_n converges to L, then the sequence $b_n = a_{n+1}$ also converges to L (because it is the same sequence, with terms shifted one to the left). Then, applying Theorem 4 to $f(x) = \sqrt{x}$, we have

$$L = \lim_{n\to\infty} a_{n+1} = \lim_{n\to\infty} \sqrt{2a_n} = \sqrt{2\lim_{n\to\infty} a_n} = \sqrt{2L}$$

Hence, $L = 2$. ■

DF TABLE 4   Recursive Sequence $a_{n+1} = \sqrt{2a_n}$

a_1	1.4142
a_2	1.6818
a_3	1.8340
a_4	1.9152
a_5	1.9571
a_6	1.9785
a_7	1.9892
a_8	1.9946

10.1 SUMMARY

- A sequence $\{a_n\}$ *converges* to a limit L if, for every $\epsilon > 0$, there is a number M such that

$$|a_n - L| < \epsilon \qquad \text{for all } n > M$$

We write $\lim_{n\to\infty} a_n = L$ or $a_n \to L$.
- If no limit exists, we say that $\{a_n\}$ *diverges*.
- In particular, if the terms increase without bound, we say that $\{a_n\}$ diverges to infinity.
- If $a_n = f(n)$ and $\lim_{x\to\infty} f(x) = L$, then $\lim_{n\to\infty} a_n = L$.
- A *geometric sequence* is a sequence $a_n = cr^n$, where c and r are nonzero. It converges to 0 for $-1 < r < 1$, converges to c for $r = 1$, and diverges otherwise.
- The Basic Limit Laws and the Squeeze Theorem apply to sequences.

- If f is continuous and $\lim_{n\to\infty} a_n = L$, then $\lim_{n\to\infty} f(a_n) = f(L)$.
- A sequence $\{a_n\}$ is
  - *bounded above* by M if $a_n \le M$ for all n.
  - *bounded below* by m if $a_n \ge m$ for all n.

If $\{a_n\}$ is bounded above and below, $\{a_n\}$ is called *bounded*.
- A sequence $\{a_n\}$ is *monotonic* if it is increasing ($a_n < a_{n+1}$) or decreasing ($a_n > a_{n+1}$).
- Bounded monotonic sequences converge (Theorem 6).

10.1 EXERCISES

Preliminary Questions

1. What is a_4 for the sequence $a_n = n^2 - n$?

2. Which of the following sequences converge to zero?

(a) $\dfrac{n^2}{n^2 + 1}$ (b) 2^n (c) $\left(\dfrac{-1}{2}\right)^n$

3. Let a_n be the nth decimal approximation to $\sqrt{2}$. That is, $a_1 = 1$, $a_2 = 1.4$, $a_3 = 1.41$, etc. What is $\lim_{n\to\infty} a_n$?

4. Which of the following sequences is defined recursively?

(a) $a_n = \sqrt{4 + n}$ (b) $b_n = \sqrt{4 + b_{n-1}}$

5. Theorem 5 says that every convergent sequence is bounded. Determine if the following statements are true or false, and if false, give a counterexample:

(a) If $\{a_n\}$ is bounded, then it converges.

(b) If $\{a_n\}$ is not bounded, then it diverges.

(c) If $\{a_n\}$ diverges, then it is not bounded.

Exercises

1. Match each sequence with its general term:

$a_1, a_2, a_3, a_4, \ldots$	General term
(a) $\frac{1}{2}, \frac{2}{3}, \frac{3}{4}, \frac{4}{5}, \ldots$	(i) $\cos \pi n$
(b) $-1, 1, -1, 1, \ldots$	(ii) $\dfrac{n!}{2^n}$
(c) $1, -1, 1, -1, \ldots$	(iii) $(-1)^{n+1}$
(d) $\frac{1}{2}, \frac{2}{4}, \frac{6}{8}, \frac{24}{16} \ldots$	(iv) $\dfrac{n}{n+1}$

2. Let $a_n = \dfrac{1}{2n - 1}$ for $n = 1, 2, 3, \ldots$. Write out the first three terms of the following sequences.

(a) $b_n = a_{n+1}$ (b) $c_n = a_{n+3}$
(c) $d_n = a_n^2$ (d) $e_n = 2a_n - a_{n+1}$

In Exercises 3–12, calculate the first four terms of the sequence, starting with $n = 1$.

3. $c_n = \dfrac{3^n}{n!}$ **4.** $b_n = \dfrac{(2n - 1)!}{n!}$

5. $a_1 = 2, \quad a_{n+1} = 2a_n^2 - 3$

6. $b_1 = 1, \quad b_n = b_{n-1} + \dfrac{1}{b_{n-1}}$

7. $b_n = 5 + \cos \pi n$ **8.** $c_n = (-1)^{2n+1}$

9. $c_n = 1 + \dfrac{1}{2} + \dfrac{1}{3} + \cdots + \dfrac{1}{n}$

10. $a_n = n + (n + 1) + (n + 2) + \cdots + (2n)$

11. $b_1 = 2, \quad b_2 = 3, \quad b_n = 2b_{n-1} + b_{n-2}$

12. $c_n = n$-place decimal approximation to e

13. Find a formula for the nth term of each sequence.

(a) $\dfrac{1}{1}, \dfrac{-1}{8}, \dfrac{1}{27}, \ldots$ (b) $\dfrac{2}{6}, \dfrac{3}{7}, \dfrac{4}{8}, \ldots$

14. Suppose that $\lim_{n\to\infty} a_n = 4$ and $\lim_{n\to\infty} b_n = 7$. Determine:

(a) $\lim_{n\to\infty} (a_n + b_n)$ (b) $\lim_{n\to\infty} a_n^3$

(c) $\lim_{n\to\infty} \cos(\pi b_n)$ (d) $\lim_{n\to\infty} (a_n^2 - 2a_n b_n)$

In Exercises 15–26, use Theorem 1 to determine the limit of the sequence or state that the sequence diverges.

15. $a_n = 12$ **16.** $a_n = 20 - \dfrac{4}{n^2}$

17. $b_n = \dfrac{5n - 1}{12n + 9}$ **18.** $a_n = \dfrac{4 + n - 3n^2}{4n^2 + 1}$

19. $c_n = -2^{-n}$ **20.** $z_n = \left(\dfrac{1}{3}\right)^n$

21. $c_n = 9^n$ **22.** $z_n = 10^{-1/n}$

23. $a_n = \dfrac{n}{\sqrt{n^2 + 1}}$ **24.** $a_n = \dfrac{n}{\sqrt{n^3 + 1}}$

25. $a_n = \ln\left(\dfrac{12n + 2}{-9 + 4n}\right)$ **26.** $r_n = \ln n - \ln(n^2 + 1)$

In Exercises 27–30, use Theorem 4 to determine the limit of the sequence.

27. $a_n = \sqrt{4 + \dfrac{1}{n}}$ **28.** $a_n = e^{4n/(3n+9)}$

29. $a_n = \cos^{-1}\left(\dfrac{n^3}{2n^3 + 1}\right)$ **30.** $a_n = \tan^{-1}(e^{-n})$

31. Let $a_n = \dfrac{n}{n + 1}$. Find a number M such that:

(a) $|a_n - 1| \le 0.001$ for $n \ge M$.

(b) $|a_n - 1| \le 0.00001$ for $n \ge M$.

Then use the limit definition to prove that $\lim_{n\to\infty} a_n = 1$.

32. Let $b_n = \left(\frac{1}{3}\right)^n$.

(a) Find a value of M such that $|b_n| \leq 10^{-5}$ for $n \geq M$.

(b) Use the limit definition to prove that $\lim_{n \to \infty} b_n = 0$.

33. Use the limit definition to prove that $\lim_{n \to \infty} n^{-2} = 0$.

34. Use the limit definition to prove that $\lim_{n \to \infty} \frac{n}{n + n^{-1}} = 1$.

In Exercises 35–62, use the appropriate limit laws and theorems to determine the limit of the sequence or show that it diverges.

35. $a_n = 10 + \left(-\frac{1}{9}\right)^n$

36. $d_n = \sqrt{n+3} - \sqrt{n}$

37. $c_n = 1.01^n$

38. $b_n = e^{1-n^2}$

39. $a_n = 2^{1/n}$

40. $b_n = n^{1/n}$

41. $c_n = \frac{9^n}{n!}$

42. $a_n = \frac{8^{2n}}{n!}$

43. $a_n = \frac{3n^2 + n + 2}{2n^2 - 3}$

44. $a_n = \frac{\sqrt{n}}{\sqrt{n} + 4}$

45. $a_n = \frac{\cos n}{n}$

46. $c_n = \frac{(-1)^n}{\sqrt{n}}$

47. $d_n = \ln 5^n - \ln n!$

48. $d_n = \ln(n^2 + 4) - \ln(n^2 - 1)$

49. $a_n = \left(2 + \frac{4}{n^2}\right)^{1/3}$

50. $b_n = \tan^{-1}\left(1 - \frac{2}{n}\right)$

51. $c_n = \ln\left(\frac{2n+1}{3n+4}\right)$

52. $c_n = \frac{n}{n + n^{1/n}}$

53. $y_n = \frac{e^n}{2^n}$

54. $a_n = \frac{n}{2^n}$

55. $y_n = \frac{e^n + (-3)^n}{5^n}$

56. $b_n = \frac{(-1)^n n^3 + 2^{-n}}{3n^3 + 4^{-n}}$

57. $a_n = n \sin \frac{\pi}{n}$

58. $b_n = \frac{n!}{\pi^n}$

59. $b_n = \frac{3 - 4^n}{2 + 7 \cdot 4^n}$

60. $a_n = \frac{3 - 4^n}{2 + 7 \cdot 3^n}$

61. $a_n = \left(1 + \frac{1}{n}\right)^n$

62. $a_n = \left(1 + \frac{1}{n^2}\right)^n$

In Exercises 63–66, find the limit of the sequence using L'Hôpital's Rule.

63. $a_n = \frac{(\ln n)^2}{n}$

64. $b_n = \sqrt{n} \ln\left(1 + \frac{1}{n}\right)$

65. $c_n = n(\sqrt{n^2 + 1} - n)$

66. $d_n = n^2(\sqrt[3]{n^3 + 1} - n)$

In Exercises 67–70, use the Squeeze Theorem to evaluate $\lim_{n \to \infty} a_n$ by verifying the given inequality.

67. $a_n = \frac{1}{\sqrt{n^4 + n^8}}$, $\quad \frac{1}{\sqrt{2n^4}} \leq a_n \leq \frac{1}{\sqrt{2n^2}}$

68. $c_n = \frac{1}{\sqrt{n^2 + 1}} + \frac{1}{\sqrt{n^2 + 2}} + \cdots + \frac{1}{\sqrt{n^2 + n}}$,

$\frac{n}{\sqrt{n^2 + n}} \leq c_n \leq \frac{n}{\sqrt{n^2 + 1}}$

69. $a_n = (2^n + 3^n)^{1/n}$, $\quad 3 \leq a_n \leq (2 \cdot 3^n)^{1/n} = 2^{1/n} \cdot 3$

70. $a_n = (n + 10^n)^{1/n}$, $\quad 10 \leq a_n \leq (2 \cdot 10^n)^{1/n}$

71. Which of the following statements is equivalent to the assertion $\lim_{n \to \infty} a_n = L$? Explain.

(a) For every $\epsilon > 0$, the interval $(L - \epsilon, L + \epsilon)$ contains at least one element of the sequence $\{a_n\}$.

(b) For every $\epsilon > 0$, the interval $(L - \epsilon, L + \epsilon)$ contains all but at most finitely many elements of the sequence $\{a_n\}$.

72. Show that $a_n = \frac{1}{2n + 1}$ is decreasing.

73. Show that $a_n = \frac{3n^2}{n^2 + 2}$ is increasing. Find an upper bound.

74. Show that $a_n = \sqrt[3]{n+1} - n$ is decreasing.

75. Give an example of a divergent sequence $\{a_n\}$ such that $\lim_{n \to \infty} |a_n|$ converges.

76. Give an example of divergent sequences $\{a_n\}$ and $\{b_n\}$ such that $\{a_n + b_n\}$ converges.

77. Using the limit definition, prove that if $\{a_n\}$ converges and $\{b_n\}$ diverges, then $\{a_n + b_n\}$ diverges.

78. Use the limit definition to prove that if $\{a_n\}$ is a convergent sequence of integers with limit L, then there exists a number M such that $a_n = L$ for all $n \geq M$.

79. Theorem 1 states that if $\lim_{x \to \infty} f(x) = L$, then the sequence $a_n = f(n)$ converges and $\lim_{n \to \infty} a_n = L$. Show that the *converse* is false. In other words, find a function f such that $a_n = f(n)$ converges but $\lim_{x \to \infty} f(x)$ does not exist.

80. Use the limit definition to prove that the limit does not change if a finite number of terms are added or removed from a convergent sequence.

81. Let $b_n = a_{n+1}$. Use the limit definition to prove that if $\{a_n\}$ converges, then $\{b_n\}$ also converges and $\lim_{n \to \infty} a_n = \lim_{n \to \infty} b_n$.

82. Let $\{a_n\}$ be a sequence such that $\lim_{n \to \infty} |a_n|$ exists and is nonzero. Show that $\lim_{n \to \infty} a_n$ exists if and only if there exists an integer M such that the sign of a_n does not change for $n > M$.

83. Proceed as in Example 13 to show that the sequence $\sqrt{3}, \sqrt{3\sqrt{3}}$, $\sqrt{3\sqrt{3\sqrt{3}}}, \ldots$ is increasing and bounded above by $M = 3$. Then prove that the limit exists and find its value.

84. Let $\{a_n\}$ be the sequence defined recursively by

$$a_0 = 0, \qquad a_{n+1} = \sqrt{2 + a_n}$$

Thus, $a_1 = \sqrt{2}$, $a_2 = \sqrt{2 + \sqrt{2}}$, $a_3 = \sqrt{2 + \sqrt{2 + \sqrt{2}}}, \ldots$.

(a) Show that if $a_n < 2$, then $a_{n+1} < 2$. Conclude by induction that $a_n < 2$ for all n.

(b) Show that if $a_n < 2$, then $a_n \leq a_{n+1}$. Conclude by induction that $\{a_n\}$ is increasing.

(c) Use (a) and (b) to conclude that $L = \lim_{n \to \infty} a_n$ exists. Then compute L by showing that $L = \sqrt{2 + L}$.

Further Insights and Challenges

85. Show that $\lim\limits_{n\to\infty} \sqrt[n]{n!} = \infty$. *Hint:* Verify that $n! \geq (n/2)^{n/2}$ by observing that half of the factors of $n!$ are greater than or equal to $n/2$.

86. Let $b_n = \dfrac{\sqrt[n]{n!}}{n}$.

(a) Show that $\ln b_n = \dfrac{1}{n} \sum\limits_{k=1}^{n} \ln \dfrac{k}{n}$.

(b) Show that $\ln b_n$ converges to $\displaystyle\int_0^1 \ln x \, dx$, and conclude that $b_n \to e^{-1}$.

87. Given positive numbers $a_1 < b_1$, define two sequences recursively by

$$a_{n+1} = \sqrt{a_n b_n}, \qquad b_{n+1} = \frac{a_n + b_n}{2}$$

(a) Show that $a_n \leq b_n$ for all n (Figure 14).

(b) Show that $\{a_n\}$ is increasing and $\{b_n\}$ is decreasing.

(c) Show that $b_{n+1} - a_{n+1} \leq \dfrac{b_n - a_n}{2}$.

(d) Prove that both $\{a_n\}$ and $\{b_n\}$ converge and have the same limit. This limit, denoted $\mathrm{AGM}(a_1, b_1)$, is called the **arithmetic-geometric mean** of a_1 and b_1.

(e) Estimate $\mathrm{AGM}(1, \sqrt{2})$ to three decimal places.

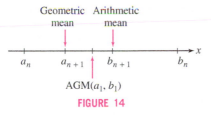

Geometric mean — Arithmetic mean

a_n a_{n+1} b_{n+1} b_n

$\mathrm{AGM}(a_1, b_1)$

FIGURE 14

88. Let $c_n = \dfrac{1}{n} + \dfrac{1}{n+1} + \dfrac{1}{n+2} + \cdots + \dfrac{1}{2n}$.

(a) Calculate c_1, c_2, c_3, c_4.

(b) Use a comparison of rectangles with the area under $y = x^{-1}$ over the interval $[n, 2n]$ to prove that

$$\int_n^{2n} \frac{dx}{x} + \frac{1}{2n} \leq c_n \leq \int_n^{2n} \frac{dx}{x} + \frac{1}{n}$$

(c) Use the Squeeze Theorem to determine $\lim\limits_{n\to\infty} c_n$.

89. Let $a_n = H_n - \ln n$, where H_n is the nth harmonic number:

$$H_n = 1 + \frac{1}{2} + \frac{1}{3} + \cdots + \frac{1}{n}$$

(a) Show that $a_n \geq 0$ for $n \geq 1$. *Hint:* Show that $H_n \geq \displaystyle\int_1^{n+1} \frac{dx}{x}$.

(b) Show that $\{a_n\}$ is decreasing by interpreting $a_n - a_{n+1}$ as an area.

(c) Prove that $\lim\limits_{n\to\infty} a_n$ exists.

This limit, denoted γ, is known as *Euler's Constant*. It appears in many areas of mathematics, including analysis and number theory, and has been calculated to more than 100 million decimal places, but it is still not known whether γ is an irrational number. The first 10 digits are $\gamma \approx 0.5772156649$.

10.2 Summing an Infinite Series

Many quantities that arise in applications cannot be computed exactly. We cannot write down an exact decimal expression for the number π or for values of the sine function such as $\sin 1$. However, sometimes these quantities can be represented as infinite sums. For example, using Taylor series (Section 10.7), we can show that

$$\sin 1 = 1 - \frac{1}{3!} + \frac{1}{5!} - \frac{1}{7!} + \frac{1}{9!} - \frac{1}{11!} + \cdots \qquad \boxed{1}$$

Infinite sums of this type are called **infinite series**. We think of them as having been obtained by adding up all of the terms in an infinite sequence.

But what precisely does Eq. (1) mean? It is impossible to add up infinitely many numbers, but what we can do is compute the **partial sums** S_N, defined as the finite sum of the terms up to and including the Nth term. Here are the first five partial sums of the infinite series for $\sin 1$:

$$S_1 = 1$$

$$S_2 = 1 - \frac{1}{3!} = 1 - \frac{1}{6} \qquad\qquad \approx 0.833$$

$$S_3 = 1 - \frac{1}{3!} + \frac{1}{5!} = 1 - \frac{1}{6} + \frac{1}{120} \qquad\qquad \approx 0.841667$$

$$S_4 = 1 - \frac{1}{6} + \frac{1}{120} - \frac{1}{5040} \qquad\qquad \approx 0.841468$$

$$S_5 = 1 - \frac{1}{6} + \frac{1}{120} - \frac{1}{5040} + \frac{1}{362{,}880} \approx \mathbf{0.8414709846}$$

Compare these values with the value obtained from a calculator:

$$\sin 1 \approx \mathbf{0.8414709848079} \qquad \text{(calculator value)}$$

We see that S_5 differs from $\sin 1$ by less than 10^{-9}. This suggests that the partial sums converge to $\sin 1$, and in fact, we will prove that

$$\sin 1 = \lim_{N \to \infty} S_N$$

(see Example 2 in Section 10.7). So although we cannot add up infinitely many numbers, it makes sense to *define* the sum of an infinite series as a limit of partial sums.

In general, an infinite series is an expression of the form

$$\sum_{n=1}^{\infty} a_n = a_1 + a_2 + a_3 + a_4 + \cdots$$

where $\{a_n\}$ is any sequence. For example,

- *Infinite series may begin with any value for the index. For example,*

$$\sum_{n=3}^{\infty} \frac{1}{n} = \frac{1}{3} + \frac{1}{4} + \frac{1}{5} + \cdots$$

When it is not necessary to specify the starting point, we write simply $\sum a_n$.
- *Any letter may be used for the index. Thus, we may write a_m, a_k, a_i, etc.*

Sequence	General term	Infinite series
$\frac{1}{3}, \frac{1}{9}, \frac{1}{27}, \dots$	$a_n = \frac{1}{3^n}$	$\displaystyle\sum_{n=1}^{\infty} \frac{1}{3^n} = \frac{1}{3} + \frac{1}{9} + \frac{1}{27} + \frac{1}{81} + \cdots$
$\frac{1}{1}, \frac{1}{4}, \frac{1}{9}, \frac{1}{16}, \dots$	$a_n = \frac{1}{n^2}$	$\displaystyle\sum_{n=1}^{\infty} \frac{1}{n^2} = \frac{1}{1} + \frac{1}{4} + \frac{1}{9} + \frac{1}{16} + \cdots$

The Nth partial sum S_N is the finite sum of the terms up to and including a_N:

$$S_N = \sum_{n=1}^{N} a_n = a_1 + a_2 + a_3 + \cdots + a_N$$

If the series begins at k, then $S_N = a_k + a_{k+1} + \cdots + a_N$.

DEFINITION Convergence of an Infinite Series   An infinite series $\displaystyle\sum_{n=k}^{\infty} a_n$ *converges* to the sum S if the sequence of its partial sums $\{S_N\}$ converge to S:

$$\lim_{N \to \infty} S_N = S$$

In this case, we write $S = \displaystyle\sum_{n=k}^{\infty} a_n$.

- If the limit does not exist, we say that the infinite series *diverges*.
- If the limit is infinite, we say that the infinite series *diverges to infinity*.

We can investigate series numerically by computing several partial sums S_N. If the sequence of partial sums shows a trend of convergence to some number S, then we have evidence (but not proof) that the series converges to S. The next example treats a **telescoping series**, where the partial sums are particularly easy to evaluate.

■ **EXAMPLE 1** **Telescoping Series** Investigate numerically:

$$S = \sum_{n=1}^{\infty} \frac{1}{n(n+1)} = \frac{1}{1(2)} + \frac{1}{2(3)} + \frac{1}{3(4)} + \frac{1}{4(5)} + \cdots$$

Then compute the sum S using the identity:

$$\frac{1}{n(n+1)} = \frac{1}{n} - \frac{1}{n+1}$$

DF TABLE 1 Partial
Sums for $\displaystyle\sum_{n=1}^{\infty} \frac{1}{n(n+1)}$

N	S_N
10	0.90909
50	0.98039
100	0.990099
200	0.995025
300	0.996678

In most cases (apart from telescoping series and the geometric series introduced below), there is no simple formula like Eq. (2) for the partial sum S_N. Therefore, we shall develop techniques that do not rely on formulas for S_N.

Solution The values of the partial sums listed in Table 1 suggest convergence to $S = 1$. To prove this, we observe that because of the identity, each partial sum collapses down to just two terms:

$$S_1 = \frac{1}{1(2)} = \frac{1}{1} - \frac{1}{2}$$

$$S_2 = \frac{1}{1(2)} + \frac{1}{2(3)} = \left(\frac{1}{1} - \frac{1}{2}\right) + \left(\frac{1}{2} - \frac{1}{3}\right) = 1 - \frac{1}{3}$$

$$S_3 = \frac{1}{1(2)} + \frac{1}{2(3)} + \frac{1}{3(4)} = \left(\frac{1}{1} - \frac{1}{2}\right) + \left(\frac{1}{2} - \frac{1}{3}\right) + \left(\frac{1}{3} - \frac{1}{4}\right) = 1 - \frac{1}{4}$$

In general,

$$S_N = \left(\frac{1}{1} - \frac{1}{2}\right) + \left(\frac{1}{2} - \frac{1}{3}\right) + \cdots + \left(\frac{1}{N-1} - \frac{1}{N}\right) + \left(\frac{1}{N} - \frac{1}{N+1}\right)$$

$$= 1 - \frac{1}{N+1}$$

2

The sum S is the limit of the sequence of partial sums:

$$S = \lim_{N\to\infty} S_N = \lim_{N\to\infty} \left(1 - \frac{1}{N+1}\right) = 1 \qquad \blacksquare$$

It is important to keep in mind the difference between a sequence $\{a_n\}$ and an infinite series $\displaystyle\sum_{n=1}^{\infty} a_n$.

■ **EXAMPLE 2** Sequences Versus Series Discuss the difference between $\{a_n\}$ and $\displaystyle\sum_{n=1}^{\infty} a_n$, where $a_n = \frac{1}{n(n+1)}$.

Make sure you understand the difference between sequences and series.

• *With a sequence, we consider the limit of the individual terms a_n.*
• *With a series, we are interested in the sum of the terms*

$$a_1 + a_2 + a_3 + \cdots$$

which is defined as the limit of the sequence of partial sums.

Solution The sequence is the list of numbers $\frac{1}{1(2)}, \frac{1}{2(3)}, \frac{1}{3(4)}, \dots$. This sequence converges to zero:

$$\lim_{n\to\infty} a_n = \lim_{n\to\infty} \frac{1}{n(n+1)} = 0$$

The infinite series is the *sum* of the numbers a_n, defined formally as the limit of the sequence of partial sums. This sum is not zero. In fact, the sum is equal to 1 by Example 1:

$$\sum_{n=1}^{\infty} a_n = \sum_{n=1}^{\infty} \frac{1}{n(n+1)} = \frac{1}{1(2)} + \frac{1}{2(3)} + \frac{1}{3(4)} + \cdots = 1 \qquad \blacksquare$$

The next theorem shows that infinite series may be added or subtracted like ordinary sums, *provided that the series converge.*

THEOREM 1 Linearity of Infinite Series If $\sum a_n$ and $\sum b_n$ converge, then $\sum (a_n \pm b_n)$ and $\sum c a_n$ also converge (c any constant), and

$$\sum a_n + \sum b_n = \sum (a_n + b_n)$$

$$\sum a_n - \sum b_n = \sum (a_n - b_n)$$

$$\sum c a_n = c \sum a_n \qquad (c \text{ any constant})$$

Proof These rules follow from the corresponding linearity rules for limits. For example,

$$\sum_{n=1}^{\infty}(a_n + b_n) = \lim_{N\to\infty}\sum_{n=1}^{N}(a_n + b_n) = \lim_{N\to\infty}\left(\sum_{n=1}^{N}a_n + \sum_{n=1}^{N}b_n\right)$$

$$= \lim_{N\to\infty}\sum_{n=1}^{N}a_n + \lim_{N\to\infty}\sum_{n=1}^{\infty}b_n = \sum_{n=1}^{\infty}a_n + \sum_{n=1}^{\infty}b_n \qquad\blacksquare$$

A main goal in this chapter is to develop techniques for determining whether a series converges or diverges. It is easy to give examples of series that diverge:

- $S = \sum_{n=1}^{\infty} 1$ diverges to infinity (the partial sums increase without bound):

$$S_1 = 1, \quad S_2 = 1 + 1 = 2, \quad S_3 = 1 + 1 + 1 = 3, \quad S_4 = 1 + 1 + 1 + 1 = 4, \quad \ldots$$

- $\sum_{n=1}^{\infty}(-1)^{n-1}$ diverges (the partial sums jump between 1 and 0):

$$S_1 = 1, \quad S_2 = 1 - 1 = 0, \quad S_3 = 1 - 1 + 1 = 1, \quad S_4 = 1 - 1 + 1 - 1 = 0, \quad \ldots$$

Next, we study geometric series, which converge or diverge depending on the common ratio r.

A **geometric series** with common ratio $r \neq 0$ is a series defined by a geometric sequence cr^n, where $c \neq 0$. If the series begins at $n = 0$, then

$$S = \sum_{n=0}^{\infty} cr^n = c + cr + cr^2 + cr^3 + cr^4 + cr^5 + \cdots$$

For $r = \frac{1}{2}$ and $c = 1$, we can visualize the geometric series starting at $n = 1$ (Figure 1):

$$S = \sum_{n=1}^{\infty}\frac{1}{2^n} = \frac{1}{2} + \frac{1}{4} + \frac{1}{8} + \frac{1}{16} + \cdots = 1$$

Adding up the terms corresponds to moving stepwise from 0 to 1, where each step is a move to the right by half of the remaining distance. Thus, $S = 1$.

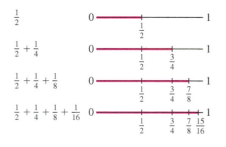

FIGURE 1 Partial sums of $\sum_{n=1}^{\infty}\frac{1}{2^n}$.

There is a simple device for computing the partial sums of a geometric series:

$$S_N = c + cr + cr^2 + cr^3 + \cdots + cr^N$$

$$rS_N = \qquad cr + cr^2 + cr^3 + \cdots + cr^N + cr^{N+1}$$

$$S_N - rS_N = c - cr^{N+1}$$

$$S_N(1 - r) = c(1 - r^{N+1})$$

If $r \neq 1$, we may divide by $(1 - r)$ to obtain

$$S_N = c + cr + cr^2 + cr^3 + \cdots + cr^N = \frac{c(1 - r^{N+1})}{1 - r} \qquad \boxed{3}$$

This formula enables us to sum the geometric series.

Geometric series are important because they:

- *arise often in applications.*
- *can be evaluated explicitly.*
- *are used to study other, nongeometric series (by comparison).*

THEOREM 2 Sum of a Geometric Series Let $c \neq 0$. If $|r| < 1$, then

$$\sum_{n=0}^{\infty} cr^n = c + cr + cr^2 + cr^3 + \cdots = \frac{c}{1-r}$$ **4**

$$\sum_{n=M}^{\infty} cr^n = cr^M + cr^{M+1} + cr^{M+2} + cr^{M+3} + \cdots = \frac{cr^M}{1-r}$$ **5**

If $|r| \geq 1$, then the geometric series diverges.

Proof If $r = 1$, then the series certainly diverges because the partial sums $S_N = Nc$ grow arbitrarily large. If $r \neq 1$, then Eq. (3) yields

$$S = \lim_{N \to \infty} S_N = \lim_{N \to \infty} \frac{c(1 - r^{N+1})}{1-r} = \frac{c}{1-r} - \frac{c}{1-r} \lim_{N \to \infty} r^{N+1}$$

If $|r| < 1$, then $\lim_{N \to \infty} r^{N+1} = 0$ and we obtain Eq. (4). If $|r| \geq 1$ and $r \neq 1$, then $\lim_{N \to \infty} r^{N+1}$ does not exist and the geometric series diverges. Finally, if the series starts with cr^M rather than cr^0, then

$$S = cr^M + cr^{M+1} + cr^{M+2} + cr^{M+3} + \cdots = r^M \sum_{n=0}^{\infty} cr^n = \frac{cr^M}{1-r}$$ ■

■ **EXAMPLE 3** Evaluate $\sum_{n=0}^{\infty} 5^{-n}$.

Solution This is a geometric series with $r = 5^{-1}$. By Eq. (4),

$$\sum_{n=0}^{\infty} 5^{-n} = 1 + \frac{1}{5} + \frac{1}{5^2} + \frac{1}{5^3} + \cdots = \frac{1}{1 - 5^{-1}} = \frac{5}{4}$$ ■

■ **EXAMPLE 4** Evaluate $\sum_{n=3}^{\infty} 7\left(-\frac{3}{4}\right)^n = 7\left(-\frac{3}{4}\right)^3 + 7\left(-\frac{3}{4}\right)^4 + 7\left(-\frac{3}{4}\right)^5 + \cdots$.

Solution This is a geometric series with $r = -\frac{3}{4}$ and $c = 7$, starting at $n = 3$. By Eq. (5),

$$\sum_{n=3}^{\infty} 7\left(-\frac{3}{4}\right)^n = \frac{7\left(-\frac{3}{4}\right)^3}{1 - \left(-\frac{3}{4}\right)} = -\frac{27}{16}$$ ■

■ **EXAMPLE 5** Find a fraction that has repeated decimal expansion $0.212121\ldots$.

Solution We can write this decimal as the series $\frac{21}{100} + \frac{21}{100^2} + \frac{21}{100^3} + \cdots$. This is a geometric series with $c = \frac{21}{100}$ and $r = \frac{1}{100}$. Thus, it converges to

$$\frac{a}{1-r} = \frac{\frac{21}{100}}{1 - \frac{1}{100}} = \frac{21}{99} = \frac{7}{33}$$ ■

■ **EXAMPLE 6** Evaluate $S = \sum_{n=0}^{\infty} \frac{2 + 3^n}{5^n}$.

Solution Write S as a sum of two geometric series. This is valid by Theorem 1 because both geometric series converge:

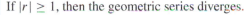
Both geometric series converge

$$\sum_{n=0}^{\infty} \frac{2 + 3^n}{5^n} = \sum_{n=0}^{\infty} \frac{2}{5^n} + \sum_{n=0}^{\infty} \frac{3^n}{5^n} = 2 \sum_{n=0}^{\infty} \frac{1}{5^n} + \sum_{n=0}^{\infty} \left(\frac{3}{5}\right)^n = 2 \cdot \frac{1}{1 - \frac{1}{5}} + \frac{1}{1 - \frac{3}{5}} = 5$$ ■

CONCEPTUAL INSIGHT Sometimes, the following *incorrect argument* is given for summing a geometric series:

$$S = \quad \frac{1}{2} + \frac{1}{4} + \frac{1}{8} + \cdots$$

$$2S = 1 + \frac{1}{2} + \frac{1}{4} + \frac{1}{8} + \cdots = 1 + S$$

Thus, $2S = 1 + S$, or $S = 1$. The answer is correct, so why is the argument wrong? It is wrong because we do not know in advance that the series converges. Observe what happens when this argument is applied to a divergent series:

$$S = 1 + 2 + 4 + 8 + 16 + \cdots$$

$$2S = \quad 2 + 4 + 8 + 16 + \cdots = S - 1$$

This would yield $2S = S - 1$, or $S = -1$, which is absurd because S diverges. We avoid such erroneous conclusions by carefully defining the sum of an infinite series as the limit of its sequence of partial sums.

The infinite series $\sum\limits_{k=1}^{\infty} 1$ diverges because the Nth partial sum $S_N = N$ diverges to infinity. It is less clear whether the following series converges or diverges:

$$\sum_{n=1}^{\infty} (-1)^{n+1} \frac{n}{n+1} = \frac{1}{2} - \frac{2}{3} + \frac{3}{4} - \frac{4}{5} + \frac{5}{6} - \cdots$$

We now introduce a useful test that allows us to conclude that this series diverges. The idea is that if the terms are not shrinking to 0 in size, then the series will not converge. This is typically the first test one applies when attempting to determine whether a series diverges.

---

*The **nth Term Divergence Test** (also known as the **Divergence Test**) is often stated as follows:*

If $\sum\limits_{n=1}^{\infty} a_n$ converges, then $\lim\limits_{n\to\infty} a_n = 0$.

In practice, we use it to prove that a given series diverges. It is important to note that it does NOT say that if $\lim\limits_{n\to\infty} a_n = 0$, then

$\sum\limits_{n=1}^{\infty} a_n$ necessarily converges. We will see

that even though $\lim\limits_{n\to\infty} \dfrac{1}{n} = 0$, the series

$\sum\limits_{n=1}^{\infty} \dfrac{1}{n}$ diverges.

THEOREM 3 **nth Term Divergence Test** If $\lim\limits_{n\to\infty} a_n \neq 0$, then the series $\sum\limits_{n=1}^{\infty} a_n$ diverges.

Proof First, note that $a_n = S_n - S_{n-1}$ because

$$S_n = (a_1 + a_2 + \cdots + a_{n-1}) + a_n = S_{n-1} + a_n$$

If $\sum\limits_{n=1}^{\infty} a_n$ converges with sum S, then

$$\lim_{n\to\infty} a_n = \lim_{n\to\infty} (S_n - S_{n-1}) = \lim_{n\to\infty} S_n - \lim_{n\to\infty} S_{n-1} = S - S = 0$$

Therefore, if a_n does not converge to zero, $\sum\limits_{n=1}^{\infty} a_n$ cannot converge. ■

■ **EXAMPLE 7** Prove the divergence of $S = \sum\limits_{n=1}^{\infty} \dfrac{n}{4n+1}$.

Solution We have

$$\lim_{n\to\infty} a_n = \lim_{n\to\infty} \frac{n}{4n+1} = \lim_{n\to\infty} \frac{1}{4 + 1/n} = \frac{1}{4}$$

The nth term a_n does not converge to zero, so the series diverges by the nth Term Divergence Test (Theorem 3). ■

■ **EXAMPLE 8** Determine the convergence or divergence of

$$S = \sum_{n=1}^{\infty} (-1)^{n-1} \frac{n}{n+1} = \frac{1}{2} - \frac{2}{3} + \frac{3}{4} - \frac{4}{5} + \cdots$$

Solution The general term $a_n = (-1)^{n-1} \frac{n}{n+1}$ does not approach a limit. Indeed, $\frac{n}{n+1}$ tends to 1, so the odd terms a_{2n+1} tend to 1, and the even terms a_{2n} tend to -1. Because $\lim_{n \to \infty} a_n$ does not exist, the series S diverges by the nth Term Divergence Test. ■

The nth Term Divergence Test tells only part of the story. If a_n does not tend to zero, then $\sum a_n$ certainly diverges. But what if a_n does tend to zero? In this case, the series may converge or it may diverge. In other words, $\lim_{n \to \infty} a_n = 0$ is a *necessary* condition of convergence, but it is *not sufficient*. As we show in the next example, it is possible for a series to diverge even though its terms tend to zero.

■ **EXAMPLE 9** **Sequence Tends to Zero, Yet the Series Diverges** Prove the divergence of

$$\sum_{n=1}^{\infty} \frac{1}{\sqrt{n}} = \frac{1}{\sqrt{1}} + \frac{1}{\sqrt{2}} + \frac{1}{\sqrt{3}} + \cdots$$

Solution The general term $1/\sqrt{n}$ tends to zero. However, because each term in the partial sum S_N is greater than or equal to $1/\sqrt{N}$, we have

$$\begin{aligned}
S_N &= \overbrace{\frac{1}{\sqrt{1}} + \frac{1}{\sqrt{2}} + \cdots + \frac{1}{\sqrt{N}}}^{N \text{ terms}} \\
&\geq \frac{1}{\sqrt{N}} + \frac{1}{\sqrt{N}} + \cdots + \frac{1}{\sqrt{N}} \\
&= N \left(\frac{1}{\sqrt{N}} \right) = \sqrt{N}
\end{aligned}$$

This shows that $S_N \geq \sqrt{N}$. But $\sqrt{N}$ increases without bound (Figure 2). Therefore, S_N also increases without bound. This proves that the series diverges. ■

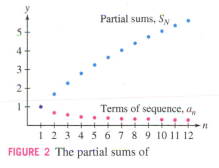

Partial sums, S_N

Terms of sequence, a_n

FIGURE 2 The partial sums of

$$\sum_{n=1}^{\infty} \frac{1}{\sqrt{n}}$$

diverge even though the terms $a_n = 1/\sqrt{n}$ tend to zero.

10.2 SUMMARY

- An *infinite series* is an expression

$$\sum_{n=1}^{\infty} a_n = a_1 + a_2 + a_3 + a_4 + \cdots$$

We call a_n the *general term* of the series. An infinite series can begin at $n = k$ for any integer k.

- The Nth *partial sum* is the finite sum of the terms up to and including the Nth term:

$$S_N = \sum_{n=1}^{N} a_n = a_1 + a_2 + a_3 + \cdots + a_N$$

- By definition, the sum of an infinite series is the limit $S = \lim_{N \to \infty} S_N$. If the limit exists, we say that the infinite series is *convergent* or *converges* to the sum S. If the limit does not exist, we say that the infinite series *diverges*.

- If the sequence of partial sums $\{S_N\}$ increase without bound, we say that S diverges to infinity.

- nth Term Divergence Test: If $\lim_{n \to \infty} a_n \neq 0$, then $\sum_{n=1}^{\infty} a_n$ diverges. However, a series may diverge even if its general term $\{a_n\}$ tends to zero.

• Partial sum of a geometric series:

$$c + cr + cr^2 + cr^3 + \cdots + cr^N = \frac{c(1 - r^{N+1})}{1 - r}$$

• Geometric series: If $|r| < 1$, then

$$\sum_{n=0}^{\infty} r^n = 1 + r + r^2 + r^3 + \cdots = \frac{1}{1 - r}$$

$$\sum_{n=M}^{\infty} cr^n = cr^M + cr^{M+1} + cr^{M+2} + \cdots = \frac{cr^M}{1 - r}$$

The geometric series diverges if $|r| \geq 1$.

Archimedes (287 BCE–212 BCE), who discovered the law of the lever, said, "Give me a place to stand on, and I can move the Earth" (quoted by Pappus of Alexandria c. 340 CE).

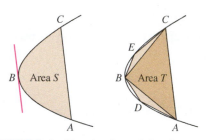

FIGURE 3 Archimedes showed that the area S of the parabolic segment is $\frac{4}{3}T$, where T is the area of $\triangle ABC$.

(Mechanics Magazine London, 1824)

HISTORICAL PERSPECTIVE

Geometric series were used as early as the third century BCE by Archimedes in a brilliant argument for determining the area S of a "parabolic segment" (shaded region in Figure 3). Given two points A and C on a parabola, there is a point B between A and C where the tangent line is parallel to $\overline{AC}$ (apparently, Archimedes was aware of the Mean Value Theorem more than 2000 years before the invention of calculus). Let T be the area of triangle $\triangle ABC$. Archimedes proved that if D is chosen in a similar fashion relative to $\overline{AB}$ and E is chosen relative to $\overline{BC}$, then

$$\frac{1}{4}T = \text{Area}(\triangle ADB) + \text{Area}(\triangle BEC) \quad \boxed{6}$$

This construction of triangles can be continued. The next step would be to construct the four triangles on the segments $\overline{AD}, \overline{DB}, \overline{BE}, \overline{EC}$, of total area $\frac{1}{4^2}T$. Then construct eight triangles of total area $\frac{1}{4^3}T$, etc. In this way, we obtain infinitely many triangles that completely fill up the parabolic segment. By the formula for the sum of a geometric series, we get

$$S = T + \frac{1}{4}T + \frac{1}{16}T + \cdots = T \sum_{n=0}^{\infty} \frac{1}{4^n} = \frac{4}{3}T$$

For this and many other achievements, Archi-

medes is ranked together with Newton and Gauss as one of the greatest scientists of all time.

The modern study of infinite series began in the seventeenth century with Newton, Leibniz, and their contemporaries. The divergence of $\sum_{n=1}^{\infty} 1/n$ (called the **harmonic series**) was known to the medieval scholar Nicole d'Oresme (1323–1382), but his proof was lost for centuries, and the result was rediscovered on more than one occasion. It was also known that the sum of the reciprocal squares $\sum_{n=1}^{\infty} 1/n^2$ converges, and in the 1640s, the Italian Pietro Mengoli put forward the challenge of finding its sum. Despite the efforts of the best mathematicians of the day, including Leibniz and the Bernoulli brothers Jakob and Johann, the problem resisted solution for nearly a century. In 1735 the great master Leonhard Euler (at the time, 28 years old) astonished his contemporaries by proving that

$$\frac{1}{1^2} + \frac{1}{2^2} + \frac{1}{3^2} + \frac{1}{4^2} + \frac{1}{5^2} + \frac{1}{6^2} + \cdots = \frac{\pi^2}{6}$$

This formula, surprising in itself, plays a role in a variety of mathematical fields. A theorem from number theory states that two whole numbers, chosen randomly, have no common factor with probability $6/\pi^2 \approx 0.6$ (the reciprocal of Euler's result). On the other hand, Euler's result and its generalizations appear in the field of statistical mechanics.

10.2 EXERCISES

Preliminary Questions

1. What role do partial sums play in defining the sum of an infinite series?

2. What is the sum of the following infinite series?

$$\frac{1}{4} + \frac{1}{8} + \frac{1}{16} + \frac{1}{32} + \frac{1}{64} + \cdots$$

3. What happens if you apply the formula for the sum of a geometric series to the following series? Is the formula valid?

$$1 + 3 + 3^2 + 3^3 + 3^4 + \cdots$$

4. A student asserts that $\sum_{n=1}^{\infty} \frac{1}{n^2} = 0$ because $\frac{1}{n^2}$ tends to zero. Is this valid reasoning?

5. A student claims that $\sum_{n=1}^{\infty} \dfrac{1}{\sqrt{n}}$ converges because

$$\lim_{n \to \infty} \dfrac{1}{\sqrt{n}} = 0$$

Is this valid reasoning?

6. Find an N such that $S_N > 25$ for the series $\sum_{n=1}^{\infty} 2$.

7. Does there exist an N such that $S_N > 25$ for the series $\sum_{n=1}^{\infty} 2^{-n}$? Explain.

8. Give an example of a divergent infinite series whose general term tends to zero.

Exercises

1. Find a formula for the general term a_n (not the partial sum) of the infinite series,

(a) $\dfrac{1}{3} + \dfrac{1}{9} + \dfrac{1}{27} + \dfrac{1}{81} + \cdots$

(b) $\dfrac{1}{1} + \dfrac{5}{2} + \dfrac{25}{4} + \dfrac{125}{8} + \cdots$

(c) $\dfrac{1}{1} - \dfrac{2^2}{2 \cdot 1} + \dfrac{3^3}{3 \cdot 2 \cdot 1} - \dfrac{4^4}{4 \cdot 3 \cdot 2 \cdot 1} + \cdots$

(d) $\dfrac{2}{1^2 + 1} + \dfrac{1}{2^2 + 1} + \dfrac{2}{3^2 + 1} + \dfrac{1}{4^2 + 1} + \cdots$

2. Write in summation notation:

(a) $1 + \dfrac{1}{4} + \dfrac{1}{9} + \dfrac{1}{16} + \cdots$

(b) $\dfrac{1}{9} + \dfrac{1}{16} + \dfrac{1}{25} + \dfrac{1}{36} + \cdots$

(c) $1 - \dfrac{1}{3} + \dfrac{1}{5} - \dfrac{1}{7} + \cdots$

(d) $\dfrac{125}{9} + \dfrac{625}{16} + \dfrac{3125}{25} + \dfrac{15{,}625}{36} + \cdots$

In Exercises 3–6, compute the partial sums S_2, S_4, and S_6.

3. $1 + \dfrac{1}{2^2} + \dfrac{1}{3^2} + \dfrac{1}{4^2} + \cdots$

4. $\sum_{k=1}^{\infty} (-1)^k k^{-1}$

5. $\dfrac{1}{1 \cdot 2} + \dfrac{1}{2 \cdot 3} + \dfrac{1}{3 \cdot 4} + \cdots$

6. $\sum_{j=1}^{\infty} \dfrac{1}{j!}$

7. The series $S = 1 + \left(\dfrac{1}{5}\right) + \left(\dfrac{1}{5}\right)^2 + \left(\dfrac{1}{5}\right)^3 + \cdots$ converges to $\dfrac{5}{4}$. Calculate S_N for $N = 1, 2, \ldots$ until you find an S_N that approximates $\dfrac{5}{4}$ with an error less than 0.0001.

8. The series $S = \dfrac{1}{0!} - \dfrac{1}{1!} + \dfrac{1}{2!} - \dfrac{1}{3!} + \cdots$ is known to converge to e^{-1} (recall that $0! = 1$). Calculate S_N for $N = 1, 2, \ldots$ until you find an S_N that approximates e^{-1} with an error less than 0.001.

In Exercises 9 and 10, use a computer algebra system to compute S_{10}, S_{100}, S_{500}, and S_{1000} for the series. Do these values suggest convergence to the given value?

9. *CAS*

$$\dfrac{\pi - 3}{4} = \dfrac{1}{2 \cdot 3 \cdot 4} - \dfrac{1}{4 \cdot 5 \cdot 6} + \dfrac{1}{6 \cdot 7 \cdot 8} - \dfrac{1}{8 \cdot 9 \cdot 10} + \cdots$$

10. *CAS*

$$\dfrac{\pi^4}{90} = 1 + \dfrac{1}{2^4} + \dfrac{1}{3^4} + \dfrac{1}{4^4} + \cdots$$

11. Calculate S_3, S_4, and S_5 and then find the sum of the telescoping series

$$S = \sum_{n=1}^{\infty} \left(\dfrac{1}{n+1} - \dfrac{1}{n+2} \right)$$

12. Write $\sum_{n=3}^{\infty} \dfrac{1}{n(n-1)}$ as a telescoping series and find its sum.

13. Calculate S_3, S_4, and S_5 and then find the sum $S = \sum_{n=1}^{\infty} \dfrac{1}{4n^2 - 1}$ using the identity

$$\dfrac{1}{4n^2 - 1} = \dfrac{1}{2} \left(\dfrac{1}{2n - 1} - \dfrac{1}{2n + 1} \right)$$

14. Use partial fractions to rewrite $\sum_{n=1}^{\infty} \dfrac{1}{n(n+3)}$ as a telescoping series and find its sum.

15. Find the sum of $\dfrac{1}{1 \cdot 3} + \dfrac{1}{3 \cdot 5} + \dfrac{1}{5 \cdot 7} + \cdots$.

16. Find a formula for the partial sum S_N of $\sum_{n=1}^{\infty} (-1)^{n-1}$ and show that the series diverges.

In Exercises 17–22, the nth Term Divergence Test (Theorem 3) to prove that the following series diverge.

17. $\sum_{n=1}^{\infty} \dfrac{n}{10n + 12}$

18. $\sum_{n=1}^{\infty} \dfrac{n}{\sqrt{n^2 + 1}}$

19. $\dfrac{0}{1} - \dfrac{1}{2} + \dfrac{2}{3} - \dfrac{3}{4} + \cdots$

20. $\sum_{n=1}^{\infty} (-1)^n n^2$

21. $\cos \dfrac{1}{2} + \cos \dfrac{1}{3} + \cos \dfrac{1}{4} + \cdots$

22. $\sum_{n=0}^{\infty} \left(\sqrt{4n^2 + 1} - n \right)$

In Exercises 23–36, use the formula for the sum of a geometric series to find the sum or state that the series diverges.

23. $\dfrac{1}{1} + \dfrac{1}{8} + \dfrac{1}{8^2} + \cdots$

24. $\dfrac{4^3}{5^3} + \dfrac{4^4}{5^4} + \dfrac{4^5}{5^5} + \cdots$

25. $\sum_{n=3}^{\infty} \left(\dfrac{3}{11} \right)^{-n}$

26. $\sum_{n=2}^{\infty} \dfrac{7 \cdot (-3)^n}{5^n}$

27. $\sum_{n=-4}^{\infty} \left(-\dfrac{4}{9} \right)^n$

28. $\sum_{n=0}^{\infty} \left(\dfrac{\pi}{e} \right)^n$

29. $\sum_{n=1}^{\infty} e^{-n}$

30. $\sum_{n=2}^{\infty} e^{3 - 2n}$

31. $\sum_{n=0}^{\infty} \dfrac{8 + 2^n}{5^n}$

32. $\sum_{n=0}^{\infty} \dfrac{3(-2)^n - 5^n}{8^n}$

33. $5 - \dfrac{5}{4} + \dfrac{5}{4^2} - \dfrac{5}{4^3} + \cdots$

34. $\dfrac{2^3}{7} + \dfrac{2^4}{7^2} + \dfrac{2^5}{7^3} + \dfrac{2^6}{7^4} + \cdots$

35. $\dfrac{7}{8} - \dfrac{49}{64} + \dfrac{343}{512} - \dfrac{2401}{4096} + \cdots$

36. $\dfrac{25}{9} + \dfrac{5}{3} + 1 + \dfrac{3}{5} + \dfrac{9}{25} + \dfrac{27}{125} + \cdots$

In Exercises 37–40, determine a reduced fraction that has this repeating decimal.

37. $0.222\ldots$

38. $0.454545\ldots$

39. $0.313131\ldots$

40. $0.217217217\ldots$

41. Determine which reduced fractions have a repeating decimal of the form $0.aaa\ldots$, where $a = 0, 1, 2, \ldots, 9$.

42. Determine which denominators are realized for reduced fractions that have a repeating decimal of the form $0.abcabcabc\ldots$, where each of a, b, and c is a one-digit number from 0 to 9.

43. Which of the following are *not* geometric series?

(a) $\displaystyle\sum_{n=0}^{\infty} \dfrac{7^n}{29^n}$

(b) $\displaystyle\sum_{n=3}^{\infty} \dfrac{1}{n^4}$

(c) $\displaystyle\sum_{n=0}^{\infty} \dfrac{n^2}{2^n}$

(d) $\displaystyle\sum_{n=5}^{\infty} \pi^{-n}$

44. Use the method of Example 9 to show that $\displaystyle\sum_{k=1}^{\infty} \dfrac{1}{k^{1/3}}$ diverges.

45. Prove that if $\displaystyle\sum_{n=1}^{\infty} a_n$ converges and $\displaystyle\sum_{n=1}^{\infty} b_n$ diverges, then $\displaystyle\sum_{n=1}^{\infty} (a_n + b_n)$ diverges. *Hint:* If not, derive a contradiction by writing

$$\sum_{n=1}^{\infty} b_n = \sum_{n=1}^{\infty} (a_n + b_n) - \sum_{n=1}^{\infty} a_n$$

46. Prove the divergence of $\displaystyle\sum_{n=0}^{\infty} \dfrac{9^n + 2^n}{5^n}$.

47. 📖 Give a counterexample to show that each of the following statements is false.

(a) If the general term a_n tends to zero, then $\displaystyle\sum_{n=1}^{\infty} a_n = 0$.

(b) The Nth partial sum of the infinite series defined by $\{a_n\}$ is a_N.

(c) If a_n tends to zero, then $\displaystyle\sum_{n=1}^{\infty} a_n$ converges.

(d) If a_n tends to L, then $\displaystyle\sum_{n=1}^{\infty} a_n = L$.

48. Suppose that $S = \displaystyle\sum_{n=1}^{\infty} a_n$ is an infinite series with partial sum $S_N = 5 - \dfrac{2}{N^2}$.

(a) What are the values of $\displaystyle\sum_{n=1}^{10} a_n$ and $\displaystyle\sum_{n=5}^{16} a_n$?

(b) What is the value of a_3?

(c) Find a general formula for a_n.

(d) Find the sum $\displaystyle\sum_{n=1}^{\infty} a_n$.

49. Compute the total area of the (infinitely many) triangles in Figure 4.

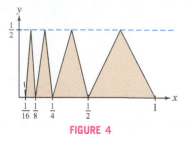

FIGURE 4

50. The winner of a lottery receives m dollars at the end of each year for N years. The present value (PV) of this prize in today's dollars is $\text{PV} = \displaystyle\sum_{i=1}^{N} m(1+r)^{-i}$, where r is the interest rate. Calculate PV if $m = \$50{,}000$, $r = 0.06$ (corresponding to 6%), and $N = 20$. What is PV if $N = \infty$?

51. If a patient takes a dose of D units of a particular drug, the amount of the dosage that remains in the patient's bloodstream after t days is De^{-kt}, where k is a positive constant depending on the particular drug.

(a) Show that if the patient takes a dose D every day for an extended period, the amount of drug in the bloodstream approaches $R = \dfrac{De^{-k}}{1 - e^{-k}}$.

(b) Show that if the patient takes a dose D once every t days for an extended period, the amount of drug in the bloodstream approaches $R = \dfrac{De^{-kt}}{1 - e^{-kt}}$.

(c) Suppose that a dosage of more than S units is considered dangerous. What is the minimal time between doses that is safe? (*Hint:* $D + R \leq S$.)

52. In economics, the multiplier effect refers to the fact that when there is an injection of money to consumers, the consumers spend a certain percentage of it. That amount recirculates through the economy and adds additional income, which comes back to the consumers and of which they spend the same percentage. This process repeats indefinitely, circulating additional money through the economy. Suppose that in order to stimulate the economy, the government institutes a tax cut of \$10 billion. If taxpayers are known to save 10% of any additional money they receive, and to spend 90%, how much total money will be circulated through the economy by that single \$10 billion tax cut?

53. Find the total length of the infinite zigzag path in Figure 5 (each zag occurs at an angle of $\frac{\pi}{4}$).

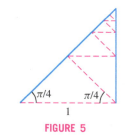

FIGURE 5

54. Evaluate $\displaystyle\sum_{n=1}^{\infty} \dfrac{1}{n(n+1)(n+2)}$. *Hint:* Find constants A, B, and C such that

$$\dfrac{1}{n(n+1)(n+2)} = \dfrac{A}{n} + \dfrac{B}{n+1} + \dfrac{C}{n+2}$$

55. Show that if a is a positive integer, then

$$\sum_{n=1}^{\infty} \frac{1}{n(n+a)} = \frac{1}{a}\left(1 + \frac{1}{2} + \cdots + \frac{1}{a}\right)$$

56. A ball dropped from a height of 10 ft begins to bounce vertically. Each time it strikes the ground, it returns to two-thirds of its previous height. What is the total vertical distance traveled by the ball if it bounces infinitely many times?

57. A ball is dropped from a height of 6 ft and begins to bounce vertically. Each time it strikes the ground, it returns to three-quarters of its previous height. What is the total vertical distance traveled by the ball if it bounces infinitely many times?

58. A unit square is cut into nine equal regions as in Figure 6(A). The central subsquare is painted red. Each of the unpainted squares is then cut into nine equal subsquares and the central square of each is painted red as in Figure 6(B). This procedure is repeated for each of the resulting unpainted squares. After continuing this process an infinite number of times, what fraction of the total area of the original square is painted?

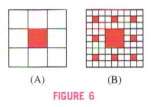

(A)   (B)

FIGURE 6

59. Let $\{b_n\}$ be a sequence and let $a_n = b_n - b_{n-1}$. Show that $\sum_{n=1}^{\infty} a_n$ converges if and only if $\lim_{n\to\infty} b_n$ exists.

60. Assumptions Matter Show, by giving counterexamples, that the assertions of Theorem 1 are not valid if the series $\sum_{n=0}^{\infty} a_n$ and $\sum_{n=0}^{\infty} b_n$ are not convergent.

Further Insights and Challenges

Exercises 61–63 use the formula

$$1 + r + r^2 + \cdots + r^{N-1} = \frac{1 - r^N}{1 - r} \qquad \boxed{7}$$

61. Professor George Andrews of Pennsylvania State University observed that we can use Eq. (7) to calculate the derivative of $f(x) = x^N$ (for $N \geq 0$). Assume that $a \neq 0$ and let $x = ra$. Show that

$$f'(a) = \lim_{x\to a} \frac{x^N - a^N}{x - a} = a^{N-1} \lim_{r\to 1} \frac{r^N - 1}{r - 1}$$

and evaluate the limit.

62. Pierre de Fermat used geometric series to compute the area under the graph of $f(x) = x^N$ over $[0, A]$. For $0 < r < 1$, let $F(r)$ be the sum of the areas of the infinitely many right-endpoint rectangles with endpoints Ar^n, as in Figure 7. As r tends to 1, the rectangles become narrower and $F(r)$ tends to the area under the graph.

(a) Show that $F(r) = A^{N+1} \dfrac{1 - r}{1 - r^{N+1}}$.

(b) Use Eq. (7) to evaluate $\displaystyle\int_0^A x^N \, dx = \lim_{r\to 1} F(r)$.

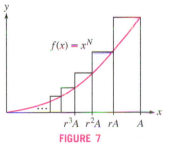

$f(x) = x^N$

$r^3A \quad r^2A \quad rA \qquad A$

FIGURE 7

63. Verify the Gregory–Leibniz formula as follows.

(a) Set $r = -x^2$ in Eq. (7) and rearrange to show that

$$\frac{1}{1+x^2} = 1 - x^2 + x^4 - \cdots + (-1)^{N-1} x^{2N-2} + \frac{(-1)^N x^{2N}}{1+x^2}$$

(b) Show, by integrating over $[0, 1]$, that

$$\frac{\pi}{4} = 1 - \frac{1}{3} + \frac{1}{5} - \frac{1}{7} + \cdots + \frac{(-1)^{N-1}}{2N-1} + (-1)^N \int_0^1 \frac{x^{2N}\, dx}{1+x^2}$$

(c) Use the Comparison Theorem for integrals to prove that

$$0 \leq \int_0^1 \frac{x^{2N}\, dx}{1+x^2} \leq \frac{1}{2N+1}$$

Hint: Observe that the integrand is $\leq x^{2N}$.

(d) Prove that

$$\frac{\pi}{4} = 1 - \frac{1}{3} + \frac{1}{5} - \frac{1}{7} + \frac{1}{9} - \cdots$$

Hint: Use (b) and (c) to show that the partial sums S_N satisfy $\left|S_N - \frac{\pi}{4}\right| \leq \frac{1}{2N+1}$, and thereby conclude that $\lim_{N\to\infty} S_N = \frac{\pi}{4}$.

64. Cantor's Disappearing Table (following Larry Knop of Hamilton College) Take a table of length L (Figure 8). At Stage 1, remove the section of length $L/4$ centered at the midpoint. Two sections remain, each with length less than $L/2$. At Stage 2, remove sections of length $L/4^2$ from each of these two sections (this stage removes $L/8$ of the table). Now four sections remain, each of length less than $L/4$. At Stage 3, remove the four central sections of length $L/4^3$, etc.

(a) Show that at the Nth stage, each remaining section has length less than $L/2^N$ and that the total amount of table removed is

$$L\left(\frac{1}{4} + \frac{1}{8} + \frac{1}{16} + \cdots + \frac{1}{2^{N+1}}\right)$$

(b) Show that in the limit as $N \to \infty$, precisely one-half of the table remains.

This result is intriguing, because there are no nonzero intervals of table left (at each stage, the remaining sections have a length less than $L/2^N$). So, the table has "disappeared." However, we can place any object longer than $L/4$ on the table. The object will not fall through because it will not fit through any of the removed sections.

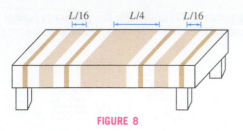

$L/16 \quad L/4 \quad L/16$

FIGURE 8

65. The **Koch snowflake** (described in 1904 by Swedish mathematician Helge von Koch) is an infinitely jagged "fractal" curve obtained as a limit of polygonal curves (it is continuous but has no tangent line at any point). Begin with an equilateral triangle (Stage 0) and produce Stage 1 by replacing each edge with four edges of one-third the length, arranged as in Figure 9. Continue the process: At the nth stage, replace each edge with four edges of one-third the length of the edge from the $(n-1)$st stage.

(a) Show that the perimeter P_n of the polygon at the nth stage satisfies $P_n = \frac{4}{3}P_{n-1}$. Prove that $\lim_{n \to \infty} P_n = \infty$. The snowflake has infinite length.

(b) Let A_0 be the area of the original equilateral triangle. Show that $(3)4^{n-1}$ new triangles are added at the nth stage, each with area $A_0/9^n$ (for $n \geq 1$). Show that the total area of the Koch snowflake is $\frac{8}{5}A_0$.

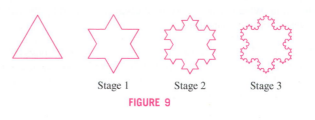

Stage 1   Stage 2   Stage 3

FIGURE 9

10.3 Convergence of Series with Positive Terms

The next three sections develop techniques for determining whether an infinite series converges or diverges. This is easier than finding the sum of an infinite series, which is possible only in special cases.

In this section, we consider **positive series** $\sum a_n$, where $a_n > 0$ for all n. We can visualize the terms of a positive series as rectangles of width 1 and height a_n (Figure 1). The partial sum

$$S_N = a_1 + a_2 + \cdots + a_N$$

is equal to the area of the first N rectangles.

The key feature of positive series is that their partial sums form an increasing sequence:

$$S_N < S_{N+1}$$

for all N. This is because S_{N+1} is obtained from S_N by adding a positive number:

$$S_{N+1} = (a_1 + a_2 + \cdots + a_N) + a_{N+1} = S_N + \underbrace{a_{N+1}}_{\text{Positive}}$$

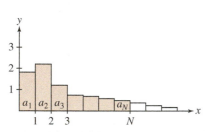

FIGURE 1 The partial sum S_N is the sum of the areas of the N shaded rectangles.

Recall that an increasing sequence converges if it is bounded above. Otherwise, it diverges (Theorem 6, Section 10.1). It follows that a positive series behaves in one of two ways.

THEOREM 1 Partial Sum Theorem for Positive Series If $S = \sum_{n=1}^{\infty} a_n$ is a positive series, then either:

(i) The partial sums S_N are bounded above. In this case, S converges. Or,

(ii) The partial sums S_N are not bounded above. In this case, S diverges.

- Theorem 1 remains true if $a_n \geq 0$. It is not necessary to assume that $a_n > 0$.
- It also remains true if $a_n > 0$ for all $n \geq M$ for some M, because the convergence or divergence of a series is not affected by the first M terms.

Assumptions Matter   The theorem does not hold for nonpositive series. Consider

$$S = \sum_{n=1}^{\infty} (-1)^{n-1} = 1 - 1 + 1 - 1 + 1 - 1 + \cdots$$

The partial sums are bounded (because $S_N = 1$ or 0), but S diverges.

Our first application of Theorem 1 is the following Integral Test. It is extremely useful because integrals are easier to evaluate than series in most cases.

The Integral Test is valid for any series

$$\sum_{n=k}^{\infty} f(n), \text{ provided that for some } M > 0,$$

f is a positive, decreasing, and continuous function of x for $x \geq M$. The convergence of the series is determined by the convergence of

$$\int_{M}^{\infty} f(x)\, dx$$

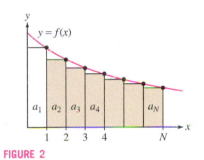

FIGURE 2

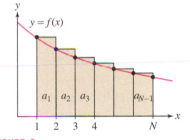

FIGURE 3

The infinite series

$$\sum_{n=1}^{\infty} \frac{1}{n}$$

is called the "harmonic series."

> **THEOREM 2 Integral Test** Let $a_n = f(n)$, where f is a positive, decreasing, and continuous function of x for $x \geq 1$.
>
> **(i)** If $\displaystyle\int_{1}^{\infty} f(x)\, dx$ converges, then $\displaystyle\sum_{n=1}^{\infty} a_n$ converges.
>
> **(ii)** If $\displaystyle\int_{1}^{\infty} f(x)\, dx$ diverges, then $\displaystyle\sum_{n=1}^{\infty} a_n$ diverges.

Proof Because f is decreasing, the shaded rectangles in Figure 2 lie below the graph of f, and therefore for all N

$$\underbrace{a_2 + \cdots + a_N}_{\text{Area of shaded rectangles in Figure 2}} \leq \int_{1}^{N} f(x)\, dx \leq \int_{1}^{\infty} f(x)\, dx$$

If the improper integral on the right converges, then the sums $a_2 + \cdots + a_N$ remain bounded. In this case, S_N also remains bounded, and the infinite series converges by the Partial Sum Theorem for Positive Series (Theorem 1). This proves (i).

On the other hand, the rectangles in Figure 3 lie above the graph of $f(x)$, so

$$\int_{1}^{N} f(x)\, dx \leq \underbrace{a_1 + a_2 + \cdots + a_{N-1}}_{\text{Area of shaded rectangles in Figure 3}} \qquad \boxed{1}$$

If $\int_{1}^{\infty} f(x)\, dx$ diverges, then $\int_{1}^{N} f(x)\, dx$ tends to ∞, and Eq. (1) shows that S_N also tends to ∞. This proves (ii). ∎

■ **EXAMPLE 1** **The Harmonic Series Diverges** Show that $\displaystyle\sum_{n=1}^{\infty} \frac{1}{n}$ diverges.

Solution Let $f(x) = \frac{1}{x}$. Then $f(n) = \frac{1}{n}$, and the Integral Test applies because f is positive, decreasing, and continuous for $x \geq 1$. The integral diverges:

$$\int_{1}^{\infty} \frac{dx}{x} = \lim_{R \to \infty} \int_{1}^{R} \frac{dx}{x} = \lim_{R \to \infty} \ln R = \infty$$

Therefore, the series $\displaystyle\sum_{n=1}^{\infty} \frac{1}{n}$ diverges. ■

■ **EXAMPLE 2** Does $\displaystyle\sum_{n=1}^{\infty} \frac{n}{(n^2 + 1)^2} = \frac{1}{2^2} + \frac{2}{5^2} + \frac{3}{10^2} + \cdots$ converge?

Solution The function $f(x) = \dfrac{x}{(x^2 + 1)^2}$ is positive and continuous for $x \geq 1$. It is decreasing because $f'(x)$ is negative:

$$f'(x) = \frac{1 - 3x^2}{(x^2 + 1)^3} < 0 \qquad \text{for } x \geq 1$$

Therefore, the Integral Test applies. Using the substitution $u = x^2 + 1$, $du = 2x\, dx$, we have

$$\int_{1}^{\infty} \frac{x}{(x^2 + 1)^2}\, dx = \lim_{R \to \infty} \int_{1}^{R} \frac{x}{(x^2 + 1)^2}\, dx = \lim_{R \to \infty} \frac{1}{2} \int_{2}^{R^2 + 1} \frac{du}{u^2}$$

$$= \lim_{R \to \infty} \frac{-1}{2u}\Big|_{2}^{R^2 + 1} = \lim_{R \to \infty} \left(\frac{1}{4} - \frac{1}{2(R^2 + 1)}\right) = \frac{1}{4}$$

Thus, the integral converges, and therefore, $\displaystyle\sum_{n=1}^{\infty} \frac{n}{(n^2 + 1)^2}$ also converges by the Integral Test. ■

The sum of the reciprocal powers n^{-p} is called a *p*-series.

THEOREM 3 Convergence of *p*-Series The infinite series $\displaystyle\sum_{n=1}^{\infty} \frac{1}{n^p}$ converges if $p > 1$ and diverges otherwise.

Proof If $p \le 0$, then the general term n^{-p} does not tend to zero, so the series diverges by the *n*th Term Divergence Test. If $p > 0$, then $f(x) = x^{-p}$ is positive and decreasing for $x \ge 1$, so the Integral Test applies. According to Theorem 1 in Section 7.7,

$$\int_1^{\infty} \frac{1}{x^p}\, dx = \begin{cases} \dfrac{1}{p-1} & \text{if } p > 1 \\ \infty & \text{if } p \le 1 \end{cases}$$

Therefore, $\displaystyle\sum_{n=1}^{\infty} \frac{1}{n^p}$ converges for $p > 1$ and diverges for $p \le 1$. ∎

Here are two examples of *p*-series:

$$p = \frac{1}{3}: \quad \sum_{n=1}^{\infty} \frac{1}{\sqrt[3]{n}} = \frac{1}{\sqrt[3]{1}} + \frac{1}{\sqrt[3]{2}} + \frac{1}{\sqrt[3]{3}} + \frac{1}{\sqrt[3]{4}} + \cdots = \infty \quad \text{diverges}$$

$$p = 2: \quad \sum_{n=1}^{\infty} \frac{1}{n^2} = \frac{1}{1} + \frac{1}{2^2} + \frac{1}{3^2} + \frac{1}{4^2} + \cdots \qquad \text{converges}$$

Another powerful method for determining convergence of positive series occurs via comparison with other series. Suppose that $0 \le a_n \le b_n$. Figure 4 suggests that if the larger sum $\sum b_n$ *converges*, then the smaller sum $\sum a_n$ also converges. Similarly, if the smaller sum *diverges*, then the larger sum also diverges.

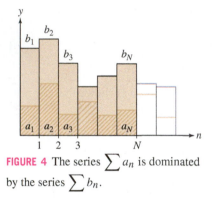

FIGURE 4 The series $\sum a_n$ is dominated by the series $\sum b_n$.

THEOREM 4 Direct Comparison Test
Assume that there exists $M > 0$ such that $0 \le a_n \le b_n$ for $n \ge M$.

(i) If $\displaystyle\sum_{n=1}^{\infty} b_n$ converges, then $\displaystyle\sum_{n=1}^{\infty} a_n$ also converges.

(ii) If $\displaystyle\sum_{n=1}^{\infty} a_n$ diverges, then $\displaystyle\sum_{n=1}^{\infty} b_n$ also diverges.

Proof We can assume, without loss of generality, that $M = 1$. If $S = \displaystyle\sum_{n=1}^{\infty} b_n$ converges, then the partial sums of $\displaystyle\sum_{n=1}^{\infty} a_n$ are bounded above by S because

$$a_1 + a_2 + \cdots + a_N \le b_1 + b_2 + \cdots + b_N \le \sum_{n=1}^{\infty} b_n = S \qquad \boxed{2}$$

Therefore, $\displaystyle\sum_{n=1}^{\infty} a_n$ converges by the Partial Sum Theorem for Positive Series (Theorem 1). This proves (i). On the other hand, if $\displaystyle\sum_{n=1}^{\infty} a_n$ diverges, then $\displaystyle\sum_{n=1}^{\infty} b_n$ must also diverge. Otherwise, we would have a contradiction to (i). ∎

■ **EXAMPLE 3** Does $\displaystyle\sum_{n=1}^{\infty} \frac{1}{\sqrt{n}\, 3^n}$ converge?

A good analogy for the Direct Comparison Test, as in Figure 5, is one balloon containing the terms for a_n, inside a bigger balloon containing the terms for b_n. As we blow air into the balloons in amounts corresponding to the subsequent terms a_n and b_n, the balloon containing the b_n terms will always be bigger than the balloon containing the a_n terms since $a_n \le b_n$. If the bigger balloon does not contain enough air to make it pop, then the smaller balloon does not pop either. Thus, if the bigger series converges, so does the smaller series. On the other hand, if the smaller balloon contains enough air to make it pop, then the bigger balloon must also pop, implying that if the smaller series diverges, the bigger series diverges as well. But in the cases that the bigger balloon pops or the smaller balloon does not pop, it says nothing about the remaining balloon.

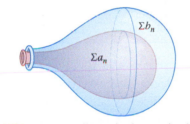

FIGURE 5 The smaller series is contained in the smaller balloon.

Solution For $n \ge 1$, we have

$$\frac{1}{\sqrt{n}\,3^n} \le \frac{1}{3^n}$$

The larger series $\displaystyle\sum_{n=1}^{\infty} \frac{1}{3^n}$ converges because it is a geometric series with $r = \frac{1}{3} < 1$. By the Direct Comparison Test, the smaller series $\displaystyle\sum_{n=1}^{\infty} \frac{1}{\sqrt{n}\,3^n}$ also converges. ∎

■ EXAMPLE 4 Does $S = \displaystyle\sum_{n=2}^{\infty} \frac{1}{(n^2+3)^{1/3}}$ converge?

Solution Let us show that

$$\frac{1}{n} \le \frac{1}{(n^2+3)^{1/3}} \qquad \text{for } n \ge 2$$

This inequality is equivalent to $(n^2+3) \le n^3$, so we must show that

$$f(x) = x^3 - (x^2+3) \ge 0 \qquad \text{for } x \ge 2$$

The function f is increasing because its derivative $f'(x) = 3x\left(x - \frac{2}{3}\right)$ is positive for $x \ge 2$. Since $f(2) = 1$, it follows that $f(x) \ge 1$ for $x \ge 2$, and our original inequality follows. We know that the smaller harmonic series $\displaystyle\sum_{n=2}^{\infty} \frac{1}{n}$ diverges. Therefore, the larger series $\displaystyle\sum_{n=2}^{\infty} \frac{1}{(n^2+1)^{1/3}}$ also diverges. ∎

■ EXAMPLE 5 **Using the Comparison Correctly** Determine the convergence of

$$\sum_{n=2}^{\infty} \frac{1}{n(\ln n)^2}$$

Solution We might be tempted to compare $\displaystyle\sum_{n=2}^{\infty} \frac{1}{n(\ln n)^2}$ to the harmonic series $\displaystyle\sum_{n=2}^{\infty} \frac{1}{n}$ using the inequality (valid for $n \ge 3$ since $\ln 3 > 1$)

$$\frac{1}{n(\ln n)^2} \le \frac{1}{n}$$

However, $\displaystyle\sum_{n=2}^{\infty} \frac{1}{n}$ diverges, and this says nothing about the *smaller* series $\displaystyle\sum \frac{1}{n(\ln n)^2}$. Fortunately, the Integral Test can be used. The substitution $u = \ln x$ yields

$$\int_2^{\infty} \frac{dx}{x(\ln x)^2} = \int_{\ln 2}^{\infty} \frac{du}{u^2} = \lim_{R \to \infty} \left(\frac{1}{\ln 2} - \frac{1}{R} \right) = \frac{1}{\ln 2} < \infty$$

The Integral Test shows that $\displaystyle\sum_{n=2}^{\infty} \frac{1}{n(\ln n)^2}$ converges. ∎

Suppose we wish to study the convergence of

$$S = \sum_{n=2}^{\infty} \frac{n^2}{n^4 - n - 1}$$

For large n, the general term is very close to $1/n^2$:

$$\frac{n^2}{n^4 - n - 1} = \frac{1}{n^2 - n^{-1} - n^{-2}} \approx \frac{1}{n^2}$$

Thus, we might try to compare S with $\sum_{n=2}^{\infty} \frac{1}{n^2}$. Unfortunately, however, the inequality goes in the wrong direction:

$$\frac{n^2}{n^4 - n - 1} > \frac{n^2}{n^4} = \frac{1}{n^2}$$

Although the smaller series $\sum_{n=2}^{\infty} \frac{1}{n^2}$ converges, we cannot use the Direct Comparison Test to say anything about our larger series. In this situation, the following variation of the Direct Comparison Test can be used.

THEOREM 5 **Limit Comparison Test** Let $\{a_n\}$ and $\{b_n\}$ be *positive* sequences. Assume that the following limit exists:

$$L = \lim_{n \to \infty} \frac{a_n}{b_n}$$

- If $L > 0$, then $\sum a_n$ converges if and only if $\sum b_n$ converges.
- If $L = \infty$ and $\sum a_n$ converges, then $\sum b_n$ converges.
- If $L = 0$ and $\sum b_n$ converges, then $\sum a_n$ converges.

Proof Assume first that L is finite (possibly zero) and that $\sum b_n$ converges. Choose a positive number $R > L$. Then $0 \le a_n/b_n \le R$ for all n sufficiently large because a_n/b_n approaches L. Therefore, $a_n \le Rb_n$. The series $\sum Rb_n$ converges because it is a multiple of the convergent series $\sum b_n$. Thus, $\sum a_n$ converges by the Direct Comparison Test.

Next, suppose that L is nonzero (positive or infinite) and that $\sum a_n$ converges. Let $L^{-1} = \lim_{n \to \infty} b_n/a_n$. Then L^{-1} is finite and we can apply the result of the previous paragraph with the roles of $\{a_n\}$ and $\{b_n\}$ reversed to conclude that $\sum b_n$ converges. ∎

CONCEPTUAL INSIGHT To remember the different cases of the Limit Comparison Test, you can think of it this way: If $L > 0$, then $a_n \approx Lb_n$ for large n. In other words, the series $\sum a_n$ and $\sum b_n$ are *roughly* multiples of each other, so one converges if and only if the other converges. If $L = \infty$, then a_n is much larger than b_n (for large n), so if $\sum a_n$ converges, $\sum b_n$ certainly converges. Finally, if $L = 0$, then b_n is much larger than a_n and the convergence of $\sum b_n$ yields the convergence of $\sum a_n$.

■ **EXAMPLE 6** Show that $\sum_{n=2}^{\infty} \frac{n^2}{n^4 - n - 1}$ converges.

Solution Let

$$a_n = \frac{n^2}{n^4 - n - 1} \quad \text{and} \quad b_n = \frac{1}{n^2}$$

We observed above that $a_n \approx b_n$ for large n. To apply the Limit Comparison Test, we observe that the limit L exists and $L > 0$:

$$L = \lim_{n \to \infty} \frac{a_n}{b_n} = \lim_{n \to \infty} \frac{n^2}{n^4 - n - 1} \cdot \frac{n^2}{1} = \lim_{n \to \infty} \frac{1}{1 - n^{-3} - n^{-4}} = 1$$

Since $\sum_{n=2}^{\infty} \frac{1}{n^2}$ converges, our series $\sum_{n=2}^{\infty} \frac{n^2}{n^4 - n - 1}$ also converges by Theorem 5. ■

■ **EXAMPLE 7** Determine whether $\displaystyle\sum_{n=3}^{\infty} \frac{1}{\sqrt{n^2+4}}$ converges.

Solution Apply the Limit Comparison Test with $a_n = \dfrac{1}{\sqrt{n^2+4}}$ and $b_n = \dfrac{1}{n}$. Then

$$L = \lim_{n\to\infty} \frac{a_n}{b_n} = \lim_{n\to\infty} \frac{n}{\sqrt{n^2+4}} = \lim_{n\to\infty} \frac{1}{\sqrt{1+4/n^2}} = 1$$

Since $\displaystyle\sum_{n=3}^{\infty} \frac{1}{n}$ diverges and $L > 0$, the series $\displaystyle\sum_{n=3}^{\infty} \frac{1}{\sqrt{n^2+4}}$ also diverges. ■

In the Limit Comparison Test, when attempting to find an appropriate b_n to compare with a_n, we typically keep only the largest power of n in the numerator and denominator of a_n, as we did in each of the previous examples.

10.3 SUMMARY

- The partial sums S_N of a positive series $S = \sum a_n$ form an increasing sequence.
- Partial Sum Theorem for Positive Series: A positive series S converges if its partial sums S_N remain bounded. Otherwise, it diverges.
- Integral Test: Assume that f is positive, decreasing, and continuous for $x > M$. Set $a_n = f(n)$. If $\displaystyle\int_M^{\infty} f(x)\, dx$ converges, then $S = \sum a_n$ converges, and if $\displaystyle\int_M^{\infty} f(x)\, dx$ diverges, then $S = \sum a_n$ diverges.
- p-Series: The series $\displaystyle\sum_{n=1}^{\infty} \frac{1}{n^p}$ converges if $p > 1$ and diverges if $p \le 1$.
- Direct Comparison Test: Assume there exists $M > 0$ such that $0 \le a_n \le b_n$ for all $n \ge M$. If $\sum b_n$ converges, then $\sum a_n$ converges, and if $\sum a_n$ diverges, then $\sum b_n$ diverges.
- Limit Comparison Test: Assume that $\{a_n\}$ and $\{b_n\}$ are positive and that the following limit exists:

$$L = \lim_{n\to\infty} \frac{a_n}{b_n}$$

  - If $L > 0$, then $\sum a_n$ converges if and only if $\sum b_n$ converges.
  - If $L = \infty$ and $\sum a_n$ converges, then $\sum b_n$ converges.
  - If $L = 0$ and $\sum b_n$ converges, then $\sum a_n$ converges.

10.3 EXERCISES

Preliminary Questions

1. Let $S = \displaystyle\sum_{n=1}^{\infty} a_n$. If the partial sums S_N are increasing, then (choose the correct conclusion):

(a) $\{a_n\}$ is an increasing sequence.

(b) $\{a_n\}$ is a positive sequence.

2. What are the hypotheses of the Integral Test?

3. Which test would you use to determine whether $\displaystyle\sum_{n=1}^{\infty} n^{-3.2}$ converges?

4. Which test would you use to determine whether $\displaystyle\sum_{n=1}^{\infty} \frac{1}{2^n + \sqrt{n}}$ converges?

5. Ralph hopes to investigate the convergence of $\displaystyle\sum_{n=1}^{\infty} \frac{e^{-n}}{n}$ by comparing it with $\displaystyle\sum_{n=1}^{\infty} \frac{1}{n}$. Is Ralph on the right track?

Exercises

In Exercises 1–14, use the Integral Test to determine whether the infinite series is convergent.

1. $\displaystyle\sum_{n=1}^{\infty} \frac{1}{n^4}$

2. $\displaystyle\sum_{n=1}^{\infty} \frac{1}{n+3}$

3. $\displaystyle\sum_{n=1}^{\infty} n^{-1/3}$

4. $\displaystyle\sum_{n=5}^{\infty} \frac{1}{\sqrt{n-4}}$

5. $\displaystyle\sum_{n=25}^{\infty} \frac{n^2}{(n^3+9)^{5/2}}$

6. $\displaystyle\sum_{n=1}^{\infty} \frac{n}{(n^2+1)^{3/5}}$

7. $\displaystyle\sum_{n=1}^{\infty} \frac{1}{n^2+1}$

8. $\displaystyle\sum_{n=4}^{\infty} \frac{1}{n^2-1}$

9. $\displaystyle\sum_{n=1}^{\infty} \frac{1}{n(n+1)}$

10. $\displaystyle\sum_{n=1}^{\infty} ne^{-n^2}$

11. $\displaystyle\sum_{n=2}^{\infty} \frac{1}{n(\ln n)^2}$

12. $\displaystyle\sum_{n=1}^{\infty} \frac{\ln n}{n^2}$

13. $\displaystyle\sum_{n=1}^{\infty} \frac{1}{2^{\ln n}}$

14. $\displaystyle\sum_{n=1}^{\infty} \frac{1}{3^{\ln n}}$

15. Show that $\displaystyle\sum_{n=1}^{\infty} \frac{1}{n^3+8n}$ converges by using the Direct Comparison Test with $\displaystyle\sum_{n=1}^{\infty} n^{-3}$.

16. Show that $\displaystyle\sum_{n=2}^{\infty} \frac{1}{\sqrt{n^2-3}}$ diverges by comparing with $\displaystyle\sum_{n=2}^{\infty} n^{-1}$.

17. Let $S = \displaystyle\sum_{n=1}^{\infty} \frac{1}{n+\sqrt{n}}$. Verify that for $n \geq 1$,

$$\frac{1}{n+\sqrt{n}} \leq \frac{1}{n}, \qquad \frac{1}{n+\sqrt{n}} \leq \frac{1}{\sqrt{n}}$$

Can either inequality be used to show that S diverges? Show that $\dfrac{1}{n+\sqrt{n}} \geq \dfrac{1}{2n}$ for $n \geq 1$ and conclude that S diverges.

18. Which of the following inequalities can be used to study the convergence of $\displaystyle\sum_{n=2}^{\infty} \frac{1}{n^2+\sqrt{n}}$? Explain.

$$\frac{1}{n^2+\sqrt{n}} \leq \frac{1}{\sqrt{n}}, \qquad \frac{1}{n^2+\sqrt{n}} \leq \frac{1}{n^2}$$

In Exercises 19–30, use the Direct Comparison Test to determine whether the infinite series is convergent.

19. $\displaystyle\sum_{n=1}^{\infty} \frac{1}{n2^n}$

20. $\displaystyle\sum_{n=1}^{\infty} \frac{n^3}{n^5+4n+1}$

21. $\displaystyle\sum_{n=1}^{\infty} \frac{1}{n^{1/3}+2^n}$

22. $\displaystyle\sum_{n=1}^{\infty} \frac{1}{\sqrt{n^3+2n-1}}$

23. $\displaystyle\sum_{m=1}^{\infty} \frac{4}{m!+4^m}$

24. $\displaystyle\sum_{n=4}^{\infty} \frac{\sqrt{n}}{n-3}$

25. $\displaystyle\sum_{k=1}^{\infty} \frac{\sin^2 k}{k^2}$

26. $\displaystyle\sum_{k=2}^{\infty} \frac{k^{1/3}}{k^{5/4}-k}$

27. $\displaystyle\sum_{n=1}^{\infty} \frac{2}{3^n+3^{-n}}$

28. $\displaystyle\sum_{k=1}^{\infty} 2^{-k^2}$

29. $\displaystyle\sum_{n=1}^{\infty} \frac{1}{(n+1)!}$

30. $\displaystyle\sum_{n=1}^{\infty} \frac{n!}{n^3}$

Exercise 31–36: For all $a > 0$ and $b > 1$, the inequalities

$$\ln n \leq n^a, \qquad n^a < b^n$$

are true for n sufficiently large (this can be proved using L'Hôpital's Rule). Use this, together with the Direct Comparison Test, to determine whether the series converges or diverges.

31. $\displaystyle\sum_{n=1}^{\infty} \frac{\ln n}{n^3}$

32. $\displaystyle\sum_{m=2}^{\infty} \frac{1}{\ln m}$

33. $\displaystyle\sum_{n=1}^{\infty} \frac{(\ln n)^{100}}{n^{1.1}}$

34. $\displaystyle\sum_{n=1}^{\infty} \frac{1}{(\ln n)^{10}}$

35. $\displaystyle\sum_{n=1}^{\infty} \frac{n}{3^n}$

36. $\displaystyle\sum_{n=1}^{\infty} \frac{n^5}{2^n}$

37. Show that $\displaystyle\sum_{n=1}^{\infty} \sin \frac{1}{n^2}$ converges. *Hint:* Use $\sin x \leq x$ for $x \geq 0$.

38. Does $\displaystyle\sum_{n=2}^{\infty} \frac{\sin(1/n)}{\ln n}$ converge? *Hint:* By Theorem 3 in Section 2.6, $\sin(1/n) > (\cos(1/n))/n$. Thus, $\sin(1/n) > 1/(2n)$ for $n > 2$ [because $\cos(1/n) > \frac{1}{2}$].

In Exercises 39–48, use the Limit Comparison Test to prove convergence or divergence of the infinite series.

39. $\displaystyle\sum_{n=2}^{\infty} \frac{n^2}{n^4-1}$

40. $\displaystyle\sum_{n=2}^{\infty} \frac{1}{n^2-\sqrt{n}}$

41. $\displaystyle\sum_{n=2}^{\infty} \frac{n}{\sqrt{n^3+1}}$

42. $\displaystyle\sum_{n=2}^{\infty} \frac{n^3}{\sqrt{n^7+2n^2+1}}$

43. $\displaystyle\sum_{n=3}^{\infty} \frac{3n+5}{n(n-1)(n-2)}$

44. $\displaystyle\sum_{n=1}^{\infty} \frac{e^n+n}{e^{2n}-n^2}$

45. $\displaystyle\sum_{n=1}^{\infty} \frac{1}{\sqrt{n} + \ln n}$

46. $\displaystyle\sum_{n=1}^{\infty} \frac{\ln(n+4)}{n^{5/2}}$

47. $\displaystyle\sum_{n=1}^{\infty} \left(1 - \cos\frac{1}{n}\right)$ *Hint: Compare with* $\displaystyle\sum_{n=1}^{\infty} n^{-2}$.

48. $\displaystyle\sum_{n=1}^{\infty} (1 - 2^{-1/n})$ *Hint: Compare with the harmonic series.*

In Exercises 49–78, determine convergence or divergence using any method covered so far.

49. $\displaystyle\sum_{n=4}^{\infty} \frac{1}{n^2 - 9}$

50. $\displaystyle\sum_{n=1}^{\infty} \frac{\cos^2 n}{n^2}$

51. $\displaystyle\sum_{n=1}^{\infty} \frac{\sqrt{n}}{4n + 9}$

52. $\displaystyle\sum_{n=1}^{\infty} \frac{n - \cos n}{n^3}$

53. $\displaystyle\sum_{n=1}^{\infty} \frac{n^2 - n}{n^5 + n}$

54. $\displaystyle\sum_{n=1}^{\infty} \frac{1}{n^2 + \sin n}$

55. $\displaystyle\sum_{n=5}^{\infty} (4/5)^{-n}$

56. $\displaystyle\sum_{n=1}^{\infty} \frac{1}{3^{n^2}}$

57. $\displaystyle\sum_{n=2}^{\infty} \frac{1}{n^{3/2} \ln n}$

58. $\displaystyle\sum_{n=2}^{\infty} \frac{(\ln n)^{12}}{n^{9/8}}$

59. $\displaystyle\sum_{k=1}^{\infty} 4^{1/k}$

60. $\displaystyle\sum_{n=1}^{\infty} \frac{4^n}{5^n - 2n}$

61. $\displaystyle\sum_{n=2}^{\infty} \frac{1}{(\ln n)^4}$

62. $\displaystyle\sum_{n=1}^{\infty} \frac{2^n}{3^n - n}$

63. $\displaystyle\sum_{n=3}^{\infty} \frac{1}{n \ln n - n}$

64. $\displaystyle\sum_{n=3}^{\infty} \frac{1}{n(\ln n)^2 - n}$

65. $\displaystyle\sum_{n=1}^{\infty} \frac{1}{n^n}$

66. $\displaystyle\sum_{n=1}^{\infty} \frac{n^2 - 4n^{3/2}}{n^3}$

67. $\displaystyle\sum_{n=1}^{\infty} \frac{1 + (-1)^n}{n}$

68. $\displaystyle\sum_{n=1}^{\infty} \frac{2 + (-1)^n}{n^{3/2}}$

69. $\displaystyle\sum_{n=1}^{\infty} \sin\frac{1}{n}$

70. $\displaystyle\sum_{n=1}^{\infty} \frac{\sin(1/n)}{\sqrt{n}}$

71. $\displaystyle\sum_{n=1}^{\infty} \frac{2n + 1}{4^n}$

72. $\displaystyle\sum_{n=3}^{\infty} \frac{1}{e^{\sqrt{n}}}$

73. $\displaystyle\sum_{n=4}^{\infty} \frac{\ln n}{n^2 - 3n}$

74. $\displaystyle\sum_{n=1}^{\infty} \frac{1}{5^{\ln n}}$

75. $\displaystyle\sum_{n=2}^{\infty} \frac{1}{n^{1/2} \ln n}$

76. $\displaystyle\sum_{n=1}^{\infty} \frac{1}{n^{3/2} - (\ln n)^4}$

77. $\displaystyle\sum_{n=2}^{\infty} \frac{4n^2 + 15n}{3n^4 - 5n^2 - 17}$

78. $\displaystyle\sum_{n=1}^{\infty} \frac{n}{4^{-n} + 5^{-n}}$

79. For which a does $\displaystyle\sum_{n=2}^{\infty} \frac{1}{n(\ln n)^a}$ converge?

80. For which a does $\displaystyle\sum_{n=2}^{\infty} \frac{1}{n^a \ln n}$ converge?

81. For which values of p does $\displaystyle\sum_{n=1}^{\infty} \frac{n^2}{(n^3 + 1)^p}$ converge?

82. For which values of p does $\displaystyle\sum_{n=1}^{\infty} \frac{e^x}{(1 + e^{2x})^p}$ converge?

Approximating Infinite Sums *In Exercises 83–85, let $a_n = f(n)$, where f is a continuous, decreasing function such that $f(x) \geq 0$ and $\int_1^{\infty} f(x)\,dx$ converges.*

83. Show that

$$\int_1^{\infty} f(x)\,dx \leq \sum_{n=1}^{\infty} a_n \leq a_1 + \int_1^{\infty} f(x)\,dx \qquad \boxed{3}$$

84. 𝐶𝐴𝑆   Using Eq. (3), show that

$$5 \leq \sum_{n=1}^{\infty} \frac{1}{n^{1.2}} \leq 6$$

This series converges slowly. Use a computer algebra system to verify that $S_N < 5$ for $N \leq 43{,}128$ and $S_{43{,}129} \approx 5.00000021$.

85. Let $S = \displaystyle\sum_{n=1}^{\infty} a_n$. Arguing as in Exercise 83, show that

$$\sum_{n=1}^{M} a_n + \int_{M+1}^{\infty} f(x)\,dx \leq S \leq \sum_{n=1}^{M+1} a_n + \int_{M+1}^{\infty} f(x)\,dx \qquad \boxed{4}$$

Conclude that

$$0 \leq S - \left(\sum_{n=1}^{M} a_n + \int_{M+1}^{\infty} f(x)\,dx\right) \leq a_{M+1} \qquad \boxed{5}$$

This provides a method for approximating S with an error of at most a_{M+1}.

86. 𝐶𝐴𝑆   Use Eq. (4) from Exercise 85 with $M = 43{,}129$ to prove that

$$5.5915810 \leq \sum_{n=1}^{\infty} \frac{1}{n^{1.2}} \leq 5.5915839$$

87. 𝐶𝐴𝑆   Apply Eq. (4) from Exercise 85 with $M = 40{,}000$ to show that

$$1.644934066 \leq \sum_{n=1}^{\infty} \frac{1}{n^2} \leq 1.644934068$$

Is this consistent with Euler's result, according to which this infinite series has sum $\pi^2/6$?

88. 𝐶𝐴𝑆   Using a CAS and Eq. (5) from Exercise 85, determine the value of $\displaystyle\sum_{n=1}^{\infty} n^{-6}$ to within an error less than 10^{-4}. Check that your result is consistent with that of Euler, who proved that the sum is equal to $\pi^6/945$.

89. 𝐶𝐴𝑆   Using a CAS and Eq. (5) from Exercise 85, determine the value of $\displaystyle\sum_{n=1}^{\infty} n^{-5}$ to within an error less than 10^{-4}.

90. How far can a stack of identical books (of mass m and unit length) extend without tipping over? The stack will not tip over if the $(n+1)$st book is placed at the bottom of the stack with its right edge located at or before the center of mass of the first n books (Figure 6). Let c_n be the center of mass of the first n books, measured along the x-axis, where we take the positive x-axis to the left of the origin as in Figure 7. Recall that if an object of mass m_1 has center of mass at x_1 and a second object of m_2 has center of mass x_2, then the center of mass of the system has x-coordinate

$$\frac{m_1 x_1 + m_2 x_2}{m_1 + m_2}$$

(a) Show that if the $(n+1)$st book is placed with its right edge at c_n, then its center of mass is located at $c_n + \frac{1}{2}$.

(b) Consider the first n books as a single object of mass nm with center of mass at c_n and the $(n+1)$st book as a second object of mass m. Show that if the $(n+1)$st book is placed with its right edge at c_n, then

$$c_{n+1} = c_n + \frac{1}{2(n+1)}.$$

(c) Prove that $\lim_{n \to \infty} c_n = \infty$. Thus, by using enough books, the stack can be extended as far as desired without tipping over.

91. The following argument proves the divergence of the harmonic series $S = \sum_{n=1}^{\infty} 1/n$ without using the Integral Test. Let

$$S_1 = 1 + \frac{1}{3} + \frac{1}{5} + \cdots, \qquad S_2 = \frac{1}{2} + \frac{1}{4} + \frac{1}{6} + \cdots$$

Show that if S converges, then

(a) S_1 and S_2 also converge and $S = S_1 + S_2$.

(b) $S_1 > S_2$ and $S_2 = \frac{1}{2} S$.

Observe that (b) contradicts (a), and conclude that S diverges.

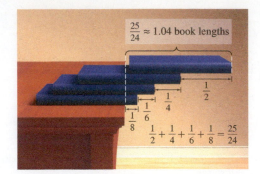

FIGURE 6

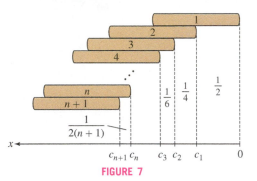

FIGURE 7

Further Insights and Challenges

92. Let $S = \sum_{n=2}^{\infty} a_n$, where $a_n = (\ln(\ln n))^{-\ln n}$.

(a) Show, by taking logarithms, that $a_n = n^{-\ln(\ln(\ln n))}$.

(b) Show that $\ln(\ln(\ln n)) \geq 2$ if $n > C$, where $C = e^{e^{e^2}}$.

(c) Show that S converges.

93. Kummer's Acceleration Method Suppose we wish to approximate $S = \sum_{n=1}^{\infty} 1/n^2$. There is a similar telescoping series whose value can be computed exactly (Example 2 in Section 10.2):

$$\sum_{n=1}^{\infty} \frac{1}{n(n+1)} = 1$$

(a) Verify that

$$S = \sum_{n=1}^{\infty} \frac{1}{n(n+1)} + \sum_{n=1}^{\infty} \left(\frac{1}{n^2} - \frac{1}{n(n+1)} \right)$$

Thus for M large,

$$S \approx 1 + \sum_{n=1}^{M} \frac{1}{n^2(n+1)} \qquad \boxed{6}$$

(b) Explain what has been gained. Why is Eq. (6) a better approximation to S than $\sum_{n=1}^{M} 1/n^2$?

(c) *CAS* Compute

$$\sum_{n=1}^{1000} \frac{1}{n^2}, \qquad 1 + \sum_{n=1}^{100} \frac{1}{n^2(n+1)}$$

Which is a better approximation to S, whose exact value is $\pi^2/6$?

94. *CAS* The series $S = \sum_{k=1}^{\infty} k^{-3}$ has been computed to more than 100 million digits. The first 30 digits are

$$S = 1.202056903159594285399738161511$$

Approximate S using Kummer's Acceleration Method of Exercise 93 with $M = 100$ and auxiliary series $R = \sum_{n=1}^{\infty} (n(n+1)(n+2))^{-1}$. According to Exercise 54 in Section 10.2, R is a telescoping series with the sum $R = \frac{1}{4}$.

10.4 Absolute and Conditional Convergence

In the previous section, we studied positive series, but we still lack the tools to analyze series with both positive and negative terms. One of the keys to understanding such series is the concept of absolute convergence.

DEFINITION Absolute Convergence The series $\sum a_n$ **converges absolutely** if $\sum |a_n|$ converges.

■ **EXAMPLE 1** Verify that the series

$$\sum_{n=1}^{\infty} \frac{(-1)^{n-1}}{n^2} = \frac{1}{1^2} - \frac{1}{2^2} + \frac{1}{3^2} - \frac{1}{4^2} + \cdots$$

converges absolutely.

Solution This series converges absolutely because the positive series (with absolute values) is a p-series with $p = 2 > 1$:

$$\sum_{n=1}^{\infty} \left| \frac{(-1)^{n-1}}{n^2} \right| = \frac{1}{1^2} + \frac{1}{2^2} + \frac{1}{3^2} + \frac{1}{4^2} + \cdots \qquad \text{(convergent } p\text{-series)} \qquad ■$$

The next theorem tells us that if the series of absolute values converges, then the original series also converges.

THEOREM 1 Absolute Convergence Implies Convergence If $\sum |a_n|$ converges, then $\sum a_n$ also converges.

Proof We have $-|a_n| \leq a_n \leq |a_n|$. By adding $|a_n|$ to all parts of the inequality, we get $0 \leq |a_n| + a_n \leq 2|a_n|$. If $\sum |a_n|$ converges, then $\sum 2|a_n|$ also converges, and therefore, $\sum (a_n + |a_n|)$ converges by the Comparison Test. Our original series converges because it is the difference of two convergent series:

$$\sum a_n = \sum (a_n + |a_n|) - \sum |a_n| \qquad\qquad ■$$

■ **EXAMPLE 2** Verify that $S = \sum_{n=1}^{\infty} \frac{(-1)^{n-1}}{n^2}$ converges.

Solution We showed that S converges absolutely in Example 1. By Theorem 1, S itself converges. ■

■ **EXAMPLE 3** Does $S = \sum_{n=1}^{\infty} \frac{(-1)^{n-1}}{\sqrt{n}} = \frac{1}{\sqrt{1}} - \frac{1}{\sqrt{2}} + \frac{1}{\sqrt{3}} - \cdots$ converge absolutely?

Solution The positive series $\sum_{n=1}^{\infty} \frac{1}{\sqrt{n}}$ is a p-series with $p = \frac{1}{2}$. It diverges because $p < 1$. Therefore, S does not converge absolutely. ■

The series in the previous example does not converge *absolutely*, but we still do not know whether or not it converges. A series $\sum a_n$ may converge without converging absolutely. In this case, we say that $\sum a_n$ is conditionally convergent.

DEFINITION Conditional Convergence An infinite series $\sum a_n$ **converges conditionally** if $\sum a_n$ converges but $\sum |a_n|$ diverges.

If a series is not absolutely convergent, how can we determine whether it is conditionally convergent? This is often a more difficult question, because we cannot use the Integral Test or the Comparison Test (they apply only to positive series). However, convergence is guaranteed in the particular case of an **alternating series**

$$S = \sum_{n=1}^{\infty}(-1)^{n-1}b_n = b_1 - b_2 + b_3 - b_4 + \cdots$$

where the terms b_n are positive and decrease to zero (Figure 1).

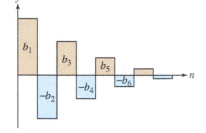

THEOREM 2 Alternating Series Test Assume that $\{b_n\}$ is a positive sequence that is decreasing and converges to 0:

$$b_1 > b_2 > b_3 > b_4 > \cdots > 0, \qquad \lim_{n \to \infty} b_n = 0$$

Then the following alternating series converges:

$$S = \sum_{n=1}^{\infty}(-1)^{n-1}b_n = b_1 - b_2 + b_3 - b_4 + \cdots$$

Furthermore,

$$0 < S < b_1 \quad \text{and} \quad S_{2N} < S < S_{2N+1}, \qquad N \geq 1$$

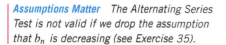

FIGURE 1 An alternating series with decreasing terms. The sum is the signed area, which is at most b_1.

Assumptions Matter The Alternating Series Test is not valid if we drop the assumption that b_n is decreasing (see Exercise 35).

As we will see, this last fact allows the estimation of such a series to any level of accuracy needed.

Notice that under the same conditions, the series

$$\sum_{n=1}^{\infty}(-1)^{n}b_n = -b_1 + b_2 - b_3 + b_4 - \cdots$$

also converges since it is just -1 times the series appearing in the theorem.

Proof We will prove that the partial sums zigzag above and below the sum S as in Figure 2. Note first that the even partial sums are increasing. Indeed, the odd-numbered terms occur with a plus sign and thus, for example,

$$S_4 + b_5 - b_6 = S_6$$

But $b_5 - b_6 > 0$ because b_n is decreasing, and therefore $S_4 < S_6$. In general,

$$S_{2N} + (b_{2N+1} - b_{2N+2}) = S_{2N+2}$$

where $b_{2n+1} - b_{2N+2} > 0$. Thus, $S_{2N} < S_{2N+2}$ and

$$0 < S_2 < S_4 < S_6 < \cdots$$

Similarly,

$$S_{2N-1} - (b_{2N} - b_{2N+1}) = S_{2N+1}$$

Therefore, $S_{2N+1} < S_{2N-1}$, and the sequence of odd partial sums is decreasing:

$$\cdots < S_7 < S_5 < S_3 < S_1$$

Finally, $S_{2N} < S_{2N} + b_{2N+1} = S_{2N+1}$. The picture is as follows:

$$0 < S_2 < S_4 < S_6 < \quad \cdots \quad < S_7 < S_5 < S_3 < S_1$$

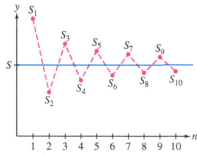

DF FIGURE 2 The partial sums of an alternating series zigzag above and below the limit. The odd partial sums decrease and the even partial sums increase.

Now, because bounded monotonic sequences converge (Theorem 6 of Section 10.1), the even and odd partial sums approach limits that are sandwiched in the middle:

$$0 < S_2 < S_4 < \cdots < \lim_{N \to \infty} S_{2N} \leq \lim_{N \to \infty} S_{2N+1} < \cdots < S_5 < S_3 < S_1 \qquad \boxed{1}$$

These two limits must have a common value L because

$$\lim_{N \to \infty} S_{2N+1} - \lim_{N \to \infty} S_{2N} = \lim_{N \to \infty} (S_{2N+1} - S_{2N}) = \lim_{N \to \infty} b_{2N+1} = 0$$

Therefore, $\lim_{N \to \infty} S_N = L$ and the infinite series converges to $S = L$. From Eq. (1) we also see that $0 < S < S_1 = b_1$ and $S_{2N} < S < S_{2N+1}$ for all N as claimed. ∎

The Alternating Series Test is the only test for conditional convergence developed in this text. Other tests, such as Abel's Criterion and the Dirichlet Test, are discussed in textbooks on analysis.

■ EXAMPLE 4 Show that $S = \displaystyle\sum_{n=1}^{\infty} \frac{(-1)^{n-1}}{\sqrt{n}} = \frac{1}{\sqrt{1}} - \frac{1}{\sqrt{2}} + \frac{1}{\sqrt{3}} - \cdots$ converges conditionally and that $0 \leq S \leq 1$.

Solution The terms $b_n = 1/\sqrt{n}$ are positive and decreasing, and $\lim_{n \to \infty} b_n = 0$. Therefore, S converges by the Alternating Series Test. Furthermore, $0 \leq S \leq 1$ because $b_1 = 1$. However, the positive series $\displaystyle\sum_{n=1}^{\infty} 1/\sqrt{n}$ diverges because it is a p-series with $p = \frac{1}{2} < 1$. Thus, S is conditionally convergent but not absolutely convergent (Figure 3). ∎

The inequality $S_{2N} < S < S_{2N+1}$ in Theorem 2 gives us important information about the error; it tells us that $|S_N - S|$ is less than $|S_N - S_{N+1}| = b_{N+1}$ for all N.

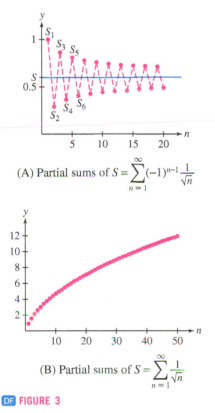

(A) Partial sums of $S = \displaystyle\sum_{n=1}^{\infty} (-1)^{n-1} \frac{1}{\sqrt{n}}$

(B) Partial sums of $S = \displaystyle\sum_{n=1}^{\infty} \frac{1}{\sqrt{n}}$

DF **FIGURE 3**

THEOREM 3 Let $S = \displaystyle\sum_{n=1}^{\infty} (-1)^{n-1} b_n$, where $\{b_n\}$ is a positive decreasing sequence that converges to 0. Then

$$\boxed{\left| S - S_N \right| < b_{N+1}} \qquad \boxed{2}$$

In other words, *when we approximate S by S_N, the error is less than the size of the first omitted term b_{N+1}.*

■ EXAMPLE 5 Alternating Harmonic Series Show that $S = \displaystyle\sum_{n=1}^{\infty} \frac{(-1)^{n-1}}{n}$ converges conditionally. Then:

(a) Show that $|S - S_6| < \frac{1}{7}$.

(b) Find an N such that S_N approximates S with an error less than 10^{-3}.

Solution The terms $b_n = 1/n$ are positive and decreasing, and $\lim_{n \to \infty} b_n = 0$. Therefore, S converges by the Alternating Series Test. The harmonic series $\displaystyle\sum_{n=1}^{\infty} 1/n$ diverges, so S converges conditionally but not absolutely. Now, applying Eq. (2), we have

$$|S - S_N| < b_{N+1} = \frac{1}{N+1}$$

For $N = 6$, we obtain $|S - S_6| < b_7 = \frac{1}{7}$. We can make the error less than 10^{-3} by choosing N so that

$$\frac{1}{N+1} \leq 10^{-3} \quad \Rightarrow \quad N + 1 \geq 10^3 \quad \Rightarrow \quad N \geq 999$$

Using a computer algebra system, we find that $S_{999} \approx 0.69365$. In Exercise 86 of Section 10.7, we will prove that $S = \ln 2 \approx 0.69314$, and thus we can verify that

$$|S - S_{999}| \approx |\ln 2 - 0.69365| \approx 0.0005 < 10^{-3} \qquad ■$$

CONCEPTUAL INSIGHT The convergence of an infinite series $\sum a_n$ depends on two factors: (1) how quickly a_n tends to zero, and (2) how much cancellation takes place among the terms. Consider

Harmonic series (diverges):	$1 + \dfrac{1}{2} + \dfrac{1}{3} + \dfrac{1}{4} + \dfrac{1}{5} + \cdots$
p-Series with $p = 2$ (converges):	$1 + \dfrac{1}{2^2} + \dfrac{1}{3^2} + \dfrac{1}{4^2} + \dfrac{1}{5^2} + \cdots$
Alternating harmonic series (converges):	$1 - \dfrac{1}{2} + \dfrac{1}{3} - \dfrac{1}{4} + \dfrac{1}{5} - \cdots$

The harmonic series diverges because reciprocals $1/n$ do not tend to zero quickly enough. By contrast, the reciprocal squares $1/n^2$ tend to zero quickly enough for the p-series with $p = 2$ to converge. The alternating harmonic series converges, but only due to the cancelation among the terms.

10.4 SUMMARY

- $\sum a_n$ converges *absolutely* if the positive series $\sum |a_n|$ converges.
- Absolute convergence implies convergence: If $\sum |a_n|$ converges, then $\sum a_n$ also converges.
- $\sum a_n$ converges *conditionally* if $\sum a_n$ converges but $\sum |a_n|$ diverges.
- Alternating Series Test: If $\{b_n\}$ is positive and decreasing and $\lim\limits_{n \to \infty} b_n = 0$, then the alternating series

$$S = \sum_{n=1}^{\infty} (-1)^{n-1} b_n = b_1 - b_2 + b_3 - b_4 + b_5 - \cdots$$

converges. Furthermore, $|S - S_N| < b_{N+1}$.
- We have developed two ways to handle nonpositive series: Show absolute convergence if possible, or use the Alternating Series Test if applicable.

10.4 EXERCISES

Preliminary Questions

1. Give an example of a series such that $\sum a_n$ converges but $\sum |a_n|$ diverges.

2. Which of the following statements is equivalent to Theorem 1?

(a) If $\displaystyle\sum_{n=0}^{\infty} |a_n|$ diverges, then $\displaystyle\sum_{n=0}^{\infty} a_n$ also diverges.

(b) If $\displaystyle\sum_{n=0}^{\infty} a_n$ diverges, then $\displaystyle\sum_{n=0}^{\infty} |a_n|$ also diverges.

(c) If $\displaystyle\sum_{n=0}^{\infty} a_n$ converges, then $\displaystyle\sum_{n=0}^{\infty} |a_n|$ also converges.

3. A student argues that $\displaystyle\sum_{n=1}^{\infty} (-1)^n \sqrt{n}$ is an alternating series and therefore converges. Is the student right?

4. Suppose that b_n is positive, decreasing, and tends to 0, and let $S = \displaystyle\sum_{n=1}^{\infty} (-1)^{n-1} b_n$. What can we say about $|S - S_{100}|$ if $a_{101} = 10^{-3}$? Is S larger or smaller than S_{100}?

Exercises

1. Show that

$$\sum_{n=0}^{\infty} \frac{(-1)^n}{2^n}$$

converges absolutely.

2. Show that the following series converges conditionally:

$$\sum_{n=1}^{\infty} (-1)^{n-1} \frac{1}{n^{2/3}} = \frac{1}{1^{2/3}} - \frac{1}{2^{2/3}} + \frac{1}{3^{2/3}} - \frac{1}{4^{2/3}} + \cdots$$

In Exercises 3–10, determine whether the series converges absolutely, conditionally, or not at all.

3. $\displaystyle\sum_{n=1}^{\infty} \frac{(-1)^{n-1}}{n^{1/3}}$

4. $\displaystyle\sum_{n=1}^{\infty} \frac{(-1)^n n^4}{n^3+1}$

5. $\displaystyle\sum_{n=0}^{\infty} \frac{(-1)^{n-1}}{(1.1)^n}$

6. $\displaystyle\sum_{n=1}^{\infty} \frac{\sin(\frac{\pi n}{4})}{n^2}$

7. $\displaystyle\sum_{n=2}^{\infty} \frac{(-1)^n}{n \ln n}$

8. $\displaystyle\sum_{n=1}^{\infty} \frac{(-1)^n}{1+\frac{1}{n}}$

9. $\displaystyle\sum_{n=2}^{\infty} \frac{\cos n\pi}{(\ln n)^2}$

10. $\displaystyle\sum_{n=1}^{\infty} \frac{\cos n}{2^n}$

11. Let $\displaystyle S = \sum_{n=1}^{\infty} (-1)^{n+1} \frac{1}{n^3}$.

(a) Calculate S_n for $1 \le n \le 10$.

(b) Use Eq. (2) to show that $0.9 \le S \le 0.902$.

12. Use Eq. (2) to approximate

$$\sum_{n=1}^{\infty} \frac{(-1)^{n+1}}{n!}$$

to four decimal places.

13. Approximate $\displaystyle\sum_{n=1}^{\infty} \frac{(-1)^{n+1}}{n^4}$ to three decimal places.

14. *CAS* Let

$$S = \sum_{n=1}^{\infty} (-1)^{n-1} \frac{n}{n^2+1}$$

Use a computer algebra system to calculate and plot the partial sums S_n for $1 \le n \le 100$. Observe that the partial sums zigzag above and below the limit.

In Exercises 15–16, find a value of N such that S_N approximates the series with an error of at most 10^{-5}. If you have a CAS, compute this value of S_N.

15. $\displaystyle\sum_{n=1}^{\infty} \frac{(-1)^{n+1}}{n(n+2)(n+3)}$

16. $\displaystyle\sum_{n=1}^{\infty} \frac{(-1)^{n+1} \ln n}{n!}$

In Exercises 17–32, determine convergence or divergence by any method.

17. $\displaystyle\sum_{n=0}^{\infty} 7^{-n}$

18. $\displaystyle\sum_{n=1}^{\infty} \frac{1}{n^{7.5}}$

19. $\displaystyle\sum_{n=1}^{\infty} \frac{1}{5^n - 3^n}$

20. $\displaystyle\sum_{n=2}^{\infty} \frac{n}{n^2-n}$

21. $\displaystyle\sum_{n=1}^{\infty} \frac{1}{3n^4+12n}$

22. $\displaystyle\sum_{n=1}^{\infty} \frac{(-1)^n}{\sqrt{n^2+1}}$

23. $\displaystyle\sum_{n=1}^{\infty} \frac{1}{\sqrt{n^2+1}}$

24. $\displaystyle\sum_{n=0}^{\infty} \frac{(-1)^n n}{\sqrt{n^2+1}}$

25. $\displaystyle\sum_{n=1}^{\infty} \frac{3^n + (-2)^n}{5^n}$

26. $\displaystyle\sum_{n=1}^{\infty} \frac{(-1)^{n+1}}{(2n+1)!}$

27. $\displaystyle\sum_{n=1}^{\infty} (-1)^n n^2 e^{-n^3/3}$

28. $\displaystyle\sum_{n=1}^{\infty} n e^{-n^3/3}$

29. $\displaystyle\sum_{n=2}^{\infty} \frac{(-1)^n}{n^{1/2}(\ln n)^2}$

30. $\displaystyle\sum_{n=2}^{\infty} \frac{1}{n(\ln n)^{1/4}}$

31. $\displaystyle\sum_{n=1}^{\infty} \frac{\ln n}{n^{1.05}}$

32. $\displaystyle\sum_{n=2}^{\infty} \frac{1}{(\ln n)^2}$

33. Show that

$$S = \frac{1}{2} - \frac{1}{2} + \frac{1}{3} - \frac{1}{3} + \frac{1}{4} - \frac{1}{4} + \cdots$$

converges by computing the partial sums. Does it converge absolutely?

34. The Alternating Series Test cannot be applied to

$$\frac{1}{2} - \frac{1}{3} + \frac{1}{2^2} - \frac{1}{3^2} + \frac{1}{2^3} - \frac{1}{3^3} + \cdots$$

Why not? Show that it converges by another method.

35. ✏️ **Assumptions Matter** Show by counterexample that the Alternating Series Test does not remain true if the sequence a_n tends to zero but is not assumed to be decreasing. *Hint:* Consider

$$R = \frac{1}{2} - \frac{1}{4} + \frac{1}{3} - \frac{1}{8} + \frac{1}{4} - \frac{1}{16} + \cdots + \left(\frac{1}{n} - \frac{1}{2^n}\right) + \cdots$$

36. Determine whether the following series converges conditionally:

$$1 - \frac{1}{3} + \frac{1}{2} - \frac{1}{5} + \frac{1}{3} - \frac{1}{7} + \frac{1}{4} - \frac{1}{9} + \frac{1}{5} - \frac{1}{11} + \cdots$$

37. Prove that if $\sum a_n$ converges absolutely, then $\sum a_n^2$ also converges. Give an example where $\sum a_n$ is only conditionally convergent and $\sum a_n^2$ diverges.

Further Insights and Challenges

38. Prove the following variant of the Alternating Series Test: If $\{b_n\}$ is a positive, decreasing sequence with $\displaystyle\lim_{n\to\infty} b_n = 0$, then the series

$$b_1 + b_2 - 2b_3 + a_4 + b_5 - 2a_6 + \cdots$$

converges. *Hint:* Show that S_{3N} is increasing and bounded by $a_1 + a_2$, and continue as in the proof of the Alternating Series Test.

39. Use Exercise 38 to show that the following series converges:

$$S = \frac{1}{\ln 2} + \frac{1}{\ln 3} - \frac{2}{\ln 4} + \frac{1}{\ln 5} + \frac{1}{\ln 6} - \frac{2}{\ln 7} + \cdots$$

40. Prove the conditional convergence of

$$R = 1 + \frac{1}{2} + \frac{1}{3} - \frac{3}{4} + \frac{1}{5} + \frac{1}{6} + \frac{1}{7} - \frac{3}{8} + \cdots$$

41. Show that the following series diverges:

$$S = 1 + \frac{1}{2} + \frac{1}{3} - \frac{2}{4} + \frac{1}{5} + \frac{1}{6} + \frac{1}{7} - \frac{2}{8} + \cdots$$

Hint: Use the result of Exercise 40 to write S as the sum of a convergent series and a divergent series.

42. Prove that

$$\sum_{n=1}^{\infty}(-1)^{n+1}\frac{(\ln n)^a}{n}$$

converges for all exponents a. *Hint:* Show that $f(x) = (\ln x)^a/x$ is decreasing for x sufficiently large.

43. We say that $\{b_n\}$ is a rearrangement of $\{a_n\}$ if $\{b_n\}$ has the same terms as $\{a_n\}$ but occurring in a different order. Show that if $\{b_n\}$ is a rearrangement of $\{a_n\}$ and $S = \sum_{n=1}^{\infty} a_n$ converges absolutely, then

$$T = \sum_{n=1}^{\infty} b_n$$

also converges absolutely. (This result does not hold if S is only conditionally convergent.) *Hint:* Prove that the partial sums

$$\sum_{n=1}^{N} |b_n|$$

are bounded. It can be shown further that $S = T$.

44. Assumptions Matter In 1829 Lejeune Dirichlet pointed out that the great French mathematician Augustin Louis Cauchy made a mistake in a published paper by improperly assuming the Limit Comparison Test to be valid for nonpositive series. Here are Dirichlet's two series:

$$\sum_{n=1}^{\infty} \frac{(-1)^n}{\sqrt{n}}, \qquad \sum_{n=1}^{\infty} \frac{(-1)^n}{\sqrt{n}}\left(1 + \frac{(-1)^n}{\sqrt{n}}\right)$$

Explain how they provide a counterexample to the Limit Comparison Test when the series are not assumed to be positive.

10.5 The Ratio and Root Tests and Strategies for Choosing Tests

Series such as

$$S = 1 + \frac{2}{1!} + \frac{2^2}{2!} + \frac{2^3}{3!} + \frac{2^4}{4!} + \cdots$$

arise in applications, but the convergence tests developed so far cannot be applied easily. Fortunately, the Ratio Test can be used for this and many other series.

> **THEOREM 1 Ratio Test** Assume that the following limit exists:
>
> $$\rho = \lim_{n\to\infty}\left|\frac{a_{n+1}}{a_n}\right|$$
>
> **(i)** If $\rho < 1$, then $\sum a_n$ converges absolutely.
>
> **(ii)** If $\rho > 1$, then $\sum a_n$ diverges.
>
> **(iii)** If $\rho = 1$, the test is inconclusive (the series may converge or diverge).

The symbol ρ is a lowercase rho, the seventeenth letter of the Greek alphabet.

Proof The idea is to compare with a geometric series. If $\rho < 1$, we may choose a number r such that $\rho < r < 1$. Since $|a_{n+1}/a_n|$ converges to ρ, there exists a number M such that $|a_{n+1}/a_n| < r$ for all $n \geq M$. Therefore,

$$|a_{M+1}| < r|a_M|$$

$$|a_{M+2}| < r|a_{M+1}| < r(r|a_M|) = r^2|a_M|$$

$$|a_{M+3}| < r|a_{M+2}| < r^3|a_M|$$

In general, $|a_{M+n}| < r^n|a_M|$, and thus,

$$\sum_{n=M}^{\infty} |a_n| = \sum_{n=0}^{\infty} |a_{M+n}| \leq \sum_{n=0}^{\infty} |a_M|r^n = |a_M|\sum_{n=0}^{\infty} r^n$$

The geometric series on the right converges because $0 < r < 1$, so $\sum_{n=M}^{\infty} |a_n|$ converges by the Direct Comparison Test and thus $\sum a_n$ converges absolutely.

If $\rho > 1$, choose r such that $1 < r < \rho$. Then there exists a number M such that $|a_{n+1}/a_n| > r$ for all $n \geq M$. Arguing as before with the inequalities reversed, we find that $|a_{M+n}| \geq r^n|a_M|$. Since r^n tends to ∞, the terms a_{M+n} do not tend to zero, and consequently, $\sum a_n$ diverges. Finally, Example 4 below shows that both convergence and divergence are possible when $\rho = 1$, so the test is inconclusive in this case. ∎

■ **EXAMPLE 1** Prove that $\displaystyle\sum_{n=1}^{\infty} \frac{2^n}{n!}$ converges.

Solution Compute the ratio and its limit with $a_n = \dfrac{2^n}{n!}$. Note that $(n+1)! = (n+1)n!$. Thus,

$$\left|\frac{a_{n+1}}{a_n}\right| = \frac{2^{n+1}}{(n+1)!} \frac{n!}{2^n} = \frac{2^{n+1}}{2^n} \frac{n!}{(n+1)!} = \frac{2}{n+1}$$

We obtain

$$\rho = \lim_{n\to\infty} \left|\frac{a_{n+1}}{a_n}\right| = \lim_{n\to\infty} \frac{2}{n+1} = 0$$

Since $\rho < 1$, the series $\displaystyle\sum_{n=1}^{\infty} \frac{2^n}{n!}$ converges by the Ratio Test. ■

■ **EXAMPLE 2** Does $\displaystyle\sum_{n=1}^{\infty} \frac{n^2}{2^n}$ converge?

Solution Apply the Ratio Test with $a_n = \dfrac{n^2}{2^n}$:

$$\left|\frac{a_{n+1}}{a_n}\right| = \frac{(n+1)^2}{2^{n+1}} \frac{2^n}{n^2} = \frac{1}{2}\left(\frac{n^2 + 2n + 1}{n^2}\right) = \frac{1}{2}\left(1 + \frac{2}{n} + \frac{1}{n^2}\right)$$

We obtain

$$\rho = \lim_{n\to\infty} \left|\frac{a_{n+1}}{a_n}\right| = \frac{1}{2} \lim_{n\to\infty}\left(1 + \frac{2}{n} + \frac{1}{n^2}\right) = \frac{1}{2}$$

Since $\rho < 1$, the series converges by the Ratio Test. ■

■ **EXAMPLE 3** Does $\displaystyle\sum_{n=0}^{\infty} (-1)^n \frac{n!}{1000^n}$ converge?

Solution This series diverges by the Ratio Test because $\rho > 1$:

$$\rho = \lim_{n\to\infty} \left|\frac{a_{n+1}}{a_n}\right| = \lim_{n\to\infty} \frac{(n+1)!}{1000^{n+1}} \frac{1000^n}{n!} = \lim_{n\to\infty} \frac{n+1}{1000} = \infty$$ ■

■ **EXAMPLE 4** **Ratio Test Inconclusive** Show that both convergence and divergence are possible when $\rho = 1$ by considering $\displaystyle\sum_{n=1}^{\infty} n^2$ and $\displaystyle\sum_{n=1}^{\infty} n^{-2}$.

Solution For $a_n = n^2$, we have

$$\rho = \lim_{n\to\infty} \left|\frac{a_{n+1}}{a_n}\right| = \lim_{n\to\infty} \frac{(n+1)^2}{n^2} = \lim_{n\to\infty} \frac{n^2 + 2n + 1}{n^2} = \lim_{n\to\infty}\left(1 + \frac{2}{n} + \frac{1}{n^2}\right) = 1$$

On the other hand, for $b_n = n^{-2}$,

$$\rho = \lim_{n\to\infty} \left|\frac{b_{n+1}}{b_n}\right| = \lim_{n\to\infty} \left|\frac{a_n}{a_{n+1}}\right| = \frac{1}{\displaystyle\lim_{n\to\infty}\left|\frac{a_{n+1}}{a_n}\right|} = 1$$

Thus, $\rho = 1$ in both cases, but, in fact, $\displaystyle\sum_{n=1}^{\infty} n^2$ diverges by the nth Term Divergence Test since $\displaystyle\lim_{n\to\infty} n^2 = \infty$, and $\displaystyle\sum_{n=1}^{\infty} n^{-2}$ converges since it is a p-series with $p = 2 > 1$. This shows that both convergence and divergence are possible when $\rho = 1$. ■

Our next test is based on the limit of the nth roots $\sqrt[n]{|a_n|}$ rather than the ratios a_{n+1}/a_n. Its proof, like that of the Ratio Test, is based on a comparison with a geometric series (see Exercise 63).

THEOREM 2 Root Test   Assume that the following limit exists:

$$L = \lim_{n \to \infty} \sqrt[n]{|a_n|}$$

(i) If $L < 1$, then $\sum a_n$ converges absolutely.

(ii) If $L > 1$, then $\sum a_n$ diverges.

(iii) If $L = 1$, the test is inconclusive (the series may converge or diverge).

■ **EXAMPLE 5** Does $\displaystyle\sum_{n=1}^{\infty} \left(\dfrac{n}{2n+3}\right)^n$ converge?

Solution We have $L = \lim_{n\to\infty} \sqrt[n]{a_n} = \lim_{n\to\infty} \dfrac{n}{2n+3} = \dfrac{1}{2}$. Since $L < 1$, the series converges by the Root Test. ■

Determining Which Test to Apply

We end this section with a brief review of all of the tests we have introduced for determining convergence so far and how one decides which test to apply.

Let $\displaystyle\sum_{n=1}^{\infty} a_n$ be given. Keep in mind that the series for which convergence or divergence

is known include the geometric series $\displaystyle\sum_{n=0}^{\infty} ar^n$, which converge for $|r| < 1$, and the p-series

$\displaystyle\sum_{n=0}^{\infty} \dfrac{1}{n^p}$, which converge for $p > 1$.

1. The nth Term Divergence Test Always check this test first. If $\lim\limits_{n\to\infty} a_n \neq 0$, then the series diverges. But if $\lim\limits_{n\to\infty} a_n = 0$, we do not know whether the series converges or diverges, and hence we move on to the next step.

2. Positive Series If all terms in the series are positive, try one of the following tests:

(a) The Direct Comparison Test Consider whether dropping terms in the numerator or denominator gives a series that we know either converges or diverges. If a larger series converges or a smaller series diverges, then the original series does the same. For

example, $\displaystyle\sum_{n=1}^{\infty} \dfrac{1}{n^2 + \sqrt{n}}$ converges because $\dfrac{1}{n^2 + \sqrt{n}} < \dfrac{1}{n^2}$, and $\displaystyle\sum_{n=1}^{\infty} \dfrac{1}{n^2}$ is a p-series

with $p = 2 > 1$ so it converges. On the other hand, this does not work for $\displaystyle\sum_{n=2}^{\infty} \dfrac{1}{n^2 - \sqrt{n}}$

since then the comparison series $\displaystyle\sum_{n=1}^{\infty} \dfrac{1}{n^2}$, while still converging, is smaller than the

original series, so the Direct Comparison Test does not apply. In this case, we can often apply the Limit Comparison Test as follows.

(b) The Limit Comparison Test Consider the dominant term in the numerator and denominator, and compare the original series to the ratio of those terms. For example,

for $\displaystyle\sum_{n=2}^{\infty} \dfrac{1}{n^2 - \sqrt{n}}$, n^2 is dominant over $\sqrt{n}$ as it grows faster as n increases. So, we

let $b_n = \dfrac{1}{n^2}$. Then

$$\lim_{n\to\infty} \dfrac{a_n}{b_n} = \lim_{n\to\infty} \dfrac{\frac{1}{n^2-\sqrt{n}}}{\frac{1}{n^2}} = \lim_{n\to\infty} \dfrac{n^2}{n^2 - \sqrt{n}} = 1$$

The limit is a positive number, so the Limit Comparison Test applies. Since $\sum_{n=1}^{\infty} \frac{1}{n^2}$ converges, so does the original series.

(c) **The Ratio Test** The Ratio Test is often effective in the presence of a factorial such as $n!$ since in the ratio, the factorial disappears after cancellation. It is also effective when there are constants to the power n, such as 2^n, since in the ratio, the power n disappears after cancelation. For example, if the series is $\sum_{n=1}^{\infty} \frac{3^n}{n!}$, then applying the Ratio Test yields

$$\lim_{n\to\infty} \left|\frac{a_{n+1}}{a_n}\right| = \lim_{n\to\infty} \frac{\frac{3^{n+1}}{(n+1)!}}{\frac{3^n}{(n)!}} = \lim_{n\to\infty} \frac{3}{n+1} = 0 < 1$$

Therefore, the series converges.

(d) **The Root Test** The Root Test is often effective when there is a term of the form $f(n)^{g(n)}$. For example, $\sum_{n=1}^{\infty} \frac{2^n}{n^{2n}}$ is a good example since applying the Root Test yields

$$\lim_{n\to\infty} |a_n|^{1/n} = \lim_{n\to\infty} \left(\frac{2^n}{n^{2n}}\right)^{1/n} = \lim_{n\to\infty} \frac{2}{n^2} = 0 < 1$$

Thus, the series converges.

(e) **The Integral Test** When the other tests fail on a positive series, consider the Integral Test. If $a_n = f(n)$ is a decreasing function, then the series converges if and only if the improper integral $\int_1^{\infty} f(x)\,dx$ converges. For example, the other tests do not easily apply to $\sum_{n=2}^{\infty} \frac{1}{n \ln n}$. However $f(x) = \frac{1}{n \ln n}$ is a decreasing function and

$$\int_2^{\infty} \frac{1}{x \ln x}\,dx = \ln(\ln x)\Big|_2^{\infty} = \infty.$$ Thus, the integral and, hence, the series diverges.

3. **Series That Are Not Positive Series**

(a) **Alternating Series Test** If the series is alternating of the form $\sum_{n=1}^{\infty} (-1)^{n-1} b_n$, show that $0 < b_n + 1 < b_n$ and $\lim_{n\to\infty} b_n = 0$. Then the Alternating Series Test shows the series converges.

(b) **Absolute Convergence** If the series is not alternating, then see if its absolute value, which is a positive series, converges using the tests for positive series. If so, the original series converges as well.

10.5 SUMMARY

- Ratio Test: Assume that $\rho = \lim_{n\to\infty} \left|\frac{a_{n+1}}{a_n}\right|$ exists. Then $\sum a_n$

  - Converges absolutely if $\rho < 1$.
  - Diverges if $\rho > 1$.
  - Inconclusive if $\rho = 1$.

- Root Test: Assume that $L = \lim_{n\to\infty} \sqrt[n]{|a_n|}$ exists. Then $\sum a_n$

  - Converges absolutely if $L < 1$.
  - Diverges if $L > 1$.
  - Inconclusive if $L = 1$.

10.5 EXERCISES

Preliminary Questions

1. In the Ratio Test, is ρ equal to $\lim\limits_{n\to\infty}\left|\dfrac{a_{n+1}}{a_n}\right|$ or $\lim\limits_{n\to\infty}\left|\dfrac{a_n}{a_{n+1}}\right|$?

2. Is the Ratio Test conclusive for $\sum\limits_{n=1}^{\infty}\dfrac{1}{2^n}$? Is it conclusive for $\sum\limits_{n=1}^{\infty}\dfrac{1}{n}$?

3. Can the Ratio Test be used to show convergence if the series is only conditionally convergent?

Exercises

In Exercises 1–20, apply the Ratio Test to determine convergence or divergence, or state that the Ratio Test is inconclusive.

1. $\sum\limits_{n=1}^{\infty}\dfrac{1}{5^n}$

2. $\sum\limits_{n=1}^{\infty}\dfrac{(-1)^{n-1}n}{5^n}$

3. $\sum\limits_{n=1}^{\infty}\dfrac{1}{n^n}$

4. $\sum\limits_{n=0}^{\infty}\dfrac{3n+2}{5n^3+1}$

5. $\sum\limits_{n=1}^{\infty}\dfrac{n}{n^2+1}$

6. $\sum\limits_{n=1}^{\infty}\dfrac{2^n}{n}$

7. $\sum\limits_{n=1}^{\infty}\dfrac{2^n}{n^{100}}$

8. $\sum\limits_{n=1}^{\infty}\dfrac{n^3}{3^{n^2}}$

9. $\sum\limits_{n=1}^{\infty}\dfrac{10^n}{2^{n^2}}$

10. $\sum\limits_{n=1}^{\infty}\dfrac{e^n}{n!}$

11. $\sum\limits_{n=1}^{\infty}\dfrac{e^n}{n^n}$

12. $\sum\limits_{n=1}^{\infty}\dfrac{n^{40}}{n!}$

13. $\sum\limits_{n=0}^{\infty}\dfrac{n!}{6^n}$

14. $\sum\limits_{n=1}^{\infty}\dfrac{n!}{n^9}$

15. $\sum\limits_{n=2}^{\infty}\dfrac{1}{n\ln n}$

16. $\sum\limits_{n=1}^{\infty}\dfrac{1}{(2n)!}$

17. $\sum\limits_{n=1}^{\infty}\dfrac{n^2}{(2n+1)!}$

18. $\sum\limits_{n=1}^{\infty}\dfrac{(n!)^3}{(3n)!}$

19. $\sum\limits_{n=2}^{\infty}\dfrac{1}{2^n+1}$

20. $\sum\limits_{n=2}^{\infty}\dfrac{1}{\ln n}$

21. Show that $\sum\limits_{n=1}^{\infty}n^k\,3^{-n}$ converges for all exponents k.

22. Show that $\sum\limits_{n=1}^{\infty}n^2 x^n$ converges if $|x|<1$.

23. Show that $\sum\limits_{n=1}^{\infty}2^n x^n$ converges if $|x|<\frac{1}{2}$.

24. Show that $\sum\limits_{n=1}^{\infty}\dfrac{r^n}{n!}$ converges for all r.

25. Show that $\sum\limits_{n=1}^{\infty}\dfrac{r^n}{n}$ converges if $|r|<1$.

26. Is there any value of k such that $\sum\limits_{n=1}^{\infty}\dfrac{2^n}{n^k}$ converges?

27. Show that $\sum\limits_{n=1}^{\infty}\dfrac{n!}{n^n}$ converges. *Hint:* Use $\lim\limits_{n\to\infty}\left(1+\dfrac{1}{n}\right)^n=e$.

In Exercises 28–33, assume that $|a_{n+1}/a_n|$ converges to $\rho=\frac{1}{3}$. What can you say about the convergence of the given series?

28. $\sum\limits_{n=1}^{\infty}na_n$

29. $\sum\limits_{n=1}^{\infty}n^3 a_n$

30. $\sum\limits_{n=1}^{\infty}2^n a_n$

31. $\sum\limits_{n=1}^{\infty}3^n a_n$

32. $\sum\limits_{n=1}^{\infty}4^n a_n$

33. $\sum\limits_{n=1}^{\infty}a_n^2$

34. Assume that $|a_{n+1}/a_n|$ converges to $\rho=4$. Does $\sum\limits_{n=1}^{\infty}a_n^{-1}$ converge (assume that $a_n\neq 0$ for all n)?

35. Is the Ratio Test conclusive for the p-series $\sum\limits_{n=1}^{\infty}\dfrac{1}{n^p}$?

In Exercises 36–41, use the Root Test to determine convergence or divergence (or state that the test is inconclusive).

36. $\sum\limits_{n=0}^{\infty}\dfrac{1}{10^n}$

37. $\sum\limits_{n=1}^{\infty}\dfrac{1}{n^n}$

38. $\sum\limits_{k=0}^{\infty}\left(\dfrac{k}{k+10}\right)^k$

39. $\sum\limits_{k=0}^{\infty}\left(\dfrac{k}{3k+1}\right)^k$

40. $\sum\limits_{n=1}^{\infty}\left(1+\dfrac{1}{n}\right)^{-n}$

41. $\sum\limits_{n=4}^{\infty}\left(1+\dfrac{1}{n}\right)^{-n^2}$

42. Prove that $\sum\limits_{n=1}^{\infty}\dfrac{2^{n^2}}{n!}$ diverges. *Hint:* Use $2^{n^2}=(2^n)^n$ and $n!\leq n^n$.

In Exercises 43–62, determine convergence or divergence using any method covered in the text so far.

43. $\sum\limits_{n=1}^{\infty}\dfrac{2^n+4^n}{7^n}$

44. $\sum\limits_{n=1}^{\infty}\dfrac{n^3}{n!}$

45. $\sum\limits_{n=1}^{\infty}\dfrac{n}{2n+1}$

46. $\sum\limits_{n=1}^{\infty}2^{1/n}$

47. $\sum\limits_{n=1}^{\infty}\dfrac{\sin n}{n^2}$

48. $\sum\limits_{n=1}^{\infty}\dfrac{n!}{(2n)!}$

49. $\sum\limits_{n=1}^{\infty}\dfrac{1}{n+\sqrt{n}}$

50. $\sum\limits_{n=2}^{\infty}\dfrac{1}{n(\ln n)^3}$

51. $\sum\limits_{n=1}^{\infty}\dfrac{n^3}{5^n}$

52. $\sum\limits_{n=2}^{\infty}\dfrac{1}{n(\ln n)^3}$

53. $\sum\limits_{n=2}^{\infty}\dfrac{1}{\sqrt{n^3-n^2}}$

54. $\sum\limits_{n=1}^{\infty}\dfrac{n^2+4n}{3n^4+9}$

55. $\sum\limits_{n=1}^{\infty}n^{-0.8}$

56. $\sum\limits_{n=1}^{\infty}(0.8)^{-n}n^{-0.8}$

57. $\sum\limits_{n=1}^{\infty}4^{-2n+1}$

58. $\sum\limits_{n=1}^{\infty}\dfrac{(-1)^{n-1}}{\sqrt{n}}$

59. $\sum\limits_{n=1}^{\infty}\sin\dfrac{1}{n^2}$

60. $\sum\limits_{n=1}^{\infty}(-1)^n\cos\dfrac{1}{n}$

61. $\sum\limits_{n=1}^{\infty}\dfrac{(-2)^n}{\sqrt{n}}$

62. $\sum\limits_{n=1}^{\infty}\left(\dfrac{n}{n+12}\right)^n$

Further Insights and Challenges

63. **Proof of the Root Test** Let $S = \sum_{n=0}^{\infty} a_n$ be a positive series, and assume that $L = \lim_{n \to \infty} \sqrt[n]{a_n}$ exists.

(a) Show that S converges if $L < 1$. *Hint:* Choose R with $L < R < 1$ and show that $a_n \le R^n$ for n sufficiently large. Then compare with the geometric series $\sum R^n$.

(b) Show that S diverges if $L > 1$.

64. Show that the Ratio Test does not apply, but verify convergence using the Direct Comparison Test for the series

$$\frac{1}{2} + \frac{1}{3^2} + \frac{1}{2^3} + \frac{1}{3^4} + \frac{1}{2^5} + \cdots$$

65. Let $S = \sum_{n=1}^{\infty} \frac{c^n n!}{n^n}$, where c is a constant.

(a) Prove that S converges absolutely if $|c| < e$ and diverges if $|c| > e$.

(b) It is known that $\lim_{n \to \infty} \frac{e^n n!}{n^{n+1/2}} = \sqrt{2\pi}$. Verify this numerically.

(c) Use the Limit Comparison Test to prove that S diverges for $c = e$.

10.6 Power Series

A **power series** with center c is an infinite series

$$F(x) = \sum_{n=0}^{\infty} a_n(x - c)^n = a_0 + a_1(x - c) + a_2(x - c)^2 + a_3(x - c)^3 + \cdots$$

where x is a variable. For example,

$$F(x) = 1 + (x - 2) + 2(x - 2)^2 + 3(x - 2)^3 + \cdots \qquad \boxed{1}$$

is a power series with center $c = 2$.

Many functions that arise in applications can be represented as power series. This includes not only the familiar trigonometric, exponential, logarithm, and root functions, but also the host of "special functions" of physics and engineering such as Bessel functions and elliptic functions.

A power series $F(x) = \sum_{n=0}^{\infty} a_n(x - c)^n$ converges for some values of x and may diverge for others. For example, if we set $x = \frac{9}{4}$ in the power series of Eq. (1), we obtain the infinite series

$$F\left(\frac{9}{4}\right) = 1 + \left(\frac{9}{4} - 2\right) + 2\left(\frac{9}{4} - 2\right)^2 + 3\left(\frac{9}{4} - 2\right)^3 + \cdots$$

$$= 1 + \left(\frac{1}{4}\right) + 2\left(\frac{1}{4}\right)^2 + 3\left(\frac{1}{4}\right)^3 + \cdots + n\left(\frac{1}{4}\right)^n + \cdots$$

This converges by the Ratio Test:

$$\lim_{n \to \infty} \left| \frac{a_{n+1}}{a_n} \right| = \lim_{n \to \infty} \left| \frac{\frac{n+1}{4^{n+1}}}{\frac{n}{4^n}} \right| = \lim_{n \to \infty} \frac{1}{4}\left(\frac{n+1}{n}\right)\frac{1/n}{1/n} = \lim_{n \to \infty} \frac{1}{4}\left(\frac{1 + 1/n}{1}\right) = \frac{1}{4}$$

On the other hand, the power series in Eq. (1) diverges for $x = 3$ by the nth Term Test:

$$F(3) = 1 + (3 - 2) + 2(3 - 2)^2 + 3(3 - 2)^3 + \cdots$$

$$= 1 + 1 + 2 + 3 + \cdots$$

There is a surprisingly simple way to describe the set of values x at which a power series $F(x)$ converges. According to our next theorem, either $F(x)$ converges absolutely for all values of x or there is a radius of convergence R such that

$F(x)$ *converges absolutely when* $|x - c| < R$ *and diverges when* $|x - c| > R$.

This means that $F(x)$ converges for x in an **interval of convergence** consisting of the open interval $(c - R, c + R)$ and possibly one or both of the endpoints $c - R$ and $c + R$ (Figure 1). Note that $F(x)$ automatically converges at $x = c$ because

$$F(c) = a_0 + a_1(c - c) + a_2(c - c)^2 + a_3(c - c)^3 + \cdots = a_0$$

We set $R = 0$ if $F(x)$ converges only for $x = c$, and we set $R = \infty$ if $F(x)$ converges for all values of x.

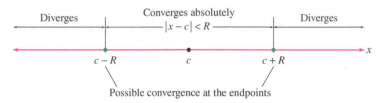

FIGURE 1 Interval of convergence of a power series.

THEOREM 1 Radius of Convergence Every power series

$$F(x) = \sum_{n=0}^{\infty} a_n(x - c)^n$$

has a radius of convergence R, which is either a nonnegative number ($R \geq 0$) or infinity ($R = \infty$). If R is finite, $F(x)$ converges absolutely when $|x - c| < R$ and diverges when $|x - c| > R$. If $R = \infty$, then $F(x)$ converges absolutely for all x.

Proof We assume that $c = 0$ to simplify the notation. If $F(x)$ converges only at $x = 0$, then $R = 0$. Otherwise, $F(x)$ converges for some nonzero value $x = B$. We claim that $F(x)$ must then converge absolutely for all $|x| < |B|$. To prove this, note that because

$$F(B) = \sum_{n=0}^{\infty} a_n B^n$$ converges, the general term $a_n B^n$ tends to zero. In particular, there exists $M > 0$ such that $|a_n B^n| < M$ for all n. Therefore,

$$\sum_{n=0}^{\infty} |a_n x^n| = \sum_{n=0}^{\infty} |a_n B^n| \left| \frac{x}{B} \right|^n < M \sum_{n=0}^{\infty} \left| \frac{x}{B} \right|^n$$

If $|x| < |B|$, then $|x/B| < 1$ and the series on the right is a convergent geometric series. By the Direct Comparison Test, the series on the left also converges. This proves that $F(x)$ converges absolutely if $|x| < |B|$.

Now let S be the set of numbers x such that $F(x)$ converges. Then S contains 0, and we have shown that if S contains a number $B \neq 0$, then S contains the open interval $(-|B|, |B|)$. If S is bounded, then S has a least upper bound $L > 0$ (see marginal note). In this case, there exist numbers $B \in S$ smaller than but arbitrarily close to L, and thus, S contains $(-B, B)$ for all $0 < B < L$. It follows that S contains the open interval $(-L, L)$. The set S cannot contain any number x with $|x| > L$, but S may contain one or both of the endpoints $x = \pm L$. So in this case, F has radius of convergence $R = L$. If S is not bounded, then S contains intervals $(-B, B)$ for B arbitrarily large. In this case, S is the entire real line **R**, and the radius of convergence is $R = \infty$. ∎

Least Upper Bound Property: If S is a set of real numbers with an upper bound M (i.e., $x \leq M$ for all $x \in S$), then S has a least upper bound L. See Appendix B.

From Theorem 1, we see that there are two steps in determining the interval of convergence of F:

Step 1. Find the radius of convergence R (using the Ratio Test, in most cases).

Step 2. Check convergence at the endpoints (if $R \neq 0$ or ∞).

■ **EXAMPLE 1** **Using the Ratio Test** Where does $F(x) = \sum_{n=0}^{\infty} \dfrac{x^n}{2^n}$ converge?

Solution

Step 1. **Find the radius of convergence.**

Let $a_n = \dfrac{x^n}{2^n}$ and compute ρ from the Ratio Test:

$$\rho = \lim_{n \to \infty} \left| \frac{a_{n+1}}{a_n} \right| = \lim_{n \to \infty} \left| \frac{x^{n+1}}{2^{n+1}} \right| \cdot \left| \frac{2^n}{x^n} \right| = \lim_{n \to \infty} \frac{1}{2} |x| = \frac{1}{2} |x|$$

We find that

$$\rho < 1 \quad \text{if} \quad \frac{1}{2} |x| < 1, \quad \text{that is, if} \quad |x| < 2$$

Thus, $F(x)$ converges if $|x| < 2$. Similarly, $\rho > 1$ if $\frac{1}{2}|x| > 1$, or $|x| > 2$. So, $F(x)$ diverges if $|x| > 2$. Therefore, the radius of convergence is $R = 2$.

Step 2. **Check the endpoints.**

The Ratio Test is inconclusive for $x = \pm 2$, so we must check these cases directly:

$$F(2) = \sum_{n=0}^{\infty} \frac{2^n}{2^n} = 1 + 1 + 1 + 1 + 1 + 1 \cdots$$

$$F(-2) = \sum_{n=0}^{\infty} \frac{(-2)^n}{2^n} = 1 - 1 + 1 - 1 + 1 - 1 \cdots$$

Both series diverge. We conclude that $F(x)$ converges only for $|x| < 2$ (Figure 2). ■

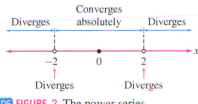

Converges
Diverges · absolutely · Diverges

$-2 \qquad 0 \qquad 2$

Diverges $\qquad$ Diverges

 FIGURE 2 The power series

$$\sum_{n=0}^{\infty} \frac{x^n}{2^n}$$

has an interval of convergence $(-2, 2)$.

■ **EXAMPLE 2** Where does $F(x) = \sum_{n=1}^{\infty} \dfrac{(-1)^n}{4^n \, n} (x - 5)^n$ converge?

Solution We compute ρ with $a_n = \dfrac{(-1)^n}{4^n \, n} (x - 5)^n$:

$$\rho = \lim_{n \to \infty} \left| \frac{a_{n+1}}{a_n} \right| = \lim_{n \to \infty} \left| \frac{(x-5)^{n+1}}{4^{n+1}(n+1)} \, \frac{4^n \, n}{(x-5)^n} \right|$$

$$= |x - 5| \lim_{n \to \infty} \left| \frac{n}{4(n+1)} \right|$$

$$= \frac{1}{4} |x - 5|$$

We find that

$$\rho < 1 \quad \text{if} \quad \frac{1}{4} |x - 5| < 1, \quad \text{that is, if} \quad |x - 5| < 4$$

Thus, $F(x)$ converges absolutely on the open interval $(1, 9)$ of radius 4 with center $c = 5$. In other words, the radius of convergence is $R = 4$. Next, we check the endpoints:

$$x = 9: \quad \sum_{n=1}^{\infty} \frac{(-1)^n}{4^n n} (9 - 5)^n = \sum_{n=1}^{\infty} \frac{(-1)^n}{n} \quad \text{converges (Alternating Series Test)}$$

$$x = 1: \quad \sum_{n=1}^{\infty} \frac{(-1)^n}{4^n n} (-4)^n = \sum_{n=1}^{\infty} \frac{1}{n} \quad \text{diverges (harmonic series)}$$

We conclude that $F(x)$ converges for x in the half-open interval $(1, 9]$ shown in Figure 3.

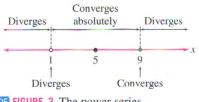

Converges
Diverges · absolutely · Diverges

$1 \qquad 5 \qquad 9$

Diverges $\qquad$ Converges

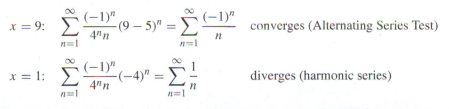 **FIGURE 3** The power series

$$\sum_{n=1}^{\infty} \frac{(-1)^n}{4^n n} (x - 5)^n$$

has an interval of convergence $(1, 9]$.

Some power series contain only even powers or only odd powers of x. The Ratio Test can still be used to find the radius of convergence.

■ **EXAMPLE 3** **An Even Power Series** Where does $\displaystyle\sum_{n=0}^{\infty} \frac{x^{2n}}{(2n)!}$ converge?

Solution Although this power series has only even powers of x, we can still apply the Ratio Test with $a_n = x^{2n}/(2n)!$. We have

$$a_{n+1} = \frac{x^{2(n+1)}}{(2(n+1))!} = \frac{x^{2n+2}}{(2n+2)!}$$

Furthermore, $(2n+2)! = (2n+2)(2n+1)(2n)!$, so

$$\rho = \lim_{n\to\infty}\left|\frac{a_{n+1}}{a_n}\right| = \lim_{n\to\infty}\frac{x^{2n+2}}{(2n+2)!}\frac{(2n)!}{x^{2n}} = |x|^2 \lim_{n\to\infty}\frac{1}{(2n+2)(2n+1)} = 0$$

Thus, $\rho = 0$ for all x, and $F(x)$ converges for all x. The radius of convergence is $R = \infty$.   ■

When a function f is represented by a power series on an interval I, we refer to it as the power series expansion *of f on I.*

Geometric series are important examples of power series. Recall the formula $\displaystyle\sum_{n=0}^{\infty} r^n = 1/(1-r)$, valid for $|r| < 1$. Writing x in place of r, we obtain a power series expansion with radius of convergence $R = 1$:

$$\boxed{\frac{1}{1-x} = \sum_{n=0}^{\infty} x^n \qquad \text{for } |x| < 1}$$ **2**

The next two examples show that we can modify this formula to find the power series expansions of other functions.

■ **EXAMPLE 4** **Geometric Series** Prove that

$$\frac{1}{1-2x} = \sum_{n=0}^{\infty} 2^n x^n \qquad \text{for } |x| < \frac{1}{2}$$

Solution Substitute $2x$ for x in Eq. (2):

$$\frac{1}{1-2x} = \sum_{n=0}^{\infty}(2x)^n = \sum_{n=0}^{\infty} 2^n x^n$$ **3**

Expansion (2) is valid for $|x| < 1$, so Eq. (3) is valid for $|2x| < 1$, or $|x| < \frac{1}{2}$.   ■

■ **EXAMPLE 5** Find a power series expansion with center $c = 0$ for

$$f(x) = \frac{1}{2 + x^2}$$

and find the interval of convergence.

Solution We need to rewrite $f(x)$ so we can use Eq. (2). We have

$$\frac{1}{2+x^2} = \frac{1}{2}\left(\frac{1}{1+\frac{1}{2}x^2}\right) = \frac{1}{2}\left(\frac{1}{1-(-\frac{1}{2}x^2)}\right) = \frac{1}{2}\left(\frac{1}{1-u}\right)$$

where $u = -\frac{1}{2}x^2$. Now substitute $u = -\frac{1}{2}x^2$ for x in Eq. (2) to obtain

$$f(x) = \frac{1}{2+x^2} = \frac{1}{2}\sum_{n=0}^{\infty}\left(-\frac{x^2}{2}\right)^n$$

$$= \sum_{n=0}^{\infty}\frac{(-1)^n x^{2n}}{2^{n+1}}$$

This expansion is valid if $|-x^2/2| < 1$, or $|x| < \sqrt{2}$. The interval of convergence is $(-\sqrt{2}, \sqrt{2})$.   ■

Our next theorem tells us that within the interval of convergence, we can treat a power series as though it were a polynomial; that is, we can differentiate and integrate term by term.

The proof of Theorem 2 is somewhat technical and is omitted. See Exercise 66 for a proof that F is continuous.

THEOREM 2 **Term-by-Term Differentiation and Integration** Assume that

$$F(x) = \sum_{n=0}^{\infty} a_n (x - c)^n$$

has radius of convergence $R > 0$. Then F is differentiable on $(c - R, c + R)$ (or for all x if $R = \infty$). Furthermore, we can integrate and differentiate term by term. For $x \in (c - R, c + R)$,

$$F'(x) = \sum_{n=1}^{\infty} n a_n (x - c)^{n-1}$$

$$\int F(x)\,dx = A + \sum_{n=0}^{\infty} \frac{a_n}{n+1}(x-c)^{n+1} \qquad (A \text{ any constant})$$

These series have the same radius of convergence R.

■ **EXAMPLE 6** **Differentiating a Power Series** Prove that for $-1 < x < 1$,

$$\frac{1}{(1-x)^2} = 1 + 2x + 3x^2 + 4x^3 + 5x^4 + \cdots$$

Solution The geometric series has radius of convergence $R = 1$:

$$\frac{1}{1-x} = 1 + x + x^2 + x^3 + x^4 + \cdots$$

By Theorem 2, we can differentiate term by term for $|x| < 1$ to obtain

$$\frac{d}{dx}\left(\frac{1}{1-x}\right) = \frac{d}{dx}(1 + x + x^2 + x^3 + x^4 + \cdots)$$

$$\frac{1}{(1-x)^2} = 1 + 2x + 3x^2 + 4x^3 + 5x^4 + \cdots \qquad ■$$

Theorem 2 is a powerful tool in the study of power series.

■ **EXAMPLE 7** **Power Series for Arctangent** Prove that for $-1 < x < 1$,

$$\tan^{-1} x = \sum_{n=0}^{\infty} \frac{(-1)^n x^{2n+1}}{2n+1} = x - \frac{x^3}{3} + \frac{x^5}{5} - \frac{x^7}{7} + \cdots \qquad \boxed{4}$$

Solution Recall that $\tan^{-1} x$ is an antiderivative of $(1 + x^2)^{-1}$. We obtain a power series expansion of this antiderivative by substituting $-x^2$ for x in the geometric series of Eq. (2):

$$\frac{1}{1+x^2} = 1 - x^2 + x^4 - x^6 + \cdots$$

This expansion is valid for $|x^2| < 1$—that is, for $|x| < 1$. By Theorem 2, we can integrate series term by term. The resulting expansion is also valid for $|x| < 1$:

$$\tan^{-1} x = \int \frac{dx}{1+x^2} = \int (1 - x^2 + x^4 - x^6 + \cdots)\,dx$$

$$= A + x - \frac{x^3}{3} + \frac{x^5}{5} - \frac{x^7}{7} + \cdots$$

Setting $x = 0$, we obtain $A = \tan^{-1} 0 = 0$. Thus, Eq. (4) is valid for $-1 < x < 1$. ■

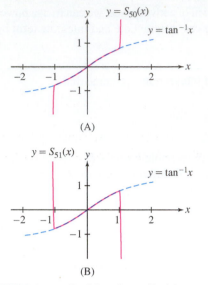

FIGURE 4 $y = S_{50}(x)$ and $y = S_{51}(x)$ are nearly indistinguishable from $y = \tan^{-1} x$ on $(-1, 1)$.

GRAPHICAL INSIGHT Let's examine the expansion of the previous example graphically. The partial sums of the power series for $f(x) = \tan^{-1} x$ are

$$S_N(x) = x - \frac{x^3}{3} + \frac{x^5}{5} - \frac{x^7}{7} + \cdots + (-1)^N \frac{x^{2N+1}}{2N + 1}$$

For large N, we can expect $S_N(x)$ to provide a good approximation to $f(x) = \tan^{-1} x$ on the interval $(-1, 1)$, where the power series expansion is valid. Figure 4 confirms this expectation: The graphs of $y = S_{50}(x)$ and $y = S_{51}(x)$ are nearly indistinguishable from the graph of $y = \tan^{-1} x$ on $(-1, 1)$. Thus, we may use the partial sums to approximate the arctangent. For example, $\tan^{-1}(0.3)$ is approximated by

$$S_4(0.3) = 0.3 - \frac{(0.3)^3}{3} + \frac{(0.3)^5}{5} - \frac{(0.3)^7}{7} + \frac{(0.3)^9}{9} \approx 0.2914569$$

Since the power series is an alternating series, the error is less than the first omitted term:

$$|\tan^{-1}(0.3) - S_4(0.3)| < \frac{(0.3)^{11}}{11} \approx 1.61 \times 10^{-7}$$

The situation changes drastically in the region $|x| > 1$, where the power series diverges and the partial sums $S_N(x)$ deviate sharply from $\tan^{-1} x$.

Power Series Solutions of Differential Equations

Power series are a basic tool in the study of differential equations. To illustrate, consider the differential equation with initial condition

$$y' = y, \qquad y(0) = 1$$

We know that $f(x) = e^x$ is the unique solution, but let's try to find a power series that satisfies this Initial Value Problem. We have

$$F(x) = \sum_{n=0}^{\infty} a_n x^n = a_0 + a_1 x + a_2 x^2 + a_3 x^3 + \cdots$$

$$F'(x) = \sum_{n=0}^{\infty} n a_n x^{n-1} = a_1 + 2a_2 x + 3a_3 x^2 + 4a_4 x^3 + \cdots$$

Therefore, $F'(x) = F(x)$ if

$$a_0 = a_1, \quad a_1 = 2a_2, \quad a_2 = 3a_3, \quad a_3 = 4a_4, \quad \ldots$$

In other words, $F'(x) = F(x)$ if $a_{n-1} = n a_n$, or

$$\boxed{a_n = \frac{a_{n-1}}{n}}$$

An equation of this type is called a *recursion relation*. It enables us to determine all of the coefficients a_n successively from the first coefficient a_0, which may be chosen arbitrarily. For example,

$$n = 1: \qquad a_1 = \frac{a_0}{1}$$

$$n = 2: \qquad a_2 = \frac{a_1}{2} = \frac{a_0}{2 \cdot 1} = \frac{a_0}{2!}$$

$$n = 3: \qquad a_3 = \frac{a_2}{3} = \frac{a_1}{3 \cdot 2} = \frac{a_0}{3 \cdot 2 \cdot 1} = \frac{a_0}{3!}$$

To obtain a general formula for a_n, apply the recursion relation n times:

$$a_n = \frac{a_{n-1}}{n} = \frac{a_{n-2}}{n(n-1)} = \frac{a_{n-3}}{n(n-1)(n-2)} = \cdots = \frac{a_0}{n!}$$

We conclude that

$$F(x) = a_0 \sum_{n=0}^{\infty} \frac{x^n}{n!}$$

In Example 3, we showed that this power series has radius of convergence $R = \infty$, so $y = F(x)$ satisfies $y' = y$ for all x. Moreover, $F(0) = a_0$, so the initial condition $y(0) = 1$ is satisfied with $a_0 = 1$.

What we have shown is that $f(x) = e^x$ and $F(x)$ with $a_0 = 1$ are both solutions of the Initial Value Problem. They must be equal because the solution is unique. This proves that for all x,

$$e^x = \sum_{n=0}^{\infty} \frac{x^n}{n!} = 1 + x + \frac{x^2}{2!} + \frac{x^3}{3!} + \frac{x^4}{4!} + \cdots$$

In this example, we knew in advance that $y = e^x$ is a solution of $y' = y$, but suppose we are given a differential equation whose solution is unknown. We can try to find a solution in the form of a power series $F(x) = \sum_{n=0}^{\infty} a_n x^n$. In favorable cases, the differential equation leads to a recursion relation that enables us to determine the coefficients a_n.

The solution in Example 8 is called the "Bessel function of order 1." The Bessel function of order n is a solution of

$$x^2 y'' + xy' + (x^2 - n^2)y = 0$$

These functions have applications in many areas of physics and engineering.

■ **EXAMPLE 8** Find a power series solution to the Initial Value Problem:

$$x^2 y'' + xy' + (x^2 - 1)y = 0, \qquad y'(0) = 1 \qquad \boxed{5}$$

Solution Assume that Eq. (5) has a power series solution $F(x) = \sum_{n=0}^{\infty} a_n x^n$. Then

$$y' = F'(x) = \sum_{n=0}^{\infty} n a_n x^{n-1} = a_1 + 2a_2 x + 3a_3 x^2 + \cdots$$

$$y'' = F''(x) = \sum_{n=0}^{\infty} n(n-1) a_n x^{n-2} = 2a_2 + 6a_3 x + 12a_4 x^2 + \cdots$$

Now substitute the series for y, y', and y'' into the differential equation (5) to determine the recursion relation satisfied by the coefficients a_n:

$$x^2 y'' + xy' + (x^2 - 1)y$$

In Eq. (6), we combine the first three series into a single series using

$$n(n-1) + n - 1 = n^2 - 1$$

and we shift the fourth series to begin at $n = 2$ rather than $n = 0$.

$$= x^2 \sum_{n=0}^{\infty} n(n-1) a_n x^{n-2} + x \sum_{n=0}^{\infty} n a_n x^{n-1} + (x^2 - 1) \sum_{n=0}^{\infty} a_n x^n$$

$$= \sum_{n=0}^{\infty} n(n-1) a_n x^n + \sum_{n=0}^{\infty} n a_n x^n - \sum_{n=0}^{\infty} a_n x^n + \sum_{n=0}^{\infty} a_n x^{n+2} \qquad \boxed{6}$$

$$= \sum_{n=0}^{\infty} (n^2 - 1) a_n x^n + \sum_{n=2}^{\infty} a_{n-2} x^n = 0$$

The differential equation is satisfied if

$$\sum_{n=0}^{\infty} (n^2 - 1) a_n x^n = -\sum_{n=2}^{\infty} a_{n-2} x^n$$

The first few terms on each side of this equation are

$$-a_0 + 0 \cdot x + 3a_2 x^2 + 8a_3 x^3 + 15a_4 x^4 + \cdots = 0 + 0 \cdot x - a_0 x^2 - a_1 x^3 - a_2 x^4 - \cdots$$

Matching up the coefficients of x^n, we find that

$$-a_0 = 0, \qquad 3a_2 = -a_0, \qquad 8a_3 = -a_1, \qquad 15a_4 = -a_2 \qquad \boxed{7}$$

In general, $(n^2 - 1)a_n = -a_{n-2}$, and this yields the recursion relation

$$\boxed{a_n = -\frac{a_{n-2}}{n^2 - 1} \qquad \text{for } n \geq 2} \qquad \boxed{8}$$

Note that $a_0 = 0$ by Eq. (7). The recursion relation forces all of the even coefficients a_2, $a_4, a_6, \ldots$ to be zero:

$$a_2 = \frac{a_0}{2^2 - 1} \text{ so } a_2 = 0, \qquad \text{and then} \qquad a_4 = \frac{a_2}{4^2 - 1} = 0 \text{ so } a_4 = 0, \qquad \text{etc.}$$

As for the odd coefficients, a_1 may be chosen arbitrarily. Because $F'(0) = a_1$, we set $a_1 = 1$ to obtain a solution $y = F(x)$ satisfying $F'(0) = 1$. Now apply Eq. (8):

$$n = 3: \qquad a_3 = -\frac{a_1}{3^2 - 1} = -\frac{1}{3^2 - 1}$$

$$n = 5: \qquad a_5 = -\frac{a_3}{5^2 - 1} = \frac{1}{(5^2 - 1)(3^2 - 1)}$$

$$n = 7: \qquad a_7 = -\frac{a_5}{7^2 - 1} = -\frac{1}{(7^2 - 1)(3^2 - 1)(5^2 - 1)}$$

This shows the general pattern of coefficients. To express the coefficients in a compact form, let $n = 2k + 1$. Then the denominator in the recursion relation (8) can be written

$$n^2 - 1 = (2k + 1)^2 - 1 = 4k^2 + 4k = 4k(k + 1)$$

and

$$a_{2k+1} = -\frac{a_{2k-1}}{4k(k + 1)}$$

Applying this recursion relation k times, we obtain the closed formula

$$a_{2k+1} = (-1)^k \left(\frac{1}{4k(k + 1)} \right) \left(\frac{1}{4(k - 1)k} \right) \cdots \left(\frac{1}{4(1)(2)} \right) = \frac{(-1)^k}{4^k \, k! \, (k + 1)!}$$

Thus, we obtain a power series representation of our solution:

$$F(x) = \sum_{k=0}^{\infty} \frac{(-1)^k}{4^k k!(k + 1)!} x^{2k+1}$$

A straightforward application of the Ratio Test shows that F has an infinite radius of convergence. Therefore, $F(x)$ is a solution of the Initial Value Problem for all x. ∎

10.6 SUMMARY

- A *power series* is an infinite series of the form

$$F(x) = \sum_{n=0}^{\infty} a_n (x - c)^n$$

The constant c is called the *center* of $F(x)$.
- Every power series $F(x)$ has a *radius of convergence* R (Figure 5) such that

  - $F(x)$ converges absolutely for $|x - c| < R$ and diverges for $|x - c| > R$.
  - $F(x)$ may converge or diverge at the endpoints $c - R$ and $c + R$.

We set $R = 0$ if $F(x)$ converges only for $x = c$ and $R = \infty$ if $F(x)$ converges for all x.

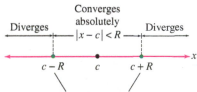

FIGURE 5 Interval of convergence of a power series.

- The *interval of convergence* of F consists of the open interval $(c - R, c + R)$ and possibly one or both endpoints $c - R$ and $c + R$.
- In many cases, the Ratio Test can be used to find the radius of convergence R. It is necessary to check convergence at the endpoints separately.
- If $R > 0$, then F is differentiable on $(c - R, c + R)$ and

$$F'(x) = \sum_{n=1}^{\infty} na_n(x - c)^{n-1}, \qquad \int F(x)\,dx = A + \sum_{n=0}^{\infty} \frac{a_n}{n + 1}(x - c)^{n+1}$$

(A is any constant). These two power series have the same radius of convergence R.

- The expansion $\dfrac{1}{1 - x} = \sum_{n=0}^{\infty} x^n$ is valid for $|x| < 1$. It can be used to derive expansions of other related functions by substitution, integration, or differentiation.

10.6 EXERCISES

Preliminary Questions

1. Suppose that $\sum a_n x^n$ converges for $x = 5$. Must it also converge for $x = 4$? What about $x = -3$?

2. Suppose that $\sum a_n (x - 6)^n$ converges for $x = 10$. At which of the points (a)–(d) must it also converge?

(a) $x = 8$ (b) $x = 11$ (c) $x = 3$ (d) $x = 0$

3. What is the radius of convergence of $F(3x)$ if $F(x)$ is a power series with radius of convergence $R = 12$?

4. The power series $F(x) = \sum_{n=1}^{\infty} nx^n$ has radius of convergence $R = 1$. What is the power series expansion of $F'(x)$ and what is its radius of convergence?

Exercises

1. Use the Ratio Test to determine the radius of convergence R of $\sum_{n=0}^{\infty} \dfrac{x^n}{2^n}$. Does it converge at the endpoints $x = \pm R$?

2. Use the Ratio Test to show that $\sum_{n=1}^{\infty} \dfrac{x^n}{\sqrt{n}2^n}$ has radius of convergence $R = 2$. Then determine whether it converges at the endpoints $R = \pm 2$.

3. Show that the power series (a)–(c) have the same radius of convergence. Then show that (a) diverges at both endpoints, (b) converges at one endpoint but diverges at the other, and (c) converges at both endpoints.

(a) $\sum_{n=1}^{\infty} \dfrac{x^n}{3^n}$ (b) $\sum_{n=1}^{\infty} \dfrac{x^n}{n3^n}$ (c) $\sum_{n=1}^{\infty} \dfrac{x^n}{n^2 3^n}$

4. Repeat Exercise 3 for the following series:

(a) $\sum_{n=1}^{\infty} \dfrac{(x - 5)^n}{9^n}$ (b) $\sum_{n=1}^{\infty} \dfrac{(x - 5)^n}{n9^n}$ (c) $\sum_{n=1}^{\infty} \dfrac{(x - 5)^n}{n^2 9^n}$

5. Show that $\sum_{n=0}^{\infty} n^n x^n$ diverges for all $x \neq 0$.

6. For which values of x does $\sum_{n=0}^{\infty} n! x^n$ converge?

7. Use the Ratio Test to show that $\sum_{n=0}^{\infty} \dfrac{x^{2n}}{3^n}$ has radius of convergence $R = \sqrt{3}$.

8. Show that $\sum_{n=0}^{\infty} \dfrac{x^{3n+1}}{64^n}$ has radius of convergence $R = 4$.

In Exercises 9–34, find the interval of convergence.

9. $\sum_{n=0}^{\infty} nx^n$

10. $\sum_{n=1}^{\infty} \dfrac{2^n}{n}x^n$

11. $\sum_{n=1}^{\infty} (-1)^n \dfrac{x^{2n+1}}{2^n n}$

12. $\sum_{n=0}^{\infty} (-1)^n \dfrac{n}{4^n}x^{2n}$

13. $\sum_{n=4}^{\infty} \dfrac{x^n}{n^5}$

14. $\sum_{n=8}^{\infty} n^7 x^n$

15. $\sum_{n=0}^{\infty} \dfrac{x^n}{(n!)^2}$

16. $\sum_{n=0}^{\infty} \dfrac{8^n}{n!}x^n$

17. $\sum_{n=0}^{\infty} \dfrac{(2n)!}{(n!)^3}x^n$

18. $\sum_{n=0}^{\infty} \dfrac{4^n}{(2n + 1)!}x^{2n-1}$

19. $\sum_{n=0}^{\infty} \dfrac{(-1)^n x^n}{\sqrt{n^2 + 1}}$

20. $\sum_{n=0}^{\infty} \dfrac{x^n}{n^4 + 2}$

21. $\sum_{n=15}^{\infty} \dfrac{x^{2n+1}}{3n + 1}$

22. $\sum_{n=9}^{\infty} \dfrac{x^n}{n - 4\ln n}$

23. $\sum_{n=2}^{\infty} \dfrac{x^n}{\ln n}$

24. $\sum_{n=2}^{\infty} \dfrac{x^{3n+2}}{\ln n}$

25. $\sum_{n=1}^{\infty} n(x - 3)^n$

26. $\sum_{n=1}^{\infty} \dfrac{(-5)^n (x - 3)^n}{n^2}$

27. $\displaystyle\sum_{n=1}^{\infty}(-1)^n n^5 (x-7)^n$

28. $\displaystyle\sum_{n=0}^{\infty} 27^n (x-1)^{3n+2}$

29. $\displaystyle\sum_{n=1}^{\infty} \frac{2^n}{3n}(x+3)^n$

30. $\displaystyle\sum_{n=0}^{\infty} \frac{(x-4)^n}{n!}$

31. $\displaystyle\sum_{n=0}^{\infty} \frac{(-5)^n}{n!}(x+10)^n$

32. $\displaystyle\sum_{n=10}^{\infty} n!\,(x+5)^n$

33. $\displaystyle\sum_{n=12}^{\infty} e^n (x-2)^n$

34. $\displaystyle\sum_{n=2}^{\infty} \frac{(x+4)^n}{(n \ln n)^2}$

In Exercises 35–40, use Eq. (2) to expand the function in a power series with center $c = 0$ and determine the interval of convergence.

35. $f(x) = \dfrac{1}{1-3x}$

36. $f(x) = \dfrac{1}{1+3x}$

37. $f(x) = \dfrac{1}{3-x}$

38. $f(x) = \dfrac{1}{4+3x}$

39. $f(x) = \dfrac{1}{1+x^2}$

40. $f(x) = \dfrac{1}{16+2x^3}$

41. Use the equalities

$$\frac{1}{1-x} = \frac{1}{-3-(x-4)} = \frac{-\frac{1}{3}}{1+\left(\frac{x-4}{3}\right)}$$

to show that for $|x-4| < 3$,

$$\frac{1}{1-x} = \sum_{n=0}^{\infty}(-1)^{n+1}\frac{(x-4)^n}{3^{n+1}}$$

42. Use the method of Exercise 41 to expand $1/(1-x)$ in power series with centers $c = 2$ and $c = -2$. Determine the interval of convergence.

43. Use the method of Exercise 41 to expand $1/(4-x)$ in a power series with center $c = 5$. Determine the interval of convergence.

44. Find a power series that converges only for x in $[2, 6)$.

45. Apply integration to the expansion

$$\frac{1}{1+x} = \sum_{n=0}^{\infty}(-1)^n x^n = 1 - x + x^2 - x^3 + \cdots$$

to prove that for $-1 < x < 1$,

$$\ln(1+x) = \sum_{n=1}^{\infty}\frac{(-1)^{n-1}x^n}{n} = x - \frac{x^2}{2} + \frac{x^3}{3} - \frac{x^4}{4} + \cdots$$

46. Use the result of Exercise 45 to prove that

$$\ln\frac{3}{2} = \frac{1}{2} - \frac{1}{2 \cdot 2^2} + \frac{1}{3 \cdot 2^3} - \frac{1}{4 \cdot 2^4} + \cdots$$

Use your knowledge of alternating series to find an N such that the partial sum S_N approximates $\ln \frac{3}{2}$ to within an error of at most 10^{-3}. Confirm using a calculator to compute both S_N and $\ln \frac{3}{2}$.

47. Let $F(x) = (x+1)\ln(1+x) - x$.

(a) Apply integration to the result of Exercise 45 to prove that for $-1 < x < 1$,

$$F(x) = \sum_{n=1}^{\infty}(-1)^{n+1}\frac{x^{n+1}}{n(n+1)}$$

(b) Evaluate at $x = \frac{1}{2}$ to prove

$$\frac{3}{2}\ln\frac{3}{2} - \frac{1}{2} = \frac{1}{1 \cdot 2 \cdot 2^2} - \frac{1}{2 \cdot 3 \cdot 2^3} + \frac{1}{3 \cdot 4 \cdot 2^4} - \frac{1}{4 \cdot 5 \cdot 2^5} + \cdots$$

(c) Use a calculator to verify that the partial sum S_4 approximates the left-hand side with an error no greater than the term a_5 of the series.

48. Prove that for $|x| < 1$,

$$\int \frac{dx}{x^4+1} = A + x - \frac{x^5}{5} + \frac{x^9}{9} - \cdots$$

Use the first two terms to approximate $\int_0^{1/2} dx/(x^4+1)$ numerically. Use the fact that you have an alternating series to show that the error in this approximation is at most 0.00022.

49. Use the result of Example 7 to show that

$$F(x) = \frac{x^2}{1 \cdot 2} - \frac{x^4}{3 \cdot 4} + \frac{x^6}{5 \cdot 6} - \frac{x^8}{7 \cdot 8} + \cdots$$

is an antiderivative of $f(x) = \tan^{-1} x$ satisfying $F(0) = 0$. What is the radius of convergence of this power series?

50. Verify that function $F(x) = x \tan^{-1} x - \frac{1}{2}\log(x^2+1)$ is an antiderivative of $f(x) = \tan^{-1} x$ satisfying $F(0) = 0$. Then use the result of Exercise 49 with $x = \frac{1}{\sqrt{3}}$ to show that

$$\frac{\pi}{6\sqrt{3}} - \frac{1}{2}\ln\frac{4}{3} = \frac{1}{1 \cdot 2(3)} - \frac{1}{3 \cdot 4(3^2)} + \frac{1}{5 \cdot 6(3^3)} - \frac{1}{7 \cdot 8(3^4)} + \cdots$$

Use a calculator to compare the value of the left-hand side with the partial sum S_4 of the series on the right.

51. Evaluate $\displaystyle\sum_{n=1}^{\infty}\frac{n}{2^n}$. *Hint:* Use differentiation to show that

$$(1-x)^{-2} = \sum_{n=1}^{\infty} nx^{n-1} \quad (\text{for } |x| < 1)$$

52. Use the power series for $(1+x^2)^{-1}$ and differentiation to prove that for $|x| < 1$,

$$\frac{2x}{(x^2+1)^2} = \sum_{n=1}^{\infty}(-1)^{n-1}(2n)x^{2n-1}$$

53. Show that the following series converges absolutely for $|x| < 1$ and compute its sum:

$$F(x) = 1 - x - x^2 + x^3 - x^4 - x^5 + x^6 - x^7 - x^8 + \cdots$$

Hint: Write $F(x)$ as a sum of three geometric series with common ratio x^3.

54. Show that for $|x| < 1$,

$$\frac{1+2x}{1+x+x^2} = 1 + x - 2x^2 + x^3 + x^4 - 2x^5 + x^6 + x^7 - 2x^8 + \cdots$$

Hint: Use the hint from Exercise 53.

55. Find all values of x such that $\displaystyle\sum_{n=1}^{\infty}\frac{x^{n^2}}{n!}$ converges.

56. Find all values of x such that the following series converges:

$$F(x) = 1 + 3x + x^2 + 27x^3 + x^4 + 243x^5 + \cdots$$

57. Find a power series $P(x) = \displaystyle\sum_{n=0}^{\infty} a_n x^n$ satisfying the differential equation $y' = -y$ with initial condition $y(0) = 1$. Then use Theorem 1 of Section 5.9 to conclude that $P(x) = e^{-x}$.

58. Let $C(x) = 1 - \dfrac{x^2}{2!} + \dfrac{x^4}{4!} - \dfrac{x^6}{6!} + \cdots$.

(a) Show that $C(x)$ has an infinite radius of convergence.

(b) Prove that $C(x)$ and $f(x) = \cos x$ are both solutions of $y'' = -y$ with initial conditions $y(0) = 1$, $y'(0) = 0$. This Initial Value Problem has a unique solution, so we have $C(x) = \cos x$ for all x.

59. Use the power series for $y = e^x$ to show that

$$\frac{1}{e} = \frac{1}{2!} - \frac{1}{3!} + \frac{1}{4!} - \cdots$$

Use your knowledge of alternating series to find an N such that the partial sum S_N approximates e^{-1} to within an error of at most 10^{-3}. Confirm this using a calculator to compute both S_N and e^{-1}.

60. Let $P(x) = \sum\limits_{n=0}^{\infty} a_n x^n$ be a power series solution to $y' = 2xy$ with initial condition $y(0) = 1$.
(a) Show that the odd coefficients a_{2k+1} are all zero.
(b) Prove that $a_{2k} = a_{2k-2}/k$ and use this result to determine the coefficients a_{2k}.

61. Find a power series $P(x)$ satisfying the differential equation

$$y'' - xy' + y = 0 \qquad \boxed{9}$$

with initial condition $y(0) = 1$, $y'(0) = 0$. What is the radius of convergence of the power series?

62. Find a power series satisfying Eq. (9) with initial condition $y(0) = 0$, $y'(0) = 1$.

63. Prove that

$$J_2(x) = \sum_{k=0}^{\infty} \frac{(-1)^k}{2^{2k+2}\, k!\,(k+3)!} x^{2k+2}$$

is a solution of the Bessel differential equation of order 2:

$$x^2 y'' + xy' + (x^2 - 4)y = 0$$

64. Why is it impossible to expand $f(x) = |x|$ as a power series that converges in an interval around $x = 0$? Explain using Theorem 2.

Further Insights and Challenges

65. Suppose that the coefficients of $F(x) = \sum\limits_{n=0}^{\infty} a_n x^n$ are *periodic*; that is, for some whole number $M > 0$, we have $a_{M+n} = a_n$. Prove that $F(x)$ converges absolutely for $|x| < 1$ and that

$$F(x) = \frac{a_0 + a_1 x + \cdots + a_{M-1} x^{M-1}}{1 - x^M}$$

Hint: Use the hint for Exercise 53.

66. Continuity of Power Series Let $F(x) = \sum\limits_{n=0}^{\infty} a_n x^n$ be a power series with radius of convergence $R > 0$.
(a) Prove the inequality

$$|x^n - y^n| \le n|x - y|(|x|^{n-1} + |y|^{n-1}) \qquad \boxed{10}$$

Hint: $x^n - y^n = (x - y)(x^{n-1} + x^{n-2}y + \cdots + y^{n-1})$.

(b) Choose R_1 with $0 < R_1 < R$. Show that the infinite series

$$M = \sum_{n=0}^{\infty} 2n|a_n|R_1^n$$

converges. *Hint:* Show that $n|a_n|R_1^n < |a_n|x^n$ for all n sufficiently large if $R_1 < x < R$.

(c) Use Eq. (10) to show that if $|x| < R_1$ and $|y| < R_1$, then $|F(x) - F(y)| \le M|x - y|$.

(d) Prove that if $|x| < R$, then F is continuous at x. *Hint:* Choose R_1 such that $|x| < R_1 < R$. Show that if $\epsilon > 0$ is given, then $|F(x) - F(y)| \le \epsilon$ for all y such that $|x - y| < \delta$, where δ is any positive number that is less than ϵ/M and $R_1 - |x|$ (see Figure 6).

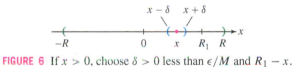

FIGURE 6 If $x > 0$, choose $\delta > 0$ less than ϵ/M and $R_1 - x$.

10.7 Taylor Series

In this section, we develop general methods for finding power series representations. Ultimately, we will see that these power series are, in fact, the extended versions of the Taylor polynomials we discussed in Section 8.4, which we used to approximate values of functions. Suppose that $f(x)$ is represented by a power series centered at $x = c$ on an interval $(c - R, c + R)$ with $R > 0$:

$$f(x) = \sum_{n=0}^{\infty} a_n(x - c)^n = a_0 + a_1(x - c) + a_2(x - c)^2 + \cdots$$

According to Theorem 2 in Section 10.6, we can compute the derivatives of f by differentiating the series expansion term by term:

$$
\begin{aligned}
f(x) &= & a_0 + & a_1(x - c) + & a_2(x - c)^2 + & a_3(x - c)^3 + \cdots \\
f'(x) &= & a_1 + & 2a_2(x - c) + & 3a_3(x - c)^2 + & 4a_4(x - c)^3 + \cdots \\
f''(x) &= & 2a_2 + & 2 \cdot 3a_3(x - c) + & 3 \cdot 4a_4(x - c)^2 + 4 \cdot 5a_5(x - c)^3 + \cdots \\
f'''(x) &= 2 \cdot 3a_3 + 2 \cdot 3 \cdot 4a_4(x - 2) + 3 \cdot 4 \cdot 5a_5(x - 2)^2 + & & & \cdots
\end{aligned}
$$

In general,

$$f^{(k)}(x) = k!a_k + \left(2 \cdot 3 \cdots (k+1)\right)a_{k+1}(x - c) + \cdots$$

Setting $x = c$ in each of these series, we find that

$$f(c) = a_0, \quad f'(c) = a_1, \quad f''(c) = 2a_2, \quad f'''(c) = 2 \cdot 3a_2, \quad \ldots, \quad f^{(k)}(c) = k!a_k, \quad \ldots$$

We see that a_k is the kth coefficient of the Taylor polynomial studied in Section 8.4:

$$a_k = \frac{f^{(k)}(c)}{k!}$$

1

Therefore, $f(x) = T(x)$, where $T(x)$ is the **Taylor series** of $f(x)$ centered at $x = c$:

$$T(x) = f(c) + f'(c)(x - c) + \frac{f''(c)}{2!}(x - c)^2 + \frac{f'''(c)}{3!}(x - c)^3 + \cdots$$

This proves the next theorem.

THEOREM 1 Taylor Series Expansion If $f(x)$ is represented by a power series centered at c in an interval $|x - c| < R$ with $R > 0$, then that power series is the Taylor series

$$T(x) = \sum_{n=0}^{\infty} \frac{f^{(n)}(c)}{n!}(x - c)^n$$

In the special case $c = 0$, $T(x)$ is also called the **Maclaurin series**:

$$f(x) = \sum_{n=0}^{\infty} \frac{f^{(n)}(0)}{n!}x^n = f(0) + f'(0)x + \frac{f''(0)}{2!}x^2 + \frac{f'''(0)}{3!}x^3 + \frac{f^{(4)}(0)}{4!}x^4 + \cdots$$

TABLE 1 Finding the Coefficients for the Taylor Series.

n	$f^{(n)}(x)$	$\dfrac{f^{(n)}(x)}{n!}$	$\dfrac{f^{(n)}(1)}{n!}$
0	x^{-3}	x^{-3}	1
1	$-3x^{-4}$	$-3x^{-4}$	-3
2	$12x^{-5}$	$6x^{-5}$	6
3	$-60x^{-6}$	$-10x^{-6}$	-10
4	$360x^{-7}$	$15x^{-7}$	15

■ **EXAMPLE 1** Find the Taylor series for $f(x) = x^{-3}$ centered at $c = 1$.

Solution It often helps to create a table, as in Table 1, to see the pattern. The derivatives of $f(x)$ are $f'(x) = -3x^{-4}$, $f''(x) = (-3)(-4)x^{-5}$, and in general,

$$f^{(n)}(x) = (-1)^n (3)(4) \cdots (n + 2)x^{-3-n}$$

Note that $(3)(4) \cdots (n + 2) = \frac{1}{2}(n + 2)!$. Therefore,

$$f^{(n)}(1) = (-1)^n \frac{1}{2}(n + 2)!$$

Noting that $(n + 2)! = (n + 2)(n + 1)n!$, we write the coefficients of the Taylor series as

$$a_n = \frac{f^{(n)}(1)}{n!} = \frac{(-1)^n \frac{1}{2}(n + 2)!}{n!} = (-1)^n \frac{(n + 2)(n + 1)}{2}$$

The Taylor series for $f(x) = x^{-3}$ centered at $c = 1$ is

$$T(x) = 1 - 3(x - 1) + 6(x - 1)^2 - 10(x - 1)^3 + \cdots$$

$$= \sum_{n=0}^{\infty} (-1)^n \frac{(n + 2)(n + 1)}{2}(x - 1)^n$$

■

Theorem 1 tells us that if we want to represent a function f by a power series centered at c, then the only candidate for the job is the Taylor series:

$$T(x) = \sum_{n=0}^{\infty} \frac{f^{(n)}(c)}{n!}(x - c)^n$$

See Exercise 94 for an example where a Taylor series $T(x)$ converges but does not converge to $f(x)$.

However, *there is no guarantee that $T(x)$ converges to $f(x)$, even if $T(x)$ converges*. To study convergence, we consider the kth partial sum, which is the Taylor polynomial of degree k:

$$T_k(x) = f(c) + f'(c)(x - c) + \frac{f''(c)}{2!}(x - c)^2 + \cdots + \frac{f^{(k)}(c)}{k!}(x - c)^k$$

In Section 8.4, we defined the remainder

$$R_k(x) = f(x) - T_k(x)$$

Since $T(x)$ is the limit of the partial sums $T_k(x)$, we see that

> *The Taylor series converges to $f(x)$ if and only if* $\lim_{k \to \infty} R_k(x) = 0$.

There is no general method for determining whether $R_k(x)$ tends to zero, but the following theorem can be applied in some important cases.

REMINDER $f(x)$ is called "infinitely differentiable" if $f^{(n)}(x)$ exists for all n.

THEOREM 2 Let $I = (c - R, c + R)$, where $R > 0$. Suppose there exists $K > 0$ such that all derivatives of f are bounded by K on I:

$$|f^{(k)}(x)| \le K \qquad \text{for all} \quad k \ge 0 \quad \text{and} \quad x \in I$$

Then f is represented by its Taylor series in I:

$$f(x) = \sum_{n=0}^{\infty} \frac{f^{(n)}(c)}{n!}(x - c)^n \qquad \text{for all} \quad x \in I$$

Proof According to the Error Bound for Taylor polynomials (Theorem 2 in Section 8.4),

$$|R_k(x)| = |f(x) - T_k(x)| \le K\frac{|x - c|^{k+1}}{(k + 1)!}$$

If $x \in I$, then $|x - c| < R$ and

$$|R_k(x)| \le K\frac{R^{k+1}}{(k + 1)!}$$

We showed in Example 9 of Section 10.1 that $R^k / k!$ tends to zero as $k \to \infty$. Therefore, $\lim_{k \to \infty} R_k(x) = 0$ for all $x \in (c - R, c + R)$, as required. ■

Taylor expansions were studied throughout the seventeenth and eighteenth centuries by Gregory, Leibniz, Newton, Maclaurin, Taylor, Euler, and others. These developments were anticipated by the great Hindu mathematician Madhava (c. 1340–1425), who discovered the expansions of sine and cosine and many other results two centuries earlier.

■ **EXAMPLE 2** **Expansions of Sine and Cosine** Show that the following Maclaurin expansions are valid for all x:

$$\sin x = \sum_{n=0}^{\infty}(-1)^n \frac{x^{2n+1}}{(2n + 1)!} = x - \frac{x^3}{3!} + \frac{x^5}{5!} - \frac{x^7}{7!} + \cdots$$

$$\cos x = \sum_{n=0}^{\infty}(-1)^n \frac{x^{2n}}{(2n)!} = 1 - \frac{x^2}{2!} + \frac{x^4}{4!} - \frac{x^6}{6!} + \cdots$$

Solution Recall that the derivatives of $f(x) = \sin x$ and their values at $x = 0$ form a repeating pattern of period 4:

$f(x)$	$f'(x)$	$f''(x)$	$f'''(x)$	$f^{(4)}(x)$	$\cdots$
$\sin x$	$\cos x$	$-\sin x$	$-\cos x$	$\sin x$	$\cdots$
0	1	0	-1	0	$\cdots$

In other words, the even derivatives are zero and the odd derivatives alternate in sign: $f^{(2n+1)}(0) = (-1)^n$. Therefore, the nonzero Taylor coefficients for $\sin x$ are

$$a_{2n+1} = \frac{(-1)^n}{(2n+1)!}$$

For $f(x) = \cos x$, the situation is reversed. The odd derivatives are zero and the even derivatives alternate in sign: $f^{(2n)}(0) = (-1)^n \cos 0 = (-1)^n$. Therefore, the nonzero Taylor coefficients for $\cos x$ are $a_{2n} = (-1)^n/(2n)!$.

We can apply Theorem 2 with $K = 1$ and any value of R because both sine and cosine satisfy $|f^{(n)}(x)| \leq 1$ for all x and n. The conclusion is that the Taylor series converges to $f(x)$ for $|x| < R$. Since R is arbitrary, the Taylor expansions hold for all x. ∎

■ **EXAMPLE 3** **Taylor Expansion of $f(x) = e^x$ at $x = c$** Find the Taylor series $T(x)$ of $f(x) = e^x$ at $x = c$.

Solution We have $f^{(n)}(c) = e^c$ for all x. Thus,

$$T(x) = \sum_{n=0}^{\infty} \frac{e^c}{n!}(x - c)^n$$

Because $f(x) = e^x$ is increasing for all $R > 0$, $|f^{(k)}(x)| \leq e^{c+R}$ for $x \in (c - R, c + R)$. Applying Theorem 2 with $K = e^{c+R}$, we conclude that $T(x)$ converges to $f(x)$ for all $x \in (c - R, c + R)$. Since R is arbitrary, the Taylor expansion holds for all x. For $c = 0$, we obtain the standard Maclaurin series

$$\boxed{e^x = 1 + x + \frac{x^2}{2!} + \frac{x^3}{3!} + \cdots}$$

∎

Shortcuts to Finding Taylor Series

There are several methods for generating new Taylor series from known ones. First of all, we can differentiate and integrate Taylor series term by term within its interval of convergence, by Theorem 2 of Section 10.6. We can also multiply two Taylor series or substitute one Taylor series into another (we omit the proofs of these facts).

In Example 4, we can also write the Maclaurin series as

$$\sum_{n=0}^{\infty} \frac{x^{n+2}}{n!}$$

■ **EXAMPLE 4** Find the Maclaurin series for $f(x) = x^2 e^x$.

Solution Multiply the known Maclaurin series for e^x by x^2:

$$x^2 e^x = x^2\left(1 + x + \frac{x^2}{2!} + \frac{x^3}{3!} + \frac{x^4}{4!} + \frac{x^5}{5!} + \cdots\right)$$

$$= x^2 + x^3 + \frac{x^4}{2!} + \frac{x^5}{3!} + \frac{x^6}{4!} + \frac{x^7}{5!} + \cdots = \sum_{n=2}^{\infty} \frac{x^n}{(n-2)!}$$

∎

■ **EXAMPLE 5** **Substitution** Find the Maclaurin series for $f(x) = e^{-x^2}$.

Solution Substitute $-x^2$ in the Maclaurin series for e^x:

$$e^{-x^2} = \sum_{n=0}^{\infty} \frac{(-x^2)^n}{n!} = \sum_{n=0}^{\infty} \frac{(-1)^n x^{2n}}{n!} = 1 - x^2 + \frac{x^4}{2!} - \frac{x^6}{3!} + \frac{x^8}{4!} - \cdots \quad \boxed{2}$$

The Taylor expansion of e^x is valid for all x, so this expansion is also valid for all x. ∎

■ **EXAMPLE 6 Integration** Find the Maclaurin series for $f(x) = \ln(1+x)$.

Solution We integrate the geometric series with common ratio $-x$ (valid for $|x| < 1$):

$$\frac{1}{1+x} = 1 - x + x^2 - x^3 + \cdots$$

$$\ln(1+x) = \int \frac{dx}{1+x} = A + x - \frac{x^2}{2} + \frac{x^3}{3} - \frac{x^4}{4} + \cdots = A + \sum_{n=1}^{\infty} (-1)^{n-1} \frac{x^n}{n}$$

The constant of integration A on the right is zero because $\ln(1+x) = 0$ for $x = 0$, so

$$\ln(1+x) = \sum_{n=1}^{\infty} (-1)^{n-1} \frac{x^n}{n}$$

This expansion is valid for $|x| < 1$. It also holds for $x = 1$ (see Exercise 86).   ■

In many cases, there is no convenient general formula for the Taylor coefficients, but we can still compute as many coefficients as desired.

■ **EXAMPLE 7 Multiplying Taylor Series** Write out the terms up to degree 5 in the Maclaurin series for $f(x) = e^x \cos x$.

Solution We multiply the fifth-order Taylor polynomials of e^x and $\cos x$ together, dropping the terms of degree greater than 5:

$$\left(1 + x + \frac{x^2}{2} + \frac{x^3}{6} + \frac{x^4}{24} + \frac{x^5}{120}\right)\left(1 - \frac{x^2}{2} + \frac{x^4}{24}\right)$$

Distributing the term on the left (and ignoring terms of degree greater than 5), we obtain

$$\left(1 + x + \frac{x^2}{2} + \frac{x^3}{6} + \frac{x^4}{24} + \frac{x^5}{120}\right) - \left(1 + x + \frac{x^2}{2} + \frac{x^3}{6}\right)\left(\frac{x^2}{2}\right) + (1+x)\left(\frac{x^4}{24}\right)$$

$$= \underbrace{1 + x - \frac{x^3}{3} - \frac{x^4}{6} - \frac{x^5}{30}}_{\text{Retain terms of degree} \leq 5}$$

We conclude that the fifth Maclaurin polynomial for $f(x) = e^x \cos x$ is

$$T_5(x) = 1 + x - \frac{x^3}{3} - \frac{x^4}{6} - \frac{x^5}{30}$$ ■

In the next example, we express the definite integral of $\sin(x^2)$ as an infinite series. This is useful because the integral cannot be evaluated explicitly. Figure 1 shows the graph of the Taylor polynomial $y = T_{12}(x)$ of the Taylor series expansion of the antiderivative.

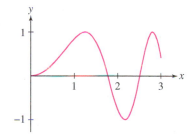

FIGURE 1 Graph of $y = T_{12}(x)$ for the power series expansion of the antiderivative

$$F(x) = \int_0^x \sin(t^2)\,dt$$

■ **EXAMPLE 8** Let $J = \int_0^1 \sin(x^2)\,dx$.

(a) Express J as an infinite series.

(b) Determine J to within an error less than 10^{-4}.

Solution

(a) The Maclaurin expansion for $f(x) = \sin x$ is valid for all x, so we have

$$\sin x = \sum_{n=0}^{\infty} \frac{(-1)^n}{(2n+1)!} x^{2n+1} \quad\Rightarrow\quad \sin(x^2) = \sum_{n=0}^{\infty} \frac{(-1)^n}{(2n+1)!} x^{4n+2}$$

We obtain an infinite series for J by integration:

$$J = \int_0^1 \sin(x^2)\,dx = \sum_{n=0}^{\infty} \frac{(-1)^n}{(2n+1)!} \int_0^1 x^{4n+2}\,dx = \sum_{n=0}^{\infty} \frac{(-1)^n}{(2n+1)!}\left(\frac{1}{4n+3}\right)$$

$$= \frac{1}{3} - \frac{1}{42} + \frac{1}{1320} - \frac{1}{75,600} + \cdots$$ 3

(b) The infinite series for J is an alternating series with decreasing terms, so the sum of the first N terms is accurate to within an error that is less than the $(N+1)$st term. The absolute value of the fourth term $1/75{,}600$ is smaller than 10^{-4}, so we obtain the desired accuracy using the first three terms of the series for J:

$$J \approx \frac{1}{3} - \frac{1}{42} + \frac{1}{1320} \approx 0.31028$$

The error satisfies

$$\left| J - \left(\frac{1}{3} - \frac{1}{42} + \frac{1}{1320} \right) \right| < \frac{1}{75{,}600} \approx 1.3 \times 10^{-5}$$

The percentage error is less than 0.005% with just three terms. ∎

Power series have additional applications.

■ **EXAMPLE 9** Determine $\lim\limits_{x \to 0} \dfrac{x - \sin x}{x^3 \cos x}$.

Solution This limit is of indeterminate form $\frac{0}{0}$, so we could use L'Hôpital's Rule repeatedly. However, instead, we will work with the Maclaurin series. We have

$$\sin x = x - \frac{x^3}{3!} + \frac{x^5}{5!} - \cdots$$

$$\cos x = 1 - \frac{x^2}{2!} + \frac{x^4}{4!} - \cdots$$

Hence, the limit becomes

$$\lim_{x \to 0} \frac{x - \sin x}{x^3 \cos x} = \lim_{x \to 0} \frac{x - (x - \frac{x^3}{3!} + \frac{x^5}{5!} - \cdots)}{x^3(1 - \frac{x^2}{2!} + \frac{x^4}{4!} - \cdots)}$$

$$= \lim_{x \to 0} \frac{\frac{x^3}{3!} - \frac{x^5}{5!} + \cdots}{x^3(1 - \frac{x^2}{2!} + \frac{x^4}{4!} - \cdots)}$$

$$= \lim_{x \to 0} \frac{x^3(\frac{1}{3!} - \frac{x^2}{5!} + \cdots)}{x^3(1 - \frac{x^2}{2!} + \frac{x^4}{4!} - \cdots)}$$

$$= \lim_{x \to 0} \frac{\frac{1}{3!} - \frac{x^2}{5!} + \cdots}{1 - \frac{x^2}{2!} + \frac{x^4}{4!} - \cdots}$$

$$= \frac{1}{3!} = \frac{1}{6} \qquad ∎$$

Binomial Series

Isaac Newton discovered an important generalization of the Binomial Theorem around 1665. For any number a (integer or not) and integer $n \geq 0$, we define the **binomial coefficient**:

$$\binom{a}{n} = \frac{a(a-1)(a-2)\cdots(a-n+1)}{n!}, \qquad \binom{a}{0} = 1$$

For example,

$$\binom{6}{3} = \frac{6 \cdot 5 \cdot 4}{3 \cdot 2 \cdot 1} = 20, \qquad \binom{\frac{4}{3}}{3} = \frac{\frac{4}{3} \cdot \frac{1}{3} \cdot (-\frac{2}{3})}{3 \cdot 2 \cdot 1} = -\frac{4}{81}$$

Let

$$f(x) = (1+x)^a$$

The **Binomial Theorem** of algebra (see Appendix C) states that for any whole number a,

$$(r+s)^a = r^a + \binom{a}{1}r^{a-1}s + \binom{a}{2}r^{a-2}s^2 + \cdots + \binom{a}{a-1}rs^{a-1} + s^a$$

Setting $r = 1$ and $s = x$, we obtain the expansion of $f(x)$:

$$(1 + x)^a = 1 + \binom{a}{1}x + \binom{a}{2}x^2 + \cdots + \binom{a}{a-1}x^{a-1} + x^a$$

We derive Newton's generalization by computing the Maclaurin series of $f(x)$ without assuming that a is a whole number. Observe that the derivatives follow a pattern:

$$f(x) = (1 + x)^a \qquad\qquad f(0) = 1$$
$$f'(x) = a(1 + x)^{a-1} \qquad\qquad f'(0) = a$$
$$f''(x) = a(a - 1)(1 + x)^{a-2} \qquad\qquad f''(0) = a(a - 1)$$
$$f'''(x) = a(a - 1)(a - 2)(1 + x)^{a-3} \qquad f'''(0) = a(a - 1)(a - 2)$$

In general, $f^{(n)}(0) = a(a - 1)(a - 2) \cdots (a - n + 1)$ and

$$\frac{f^{(n)}(0)}{n!} = \frac{a(a - 1)(a - 2) \cdots (a - n + 1)}{n!} = \binom{a}{n}$$

When a is a positive whole number, $\binom{a}{n}$ is zero for $n > a$, and in this case, the binomial series breaks off at degree n. The binomial series is an infinite series when a is not a positive whole number.

Hence, the Maclaurin series for $f(x) = (1 + x)^a$ is the binomial series

$$\sum_{n=0}^{\infty} \binom{a}{n}x^n = 1 + ax + \frac{a(a - 1)}{2!}x^2 + \frac{a(a - 1)(a - 2)}{3!}x^3 + \cdots + \binom{a}{n}x^n + \cdots$$

The Ratio Test shows that this series has radius of convergence $R = 1$ (Exercise 88), and an additional argument (developed in Exercise 89) shows that it converges to $(1 + x)^a$ for $|x| < 1$.

> **THEOREM 3** **The Binomial Series** For any exponent a and for $|x| < 1$,
>
> $$(1 + x)^a = 1 + \frac{a}{1!}x + \frac{a(a - 1)}{2!}x^2 + \frac{a(a - 1)(a - 2)}{3!}x^3 + \cdots + \binom{a}{n}x^n + \cdots$$

■ **EXAMPLE 10** Find the terms through degree 4 in the Maclaurin expansion of

$$f(x) = (1 + x)^{4/3}$$

Solution The binomial coefficients $\binom{a}{n}$ for $a = \frac{4}{3}$ for $0 < n < 4$ are

$$1, \quad \frac{\frac{4}{3}}{1!} = \frac{4}{3}, \quad \frac{\frac{4}{3}\left(\frac{1}{3}\right)}{2!} = \frac{2}{9}, \quad \frac{\frac{4}{3}\left(\frac{1}{3}\right)\left(-\frac{2}{3}\right)}{3!} = -\frac{4}{81}, \quad \frac{\frac{4}{3}\left(\frac{1}{3}\right)\left(-\frac{2}{3}\right)\left(-\frac{5}{3}\right)}{4!} = \frac{5}{243}$$

Therefore, $(1 + x)^{4/3} \approx 1 + \frac{4}{3}x + \frac{2}{9}x^2 - \frac{4}{81}x^3 + \frac{5}{243}x^4 + \cdots$. ■

■ **EXAMPLE 11** Find the Maclaurin series for

$$f(x) = \frac{1}{\sqrt{1 - x^2}}$$

Solution First, let's find the coefficients in the binomial series for $(1 + x)^{-1/2}$:

$$1, \quad \frac{-\frac{1}{2}}{1!} = -\frac{1}{2}, \quad \frac{-\frac{1}{2}\left(-\frac{3}{2}\right)}{1 \cdot 2} = \frac{1 \cdot 3}{2 \cdot 4}, \quad \frac{-\frac{1}{2}\left(-\frac{3}{2}\right)\left(-\frac{5}{2}\right)}{1 \cdot 2 \cdot 3} = \frac{1 \cdot 3 \cdot 5}{2 \cdot 4 \cdot 6}$$

The general pattern is

$$\binom{-\frac{1}{2}}{n} = \frac{-\frac{1}{2}\left(-\frac{3}{2}\right)\left(-\frac{5}{2}\right) \cdots \left(-\frac{2n-1}{2}\right)}{1 \cdot 2 \cdot 3 \cdots n} = (-1)^n \frac{1 \cdot 3 \cdot 5 \cdots (2n - 1)}{2 \cdot 4 \cdot 6 \cdot 2n}$$

Thus, the following binomial expansion is valid for $|x| < 1$:

$$\frac{1}{\sqrt{1 + x}} = 1 + \sum_{n=1}^{\infty} (-1)^n \frac{1 \cdot 3 \cdot 5 \cdots (2n - 1)}{2 \cdot 4 \cdot 6 \cdots (2n)}x^n = 1 - \frac{1}{2}x + \frac{1 \cdot 3}{2 \cdot 4}x^2 - \cdots$$

If $|x| < 1$, then $|x|^2 < 1$, and we can substitute $-x^2$ for x to obtain

$$\frac{1}{\sqrt{1-x^2}} = 1 + \sum_{n=1}^{\infty} \frac{1 \cdot 3 \cdot 5 \cdots (2n-1)}{2 \cdot 4 \cdot 6 \cdots 2n} x^{2n} = 1 + \frac{1}{2}x^2 + \frac{1 \cdot 3}{2 \cdot 4}x^4 + \cdots \qquad \boxed{4}$$

FIGURE 2 Pendulum released at an angle θ.

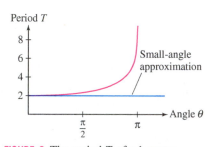

FIGURE 3 The period T of a 1-meter pendulum as a function of the angle θ at which it is released.

Taylor series are particularly useful for studying the so-called *special functions* (such as Bessel and hypergeometric functions) that appear in a wide range of physics and engineering applications. One example is the following **elliptic integral of the first kind**, defined for $|k| < 1$:

$$E(k) = \int_0^{\pi/2} \frac{dt}{\sqrt{1 - k^2 \sin^2 t}}$$

This function is used in physics to compute the period T of pendulum of length L released from an angle θ (Figure 2). We can use the "small-angle approximation" $T \approx 2\pi \sqrt{L/g}$ when θ is small (where $g = 9.8$ m/s^2), but this approximation breaks down for large angles (Figure 3). The exact value of the period is $T = 4\sqrt{L/g}\, E(k)$, where $k = \sin \frac{1}{2}\theta$.

■ **EXAMPLE 12 Elliptic Function** Find the Maclaurin series for $E(k)$ and estimate $E(k)$ for $k = \sin\frac{\pi}{6}$.

Solution Substitute $x = k \sin t$ in the Taylor expansion (4):

$$\frac{1}{\sqrt{1 - k^2 \sin^2 t}} = 1 + \frac{1}{2}k^2 \sin^2 t + \frac{1 \cdot 3}{2 \cdot 4}k^4 \sin^4 t + \frac{1 \cdot 3 \cdot 5}{2 \cdot 4 \cdot 6}k^6 \sin^6 t + \cdots$$

This expansion is valid because $|k| < 1$ and hence $|x| = |k \sin t| < 1$. Thus, $E(k)$ is equal to

$$\int_0^{\pi/2} \frac{dt}{\sqrt{1 - k^2 \sin^2 t}} = \int_0^{\pi/2} dt + \sum_{n=1}^{\infty} \frac{1 \cdot 3 \cdots (2n-1)}{2 \cdot 4 \cdot (2n)} \left(\int_0^{\pi/2} \sin^{2n} t\, dt \right) k^{2n}$$

According to Exercise 80 in Section 7.2,

$$\int_0^{\pi/2} \sin^{2n} t\, dt = \left(\frac{1 \cdot 3 \cdots (2n-1)}{2 \cdot 4 \cdot (2n)} \right) \frac{\pi}{2}$$

This yields

$$E(k) = \frac{\pi}{2} + \frac{\pi}{2} \sum_{n=1}^{\infty} \left(\frac{1 \cdot 3 \cdots (2n-1)^2}{2 \cdot 4 \cdots (2n)} \right)^2 k^{2n}$$

We approximate $E(k)$ for $k = \sin\left(\frac{\pi}{6}\right) = \frac{1}{2}$ using the first five terms:

$$E\left(\frac{1}{2}\right) \approx \frac{\pi}{2} \left(1 + \left(\frac{1}{2}\right)^2 \left(\frac{1}{2}\right)^2 + \left(\frac{1 \cdot 3}{2 \cdot 4}\right)^2 \left(\frac{1}{2}\right)^4 \right.$$

$$\left. + \left(\frac{1 \cdot 3 \cdot 5}{2 \cdot 4 \cdot 6}\right)^2 \left(\frac{1}{2}\right)^6 + \left(\frac{1 \cdot 3 \cdot 5 \cdot 7}{2 \cdot 4 \cdot 6 \cdot 8}\right)^2 \left(\frac{1}{2}\right)^8 \right)$$

$$\approx 1.68517$$

The value given by a computer algebra system to seven places is $E\left(\frac{1}{2}\right) \approx 1.6856325$. ■

In Table 2, we provide a list of useful Taylor series centered at $x = 0$ (also called a Maclaurin series) and the values of x for which they converge.

TABLE 2

$f(x)$	Maclaurin series	Converges to $f(x)$ for
e^x	$\sum_{n=0}^{\infty} \frac{x^n}{n!} = 1 + x + \frac{x^2}{2!} + \frac{x^3}{3!} + \frac{x^4}{4!} + \cdots$	All x
$\sin x$	$\sum_{n=0}^{\infty} \frac{(-1)^n x^{2n+1}}{(2n+1)!} = x - \frac{x^3}{3!} + \frac{x^5}{5!} - \frac{x^7}{7!} + \cdots$	All x
$\cos x$	$\sum_{n=0}^{\infty} \frac{(-1)^n x^{2n}}{(2n)!} = 1 - \frac{x^2}{2!} + \frac{x^4}{4!} - \frac{x^6}{6!} + \cdots$	All x
$\frac{1}{1-x}$	$\sum_{n=0}^{\infty} x^n = 1 + x + x^2 + x^3 + x^4 + \cdots$	$|x| < 1$
$\frac{1}{1+x}$	$\sum_{n=0}^{\infty} (-1)^n x^n = 1 - x + x^2 - x^3 + x^4 - \cdots$	$|x| < 1$
$\ln(1+x)$	$\sum_{n=1}^{\infty} \frac{(-1)^{n-1} x^n}{n} = x - \frac{x^2}{2} + \frac{x^3}{3} - \frac{x^4}{4} + \cdots$	$|x| < 1$ and $x = 1$
$\tan^{-1} x$	$\sum_{n=0}^{\infty} \frac{(-1)^n x^{2n+1}}{2n+1} = x - \frac{x^3}{3} + \frac{x^5}{5} - \frac{x^7}{7} + \cdots$	$|x| \leq 1$
$(1+x)^a$	$\sum_{n=0}^{\infty} \binom{a}{n} x^n = 1 + ax + \frac{a(a-1)}{2!} x^2 + \frac{a(a-1)(a-2)}{3!} x^3 + \cdots$	$|x| < 1$

10.7 SUMMARY

- *Taylor series of $f(x)$ centered at $x = c$:*

$$T(x) = \sum_{n=0}^{\infty} \frac{f^{(n)}(c)}{n!} (x-c)^n$$

The partial sum $T_k(x)$ is the kth Taylor polynomial.
- *Maclaurin series ($c = 0$):*

$$T(x) = \sum_{n=0}^{\infty} \frac{f^{(n)}(0)}{n!} x^n$$

- If $f(x)$ is represented by a power series $\sum_{n=0}^{\infty} a_n(x-c)^n$ for $|x-c| < R$ with $R > 0$, then this power series is necessarily the Taylor series centered at $x = c$.
- A function f is represented by its Taylor series $T(x)$ if and only if the remainder $R_k(x) = f(x) - T_k(x)$ tends to zero as $k \to \infty$.
- Let $I = (c - R, c + R)$ with $R > 0$. Suppose that there exists $K > 0$ such that $|f^{(k)}(x)| < K$ for all $x \in I$ and all k. Then f is represented by its Taylor series on I; that is, $f(x) = T(x)$ for $x \in I$.
- A good way to find the Taylor series of a function is to start with known Taylor series and apply one of the following operations: differentiation, integration, multiplication, or substitution.
- For any exponent a, the binomial expansion is valid for $|x| < 1$:

$$(1+x)^a = 1 + ax + \frac{a(a-1)}{2!} x^2 + \frac{a(a-1)(a-2)}{3!} x^3 + \cdots + \binom{a}{n} x^n + \cdots$$

10.7 EXERCISES

Preliminary Questions

1. Determine $f(0)$ and $f'''(0)$ for a function f with Maclaurin series

$$T(x) = 3 + 2x + 12x^2 + 5x^3 + \cdots$$

2. Determine $f(-2)$ and $f^{(4)}(-2)$ for a function with Taylor series

$$T(x) = 3(x + 2) + (x + 2)^2 - 4(x + 2)^3 + 2(x + 2)^4 + \cdots$$

3. What is the easiest way to find the Maclaurin series for the function $f(x) = \sin(x^2)$?

4. Find the Taylor series for f centered at $c = 3$ if $f(3) = 4$ and $f'(x)$ has a Taylor expansion

$$f'(x) = \sum_{n=1}^{\infty} \frac{(x - 3)^n}{n}$$

5. Let $T(x)$ be the Maclaurin series of $f(x)$. Which of the following guarantees that $f(2) = T(2)$?

(a) $T(x)$ converges for $x = 2$.

(b) The remainder $R_k(2)$ approaches a limit as $k \to \infty$.

(c) The remainder $R_k(2)$ approaches zero as $k \to \infty$.

Exercises

1. Write out the first four terms of the Maclaurin series of $f(x)$ if

$$f(0) = 2, \quad f'(0) = 3, \quad f''(0) = 4, \quad f'''(0) = 12$$

2. Write out the first four terms of the Taylor series of $f(x)$ centered at $c = 3$ if

$$f(3) = 1, \quad f'(3) = 2, \quad f''(3) = 12, \quad f'''(3) = 3$$

In Exercises 3–18, find the Maclaurin series and find the interval on which the expansion is valid.

3. $f(x) = \dfrac{1}{1 - 2x}$

4. $f(x) = \dfrac{x}{1 - x^4}$

5. $f(x) = \cos 3x$

6. $f(x) = \sin(2x)$

7. $f(x) = \sin(x^2)$

8. $f(x) = e^{4x}$

9. $f(x) = \ln(1 - x^2)$

10. $f(x) = (1 - x)^{-1/2}$

11. $f(x) = \tan^{-1}(x^2)$

12. $f(x) = x^2 e^{x^2}$

13. $f(x) = e^{x-2}$

14. $f(x) = \dfrac{1 - \cos x}{x}$

15. $f(x) = \ln(1 - 5x)$

16. $f(x) = (x^2 + 2x)e^x$

17. $f(x) = \sinh x$

18. $f(x) = \cosh x$

In Exercises 19–28, find the terms through degree 4 of the Maclaurin series of $f(x)$. Use multiplication and substitution as necessary.

19. $f(x) = e^x \sin x$

20. $f(x) = e^x \ln(1 - x)$

21. $f(x) = \dfrac{\sin x}{1 - x}$

22. $f(x) = \dfrac{1}{1 + \sin x}$

23. $f(x) = (1 + x)^{1/4}$

24. $f(x) = (1 + x)^{-3/2}$

25. $f(x) = e^x \tan^{-1} x$

26. $f(x) = \sin(x^3 - x)$

27. $f(x) = e^{\sin x}$

28. $f(x) = e^{(e^x)}$

In Exercises 29–38, find the Taylor series centered at c and the interval on which the expansion is valid.

29. $f(x) = \dfrac{1}{x}, \quad c = 1$

30. $f(x) = e^{3x}, \quad c = -1$

31. $f(x) = \dfrac{1}{1 - x}, \quad c = 5$

32. $f(x) = \sin x, \quad c = \dfrac{\pi}{2}$

33. $f(x) = x^4 + 3x - 1, \quad c = 2$

34. $f(x) = x^4 + 3x - 1, \quad c = 0$

35. $f(x) = \dfrac{1}{x^2}, \quad c = 4$

36. $f(x) = \sqrt{x}, \quad c = 4$

37. $f(x) = \dfrac{1}{1 - x^2}, \quad c = 3$

38. $f(x) = \dfrac{1}{3x - 2}, \quad c = -1$

39. Use the identity $\cos^2 x = \frac{1}{2}(1 + \cos 2x)$ to find the Maclaurin series for $f(x) = \cos^2 x$.

40. Show that for $|x| < 1$,

$$\tanh^{-1} x = x + \frac{x^3}{3} + \frac{x^5}{5} + \cdots$$

Hint: Recall that $\dfrac{d}{dx} \tanh^{-1} x = \dfrac{1}{1 - x^2}$.

41. Use the Maclaurin series for $\ln(1 + x)$ and $\ln(1 - x)$ to show that

$$\frac{1}{2} \ln\left(\frac{1 + x}{1 - x}\right) = x + \frac{x^3}{3} + \frac{x^5}{5} + \cdots$$

for $|x| < 1$. What can you conclude by comparing this result with that of Exercise 40?

42. Differentiate the Maclaurin series for $\dfrac{1}{1 - x}$ twice to find the Maclaurin series of $\dfrac{1}{(1 - x)^3}$.

43. Show, by integrating the Maclaurin series for $f(x) = \dfrac{1}{\sqrt{1 - x^2}}$, that for $|x| < 1$,

$$\sin^{-1} x = x + \sum_{n=1}^{\infty} \frac{1 \cdot 3 \cdot 5 \cdots (2n - 1)}{2 \cdot 4 \cdot 6 \cdots (2n)} \frac{x^{2n+1}}{2n + 1}$$

44. Use the first five terms of the Maclaurin series in Exercise 43 to approximate $\sin^{-1} \frac{1}{2}$. Compare the result with the calculator value.

45. How many terms of the Maclaurin series of $f(x) = \ln(1 + x)$ are needed to compute $\ln 1.2$ to within an error of at most 0.0001? Make the computation and compare the result with the calculator value.

46. Show that

$$\pi - \frac{\pi^3}{3!} + \frac{\pi^5}{5!} - \frac{\pi^7}{7!} + \cdots$$

converges to zero. How many terms must be computed to get within 0.01 of zero?

47. Use the Maclaurin expansion for e^{-t^2} to express the function $F(x) = \int_0^x e^{-t^2} dt$ as an alternating power series in x (Figure 4).

(a) How many terms of the Maclaurin series are needed to approximate the integral for $x = 1$ to within an error of at most 0.001?

(b) `CAS` Carry out the computation and check your answer using a computer algebra system.

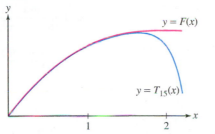

FIGURE 4 The Maclaurin polynomial $T_{15}(x)$ for $F(t) = \int_0^x e^{-t^2}\, dt$.

48. Let $F(x) = \int_0^x \dfrac{\sin t\, dt}{t}$. Show that

$$F(x) = x - \frac{x^3}{3 \cdot 3!} + \frac{x^5}{5 \cdot 5!} - \frac{x^7}{7 \cdot 7!} + \cdots$$

Evaluate $F(1)$ to three decimal places.

In Exercises 49–52, express the definite integral as an infinite series and find its value to within an error of at most 10^{-4}.

49. $\displaystyle\int_0^1 \cos(x^2)\, dx$

50. $\displaystyle\int_0^1 \tan^{-1}(x^2)\, dx$

51. $\displaystyle\int_0^1 e^{-x^3}\, dx$

52. $\displaystyle\int_0^1 \frac{dx}{\sqrt{x^4 + 1}}$

In Exercises 53–56, express the integral as an infinite series.

53. $\displaystyle\int_0^x \frac{1 - \cos t}{t}\, dt$, for all x

54. $\displaystyle\int_0^x \frac{t - \sin t}{t}\, dt$, for all x

55. $\displaystyle\int_0^x \ln(1 + t^2)\, dt$, for $|x| < 1$

56. $\displaystyle\int_0^x \frac{dt}{\sqrt{1 - t^4}}$, for $|x| < 1$

57. Which function has Maclaurin series $\displaystyle\sum_{n=0}^{\infty} (-1)^n 2^n x^n$?

58. Which function has Maclaurin series

$$\sum_{k=0}^{\infty} \frac{(-1)^k}{3^{k+1}} (x - 3)^k ?$$

For which values of x is the expansion valid?

59. Using Maclaurin series, determine to exactly what value the following series converges:

$$\sum_{n=0}^{\infty} (-1)^n \frac{(\pi)^{2n}}{(2n)!}$$

60. Using Maclaurin series, determine to exactly what value the following series converges:

$$\sum_{n=0}^{\infty} \frac{(\ln 5)^n}{n!}$$

In Exercises 61–64, use Theorem 2 to prove that the $f(x)$ is represented by its Maclaurin series for all x.

61. $f(x) = \sin(x/2) + \cos(x/3)$

62. $f(x) = e^{-x}$

63. $f(x) = \sinh x$

64. $f(x) = (1 + x)^{100}$

In Exercises 65–68, find the functions with the following Maclaurin series (refer to Table 2 on page 571).

65. $1 + x^3 + \dfrac{x^6}{2!} + \dfrac{x^9}{3!} + \dfrac{x^{12}}{4!} + \cdots$

66. $1 - 4x + 4^2 x^2 - 4^3 x^3 + 4^4 x^4 - 4^5 x^5 + \cdots$

67. $1 - \dfrac{5^3 x^3}{3!} + \dfrac{5^5 x^5}{5!} - \dfrac{5^7 x^7}{7!} + \cdots$

68. $x^4 - \dfrac{x^{12}}{3} + \dfrac{x^{20}}{5} - \dfrac{x^{28}}{7} + \cdots$

In Exercises 69 and 70, let

$$f(x) = \frac{1}{(1 - x)(1 - 2x)}$$

69. Find the Maclaurin series of $f(x)$ using the identity

$$f(x) = \frac{2}{1 - 2x} - \frac{1}{1 - x}$$

70. Find the Taylor series for $f(x)$ at $c = 2$. *Hint:* Rewrite the identity of Exercise 69 as

$$f(x) = \frac{2}{-3 - 2(x - 2)} - \frac{1}{-1 - (x - 2)}$$

71. When a voltage V is applied to a series circuit consisting of a resistor R and an inductor L, the current at time t is

$$I(t) = \left(\frac{V}{R}\right)\left(1 - e^{-Rt/L}\right)$$

Expand $I(t)$ in a Maclaurin series. Show that $I(t) \approx \dfrac{Vt}{L}$ for small t.

72. Use the result of Exercise 71 and your knowledge of alternating series to show that

$$\frac{Vt}{L}\left(1 - \frac{R}{2L}t\right) \le I(t) \le \frac{Vt}{L} \qquad \text{(for all } t\text{)}$$

73. Find the Maclaurin series for $f(x) = \cos(x^3)$ and use it to determine $f^{(6)}(0)$.

74. Find $f^{(7)}(0)$ and $f^{(8)}(0)$ for $f(x) = \tan^{-1} x$ using the Maclaurin series.

75. ✏ Use substitution to find the first three terms of the Maclaurin series for $f(x) = e^{x^{20}}$. How does the result show that $f^{(k)}(0) = 0$ for $1 \le k \le 19$?

76. Use the binomial series to find $f^{(8)}(0)$ for $f(x) = \sqrt{1 - x^2}$.

77. Does the Maclaurin series for $f(x) = (1 + x)^{3/4}$ converge to $f(x)$ at $x = 2$? Give numerical evidence to support your answer.

78. ✏ Explain the steps required to verify that the Maclaurin series for $f(x) = e^x$ converges to $f(x)$ for all x.

79. `GU` Let $f(x) = \sqrt{1 + x}$.

(a) Use a graphing calculator to compare the graph of f with the graphs of the first five Taylor polynomials for f. What do they suggest about the interval of convergence of the Taylor series?

(b) Investigate numerically whether or not the Taylor expansion for f is valid for $x = 1$ and $x = -1$.

80. Use the first five terms of the Maclaurin series for the elliptic integral $E(k)$ to estimate the period T of a 1-meter pendulum released at an angle $\theta = \frac{\pi}{4}$ (see Example 12).

81. Use Example 12 and the approximation $\sin x \approx x$ to show that the period T of a pendulum released at an angle θ has the following second-order approximation:

$$T \approx 2\pi \sqrt{\frac{L}{g}} \left(1 + \frac{\theta^2}{16}\right)$$

In Exercises 82–85, find the Maclaurin series of the function and use it to calculate the limit.

82. $\lim\limits_{x \to 0} \dfrac{\cos x - 1 + \frac{x^2}{2}}{x^4}$

83. $\lim\limits_{x \to 0} \dfrac{\sin x - x + \frac{x^3}{6}}{x^5}$

84. $\lim\limits_{x \to 0} \dfrac{\tan^{-1} x - x \cos x - \frac{1}{6}x^3}{x^5}$

85. $\lim\limits_{x \to 0} \left(\dfrac{\sin(x^2)}{x^4} - \dfrac{\cos x}{x^2}\right)$

Further Insights and Challenges

86. In this exercise, we show that the Maclaurin expansion of $f(x) = \ln(1 + x)$ is valid for $x = 1$.

(a) Show that for all $x \neq -1$,

$$\frac{1}{1+x} = \sum_{n=0}^{N} (-1)^n x^n + \frac{(-1)^{N+1} x^{N+1}}{1+x}$$

(b) Integrate from 0 to 1 to obtain

$$\ln 2 = \sum_{n=1}^{N} \frac{(-1)^{n-1}}{n} + (-1)^{N+1} \int_0^1 \frac{x^{N+1}\, dx}{1+x}$$

(c) Verify that the integral on the right tends to zero as $N \to \infty$ by showing that it is smaller than $\int_0^1 x^{N+1} dx$.

(d) Prove the formula

$$\ln 2 = 1 - \frac{1}{2} + \frac{1}{3} - \frac{1}{4} + \cdots$$

87. Let $g(t) = \dfrac{1}{1+t^2} - \dfrac{t}{1+t^2}$.

(a) Show that $\displaystyle\int_0^1 g(t)\, dt = \dfrac{\pi}{4} - \dfrac{1}{2}\ln 2$.

(b) Show that $g(t) = 1 - t - t^2 + t^3 + t^4 - t^5 - t^6 + \cdots$.

(c) Evaluate $S = 1 - \frac{1}{2} - \frac{1}{3} + \frac{1}{4} + \frac{1}{5} - \frac{1}{6} - \frac{1}{7} + \cdots$.

In Exercises 88 and 89, we investigate the convergence of the binomial series

$$T_a(x) = \sum_{n=0}^{\infty} \binom{a}{n} x^n$$

88. Prove that $T_a(x)$ has radius of convergence $R = 1$ if a is not a whole number. What is the radius of convergence if a is a whole number?

89. By Exercise 88, $T_a(x)$ converges for $|x| < 1$, but we do not yet know whether $T_a(x) = (1 + x)^a$.

(a) Verify the identity

$$a\binom{a}{n} = n\binom{a}{n} + (n+1)\binom{a}{n+1}$$

(b) Use (a) to show that $y = T_a(x)$ satisfies the differential equation $(1 + x)y' = ay$ with initial condition $y(0) = 1$.

(c) Prove $T_a(x) = (1 + x)^a$ for $|x| < 1$ by showing that the derivative of the ratio $\dfrac{T_a(x)}{(1+x)^a}$ is zero.

90. The function $G(k) = \displaystyle\int_0^{\pi/2} \sqrt{1 - k^2 \sin^2 t}\, dt$ is called an **elliptic integral of the second kind**. Prove that for $|k| < 1$,

$$G(k) = \frac{\pi}{2} - \frac{\pi}{2} \sum_{n=1}^{\infty} \left(\frac{1 \cdot 3 \cdots (2n-1)}{2 \cdots 4 \cdot (2n)}\right)^2 \frac{k^{2n}}{2n-1}$$

91. Assume that $a < b$ and let L be the arc length (circumference) of the ellipse $\left(\frac{x}{a}\right)^2 + \left(\frac{y}{b}\right)^2 = 1$ shown in Figure 5. There is no explicit formula for L, but it is known that $L = 4bG(k)$, with $G(k)$ as in Exercise 90 and $k = \sqrt{1 - a^2/b^2}$. Use the first three terms of the expansion of Exercise 90 to estimate L when $a = 4$ and $b = 5$.

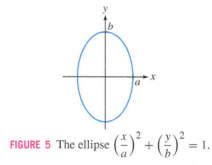

FIGURE 5 The ellipse $\left(\dfrac{x}{a}\right)^2 + \left(\dfrac{y}{b}\right)^2 = 1$.

92. Use Exercise 90 to prove that if $a < b$ and a/b is near 1 (a nearly circular ellipse), then

$$L \approx \frac{\pi}{2}\left(3b + \frac{a^2}{b}\right)$$

Hint: Use the first two terms of the series for $G(k)$.

93. Irrationality of e Prove that e is an irrational number using the following argument by contradiction. Suppose that $e = M/N$, where M, N are nonzero integers.

(a) Show that $M! e^{-1}$ is a whole number.

(b) Use the power series for $f(x) = e^x$ at $x = -1$ to show that there is an integer B such that $M! e^{-1}$ equals

$$B + (-1)^{M+1}\left(\frac{1}{M+1} - \frac{1}{(M+1)(M+2)} + \cdots\right)$$

(c) Use your knowledge of alternating series with decreasing terms to conclude that $0 < |M! e^{-1} - B| < 1$ and observe that this contradicts (a). Hence, e is not equal to M/N.

94. Use the result of Exercise 75 in Section 4.5 to show that the Maclaurin series of the function

$$f(x) = \begin{cases} e^{-1/x^2} & \text{for } x \neq 0 \\ 0 & \text{for } x = 0 \end{cases}$$

is $T(x) = 0$. This provides an example of a function f whose Maclaurin series converges but does not converge to $f(x)$ (except at $x = 0$).

CHAPTER REVIEW EXERCISES

1. Let $a_n = \dfrac{n-3}{n!}$ and $b_n = a_{n+3}$. Calculate the first three terms in each sequence.

(a) a_n^2

(b) b_n

(c) $a_n b_n$

(d) $2a_{n+1} - 3a_n$

2. Prove that $\lim\limits_{n\to\infty} \dfrac{2n-1}{3n+2} = \dfrac{2}{3}$ using the limit definition.

In Exercises 3–8, compute the limit (or state that it does not exist) assuming that $\lim\limits_{n\to\infty} a_n = 2$.

3. $\lim\limits_{n\to\infty} (5a_n - 2a_n^2)$

4. $\lim\limits_{n\to\infty} \dfrac{1}{a_n}$

5. $\lim\limits_{n\to\infty} e^{a_n}$

6. $\lim\limits_{n\to\infty} \cos(\pi a_n)$

7. $\lim\limits_{n\to\infty} (-1)^n a_n$

8. $\lim\limits_{n\to\infty} \dfrac{a_n + n}{a_n + n^2}$

In Exercises 9–22, determine the limit of the sequence or show that the sequence diverges.

9. $a_n = \sqrt{n+5} - \sqrt{n+2}$

10. $a_n = \dfrac{3n^3 - n}{1 - 2n^3}$

11. $a_n = 2^{1/n^2}$

12. $a_n = \dfrac{10^n}{n!}$

13. $b_m = 1 + (-1)^m$

14. $b_m = \dfrac{1 + (-1)^m}{m}$

15. $b_n = \tan^{-1}\left(\dfrac{n+2}{n+5}\right)$

16. $a_n = \dfrac{100^n}{n!} - \dfrac{3 + \pi^n}{5^n}$

17. $b_n = \sqrt{n^2 + n} - \sqrt{n^2 + 1}$

18. $c_n = \sqrt{n^2 + n} - \sqrt{n^2 - n}$

19. $b_m = \left(1 + \dfrac{1}{m}\right)^{3m}$

20. $c_n = \left(1 + \dfrac{3}{n}\right)^n$

21. $b_n = n\big(\ln(n+1) - \ln n\big)$

22. $c_n = \dfrac{\ln(n^2 + 1)}{\ln(n^3 + 1)}$

23. Use the Squeeze Theorem to show that $\lim\limits_{n\to\infty} \dfrac{\arctan(n^2)}{\sqrt{n}} = 0$.

24. Give an example of a divergent sequence $\{a_n\}$ such that $\{\sin a_n\}$ is convergent.

25. Calculate $\lim\limits_{n\to\infty} \dfrac{a_{n+1}}{a_n}$, where $a_n = \dfrac{1}{2}3^n - \dfrac{1}{3}2^n$.

26. Define $a_{n+1} = \sqrt{a_n + 6}$ with $a_1 = 2$.

(a) Compute a_n for $n = 2, 3, 4, 5$.

(b) Show that $\{a_n\}$ is increasing and is bounded by 3.

(c) Prove that $\lim\limits_{n\to\infty} a_n$ exists and find its value.

27. Calculate the partial sums S_4 and S_7 of the series $\sum\limits_{n=1}^{\infty} \dfrac{n-2}{n^2 + 2n}$.

28. Find the sum $1 - \dfrac{1}{4} + \dfrac{1}{4^2} - \dfrac{1}{4^3} + \cdots$.

29. Find the sum $\dfrac{4}{9} + \dfrac{8}{27} + \dfrac{16}{81} + \dfrac{32}{243} + \cdots$.

30. Use series to determine a reduced fraction that has decimal expansion $0.121212\cdots$.

31. Use series to determine a reduced fraction that has decimal expansion $0.108108108\cdots$.

32. Find the sum $\sum\limits_{n=2}^{\infty} \left(\dfrac{2}{e}\right)^n$.

33. Find the sum $\sum\limits_{n=-1}^{\infty} \dfrac{2^{n+3}}{3^n}$.

34. Show that $\sum\limits_{n=1}^{\infty} \left(b - \tan^{-1} n^2\right)$ diverges if $b \neq \dfrac{\pi}{2}$.

35. Give an example of divergent series $\sum\limits_{n=1}^{\infty} a_n$ and $\sum\limits_{n=1}^{\infty} b_n$ such that $\sum\limits_{n=1}^{\infty} (a_n + b_n) = 1$.

36. Let $S = \sum\limits_{n=1}^{\infty} \left(\dfrac{1}{n} - \dfrac{1}{n+2}\right)$. Compute S_N for $N = 1, 2, 3, 4$. Find S by showing that

$$S_N = \dfrac{3}{2} - \dfrac{1}{N+1} - \dfrac{1}{N+2}$$

37. Evaluate $S = \sum\limits_{n=3}^{\infty} \dfrac{1}{n(n+3)}$.

38. Find the total area of the infinitely many circles on the interval $[0, 1]$ in Figure 1.

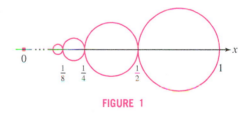

FIGURE 1

In Exercises 39–42, use the Integral Test to determine whether the infinite series converges.

39. $\sum\limits_{n=1}^{\infty} \dfrac{n^2}{n^3 + 1}$

40. $\sum\limits_{n=1}^{\infty} \dfrac{n^2}{(n^3 + 1)^{1.01}}$

41. $\sum\limits_{n=1}^{\infty} \dfrac{1}{(n+2)(\ln(n+2))^3}$

42. $\sum\limits_{n=1}^{\infty} \dfrac{n^3}{e^{n^4}}$

In Exercises 43–50, use the Direct Comparison or Limit Comparison Test to determine whether the infinite series converges.

43. $\sum\limits_{n=1}^{\infty} \dfrac{1}{(n+1)^2}$

44. $\sum\limits_{n=1}^{\infty} \dfrac{1}{\sqrt{n}+n}$

45. $\sum\limits_{n=2}^{\infty} \dfrac{n^2 + 1}{n^{3.5} - 2}$

46. $\sum\limits_{n=1}^{\infty} \dfrac{1}{n - \ln n}$

47. $\sum\limits_{n=2}^{\infty} \dfrac{n}{\sqrt{n^5 + 5}}$

48. $\sum\limits_{n=1}^{\infty} \dfrac{1}{3^n - 2^n}$

49. $\displaystyle\sum_{n=1}^{\infty} \frac{n^{10} + 10^n}{n^{11} + 11^n}$

50. $\displaystyle\sum_{n=1}^{\infty} \frac{n^{20} + 21^n}{n^{21} + 20^n}$

51. Determine the convergence of $\displaystyle\sum_{n=1}^{\infty} \frac{2^n + n}{3^n - 2}$ using the Limit Comparison Test with $b_n = \left(\frac{2}{3}\right)^n$.

52. Determine the convergence of $\displaystyle\sum_{n=1}^{\infty} \frac{\ln n}{1.5^n}$ using the Limit Comparison Test with $b_n = \frac{1}{1.4^n}$.

53. Let $a_n = 1 - \sqrt{1 - \frac{1}{n}}$. Show that $\lim_{n\to\infty} a_n = 0$ and that $\displaystyle\sum_{n=1}^{\infty} a_n$ diverges. *Hint:* Show that $a_n \geq \frac{1}{2n}$.

54. Determine whether $\displaystyle\sum_{n=2}^{\infty} \left(1 - \sqrt{1 - \frac{1}{n^2}}\right)$ converges.

55. Let $S = \displaystyle\sum_{n=1}^{\infty} \frac{n}{(n^2 + 1)^2}$.

(a) Show that S converges.

(b) *CAS* Use Eq. (4) in Exercise 85 of Section 10.3 with $M = 99$ to approximate S. What is the maximum size of the error?

In Exercises 56–59, determine whether the series converges absolutely. If it does not, determine whether it converges conditionally.

56. $\displaystyle\sum_{n=1}^{\infty} \frac{(-1)^n}{\sqrt[3]{n} + 2n}$

57. $\displaystyle\sum_{n=1}^{\infty} \frac{(-1)^n}{n^{1.1} \ln(n + 1)}$

58. $\displaystyle\sum_{n=1}^{\infty} \frac{\cos\left(\frac{\pi}{4} + \pi n\right)}{\sqrt{n}}$

59. $\displaystyle\sum_{n=1}^{\infty} \frac{\cos\left(\frac{\pi}{4} + 2\pi n\right)}{\sqrt{n}}$

60. *CAS* Use a computer algebra system to approximate $\displaystyle\sum_{n=1}^{\infty} \frac{(-1)^n}{n^3 + \sqrt{n}}$ to within an error of at most 10^{-5}.

61. Catalan's constant is defined by $K = \displaystyle\sum_{k=0}^{\infty} \frac{(-1)^k}{(2k + 1)^2}$.

(a) How many terms of the series are needed to calculate K with an error of less than 10^{-6}?

(b) *CAS* Carry out the calculation.

62. Give an example of conditionally convergent series $\displaystyle\sum_{n=1}^{\infty} a_n$ and $\displaystyle\sum_{n=1}^{\infty} b_n$ such that $\displaystyle\sum_{n=1}^{\infty}(a_n + b_n)$ converges absolutely.

63. Let $\displaystyle\sum_{n=1}^{\infty} a_n$ be an absolutely convergent series. Determine whether the following series are convergent or divergent:

(a) $\displaystyle\sum_{n=1}^{\infty} \left(a_n + \frac{1}{n^2}\right)$

(b) $\displaystyle\sum_{n=1}^{\infty}(-1)^n a_n$

(c) $\displaystyle\sum_{n=1}^{\infty} \frac{1}{1 + a_n^2}$

(d) $\displaystyle\sum_{n=1}^{\infty} \frac{|a_n|}{n}$

64. Let $\{a_n\}$ be a positive sequence such that $\lim_{n\to\infty} \sqrt[n]{a_n} = \frac{1}{2}$. Determine whether the following series converge or diverge:

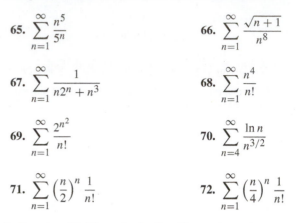

(a) $\displaystyle\sum_{n=1}^{\infty} 2a_n$

(b) $\displaystyle\sum_{n=1}^{\infty} 3^n a_n$

(c) $\displaystyle\sum_{n=1}^{\infty} \sqrt{a_n}$

In Exercises 65–72, apply the Ratio Test to determine convergence or divergence, or state that the Ratio Test is inconclusive.

65. $\displaystyle\sum_{n=1}^{\infty} \frac{n^5}{5^n}$

66. $\displaystyle\sum_{n=1}^{\infty} \frac{\sqrt{n + 1}}{n^8}$

67. $\displaystyle\sum_{n=1}^{\infty} \frac{1}{n2^n + n^3}$

68. $\displaystyle\sum_{n=1}^{\infty} \frac{n^4}{n!}$

69. $\displaystyle\sum_{n=1}^{\infty} \frac{2^{n^2}}{n!}$

70. $\displaystyle\sum_{n=4}^{\infty} \frac{\ln n}{n^{3/2}}$

71. $\displaystyle\sum_{n=1}^{\infty} \left(\frac{n}{2}\right)^n \frac{1}{n!}$

72. $\displaystyle\sum_{n=1}^{\infty} \left(\frac{n}{4}\right)^n \frac{1}{n!}$

In Exercises 73–76, apply the Root Test to determine convergence or divergence, or state that the Root Test is inconclusive.

73. $\displaystyle\sum_{n=1}^{\infty} \frac{1}{4^n}$

74. $\displaystyle\sum_{n=1}^{\infty} \left(\frac{2}{n}\right)^n$

75. $\displaystyle\sum_{n=1}^{\infty} \left(\frac{3}{4n}\right)^n$

76. $\displaystyle\sum_{n=1}^{\infty} \left(\cos\frac{1}{n}\right)^{n^3}$

In Exercises 77–100, determine convergence or divergence using any method covered in the text.

77. $\displaystyle\sum_{n=1}^{\infty} \left(\frac{2}{3}\right)^n$

78. $\displaystyle\sum_{n=1}^{\infty} \frac{\pi^{7n}}{e^{8n}}$

79. $\displaystyle\sum_{n=1}^{\infty} e^{-0.02n}$

80. $\displaystyle\sum_{n=1}^{\infty} ne^{-0.02n}$

81. $\displaystyle\sum_{n=1}^{\infty} \frac{(-1)^{n-1}}{\sqrt{n} + \sqrt{n + 1}}$

82. $\displaystyle\sum_{n=10}^{\infty} \frac{1}{n(\ln n)^{3/2}}$

83. $\displaystyle\sum_{n=2}^{\infty} \frac{(-1)^n}{\ln n}$

84. $\displaystyle\sum_{n=1}^{\infty} \frac{n!}{(2n)!}$

85. $\displaystyle\sum_{n=2}^{\infty} \frac{n}{1 + 100n}$

86. $\displaystyle\sum_{n=2}^{\infty} \frac{n^3 - 2n^2 + n - 4}{2n^4 + 3n^3 - 4n^2 - 1}$

87. $\displaystyle\sum_{n=1}^{\infty} \frac{\cos n}{n^{3/2}}$

88. $\displaystyle\sum_{n=1}^{\infty} \frac{n}{\sqrt{n^{3/2} + 1}}$

89. $\displaystyle\sum_{n=1}^{\infty} \left(\frac{n}{5n + 2}\right)^n$

90. $\displaystyle\sum_{n=1}^{\infty} \frac{e^n}{n!}$

91. $\displaystyle\sum_{n=1}^{\infty} \frac{1}{n\sqrt{n} + \ln n}$

92. $\displaystyle\sum_{n=1}^{\infty} \frac{1}{\sqrt[3]{n}(1 + \sqrt{n})}$

93. $\displaystyle\sum_{n=1}^{\infty} \left(\frac{1}{\sqrt{n}} - \frac{1}{\sqrt{n + 1}}\right)$

94. $\displaystyle\sum_{n=1}^{\infty} \left(\ln n - \ln(n + 1)\right)$

95. $\displaystyle\sum_{n=1}^{\infty} \frac{1}{n + \sqrt{n}}$

96. $\displaystyle\sum_{n=2}^{\infty} \frac{\cos(\pi n)}{n^{2/3}}$

97. $\displaystyle\sum_{n=2}^{\infty} \frac{1}{n \ln n}$

98. $\displaystyle\sum_{n=2}^{\infty} \frac{1}{\ln^3 n}$

99. $\displaystyle\sum_{n=1}^{\infty} \sin^2 \frac{\pi}{n}$

100. $\displaystyle\sum_{n=0}^{\infty} \frac{2^{2n}}{n!}$

In Exercises 101–106, find the interval of convergence of the power series.

101. $\displaystyle\sum_{n=0}^{\infty} \frac{2^n x^n}{n!}$

102. $\displaystyle\sum_{n=0}^{\infty} \frac{x^n}{n + 1}$

103. $\displaystyle\sum_{n=0}^{\infty} \frac{n^6}{n^8 + 1}(x - 3)^n$

104. $\displaystyle\sum_{n=0}^{\infty} n x^n$

105. $\displaystyle\sum_{n=0}^{\infty} (nx)^n$

106. $\displaystyle\sum_{n=2}^{\infty} \frac{(2x - 3)^n}{n \ln n}$

107. Expand $f(x) = \dfrac{2}{4 - 3x}$ as a power series centered at $c = 0$. Determine the values of x for which the series converges.

108. Prove that

$$\sum_{n=0}^{\infty} n e^{-nx} = \frac{e^{-x}}{(1 - e^{-x})^2}$$

Hint: Express the left-hand side as the derivative of a geometric series.

109. Let $F(x) = \displaystyle\sum_{k=0}^{\infty} \frac{x^{2k}}{2^k \cdot k!}$.

(a) Show that $F(x)$ has infinite radius of convergence.

(b) Show that $y = F(x)$ is a solution of

$$y'' = xy' + y, \qquad y(0) = 1, \qquad y'(0) = 0$$

(c) CAS   Plot the partial sums S_N for $N = 1, 3, 5, 7$ on the same set of axes.

110. Find a power series $P(x) = \displaystyle\sum_{n=0}^{\infty} a_n x^n$ that satisfies the Laguerre differential equation

$$xy'' + (1 - x)y' - y = 0$$

with initial condition satisfying $P(0) = 1$.

111. Use power series to evaluate $\displaystyle\lim_{x \to 0} \frac{x^2 e^x}{\cos x - 1}$.

112. Use power series to evaluate $\displaystyle\lim_{x \to 0} \frac{x^2(1 - \ln(x + 1))}{\sin x - x}$.

In Exercises 113–122, find the Taylor series centered at c.

113. $f(x) = e^{4x}, \quad c = 0$

114. $f(x) = e^{2x}, \quad c = -1$

115. $f(x) = x^4, \quad c = 2$

116. $f(x) = x^3 - x, \quad c = -2$

117. $f(x) = \sin x, \quad c = \pi$

118. $f(x) = e^{x-1}, \quad c = -1$

119. $f(x) = \dfrac{1}{1 - 2x}, \quad c = -2$

120. $f(x) = \dfrac{1}{(1 - 2x)^2}, \quad c = -2$

121. $f(x) = \ln \dfrac{x}{2}, \quad c = 2$

122. $f(x) = x \ln\left(1 + \dfrac{x}{2}\right), \quad c = 0$

In Exercises 123–126, find the first three terms of the Maclaurin series of $f(x)$ and use it to calculate $f^{(3)}(0)$.

123. $f(x) = (x^2 - x)e^{x^2}$

124. $f(x) = \tan^{-1}(x^2 - x)$

125. $f(x) = \dfrac{1}{1 + \tan x}$

126. $f(x) = (\sin x)\sqrt{1 + x}$

127. Calculate $\dfrac{\pi}{2} - \dfrac{\pi^3}{2^3 3!} + \dfrac{\pi^5}{2^5 5!} - \dfrac{\pi^7}{2^7 7!} + \cdots$.

128. Find the Maclaurin series of the function $F(x) = \displaystyle\int_0^x \frac{e^t - 1}{t}\, dt$.

Ellipses, which are a type of conic section, and parametric equations are used to track satellites as they orbit Earth. (*Chad Baker/Getty Images*)

11 PARAMETRIC EQUATIONS, POLAR COORDINATES, AND CONIC SECTIONS

This chapter introduces two important new tools. First, we consider parametric equations, which describe curves in a form that is particularly useful for analyzing motion and is indispensable in fields such as computer graphics and computer-aided design. We then study polar coordinates, an alternative to rectangular coordinates that simplifies computations in many applications. The chapter closes with a discussion of the conic sections (ellipses, hyperbolas, and parabolas).

11.1 Parametric Equations

We use the term "particle" when we treat an object as a moving point, ignoring its internal structure.

Imagine a particle moving along a curve C in the plane as in Figure 1. We would like to be able to describe the particle's motion; however, the curve C is not the graph of a function $y = h(x)$ since it fails the vertical line test. Instead, we can describe the particle's motion by specifying its coordinates as functions of time t:

$$x = f(t), \qquad y = g(t) \qquad \boxed{1}$$

In other words, at time t, the particle is located at the point

$$c(t) = (f(t), g(t))$$

The equations (1) are called **parametric equations**, and C is called a **parametric curve**. We refer to $c(t)$ as a **parametrization** with **parameter** t.

Because x and y are functions of t, we often write $c(t) = (x(t), y(t))$ instead of $(f(t), g(t))$. Of course, we are free to use any variable for the parameter (such as s or θ). In plots of parametric curves, the direction of motion is often indicated by an arrow as in Figure 1.

DF FIGURE 1 Particle moving along a curve C in the plane.

■ **EXAMPLE 1** Sketch the curve with parametric equations

$$x = 2t - 4, \qquad y = 3 + t^2 \qquad \boxed{2}$$

Solution First compute the x- and y-coordinates for several values of t as in Table 1, and plot the corresponding points (x, y) as in Figure 2. Then join the points by a smooth curve, indicating the direction of motion (direction of increasing t) with an arrow. ■

TABLE 1

t	$x = 2t - 4$	$y = 3 + t^2$
-2	-8	7
0	-4	3
2	0	7
4	4	19

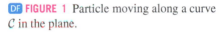

DF FIGURE 2 The parametric curve $x = 2t - 4$, $y = 3 + t^2$.

579

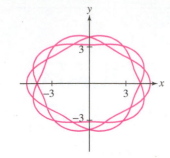

FIGURE 3 The parametric curve
$x = 5\cos(3t)\cos\left(\frac{2}{3}\sin(5t)\right)$,
$y = 4\sin(3t)\cos\left(\frac{2}{3}\sin(5t)\right)$.

CONCEPTUAL INSIGHT The graph of $y = x^2$ can be parametrized in a simple way. We take $x = t$, and we take $y = t^2$. Then, clearly this generates the right curve, since eliminating the variable t returns us to the original equation $y = x^2$. Thus, the parabola $y = x^2$ is parametrized by $c(t) = (t, t^2)$. More generally, we can parametrize the graph of $y = f(x)$ by taking $x = t$ and $y = f(t)$. Therefore, $c(t) = (t, f(t))$ parametrizes the graph. For another example, the graph of $y = e^x$ is parametrized by $c(t) = (t, e^t)$. An advantage of parametric equations is that they enable us to describe curves that are not graphs of functions. For example, the curve in Figure 3 is not of the form $y = f(x)$ but it can be expressed parametrically.

As we have just noted, a parametric curve $c(t)$ need not be the graph of a function. If it is, however, it may be possible to find the function f by "eliminating the parameter" as in the next example.

■ **EXAMPLE 2** **Eliminating the Parameter** Describe the parametric curve

$$c(t) = (2t - 4, 3 + t^2)$$

of the previous example in the form $y = f(x)$.

Solution We "eliminate the parameter" by solving for y as a function of x. First, express t in terms of x: Since $x = 2t - 4$, we have $t = \frac{1}{2}x + 2$. Then substitute

$$y = 3 + t^2 = 3 + \left(\frac{1}{2}x + 2\right)^2 = 7 + 2x + \frac{1}{4}x^2$$

Thus, $c(t)$ traces out the graph of $f(x) = 7 + 2x + \frac{1}{4}x^2$ shown in Figure 2. ■

■ **EXAMPLE 3** A model rocket follows the trajectory

$$c(t) = (80t, 200t - 4.9t^2)$$

until it hits the ground, with t in seconds and distance in meters (Figure 4). Find:

(a) The rocket's height at $t = 5$ s. **(b)** Its maximum height.

Solution The height of the rocket at time t is $y(t) = 200t - 4.9t^2$.

(a) The height at $t = 5$ s is

$$y(5) = 200(5) - 4.9(5^2) = 877.5 \text{ m}$$

(b) The maximum height occurs at the critical point of $y(t)$:

$$y'(t) = \frac{d}{dt}(200t - 4.9t^2) = 200 - 9.8t = 0 \quad \Rightarrow \quad t = \frac{200}{9.8} \approx 20.4 \text{ s}$$

The rocket's maximum height is $y(20.4) = 200(20.4) - 4.9(20.4)^2 \approx 2041$ m. ■

We now discuss parametrizations of lines and circles. They will appear frequently in later chapters.

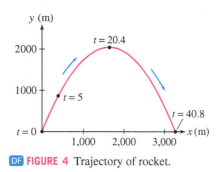

DF FIGURE 4 Trajectory of rocket.

CAUTION *The graph of height versus time for an object tossed in the air is a parabola (by Galileo's formula). But keep in mind that Figure 4 is **not** a graph of height versus time. It shows the actual path of the rocket (which has both a vertical and a horizontal displacement).*

THEOREM 1 Parametrization of a Line

(a) The line through $P = (a, b)$ of slope m is parametrized by

$$\boxed{x = a + rt, \qquad y = b + st \qquad -\infty < t < \infty} \qquad \boxed{3}$$

for any r and s (with $r \neq 0$) such that $m = s/r$.

(b) The line through $P = (a, b)$ and $Q = (c, d)$ has parametrization

$$\boxed{x = a + t(c - a), \qquad y = b + t(d - b) \qquad -\infty < t < \infty} \qquad \boxed{4}$$

The segment from P to Q corresponds to $0 \leq t \leq 1$.

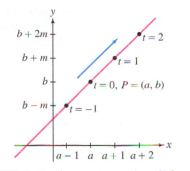

FIGURE 5 The line $y - a = m(x - b)$ has parametrization $c(t) = (a + t, b + mt)$. This corresponds to $r = 1$, $s = m$ in Eq. (3).

Proof **(a)** Solve $x = a + rt$ for t in terms of x to obtain $t = (x - a)/r$. Then

$$y = b + st = b + s\left(\frac{x - a}{r}\right) = b + m(x - a) \quad \text{or} \quad y - b = m(x - a)$$

This is the equation of the line through $P = (a, b)$ of slope m. The choice $r = 1$ and $s = m$ yields the parametrization in Figure 5.

The parametrization in **(b)** defines a line that satisfies $(x(0), y(0)) = (a, b)$ and $(x(1), y(1)) = (c, d)$. Thus, it parametrizes the line through P and Q and traces the segment from P to Q as t varies from 0 to 1. ∎

■ **EXAMPLE 4** **Parametrization of a Line** Find parametric equations for the line through $P = (3, -1)$ of slope $m = 4$.

Solution We can parametrize the line by taking $r = 1$ and $s = 4$ in Eq. (3):

$$x = 3 + t, \qquad y = -1 + 4t$$

This is also written as $c(t) = (3 + t, -1 + 4t)$. Another parametrization of the line is $c(t) = (3 + 5t, -1 + 20t)$, corresponding to $r = 5$ and $s = 20$ in Eq. (3). ∎

The circle of radius R centered at the origin has the parametrization

$$x = R \cos\theta, \qquad y = R \sin\theta$$

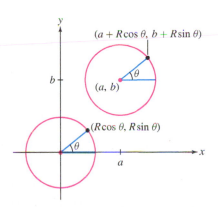

FIGURE 6 Parametrization of a circle of radius R with center (a, b).

The parameter θ represents the angle corresponding to the point (x, y) on the circle (Figure 6). The circle is traversed once in the counterclockwise direction as θ varies over a half-open interval of length 2π such as $[0, 2\pi)$ or $[-\pi, \pi)$.

More generally, the circle of radius R with center (a, b) has parametrization (Figure 6)

$$\boxed{x = a + R \cos\theta, \qquad y = b + R \sin\theta} \qquad \boxed{5}$$

As a check, let's verify that a point (x, y) given by Eq. (5) satisfies the equation of the circle of radius R centered at (a, b):

$$(x - a)^2 + (y - b)^2 = (a + R \cos\theta - a)^2 + (b + R \sin\theta - b)^2$$
$$= R^2 \cos^2\theta + R^2 \sin^2\theta = R^2$$

In general, to **translate** (meaning "to move") a parametric curve horizontally a units and vertically b units, replace $c(t) = (x(t), y(t))$ by $c(t) = (a + x(t), b + y(t))$.

Suppose we have a parametrization $c(t) = (x(t), y(t))$, where $x(t)$ is an even function and $y(t)$ is an odd function, that is, $x(-t) = x(t)$ and $y(-t) = -y(t)$. In this case, $c(-t)$ is the *reflection* of $c(t)$ across the x-axis:

$$c(-t) = (x(-t), y(-t)) = (x(t), -y(t))$$

The curve, therefore, is *symmetric* with respect to the x-axis. We apply this remark in the next example and in Example 7 below.

■ **EXAMPLE 5** **Parametrization of an Ellipse** Verify that the ellipse with equation $\left(\frac{x}{a}\right)^2 + \left(\frac{y}{b}\right)^2 = 1$ is parametrized by

$$\boxed{c(t) = (a \cos t, b \sin t) \qquad (\text{for } -\pi \leq t < \pi)}$$

Plot the case $a = 4$, $b = 2$.

Solution To verify that $c(t)$ parametrizes the ellipse, show that the equation of the ellipse is satisfied with $x = a \cos t$, $y = b \sin t$:

$$\left(\frac{x}{a}\right)^2 + \left(\frac{y}{b}\right)^2 = \left(\frac{a \cos t}{a}\right)^2 + \left(\frac{b \sin t}{b}\right)^2 = \cos^2 t + \sin^2 t = 1$$

NOTATION We use $x(t)$ and $y(t)$ to represent functions of t, where $x(t)$ and $y(t)$ correspond to the x-coordinate and y-coordinate of a parametric curve.

TABLE 2

t	$x(t) = 4\cos t$	$y(t) = 2\sin t$
0	4	0
$\dfrac{\pi}{6}$	$2\sqrt{3}$	1
$\dfrac{\pi}{3}$	2	$\sqrt{3}$
$\dfrac{\pi}{2}$	0	2
$\dfrac{2\pi}{3}$	-2	$\sqrt{3}$
$\dfrac{5\pi}{6}$	$-2\sqrt{3}$	1
π	-4	0

To plot the case $a = 4$, $b = 2$, we connect the points for the t-values in Table 2 [see Figure 7(A)]. This gives us the top half of the ellipse for $0 \le t \le \pi$. Then we observe that $x(t) = 4\cos t$ is even and $y(t) = 2\sin t$ is odd. As noted above, this tells us that the bottom half of the ellipse is obtained by symmetry with respect to the x-axis, as in Figure 7(B). Alternatively, we could also evaluate $x(t)$ and $y(t)$ for negative values of t between $-\pi$ and 0 to determine the bottom portion of the ellipse. ∎

A parametric curve $c(t)$ is also called a **path**. This term emphasizes that $c(t)$ describes not just an underlying curve $\mathcal{C}$, but a particular way of moving along the curve.

> **CONCEPTUAL INSIGHT** The parametric equations for the ellipse in Example 5 illustrate a key difference between the path $c(t)$ and its underlying curve $\mathcal{C}$. The curve $\mathcal{C}$ is an ellipse in the plane, whereas $c(t)$ describes a particular, counterclockwise motion of a particle along the ellipse. If we let t vary from 0 to 4π, then the particle goes around the ellipse twice.
>
> A key feature of parametrizations is that they are not unique. In fact, every curve can be parametrized in infinitely many different ways. For instance, the parabola $y = x^2$ is parametrized not only by (t, t^2) but also by (t^3, t^6), or (t^5, t^{10}), and so on.

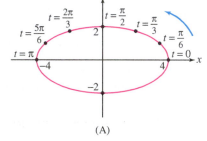

(A)

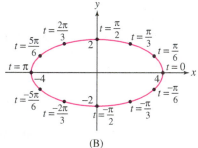

(B)

DF FIGURE 7 Ellipse with parametric equations $x = 4\cos t$, $y = 2\sin t$.

■ **EXAMPLE 6**   **Different Parametrizations of the Same Curve** Describe the motion of a particle moving along each of the following paths:

(a) $c_1(t) = (t^3, t^6)$ **(b)** $c_2(t) = (t^2, t^4)$ **(c)** $c_3(t) = (\cos t, \cos^2 t)$

Solution Each of these parametrizations satisfies $y = x^2$, so all three parametrize portions of the parabola $y = x^2$.

(a) As t varies from $-\infty$ to ∞, t^3 also varies from $-\infty$ to ∞. Therefore, $c_1(t) = (t^3, t^6)$ traces the entire parabola $y = x^2$, moving from left to right and passing through each point once [Figure 8(A)].

(b) Since $x = t^2 \ge 0$, the path $c_2(t) = (t^2, t^4)$ traces only the right half of the parabola. The particle comes in toward the origin as t varies from $-\infty$ to 0, and it goes back out to the right as t varies from 0 to ∞ [Figure 8(B)].

(c) As t varies from $-\infty$ and ∞, $\cos t$ oscillates between 1 and -1. Thus, a particle following the path $c_3(t) = (\cos t, \cos^2 t)$ oscillates back and forth between the points $(1, 1)$ and $(-1, 1)$ on the parabola [Figure 8(C)]. ∎

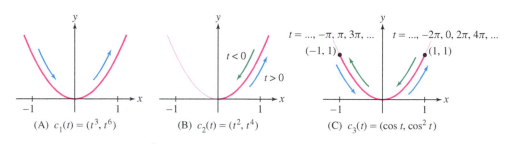

FIGURE 8 Three parametrizations of portions of the parabola.

(A) $c_1(t) = (t^3, t^6)$ (B) $c_2(t) = (t^2, t^4)$ (C) $c_3(t) = (\cos t, \cos^2 t)$

■ **EXAMPLE 7**   **Using Symmetry to Sketch a Loop** Sketch the parametric curve

$$c(t) = (t^2 + 1, t^3 - 4t)$$

Label the points corresponding to $t = 0, \pm 1, \pm 2, \pm 2.5$.

Solution

Step 1. **Use symmetry.**

Observe that $x(t) = t^2 + 1$ is an even function and that $y(t) = t^3 - 4t$ is an odd function. As noted before Example 5, this tells us that $c(t)$ is symmetric with respect to the x-axis. Therefore, we will plot the curve for $t \ge 0$ and reflect across the x-axis to obtain the part for $t \le 0$.

Step 2. Analyze $x(t)$, $y(t)$ as functions of t.

We have $x(t) = t^2 + 1$ and $y(t) = t^3 - 4t$. The x-coordinate $x(t) = t^2 + 1$ increases to ∞ as $t \to \infty$. To analyze the y-coordinate, we graph $y(t) = t^3 - 4t = t(t-2)(t+2)$ as a function of t (*not* as a function of x). Since $y(t)$ is the height above the x-axis, Figure 9(A) shows that

$$y(t) < 0 \quad \text{for} \quad 0 < t < 2, \quad \Rightarrow \quad \text{curve below } x\text{-axis}$$

$$y(t) > 0 \quad \text{for} \quad t > 2, \quad \Rightarrow \quad \text{curve above } x\text{-axis}$$

So, the curve starts at $c(0) = (1, 0)$, dips below the x-axis, and returns to the x-axis at $t = 2$. Both $x(t)$ and $y(t)$ tend to ∞ as $t \to \infty$. The curve is concave up because $y(t)$ increases more rapidly than $x(t)$.

Step 3. Plot points and join by an arc.

The points $c(0), c(1), c(2), c(2.5)$ tabulated in Table 3 are plotted and joined by an arc to create the sketch for $t \geq 0$ as in Figure 9(B). The sketch is completed by reflecting across the x-axis as in Figure 9(C). ■

TABLE 3

t	$x = t^2 + 1$	$y = t^3 - 4t$
0	1	0
1	2	-3
2	5	0
2.5	7.25	5.625

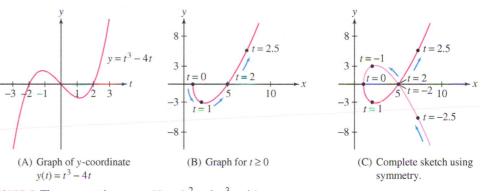

(A) Graph of y-coordinate
$y(t) = t^3 - 4t$

(B) Graph for $t \geq 0$

(C) Complete sketch using symmetry.

FIGURE 9 The parametric curve $c(t) = (t^2 + 1, t^3 - 4t)$.

A **cycloid** is a curve traced by a point on the circumference of a rolling wheel as in Figure 10. Cycloids are famous for their "brachistochrone property" (see the marginal note below).

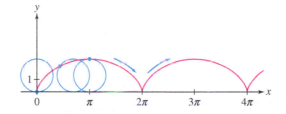

FIGURE 10 A cycloid.

A stellar cast of mathematicians (including Galileo, Pascal, Newton, Leibniz, Huygens, and Bernoulli) studied the cycloid and discovered many of its remarkable properties. A slide designed so that an object sliding down (without friction) reaches the bottom in the least time must have the shape of an inverted cycloid. This is the brachistochrone property, a term derived from the Greek brachistos, "shortest," and chronos, "time."

■ **EXAMPLE 8** **Parametrizing the Cycloid** Find parametric equations for the cycloid generated by a point P on the unit circle.

Solution The point P is located at the origin at $t = 0$. At time t, the circle has rolled t radians along the x-axis and the center C of the circle then has coordinates $(t, 1)$ as in Figure 11(A). Figure 11(B) shows that we get from C to P by moving down $\cos t$ units and to the left $\sin t$ units, giving us the parametric equations

$$x(t) = t - \sin t, \qquad y(t) = 1 - \cos t \qquad \boxed{6}$$

■

The argument in Example 8 shows in a similar fashion that the cycloid generated by a circle of radius R has parametric equations

$$x = Rt - R\sin t, \qquad y = R - R\cos t \qquad \boxed{7}$$

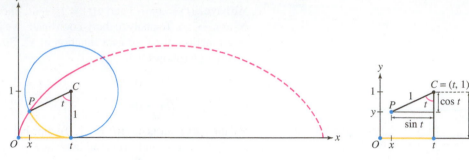

(A) Position of P at time t

(B) P has coordinates
$x = t - \sin t, \; y = 1 - \cos t$

DF FIGURE 11

Tangent Lines to Parametric Curves

Just as we use tangent lines to the graph of $y = f(x)$ to determine the rate of change of the function f, we would like to be able to determine how y changes with x when the curve is described by parametric equations. The slope of the tangent line is the derivative dy/dx, but we have to use the Chain Rule to compute it because y is not given explicitly as a function of x. Write $x = f(t)$, $y = g(t)$. Then, by the Chain Rule,

$$g'(t) = \frac{dy}{dt} = \frac{dy}{dx}\frac{dx}{dt} = \frac{dy}{dx} f'(t)$$

> **NOTATION** In this section, we write $f'(t)$, $x'(t)$, $y'(t)$, and so on to denote the derivative with respect to t.

If $f'(t) \neq 0$, we can divide by $f'(t)$ to obtain

$$\frac{dy}{dx} = \frac{g'(t)}{f'(t)}$$

This calculation is valid if $f(t)$ and $g(t)$ are differentiable, $f'(t)$ is continuous, and $f'(t) \neq 0$. In this case, the inverse $t = f^{-1}(x)$ exists, and the composite $y = g(f^{-1}(x))$ is a differentiable function of x.

> **CAUTION** Do not confuse dy/dx with the derivatives dx/dt and dy/dt, which are derivatives with respect to the parameter t. Only dy/dx is the slope of the tangent line.

> **THEOREM 2 Slope of the Tangent Line** Let $c(t) = (x(t), y(t))$, where $x(t)$ and $y(t)$ are differentiable. Assume that $x'(t)$ is continuous and $x'(t) \neq 0$. Then
>
> $$\frac{dy}{dx} = \frac{dy/dt}{dx/dt} = \frac{y'(t)}{x'(t)} \qquad \boxed{8}$$

■ **EXAMPLE 9** Let $c(t) = (t^2 + 1, t^3 - 4t)$. Find:

(a) An equation of the tangent line at $t = 3$

(b) The points where the tangent is horizontal (Figure 12).

Solution We have

$$\frac{dy}{dx} = \frac{y'(t)}{x'(t)} = \frac{(t^3 - 4t)'}{(t^2 + 1)'} = \frac{3t^2 - 4}{2t}$$

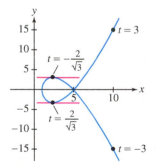

FIGURE 12 Horizontal tangent lines on $c(t) = (t^2 + 1, t^3 - 4t)$.

(a) The slope at $t = 3$ is

$$\left.\frac{dy}{dx} = \frac{3t^2 - 4}{2t}\right|_{t=3} = \frac{3(3)^2 - 4}{2(3)} = \frac{23}{6}$$

Since $c(3) = (10, 15)$, the equation of the tangent line in point-slope form is

$$y - 15 = \frac{23}{6}(x - 10)$$

(b) The slope dy/dx is zero if $y'(t) = 0$ and $x'(t) \neq 0$. We have $y'(t) = 3t^2 - 4 = 0$ for $t = \pm 2/\sqrt{3}$ (and $x'(t) = 2t \neq 0$ for these values of t). Therefore, the tangent line is horizontal at the points

$$c\left(-\frac{2}{\sqrt{3}}\right) = \left(\frac{7}{3}, \frac{16}{3\sqrt{3}}\right), \qquad c\left(\frac{2}{\sqrt{3}}\right) = \left(\frac{7}{3}, -\frac{16}{3\sqrt{3}}\right) \qquad ∎$$

Parametric curves are widely used in the field of computer graphics. A particularly important class of curves are **Bézier curves**, which we discuss here briefly in the cubic case. Given four "control points" (Figure 13):

$$P_0 = (a_0, b_0), \qquad P_1 = (a_1, b_1), \qquad P_2 = (a_2, b_2), \qquad P_3 = (a_3, b_3)$$

the Bézier curve $c(t) = (x(t), y(t))$ is defined for $0 \leq t \leq 1$ by

$$x(t) = a_0(1-t)^3 + 3a_1 t(1-t)^2 + 3a_2 t^2(1-t) + a_3 t^3 \qquad \boxed{9}$$

$$y(t) = b_0(1-t)^3 + 3b_1 t(1-t)^2 + 3b_2 t^2(1-t) + b_3 t^3 \qquad \boxed{10}$$

Bézier curves were invented in the 1960s by the French engineer Pierre Bézier (1910–1999), who worked for the Renault car company. They are based on the properties of Bernstein polynomials, introduced 50 years earlier by the Russian mathematician Sergei Bernstein to study the approximation of continuous functions by polynomials. Today, Bézier curves are used in standard graphics programs, such as Adobe Illustrator™ and Corel Draw™, and in the construction and storage of computer fonts such as TrueType™ and PostScript™ fonts.

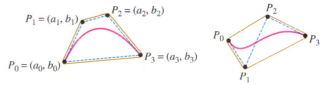

DF FIGURE 13 Cubic Bézier curves specified by four control points.

Note that $c(0) = (a_0, b_0)$ and $c(1) = (a_3, b_3)$, so the Bézier curve begins at P_0 and ends at P_3 (Figure 13). It can also be shown that the Bézier curve is contained within the quadrilateral (shown in blue) with vertices P_0, P_1, P_2, P_3. However, $c(t)$ does not pass through P_1 and P_2. Instead, these intermediate control points determine the slopes of the tangent lines at P_0 and P_3, as we show in the next example (also, see Exercises 67–70).

■ **EXAMPLE 10** Show that the Bézier curve is tangent to segment $\overline{P_0 P_1}$ at P_0.

Solution The Bézier curve passes through P_0 at $t = 0$, so we must show that the slope of the tangent line at $t = 0$ is equal to the slope of $\overline{P_0 P_1}$. To find the slope, we compute the derivatives:

$$x'(t) = -3a_0(1-t)^2 + 3a_1(1 - 4t + 3t^2) + a_2(2t - 3t^2) + 3a_3 t^2$$

$$y'(t) = -3b_0(1-t)^2 + 3b_1(1 - 4t + 3t^2) + b_2(2t - 3t^2) + 3b_3 t^2$$

Evaluating at $t = 0$, we obtain $x'(0) = 3(a_1 - a_0)$, $y'(0) = 3(b_1 - b_0)$, and

$$\left.\frac{dy}{dx}\right|_{t=0} = \frac{y'(0)}{x'(0)} = \frac{3(b_1 - b_0)}{3(a_1 - a_0)} = \frac{b_1 - b_0}{a_1 - a_0}$$

This is equal to the slope of the line through $P_0 = (a_0, b_0)$ and $P_1 = (a_1, b_1)$ as claimed (provided that $a_1 \neq a_0$). ∎

Area Under a Parametric Curve

As we know, the area under a curve given by $y = h(x)$ when $h(x) \geq 0$ for $a \leq x \leq b$ is given by

$$A = \int_a^b h(x)\, dx$$

When the curve $y = h(x)$ is traced once by a parametric curve $c(t) = (x(t), y(t))$ as in Figure 14, where $x(t_0) = a$ and $x(t_1) = b$, then we can substitute, replacing $y = h(x)$ by $y(t)$ and dx by $x'(t)dt$, yielding a formula for the area A under the curve:

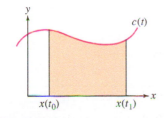

FIGURE 14 Finding area under a parametric curve $c(t)$.

$$\boxed{A = \int_{t_0}^{t_1} y(t) x'(t)\, dt} \qquad \boxed{11}$$

■ **EXAMPLE 11** Find the area inside the ellipse of Example 5.

Solution The ellipse was found to be parametrized by $c(t) = (a \cos t, b \sin t)$ for $-\pi \le t \le \pi$. The top half of the ellipse corresponds to $0 \le t \le \pi$, and since this yields the graph of a function that is nonnegative, we can find the area under the curve using formula (10). Notice, however, that $t_0 = \pi$ and $t_1 = 0$, so $x(t_0) = -a$ and $x(t_1) = a$. Thus, the area of the ellipse is given by

$$A = 2 \int_\pi^0 \underbrace{b \sin t}_{y(t)} \underbrace{(-a \sin t)}_{x'(t)} \, dt = 2ab \int_0^\pi \sin^2 t \, dt$$

$$= 2ab \int_0^\pi \frac{1 - \cos 2t}{2} \, dt = 2ab \left(\frac{t}{2} - \frac{\sin 2t}{4} \right) \bigg|_0^\pi = \pi ab$$ ■

11.1 SUMMARY

- A parametric curve $c(t) = (f(t), g(t))$ describes the path of a particle moving along a curve as a function of the parameter t.
- Parametrizations are not unique: Every curve $\mathcal{C}$ can be parametrized in infinitely many ways. Furthermore, the path $c(t)$ may traverse all or part of $\mathcal{C}$ more than once.
- Slope of the tangent line at $c(t)$:

$$\frac{dy}{dx} = \frac{dy/dt}{dx/dt} = \frac{y'(t)}{x'(t)} \qquad \text{[valid if } x'(t) \ne 0]$$

- Do not confuse the slope of the tangent line dy/dx with the derivatives dy/dt and dx/dt, with respect to t.
- Standard parametrizations:

  - Line of slope $m = s/r$ through $P = (a, b)$: $c(t) = (a + rt, b + st)$
  - Circle of radius R centered at $P = (a, b)$: $c(t) = (a + R \cos t, b + R \sin t)$
  - Cycloid generated by a circle of radius R: $c(t) = (R(t - \sin t), R(1 - \cos t))$

- Area under a parametric curve $c(t) = (x(t), y(t))$ that does not dip below the x-axis and that traces once the graph of a function is given by $A = \displaystyle\int_{t_0}^{t_1} y(t) x'(t) \, dt$.

11.1 EXERCISES

Preliminary Questions

1. Describe the shape of the curve $x = 3 \cos t$, $y = 3 \sin t$.

2. How does $x = 4 + 3 \cos t$, $y = 5 + 3 \sin t$ differ from the curve in the previous question?

3. What is the maximum height of a particle whose path has parametric equations $x = t^9$, $y = 4 - t^2$?

4. Can the parametric curve $(t, \sin t)$ be represented as a graph $y = f(x)$? What about $(\sin t, t)$?

5. (a) Describe the path of an ant that is crawling along the plane according to $c_1(t) = (f(t), f(t))$, where $f(t)$ is an increasing function.

(b) Compare that path to the path of a second ant crawling according to $c_2(t) = f(2t), f(2t))$.

6. Find three different parametrizations of the graph of $y = x^3$.

7. Match the derivatives with a verbal description:

(a) $\dfrac{dx}{dt}$ **(b)** $\dfrac{dy}{dt}$ **(c)** $\dfrac{dy}{dx}$

(i) Slope of the tangent line to the curve

(ii) Vertical rate of change with respect to time

(iii) Horizontal rate of change with respect to time

Exercises

1. Find the coordinates at times $t = 0, 2, 4$ of a particle following the path $x = 1 + t^3$, $y = 9 - 3t^2$.

2. Find the coordinates at $t = 0, \frac{\pi}{4}, \pi$ of a particle moving along the path $c(t) = (\cos 2t, \sin^2 t)$.

3. Show that the path traced by the rocket in Example 3 is a parabola by eliminating the parameter.

4. Use the table of values to sketch the parametric curve $(x(t), y(t))$, indicating the direction of motion.

t	-3	-2	-1	0	1	2	3
x	-15	0	3	0	-3	0	15
y	5	0	-3	-4	-3	0	5

5. Graph the parametric curves. Include arrows indicating the direction of motion.

(a) (t, t), $\quad -\infty < t < \infty$

(b) $(\sin t, \sin t)$, $\quad 0 \le t \le 2\pi$

(c) (e^t, e^t), $\quad -\infty < t < \infty$

(d) (t^3, t^3), $\quad -1 \le t \le 1$

6. Give two different parametrizations of the line through $(4, 1)$ with slope 2.

In Exercises 7–14, express in the form $y = f(x)$ by eliminating the parameter.

7. $x = t + 3$, $\quad y = 4t$

8. $x = t^{-1}$, $\quad y = t^{-2}$

9. $x = t$, $\quad y = \tan^{-1}(t^3 + e^t)$

10. $x = t^2$, $\quad y = t^3 + 1$

11. $x = e^{-2t}$, $\quad y = 6e^{4t}$

12. $x = 1 + t^{-1}$, $\quad y = t^2$

13. $x = \ln t$, $\quad y = 2 - t$

14. $x = \cos t$, $\quad y = \tan t$

In Exercises 15–18, graph the curve and draw an arrow specifying the direction corresponding to motion.

15. $x = \frac{1}{2}t$, $\quad y = 2t^2$

16. $x = 2 + 4t$, $\quad y = 3 + 2t$

17. $x = \pi t$, $\quad y = \sin t$

18. $x = t^2$, $\quad y = t^3$

19. Match the parametrizations (a)–(d) below with their plots in Figure 15, and draw an arrow indicating the direction of motion.

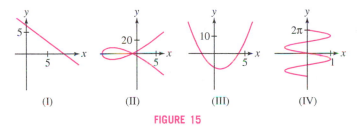

(I)   (II)   (III)   (IV)

FIGURE 15

(a) $c(t) = (\sin t, -t)$

(b) $c(t) = (t^2 - 9, 8t - t^3)$

(c) $c(t) = (1 - t, t^2 - 9)$

(d) $c(t) = (4t + 2, 5 - 3t)$

20. Find an interval of t-values such that $c(t) = (\cos t, \sin t)$ traces the lower half of the unit circle.

21. A particle follows the trajectory

$$x(t) = \frac{1}{4}t^3 + 2t, \qquad y(t) = 20t - t^2$$

with t in seconds and distance in centimeters.

(a) What is the particle's maximum height?

(b) When does the particle hit the ground and how far from the origin does it land?

22. Find an interval of t-values such that $c(t) = (2t + 1, 4t - 5)$ parametrizes the segment from $(0, -7)$ to $(7, 7)$.

In Exercises 23–38, find parametric equations for the given curve.

23. $y = 9 - 4x$

24. $y = 8x^2 - 3x$

25. $4x - y^2 = 5$

26. $x^2 + y^2 = 49$

27. $(x + 9)^2 + (y - 4)^2 = 49$

28. $\left(\frac{x}{5}\right)^2 + \left(\frac{y}{12}\right)^2 = 1$

29. Line of slope 8 through $(-4, 9)$

30. Line through $(2, 5)$ perpendicular to $y = 3x$

31. Line through $(3, 1)$ and $(-5, 4)$

32. Line through $\left(\frac{1}{3}, \frac{1}{6}\right)$ and $\left(-\frac{7}{6}, \frac{5}{3}\right)$

33. Segment joining $(1, 1)$ and $(2, 3)$

34. Segment joining $(-3, 0)$ and $(0, 4)$

35. Circle of radius 4 with center $(3, 9)$

36. Ellipse of Exercise 28, with its center translated to $(7, 4)$

37. $y = x^2$, translated so that the minimum occurs at $(-4, -8)$

38. $y = \cos x$, translated so that a maximum occurs at $(3, 5)$

In Exercises 39–42, find a parametrization $c(t)$ of the curve satisfying the given condition.

39. $y = 3x - 4$, $\quad c(0) = (2, 2)$

40. $y = 3x - 4$, $\quad c(3) = (2, 2)$

41. $y = x^2$, $\quad c(0) = (3, 9)$

42. $x^2 + y^2 = 4$, $\quad c(0) = (1, \sqrt{3})$

43. Describe $c(t) = (\sec t, \tan t)$ for $0 \le t < \frac{\pi}{2}$ in the form $y = f(x)$. Specify the domain of x.

44. Find a parametrization of the right branch $(x > 0)$ of the hyperbola

$$\left(\frac{x}{a}\right)^2 - \left(\frac{y}{b}\right)^2 = 1$$

using $\cosh t$ and $\sinh t$. How can you parametrize the branch $x < 0$?

45. The graphs of $x(t)$ and $y(t)$ as functions of t are shown in Figure 16(A). Which of (I)–(III) is the plot of $c(t) = (x(t), y(t))$? Explain.

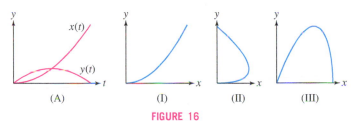

(A)   (I)   (II)   (III)

FIGURE 16

46. Which graph, (I) or (II), is the graph of $x(t)$ and which is the graph of $y(t)$ for the parametric curve in Figure 17(A)?

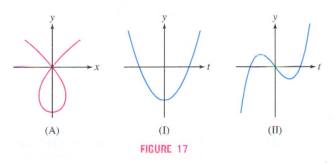

(A)   (I)   (II)

FIGURE 17

47. Sketch $c(t) = (t^3 - 4t, t^2)$ following the steps in Example 7.

48. Sketch $c(t) = (t^2 - 4t, 9 - t^2)$ for $-4 \le t \le 10$.

In Exercises 49–54, use Eq. (8) to find dy/dx at the given point.

49. $(t^3, t^2 - 1)$, $t = -4$

50. $(2t + 9, 7t - 9)$, $t = 1$

51. $(s^{-1} - 3s, s^3)$, $s = -1$

52. $(\sin 2\theta, \cos 3\theta)$, $\theta = \frac{\pi}{6}$

53. $(\sin^3 \theta, \cos \theta)$, $\theta = \frac{\pi}{4}$

54. (e^t, t^2), $t = 1$

In Exercises 55–58, find an equation $y = f(x)$ for the parametric curve and compute dy/dx in two ways: using Eq. (8) and by differentiating $f(x)$.

55. $c(t) = (2t + 1, 1 - 9t)$

56. $c(t) = \left(\frac{1}{2}t, \frac{1}{4}t^2 - t\right)$

57. $x = s^3$, $y = s^6 + s^{-3}$

58. $x = \cos \theta$, $y = \cos \theta + \sin^2 \theta$

59. Find the points on the parametric curve $c(t) = (3t^2 - 2t, t^3 - 6t)$ where the tangent line has slope 3.

60. Find the equation of the tangent line to the cycloid generated by a circle of radius 4 at $t = \frac{\pi}{2}$.

In Exercises 61–64, let $c(t) = (t^2 - 9, t^2 - 8t)$ (see Figure 18).

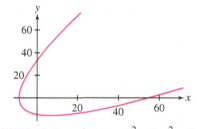

FIGURE 18 Plot of $c(t) = (t^2 - 9, t^2 - 8t)$.

61. Draw an arrow indicating the direction of motion, and determine the interval of t-values corresponding to the portion of the curve in each of the four quadrants.

62. Find the equation of the tangent line at $t = 4$.

63. Find the points where the tangent has slope $\frac{1}{2}$.

64. Find the points where the tangent is horizontal or vertical.

65. Let A and B be the points where the ray of angle θ intersects the two concentric circles of radii $r < R$ centered at the origin (Figure 19). Let P be the point of intersection of the horizontal line through A and the vertical line through B. Express the coordinates of P as a function of θ and describe the curve traced by P for $0 \leq \theta \leq 2\pi$.

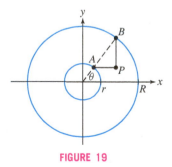

FIGURE 19

66. A 10-ft ladder slides down a wall as its bottom B is pulled away from the wall (Figure 20). Using the angle θ as a parameter, find the parametric equations for the path followed by (a) the top of the ladder A, (b) the bottom of the ladder B, and (c) the point P located 4 ft from the top of the ladder. Show that P describes an ellipse.

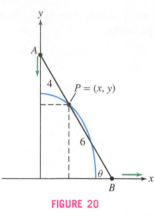

FIGURE 20

In Exercises 67–70, refer to the Bézier curve defined by Eqs. (9) and (10).

67. Show that the Bézier curve with control points
$$P_0 = (1, 4), \quad P_1 = (3, 12), \quad P_2 = (6, 15), \quad P_3 = (7, 4)$$
has parametrization
$$c(t) = (1 + 6t + 3t^2 - 3t^3, 4 + 24t - 15t^2 - 9t^3)$$
Verify that the slope at $t = 0$ is equal to the slope of the segment $\overline{P_0 P_1}$.

68. Find an equation of the tangent line to the Bézier curve in Exercise 67 at $t = \frac{1}{3}$.

69. CAS Find and plot the Bézier curve $c(t)$ with control points
$$P_0 = (3, 2), \quad P_1 = (0, 2), \quad P_2 = (5, 4), \quad P_3 = (2, 4)$$

70. Show that a cubic Bézier curve is tangent to the segment $\overline{P_2 P_3}$ at P_3.

71. A bullet fired from a gun follows the trajectory
$$x = at, \qquad y = bt - 16t^2 \quad (a, b > 0)$$
Show that the bullet leaves the gun at an angle $\theta = \tan^{-1}\left(\frac{b}{a}\right)$ and lands at a distance $\dfrac{ab}{16}$ from the origin.

72. CAS Plot $c(t) = (t^3 - 4t, t^4 - 12t^2 + 48)$ for $-3 \leq t \leq 3$. Find the points where the tangent line is horizontal or vertical.

73. CAS Plot the astroid $x = \cos^3 \theta$, $y = \sin^3 \theta$ and find the equation of the tangent line at $\theta = \frac{\pi}{3}$.

74. Find the equation of the tangent line at $t = \frac{\pi}{4}$ to the cycloid generated by the unit circle with parametric equation (6).

75. Find the points with a horizontal tangent line on the cycloid with parametric equation (6).

76. Property of the Cycloid Prove that the tangent line at a point P on the cycloid always passes through the top point on the rolling circle as indicated in Figure 21. Assume the generating circle of the cycloid has radius 1.

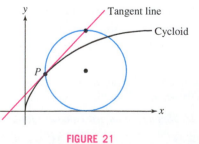

FIGURE 21

77. A *curtate cycloid* (Figure 22) is the curve traced by a point at a distance h from the center of a circle of radius R rolling along the x-axis where $h < R$. Show that this curve has parametric equations $x = Rt - h \sin t$, $y = R - h \cos t$.

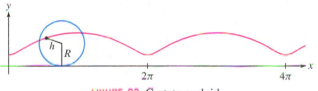

FIGURE 22 Curtate cycloid.

78. $\boxed{CAS}$ Use a computer algebra system to explore what happens when $h > R$ in the parametric equations of Exercise 77. Describe the result.

79. Show that the line of slope t through $(-1, 0)$ intersects the unit circle in the point with coordinates

$$x = \frac{1 - t^2}{t^2 + 1}, \qquad y = \frac{2t}{t^2 + 1} \qquad \boxed{12}$$

Conclude that these equations parametrize the unit circle with the point $(-1, 0)$ excluded (Figure 23). Show further that $t = y/(x + 1)$.

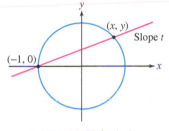

FIGURE 23 Unit circle.

80. The **folium of Descartes** is the curve with equation $x^3 + y^3 = 3axy$, where $a \neq 0$ is a constant (Figure 24).

(a) Show that the line $y = tx$ intersects the folium at the origin and at one other point P for all $t \neq -1, 0$. Express the coordinates of P in terms of t to obtain a parametrization of the folium. Indicate the direction of the parametrization on the graph.

(b) Describe the interval of t-values parametrizing the parts of the curve in quadrants I, II, and IV. Note that $t = -1$ is a point of discontinuity of the parametrization.

(c) Calculate dy/dx as a function of t and find the points with horizontal or vertical tangent.

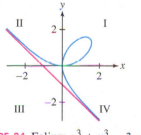

FIGURE 24 Folium $x^3 + y^3 = 3axy$.

81. Use the results of Exercise 80 to show that the asymptote of the folium is the line $x + y = -a$. *Hint:* Show that $\lim_{t \to -1} (x + y) = -a$.

82. Find a parametrization of $x^{2n+1} + y^{2n+1} = ax^n y^n$, where a and n are constants.

83. Second Derivative for a Parametrized Curve Given a parametrized curve $c(t) = (x(t), y(t))$, show that

$$\frac{d}{dt}\left(\frac{dy}{dx}\right) = \frac{x'(t)y''(t) - y'(t)x''(t)}{x'(t)^2}$$

Use this to prove the formula

$$\boxed{\frac{d^2y}{dx^2} = \frac{x'(t)y''(t) - y'(t)x''(t)}{x'(t)^3}} \qquad \boxed{13}$$

84. The second derivative of $y = x^2$ is $dy^2/d^2x = 2$. Verify that Eq. (13) applied to $c(t) = (t, t^2)$ yields $dy^2/d^2x = 2$. In fact, any parametrization may be used. Check that $c(t) = (t^3, t^6)$ and $c(t) = (\tan t, \tan^2 t)$ also yield $dy^2/d^2x = 2$.

In Exercises 85–88, use Eq. (13) to find d^2y/dx^2.

85. $x = t^3 + t^2$, $\quad y = 7t^2 - 4$, $\quad t = 2$

86. $x = s^{-1} + s$, $\quad y = 4 - s^{-2}$, $\quad s = 1$

87. $x = 8t + 9$, $\quad y = 1 - 4t$, $\quad t = -3$

88. $x = \cos\theta$, $\quad y = \sin\theta$, $\quad \theta = \frac{\pi}{4}$

89. Use Eq. (13) to find the t-intervals on which $c(t) = (t^2, t^3 - 4t)$ is concave up.

90. Use Eq. (13) to find the t-intervals on which $c(t) = (t^2, t^4 - 4t)$ is concave up.

91. Calculate the area under $y = x^2$ over $[0, 1]$ using Eq. (11) with the parametrizations (t^3, t^6) and (t^2, t^4).

92. What does Eq. (11) say if $c(t) = (t, f(t))$?

93. Consider the curve $c(t) = (t^2, t^3)$ for $0 \leq t \leq 1$.
(a) Find the area under the curve using Eq. (11).
(b) Find the area under the curve by expressing y as a function of x and finding the area using the standard method.

94. Compute the area under the parametrized curve $c(t) = (e^t, t)$ for $0 \leq t \leq 1$ using Eq. (11).

95. Compute the area under the parametrized curve given by $c(t) = (\sin t, \cos^2 t)$ for $0 \leq t \leq \pi/2$ using Eq. (11).

96. Sketch the graph of $c(t) = (\ln t, 2 - t)$ for $1 \leq t \leq 2$ and compute the area under the graph using Eq. (11).

97. Galileo tried unsuccessfully to find the area under a cycloid. Around 1630, Gilles de Roberval proved that the area under one arch of the cycloid $c(t) = (Rt - R\sin t, R - R\cos t)$ generated by a circle of radius R is equal to three times the area of the circle (Figure 25). Verify Roberval's result using Eq. (11).

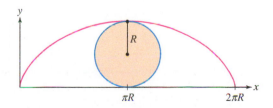
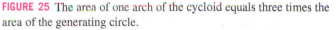

FIGURE 25 The area of one arch of the cycloid equals three times the area of the generating circle.

Further Insights and Challenges

98. Prove the following generalization of Exercise 97: For all $t > 0$, the area of the cycloidal sector OPC is equal to three times the area of the circular segment cut by the chord PC in Figure 26.

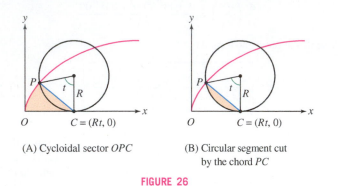

(A) Cycloidal sector OPC

(B) Circular segment cut by the chord PC

FIGURE 26

99. Derive the formula for the slope of the tangent line to a parametric curve $c(t) = (x(t), y(t))$ using a method different from that presented in the text. Assume that $x'(t_0)$ and $y'(t_0)$ exist and $x'(t_0) \neq 0$. Show that

$$\lim_{h \to 0} \frac{y(t_0 + h) - y(t_0)}{x(t_0 + h) - x(t_0)} = \frac{y'(t_0)}{x'(t_0)}$$

Then explain why this limit is equal to the slope dy/dx. Draw a diagram showing that the ratio in the limit is the slope of a secant line.

100. Verify that the **tractrix** curve ($\ell > 0$)

$$c(t) = \left(t - \ell \tanh \frac{t}{\ell}, \ell \operatorname{sech} \frac{t}{\ell}\right)$$

has the following property: For all t, the segment from $c(t)$ to $(t, 0)$ is tangent to the curve and has length ℓ (Figure 27).

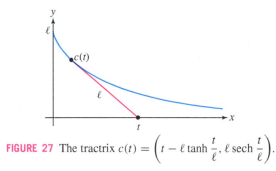

FIGURE 27 The tractrix $c(t) = \left(t - \ell \tanh \dfrac{t}{\ell}, \ell \operatorname{sech} \dfrac{t}{\ell}\right)$.

101. In Exercise 59 of Section 9.1, we described the tractrix by the differential equation

$$\frac{dy}{dx} = -\frac{y}{\sqrt{\ell^2 - y^2}}$$

Show that the parametric curve $c(t)$ identified as the tractrix in Exercise 100 satisfies this differential equation. Note that the derivative on the left is taken with respect to x, not t.

In Exercises 102 and 103, refer to Figure 28.

102. In the parametrization $c(t) = (a \cos t, b \sin t)$ of an ellipse, t is *not* an angular parameter unless $a = b$ (in which case, the ellipse is a circle). However, t can be interpreted in terms of area: Show that if $c(t) = (x, y)$, then $t = (2/ab)A$, where A is the area of the shaded region in Figure 28. *Hint:* Use Eq. (11).

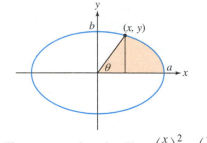

FIGURE 28 The parameter θ on the ellipse $\left(\dfrac{x}{a}\right)^2 + \left(\dfrac{y}{b}\right)^2 = 1$.

103. Show that the parametrization of the ellipse by the angle θ is

$$x = \frac{ab \cos \theta}{\sqrt{a^2 \sin^2 \theta + b^2 \cos^2 \theta}}$$

$$y = \frac{ab \sin \theta}{\sqrt{a^2 \sin^2 \theta + b^2 \cos^2 \theta}}$$

11.2 Arc Length and Speed

We now derive a formula for the arc length of a curve in parametric form. Recall that in Section 8.1, arc length was defined as the limit of the lengths of polygonal approximations (Figure 1).

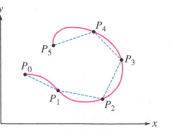

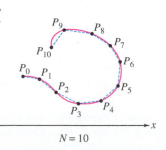

DF **FIGURE 1** Polygonal approximations for $N = 5$ and $N = 10$.

$N = 5$

$N = 10$

Given a parametrization $c(t) = (x(t), y(t))$ for $a \leq t \leq b$, we construct a polygonal approximation L consisting of the N segments by joining points

$$P_0 = c(t_0), \quad P_1 = c(t_1), \quad \ldots, \quad P_N = c(t_N)$$

corresponding to a choice of values $t_0 = a < t_1 < t_2 < \cdots < t_N = b$. By the distance formula,

$$P_{i-1}P_i = \sqrt{\left(x(t_i) - x(t_{i-1})\right)^2 + \left(y(t_i) - y(t_{i-1})\right)^2} \qquad \boxed{1}$$

Now assume that $x(t)$ and $y(t)$ are differentiable. According to the Mean Value Theorem, there are values t_i^* and t_i^{**} in the interval $[t_{i-1}, t_i]$ such that

$$x(t_i) - x(t_{i-1}) = x'(t_i^*)\Delta t_i, \qquad y(t_i) - y(t_{i-1}) = y'(t_i^{**})\Delta t_i$$

where $\Delta t_i = t_i - t_{i-1}$, and therefore,

$$P_{i-1}P_i = \sqrt{x'(t_i^*)^2 \Delta t_i^2 + y'(t_i^{**})^2 \Delta t_i^2} = \sqrt{x'(t_i^*)^2 + y'(t_i^{**})^2}\, \Delta t_i$$

The length of the polygonal approximation L is equal to the sum

$$\sum_{i=1}^{N} P_{i-1}P_i = \sum_{i=1}^{N} \sqrt{x'(t_i^*)^2 + y'(t_i^{**})^2}\, \Delta t_i \qquad \boxed{2}$$

This is *nearly* a Riemann sum for the function $\sqrt{x'(t)^2 + y'(t)^2}$. It would be a true Riemann sum if the intermediate values t_i^* and t_i^{**} were equal. Although they are not necessarily equal, it can be shown (and we will take for granted) that if $x'(t)$ and $y'(t)$ are continuous, then the sum in Eq. (2) still approaches the integral as the widths Δt_i tend to 0. Thus,

$$s = \lim_{\Delta t_i \to 0} \sum_{i=1}^{N} P_{i-1}P_i = \int_a^b \sqrt{x'(t)^2 + y'(t)^2}\, dt$$

Because of the square root, the arc length integral cannot be evaluated explicitly except in special cases, but we can always approximate it numerically.

THEOREM 1 Arc Length Let $c(t) = (x(t), y(t))$, where $x'(t)$ and $y'(t)$ exist and are continuous. Then the arc length s of $c(t)$ for $a \leq t \leq b$ is equal to

$$s = \int_a^b \sqrt{x'(t)^2 + y'(t)^2}\, dt \qquad \boxed{3}$$

The graph of a function $y = f(x)$ has parametrization $c(t) = (t, f(t))$. In this case,

$$\sqrt{x'(t)^2 + y'(t)^2} = \sqrt{1 + f'(t)^2}$$

and Eq. (3) reduces to the arc length formula derived in Section 8.1.

As mentioned above, the arc length integral can be evaluated explicitly only in special cases. The circle and the cycloid are two such cases.

■ **EXAMPLE 1** Use Eq. (3) to calculate the arc length of a circle of radius R.

Solution With the parametrization $x = R\cos\theta$, $y = R\sin\theta$,

$$x'(\theta)^2 + y'(\theta)^2 = (-R\sin\theta)^2 + (R\cos\theta)^2 = R^2(\sin^2\theta + \cos^2\theta) = R^2$$

We obtain the expected result:

$$s = \int_0^{2\pi} \sqrt{x'(\theta)^2 + y'(\theta)^2}\, d\theta = \int_0^{2\pi} R\, d\theta = 2\pi R \qquad \blacksquare$$

■ **EXAMPLE 2** Find the arc length of the curve given in parametric form by $c(t) = (t^2, t^3)$ for $0 \leq t \leq 1$.

Solution The arc length of this curve is given by

$$s = \int_0^1 \sqrt{x'(t)^2 + y'(t)^2}\, dt = \int_0^1 \sqrt{(2t)^2 + (3t^2)^2}\, dt$$

$$= \int_0^1 t\sqrt{4 + 9t^2}\, dt$$

Letting $u = 4 + 9t^2$, and therefore $du = 18t\, dt$, we obtain

$$s = \frac{1}{18}\int_4^{13} \sqrt{u}\, du = \frac{2}{3}\frac{u^{\frac{3}{2}}}{18}\Big|_4^{13} = \frac{1}{27}(13^{\frac{3}{2}} - (4)^{\frac{3}{2}}) \approx 1.4397$$ ■

■ **EXAMPLE 3** **Length of the Cycloid** Calculate the length s of one arch of the cycloid generated by a circle of radius $R = 2$ (Figure 2).

Solution We use the parametrization of the cycloid in Eq. (7) of Section 11.1:

$$x(t) = 2(t - \sin t), \qquad y(t) = 2(1 - \cos t)$$
$$x'(t) = 2(1 - \cos t), \qquad y'(t) = 2 \sin t$$

Thus,

$$x'(t)^2 + y'(t)^2 = 2^2(1 - \cos t)^2 + 2^2 \sin^2 t$$

$$= 4 - 8\cos t + 4\cos^2 t + 4\sin^2 t$$

$$= 8 - 8\cos t$$

$$= 16\sin^2 \frac{t}{2} \qquad \text{(Use the identity recalled in the margin.)}$$

One arch of the cycloid is traced as t varies from 0 to 2π, so

$$s = \int_0^{2\pi} \sqrt{x'(t)^2 + y'(t)^2}\, dt = \int_0^{2\pi} 4\sin\frac{t}{2}\, dt = -8\cos\frac{t}{2}\Big|_0^{2\pi} = -8(-1) + 8 = 16$$

Note that because $\sin\frac{t}{2} \geq 0$ for $0 \leq t \leq 2\pi$, we did not need an absolute value when taking the square root of $16\sin^2\frac{t}{2}$. ■

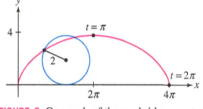

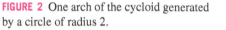

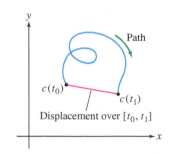

FIGURE 2 One arch of the cycloid generated by a circle of radius 2.

◄·· **REMINDER**

$$\frac{1 - \cos t}{2} = \sin^2 \frac{t}{2}$$

In Chapter 13, we will discuss not just the speed but also the velocity of a particle moving along a curved path. Velocity is "speed plus direction" and is represented by a vector.

Now consider a particle moving along a path $c(t)$. The distance traveled by the particle over the time interval $[t_0, t]$ is given by the arc length integral:

$$s(t) = \int_{t_0}^t \sqrt{x'(u)^2 + y'(u)^2}\, du$$

On the other hand, speed is defined as the rate of change of distance traveled with respect to time, so by the Fundamental Theorem of Calculus,

$$\text{Speed} = \frac{ds}{dt} = \frac{d}{dt}\int_{t_0}^t \sqrt{x'(u)^2 + y'(u)^2}\, du = \sqrt{x'(t)^2 + y'(t)^2}$$

THEOREM 2 **Speed Along a Parametrized Path** The speed of $c(t) = (x(t), y(t))$ is

$$\boxed{\text{Speed} = \frac{ds}{dt} = \sqrt{x'(t)^2 + y'(t)^2}}$$

FIGURE 3 The distance along the path is greater than or equal to the displacement.

The next example illustrates the difference between distance traveled along a path and **displacement** (also called the net change in position). The displacement along a path is the distance between the initial point $c(t_0)$ and the endpoint $c(t_1)$. The distance traveled is greater than the displacement unless the particle happens to move in a straight line (Figure 3).

■ **EXAMPLE 4**   A particle travels along the path $c(t) = (2t, 1 + t^{3/2})$. Find:

(a) The particle's speed at $t = 1$ (assume units of meters and minutes).

(b) The distance traveled s and displacement d during the interval $0 \le t \le 4$.

Solution   We have

$$x'(t) = 2, \qquad y'(t) = \frac{3}{2}t^{1/2}$$

The speed at time t is

$$s'(t) = \sqrt{x'(t)^2 + y'(t)^2} = \sqrt{4 + \frac{9}{4}t} \quad \text{m/min}$$

(a) The particle's speed at $t = 1$ is $s'(1) = \sqrt{4 + \frac{9}{4}} = 2.5$ m/min.

(b) The distance traveled in the first 4 min is

$$s = \int_0^4 \sqrt{4 + \frac{9}{4}t}\, dt = \frac{8}{27}\left(4 + \frac{9}{4}t\right)^{3/2}\Big|_0^4 = \frac{8}{27}(13^{3/2} - 8) \approx 11.52 \text{ m}$$

The displacement d is the distance from the initial point $c(0) = (0, 1)$ to the endpoint $c(4) = (8, 1 + 4^{3/2}) = (8, 9)$ (see Figure 4):

$$d = \sqrt{(8 - 0)^2 + (9 - 1)^2} = 8\sqrt{2} \approx 11.31 \text{ m} \qquad ■$$

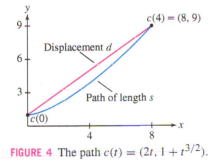

FIGURE 4 The path $c(t) = (2t, 1 + t^{3/2})$.

In physics, we often describe the path of a particle moving with constant speed along a circle of radius R in terms of a constant ω (lowercase Greek omega) as follows:

$$c(t) = (R \cos \omega t, R \sin \omega t)$$

The constant ω, called the *angular velocity*, is the rate of change with respect to time of the particle's angle θ (Figure 5).

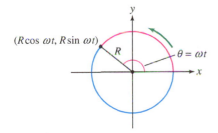

FIGURE 5 A particle moving on a circle of radius R with angular velocity ω has speed $|\omega|R$.

■ **EXAMPLE 5   Angular Velocity**   Calculate the speed of the circular path of radius R and angular velocity ω. What is the speed if $R = 3$ m and $\omega = 4$ radians per second (rad/s)?

Solution   We have $x = R \cos \omega t$ and $y = R \sin \omega t$, and

$$x'(t) = -\omega R \sin \omega t, \qquad y'(t) = \omega R \cos \omega t$$

The particle's speed is

$$\frac{ds}{dt} = \sqrt{x'(t)^2 + y'(t)^2} = \sqrt{(-\omega R \sin \omega t)^2 + (\omega R \cos \omega t)^2}$$

$$= \sqrt{\omega^2 R^2(\sin^2 \omega t + \cos^2 \omega t)} = |\omega|R$$

Thus, the speed is constant with value $|\omega|R$. If $R = 3$ m and $\omega = 4$ rad/s, then the speed is $|\omega|R = 3(4) = 12$ m/s. ■

Consider the surface obtained by rotating a parametric curve $c(t) = (x(t), y(t))$ about the x-axis. The surface area is given by Eq. (4) in the next theorem. It can be derived in much the same way as the formula for a surface of revolution of a graph $y = f(x)$ in Section 8.1. In this theorem, we assume that $y(t) \ge 0$ so the parametric curve $c(t)$ lies above the x-axis, and that $x(t)$ is increasing so the curve does not reverse direction.

THEOREM 3   Surface Area   Let $c(t) = (x(t), y(t))$, where $y(t) \ge 0$, $x(t)$ is increasing, and $x'(t)$ and $y'(t)$ are continuous. Then the surface obtained by rotating $c(t)$ about the x-axis for $a \le t \le b$ has surface area

$$S = 2\pi \int_a^b y(t)\sqrt{x'(t)^2 + y'(t)^2}\, dt \qquad \boxed{4}$$

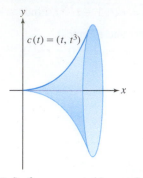

FIGURE 6 Surface generated by revolving the curve about the x-axis.

■ **EXAMPLE 6** Calculate the surface area of the surface obtained by rotating the parametric curve $c(t) = (t, t^3)$ about the x-axis for $0 \leq t \leq 1$. The surface appears as in Figure 6.

Solution We have $x'(t) = 1$ and $y'(t) = 3t^2$.
Therefore,

$$S = 2\pi \int_0^1 t^3 \sqrt{1 + (3t^2)^2}\, dt = 2\pi \int_0^1 t^3 \sqrt{1 + 9t^4}\, dt$$

With the substitution $u = 1 + 9t^4$ and $du = 36t^3\, dt$, we obtain

$$S = 2\pi \frac{1}{36} \int_1^{10} \sqrt{u}\, du = \frac{\pi}{18} \left(\frac{2}{3}\right) u^{\frac{3}{2}} \Big|_1^{10} = \frac{\pi}{27} (10^{\frac{3}{2}} - 1) \approx 3.5631 \qquad ■$$

11.2 SUMMARY

- Arc length of $c(t) = (x(t), y(t))$ for $a \leq t \leq b$:

$$s = \text{arc length} = \int_a^b \sqrt{x'(t)^2 + y'(t)^2}\, dt$$

- The arc length is the distance along the path $c(t)$. The *displacement* is the distance from the starting point $c(a)$ to the endpoint $c(b)$.
- Arc length integral as a function of t:

$$s(t) = \int_{t_0}^t \sqrt{x'(u)^2 + y'(u)^2}\, du$$

- Speed at time t:

$$\frac{ds}{dt} = \sqrt{x'(t)^2 + y'(t)^2}$$

- Surface area of the surface obtained by rotating $c(t) = (x(t), y(t))$ about the x-axis for $a \leq t \leq b$:

$$S = 2\pi \int_a^b y(t) \sqrt{x'(t)^2 + y'(t)^2}\, dt$$

11.2 EXERCISES

Preliminary Questions

1. What is the definition of arc length?

2. Can the distance traveled by a particle ever be less than its displacement? When are they equal?

3. What is the interpretation of $\sqrt{x'(t)^2 + y'(t)^2}$ for a particle following the trajectory $(x(t), y(t))$?

4. A particle travels along a path from $(0, 0)$ to $(3, 4)$. What is the displacement? Can the distance traveled be determined from the information given?

5. A particle traverses the parabola $y = x^2$ with constant speed 3 cm/s. What is the distance traveled during the first minute? *Hint:* Only simple computation is necessary.

6. If the straight line segment given by $c(t) = (3, t)$ for $0 \leq t \leq 2$ is rotated around the x-axis, what surface area results? *Hint:* Only simple computation is necessary.

Exercises

In Exercises 1–10, use Eq. (3) to find the length of the path over the given interval.

1. $(3t + 1, 9 - 4t)$, $0 \leq t \leq 2$

2. $(1 + 2t, 2 + 4t)$, $1 \leq t \leq 4$

3. $(2t^2, 3t^2 - 1)$, $0 \leq t \leq 4$

4. $(3t, 4t^{3/2})$, $0 \leq t \leq 1$

5. $(3t^2, 4t^3)$, $1 \leq t \leq 4$

6. $(t^3 + 1, t^2 - 3)$, $0 \leq t \leq 1$

7. $(\sin 3t, \cos 3t)$, $0 \leq t \leq \pi$

8. $(\sin\theta - \theta\cos\theta, \cos\theta + \theta\sin\theta)$, $0 \leq \theta \leq 2$

In Exercises 9 and 10, use the identity

$$\frac{1 - \cos t}{2} = \sin^2 \frac{t}{2}$$

9. $(2\cos t - \cos 2t,\ 2\sin t - \sin 2t),\quad 0 \le t \le \frac{\pi}{2}$

10. $(5(\theta - \sin\theta),\ 5(1 - \cos\theta)),\quad 0 \le \theta \le 2\pi$

11. Show that one arch of a cycloid generated by a circle of radius R has length $8R$.

12. Find the length of the spiral $c(t) = (t\cos t,\ t\sin t)$ for $0 \le t \le 2\pi$ to three decimal places (Figure 7). *Hint:* Use the formula

$$\int \sqrt{1 + t^2}\, dt = \frac{1}{2} t\sqrt{1 + t^2} + \frac{1}{2}\ln\left(t + \sqrt{1 + t^2}\right)$$

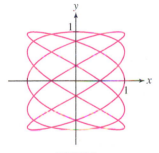

FIGURE 7 The spiral $c(t) = (t\cos t,\ t\sin t)$.

13. Find the length of the parabola given by $c(t) = (t, t^2)$ for $0 \le t \le 1$. See the hint for Exercise 12.

14. *CAS* Find a numerical approximation to the length of $c(t) = (\cos 5t,\ \sin 3t)$ for $0 \le t \le 2\pi$ (Figure 8).

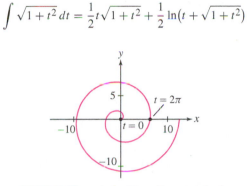

FIGURE 8

In Exercises 15–20, determine the speed $\frac{ds}{dt}$ at time t (assume units of meters and seconds).

15. $(t^3, t^2),\quad t = 2$

16. $(3\sin 5t,\ 8\cos 5t),\quad t = \frac{\pi}{4}$

17. $(5t + 1,\ 4t - 3),\quad t = 9$

18. $(\ln(t^2 + 1),\ t^3),\quad t = 1$

19. $(t^2, e^t),\quad t = 0$

20. $(\sin^{-1} t,\ \tan^{-1} t),\quad t = 0$

21. Find the minimum speed of a particle with trajectory $c(t) = (t^3 - 4t,\ t^2 + 1)$ for $t \ge 0$. *Hint:* It is easier to find the minimum of the square of the speed.

22. Find the minimum speed of a particle with trajectory $c(t) = (t^3, t^{-2})$ for $t \ge 0.5$.

23. Find the speed of the cycloid $c(t) = (4t - 4\sin t,\ 4 - 4\cos t)$ at points where the tangent line is horizontal.

24. Calculate the arc length integral $s(t)$ for the *logarithmic spiral* $c(t) = (e^t \cos t,\ e^t \sin t)$.

CAS *In Exercises 25–28, plot the curve and use the Midpoint Rule with $N = 10, 20, 30$, and 50 to approximate its length.*

25. $c(t) = (\cos t,\ e^{\sin t})$ for $0 \le t \le 2\pi$

26. $c(t) = (t - \sin 2t,\ 1 - \cos 2t)$ for $0 \le t \le 2\pi$

27. The ellipse $\left(\frac{x}{5}\right)^2 + \left(\frac{y}{3}\right)^2 = 1$

28. $x = \sin 2t,\quad y = \sin 3t$ for $0 \le t \le 2\pi$

29. If you unwind thread from a stationary circular spool, keeping the thread taut at all times, then the endpoint traces a curve $\mathcal{C}$ called the **involute** of the circle (Figure 9). Observe that $\overline{PQ}$ has length $R\theta$. Show that $\mathcal{C}$ is parametrized by

$$c(\theta) = \left(R(\cos\theta + \theta\sin\theta),\qquad R(\sin\theta - \theta\cos\theta)\right)$$

Then find the length of the involute for $0 \le \theta \le 2\pi$.

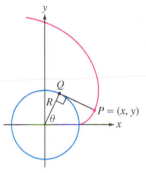

FIGURE 9 Involute of a circle.

30. Let $a > b$ and set

$$k = \sqrt{1 - \frac{b^2}{a^2}}$$

Use a parametric representation to show that the ellipse $\left(\frac{x}{a}\right)^2 + \left(\frac{y}{b}\right)^2 = 1$ has length $L = 4aG\left(\frac{\pi}{2}, k\right)$, where

$$G(\theta, k) = \int_0^\theta \sqrt{1 - k^2 \sin^2 t}\, dt$$

is the *elliptic integral of the second kind*.

In Exercises 31–38, use Eq. (4) to compute the surface area of the given surface.

31. The cone generated by revolving $c(t) = (t, mt)$ about the x-axis for $0 \le t \le A$.

32. A sphere of radius R.

33. The surface generated by revolving the curve $c(t) = (t^2, t)$ about the x-axis for $0 \le t \le 1$.

34. The surface generated by revolving the curve $c(t) = (t, e^t)$ about the x-axis for $0 \le t \le 1$.

35. The surface generated by revolving the curve $c(t) = (\sin^2 t,\ \cos^2 t)$ about the x-axis for $0 \le t \le \frac{\pi}{2}$.

36. The surface generated by revolving the curve $c(t) = (t, \sin t)$ about the x-axis for $0 \le t \le 2\pi$.

37. The surface generated by revolving one arch of the cycloid $c(t) = (t - \sin t, 1 - \cos t)$ about the x-axis

38. The surface generated by revolving the astroid $c(t) = (\cos^3 t, \sin^3 t)$ about the x-axis for $0 \le t \le \frac{\pi}{2}$

Further Insights and Challenges

39. $\mathcal{CAS}$ Let $b(t)$ be the "Butterfly Curve":

$$x(t) = \sin t \left(e^{\cos t} - 2 \cos 4t - \sin\left(\frac{t}{12}\right)^5 \right)$$

$$y(t) = \cos t \left(e^{\cos t} - 2 \cos 4t - \sin\left(\frac{t}{12}\right)^5 \right)$$

(a) Use a computer algebra system to plot $b(t)$ and the speed $s'(t)$ for $0 \le t \le 12\pi$.

(b) Approximate the length $b(t)$ for $0 \le t \le 10\pi$.

40. $\mathcal{CAS}$ Let $a \ge b > 0$ and set $k = \dfrac{2\sqrt{ab}}{a - b}$. Show that the **trochoid**

$$x = at - b \sin t, \qquad y = a - b \cos t, \quad 0 \le t \le T$$

has length $2(a - b)G\left(\frac{T}{2}, k\right)$, with $G(\theta, k)$ as in Exercise 30.

41. A satellite orbiting at a distance R from the center of the earth follows the circular path $x(t) = R \cos \omega t$, $y(t) = R \sin \omega t$.

(a) Show that the period T (the time of one revolution) is $T = 2\pi/\omega$.

(b) According to Newton's Laws of Motion and Gravity,

$$x''(t) = -Gm_e \frac{x}{R^3}, \qquad y''(t) = -Gm_e \frac{y}{R^3}$$

where G is the universal gravitational constant and m_e is the mass of the earth. Prove that $R^3/T^2 = Gm_e/4\pi^2$. Thus, R^3/T^2 has the same value for all orbits (a special case of Kepler's Third Law).

42. The acceleration due to gravity on the surface of the earth is

$$g = \frac{Gm_e}{R_e^2} = 9.8 \text{ m/s}^2, \quad \text{where } R_e = 6378 \text{ km}$$

Use Exercise 41(b) to show that a satellite orbiting at the earth's surface would have period $T_e = 2\pi \sqrt{R_e/g} \approx 84.5$ min. Then estimate the distance R_m from the moon to the center of the earth. Assume that the period of the moon (sidereal month) is $T_m \approx 27.43$ days.

Polar coordinates are appropriate when distance from the origin or angle plays a role. For example, the gravitational force exerted on a planet by the sun depends only on the distance r from the sun and is conveniently described in polar coordinates.

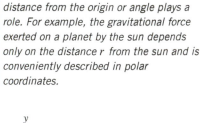
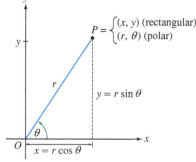

 FIGURE 1

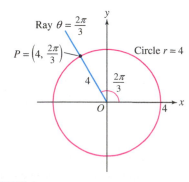

FIGURE 2

11.3 Polar Coordinates

The rectangular coordinates that we have utilized up to now provide a useful way to represent points in the plane. However, there are a variety of situations where a different coordinate system is more natural. In polar coordinates, we label a point P by coordinates (r, θ), where r is the distance to the origin O and θ is the angle between $\overline{OP}$ and the positive x-axis (Figure 1). By convention, an angle is positive if the corresponding rotation is counterclockwise. We call r the **radial coordinate** and θ the **angular coordinate**.

The point P in Figure 2 has polar coordinates $(r, \theta) = \left(4, \frac{2\pi}{3}\right)$. It is located at distance $r = 4$ from the origin (so it lies on the circle of radius 4), and it lies on the ray of angle $\theta = \frac{2\pi}{3}$. Notice that it can also be described by $(r, \theta) = \left(4, \frac{-4\pi}{3}\right)$. Unlike Cartesian coordinates, polar coordinates are not unique, as we will discuss in more detail shortly.

Figure 3 shows the two families of **grid lines** in polar coordinates:

$$\text{Circle centered at } O \quad \longleftrightarrow \quad r = \text{constant}$$

$$\text{Ray starting at } O \quad \longleftrightarrow \quad \theta = \text{constant}$$

Every point in the plane other than the origin lies at the intersection of the two grid lines and these two grid lines determine its polar coordinates. For example, point Q in Figure 3 lies on the circle $r = 3$ and the ray $\theta = \frac{5\pi}{6}$, so $Q = \left(3, \frac{5\pi}{6}\right)$ in polar coordinates.

Figure 1 shows that polar and rectangular coordinates are related by the equations $x = r \cos \theta$ and $y = r \sin \theta$. On the other hand, $r^2 = x^2 + y^2$ by the distance formula, and $\tan \theta = y/x$ if $x \ne 0$. This yields the conversion formulas:

Polar to Rectangular	Rectangular to Polar
$x = r \cos \theta$	$r = \sqrt{x^2 + y^2}$
$y = r \sin \theta$	$\tan \theta = \dfrac{y}{x} \quad (x \ne 0)$

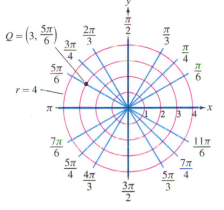

$Q = \left(3, \frac{5\pi}{6}\right)$

$r = 4$

FIGURE 3 Grid lines in polar coordinates.

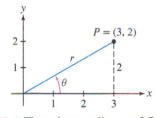

FIGURE 4 The polar coordinates of P satisfy $r = \sqrt{3^2 + 2^2}$ and $\tan\theta = \frac{2}{3}$.

By definition,

$$-\frac{\pi}{2} < \tan^{-1}x < \frac{\pi}{2}$$

If $r > 0$, a coordinate θ of $P = (x, y)$ is

$$\theta = \begin{cases} \tan^{-1}\dfrac{y}{x} & \text{if } x > 0 \\[2mm] \tan^{-1}\dfrac{y}{x} + \pi & \text{if } x < 0 \\[2mm] \pm\dfrac{\pi}{2} & \text{if } x = 0 \end{cases}$$

■ **EXAMPLE 1** **From Polar to Rectangular Coordinates** Find the rectangular coordinates of point Q in Figure 3.

Solution The point $Q = (r, \theta) = \left(3, \frac{5\pi}{6}\right)$ has rectangular coordinates:

$$x = r\cos\theta = 3\cos\left(\frac{5\pi}{6}\right) = 3\left(-\frac{\sqrt{3}}{2}\right) = -\frac{3\sqrt{3}}{2}$$

$$y = r\sin\theta = 3\sin\left(\frac{5\pi}{6}\right) = 3\left(\frac{1}{2}\right) = \frac{3}{2}$$

■ **EXAMPLE 2** **From Rectangular to Polar Coordinates** Find the polar coordinates of point P in Figure 4.

Solution Since $P = (x, y) = (3, 2)$,

$$r = \sqrt{x^2 + y^2} = \sqrt{3^2 + 2^2} = \sqrt{13} \approx 3.6$$

$$\tan\theta = \frac{y}{x} = \frac{2}{3}$$

and because P lies in the first quadrant,

$$\theta = \tan^{-1}\frac{y}{x} = \tan^{-1}\frac{2}{3} \approx 0.588$$

Thus, P has polar coordinates $(r, \theta) \approx (3.6, 0.588)$. ■

A few remarks are in order before proceeding:

- The angular coordinate is not unique because (r, θ) and $(r, \theta + 2\pi n)$ *label the same point* for any integer n. For instance, point P in Figure 5 has radial coordinate $r = 2$, but its angular coordinate can be any one of $\frac{\pi}{2}, \frac{5\pi}{2}, \ldots$ or $-\frac{3\pi}{2}, -\frac{7\pi}{2}, \ldots$.
- The origin O has no well-defined angular coordinate, so we assign to O the polar coordinates $(0, \theta)$ for any angle θ.

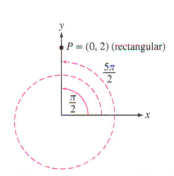

FIGURE 5 The angular coordinate of $P = (0, 2)$ is $\frac{\pi}{2}$ or any angle $\frac{\pi}{2} + 2\pi n$, where n is an integer.

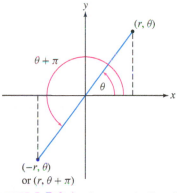

FIGURE 6 Relation between (r, θ) and $(-r, \theta)$.

- By convention, we allow *negative* radial coordinates. By definition, $(-r, \theta)$ is the reflection of (r, θ) through the origin (Figure 6). With this convention, $(-r, \theta)$ and $(r, \theta + \pi)$ represent the same point.
- We may specify unique polar coordinates for points other than the origin by placing restrictions on r and θ. We commonly choose $r > 0$ and $0 \le \theta < 2\pi$, but other choices are sometimes made.

When determining the angular coordinate of a point $P = (x, y)$, remember that there are two angles between 0 and 2π satisfying $\tan\theta = y/x$. You must choose θ so that (r, θ) lies in the quadrant containing P if you use $r > 0$ and in the opposite quadrant if you use $r < 0$ (Figure 7).

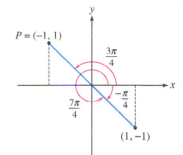

FIGURE 7

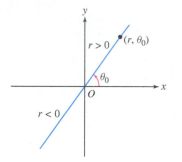

FIGURE 8 Lines through O with polar equation $\theta = \theta_0$.

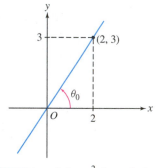

FIGURE 9 Line of slope $\frac{3}{2}$ through the origin.

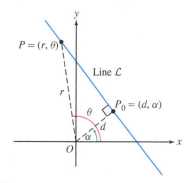

DF FIGURE 10 P_0 is the point on $\mathcal{L}$ closest to the origin.

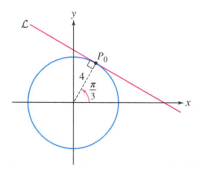

FIGURE 11 The tangent line has equation $r = 4\sec\left(\theta - \frac{\pi}{3}\right)$.

■ **EXAMPLE 3 Choosing θ Correctly** Find two polar representations of $P = (-1, 1)$, one with $r > 0$ and one with $r < 0$.

Solution The point $P = (x, y) = (-1, 1)$ has polar coordinates (r, θ), where

$$r = \sqrt{(-1)^2 + 1^2} = \sqrt{2}, \qquad \tan\theta = \tan\frac{y}{x} = -1$$

However, θ is not given by

$$\tan^{-1}\frac{y}{x} = \tan^{-1}\left(\frac{1}{-1}\right) = -\frac{\pi}{4}$$

Because $\theta = -\frac{\pi}{4}$, this would place P in the fourth quadrant (Figure 7). Since P is in the second quadrant, the correct angle is

$$\theta = \tan^{-1}\frac{y}{x} + \pi = -\frac{\pi}{4} + \pi = \frac{3\pi}{4}$$

If we wish to use the negative radial coordinate $r = -\sqrt{2}$, then the angle becomes $\theta = -\frac{\pi}{4}$ or $\frac{7\pi}{4}$. Thus,

$$P = \left(\sqrt{2}, \frac{3\pi}{4}\right) \qquad \text{or} \qquad \left(-\sqrt{2}, \frac{7\pi}{4}\right) \qquad \blacksquare$$

A curve is described in polar coordinates by an equation involving r and θ, which we call a **polar equation**. By convention, we allow solutions with $r < 0$.

A line through the origin O has the simple equation $\theta = \theta_0$, where θ_0 is the angle between the line and the x-axis (Figure 8). Indeed, the points with $\theta = \theta_0$ are (r, θ_0), where r is arbitrary (positive, negative, or zero).

■ **EXAMPLE 4 Line Through the Origin** Find the polar equation of the line through the origin of slope $\frac{3}{2}$ (Figure 9).

Solution A line of slope m makes an angle θ_0 with the x-axis, where $m = \tan\theta_0$. In our case, $\theta_0 = \tan^{-1}\frac{3}{2} \approx 0.98$. The equation of the line is $\theta = \tan^{-1}\frac{3}{2}$ or $\theta \approx 0.98$. ■

To describe lines that do not pass through the origin, we note that any such line has a unique point P_0 that is *closest* to the origin. The next example shows how to write down the polar equation of the line in terms of P_0 (Figure 10).

■ **EXAMPLE 5 Line Not Passing Through the Origin** Show that

$$\boxed{r = d\sec(\theta - \alpha)} \qquad \boxed{1}$$

is the polar equation of the line $\mathcal{L}$ whose point closest to the origin is $P_0 = (d, \alpha)$.

Solution The point P_0 is obtained by dropping a perpendicular from the origin to $\mathcal{L}$ (Figure 10), and if $P = (r, \theta)$ is any point on $\mathcal{L}$ other than P_0, then $\triangle OPP_0$ is a right triangle. Therefore, $d/r = \cos(\theta - \alpha)$, or $r = d\sec(\theta - \alpha)$, as claimed. ■

■ **EXAMPLE 6** Find the polar equation of the line $\mathcal{L}$ tangent to the circle $r = 4$ at the point with polar coordinates $P_0 = \left(4, \frac{\pi}{3}\right)$.

Solution The point on $\mathcal{L}$ closest to the origin is P_0 itself (Figure 11). Therefore, we take $(d, \alpha) = \left(4, \frac{\pi}{3}\right)$ in Eq. (1) to obtain the equation $r = 4\sec\left(\theta - \frac{\pi}{3}\right)$. ■

■ **EXAMPLE 7** Sketch the curve corresponding to $r = 1 + \sin\theta$.

Solution If we let θ vary from 0 to 2π, we see all possible values of the function, and then it will repeat. So, we consider values in the range from 0 to 2π.

	A	B	C	D	E	F	G	H
θ	0	$\dfrac{\pi}{4}$	$\dfrac{\pi}{2}$	$\dfrac{3\pi}{4}$	π	$\dfrac{5\pi}{4}$	$\dfrac{3\pi}{2}$	$\dfrac{7\pi}{4}$
$r = 1 + \sin\theta$	1	1.707	2	1.707	1	0.293	0	0.293

For each of the given angles, we plot the point as in Figure 12, and then we connect the points with a smooth curve. The resulting curve is called a *cardioid*, which is Greek for the "heart" that it resembles. ■

Often, it is hard to guess the shape of a graph of a polar equation. In some cases, it is helpful to rewrite the equation in rectangular coordinates.

■ **EXAMPLE 8** **Converting to Rectangular Coordinates** Identify the curve with polar equation $r = 2a\cos\theta$ (a a positive constant).

Solution Multiply the equation by r to obtain $r^2 = 2ar\cos\theta$. Because $r^2 = x^2 + y^2$ and $x = r\cos\theta$, this equation becomes

$$x^2 + y^2 = 2ax \qquad \text{or} \qquad x^2 - 2ax + y^2 = 0$$

Then complete the square to obtain $(x - a)^2 + y^2 = a^2$. This is the equation of the circle of radius a and center $(a, 0)$ (Figure 13). ■

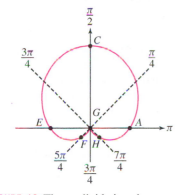

FIGURE 12 The cardioid given by $r = 1 + \sin\theta$.

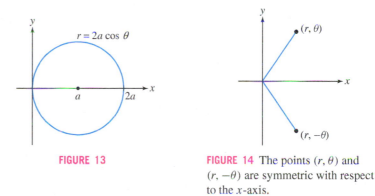

FIGURE 13

FIGURE 14 The points (r, θ) and $(r, -\theta)$ are symmetric with respect to the x-axis.

A similar calculation shows that the circle $x^2 + (y - a)^2 = a^2$ of radius a and center $(0, a)$ has polar equation $r = 2a\sin\theta$. In the next example, we make use of symmetry. Note that the points (r, θ) and $(r, -\theta)$ are symmetric with respect to the x-axis (Figure 14).

■ **EXAMPLE 9** **Symmetry About the x-Axis** Sketch the *limaçon* curve $r = 2\cos\theta - 1$.

Solution Since $f(\theta) = \cos\theta$ is periodic, it suffices to plot points for $-\pi \le \theta \le \pi$.

Step 1. Plot points.
To get started, we plot points A–G on a grid and join them by a smooth curve (Figure 15).

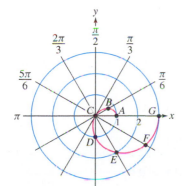

DF FIGURE 15 Plotting $r = 2\cos\theta - 1$ using a grid.

	A	B	C	D	E	F	G
θ	0	$\dfrac{\pi}{6}$	$\dfrac{\pi}{3}$	$\dfrac{\pi}{2}$	$\dfrac{2\pi}{3}$	$\dfrac{5\pi}{6}$	π
$r = 2\cos\theta - 1$	1	0.73	0	-1	-2	-2.73	-3

Step 2. **Analyze r as a function of θ.**

For a better understanding, it is helpful to graph r as a function of θ in rectangular coordinates. Figure 16(A) shows that

As θ varies from 0 to $\frac{\pi}{3}$, r varies from 1 to 0.

As θ varies from $\frac{\pi}{3}$ to π, r is *negative* and varies from 0 to -3.

We conclude:

- The graph begins at point A in Figure 16(B) and moves in toward the origin as θ varies from 0 to $\frac{\pi}{3}$.
- Since r is negative for $\frac{\pi}{3} \leq \theta \leq \pi$, the curve continues into the third and fourth quadrants (rather than into the first and second quadrants), moving toward the point $G = (-3, \pi)$ in Figure 16(C).

Step 3. **Use symmetry.**

Since $r(\theta) = r(-\theta)$, the curve is symmetric with respect to the x-axis. So, the part of the curve with $-\pi \leq \theta \leq 0$ is obtained by reflection through the x-axis as in Figure 16(D). ∎

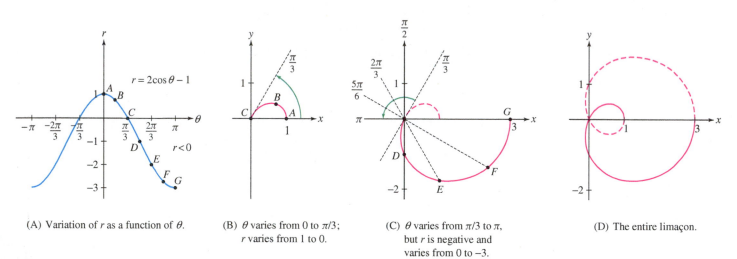

(A) Variation of r as a function of θ.

(B) θ varies from 0 to $\pi/3$; r varies from 1 to 0.

(C) θ varies from $\pi/3$ to π, but r is negative and varies from 0 to -3.

(D) The entire limaçon.

DF **FIGURE 16** The curve $r = 2\cos\theta - 1$ is called the *limaçon*, from the Latin word for "snail." It was first described in 1525 by the German artist Albrecht Dürer.

11.3 SUMMARY

- A point $P = (x, y)$ has polar coordinates (r, θ), where r is the distance to the origin and θ is the angle between the positive x-axis and the segment $\overline{OP}$, measured in the counterclockwise direction:

$$x = r\cos\theta, \qquad r = \sqrt{x^2 + y^2}$$

$$y = r\sin\theta, \qquad \tan\theta = \frac{y}{x} \quad (x \neq 0)$$

- The angular coordinate θ must be chosen so that (r, θ) lies in the proper quadrant. If $r > 0$, then

$$\theta = \begin{cases} \tan^{-1}\dfrac{y}{x} & \text{if } x > 0 \\[2mm] \tan^{-1}\dfrac{y}{x} + \pi & \text{if } x < 0 \\[2mm] \pm\dfrac{\pi}{2} & \text{if } x = 0 \end{cases}$$

- Nonuniqueness: (r, θ) and $(r, \theta + 2n\pi)$ represent the same point for all integers n. The origin O has polar coordinates $(0, \theta)$ for any θ.

• Negative radial coordinates: $(-r, \theta)$ and $(r, \theta + \pi)$ represent the same point.
• Polar equations:

Curve	Polar equation
Circle of radius R, center at the origin	$r = R$
Line through origin of slope $m = \tan\theta_0$	$\theta = \theta_0$
Line on which $P_0 = (d, \alpha)$ is the point closest to the origin	$r = d\sec(\theta - \alpha)$
Circle of radius a, center at $(a, 0)$ $(x-a)^2 + y^2 = a^2$	$r = 2a\cos\theta$
Circle of radius a, center at $(0, a)$ $x^2 + (y-a)^2 = a^2$	$r = 2a\sin\theta$

11.3 EXERCISES

Preliminary Questions

1. Points P and Q with the same radial coordinate (choose the correct answer):

(a) lie on the same circle with the center at the origin.

(b) lie on the same ray based at the origin.

2. Give two polar representations for the point $(x, y) = (0, 1)$, one with negative r and one with positive r.

3. Describe each of the following curves:

(a) $r = 2$ **(b)** $r^2 = 2$ **(c)** $r\cos\theta = 2$

4. If $f(-\theta) = f(\theta)$, then the curve $r = f(\theta)$ is symmetric with respect to the (choose the correct answer):

(a) x-axis.   **(b)** y-axis.   **(c)** origin.

Exercises

1. Find polar coordinates for each of the seven points plotted in Figure 17. (Choose $r \geq 0$ and θ in $[0, 2\pi)$.)

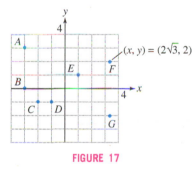

FIGURE 17

(c) $\left(-2, -\dfrac{3\pi}{2}\right)$ **(d)** $\left(-2, \dfrac{7\pi}{2}\right)$

(e) $\left(-2, -\dfrac{\pi}{2}\right)$ **(f)** $\left(2, -\dfrac{7\pi}{2}\right)$

7. Describe each shaded sector in Figure 18 by inequalities in r and θ.

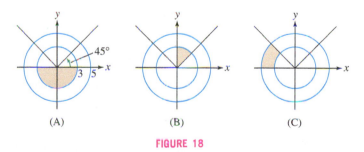

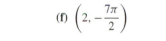

(A)   **(B)**   **(C)**

FIGURE 18

2. Plot the points with polar coordinates:

(a) $\left(2, \dfrac{\pi}{6}\right)$ **(b)** $\left(4, \dfrac{3\pi}{4}\right)$ **(c)** $\left(3, -\dfrac{\pi}{2}\right)$ **(d)** $\left(0, \dfrac{\pi}{6}\right)$

3. Convert from rectangular to polar coordinates:

(a) $(1, 0)$ **(b)** $(3, \sqrt{3})$ **(c)** $(-2, 2)$ **(d)** $(-1, \sqrt{3})$

4. Convert from rectangular to polar coordinates using a calculator (make sure your choice of θ gives the correct quadrant):

(a) $(2, 3)$ **(b)** $(4, -7)$ **(c)** $(-3, -8)$ **(d)** $(-5, 2)$

5. Convert from polar to rectangular coordinates:

(a) $\left(3, \dfrac{\pi}{6}\right)$ **(b)** $\left(6, \dfrac{3\pi}{4}\right)$ **(c)** $\left(0, \dfrac{\pi}{5}\right)$ **(d)** $\left(5, -\dfrac{\pi}{2}\right)$

6. Which of the following are possible polar coordinates for the point P with rectangular coordinates $(0, -2)$?

(a) $\left(2, \dfrac{\pi}{2}\right)$ **(b)** $\left(2, \dfrac{7\pi}{2}\right)$

8. Find the equation in polar coordinates of the line through the origin with slope $\frac{1}{2}$.

9. What is the slope of the line $\theta = \dfrac{3\pi}{5}$?

10. Which of $r = 2\sec\theta$ and $r = 2\csc\theta$ defines a horizontal line?

In Exercises 11–16, convert to an equation in rectangular coordinates.

11. $r = 7$ **12.** $r = \sin\theta$

13. $r = 2\sin\theta$ **14.** $r = 2\csc\theta$

15. $r = \dfrac{1}{\cos\theta - \sin\theta}$ **16.** $r = \dfrac{1}{2 - \cos\theta}$

In Exercises 17–22, convert to an equation in polar coordinates of the form $r = f(\theta)$.

17. $x^2 + y^2 = 5$

18. $x = 5$

19. $y = x^2$

20. $xy = 1$

21. $e^{\sqrt{x^2+y^2}} = 1$

22. $\ln x = 1$

23. Match each equation with its description:

(a) $r = 2$ (i) Vertical line
(b) $\theta = 2$ (ii) Horizontal line
(c) $r = 2 \sec \theta$ (iii) Circle
(d) $r = 2 \csc \theta$ (iv) Line through origin

24. Suppose that $P = (x, y)$ has polar coordinates (r, θ). Find the polar coordinates for the points:

(a) $(x, -y)$ (b) $(-x, -y)$ (c) $(-x, y)$ (d) (y, x)

25. Find the values of θ in the plot of $r = 4 \cos \theta$ corresponding to points A, B, C, D in Figure 19. Then indicate the portion of the graph traced out as θ varies in the following intervals:

(a) $0 \le \theta \le \frac{\pi}{2}$ (b) $\frac{\pi}{2} \le \theta \le \pi$ (c) $\pi \le \theta \le \frac{3\pi}{2}$

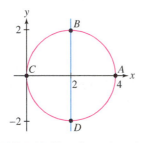

FIGURE 19 Plot of $r = 4 \cos \theta$.

26. Match each equation in rectangular coordinates with its equation in polar coordinates:

(a) $x^2 + y^2 = 4$ (i) $r^2(1 - 2\sin^2 \theta) = 4$
(b) $x^2 + (y - 1)^2 = 1$ (ii) $r(\cos \theta + \sin \theta) = 4$
(c) $x^2 - y^2 = 4$ (iii) $r = 2 \sin \theta$
(d) $x + y = 4$ (iv) $r = 2$

27. What are the polar equations of the lines parallel to the line $r \cos\left(\theta - \frac{\pi}{3}\right) = 1$?

28. Show that the circle with its center at $\left(\frac{1}{2}, \frac{1}{2}\right)$ in Figure 20 has polar equation $r = \sin \theta + \cos \theta$ and find the values of θ between 0 and π corresponding to points A, B, C, and D.

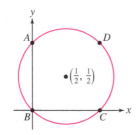

FIGURE 20 Plot of $r = \sin \theta + \cos \theta$.

29. Sketch the curve $r = \frac{1}{2}\theta$ (the spiral of Archimedes) for θ between 0 and 2π by plotting the points for $\theta = 0, \frac{\pi}{4}, \frac{\pi}{2}, \ldots, 2\pi$.

30. Sketch $r = 3 \cos \theta - 1$ (see Example 9).

31. Sketch the cardioid curve $r = 1 + \cos \theta$.

32. Show that the cardioid of Exercise 31 has equation

$$(x^2 + y^2 - x)^2 = x^2 + y^2$$

in rectangular coordinates.

33. Figure 21 displays the graphs of $r = \sin 2\theta$ in rectangular coordinates and in polar coordinates, where it is a "rose with four petals." Identify:

(a) The points in (B) corresponding to points A–I in (A).

(b) The parts of the curve in (B) corresponding to the angle intervals $\left[0, \frac{\pi}{2}\right]$, $\left[\frac{\pi}{2}, \pi\right]$, $\left[\pi, \frac{3\pi}{2}\right]$, and $\left[\frac{3\pi}{2}, 2\pi\right]$.

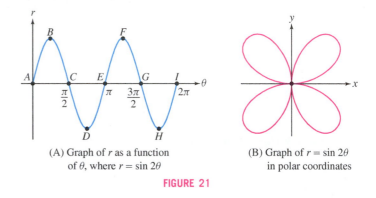

(A) Graph of r as a function of θ, where $r = \sin 2\theta$

(B) Graph of $r = \sin 2\theta$ in polar coordinates

FIGURE 21

34. Sketch the curve $r = \sin 3\theta$. First fill in the table of r-values below and plot the corresponding points of the curve. Notice that the three petals of the curve correspond to the angle intervals $\left[0, \frac{\pi}{3}\right]$, $\left[\frac{\pi}{3}, \frac{2\pi}{3}\right]$, and $\left[\frac{\pi}{3}, \pi\right]$. Then plot $r = \sin 3\theta$ in rectangular coordinates and label the points on this graph corresponding to (r, θ) in the table.

θ	0	$\frac{\pi}{12}$	$\frac{\pi}{6}$	$\frac{\pi}{4}$	$\frac{\pi}{3}$	$\frac{5\pi}{12}$	$\cdots$	$\frac{11\pi}{12}$	π
r									

35. *CAS* Plot the **cissoid** $r = 2 \sin \theta \tan \theta$ and show that its equation in rectangular coordinates is

$$y^2 = \frac{x^3}{2 - x}$$

36. Prove that $r = 2a \cos \theta$ is the equation of the circle in Figure 22 using only the fact that a triangle inscribed in a circle with one side a diameter is a right triangle.

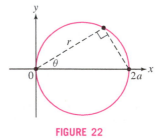

FIGURE 22

37. Show that

$$r = a \cos \theta + b \sin \theta$$

is the equation of a circle passing through the origin. Express the radius and center (in rectangular coordinates) in terms of a and b and write down the equation in rectangular coordinates.

38. Use the previous exercise to write the equation of the circle of radius 5 and center $(3, 4)$ in the form $r = a \cos \theta + b \sin \theta$.

39. Use the identity $\cos 2\theta = \cos^2 \theta - \sin^2 \theta$ to find a polar equation of the hyperbola $x^2 - y^2 = 1$.

40. Find an equation in rectangular coordinates for the curve $r^2 = \cos 2\theta$.

41. Show that $\cos 3\theta = \cos^3 \theta - 3 \cos \theta \sin^2 \theta$ and use this identity to find an equation in rectangular coordinates for the curve $r = \cos 3\theta$.

42. Use the addition formula for the cosine to show that the line $\mathcal{L}$ with polar equation $r \cos(\theta - \alpha) = d$ has the equation in rectangular coordinates $(\cos \alpha)x + (\sin \alpha)y = d$. Show that $\mathcal{L}$ has slope $m = -\cot \alpha$ and y-intercept $d/\sin \alpha$.

In Exercises 43–46, find an equation in polar coordinates of the line $\mathcal{L}$ with the given description.

43. The point on $\mathcal{L}$ closest to the origin has polar coordinates $\left(2, \frac{\pi}{9}\right)$.

44. The point on $\mathcal{L}$ closest to the origin has rectangular coordinates $(-2, 2)$.

45. $\mathcal{L}$ is tangent to the circle $r = 2\sqrt{10}$ at the point with rectangular coordinates $(-2, -6)$.

46. $\mathcal{L}$ has slope 3 and is tangent to the unit circle in the fourth quadrant.

47. Show that every line that does not pass through the origin has a polar equation of the form

$$r = \frac{b}{\sin \theta - a \cos \theta}$$

where $b \neq 0$.

48. By the Law of Cosines, the distance d between two points (Figure 23) with polar coordinates (r, θ) and (r_0, θ_0) is

$$d^2 = r^2 + r_0^2 - 2rr_0 \cos(\theta - \theta_0)$$

Use this distance formula to show that

$$r^2 - 10r \cos\left(\theta - \frac{\pi}{4}\right) = 56$$

is the equation of the circle of radius 9 whose center has polar coordinates $\left(5, \frac{\pi}{4}\right)$.

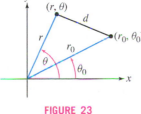

FIGURE 23

49. For $a > 0$, a **lemniscate curve** is the set of points P such that the product of the distances from P to $(a, 0)$ and $(-a, 0)$ is a^2. Show that the equation of the lemniscate is

$$(x^2 + y^2)^2 = 2a^2(x^2 - y^2)$$

Then find the equation in polar coordinates. To obtain the simplest form of the equation, use the identity $\cos 2\theta = \cos^2 \theta - \sin^2 \theta$. Plot the lemniscate for $a = 2$ if you have a computer algebra system.

50. Let c be a fixed constant. Explain the relationship between the graphs of:

(a) $y = f(x + c)$ and $y = f(x)$ (rectangular).

(b) $r = f(\theta + c)$ and $r = f(\theta)$ (polar).

(c) $y = f(x) + c$ and $y = f(x)$ (rectangular).

(d) $r = f(\theta) + c$ and $r = f(\theta)$ (polar).

51. The Derivative in Polar Coordinates Show that a polar curve $r = f(\theta)$ has parametric equations

$$x = f(\theta) \cos \theta, \qquad y = f(\theta) \sin \theta$$

Then apply Theorem 2 of Section 11.1 to prove

$$\frac{dy}{dx} = \frac{f(\theta) \cos \theta + f'(\theta) \sin \theta}{-f(\theta) \sin \theta + f'(\theta) \cos \theta} \qquad \boxed{2}$$

where $f'(\theta) = df/d\theta$.

52. Use Eq. (2) to find the slope of the tangent line to $r = \sin \theta$ at $\theta = \frac{\pi}{3}$.

53. Use Eq. (2) to find the slope of the tangent line to $r = \theta$ at $\theta = \frac{\pi}{2}$ and $\theta = \pi$.

54. Find the equation in rectangular coordinates of the tangent line to $r = 4 \cos 3\theta$ at $\theta = \frac{\pi}{6}$.

55. Find the polar coordinates of the points on the lemniscate $r^2 = \cos 2\theta$ in Figure 24 where the tangent line is horizontal.

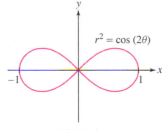

$r^2 = \cos(2\theta)$

FIGURE 24

56. Find the polar coordinates of the points on the cardioid $r = 1 + \cos \theta$ where the tangent line is horizontal (see Figure 25(A)).

57. Use Eq. (2) to show that for $r = \sin \theta + \cos \theta$,

$$\frac{dy}{dx} = \frac{\cos 2\theta + \sin 2\theta}{\cos 2\theta - \sin 2\theta}$$

Then calculate the slopes of the tangent lines at points A, B, C in Figure 20.

Further Insights and Challenges

58. Let $y = f(x)$ be a periodic function of period 2π—that is, $f(x) = f(x + 2\pi)$. Explain how this periodicity is reflected in the graph of:

(a) $y = f(x)$ in rectangular coordinates.

(b) $r = f(\theta)$ in polar coordinates.

59. GU Use a graphing utility to convince yourself that the polar equations $r = f_1(\theta) = 2\cos\theta - 1$ and $r = f_2(\theta) = 2\cos\theta + 1$ have the same graph. Then explain why. *Hint:* Show that the points $(f_1(\theta + \pi), \theta + \pi)$ and $(f_2(\theta), \theta)$ coincide.

60. CAS We investigate how the shape of the limaçon curve $r = b + \cos\theta$ depends on the constant b (see Figure 25).

(a) Argue as in Exercise 59 to show that the constants b and $-b$ yield the same curve.

(b) Plot the limaçon for $b = 0, 0.2, 0.5, 0.8, 1$ and describe how the curve changes.

(c) Plot the limaçon for $b = 1.2, 1.5, 1.8, 2, 2.4$ and describe how the curve changes.

(d) Use Eq. (2) to show that

$$\frac{dy}{dx} = -\left(\frac{b\cos\theta + \cos 2\theta}{b + 2\cos\theta}\right)\csc\theta$$

(e) Find the points where the tangent line is vertical. Note that there are three cases: $0 \le b < 2$, $b = 2$, and $b > 2$. Do the plots constructed in (b) and (c) reflect your results?

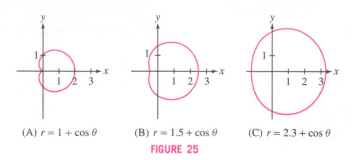

(A) $r = 1 + \cos\theta$ (B) $r = 1.5 + \cos\theta$ (C) $r = 2.3 + \cos\theta$

FIGURE 25

11.4 Area and Arc Length in Polar Coordinates

Integration in polar coordinates involves finding not the area *underneath* a curve, but rather the area of a sector bounded by a curve as in Figure 1(A). Consider the region bounded by the curve $r = f(\theta)$ where $f(\theta) \ge 0$ and the two rays $\theta = \alpha$ and $\theta = \beta$ with $\alpha < \beta$. To derive a formula for the area, divide the region into N narrow sectors of angle $\Delta\theta = (\beta - \alpha)/N$ corresponding to a partition of the interval $[\alpha, \beta]$ as in Figure 1(B):

$$\theta_0 = \alpha < \theta_1 < \theta_2 < \cdots < \theta_N = \beta$$

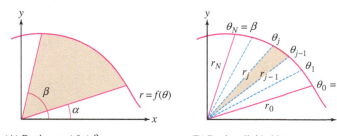

DF **FIGURE 1** Area bounded by the curve $r = f(\theta)$ and the two rays $\theta = \alpha$ and $\theta = \beta$.

(A) Region $\alpha \le \theta \le \beta$ (B) Region divided into narrow sectors

Recall that a circular sector of angle $\Delta\theta$ and radius r has area $\frac{1}{2}r^2\Delta\theta$ (Figure 2). If $\Delta\theta$ is small, the jth narrow sector (Figure 3) is nearly a circular sector of radius $r_j = f(\theta_j)$, so its area is *approximately* $\frac{1}{2}r_j^2\Delta\theta$. The total area is approximated by the sum:

$$\text{Area of region} \approx \sum_{j=1}^{N} \frac{1}{2}r_j^2\Delta\theta = \frac{1}{2}\sum_{j=1}^{N} f(\theta_j)^2\Delta\theta \qquad \boxed{1}$$

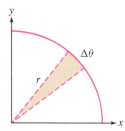

FIGURE 2 The area of a circular sector is the fraction of the total area of the circle given by the angle divided by 2π, that is, $\frac{\Delta\theta}{2\pi}\pi r^2 = \frac{1}{2}r^2\Delta\theta$.

This is a Riemann sum for the integral $\frac{1}{2}\int_\alpha^\beta f(\theta)^2\,d\theta$. If f is continuous, then the sum approaches the integral as $N \to \infty$, and we obtain the following formula.

> **THEOREM 1 Area in Polar Coordinates** If f is a continuous function, then the area bounded by a curve in polar form $r = f(\theta)$ and the rays $\theta = \alpha$ and $\theta = \beta$ (with $\alpha < \beta$) is equal to
>
> $$\frac{1}{2}\int_\alpha^\beta r^2\,d\theta = \frac{1}{2}\int_\alpha^\beta f(\theta)^2\,d\theta \qquad \boxed{2}$$

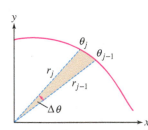

DF **FIGURE 3** The area of the jth sector is approximately $\frac{1}{2}r_j^2\Delta\theta$.

We know that $r = R$ defines a circle of radius R. By Eq. (2), the area is equal to $\frac{1}{2}\int_0^{2\pi} R^2\,d\theta = \frac{1}{2}R^2(2\pi) = \pi R^2$, as expected.

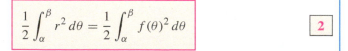

CAUTION Keep in mind that the integral $\frac{1}{2} \int_{\alpha}^{\beta} r^2 \, d\theta$ does **not** compute the area **under** a curve as in Figure 4(B), but rather computes the area "swept out" by a radial segment as θ varies from α to β, as in Figure 4(A).

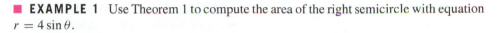

■ **EXAMPLE 1** Use Theorem 1 to compute the area of the right semicircle with equation $r = 4 \sin \theta$.

Solution The equation $r = 4 \sin \theta$ defines a circle of radius 2 tangent to the x-axis at the origin. The right semicircle is "swept out" as θ varies from 0 to $\frac{\pi}{2}$ as in Figure 4(A). By Eq. (2), the area of the right semicircle is

$$\frac{1}{2} \int_0^{\pi/2} r^2 \, d\theta = \frac{1}{2} \int_0^{\pi/2} (4 \sin \theta)^2 \, d\theta = 8 \int_0^{\pi/2} \sin^2 \theta \, d\theta \qquad \boxed{3}$$

←·· **REMINDER** In Eq. (3), we use the identity

$$\sin^2 \theta = \frac{1}{2}(1 - \cos 2\theta) \qquad \boxed{4}$$

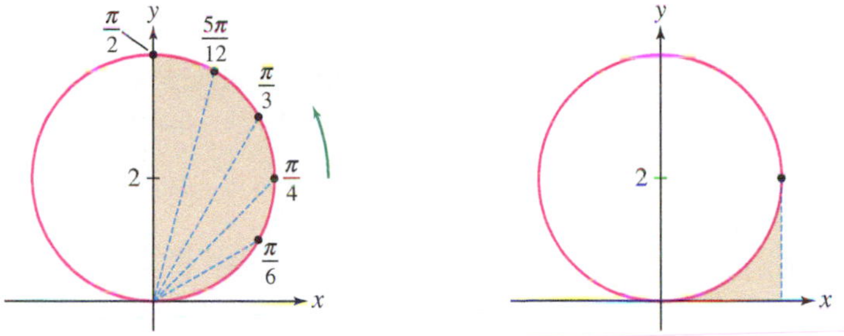

$$= 8 \int_0^{\pi/2} \frac{1}{2}(1 - \cos 2\theta) \, d\theta$$

$$= (4\theta - 2 \sin 2\theta) \Big|_0^{\pi/2} = 4 \left(\frac{\pi}{2} \right) - 0 = 2\pi \qquad ■$$

(A) The polar integral computes the area swept out by a radial segment.

(B) The ordinary integral in rectangular coordinates computes the area underneath a curve.

FIGURE 4

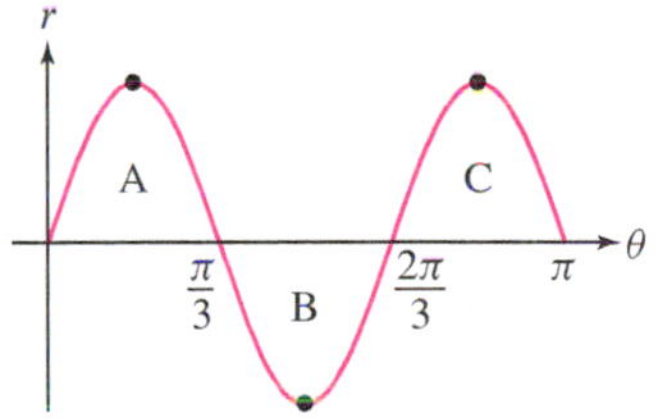

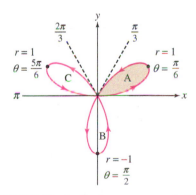

FIGURE 5 Graph of $r = \sin 3\theta$ as a function of θ.

■ **EXAMPLE 2** Sketch $r = \sin 3\theta$ and compute the area of one "petal."

Solution To sketch the curve, we first graph $r = \sin 3\theta$ in rectangular coordinates. Figure 5 shows that the radius r varies from 0 to 1 and back to 0 as θ varies from 0 to $\frac{\pi}{3}$. This gives petal A in Figure 6. Petal B is traced as θ varies from $\frac{\pi}{3}$ to $\frac{2\pi}{3}$ (with $r \leq 0$), and petal C is traced for $\frac{2\pi}{3} \leq \theta \leq \pi$. We find that the area of petal A [using Eq. (4) in the margin to evaluate the integral] is equal to

$$\frac{1}{2} \int_0^{\pi/3} (\sin 3\theta)^2 \, d\theta = \frac{1}{2} \int_0^{\pi/3} \left(\frac{1 - \cos 6\theta}{2} \right) \, d\theta = \left(\frac{1}{4}\theta - \frac{1}{24} \sin 6\theta \right) \Big|_0^{\pi/3} = \frac{\pi}{12} \qquad ■$$

The region sandwiched between two polar curves $r = f_1(\theta)$ and $r = f_2(\theta)$ with $f_2(\theta) \geq f_1(\theta) > 0$, for $\alpha \leq \theta \leq \beta$, is called a **radially simple region**. It has the property that every radial line through the origin that intersects the region does so in either a single point or a single line segment that begins on the curve $r = f_1(\theta)$ and ends on the curve $r = f_2(\theta)$ (Figure 7). For such a radially simple region, we have the following formula for its area:

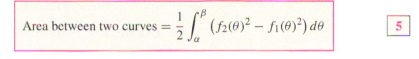

$$\text{Area between two curves} = \frac{1}{2} \int_{\alpha}^{\beta} \left(f_2(\theta)^2 - f_1(\theta)^2 \right) \, d\theta \qquad \boxed{5}$$

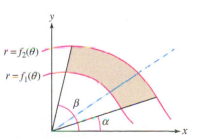

FIGURE 6 Graph of polar curve $r = \sin 3\theta$, a "rose with three petals."

FIGURE 7 A radially simple region.

■ **EXAMPLE 3** Area Between Two Curves Find the area of the region inside the circle $r = 2 \cos \theta$ but outside the circle $r = 1$ [Figure 8(A)].

Solution The two circles intersect at the points where $(r, 2 \cos \theta) = (r, 1)$ or, in other words, when $2 \cos \theta = 1$. This yields $\cos \theta = \frac{1}{2}$, which has solutions $\theta = \pm \frac{\pi}{3}$.

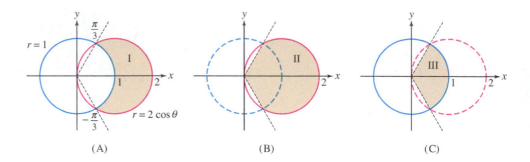

(A) (B) (C)

FIGURE 8 Region I is the difference of regions II and III.

We see in Figure 8 that region I is the difference of regions II and III in Figures 8(B) and (C). Therefore,

$$\text{Area of I} = \text{Area of II} - \text{Area of III}$$

◀·· *REMINDER In Eq. (6), we use the identity*

$$\cos^2 \theta = \frac{1}{2}(1 + \cos 2\theta)$$

$$= \frac{1}{2} \int_{-\pi/3}^{\pi/3} (2 \cos \theta)^2 \, d\theta - \frac{1}{2} \int_{-\pi/3}^{\pi/3} (1)^2 \, d\theta$$

$$= \frac{1}{2} \int_{-\pi/3}^{\pi/3} (4 \cos^2 \theta - 1) \, d\theta = \frac{1}{2} \int_{-\pi/3}^{\pi/3} (2 \cos 2\theta + 1) \, d\theta \qquad \boxed{6}$$

$$= \frac{1}{2}(\sin 2\theta + \theta)\Big|_{-\pi/3}^{\pi/3} = \frac{\sqrt{3}}{2} + \frac{\pi}{3} \approx 1.91 \qquad ■$$

We close this section by deriving a formula for arc length in polar coordinates. Observe that a polar curve $r = f(\theta)$ has a parametrization with θ as a parameter:

$$x = r \cos \theta = f(\theta) \cos \theta, \qquad y = r \sin \theta = f(\theta) \sin \theta$$

Using a prime to denote the derivative with respect to θ, we have

$$x'(\theta) = \frac{dx}{d\theta} = -f(\theta) \sin \theta + f'(\theta) \cos \theta$$

$$y'(\theta) = \frac{dy}{d\theta} = f(\theta) \cos \theta + f'(\theta) \sin \theta$$

Recall from Section 11.2 that arc length is obtained by integrating $\sqrt{x'(\theta)^2 + y'(\theta)^2}$. Straightforward algebra shows that $x'(\theta)^2 + y'(\theta)^2 = f(\theta)^2 + f'(\theta)^2$; thus,

$$\boxed{\text{Arc length } s = \int_{\alpha}^{\beta} \sqrt{f(\theta)^2 + f'(\theta)^2} \, d\theta} \qquad \boxed{7}$$

■ **EXAMPLE 4** Find the total length of the circle $r = 2a \cos \theta$ for $a > 0$.

Solution In this case, $f(\theta) = 2a \cos \theta$ and

$$f(\theta)^2 + f'(\theta)^2 = 4a^2 \cos^2 \theta + 4a^2 \sin^2 \theta = 4a^2$$

The total length of this circle of radius a has the expected value:

$$\int_0^{\pi} \sqrt{f(\theta)^2 + f'(\theta)^2} \, d\theta = \int_0^{\pi} (2a) \, d\theta = 2\pi a$$

FIGURE 9 Graph of $r = 2a \cos \theta$.

Note that the upper limit of integration is π rather than 2π because the entire circle is traced out as θ varies from 0 to π (see Figure 9). ■

11.4 SUMMARY

- Area of the sector bounded by a polar curve $r = f(\theta)$ and two rays $\theta = \alpha$ and $\theta = \beta$ (Figure 10):

$$\text{Area} = \frac{1}{2} \int_{\alpha}^{\beta} f(\theta)^2 \, d\theta$$

- Area of a radially simple region, which is between $r = f_1(\theta)$ and $r = f_2(\theta)$, where $f_2(\theta) \geq f_1(\theta)$ (Figure 11):

$$\text{Area} = \frac{1}{2} \int_{\alpha}^{\beta} \left(f_2(\theta)^2 - f_1(\theta)^2 \right) d\theta$$

- Arc length of the polar curve $r = f(\theta)$ for $\alpha \leq \theta \leq \beta$:

$$\text{Arc length} = \int_{\alpha}^{\beta} \sqrt{f(\theta)^2 + f'(\theta)^2} \, d\theta$$

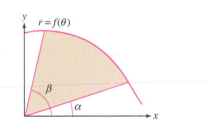

FIGURE 10 Region bounded by the polar curve $r = f(\theta)$ and the rays $\theta = \alpha, \theta = \beta$.

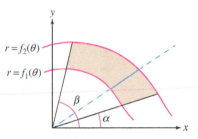

FIGURE 11 Region between two polar curves.

11.4 EXERCISES

Preliminary Questions

1. Polar coordinates are suited to finding the area (choose one):

(a) under a curve between $x = a$ and $x = b$.

(b) bounded by a curve and two rays through the origin.

2. Is the formula for area in polar coordinates valid if $f(\theta)$ takes negative values?

3. The horizontal line $y = 1$ has polar equation $r = \csc \theta$. Which area is represented by the integral $\dfrac{1}{2} \displaystyle\int_{\pi/6}^{\pi/2} \csc^2 \theta \, d\theta$ (Figure 12)?

(a) $\square ABCD$ (b) $\triangle ABC$ (c) $\triangle ACD$

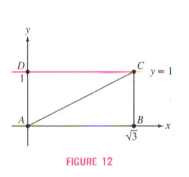

FIGURE 12

Exercises

1. Sketch the area bounded by the circle $r = 5$ and the rays $\theta = \frac{\pi}{2}$ and $\theta = \pi$, and compute its area as an integral in polar coordinates.

2. Sketch the region bounded by the line $r = \sec \theta$ and the rays $\theta = 0$ and $\theta = \frac{\pi}{3}$. Compute its area in two ways: as an integral in polar coordinates and using geometry.

3. Calculate the area of the circle $r = 4 \sin \theta$ as an integral in polar coordinates (see Figure 4). Be careful to choose the correct limits of integration.

4. Find the area of the shaded triangle in Figure 13 as an integral in polar coordinates. Then find the rectangular coordinates of P and Q, and compute the area via geometry.

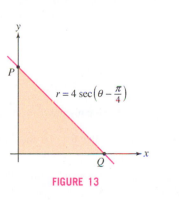

FIGURE 13

5. Find the area of the shaded region in Figure 14. Note that θ varies from 0 to $\frac{\pi}{2}$.

6. Which interval of θ-values corresponds to the the shaded region in Figure 15? Find the area of the region.

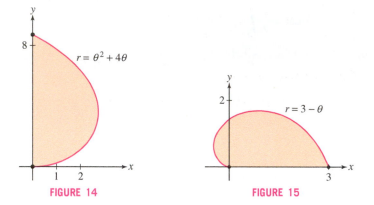

FIGURE 14   FIGURE 15

7. Find the total area enclosed by the cardioid in Figure 16.

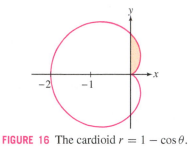

FIGURE 16 The cardioid $r = 1 - \cos \theta$.

8. Find the area of the shaded region in Figure 16.

9. Find the area of one leaf of the "four-petaled rose" $r = \sin 2\theta$ (Figure 17). Then prove that the total area of the rose is equal to one-half the area of the circumscribed circle.

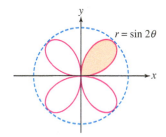

FIGURE 17 Four-petaled rose $r = \sin 2\theta$.

10. Find the area enclosed by one loop of the lemniscate with equation $r^2 = \cos 2\theta$ (Figure 18). Choose your limits of integration carefully.

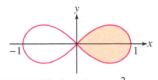

FIGURE 18 The lemniscate $r^2 = \cos 2\theta$.

11. Sketch the spiral $r = \theta$ for $0 \le \theta \le 2\pi$ and find the area bounded by the curve and the first quadrant.

12. Find the area of the intersection of the circles $r = \sin \theta$ and $r = \cos \theta$.

13. Find the area of region A in Figure 19.

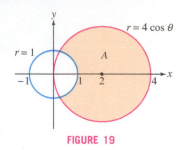

FIGURE 19

14. Find the area of the shaded region in Figure 20 enclosed by the circle $r = \frac{1}{2}$ and a petal of the curve $r = \cos 3\theta$. *Hint:* Compute the area of both the petal and the region inside the petal and outside the circle.

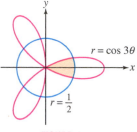

FIGURE 20

15. Find the area of the inner loop of the limaçon with polar equation $r = 2 \cos \theta - 1$ (Figure 21).

16. Find the area of the shaded region in Figure 21 between the inner and outer loop of the limaçon $r = 2 \cos \theta - 1$.

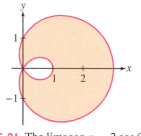

FIGURE 21 The limaçon $r = 2 \cos \theta - 1$.

17. Find the area of the part of the circle $r = \sin \theta + \cos \theta$ in the fourth quadrant (see Exercise 28 in Section 11.3).

18. Find the area of the region inside the circle $r = 2 \sin \left(\theta + \frac{\pi}{4} \right)$ and above the line $r = \sec \left(\theta - \frac{\pi}{4} \right)$.

19. Find the area between the two curves in Figure 22(A).

20. Find the area between the two curves in Figure 22(B).

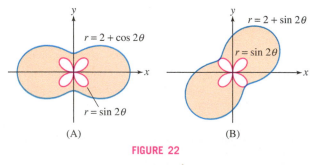

(A)   (B)

FIGURE 22

21. Find the area inside both curves in Figure 23.

22. Find the area of the region that lies inside one but not both of the curves in Figure 23.

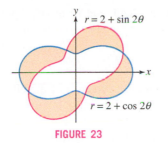

FIGURE 23

23. Calculate the total length of the circle $r = 4\sin\theta$ as an integral in polar coordinates.

24. Sketch the segment $r = \sec\theta$ for $0 \le \theta \le A$. Then compute its length in two ways: as an integral in polar coordinates and using trigonometry.

In Exercises 25–32, compute the length of the polar curve.

25. The curve of $r = \theta^2$ for $0 \le \theta \le \pi$

26. The spiral $r = \theta$ for $0 \le \theta \le A$

27. The curve $r = \sin\theta$ for $0 \le \theta \le \pi$

28. The equiangular spiral $r = e^\theta$ for $0 \le \theta \le 2\pi$

29. $r = \sqrt{1 + \sin 2\theta}$ for $0 \le \theta \le \pi/4$

30. The cardioid $r = 1 - \cos\theta$ in Figure 16

31. $r = \cos^2\theta$

32. $r = 1 + \theta$ for $0 \le \theta \le \pi/2$

In Exercises 33–36, express the length of the curve as an integral but do not evaluate it.

33. $r = e^\theta + 1, \quad 0 \le \theta \le \pi/2$.

34. $r = (2 - \cos\theta)^{-1}, \quad 0 \le \theta \le 2\pi$

35. $r = \sin^3\theta, \quad 0 \le \theta \le 2\pi$

36. $r = \sin\theta\cos\theta, \quad 0 \le \theta \le \pi$

In Exercises 37–40, use a computer algebra system to calculate the total length to two decimal places.

37. $\boxed{\text{CAS}}$ The three-petal rose $r = \cos 3\theta$ in Figure 20

38. $\boxed{\text{CAS}}$ The curve $r = 2 + \sin 2\theta$ in Figure 23

39. $\boxed{\text{CAS}}$ The curve $r = \theta\sin\theta$ in Figure 24 for $0 \le \theta \le 4\pi$

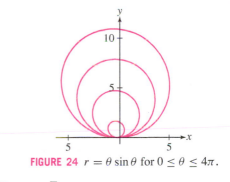

FIGURE 24 $r = \theta\sin\theta$ for $0 \le \theta \le 4\pi$.

40. $\boxed{\text{CAS}}$ $r = \sqrt{\theta}, \quad 0 \le \theta \le 4\pi$

Further Insights and Challenges

41. Suppose that the polar coordinates of a moving particle at time t are $(r(t), \theta(t))$. Prove that the particle's speed is equal to
$$\sqrt{(dr/dt)^2 + r^2(d\theta/dt)^2}.$$

42. $\boxed{}$ Compute the speed at time $t = 1$ of a particle whose polar coordinates at time t are $r = t$, $\theta = t$ (use Exercise 41). What would the speed be if the particle's rectangular coordinates were $x = t, y = t$? Why is the speed increasing in one case and constant in the other?

11.5 Conic Sections

The conics were first studied by the ancient Greek mathematicians, beginning possibly with Menaechmus (c. 380–320 BCE) and including Archimedes (287–212 BCE) and Apollonius (c. 262–190 BCE).

Three familiar families of curves—ellipses, hyperbolas, and parabolas—appear throughout mathematics and its applications. They are called **conic sections** because they are obtained as the intersection of a cone with a suitable plane (Figure 1). Our goal in this section is to derive equations for the conic sections from their geometric definitions as curves in the plane.

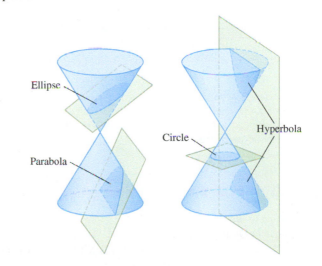

DF FIGURE 1 The conic sections are obtained by intersecting a plane and a cone.

An **ellipse** is an oval-shaped curve [Figure 2(A)] consisting of all points P such that the sum of the distances to two fixed points F_1 and F_2 is a constant $K > 0$:

$$PF_1 + PF_2 = K \qquad \boxed{1}$$

We assume always that K is greater than the distance F_1F_2 between the foci, because the ellipse reduces to the line segment $\overline{F_1F_2}$ if $K = F_1F_2$, and it has no points at all if $K < F_1F_2$.

The points F_1 and F_2 are called the **foci** (plural of "focus") of the ellipse. Note that if the foci coincide, then Eq. (1) reduces to $2PF_1 = K$ and we obtain a circle of radius $\frac{1}{2}K$ centered at F_1.

We use the following terminology:

• The midpoint of $\overline{F_1F_2}$ is the **center** of the ellipse.
• The line through the foci is the **focal axis**.
• The line through the center perpendicular to the focal axis is the **conjugate axis**.

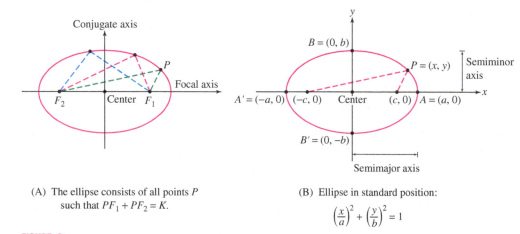

(A) The ellipse consists of all points P
such that $PF_1 + PF_2 = K$.

(B) Ellipse in standard position:

$$\left(\frac{x}{a}\right)^2 + \left(\frac{y}{b}\right)^2 = 1$$

FIGURE 2

The ellipse is said to be in **standard position** if the focal and conjugate axes are the x- and y-axes, as shown in Figure 2(B). In this case, the foci have coordinates $F_1 = (c, 0)$ and $F_2 = (-c, 0)$ for some $c > 0$. Let us prove that the equation of this ellipse has the particularly simple form

$$\left(\frac{x}{a}\right)^2 + \left(\frac{y}{b}\right)^2 = 1 \qquad \boxed{2}$$

where $a = K/2$ and $b = \sqrt{a^2 - c^2}$.

By the distance formula, $P = (x, y)$ lies on the ellipse in Figure 2(B) if

$$PF_1 + PF_2 = \sqrt{(x-c)^2 + y^2} + \sqrt{(x+c)^2 + y^2} = 2a \qquad \boxed{3}$$

Move the first term on the left over to the right and square both sides:

$$(x+c)^2 + y^2 = 4a^2 - 4a\sqrt{(x-c)^2 + y^2} + (x-c)^2 + y^2$$

$$4a\sqrt{(x-c)^2 + y^2} = 4a^2 + (x-c)^2 - (x+c)^2 = 4a^2 - 4cx$$

Strictly speaking, it is necessary to show that if $P = (x, y)$ satisfies Eq. (4), then it also satisfies Eq. (3). When we begin with Eq. (4) and reverse the algebraic steps, the process of taking square roots leads to the relation

$$\sqrt{(x-c)^2 + y^2} \pm \sqrt{(x+c)^2 + y^2} = \pm 2a$$

However, this equation has no solutions unless both signs are positive because $a > c$.

Now divide by 4, square, and simplify:

$$a^2(x^2 - 2cx + c^2 + y^2) = a^4 - 2a^2cx + c^2x^2$$

$$(a^2 - c^2)x^2 + a^2y^2 = a^4 - a^2c^2 = a^2(a^2 - c^2)$$

$$\frac{x^2}{a^2} + \frac{y^2}{a^2 - c^2} = 1 \qquad \boxed{4}$$

This is Eq. (2) with $b^2 = a^2 - c^2$ as claimed.

The ellipse intersects the axes in four points A, A', B, B' called **vertices**. Vertices A and A' along the focal axis are called the **focal vertices**. Following common usage, the numbers a and b are referred to as the **semimajor axis** and the **semiminor axis** (even though they are numbers rather than axes).

THEOREM 1 Ellipse in Standard Position Let $a > b > 0$, and set $c = \sqrt{a^2 - b^2}$. The ellipse $PF_1 + PF_2 = 2a$ with foci $F_1 = (c, 0)$ and $F_2 = (-c, 0)$ has equation

$$\left(\frac{x}{a}\right)^2 + \left(\frac{y}{b}\right)^2 = 1 \qquad \boxed{5}$$

Furthermore, the ellipse has:

- semimajor axis a, semiminor axis b.
- focal vertices $(\pm a, 0)$, minor vertices $(0, \pm b)$.

If $b > a > 0$, then Eq. (5) defines an ellipse with foci $(0, \pm c)$, where $c = \sqrt{b^2 - a^2}$.

■ **EXAMPLE 1** Find the equation of the ellipse with foci $(\pm\sqrt{11}, 0)$ and semimajor axis $a = 6$. Then find the semiminor axis and sketch the graph.

Solution The foci are $(\pm c, 0)$ with $c = \sqrt{11}$, and the semimajor axis is $a = 6$, so we can use the relation $c = \sqrt{a^2 - b^2}$ to find b:

$$b^2 = a^2 - c^2 = 6^2 - (\sqrt{11})^2 = 25 \quad \Rightarrow \quad b = 5$$

Thus, the semiminor axis is $b = 5$ and the ellipse has equation $\left(\frac{x}{6}\right)^2 + \left(\frac{y}{5}\right)^2 = 1$. To sketch this ellipse, plot the vertices $(\pm 6, 0)$ and $(0, \pm 5)$ and connect them as in Figure 3. ■

FIGURE 3

To write down the equation of an ellipse with axes parallel to the x- and y-axes and center translated to the point $C = (h, k)$, replace x by $x - h$ and y by $y - k$ in the equation (Figure 4):

$$\left(\frac{x - h}{a}\right)^2 + \left(\frac{y - k}{b}\right)^2 = 1$$

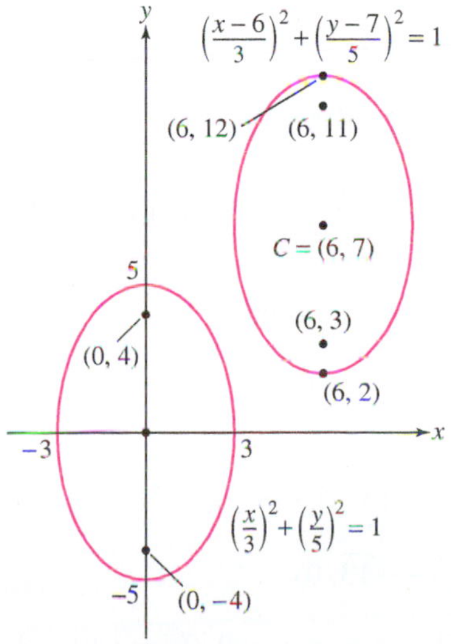

■ **EXAMPLE 2 Translating an Ellipse** Find an equation of the ellipse with center $C = (6, 7)$, vertical focal axis, semimajor axis 5, and semiminor axis 3. Where are the foci located?

Solution Since the focal axis is vertical, we have $a = 3$ and $b = 5$, so that $a < b$ (Figure 4). The ellipse centered at the origin would have equation $\left(\frac{x}{3}\right)^2 + \left(\frac{y}{5}\right)^2 = 1$. When the center is translated to $(h, k) = (6, 7)$, the equation becomes

$$\left(\frac{x - 6}{3}\right)^2 + \left(\frac{y - 7}{5}\right)^2 = 1$$

Furthermore, $c = \sqrt{b^2 - a^2} = \sqrt{5^2 - 3^2} = 4$, so the foci are located ± 4 vertical units above and below the center—that is, $F_1 = (6, 11)$ and $F_2 = (6, 3)$. ■

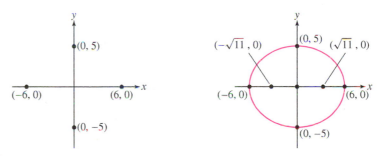

FIGURE 4 An ellipse with vertical major axis and its translate with center $C = (6, 7)$.

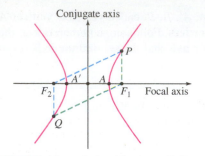

FIGURE 5 A hyperbola with center $(0, 0)$.

A **hyperbola** is the set of all points P such that the difference of the distances from P to two foci F_1 and F_2 is $\pm K$:

$$PF_1 - PF_2 = \pm K \qquad \boxed{6}$$

We assume that K is less than the distance $\overline{F_1 F_2}$ between the foci (the hyperbola has no points if $K > \overline{F_1 F_2}$). Note that a hyperbola consists of two branches corresponding to the choices of sign $\pm$ (Figure 5).

As before, the midpoint of $\overline{F_1 F_2}$ is the **center** of the hyperbola, the line through F_1 and F_2 is called the **focal axis**, and the line through the center perpendicular to the focal axis is called the **conjugate axis**. The **vertices** are the points where the focal axis intersects the hyperbola; they are labeled A and A' in Figure 5. The hyperbola is said to be in standard position when the focal and conjugate axes are the x- and y-axes, respectively, as in Figure 6. The next theorem can be verified in much the same way as Theorem 1.

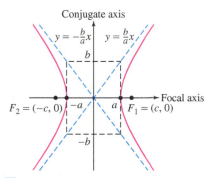

DF FIGURE 6 Hyperbola in standard position.

> **THEOREM 2 Hyperbola in Standard Position** Let $a > 0$ and $b > 0$, and set $c = \sqrt{a^2 + b^2}$. The hyperbola $PF_1 - PF_2 = \pm 2a$ with foci $F_1 = (c, 0)$ and $F_2 = (-c, 0)$ has equation
>
> $$\left(\frac{x}{a}\right)^2 - \left(\frac{y}{b}\right)^2 = 1 \qquad \boxed{7}$$

A hyperbola has two **asymptotes** $y = \pm\frac{b}{a}x$ that are, we claim, diagonals of the rectangle whose sides pass through $(\pm a, 0)$ and $(0, \pm b)$ as in Figure 6. To prove this, consider a point (x, y) on the hyperbola in the first quadrant. By Eq. (7),

$$y = \sqrt{\frac{b^2}{a^2}x^2 - b^2} = \frac{b}{a}\sqrt{x^2 - a^2}$$

The following limit shows that a point (x, y) on the hyperbola approaches the line $y = \frac{b}{a}x$ as $x \to \infty$:

$$\lim_{x \to \infty}\left(y - \frac{b}{a}x\right) = \frac{b}{a}\lim_{x \to \infty}\left(\sqrt{x^2 - a^2} - x\right)$$

$$= \frac{b}{a}\lim_{x \to \infty}\left(\sqrt{x^2 - a^2} - x\right)\left(\frac{\sqrt{x^2 - a^2} + x}{\sqrt{x^2 - a^2} + x}\right)$$

$$= \frac{b}{a}\lim_{x \to \infty}\left(\frac{-a^2}{\sqrt{x^2 - a^2} + x}\right) = 0$$

The asymptotic behavior in the remaining quadrants is similar.

■ **EXAMPLE 3** Find the foci of the hyperbola $9x^2 - 4y^2 = 36$. Sketch its graph and asymptotes.

Solution First divide by 36 to write the equation in standard form:

$$\frac{x^2}{4} - \frac{y^2}{9} = 1 \qquad \text{or} \qquad \left(\frac{x}{2}\right)^2 - \left(\frac{y}{3}\right)^2 = 1$$

Thus, $a = 2$, $b = 3$, and $c = \sqrt{a^2 + b^2} = \sqrt{4 + 9} = \sqrt{13}$. The foci are

$$F_1 = (\sqrt{13}, 0), \qquad F_2 = (-\sqrt{13}, 0)$$

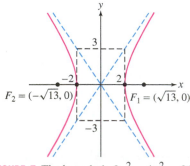

FIGURE 7 The hyperbola $9x^2 - 4y^2 = 36$.

To sketch the graph, we draw the rectangle through the points $(\pm 2, 0)$ and $(0, \pm 3)$ as in Figure 7. The diagonals of the rectangle are the asymptotes $y = \pm\frac{3}{2}x$. The hyperbola passes through the vertices $(\pm 2, 0)$ and approaches the asymptotes. ■

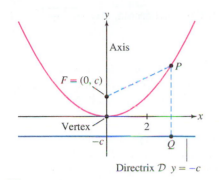

FIGURE 8 Parabola with focus $(0, c)$ and directrix $y = -c$.

Unlike the ellipse and hyperbola, which are defined in terms of two foci, a **parabola** is the set of points P equidistant from a focus F and a line $\mathcal{D}$ called the **directrix**:

$$PF = PD \qquad \boxed{8}$$

Here, when we speak of the *distance* from a point P to a line $\mathcal{D}$, we mean the distance from P to the point Q on $\mathcal{D}$ closest to P, obtained by dropping a perpendicular from P to $\mathcal{D}$ (Figure 8). We denote this distance by PD.

The line through the focus F perpendicular to $\mathcal{D}$ is called the **axis** of the parabola. The **vertex** is the point where the parabola intersects its axis. We say that the parabola is in standard position if, for some c, the focus is $F = (0, c)$ and the directrix is $y = -c$, as shown in Figure 8. We verify in Exercise 73 that the vertex is then located at the origin and the equation of the parabola is $y = x^2/4c$. If $c < 0$, then the parabola opens downward.

THEOREM 3 Parabola in Standard Position Let $c \neq 0$. The parabola with focus $F = (0, c)$ and directrix $y = -c$ has equation

$$y = \frac{1}{4c}x^2 \qquad \boxed{9}$$

The vertex is located at the origin. The parabola opens upward if $c > 0$ and downward if $c < 0$.

■ **EXAMPLE 4** The standard parabola with directrix $y = -2$ is translated so that its vertex is located at $(2, 8)$. Find its equation, directrix, and focus.

Solution By Eq. (9) with $c = 2$, the standard parabola with directrix $y = -2$ has equation $y = \frac{1}{8}x^2$ (Figure 9). The focus of this standard parabola is $(0, c) = (0, 2)$, which is 2 units above the vertex $(0, 0)$.

To obtain the equation when the parabola is translated with vertex at $(2, 8)$, we replace x by $x - 2$ and y by $y - 8$:

$$y - 8 = \frac{1}{8}(x - 2)^2 \qquad \text{or} \qquad y = \frac{1}{8}x^2 - \frac{1}{2}x + \frac{17}{2}$$

The vertex has moved up 8 units, so the directrix also moves up 8 units to become $y = 6$. The new focus is 2 units above the new vertex $(2, 8)$, so the new focus is $(2, 10)$. ■

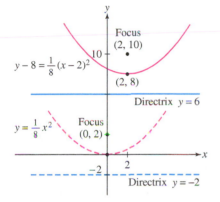

FIGURE 9 A parabola and its translate.

Eccentricity

Some ellipses are flatter than others, just as some hyperbolas are steeper. The "shape" of a conic section is measured by a number e called the **eccentricity**. For an ellipse or hyperbola,

$$e = \frac{\text{distance betweeen foci}}{\text{distance between vertices on focal axis}}$$

A parabola is defined to have eccentricity $e = 1$.

◄·· REMINDER

Standard ellipse:

$$\left(\frac{x}{a}\right)^2 + \left(\frac{y}{b}\right)^2 = 1, \quad c = \sqrt{a^2 - b^2}$$

Standard hyperbola:

$$\left(\frac{x}{a}\right)^2 - \left(\frac{y}{b}\right)^2 = 1, \quad c = \sqrt{a^2 + b^2}$$

THEOREM 4 For ellipses and hyperbolas in standard position,

$$e = \frac{c}{a}$$

1. An ellipse has eccentricity $0 \leq e < 1$.
2. A hyperbola has eccentricity $e > 1$.

Proof The foci are located at $(\pm c, 0)$ and the vertices are on the focal axis at $(\pm a, 0)$. Therefore,

$$e = \frac{\text{distance between foci}}{\text{distance between vertices on focal axis}} = \frac{2c}{2a} = \frac{c}{a}$$

For an ellipse, $c = \sqrt{a^2 - b^2}$ and so $e = c/a < 1$. For a hyperbola, $c = \sqrt{a^2 + b^2}$ and thus $e = c/a > 1$.   ∎

How does eccentricity determine the shape of a conic [Figure 10(A)]? Consider the ratio b/a of the semiminor axis to the semimajor axis of an ellipse. The ellipse is nearly circular if b/a is close to 1, whereas it is elongated and flat if b/a is small. Now

$$\frac{b}{a} = \frac{\sqrt{a^2 - c^2}}{a} = \sqrt{1 - \frac{c^2}{a^2}} = \sqrt{1 - e^2}$$

This shows that b/a gets smaller (and the ellipse gets flatter) as $e \to 1$ [Figure 10(B)]. The "roundest" ellipse is the circle, with $e = 0$.

Similarly, for a hyperbola,

$$\frac{b}{a} = \sqrt{1 + e^2}$$

The ratios $\pm b/a$ are the slopes of the asymptotes, so the asymptotes get steeper as $e \to \infty$ [Figure 10(C)].

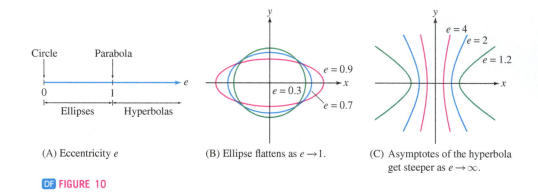

(A) Eccentricity e

(B) Ellipse flattens as $e \to 1$.

(C) Asymptotes of the hyperbola get steeper as $e \to \infty$.

DF FIGURE 10

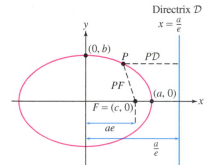

DF FIGURE 11 The ellipse consists of points P such that $PF = ePD$.

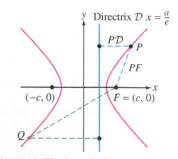

FIGURE 12 The hyperbola consists of points P such that $PF = ePD$.

CONCEPTUAL INSIGHT   There is a more precise way to explain how eccentricity determines the shape of a conic. We can prove that if two conics C_1 and C_2 have the same eccentricity e, then there is a change of scale that makes C_1 *congruent* to C_2. Changing the scale means changing the units along the x- and y-axes by a common positive factor. A curve scaled by a factor of 10 has the same shape but is ten times as large. This corresponds, for example, to changing units from centimeters to millimeters (smaller units make for a larger figure). By "congruent" we mean that after scaling, it is possible to move C_1 by a rigid motion (involving rotation and translation, but no stretching or bending) so that it lies directly on top of C_2.

All circles ($e = 0$) have the same shape because scaling by a factor $r > 0$ transforms a circle of radius R into a circle of radius rR. Similarly, any two parabolas ($e = 1$) become congruent after suitable scaling. However, an ellipse of eccentricity $e = 0.5$ cannot be made congruent to an ellipse of eccentricity $e = 0.8$ by scaling (see Exercise 74).

Eccentricity can be used to give a unified focus-directrix definition of the conic sections. Given a point F (the focus), a line $\mathcal{D}$ (the directrix), and a number $e > 0$, we consider the set of all points P such that

$$\boxed{PF = ePD} \qquad \textbf{10}$$

For $e = 1$, this is our definition of a parabola. According to the next theorem, Eq. (10) defines a conic section of eccentricity e for all $e > 0$ (Figures 11 and 12). Note, however, that there is no focus-directrix definition for circles ($e = 0$).

THEOREM 5 Focus-Directrix Definition   For all $e > 0$, the set of points satisfying Eq. (10) is a conic section of eccentricity e. Furthermore,

- **Ellipse:** Let $a > b > 0$ and $c = \sqrt{a^2 - b^2}$. The ellipse

$$\left(\frac{x}{a}\right)^2 + \left(\frac{y}{b}\right)^2 = 1$$

satisfies Eq. (10) with $F = (c, 0)$, $e = \frac{c}{a}$, and vertical directrix $x = \frac{a}{e}$.

- **Hyperbola:** Let $a, b > 0$ and $c = \sqrt{a^2 + b^2}$. The hyperbola

$$\left(\frac{x}{a}\right)^2 - \left(\frac{y}{b}\right)^2 = 1$$

satisfies Eq. (10) with $F = (c, 0)$, $e = \frac{c}{a}$, and vertical directrix $x = \frac{a}{e}$.

Proof   Assume that $e > 1$ (the case $e < 1$ is similar, see Exercise 66). We may choose our coordinate axes so that the focus F lies on the x-axis and the directrix is vertical, lying to the left of F, as in Figure 13. Anticipating the final result, we let d be the distance from the focus F to the directrix $\mathcal{D}$ and set

$$c = \frac{d}{1 - e^{-2}}, \qquad a = \frac{c}{e}, \qquad b = \sqrt{c^2 - a^2}$$

Since we are free to shift the y-axis, let us choose the y-axis so that the focus has coordinates $F = (c, 0)$. Then the directrix is the line

$$x = c - d = c - c(1 - e^{-2})$$

$$= c e^{-2} = \frac{a}{e}$$

Now, the equation $PF = ePD$ for a point $P = (x, y)$ may be written

$$\underbrace{\sqrt{(x - c)^2 + y^2}}_{PF} = \underbrace{e\sqrt{\left(x - (a/e)\right)^2}}_{PD}$$

Algebraic manipulation yields

$$(x - c)^2 + y^2 = e^2\left(x - (a/e)\right)^2 \qquad \text{(square)}$$

$$x^2 - 2cx + c^2 + y^2 = e^2 x^2 - 2aex + a^2$$

$$x^2 - \cancel{2aex} + a^2 e^2 + y^2 = e^2 x^2 - \cancel{2aex} + a^2 \qquad \text{(use } c = ae\text{)}$$

$$(e^2 - 1)x^2 - y^2 = a^2(e^2 - 1) \qquad \text{(rearrange)}$$

$$\frac{x^2}{a^2} - \frac{y^2}{a^2(e^2 - 1)} = 1 \qquad \text{(divide)}$$

This is the desired equation because $a^2(e^2 - 1) = c^2 - a^2 = b^2$. ■

EXAMPLE 5   Find the equation, foci, and directrix of the standard ellipse with eccentricity $e = 0.8$ and focal vertices $(\pm 10, 0)$.

Solution   The vertices are $(\pm a, 0)$ with $a = 10$ (Figure 14). By Theorem 5,

$$c = ae = 10(0.8) = 8, \qquad b = \sqrt{a^2 - c^2} = \sqrt{10^2 - 8^2} = 6$$

Thus, our ellipse has equation

$$\left(\frac{x}{10}\right)^2 + \left(\frac{y}{6}\right)^2 = 1$$

The foci are $(\pm c, 0) = (\pm 8, 0)$ and the directrix is $x = \frac{a}{e} = \frac{10}{0.8} = 12.5$. ■

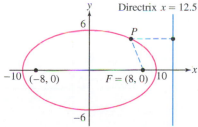

FIGURE 13

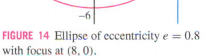

FIGURE 14 Ellipse of eccentricity $e = 0.8$ with focus at $(8, 0)$.

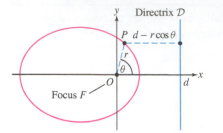

FIGURE 15 Focus-directrix definition of the ellipse in polar coordinates.

In Section 13.6, we discuss the famous law of Johannes Kepler stating that the orbit of a planet around the sun is an ellipse with one focus at the sun. In this discussion, we will need to write the equation of an ellipse in polar coordinates. To derive the polar equations of the conic sections, it is convenient to use the focus-directrix definition with focus F at the origin O and vertical line $x = d$ as directrix $\mathcal{D}$ (Figure 15). Note from the figure that if $P = (r, \theta)$, then

$$PF = r, \qquad PD = d - r\cos\theta$$

Thus, the focus-directrix equation of the ellipse $PF = e\,PD$ becomes $r = e(d - r\cos\theta)$, or $r(1 + e\cos\theta) = ed$. This proves the following result, which is also valid for the hyperbola and parabola (see Exercise 67).

> **THEOREM 6 Polar Equation of a Conic Section** The conic section of eccentricity $e > 0$ with focus at the origin and directrix $x = d$ has polar equation
>
> $$r = \frac{ed}{1 + e\cos\theta} \qquad \boxed{11}$$

■ **EXAMPLE 6** Find the eccentricity, directrix, and focus of the conic section

$$r = \frac{24}{4 + 3\cos\theta}$$

Solution First, we write the equation in the standard form

$$r = \frac{24}{4 + 3\cos\theta} = \frac{6}{1 + \frac{3}{4}\cos\theta}$$

Comparing with Eq. (11), we see that $e = \frac{3}{4}$ and $ed = 6$. Therefore, $d = 8$. Since $e < 1$, the conic is an ellipse. By Theorem 6, the directrix is the line $x = 8$ and the focus is the origin. ■

Reflective Properties of Conic Sections

The conic sections have numerous geometric properties. Especially important are the *reflective properties*, which are used in optics and communications (e.g., in antenna and telescope design; Figure 16). We describe these properties here briefly without proof (but see Exercises 68–70 and Exercise 71 for proofs of the reflective property of ellipses).

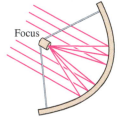

- **Ellipse:** The segments $F_1 P$ and $F_2 P$ make equal angles with the tangent line at a point P on the ellipse. Therefore, a beam of light originating at focus F_1 is reflected off the ellipse toward the second focus F_2 [Figure 17(A)]. See also Figure 18.
- **Hyperbola:** The tangent line at a point P on the hyperbola bisects the angle formed by the segments $F_1 P$ and $F_2 P$. Therefore, a beam of light directed toward F_2 is reflected off the hyperbola toward the second focus F_1 [Figure 17(B)].
- **Parabola:** The segment FP and the line through P parallel to the axis make equal angles with the tangent line at a point P on the parabola [Figure 17(C)]. Therefore, a beam of light approaching P from above in the axial direction is reflected off the parabola toward the focus F.

FIGURE 16 The paraboloid shape of this radio telescope directs the incoming signal to the focus. *(Stockbyte)*

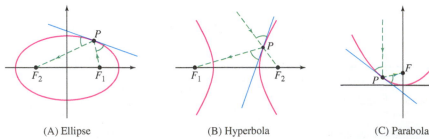

(A) Ellipse   (B) Hyperbola   (C) Parabola

FIGURE 17

FIGURE 18 The ellipsoidal dome of the National Statuary Hall in the U.S. Capitol Building creates a "whisper chamber." Legend has it that John Quincy Adams would locate at one focus in order to eavesdrop on conversations taking place at the other focus. *(Architect of the Capitol)*

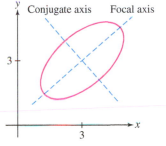

FIGURE 19 The ellipse with equation $6x^2 - 8xy + 8y^2 - 12x - 24y + 38 = 0$.

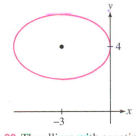

FIGURE 20 The ellipse with equation $4x^2 + 9y^2 + 24x - 72y + 144 = 0$.

If $(\tilde{x}, \tilde{y})$ are coordinates relative to axes rotated by an angle θ as in Figure 21, then

$$x = \tilde{x}\cos\theta - \tilde{y}\sin\theta \qquad \boxed{13}$$

$$y = \tilde{x}\sin\theta + \tilde{y}\cos\theta \qquad \boxed{14}$$

See Exercise 75. In Exercise 76, we show that the cross term disappears when Eq. (12) is rewritten in terms of $\tilde{x}$ and $\tilde{y}$ for the angle

$$\theta = \frac{1}{2}\cot^{-1}\frac{a - c}{b} \qquad \boxed{15}$$

General Equations of Degree 2

The equations of the standard conic sections are special cases of the general equation of degree 2 in x and y:

$$\boxed{ax^2 + bxy + cy^2 + dx + ey + f = 0} \qquad \boxed{12}$$

Here, a, b, c, d, e, f are constants with a, b, c not all zero. It turns out that this general equation of degree 2 does not give rise to any new types of curves. Apart from certain "degenerate cases," Eq. (12) defines a conic section that is not necessarily in standard position: It need not be centered at the origin, and its focal and conjugate axes may be rotated relative to the coordinate axes. For example, the equation

$$6x^2 - 8xy + 8y^2 - 12x - 24y + 38 = 0$$

defines an ellipse with its center at $(3, 3)$ whose axes are rotated (Figure 19).

We say that Eq. (12) is **degenerate** if the set of solutions is a pair of intersecting lines, a pair of parallel lines, a single line, a point, or the empty set. For example:

- $x^2 - y^2 = 0$ defines a pair of intersecting lines $y = x$ and $y = -x$.
- $x^2 - x = 0$ defines a pair of parallel lines $x = 0$ and $x = 1$.
- $x^2 = 0$ defines a single line (the y-axis).
- $x^2 + y^2 = 0$ has just one solution $(0, 0)$.
- $x^2 + y^2 = -1$ has no solutions.

Now assume that Eq. (12) is nondegenerate. The term bxy is called the *cross term*. When the cross term is zero (i.e., when $b = 0$), we can "complete the square" to show that Eq. (12) defines a translate of the conic in standard position. In other words, the axes of the conic are parallel to the coordinate axes. This is illustrated in the next example.

■ **EXAMPLE 7 Completing the Square** Show that

$$4x^2 + 9y^2 + 24x - 72y + 144 = 0$$

defines a translate of a conic section in standard position (Figure 20).

Solution Since there is no cross term, we may complete the square of the terms involving x and y separately:

$$4x^2 + 9y^2 + 24x - 72y + 144 = 0$$

$$4(x^2 + 6x + 9 - 9) + 9(y^2 - 8y + 16 - 16) + 144 = 0$$

$$4(x + 3)^2 - 4(9) + 9(y - 4)^2 - 9(16) + 144 = 0$$

$$4(x + 3)^2 + 9(y - 4)^2 = 36$$

Therefore, this quadratic equation can be rewritten in the form

$$\left(\frac{x + 3}{3}\right)^2 + \left(\frac{y - 4}{2}\right)^2 = 1 \qquad ■$$

When the cross term bxy is nonzero, Eq. (12) defines a conic whose axes are rotated relative to the coordinate axes. The marginal note describes how this may be verified in general. We illustrate with the following example.

■ **EXAMPLE 8** Show that $2xy = 1$ defines a conic section whose focal and conjugate axes are rotated relative to the coordinate axes.

Solution Figure 22(A) shows axes labeled $\tilde{x}$ and $\tilde{y}$ that are rotated by 45° relative to the standard coordinate axes. A point P with coordinates (x, y) may also be described by

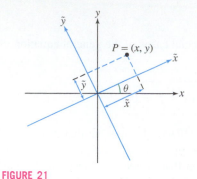

FIGURE 21

coordinates $(\tilde{x}, \tilde{y})$ relative to these rotated axes. Applying Eqs. (13) and (14) with $\theta = \frac{\pi}{4}$, we find that (x, y) and $(\tilde{x}, \tilde{y})$ are related by the formulas

$$x = \frac{\tilde{x} - \tilde{y}}{\sqrt{2}}, \qquad y = \frac{\tilde{x} + \tilde{y}}{\sqrt{2}}$$

Therefore, if $P = (x, y)$ lies on the hyperbola — that is, if $2xy = 1$ — then

$$2xy = 2 \left(\frac{\tilde{x} - \tilde{y}}{\sqrt{2}} \right) \left(\frac{\tilde{x} + \tilde{y}}{\sqrt{2}} \right) = \tilde{x}^2 - \tilde{y}^2 = 1$$

Thus, the coordinates $(\tilde{x}, \tilde{y})$ satisfy the equation of the standard hyperbola $\tilde{x}^2 - \tilde{y}^2 = 1$ whose focal and conjugate axes are the $\tilde{x}$- and $\tilde{y}$-axes, respectively. ∎

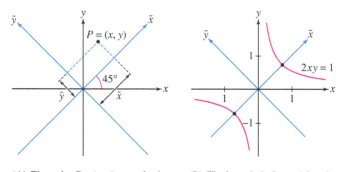

FIGURE 22 The $\tilde{x}$- and $\tilde{y}$-axes are rotated at a 45° angle relative to the x- and y-axes.

(A) The point $P = (x, y)$ may also be described by coordinates $(\tilde{x}, \tilde{y})$ relative to the rotated axis.

(B) The hyperbola $2xy = 1$ has the standard form $\tilde{x}^2 - \tilde{y}^2 = 1$ relative to the $\tilde{x}, \tilde{y}$ axes.

We conclude our discussion of conics by stating the Discriminant Test. Suppose that the equation

$$ax^2 + bxy + cy^2 + dx + ey + f = 0$$

is nondegenerate and thus defines a conic section. According to the Discriminant Test, the type of conic is determined by the **discriminant** D:

$$\boxed{D = b^2 - 4ac}$$

We have the following cases:

- $D < 0$: Ellipse or circle
- $D > 0$: Hyperbola
- $D = 0$: Parabola

For example, the discriminant of the equation $2xy = 1$ is

$$D = b^2 - 4ac = 2^2 - 0 = 4 > 0$$

According to the Discriminant Test, $2xy = 1$ defines a hyperbola. This agrees with our conclusion in Example 8.

11.5 SUMMARY

- An *ellipse* with foci F_1 and F_2 is the set of points P such that $PF_1 + PF_2 = K$, where K is a constant such that $K > F_1 F_2$. The equation in standard position is

$$\left(\frac{x}{a} \right)^2 + \left(\frac{y}{b} \right)^2 = 1$$

The vertices of the ellipse are $(\pm a, 0)$ and $(0, \pm b)$.

	Focal axis	Foci	Focal vertices
$a > b$	x-axis	$(\pm c, 0)$ with $c = \sqrt{a^2 - b^2}$	$(\pm a, 0)$
$a < b$	y-axis	$(0, \pm c)$ with $c = \sqrt{b^2 - a^2}$	$(0, \pm b)$

Eccentricity: $e = \frac{c}{a}$ $(0 \le e < 1)$. Directrix: $x = \frac{a}{e}$ (if $a > b$).

- A *hyperbola* with foci F_1 and F_2 is the set of points P such that

$$PF_1 - PF_2 = \pm K$$

where K is a constant such that $0 < K < F_1 F_2$. The equation in standard position is

$$\left(\frac{x}{a}\right)^2 - \left(\frac{y}{b}\right)^2 = 1$$

Focal axis	Foci	Vertices	Asymptotes
x-axis	$(\pm c, 0)$ with $c = \sqrt{a^2 + b^2}$	$(\pm a, 0)$	$y = \pm \frac{b}{a} x$

Eccentricity: $e = \frac{c}{a}$ $(e > 1)$. Directrix: $x = \frac{a}{e}$.

- A *parabola* with focus F and directrix $\mathcal{D}$ is the set of points P such that $PF = P\mathcal{D}$. The equation in standard position is

$$y = \frac{1}{4c} x^2$$

Focus $F = (0, c)$, directrix $y = -c$, and vertex at the origin $(0, 0)$.
- *Focus-directrix definition* of conic with focus F and directrix $\mathcal{D}$: $PF = eP\mathcal{D}$.
- To translate a conic section h units horizontally and k units vertically, replace x by $x - h$ and y by $y - k$ in the equation.
- Polar equation of conic of eccentricity $e > 0$, focus at the origin, directrix $x = d$:

$$r = \frac{ed}{1 + e\cos\theta}$$

11.5 EXERCISES

Preliminary Questions

1. Decide if the equation defines an ellipse, a hyperbola, a parabola, or no conic section at all.

(a) $4x^2 - 9y^2 = 12$ (b) $-4x + 9y^2 = 0$

(c) $4y^2 + 9x^2 = 12$ (d) $4x^3 + 9y^3 = 12$

2. For which conic sections do the vertices lie between the foci?

3. What are the foci of

$$\left(\frac{x}{a}\right)^2 + \left(\frac{y}{b}\right)^2 = 1 \quad \text{if } a < b?$$

4. What is the geometric interpretation of b/a in the equation of a hyperbola in standard position?

Exercises

In Exercises 1–6, find the vertices and foci of the conic section.

1. $\left(\frac{x}{9}\right)^2 + \left(\frac{y}{4}\right)^2 = 1$ **2.** $\dfrac{x^2}{9} + \dfrac{y^2}{4} = 1$

3. $\left(\frac{x}{4}\right)^2 - \left(\frac{y}{9}\right)^2 = 1$ **4.** $\dfrac{x^2}{4} - \dfrac{y^2}{9} = 36$

5. $\left(\frac{x-3}{7}\right)^2 - \left(\frac{y+1}{4}\right)^2 = 1$

6. $\left(\frac{x-3}{4}\right)^2 + \left(\frac{y+1}{7}\right)^2 = 1$

In Exercises 7–10, find the equation of the ellipse obtained by translating (as indicated) the ellipse

$$\left(\frac{x-8}{6}\right)^2 + \left(\frac{y+4}{3}\right)^2 = 1$$

7. Translated with center at the origin

8. Translated with center at $(-2, -12)$

9. Translated to the right 6 units

10. Translated down 4 units

In Exercises 11–14, find the equation of the given ellipse.

11. Vertices $(\pm 5, 0)$ and $(0, \pm 7)$

12. Foci $(\pm 6, 0)$ and focal vertices $(\pm 10, 0)$

13. Foci $(0, \pm 10)$ and eccentricity $e = \frac{3}{5}$

14. Vertices $(4, 0)$, $(28, 0)$ and eccentricity $e = \frac{2}{3}$

In Exercises 15–20, find the equation of the given hyperbola.

15. Vertices $(\pm 3, 0)$ and foci $(\pm 5, 0)$

16. Vertices $(\pm 3, 0)$ and asymptotes $y = \pm \frac{1}{2}x$

17. Foci $(\pm 4, 0)$ and eccentricity $e = 2$

18. Vertices $(0, \pm 6)$ and eccentricity $e = 3$

19. Vertices $(-3, 0)$, $(7, 0)$ and eccentricity $e = 3$

20. Vertices $(0, -6)$, $(0, 4)$ and foci $(0, -9)$, $(0, 7)$

In Exercises 21–28, find the equation of the parabola with the given properties.

21. Vertex $(0, 0)$, focus $\left(\frac{1}{12}, 0\right)$

22. Vertex $(0, 0)$, focus $(0, 2)$

23. Vertex $(0, 0)$, directrix $y = -5$

24. Vertex $(3, 4)$, directrix $y = -2$

25. Focus $(0, 4)$, directrix $y = -4$

26. Focus $(0, -4)$, directrix $y = 4$

27. Focus $(2, 0)$, directrix $x = -2$

28. Focus $(-2, 0)$, vertex $(2, 0)$

In Exercises 29–38, find the vertices, foci, center (if an ellipse or a hyperbola), and asymptotes (if a hyperbola).

29. $x^2 + 4y^2 = 16$ **30.** $4x^2 + y^2 = 16$

31. $\left(\frac{x-3}{4}\right)^2 - \left(\frac{y+5}{7}\right)^2 = 1$ **32.** $3x^2 - 27y^2 = 12$

33. $4x^2 - 3y^2 + 8x + 30y = 215$

34. $y = 4x^2$ **35.** $y = 4(x-4)^2$

36. $8y^2 + 6x^2 - 36x - 64y + 134 = 0$

37. $4x^2 + 25y^2 - 8x - 10y = 20$

38. $16x^2 + 25y^2 - 64x - 200y + 64 = 0$

In Exercises 39–42, use the Discriminant Test to determine the type of the conic section (in each case, the equation is nondegenerate). Plot the curve if you have a computer algebra system.

39. $4x^2 + 5xy + 7y^2 = 24$

40. $x^2 - 2xy + y^2 + 24x - 8 = 0$

41. $2x^2 - 8xy + 3y^2 - 4 = 0$

42. $2x^2 - 3xy + 5y^2 - 4 = 0$

43. Show that the "conic" $x^2 + 3y^2 - 6x + 12y + 23 = 0$ has no points.

44. For which values of a does the conic $3x^2 + 2y^2 - 16y + 12x = a$ have at least one point?

45. Show that $\dfrac{b}{a} = \sqrt{1 - e^2}$ for a standard ellipse of eccentricity e.

46. Show that the eccentricity of a hyperbola in standard position is $e = \sqrt{1 + m^2}$, where $\pm m$ are the slopes of the asymptotes.

47. Explain why the dots in Figure 23 lie on a parabola. Where are the focus and directrix located?

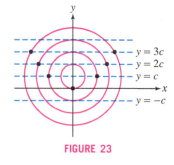

FIGURE 23

48. Find the equation of the ellipse consisting of points P such that $PF_1 + PF_2 = 12$, where $F_1 = (4, 0)$ and $F_2 = (-2, 0)$.

49. A **latus rectum** of a conic section is a chord through a focus parallel to the directrix. Find the area bounded by the parabola $y = x^2/(4c)$ and its latus rectum (refer to Figure 8).

50. Show that the tangent line at a point $P = (x_0, y_0)$ on the hyperbola $\left(\frac{x}{a}\right)^2 - \left(\frac{y}{b}\right)^2 = 1$ has equation

$$Ax - By = 1$$

where $A = \dfrac{x_0}{a^2}$ and $B = \dfrac{y_0}{b^2}$.

In Exercises 51–54, find the polar equation of the conic with the given eccentricity and directrix, and focus at the origin.

51. $e = \frac{1}{2}$, $x = 3$ **52.** $e = \frac{1}{2}$, $x = -3$

53. $e = 1$, $x = 4$ **54.** $e = \frac{3}{2}$, $x = -4$

In Exercises 55–58, identify the type of conic, the eccentricity, and the equation of the directrix.

55. $r = \dfrac{8}{1 + 4\cos\theta}$ **56.** $r = \dfrac{8}{4 + \cos\theta}$

57. $r = \dfrac{8}{4 + 3\cos\theta}$ **58.** $r = \dfrac{12}{4 + 3\cos\theta}$

59. Find a polar equation for the hyperbola with focus at the origin, directrix $x = -2$, and eccentricity $e = 1.2$.

60. Let $\mathcal{C}$ be the ellipse $r = de/(1 + e\cos\theta)$, where $e < 1$. Show that the x-coordinates of the points in Figure 24 are as follows:

Point	A	C	F_2	A'
x-coordinate	$\dfrac{de}{e+1}$	$-\dfrac{de^2}{1-e^2}$	$\dfrac{2de^2}{1-e^2}$	$\dfrac{de}{1-e}$

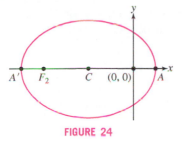

FIGURE 24

61. Find an equation in rectangular coordinates of the conic

$$r = \frac{16}{5 + 3\cos\theta}$$

Hint: Use the results of Exercise 60.

62. Let $e > 1$. Show that the vertices of the hyperbola $r = \dfrac{de}{1 + e\cos\theta}$ have x-coordinates $\dfrac{ed}{e+1}$ and $\dfrac{ed}{e-1}$.

63. Kepler's First Law states that planetary orbits are ellipses with the sun at one focus. The orbit of Pluto has eccentricity $e \approx 0.25$. Its **perihelion** (closest distance to the sun) is approximately 2.7 billion miles. Find the **aphelion** (farthest distance from the sun).

64. Kepler's Third Law states that the ratio $T/a^{3/2}$ is equal to a constant C for all planetary orbits around the sun, where T is the period (time for a complete orbit) and a is the semimajor axis.

(a) Compute C in units of days and kilometers, given that the semimajor axis of the earth's orbit is 150×10^6 km.

(b) Compute the period of Saturn's orbit, given that its semimajor axis is approximately 1.43×10^9 km.

(c) Saturn's orbit has eccentricity $e = 0.056$. Find the perihelion and aphelion of Saturn (see Exercise 63).

Further Insights and Challenges

65. Verify Theorem 2.

66. Verify Theorem 5 in the case $0 < e < 1$. *Hint:* Repeat the proof of Theorem 5, but set $c = d/(e^{-2} - 1)$.

67. Verify that if $e > 1$, then Eq. (11) defines a hyperbola of eccentricity e, with its focus at the origin and directrix at $x = d$.

Reflective Property of the Ellipse   In Exercises 68–70, we prove that the focal radii at a point on an ellipse make equal angles with the tangent line $\mathcal{L}$. Let $P = (x_0, y_0)$ be a point on the ellipse in Figure 25 with foci $F_1 = (-c, 0)$ and $F_2 = (c, 0)$, and eccentricity $e = c/a$.

68. Show that the equation of the tangent line at P is $Ax + By = 1$, where $A = \dfrac{x_0}{a^2}$ and $B = \dfrac{y_0}{b^2}$.

69. Points R_1 and R_2 in Figure 25 are defined so that $\overline{F_1 R_1}$ and $\overline{F_2 R_2}$ are perpendicular to the tangent line.

70. (a) Prove that $PF_1 = a + x_0 e$ and $PF_2 = a - x_0 e$. *Hint:* Show that $PF_1{}^2 - PF_2{}^2 = 4x_0 c$. Then use the defining property $PF_1 + PF_2 = 2a$ and the relation $e = c/a$.

(b) Verify that $\dfrac{F_1 R_1}{PF_1} = \dfrac{F_2 R_2}{PF_2}$.

(c) Show that $\sin\theta_1 = \sin\theta_2$. Conclude that $\theta_1 = \theta_2$.

71. 📖   Here is another proof of the Reflective Property.

(a) Figure 25 suggests that $\mathcal{L}$ is the unique line that intersects the ellipse only in the point P. Assuming this, prove that $QF_1 + QF_2 > PF_1 + PF_2$ for all points Q on the tangent line other than P.

(b) Use the Principle of Least Distance (Example 6 in Section 4.7) to prove that $\theta_1 = \theta_2$.

72. Show that the length QR in Figure 26 is independent of the point P.

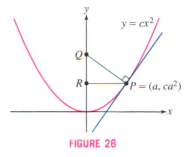

FIGURE 26

73. Show that $y = x^2/4c$ is the equation of a parabola with directrix $y = -c$, focus $(0, c)$, and the vertex at the origin, as stated in Theorem 3.

74. Consider two ellipses in standard position:

$$E_1 : \quad \left(\frac{x}{a_1}\right)^2 + \left(\frac{y}{b_1}\right)^2 = 1$$

$$E_2 : \quad \left(\frac{x}{a_2}\right)^2 + \left(\frac{y}{b_2}\right)^2 = 1$$

We say that E_1 is similar to E_2 under scaling if there exists a factor $r > 0$ such that for all (x, y) on E_1, the point (rx, ry) lies on E_2. Show that E_1 and E_2 are similar under scaling if and only if they have the same eccentricity. Show that any two circles are similar under scaling.

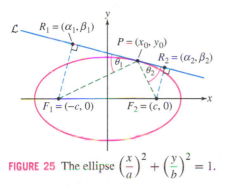

FIGURE 25   The ellipse $\left(\dfrac{x}{a}\right)^2 + \left(\dfrac{y}{b}\right)^2 = 1.$

(a) Show, with A and B as in Exercise 68, that

$$\frac{\alpha_1 + c}{\beta_1} = \frac{\alpha_2 - c}{\beta_2} = \frac{A}{B}$$

(b) Use (a) and the distance formula to show that

$$\frac{F_1 R_1}{F_2 R_2} = \frac{\beta_1}{\beta_2}$$

(c) Use (a) and the equation of the tangent line in Exercise 68 to show that

$$\beta_1 = \frac{B(1 + Ac)}{A^2 + B^2}, \qquad \beta_2 = \frac{B(1 - Ac)}{A^2 + B^2}$$

75. Derive Equations (13) and (14) in the text as follows. Write the coordinates of P with respect to the rotated axes in Figure 21 in polar form $\tilde{x} = r \cos\alpha$, $\tilde{y} = r \sin\alpha$. Explain why P has polar coordinates $(r, \alpha + \theta)$ with respect to the standard x- and y-axes, and derive (13) and (14) using the addition formulas for cosine and sine.

76. If we rewrite the general equation of degree 2 (Eq. 12) in terms of variables $\tilde{x}$ and $\tilde{y}$ that are related to x and y by Eqs. (13) and (14), we obtain a new equation of degree 2 in $\tilde{x}$ and $\tilde{y}$ of the same form but with different coefficients:

$$a'\tilde{x}^2 + b'\tilde{x}\tilde{y} + c'\tilde{y}^2 + d'\tilde{x} + e'\tilde{y} + f' = 0$$

(a) Show that $b' = b\cos 2\theta + (c - a)\sin 2\theta$.

(b) Show that if $b \neq 0$, then we obtain $b' = 0$ for

$$\theta = \frac{1}{2}\cot^{-1}\frac{a - c}{b}$$

This proves that it is always possible to eliminate the cross term bxy by rotating the axes through a suitable angle.

CHAPTER REVIEW EXERCISES

1. Which of the following curves pass through the point $(1, 4)$?

(a) $c(t) = (t^2, t + 3)$ **(b)** $c(t) = (t^2, t - 3)$

(c) $c(t) = (t^2, 3 - t)$ **(d)** $c(t) = (t - 3, t^2)$

2. Find parametric equations for the line through $P = (2, 5)$ perpendicular to the line $y = 4x - 3$.

3. Find parametric equations for the circle of radius 2 with center $(1, 1)$. Use the equations to find the points of intersection of the circle with the x- and y-axes.

4. Find a parametrization $c(t)$ of the line $y = 5 - 2x$ such that $c(0) = (2, 1)$.

5. Find a parametrization $c(\theta)$ of the unit circle such that $c(0) = (-1, 0)$.

6. Find a path $c(t)$ that traces the parabolic arc $y = x^2$ from $(0, 0)$ to $(3, 9)$ for $0 \leq t \leq 1$.

7. Find a path $c(t)$ that traces the line $y = 2x + 1$ from $(1, 3)$ to $(3, 7)$ for $0 \leq t \leq 1$.

8. Sketch the graph $c(t) = (1 + \cos t, \sin 2t)$ for $0 \leq t \leq 2\pi$ and draw arrows specifying the direction of motion.

In Exercises 9–12, express the parametric curve in the form $y = f(x)$.

9. $c(t) = (4t - 3, 10 - t)$ **10.** $c(t) = (t^3 + 1, t^2 - 4)$

11. $c(t) = \left(3 - \frac{2}{t}, t^3 + \frac{1}{t}\right)$ **12.** $x = \tan t$, $y = \sec t$

In Exercises 13–16, calculate dy/dx at the point indicated.

13. $c(t) = (t^3 + t, t^2 - 1)$, $t = 3$

14. $c(\theta) = (\tan^2\theta, \cos\theta)$, $\theta = \frac{\pi}{4}$

15. $c(t) = (e^t - 1, \sin t)$, $t = 20$

16. $c(t) = (\ln t, 3t^2 - t)$, $P = (0, 2)$

17. [CAS] Find the point on the cycloid $c(t) = (t - \sin t, 1 - \cos t)$ where the tangent line has slope $\frac{1}{2}$.

18. Find the points on $(t + \sin t, t - 2\sin t)$ where the tangent is vertical or horizontal.

19. Find the equation of the Bézier curve with control points

$$P_0 = (-1, -1), \quad P_1 = (-1, 1), \quad P_2 = (1, 1), \quad P_3(1, -1)$$

20. Find the speed at $t = \frac{\pi}{4}$ of a particle whose position at time t seconds is $c(t) = (\sin 4t, \cos 3t)$.

21. Find the speed (as a function of t) of a particle whose position at time t seconds is $c(t) = (\sin t + t, \cos t + t)$. What is the particle's maximal speed?

22. Find the length of $(3e^t - 3, 4e^t + 7)$ for $0 \leq t \leq 1$.

In Exercises 23 and 24, let $c(t) = (e^{-t}\cos t, e^{-t}\sin t)$.

23. Show that $c(t)$ for $0 \leq t < \infty$ has finite length and calculate its value.

24. Find the first positive value of t_0 such that the tangent line to $c(t_0)$ is vertical, and calculate the speed at $t = t_0$.

25. [CAS] Plot $c(t) = (\sin 2t, 2\cos t)$ for $0 \leq t \leq \pi$. Express the length of the curve as a definite integral, and approximate it using a computer algebra system.

26. Convert the points $(x, y) = (1, -3), (3, -1)$ from rectangular to polar coordinates.

27. Convert the points $(r, \theta) = \left(1, \frac{\pi}{6}\right), \left(3, \frac{5\pi}{4}\right)$ from polar to rectangular coordinates.

28. Write $(x + y)^2 = xy + 6$ as an equation in polar coordinates.

29. Write $r = \frac{2\cos\theta}{\cos\theta - \sin\theta}$ as an equation in rectangular coordinates.

30. Show that $r = \frac{4}{7\cos\theta - \sin\theta}$ is the polar equation of a line.

31. [GU] Convert the equation

$$9(x^2 + y^2) = (x^2 + y^2 - 2y)^2$$

to polar coordinates, and plot it with a graphing utility.

32. Calculate the area of the circle $r = 3\sin\theta$ bounded by the rays $\theta = \frac{\pi}{3}$ and $\theta = \frac{2\pi}{3}$.

33. Calculate the area of one petal of $r = \sin 4\theta$ (see Figure 1).

34. The equation $r = \sin(n\theta)$, where $n \geq 2$ is even, is a "rose" of $2n$ petals (Figure 1). Compute the total area of the flower, and show that it does not depend on n.

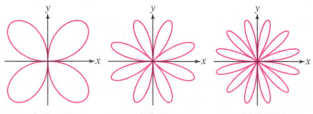

$n = 2$ (4 petals) $n = 4$ (8 petals) $n = 6$ (12 petals)

FIGURE 1 Plot of $r = \sin(n\theta)$.

35. Calculate the total area enclosed by the curve $r^2 = \cos\theta\, e^{\sin\theta}$ (Figure 2).

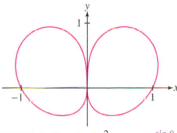

FIGURE 2 Graph of $r^2 = \cos\theta\, e^{\sin\theta}$.

36. Find the shaded area in Figure 3.

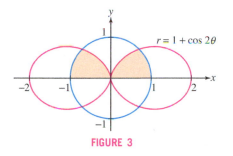

$r = 1 + \cos 2\theta$

FIGURE 3

37. Find the area enclosed by the cardioid $r = a(1 + \cos\theta)$, where $a > 0$.

38. Calculate the length of the curve with polar equation $r = \theta$ in Figure 4.

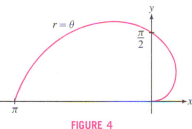

$r = \theta$

FIGURE 4

39. ⌐ℛ⌐ Figure 5 shows the graph of $r = e^{0.5\theta}\sin\theta$ for $0 \le \theta \le 2\pi$. Use a computer algebra system to approximate the difference in length between the outer and inner loops.

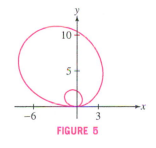

FIGURE 5

40. 📖 Show that $r = f_1(\theta)$ and $r = f_2(\theta)$ define the same curves in polar coordinates if $f_1(\theta) = -f_2(\theta + \pi)$. Use this to show that the following define the same conic section:

$$r = \frac{de}{1 - e\cos\theta}, \qquad r = \frac{-de}{1 + e\cos\theta}$$

In Exercises 41–44, identify the conic section. Find the vertices and foci.

41. $\left(\dfrac{x}{3}\right)^2 + \left(\dfrac{y}{2}\right)^2 = 1$

42. $x^2 - 2y^2 = 4$

43. $\left(2x + \tfrac{1}{2}y\right)^2 = 4 - (x - y)^2$

44. $(y - 3)^2 = 2x^2 - 1$

In Exercises 45–50, find the equation of the conic section indicated.

45. Ellipse with vertices $(\pm 8, 0)$, foci $(\pm\sqrt{3}, 0)$

46. Ellipse with foci $(\pm 8, 0)$, eccentricity $\tfrac{1}{8}$

47. Hyperbola with vertices $(\pm 8, 0)$, asymptotes $y = \pm\tfrac{3}{4}x$

48. Hyperbola with foci $(2, 0)$ and $(10, 0)$, eccentricity $e = 4$

49. Parabola with focus $(8, 0)$, directrix $x = -8$

50. Parabola with vertex $(4, -1)$, directrix $x = 15$

51. Find the asymptotes of the hyperbola $3x^2 + 6x - y^2 - 10y = 1$.

52. Show that the "conic section" with equation $x^2 - 4x + y^2 + 5 = 0$ has no points.

53. Show that the relation $\dfrac{dy}{dx} = (e^2 - 1)\dfrac{x}{y}$ holds on a standard ellipse or hyperbola of eccentricity e.

54. The orbit of Jupiter is an ellipse with the sun at a focus. Find the eccentricity of the orbit if the perihelion (closest distance to the sun) equals 740×10^6 km and the aphelion (farthest distance from the sun) equals 816×10^6 km.

55. Refer to Figure 25 in Section 11.5. Prove that the product of the perpendicular distances F_1R_1 and F_2R_2 from the foci to a tangent line of an ellipse is equal to the square b^2 of the semiminor axes.

A THE LANGUAGE OF MATHEMATICS

One of the challenges in learning calculus is growing accustomed to its precise language and terminology, especially in the statements of theorems. In this section, we analyze a few details of logic that are helpful, and indeed essential, in understanding and applying theorems properly.

Many theorems in mathematics involve an **implication**. If A and B are statements, then the implication $A \implies B$ is the assertion that A implies B:

$$A \implies B: \qquad \textit{If A is true, then B is true.}$$

Statement A is called the **hypothesis** (or premise) and statement B the **conclusion** of the implication. Here is an example: *If m and n are even integers, then m + n is an even integer.* This statement may be divided into a hypothesis and conclusion:

$$\underbrace{m \text{ and } n \text{ are even integers}}_{A} \implies \underbrace{m + n \text{ is an even integer}}_{B}$$

In everyday speech, implications are often used in a less precise way. An example is: *If you work hard, then you will succeed.* Furthermore, some statements that do not initially have the form $A \implies B$ may be restated as implications. For example, the statement, "Cats are mammals," can be rephrased as follows:

$$\text{Let } X \text{ be an animal.} \quad \underbrace{X \text{ is a cat}}_{A} \implies \underbrace{X \text{ is a mammal}}_{B}$$

When we say that an implication $A \implies B$ is true, we do not claim that A or B is necessarily true. Rather, we are making the conditional statement that *if* A happens to be true, *then* B is also true. In the above, if X does not happen to be a cat, the implication tells us nothing.

The **negation** of a statement A is the assertion that A is false and is denoted $\neg A$.

Statement A	Negation $\neg A$
X lives in California.	X does not live in California.
$\triangle ABC$ is a right triangle.	$\triangle ABC$ is not a right triangle.

The negation of the negation is the original statement: $\neg(\neg A) = A$. To say that X does *not not live in California* is the same as saying that X *lives in California*.

■ **EXAMPLE 1** State the negation of each statement.

(a) The door is open and the dog is barking.

(b) The door is open or the dog is barking (or both).

Solution

(a) The first statement is true if two conditions are satisfied (door open and dog barking), and it is false if at least one of these conditions is not satisfied. So the negation is

Either the door is not open *OR* the dog is not barking *(or both)*.

(b) The second statement is true if at least one of the conditions (door open or dog barking) is satisfied, and it is false if neither condition is satisfied. So the negation is

The door is not open *AND* the dog is not barking. ■

Contrapositive and Converse

Two important operations are the formation of the contrapositive and the formation of the converse of a statement. The **contrapositive** of $A \implies B$ is the statement "If B is false, then A is false":

> The contrapositive of $A \implies B$ is $\neg B \implies \neg A$.

Keep in mind that when we form the contrapositive, we reverse the order of A and B. The contrapositive of $A \implies B$ is NOT $\neg A \implies \neg B$.

Here are some examples:

Statement	Contrapositive
If X is a cat, then X is a mammal.	If X is not a mammal, then X is not a cat.
If you work hard, then you will succeed.	If you did not succeed, then you did not work hard.
If m and n are both even, then $m + n$ is even.	If $m + n$ is not even, then m and n are not both even.

A key observation is this:

The contrapositive and the original implication are equivalent.

The fact that $A \implies B$ is equivalent to its contrapositive $\neg B \implies \neg A$ is a general rule of logic that does not depend on what A and B happen to mean. This rule belongs to the subject of "formal logic," which deals with logical relations between statements without concern for the actual content of these statements.

In other words, if an implication is true, then its contrapositive is automatically true, and vice versa. In essence, an implication and its contrapositive are two ways of saying the same thing. For example, the contrapositive, "If X is not a mammal, then X is not a cat," is a roundabout way of saying that cats are mammals.

The **converse** of $A \implies B$ is the *reverse* implication $B \implies A$:

Implication: $A \implies B$	Converse $B \implies A$
If A is true, then B is true.	If B is true, then A is true.

The converse plays a very different role than the contrapositive because *the converse is NOT equivalent to the original implication*. The converse may be true or false, even if the original implication is true. Here are some examples:

True Statement	Converse	Converse True or False?
If X is a cat, then X is a mammal.	If X is a mammal, then X is a cat.	False
If m is even, then m^2 is even.	If m^2 is even, then m is even.	True

A counterexample is an example that satisfies the hypothesis but not the conclusion of a statement. If a single counterexample exists, then the statement is false. However, we cannot prove that a statement is true merely by giving an example.

■ **EXAMPLE 2 An Example Where the Converse Is False** Show that the converse of, "If m and n are even, then $m + n$ is even," is false.

Solution The converse is, "If $m + n$ is even, then m and n are even." To show that the converse is false, we display a counterexample. Take $m = 1$ and $n = 3$ (or any other pair of odd numbers). The sum is even (since $1 + 3 = 4$) but neither 1 nor 3 is even. Therefore, the converse is false. ■

■ **EXAMPLE 3 An Example Where the Converse Is True** State the contrapositive and converse of the Pythagorean Theorem. Are either or both of these true?

Solution Consider a triangle with sides a, b, and c, and let θ be the angle opposite the side of length c, as in Figure 1. The Pythagorean Theorem states that if $\theta = 90°$, then $a^2 + b^2 = c^2$. Here are the contrapositive and converse:

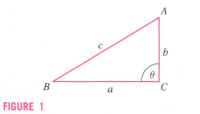

FIGURE 1

Pythagorean Theorem	$\theta = 90° \implies a^2 + b^2 = c^2$	True
Contrapositive	$a^2 + b^2 \neq c^2 \implies \theta \neq 90°$	Automatically true
Converse	$a^2 + b^2 = c^2 \implies \theta = 90°$	True (but not automatic)

The contrapositive is automatically true because it is just another way of stating the original theorem. The converse is not automatically true since there could conceivably exist a nonright triangle that satisfies $a^2 + b^2 = c^2$. However, the converse of the Pythagorean Theorem is, in fact, true. This follows from the Law of Cosines (see Exercise 38). ∎

When both a statement $A \implies B$ and its converse $B \implies A$ are true, we write $A \iff B$. In this case, A and B are **equivalent**. We often express this with the phrase

$$A \iff B \qquad A \text{ is true } \textit{if and only if } B \text{ is true.}$$

For example,

$$a^2 + b^2 = c^2 \qquad \text{if and only if} \qquad \theta = 90°$$

$$\text{It is morning} \qquad \text{if and only if} \qquad \text{the sun is rising.}$$

We mention the following variations of terminology involving implications that you may come across:

Statement	Is Another Way of Saying
A is true <u>if</u> B is true.	$B \implies A$
A is true <u>only if</u> B is true.	$A \implies B$ (A cannot be true unless B is also true.)
For A to be true, <u>it is necessary</u> that B be true.	$A \implies B$ (A cannot be true unless B is also true.)
For A to be true, <u>it is sufficient</u> that B be true.	$B \implies A$
For A to be true, it is <u>necessary and sufficient</u> that B be true.	$B \iff A$

Analyzing a Theorem

To see how these rules of logic arise in calculus, consider the following result from Section 4.2:

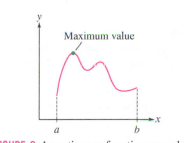
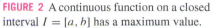

FIGURE 2 A continuous function on a closed interval $I = [a, b]$ has a maximum value.

> **THEOREM 1 Existence of Extrema on a Closed Interval** A continuous function f on a closed (bounded) interval $I = [a, b]$ takes on both a minimum and a maximum value on I (Figure 2).

To analyze this theorem, let's write out the hypotheses and conclusion separately:

Hypotheses A: f is continuous and I is closed.

Conclusion B: f takes on a minimum and a maximum value on I.

A first question to ask is: "Are the hypotheses necessary?" Is the conclusion still true if we drop one or both assumptions? To show that both hypotheses are necessary, we provide counterexamples:

- **The continuity of f is a necessary hypothesis.** Figure 3(A) shows the graph of a function on a closed interval $[a, b]$ that is not continuous. This function has no maximum value on $[a, b]$, which shows that the conclusion may fail if the continuity hypothesis is not satisfied.
- **The hypothesis that I is closed is necessary.** Figure 3(B) shows the graph of a continuous function on an *open interval* (a, b). This function has no maximum value, which shows that the conclusion may fail if the interval is not closed.

We see that both hypotheses in Theorem 1 are necessary. In stating this, we do not claim that the conclusion *always* fails when one or both of the hypotheses are not satisfied. We claim only that the conclusion *may* fail when the hypotheses are not satisfied. Next, let's analyze the contrapositive and converse:

- **Contrapositive $\neg B \implies \neg A$ (automatically true):** If f does not have a minimum and a maximum value on I, then either f is not continuous or I is not closed (or both).
- **Converse $B \implies A$ (in this case, false):** If f has a minimum and a maximum value on I, then f is continuous and I is closed. We prove this statement false with a counterexample [Figure 3(C)].

As we know, the contrapositive is merely a way of restating the theorem, so it is automatically true. The converse is not automatically true, and in fact, in this case it is false. The function in Figure 3(C) provides a counterexample to the converse: f has a maximum value on $I = (a, b)$, but f is not continuous and I is not closed.

The technique of proof by contradiction is also known by its Latin name reductio ad absurdum or "reduction to the absurd." The ancient Greek mathematicians used proof by contradiction as early as the fifth century BCE, and Euclid (325–265 BCE) employed it in his classic treatise on geometry entitled The Elements. A famous example is the proof that $\sqrt{2}$ is irrational in Example 4. The philosopher Plato (427–347 BCE) wrote: "He is unworthy of the name of man who is ignorant of the fact that the diagonal of a square is incommensurable with its side."

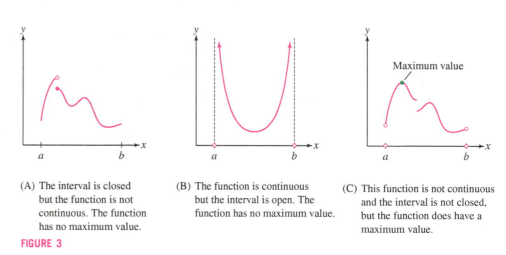

(A) The interval is closed but the function is not continuous. The function has no maximum value.

(B) The function is continuous but the interval is open. The function has no maximum value.

(C) This function is not continuous and the interval is not closed, but the function does have a maximum value.

FIGURE 3

Mathematicians have devised various general strategies and methods for proving theorems. The method of proof by induction is discussed in Appendix C. Another important method is **proof by contradiction**, also called **indirect proof**. Suppose our goal is to prove statement A. In a proof by contradiction, we start by assuming that A is false, and then show that this leads to a contradiction. Therefore, A must be true (to avoid the contradiction).

FIGURE 4 The diagonal of the unit square has length $\sqrt{2}$.

■ **EXAMPLE 4 Proof by Contradiction** The number $\sqrt{2}$ is irrational (Figure 4).

Solution Assume that the theorem is false, namely that $\sqrt{2} = p/q$, where p and q are whole numbers. We may assume that p/q is in lowest terms, and therefore, at most one of p and q is even. Note that if the square m^2 of a whole number is even, then m itself must be even.

The relation $\sqrt{2} = p/q$ implies that $2 = p^2/q^2$ or $p^2 = 2q^2$. This shows that p must be even. But if p is even, then $p = 2m$ for some whole number m, and $p^2 = 4m^2$. Because $p^2 = 2q^2$, we obtain $4m^2 = 2q^2$, or $q^2 = 2m^2$. This shows that q is also even. But we chose p and q so that at most one of them is even. This contradiction shows that our original assumption, that $\sqrt{2} = p/q$, must be false. Therefore, $\sqrt{2}$ is irrational. ■

One of the most famous problems in mathematics is known as "Fermat's Last Theorem." It states that the equation

$$x^n + y^n = z^n$$

has no solutions in positive integers if $n \geq 3$. In a marginal note written around 1630, Fermat claimed to have a proof, and over the centuries, that assertion was verified for many values of the exponent n. However, only in 1994 did the British-American mathematician Andrew Wiles, working at Princeton University, find a complete proof.

CONCEPTUAL INSIGHT   The hallmark of mathematics is precision and rigor. A theorem is established, not through observation or experimentation, but by a proof that consists of a chain of reasoning with no gaps.

This approach to mathematics comes down to us from the ancient Greek mathematicians, especially Euclid, and it remains the standard in contemporary research. In recent decades, the computer has become a powerful tool for mathematical experimentation and data analysis. Researchers may use experimental data to discover potential new mathematical facts, but the title "theorem" is not bestowed until someone writes down a proof.

This insistence on theorems and proofs distinguishes mathematics from the other sciences. In the natural sciences, facts are established through experiment and are subject to change or modification as more knowledge is acquired. In mathematics, theories are also developed and expanded, but previous results are not invalidated. The Pythagorean Theorem was discovered in antiquity and is a cornerstone of plane geometry. In the nineteenth century, mathematicians began to study more general types of geometry (of the type that eventually led to Einstein's four-dimensional space-time geometry in the Theory of Relativity). The Pythagorean Theorem does not hold in these more general geometries, but its status in plane geometry is unchanged.

A. SUMMARY

- The implication $A \Longrightarrow B$ is the assertion, "If A is true, then B is true."
- The *contrapositive* of $A \Longrightarrow B$ is the implication $\neg B \Longrightarrow \neg A$, which says, "If B is false, then A is false." An implication and its contrapositive are equivalent (one is true if and only if the other is true).
- The *converse* of $A \Longrightarrow B$ is $B \Longrightarrow A$. An implication and its converse are not necessarily equivalent. One may be true and the other false.
- A and B are *equivalent* if $A \Longrightarrow B$ and $B \Longrightarrow A$ are both true.
- In a proof by contradiction (in which the goal is to prove statement A), we start by assuming that A is false and show that this assumption leads to a contradiction.

A. EXERCISES

Preliminary Questions

1. Which is the contrapositive of $A \Longrightarrow B$?

(a) $B \Longrightarrow A$ (b) $\neg B \Longrightarrow A$

(c) $\neg B \Longrightarrow \neg A$ (d) $\neg A \Longrightarrow \neg B$

2. Which of the choices in Question 1 is the converse of $A \Longrightarrow B$?

3. Suppose that $A \Longrightarrow B$ is true. Which is then automatically true, the converse or the contrapositive?

4. Restate as an implication: "A triangle is a polygon."

Exercises

1. Which is the negation of the statement, "The car and the shirt are both blue"?

(a) Neither the car nor the shirt is blue.

(b) The car is not blue and/or the shirt is not blue.

2. Which is the contrapositive of the implication, "If the car has gas, then it will run"?

(a) If the car has no gas, then it will not run.

(b) If the car will not run, then it has no gas.

In Exercises 3–8, state the negation.

3. The time is 4 o'clock.

4. $\triangle ABC$ is an isosceles triangle.

5. m and n are odd integers.

6. Either m is odd or n is odd.

7. x is a real number and y is an integer.

8. f is a linear function.

In Exercises 9–14, state the contrapositive and converse.

9. If m and n are odd integers, then mn is odd.

10. If today is Tuesday, then we are in Belgium.

11. If today is Tuesday, then we are not in Belgium.

12. If $x > 4$, then $x^2 > 16$.

13. If m^2 is divisible by 3, then m is divisible by 3.

14. If $x^2 = 2$, then x is irrational.

In Exercise 15–18, give a counterexample to show that the converse of the statement is false.

15. If m is odd, then $2m + 1$ is also odd.

16. If $\triangle ABC$ is equilateral, then it is an isosceles triangle.

17. If m is divisible by 9 and 4, then m is divisible by 12.

18. If m is odd, then $m^3 - m$ is divisible by 3.

In Exercise 19–22, determine whether the converse of the statement is false.

19. If $x > 4$ and $y > 4$, then $x + y > 8$.

20. If $x > 4$, then $x^2 > 16$.

21. If $|x| > 4$, then $x^2 > 16$.

22. If m and n are even, then mn is even.

In Exercises 23 and 24, state the contrapositive and converse (it is not necessary to know what these statements mean).

23. If f and g are differentiable, then fg is differentiable.

24. If the force field is radial and decreases as the inverse square of the distance, then all closed orbits are ellipses.

*In Exercises 25–28, the **inverse** of $A \Longrightarrow B$ is the implication $\neg A \Longrightarrow \neg B$.*

25. Which of the following is the inverse of the implication, "If she jumped in the lake, then she got wet"?

(a) If she did not get wet, then she did not jump in the lake.

(b) If she did not jump in the lake, then she did not get wet.

Is the inverse true?

26. State the inverses of these implications:

(a) If X is a mouse, then X is a rodent.

(b) If you sleep late, you will miss class.

(c) If a star revolves around the sun, then it's a planet.

27. Explain why the inverse is equivalent to the converse.

28. State the inverse of the Pythagorean Theorem. Is it true?

29. Theorem 1 in Section 2.4 states the following: "If f and g are continuous functions, then $f + g$ is continuous." Does it follow logically that if f and g are not continuous, then $f + g$ is not continuous?

30. Write out a proof by contradiction for this fact: There is no smallest positive rational number. Base your proof on the fact that if $r > 0$, then $0 < r/2 < r$.

31. Use proof by contradiction to prove that if $x + y > 2$, then $x > 1$ or $y > 1$ (or both).

In Exercises 32–35, use proof by contradiction to show that the number is irrational.

32. $\sqrt{\frac{1}{2}}$ **33.** $\sqrt{3}$ **34.** $\sqrt[3]{2}$ **35.** $\sqrt[4]{11}$

36. An isosceles triangle is a triangle with two equal sides. The following theorem holds: If $\triangle$ is a triangle with two equal angles, then $\triangle$ is an isosceles triangle.

(a) What is the hypothesis?

(b) Show by providing a counterexample that the hypothesis is necessary.

(c) What is the contrapositive?

(d) What is the converse? Is it true?

37. Consider the following theorem: Let f be a quadratic polynomial with a positive leading coefficient. Then f has a minimum value.

(a) What are the hypotheses?

(b) What is the contrapositive?

(c) What is the converse? Is it true?

Further Insights and Challenges

38. Let a, b, and c be the sides of a triangle and let θ be the angle opposite c. Use the Law of Cosines (Theorem 1 in Section 1.4) to prove the converse of the Pythagorean Theorem.

39. Carry out the details of the following proof by contradiction that $\sqrt{2}$ is irrational (this proof is due to R. Palais). If $\sqrt{2}$ is rational, then $n\sqrt{2}$ is a whole number for some whole number n. Let n be the smallest such whole number and let $m = n\sqrt{2} - n$.

(a) Prove that $m < n$.

(b) Prove that $m\sqrt{2}$ is a whole number.

Explain why (a) and (b) imply that $\sqrt{2}$ is irrational.

40. Generalize the argument of Exercise 39 to prove that $\sqrt{A}$ is irrational if A is a whole number but not a perfect square. *Hint:* Choose n

as before and let $m = n\sqrt{A} - n\lfloor \sqrt{A} \rfloor$, where $\lfloor x \rfloor$ is the greatest integer function.

41. Generalize further and show that for any whole number r, the rth root $\sqrt[r]{A}$ is irrational unless A is an rth power. *Hint:* Let $x = \sqrt[r]{A}$. Show that if x is rational, then we may choose a smallest whole number n such that nx^j is a whole number for $j = 1, \dots, r - 1$. Then consider $m = nx - n[x]$ as before.

42. Given a finite list of prime numbers $p_1, \dots, p_N$, let $M = p_1 \cdot p_2 \cdots p_N + 1$. Show that M is not divisible by any of the primes $p_1, \dots, p_N$. Use this and the fact that every number has a prime factorization to prove that there exist infinitely many prime numbers. This argument was advanced by Euclid in *The Elements*.

B PROPERTIES OF REAL NUMBERS

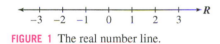

FIGURE 1 The real number line.

In this appendix, we discuss the basic properties of real numbers. First, let us recall that a real number is a number that may be represented by a finite or infinite decimal (also called a decimal expansion). The set of all real numbers is denoted **R** and is often visualized as the "number line" (Figure 1).

Thus, a real number a is represented as

$$a = \pm n.a_1 a_2 a_3 a_4 \ldots,$$

where n is any whole number and each digit a_j is a whole number between 0 and 9. For example, $10\pi = 31.41592\ldots$. Recall that a is rational if its expansion is finite or repeating, and is irrational if its expansion is nonrepeating. Furthermore, the decimal expansion is unique apart from the following exception: Every finite expansion is equal to an expansion in which the digit 9 repeats. For example, $0.5 = 0.4999\cdots = 0.49\bar{9}$.

We shall take for granted that the operations of addition and multiplication are defined on **R**—that is, on the set of all decimals. Roughly speaking, addition and multiplication of infinite decimals are defined in terms of finite decimals. For $d \geq 1$, define the dth truncation of $a = n.a_1 a_2 a_3 a_4 \ldots$ to be the finite decimal $a(d) = a.a_1 a_2 \ldots a_d$ obtained by truncating at the dth place. To form the sum $a + b$, assume that both a and b are infinite (possibly ending with repeated nines). This eliminates any possible ambiguity in the expansion. Then the nth digit of $a + b$ is equal to the nth digit of $a(d) + b(d)$ for d sufficiently large [from a certain point onward, the nth digit of $a(d) + b(d)$ no longer changes, and this value is the nth digit of $a + b$]. Multiplication is defined similarly. Furthermore, the Commutative, Associative, and Distributive Laws hold (Table 1).

TABLE 1	Algebraic Laws
Commutative Laws:	$a + b = b + a, \quad ab = ba$
Associative Laws:	$(a + b) + c = a + (b + c), \quad (ab)c = a(bc)$
Distributive Law:	$a(b + c) = ab + ac$

Every real number x has an additive inverse $-x$ such that $x + (-x) = 0$, and every nonzero real number x has a multiplicative inverse x^{-1} such that $x(x^{-1}) = 1$. We do not regard subtraction and division as separate algebraic operations because they are defined in terms of inverses. By definition, the difference $x - y$ is equal to $x + (-y)$, and the quotient x/y is equal to $x(y^{-1})$ for $y \neq 0$.

In addition to the algebraic operations, there is an **order relation** on **R**: For any two real numbers a and b, precisely one of the following is true:

$$\text{Either} \quad a = b, \quad \text{or} \quad a < b, \quad \text{or} \quad a > b$$

To distinguish between the conditions $a \leq b$ and $a < b$, we often refer to $a < b$ as a **strict inequality**. Similar conventions hold for $>$ and $\geq$. The rules given in Table 2 allow us to manipulate inequalities. The last order property says that an inequality reverses direction when multiplied by a negative number c. For example,

$$-2 < 5 \quad \text{but} \quad (-3)(-2) > (-3)5$$

TABLE 2 Order Properties

If $a < b$ and $b < c$, then $a < c$.
If $a < b$ and $c < d$, then $a + c < b + d$.
If $a < b$ and $c > 0$, then $ac < bc$.
If $a < b$ and $c < 0$, then $ac > bc$.

The algebraic and order properties of real numbers are certainly familiar. We now discuss the less familiar **Least Upper Bound (LUB) Property** of the real numbers. This property is one way of expressing the so-called **completeness** of the real numbers. There are other ways of formulating completeness (such as the so-called nested interval property discussed in any book on analysis) that are equivalent to the LUB Property and serve the same purpose. Completeness is used in calculus to construct rigorous proofs of basic theorems about continuous functions, such as the Intermediate Value Theorem, (IVT) or the existence of extreme values on a closed interval. The underlying idea is that the real number line "has no holes." We elaborate on this idea below. First, we introduce the necessary definitions.

Suppose that S is a nonempty set of real numbers. A number M is called an **upper bound** for S if

$$x \leq M \qquad \text{for all } x \in S$$

If S has an upper bound, we say that S is **bounded above**. A **least upper bound** L is an upper bound for S such that every other upper bound M satisfies $M \geq L$. For example (Figure 2),

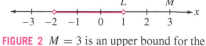

FIGURE 2 $M = 3$ is an upper bound for the set $S = (-2, 1)$. The LUB is $L = 1$.

- $M = 3$ is an upper bound for the open interval $S = (-2, 1)$.
- $L = 1$ is the LUB for $S = (-2, 1)$.

We now state the LUB Property of the real numbers.

THEOREM 1 Existence of a Least Upper Bound Let S be a nonempty set of real numbers that is bounded above. Then S has an LUB.

In a similar fashion, we say that a number B is a **lower bound** for S if $x \geq B$ for all $x \in S$. We say that S is **bounded below** if S has a lower bound. A **greatest lower bound** (GLB) is a lower bound M such that every other lower bound B satisfies $B \leq M$. The set of real numbers also has the GLB Property: If S is a nonempty set of real numbers that is bounded below, then S has a GLB. This may be deduced immediately from Theorem 1. For any nonempty set of real numbers S, let $-S$ be the set of numbers of the form $-x$ for $x \in S$. Then $-S$ has an upper bound if S has a lower bound. Consequently, $-S$ has an LUB L by Theorem 1, and $-L$ is a GLB for S.

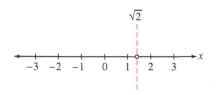

FIGURE 3 The rational numbers have a "hole" at the location $\sqrt{2}$.

CONCEPTUAL INSIGHT Theorem 1 may appear quite reasonable, but perhaps it is not clear why it is useful. We suggested above that the LUB Property expresses the idea that **R** is "complete" or "has no holes." To illustrate this idea, let's compare **R** to the set of rational numbers, denoted **Q**. Intuitively, **Q** is not complete because the irrational numbers are missing. For example, **Q** has a "hole" where the irrational number $\sqrt{2}$ should be located (Figure 3). This hole divides **Q** into two halves that are not connected to each other (the half to the left and the half to the right of $\sqrt{2}$). Furthermore, the half on the left is bounded above but no rational number is an LUB, and the half on the right is bounded below but no rational number is a GLB. The LUB and GLB are both equal to the irrational number $\sqrt{2}$, which exists in only **R** but not **Q**. So unlike **R**, the rational numbers **Q** do not have the LUB property.

■ **EXAMPLE 1** Show that 2 has a square root by applying the LUB Property to the set

$$S = \{x : x^2 < 2\}$$

Solution First, we note that S is bounded with the upper bound $M = 2$. Indeed, if $x > 2$, then x satisfies $x^2 > 4$, and hence x does not belong to S. By the LUB Property, S has a least upper bound. Call it L. We claim that $L = \sqrt{2}$, or, equivalently, that $L^2 = 2$. We prove this by showing that $L^2 \geq 2$ and $L^2 \leq 2$.

If $L^2 < 2$, let $b = L + h$, where $h > 0$. Then

$$b^2 = L^2 + 2Lh + h^2 = L^2 + h(2L + h) \qquad \boxed{1}$$

We can make the quantity $h(2L + h)$ as small as desired by choosing $h > 0$ small enough. In particular, we may choose a positive h so that $h(2L + h) < 2 - L^2$. For this choice, $b^2 < L^2 + (2 - L^2) = 2$ by Eq. (1). Therefore, $b \in S$. But $b > L$ since $h > 0$, and thus

L is not an upper bound for S, in contradiction to our hypothesis on L. We conclude that $L^2 \geq 2$.

If $L^2 > 2$, let $b = L - h$, where $h > 0$. Then

$$b^2 = L^2 - 2Lh + h^2 = L^2 - h(2L - h)$$

Now choose h positive but small enough so that $0 < h(2L - h) < L^2 - 2$. Then $b^2 > L^2 - (L^2 - 2) = 2$. But $b < L$, so b is a smaller lower bound for S. Indeed, if $x \geq b$, then $x^2 \geq b^2 > 2$, and x does not belong to S. This contradicts our hypothesis that L is the LUB. We conclude that $L^2 \leq 2$, and since we have already shown that $L^2 \geq 2$, we have $L^2 = 2$ as claimed. ∎

We now prove three important theorems, the third of which is used in the proof of the LUB Property below.

THEOREM 2 Bolzano–Weierstrass Theorem Let S be a bounded, infinite set of real numbers. Then there exists a sequence of distinct elements $\{a_n\}$ in S such that the limit $L = \lim\limits_{n \to \infty} a_n$ exists.

Proof For simplicity of notation, we assume that S is contained in the unit interval $[0, 1]$ (a similar proof works in general). If $k_1, k_2, \ldots, k_n$ is a sequence of n digits (i.e., each k_j is a whole number and $0 \leq k_j \leq 9$), let

$$S(k_1, k_2, \ldots, k_n)$$

be the set of $x \in S$ whose decimal expansion begins $0.k_1k_2 \ldots k_n$. The set S is the union of the subsets $S(0), S(1), \ldots, S(9)$, and since S is infinite, at least one of these subsets must be infinite. Therefore, we may choose k_1 so that $S(k_1)$ is infinite. In a similar fashion, at least one of the set $S(k_1, 0), S(k_2, 1), \ldots, S(k_1, 9)$ must be infinite, so we may choose k_2 so that $S(k_1, k_2)$ is infinite. Continuing in this way, we obtain an infinite sequence $\{k_n\}$ such that $S(k_1, k_2, \ldots, k_n)$ is infinite for all n. We may choose a sequence of elements $a_n \in S(k_1, k_2, \ldots, k_n)$ with the property that a_n differs from $a_1, \ldots, a_{n-1}$ for all n. Let L be the infinite decimal $0.k_1k_2k_3 \ldots$. Then $\lim\limits_{n \to \infty} a_n = L$ since $|L - a_n| < 10^{-n}$ for all n. ∎

We use the Bolzano–Weierstrass Theorem to prove two important results about sequences $\{a_n\}$. Recall that an upper bound for $\{a_n\}$ is a number M such that $a_j \leq M$ for all j. If an upper bound exists, $\{a_n\}$ is said to be bounded from above. Lower bounds are defined similarly and $\{a_n\}$ is said to be bounded from below if a lower bound exists. A sequence is bounded if it is bounded from above and below. A **subsequence** of $\{a_n\}$ is a sequence of elements $a_{n_1}, a_{n_2}, a_{n_3}, \ldots$, where $n_1 < n_2 < n_3 < \cdots$.

Now consider a bounded sequence $\{a_n\}$. If infinitely many of the a_n are distinct, the Bolzano–Weierstrass Theorem implies that there exists a subsequence $\{a_{n_1}, a_{n_2}, \ldots\}$ such that $\lim\limits_{n \to \infty} a_{n_k}$ exists. Otherwise, infinitely many of the a_n must coincide, and these terms form a convergent subsequence. This proves the next result.

I Section 10.1

THEOREM 3 Every bounded sequence has a convergent subsequence.

THEOREM 4 Bounded Monotonic Sequences Converge

- If $\{a_n\}$ is increasing and $a_n \leq M$ for all n, then $\{a_n\}$ converges and $\lim\limits_{n \to \infty} a_n \leq M$.
- If $\{a_n\}$ is decreasing and $a_n \geq M$ for all n, then $\{a_n\}$ converges and $\lim\limits_{n \to \infty} a_n \geq M$.

Proof Suppose that $\{a_n\}$ is increasing and bounded above by M. Then $\{a_n\}$ is automatically bounded below by $m = a_1$ since $a_1 \leq a_2 \leq a_3 \cdots$. Hence, $\{a_n\}$ is bounded, and by

Theorem 3, we may choose a convergent subsequence $a_{n_1}, a_{n_2}, \ldots$. Let

$$L = \lim_{k \to \infty} a_{n_k}$$

Observe that $a_n \leq L$ for all n. For if not, then $a_n > L$ for some n and then $a_{n_k} \geq a_n > L$ for all k such that $n_k \geq n$. But this contradicts that $a_{n_k} \to L$. Now, by definition, for any $\epsilon > 0$, there exists $N_\epsilon > 0$ such that

$$|a_{n_k} - L| < \epsilon \qquad \text{if } n_k > N_\epsilon$$

Choose m such that $n_m > N_\epsilon$. If $n \geq n_m$, then $a_{n_m} \leq a_n \leq L$, and therefore,

$$|a_n - L| \leq |a_{n_m} - L| < \epsilon \qquad \text{for all } n \geq n_m$$

This proves that $\lim_{n \to \infty} a_n = L$, as desired. It remains to prove that $L \leq M$. If $L > M$, let $\epsilon = (L - M)/2$ and choose N so that

$$|a_n - L| < \epsilon \qquad \text{if } k > N$$

Then $a_n > L - \epsilon = M + \epsilon$. This contradicts our assumption that M is an upper bound for $\{a_n\}$. Therefore, $L \leq M$ as claimed. ■

Proof of Theorem 1 We now use Theorem 4 to prove the LUB Property (Theorem 1). As above, if x is a real number, let $x(d)$ be the truncation of x of length d. For example,

$$\text{If } x = 1.41569, \text{ then } x(3) = 1.415.$$

We say that x is a *decimal of length d* if $x = x(d)$. Any two distinct decimals of length d differ by at least 10^{-d}. It follows that for any two real numbers $A < B$, there are at most finitely many decimals of length d between A and B.

Now let S be a nonempty set of real numbers with an upper bound M. We shall prove that S has an LUB. Let $S(d)$ be the set of truncations of length d:

$$S(d) = \{x(d) : x \in S\}$$

We claim that $S(d)$ has a maximum element. To verify this, choose any $a \in S$. If $x \in S$ and $x(d) > a(d)$, then

$$a(d) \leq x(d) \leq M$$

Thus, by the remark of the previous paragraph, there are at most finitely many values of $x(d)$ in $S(d)$ larger than $a(d)$. The largest of these is the maximum element in $S(d)$.

For $d = 1, 2, \ldots$, choose an element x_d such that $x_d(d)$ is the maximum element in $S(d)$. By construction, $\{x_d(d)\}$ is an increasing sequence (since the largest dth truncation cannot get smaller as d increases). Furthermore, $x_d(d) \leq M$ for all d. We now apply Theorem 4 to conclude that $\{x_d(d)\}$ converges to a limit L. We claim that L is the LUB of S. Observe first that L is an upper bound for S. Indeed, if $x \in S$, then $x(d) \leq L$ for all d and thus $x \leq L$. To show that L is the LUB, suppose that M is an upper bound such that $M < L$. Then $x_d \leq M$ for all d and hence $x_d(d) \leq M$ for all d. But then

$$L = \lim_{d \to \infty} x_d(d) \leq M$$

This is a contradiction since $M < L$. Therefore, L is the LUB of S. ■

As mentioned above, the LUB Property is used in calculus to establish certain basic theorems about continuous functions. As an example, we prove the IVT. Another example is the theorem on the existence of extrema on a closed interval (see Appendix D).

THEOREM 5 Intermediate Value Theorem If f is continuous on a closed interval $[a, b]$ and $f(a) \neq f(b)$, then for every value M between $f(a)$ and $f(b)$, there exists at least one value $c \in (a, b)$ such that $f(c) = M$.

Proof Assume first that $M = 0$. Replacing $f(x)$ by $-f(x)$ if necessary, we may assume that $f(a) < 0$ and $f(b) > 0$. Now let

$$S = \{x \in [a, b] : f(x) < 0\}$$

Then $a \in S$ since $f(a) < 0$ and thus S is nonempty. Clearly, b is an upper bound for S. Therefore, by the LUB Property, S has an LUB L. We claim that $f(L) = 0$. If not, set $r = f(L)$. Assume first that $r > 0$.

Since f is continuous, there exists a number $\delta > 0$ such that

$$\text{if } |x - L| < \delta, \text{ then } |f(x) - f(L)| = |f(x) - r| < \frac{1}{2}r.$$

Equivalently,

$$\text{if } |x - L| < \delta, \text{ then } \frac{1}{2}r < f(x) < \frac{3}{2}r.$$

The number $\frac{1}{2}r$ is positive, so we conclude that

$$\text{if } L - \delta < x < L + \delta, \text{ then } f(x) > 0.$$

By definition of L, $f(x) \geq 0$ for all $x \in [a, b]$ such that $x > L$, and thus $f(x) \geq 0$ for all $x \in [a, b]$ such that $x > L - \delta$. Thus, $L - \delta$ is an upper bound for S. This is a contradiction since L is the LUB of S, and it follows that $r = f(L)$ cannot satisfy $r > 0$. Similarly, r cannot satisfy $r < 0$. We conclude that $f(L) = 0$ as claimed.

Now, if M is nonzero, let $g(x) = f(x) - M$. Then 0 lies between $g(a)$ and $g(b)$, and by what we have proved, there exists $c \in (a, b)$ such that $g(c) = 0$. But then $f(c) = g(c) + M = M$, as desired. ∎

C INDUCTION AND THE BINOMIAL THEOREM

The Principle of Induction is a method of proof that is widely used to prove that a given statement $P(n)$ is valid for all natural numbers $n = 1, 2, 3, \ldots$. Here are two statements of this kind:

- $P(n)$: The sum of the first n odd numbers is equal to n^2.
- $P(n)$: $\dfrac{d}{dx}x^n = nx^{n-1}$.

The first statement claims that for all natural numbers n,

$$\underbrace{1 + 3 + \cdots + (2n - 1)}_{\text{Sum of first } n \text{ odd numbers}} = n^2 \qquad \boxed{1}$$

We can check directly that $P(n)$ is true for the first few values of n:

$$P(1) \text{ is the equality:} \qquad 1 = 1^2 \quad \text{(true)}$$
$$P(2) \text{ is the equality:} \qquad 1 + 3 = 2^2 \quad \text{(true)}$$
$$P(3) \text{ is the equality:} \qquad 1 + 3 + 5 = 3^2 \quad \text{(true)}$$

The Principle of Induction may be used to establish $P(n)$ for all n.

The Principle of Induction applies if $P(n)$ is an assertion defined for $n \geq n_0$, where n_0 is a fixed integer. Assume that

(i) Initial step: $P(n_0)$ is true.

(ii) Induction step: If $P(n)$ is true for $n = k$, then $P(n)$ is also true for $n = k + 1$.

Then $P(n)$ is true for all $n \geq n_0$.

> **THEOREM 1 Principle of Induction**   Let $P(n)$ be an assertion that depends on a natural number n. Assume that
>
> **(i) Initial step:** $P(1)$ is true.
> **(ii) Induction step:** If $P(n)$ is true for $n = k$, then $P(n)$ is also true for $n = k + 1$.
>
> Then $P(n)$ is true for all natural numbers $n = 1, 2, 3, \ldots$.

■ **EXAMPLE 1**   Prove that $1 + 3 + \cdots + (2n - 1) = n^2$ for all natural numbers n.

Solution   As above, we let $P(n)$ denote the equality

$$P(n): \qquad 1 + 3 + \cdots + (2n - 1) = n^2$$

Step 1. **Initial step: Show that $P(1)$ is true.**
We checked this above. $P(1)$ is the equality $1 = 1^2$.

Step 2. **Induction step: Show that if $P(n)$ is true for $n = k$, then $P(n)$ is also true for $n = k + 1$.**
Assume that $P(k)$ is true. Then

$$1 + 3 + \cdots + (2k - 1) = k^2$$

Add $2k + 1$ to both sides:

$$\left[1 + 3 + \cdots + (2k - 1)\right] + (2k + 1) = k^2 + 2k + 1 = (k + 1)^2$$
$$1 + 3 + \cdots + (2k + 1) = (k + 1)^2$$

This is precisely the statement $P(k + 1)$. Thus, $P(k + 1)$ is true whenever $P(k)$ is true. By the Principle of Induction, $P(k)$ is true for all k. ■

The intuition behind the Principle of Induction is the following. If $P(n)$ were not true for all n, then there would exist a smallest natural number k such that $P(k)$ is false. Furthermore, $k > 1$ since $P(1)$ is true. Thus, $P(k-1)$ is true [otherwise, $P(k)$ would not be the smallest "counterexample"]. On the other hand, if $P(k-1)$ is true, then $P(k)$ is also true by the induction step. This is a contradiction. So $P(k)$ must be true for all k.

■ **EXAMPLE 2** Use Induction and the Product Rule to prove that for all whole numbers n,

$$\frac{d}{dx}x^n = nx^{n-1}$$

Solution Let $P(n)$ be the formula $\dfrac{d}{dx}x^n = nx^{n-1}$.

Step 1. **Initial step: Show that $P(1)$ is true.**
We use the limit definition to verify $P(1)$:

$$\frac{d}{dx}x = \lim_{h\to 0}\frac{(x+h)-x}{h} = \lim_{h\to 0}\frac{h}{h} = \lim_{h\to 0} 1 = 1$$

Step 2. **Induction step: Show that if $P(n)$ is true for $n = k$, then $P(n)$ is also true for $n = k+1$.**

To carry out the induction step, assume that $\dfrac{d}{dx}x^k = kx^{k-1}$, where $k \geq 1$. Then, by the Product Rule,

$$\frac{d}{dx}x^{k+1} = \frac{d}{dx}(x \cdot x^k) = x\frac{d}{dx}x^k + x^k\frac{d}{dx}x = x(kx^{k-1}) + x^k$$
$$= kx^k + x^k = (k+1)x^k$$

This shows that $P(k+1)$ is true.

By the Principle of Induction, $P(n)$ is true for all $n \geq 1$.   ■

As another application of induction, we prove the Binomial Theorem, which describes the expansion of the binomial $(a+b)^n$. The first few expansions are familiar:

$$(a+b)^1 = a + b$$
$$(a+b)^2 = a^2 + 2ab + b^2$$
$$(a+b)^3 = a^3 + 3a^2b + 3ab^2 + b^3$$

In general, we have an expansion

$$(a+b)^n = a^n + \binom{n}{1}a^{n-1}b + \binom{n}{2}a^{n-2}b^2 + \binom{n}{3}a^{n-3}b^3$$

$$\boxed{2}$$

$$+ \cdots + \binom{n}{n-1}ab^{n-1} + b^n$$

where the coefficient of $x^{n-k}x^k$, denoted $\dbinom{n}{k}$, is called the **binomial coefficient**. Note that the first term in Eq. (2) corresponds to $k=0$ and the last term to $k=n$; thus, $\dbinom{n}{0} = \dbinom{n}{n} = 1$. In summation notation,

$$(a+b)^n = \sum_{k=0}^{n}\binom{n}{k}a^k b^{n-k}$$

In Pascal's Triangle, the nth row displays the coefficients in the expansion of $(a+b)^n$:

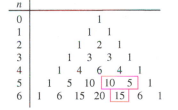

n												
0						1						
1					1		1					
2				1		2		1				
3			1		3		3		1			
4		1		4		6		4		1		
5	1		5		10		10		5		1	
6	1	6		15		20		15		6		1

The triangle is constructed as follows: Each entry is the sum of the two entries above it in the previous line. For example, the entry 15 in line $n = 6$ is the sum $10 + 5$ of the entries above it in line $n = 5$. The recursion relation guarantees that the entries in the triangle are the binomial coefficients.

Pascal's Triangle (described in the marginal note on page A14) can be used to compute binomial coefficients if n and k are not too large. The Binomial Theorem provides the following general formula:

$$\binom{n}{k} = \frac{n!}{k!\,(n-k)!} = \frac{n(n-1)(n-2)\cdots(n-k+1)}{k(k-1)(k-2)\cdots 2 \cdot 1} \qquad \boxed{3}$$

Before proving this formula, we prove a recursion relation for binomial coefficients. Note, however, that Eq. (3) is certainly correct for $k=0$ and $k=n$ (recall that by convention, $0! = 1$):

$$\binom{n}{0} = \frac{n!}{(n-0)!\,0!} = \frac{n!}{n!} = 1, \qquad \binom{n}{n} = \frac{n!}{(n-n)!\,n!} = \frac{n!}{n!} = 1$$

THEOREM 2   Recursion Relation for Binomial Coefficients

$$\binom{n}{k} = \binom{n-1}{k} + \binom{n-1}{k-1} \qquad \text{for } 1 \le k \le n-1$$

Proof We write $(a+b)^n$ as $(a+b)(a+b)^{n-1}$ and expand in terms of binomial coefficients:

$$(a+b)^n = (a+b)(a+b)^{n-1}$$

$$\sum_{k=0}^{n} \binom{n}{k} a^{n-k} b^k = (a+b) \sum_{k=0}^{n-1} \binom{n-1}{k} a^{n-1-k} b^k$$

$$= a \sum_{k=0}^{n-1} \binom{n-1}{k} a^{n-1-k} b^k + b \sum_{k=0}^{n-1} \binom{n-1}{k} a^{n-1-k} b^k$$

$$= \sum_{k=0}^{n-1} \binom{n-1}{k} a^{n-k} b^k + \sum_{k=0}^{n-1} \binom{n-1}{k} a^{n-(k+1)} b^{k+1}$$

Replacing k by $k-1$ in the second sum, we obtain

$$\sum_{k=0}^{n} \binom{n}{k} a^{n-k} b^k = \sum_{k=0}^{n-1} \binom{n-1}{k} a^{n-k} b^k + \sum_{k=1}^{n} \binom{n-1}{k-1} a^{n-k} b^k$$

On the right-hand side, the first term in the first sum is a^n and the last term in the second sum is b^n. Thus, we have

$$\sum_{k=0}^{n} \binom{n}{k} a^{n-k} b^k = a^n + \left(\sum_{k=1}^{n-1} \left(\binom{n-1}{k} + \binom{n-1}{k-1} \right) a^{n-k} b^k \right) + b^n$$

The recursion relation follows because the coefficients of $a^{n-k} b^k$ on the two sides of the equation must be equal. ∎

We now use induction to prove Eq. (3). Let $P(n)$ be the claim

$$\binom{n}{k} = \frac{n!}{k!\,(n-k)!} \qquad \text{for } 0 \le k \le n$$

We have $\binom{1}{0} = \binom{1}{1} = 1$ since $(a+b)^1 = a+b$, so $P(1)$ is true. Furthermore, $\binom{n}{n} = \binom{n}{0} = 1$ as observed above, since a^n and b^n have coefficient 1 in the expansion of $(a + b)^n$. For the inductive step, assume that $P(n)$ is true. By the recursion relation, for $1 \leq k \leq n$, we have

$$\binom{n+1}{k} = \binom{n}{k} + \binom{n}{k-1} = \frac{n!}{k!\,(n-k)!} + \frac{n!}{(k-1)!\,(n-k+1)!}$$

$$= n!\left(\frac{n+1-k}{k!\,(n+1-k)!} + \frac{k}{k!\,(n+1-k)!}\right) = n!\left(\frac{n+1}{k!\,(n+1-k)!}\right)$$

$$= \frac{(n+1)!}{k!\,(n+1-k)!}$$

Thus, $P(n + 1)$ is also true and the Binomial Theorem follows by induction.

■ **EXAMPLE 3** Use the Binomial Theorem to expand $(x + y)^5$ and $(x + 2)^3$.

Solution The fifth row in Pascal's Triangle yields

$$(x + y)^5 = x^5 + 5x^4 y + 10x^3 y^2 + 10x^2 y^3 + 5xy^4 + y^5$$

The third row in Pascal's Triangle yields

$$(x + 2)^3 = x^3 + 3x^2(2) + 3x(2)^2 + 2^3 = x^3 + 6x^2 + 12x + 8 \qquad ■$$

C. EXERCISES

In Exercises 1–4, use the Principle of Induction to prove the formula for all natural numbers n.

1. $1 + 2 + 3 + \cdots + n = \dfrac{n(n+1)}{2}$

2. $1^3 + 2^3 + 3^3 + \cdots + n^3 = \dfrac{n^2(n+1)^2}{4}$

3. $\dfrac{1}{1 \cdot 2} + \dfrac{1}{2 \cdot 3} + \cdots + \dfrac{1}{n(n+1)} = \dfrac{n}{n+1}$

4. $1 + x + x^2 + \cdots + x^n = \dfrac{1 - x^{n+1}}{1 - x}$ for any $x \neq 1$

5. Let $P(n)$ be the statement $2^n > n$.

(a) Show that $P(1)$ is true.

(b) Observe that if $2^n > n$, then $2^n + 2^n > 2n$. Use this to show that if $P(n)$ is true for $n = k$, then $P(n)$ is true for $n = k + 1$. Conclude that $P(n)$ is true for all n.

6. Use induction to prove that $n! > 2^n$ for $n \geq 4$.

Let $\{F_n\}$ be the Fibonacci sequence, defined by the recursion formula

$$F_n = F_{n-1} + F_{n-2}, \qquad F_1 = F_2 = 1$$

The first few terms are $1, 1, 2, 3, 5, 8, 13, \ldots$. In Exercises 7–10, use induction to prove the identity.

7. $F_1 + F_2 + \cdots + F_n = F_{n+2} - 1$

8. $F_1^2 + F_2^2 + \cdots + F_n^2 = F_{n+1} F_n$

9. $F_n = \dfrac{R_+^n - R_-^n}{\sqrt{5}}$, where $R_\pm = \dfrac{1 \pm \sqrt{5}}{2}$

10. $F_{n+1} F_{n-1} = F_n^2 + (-1)^n$. *Hint:* For the induction step, show that

$$F_{n+2} F_n = F_{n+1} F_n + F_n^2$$

$$F_{n+1}^2 = F_{n+1} F_n + F_{n+1} F_{n-1}$$

11. Use induction to prove that $f(n) = 8^n - 1$ is divisible by 7 for all natural numbers n. *Hint:* For the induction step, show that

$$8^{k+1} - 1 = 7 \cdot 8^k + (8^k - 1)$$

12. Use induction to prove that $n^3 - n$ is divisible by 3 for all natural numbers n.

13. Use induction to prove that $5^{2n} - 4^n$ is divisible by 7 for all natural numbers n.

14. Use Pascal's Triangle to write out the expansions of $(a + b)^6$ and $(a - b)^4$.

15. Expand $(x + x^{-1})^4$.

16. What is the coefficient of x^9 in $(x^3 + x)^5$?

17. Let $S(n) = \displaystyle\sum_{k=0}^{n} \binom{n}{k}$.

(a) Use Pascal's Triangle to compute $S(n)$ for $n = 1, 2, 3, 4$.

(b) Prove that $S(n) = 2^n$ for all $n \geq 1$. *Hint:* Expand $(a + b)^n$ and evaluate at $a = b = 1$.

18. Let $T(n) = \displaystyle\sum_{k=0}^{n} (-1)^k \binom{n}{k}$.

(a) Use Pascal's Triangle to compute $T(n)$ for $n = 1, 2, 3, 4$.

(b) Prove that $T(n) = 0$ for all $n \geq 1$. *Hint:* Expand $(a + b)^n$ and evaluate at $a = 1, b = -1$.

D ADDITIONAL PROOFS

In this appendix, we provide proofs of several theorems that were stated or used in the text.

I Section 2.3

> **THEOREM 1** **Basic Limit Laws** Assume that $\lim\limits_{x \to c} f(x)$ and $\lim\limits_{x \to c} g(x)$ exist. Then:
>
> **(i)** $\lim\limits_{x \to c} \big(f(x) + g(x)\big) = \lim\limits_{x \to c} f(x) + \lim\limits_{x \to c} g(x)$
>
> **(ii)** For any number k, $\lim\limits_{x \to c} kf(x) = k \lim\limits_{x \to c} f(x)$
>
> **(iii)** $\lim\limits_{x \to c} f(x)g(x) = \Big(\lim\limits_{x \to c} f(x) \Big) \Big(\lim\limits_{x \to c} g(x) \Big)$
>
> **(iv)** If $\lim\limits_{x \to c} g(x) \neq 0$, then
>
> $$\lim_{x \to c} \frac{f(x)}{g(x)} = \frac{\lim\limits_{x \to c} f(x)}{\lim\limits_{x \to c} g(x)}$$

Proof Let $L = \lim\limits_{x \to c} f(x)$ and $M = \lim\limits_{x \to c} g(x)$. The Sum Law (i) was proved in Section 2.9. Observe that (ii) is a special case of (iii), where $g(x) = k$ is a constant function. Thus, it will suffice to prove the Product Law (iii). We write

$$f(x)g(x) - LM = f(x)(g(x) - M) + M(f(x) - L)$$

and apply the Triangle Inequality to obtain

$$|f(x)g(x) - LM| \leq |f(x)(g(x) - M)| + |M(f(x) - L)| \qquad \boxed{1}$$

By the limit definition, we may choose $\delta > 0$ so that

$$\text{if } 0 < |x - c| < \delta, \text{ then } |f(x) - L| < 1.$$

If follows that $|f(x)| < |L| + 1$ for $0 < |x - c| < \delta$. Now choose any number $\epsilon > 0$. Applying the limit definition again, we see that by choosing a smaller δ if necessary, we may also ensure that if $0 < |x - c| < \delta$, then

$$|f(x) - L| \leq \frac{\epsilon}{2(|M| + 1)} \qquad \text{and} \qquad |g(x) - M| \leq \frac{\epsilon}{2(|L| + 1)}$$

Using Eq. (1), we see that if $0 < |x - c| < \delta$, then

$$|f(x)g(x) - LM| \leq |f(x)| \, |g(x) - M| + |M| \, |f(x) - L|$$

$$\leq (|L| + 1) \frac{\epsilon}{2(|L| + 1)} + |M| \frac{\epsilon}{2(|M| + 1)}$$

$$\leq \frac{\epsilon}{2} + \frac{\epsilon}{2} = \epsilon$$

Since ϵ is arbitrary, this proves that $\lim\limits_{x \to c} f(x)g(x) = LM$. To prove the Quotient Law (iv), it suffices to verify that if $M \neq 0$, then

$$\lim_{x \to c} \frac{1}{g(x)} = \frac{1}{M} \qquad \boxed{2}$$

For if Eq. (2) holds, then we may apply the Product Law to $f(x)$ and $g(x)^{-1}$ to obtain the Quotient Law:

$$\lim_{x \to c} \frac{f(x)}{g(x)} = \lim_{x \to c} f(x) \frac{1}{g(x)} = \Big(\lim_{x \to c} f(x) \Big) \Big(\lim_{x \to c} \frac{1}{g(x)} \Big)$$

$$= L \Big(\frac{1}{M} \Big) = \frac{L}{M}$$

We now verify Eq. (2). Since $g(x)$ approaches M and $M \neq 0$, we may choose $\delta > 0$ so that if $0 < |x - c| < \delta$, then $|g(x)| \geq |M|/2$. Now choose any number $\epsilon > 0$. By choosing a smaller δ if necessary, we may also ensure that

$$\text{for } 0 < |x - c| < \delta, \text{ then } |M - g(x)| < \epsilon |M| \left(\frac{|M|}{2} \right).$$

Then

$$\left| \frac{1}{g(x)} - \frac{1}{M} \right| = \left| \frac{M - g(x)}{Mg(x)} \right| \leq \left| \frac{M - g(x)}{M(M/2)} \right| \leq \frac{\epsilon |M|(|M|/2)}{|M|(|M|/2)} = \epsilon$$

Since ϵ is arbitrary, the limit in Eq. (2) is proved. ∎

The following result was used in the text.

THEOREM 2 Limits Preserve Inequalities Let (a, b) be an open interval and let $c \in (a, b)$. Suppose that $f(x)$ and $g(x)$ are defined on (a, b), except possibly at c. Assume that

$$f(x) \leq g(x) \qquad \text{for } x \in (a, b), \quad x \neq c,$$

and that the limits $\lim_{x \to c} f(x)$ and $\lim_{x \to c} g(x)$ exist. Then

$$\lim_{x \to c} f(x) \leq \lim_{x \to c} g(x)$$

Proof Let $L = \lim_{x \to c} f(x)$ and $M = \lim_{x \to c} g(x)$. To show that $L \leq M$, we use proof by contradiction. If $L > M$, let $\epsilon = \frac{1}{2}(L - M)$. By the formal definition of limits, we may choose $\delta > 0$ so that the following two conditions are satisfied:

$$\text{If } |x - c| < \delta, \text{ then } |M - g(x)| < \epsilon.$$

$$\text{If } |x - c| < \delta, \text{ then } |L - f(x)| < \epsilon.$$

But then

$$f(x) > L - \epsilon = M + \epsilon > g(x)$$

This is a contradiction since $f(x) \leq g(x)$. We conclude that $L \leq M$. ∎

THEOREM 3 Limit of a Composite Function Assume that the following limits exist:

$$L = \lim_{x \to c} g(x) \qquad \text{and} \qquad M = \lim_{x \to L} f(x)$$

Then $\lim_{x \to c} f(g(x)) = M$.

Proof Let $\epsilon > 0$ be given. By the limit definition, there exists $\delta_1 > 0$ such that

$$\text{if } 0 < |x - L| < \delta_1, \text{ then } |f(x) - M| < \epsilon. \qquad \boxed{3}$$

Similarly, there exists $\delta > 0$ such that

$$\text{if } 0 < |x - c| < \delta, \text{ then } |g(x) - L| < \delta_1. \qquad \boxed{4}$$

We replace x by $g(x)$ in Eq. (3) and apply Eq. (4) to obtain:

$$\text{If } 0 < |x - c| < \delta, \text{ then } |f(g(x)) - M| < \epsilon.$$

Since ϵ is arbitrary, this proves that $\lim_{x \to c} f(g(x)) = M$. ∎

| Section 2.4

> **THEOREM 4 Continuity of Composite Functions** Let $F(x) = f(g(x))$ be a composite function. If g is continuous at $x = c$ and f is continuous at $x = g(c)$, then F is continuous at $x = c$.

Proof By definition of continuity,

$$\lim_{x \to c} g(x) = g(c) \qquad \text{and} \qquad \lim_{x \to g(c)} f(x) = f(g(c))$$

Therefore, we may apply Theorem 3 to obtain

$$\lim_{x \to c} f(g(x)) = f(g(c))$$

This proves that $F(x) = f(g(x))$ is continuous at $x = c$. ∎

| Section 2.6

> **THEOREM 5 Squeeze Theorem** Assume that for $x \neq c$ (in some open interval containing c),
>
> $$l(x) \leq f(x) \leq u(x) \qquad \text{and} \qquad \lim_{x \to c} l(x) = \lim_{x \to c} u(x) = L$$
>
> Then $\lim_{x \to c} f(x)$ exists and
>
> $$\lim_{x \to c} f(x) = L$$

Proof Let $\epsilon > 0$ be given. We may choose $\delta > 0$ such that

$$\text{if } 0 < |x - c| < \delta, \text{ then } |l(x) - L| < \epsilon \text{ and } |u(x) - L| < \epsilon.$$

In principle, a different δ may be required to obtain the two inequalities for $l(x)$ and $u(x)$, but we may choose the smaller of the two deltas. Thus, if $0 < |x - c| < \delta$, we have

$$L - \epsilon < l(x) < L + \epsilon$$

and

$$L - \epsilon < u(x) < L + \epsilon$$

Since $f(x)$ lies between $l(x)$ and $u(x)$, it follows that

$$L - \epsilon < l(x) \leq f(x) \leq u(x) < L + \epsilon$$

and therefore $|f(x) - L| < \epsilon$ if $0 < |x - c| < \delta$. Since ϵ is arbitrary, this proves that $\lim_{x \to c} f(x) = L$, as desired. ∎

| Section 4.2

> **THEOREM 6 Existence of Extrema on a Closed Interval** A continuous function f on a closed (bounded) interval $I = [a, b]$ takes on both a minimum and a maximum value on I.

Proof We prove that f takes on a maximum value in two steps (the case of a minimum is similar).

***Step 1.* Prove that f is bounded from above.**

We use proof by contradiction. If f is not bounded from above, then there exist points $a_n \in [a, b]$ such that $f(a_n) \geq n$ for $n = 1, 2, \ldots$. By Theorem 3 in Appendix B, we may choose a subsequence of elements $a_{n_1}, a_{n_2}, \ldots$ that converges to a limit in $[a, b]$—say, $\lim_{k \to \infty} a_{n_k} = L$. Since f is continuous, there exists $\delta > 0$ such that

$$\text{if } x \in [a, b] \text{ and } |x - L| < \delta, \text{ then } |f(x) - f(L)| < 1.$$

Therefore,

$$\text{if } x \in [a, b] \text{ and } x \in (L - \delta, L + \delta), \text{ then } f(x) < f(L) + 1. \qquad \boxed{5}$$

For k sufficiently large, a_{n_k} lies in $(L - \delta, L + \delta)$ because $\lim_{k \to \infty} a_{n_k} = L$. By Eq. (5), $f(a_{n_k})$ is bounded by $f(L) + 1$. However, $f(a_{n_k}) = n_k$ tends to infinity as $k \to \infty$. This is a contradiction. Hence, our assumption that f is not bounded from above is false.

Step 2. **Prove that f takes on a maximum value.**

The range of f on $I = [a, b]$ is the set

$$S = \{f(x) : x \in [a, b]\}$$

By the previous step, S is bounded from above and therefore has a least upper bound M by the LUB Property. Thus, $f(x) \leq M$ for all $x \in [a, b]$. To complete the proof, we show that $f(c) = M$ for some $c \in [a, b]$. This will show that f attains the maximum value M on $[a, b]$.

By definition, $M - 1/n$ is not an upper bound for $n \geq 1$, and therefore, we may choose a point b_n in $[a, b]$ such that

$$M - \frac{1}{n} \leq f(b_n) \leq M$$

Again by Theorem 3 in Appendix B, there exists a subsequence of elements $\{b_{n_1}, b_{n_2}, \dots\}$ in $\{b_1, b_2, \dots\}$ that converges to a limit—say,

$$\lim_{k \to \infty} b_{n_k} = c$$

Furthermore, this limit c belongs to $[a, b]$ because $[a, b]$ is closed. Let $\epsilon > 0$. Since f is continuous, we may choose k so large that the following two conditions are satisfied: $|f(c) - f(b_{n_k})| < \epsilon/2$ and $n_k > 2/\epsilon$. Then

$$|f(c) - M| \leq |f(c) - f(b_{n_k})| + |f(b_{n_k}) - M| \leq \frac{\epsilon}{2} + \frac{1}{n_k} \leq \frac{\epsilon}{2} + \frac{\epsilon}{2} = \epsilon$$

Thus, $|f(c) - M|$ is smaller than ϵ for all positive numbers ϵ. But this is not possible unless $|f(c) - M| = 0$. Thus, $f(c) = M$, as desired.   ■

| *Section 5.2*

> **THEOREM 7** **Continuous Functions Are Integrable**   If f is continuous on $[a, b]$, then f is integrable over $[a, b]$.

Proof We shall make the simplifying assumption that f is differentiable and that its derivative f' is bounded. In other words, we assume that $|f'(x)| \leq K$ for some constant K. This assumption is used to show that f cannot vary too much in a small interval. More precisely, let us prove that if $[a_0, b_0]$ is any closed interval contained in $[a, b]$ and if m and M are the minimum and maximum values of f on $[a_0, b_0]$, then

$$|M - m| \leq K|b_0 - a_0| \qquad \boxed{6}$$

Figure 1 illustrates the idea behind this inequality. Suppose that $f(x_1) = m$ and $f(x_2) = M$, where x_1 and x_2 lie in $[a_0, b_0]$. If $x_1 \neq x_2$, then by the Mean Value Theorem (MVT), there is a point c between x_1 and x_2 such that

$$\frac{M - m}{x_2 - x_1} = \frac{f(x_2) - f(x_1)}{x_2 - x_1} = f'(c)$$

Since x_1, x_2 lie in $[a_0, b_0]$, we have $|x_2 - x_1| \leq |b_0 - a_0|$, and thus,

$$|M - m| = |f'(c)| \, |x_2 - x_1| \leq K|b_0 - a_0|$$

This proves Eq. (6).

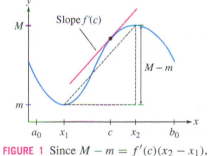

FIGURE 1 Since $M - m = f'(c)(x_2 - x_1)$, we conclude that $M - m \leq K(b_0 - a_0)$.

We divide the rest of the proof into two steps. Consider a partition P:

$$P: \qquad x_0 = a < x_1 < \quad \cdots \quad < x_{N-1} < x_N = b$$

Let m_i be the minimum value of f on $[x_{i-1}, x_i]$ and M_i the maximum on $[x_{i-1}, x_i]$. We define the *lower* and *upper* Riemann sums

$$L(f, P) = \sum_{i=1}^{N} m_i \, \Delta x_i, \qquad U(f, P) = \sum_{i=1}^{N} M_i \, \Delta x_i$$

These are the particular Riemann sums in which the intermediate point in $[x_{i-1}, x_i]$ is the point where f takes on its minimum or maximum on $[x_{i-1}, x_i]$. Figure 2 illustrates the case $N = 4$.

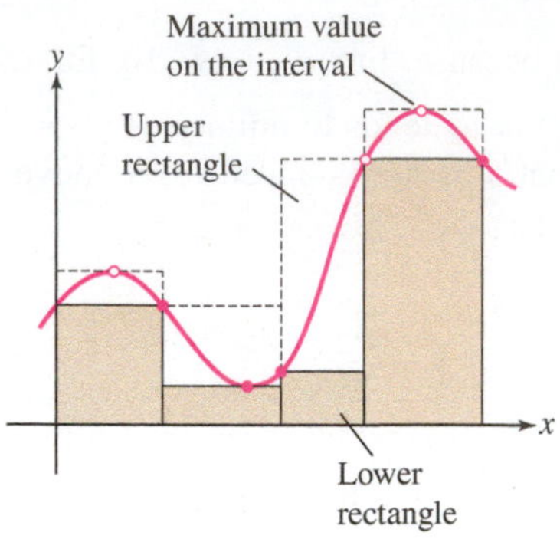

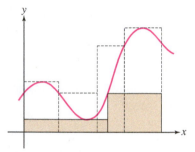

Maximum value on the interval

Upper rectangle

Lower rectangle

FIGURE 2 Lower and upper rectangles for a partition of length $N = 4$.

Step 1. Prove that the lower and upper sums approach a limit.
We observe that

$$L(f, P_1) \le U(f, P_2) \quad \text{for any two partitions } P_1 \text{ and } P_2 \qquad \boxed{7}$$

Indeed, if a subinterval I_1 of P_1 overlaps with a subinterval I_2 of P_2, then the minimum of f on I_1 is less than or equal to the maximum of f on I_2 (Figure 3). In particular, the lower sums are bounded above by $U(f, P)$ for all partitions P. Let L be the least upper bound of the lower sums. Then for all partitions P,

$$L(f, P) \le L \le U(f, P) \qquad \boxed{8}$$

According to Eq. (6), $|M_i - m_i| \le K \, \Delta x_i$ for all i. Since $\|P\|$ is the largest of the widths Δx_i, we see that $|M_i - m_i| \le K \|P\|$ and

$$|U(f, P) - L(f, P)| \le \sum_{i=1}^{N} |M_i - m_i| \, \Delta x_i$$

$$\le K \|P\| \sum_{i=1}^{N} \Delta x_i = K \|P\| \, |b - a| \qquad \boxed{9}$$

FIGURE 3 The lower rectangles always lie below the upper rectangles, even when the partitions are different.

Let $c = K \, |b - a|$. Using Eq. (8) and Eq. (9), we obtain

$$|L - U(f, P)| \le |U(f, P) - L(f, P)| \le c \|P\|$$

We conclude that $\lim_{\|P\| \to 0} |L - U(f, P)| = 0$. Similarly,

$$|L - L(f, P)| \le c \|P\|$$

and

$$\lim_{\|P\| \to 0} |L - L(f, P)| = 0$$

Thus, we have

$$\lim_{\|P\| \to 0} U(f, P) = \lim_{\|P\| \to 0} L(f, P) = L$$

Step 2. Prove that $\displaystyle\int_a^b f(x) \, dx$ exists and has value L.
Recall that for any choice C of intermediate points $c_i \in [x_{i-1}, x_i]$, we define the Riemann sum

$$R(f, P, C) = \sum_{i=1}^{N} f(c_i) \Delta x_i$$

We have

$$L(f, P) \le R(f, P, C) \le U(f, P)$$

Indeed, since $c_i \in [x_{i-1}, x_i]$, we have $m_i \leq f(c_i) \leq M_i$ for all i and

$$\sum_{i=1}^{N} m_i \, \Delta x_i \leq \sum_{i=1}^{N} f(c_i) \, \Delta x_i \leq \sum_{i=1}^{N} M_i \, \Delta x_i$$

It follows that

$$|L - R(f, P, C)| \leq |U(f, P) - L(f, P)| \leq c \|P\|$$

This shows that $R(f, P, C)$ converges to L as $\|P\| \to 0$. ∎

| Section 10.1

THEOREM 8   If f is continuous and $\{a_n\}$ is a sequence such that the limit $\lim_{n \to \infty} a_n = L$ exists, then

$$\lim_{n \to \infty} f(a_n) = f(L)$$

Proof Choose any $\epsilon > 0$. Since f is continuous, there exists $\delta > 0$ such that

$$\text{if } 0 < |x - L| < \delta, \text{ then } |f(x) - f(L)| < \epsilon.$$

Since $\lim_{n \to \infty} a_n = L$, there exists $N > 0$ such that $|a_n - L| < \delta$ for $n > N$. Thus,

$$|f(a_n) - f(L)| < \epsilon \qquad \text{for } n > N$$

It follows that $\lim_{n \to \infty} f(a_n) = f(L)$. ∎

| Section 14.3

THEOREM 9 Clairaut's Theorem   If f_{xy} and f_{yx} are both continuous functions on a disk D, then $f_{xy}(a, b) = f_{yx}(a, b)$ for all $(a, b) \in D$.

Proof We prove that both $f_{xy}(a, b)$ and $f_{yx}(a, b)$ are equal to the limit

$$L = \lim_{h \to 0} \frac{f(a+h, b+h) - f(a+h, b) - f(a, b+h) + f(a, b)}{h^2}$$

Let $F(x) = f(x, b+h) - f(x, b)$. The numerator in the limit is equal to

$$F(a+h) - F(a)$$

and $F'(x) = f_x(x, b+h) - f_x(x, b)$. By the MVT, there exists a_1 between a and $a+h$ such that

$$F(a+h) - F(a) = hF'(a_1) = h(f_x(a_1, b+h) - f_x(a_1, b))$$

By the MVT applied to f_x, there exists b_1 between b and $b+h$ such that

$$f_x(a_1, b+h) - f_x(a_1, b) = hf_{xy}(a_1, b_1)$$

Thus,

$$F(a+h) - F(a) = h^2 f_{xy}(a_1, b_1)$$

and

$$L = \lim_{h \to 0} \frac{h^2 f_{xy}(a_1, b_1)}{h^2} = \lim_{h \to 0} f_{xy}(a_1, b_1) = f_{xy}(a, b)$$

The last equality follows from the continuity of f_{xy} since (a_1, b_1) approaches (a, b) as $h \to 0$. To prove that $L = f_{yx}(a, b)$, repeat the argument using the function $F(y) = f(a+h, y) - f(a, y)$, with the roles of x and y reversed. ∎

> **THEOREM 10 Criterion for Differentiability** If $f_x(x, y)$ and $f_y(x, y)$ exist and are continuous on an open disk D, then $f(x, y)$ is differentiable on D.

Proof Let $(a, b) \in D$ and set

$$L(x, y) = f(a, b) + f_x(a, b)(x - a) + f_y(a, b)(y - b)$$

It is convenient to switch to the variables h and k, where $x = a + h$ and $y = b + k$. Set

$$\Delta f = f(a + h, b + k) - f(a, b)$$

Then

$$L(x, y) = f(a, b) + f_x(a, b)h + f_y(a, b)k$$

and we may define the function

$$e(h, k) = f(x, y) - L(x, y) = \Delta f - (f_x(a, b)h + f_y(a, b)k)$$

To prove that $f(x, y)$ is differentiable, we must show that

$$\lim_{(h,k) \to (0,0)} \frac{e(h, k)}{\sqrt{h^2 + k^2}} = 0$$

To do this, we write Δf as a sum of two terms:

$$\Delta f = (f(a + h, b + k) - f(a, b + k)) + (f(a, b + k) - f(a, b))$$

and apply the MVT to each term separately. We find that there exist a_1 between a and $a + h$, and b_1 between b and $b + k$, such that

$$f(a + h, b + k) - f(a, b + k) = h f_x(a_1, b + k)$$

$$f(a, b + k) - f(a, b) = k f_y(a, b_1)$$

Therefore,

$$e(h, k) = h(f_x(a_1, b + k) - f_x(a, b)) + k(f_y(a, b_1) - f_y(a, b))$$

and for $(h, k) \neq (0, 0)$,

$$\left| \frac{e(h, k)}{\sqrt{h^2 + k^2}} \right| = \left| \frac{h(f_x(a_1, b + k) - f_x(a, b)) + k(f_y(a, b_1) - f_y(a, b))}{\sqrt{h^2 + k^2}} \right|$$

$$\leq \left| \frac{h(f_x(a_1, b + k) - f_x(a, b))}{\sqrt{h^2 + k^2}} \right| + \left| \frac{k(f_y(a, b_1) - f_y(a, b))}{\sqrt{h^2 + k^2}} \right|$$

$$= |f_x(a_1, b + k) - f_x(a, b)| + |f_y(a, b_1) - f_y(a, b)|$$

In the second line, we use the Triangle Inequality [see Eq. (1) in Section 1.1], and we may pass to the third line because $\left| h/\sqrt{h^2 + k^2} \right|$ and $\left| k/\sqrt{h^2 + k^2} \right|$ are both less than 1. Both terms in the last line tend to zero as $(h, k) \to (0, 0)$ because f_x and f_y are assumed to be continuous. This completes the proof that $f(x, y)$ is differentiable. ■

ANSWERS TO ODD-NUMBERED EXERCISES

Chapter 1

Section 1.1 Preliminary Questions

1. $a = -3$ and $b = 1$

2. The numbers $a \geq 0$ satisfy $|a| = a$ and $|-a| = a$. The numbers $a \leq 0$ satisfy $|a| = -a$.

3. $a = -3$ and $b = 1$ **4.** No **5.** $(9, -4)$

6. **(a)** First quadrant. **(b)** Second quadrant. **(c)** Fourth quadrant. **(d)** Third quadrant.

7. 3 **8.** (b) **9.** Symmetry with respect to the origin

10. The only function that is both even and odd is the constant function $f(x) = 0$.

Section 1.1 Exercises

1. $r = \frac{12337}{1250}$ **3.** $|x| \leq 2$ **5.** $|x - 2| < 2$ **7.** $|x - 3| \leq 2$
9. $-8 < x < 8$ **11.** $-3 < x < 2$ **13.** $(-4, 4)$ **15.** $(2, 6)$
17. $\left[-\frac{7}{4}, \frac{9}{4}\right]$ **19.** $(-\infty, 2) \cup (6, \infty)$
21. $(-\infty, -\sqrt{3}) \cup (\sqrt{3}, \infty)$

23. **(a)** **(i)** **(b)** **(iii)** **(c)** **(v)** **(d)** **(vi)** **(e)** **(ii)** **(f)** **(iv)**

25. $(-3, 1)$

29. $|a + b - 13| = |(a - 5) + (b - 8)| \leq |a - 5| + |b - 8| < \frac{1}{2} + \frac{1}{2} = 1$

31. **(a)** 11 **(b)** 1

33. $r_1 = \frac{3}{11}$ and $r_2 = \frac{4}{15}$

35. Let $a = 1$ and $b = .\overline{9}$ (see the discussion before Example 1). The decimal expansions of a and b do not agree, but $|1 - .\overline{9}| < 10^{-k}$ for all k.

37. **(a)** $(x - 2)^2 + (y - 4)^2 = 9$

(b) $(x - 2)^2 + (y - 4)^2 = 26$

39. $D = \{r, s, t, u\}; R = \{A, B, E\}$ **41.** D : all reals; R : all reals
43. D : all reals; R : all reals **45.** D : all reals; $R : \{y : y \geq 0\}$
47. $D : \{x : x \neq 0\}; R : \{y : y > 0\}$ **49.** On the interval $(-1, \infty)$
51. On the interval $(0, \infty)$

53. Zeros: ± 2; Increasing: $x > 0$; Decreasing: $x < 0$; Symmetry: $f(-x) = f(x)$, so y-axis symmetry.

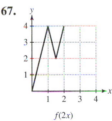

55. Zeros: $0, \pm 2$; Symmetry: $f(-x) = -f(x)$, so origin symmetry.

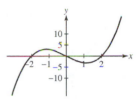

57. This is an x-axis reflection of $y = x^3$ translated up 2 units. There is one zero at $x = \sqrt[3]{2}$.

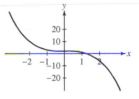

59. (B)

61. **(a)** Odd **(b)** Odd **(c)** Neither odd nor even **(d)** Even

65. $D : [0, 4]; R : [0, 4]$

67.

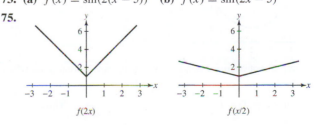

$f(2x)$ $f(x/2)$ $2f(x)$

69.

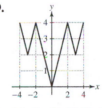

71. **(a)** $D : [4, 8], R : [5, 9]$. **(b)** $D : [1, 5], R : [2, 6]$.
(c) $D : \left[\frac{4}{3}, \frac{8}{3}\right], R : [2, 6]$. **(d)** $D : [4, 8], R : [6, 18]$.

73. **(a)** $f(x) = \sin(2(x - 5))$ **(b)** $f(x) = \sin(2x - 5)$

75.

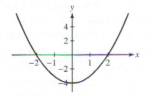

$f(2x)$ $f(x/2)$

77.

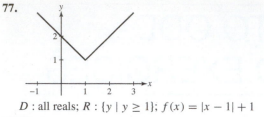

D : all reals; R : $\{y \mid y \geq 1\}$; $f(x) = |x - 1| + 1$

79. Even:

$(f + g)(-x) = f(-x) + g(-x) \overset{\text{even}}{=} f(x) + g(x) = (f + g)(x)$

Odd: $(f + g)(-x) = f(-x) + g(-x) \overset{\text{odd}}{=} -f(x) + -g(x) = -(f + g)(x)$

85. (a) There are many possibilities, one of which is

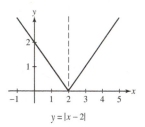

$y = |x - 2|$

(b) Let $g(x) = f(x + a)$. Then

$g(-x) = f(-x + a) = f(a - x) = f(a + x) = g(x)$

Section 1.2 Preliminary Questions

1. -4 **2.** No.

3. Parallel to the y-axis when $b = 0$; parallel to the x-axis when $a = 0$

4. $\Delta y = 9$ **5.** -4 **6.** $(x - 0)^2 + 1$

Section 1.2 Exercises

1. $m = 3$; $y = 12$; $x = -4$ **3.** $m = -\frac{4}{9}$; $y = \frac{1}{3}$; $x = \frac{3}{4}$

5. $m = 3$ **7.** $m = -\frac{3}{4}$ **9.** $y = 3x + 8$ **11.** $y = 3x - 12$

13. $y = -2$ **15.** $y = 3x - 2$ **17.** $3x + 5y = 21$ **19.** $y = 4$

21. $y = -2x + 9$ **23.** $3x + 4y = 12$

25. (a) $c = -\frac{1}{4}$ **(b)** $c = -2$

(c) No value for c that will make this slope equal to 0 **(d)** $c = 0$

27. $N(P) = -5P + 15,000$; $\Delta N = -500$

29. (a) 40.0248 cm **(b)** 64.9597 in.

(c) $L = 65(1 + \alpha(T - 100))$

31. $b = 4$

33. No, because the slopes between consecutive data points are not equal.

35. (a) 1 or $-\frac{1}{4}$ **(b)** $1 \pm \sqrt{2}$

37. Minimum value is 0 **39.** Minimum value is -7

41. Maximum value is $\frac{137}{16}$ **43.** Maximum value is $\frac{1}{3}$

45.

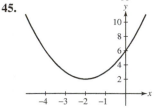

47. A double root occurs when $c = \pm 2$. There are no real roots when $-2 < c < 2$.

49. For all $x \geq 0$, $0 \leq \left(x^{1/2} - x^{-1/2}\right)^2 = x - 2 + \frac{1}{x}$. Thus,

$x + \frac{1}{x} \geq 2$.

53. $4 \pm \sqrt{8}$ **57.** For x^2, $\frac{\Delta y}{\Delta x} = \frac{x_2^2 - x_1^2}{x_2 - x_1} = x_2 + x_1$.

61.

$(x - \alpha)(x - \beta) = x^2 - \alpha x - \beta x + \alpha \beta = x^2 + (-\alpha - \beta)x + \alpha \beta$

Section 1.3 Preliminary Questions

1. One example is $f(x) = \frac{3x^2 - 2}{7x^3 + x - 1}$.

2. $y = |x|$ is not a polynomial; $y = |x^2 + 1|$ is a polynomial.

3. The domain of $f \circ g$ is the empty set.

4. Decreasing **5.** One possibility is $f(x) = e^x - \sin x$.

Section 1.3 Exercises

1. $x \geq 0$ **3.** All reals **5.** $t \neq -2$ **7.** $u \neq \pm 2$ **9.** $x \neq 0, 1$

11. $y > 0$ **13.** Polynomial **15.** Algebraic **17.** Transcendental

19. Rational **21.** Transcendental **23.** Rational **25.** Yes

27. $f(g(x)) = \sqrt{x + 1}$; D: $x \geq -1$, $g(f(x)) = \sqrt{x} + 1$; D: $x \geq 0$

29. $f(g(x)) = 2^{x^2}$; D: **R**, $g(f(x)) = (2^x)^2 = 2^{2x}$; D: **R**

31. $f(g(x)) = \cos(x^3 + x^2)$; D: **R**, $g(f(\theta)) = \cos^3 \theta + \cos^2 \theta$; D: **R**

33. $f(g(t)) = \frac{1}{\sqrt{-t^2}}$; D: Not valid for any t,

$g(f(t)) = -\left(\frac{1}{\sqrt{t}}\right)^2 = -\frac{1}{t}$; D: $t > 0$

35.

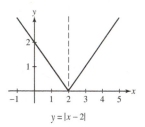

37.

39.

$P(t + 10) = 30 \cdot 2^{0.1(t+10)} = 30 \cdot 2^{0.1t+1} = 2(30 \cdot 2^{0.1t}) = 2P(t)$;

$g\left(t + \frac{1}{k}\right) = a2^{k(t+1/k)} = a2^{kt+1} = 2a2^{kt} = 2g(t)$

41. $f(x) = x^2$:

$\delta f(x) = f(x + 1) - f(x) = (x + 1)^2 - x^2 = 2x + 1$

$f(x) = x$: $\delta f(x) = x + 1 - x = 1$

$f(x) = x^3$: $\delta f(x) = (x + 1)^3 - x^3 = 3x^2 + 3x + 1$

43.

$\delta(f + g) = (f(x + 1) + g(x + 1)) - (f(x) - g(x))$

$= (f(x + 1) - f(x)) + (g(x + 1) - g(x)) = \delta f(x) + \delta g(x)$

$\delta(cf) = cf(x + 1) - cf(x) = c(f(x + 1) - f(x)) = c\delta f(x)$.

Section 1.4 Preliminary Questions

1. It is possible if the rotations differ by an integer multiple of 2π.

2. $\frac{9\pi}{4}$ and $\frac{41\pi}{4}$ **3.** $-\frac{5\pi}{3}$ **4. (a)**

5. Let O denote the center of the unit circle, and let P be a point on the unit circle such that the radius $\overline{OP}$ makes an angle θ with the positive x-axis. Then, $\sin \theta$ is the y-coordinate of the point P.

6. Let O denote the center of the unit circle, and let P be a point on the unit circle such that the radius $\overline{OP}$ makes an angle θ with the positive x-axis. The angle $\theta + 2\pi$ is obtained from the angle θ by making one full revolution around the circle. The angle $\theta + 2\pi$ will therefore have the radius $\overline{OP}$ as its terminal side.

Section 1.4 Exercises

1. $5\pi/4$

3. (a) $\frac{180°}{\pi} \approx 57.3°$ (b) $60°$ (c) $\frac{75°}{\pi} \approx 23.87°$ (d) $225°$

5. $s = r\theta = 3.6$; $s = r\phi = 8$

7.

θ	$(\cos\theta, \sin\theta)$	θ	$(\cos\theta, \sin\theta)$
$\frac{\pi}{2}$	$(0, 1)$	$\frac{5\pi}{4}$	$\left(\frac{-\sqrt{2}}{2}, \frac{-\sqrt{2}}{2}\right)$
$\frac{2\pi}{3}$	$\left(\frac{-1}{2}, \frac{\sqrt{3}}{2}\right)$	$\frac{4\pi}{3}$	$\left(\frac{-1}{2}, \frac{-\sqrt{3}}{2}\right)$
$\frac{3\pi}{4}$	$\left(\frac{-\sqrt{2}}{2}, \frac{\sqrt{2}}{2}\right)$	$\frac{3\pi}{2}$	$(0, -1)$
$\frac{5\pi}{6}$	$\left(\frac{-\sqrt{3}}{2}, \frac{1}{2}\right)$	$\frac{5\pi}{3}$	$\left(\frac{1}{2}, \frac{-\sqrt{3}}{2}\right)$
π	$(-1, 0)$	$\frac{7\pi}{4}$	$\left(\frac{\sqrt{2}}{2}, \frac{-\sqrt{2}}{2}\right)$
$\frac{7\pi}{6}$	$\left(\frac{-\sqrt{3}}{2}, \frac{-1}{2}\right)$	$\frac{11\pi}{6}$	$\left(\frac{\sqrt{3}}{2}, \frac{-1}{2}\right)$

9. $\theta = \frac{\pi}{3}, \frac{5\pi}{3}$ **11.** $\theta = \frac{3\pi}{4}, \frac{7\pi}{4}$ **13.** $x = \frac{\pi}{3}, \frac{2\pi}{3}$

15.

θ	$\frac{\pi}{6}$	$\frac{\pi}{4}$	$\frac{\pi}{3}$	$\frac{\pi}{2}$	$\frac{2\pi}{3}$	$\frac{3\pi}{4}$	$\frac{5\pi}{6}$
$\tan\theta$	$\frac{1}{\sqrt{3}}$	1	$\sqrt{3}$	und	$-\sqrt{3}$	-1	$-\frac{1}{\sqrt{3}}$
$\sec\theta$	$\frac{2}{\sqrt{3}}$	$\sqrt{2}$	2	und	-2	$-\sqrt{2}$	$-\frac{2}{\sqrt{3}}$

17. The hypotenuse of the triangle will have length $\sqrt{1 + c^2}$.

19. $\sin\theta = \frac{12}{13}$ and $\tan\theta = \frac{12}{5}$

21. $\sin\theta = \frac{2}{\sqrt{53}}$, $\sec\theta = \frac{\sqrt{53}}{7}$ and $\cot\theta = \frac{7}{2}$ **23.** $23/25$

25. $\cos\theta = -\frac{\sqrt{21}}{5}$ and $\tan\theta = -\frac{2\sqrt{21}}{21}$

27. $\cos\theta = -\frac{4}{5}$

29. Let's start with the four points in Figure 23(A).

- The point in the first quadrant:

$$\sin\theta = 0.918, \quad \cos\theta = 0.3965, \quad \text{and} \tan\theta = \frac{0.918}{0.3965} = 2.3153$$

- The point in the second quadrant:

$$\sin\theta = 0.3965, \quad \cos\theta = -0.918, \quad \text{and}$$

$$\tan\theta = \frac{0.3965}{-0.918} = -0.4319$$

- The point in the third quadrant:

$$\sin\theta = -0.918, \quad \cos\theta = -0.3965, \quad \text{and}$$

$$\tan\theta = \frac{-0.918}{-0.3965} = 2.3153$$

- The point in the fourth quadrant:

$$\sin\theta = -0.3965, \quad \cos\theta = 0.918, \quad \text{and}$$

$$\tan\theta = \frac{-0.3965}{0.918} = -0.4319$$

Now consider the four points in Figure 23(B).

- The point in the first quadrant:

$$\sin\theta = 0.918, \quad \cos\theta = 0.3965, \quad \text{and}$$

$$\tan\theta = \frac{0.918}{0.3965} = 2.3153$$

- The point in the second quadrant:

$$\sin\theta = 0.918, \quad \cos\theta = -0.3965, \quad \text{and}$$

$$\tan\theta = \frac{0.918}{0.3965} = -2.3153$$

- The point in the third quadrant:

$$\sin\theta = -0.918, \quad \cos\theta = -0.3965, \quad \text{and}$$

$$\tan\theta = \frac{-0.918}{-0.3965} = 2.3153$$

- The point in the fourth quadrant:

$$\sin\theta = -0.918, \quad \cos\theta = 0.3965, \quad \text{and}$$

$$\tan\theta = \frac{-0.918}{0.3965} = -2.3153$$

31. $\cos\psi = 0.3$, $\sin\psi = \sqrt{0.91}$, $\cot\psi = \frac{0.3}{\sqrt{0.91}}$ and $\csc\psi = \frac{1}{\sqrt{0.91}}$

33. $\cos\left(\frac{\pi}{3} + \frac{\pi}{4}\right) = \frac{\sqrt{2}-\sqrt{6}}{4}$

35.

37.

39. $3\cos(\theta/2)$; period 4π; amplitude 3

41. If $|c| > 1$, no points of intersection; if $|c| = 1$, one point of intersection; if $|c| < 1$, two points of intersection.

43. $\theta = 0, \frac{2\pi}{5}, \frac{4\pi}{5}, \pi, \frac{6\pi}{5}, \frac{8\pi}{5}$ **45.** $\theta = \frac{\pi}{6}, \frac{\pi}{2}, \frac{5\pi}{6}, \frac{7\pi}{6}, \frac{3\pi}{2}, \frac{11\pi}{6}$

47. $\cos 2\theta = \cos(\theta + \theta) = \cos\theta\cos\theta - \sin\theta\sin\theta = \cos^2\theta - \sin^2\theta = 2\cos^2\theta - 1$

49. Substitute $x = \theta/2$ into the double angle formula for sine, $\sin^2 x = \frac{1}{2}(1 - \cos 2x)$, then take the square root of both sides.

51. $\cos(\theta + \pi) = \cos\theta\cos\pi - \sin\theta\sin\pi = \cos\theta(-1) = -\cos\theta$

53. $\tan(\pi - \theta) = \frac{\sin(\pi-\theta)}{\cos(\pi-\theta)} = \frac{-\sin(-\theta)}{-\cos(-\theta)} = \frac{\sin\theta}{-\cos\theta} = -\tan\theta$.

55. $\frac{\sin 2x}{1+\cos 2x} = \frac{2\sin x\cos x}{1+2\cos^2 x-1} = \frac{2\sin x\cos x}{2\cos^2 x} = \frac{\sin x}{\cos x} = \tan x$

59. 16.928

Section 1.5 Preliminary Questions

1. (a), (b), (f)

2. Many different teenagers will have the same last name, so this function will not be one-to-one.

3. This function is one-to-one, and $f^{-1}(6:27) = $ Hamilton Township.

4. The graph of the inverse function is the reflection of the graph of $y = f(x)$ through the line $y = x$.

5. (b) and (c)   **6.** $\theta = 3\pi$; No

Section 1.5 Exercises

1. $f^{-1}(x) = \frac{x+4}{7}$ **3.** $[-\pi/2, \pi/2]$

5.
- $f(g(x)) = \left((x - 3)^{1/3}\right)^3 + 3 = x - 3 + 3 = x$.
- $g(f(x)) = \left(x^3 + 3 - 3\right)^{1/3} = \left(x^3\right)^{1/3} = x$.

7. $R(v) = \frac{2GM}{v^2}$

9. $f^{-1}(x) = 4 - x$.

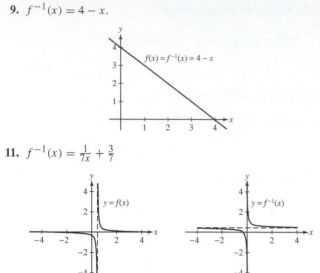

$f(x) = f^{-1}(x) = 4 - x$

11. $f^{-1}(x) = \frac{1}{7x} + \frac{3}{7}$

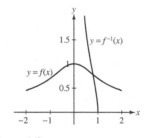

$y = f(x)$　　　$y = f^{-1}(x)$

13. Domain $\{x : x \geq 0\}$: $f^{-1}(x) = \frac{\sqrt{1-x^2}}{x}$; domain $\{x : x \leq 0\}$:
$f^{-1}(x) = -\frac{\sqrt{1-x^2}}{x}$

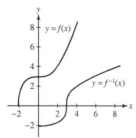

$y = f^{-1}(x)$

$y = f(x)$

15. $f^{-1}(x) = (x^2 - 9)^{1/3}$

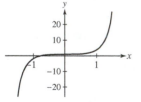

$y = f(x)$

$y = f^{-1}(x)$

17. Figures (B) and (C)

19. (a)

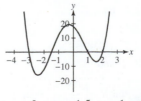

(b) $(-\infty, \infty)$　**(c)** $f^{-1}(3) = 1$

21. Domain $x \leq 1$: $f^{-1}(x) = 1 - \sqrt{x+1}$; domain $x \geq 1$:
$f^{-1}(x) = 1 + \sqrt{x+1}$

23. $f^{-1}(x) = \begin{cases} x & \text{when } x < 0 \\ \frac{1}{2}x & \text{when } x \geq 0 \end{cases}$

25. f is not one-to-one.

27. 0　**29.** $\frac{\pi}{4}$　**31.** $\frac{\pi}{3}$　**33.** $\frac{\pi}{3}$　**35.** $\frac{\pi}{2}$　**37.** $-\frac{\pi}{4}$　**39.** π

41. not defined　**43.** $\frac{\sqrt{1-x^2}}{x}$　**45.** $\frac{1}{\sqrt{x^2-1}}$　**47.** $\frac{\sqrt{5}}{3}$　**49.** $\frac{4}{3}$

51. $\sqrt{3}$　**53.** $\frac{1}{20}$

Section 1.6 Preliminary Questions

1. (a) Correct　**(b)** Correct　**(c)** Incorrect　**(d)** Correct

2. $\log_{b^2}(b^4) = 2$　**3.** For $0 < x < 1$　**4.** $\ln(-3)$ is not defined

5. This phrase is a verbal description of the general property of
logarithms that states $\log(ab) = \log a + \log b$.

6. D: $x > 0$; R: real numbers

7. $f(x) = \cosh x$ and $f(x) = \operatorname{sech} x$

8. $f(x) = \sinh x$ and $f(x) = \tanh x$

9. Both types of functions have the same parity, they share similar
identities, and values of trigonometric functions lie on a circle while
those of hyperbolic functions lie on a hyperbola.

Section 1.6 Exercises

1. (a) 1　**(b)** 29　**(c)** 1　**(d)** 81　**(e)** 16　**(f)** 0

3. $x = 1$　**5.** $x = -1/2$　**7.** $x = -1/3$　**9.** $k = 9$　**11.** 3　**13.** 0

15. $\frac{5}{3}$　**17.** $\frac{1}{3}$　**19.** $\frac{5}{6}$　**21.** 1　**23.** 7　**25.** 29

27. (a) $\ln 1600$ **(b)** $\ln(9x^{7/2})$　**29.** $t = \frac{1}{5}\ln\left(\frac{100}{7}\right)$

31. $x = -1$ or $x = 3$　**33.** $x = e$　**35.** $y = (3 + \ln x)/2$

37.

x	-3	0	5
$\sinh x = \dfrac{e^x - e^{-x}}{2}$	-10.0179	0	74.203
$\cosh x = \dfrac{e^x + e^{-x}}{2}$	10.0677	1	74.210

39. $\ln(2 \cdot 1) \neq (\ln 2)(\ln 1)$

41. $\tanh(-x) = \dfrac{\sinh(-x)}{\cosh(-x)} = -\dfrac{\sinh(x)}{\cosh(x)} = -\tanh(x)$

43. $a = 8$; 1000 earthquakes

49. (a) By Galileo's Law, $w = 500 + 10 = 510$ m/s. Using
Einstein's Law, $w = c \cdot \tanh(1.7 \times 10^{-6}) \approx 510$ m/s.

(b) By Galileo's Law, $u + v = 10^7 + 10^6 = 1.1 \times 10^7$ m/s. By
Einstein's Law, $w \approx c \cdot \tanh(0.036679) \approx 1.09988 \times 10^7$ m/s.

51. Let $y = \log_b x$. Then $x = b^y$ and $\log_a x = \log_a b^y = y \log_a b$.
Thus, $y = \dfrac{\log_a x}{\log_a b}$.

53. $13 \cosh x - 3 \sinh x$

Section 1.7 Preliminary Questions

1. It is best to experiment.

2. (a) The screen will display nothing.

(b) The screen will display the portion of the parabola between the
points $(0, 3)$ and $(1, 4)$.

3. No

4. Experiment with the viewing window to zoom in on the lowest
point on the graph of the function. The y-coordinate of the lowest
point on the graph is the minimum value of the function.

Section 1.7 Exercises

1.

$x = -3, x = -1.5, x = 1,$ and $x = 2$

3. Two positive solutions   **5.** There are no solutions.

7. Nothing. An appropriate viewing window: [50, 150] by [1000, 2000]

9.

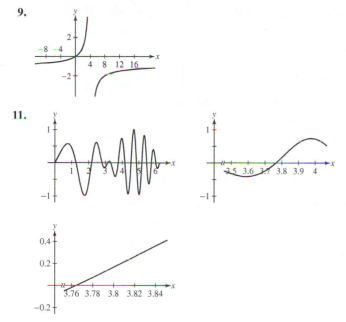

11.

13. The table and graphs below suggest that as n gets large, $n^{1/n}$ approaches 1.

n	$n^{1/n}$
10	1.258925412
10^2	1.047128548
10^3	1.006931669
10^4	1.000921458
10^5	1.000115136
10^6	1.000013816

15. The table and graphs below suggest that as n gets large, $f(n)$ tends toward ∞.

n	$\left(1 + \frac{1}{n}\right)^{n^2}$
10	13780.61234
10^2	$1.635828711 \times 10^{43}$
10^3	$1.195306603 \times 10^{434}$
10^4	$5.341783312 \times 10^{4342}$
10^5	$1.702333054 \times 10^{43429}$
10^6	$1.839738749 \times 10^{434294}$

17. The table and graphs below suggest that as x gets large, $f(x)$ approaches 1.

x	$\left(x \tan \frac{1}{x}\right)^x$
10	1.033975758
10^2	1.003338973
10^3	1.000333389
10^4	1.000033334
10^5	1.000003333
10^6	1.000000333

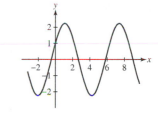

19.

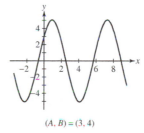

$(A, B) = (1, 1)$ $(A, B) = (1, 2)$

$(A, B) = (3, 4)$

21. $x \in (-2, 0) \cup (3, \infty)$

23.

$$f_3(x) = \frac{1}{2}\left(\frac{1}{2}(x + 1) + \frac{x}{\frac{1}{2}(x + 1)}\right) = \frac{x^2 + 6x + 1}{4(x + 1)}$$

$$f_4(x) = \frac{1}{2}\left(\frac{x^2 + 6x + 1}{4(x + 1)} + \frac{x}{\frac{x^2+6x+1}{4(x+1)}}\right) = \frac{x^4 + 28x^3 + 70x^2 + 28x + 1}{8(1 + x)(1 + 6x + x^2)}$$

and

$$f_5(x) = \frac{1 + 120x + 1820x^2 + 8008x^3 + 12870x^4 + 8008x^5 + 1820x^6 + 120x^7 + x^8}{16(1 + x)(1 + 6x + x^2)(1 + 28x + 70x^2 + 28x^3 + x^4)}$$

It appears as if the f_n are asymptotic to $\sqrt{x}$.

Chapter 1 Review

1. $\{x : |x - 7| < 3\}$ **3.** $[-5, -1] \cup [3, 7]$

5. $(x, 0)$ with $x \geq 0$; $(0, y)$ with $y < 0$

7.

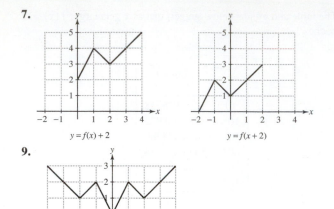

$y = f(x) + 2$ $y = f(x+2)$

9.

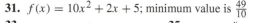

11. $D : \{x : x \geq -1\}; R : \{y : y \geq 0\}$

13. $D : \{x : x \neq 3\}; R : \{y : y \neq 0\}$

15. **(a)** Decreasing   **(b)** Neither   **(c)** Neither   **(d)** Increasing

17. $2x - 3y = -14$ **19.** $6x - y = 53$

21. $x + 3y = 5$ **23.** $x + y = 5$ **25.** Yes

27. $C(P) = -2250P + 1,225,000; 225,000$ customers

29. Roots: $x = -2, x = 0$ and $x = 2$; decreasing: $x < -1.4$ and $0 < x < 1.4$

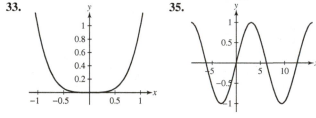

31. $f(x) = 10x^2 + 2x + 5$; minimum value is $\frac{49}{10}$

33.   **35.**

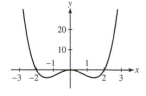

37.

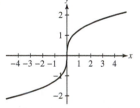

39. Let $g(x) = f\left(\frac{1}{3}x\right)$. Then
$g(x - 3b) = f\left(\frac{1}{3}(x - 3b)\right) = f\left(\frac{1}{3}x - b\right)$. The graph of
$y = \left|\frac{1}{3}x - 4\right|$:

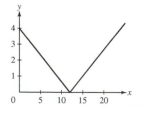

41. $f(t) = t^4$ and $g(t) = 12t + 9$

43. **(a)** π **(b)** 4π **(c)** 4π

45. **(a)** $a = b = \pi/2$ **(b)** $a = \pi$

47. $x = \pi/2, x = 7\pi/6, x = 3\pi/2$ and $x = 11\pi/6$

49. There are no solutions

51. **(a)** No match.   **(b)** No match.   **(c)** **(i)**   **(d)** **(iii)**

53. $f^{-1}(x) = \sqrt[3]{x^2 + 8}; D : \{x : x \geq 0\}; R : \{y : y \geq 2\}$

55. For $\{t : t \leq 3\}, h^{-1}(t) = 3 - \sqrt{t}$. For $t \geq 3, h^{-1}(t) = 3 + \sqrt{t}$.

57. **(a)** Yes   **(b)** Yes $f^{-1}(x) = \begin{cases} -\sqrt{-x} & \text{when } x < 0 \\ x & \text{when } x \geq 0 \end{cases}$

59. **(a)** **(iii)**   **(b)** **(iv)**   **(c)** **(ii)**   **(d)** **(i)**

Chapter 2

Section 2.1 Preliminary Questions

1. The graph of position as a function of time

2. No. Instantaneous velocity is defined as the limit of average velocity as time elapsed shrinks to zero.

3. The slope of the line tangent to the graph of position as a function of time at $t = t_0$

4. The slope of the secant line over the interval $[x_0, x_1]$ approaches the slope of the tangent line at $x = x_0$.

5. The graph of atmospheric temperature as a function of altitude. Possible units for this rate of change are °F/ft or °C/m.

Section 2.1 Exercises

1. **(a)** 11.025 m   **(b)** 22.05 m/s

(c)

time interval	[2, 2.01]	[2, 2.005]	[2, 2.001]	[2, 2.00001]
average velocity	19.649	19.6245	19.6049	19.600049

The instantaneous velocity at $t = 2$ is 19.6 m/s.

3. 0.559017 mg/°C

5. 0.3 m/s

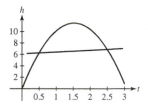

7. **(a)** Dollars/year   **(b)** [0, 0.5]: 7.8461; [0, 1]: 8

(c) Approximately $8/year

9. **(a)** Approximately 5% per year.

(b) Increases

(c) Approximately 5.5% per year.

(d) Greater at $t = 2012$; less at $t = 2008$.

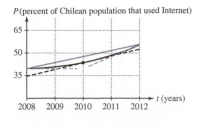

11. 12   **13.** −0.0625   **15.** 1.00   **17.** 0.333

19. (a) $[0, 0.1]$: -144.721 cm/s; $[3, 3.5]$: 0 cm/s   **(b)** 0 cm/s

21. (a) 3000 cells per hour

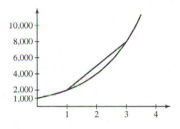

(b) 1250 cells per hour. The quantity m represents the instantaneous rate of change of $P(t)$ at $t = 1$.

23. Sales decline more slowly as time increases.

25.   • In graph (A), the particle is (c) slowing down.
   • In graph (B), the particle is (b) speeding up and then slowing down.
   • In graph (C), the particle is (d) slowing down and then speeding up.
   • In graph (D), the particle is (a) speeding up.

27. (a) Percent/day; measures how quickly the population of flax plants is becoming infected.

(b) $[40, 52]$, $[0, 12]$, $[20, 32]$

(c) The average rates of infection over the intervals $[30, 40]$, $[40, 50]$, $[30, 50]$ are $.9$, $.5$, $.7$ %/d, respectively.

(d) 0.55%/day

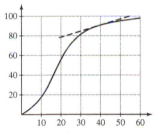

31. (B)

33. Interval $[1, t]$: average rate of change is $t + 1$, instantaneous rate of change is $1 + 1 = 2$; interval $[2, t]$: average rate of change is $t + 2$, instantaneous rate of change is $2 + 2 = 4$.

35. $x^2 + 2x + 4$; instantaneous rate of change is $2^2 + 2*2 + 4 = 12$.

Section 2.2 Preliminary Questions

1. 1   **2.** π **3.** 20   **4.** Yes; $f(x) = \frac{x^2-9}{x-3}$ at $c = 3$ and $c = 3$.

5. $\lim\limits_{x \to 1^-} f(x) = \infty$ and $\lim\limits_{x \to 1^+} f(x) = 3$

6. No because $\lim\limits_{x \to 5^-} f(x)$ may not be equal to $f(5)$.

7. Yes

Section 2.2 Exercises

1.

x	0.998	0.999	0.9995	0.99999
$f(x)$	1.498501	1.499250	1.499625	1.499993

x	1.00001	1.0005	1.001	1.002
$f(x)$	1.500008	1.500375	1.500750	1.501500

The limit as $x \to 1$ is $\frac{3}{2}$.

3.

y	1.998	1.999	1.9999
$f(y)$	0.59984	0.59992	0.599992

y	2.0001	2.001	2.002
$f(y)$	0.600008	0.60008	0.60016

The limit as $y \to 2$ is $\frac{3}{5}$.

5. 1.5   **7.** 21   **9.** $|3x - 12| = 3|x - 4|$

11. $|(5x + 2) - 17| = |5x - 15| = 5|x - 3|$

13. Suppose $|x| < 1$ so that
$|x^2 - 0| = |x + 0||x - 0| = |x||x| < |x|$.

15. If $|x| < 1$, $|4x + 2|$ can be no bigger than 6, so
$|4x^2 + 2x + 5 - 5| = |4x^2 + 2x| = |x||4x + 2| < 6|x|$.

17. $\frac{1}{2}$ **19.** $\frac{5}{3}$ **21.** 2   **23.** 0

25. As $x \to 4^-$, $f(x) \to -\infty$; similarly, as $x \to 4^+$, $f(x) \to \infty$

27. $-1/5$ **29.** $-\infty$ **31.** 0   **33.** 1

35. 2.718 (The exact answer is e.)   **37.** ∞

39.

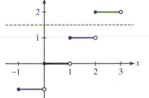

(a) $c - 1$ **(b)** c **(c)** 2

41. $\lim\limits_{x \to 0^-} f(x) = -1$, $\lim\limits_{x \to 0^+} f(x) = 1$

43. $\lim\limits_{x \to 0^-} f(x) = \infty$, $\lim\limits_{x \to 0^+} f(x) = \frac{1}{6}$

45. $\lim\limits_{x \to -2^-} \frac{4x^2 + 7}{x^3 + 8} = -\infty$, $\lim\limits_{x \to -2^+} \frac{4x^2 + 7}{x^3 + 8} = \infty$

47. $\lim\limits_{x \to 1^\pm} \frac{x^5 + x - 2}{x^2 + x - 2} = 2$

49.   • $\lim\limits_{x \to 2^-} f(x) = \infty$ and $\lim\limits_{x \to 2^+} f(x) = \infty$.
   • $\lim\limits_{x \to 4^-} f(x) = -\infty$ and $\lim\limits_{x \to 4^+} f(x) = 10$.
The vertical asymptotes are the vertical lines $x = 2$ and $x = 4$.

51.   **53.**

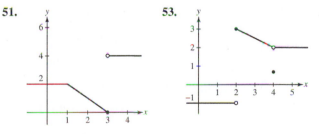

55.   • $\lim\limits_{x \to 1^-} f(x) = \lim\limits_{x \to 1^+} f(x) = 3$
   • $\lim\limits_{x \to 3^-} f(x) = -\infty$
   • $\lim\limits_{x \to 3^+} f(x) = 4$
   • $\lim\limits_{x \to 5^-} f(x) = 2$
   • $\lim\limits_{x \to 5^+} f(x) = -3$
   • $\lim\limits_{x \to 6^-} f(x) = \lim\limits_{x \to 6^+} f(x) = \infty$

57. $\frac{5}{2}$

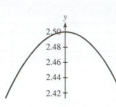

59. 0.693 (The exact answer is ln 2.)

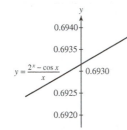

61. -12

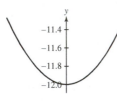

63. For n even

65. (a) No **(b)** $f(\frac{1}{2n}) = 1$ for all integers n.

(c) At $x = 1, \frac{1}{3}, \frac{1}{5}, \ldots$, the value of $f(x)$ is always -1.

67. $\lim\limits_{\theta \to 0} \dfrac{\sin n\theta}{\theta} = n$ **69.** $\frac{1}{2}, 2, \frac{3}{2}, \frac{2}{3}$; $\lim\limits_{x \to 1} \dfrac{x^n - 1}{x^m - 1} = \dfrac{n}{m}$

71. (a)

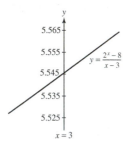

(b) $L = 5.545$.

Section 2.3 Preliminary Questions

1. Suppose $\lim_{x \to c} f(x)$ and $\lim_{x \to c} g(x)$ both exist. The Sum Law states that

$$\lim_{x \to c} (f(x) + g(x)) = \lim_{x \to c} f(x) + \lim_{x \to c} g(x)$$

Provided $\lim_{x \to c} g(x) \neq 0$, the Quotient Law states that

$$\lim_{x \to c} \frac{f(x)}{g(x)} = \frac{\lim_{x \to c} f(x)}{\lim_{x \to c} g(x)}$$

2. (b) **3. (a)**

Section 2.3 Exercises

1. 9 **3.** $\frac{1}{16}$ **5.** $\frac{1}{2}$ **7.** 4.6 **9.** 1 **11.** 9 **13.** $-\frac{2}{5}$ **15.** 10
17. $\frac{1}{5}$ **19.** $\frac{1}{5}$ **21.** $\frac{2}{5}$ **23.** 64 **27.** 3 **29.** $\frac{1}{16}$ **31.** No
33. $f(x) = 1/x$ and $g(x) = -1/x$ **37.** Write $g(t) = \frac{tg(t)}{t}$.
39. (b)

Section 2.4 Preliminary Questions

1. Continuity **2.** $f(3) = \frac{1}{2}$ **3.** No **4.** No; Yes
5. (a) False. The correct statement is, "f is continuous at $x = a$ if the left- and right-hand limits of $f(x)$ as $x \to a$ exist and equal $f(a)$."
(b) True.
(c) False. The correct statement is, "If the left- and right-hand limits of $f(x)$ as $x \to a$ are equal but not equal to $f(a)$, then f has a removable discontinuity at $x = a$."
(d) True.
(e) False. The correct statement is, "If f and g are continuous at $x = a$ and $g(a) \neq 0$, then f/g is continuous at $x = a$."

Section 2.4 Exercises

1. • The function f is discontinuous at $x = 1$; it is right-continuous there.
  • The function f is discontinuous at $x = 3$; it is neither left-continuous nor right-continuous there.
  • The function f is discontinuous at $x = 5$; it is left-continuous there.
  None of these discontinuities is removable.

3. $x = 3$; redefine $g(3) = 4$

5. The function f is discontinuous at $x = 0$, at which $\lim\limits_{x \to 0^-} f(x) = \infty$ and $\lim\limits_{x \to 0^+} f(x) = 2$. The function f is also discontinuous at $x = 2$, at which $\lim\limits_{x \to 2^-} f(x) = 6$ and $\lim\limits_{x \to 2^+} f(x) = 6$. The discontinuity at $x = 2$ is removable. Assigning $f(2) = 6$ makes f continuous at $x = 2$.

7. $y = x$ and $y = \sin x$ are continuous, so is $f(x) = x + \sin x$ by Continuity Law (i).

9. Since $y = x$ and $y = \sin x$ are continuous, so are $y = 3x$ and $y = 4 \sin x$ by Continuity Law (ii). Thus, $f(x) = 3x + 4 \sin x$ is continuous by Continuity Law (i).

11. Since $y = x$ is continuous, so is $y = x^2$ by Continuity Law (iii). Recall that constant functions, such as 1, are continuous. Thus, $y = x^2 + 1$ is continuous by Continuity Law (i). Finally,
$$f(x) = \frac{1}{x^2 + 1}$$
is continuous by Continuity Law (iv) because $x^2 + 1$ is never 0.

13. The function f is a composite of two continuous functions: $y = \cos x$ and $y = x^2$, so f is continuous by Theorem 5.

15. $y = e^x$ and $y = \cos 3x$ are continuous, so $f(x) = e^x \cos 3x$ is continuous by Continuity Law (iii).

17. Discontinuous at $x = 0$, at which there is an infinite discontinuity. The function is neither left- nor right-continuous at $x = 0$.

19. Discontinuous at $x = 1$, at which there is an infinite discontinuity. The function is neither left- nor right-continuous at $x = 1$.

21. Discontinuous at even integers, at which there are jump discontinuities. Function is right-continuous at the even integers but not left-continuous.

23. Discontinuous at $x = \frac{1}{2}$, at which there is an infinite discontinuity. The function is neither left- nor right-continuous at $x = \frac{1}{2}$.

25. Continuous for all x

27. Jump discontinuity at $x = 2$. Function is left-continuous at $x = 2$ but not right-continuous.

29. Discontinuous whenever $t = \frac{(2n+1)\pi}{4}$, where n is an integer. At every such value of t there is an infinite discontinuity. The function is neither left- nor right-continuous at any of these points of discontinuity.

31. Continuous everywhere

33. Discontinuous at $x = 0$, at which there is an infinite discontinuity. The function is neither left- nor right-continuous at $x = 0$.

35. The domain is all real numbers. Both $y = \sin x$ and $y = \cos x$ are continuous on this domain, so $f(x) = 2\sin x + 3\cos x$ is continuous by Continuity Laws (i) and (ii).

37. Domain is $x \geq 0$. Since $y = \sqrt{x}$ and $y = \sin x$ are continuous, so is $f(x) = \sqrt{x}\sin x$ by Continuity Law (iii).

39. Domain is all real numbers. Both $y = x^{2/3}$ and $y = 2^x$ are continuous on this domain, so $f(x) = x^{2/3}2^x$ is continuous by Continuity Law (iii).

41. Domain is $x \neq 0$. Because the function $y = x^{4/3}$ is continuous and not equal to zero for $x \neq 0$, $f(x) = x^{-4/3}$ is continuous for $x \neq 0$ by Continuity Law (iv).

43. Domain is all $x \neq \pm(2n-1)\pi/2$, where n is a positive integer. Because $y = \tan x$ is continuous on this domain, it follows from Continuity Law (iii) that $f(x) = \tan^2 x$ is also continuous on this domain.

45. Domain of $f(x) = (x^4 + 1)^{3/2}$ is all real numbers. Because $y = x^{3/2}$ and the polynomial $y = x^4 + 1$ are both continuous, so is the composite function $f(x) = (x^4 + 1)^{3/2}$.

47. Domain is all $x \neq \pm 1$. Because the functions $y = \cos x$ and $y = x^2$ are continuous on this domain, so is the composite function $y = \cos(x^2)$. Finally, because the polynomial $y = x^2 - 1$ is continuous and not equal to zero for $x \neq \pm 1$, the function $f(x) = \frac{\cos(x^2)}{x^2-1}$ is continuous by Continuity Law (iv).

49. f is right-continuous at $x = 1$; f is continuous at $x = 2$.

51. The function f is continuous everywhere.

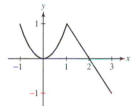

53. The function f is neither left- nor right-continuous at $x = 2$.

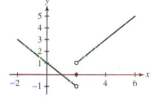

55. $\lim\limits_{x\to 4} \frac{x^2-16}{x-4} = \lim\limits_{x\to 4}(x+4) = 8 \neq 10 = f(4)$

57. $c = \frac{5}{3}$ **59.** $a = 2$ and $b = 1$

61. (a) No   (b) $g(1) = -\frac{\pi}{2}$

63.   **65.**

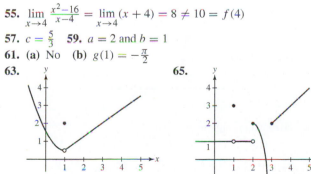

67. -6 **69.** $\frac{1}{3}$ **71.** -1 **73.** $\frac{1}{32}$ **75.** 27 **77.** 1000 **79.** $\frac{\pi}{2}$

81. No. Take $f(x) = -x^{-1}$ and $g(x) = x^{-1}$

83. $f(x) = |g(x)|$ is a composition of the continuous functions g and $y = |x|$

85. No.

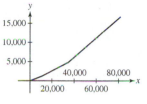

87. $f(x) = 3$ and $g(x) = [x]$

89. In this case, $y = f(x)^2$ is the constant function 1.

Section 2.5 Preliminary Questions

1. $\frac{x^2-1}{\sqrt{x+3}-2}$

2. (a) $f(x) = \frac{x^2-1}{x-1}$ (b) $f(x) = \frac{x^2-1}{x-1}$ (c) $f(x) = \frac{1}{x}$

3. The "simplify and plug-in" strategy is based on simplifying a function that is indeterminate to a continuous function. Once the simplification has been made, the limit of the remaining continuous function is obtained by evaluation.

Section 2.5 Exercises

1. $\lim\limits_{x\to 6}\frac{x^2-36}{x-6} = \lim\limits_{x\to 6}\frac{(x-6)(x+6)}{x-6} = \lim\limits_{x\to 6}(x+6) = 12$

3. 0 **5.** $\frac{1}{14}$ **7.** -1 **9.** $\frac{11}{10}$ **11.** 2 **13.** 1 **15.** 2 **17.** $\frac{1}{8}$

19. $-\frac{1}{4}$

21. Limit does not exist.

• As $h \to 0+$, $\dfrac{\sqrt{h+2}-2}{h} \to -\infty$.

• As $h \to 0-$, $\dfrac{\sqrt{h+2}-2}{h} \to \infty$.

23. 2 **25.** $\frac{1}{4}$ **27.** 1 **29.** $-\frac{1}{2}$ **31.** 9 **33.** $\frac{1}{2}$

35. $\lim\limits_{x\to 4} f(x) \approx 2.00$; to two decimal places, this matches the value of 2 obtained in Exercise 23.

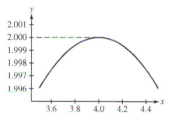

37. 12 **39.** -1 **41.** $\frac{4}{3}$ **43.** $\frac{1}{4}$ **45.** $2a$ **47.** $-4+5a$ **49.** $4a$

51. $\frac{1}{2\sqrt{a}}$ **53.** $3a^2$ **55.** $c = -1$ and $c = 6$ **57.** $c = 3$ **59.** $+$

Section 2.6 Preliminary Questions

1. $\lim_{x\to 0} f(x) = 0$; No

2. Assume that for $x \neq c$ (in some open interval containing c),

$$l(x) \leq f(x) \leq u(x)$$

and that $\lim\limits_{x\to c} l(x) = \lim\limits_{x\to c} u(x) = L$. Then $\lim\limits_{x\to c} f(x)$ exists and

$$\lim\limits_{x\to c} f(x) = L$$

3. (a)

Section 2.6 Exercises

1. For all $x \neq 1$ on the open interval $(0, 2)$ containing $x = 1$, $\ell(x) \leq f(x) \leq u(x)$. Moreover,

$$\lim_{x \to 1} \ell(x) = \lim_{x \to 1} u(x) = 2$$

Therefore, by the Squeeze Theorem,

$$\lim_{x \to 1} f(x) = 2$$

3. $\lim_{x \to 7} f(x) = 6$; No

5. **(a)** *not* sufficient information   **(b)** $\lim_{x \to 1} f(x) = 1$
(c) $\lim_{x \to 1} f(x) = 3$

7. $\lim_{x \to 0} x^2 \cos \frac{1}{x} = 0$ **9.** $\lim_{x \to 1} (x - 1) \sin \frac{\pi}{x - 1} = 0$

11. $\lim_{t \to 0} (2^t - 1) \cos \frac{1}{t} = 0$ **13.** $\lim_{t \to 2} (t^2 - 4) \cos \frac{1}{t - 2} = 0$

15. $\lim_{\theta \to \frac{\pi}{2}} \cos \theta \cos(\tan \theta) = 0$ **17.** 1   **19.** 3   **21.** 1   **23.** 0

25. $\frac{2\sqrt{2}}{\pi}$ **27. (b)** $L = 14$ **29.** 9   **31.** $\frac{1}{5}$ **33.** $\frac{7}{3}$ **35.** $\frac{1}{25}$

37. 6   **39.** $-\frac{3}{4}$ **41.** $\frac{1}{2}$ **43.** $\frac{6}{5}$ **45.** 0   **47.** 0   **49.** -1

53. $-\frac{9}{2}$

55. $\lim_{t \to 0^+} \frac{\sqrt{1 - \cos t}}{t} = \frac{\sqrt{2}}{2}$; $\lim_{t \to 0^-} \frac{\sqrt{1 - \cos t}}{t} = -\frac{\sqrt{2}}{2}$; does not exist.

59. **(a)**

x	$c - .01$	$c - .001$	$c + .001$	$c + .01$
$\dfrac{\sin x - \sin c}{x - c}$	.999983	.99999983	.99999983	.999983

Here, $c = 0$ and $\cos c = 1$.

x	$c - .01$	$c - .001$	$c + .001$	$c + .01$
$\dfrac{\sin x - \sin c}{x - c}$	.868511	.866275	.865775	.863511

Here, $c = \frac{\pi}{6}$ and $\cos c = \frac{\sqrt{3}}{2} \approx .866025$.

x	$c - .01$	$c - .001$	$c + .001$	$c + .01$
$\dfrac{\sin x - \sin c}{x - c}$	.504322	.500433	.499567	.495662

Here, $c = \frac{\pi}{3}$ and $\cos c = \frac{1}{2}$.

x	$c - .01$	$c - .001$	$c + .001$	$c + .01$
$\dfrac{\sin x - \sin c}{x - c}$	.710631	.707460	.706753	.703559

Here, $c = \frac{\pi}{4}$ and $\cos c = \frac{\sqrt{2}}{2} \approx 0.707107$.

x	$c - .01$	$c - .001$	$c + .001$	$c + .01$
$\dfrac{\sin x - \sin c}{x - c}$	.005000	.000500	$-.000500$	$-.005000$

Here, $c = \frac{\pi}{2}$ and $\cos c = 0$.

(b) $\lim_{x \to c} \dfrac{\sin x - \sin c}{x - c} = \cos c$.
(c)

x	$c - .01$	$c - .001$	$c + .001$	$c + .01$
$\dfrac{\sin x - \sin c}{x - c}$	$-.411593$	$-.415692$	$-.416601$	$-.420686$

Here, $c = 2$ and $\cos c = \cos 2 \approx -.416147$.

x	$c - .01$	$c - .001$	$c + .001$	$c + .01$
$\dfrac{\sin x - \sin c}{x - c}$	.863511	.865775	.866275	.868511

Here, $c = -\frac{\pi}{6}$ and $\cos c = \frac{\sqrt{3}}{2} \approx .866025$.

Section 2.7 Preliminary Questions

1. **(a)** Correct   **(b)** Not correct   **(c)** Not correct   **(d)** Correct
2. **(a)** $\lim_{x \to \infty} x^3 = \infty$ **(b)** $\lim_{x \to -\infty} x^3 = -\infty$
(c) $\lim_{x \to -\infty} x^4 = \infty$
3.

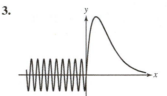

4. Negative   **5.** Negative
6. As $x \to \infty$, $\frac{1}{x} \to 0$, so

$$\lim_{x \to \infty} \sin \frac{1}{x} = \sin 0 = 0$$

On the other hand, $\frac{1}{x} \to \pm\infty$ as $x \to 0$, and as $\frac{1}{x} \to \pm\infty$, $\sin \frac{1}{x}$ oscillates infinitely often.

Section 2.7 Exercises

1. $y = 1$ and $y = 2$
3.

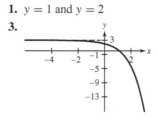

5. **(a)** From the table below, it appears that

$$\lim_{x \to \pm\infty} \frac{x^3}{x^3 + x} = 1$$

x	± 50	± 100	± 500	± 1000
$f(x)$	.999600	.999900	.999996	.999999

(b) From the graph below, it also appears that

$$\lim_{x \to \pm\infty} \frac{x^3}{x^3 + x} = 1$$

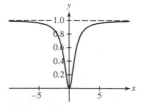

(c) The horizontal asymptote of f is $y = 1$.
7. 1   **9.** 0   **11.** $\frac{7}{4}$ **13.** $-\infty$ **15.** ∞ **17.** $y = \frac{1}{4}$ **19.** $y = \frac{2}{3}$
and $y = -\frac{2}{3}$ **21.** $y = 0$ **23.** 0   **25.** 2   **27.** $\frac{1}{16}$ **29.** 0

31. $\frac{\pi}{2}$; the graph of $y = \tan^{-1} x$ has a horizontal asymptote at $y = \frac{\pi}{2}$.

33. (a) $\lim\limits_{s\to\infty} R(s) = \lim\limits_{s\to\infty} \dfrac{As}{K+s} = \lim\limits_{s\to\infty} \dfrac{A}{1+\frac{K}{s}} = A.$

(b) $R(K) = \dfrac{AK}{K+K} = \dfrac{AK}{2K} = \dfrac{A}{2}$ half of the limiting value.

(c) 3.75 mM

35. 0   **37.** ∞ **39.** $\ln\frac{3}{2}$ **41.** $-\frac{\pi}{2}$

45. $\lim\limits_{x\to\infty} \dfrac{3x^2 - x}{2x^2 + 5} = \lim\limits_{t\to 0+} \dfrac{3-t}{2+5t^2} = \dfrac{3}{2}$

47. • $b = 0.2$:

x	5	10	50	100
$G(x)$	1.000064	1.000000	1.000000	1.000000

It appears that $G(0.2) = 1$.
  • $b = 0.8$:

x	5	10	50	100
$G(x)$	1.058324	1.010251	1.000000	1.000000

It appears that $G(0.8) = 1$.
  • $b = 2$:

x	5	10	50	100
$G(x)$	2.012347	2.000195	2.000000	2.000000

It appears that $G(2) = 2$.
  • $b = 3$:

x	5	10	50	100
$G(x)$	3.002465	3.000005	3.000000	3.000000

It appears that $G(3) = 3$.
  • $b = 5$:

x	5	10	50	100
$G(x)$	5.000320	5.000000	5.000000	5.000000

It appears that $G(5) = 5$.
Based on these observations, we conjecture that $G(b) = 1$ if $0 \le b \le 1$ and $G(b) = b$ for $b > 1$. The graph of $y = G(b)$ is shown below; the graph does appear to be continuous.

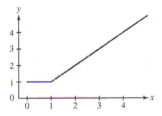

Section 2.8 Preliminary Questions

1. Observe that $f(x) = x^2$ is continuous on $[0, 1]$ with $f(0) = 0$ and $f(1) = 1$. Because $f(0) < 0.5 < f(1)$, the Intermediate Value Theorem guarantees there is a $c \in [0, 1]$ such that $f(c) = 0.5$.

2. We must assume that temperature is a continuous function of time.

3. If f is continuous on $[a, b]$, then the horizontal line $y = k$ for every k between $f(a)$ and $f(b)$ intersects the graph of $y = f(x)$ at least once.

4.

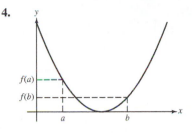

5. (a) Sometimes true.   **(b)** Always true.   **(c)** Never true.
(d) Sometimes true.

Section 2.8 Exercises

1. Observe that $f(1) = 2$ and $f(2) = 10$. Since f is a polynomial, it is continuous everywhere; in particular on $[1, 2]$. Therefore, by the IVT, there is a $c \in [1, 2]$ such that $f(c) = 9$.

3. $g(0) = 0$ and $g(\frac{\pi}{4}) = \frac{\pi^2}{16}$. g is continuous for all t between 0 and $\frac{\pi}{4}$, and $0 < \frac{1}{2} < \frac{\pi^2}{16}$; therefore, by the IVT, there is a $c \in [0, \frac{\pi}{4}]$ such that $g(c) = \frac{1}{2}$.

5. Let $f(x) = x - \cos x$. Observe that f is continuous with $f(0) = -1$ and $f(1) = 1 - \cos 1 \approx .46$. Therefore, by the IVT, there is a $c \in [0, 1]$ such that $f(c) = c - \cos c = 0$.

7. Let $f(x) = \sqrt{x} + \sqrt{x+2} - 3$. Note that f is continuous on $\left[\frac{1}{4}, 2\right]$ with $f(\frac{1}{4}) = -1$ and $f(2) = \sqrt{2} - 1 \approx .41$. Therefore, by the IVT, there is a $c \in \left[\frac{1}{4}, 2\right]$ such that $f(c) = \sqrt{c} + \sqrt{c+2} - 3 = 0$.

9. Let $f(x) = x^2$. Observe that f is continuous with $f(1) = 1$ and $f(2) = 4$. Therefore, by the IVT, there is a $c \in [1, 2]$ such that $f(c) = c^2 = 2$.

11. For each positive integer k, let $f(x) = x^k - \cos x$. Observe that f is continuous on $\left[0, \frac{\pi}{2}\right]$ with $f(0) = -1$ and $f(\frac{\pi}{2}) = \left(\frac{\pi}{2}\right)^k > 0$. Therefore, by the IVT, there is a $c \in \left[0, \frac{\pi}{2}\right]$ such that $f(c) = c^k - \cos(c) = 0$.

13. Let $f(x) = 2^x + 3^x - 4^x$. Observe that f is continuous on $[0, 2]$ with $f(0) = 1 > 0$ and $f(2) = -3 < 0$. Therefore, by the IVT, there is a $c \in (0, 2)$ such that $f(c) = 2^c + 3^c - 4^c = 0$.

15. Let $f(x) = e^x + \ln x$. Observe that f is continuous on $[e^{-2}, 1]$ with $f(e^{-2}) = e^{e^{-2}} - 2 < 0$ and $f(1) = e > 0$. Therefore, by the IVT, there is a $c \in (e^{-2}, 1) \subset (0, 1)$ such that $f(c) = e^c + \ln c = 0$.

17. Apply Corollary 2 to the Intermediate Value Theorem. f is a polynomial and continuous everywhere. $f(-3) = 170$, $f(-2) = -25$, $f(-1) = 2$, $f(0) = -1$, $f(1) = 2$, $f(2) = -25$, and $f(3) = 170$. Thus, f has a zero in each of these intervals: $(-3, -2)$, $(-2, -1)$, $(-1, 0)$, $(0, 1)$, $(1, 2)$, $(2, 3)$, and f must have six distinct solutions.

19. The IVT does not apply since g is not continuous on the interval $[-1, 1]$. However, this function does take on all values between $g(-1) = 1$ and $g(1) = 2$.

21. (a) $f(1) = 1$, $f(1.5) = 2^{1.5} - (1.5)^3 < 3 - 3.375 < 0$. Hence, $f(x) = 0$ for some x between 1 and 1.5.

(b) $f(1.25) \approx 0.4253 > 0$ and $f(1.5) < 0$. Hence, $f(x) = 0$ for some x between 1.25 and 1.5.

(c) $f(1.375) \approx -0.0059$. Hence, $f(x) = 0$ for some x between 1.25 and 1.375.

23. $[0, .25]$

25. 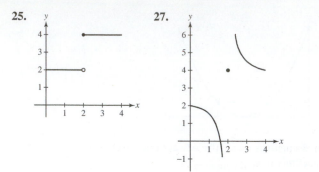 **27.**

15.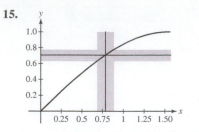

29. No; No

Section 2.9 Preliminary Questions

1. (c)

2. (b) and (d) are true.

Section 2.9 Exercises

1. $L = 4$, $\epsilon = .8$, and $\delta = .1$

3. (a)
$$|f(x) - 35| = |8x + 3 - 35| = |8x - 32| = |8(x - 4)| = 8\,|x - 4|$$
(b) Let $\epsilon > 0$. Let $\delta = \epsilon/8$ and suppose $|x - 4| < \delta$. By part (a),
$|f(x) - 35| = 8|x - 4| < 8\delta$. Substituting $\delta = \epsilon/8$, we see
$|f(x) - 35| < 8\epsilon/8 = \epsilon$.

5. (a) If $0 < |x - 2| < \delta = .01$, then $|x| < 3$ and
$\left|x^2 - 4\right| = |x - 2||x + 2| \le |x - 2|\,(|x| + 2) < 5|x - 2| < .05$.
(b) If $0 < |x - 2| < \delta = .0002$, then $|x| < 2.0002$ and
$$\left|x^2 - 4\right| = |x - 2||x + 2| \le |x - 2|\,(|x| + 2) < 4.0002|x - 2|$$
$$< .00080004 < .0009$$

(c) $\delta = 10^{-5}$

7. $\delta = 6 \times 10^{-4}$

9. $\delta = 0.25$ **11.** $\delta = 0.02$

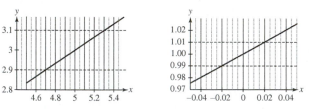

13. (a) Since $|x - 2| < 1$, it follows that $1 < x < 3$, in particular
that $x > 1$. Because $x > 1$, then $\dfrac{1}{x} < 1$ and
$$\left|\frac{1}{x} - \frac{1}{2}\right| = \left|\frac{2 - x}{2x}\right| = \frac{|x - 2|}{2x} < \frac{1}{2}|x - 2|$$

(b) Let $\delta = \min\{1, 2\epsilon\}$ and suppose that $|x - 2| < \delta$. Then by part
(a) we have
$$\left|\frac{1}{x} - \frac{1}{2}\right| < \frac{1}{2}|x - 2| < \frac{1}{2}\delta < \frac{1}{2} \cdot 2\epsilon = \epsilon$$

(c) Choose $\delta = .02$.

(d) Let $\epsilon > 0$ be given. Then whenever
$0 < |x - 2| < \delta = \min\{1, 2\epsilon\}$, we have
$$\left|\frac{1}{x} - \frac{1}{2}\right| < \frac{1}{2}\delta \le \epsilon$$

17. Given $\epsilon > 0$, we let
$$\delta = \min\left\{|c|, \frac{\epsilon}{3|c|}\right\}$$

Then, for $|x - c| < \delta$, we have
$$|x^2 - c^2| = |x - c|\,|x + c| < 3|c|\delta < 3|c|\frac{\epsilon}{3|c|} = \epsilon$$

19. Let $\epsilon > 0$ be given. Let $\delta = \min(1, 3\epsilon)$. If $|x - 4| < \delta$,
$$\left|\sqrt{x} - 2\right| = |x - 4|\left|\frac{1}{\sqrt{x} + 2}\right| < |x - 4|\frac{1}{3} < \delta\frac{1}{3} < 3\epsilon\frac{1}{3} = \epsilon$$

21. Let $\epsilon > 0$ be given. Let $\delta = \min(1, \frac{\epsilon}{7})$, and assume $|x - 1| < \delta$.
Since $\delta < 1$, $0 < x < 2$. Since $x^2 + x + 1$ increases as x increases for
$x > 0$, $x^2 + x + 1 < 7$ for $0 < x < 2$, so
$$\left|x^3 - 1\right| = |x - 1|\left|x^2 + x + 1\right| < 7|x - 1| < 7\frac{\epsilon}{7} = \epsilon$$

23. Let $\epsilon > 0$ be given. Let $\delta = \min(1, \frac{4}{5}\epsilon)$, and suppose
$|x - 2| < \delta$. Since $\delta < 1$, $|x - 2| < 1$, so $1 < x < 3$. This means that
$4x^2 > 4$ and $|2 + x| < 5$ so that $\frac{2+x}{4x^2} < \frac{5}{4}$. We get
$$\left|x^{-2} - \frac{1}{4}\right| = |2 - x|\left|\frac{2 + x}{4x^2}\right| < \frac{5}{4}|x - 2| < \frac{5}{4} \cdot \frac{4}{5}\epsilon = \epsilon$$

25. Let L be any real number. Let $\delta > 0$ be any small positive
number. Let $x = \frac{\delta}{2}$, which satisfies $|x| < \delta$, and $f(x) = 1$. We
consider two cases:

• $(|f(x) - L| \ge \frac{1}{2})$: we are done.
• $(|f(x) - L| < \frac{1}{2})$: This means $\frac{1}{2} < L < \frac{3}{2}$. In this case, let
  $x = -\frac{\delta}{2}$. $f(x) = -1$ so $\frac{3}{2} < L - f(x)$.

In either case, there exists an x such that $|x| < \frac{\delta}{2}$, but $|f(x) - L| \ge \frac{1}{2}$.

27. Let $\epsilon > 0$ and let $\delta = \min(1, \frac{\epsilon}{2})$. Then, whenever $|x - 1| < \delta$, it
follows that $0 < x < 2$. If $1 < x < 2$, then $\min(x, x^2) = x$ and
$$|f(x) - 1| = |x - 1| < \delta < \frac{\epsilon}{2} < \epsilon$$

On the other hand, if $0 < x < 1$, then $\min(x, x^2) = x^2$, $|x + 1| < 2$
and
$$|f(x) - 1| = |x^2 - 1| = |x - 1|\,|x + 1| < 2\delta < \epsilon$$

Thus, whenever $|x - 1| < \delta$, $|f(x) - 1| < \epsilon$.

31. Suppose that $\lim_{x \to c} f(x) = L$. Let $\epsilon > 0$ be given. Since
$\lim_{x \to c} f(x) = L$, we know there is a $\delta > 0$ such that $|x - c| < \delta$ forces
$|f(x) - L| < \epsilon/|a|$. Suppose $|x - c| < \delta$. Then
$|af(x) - aL| = |a|\,|f(x) - aL| < |a|(\epsilon/|a|) = \epsilon$.

Chapter 2 Review

1. Average velocity approximately 0.954 m/s; instantaneous velocity approximately 0.894 m/s

3. $\frac{200}{9}$ **5.** 1.50   **7.** 1.69   **9.** 2.00

11. 5   **13.** $-\frac{1}{2}$ **15.** $\frac{1}{6}$ **17.** 2

19. Does not exist;

$$\lim_{t \to 9-} \frac{t-6}{\sqrt{t}-3} = -\infty \quad \text{and} \quad \lim_{t \to 9+} \frac{t-6}{\sqrt{t}-3} = \infty$$

21. ∞

23. Does not exist;

$$\lim_{x \to 1-} \frac{x^3 - 2x}{x-1} = \infty \quad \text{and} \quad \lim_{x \to 1+} \frac{x^3 - 2x}{x-1} = -\infty$$

25. 2   **27.** $\frac{2}{3}$ **29.** $-\frac{1}{2}$ **31.** $3b^2$ **33.** $\frac{1}{9}$ **35.** ∞

37. Does not exist;

$$\lim_{\theta \to \frac{\pi}{2}-} \theta \sec \theta = \infty \quad \text{and} \quad \lim_{\theta \to \frac{\pi}{2}+} \theta \sec \theta = -\infty$$

39. Does not exist;

$$\lim_{\theta \to 0-} \frac{\cos \theta - 2}{\theta} = \infty \quad \text{and} \quad \lim_{\theta \to 0+} \frac{\cos \theta - 2}{\theta} = -\infty$$

41. ∞ **43.** ∞

45. Does not exist;

$$\lim_{x \to \frac{\pi}{2}-} \tan x = \infty \quad \text{and} \quad \lim_{x \to \frac{\pi}{2}+} \tan x = -\infty$$

47. 0   **49.** 0

51. According to the graph of f,

$$\lim_{x \to 0-} f(x) = \lim_{x \to 0+} f(x) = 1$$

$$\lim_{x \to 2-} f(x) = \lim_{x \to 2+} f(x) = \infty$$

$$\lim_{x \to 4-} f(x) = -\infty$$

$$\lim_{x \to 4+} f(x) = \infty$$

The function is both left- and right-continuous at $x = 0$ and neither left- nor right-continuous at $x = 2$ and $x = 4$.

53. At $x = 0$, the function has an infinite discontinuity but is left-continuous.

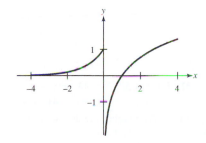

55. g has a jump discontinuity at $x = -1$; g is left-continuous at $x = -1$.

57. $b = 7$; h has a jump discontinuity at $x = -2$.

59. Does not have any horizontal asymptotes

61. $y = 2$ **63.** $y = 1$

65.

$$B = B \cdot 1 = B \cdot L =$$

$$\lim_{x \to a} g(x) \cdot \lim_{x \to a} \frac{f(x)}{g(x)} = \lim_{x \to a} g(x) \frac{f(x)}{g(x)} = \lim_{x \to a} f(x) = A$$

67. $f(x) = \dfrac{1}{(x-a)^3}$ and $g(x) = \dfrac{1}{(x-a)^5}$

71. Let $f(x) = x^2 - \cos x$. Now, f is continuous over the interval $[0, \frac{\pi}{2}]$, $f(0) = -1 < 0$ and $f(\frac{\pi}{2}) = \frac{\pi^2}{4} > 0$. Therefore, by the Intermediate Value Theorem, there exists a $c \in (0, \frac{\pi}{2})$ such that $f(c) = 0$; consequently, the curves $y = x^2$ and $y = \cos x$ intersect.

73. Let $f(x) = e^{-x^2} - x$. Observe that f is continuous on $[0, 1]$ with $f(0) = e^0 - 0 = 1 > 0$ and $f(1) = e^{-1} - 1 < 0$. Therefore, the IVT guarantees there exists a $c \in (0, 1)$ such that $f(c) = e^{-c^2} - c = 0$.

75. $g(x) = \lfloor x \rfloor$; On the interval

$$x \in \left[\frac{a}{2 + 2\pi a}, \frac{a}{2} \right] \subset [-a, a]$$

$\frac{1}{x}$ runs from $\frac{2}{a}$ to $\frac{2}{a} + 2\pi$, so the sine function covers one full period and clearly takes on every value from $-\sin a$ through $\sin a$.

77. $\delta = 0.55$;

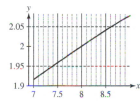

79. Let $\epsilon > 0$ and take $\delta = \epsilon/8$. Then, whenever $|x - (-1)| = |x + 1| < \delta$,

$$|f(x) - (-4)| = |4 + 8x + 4| = 8|x + 1| < 8\delta = \epsilon$$

Chapter 3

Section 3.1 Preliminary Questions

1. B and D

2. $\dfrac{f(x) - f(a)}{x - a}$ and $\dfrac{f(a + h) - f(a)}{h}$

3. $a = 3$ and $h = 2$

4. Derivative of the function $f(x) = \tan x$ at $x = \frac{\pi}{4}$

5. (a) The difference in height between the points $(0.9, \sin 0.9)$ and $(1.3, \sin 1.3)$

(b) The slope of the secant line between the points $(0.9, \sin 0.9)$ and $(1.3, \sin 1.3)$

(c) The slope of the tangent line to the graph at $x = 0.9$

Section 3.1 Exercises

1. $f'(3) = 30$ **3.** $f'(0) = 9$ **5.** $f'(-1) = -2$

7. $f'(1) = 5$

9. Slope of the secant line $= 1$; the secant line through $(2, f(2))$ and $(2.5, f(2.5))$ has a larger slope than the tangent line at $x = 2$.

11. $f'(1) \approx 0$; $f'(2) \approx 0.8$

13. $f'(1) = f'(2) = 0$; $f'(4) = \frac{1}{2}$; $f'(7) = 0$

15. $f'(5.5)$ **17.** $f'(x) = 7$ **19.** $g'(t) = -3$ **21.** $y = 2x - 1$

23. The tangent line at any point is the line itself.

25. $f(-2 + h) = \dfrac{1}{-2 + h}$; $-\dfrac{1}{3}$ **27.** $f'(5) = -\dfrac{1}{10\sqrt{5}}$

29. $f'(3) = 22$; $y = 22x - 18$ **31.** $f'(3) = -11$; $y = -11t + 18$

33. $f'(0) = 1$; $y = x$ **35.** $f'(8) = -\dfrac{1}{64}$; $y = -\dfrac{1}{64}x + \dfrac{1}{4}$

37. $f'(-2) = -1$; $y = -x - 1$

39. $f'(1) = \dfrac{1}{2\sqrt{5}}$; $y = \dfrac{1}{2\sqrt{5}}x + \dfrac{9}{2\sqrt{5}}$

41. $f'(4) = -\dfrac{1}{16}$; $y = -\dfrac{1}{16}x + \dfrac{3}{4}$

43. $f'(3) = \dfrac{3}{\sqrt{10}}$; $y = \dfrac{3}{\sqrt{10}}t + \dfrac{1}{\sqrt{10}}$ **45.** $f'(0) = 0$; $y = 1$

47. $W'(4) \approx 0.9$ kg/year; slope of the tangent is zero at $t = 10$ and at $t = 11.6$; slope of the tangent line is negative for $10 < t < 11.6$.

49. **(a)** $f'(0) \approx -0.68$

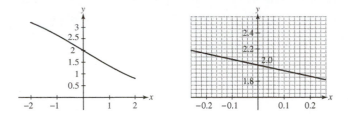

(b) $y = -0.68x + 2$

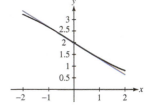

51. For $1 < x < 2.5$ and for $x > 3.5$ **53.** $f(x) = x^3$ and $a = 5$

55. $f(x) = \sin x$ and $a = \dfrac{\pi}{6}$ **57.** $f(x) = 5^x$ and $a = 2$

59. $f'\left(\dfrac{\pi}{4}\right) \approx 0.7071$

61. • On curve (A), $f'(1)$ is larger than
$$\frac{f(1+h) - f(1)}{h}$$
The curve is bending downward, so the secant line to the right is at a lower angle than the tangent line.
• On curve (B), $f'(1)$ is smaller than
$$\frac{f(1+h) - f(1)}{h}$$
The curve is bending upward, so the secant line to the right is at a steeper angle than the tangent line.

63. **(b)** $f'(4) \approx 20.0000$

(c) $y = 20x - 48$

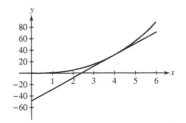

65. $c \approx 0.37$.

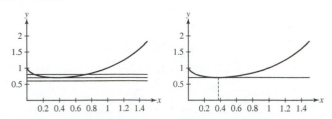

67.

$$P'(303) \approx \frac{P(313) - P(293)}{20} = \frac{0.0808 - 0.0278}{20} = 0.00265 \text{ atm/K};$$

$$P'(313) \approx \frac{P(323) - P(303)}{20} = \frac{0.1311 - 0.0482}{20} = 0.004145 \text{ atm/K};$$

$$P'(323) \approx \frac{P(333) - P(313)}{20} = \frac{0.2067 - 0.0808}{20} = 0.006295 \text{ atm/K};$$

$$P'(333) \approx \frac{P(343) - P(323)}{20} = \frac{0.3173 - 0.1311}{20} = 0.00931 \text{ atm/K};$$

$$P'(343) \approx \frac{P(353) - P(333)}{20} = \frac{0.4754 - 0.2067}{20} = 0.013435 \text{ atm/K}$$

69. -0.375 kph·km/car **71.** $i(3) = 0.06$ amperes

73. $v'(4) \approx 160$; $C \approx 0.2$ farads

75. It is the slope of the secant line connecting the points $(a - h, f(a - h))$ and $(a + h, f(a + h))$ on the graph of f.

77. Let $f(x) = px^2 + qx + r$; SDQ: $\big(f(a+h) - f(a-h)\big)/2h = 2h(2pa + q)/2h = 2pa + q$; $f'(a) = \lim_{h \to 0} \big(f(a+h) - f(a)\big)/h = 2pa + q$

Section 3.2 Preliminary Questions

1. 8 **2.** $(f - g)'(1) = -2$ and $(3f + 2g)'(1) = 19$

3. (a), (b), (c), and (f) **4.** (b)

5. The line tangent to $f(x) = e^x$ at $x = 0$ has slope equal to 1.

Section 3.2 Exercises

1. $f'(x) = 3$ **3.** $f'(x) = 3x^2$ **5.** $f'(x) = 1 - \dfrac{1}{2\sqrt{x}}$

7. $\dfrac{d}{dx}x^4\bigg|_{x=-2} = 4(-2)^3 = -32$

9. $\dfrac{d}{dt}t^{2/3}\bigg|_{t=8} = \dfrac{2}{3}(8)^{-1/3} = \dfrac{1}{3}$

11. $0.35x^{-0.65}$ **13.** $\sqrt{17}t^{\sqrt{17}-1}$

15. $f'(x) = 4x^3$; $y = 32x - 48$

17. $f'(x) = 5 - 16x^{-1/2}$; $y = -3x - 32$

19. **(a)** $\dfrac{d}{dx}12e^x = 12e^x$ **(b)** $\dfrac{d}{dt}(25t - 8e^t) = 25 - 8e^t$

(c) $\dfrac{d}{dt}e^{t-3} = e^{t-3}$

21. $f'(x) = 6x^2 - 6x$ **23.** $f'(x) = \dfrac{20}{3}x^{2/3} + 6x^{-3}$

25. $g'(z) = -\dfrac{5}{2}z^{-19/14} - 5z^{-6}$ **27.** $f'(s) = \dfrac{1}{4}s^{-3/4} + \dfrac{1}{3}s^{-2/3}$

29. $g'(x) = 0$ **31.** $h'(t) = 5e^{t-3}$ **33.** $P'(s) = 32s - 24$

35. $g'(x) = -6x^{-5/2}$ **37.** 1 **39.** -60 **41.** $1 - e^4$

43. • The graph in (A) matches the derivative in (III).
• The graph in (B) matches the derivative in (I).
• The graph in (C) matches the derivative in (II).
• The graph in (D) matches the derivative in (III).
(A) and (D) have the same derivative because the graph in (D) is just a vertical translation of the graph in (A).

45. Label the graph in (A) as f, the graph in (B) as h, and the graph in (C) as g.

47. (B) might be the graph of the derivative of f.

49. **(a)** $\dfrac{d}{dt}ct^3 = 3ct^2$ **(b)** $\dfrac{d}{dz}(5z + 4cz^2) = 5 + 8cz$

(c) $\dfrac{d}{dy}(9c^2y^3 - 24c) = 27c^2y^2$

51. $x = \frac{1}{2}$ **53.** $a = 2$ and $b = -3$

55.   • $f'(x) = 3x^2 - 3 \geq -3$ since $3x^2$ is nonnegative.
 • The two parallel tangent lines with slope 2 are shown with the graph of f here.

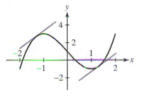

57. $f'(x) = \frac{3}{2}x^{1/2}$

59. $f'(0) = 1$; $y = x$

61. Decreasing; $y = -0.63216(m - 33) + 83.445$;
$y = -0.25606(m - 68) + 69.647$

63.

$$P'(303) \approx \frac{P(313) - P(293)}{20} = \frac{0.0808 - 0.0278}{20} = 0.00265 \text{ atm/K};$$

$$P'(313) \approx \frac{P(323) - P(303)}{20} = \frac{0.1311 - 0.0482}{20} = 0.004145 \text{ atm/K};$$

$$P'(323) \approx \frac{P(333) - P(313)}{20} = \frac{0.2067 - 0.0808}{20} = 0.006295 \text{ atm/K};$$

$$P'(333) \approx \frac{P(343) - P(323)}{20} = \frac{0.3173 - 0.1311}{20} = 0.00931 \text{ atm/K};$$

$$P'(343) \approx \frac{P(353) - P(333)}{20} = \frac{0.4754 - 0.2067}{20} = 0.013435 \text{ atm/K}$$

$\frac{T^2}{P} \frac{dP}{dT}$ is roughly constant, suggesting that the Clausius–Clapeyron law is valid, and that $k \approx 5000$.

67.

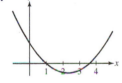

69. For $x < 0$, $f(x) = -x^2$, and $f'(x) = -2x$. For $x > 0$, $f(x) = x^2$, and $f'(x) = 2x$. Thus, $f'(0) = 0$.

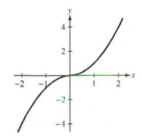

71. $c = 1$ **73.** $c = 0$ **75.** $c = \pm 1$

77. It appears that f is not differentiable at $a = 0$. Moreover, the tangent line does not exist at this point.

79. It appears that f is not differentiable at $a = 3$. Moreover, the tangent line appears to be vertical.

81. It appears that f is not differentiable at $a = 0$. Moreover, the tangent line does not exist at this point.

83. $(1, 8)$ **85.** $\frac{10}{7}$

87. The normal line intersects the x-axis at the point T with coordinates $(x + f(x)f'(x), 0)$. The point R has coordinates $(x, 0)$, so the subnormal is $|x + f(x)f'(x) - x| = |f(x)f'(x)|$.

89. The tangent line to f at $x = a$ is $y = 2ax - a^2$. The x-intercept of this line is $\frac{a}{2}$, so the subtangent is $a - a/2 = a/2$.

91. The subtangent is $\frac{1}{n}a$.   **93.** $r \leq \frac{1}{2}$

Section 3.3 Preliminary Questions

1. (a) False. The notation fg denotes the function whose value at x is $f(x)g(x)$.

(b) True.

(c) False. The derivative of a product fg is $f'(x)g(x) + f(x)g'(x)$.

(d) False. $\frac{d}{dx}(fg)\Big|_{x=4} = f(4)g'(4) + g(4)f'(4)$.

(e) True.

2. -1 **3.** 5

Section 3.3 Exercises

1. $f'(x) = 10x^4 + 3x^2$ **3.** $f'(x) = e^x(x^2 + 2x)$

5. $\frac{dh}{ds} = -\frac{7}{2}s^{-3/2} + \frac{3}{2}s^{-5/2} + 14$; $\frac{dh}{ds}\Big|_{s=4} = \frac{871}{64}$

7. $f'(x) = \frac{-2}{(x - 2)^2}$ **9.** $\frac{dg}{dt} = -\frac{4t}{(t^2 - 1)^2}$; $\frac{dg}{dt}\Big|_{t=-2} = \frac{8}{9}$

11. $g'(x) = -\frac{e^x}{(1 + e^x)^2}$ **13.** $f'(t) = 6t^2 + 2t - 4$

15. $h'(t) = 1$ for $t \neq 1$ **17.** $f'(x) = 6x^5 + 4x^3 + 18x^2 + 5$

19. $\frac{dy}{dx} = -\frac{1}{(x + 10)^2}$; $\frac{dy}{dx}\Big|_{x=3} = -\frac{1}{169}$

21. $f'(x) = 1$ for $x \geq 0$

23. $\frac{dy}{dx} = \frac{2x^5 - 20x^3 + 8x}{(x^2 - 5)^2}$; $\frac{dy}{dx}\Big|_{x=2} = -80$

25. $\frac{dz}{dx} = -\frac{3x^2}{(x^3 + 1)^2}$; $\frac{dz}{dx}\Big|_{x=1} = -\frac{3}{4}$

27. $h'(t) = \frac{-2t^3 - t^2 + 1}{(t^3 + t^2 + t + 1)^2}$

29. $f'(t) = 0$ **31.** $f'(x) = 3x^2 - 6x - 13$

33. $f'(x) = \frac{xe^x}{(x + 1)^2}$ **35.** For $z \neq -2$ and $z \neq 1$, $g'(z) = 2z - 1$

37. $f'(t) = \frac{-xt^2 + 8t - x^2}{(t^2 - x)^2}$

39. $(fg)'(4) = -20$ and $(f/g)'(4) = 0$

41. $G'(4) = -10$ **43.** $F'(0) = -7$ **45.** $\frac{d}{dx}e^{2x} = 2e^{2x}$

47. From the plot of f shown below, we see that f is decreasing on its domain $\{x : x \neq \pm 1\}$. Consequently, $f'(x)$ must be negative. Using the quotient rule, we find

$$f'(x) = \frac{(x^2 - 1)(1) - x(2x)}{(x^2 - 1)^2} = -\frac{x^2 + 1}{(x^2 - 1)^2}$$

which is negative for all $x \neq \pm 1$.

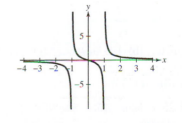

49. $a = 1$

51. (a) Given $R(t) = N(t)S(t)$, it follows that

$$\frac{dR}{dt} = N(t)S'(t) + S(t)N'(t)$$

(b) $\left.\dfrac{dR}{dt}\right|_{t=0} = 1,250,000$

(c) The term $5S(0)$ is larger than the term $10,000N(0)$. Thus, if only one leg of the campaign can be implemented, it should be part A: increase the number of stores by 5 per month.

53.   • At $x = -1$, the tangent line is $y = \dfrac{1}{2}x + 1$.

   • At $x = 1$, the tangent line is $y = -\dfrac{1}{2}x + 1$.

55. Let $g = f^2 = ff$. Then
$g' = \left(f^2\right)' = (ff)' = ff' + ff' = 2ff'$.

57. Let $p = fgh$. Then
$p' = (fgh)' = f\left(gh' + hg'\right) + ghf' = f'gh + fg'h + fgh'$.

61.

$$\frac{d}{dx}(xf(x)) = \lim_{h \to 0} \frac{(x+h)f(x+h) - f(x)}{h}$$

$$= \lim_{h \to 0}\left(x\frac{f(x+h) - f(x)}{h} + f(x+h)\right)$$

$$= x\lim_{h \to 0}\frac{f(x+h) - f(x)}{h} + \lim_{h \to 0} f(x+h)$$

$$= xf'(x) + f(x)$$

65. (a) Is a multiple root   **(b)** Not a multiple root

67.

$$m(ab)(ab)^x = \frac{d}{dx}(ab)^x = \frac{d}{dx}\left(a^x b^x\right)$$

$$= a^x\frac{d}{dx}b^x + b^x\frac{d}{dx}a^x$$

$$= m(b)a^x b^x + m(a)a^x b^x = (m(a) + m(b))(ab)^x$$

Section 3.4 Preliminary Questions

1. (a) atmospheres/meter   **(b)** moles/(liter·hour)

2. 90 mph   **3.** $f(26) \approx 43.75$

4. (a) $P'(2009)$ measures the rate of change of the population of Freedonia in the year 2009.

(b) $P(2010) \approx 5.2$ million.

Section 3.4 Exercises

1. 10 square units per unit increase

3.

c	ROC of $f(x)$ with respect to x at $x = c$
1	$f'(1) = \frac{1}{3}$
8	$f'(8) = \frac{1}{12}$
27	$f'(27) = \frac{1}{27}$

5. $d' = 2$ **7.** $dV/dr = 3\pi r^2$

9. (a) 100 km/hour   **(b)** 100 km/hour   **(c)** 0 km/hour
(d) -50 km/hour

11. (a) (i)   **(b) (ii)**   **(c) (iii)**

13. $\dfrac{dT}{dt} \approx -1.5625°C/hour$ **15.** -8×10^{-6} 1/s

17. $\left.\dfrac{dT}{dh}\right|_{h=30} \approx 2.94°C/km$; $\left.\dfrac{dT}{dh}\right|_{h=70} \approx -3.33°C/km$; $\dfrac{dT}{dh} = 0$

over the interval $[13, 23]$, and near the points $h = 50$ and $h = 90$

19. $v'_{esc}(r) = -1.41 \times 10^7 r^{-3/2}$ **21.** $t = \frac{5}{2}$ s

23. The particle passes through the origin when $t = 0$ s and when $t = 3\sqrt{2} \approx 4.24$ s. The particle is instantaneously motionless when $t = 0$ s and when $t = 3$ s.

25. Maximum velocity: 200 m/s; maximum height: 2040.82 m

27. Initial velocity: $v_0 = 19.6$ m/s; maximum height: 19.6 m

31. (a) $\dfrac{dV}{dv} = -1$ **(b)** -4

35. Rate of change of BSA with respect to mass: $\dfrac{\sqrt{5}}{20\sqrt{m}}$; $m = 70$ kg,

rate of change is $\approx 0.0133631\frac{m^2}{kg}$; $m = 80$ kg, rate of change is

$\frac{1}{80}\frac{m^2}{kg}$; BSA increases more rapidly at lower body mass.

37. 2

39. $\sqrt{2} - \sqrt{1} \approx \frac{1}{2}$; the actual value, to six decimal places, is 0.414214. $\sqrt{101} - \sqrt{100} \approx .05$; the actual value, to six decimal places, is 0.0498756.

41.   • $F(65) = 282.75$ ft
   • Increasing speed from 65 to 66 therefore increases stopping distance by approximately 7.6 ft.
   • The actual increase in stopping distance when speed increases from 65 mph to 66 mph is
   $F(66) - F(65) = 290.4 - 282.75 = 7.65$ ft, which differs by less than 1% from the estimate found using the derivative.

43. The cost of producing 2000 bagels is $796. The cost of the 2001st bagel is approximately $0.244, which is indistinguishable from the estimated cost.

45. An increase in oil prices of a dollar leads to a decrease in demand of 0.5625 barrels a year, and a decrease of a dollar leads to an *increase* in demand of 0.5625 barrels a year.

47. $\dfrac{dB}{dI} = \dfrac{2k}{3I^{1/3}}$; $\dfrac{dH}{dW} = \dfrac{3k}{2}W^{1/2}$

(a) As I increases, $\frac{dB}{dI}$ shrinks, so the rate of change of perceived intensity decreases as the actual intensity increases.

(b) As W increases, $\frac{dH}{dW}$ increases as well, so the rate of change of perceived weight increases as weight increases.

49. (a) The average income among households in the bottom rth part is

$$\frac{F(r)T}{rN} = \frac{F(r)}{r}\cdot\frac{T}{N} = \frac{F(r)}{r}A$$

(b) The average income of households belonging to an interval $[r, r + \Delta r]$ is equal to

$$\frac{F(r + \Delta r)T - F(r)T}{\Delta r N} = \frac{F(r + \Delta r) - F(r)}{\Delta r}\cdot\frac{T}{N}$$

$$= \frac{F(r + \Delta r) - F(r)}{\Delta r}A$$

(c) Take the result from part (b) and let $\Delta r \to 0$. Because

$$\lim_{\Delta r \to 0}\frac{F(r + \Delta r) - F(r)}{\Delta r} = F'(r)$$

we find that a household in the $100r$th percentile has income $F'(r)A$.

(d) The point P in Figure 14(B) has an r-coordinate of 0.6, while the point Q has an r-coordinate of roughly 0.75. Thus, on curve L_1, 40% of households have $F'(r) > 1$ and therefore have above-average income. On curve L_2, roughly 25% of households have above-average income.

53. By definition, the slope of the line through $(0, 0)$ and $(x, C(x))$ is

$$\frac{C(x) - 0}{x - 0} = \frac{C(x)}{x} = C_{\text{avg}}(x)$$

- At point A, average cost is greater than marginal cost.
- At point B, average cost is greater than marginal cost.
- At point C, average cost and marginal cost are nearly the same.
- At point D, average cost is less than marginal cost.

Section 3.5 Preliminary Questions

1. The first derivative of stock prices must be positive, while the second derivative must be negative.

2. True   **3.** All quadratic polynomials   **4.** e^x

Section 3.5 Exercises

1. $y'' = 28$ and $y''' = 0$ **3.** $y'' = 12x^2 - 50$ and $y''' = 24x$

5. $y'' = 8\pi r$ and $y''' = 8\pi$

7. $y'' = -\frac{16}{5}t^{-6/5} + \frac{4}{3}t^{-4/3}$ and $y''' = \frac{96}{25}t^{-11/5} - \frac{16}{9}t^{-7/3}$

9. $y'' = -8z^{-3}$ and $y''' = 24z^{-4}$

11. $y'' = 12\theta + 14$ and $y''' = 12$

13. $y'' = -8x^{-3}$ and $y''' = 24x^{-4}$

15. $y'' = (x^5 + 10x^4 + 20x^3)e^x$ and $y''' = (x^5 + 15x^4 + 60x^3 + 60x^2)e^x$

17. $f^{(4)}(1) = 24$ **19.** $\left.\dfrac{d^2y}{dt^2}\right|_{t=1} = 54$

21. $\left.\dfrac{d^4x}{dt^4}\right|_{t=16} = \dfrac{3465}{134217728}$ **23.** $f'''(-3) = 4e^{-3} - 6$

25. $h''(1) = \frac{7}{4}e$

27. $y^{(0)}(0) = d$, $y^{(1)}(0) = c$, $y^{(2)}(0) = 2b$, $y^{(3)}(0) = 6a$, $y^{(4)}(0) = 24$, and $y^{(5)}(0) = 0$

29. $\dfrac{d^6}{dx^6}x^{-1} = 720x^{-7}$ **31.** $f^{(n)}(x) = (-1)^n(n + 1)!x^{-(n+2)}$

33. $f^{(n)}(x) = (-1)^n \dfrac{(2n-1)\times(2n-3)\times\cdots\times 1}{2^n} x^{-(2n+1)/2}$

35. $f^{(n)}(x) = (-1)^n(x - n)e^{-x}$

37. (a) $a(5) = -120$ m/min^2

(b) The acceleration of the helicopter for $0 \leq t \leq 6$ is shown in the figure below. As the acceleration of the helicopter is negative, the velocity of the helicopter must be decreasing. Because the velocity is positive for $0^- \leq t \leq 5$, the helicopter is slowing down between 0 and 5 minutes and speeding up between 5 and 6 minutes.

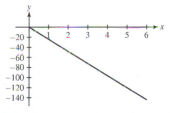

39. (a) f'' **(b)** f' **(c)** f

41. Roughly from time 10 to time 20 and from time 30 to time 40

43. $n = 4, -1$

45. (a) $v(t) = -5.12$ m/s   **(b)** $v(t) = -7.25$ m/s

47. A possible plot of the drill bit's vertical velocity follows:

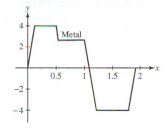

A graph of the acceleration is extracted from this graph:

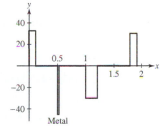

49. (a) Traffic speed must be reduced when the road gets more crowded, so we expect $\frac{dS}{dQ}$ to be negative.

(b) The decrease in speed due to a 1-unit increase in density is approximately $\frac{dS}{dQ}$ (a negative number). Since $\frac{d^2S}{dQ^2} = 5764Q^{-3} > 0$ is positive, this tells us that $\frac{dS}{dQ}$ gets larger as Q increases.

(c) dS/dQ is plotted below. The fact that this graph is increasing shows that $d^2S/dQ^2 > 0$.

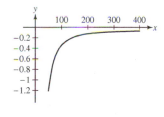

51.

$$f'(x) = -\frac{3}{(x - 1)^2} = (-1)^1 \frac{3 \cdot 1}{(x - 1)^{1+1}};$$

$$f''(x) = \frac{6}{(x - 1)^3} = (-1)^2 \frac{3 \cdot 2 \cdot 1}{(x - 1)^{2+1}};$$

$$f'''(x) = -\frac{18}{(x - 1)^4} = (-1)^3 \frac{3 \cdot 3!}{(x - 1)^{3+1}}; \text{ and}$$

$$f^{(4)}(x) = \frac{72}{(x - 1)^5} = (-1)^4 \frac{3 \cdot 4!}{(x - 1)^{4+1}}$$

From the pattern observed above, we conjecture

$$f^{(k)}(x) = (-1)^k \frac{3 \cdot k!}{(x - 1)^{k+1}}$$

53. 99!

55. $(fg)''' = f'''g + 3f''g' + 3f'g'' + fg'''$;

$$(fg)^{(n)} = \sum_{k=0}^{n} \binom{n}{k} f^{(n-k)} g^{(k)}$$

Section 3.6 Preliminary Questions

1. (a) $\dfrac{d}{dx}(\sin x + \cos x) = -\sin x + \cos x$

(b) $\dfrac{d}{dx}\sec x = \sec x \tan x$ **(c)** $\dfrac{d}{dx}\cot x = -\csc^2 x$

2. (a) This function can be differentiated using the Product Rule.

(b) We have not yet discussed how to differentiate a function like this.

(c) This function can be differentiated using the Product Rule.

3. 0

4. The difference quotient for the function $f(x) = \sin x$ involves the expression $\sin(x + h)$. The addition formula for the sine function is used to expand this expression as

$\sin(x + h) = \sin x \cos h + \sin h \cos x$.

Section 3.6 Exercises

1. $y = \dfrac{\sqrt{2}}{2} x + \dfrac{\sqrt{2}}{2}\left(1 - \dfrac{\pi}{4}\right)$ **3.** $y = 2x + 1 - \dfrac{\pi}{2}$

5. $f'(x) = \cos^2 x - \sin^2 x$ **7.** $f'(x) = 2 \sin x \cos x$

9. $H'(t) = \sec t + 2 \sin t \sec^2 t \tan t$

11. $f'(\theta) = \sec \theta \left(\sec^2 \theta + \tan^2 \theta \right)$

13. $f'(x) = (8x^3 + 4x^{-2}) \sec x + (2x^4 - 4x^{-1}) \sec x \tan x$

15. $y' = \dfrac{\theta \sec \theta \tan \theta - \sec \theta}{\theta^2}$ **17.** $R'(y) = \dfrac{4 \cos y - 3}{\sin^2 y}$

19. $f'(x) = \dfrac{2 \sec^2 x}{(1 - \tan x)^2}$ **21.** $f'(x) = e^x (\cos x + \sin x)$

23. $f'(\theta) = e^\theta (5 \sin \theta + 5 \cos \theta - 4 \tan \theta - 4 \sec^2 \theta)$

25. $y = 1$ **27.** $y = \frac{2}{3}t + (3\sqrt{3} - 2\pi)/9$

29. $y = (1 - \sqrt{3})\left(\theta - \dfrac{\pi}{3}\right) + 1 + \sqrt{3}$

31. $y = x + 1$ **33.** $y = 2e^{\pi/2}\left(t - \dfrac{\pi}{2}\right) + e^{\pi/2}$

35. $\cot x = \frac{\cos x}{\sin x}$; use the Quotient Rule.

37. $\csc x = \frac{1}{\sin x}$; use the Quotient Rule.

39. $f''(\theta) = 2 \cos \theta - \theta \sin \theta$

41. $y'' = 2 \sec^2 x \tan x$
$y''' = 2 \sec^4 x + 4 \sec^2 x \tan^2 x$

43. • Then $f'(x) = -\sin x$, $f''(x) = -\cos x$, $f'''(x) = \sin x$, $f^{(4)}(x) = \cos x$, and $f^{(5)}(x) = -\sin x$.

• Accordingly, the successive derivatives of f cycle among

$$\{-\sin x, -\cos x, \sin x, \cos x\}$$

in that order. Since 8 is a multiple of 4, we have $f^{(8)}(x) = \cos x$.

• Since 36 is a multiple of 4, we have $f^{(36)}(x) = \cos x$. Therefore, $f^{(37)}(x) = -\sin x$.

45. $x = \frac{\pi}{4}, \frac{3\pi}{4}, \frac{5\pi}{4}, \frac{7\pi}{4}$

47. (a)

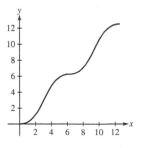

(b) Since $g'(t) = 1 - \cos t \geq 0$ for all t, the slope of the tangent line to g is always nonnegative.

(c) $t = 0, 2\pi, 4\pi$

49. $f'(x) = \sec^2 x = \frac{1}{\cos^2 x}$. Note that $f'(x) = \frac{1}{\cos^2 x}$ has numerator 1; the equation $f'(x) = 0$ therefore has no solution. The least slope for a tangent line to $\tan x$ is 1. Here is a graph of f'.

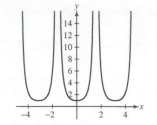

51. $\dfrac{dR}{d\theta} = (v_0^2/9.8)(-\sin^2 \theta + \cos^2 \theta)$; if $\theta = 7\pi/24$, increasing the angle will *decrease* the range.

53.

$$f'(x) = \lim_{h \to 0} \frac{\cos(x + h) - \cos x}{h} = \lim_{h \to 0} \frac{\cos x \cos h - \sin x \sin h - \cos x}{h}$$

$$= \lim_{h \to 0} \left((-\sin x) \frac{\sin h}{h} + (\cos x) \frac{\cos h - 1}{h} \right)$$

$$= (-\sin x) \cdot 1 + (\cos x) \cdot 0 = -\sin x$$

Section 3.7 Preliminary Questions

1. (a) The outer function is $\sqrt{x}$, and the inner function is $4x + 9x^2$.

(b) The outer function is $\tan x$, and the inner function is $x^2 + 1$.

(c) The outer function is x^5, and the inner function is $\sec x$.

(d) The outer function is x^4, and the inner function is $1 + e^x$.

2. The function $\frac{x}{x+1}$ can be differentiated using the Quotient Rule, and the functions $\sqrt{x} \cdot \sec x$ and xe^x can be differentiated using the Product Rule. The functions $\tan(7x^2 + 2)$, $\sqrt{x \cos x}$ and $e^{\sin x}$ require the Chain Rule.

3. (b)

4. We do not have enough information to compute $F'(4)$. We are missing the value of $f'(1)$.

Section 3.7 Exercises

1.

$f(g(x))$	$f'(u)$	$f'(g(x))$	$g'(x)$	$(f \circ g)'$
$(x^4 + 1)^{3/2}$	$\frac{3}{2}u^{1/2}$	$\frac{3}{2}(x^4 + 1)^{1/2}$	$4x^3$	$6x^3(x^4 + 1)^{1/2}$

3.

$f(g(x))$	$f'(u)$	$f'(g(x))$	$g'(x)$	$(f \circ g)'$
$\tan(x^4)$	$\sec^2 u$	$\sec^2(x^4)$	$4x^3$	$4x^3 \sec^2(x^4)$

5. $4(x + \sin x)^3 (1 + \cos x)$

7. (a) $2x \sin(9 - x^2)$ **(b)** $\dfrac{\sin(x^{-1})}{x^2}$ **(c)** $-\sec^2 x \sin(\tan x)$

9. 12 **11.** $12x^3(x^4 + 5)^2$ **13.** $\dfrac{7}{2\sqrt{7x - 3}}$

15. $-2(x^2 + 9x)^{-3}(2x + 9)$ **17.** $-4 \cos^3 \theta \sin \theta$

19. $9(2 \cos \theta + 5 \sin \theta)^8 (5 \cos \theta - 2 \sin \theta)$

21. e^{x-12} **23.** $2 \cos(2x + 1)$ **25.** $e^{x + x^{-1}}\left(1 - x^{-2}\right)$

27.

$$\frac{d}{dx} f(g(x)) = -\sin(x^2 + 1)(2x) = -2x \sin(x^2 + 1)$$

$$\frac{d}{dx} g(f(x)) = -2 \sin x \cos x$$

29. $2x \cos\left(x^2\right)$ **31.** $\dfrac{t}{\sqrt{t^2 + 9}}$

33. $\dfrac{2}{3}\left(x^4 - x^3 - 1\right)^{-1/3}\left(4x^3 - 3x^2\right)$

35. $\dfrac{8(1+x)^3}{(1-x)^5}$ **37.** $-\dfrac{\sec(1/x)\tan(1/x)}{x^2}$

39. $(1-\sin\theta)\sec^2(\theta+\cos\theta)$ **41.** $-18te^{2-9t^2}$

43. $(2x+4)\sec^2(x^2+4x)$ **45.** $\cos(1-3x)+3x\sin(1-3x)$

47. $2(4t+9)^{-1/2}$ **49.** $4(\sin x-3x^2)(x^3+\cos x)^{-5}$

51. $\dfrac{\cos 2x}{\sqrt{2\sin 2x}}$ **53.** $\dfrac{x\cos(x^2)-3\sin 6x}{\sqrt{\cos 6x+\sin(x^2)}}$

55. $3(\tan^2 x\sec^2 x+x^2\sec^2(x^3))$ **57.** $\dfrac{-1}{\sqrt{z+1}\,(z-1)^{3/2}}$

59. $\dfrac{\sin(-1)-\sin(1+x)}{(1+\cos x)^2}$ **61.** $-35x^4\cot^6\left(x^5\right)\csc^2\left(x^5\right)$

63. $-180x^3\cot^4\left(x^4+1\right)\csc^2\left(x^4+1\right)\left(1+\cot^5\left(x^4+1\right)\right)^8$

65. $24(2e^{3x}+3e^{-2x})^3(e^{3x}-e^{-2x})$

67. $4(x+1)(x^2+2x+3)e^{(x^2+2x+3)^2}$

69. $\dfrac{1}{8\sqrt{x}\sqrt{1+\sqrt{x}}\sqrt{1+\sqrt{1+\sqrt{x}}}}$

71. $-\dfrac{k}{3}(kx+b)^{-4/3}$ **73.** $2\cos\left(x^2\right)-4x^2\sin\left(x^2\right)$

75. $-336(9-x)^5$ **77.** $\left.\dfrac{dv}{dP}\right|_{P=1.5}=\dfrac{290\sqrt{3}}{3}\dfrac{\text{m}}{\text{s}\cdot\text{atm}}$

79. (a) When $r=3$, $\dfrac{dV}{dt}=1.6\pi(3)^2\approx 45.24$ cm²/s.

(b) When $t=3$, we have $r=1.2$. Hence, $\dfrac{dV}{dt}=1.6\pi(1.2)^2\approx 7.24$ cm/s.

81. $W'(10)\approx 0.3566$ kg/year   **83. (a)** $\dfrac{\pi}{360}$ **(b)** $1+\dfrac{\pi}{90}$

85. $5\sqrt{3}$ **87.** 12   **89.** $\dfrac{1}{16}$ **91.** $\left.\dfrac{dP}{dt}\right|_{t=3}=-0.727\dfrac{\text{dollars}}{\text{year}}$

93. $\dfrac{dP}{dh}=-4.03366\times 10^{-16}(288.14-0.000649\,h)^{4.256}$; for each additional meter of altitude, $\Delta P\approx -1.15\times 10^{-2}$ Pa

95. 0.0973 kelvins/year   **97.** $f''(g(x))(g'(x))^2+f'(g(x))g''(x)$

99. Let $u=h(x)$, $v=g(u)$, and $w=f(v)$. Then

$$\dfrac{dw}{dx}=\dfrac{df}{dv}\dfrac{dv}{dx}=\dfrac{df}{dv}\dfrac{dv}{du}\dfrac{du}{dx}=f'(g(h(x))g'(h(x))h'(x)$$

103. For $n=1$, we find

$$\dfrac{d}{dx}\sin x=\cos x=\sin\left(x+\dfrac{\pi}{2}\right)$$

as required. Now, suppose that for some positive integer k,

$$\dfrac{d^k}{dx^k}\sin x=\sin\left(x+\dfrac{k\pi}{2}\right)$$

Then

$$\dfrac{d^{k+1}}{dx^{k+1}}\sin x=\dfrac{d}{dx}\sin\left(x+\dfrac{k\pi}{2}\right)$$
$$=\cos\left(x+\dfrac{k\pi}{2}\right)=\sin\left(x+\dfrac{(k+1)\pi}{2}\right)$$

Section 3.8 Preliminary Questions

1. The Chain Rule

2. (a) This is correct.   **(b)** This is correct.

(c) This is incorrect. Because the differentiation is with respect to the variable x, the Chain Rule is needed to obtain

$$\dfrac{d}{dx}\sin(y^2)=2y\cos(y^2)\dfrac{dy}{dx}$$

3. There are two mistakes in Jason's answer. First, Jason should have applied the product rule to the second term to obtain

$$\dfrac{d}{dx}(2xy)=2x\dfrac{dy}{dx}+2y$$

Second, he should have applied the general power rule to the third term to obtain

$$\dfrac{d}{dx}y^3=3y^2\dfrac{dy}{dx}$$

4. (b)   **5.** $g(x)=\tan^{-1}x$

6. The derivatives of $\sin^{-1}x$ and $\cos^{-1}x$ are negatives of each other.

Section 3.8 Exercises

1. $(2,1)$, $\dfrac{dy}{dx}=-\dfrac{2}{3}$ **3.** $\dfrac{d}{dx}\left(x^2y^3\right)=3x^2y^2y'+2xy^3$

5. $\dfrac{d}{dx}\left(\left(x^2+y^2\right)^{3/2}\right)=3\left(x+yy'\right)\sqrt{x^2+y^2}$

7. $\dfrac{d}{dx}\dfrac{y}{y+1}=\dfrac{y'}{(y+1)^2}$ **9.** $y'=-\dfrac{2x}{9y^2}$

11. $y'=\dfrac{2xy+6x^2y-1}{1-x^2-2x^3}$ **13.** $R'=-\dfrac{3R}{5x}$

15. $y'=\dfrac{y(y^2-x^2)}{x(y^2-x^2-2xy^2)}$ **17.** $y'=\dfrac{9}{4}x^{1/2}y^{5/3}$

19. $y'=\dfrac{(2x+1)y^2}{y^2-1}$ **21.** $y'=\dfrac{1-\cos(x+y)}{\cos(x+y)+\sin y}$

23. $y'=\dfrac{e^y-2y}{2x+3y^2-xe^y}$ **25.** $y'=\dfrac{xy-y}{xy+x}$ **27.** $5/4$

29. $\dfrac{1}{4\sqrt{15}}$ **31.** $\dfrac{7}{\sqrt{1-49x^2}}$ **33.** $\dfrac{-2x}{\sqrt{1-x^4}}$

35. $\tan^{-1}x+\dfrac{x}{x^2+1}$ **37.** $\dfrac{e^x}{\sqrt{1-e^{2x}}}$ **39.** $\dfrac{1-t}{\sqrt{1-t^2}}$

41. $\dfrac{3(\tan^{-1}x)^2}{x^2+1}$ **43.** 0

45. $\cos y=x$, so implicit differentiation gives $-\sin(y)y'=1$, or $\dfrac{dy}{dx}=-\csc y=\dfrac{-1}{\sqrt{1-x^2}}$.

47. Since $1+\tan^2 y=\sec^2 y$, and $\sec y=x$, $\Rightarrow\tan y=\pm\sqrt{x^2-1}$. If $x\geq 1$, $\text{arcsec}\,x=y\in\left[0,\frac{\pi}{2}\right)\Rightarrow\tan y>0$. If $x\leq -1$, $\text{arcsec}\,x=y\in\left(-\frac{\pi}{2},0\right]\Rightarrow\tan y=-\sqrt{x^2-1}<0$.

49. Multiplying both sides of $x+yx^{-1}=1$ by x gives $x^2+y=x$, so the two define the same curve except when $x=0$. Since $y=x-x^2$, differentiating the first form gives $y'=\dfrac{y}{x}-x=\dfrac{x-x^2}{x}-x=1-2x$.

51. $\dfrac{1}{4}$ **53.** $y=-\dfrac{1}{2}x+2$ **55.** $y=-2x+2$

57. $y=-\dfrac{12}{5}x+\dfrac{32}{5}$ **59.** $y=\dfrac{4}{3}x+\dfrac{4}{3}$

61. The tangent is horizontal at the points $(-1,\sqrt{3})$ and $(-1,-\sqrt{3})$.

63. The tangent line is horizontal at

$$\left(\dfrac{2\sqrt{78}}{13},-\dfrac{4\sqrt{78}}{13}\right)\quad\text{and}\quad\left(-\dfrac{2\sqrt{78}}{13},\dfrac{4\sqrt{78}}{13}\right)$$

65. The slopes corresponding to $x=0$ are $\dfrac{-1\pm\sqrt[4]{2}}{4\sqrt[4]{8}}$. At $(1,1)$, the tangent line has equation $x-5y=-4$.

67. $(2^{1/3},2^{2/3})$ **69.** $x=\dfrac{1}{2},1\pm\sqrt{2}$

71.   • At $(1,2)$, $y'=\dfrac{1}{3}$

- At $(1, -2)$, $y' = -\dfrac{1}{3}$

- At $(1, \frac{1}{2})$, $y' = \dfrac{11}{12}$

- At $(1, -\frac{1}{2})$, $y' = -\dfrac{11}{12}$

73. There are vertical tangent lines at six points $(\pm 1, 0)$ and $\left(\pm \frac{\sqrt{3}}{2}, \pm \frac{\sqrt{2}}{2} \right)$.

75. $\dfrac{dx}{dy} = \dfrac{2y}{3x^2 - 4}$; it follows that $\dfrac{dx}{dy} = 0$ when $y = 0$, so the tangent line to this curve is vertical at the points where the curve intersects the x-axis.

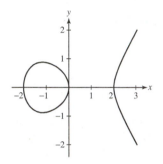

77. $y'' = (y^2 - 2xyy')/y^4 = (y^2 - 2xyx/y^2)/y^4 = (y^3 - 2x^2)/y^5$

79. (a) $y' = -y^2/(2xy + 1)$; $y'|(1, 1) = -1/3$
(b) $y'' = (2y^3 - 2xy^2y' - 2yy')/(2xy + 1)^2$; $y''|(1, 1) = 10/27$

81. The product rule gives us $\dfrac{dx}{dt} y + x \dfrac{dy}{dt} = 0$, or $\dfrac{dy}{dt} = -\dfrac{y}{x} \dfrac{dx}{dt}$

83. (a) $\dfrac{dy}{dt} = \dfrac{x^2}{y^2} \dfrac{dx}{dt}$ **(b)** $\dfrac{dy}{dt} = -\dfrac{x+y}{2y^3 + x} \dfrac{dx}{dt}$

85. The derivatives of these two curves are $y' = x/y$ and $y' = -y/x$, respectively. So at any point (x, y) satisfying both equations, the slopes of one tangent line will be the negative reciprocal of the other slope. That is, the tangent lines will be perpendicular.

87. • Upper branch:

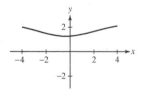

• Lower part of lower left curve:

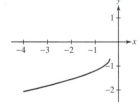

• Upper part of lower left curve:

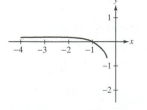

• Upper part of lower right curve:

• Lower part of lower right curve:

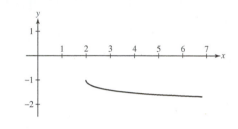

Section 3.9 Preliminary Questions

1. $\ln 4$ **2.** $\dfrac{1}{10}$ **3.** e^2 **4.** e^3

5. $y^{(100)} = \cosh x$ and $y^{(101)} = \sinh x$

Section 3.9 Exercises

1. $\dfrac{d}{dx} x \ln x = \ln x + 1$ **3.** $\dfrac{d}{dx} 2^{x^3} = 2^{x^3} * 3x^2 * \ln 2$

5. $\dfrac{d}{dx} \ln(9x^2 - 8) = \dfrac{18x}{9x^2 - 8}$ **7.** $\dfrac{d}{dx} (\ln x)^2 = \dfrac{2}{x} \ln x$

9. $\dfrac{d}{dx} e^{(\ln x)^2} = \dfrac{2}{x} \ln x * e^{(\ln x)^2}$ **11.** $\dfrac{d}{dx} \ln(\ln x) = \dfrac{1}{x \ln x}$

13. $\dfrac{d}{dx} (\ln(\ln x))^3 = \dfrac{3(\ln(\ln x))^2}{x \ln x}$

15. $\dfrac{d}{dx} \ln((x + 1)(2x + 9)) = \dfrac{4x + 11}{(x + 1)(2x + 9)}$

17. $\dfrac{d}{dx} 11^x = \ln 11 \cdot 11^x$

19. $\dfrac{d}{dx} \dfrac{2^x - 3^{-x}}{x} = \dfrac{x(2^x \ln 2 + 3^{-x} \ln 3) - (2^x - 3^{-x})}{x^2}$

21. $f'(x) = \dfrac{1}{x} \cdot \dfrac{1}{\ln 2}$ **23.** $\dfrac{d}{dt} \log_3(\sin t) = \dfrac{\cot t}{\ln 3}$

25. $y = 36 \ln 6(x - 2) + 36$ **27.** $y = 3^{20} \ln 3(t - 2) + 3^{18}$

29. $y = 5^{-1}$ **31.** $y = -1(t - 1) + \ln 4$

33. $y = \dfrac{12}{25 \ln 5} (z - 3) + 2$ **35.** $y = \dfrac{8}{\ln 2} \left(w - \dfrac{1}{8} \right) - 3$

37. $y' = 2x + 14$ **39.** $y' = 3x^2 - 12x - 79$

41. $y' = \dfrac{x(x^2 + 1)}{\sqrt{x + 1}} \left(\dfrac{1}{x} + \dfrac{2x}{x^2 + 1} - \dfrac{1}{2(x + 1)} \right)$

43.

$$y' = \dfrac{1}{2} \sqrt{\dfrac{x(x + 2)}{(2x + 1)(3x + 2)}} \cdot \left(\dfrac{1}{x} + \dfrac{1}{x + 2} - \dfrac{2}{2x + 1} - \dfrac{3}{3x + 2} \right)$$

45. $\dfrac{d}{dx} x^{3x} = x^{3x}(3 + 3 \ln x)$ **47.** $\dfrac{d}{dx} x^{e^x} = x^{e^x} \left(\dfrac{e^x}{x} + e^x \ln x \right)$

49. $y' = x^{\cos x} ((\cos x)/x - \sin x \ln x)$

51. $\dfrac{d}{dx} \sinh(9x) = 9 \cosh(9x)$

53. $\dfrac{d}{dt} \cosh^2(9 - 3t) = -6 \cosh(9 - 3t) \sinh(9 - 3t)$

55. $\dfrac{d}{dx}\sqrt{\cosh x + 1} = \dfrac{1}{2}(\cosh x + 1)^{-1/2}\sinh x$

57. $\dfrac{dy}{dt} = -\dfrac{\operatorname{csch} t\,(\operatorname{csch} t + 2\operatorname{sech} t)}{(1 + \tanh t)^2}$

59. $\dfrac{d}{dx}\sinh(\ln x) = \dfrac{\cosh(\ln x)}{x}$

61. $\dfrac{d}{dx}\tanh(e^x) = e^x\operatorname{sech}^2(e^x)$

63. $\dfrac{d}{dx}\operatorname{sech}(\sqrt{x}) = -\dfrac{1}{2}x^{-1/2}\operatorname{sech}\sqrt{x}\tanh\sqrt{x}$

65. $\dfrac{d}{dx}\operatorname{sech} x\coth x = -\operatorname{csch} x\coth x$

67. $\dfrac{d}{dx}\cosh^{-1}(3x) = \dfrac{3}{\sqrt{9x^2 - 1}}$

69. $\dfrac{d}{dx}(\sinh^{-1}(x^2))^3 = 3(\sinh^{-1}(x^2))^2\dfrac{2x}{\sqrt{x^4 + 1}}$

71. $\dfrac{d}{dx}e^{\cosh^{-1}x} = e^{\cosh^{-1}x}\left(\dfrac{1}{\sqrt{x^2 - 1}}\right)$

73. $\dfrac{d}{dt}\tanh^{-1}(\ln t) = \dfrac{1}{t(1 - (\ln t)^2)}$

75. $\dfrac{d}{dx}\coth x = \dfrac{d}{dx}\dfrac{\cosh x}{\sinh x} = \dfrac{\sinh^2 x - \cosh^2 x}{\sinh^2 x} = \dfrac{-1}{\sinh^2 x} = -\operatorname{csch}^2 x$

79. 1.22 cents per year

83. (a) $\dfrac{dP}{dT} = -\dfrac{1}{T\ln 10}$ (b) $\Delta P \approx -0.054$

85. $\dfrac{d}{dx}\log_b x = \dfrac{d}{dx}\dfrac{\ln x}{\ln b} = \dfrac{1}{(\ln b)x}$

Section 3.10 Preliminary Questions

1. Let s and V denote the length of the side and the corresponding volume of a cube, respectively. Determine $\dfrac{dV}{dt}$ if $\dfrac{ds}{dt} = 0.5$ cm/s.

2. $\dfrac{dV}{dt} = 4\pi r^2\dfrac{dr}{dt}$ **3.** Determine $\dfrac{dh}{dt}$ if $\dfrac{dV}{dt} = 2$ cm³/min

4. Determine $\dfrac{dV}{dt}$ if $\dfrac{dh}{dt} = 1$ cm/min

Section 3.10 Exercises

1. 0.039 ft/min

3. (a) $100\pi \approx 314.16$ m²/min   (b) $24\pi \approx 75.40$ m²/min

5. 27000π cm³/min   **7.** 9600π cm²/min

9. -0.632 m/s   **11.** $x = 4.737$ m; $\dfrac{dx}{dt} \approx 0.405$ m/s

13. $\dfrac{9}{8\pi} \approx 0.36$ m/min   **15.** $\dfrac{1000\pi}{3} \approx 1047.20$ cm³/s

17. 0.675 m/s

19. (a) 594.6 km/hr   (b) 0 km/h

21. 1.22 km/min   **23.** $\dfrac{1200}{241} \approx 4.98$ rad/h

25. (a) $\dfrac{100\sqrt{13}}{13} \approx 27.735$ km/h   (b) 112.962 km/h

27. $\sqrt{16.2} \approx 4.025$ m   **29.** $\dfrac{5}{3}$ m/s   **31.** -1.92 kPa/min

33. $-\dfrac{1}{8}$ rad/s

35. (b): when $x = 1$, $L'(t) = 0$; when $x = 2$, $L'(t) = \dfrac{16}{3}$

37. $-4\sqrt{5} \approx -8.94$ ft/s   **39.** -0.79 m/min

41. Let the equation $y = f(x)$ describe the shape of the roller coaster track. Taking $\dfrac{d}{dt}$ of both sides of this equation yields $\dfrac{dy}{dt} = f'(x)\dfrac{dx}{dt}$.

43. (a) The distance formula gives

$$L = \sqrt{(x - r\cos\theta)^2 + (-r\sin\theta)^2}$$

Thus,

$$L^2 = (x - r\cos\theta)^2 + r^2\sin^2\theta$$

(b) From (a), we have

$$0 = 2(x - r\cos\theta)\left(\dfrac{dx}{dt} + r\sin\theta\dfrac{d\theta}{dt}\right) + 2r^2\sin\theta\cos\theta\dfrac{d\theta}{dt}$$

(c) $-80\pi \approx -251.33$ cm/min

45. (c): $\dfrac{3\sqrt{5}}{250} \approx 0.027$ m/min

Chapter 3 Review

1. 3; the slope of the secant line through the points $(2, 7)$ and $(0, 1)$ on the graph of $f(x)$

3. $\dfrac{8}{3}$; the value of the difference quotient should be larger than the value of the derivative

5. $f'(1) = 1$; $y = x - 1$ **7.** $f'(4) = -\dfrac{1}{16}$; $y = -\dfrac{1}{16}x + \dfrac{1}{2}$

9. $-2x$ **11.** $\dfrac{1}{(2 - x)^2}$ **13.** $f'(1)$, where $f(x) = \sqrt{x}$

15. $f'(\pi)$, where $f(t) = \sin t\cos t$ **17.** $f(4) = -2$; $f'(4) = 3$

19. (C) is the graph of $f'(x)$.

21.

23. (a) 8.05 cm/year   (b) Larger over the first half

(c) $h'(3) \approx 7.8$ cm/year; $h'(8) \approx 6.0$ cm/year

25. $A'(t)$ measures the rate of change in automobile production in the United States; $A'(1971) \approx 0.25$ million automobiles/year; $A'(1974)$ would be negative.

27. (b)   **29.** $15x^4 - 14x$ **31.** $-7.3t^{-8.3}$ **33.** $\dfrac{1 - 2x - x^2}{(x^2 + 1)^2}$

35. $6(4x^3 - 9)(x^4 - 9x)^5$ **37.** $27x(2 + 9x^2)^{1/2}$

39. $\dfrac{2 - z}{2(1 - z)^{3/2}}$ **41.** $2x - \dfrac{3}{2}x^{-5/2}$

43.

$$\dfrac{1}{2}\left(x + \sqrt{x + \sqrt{x}}\right)^{-1/2}\left(1 + \dfrac{1}{2}\left(x + \sqrt{x}\right)^{-1/2}\left(1 + \dfrac{1}{2}x^{-1/2}\right)\right)$$

45. $-3t^{-4}\sec^2(t^{-3})$ **47.** $-6\sin^2 x\cos^2 x + 2\cos^4 x$

49. $\dfrac{1 + \sec t - t\sec t\tan t}{(1 + \sec t)^2}$ **51.** $\dfrac{8\csc^2\theta}{(1 + \cot\theta)^2}$

53. $-\dfrac{\sec^2(\sqrt{1 + \csc\theta})\csc\theta\cot\theta}{2(\sqrt{1 + \csc\theta})}$ **55.** $-36e^{-4x}$

57. $(4 - 2t)e^{4t - t^2}$ **59.** $\dfrac{8x}{4x^2 + 1}$ **61.** $\dfrac{2\ln s}{s}$

63. $\cot\theta$ **65.** $\sec(z + \ln z)\tan(z + \ln z)\left(1 + \dfrac{1}{z}\right)$

67. $-2(\ln 7)(7^{-2x})$ **69.** $\dfrac{1}{1 + (\ln x)^2}\cdot\dfrac{1}{x}$

71. $-\dfrac{1}{|x|\sqrt{x^2 - 1}\csc^{-1}x}$ **73.** $\dfrac{2\ln s}{s}s^{\ln s}$

75. $2(\sin^2 t)^t(t \cot t + \ln \sin t)$ **77.** $2t \cosh(t^2)$ **79.** $\dfrac{e^x}{1 - e^{2x}}$

81. $\alpha = 0$ and $\alpha > 1$

83. Let $f(x) = xe^{-x}$. Then $f'(x) = e^{-x}(1 - x)$. On $[1, \infty)$, $f'(x) < 0$, so $f(x)$ is decreasing and therefore one-to-one. The domain of $g(x)$ is $(0, e^{-1}]$, and the range is $[1, \infty)$. $g'(2e^{-2}) = -e^2$.

85. -27 **87.** $-\dfrac{57}{16}$ **89.** -18 **91.** $(-1, -1)$ and $(3, 7)$

93. $a = \dfrac{1}{6}$ **95.** $72x - 10$ **97.** $-(2x + 3)^{-3/2}$

99. $8x^2 \sec^2(x^2) \tan(x^2) + 2 \sec^2(x^2)$ **101.** $\dfrac{dy}{dx} = \dfrac{x^2}{y^2}$

103. $\dfrac{dy}{dx} = \dfrac{y^2 + 4x}{1 - 2xy}$ **105.** $\dfrac{dy}{dx} = \dfrac{\cos(x + y)}{1 - \cos(x + y)}$

107. For the plot on the left, the red, green, and blue curves, respectively, are the graphs of f, f', and f''. For the plot on the right, the green, red, and blue curves, respectively, are the graphs of f, f', and f''.

109. $\dfrac{(x + 1)^3}{(4x - 2)^2} \left(\dfrac{3}{x + 1} - \dfrac{4}{2x - 1} \right)$ **111.** $4e^{(x-1)^2} e^{(x-3)^2}(x - 2)$

113. $\dfrac{e^{3x}(x - 2)^2}{(x + 1)^2} \left(3 + \dfrac{2}{x - 2} - \dfrac{2}{x + 1} \right)$

115. $\dfrac{dR}{dp} = p\dfrac{dq}{dp} + q = q\dfrac{p}{q}\dfrac{dq}{dp} + q = q(E + 1)$

117. $\dfrac{dh}{dt} = \dfrac{1}{15}$ m/min   **119.** $\dfrac{-11\pi}{9\sqrt{89}} \approx -0.407$ cm/min

121. $\dfrac{640}{(336)^2} \approx 0.00567$ cm/s   **123.** 0.284 m/s

Chapter 4

Section 4.1 Preliminary Questions

1. True   **2.** $g(1.2) - g(1) \approx 0.8$ **3.** $f(2.1) \approx 1.3$

4. The Linear Approximation tells us that up to a small error, the change in output Δf is directly proportional to the change in input Δx when Δx is small.

Section 4.1 Exercises

1. $\Delta f \approx 0.12$ **3.** $\Delta f \approx -0.00222$ **5.** $\Delta f \approx 0.003333$

7. $\Delta f \approx 0.0074074$

9. $\Delta f \approx 0.05$; error is 0.000610; percentage error is 1.24%.

11. $\Delta f \approx -0.03$; error is 0.0054717; percentage error is 22.31%.

13. $\Delta y \approx -0.007$ **15.** $\Delta y \approx -0.026667$

17. $\Delta f \approx 0.1$; error is 0.00098.

19. $\Delta f \approx -0.0005$; error is 3.71902×10^{-6}.

21. $\Delta f \approx 0.083333$; error is 3.25×10^{-3}.

23. $\Delta f \approx -0.1$; error is 4.84×10^{-3}.   **25.** $f(4.03) \approx 2.01$

27. $\sqrt{2.1} - \sqrt{2}$ is larger than $\sqrt{9.1} - \sqrt{9}$.

29. $R(9) = 25110$ euros; if p is raised by 0.5 euros, then $\Delta R \approx 585$ euros; on the other hand, if p is lowered by 0.5 euros, then $\Delta R \approx -585$ euros.

31. $\Delta L \approx -0.00171$ cm

33. (a) $\Delta P \approx -0.434906$ kilopascals (kPa)

(b) The actual change in pressure is -0.418274 kPa; the percentage error is 3.98%.

35. (a) $\Delta W \approx W'(R)\Delta x = -\dfrac{2wR^2}{R^3}h = -\dfrac{2wh}{R} \approx -0.0005wh$

(b) $\Delta W \approx -0.7$ pounds (lb)

37. (a) $\Delta h \approx 0.71$ cm   **(b)** $\Delta h \approx 1.02$ cm

(c) There is a bigger effect at higher velocities.

39. (a) If $\theta = 34°$ (i.e., $t = \frac{17}{90}\pi$), then

$$\Delta s \approx s'(t)\Delta t = \frac{625}{16} \cos\left(\frac{17}{45}\pi\right) \Delta t$$

$$= \frac{625}{16} \cos\left(\frac{17}{45}\pi\right) \Delta\theta \cdot \frac{\pi}{180} \approx 0.255\Delta\theta$$

(b) If $\Delta\theta = 2°$, this gives $\Delta s \approx 0.51$ ft, in which case the shot would not have been successful, having been off half a foot.

41. $\Delta V \approx 4\pi(25)^2(0.5) \approx 3927$ cm^3; $\Delta S \approx 8\pi(25)(0.5) \approx 314.2$ cm^2

43. $P = 6$ atm; $\Delta P \approx \pm 0.45$ atm

45. $L(x) = 4x - 3$ **47.** $L(\theta) = \theta - \dfrac{\pi}{4} + \dfrac{1}{2}$

49. $L(x) = -\dfrac{1}{2}x + 1$ **51.** $L(x) = 1$ **53.** $L(x) = \dfrac{1}{2}e(x + 1)$

55. $f(2) = 8$

57. $\sqrt{16.2} \approx L(16.2) = 4.025$. Graphs of f and L are shown below. Because the graph of L lies above the graph of f, we expect that the estimate from the Linear Approximation is too large.

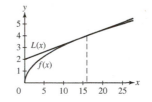

59. $\dfrac{1}{\sqrt{17}} \approx L(17) = 0.24219$; the percentage error is 0.14%.

61. $\dfrac{1}{(10.03)^2} \approx L(10.03) = 0.00994$; the percentage error is 0.0027%.

63. $(64.1)^{1/3} \approx L(64.1) = 4.002083$; the percentage error is 0.000019%.

65. $\cos^{-1}(0.52) \approx L(0.02) = 1.024104$; the percentage error is 0.015%.

67. $e^{-0.012} \approx L(-0.012) = 0.988$; the percentage error is 0.0073%.

69. Let $f(x) = \sqrt{x}$. Then $f(9) = 3$, $f'(x) = \frac{1}{2}x^{-1/2}$ and $f'(9) = \frac{1}{6}$. Therefore, by the Linear Approximation,

$$f(9 + h) - f(9) = \sqrt{9 + h} - 3 \approx \frac{1}{6}h$$

Moreover, $f''(x) = -\frac{1}{4}x^{-3/2}$, so $|f''(x)| = \frac{1}{4}x^{-3/2}$. Because this is a decreasing function, it follows that for $x \geq 9$,

$$K = \max|f''(x)| \leq |f''(9)| = \frac{1}{108} < 0.01$$

From the following table, we see that for $h = 10^{-n}$, $1 \leq n \leq 4$, $E \leq \frac{1}{2}Kh^2$.

h	$E = \left|\sqrt{9 + h} - 3 - \frac{1}{6}h\right|$	$\frac{1}{2}Kh^2$
10^{-1}	4.604×10^{-5}	5.00×10^{-5}
10^{-2}	4.627×10^{-7}	5.00×10^{-7}
10^{-3}	4.629×10^{-9}	5.00×10^{-9}
10^{-4}	4.627×10^{-11}	5.00×10^{-11}

71. $\dfrac{dy}{dx}\bigg|_{(2,1)} = -\dfrac{1}{3}$; $y \approx L(2.1) = 0.967$

73. $L(x) = -\dfrac{14}{25}x + \dfrac{36}{25}$; $y \approx L(-1.1) = 2.056$

75. Let $f(x) = x^2$. Then

$$\Delta f = f(5+h) - f(5) = (5+h)^2 - 5^2 = h^2 + 10h$$

and

$$E = |\Delta f - f'(5)h| = |h^2 + 10h - 10h| = h^2 = \frac{1}{2}(2)h^2 = \frac{1}{2}Kh^2$$

Section 4.2 Preliminary Questions

1. A critical point is a value of the independent variable x in the domain of a function f at which either $f'(x) = 0$ or $f'(x)$ does not exist.

2. (b)   **3.** (b)

4. Fermat's Theorem claims: If $f(c)$ is a local extreme value, then either $f'(c) = 0$ or $f'(c)$ does not exist.

Section 4.2 Exercises

1. (a) 3   (b) 6   (c) Local maximum of 5 at $x = 5$

(d) Answers may vary. One example is the interval $[4, 8]$. Another is $[2, 6]$.

(e) Answers may vary. One example is $[0, 2]$.

3. $x = 1$ **5.** $x = -3$ and $x = 6$ **7.** $x = 2$ **9.** $x = \pm 1$

11. $t = 3$ and $t = -1$ **13.** $x = -\frac{1}{2}$

15. $\theta = \dfrac{n\pi}{2}$ **17.** $x = \dfrac{1}{e}$ **19.** $x = \pm\dfrac{\sqrt{3}}{2}, \pm 1$

21. (a) $c = 2$ and $f(c) = -3$ (b) $f(0) = f(4) = 1$

(c) Maximum value: 1; minimum value: -3.

(d) Maximum value: 1; minimum value: -2.

23. $x = \dfrac{\pi}{4}$; Maximum value: $\sqrt{2}$; minimum value: 1

25. Maximum value: 1

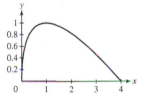

27. Critical point: $x \approx 0.652185$; maximum value: approximately 0.561096

29. Minimum: $f(-1) = 3$, maximum: $f(2) = 21$

31. Minimum: $f(0) = 0$, maximum: $f(3) = 9$

33. Minimum: $f(4) = -24$, maximum: $f(6) = 8$

35. Minimum: $f(1) = 5$, maximum: $f(2) = 28$

37. Minimum: $f(2) = -128$, maximum: $f(-2) = 128$

39. Minimum: $f(6) = 18.5$, maximum: $f(5) = 26$

41. Minimum: $f(1) = -1$, maximum: $f(0) = f(3) = 0$

43. Minimum: $f(0) = 2\sqrt{6} \approx 4.9$, maximum: $f(2) = 4\sqrt{2} \approx 5.66$

45. Minimum: $f\left(\dfrac{\sqrt{3}}{2}\right) \approx -0.589980$, maximum:

$f(4) \approx 0.472136$

47. Minimum: $f(0) = f\left(\dfrac{\pi}{2}\right) = 0$, maximum: $f\left(\dfrac{\pi}{4}\right) = \dfrac{1}{2}$

49. Minimum: $f(0) = -1$, maximum:

$f\left(\dfrac{\pi}{4}\right) = \sqrt{2}\left(\dfrac{\pi}{4} - 1\right) \approx -0.303493$

51. Minimum: $g\left(\dfrac{\pi}{3}\right) = \dfrac{\pi}{3} - \sqrt{3} \approx -0.685$, maximum:

$g\left(\dfrac{5}{3}\pi\right) = \dfrac{5}{3}\pi + \sqrt{3} \approx 6.968$

53. Minimum: $f\left(\dfrac{\pi}{4}\right) = 1 - \dfrac{\pi}{2} \approx -0.570796$, maximum: $f(0) = 0$

55. Minimum: $f(1) = 0$, maximum is $f(e) = e^{-1} \approx 0.367879$

57. Minimum: $f(1) = 3e - e^2 \approx 0.765789$, maximum:

$f\left(\ln\left(\dfrac{3}{2}\right)\right) = \dfrac{9}{4}$

59. (d) $\dfrac{\pi}{6}, \dfrac{\pi}{2}, \dfrac{5\pi}{6}, \dfrac{7\pi}{6}, \dfrac{3\pi}{2}$, and $\dfrac{11\pi}{6}$; the maximum value is

$f(\frac{\pi}{6}) = f(\frac{7\pi}{6}) = \dfrac{3\sqrt{3}}{2}$ and the minimum value is

$f(\frac{5\pi}{6}) = f(\frac{11\pi}{6}) = -\dfrac{3\sqrt{3}}{2}$.

(e) We can see that there are six flat points on the graph between 0 and 2π, as predicted. There are four local extrema, and two points at $(\frac{\pi}{2}, 0)$ and $(\frac{3\pi}{2}, 0)$ where the graph has neither a local maximum nor a local minimum.

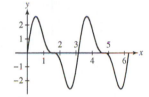

61. Critical point: $x = 2$; minimum value: $f(2) = 0$, maximum: $f(0) = f(4) = 2$

63. Critical point: $x = 2$; minimum value: $f(2) = 0$, maximum: $f(4) = 20$

65. $c = 1$ **67.** $c = \dfrac{15}{4}$

69. $f(0) < 0$ and $f(2) > 0$, so there is at least one root by the Intermediate Value Theorem; there cannot be another root because $f'(x) \geq 4$ for all x.

71. There cannot be a root $c > 0$ because $f'(x) > 4$ for all $x > 0$.

75. $b \approx 2.86$

77. (a) $F = \dfrac{1}{2}\left(1 - \dfrac{v_2^2}{v_1^2}\right)\left(1 + \dfrac{v_2}{v_1}\right)$

(b) $F(r)$ achieves its maximum value when $r = 1/3$.

(c) If v_2 were 0, then no air would be passing through the turbine, which is not realistic.

81.   • The maximum value of f on $[0, 1]$ is

$$f\left(\left(\dfrac{a}{b}\right)^{1/(b-a)}\right) = \left(\dfrac{a}{b}\right)^{a/(b-a)} - \left(\dfrac{a}{b}\right)^{b/(b-a)}$$

   • $\dfrac{1}{4}$

83. Critical points: $x = 1$, $x = 4$, and $x = \dfrac{5}{2}$; maximum value: $f(1) = f(4) = \dfrac{5}{4}$, minimum value: $f(-5) = \dfrac{17}{70}$

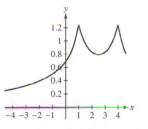

85. (a) There are therefore four points at which the derivative is zero:

$(-1, -\sqrt{2}), (-1, \sqrt{2}), (1, -\sqrt{2}), (1, \sqrt{2})$

There are also critical points where the derivative does not exist:

$$(0, 0), (\pm\sqrt[4]{27}, 0)$$

(b) The curve $27x^2 = (x^2 + y^2)^3$ and its horizontal tangents are plotted below.

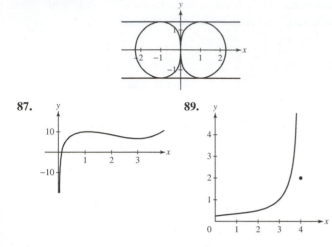

87.

89.

91. If $f(x) = a \sin x + b \cos x$, then $f'(x) = a \cos x - b \sin x$, so $f'(x) = 0$ implies $a \cos x - b \sin x = 0$. This implies $\tan x = \frac{a}{b}$. Then

$$\sin x = \frac{\pm a}{\sqrt{a^2 + b^2}} \quad \text{and} \quad \cos x = \frac{\pm b}{\sqrt{a^2 + b^2}}$$

Therefore,

$$f(x) = a \sin x + b \cos x = a \frac{\pm a}{\sqrt{a^2 + b^2}} + b \frac{\pm b}{\sqrt{a^2 + b^2}}$$

$$= \pm \frac{a^2 + b^2}{\sqrt{a^2 + b^2}} = \pm\sqrt{a^2 + b^2}$$

93. Let $f(x) = x^2 + rx + s$ and suppose that $f(x)$ takes on both positive and negative values. This will guarantee that f has two real roots. By the quadratic formula, the roots of f are

$$x = \frac{-r \pm \sqrt{r^2 - 4s}}{2}$$

Observe that the midpoint between these roots is

$$\frac{1}{2} \left(\frac{-r + \sqrt{r^2 - 4s}}{2} + \frac{-r - \sqrt{r^2 - 4s}}{2} \right) = -\frac{r}{2}$$

Next, $f'(x) = 2x + r = 0$ when $x = -\frac{r}{2}$ and, because the graph of f is an upward-opening parabola, it follows that $f(-\frac{r}{2})$ is a minimum.

95. $b > \frac{1}{4}a^2$

97.
 • Let f be a continuous function with $f(a)$ and $f(b)$ local minima on the interval $[a, b]$. By Theorem 1, $f(x)$ must take on both a minimum and a maximum on $[a, b]$. Since local minima occur at $f(a)$ and $f(b)$, the maximum must occur at some other point in the interval, call it c, where $f(c)$ is a local maximum.
 • The function graphed here is discontinuous at $x = 0$.

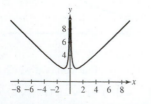

1. $m = 3$ **2. (c)**
3. Yes. The figure below displays a function that takes on only negative values but has a positive derivative.

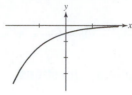

4. (a) $f(c)$ must be a local maximum.   **(b)** No.

1. $c = 4$ **3.** $c = \dfrac{3\pi}{4}$ or $\dfrac{7\pi}{4}$ **5.** $c = \pm\sqrt{7}$

7. $c = -\dfrac{1}{2} \ln \left(\dfrac{1 - e^{-6}}{6} \right)$

9. The slope of the secant line between $x = 0$ and $x = 1$ is

$$\frac{f(1) - f(0)}{1 - 0} = 1$$

Since $f'(x) = 2x$, solving $2c = 1$ gives $c = \dfrac{1}{2}$. A graph of f and the tangent line appears below:

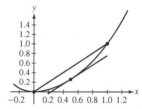

11. The slope of the secant line between $x = 0$ and $x = 1$ is

$$\frac{f(1) - f(0)}{1 - 0} = e - 1$$

Since $f'(x) = e^x$, solving $e^c = e - 1$ gives $c = \ln(e - 1)$. A graph of f and the tangent line is given below:

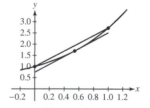

13. The slope of the secant line between $x = 0$ and $x = 1$ is

$$\frac{f(1) - f(0)}{1 - 0} = \frac{2 - 0}{1} = 2$$

It appears that the x-coordinate of the point of tangency is approximately 0.62.

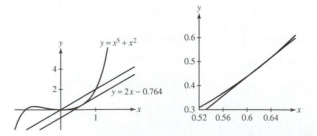

15. The derivative is positive on the intervals $(-\infty, 1) \cup (3, 5)$ and negative on the intervals $(1, 3) \cup (5, 6)$.

17. $f(2)$ is a local maximum; $f(4)$ is a local minimum

19. **21.**

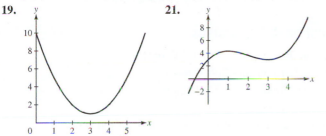

23. Critical point: $c = 3$; since the derivative changes sign from $+$ to $-$, this is a point of local maximum.

25. Critical points: $c = -2$ and $c = 0$. Since $f'(x)$ changes sign from $+$ to $-$ at $c = -2$, this is a point of local maximum. Since $f'(x)$ changes sign from $-$ to $+$ at $c = 0$, this is a point of local minimum.

27. $c = \frac{7}{2}$

x	$\left(-\infty, \frac{7}{2}\right)$	$7/2$	$\left(\frac{7}{2}, \infty\right)$
f'	$+$	0	$-$
f	$\nearrow$	M	$\searrow$

29. $c = 0, 8$

x	$(-\infty, 0)$	0	$(0, 8)$	8	$(8, \infty)$
f'	$+$	0	$-$	0	$+$
f	$\nearrow$	M	$\searrow$	m	$\nearrow$

31. $c = -2, -1, 1$

x	$(-\infty, -2)$	-2	$(-2, -1)$	-1	$(-1, 1)$	1	$(1, \infty)$
f'	$-$	0	$+$	0	$-$	0	$+$
f	$\searrow$	m	$\nearrow$	M	$\searrow$	m	$\nearrow$

33. $c = -2, -1$

x	$(-\infty, -2)$	-2	$(-2, -1)$	-1	$(-1, \infty)$
f'	$+$	0	$-$	0	$+$
f	$\nearrow$	M	$\searrow$	m	$\nearrow$

35. $c = 0$

x	$(-\infty, 0)$	0	$(0, \infty)$
f'	$+$	0	$+$
f	$\nearrow$	$\neq$	$\nearrow$

37. $c = \left(\frac{3}{2}\right)^{2/5}$

x	$\left(0, \left(\frac{3}{2}\right)^{2/5}\right)$	$\frac{3}{2}^{2/5}$	$\left(\left(\frac{3}{2}\right)^{2/5}, \infty\right)$
f'	$-$	0	$+$
f	$\searrow$	m	$\nearrow$

39. $c = 1$

x	$(0, 1)$	1	$(1, \infty)$
f'	$-$	0	$+$
f	$\searrow$	m	$\nearrow$

41. $c = 0$

x	$(-\infty, 0)$	0	$(0, \infty)$
f'	$+$	0	$-$
f	$\nearrow$	M	$\searrow$

43. $c = 0$

x	$(-\infty, 0)$	0	$(0, \infty)$
f'	$+$	0	$+$
f	$\nearrow$	$\neg$	$\nearrow$

45. $c = \frac{\pi}{2}$ and $c = \pi$

x	$\left(0, \frac{\pi}{2}\right)$	$\frac{\pi}{2}$	$\left(\frac{\pi}{2}, \pi\right)$	π	$(\pi, 2\pi)$
f'	$+$	0	$-$	0	$+$
f	$\nearrow$	M	$\searrow$	m	$\nearrow$

47. $c = \frac{\pi}{2}, \frac{7\pi}{6}, \frac{3\pi}{2}$, and $\frac{11\pi}{6}$

x	$\left(0, \frac{\pi}{2}\right)$	$\frac{\pi}{2}$	$\left(\frac{\pi}{2}, \frac{7\pi}{6}\right)$	$\frac{7\pi}{6}$	$\left(\frac{7\pi}{6}, \frac{3\pi}{2}\right)$
f'	$+$	0	$-$	0	$+$
f	$\nearrow$	M	$\searrow$	m	$\nearrow$

x	$\frac{3\pi}{2}$	$\left(\frac{3\pi}{2}, \frac{11\pi}{6}\right)$	$\frac{11\pi}{6}$	$\left(\frac{11\pi}{6}, 2\pi\right)$
f'	0	$-$	0	$+$
f	M	$\searrow$	m	$\nearrow$

49. $c = 0$

x	$(-\infty, 0)$	0	$(0, \infty)$
f'	$-$	0	$+$
f	$\searrow$	m	$\nearrow$

51. $c = -\frac{\pi}{4}$

x	$\left[-\frac{\pi}{2}, -\frac{\pi}{4}\right)$	$-\frac{\pi}{4}$	$\left(-\frac{\pi}{4}, \frac{\pi}{2}\right]$
f'	$+$	0	$-$
f	$\nearrow$	M	$\searrow$

53. $c = \pm 1$

x	$(-\infty, -1)$	-1	$(-1, 1)$	1	$(1, \infty)$
f'	$-$	0	$+$	0	$-$
f	$\searrow$	m	$\nearrow$	M	$\searrow$

55. $c = 1$

x	$(0, 1)$	1	$(1, \infty)$
f'	$-$	0	$+$
f	$\searrow$	m	$\nearrow$

57. $c = 0$; f' is positive on $(-\infty, 0)$ and on $(0, \infty)$ and is undefined at 0; f is increasing on $(-\infty, 0)$ and $(0, \infty)$; $x = 0$ is not a local minimum or maximum.

x	$(-\infty, 0)$	0	$(0, \infty)$
f'	$+$	0	$+$
f	$\nearrow$	$-$	$\nearrow$

59. $\left(\dfrac{1}{e}\right)^{1/e} \approx 0.692201$ **61.** $f'(x) > 0$ for all x

63. The graph of h is shown below at the left. Because $h(x)$ is negative for $x < -1$ and for $0 < x < 1$, it follows that f is decreasing for $x < -1$ and for $0 < x < 1$. Similarly, f is increasing for $-1 < x < 0$ and for $x > 1$ because $h(x)$ is positive on these intervals. Moreover, $f(x)$ has local minima at $x = -1$ and $x = 1$ and a local maximum at $x = 0$. A plausible graph for f is shown below at the right.

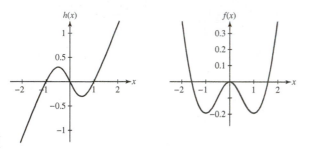

65. $f'(x) < 0$ as long as $x < 500$, so
$800^2 + 200^2 = f(200) > f(400) = 600^2 + 400^2$.

67. *every point* $c \in (a, b)$

75. **(a)** Let $g(x) = \cos x$ and $f(x) = 1 - \frac{1}{2}x^2$. Then
$f(0) = g(0) = 1$ and $g'(x) = -\sin x \geq -x = f'(x)$ for $x \geq 0$ by Exercise 73. Now apply Exercise 73 to conclude that
$\cos x \geq 1 - \frac{1}{2}x^2$ for $x \geq 0$.

(b) Let $g(x) = \sin x$ and $f(x) = x - \frac{1}{6}x^3$. Then $f(0) = g(0) = 0$ and $g'(x) = \cos x \geq 1 - \frac{1}{2}x^2 = f'(x)$ for $x \geq 0$ by part (a). Now apply Exercise 73 to conclude that $\sin x \geq x - \frac{1}{6}x^3$ for $x \geq 0$.

(c) Let $g(x) = 1 - \frac{1}{2}x^2 + \frac{1}{24}x^4$ and $f(x) = \cos x$. Then
$f(0) = g(0) = 1$ and $g'(x) = -x + \frac{1}{6}x^3 \geq -\sin x = f'(x)$ for $x \geq 0$ by part (b). Now apply Exercise 73 to conclude that
$\cos x \leq 1 - \frac{1}{2}x^2 + \frac{1}{24}x^4$ for $x \geq 0$.

(d) The next inequality in the series is $\sin x \leq x - \frac{1}{6}x^3 + \frac{1}{120}x^5$, valid for $x \geq 0$.

77.   • Let $f''(x) = 0$ for all x. Then $f'(x) = $ constant for all x. Since $f'(0) = m$, we conclude that $f'(x) = m$ for all x.
   • Let $g(x) = f(x) - mx$. Then
$g'(x) = f'(x) - m = m - m = 0$, which implies that
$g(x) = $ constant for all x and, consequently,
$f(x) - mx = $ constant for all x. Rearranging the statement,
$f(x) = mx + $ constant. Since $f(0) = b$, we conclude that
$f(x) = mx + b$ for all x.

79. **(a)** Let $g(x) = f(x)^2 + f'(x)^2$. Then

$$g'(x) = 2f(x)f'(x) + 2f'(x)f''(x)$$

$$= 2f(x)f'(x) + 2f'(x)(-f(x)) = 0$$

Because $g'(0) = 0$ for all x, $g(x) = f(x)^2 + f'(x)^2$ must be a constant function. To determine the value of C, we can substitute any number for x. In particular, for this problem, we want to substitute $x = 0$ and find $C = f(0)^2 + f'(0)^2$. Hence,

$$f(x)^2 + f'(x)^2 = f(0)^2 + f'(0)^2$$

(b) Let $f(x) = \sin x$. Then $f'(x) = \cos x$ and $f''(x) = -\sin x$, so $f''(x) = -f(x)$. Finally, if we take $f(x) = \sin x$, the result from part (a) guarantees that

$$\sin^2 x + \cos^2 x = \sin^2 0 + \cos^2 0 = 0 + 1 = 1$$

Section 4.4 Preliminary Questions

1. **(a)** increasing   **2.** $f(c)$ is a local maximum.
3. False   **4.** False; $f(x)$ may undefined at $x = c$.

Section 4.4 Exercises

1. **(a)** In C, we have $f''(x) < 0$ for all x.
(b) In A, $f''(x)$ goes from $+$ to $-$.
(c) In B, we have $f''(x) > 0$ for all x.
(d) In D, $f''(x)$ goes from $-$ to $+$.

3. Concave up everywhere; no points of inflection

5. Concave up for $x < -\sqrt{3}$ and for $0 < x < \sqrt{3}$; concave down for $-\sqrt{3} < x < 0$ and for $x > \sqrt{3}$; point of inflection at $x = 0$ and at $x = \pm\sqrt{3}$

7. Concave up for $0 < \theta < \pi$; concave down for $\pi < \theta < 2\pi$; point of inflection at $\theta = \pi$

9. Concave down for $0 < x < 9$; concave up for $x > 9$; point of inflection when $x = 9$

11. Concave up on $(0, 1)$; concave down on $(-\infty, 0) \cup (1, \infty)$; point of inflection at both $x = 0$ and $x = 1$

13. Concave up for $|x| > 1$; concave down for $|x| < 1$; point of inflection at both $x = -1$ and $x = 1$

15. Concave down for $x < \frac{2}{3}$; concave up for $x > \frac{2}{3}$; point of inflection at $x = \frac{2}{3}$

17. Concave down for $x < \frac{1}{2}$; concave up for $x > \frac{1}{2}$; point of inflection at $x = \frac{1}{2}$

19. **(a)** Starts 100 km away traveling towards us at slower and slower speeds, stops when it gets to us (after 2 h), then turns around and goes back to 100 km away, traveling at increasing speeds.
(b) The velocity is always increasing. When the ambulance is moving towards us (negative velocity), that means the speed is decreasing; when it is moving away, it means the speed is increasing.

21. Near the point of inflection, the curve is roughly a straight line going through $(55, 200)$ and $(35, 100)$, so the rate of change is roughly $\frac{200-100}{55-35} = 5$ cm/day. So when the growth rate starts to slow down, the height is growing at about 5 cm/day. Plots of the first and second derivatives are

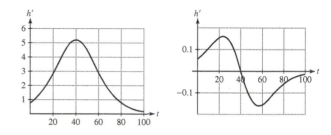

23. Points of inflection are a, d, and f. The function is concave down on $[0, a) \cup (d, f)$.

25. **(a)** f is increasing on $(0, 0.4)$.
(b) f is decreasing on $(-\infty, 0) \cup (0, 4, 1) \cup (1, 1, 2)$.
(c) f is concave up on $(0, 0.17) \cup (0.64, 1)$.
(d) f is concave down on $(0.17, 0.64) \cup (1, 1.2)$.

27. Critical points are $x = 3$ and $x = 5$; $f(3) = 54$ is a local maximum, and $f(5) = 50$ is a local minimum.

29. Critical points are $x = 0$ and $x = 1$; $f(0) = 0$ is a local minimum, Second derivative test is inconclusive at $x = 1$.

31. Critical points are $x = -4$ and $x = 2$; $f(-4) = -16$ is a local maximum, and $f(2) = -4$ is a local minimum.

33. Critical points are $x = 0$ and $x = \frac{2}{9}$; $f\left(\frac{2}{9}\right)$ is a local minimum; $f''(x)$ is undefined at $x = 0$, so the Second Derivative Test cannot be applied there.

35. Critical points are $x = 0$, $x = \frac{\pi}{3}$ and $x = \pi$; $f(0)$ is a local minimum, $f\left(\frac{\pi}{3}\right)$ is a local maximum, and $f(\pi)$ is a local minimum.

37. Critical points are $x = \pm\frac{\sqrt{2}}{2}$; $f\left(\frac{\sqrt{2}}{2}\right)$ is a local maximum and $f\left(-\frac{\sqrt{2}}{2}\right)$ is a local minimum.

39. Critical point is $x = e^{-1/3}$; $f\left(e^{-1/3}\right)$ is a local minimum.

41.

x	$\left(-\infty, \frac{1}{3}\right)$	$\frac{1}{3}$	$\left(\frac{1}{3}, 1\right)$	1	$(1, \infty)$
f'	$+$	0	$-$	0	$+$
f	↗	M	↘	m	↗

x	$\left(-\infty, \frac{2}{3}\right)$	$\frac{2}{3}$	$\left(\frac{2}{3}, \infty\right)$
f''	$-$	0	$+$
f	⌢	I	⌣

43.

t	$(-\infty, 0)$	0	$\left(0, \frac{2}{3}\right)$	$\frac{2}{3}$	$\left(\frac{2}{3}, \infty\right)$
f'	$-$	0	$+$	0	$-$
f	↘	m	↗	M	↘

t	$\left(-\infty, \frac{1}{3}\right)$	$\frac{1}{3}$	$\left(\frac{1}{3}, \infty\right)$
f''	$+$	0	$-$
f	⌣	I	⌢

45. $f''(x) > 0$ for all $x \geq 0$, which means there are no inflection points.

x	0	$\left(0, (2)^{2/3}\right)$	$(2)^{2/3}$	$\left((2)^{2/3}, \infty\right)$
f'	U	$-$	0	$+$
f	M	↘	m	↗

47.

x	$\left(-\infty, -3\sqrt{3}\right)$	$-3\sqrt{3}$	$\left(-3\sqrt{3}, 3\sqrt{3}\right)$	$3\sqrt{3}$	$\left(3\sqrt{3}, \infty\right)$
f'	$-$	0	$+$	0	$-$
f	↘	m	↗	M	↘

x	$(-\infty, -9)$	-9	$(-9, 0)$	0	$(0, 9)$	9	$(9, \infty)$
f''	$-$	0	$+$	0	$-$	0	$+$
f	⌢	I	⌣	I	⌢	I	⌣

49.

x	$\left(-\infty, -\left(\frac{3}{5}\right)^{3/2}\right)$	$-\left(\frac{3}{5}\right)^{3/2}$	$\left(-\left(\frac{3}{5}\right)^{3/2}, 0\right)$	0	$\left(0, \left(\frac{3}{5}\right)^{3/2}\right)$	$\left(\frac{3}{5}\right)^{3/2}$	$\left(\left(\frac{3}{5}\right)^{3/2}, \infty\right)$
f''	$-$	$-$	$-$	undef	$+$	$+$	$+$

51.

θ	$(0, \pi)$	π	$(\pi, 2\pi)$
f'	$+$	0	$+$
f	↗	¬	↗

θ	0	$(0, \pi)$	π	$(\pi, 2\pi)$	2π
f''	0	$-$	0	$+$	0
f	¬	⌢	I	⌣	¬

53.

x	$\left(-\frac{\pi}{2}, \frac{\pi}{2}\right)$
f'	$+$
f	↗

x	$\left(-\frac{\pi}{2}, 0\right)$	0	$\left(0, \frac{\pi}{2}\right)$
f''	$-$	0	$+$
f	⌢	I	⌣

55.

x	$\left(0, 1 + \sqrt{3}\right)$	$1 + \sqrt{3}$	$\left(1 + \sqrt{3}, \infty\right)$
f'	$+$	0	$-$
f	↗	M	↘

x	$(0, 4)$	4	$(4, \infty)$
f''	$-$	0	$+$
f	⌢	I	⌣

57.

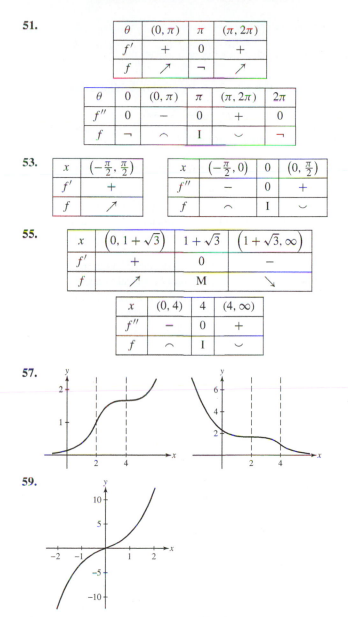

59.

61. (a) Near the beginning of the epidemic, the graph of R is concave up. Near the epidemic's end, R is concave down.

(b) "Epidemic subsiding: number of new cases declining."

63. The point of inflection should occur when the water level is equal to the radius of the sphere. A possible graph of V is shown below.

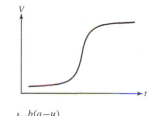

65. (a) $f'(u) = \dfrac{be^{b(a-u)}}{\left(1 + e^{b(a-u)}\right)^2} > 0$

(b) $u = a$

67. (a) From the definition of the derivative, we have

$$f''(c) = \lim_{h \to 0} \frac{f'(c+h) - f'(c)}{h} = \lim_{h \to 0} \frac{f'(c+h)}{h}$$

(b) We are given that $f''(c) > 0$. By part (a), it follows that

$$\lim_{h \to 0} \frac{f'(c+h)}{h} > 0$$

In other words, for sufficiently small h,

$$\frac{f'(c+h)}{h} > 0$$

Now, if h is sufficiently small but negative, then $f'(c+h)$ must also be negative [so that the ratio $f'(c+h)/h$ will be positive] and $c + h < c$. On the other hand, if h is sufficiently small but positive, then $f'(c+h)$ must also be positive and $c + h > c$. Thus, there exists an open interval (a, b) containing c such that $f'(x) < 0$ for $a < x < c$ and $f'(c) > 0$ for $c < x < b$. Finally, because $f'(x)$ changes from negative to positive at $x = c$, $f(c)$ must be a local minimum.

69. (b) $f(x)$ has a point of inflection at $x = 0$ and at $x = \pm 1$. The figure below shows the graph of $y = f(x)$ and its tangent lines at each of the points of inflection. It is clear that each tangent line crosses the graph of f at the inflection point.

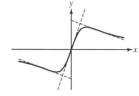

71. Let $f(x) = a_n x^n + a_{n-1} x^{n-1} + \cdots + a_1 x + a_0$ be a polynomial of degree n. Then

$$f'(x) = na_n x^{n-1} + (n-1)a_{n-1} x^{n-2} + \cdots + 2a_2 x + a_1$$

and

$$f''(x) = n(n-1)a_n x^{n-2}$$

$$+ (n-1)(n-2)a_{n-1} x^{n-3} + \cdots + 6a_3 x + 2a_2$$

If $n \geq 3$ and is odd, then $n - 2$ is also odd and f'' is a polynomial of odd degree. Therefore, f'' must take on both positive and negative values. It follows that $f''(x)$ has at least one root c such that $f''(x)$ changes sign at c. The function f will then have a point of inflection at $x = c$. On the other hand, the functions $f(x) = x^2$, x^4, and x^8 are polynomials of even degree that do not have any points of inflection.

Section 4.5 Preliminary Questions

1. Not of the form $\frac{0}{0}$ or $\frac{\infty}{\infty}$ **2.** No

Section 4.5 Exercises

1. L'Hôpital's Rule does not apply.

3. L'Hôpital's Rule does not apply.

5. L'Hôpital's Rule does not apply.

7. L'Hôpital's Rule does not apply.

9. 0   **11.** Quotient of the form $\frac{\infty}{\infty}$; $-\dfrac{9}{2}$

13. Quotient of the form $\frac{\infty}{\infty}$; 0   **15.** Quotient of the form $\frac{\infty}{\infty}$; 0

17. $\dfrac{5}{6}$ **19.** $-\dfrac{3}{5}$ **21.** $-\dfrac{7}{3}$ **23.** $\dfrac{9}{7}$ **25.** $\dfrac{2}{7}$ **27.** 1   **29.** 2

31. -1 **33.** $\dfrac{1}{2}$ **35.** 0   **37.** $-\dfrac{2}{\pi}$ **39.** 1   **41.** Does not exist

43. 0   **45.** $\ln a$ **47.** e **49.** $e^{-3/2}$ **51.** 1   **53.** $\dfrac{1}{\pi}$

55.

$$\lim_{x \to \pi/2} \frac{\cos mx}{\cos nx} = \begin{cases} (-1)^{(m-n)/2}, & m, n \text{ even} \\ \text{does not exist}, & m \text{ even}, n \text{ odd} \\ 0 & m \text{ odd}, n \text{ even} \\ (-1)^{(m-n)/2} \dfrac{m}{n}, & m, n \text{ odd} \end{cases}$$

57. $\displaystyle \lim_{x \to 0} \ln\left((1+x)^{1/x}\right) = \lim_{x \to 0} \frac{1}{x} \ln(1+x) = \lim_{x \to 0} \frac{\ln(1+x)}{x} = 1$

Thus, $\displaystyle \lim_{x \to 0} (1+x)^{1/x} = e^1 = e$; $x = 0.0005$.

61. (a) $\lim_{x \to 0+} f(x) = 0$; $\lim_{x \to \infty} f(x) = e^0 = 1$

(b) f is increasing for $0 < x < e$, is decreasing for $x > e$, and has a maximum at $x = e$. The maximum value is $f(e) = e^{1/e} \approx 1.444668$.

63. Neither

65. $\displaystyle \lim_{x \to \infty} \frac{\ln x}{x^a} = \lim_{x \to \infty} \frac{x^{-1}}{ax^{a-1}} = \lim_{x \to \infty} \frac{1}{a} x^{-a} = 0$

69. (a) $1 \leq 2 + \sin x \leq 3$, so

$$\frac{x}{x^2 + 1} \leq \frac{x(2 + \sin x)}{x^2 + 1} \leq \frac{3x}{x^2 + 1}$$

It follows by the Squeeze Theorem that

$$\lim_{x \to \infty} \frac{x(2 + \sin x)}{x^2 + 1} = 0$$

(b) $\displaystyle \lim_{x \to \infty} f(x) = \lim_{x \to \infty} x(2 + \sin x) \geq \lim_{x \to \infty} x = \infty$ and $\displaystyle \lim_{x \to \infty} g(x) = \lim_{x \to \infty} (x^2 + 1) = \infty$, but

$$\lim_{x \to \infty} \frac{f'(x)}{g'(x)} = \lim_{x \to \infty} \frac{x(\cos x) + (2 + \sin x)}{2x}$$

does not exist since $\cos x$ oscillates. This does not violate L'Hôpital's Rule since the theorem clearly states

$$\lim_{x \to \infty} \frac{f(x)}{g(x)} = \lim_{x \to \infty} \frac{f'(x)}{g'(x)}$$

"provided the limit on the right exists."

71. (a) Using Exercise 70, we see that $G(b) = e^{H(b)}$. Thus, $G(b) = 1$ if $0 \leq b \leq 1$ and $G(b) = b$ if $b > 1$.

(b)

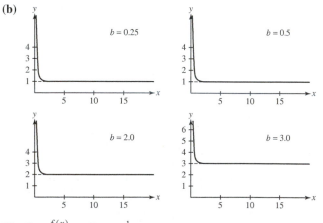

73. $\displaystyle \lim_{x \to 0} \frac{f(x)}{x^k} = \lim_{x \to 0} \frac{1}{x^k e^{1/x^2}}$. Let $t = 1/x$. As $x \to 0$, $t \to \infty$. Thus,

$$\lim_{x \to 0} \frac{1}{x^k e^{1/x^2}} = \lim_{t \to \infty} \frac{t^k}{e^{t^2}} = 0$$

by Exercise 72.

75. For $x \neq 0$, $f'(x) = e^{-1/x^2}\left(\frac{2}{x^3}\right)$. Here, $P(x) = 2$ and $r = 3$.

Assume $f^{(k)}(x) = \frac{P(x)e^{-1/x^2}}{x^r}$. Then

$$f^{(k+1)}(x) = e^{-1/x^2}\left(\frac{x^3 P'(x) + (2 - rx^2)P(x)}{x^{r+3}}\right)$$

which is of the form desired.

Moreover, from Exercise 74, $f'(0) = 0$. Suppose $f^{(k)}(0) = 0$. Then

$$f^{(k+1)}(0) = \lim_{x \to 0} \frac{f^{(k)}(x) - f^{(k)}(0)}{x - 0} = \lim_{x \to 0} \frac{P(x)e^{-1/x^2}}{x^{r+1}}$$

$$= P(0) \lim_{x \to 0} \frac{f(x)}{x^{r+1}} = 0$$

79. $\lim\limits_{x \to 0} \frac{\sin x}{x} = \lim\limits_{x \to 0} \frac{\cos x}{1} = 1$. To use L'Hôpital's Rule to evaluate $\lim_{x \to 0} \frac{\sin x}{x}$, we must know that the derivative of $\sin x$ is $\cos x$, but to determine the derivative of $\sin x$, we must be able to evaluate $\lim_{x \to 0} \frac{\sin x}{x}$.

81. **(a)** $e^{-1/6} \approx 0.846481724$

x	1	0.1	0.01
$\left(\dfrac{\sin x}{x}\right)^{1/x^2}$	0.841471	0.846435	0.846481

(b) 1/3

x	± 1	± 0.1	± 0.01
$\dfrac{1}{\sin^2 x} - \dfrac{1}{x^2}$	0.412283	0.334001	0.333340

Section 4.6 Preliminary Questions

1. An arc with the sign combination $++$ (increasing, concave up) is shown below at the left. An arc with the sign combination $-+$ (decreasing, concave up) is shown below at the right.

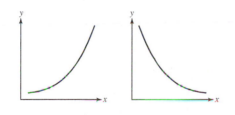

2. **(c)**

3. $x = 4$ is not in the domain of f.

Section 4.6 Exercises

1. • In A, f is decreasing and concave up, so $f' < 0$ and $f'' > 0$.
  • In B, f is increasing and concave up, so $f' > 0$ and $f'' > 0$.
  • In C, f is increasing and concave down, so $f' > 0$ and $f'' < 0$.
  • In D, f is decreasing and concave down, so $f' < 0$ and $f'' < 0$.
  • In E, f is decreasing and concave up, so $f' < 0$ and $f'' > 0$.
  • In F, f is increasing and concave up, so $f' > 0$ and $f'' > 0$.
  • In G, f is increasing and concave down, so $f' > 0$ and $f'' < 0$.

3. This function changes from concave up to concave down at $x = -1$ and from increasing to decreasing at $x = 0$.

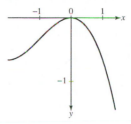

5. The function is decreasing everywhere and changes from concave up to concave down at $x = -1$ and from concave down to concave up at $x = -\frac{1}{2}$.

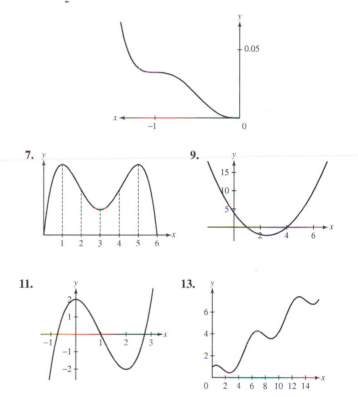

7.

9.

11.

13.

15. Local maximum at $x = -16$, a local minimum at $x = 0$, and an inflection point at $x = -8$

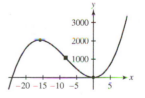

17. $f(0)$ is a local minimum, $f(\frac{1}{6})$ is a local maximum, and there is a point of inflection at $x = \frac{1}{12}$.

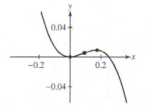

19. f has local minima at $x = \pm\sqrt{6}$, a local maximum at $x = 0$, and inflection points at $x = \pm\sqrt{2}$.

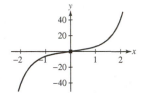

21. Graph has no critical points and is always increasing, inflection point at $(0, 0)$.

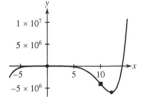

23. $f\left(\frac{1-\sqrt{33}}{8}\right)$ and $f(2)$ are local minima, and $f\left(\frac{1+\sqrt{33}}{8}\right)$ is a local maximum; points of inflection both at $x = 0$ and $x = \frac{3}{2}$.

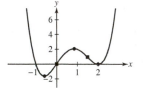

25. $f(0)$ is a local maximum, $f(12)$ is a local minimum, and there is a point of inflection at $x = 10$.

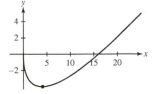

27. $f(4)$ is a local minimum, and the graph is always concave up.

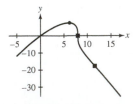

29. f has a local maximum at $x = 6$ and inflection points at $x = 8$ and $x = 12$.

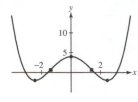

31. f has a local minimum at $x = -\frac{\sqrt{2}}{2}$, a local maximum at $x = \frac{\sqrt{2}}{2}$, inflection points at $x = 0$ and at $x = \pm\sqrt{\frac{3}{2}}$, and a horizontal asymptote at $y = 0$.

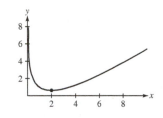

33. $f(2)$ is a local minimum and the graph is always concave up.

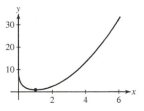

35. f has a local minimum at $x = 1$ and no inflection points. It is concave up everywhere. It has a vertical asymptote at $x = 0$.

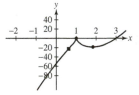

37. Graph has an inflection point at $x = \frac{3}{5}$, a local maximum at $x = 1$ (at which the graph has a cusp), and a local minimum at $x = \frac{9}{5}$.

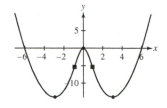

39. f has a local maximum at $x = 0$, local minima at $x = \pm 3$, and points of inflection at $x = \pm\sqrt{-6 + 3\sqrt{5}}$.

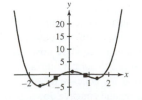

41. f has local minima at $x = -1.473$ and $x = 1.347$, a local maximum at $x = 0.126$, and points of inflection at $x = \pm\sqrt{\frac{2}{3}}$.

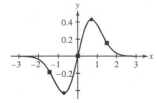

43. Graph has an inflection point at $x = \pi$, and no local maxima or minima.

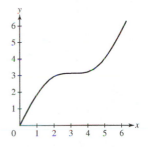

45. Local maximum at $x = \frac{\pi}{2}$, a local minimum at $x = \frac{3\pi}{2}$, and inflection points at $x = \frac{\pi}{6}$ and $x = \frac{5\pi}{6}$.

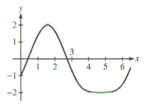

47. Local maximum at $x = \frac{\pi}{6}$ and a point of inflection at $x = \frac{2\pi}{3}$.

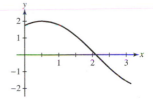

49. In both cases, there is a point where f is not differentiable at the transition from increasing to decreasing or decreasing to increasing.

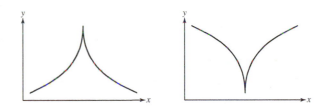

51. Graph (B) cannot be the graph of a polynomial.

53. (B) is the graph of $f(x) = \dfrac{3x^2}{x^2 - 1}$; (A) is the graph of

$f(x) = \dfrac{3x}{x^2 - 1}$.

55. f is decreasing for all $x \neq \frac{1}{3}$, concave up for $x > \frac{1}{3}$, concave down for $x < \frac{1}{3}$, has a horizontal asymptote at $y = 0$ and a vertical asymptote at $x = \frac{1}{3}$.

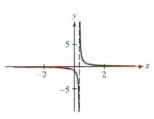

57. f is decreasing for all $x \neq 2$, concave up for $x > 2$, concave down for $x < 2$, has a horizontal asymptote at $y = 1$ and a vertical asymptote at $x = 2$.

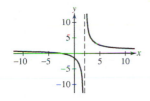

59. f is decreasing for all $x \neq 0, 1$, concave up for $0 < x < \frac{1}{2}$ and $x > 1$, concave down for $x < 0$ and $\frac{1}{2} < x < 1$, has a horizontal asymptote at $y = 0$ and vertical asymptotes at $x = 0$ and $x = 1$.

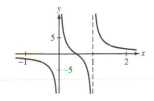

61. f is increasing for $x < 0$ and $0 < x < 1$ and decreasing for $1 < x < 2$ and $x > 2$; f is concave up for $x < 0$ and $x > 2$ and concave down for $0 < x < 2$; f has a horizontal asymptote at $y = 0$ and vertical asymptotes at $x = 0$ and $x = 2$.

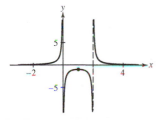

63. f is increasing for $x < 2$ and for $2 < x < 3$, is decreasing for $3 < x < 4$ and for $x > 4$, and has a local maximum at $x = 3$; f is concave up for $x < 2$ and for $x > 4$ and is concave down for $2 < x < 4$; f has a horizontal asymptote at $y = 0$ and vertical asymptotes at $x = 2$ and $x = 4$.

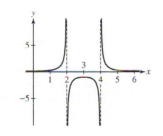

65. f is increasing for $|x| > 2$ and decreasing for $-2 < x < 0$ and for $0 < x < 2$; f is concave down for $-2\sqrt{2} < x < 0$ and for $x > 2\sqrt{2}$ and concave up for $x < -2\sqrt{2}$ and for $0 < x < 2\sqrt{2}$; f has a horizontal asymptote at $y = 1$ and a vertical asymptote at $x = 0$.

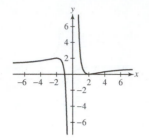

67. f is increasing for $x < 0$ and for $x > 2$ and decreasing for $0 < x < 2$; f is concave up for $x < 0$ and for $0 < x < 1$, is concave down for $1 < x < 2$ and for $x > 2$, and has a point of inflection at $x = 1$; f has a horizontal asymptote at $y = 0$ and vertical asymptotes at $x = 0$ and $x = 2$.

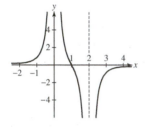

69. f is increasing for $x < 0$, decreasing for $x > 0$, and has a local maximum at $x = 0$; f is concave up for $|x| > 1/\sqrt{5}$, is concave down for $|x| < 1/\sqrt{5}$, and has points of inflection at $x = \pm 1/\sqrt{5}$; f has a horizontal asymptote at $y = 0$ and no vertical asymptotes.

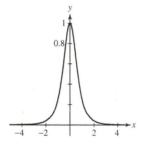

71. f is increasing for $x < 0$ and decreasing for $x > 0$; f is concave down for $|x| < \frac{\sqrt{2}}{2}$ and concave up for $|x| > \frac{\sqrt{2}}{2}$; f has a horizontal asymptote at $y = 0$ and no vertical asymptotes.

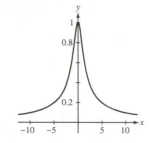

75. f is increasing for $x < -2$ and for $x > 0$, is decreasing for $-2 < x < -1$ and for $-1 < x < 0$, has a local minimum at $x = 0$, has a local maximum at $x = -2$, is concave down on $(-\infty, -1)$ and concave up on $(-1, \infty)$; f has a vertical asymptote at $x = -1$; by polynomial division, $f(x) = x - 1 + \frac{1}{x+1}$ and

$$\lim_{x \to \pm\infty} \left(x - 1 + \frac{1}{x+1} - (x-1) \right) = 0$$

which implies that the slant asymptote is $y = x - 1$.

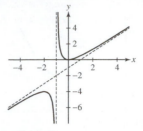

77. $y = x + 2$ is the slant asymptote of f; local minimum at $x = 2 + \sqrt{3}$, a local maximum at $x = 2 - \sqrt{3}$, and f is concave down on $(-\infty, 2)$ and concave up on $(2, \infty)$; vertical asymptote at $x = 2$.

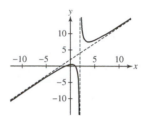

Section 4.7 Preliminary Questions

1. $b + h + \sqrt{b^2 + h^2} = 10$

2. If the function tends to infinity at the endpoints of the interval, then the function must take on a minimum value at a critical point.

3. No

Section 4.7 Exercises

1. **(a)** $y = \frac{3}{2} - x$ **(b)** $A = x(\frac{3}{2} - x) = \frac{3}{2}x - x^2$

(c) Closed interval $\left[0, \frac{3}{2}\right]$

(d) The maximum area 0.5625 m^2 is achieved with $x = y = \frac{3}{4}$ m.

3. One side of length 6 and two of length 3 **5.** 4 and 32

7. Allot approximately 5.28 m of the wire to the circle.

9. 20 and 20

11. Not possible. y can be arbitrarily large; then x is positive but close to zero and $x + 2y \approx 2y$ is arbitrarily large.

13. **(a)** The box should be a cube with side length $12^{1/3}$

(b) The box should be a cube with side length $\dfrac{\sqrt{30}}{3}$.

15. The corral of maximum area has dimensions

$$x = \frac{300}{1 + \pi/4} \text{ m} \quad \text{and} \quad y = \frac{150}{1 + \pi/4} \text{ m}$$

where x is the width of the corral and therefore the diameter of the semicircle and y is the height of the rectangular section.

17. Square of side length $4\sqrt{2}$ **19.** About 1.43 m **21.** $\left(\dfrac{1}{2}, \dfrac{1}{2}\right)$

23. $(0.632784, -1.090410)$ **25.** $\theta = \dfrac{\pi}{2}$ **27.** $\dfrac{3\sqrt{3}}{4}r^2$

29. 60 cm wide by 100 cm high for the full poster (48 cm by 80 cm for the printed matter)

31. Radius: $\sqrt{\frac{2}{3}}R$; half-height: $\dfrac{R}{\sqrt{3}}$

33. $x = 10\sqrt{5} \approx 22.36$ m and $y = 20\sqrt{5} \approx 44.72$ m, where x is the length of the brick wall and y is the length of an adjacent side

35. 1.0718   **37.** $LH + \frac{1}{2}(L^2 + H^2)$ **39.** $y = -3x + 24$

43. $s = 3\sqrt[3]{4}$ m and $h = 2\sqrt[3]{4}$ m, where s is the length of the side of the square bottom of the box and h is the height of the box

45. **(a)** Each compartment has length of 600 m and width of 400 m.
(b) 240,000 m^2

47. $N \approx 58.14$ lb and $P \approx 77.33$ lb   **49.** \$990

51. 1.2 million euros in equipment and 600,000 euros in labor

53. Brandon swims diagonally to a point located 20.2 m downstream and then runs the rest of the way.

57. $A = B = 30$ cm

59. $x = \dfrac{x_1 + x_2 + \cdots + x_n}{n}$

63. **(a)** 900 m^2 when $x = -10$ **(b)** $[0, 20]$; 800 m^2 when $x = 0$

65. $\tan \theta = \frac{5}{2}$

67. **(b)** $\left(\dfrac{17}{0.003}\right)^{1/4} \approx 8.676247$

(d) $v_d \approx 11.418583$; $D(v_d) \approx 191.741$ km

69. $s = \left(\dfrac{b^{2/3}}{2^{2/3}} + h^{2/3}\right)^{3/2}$ **71.** $\left(a^{2/3} + b^{2/3}\right)^{3/2}$

73. **(a)** $\alpha = 0$ corresponds to shooting the ball directly at the basket, while $\alpha = \pi/2$ corresponds to shooting the ball directly upward. In neither case is it possible for the ball to go into the basket. If the angle α is extremely close to 0, the ball is shot almost directly at the basket; on the other hand, if the angle α is extremely close to $\pi/2$, the ball is launched almost vertically. In either one of these cases, the ball has to travel at an enormous speed.

(b) The minimum clearly occurs where $\theta = \pi/3$.

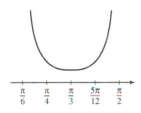

(c) $v^2 = \dfrac{16d}{F(\theta)}$; hence, v^2 is smallest whenever $F(\theta)$ is greatest.

(d) A critical point of F occurs where $\cos(\alpha - 2\theta) = 0$ so that $\alpha - 2\theta = -\frac{\pi}{2}$ (negative because $2\theta > \theta > \alpha$), and this gives us $\theta = \alpha/2 + \pi/4$. The minimum value $F(\theta_0)$ takes place at $\theta_0 = \alpha/2 + \pi/4$.

(e) Plug in $\theta_0 = \alpha/2 + \pi/4$. From Figure 34, we see that

$$\cos\alpha = \frac{d}{\sqrt{d^2 + h^2}} \quad \text{and} \quad \sin\alpha = \frac{h}{\sqrt{d^2 + h^2}}$$

(f) This shows that the minimum velocity required to launch the ball to the basket drops as shooter height increases. This shows one of the ways height is an advantage in free throws; a taller shooter need not shoot the ball as hard to reach the basket.

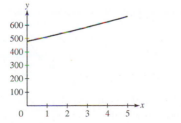

75. **(a)** From the figure, we see that

$$\theta(x) = \tan^{-1}\frac{c - f(x)}{x} - \tan^{-1}\frac{b - f(x)}{x}$$

Then

$\theta'(x)$

$$= \frac{b - (f(x) - xf'(x))}{x^2 + (b - f(x))^2} - \frac{c - (f(x) - xf'(x))}{x^2 + (c - f(x))^2}$$

$$= (b - c)\frac{x^2 - bc + (b + c)(f(x) - xf'(x)) - (f(x))^2 + 2xf(x)f'(x)}{(x^2 + (b - f(x))^2)(x^2 + (c - f(x))^2)}$$

$$= (b - c)\frac{(x^2 + (xf'(x))^2) - (bc - (b + c)(f(x) - xf'(x)) + (f(x) - xf'(x))^2)}{(x^2 + (b - f(x))^2)(x^2 + (c - f(x))^2)}$$

$$= (b - c)\frac{(x^2 + (xf'(x))^2) - (b - (f(x) - xf'(x)))(c - (f(x) - xf'(x)))}{(x^2 + (b - f(x))^2)(x^2 + (c - f(x))^2)}$$

(b) The point Q is the y-intercept of the line tangent to the graph of f at point P. The equation of this tangent line is

$$Y - f(x) = f'(x)(X - x)$$

The y-coordinate of Q is then $f(x) - xf'(x)$.
(c) From the figure, we see that

$$BQ = b - (f(x) - xf'(x)),$$

$$CQ = c - (f(x) - xf'(x))$$

and

$$PQ = \sqrt{x^2 + (f(x) - (f(x) - xf'(x)))^2} = \sqrt{x^2 + (xf'(x))^2}$$

Comparing these expressions with the numerator of $d\theta/dx$, it follows that $\dfrac{d\theta}{dx} = 0$ is equivalent to

$$PQ^2 = BQ \cdot CQ$$

(d) The equation $PQ^2 = BQ \cdot CQ$ is equivalent to

$$\frac{PQ}{BQ} = \frac{CQ}{PQ}$$

In other words, the sides CQ and PQ from the triangle $\triangle QCP$ are proportional in length to the sides PQ and BQ from the triangle $\triangle QPB$. As $\angle PQB = \angle CQP$, it follows that triangles $\triangle QCP$ and $\triangle QPB$ are similar.

Section 4.8 Preliminary Questions

1. One

2. Every term in the Newton's Method sequence will remain x_0.

3. Newton's Method will fail.

4. Yes, that is a reasonable description. The iteration formula for Newton's Method was derived by solving the equation of the tangent line to $y = f(x)$ at x_0 for its x-intercept.

Section 4.8 Exercises

1.

n	1	2	3
x_n	2.5	2.45	2.44948980

3.

n	1	2	3
x_n	2.16666667	2.15450362	2.15443469

5.

n	1	2	3
x_n	0.28540361	0.24288009	0.24267469

7. We take $x_0 = -1.4$, based on the figure, and then calculate

n	1	2	3
x_n	-1.330964467	-1.328272820	-1.328268856

9. $r_1 \approx 0.25917$ and $r_2 \approx 2.54264$

11. $\sqrt{11} \approx 3.317$; a calculator yields 3.31662479.

13. $2^{7/3} \approx 5.040$; a calculator yields 5.0396842.

15. 2.093064358   **17.** -2.225 **19.** 1.749

21. $x = 4.49341$, which is approximately 1.4303π

23. $(2.7984, -0.941684)$

25. (a) $P \approx \$156.69$

(b) $b \approx 1.02121$; the interest rate is around 25.45%.

27. (a) The sector SAB is the slice OAB with the triangle OBS removed. OAB is a central sector with arc θ and radius $\overline{OA} = a$, and therefore has area $\frac{a^2\theta}{2}$. OBS is a triangle with height $a\sin\theta$ and base length $\overline{OS} = ea$. Hence, the area of the sector is

$$\frac{a^2}{2}\theta - \frac{1}{2}ea^2\sin\theta = \frac{a^2}{2}(\theta - e\sin\theta)$$

(b) Since Kepler's Second Law indicates that the area of the sector is proportional to the time t since the planet passed point A, we get

$$\pi a^2(t/T) = a^2/2\,(\theta - e\sin\theta)$$

$$2\pi\frac{t}{T} = \theta - e\sin\theta$$

(c) From the point of view of the Sun, Mercury has traversed an angle of approximately 1.76696 radians $= 101.24°$. Mercury has therefore traveled more than one fourth of the way around (from the point of view of central angle) during this time.

29. The sequence of iterates diverges spectacularly, since $x_n = (-2)^n x_0$.

31. (a) Let $f(x) = \frac{1}{x} - c$. Then

$$x - \frac{f(x)}{f'(x)} = x - \frac{\frac{1}{x} - c}{-x^{-2}} = 2x - cx^2$$

(b) For $c = 10.3$, we have $f(x) = \frac{1}{x} - 10.3$ and thus $x_{n+1} = 2x_n - 10.3x_n^2$

- Take $x_0 = 0.1$.

n	1	2	3
x_n	0.097	0.0970873	0.09708738

- Take $x_0 = 0.5$.

n	1	2	3
x_n	-1.575	-28.7004375	-8541.66654

(c) The graph is disconnected. If $x_0 = .5$, $(x_1, f(x_1))$ is on the other portion of the graph, which will never converge to any point under Newton's Method.

33. $\theta \approx 1.2757$; hence, $h = L\dfrac{1 - \cos\theta}{2\sin\theta} \approx 1.11181$

35. (a) $a = 46.95$ (b) $s = 29.24$

37. (a) $a \approx 28.46$

(b) $\Delta L = 1$ ft yields $\Delta s \approx 0.61$; $\Delta L = 5$ yields $\Delta s \approx 3.05$.

(c) $s(161) - s(160) = 0.62$, very close to the approximation obtained from the Linear Approximation; $s(165) - s(160) = 3.02$, again very close to the approximation obtained from the Linear Approximation.

Chapter 4 Review

1. $8.1^{1/3} - 2 \approx 0.00833333$; error is 3.445×10^{-5}.

3. $625^{1/4} - 624^{1/4} \approx 0.002$; error is 1.201×10^{-6}.

5. $\frac{1}{1.02} \approx 0.98$; error is 3.922×10^{-4}.

7. $L(x) = 5 + \dfrac{1}{10}(x - 25)$ **9.** $L(r) = 36\pi(r - 2)$

11. $L(x) = \dfrac{1}{\sqrt{e}}(2 - x)$ **13.** $\Delta s \approx 0.632$

15. (a) An increase of \$1500 in revenue.

(b) A small increase in price would result in a decrease in revenue.

17. 9%   **21.** $c = \dfrac{3}{\ln 4} \approx 2.164 \in (1, 4)$

23. Let $x > 0$. Because f is continuous on $[0, x]$ and differentiable on $(0, x)$, the Mean Value Theorem guarantees there exists a $c \in (0, x)$ such that

$$f'(c) = \frac{f(x) - f(0)}{x - 0} \qquad \text{or} \qquad f(x) = f(0) + xf'(c)$$

Now, we are given $f(0) = 4$ and $f'(x) \le 2$ for $x > 0$. Therefore, for all $x \ge 0$,

$$f(x) \le 4 + x(2) = 2x + 4$$

25. $x = \frac{2}{3}$ and $x = 2$ are critical points; $f\left(\frac{2}{3}\right)$ is a local maximum, while $f(2)$ is a local minimum.

27. $x = 0$, $x = -2$ and $x = -\frac{4}{5}$ are critical points; $f(-2)$ is neither a local maximum nor a local minimum, $f\left(-\frac{4}{5}\right)$ is a local maximum and $f(0)$ is a local minimum.

29. $\theta = \dfrac{3\pi}{4} + n\pi$ is a critical point for all integers n; $g\left(\dfrac{3\pi}{4} + n\pi\right)$ is neither a local maximum nor a local minimum for any integer n.

31. Maximum value is 21; minimum value is -11.

33. Minimum value is -1; maximum value is $\dfrac{5}{4}$.

35. Minimum value is -1; maximum value is 3.

37. Minimum value is $12 - 12\ln 12 \approx -17.818880$; maximum value is $40 - 12\ln 40 \approx -4.266553$.

39. Critical points are $x = 1$ and $x = 3$. The minimum value is 2, occurring at $x = 3$, and the maximum value is 17, occurring at the right endpoint $x = 8$.

41. $x = \dfrac{4}{3}$ **43.** $x = \pm\dfrac{2}{\sqrt{3}}$ **45.** $x = 1$ and $x = 4$

47. No horizontal asymptotes; no vertical asymptotes

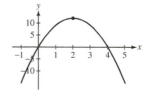

49. No horizontal asymptotes; no vertical asymptotes

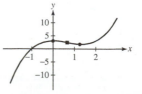

51. $y = 0$ is a horizontal asymptote; $x = -1$ is a vertical asymptote.

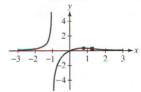

53. Horizontal asymptote of $y = 0$; no vertical asymptotes

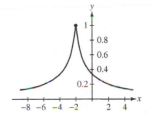

55.

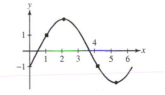

57.

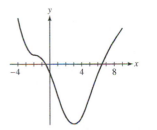

59. $b = \sqrt[3]{12}$ m and $h = \frac{1}{3}\sqrt[3]{12}$ m   **63.** $\frac{16}{9}\pi$ **69.** $\sqrt[3]{25} = 2.9240$

71. $\left(0, \frac{2}{e}\right)$ is a local minimum.

73. Local minimum at $x = e^{-1}$; no points of inflection;
$\lim_{x \to 0+} x \ln x = 0$; $\lim_{x \to \infty} x \ln x = \infty$

75. Local maximum at $x = e^{-2}$ and a local minimum at $x = 1$; point of inflection at $x = e^{-1}$; $\lim_{x \to 0+} x(\ln x)^2 = 0$;
$\lim_{x \to \infty} x(\ln x)^2 = \infty$

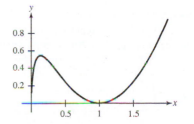

77. As $x \to \infty$, both $2x - \sin x$ and $3x + \cos 2x$ tend toward infinity, so L'Hôpital's Rule applies to $\lim_{x \to \infty} \dfrac{2x - \sin x}{3x + \cos 2x}$; however, the resulting limit, $\lim_{x \to \infty} \dfrac{2 - \cos x}{3 - 2 \sin 2x}$, does not exist due to the oscillation of $\sin x$ and $\cos x$. To evaluate the limit, we note

$$\lim_{x \to \infty} \frac{2x - \sin x}{3x + \cos 2x} = \lim_{x \to \infty} \frac{2 - \frac{\sin x}{x}}{3 + \frac{\cos 2x}{x}} = \frac{2}{3}$$

79. 4   **81.** 0   **83.** 3   **85.** $\ln 2$ **87.** $\frac{1}{6}$ **89.** 2

Chapter 5

Section 5.1 Preliminary Questions

1. The right endpoints of the subintervals are then $\frac{5}{2}, 3, \frac{7}{2}, 4, \frac{9}{2}, 5$, while the left endpoints are $2, \frac{5}{2}, 3, \frac{7}{2}, 4, \frac{9}{2}$.

2. (a) $\frac{9}{2}$ (b) $\frac{3}{2}$ and 2

3. (a) *Are* the same   (b) *Not* the same

(c) *Are* the same   (d) *Are* the same

4. The first term in the sum $\sum_{j=0}^{100} j$ is equal to zero, so it may be dropped; on the other hand, the first term in $\sum_{j=0}^{100} 1$ is not zero.

5. On $[3, 7]$, the function $f(x) = x^{-2}$ is a decreasing function.

Section 5.1 Exercises

1. Over the interval $[0, 3]$: 0.96 km; over the interval $[1, 2.5]$: 0.5 km

3. 28.5 cm; The figure below is a graph of the rainfall as a function of time. The area of the shaded region represents the total rainfall.

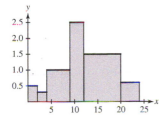

5. $L_5 = 46$; $R_5 = 44$

7. (a) $L_6 = 16.5$; $R_6 = 19.5$

(b) Via geometry (see figure below), the exact area is $A = 18$. Thus, L_6 underestimates the true area ($L_6 - A = -1.5$), while R_6 overestimates the true area ($R_6 - A = +1.5$).

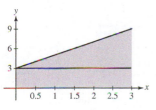

9. $R_3 = 32$; $L_3 = 20$; the area under the graph is larger than L_3 but smaller than R_3.

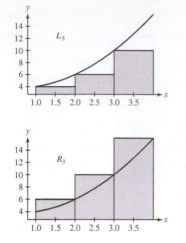

11. $R_3 = 2.5$; $M_3 = 2.875$; $L_6 = 3.4375$

13. (a) $L_4 = 1.75$ **(b)** $R_4 = 3.75$ **(c)** The actual area A under the curve $f(x) = x^2$ over the interval $[0, 2]$ satisfies $L_4 < A < R_4$.

15. $R_3 = \dfrac{16}{3}$ **17.** $M_6 = 87$ **19.** $M_5 \approx 1.30$

21. $L_4 \approx 0.410236$ **23.** $\displaystyle\sum_{k=4}^{8} k^7$ **25.** $\displaystyle\sum_{k=2}^{5} (2^k + 2)$

27. $\displaystyle\sum_{i=1}^{n} \dfrac{i}{(i+1)(i+2)}$

29. (a) 45 **(b)** 24 **(c)** 99

31. (a) -1 **(b)** 13 **(c)** 12

33. 15,050 **35.** 352,800 **37.** 1,093,350 **39.** 41,650

41. $-123,165$ **43.** $\dfrac{1}{2}$ **45.** $\dfrac{1}{3}$

47. 18; the region under the graph is a triangle with base 2 and height 18.

49. 12; the region under the curve is a trapezoid with base width 4 and heights 2 and 4.

51. 2; the region under the curve over $[0, 2]$ is a triangle with base and height 2.

53. $\lim_{N\to\infty} R_N = 16$ **55.** $R_N = \dfrac{1}{3} + \dfrac{1}{2N} + \dfrac{1}{6N^2}; \dfrac{1}{3}$

57. $R_N = 222 + \dfrac{189}{N} + \dfrac{27}{N^2}; 222$ **59.** $R_N = 2 + \dfrac{6}{N} + \dfrac{8}{N^2}; 2$

61. $R_N = (b-a)(2a+1) + (b-a)^2 + \dfrac{(b-a)^2}{N};$
$(b^2 + b) - (a^2 + a)$

63. The area between the graph of $f(x) = x^4$ and the x-axis over the interval $[0, 1]$

65. The area between the graph of $y = e^x$ and the x-axis over the interval $[-2, 3]$

67. $\displaystyle\lim_{N\to\infty} R_N = \lim_{N\to\infty} \dfrac{\pi}{N} \sum_{k=1}^{N} \sin\left(\dfrac{k\pi}{N}\right)$

69. $\displaystyle\lim_{N\to\infty} L_N = \lim_{N\to\infty} \dfrac{4}{N} \sum_{j=0}^{N-1} \sqrt{15 + \dfrac{8j}{N}}$

71. $\displaystyle\lim_{N\to\infty} M_N = \lim_{N\to\infty} \dfrac{1}{2N} \sum_{j=1}^{N} \tan\left(\dfrac{1}{2} + \dfrac{1}{2N}\left(j - \dfrac{1}{2}\right)\right)$

73. Represents the area between the graph of $y = f(x) = \sqrt{1 - x^2}$ and the x-axis over the interval $[0, 1]$. This is the portion of the circular disk $x^2 + y^2 \le 1$ that lies in the first quadrant. Accordingly, its area is $\dfrac{\pi}{4}$.

75. Of the three approximations, R_N is the least accurate, and then L_N and finally M_N are the most accurate.

77. The area A under the curve is somewhere between $L_4 \approx 0.518$ and $R_4 \approx 0.768$.

79. f is increasing over the interval $[0, \pi/2]$, so $0.79 \approx L_4 \le A \le R_4 \approx 1.18$.

81. $L_{100} = 0.793988$; $R_{100} = 0.80399$; $L_{200} = 0.797074$; $R_{200} = 0.802075$; thus, $A = 0.80$ to two decimal places.

83. (a) Let $f(x) = e^x$ on $[0, 1]$. With $n = N$, $\Delta x = (1 - 0)/N = 1/N$ and

$$x_j = a + j\Delta x = \dfrac{j}{N}$$

for $j = 0, 1, 2, \ldots, N$. Therefore,

$$L_N = \Delta x \sum_{j=0}^{N-1} f(x_j) = \dfrac{1}{N} \sum_{j=0}^{N-1} e^{j/N}$$

(b) Applying Eq. (8) with $r = e^{1/N}$, we have

$$L_N = \dfrac{1}{N} \dfrac{(e^{1/N})^N - 1}{e^{1/N} - 1} = \dfrac{e - 1}{N(e^{1/N} - 1)}$$

(c) $A = e - 1$

85.

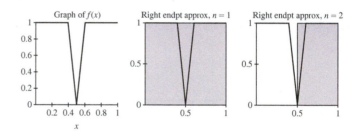

87. When f' is large, the graph of f is steeper and hence there is more gap between f and L_N or R_N.

91. $N > 30,000$

Section 5.2 Preliminary Questions

1. 2

2. (a) False. $\int_a^b f(x)\,dx$ is the *signed* area between the graph and the x-axis.

(b) True **(c)** True

3. Because $\cos(\pi - x) = -\cos x$, the "negative" area between the graph of $y = \cos x$ and the x-axis over $[\frac{\pi}{2}, \pi]$ exactly cancels the "positive" area between the graph and the x-axis over $[0, \frac{\pi}{2}]$.

4. $\displaystyle\int_{-1}^{-5} 8\,dx$

Section 5.2 Exercises

1. The region bounded by the graph of $y = 2x$ and the x-axis over the interval $[-3, 3]$ consists of two right triangles. One has area $\frac{1}{2}(3)(6) = 9$ below the axis, and the other has area $\frac{1}{2}(3)(6) = 9$ above the axis. Hence,

$$\int_{-3}^{3} 2x\,dx = 9 - 9 = 0$$

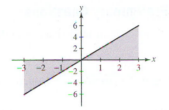

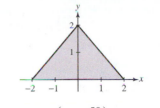

3. The region bounded by the graph of $y = 3x + 4$ and the x-axis over the interval $[-2, 1]$ consists of two right triangles. One has area $\frac{1}{2}(\frac{2}{3})(2) = \frac{2}{3}$ below the axis, and the other has area $\frac{1}{2}(\frac{7}{3})(7) = \frac{49}{6}$ above the axis. Hence,

$$\int_{-2}^{1} (3x + 4)\, dx = \frac{49}{6} - \frac{2}{3} = \frac{15}{2}$$

11. (a) $\displaystyle\lim_{N \to \infty} R_N = \lim_{N \to \infty} \left(30 - \frac{50}{N} \right) = 30$

(b) The region bounded by the graph of $y = 8 - x$ and the x-axis over the interval $[0, 10]$ consists of two right triangles. One triangle has area $\frac{1}{2}(8)(8) = 32$ above the axis, and the other has area $\frac{1}{2}(2)(2) = 2$ below the axis. Hence,

$$\int_{0}^{10} (8 - x)\, dx = 32 - 2 = 30$$

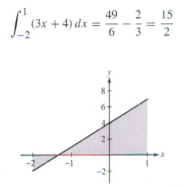

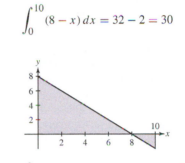

5. The region bounded by the graph of $y = 7 - x$ and the x-axis over the interval $[6, 8]$ consists of two right triangles. One triangle has area $\frac{1}{2}(1)(1) = \frac{1}{2}$ above the axis, and the other has area $\frac{1}{2}(1)(1) = \frac{1}{2}$ below the axis. Hence,

$$\int_{6}^{8} (7 - x)\, dx = \frac{1}{2} - \frac{1}{2} = 0$$

13. (a) $-\dfrac{\pi}{2}$ **(b)** $\dfrac{3\pi}{2}$

15. $\displaystyle\int_{0}^{3} g(t)\, dt = \frac{3}{2}; \int_{3}^{5} g(t)\, dt = 0$

17. The partition P is defined by

$$x_0 = 0 \quad < \quad x_1 = 1 \quad < \quad x_2 = 2.5 \quad < \quad x_3 = 3.2 \quad < \quad x_4 = 5$$

The set of sample points is given by
$C = \{c_1 = 0.5, c_2 = 2, c_3 = 3, c_4 = 4.5\}$. Finally, the value of the Riemann sum is

$$34.25(1 - 0) + 20(2.5 - 1) + 8(3.2 - 2.5) + 15(5 - 3.2) = 96.85$$

19. $R(f, P, C) = 1.59$; here is a sketch of the graph of f and the rectangles.

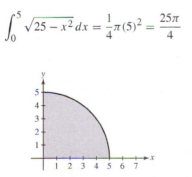

7. The region bounded by the graph of $y = \sqrt{25 - x^2}$ and the x-axis over the interval $[0, 5]$ is one-quarter of a circle of radius 5. Hence,

$$\int_{0}^{5} \sqrt{25 - x^2}\, dx = \frac{1}{4}\pi (5)^2 = \frac{25\pi}{4}$$

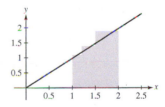

21. $R(f, P, C) = 44.625$; here is a sketch of the graph of f and the rectangles.

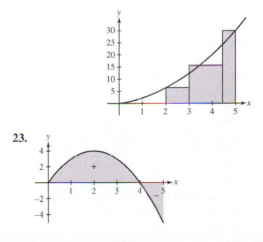

9. The region bounded by the graph of $y = 2 - |x|$ and the x-axis over the interval $[-2, 2]$ is a triangle above the axis with base 4 and height 2. Consequently,

$$\int_{-2}^{2} (2 - |x|)\, dx = \frac{1}{2}(2)(4) = 4$$

23.

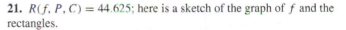

25.

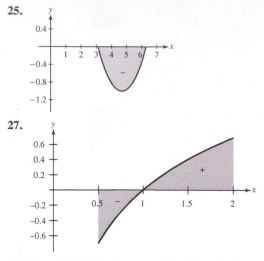

27.

29. The integrand is always positive. The integral must therefore be positive, since the signed area has only a positive part.

31. Since $y = x$ is an increasing function on $[0, 2\pi]$, and $\sin x \geq 0$ on $[0, \pi]$ but $\sin x \leq 0$ on $[\pi, 2\pi]$, it follows that the area below the x-axis is enclosed by $y = x \sin x$ on the interval $[0, \pi]$. Hence, the total area on $[0, 2\pi]$ will be negative, so the definite integral will be negative as well.

33. 36   **35.** 243   **37.** $-\dfrac{2}{3}$ **39.** $\dfrac{196}{3}$ **41.** $\dfrac{1}{3}a^3 - \dfrac{1}{2}a^2 + \dfrac{5}{6}$

43. 17   **45.** -12 **47.** No.   **49.** $\dfrac{81}{4}$ **51.** $-\dfrac{63}{4}$ **53.** 7   **55.** 8

57. -7 **59.** $\displaystyle\int_0^7 f(x)\,dx$ **61.** $\displaystyle\int_5^9 f(x)\,dx$ **63.** $\dfrac{4}{5}$ **65.** $-\dfrac{35}{2}$

67. When $f(x)$ takes on both positive and negative values on $[a, b]$, $\int_a^b f(x)\,dx$ represents the signed area between f and the x-axis, whereas $\int_a^b |f(x)|\,dx$ represents the total (unsigned) area between f and the x-axis. Any negatively signed areas that were part of $\int_a^b f(x)\,dx$ are regarded as positive areas in $\int_a^b |f(x)|\,dx$.

69. $[-1, \sqrt{2}]$ or $[-\sqrt{2}, 1]$ **71.** 9   **73.** $\dfrac{1}{2}$

75. On the interval $[0, 1]$, $x^5 \leq x^4$; on the other hand, $x^4 \leq x^5$ for $x \in [1, 2]$.

77. $y = \sin x$ is increasing on $[0.2, 0.3]$. Accordingly, for $0.2 \leq x \leq 0.3$, we have

$$m = 0.198 \leq 0.19867 \approx \sin 0.2 \leq \sin x \leq \sin 0.3$$

$$\approx 0.29552 \leq 0.296 = M$$

Therefore, by the Comparison Theorem, we have

$$0.0198 = m(0.3 - 0.2) = \int_{0.2}^{0.3} m\,dx \leq \int_{0.2}^{0.3} \sin x\,dx \leq \int_{0.2}^{0.3} M\,dx$$

$$= M(0.3 - 0.2) = 0.0296$$

79. f is decreasing and nonnegative on the interval $[\pi/4, \pi/2]$. Therefore, $0 \leq f(x) \leq f(\pi/4) = \frac{2\sqrt{2}}{\pi}$ for all x in $[\pi/4, \pi/2]$.

81. The assertion $f'(x) \leq g'(x)$ is false. Consider $a = 0, b = 1$, $f(x) = x, g(x) = 2$. $f(x) \leq g(x)$ for all x in the interval $[0, 1]$, but $f'(x) = 1$, while $g'(x) = 0$ for all x.

83. If f is an odd function, then $f(-x) = -f(x)$ for all x. Accordingly, for every positively signed area in the right half-plane where f is above the x-axis, there is a corresponding negatively signed area in the left half-plane where f is below the x-axis. Similarly, for every negatively signed area in the right half-plane where f is below the x-axis, there is a corresponding positively signed area in the left half-plane where f is above the x-axis.

Section 5.3 Preliminary Questions

1. Any constant function is an antiderivative for the function $f(x) = 0$.

2. No difference   **3.** No

4. (a) False. Even if $f(x) = g(x)$, the antiderivatives F and G may differ by an additive constant.

(b) True. This follows from the fact that the derivative of any constant is 0.

(c) False. If the functions f and g are different, then the antiderivatives F and G differ by a linear function: $F(x) - G(x) = ax + b$ for some constants a and b.

5. No

Section 5.3 Exercises

1. $6x^3 + C$ **3.** $\dfrac{2}{5}x^5 - 8x^3 + 12 \ln |x| + C$

5. $2 \sin x + 9 \cos x + C$ **7.** $12e^x + 5x^{-1} + C$

9. (a) (ii)   (b) (iii)   (c) (i)   (d) (iv)

11. $4x - 9x^2 + C$ **13.** $\dfrac{11}{5}t^{5/11} + C$ **15.** $3t^6 - 2t^5 - 14t^2 + C$

17. $5z^{1/5} - \dfrac{3}{5}z^{5/3} + \dfrac{4}{9}z^{9/4} + C$ **19.** $\dfrac{3}{2}x^{2/3} + C$ **21.** $-\dfrac{18}{t^2} + C$

23. $\dfrac{2}{5}t^{5/2} + \dfrac{1}{2}t^2 + \dfrac{2}{3}t^{3/2} + t + C$

25. $\dfrac{1}{2}x^2 + 3 \ln |x| + 4x^{-1} + C$ **27.** $12 \sec x + C$

29. $-\csc t + C$ **31.** $\dfrac{1}{3}\tan(3\theta) + C$ **33.** $\dfrac{25}{3}\tan(3z) + C$

35. $\dfrac{1}{3}\sin(3\theta) - 2\tan\left(\dfrac{\theta}{4}\right) + C$ **37.** $\dfrac{3}{5}e^{5x} + C$

39. $4x^2 + 2e^{5-2x} + C$

41. Graph (B) does not have the same local extrema as indicated by $y = f(x)$ and therefore is *not* an antiderivative of $y = f(x)$.

43. $\dfrac{d}{dx}\left(\dfrac{1}{7}(x + 13)^7 + C\right) = (x + 13)^6$

45. $\dfrac{d}{dx}\left(\dfrac{1}{12}(4x + 13)^3 + C\right) = \dfrac{1}{4}(4x + 13)^2(4) = (4x + 13)^2$

47. $y = \dfrac{1}{4}x^4 + 4$ **49.** $y = t^2 + 3t^3 - 2$ **51.** $y = \dfrac{2}{3}t^{3/2} + \dfrac{1}{3}$

53. $y = \dfrac{1}{12}(3x + 2)^4 - \dfrac{1}{3}$ **55.** $y = 1 - \cos x$

57. $y = 3 + \dfrac{1}{5}\sin 5x$ **59.** $y = e^x - e^2$ **61.** $y = -3e^{12-3t} + 10$

63. $f'(x) = 6x^2 + 1$; $f(x) = 2x^3 + x + 2$

65. $f'(x) = \frac{1}{4}x^4 - x^2 + x + 1$; $f(x) = \dfrac{1}{20}x^5 - \dfrac{1}{3}x^3 + \dfrac{1}{2}x^2 + x$

67. $f'(t) = -2t^{-1/2} + 2$; $f(t) = -4t^{1/2} + 2t + 4$

69. $f'(t) = \dfrac{1}{2}t^2 - \sin t + 2$; $f(t) = \dfrac{1}{6}t^3 + \cos t + 2t - 3$

71. The differential equation satisfied by $s(t)$ is

$$\frac{ds}{dt} = v(t) = 6t^2 - t$$

and the associated initial condition is $s(1) = 0$;

$$s(t) = 2t^3 - \frac{1}{2}t^2 - \frac{3}{2}.$$

73. $v_y = -49$ m/s

75. The differential equation satisfied by $s(t)$ is

$$\frac{ds}{dt} = v(t) = \sin(\pi t/2)$$

and the associated initial condition is $s(0) = 0$;

$$s(t) = \frac{2}{\pi}(1 - \cos(\pi t/2)).$$

77. 6.25 s; 78.125 m   **79.** 300 m/s   **83.** $c_1 = 1$ and $c_2 = -1$

85. (a) By the Chain Rule, we have

$$\frac{d}{dx}\left(\frac{1}{2}F(2x)\right) = \frac{1}{2}F'(2x)\cdot 2 = F'(2x) = f(2x)$$

Thus, $y = \frac{1}{2}F(2x)$ is an antiderivative of $y = f(2x)$.

(b) $\dfrac{1}{k}F(kx) + C$

Section 5.4 Preliminary Questions

1. (a) 4   (b) The signed area between $y = f(x)$ and the x-axis

2. 3

3. (a) False. The FTC I is valid for continuous functions.

(b) False. The FTC I works for any antiderivative of the integrand.

(c) False. If you cannot find an antiderivative of the integrand, you cannot use the FTC I to evaluate the definite integral, but the definite integral may still exist.

4. 0

Section 5.4 Exercises

1. $A = \dfrac{1}{3}$

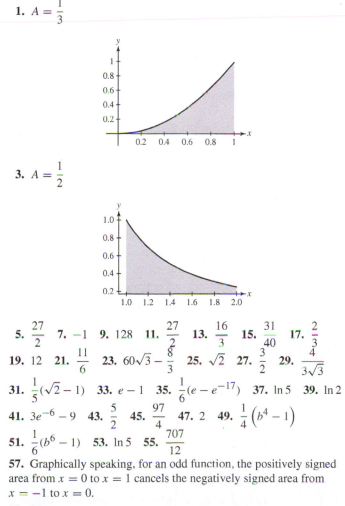

3. $A = \dfrac{1}{2}$

5. $\dfrac{27}{2}$ **7.** -1 **9.** 128   **11.** $\dfrac{27}{2}$ **13.** $\dfrac{16}{3}$ **15.** $\dfrac{31}{40}$ **17.** $\dfrac{2}{3}$

19. 12   **21.** $\dfrac{11}{6}$ **23.** $60\sqrt{3} - \dfrac{8}{3}$ **25.** $\sqrt{2}$ **27.** $\dfrac{3}{2}$ **29.** $\dfrac{4}{3\sqrt{3}}$

31. $\dfrac{1}{5}(\sqrt{2} - 1)$ **33.** $e - 1$ **35.** $\dfrac{1}{6}(e - e^{-17})$ **37.** $\ln 5$ **39.** $\ln 2$

41. $3e^{-6} - 9$ **43.** $\dfrac{5}{2}$ **45.** $\dfrac{97}{4}$ **47.** 2   **49.** $\dfrac{1}{4}\left(b^4 - 1\right)$

51. $\dfrac{1}{6}(b^6 - 1)$ **53.** $\ln 5$ **55.** $\dfrac{707}{12}$

57. Graphically speaking, for an odd function, the positively signed area from $x = 0$ to $x = 1$ cancels the negatively signed area from $x = -1$ to $x = 0$.

59. 24

61. $\int_0^1 x^n\, dx$ represents the area between the positive curve $f(x) = x^n$ and the x-axis over the interval $[0, 1]$. This area gets smaller as n gets larger, as is readily evident in the following graph, which shows curves for several values of n.

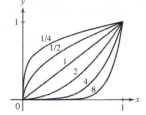

67. Let $a > b$ be real numbers, and let $f(x)$ be such that $|f'(x)| \leq K$ for $x \in [a, b]$. By FTC,

$$\int_a^x f'(t)\, dt = f(x) - f(a)$$

Since $f'(x) \geq -K$ for all $x \in [a, b]$, we get

$$f(x) - f(a) = \int_a^x f'(t)\, dt \geq -K(x - a)$$

Since $f'(x) \leq K$ for all $x \in [a, b]$, we get

$$f(x) - f(a) = \int_a^x f'(t)\, dt \leq K(x - a)$$

Combining these two inequalities yields

$$-K(x - a) \leq f(x) - f(a) \leq K(x - a)$$

so that, by definition,

$$|f(x) - f(a)| \leq K|x - a|$$

Section 5.5 Preliminary Questions

1. (a) No   (b) Yes

2. (c)

3. Yes. All continuous functions have an antiderivative, namely,

$$\int_a^x f(t)\, dt.$$

4. (b), (e), and (f)

Section 5.5 Exercises

1. $A(x) = \displaystyle\int_{-2}^x (2t + 4)\, dt = (x + 2)^2$

3. $G(1) = 0;\ G'(1) = -1$ and $G'(2) = 2;\ G(x) = \dfrac{1}{3}x^3 - 2x + \dfrac{5}{3}$

5. $G(1) = 0;\ G'(0) = 0$ and $G'(\frac{\pi}{4}) = 1$

7. $\dfrac{1}{5}x^5 - \dfrac{32}{5}$ **9.** $1 - \cos x$ **11.** $\dfrac{1}{3}e^{3x} - \dfrac{1}{3}e^{12}$ **13.** $\dfrac{1}{2}x^4 - \dfrac{1}{2}$

15. $-e^{-9x-2} + e^{-3x}$ **17.** $F(x) = \displaystyle\int_5^x \sqrt{t^3 + 1}\, dt$

19. $F(x) = \displaystyle\int_0^x \sec t\, dt$ **21.** $x^5 - 9x^3$ **23.** $\sec(5t - 9)$

25. (a) $A(2) = 4;\ A(3) = 6.5;\ A'(2) = 2$ and $A'(3) = 3$
(b)

$$A(x) = \begin{cases} 2x, & 0 \leq x < 2 \\ \frac{1}{2}x^2 + 2, & 2 \leq x \leq 4 \end{cases}$$

29. $\dfrac{2x^3}{x^2 + 1}$ **31.** $-\cos^4 s \sin s$ **33.** $2x \tan(x^2) - \dfrac{\tan(\sqrt{x})}{2\sqrt{x}}$

35. The minimum value of $A(x)$ is $A(1.5) = -1.25$; the maximum value of A is $A(4.5) = 1.25$.

37. $A(x) = (x - 2) - 1$ and $B(x) = (x - 2)$

39. **(a)** A does not have a local maximum at P.

(b) A has a local minimum at R.

(c) A has a local maximum at S.

(d) True

41. $g(x) = 2x + 1$; $c = 2$ or $c = -3$

43. **(a)** If $x = c$ is an inflection point of A, then $A''(c) = f'(c) = 0$.

(b) If A is concave up, then $A''(x) > 0$. Since A is the area function associated with f, $A'(x) = f(x)$ by FTC II, so $A''(x) = f'(x)$. Therefore, $f'(x) > 0$, so f is increasing.

(c) If A is concave down, then $A''(x) < 0$. Since A is the area function associated with $f(x)$, $A'(x) = f(x)$ by FTC II, so $A''(x) = f'(x)$. Therefore, $f'(x) < 0$, so f is decreasing.

45. **(a)** A is increasing on the intervals $(0, 4)$ and $(8, 12)$ and is decreasing on the intervals $(4, 8)$ and $(12, \infty)$.

(b) Local minimum: $x = 8$; local maximum: $x = 4$ and $x = 12$

(c) A has inflection points at $x = 2$, $x = 6$, and $x = 10$.

(d) A is concave up on the intervals $(0, 2)$ and $(6, 10)$ and is concave down on the intervals $(2, 6)$ and $(10, \infty)$.

47. The graph of one such function is

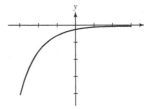

49. Smallest positive critical point: $x = (\pi/2)^{2/3}$ corresponds to a local maximum; smallest positive inflection point: $x = \pi^{2/3}$, $y = F(x)$ changes from concave down to concave up.

51. **(a)** Then by the FTC, Part II, $A'(x) = f(x)$ and thus $y = A(x)$ and $y = F(x)$ are both antiderivatives of $y = f(x)$. Hence, $F(x) = A(x) + C$ for some constant C.

(b)

$$F(b) - F(a) = (A(b) + C) - (A(a) + C) = A(b) - A(a)$$

$$= \int_a^b f(t)\,dt - \int_a^a f(t)\,dt$$

$$= \int_a^b f(t)\,dt - 0 = \int_a^b f(t)\,dt$$

which proves the FTC, Part I.

53. Write

$$\int_{u(x)}^{v(x)} f(x)\,dx = \int_{u(x)}^0 f(x)\,dx + \int_0^{v(x)} f(x)\,dx$$

$$= \int_0^{v(x)} \cdot f(x)\,dx - \int_0^{u(x)} f(x)\,dx$$

Then, by the Chain Rule and the FTC,

$$\frac{d}{dx} \int_{u(x)}^{v(x)} f(x)\,dx = \frac{d}{dx} \int_0^{v(x)} f(x)\,dx - \frac{d}{dx} \int_0^{u(x)} f(x)\,dx$$

$$= f(v(x))v'(x) - f(u(x))u'(x)$$

Section 5.6 Preliminary Questions

1. The total drop in temperature of the metal object in the first T minutes after being submerged in the cold water

2. 560 km

3. Quantities **(a)** and **(c)** would naturally be represented as derivatives; quantities **(b)** and **(d)** would naturally be represented as integrals.

Section 5.6 Exercises

1. 15,250 gal   **3.** 3,660,000   **5.** 33 m   **7.** 3.675 m

9. Displacement: 10 m; distance: 26 m

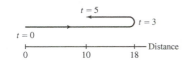

11. Displacement: 0 m; distance: 1 m

13. 39 m/s   **15.** 9200 cars

17. Total cost: \$650; average cost of first 10: \$37.50; average cost of last 10: \$27.50

19. 112.5 ft

21. The area under the graph in Figure 5 represents the total power consumption over one day in California; 3.627×10^{14} joules

23. **(a)** 2.916×10^{10}

(b) Approximately 240,526 asteroids of diameter 50 km

25. $\int_0^{365} R(t)\,dt \approx 605.05$ billion ft^3

27. $100 \le t \le 150$: 404.968 families; $350 \le t \le 400$: 245.812 families

29. The particle's velocity is $v(t) = s'(t) = t^{-2}$, an antiderivative for which is $F(t) = -t^{-1}$. Hence, the particle's position at time t is

$$s(t) = \int_1^t s'(u)\,du = F(u)\Big|_1^t = F(t) - F(1) = 1 - \frac{1}{t} < 1$$

for all $t \ge 1$. Thus, the particle will never pass $x = 1$, which implies it will never pass $x = 2$ either.

31. **(a)** $CS = \int_0^{q^*} [D(q) - p^*]\,dq$ **(b)** $PS = \int_0^{q^*} [p^* - S(q)]\,dq$

Section 5.7 Preliminary Questions

1. **(a)** and **(b)**

2. **(a)** $u(x) = x^2 + 9$ **(b)** $u(x) = x^3$ **(c)** $u(x) = \cos x$

3. **(c)**

Section 5.7 Exercises

1. $du = (3x^2 - 2x)\,dx$ **3.** $du = -2x \sin(x^2)\,dx$

5. $du = 4e^{4x+1}\,dx$

7. $\int (x - 7)^3\,dx = \int u^3\,du = \frac{1}{4}u^4 + C = \frac{1}{4}(x - 7)^4 + C$

9. $\int (3t - 4)^5\,dt = \int \frac{1}{3}u^5\,du = \frac{1}{18}u^6 + C = \frac{1}{18}(3t - 5)^6 + C$

11.

$\int t\sqrt{t^2 + 1}\,dt = \frac{1}{2}\int u^{1/2}\,du = \frac{1}{3}u^{3/2} + C = \frac{1}{3}(t^2 + 1)^{3/2} + C$

13. $\int \dfrac{t^3}{(4-2t^4)^{11}} dt = -\dfrac{1}{8}\int u^{-11} du = \dfrac{1}{80}u^{-10} + C = \dfrac{1}{80}(4-2t^4)^{-10} + C$

15.

$$\int x(x+1)^9 dx = \int (u-1)u^9 du = \int (u^{10} - u^9) du$$

$$= \dfrac{1}{11}u^{11} - \dfrac{1}{10}u^{10} + C = \dfrac{1}{11}(x+1)^{11} - \dfrac{1}{10}(x+1)^{10} + C$$

17.

$$\int x^2\sqrt{x+1}\,dx = \int (u-1)^2 u^{1/2}\,du = \int (u^{5/2} - 2u^{3/2} + u^{1/2})\,du$$

$$= \dfrac{2}{7}u^{7/2} - \dfrac{4}{5}u^{5/2} + \dfrac{2}{3}u^{3/2} + C$$

$$= \dfrac{2}{7}(x+1)^{7/2} - \dfrac{4}{5}(x+1)^{5/2} + \dfrac{2}{3}(x+1)^{3/2} + C$$

19. $\int \sin^2\theta\cos\theta\,d\theta = \int u^2\,du = \dfrac{1}{3}u^3 + C = \dfrac{1}{3}\sin^3\theta + C$

21. $\int xe^{-x^2}\,dx = -\dfrac{1}{2}\int e^u\,du = -\dfrac{1}{2}e^u + C = -\dfrac{1}{2}e^{-x^2} + C$

23. $\int \dfrac{(\ln x)^2}{x}\,dx = \int u^2\,du = \dfrac{1}{3}u^3 + C = \dfrac{1}{3}(\ln x)^3 + C$

25. $u = x^4;\ \dfrac{1}{4}\sin(x^4) + C$ **27.** $u = x^{3/2};\ \dfrac{2}{3}\sin(x^{3/2}) + C$

29. $\dfrac{1}{40}(4x+5)^{10} + C$ **31.** $2\sqrt{t+12} + C$

33. $-\dfrac{1}{4(x^2+2x)^2} + C$ **35.** $\sqrt{x^2+9} + C$ **37.** $\dfrac{1}{3}(x^3+x)^3 + C$

39. $\dfrac{1}{36}(3x+8)^{12} + C$ **41.** $\dfrac{2}{9}(x^3+1)^{3/2} + C$

43. $-\dfrac{1}{2}(x+5)^{-2} + C$ **45.** $\dfrac{1}{39}(z^3+1)^{13} + C$

47. $\dfrac{4}{9}(x+1)^{9/4} + \dfrac{4}{5}(x+1)^{5/4} + C$ **49.** $\dfrac{1}{3}\cos(8-3\theta) + C$

51. $2\sin\sqrt{t} + C$ **53.** $\dfrac{1}{4}\ln|\sec(4\theta+9)| + C$ **55.** $\ln|\sin x| + C$

57. $\dfrac{1}{4}\tan(4x+9) + C$ **59.** $2\tan(\sqrt{x}) + C$

61. $-\dfrac{1}{6}(\cos 4x + 1)^{3/2} + C$ **63.** $\dfrac{1}{2}(\sec\theta - 1)^2 + C$

65. $\dfrac{1}{14}e^{14x-7} + C$ **67.** $-\dfrac{1}{3(e^x+1)^3} + C$ **69.** $-\dfrac{1}{e^t+1} + C$

71. $\dfrac{1}{5}(\ln x)^5 + C$ **73.** $-\ln|\cos(\ln x)| + C$

75. $-\dfrac{2}{1+\sqrt{x}} + \dfrac{1}{(1+\sqrt{x})^2} + C$

77. With $u = \sin x$, $\dfrac{1}{2}\sin^2 x + C_1$; with $u = \cos x$, $-\dfrac{1}{2}\cos^2 x + C_2$; the two results differ by a constant.

79. $u = \pi$ and $u = 4\pi$ **81.** 136   **83.** $3 - \sqrt{5}$ **85.** $\dfrac{3}{16}$ **87.** $\dfrac{98}{3}$

89. $\dfrac{243}{4}$ **91.** $\dfrac{1}{2}$ **93.** $\dfrac{1}{2}\ln(\sec 1)$ **95.** $\dfrac{1}{4}$ **97.** $\dfrac{20}{3}\sqrt{5} - \dfrac{32}{5}\sqrt{3}$

99. (a) The probability that $v \in [0, b]$ is

$$\int_0^b \dfrac{1}{32}ve^{-v^2/64}\,dv$$

Let $u = -v^2/64$. Then $du = -v/32\,dv$ and

$$\int_0^b \dfrac{1}{32}ve^{-v^2/64}\,dv = -\int_0^{-b^2/64} e^u\,du$$

$$= -e^u\Big|_0^{-b^2/64} = -e^{-b^2/64} + 1$$

(b) $e^{-1/16} - e^{-25/64}$

101. $\dfrac{1}{4}f(x)^4 + C$ **103.** $\ln|f(x)| + C$

105. Let $u = \sin\theta$. Then $u(\pi/6) = 1/2$ and $u(0) = 0$, as required. Furthermore, $du = \cos\theta\,d\theta$, so

$$d\theta = \dfrac{du}{\cos\theta}$$

If $\sin\theta = u$, then $u^2 + \cos^2\theta = 1$, so $\cos\theta = \sqrt{1-u^2}$. Therefore, $d\theta = du/\sqrt{1-u^2}$. This gives

$$\int_0^{\pi/6} f(\sin\theta)\,d\theta = \int_0^{1/2} f(u)\dfrac{1}{\sqrt{1-u^2}}\,du$$

107. $I = \pi/4$

Section 5.8 Preliminary Questions

1. (a) $b = 3$ **(b)** $b = e^3$

2. $b = \sqrt{3}$ **3. (b)**   **4.** $x = 4u$

Section 5.8 Exercises

1. $\ln 9$ **3.** 3   **5.** $\dfrac{1}{3}\ln 4$ **7.** $\dfrac{\pi}{12}$ **9.** $\dfrac{\pi}{6}$

11. Let $u = x/3$. Then $x = 3u$, $dx = 3\,du$, $9 + x^2 = 9(1+u^2)$, and

$$\int \dfrac{dx}{9+x^2} = \int \dfrac{3\,du}{9(1+u^2)} = \dfrac{1}{3}\int \dfrac{du}{1+u^2}$$

$$= \dfrac{1}{3}\tan^{-1}u + C = \dfrac{1}{3}\tan^{-1}\dfrac{x}{3} + C$$

13. $\dfrac{\pi}{3\sqrt{3}}$ **15.** $\dfrac{1}{4}\sin^{-1}(4t) + C$ **17.** $\dfrac{1}{\sqrt{3}}\sin^{-1}\sqrt{\dfrac{3}{5}}t + C$

19. $\dfrac{1}{\sqrt{3}}\sec^{-1}(2x) + C$ **21.** $\dfrac{1}{2}\sec^{-1}x^2 + C$

23. $\dfrac{\pi}{4} - \tan^{-1}(1/2)$ **25.** $\dfrac{(\tan^{-1}x)^2}{2} + C$ **27.** $\dfrac{2}{\ln 3}$ **29.** $\dfrac{1}{\ln 2}$

31. $-\dfrac{1}{\ln 9}\cos(9^x) + C$ **33.** $\dfrac{1}{2}e^{y^2} + C$ **35.** $\dfrac{1}{4}\sqrt{4x^2+9} + C$

37. $-\dfrac{7^{-x}}{\ln 7} + C$ **39.** $\dfrac{1}{8}\tan^8\theta + C$ **41.** $-\sqrt{7-t^2} + C$

43. $\dfrac{3}{2}\ln(x^2+4) + \tan^{-1}(x/2) + C$ **45.** $\dfrac{1}{2}\sin^{-1}\left(\dfrac{2}{3}x\right) + C$

47. $-e^{-x} - 2x^2 + C$ **49.** $e^x - \dfrac{e^{3x}}{3} + C$

51. $-\sqrt{4-x^2} + 5\sin^{-1}(x/2) + C$ **53.** $\sin(e^x) + C$

55. $\dfrac{1}{4}\sin^{-1}\left(\dfrac{4x}{3}\right) + C$ **57.** $\dfrac{e^{7x}}{7} + \dfrac{3e^{5x}}{5} + e^{3x} + e^x + C$

59. $\dfrac{1}{3}\ln|x^3 + 2| + C$ **61.** $\ln|\sin x| + C$ **63.** $\dfrac{1}{8}(4\ln x + 5)^2 + C$

65. $\dfrac{3^{x^2}}{2\ln 3} + C$ **67.** $\dfrac{(\ln(\sin x))^2}{2} + C$

69. $\dfrac{2}{7}(t-3)^{7/2} + \dfrac{12}{5}(t-3)^{5/2} + 6(t-3)^{3/2} + C$

71. The definite integral $\int_0^x \sqrt{1-t^2}\,dt$ represents the area of the region under the upper half of the unit circle from 0 to x. The region consists of a sector of the circle and a right triangle. The sector has a central angle of $\dfrac{\pi}{2} - \theta$, where $\cos\theta = x$, and the right triangle has a base of length x and a height of $\sqrt{1-x^2}$.

73. Show that $\dfrac{d}{dt}\left(\sqrt{1-t^2} + t\sin^{-1}t\right) = \sin^{-1}t$.

75. Integrating both sides of the inequality $e^t \geq 1$ yields

$$\int_0^x e^t \, dt = e^x - 1 \geq x \quad \text{or} \quad e^x \geq 1 + x$$

Integrating both sides of this new inequality then gives

$$\int_0^x e^t \, dt = e^x - 1 \geq x + x^2/2 \quad \text{or} \quad e^x \geq 1 + x + x^2/2$$

Finally, integrating both sides again gives

$$\int_0^x e^t \, dt = e^x - 1 \geq x + x^2/2 + x^3/6$$

or

$$e^x \geq 1 + x + x^2/2 + x^3/6$$

as requested.

77. By Exercise 76, $e^x \geq 1 + x + \frac{x^2}{2} + \frac{x^3}{6}$. Thus,

$$\frac{e^x}{x^2} \geq \frac{1}{x^2} + \frac{1}{x} + \frac{1}{2} + \frac{x}{6} \geq \frac{x}{6}$$

Since $\lim_{x \to \infty} x/6 = \infty$, $\lim_{x \to \infty} e^x/x^2 = \infty$. More generally, by Exercise 75,

$$e^x \geq 1 + \frac{x^2}{2} + \cdots + \frac{x^{n+1}}{(n+1)!}$$

Thus,

$$\frac{e^x}{x^n} \geq \frac{1}{x^n} + \cdots + \frac{x}{(n+1)!} \geq \frac{x}{(n+1)!}$$

Since $\lim_{x \to \infty} \frac{x}{(n+1)!} = \infty$, $\lim_{x \to \infty} \frac{e^x}{x^n} = \infty$.

79. (a) The domain of G is $x > 0$ and, by part (i) of the previous exercise, the range of G is **R**. Now,

$$G'(x) = \frac{1}{x} > 0$$

for all $x > 0$. Thus, G is increasing on its domain, which implies that G has an inverse. The domain of the inverse is **R** and the range is $\{x : x > 0\}$. Let F denote the inverse of G.

(b) Let x and y be real numbers and suppose that $x = G(w)$ and $y = G(z)$ for some positive real numbers w and z. Then, using part (b) of the previous exercise,

$$F(x + y) = F(G(w) + G(z)) = F(G(wz)) = wz = F(x) + F(y)$$

(c) Let r be any real number. By part (k) of the previous exercise, $G(E^r) = r$. By definition of an inverse function, it then follows that $F(r) = E^r$.

(d) By the formula for the derivative of an inverse function,

$$F'(x) = \frac{1}{G'(F(x))} = \frac{1}{1/F(x)} = F(x)$$

81.

$$\lim_{n \to -1} \int_1^x t^n \, dt = \lim_{n \to -1} \frac{t^{n+1}}{n+1} \Big|_1^x = \lim_{n \to -1} \left(\frac{x^{n+1}}{n+1} - \frac{1^{n+1}}{n+1} \right)$$

$$= \lim_{n \to -1} \frac{x^{n+1} - 1}{n+1} = \lim_{n \to -1} (x^{n+1}) \ln x$$

$$= \ln x = \int_1^x t^{-1} \, dt$$

83. (a) Interpreting the graph with y as the independent variable, we see that the function is $x = e^y$. Integrating in y then gives the area of the shaded region as $\int_0^{\ln a} e^y \, dy$.

(b) We can obtain the area under the graph of $y = \ln x$ from $x = 1$ to $x = a$ by computing the area of the rectangle extending from $x = 0$ to $x = a$ horizontally and from $y = 0$ to $y = \ln a$ vertically and then subtracting the area of the shaded region. This yields

$$\int_1^a \ln x \, dx = a \ln a - \int_0^{\ln a} e^y \, dy$$

(c) By direct calculation,

$$\int_0^{\ln a} e^y \, dy = e^y \Big|_0^{\ln a} = a - 1$$

Thus,

$$\int_1^a \ln x \, dx = a \ln a - (a - 1) = a \ln a - a + 1$$

(d) Based on these results, it appears that

$$\int \ln x \, dx = x \ln x - x + C$$

Section 5.9 Preliminary Questions

1. Doubling time is inversely proportional to the growth constant. Consequently, the quantity with $k = 3.4$ doubles more rapidly.
2. It takes longer for the population to increase from one cell to two cells.
3. $\frac{dS}{dn} = -\ln 2 S(n)$ **4. (b)**
5. If the interest rate goes up, the present value of \$1 a year from now will decrease.

Section 5.9 Exercises

1. (a) 2000 bacteria initially   **(b)** $t = \frac{1}{1.3} \ln 5 \approx 1.24$ h
3. $f(t) = 5e^{t \ln 7}$
5. $N'(t) = \frac{\ln 2}{3} N(t)$; 1,048,576 molecules after 1 h
7. $y(t) = Ce^{-5t}$ for some constant C; $y(t) = 3.4e^{-5t}$
9. $y(t) = 1000e^{3(t-2)}$ **11.** 5.33 years
13. $k \approx 0.023$ h^{-1}; $P_0 \approx 332$
15. Double: 11.55 years; triple: 18.31 years; seven-fold: 32.43 years
17. One-half: 1.98 days; one-third: 3.14 days; one-tenth: 6.58 days
19. Set I   **21. (a)** 26.39 years   **(b)** 1969
23. 7600 years   **25.** 2.34×10^{-13} to 2.98×10^{-13} **27.** 2.55 h
29. (a) Yes, the graph looks like an exponential graph especially toward the latter years; $k \approx 0.369$ year^{-1}.
(b)

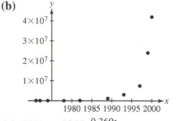

(c) $N(t) = 2250e^{0.369t}$
(d) The doubling time is $\ln 2/0.369 \approx 1.88$ years.
(e) $\approx 2.53 \times 10^{10}$ transistors
(f) No, you can't make a microchip smaller than an atom.
31. With $t_0 = 10$, the doubling time is then 24; with $t_0 = 20$, the doubling time is 44.
33. (a) $P(10) = \$4870.38$ **(b)** $P(10) = \$4902.71$
(c) $P(10) = \$4919.21$
35. (a) 1.0508   **(b)** 1.0513

37. $12,752.56

39. In 3 years:

(a) $PV = \$4176.35$

(b) $PV = \$3594.62$

In 5 years:

(a) $PV = \$3704.09$

(b) $PV = \$2884.75$

41. 9.16%

43. (a) The present value of the reduced labor costs is

$$7000(e^{-0.08} + e^{-0.16} + e^{-0.24} + e^{-0.32} + e^{-0.4}) = \$27,708.50$$

This is more than the $25,000 cost of the computer system, so the computer system should be purchased.

(b) The present value of the savings is

$$\$27,708.50 - \$25,000 = \$2708.50$$

45. $39,346.93   **47.** $41,906.75   **51.** $R = \$1200$

53. $71,460.53   **55.** $T = -\dfrac{1}{k}\ln\left(1 - \dfrac{d}{L}\right)$

57. $P(t) = 204e^{ae^{0.15t}}$ with $a \approx -0.02$; 136 rats after 20 months

59. For m-fold growth, $P(t) = mP_0$ for some t. Solving $mP_0 = P_0e^{kt}$ for t, we find $t = \dfrac{\ln m}{k}$.

61. Start by expressing

$$\ln\left(1 + \frac{x}{n}\right) = \int_1^{1+x/n} \frac{dt}{t}$$

Following the proof in the text, we note that

$$\frac{x}{n+x} \le \ln\left(1 + \frac{x}{n}\right) \le \frac{x}{n}$$

provided $x > 0$, while

$$\frac{x}{n} \le \ln\left(1 + \frac{x}{n}\right) \le \frac{x}{n+x}$$

when $x < 0$. Multiplying both sets of inequalities by n and passing to the limit as $n \to \infty$, the squeeze theorem guarantees that

$$\lim_{n\to\infty}\left(\ln\left(1 + \frac{x}{n}\right)\right)^n = x$$

Finally,

$$\lim_{n\to\infty}\left(1 + \frac{x}{n}\right)^n = e^x$$

63. (a) 9.38%

(b) In general,

$$P_0(1 + r/M)^{Mt} = P_0(1 + r_e)^t$$

so $(1 + r/M)^{Mt} = (1 + r_e)^t$ or $r_e = (1 + r/M)^M - 1$. If interest is compounded continuously, then $P_0e^{rt} = P_0(1 + r_e)^t$, so $e^{rt} = (1 + r_e)^t$ or $r_e = e^r - 1$.

(c) 11.63%

(d) 18.26%

Chapter 5 Review

1. $L_4 = \dfrac{23}{4}$; $M_4 = 7$

3. In general, R_N is larger than $\int_a^b f(x)\,dx$ on any interval $[a, b]$ over which $f(x)$ is increasing. Given the graph of f, we may take $[a, b] = [0, 2]$. In order for L_4 to be larger than $\int_a^b f(x)\,dx$, f must be decreasing over the interval $[a, b]$. We may therefore take $[a, b] = [2, 3]$.

5. $R_6 = \dfrac{625}{8}$

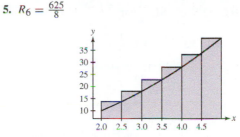

$M_6 = \dfrac{1127}{16}$

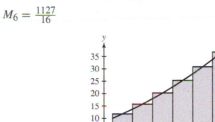

$L_6 = \dfrac{505}{8}$; The rectangles corresponding to this approximation are shown below.

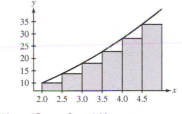

7. $R_N = \dfrac{141}{2} + \dfrac{45}{N} + \dfrac{9}{2N^2}$; $\dfrac{141}{2}$

9. $R_5 \approx 0.733732$; $M_5 \approx 0.786231$; $L_5 \approx 0.833732$

11. The area represented by the shaded rectangles is R_5; $R_5 = 90$; $L_5 = 90$.

13. $\displaystyle\lim_{N\to\infty} \frac{\pi}{6N}\sum_{j=1}^{N}\sin\left(\frac{\pi}{3} + \frac{\pi j}{6N}\right) = \int_{\pi/3}^{\pi/2}\sin x\,dx = \frac{1}{2}$

15. $\displaystyle\lim_{N\to\infty} \frac{5}{N}\sum_{j=1}^{N}\sqrt{4 + 5j/N} = \int_{4}^{9}\sqrt{x}\,dx = \frac{38}{3}$

17. $x^4 - \dfrac{2}{3}x^3 + C$ **19.** $-\cos(\theta - 8) + C$

21. $-2t^{-2} + 4t^{-3} + C$ **23.** $\tan x + C$ **25.** $\dfrac{1}{5}(y+2)^5 + C$

27. $e^x - \dfrac{x^2}{2} + C$ **29.** $4\ln|x| + C$ **31.** $y(x) = x^4 + 3$

33. $y(x) = 2\sqrt{x} - 1$ **35.** $y(x) = -e^{-x} + 4$

37. $f(t) = \dfrac{t^2}{2} - \dfrac{t^3}{3} - t + 2$

39. $\dfrac{1}{4}\ln\dfrac{5}{3}$ **41.** $\dfrac{1}{5}\left(1 - \dfrac{9\sqrt{3}}{32}\right)$

43. $4x^5 - \dfrac{9}{4}x^4 - x^2 + C$ **45.** $\dfrac{4}{5}x^5 - 3x^4 + 3x^3 + C$

47. $\dfrac{1}{4}x^4 + x^3 + C$ **49.** $\dfrac{46}{3}$ **51.** 3

53. $\dfrac{1}{150}(10t - 7)^{15} + C$ **55.** $-\dfrac{1}{24}(3x^4 + 9x^2)^{-4} + C$ **57.** 506

59. $-\dfrac{3\sqrt{3}}{2\pi}$ **61.** $\dfrac{1}{27}\tan(9t^3 + 1) + C$ **63.** $\dfrac{1}{2}\cot(9 - 2\theta) + C$

65. $3 - \dfrac{3\sqrt[3]{4}}{2}$ **67.** $-\dfrac{1}{2}e^{9-2x} + C$ **69.** $\dfrac{1}{3}e^{x^3} + C$

71. $\dfrac{10^x e^x}{\ln 10 + 1} + C$ **73.** $\dfrac{1}{2(e^{-x} + 2)^2} + C$ **75.** $\dfrac{1}{2}\ln 2$

77. $\tan^{-1}(\ln t) + C$ **79.** $\frac{1}{2}$ **81.** $\frac{1}{6}\tan^{-1}\left(\frac{2x}{3}\right) + C$

83. $\sec^{-1}12 - \sec^{-1}4$ **85.** $\frac{\pi}{12}$ **87.** $\frac{1}{2}\sin^{-1}(x^2) + C$

89. $\frac{1}{\sqrt{2}}\tan^{-1}(4\sqrt{2})$ **91.** $\frac{\pi^4}{1024}$ **93.** $\int_{-2}^{6} f(x)\,dx$

95. Local minimum at $x = 0$, no local maxima, inflection points at $x = \pm 1$

97. Daily consumption: 9.312 million gal; from 6 PM to midnight: 1.68 million gal

99. $208,245 **101.** 0

105. The function $f(x) = 2^x$ is increasing, so $1 \le x \le 2$ implies that $2 = 2^1 \le 2^x \le 2^2 = 4$. Consequently,

$$2 = \int_1^2 2\,dx \le \int_1^2 2^x\,dx \le \int_1^2 4\,dx = 4$$

On the other hand, the function $f(x) = 3^{-x}$ is decreasing, so $1 \le x \le 2$ implies that

$$\frac{1}{9} = 3^{-2} \le 3^{-x} \le 3^{-1} = \frac{1}{3}$$

It then follows that

$$\frac{1}{9} = \int_1^2 \frac{1}{9}\,dx \le \int_1^2 3^{-x}\,dx \le \int_1^2 \frac{1}{3}\,dx = \frac{1}{3}$$

107. $\frac{4}{3} \le \int_0^1 f(x)\,dx \le \frac{5}{3}$ **109.** $-\frac{1}{1+\pi}$

111. $\sin^3 x \cos x$ **113.** -2

115. Consider the figure below, which displays a portion of the graph of a linear function.

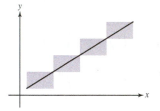

The shaded rectangles represent the differences between the right-endpoint approximation R_N and the left-endpoint approximation L_N. Because the graph of $y = f(x)$ is a line, the lower portion of each shaded rectangle is exactly the same size as the upper portion. Therefore, if we average L_N and R_N, the error in the two approximations will exactly cancel, leaving

$$\frac{1}{2}(R_N + L_N) = \int_a^b f(x)\,dx$$

117. Let

$$F(x) = x\sqrt{x^2 - 1} - 2\int_1^x \sqrt{t^2 - 1}\,dt$$

Then

$$\frac{dF}{dx} = \sqrt{x^2 - 1} + \frac{x^2}{\sqrt{x^2 - 1}} - 2\sqrt{x^2 - 1}$$

$$= \frac{x^2}{\sqrt{x^2 - 1}} - \sqrt{x^2 - 1} = \frac{1}{\sqrt{x^2 - 1}}$$

Also, $\frac{d}{dx}(\cosh^{-1}x) = \frac{1}{\sqrt{x^2-1}}$; therefore, $y = F(x)$ and $y = \cosh^{-1} x$ have the same derivative. We conclude that $y = F(x)$ and $y = \cosh^{-1} x$ differ by a constant:

$$F(x) = \cosh^{-1}x + C$$

Now, let $x = 1$. Because $F(1) = 0$ and $\cosh^{-1} 1 = 0$, it follows that $C = 0$. Therefore,

$$F(x) = \cosh^{-1}x$$

121. Approximately 6065.9 years **123.** 5.03% **125.** $17,979.10

Chapter 6

Section 6.1 Preliminary Questions

1. Area of the region between the graphs of $y = f(x)$ and $y = g(x)$, bounded on the left by the vertical line $x = a$ and on the right by the vertical line $x = b$

2. Yes **3.** $\int_0^3 (f(x) - g(x))\,dx - \int_3^5 (g(x) - f(x))\,dx$

4. Negative

5. The area of the region bounded by the graphs of the functions f and g, and the vertical lines $x = a$ and $x = b$

6.

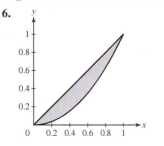

Section 6.1 Exercises

1. 102 **3.** $\frac{32}{3}$

5. $\sqrt{2} - 1$

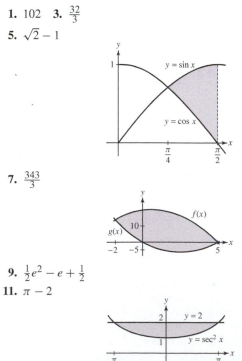

7. $\frac{343}{3}$

9. $\frac{1}{2}e^2 - e + \frac{1}{2}$

11. $\pi - 2$

13. Horizontally simple   **15.** Neither   **17.** $\frac{160}{3}$

19. $\frac{12\sqrt{3}-12+\left(\sqrt{3}-2\right)\pi}{24}$ **21.** $2-\frac{\pi}{2}$ **23.** $\frac{1331}{6}$ **25.** 256

27. $\frac{32}{3}$ **29.** $\frac{64}{3}$

31. $\frac{64}{3}$

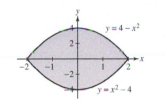

33. 2

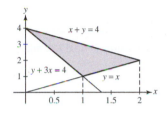

35. $\frac{128}{3}$

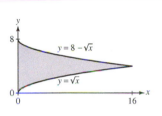

37. $\frac{1}{2}$

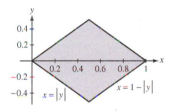

39. $\frac{1225}{8}$

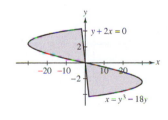

41. $\frac{32}{3}$

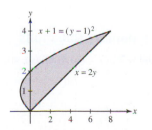

43. $\frac{3\sqrt{3}}{4}$

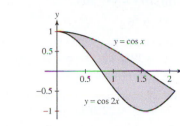

45. $\frac{2-\sqrt{2}}{2}$

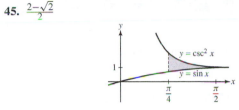

47. $4\ln 2 - 2 \approx 0.77259$

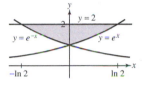

49. ≈ 0.7567130951

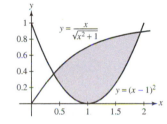

51. (a) (ii)   (b) No   (c) At 10 s, athlete 1; at 25 s, athlete 2

53. $\frac{8}{3}c^{3/2}$; $c = \frac{9^{1/3}}{4} \approx 0.520021$

55. $\int_{-\sqrt{(-1+\sqrt{5})/2}}^{\sqrt{(-1+\sqrt{5})/2}} \left[(1+x^2)^{-1} - x^2\right]\, dx$

57. 0.8009772242   **59.** 214.75 in.^{2}

61. (b) $\frac{1}{3}$ (c) 0   (d) 1

63. $m = 1 - \left(\frac{1}{2}\right)^{1/3} \approx 0.206299$

Section 6.2 Preliminary Questions

1. 3   **2.** 15

3. Flow rate is the volume of fluid that passes through a cross-sectional area at a given point per unit time.

4. The fluid velocity depended only on the radial distance from the center of the tube.

5. 15

Section 6.2 Exercises

1. (a) $\frac{4}{25}(20-y)^2$ (b) $\frac{1280}{3}$

3. $\frac{\pi R^2 h}{3}$ **5.** $\pi\left(Rh^2 - \frac{h^3}{3}\right)$ **7.** $\frac{1}{6}abc$ **9.** $\frac{8}{3}$ **11.** 36   **13.** 18

15. $\frac{\pi}{3}$ **17.** 96π

21. (a) $2\sqrt{r^2 - y^2}$ (b) $4(r^2 - y^2)$ (c) $\frac{16}{3}r^3$

23. 160π **25.** 5 kg **27.** 0.36 g

29. $P \approx 4423.59$ thousand **31.** $L_{10} = 233.86$, $R_6 = 290.56$
33. $P \approx 61$ deer **35.** $Q = 128\pi$ cm³/s **37.** $Q = \frac{8\pi}{3}$ cm³/s
39. 16 **41.** $\frac{3}{\pi}$ **43.** $\frac{1}{10}$ **45.** -4 **47.** $\frac{1}{n+1}$

49. Over $[0,24]$, the average temperature is 20; over $[2,6]$, the average temperature is $20 + \frac{15}{2\pi} \approx 22.387325$.

51. $\approx 79.56°$F **53.** $\frac{100}{\pi}$ **55.** $\frac{17}{2}$ m/s **57.** 159.033 m/s
59. $\frac{3}{5^{1/4}} \approx 2.006221$

61. Mean Value Theorem for Integrals; $c = \frac{A}{\sqrt[3]{4}}$

63. Over $[0, 1]$, $f(x)$; over $[1, 2]$, g

65. Many solutions exist. One could be

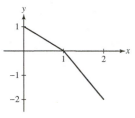

67. $v_0/2$

Section 6.3 Preliminary Questions

1. (a), (c) **2.** True

3. False, the cross sections will be washers.

4. (b)

Section 6.3 Exercises

1. (a)

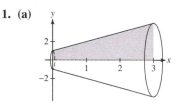

(b) Disk with radius $x + 1$

(c) $V = 21\pi$

3. (a)

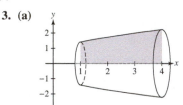

(b) Disk with radius $\sqrt{x + 1}$

(c) $V = \frac{21\pi}{2}$

5. $V = \frac{81\pi}{10}$ **7.** $V = \frac{24{,}573\pi}{13}$ **9.** $V = \pi$

11. $V = \frac{\pi}{2}\left(e^2 - 1\right)$ **13.** (ii)

15. (a)

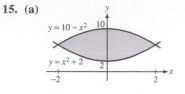

(b) A washer with outer radius $R = 10 - x^2$ and inner radius $r = x^2 + 2$

(c) $V = 256\pi$

17. (a)

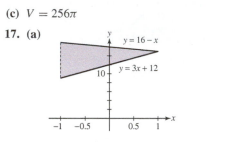

(b) A washer with outer radius $R = 16 - x$ and inner radius $r = 3x + 12$

(c) $V = \frac{656\pi}{3}$

19. (a)

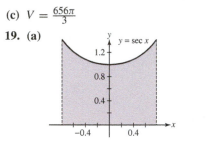

(b) A circular disk with radius $R = \sec x$ (c) $V = 2\pi$

21. $V = \frac{15\pi}{2}$ **23.** $V = \frac{3\pi}{10}$ **25.** $V = 32\pi$ **27.** $V = \frac{704\pi}{15}$
29. $V = \frac{128\pi}{5}$ **31.** $V = 40\pi$ **33.** $V = \frac{376\pi}{15}$ **35.** $V = \frac{824\pi}{15}$
37. $V = \frac{32\pi}{3}$ **39.** $V = \frac{1872\pi}{5}$ **41.** $V = \frac{1400\pi}{3}$
43. $V = \pi\left(\frac{7\pi}{9} - \sqrt{3}\right)$ **45.** $V = \frac{96\pi}{5}$ **47.** $V = \frac{16\pi}{35}$
49. $V = \frac{1184\pi}{15}$ **51.** $V = 7\pi\left(1 - \ln 2\right)$
53. $V \approx 12{,}120\pi \approx 38{,}076.1$ cm³ **55.** $V = \frac{1}{3}\pi r^2 h$
57. $V = \frac{32\pi}{105}$

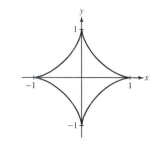

59. $V = 4\pi\sqrt{3}$ **61.** $V = \frac{4}{3}\pi a^2 b$

Section 6.4 Preliminary Questions

1. (a) Radius h and height r (b) Radius r and height h

2. (a) With respect to x (b) With respect to y

3. $V = 2\pi \displaystyle\int_0^8 y \cdot 1 \, dy = 64\pi$

Section 6.4 Exercises

1. $V = \frac{2}{5}\pi$

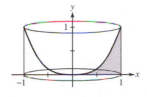

3. $V = 4\pi$

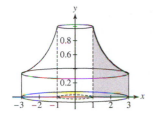

5. $V = 18\pi\left(2\sqrt{2} - 1\right)$

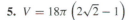
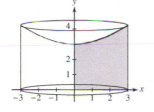

7. $V = \frac{32\pi}{3}$ **9.** $V = 16\pi$ **11.** $V = \frac{32\pi}{5}$

13. The point of intersection is $x = 1.376769504$; $V = 1.321975576$.

15. $V = \frac{3\pi}{5}$

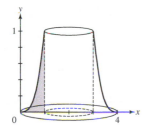

17. $V = \frac{280\pi}{81}$

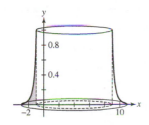

19. $V = \frac{1}{3}\pi a^3 + \pi a^2$

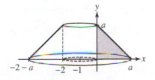

21. $V = \frac{\pi}{3}$

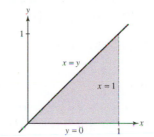

23. $V = \frac{128\pi}{3}$

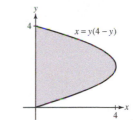

25. $V = \frac{256\pi}{15}$

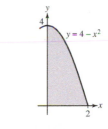

27. (b)

29. (a) $V = \frac{576\pi}{7}$ **(b)** $V = \frac{96\pi}{5}$

31. (a) $\overline{AB}$ generates a disk with radius $R = h(y)$; $\overline{CB}$ generates a shell with radius x and height $f(x)$.

(b) Shell, $V = 2\pi\int_0^2 xf(x)\,dx$; Disk, $V = \pi\int_0^{1.3}(h(y))^2\,dy$

33. $V = \frac{602\pi}{5}$ **35.** $V = 8\pi$ **37.** $V = \frac{40\pi}{3}$ **39.** $V = \frac{1024\pi}{15}$

41. $V = 16\pi$ **43.** $V = \frac{32\pi}{3}$ **45.** $V = \frac{776\pi}{15}$ **47.** $V = \frac{625\pi}{6}$

49. $\frac{3\pi}{10}$ **51.** $V = \frac{121\pi}{525}$ **53.** $V = \frac{563\pi}{30}$

55. $V = \frac{4}{3}\pi r^3$ **57.** $V = 2\pi^2 ab^2$

59. $x^2 + y^2 = 1$ over $[0, 1]$; $y = 1 - x$ over $[0, 1]$

Section 6.5 Preliminary Questions

1. Because the required force is not constant through the stretching process

2. The force involved in lifting the tank is the weight of the tank, which is constant.

3. $\frac{1}{2}kx^2$

4. When the force applied is in the opposite direction to the motion

Section 6.5 Exercises

1. $W = 627.2$ joules   **3.** $W = 5.76$ joules   **5.** $W = 8$ joules

7. $W = 11.25$ joules   **9.** $W = 3.800$ joules

11. $W = 105,840$ joules

13. $W = \frac{56,448\pi}{5}$ joules $\approx 3.547 \times 10^4$ joules

15. $W \approx 1.842 \times 10^{12}$ joules   **17.** $W = 3.92 \times 10^6$ joules

19. $W \approx 1.18 \times 10^8$ joules   **21.** $W = 9800\pi \ell r^3$ joules

23. $W = 2.94 \times 10^6$ joules   **25.** $W \approx 3.79 \times 10^6$ joules

27. $W = 3920$ joules   **29.** $W = 529.2$ joules

31. $W = 1470$ joules   **33.** $W = 374.85$ joules

37. $W \approx 5.16 \times 10^9$ joules   **41.** $\sqrt{2GM_e \left(\frac{1}{R_e} - \frac{1}{r+R_e} \right)}$ m/s

43. $v_{\text{esc}} = \sqrt{\frac{2GM_e}{R_e}}$ m/s

Chapter 6 Review

1. $\frac{32}{3}$ **3.** $\frac{1}{2}$ **5.** 24   **7.** $\frac{1}{2}$ **9.** $3\sqrt{2} - 1$ **11.** $e - \frac{3}{2}$

13. Intersection points $x = 0$, $x = 0.7145563847$;
Area $= 0.08235024596$

15. $V = 4\pi$ **17.** 2.7552 kg   **19.** $\frac{9}{4}$ **21.** $\frac{1}{2}\sinh 1$ **23.** $\frac{3\pi}{4}$

25. 27   **27.** $\frac{2\pi m^5}{15}$ **29.** $V = \frac{162\pi}{5}$ **31.** $V = 64\pi$ **33.** $V = 8\pi$

35. $V = \frac{56\pi}{15}$ **37.** $V = \frac{128\pi}{15}$ **39.** $V = 4\pi \left(1 - \frac{1}{\sqrt{e}}\right)$

41. $V = 2\pi \left(c + \frac{c^3}{3}\right)$ **43.** $V = c\pi$

45. (a) $\int_0^1 \left(\sqrt{1 - (x-1)^2} - \left(1 - \sqrt{1-x^2}\right) \right) dx$

(b) $\pi \int_0^1 \left[(1 - (x-1)^2) - (1 - \sqrt{1-x^2})^2 \right] dx$

47. $V = \pi \int_0^{1/a} \left(a\sqrt{x} - ax^2 \right)^2 dx = \frac{\pi}{6}$ **49.** $W = 1.6$ joules

51. $W = 1.93 \times 10^{10}$ joules   **53.** $9800\pi \left(h^3 + 2h^2 - \frac{1}{4}h^4 \right)$ joules

Chapter 7

Section 7.1 Preliminary Questions

1. The Integration by Parts formula is derived from the Product Rule.

3. Transforming $v' = x$ into $v = \frac{1}{2}x^2$ increases the power of x and makes the new integral harder than the original.

Section 7.1 Exercises

1. $-x \cos x + \sin x + C$ **3.** $e^x (2x + 7) + C$

5. $\frac{x^4}{16}(4 \ln x - 1) + C$ **7.** $-e^{-x}(4x + 1) + C$

9. $\frac{1}{25}(5x - 1)e^{5x+2} + C$ **11.** $\frac{1}{2}x \sin 2x + \frac{1}{4}\cos 2x + C$

13. $-x^2 \cos x + 2x \sin x + 2 \cos x + C$

15. $-\frac{1}{2}e^{-x}(\sin x + \cos x) + C$

17. $-\frac{1}{26}e^{-5x}(\cos(x) + 5\sin(x)) + C$

19. $\frac{1}{4}x^2(2\ln x - 1) + C$ **21.** $\frac{x^3}{3}\left(\ln x - \frac{1}{3}\right) + C$

23. $x\left[(\ln x)^2 - 2\ln x + 2\right] + C$ **25.** $x \cos^{-1} x - \sqrt{1 - x^2} + C$

27. $x \sec^{-1} x - \ln|x + \sqrt{x^2 - 1}| + C$

29. $\dfrac{3^x (\sin x + \ln 3 \cos x)}{1 + (\ln 3)^2} + C$

31. $(x^2 + 2)\sinh x - 2x \cosh x + C$

33. $x \tanh^{-1} 4x + \frac{1}{8}\ln|1 - 16x^2| + C$ **35.** $2e^{\sqrt{x}}(\sqrt{x} - 1) + C$

37. $\frac{1}{4}x \sin 4x + \frac{1}{16}\cos 4x + C$

39. $\frac{2}{3}(x + 1)^{3/2} - 2(x + 1)^{1/2} + C$

41. $\sin x \ln(\sin x) - \sin x + C$

43. $2xe^{\sqrt{x}} - 4\sqrt{x}e^{\sqrt{x}} + 4e^{\sqrt{x}} + C$

45. $\frac{1}{4}(\ln x)^2[2\ln(\ln x) - 1] + C$

47. $\frac{1}{16}(11e^{12} + 1)$

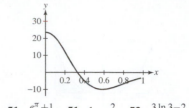

49. $2\ln 2 - \frac{3}{4}$ **51.** $\frac{e^\pi + 1}{2}$ **51.** $1 - \frac{2}{e}$ **53.** $\frac{3\ln 3 - 2}{\ln^2 3}$ **55.** $\frac{e^\pi + 1}{2}$

57. $-\frac{1}{5}\sin^4 x \cos x - \frac{4}{15}\sin^2 x \cos x - \frac{8}{15}\cos x + C$

59. $\frac{1}{3}\cos^2 x \sin x + \frac{2}{3}\sin x + C$

61. $e^x(x^3 - 3x^2 + 6x - 6) + C$

63. $\int x^n e^{-x}\, dx = -x^n e^{-x} + n \int x^{n-1} e^{-x}\, dx$ **65.** $\pi \left(\pi^2 - 4\right)$

67. Use Integration by Parts, with $u = \ln x$ and $dv = \sqrt{x}\, dx$.

69. Use substitution, followed by algebraic manipulation, with $u = 4 - x^2$ and $du = -2x\, dx$.

71. Use substitution with $u = x^2 + 4x + 3$, $\frac{du}{2} = (x + 2)\, dx$.

73. Use Integration by Parts, with $u = x$ and $dv = \sin(3x + 4)\, dx$.

75. $x(\sin^{-1} x)^2 + 2\sqrt{1 - x^2}\sin^{-1} x - 2x + C$

77. $\frac{1}{4}x^4 \sin(x^4) + \frac{1}{4}\cos(x^4) + C$

79. $2\pi(e^2 + 1)$ **81.** \$42,995.10

83. For $k = 2$: $x(\ln x)^2 - 2x \ln x + 2x + C$; for $k = 3$: $x(\ln x)^3 - 3x(\ln x)^2 + 6x \ln x - 6x + C$

85. Use Integration by Parts with $u = x$ and $v' = b^x$.

87. (b) $V(x) = \frac{1}{2}x^2 + \frac{1}{2}$ is simpler and yields $\frac{1}{2}(x^2 \tan^{-1} x - x + \tan^{-1} x) + C$

89. An example of a function satisfying these properties for some λ is $f(x) = \sin \pi x$.

91. (a) $I_n = \frac{1}{2}x^{n-1}\sin(x^2) - \frac{n-1}{2}J_{n-2}$;
(c) $\frac{1}{2}x^2 \sin(x^2) + \frac{1}{2}\cos(x^2) + C$

Section 7.2 Preliminary Questions

1. Rewrite $\sin^5 x = \sin x \sin^4 x = \sin x(1 - \cos^2 x)^2$ and then substitute $u = \cos x$.

3. No, a reduction formula is not needed because the sine function is raised to an odd power.

5. The second integral requires the use of reduction formulas and therefore more work.

Section 7.2 Exercises

1. $\sin x - \frac{1}{3}\sin^3 x + C$ **3.** $-\frac{1}{3}\cos^3 \theta + \frac{1}{5}\cos^5 \theta + C$

5. $-\frac{1}{4}\cos^4 t + \frac{1}{6}\cos^6 t + C$ **7.** 2

9. $\frac{3y}{8} + \frac{1}{4}\sin(2y) + \frac{1}{32}\sin(4y) + C$

11. $\frac{1}{6}\sin^5 x \cos x - \frac{1}{24}\sin^3 x \cos x - \frac{1}{16}\sin x \cos x + \frac{1}{16}x + C$

13. $\frac{1}{5}\sin^4 x \cos x - \frac{1}{15}\sin^2 x \cos x - \frac{2}{15}\cos x + C$

15. $\frac{1}{3}\sec^3 x - \sec x + C$

17. $\frac{1}{5}\tan x \sec^4 x - \frac{1}{15}\tan(x)\sec^2 x - \frac{2}{15}\tan x + C$

19. $-\frac{1}{2}\cot^2 x + \ln|\csc x| + C$ **21.** $-\frac{1}{6}\cot^6 x + C$

23. $-\frac{1}{6}\cos^6 x + C$

25. $\frac{1}{12}\cos^3(3x + 2)\sin(3x + 2) + \frac{1}{8}(3x + 2) + \frac{1}{16}\sin(6x + 4) + C$

27. $\frac{1}{5\pi}\sin^5(\pi\theta) - \frac{1}{7\pi}\sin^7(\pi\theta) + C$

29. $-\frac{1}{12}\sin^3(3x)\cos(3x) - \frac{1}{8}\sin(3x)\cos(3x) + \frac{3}{8}x + C$

31. $\frac{1}{2}\sin^2 x - \frac{1}{2\sin^2 x} - 2\ln|\sin x| + C$

33. $\frac{1}{2}\cot(3-2x) + C$ 35. $\frac{1}{2}\tan^2 x + C$

37. $\frac{1}{8}\sec^8 x - \frac{1}{3}\sec^6 x + \frac{1}{4}\sec^4 x + C$

39. $\frac{1}{9}\tan^9 x + \frac{1}{7}\tan^7 x + C$

41. $-\frac{1}{9}\csc^9 x + \frac{2}{7}\csc^7 x - \frac{1}{5}\csc^5 x + C$ 43. $\frac{1}{4}\sin^2 2x + C$

45. $\frac{1}{6}\cos^2(t^2)\sin(t^2) + \frac{1}{3}\sin(t^2) + C$

47. $\frac{1}{2}\cos(\sin t)\sin(\sin t) + \frac{1}{2}\sin t + C$ 49. π 51. $\frac{8}{15}$

53. $\ln\left(\sqrt{2}+1\right)$ 55. $\ln 2$ 57. $\frac{8}{3}$

59. $-\frac{6}{7}$ 61. $\frac{1}{24}$

63. First, observe $\sin 4x = 2\sin 2x\cos 2x = 2\sin 2x(1-2\sin^2 x) = 2\sin 2x - 4\sin 2x\sin^2 x = 2\sin 2x - 8\sin^3 x\cos x$. Then,
$\frac{1}{32}(12x - 8\sin 2x + \sin 4x) + C = \frac{3}{8}x - \frac{3}{16}\sin 2x - \frac{1}{4}\sin^3 x\cos x + C = \frac{3}{8}x - \frac{3}{8}\sin x\cos x - \frac{1}{4}\sin^3 x\cos x + C$.

65. $\frac{\pi^2}{2}$ 67. $\frac{1}{8}x - \frac{1}{16}\sin 2x\cos 2x + C$

69. $\frac{1}{16}x - \frac{1}{48}\sin 2x - \frac{1}{32}\sin 2x\cos 2x + \frac{1}{48}\cos^2 2x\sin 2x + C$

71. Use the identity $\tan^2 x = \sec^2 x - 1$ and the substitution $u = \tan x, du = \sec^2 x\,dx$.

73. (a) $I_0 = \int_0^{\pi/2}\sin^0 x\,dx = \frac{\pi}{2}; I_1 = \int_0^{\pi/2}\sin x\,dx = 1$

(b) $\frac{m-1}{m}\int_0^{\pi/2}\sin^{m-2}x\,dx$

(c) $I_2 = \frac{\pi}{4}; I_3 = \frac{2}{3}; I_4 = \frac{3\pi}{16}; I_5 = \frac{8}{15}$

75. $\cos(x) - \cos(x)\ln(\sin(x)) + \ln|\csc(x) - \cot(x)| + C$

79. Use Integration by Parts with $u = \sec^{m-2}x$ and $v' = \sec^2 x$.

Section 7.3 Preliminary Questions

1. (a) $x = 3\sin\theta$ (b) $x = 4\sec\theta$ (c) $x = 4\tan\theta$
(d) $x = \sqrt{5}\sec\theta$

3. $2x\sqrt{1-x^2}$

Section 7.3 Exercises

1. (a) $\theta + C$ (b) $\sin^{-1}\left(\frac{x}{3}\right) + C$

3. (a) $\int \frac{dx}{\sqrt{4x^2+9}} = \frac{1}{2}\int\sec\theta\,d\theta$

(b) $\frac{1}{2}\ln|\sec\theta + \tan\theta| + C$

(c) $\frac{1}{2}\ln\left|\sqrt{4x^2+9} + 2x\right| + C$

5. $\frac{8}{\sqrt{5}}\arcsin\left(\frac{x\sqrt{5}}{4}\right) + \frac{1}{2}x\sqrt{16-5x^2} + C$ 7. $\frac{1}{3}\sec^{-1}\left(\frac{x}{3}\right) + C$

9. $\frac{-x}{4\sqrt{x^2-4}} + C$ 11. $\sqrt{x^2-4} + C$

13. (a) $-\sqrt{1-x^2}$ (b) $\frac{1}{8}(\arcsin x - x\sqrt{1-x^2}(1-2x^2))$

(c) $-\frac{1}{3}(1-x^2)^{\frac{3}{2}} + \frac{1}{5}(1-x^2)^{\frac{5}{2}}$

(d) $\sqrt{1-x^2}\left(-\frac{x^3}{4} - \frac{3x}{8}\right) + \frac{3}{8}\arcsin(x)$

15. $\frac{9}{2}\sin^{-1}\left(\frac{x}{3}\right) - \frac{1}{2}x\sqrt{9-x^2} + C$

17. $\frac{1}{4}\ln\left|\frac{\sqrt{x^2+16}-4}{x}\right| + C$ 19. $\ln\left|x + \sqrt{x^2-9}\right| + C$

21. $-\frac{\sqrt{5-y^2}}{5y} + C$ 23. $\frac{1}{5}\ln\left|5x + \sqrt{25x^2+2}\right| + C$

25. $\frac{1}{16}\sec^{-1}\left(\frac{z}{2}\right) + \frac{\sqrt{z^2-4}}{8z^2} + C$

27. $\frac{1}{12}x\sqrt{6x^2-49} + \frac{49\ln\left(x\sqrt{6}+\sqrt{6x^2-49}\right)}{12\sqrt{6}} + C$

29. $\frac{1}{54}\tan^{-1}\left(\frac{x}{3}\right) + \frac{1}{180}$ 31. $-\frac{x}{\sqrt{x^2-1}} + \ln\left|x + \sqrt{x^2+1}\right| + C$

33. Use the substitution $x = \sqrt{a}\,u$.

35. (a) $x^2 - 4x + 8 = x^2 - 4x + 4 + 4 = (x-2)^2 + 4$

(b) $\ln\left|\sqrt{u^2+4} + u\right| + C$

(c) $\ln\left|\sqrt{(x-2)^2+4} + x - 2\right| + C$

37. $\ln\left|\sqrt{x^2+4x+13} + x + 2\right| + C$

39. $\frac{1}{\sqrt{6}}\ln\left|12x + 1 + 2\sqrt{6}\sqrt{x+6x^2}\right| + C$

41. $\frac{1}{2}(x-2)\sqrt{x^2-4x+3} - \frac{1}{2}\ln\left|x-2+\sqrt{x^2-4x+3}\right| + C$

43. $x\sec^{-1}x - \ln\left|x + \sqrt{x^2-1}\right| + C$

45. $x(\ln(x^2+1) - 2) + 2\tan^{-1}x + C$

47. $\frac{\pi}{4}$ 49. $4\pi\left[\sqrt{3} - \ln\left|2 + \sqrt{3}\right|\right]$ 51. $\frac{1}{2}\ln\left(\frac{x-1}{x+1}\right) + C$

53. (a) $1.789 \times 10^6\ \frac{V}{m}$ (b) $3.526 \times 10^6\ \frac{V}{m}$

Section 7.4 Preliminary Questions

1. (a) $x = \sinh t$ (b) $x = 3\sinh t$ (c) $3x = \sinh t$

3. $\frac{1}{2}\ln\left|\frac{1+x}{1-x}\right|$

Section 7.4 Exercises

1. $\frac{1}{3}\sinh(3x) + C$ 3. $\frac{1}{2}\cosh(x^2+1) + C$

5. $-\frac{1}{2}\tanh(1-2x) + C$ 7. $\frac{\tanh^2 x}{2} + C$ 9. $\ln\cosh x + C$

11. $\ln|\sinh x| + C$ 13. $\frac{1}{16}\sinh(8x-18) - \frac{1}{2}x + C$

15. $\frac{1}{32}\sinh 4x - \frac{1}{8}x + C$ 17. $\cosh^{-1}x + C$

19. $\frac{1}{5}\sinh^{-1}\left(\frac{5x}{4}\right) + C$ 21. $\frac{1}{2}x\sqrt{x^2-1} - \frac{1}{2}\cosh^{-1}x + C$

23. $2\tanh^{-1}\left(\frac{1}{2}\right)$ 25. $\sinh^{-1}1$

27. $\frac{1}{4}\left(\operatorname{csch}^{-1}\left(-\frac{1}{4}\right) - \operatorname{csch}^{-1}\left(-\frac{3}{4}\right)\right)$

29. $\cosh^{-1}x - \frac{\sqrt{x^2-1}}{x} + C$

31. Let $x = \sinh t$ for the first formula and $x = \cosh t$ for the second.

33. $\frac{1}{2}x\sqrt{x^2+16} + 8\ln\left|\frac{x}{4} + \sqrt{\left(\frac{x}{4}\right)^2+1}\right| + C$

35. Using Integration by Parts with $u = \cosh^{n-1}x$ and $v' = \cosh x$ to begin proof

37. $-\frac{1}{2}\left(\tanh^{-1}x\right)^2 + C$ 39. $x\tanh^{-1}x + \frac{1}{2}\ln|1-x^2| + C$

41. $u = \sqrt{\frac{\cosh x-1}{\cosh x+1}}$. From this it follows that $\cosh x = \frac{1+u^2}{1-u^2}$, $\sinh x = \frac{2u}{1-u^2}$, and $dx = \frac{2du}{1-u^2}$.

43. $\int du = u + C = \tanh\frac{x}{2} + C$

45. Let $gd(y) = \tan^{-1}(\sinh y)$. Then

$$\frac{d}{dy}gd(y) = \frac{1}{1+\sinh^2 y}\cosh y = \frac{1}{\cosh y} = \operatorname{sech} y$$

where we have used the identity $1 + \sinh^2 y = \cosh^2 y$.

47. Let $x = gd(y) = \tan^{-1}(\sinh y)$. Solving for y yields $y = \sinh^{-1}(\tan x)$. Therefore, $gd^{-1}(y) = \sinh^{-1}(\tan y)$.

49. Let $x = it$. Then $\cosh^2 x = (\cosh(it))^2 = \cos^2 t$ and $\sinh^2 x = (\sinh(it))^2 = i^2\sin^2 t = -\sin^2 t$. Thus, $1 = \cosh^2(it) - \sinh^2(it) = \cos^2 t - (-\sin^2 t) = \cos^2 t + \sin^2 t$, as desired.

Section 7.5 Preliminary Questions

1. No, f cannot be a rational function because the integral of a rational function cannot contain a term with a noninteger exponent such as $\sqrt{x+1}$.

3. (a) Square is already completed; irreducible.

(b) Square is already completed; factors as $(x - \sqrt{5})(x + \sqrt{5})$.

(c) $x^2 + 4x + 6 = (x + 2)^2 + 2$; irreducible.

(d) $x^2 + 4x + 2 = (x + 2)^2 - 2$; factors as $(x + 2 - \sqrt{2})(x + 2 + \sqrt{2})$.

Section 7.5 Exercises

1. (a) $\dfrac{x^2 + 4x + 12}{(x+2)(x^2+4)} = \dfrac{1}{x+2} + \dfrac{4}{x^2+4}$.

(b) $\dfrac{2x^2 + 8x + 24}{(x+2)^2(x^2+4)} = \dfrac{1}{x+2} + \dfrac{2}{(x+2)^2} + \dfrac{-x+2}{x^2+4}$.

(c) $\dfrac{x^2 - 4x + 8}{(x-1)^2(x-2)^2} = \dfrac{-8}{x-2} + \dfrac{4}{(x-2)^2} + \dfrac{8}{x-1} + \dfrac{5}{(x-1)^2}$.

(d) $\dfrac{x^4 - 4x + 8}{(x+2)(x^2+4)} = x - 2 + \dfrac{4}{x+2} - \dfrac{4x-4}{x^2+4}$.

3. -2 **5.** $\frac{1}{9}(3x + 4\ln(3x - 4)) + C$

7. $\frac{x^3}{3} + \ln|x + 2| + C$ **9.** $-\frac{1}{2}\ln|x - 2| + \frac{1}{2}\ln|x - 4| + C$

11. $\ln|x| - \ln|2x + 1| + C$ **13.** $x - 3\arctan\left(\frac{x}{3}\right) + C$

15. $2\ln|x + 3| - \ln|x + 5| - \frac{2}{3}\ln|3x - 2| + C$

17. $3\ln|x - 1| - 2\ln|x + 1| - \frac{5}{x+1} + C$

19. $2\ln|x - 1| - \frac{1}{x-1} - 2\ln|x - 2| - \frac{1}{x-2} + C$

21. $\ln|x| - \ln|x + 2| + \frac{2}{x+2} + \frac{2}{(x+2)^2} + C$

23. $\frac{1}{2\sqrt{6}}\ln\left|\sqrt{2}x - \sqrt{3}\right| - \frac{1}{2\sqrt{6}}\ln\left|\sqrt{2}x + \sqrt{3}\right| + C$

25. $\frac{1}{2(x+1)} + \frac{1}{4}\ln|x - 1| - \frac{1}{4}\ln|x + 1| + C$

27. $\frac{5}{2x+5} - \frac{5}{4(2x+5)^2} + \frac{1}{2}\ln|2x + 5| + C$

29. $-\ln|x| + \ln|x - 1| + \frac{1}{x-1} - \frac{1}{2(x-1)^2} + C$

31. $x + \ln|x| - 3\ln|x + 1| + C$

33. $2\ln|x - 1| + \frac{1}{2}\ln|x^2 + 1| - 3\tan^{-1}x + C$

35. $\frac{1}{25}\ln|x| - \frac{1}{50}\ln|x^2 + 25| + C$

37. $6x - 14\ln|x + 3| + 2\ln|x - 1| + C$

39. $-\frac{1}{5}\ln|x - 1| - \frac{1}{x-1} + \frac{1}{10}\ln|x^2 + 9| - \frac{4}{15}\tan^{-1}\left(\frac{x}{3}\right) + C$

41. $\frac{1}{64}\ln|x| - \frac{1}{128}\ln|x^2 + 8| + \frac{1}{16(x^2+8)} + C$

43. $\frac{1}{6}\ln|x + 2| - \frac{1}{12}\ln|x^2 + 4x + 10| + C$

45. $\ln|x| - \frac{1}{2}\left|x^2 + 2x + 5\right| + \frac{15-5x}{8(x^2+2x+5)} - \frac{13}{16}\tan^{-1}\left(\frac{x+1}{2}\right) + C$

47. $\frac{1}{2}\arctan(x^2) + C$ **49.** $\ln|e^x - 1| - x + C$

51. $2\sqrt{x} + \ln|\sqrt{x} - 1| - \ln|\sqrt{x} + 1| + C$

53. $\ln\left|x^{1/4} - 2\right| - \ln\left|x^{1/4} + 2\right| + C$

55. $\ln\left|\dfrac{x}{\sqrt{x^2-1}} - \dfrac{1}{\sqrt{x^2-1}}\right| + C = \ln\left|\dfrac{x-1}{\sqrt{x^2-1}}\right| + C$

57. If $\theta = 2\tan^{-1}t$, then $d\theta = 2\,dt/(1 + t^2)$. We also have $\cos\left(\frac{\theta}{2}\right) = 1/\sqrt{1 + t^2}$ and $\sin\left(\frac{\theta}{2}\right) = t/\sqrt{1 + t^2}$. To find $\cos\theta$, we use the double angle identity $\cos\theta = 1 - 2\sin^2\left(\frac{\theta}{2}\right)$. This gives us $\cos\theta = \frac{1-t^2}{1+t^2}$. To find $\sin\theta$, we use the double angle identity $\sin\theta = 2\sin\left(\frac{\theta}{2}\right)\cos\left(\frac{\theta}{2}\right)$. This gives us $\sin\theta = \frac{2t}{1+t^2}$. It follows then

that $\displaystyle\int \frac{d\theta}{\cos\theta + \frac{3}{4}\sin\theta} =$

$-\dfrac{4}{5}\ln\left|2 - \tan\left(\dfrac{\theta}{2}\right)\right| + \dfrac{4}{5}\ln\left|1 + 2\tan\left(\dfrac{\theta}{2}\right)\right| + C.$

59. Partial fraction decomposition shows $\dfrac{1}{(x-a)(x-b)} = \dfrac{\frac{1}{a-b}}{x-a} + \dfrac{\frac{1}{b-a}}{x-b}$.

This can be used to show $\int \dfrac{dx}{(x-a)(x-b)} = \dfrac{1}{a-b}\ln\left|\dfrac{x-a}{x-b}\right| + C.$

61. $\dfrac{2}{x-6} + \dfrac{1}{x+2}$

Section 7.6 Preliminary Questions

1. Integration by parts with $u = x$, $dv = \sin x\,dx$

2. Trigonometric substitution $x = \tan\theta$, $dx = \sec^2\theta\,d\theta$

3. Partial fractions

4. Substitute $u = \cos x$, $du = -\sin x\,dx$.

5. Integration by parts with $u = \ln x$, $dv = x\,dx$

6. Trigonometric substitution $x = \sin\theta$, $dx = \cos\theta\,d\theta$

7. **Trig Method.** Rewrite as $\int\left(1 - \sin^2 x\right)\cos^2 x \sin x\,dx$. Then use u-substitution with $u = \cos x$ and $du = -\sin x\,dx$.

8.
$$\int \frac{u^2\,du}{a + bu} = \frac{1}{2b^3}\left[(a + bu)^2 - 4a(a + bu) + 2a^2\ln|a + bu|\right] + C$$

9. Trig substitution with $x = \frac{5}{4}\tan\theta$ and $dx = \frac{5}{4}\sec^2\theta\,d\theta$

10. $\int \sec^3 u\,du = \frac{1}{2}\sec u\tan u + \frac{1}{2}\ln|\sec u + \tan u| + C$

11. Complete the square, followed by u-substitution with $u = x + 1$ and $du = dx$. Then use trig substitution with $u = 2\tan\theta$ and $du = 2\sec^2\theta\,d\theta$.

Section 7.6 Exercises

1. Complete the square to get $\int \dfrac{x\,dx}{\sqrt{21 - (x+3)^2}}$ and then use the substitution $x + 3 = \sqrt{21}\sin u$ so that $dx = \sqrt{21}\cos u\,du$.

3. **Trig Method.** Rewrite as $\int \sin^3 x\left(1 - \sin^2 x\right)\cos x\,dx$ followed by u-substitution with $u = \sin x$ and $du = \cos x\,dx$.

5. Trig substitution $x = 3\sin\theta$ and $dx = 3\cos\theta\,d\theta$

7. None   **9.** Partial fractions   **11.** $-\dfrac{\sqrt{4-x^2}}{4x} + C$

13. $\dfrac{x}{2} + \dfrac{1}{8}\sin 4x\cos 4x + C$ **15.** $\dfrac{x}{18(9+x^2)} + \dfrac{1}{54}\tan^{-1}\left(\frac{x}{3}\right) + C$

17. $\sec x - \frac{2}{3}\sec^3 x + \frac{1}{5}\sec^5 x + C$

19. $x\ln\left|x^4 + 1\right| - 4x + 2\tan^{-1}x - \ln|x - 1| + \ln|x + 1| + C$

21. $\ln\left|\sqrt{x^2 - 1} + x\right| - \dfrac{x}{\sqrt{x^2-1}} + C$

23. $-\dfrac{x+6}{8(x^2+4x+8)} - \dfrac{1}{16}\tan^{-1}\left(\frac{x+2}{2}\right) + C$

25. $6\tan^{-1}\left(x^{1/6}\right) - 6x^{1/6} + 2\sqrt{x} - \frac{6}{5}x^{5/6} + \frac{6}{7}x^{7/6} + C$

27. $\frac{1}{3}\ln\left|3e^t + 1\right| + C$ **29.** $\frac{1}{3}x^3\ln x - \frac{x^3}{9} + C$

31. $\frac{1}{2}\tan^{-1}\left(\frac{x+1}{2}\right) + C$ **33.** $\frac{2}{9}\left(1 + x^3\right)^{3/2} + C$

35. $x - \ln\left(1 + e^x\right) + C$

37. $2\sqrt{x} + x + \frac{2}{3}x^{3/2} + 2\ln\left|\sqrt{x} - 1\right| + C$

39. $\frac{2}{5}(x + 1)^{5/2} - \frac{4}{3}(x + 1)^{3/2} + 2\sqrt{x + 1} + C$

41. $x + \cos x - \frac{1}{4}\sin 2x - \frac{1}{3}\cos 3x + \frac{1}{8}\sin 4x + C$

43. $\frac{1}{2}\cos^2 x - \ln|\cos x| + C$

45. $x\ln\left(x^2 + 9\right) + 6\tan^{-1}\left(\frac{x}{3}\right) - 2x + C$

47. $-\frac{1}{3}\cos^3 x + \frac{2}{5}\cos^5 x - \frac{1}{7}\cos^7 x + C$

49. $\sin x - \sin^3 x + \frac{3}{5}\sin^5 x - \frac{1}{7}\sin^7 x + C$

51. $\frac{1}{2}x^2 - \frac{1}{4}\ln\left(x^2 + 1\right) + \frac{1}{4}\ln|x - 1| + \frac{1}{4}\ln|x + 1| + C$

53. $\frac{1}{4}\sin 2x + 9\tan x - \frac{11}{2}x + C$

55. $\frac{1}{32}x^4 + \frac{1}{4}x^4(\ln x)^2 - \frac{1}{8}x^4 \ln x + C$ **57.** $\cosh^{-1}\left(\frac{x}{6}\right) + C$

Section 7.7 Preliminary Questions

1. (a) The integral converges.

(b) The integral diverges.

(c) The integral diverges.

(d) The integral converges.

3. Any value of b satisfying $|b| \geq 2$ will make this an improper integral.

5. Knowing that an integral is smaller than a divergent integral does not allow us to draw any conclusions using the comparison test.

Section 7.7 Exercises

1. (a) Improper. The function $y = x^{-1/3}$ is infinite at 0.

(b) Improper. Infinite interval of integration.

(c) Improper. Infinite interval of integration.

(d) Proper. The function $y = e^{-x}$ is continuous on the finite interval [0, 1].

(e) Improper. The function $y = \sec x$ is infinite at $\frac{\pi}{2}$.

(f) Improper. Infinite interval of integration.

(g) Proper. The function $y = \sin x$ is continuous on the finite interval [0, 1].

(h) Proper. The function $y = 1/\sqrt{3 - x^2}$ is continuous on the finite interval [0, 1].

(i) Improper. Infinite interval of integration.

(j) Improper. The function $y = \ln x$ is infinite at 0.

3. $\int_1^\infty x^{-2/3}\,dx = \lim_{R\to\infty} \int_1^R x^{-2/3}\,dx = \lim_{R\to\infty} 3\left(R^{1/3} - 1\right) = \infty$

5. The integral does not converge.

7. The integral converges; $I = 10{,}000e^{0.0004}$.

9. The integral does not converge.

11. The integral converges; $I = 4$.

13. The integral converges; $I = \frac{1}{8}$.

15. The integral converges; $I = 2$.

17. The integral converges; $I = 0$.

19. The integral converges; $I = \frac{1}{3e^{12}}$.

21. The integral converges; $I = \frac{1}{3}$.

23. The integral converges; $I = 2\sqrt{2}$.

25. The integral does not converge.

27. The integral converges; $I = \frac{1}{2}$.

29. The integral converges; $I = \frac{1}{2}$.

31. The integral converges; $I = \frac{\pi}{2}$.

33. The integral does not converge.

35. The integral does not converge.

37. The integral converges; $I = -1$.

39. The integral does not converge.

41. (a) Partial fractions yield $\frac{dx}{(x-2)(x-3)} = \frac{dx}{x-3} - \frac{dx}{x-2}$. This yields $\int_4^R \frac{dx}{(x-2)(x-3)} = \ln\left|\frac{R-3}{R-2}\right| - \ln\frac{1}{2}$.

(b) $I = \lim_{R\to\infty}\left(\ln\left|\frac{R-3}{R-2}\right| - \ln\frac{1}{2}\right) = \ln 1 - \ln\frac{1}{2} = \ln 2$

43. The integral does not converge.

45. The integral does not converge.

47. The integral converges; $I = 0$.

49. $\int_{-1}^1 \frac{dx}{x^{1/3}} = \int_{-1}^0 \frac{dx}{x^{1/3}} + \int_0^1 \frac{dx}{x^{1/3}} = 0$

51. The integral converges for $a < 0$.

53. $\int_{-\infty}^\infty \frac{dx}{1+x^2} = \pi$.

55. $\frac{1}{x^3+4} \leq \frac{1}{x^3}$. Therefore, by the Comparison Test, the integral converges.

57. For $x \geq 1$, $x^2 \geq x$, so $-x^2 \leq -x$ and $e^{-x^2} \leq e^{-x}$. Now $\int_1^\infty e^{-x}\,dx$ converges, so $\int_1^\infty e^{-x^2}\,dx$ converges by the Comparison Test. We conclude that our integral converges by writing it as a sum: $\int_0^\infty e^{-x^2}\,dx = \int_0^1 e^{-x^2}\,dx + \int_1^\infty e^{-x^2}\,dx$.

59. Let $f(x) = \frac{1 - \sin x}{x^2}$. Since $f(x) \leq \frac{2}{x^2}$ and $\int_1^\infty 2x^{-2}\,dx = 2$, it follows that $\int_1^\infty \frac{1 - \sin x}{x^2}\,dx$ converges by the Comparison Test.

61. The integral converges.

63. The integral does not converge.

65. The integral converges.

67. The integral does not converge.

69. The integral converges.

71. The integral converges.

73. The integral converges.

75. The integral does not converge.

77. $\int_0^1 \frac{dx}{x^{1/2}(x + 1)}$ and $\int_1^\infty \frac{dx}{x^{1/2}(x + 1)}$ both converge; therefore, J converges.

79. $\$\frac{250}{0.07}$ **81.** $\$2{,}000{,}000$

83. $W = \lim_{T\to\infty} CV^2\left(\frac{1}{2} - e^{-T/RC} + \frac{1}{2}e^{-2T/RC}\right) = CV^2\left(\frac{1}{2} - 0 + 0\right) = \frac{1}{2}CV^2$

85. 2π

87. The integrand is infinite at the upper limit of integration, $x = \sqrt{2E/k}$, so the integral is improper. $T = \lim_{R\to\sqrt{2E/k}} T(R) = 4\sqrt{\frac{m}{k}}\sin^{-1}(1) = 2\pi\sqrt{\frac{m}{k}}$

89. $Lf(s) = \frac{-1}{s^2 + \alpha^2} \lim_{t\to\infty} e^{-st}(s\sin(\alpha t) + \alpha\cos(\alpha t)) - \alpha$.

91. $\frac{s}{s^2+\alpha^2}$ **93.** $J_n = \frac{n}{\alpha}J_{n-1} = \frac{n}{\alpha}\cdot\frac{(n-1)!}{\alpha^n} = \frac{n!}{\alpha^{n+1}}$

95. $E = \frac{8\pi h}{c^3}\int_0^\infty \frac{\nu^3}{e^{\alpha\nu} - 1}\,d\nu$. Because $\alpha > 0$ and $8\pi h/c^3$ is a constant, we know E is finite by Exercise 92.

97. Because $t > \ln t$ for $t > 2$, $F(x) = \int_2^x \frac{dt}{\ln t} > \int_2^x \frac{dt}{t} > \ln x$. Thus, $F(x) \to \infty$ as $x \to \infty$. Moreover, $\lim_{x\to\infty} G(x) = \lim_{x\to\infty} \frac{1}{1/x} = \lim_{x\to\infty} x = \infty$. Thus, $\lim_{x\to\infty} \frac{F(x)}{G(x)}$ is of the form ∞/∞, and L'Hôpital's Rule applies. Finally, $L = \lim_{x\to\infty} \frac{F(x)}{G(x)} = \lim_{x\to\infty} \frac{\frac{1}{\ln x}}{\frac{\ln x - 1}{(\ln x)^2}} = \lim_{x\to\infty} \frac{\ln x}{\ln x - 1} = 1$.

99. The integral is absolutely convergent. Use the Comparison Test with $\frac{1}{x^2}$.

Section 7.8 Preliminary Questions

1. No, $p(x) \geq 0$ fails. **3.** $p(x) = 4e^{-4x}$

Section 7.8 Exercises

1. $C = 2$; $P(0 \leq X \leq 1) = \frac{3}{4}$

3. $C = \frac{1}{\pi}$; $P\left(-\frac{1}{2} \leq X \leq \frac{1}{2}\right) = \frac{1}{3}$

5. $C = \frac{2}{\pi}$; $P\left(-\frac{1}{2} \leq X \leq 1\right) = \frac{2}{3} + \frac{\sqrt{3}}{4\pi}$

7. $\int_1^\infty 3x^{-4} = 1$; $\mu = \frac{3}{2}$

9. Integration confirms $\int_0^\infty \frac{1}{50} e^{-t/50} = 1$.

11. $e^{-\frac{3}{2}} \approx 0.2231$ **13.** $\frac{1}{2}\left(2 - 10e^{-2}\right) \approx 0.32$

15. $F(-\frac{2}{3}) - F(-\frac{13}{6}) \approx 0.2374$

17. **(a)** ≈ 0.8849 **(b)** ≈ 0.6554

19. $1 - F(z)$ and $F(-z)$ are the same area on opposite tails of the distribution function. Simple algebra with the standard normal cumulative distribution function shows
$P(\mu - r\sigma \leq X \leq \mu + r\sigma) = 2F(r) - 1$.

21. ≈ 0.0062 **25.** $\mu = \frac{5}{3}$; $\sigma = \frac{2\sqrt{5}}{3}$

27. $\mu = 3$; $\sigma = 3$

29. **(a)** $f(t)$ is the fraction of initial atoms present at time t. Therefore, the fraction of atoms that decay is going to be the rate of change of the total number of atoms. Over a small interval, this is simply $-f'(t)\Delta t$.

(b) The fraction of atoms that decay over an arbitrarily small interval is equivalent to the probability that an individual atom will decay over that same interval. Thus, the probability density function becomes $-f'(t)$.

(c) $\int_0^\infty -tf'(t)\,dt = \frac{1}{k}$

Section 7.9 Preliminary Questions

1. $T_1 = 6$; $T_2 = 7$

3. The Trapezoidal Rule integrates linear functions exactly, so the error will be zero.

5. The two graphical interpretations of the Midpoint Rule are the sum of the areas of the midpoint rectangles and the sum of the areas of the tangential trapezoids.

Section 7.9 Exercises

1. $T_4 = 2.75$; $M_4 = 2.625$ **3.** $T_6 = 64.6875$; $M_6 \approx 63.2813$

5. $T_6 \approx 1.4054$; $M_6 \approx 1.3769$ **7.** $T_6 \approx 1.1703$; $M_6 \approx 1.2063$

9. $T_5 \approx 0.3846$; $M_5 \approx 0.3871$ **11.** $T_5 \approx 0.7444$; $M_5 \approx 0.7481$

13. $S_4 \approx 5.2522$ **15.** $S_6 \approx 1.1090$ **17.** $S_4 \approx 0.7469$

19. $S_8 \approx 2.5450$ **21.** $S_{10} \approx 0.3466$ **23.** ≈ 2.4674

25. ≈ 1.8769 **27.** ≈ 608.611

29. **(a)** Assuming the speed of the tsunami is a continuous function, at x miles from the shore, the speed is $\sqrt{15 f(x)}$. Covering an infinitesimally small distance, dx, the time T required for the tsunami to cover that distance becomes $\dfrac{dx}{\sqrt{15 f(x)}}$. It follows from this that
$T = \int_0^M \frac{dx}{\sqrt{15 f(x)}}$.

(b) ≈ 3.347 h.

31. **(a)** Since x^3 is concave up on $[0, 2]$, T_6 is too large.

(b) We have $f'(x) = 3x^2$ and $f''(x) = 6x$. Since $|f''(x)| = |6x|$ is *increasing* on $[0, 2]$, its maximum value occurs at $x = 2$ and we may take $K_2 = |f''(2)| = 12$. Thus, Error$(T_6) \leq \frac{2}{9}$.

(c) Error$(T_6) \approx 0.1111 < \frac{2}{9}$

33. T_{10} will overestimate the integral. Error$(T_{10}) \leq 0.045$.

35. M_{10} will overestimate the integral. Error$(M_{10}) \leq 0.0113$

37. $N \geq 10^3$; Error $\approx 3.333 \times 10^{-7}$

39. $N \geq 750$; Error $\approx 2.805 \times 10^{-7}$

41. Error$(T_{10}) \leq 0.0225$; Error$(M_{10}) \leq 0.01125$

43. $S_8 \approx 4{,}0467$; Error $(S_8) \leq 0.00833$; $N \geq 78$

45. Error$(S_{40}) \leq 1.017 \times 10^{-4}$.   **47.** $N = 306$ **49.** $N = 186$

51. **(a)** The maximum value of $|f^{(4)}(x)|$ on the interval $[0, 1]$ is 24.

(b) $N = 20$; $S_{20} \approx 0.785398$; $\left|0.785398 - \frac{\pi}{4}\right| \approx 1.55 \times 10^{-10}$

53. **(a)** Notice $|f''(x)| = |2\cos(x^2) - 4x^2 \sin(x^2)|$; the proof follows.

(b) When $K_2 = 6$, Error$(M_N) \leq \frac{1}{4N^2}$.

(c) $N \geq 16$

55. Error$(T_4) \approx 0.1039$; Error$(T_8) \approx 0.0258$; Error$(T_{16}) \approx 0.0064$; Error$(T_{32}) \approx 0.0016$; Error$(T_{64}) \approx 0.0004$. These are about twice as large as the error in M_N.

57. $S_2 = \frac{1}{4}$. This is the exact value of the integral.

59. $T_N = \dfrac{r(b^2 - a^2)}{2} + s(b - a) = \displaystyle\int_a^b f(x)\,dx$

61. **(a)** This result follows because the even-numbered interior endpoints overlap:

$$\sum_{i=0}^{(N-2)/2} S_2^{2j} = \frac{b-a}{6}\left[(y_0 + 4y_1 + y_2) + (y_2 + 4y_3 + y_4) + \cdots\right]$$

$$= \frac{b-a}{6}\left[y_0 + 4y_1 + 2y_2 + 4y_3 + 2y_4 + \cdots + 4y_{N-1} + y_N\right] = S_N$$

(b) If $f(x)$ is a quadratic polynomial, then by part (a), we have

$$S_N = S_2^0 + S_2^2 + \cdots + S_2^{N-2} = \int_a^b f(x)\,dx$$

63. Let $f(x) = ax^3 + bx^2 + cx + d$, with $a \neq 0$, be any cubic polynomial. Then $f^{(4)}(x) = 0$, so we can take $K_4 = 0$. This yields Error$(S_N) \leq \dfrac{0}{180N^4} = 0$. In other words, S_N is exact for all cubic polynomials for all N.

Chapter 7 Review

1. **(a)** (v) **(b)** (iv) **(c)** (iii) **(d)** (i) **(e)** (ii)

3. $\dfrac{\sin^9\theta}{9} - \dfrac{\sin^{11}\theta}{11} + C$

5.
$\dfrac{\tan\theta\sec^5\theta}{6} - \dfrac{7\tan\theta\sec^3\theta}{24} + \dfrac{\tan\theta\sec\theta}{16} + \dfrac{1}{16}\ln|\sec\theta + \tan\theta| + C$

7. $-\dfrac{1}{\sqrt{x^2-1}} - \sec^{-1}x + C$ **9.** $2\tan^{-1}\sqrt{x} + C$

11. $-\dfrac{\tan^{-1}x}{x} + \ln|x| - \dfrac{1}{2}\ln\left(1 + x^2\right) + C$

13. $\dfrac{5}{32}e^4 - \dfrac{1}{32} \approx 8.50$ **15.** $\dfrac{\cos^{12}6\theta}{72} - \dfrac{\cos^{10}6\theta}{60} + C$

17. $5\ln|x - 1| + \ln|x + 1| + C$

19. $\dfrac{\tan^3\theta}{3} + \tan\theta + C$ **21.** $\ln 8 - 1$

23. $-\dfrac{\cos^5\theta}{5} + \dfrac{2\cos^3\theta}{3} - \cos\theta + C$ **25.** $-\dfrac{1}{4}$

27. $\dfrac{2}{3}(\tan x)^{3/2} + C$

29. $\dfrac{\sin^6\theta}{6} - \dfrac{\sin^8\theta}{8} + C$ **31.** $-\dfrac{1}{3}u^3 + C = -\dfrac{1}{3}\cot^3 x + C$

33.
$\displaystyle\int_{\pi/4}^{\pi/3} \cot^2(x)\csc^3(x)\,dx = \frac{1}{16}\ln 3 + \frac{1}{8}\ln\left(\sqrt{2} - 1\right) + \frac{3}{8}\sqrt{2} - \frac{5}{56}$

35. $\dfrac{1}{49}\ln\left|\dfrac{t+4}{t-3}\right| - \dfrac{1}{7}\cdot\dfrac{1}{t-3} + C$ **37.** $\dfrac{1}{2}\sec^{-1}\dfrac{x}{2} + C$

39. $\displaystyle\int \frac{dx}{x^{3/2} + ax^{1/2}} = \begin{cases} \frac{2}{\sqrt{a}}\tan^{-1}\sqrt{\frac{x}{a}} + C & a > 0 \\[2mm] \frac{1}{\sqrt{-a}}\ln\left|\frac{\sqrt{x}-\sqrt{-a}}{\sqrt{x}+\sqrt{-a}}\right| + C & a < 0 \\[2mm] -\frac{2}{\sqrt{x}} + C & a = 0 \end{cases}$

41. $\ln|x+2| + 5/(x+2) - 3/(x+2)^2 + C$

43. $-\ln|x-2| - 2\frac{1}{x-2} + \frac{1}{2}\ln\left(x^2+4\right) + C$

45. $\frac{1}{3}\tan^{-1}\left(\frac{x+4}{3}\right) + C$ **47.** $\frac{1}{2}\ln|x-1| - \frac{1}{2}\ln|x+1| + C$

49. $-\frac{(x^2+4)^{3/2}}{48x^3} + \frac{\sqrt{x^2+4}}{16x} + C$ **51.** $-\frac{1}{9}e^{4-3x}(3x+4) + C$

53. $\frac{1}{2}x^2\sin x^2 + \frac{1}{2}\cos x^2 + C$

55. $\frac{x^2}{2}\tanh^{-1}x + \frac{x}{2} - \frac{1}{4}\ln\left|\frac{1+x}{1-x}\right| + C$

57. $x\ln\left(x^2+9\right) - 2x + 6\tan^{-1}\left(\frac{x}{3}\right) + C$

59. $\frac{1}{2}\sinh 2$ **61.** $t + \frac{1}{4}\coth(1-4t) + C$ **63.** $\frac{\pi}{3}$

65. $\tan^{-1}(\tanh x) + C$

67. (a) $I_n = \displaystyle\int \frac{x^n}{x^2+1}\,dx = \int \frac{x^{n-2}(x^2+1-1)}{x^2+1}\,dx =$

$\displaystyle\int x^{n-2}\,dx - \int \frac{x^{n-2}}{x^2+1}\,dx = \frac{x^{n-1}}{n-1} - I_{n-2}$

(b) $I_0 = \tan^{-1}x + C$; $I_1 = \frac{1}{2}\ln\left(x^2+1\right) + C$;

$I_2 = x - \tan^{-1}x + C$; $I_3 = \frac{x^2}{2} - \frac{1}{2}\ln\left(x^2+1\right) + C$;

$I_4 = \frac{x^3}{3} - x + \tan^{-1}x + C$; $I_5 = \frac{x^4}{4} - \frac{x^2}{2} + \frac{1}{2}\ln\left(x^2+1\right) + C$

(c) Prove by induction; show it works for $n = 1$, then assume it works for $n = k$ and use that to show it works for $n = k + 1$.

69. $\frac{3}{4}$ **71.** $C = 2$; $p(0 \le X \le 1) = 1 - \frac{2}{e}$

73. (a) 0.1587   (b) 0.49997

75. Integral converges; $I = \frac{1}{2}$.   **77.** Integral converges; $I = 3\sqrt[3]{4}$.

79. Integral converges; $I = \frac{\pi}{2}$.

81. The integral does not converge.

83. The integral does not converge.   **85.** The integral converges.

87. The integral converges.   **89.** The integral converges.

91. π **95.** $\frac{2}{(s-\alpha)^3}$

97. (a) T_N is smaller and M_N is larger than the integral.
(b) M_N is smaller and T_N is larger than the integral.
(c) M_N is smaller and T_N is larger than the integral.
(d) T_N is smaller and M_N is larger than the integral.

99. $M_5 \approx 0.7481$ **101.** $M_4 \approx 0.7450$ **103.** $S_4 \approx 0.7469$

105. $V \approx T_9 \approx 20$ acre-ft $= 871{,}200$ ft^3 **107.** Error $\le \frac{3}{128}$.

109. $N \ge 29$

Chapter 8

Section 8.1 Preliminary Questions

1. $\int_0^\pi \sqrt{1+\sin^2 x}\,dx$ **2.** $\int_0^h 2\pi r\,dx = 2\pi rh$ **3.** Yes

4. The graph of $y = f(x) + C$ is a vertical translation of the graph of $y = f(x)$; hence, the two graphs should have the same arc length. We can explicitly establish this as follows:

$$\text{Length of } y = f(x) + C = \int_a^b \sqrt{1 + \left[\frac{d}{dx}(f(x)+C)\right]^2}\,dx$$

$$= \int_a^b \sqrt{1 + [f'(x)]^2}\,dx$$

$$= \text{length of } y = f(x)$$

5. Since $\sqrt{1 + f'(x)^2} \ge 1$ for any function f, we have

$$\text{Length of graph of } y = f(x) \text{ over } [1,4] = \int_1^4 \sqrt{1 + f'(x)^2}\,dx$$

$$\ge \int_1^4 1\,dx = 3$$

Section 8.1 Exercises

1. $L = \int_2^6 \sqrt{1+16x^6}\,dx$ **3.** $\frac{13}{12}$ **5.** $3\sqrt{10}$

7. $\frac{1}{27}(22\sqrt{22} - 13\sqrt{13})$ **9.** $e^2 + \frac{\ln 2}{2} + \frac{1}{4}$

11. $\int_1^2 \sqrt{1+x^6}\,dx \approx 3.957736$ **13.** $\int_1^2 \sqrt{1+\frac{1}{x^4}}\,dx \approx 1.132123$

15. $\int_1^3 \sqrt{1+\frac{1}{x^2}}\,dx = 2.29896$ **17.** 6

21. $a = \sinh^{-1}(5) = \ln(5 + \sqrt{26})$

25. Let s denote the arc length. Then $s = \frac{a}{2}\sqrt{1+4a^2} + \frac{1}{4}\ln|\sqrt{1+4a^2} + 2a|$. Thus, when $a = 1$, $s = \frac{1}{2}\sqrt{5} + \frac{1}{4}\ln(\sqrt{5} + 2) \approx 1.478943$.

27. $\sqrt{1+e^{2a}} + \frac{1}{2}\ln\frac{\sqrt{1+e^{2a}}-1}{\sqrt{1+e^{2a}}+1} - \sqrt{2} + \frac{1}{2}\ln\frac{1+\sqrt{2}}{\sqrt{2}-1}$

29. $\ln(1+\sqrt{2})$ **33.** 1.552248   **35.** $16\pi\sqrt{2}$

37. $\frac{\pi}{27}(145^{3/2} - 1)$ **39.** $\frac{384\pi}{5}$ **41.** $\frac{\pi}{16}(e^4 - 9)$

43. $2\pi\int_1^3 x^{-1}\sqrt{1+x^{-4}}\,dx \approx 7.60306$

45. $2\pi\int_0^2 e^{-x^2/2}\sqrt{1+x^2e^{-x^2}}\,dx \approx 8.222696$

47. $2\pi\ln 2 + \frac{15\pi}{8}$ **49.** $4\pi^2 br$

51. $2\pi b^2 + \frac{2\pi ba^2}{\sqrt{a^2-b^2}}\sin^{-1}\left(\frac{\sqrt{a^2-b^2}}{a}\right)$

Section 8.2 Preliminary Questions

1. Pressure is defined as force per unit area.

2. The factor of proportionality is the weight density of the fluid, $w = \rho g$.

3. Fluid force acts in the direction perpendicular to the side of the submerged object.

4. Pressure depends only on depth and does not change horizontally at a given depth.

5. When a plate is submerged vertically, the pressure is not constant along the plate, so the fluid force is not equal to the pressure times the area.

Section 8.2 Exercises

1. (a) Top: $F = 176{,}400$ newtons; bottom: $F = 705{,}600$ newtons

(b) $F \approx \sum_{j=1}^N \rho g 3 y_j\,\Delta y$ (c) $F = \int_2^8 \rho g 3 y\,dy$

(d) $F = 882{,}000$ newtons

3. Difference = force on lower plate − force on upper plate = 19,600 newtons

5. (a) The width of the triangle varies linearly from 0 at a depth of $y = 3$ m to 1 at a depth of $y = 5$ m. Thus, $f(y) = \frac{1}{2}(y-3)$.

(b) The area of the strip at depth y is $\frac{1}{2}(y-3)\Delta y$, and the pressure at depth y is $\rho g y$, where $\rho = 10^3$ kg/m^3 and $g = 9.8$. Thus, the fluid force acting on the strip at depth y is approximately equal to $\rho g \frac{1}{2} y(y-3)\Delta y$.

(c) $F \approx \sum\limits_{j=1}^{N} \rho g \frac{1}{2} y_j (y_j - 3) \, \Delta y \to \int_3^5 \rho g \frac{1}{2} y (y - 3) \, dy$

(d) $F = \frac{127,400}{3}$ newtons

7. (b) $F = \frac{19,600}{3} r^3$ newtons

9. $F = \frac{19,600}{3} r^3 + 4900 \pi m r^2$ newtons   11. $F \approx 328,224,000$ lb

13. $F = \frac{815,360}{3}$ newtons   15. $F \approx 6153.18$ newtons

17. $F \approx 5652.4$ newtons   19. $F = 940,800$ newtons

21. 5.4604×10^{11} newtons   23. $F = (15b + 30a)h^2$ lb

25. Front and back: $F = \frac{62.5\sqrt{3}}{9} H^3$; slanted sides: $F = \frac{62.5\sqrt{3}}{3} \ell H^2$

Section 8.3 Preliminary Questions

1. $M_x = M_y = 0$ 2. $M_x = 21$ 3. $M_x = 5; M_y = 10$

4. Because a rectangle is symmetric with respect to both the vertical line and the horizontal line through the center of the rectangle, the Symmetry Principle guarantees that the centroid of the rectangle must lie along both these lines. The only point in common to both lines of symmetry is the center of the rectangle, so the centroid of the rectangle must be the center of the rectangle.

5. If the plate looks like a ring, then the center of mass doesn't occur at any point on the plate.

Section 8.3 Exercises

1. (a) $M_x = 4m; M_y = 9m$; center of mass: $\left(\frac{9}{4}, 1\right)$

(b) $\left(\frac{46}{17}, \frac{14}{17}\right)$

5. A sketch of the lamina is shown here.

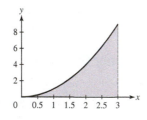

(a) $M_x = \frac{729}{10}; M_y = \frac{243}{4}$

(b) Area $= 9$ cm²; center of mass: $\left(\frac{9}{4}, \frac{27}{10}\right)$

7. $M_x = \frac{64\rho}{7}; M_y = \frac{32\rho}{5}$; center of mass: $\left(\frac{8}{5}, \frac{16}{7}\right)$

9. (a) $M_x = 24$

(b) $M = 12$, so $y_{cm} = 2$; center of mass: $(0, 2)$

11. $\left(\frac{93}{35}, \frac{45}{56}\right)$ 13. $\left(\frac{9}{8}, \frac{18}{5}\right)$

15. $\left(\frac{1-5e^{-4}}{1-e^{-4}}, \frac{1-e^{-8}}{4(1-e^{-4})}\right)$ 17. $\left(\frac{\pi}{2}, \frac{\pi}{8}\right)$

19. A sketch of the region is shown here.

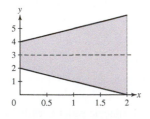

The region is clearly symmetric about the line $y = 3$, so we expect the centroid of the region to lie along this line. We find $M_x = 24$, $M_y = \frac{28}{3}$, centroid: $\left(\frac{7}{6}, 3\right)$.

21. $\left(\frac{9}{20}, \frac{9}{20}\right)$ 23. $\left(\frac{1}{2(e-2)}, \frac{e^2-3}{4(e-2)}\right)$ 25. $\left(\frac{\pi\sqrt{2}-4}{4(\sqrt{2}-1)}, \frac{1}{4(\sqrt{2}-1)}\right)$

27. A sketch of the region is shown here. Centroid: $\left(0, \frac{2}{7}\right)$

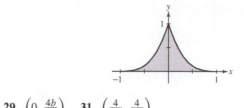

29. $\left(0, \frac{4b}{3\pi}\right)$ 31. $\left(\frac{4}{3\pi}, \frac{4}{3\pi}\right)$

33. $\left(0, \frac{\frac{2}{3}(r^2-h^2)^{3/2}}{r^2 \sin^{-1}\sqrt{1-h^2/r^2}-h\sqrt{r^2-h^2}}\right)$; with $r = 1$ and $h = \frac{1}{2}$:

$\left(0, \frac{3\sqrt{3}}{4\pi-3\sqrt{3}}\right) \approx (0, 0.71)$

35. $V = \frac{\pi r^2 H}{3}$ 37. $\left(0, \frac{49}{24}\right)$ 39. $\left(-\frac{4}{9\pi}, \frac{4}{9\pi}\right)$

41. For the square on the left: $(4, 4)$; for the square on the right: $\left(4, \frac{25}{7}\right)$

Section 8.4 Preliminary Questions

1. $T_3(x) = 9 + 8(x - 3) + 2(x - 3)^2 + 2(x - 3)^3$

2. The polynomial graphed on the right is a Maclaurin polynomial.

3. A Maclaurin polynomial gives the value of $f(0)$ exactly.

4. The correct statement is (b): $|T_3(2) - f(2)| \le \frac{2}{3}$.

Section 8.4 Exercises

1. $T_2(x) = x; T_3(x) = x - \frac{x^3}{6}$

3. $T_2(x) = \frac{1}{3} - \frac{1}{9}(x - 2) + \frac{1}{27}(x - 2)^2$;
$T_3(x) = \frac{1}{3} - \frac{1}{9}(x - 2) + \frac{1}{27}(x - 2)^2 - \frac{1}{81}(x - 2)^3$

5. $T_2(x) = 75 + 106(x - 3) + 54(x - 3)^2$;
$T_3(x) = 75 + 106(x - 3) + 54(x - 3)^2 + 12(x - 3)^3$

7. $T_2(x) = x; T_3(x) = x + \frac{x^3}{3}$

9. $T_2(x) = 2 - 3x + \frac{5x^2}{2}; T_3(x) = 2 - 3x + \frac{5x^2}{2} - \frac{3x^3}{2}$

11. $T_2(x) = \frac{1}{e} + \frac{1}{e}(x - 1) - \frac{1}{2e}(x - 1)^2$;
$T_3(x) = \frac{1}{e} + \frac{1}{e}(x - 1) - \frac{1}{2e}(x - 1)^2 - \frac{1}{6e}(x - 1)^3$

13. $T_2(x) = (x - 1) - \frac{3(x-1)^2}{2}$;
$T_3(x) = (x - 1) - \frac{3(x-1)^2}{2} + \frac{11(x-1)^3}{6}$

15. Let $f(x) = e^x$. Then, for all n,

$$f^{(n)}(x) = e^x \quad \text{and} \quad f^{(n)}(0) = 1$$

It follows that

$$T_n(x) = 1 + \frac{x}{1!} + \frac{x^2}{2!} + \cdots + \frac{x^n}{n!}$$

19. $T_n(x) = 1 - x + x^2 - x^3 + \cdots + (-1)^n x^n$

21. $T_n(x) = e + e(x - 1) + \frac{e(x-1)^2}{2!} + \cdots + \frac{e(x-1)^n}{n!}$

23.
$T_n(x) = \frac{1}{\sqrt{2}} - \frac{1}{\sqrt{2}}\left(x - \frac{\pi}{4}\right) - \frac{1}{2\sqrt{2}}\left(x - \frac{\pi}{4}\right)^2 + \frac{1}{6\sqrt{2}}\left(x - \frac{\pi}{4}\right)^3 \cdots$
In general, the coefficient of $(x - \pi/4)^n$ is

$$\pm \frac{1}{(\sqrt{2})n!}$$

with the pattern of signs $+, -, -, +, +, -, -, \dots$.

25. $T_2(x) = 1 + x + \frac{x^2}{2}; |T_2(-0.5) - f(-0.5)| \approx 0.018469$

27. $T_2(x) = 1 - \frac{2}{3}(x-1) + \frac{5}{9}(x-1)^2$;
$|f(1.2) - T_2(1.2)| \approx 0.00334008$

29. $T_3(x) = 1 + \frac{1}{2}(x-1) - \frac{1}{8}(x-1)^2 + \frac{1}{16}(x-1)^3$

31. $\frac{e^{1.1}|1.1|^4}{4!}$

33. $T_5(x) = 1 - \frac{x^2}{2} + \frac{x^4}{24}$; maximum error $= \frac{(0.25)^6}{6!}$

35. $T_3(x) = \frac{1}{2} - \frac{1}{16}(x-4) + \frac{3}{256}(x-4)^2 - \frac{5}{2048}(x-4)^3$;
maximum error $= \frac{35(0.3)^4}{65,536}$

37. $T_3(x) = x - \frac{x^3}{3}$; $T_3\left(\frac{1}{2}\right) = \frac{11}{24}$. With $K = 5$,

$$\left| T_3\left(\tfrac{1}{2}\right) - \tan^{-1}\tfrac{1}{2} \right| \le \frac{5\left(\tfrac{1}{2}\right)^4}{4!} = \frac{5}{384}$$

39. $K = 6.25$ is acceptable.   **41.** $n = 4$ **43.** $n = 6$ **47.** $n = 4$

51. $T_{4n}(x) = 1 - \frac{x^4}{2} + \frac{x^8}{4!} + \cdots + (-1)^n \frac{x^{4n}}{(2n)!}$

53. At $a = 0$,

$$T_1(x) = -4 - x$$
$$T_2(x) = -4 - x + 2x^2$$
$$T_3(x) = -4 - x + 2x^2 + 3x^3 = f(x)$$
$$T_4(x) = T_3(x)$$
$$T_5(x) = T_3(x)$$

At $a = 1$,

$$T_1(x) = 12(x-1)$$
$$T_2(x) = 12(x-1) + 11(x-1)^2$$
$$T_3(x) = 12(x-1) + 11(x-1)^2 + 3(x-1)^3$$
$$= -4 - x + 2x^2 + 3x^3 = f(x)$$
$$T_4(x) = T_3(x)$$
$$T_5(x) = T_3(x)$$

55. $T_2(t) = 60 + 24t - \frac{3}{2}t^2$; truck's distance from the intersection after 4 s is ≈ 132 m.

57. (a) $T_3(x) = -\frac{k}{R^3}x + \frac{3k}{2R^5}x^3$

65. $T_4(x) = 1 - x^2 + \frac{1}{2}x^4$; the error is approximately
$|0.461458 - 0.461281| = 0.000177$.

67. (b) $\int_0^{1/2} T_4(x)\,dx = \frac{1841}{3840}$; error bound:

$$\left| \int_0^{1/2} \cos x\,dx - \int_0^{1/2} T_4(x)\,dx \right| < \frac{\left(\tfrac{1}{2}\right)^7}{6!}$$

69. (a) $T_6(x) = x^2 - \frac{1}{6}x^6$

Chapter 8 Review

1. $\frac{779}{240}$ **3.** $4\sqrt{17}$ **7.** $24\pi\sqrt{2}$ **9.** $\frac{67\pi}{36}$

11. $12\pi + 4\pi^2$ **13.** 176,400 newtons

15. Fluid force on a triangular face: $183,750\sqrt{3} + 306,250$ newtons; fluid force on a slanted rectangular edge: $122,500\sqrt{3} + 294,000$ newtons

17. $M_x = 20,480$; $M_y = 25,600$; center of mass: $\left(2, \frac{8}{5}\right)$

19. $\left(0, \frac{2}{\pi}\right)$ **21.** $\frac{3\pi}{10}$

23. $T_3(x) = 1 + 3(x-1) + 3(x-2)^2 + (x-1)^3$

25. $T_4(x) = (x-1) + \frac{1}{2}(x-1)^2 - \frac{1}{6}(x-1)^3 + \frac{1}{12}(x-1)^4$

27. $T_4(x) = x - x^3$

29. $T_n(x) = 1 + 3x + \frac{1}{2!}(3x)^2 + \frac{1}{3!}(3x)^3 + \cdots + \frac{1}{n!}(3x)^n$

31. $T_3(1.1) = 0.832981496$; $\left| T_3(1.1) - \tan^{-1}1.1 \right| = 2.301 \times 10^{-7}$

33. $n = 11$ is sufficient.

35. The nth Maclaurin polynomial for $g(x) = \frac{1}{1+x}$ is
$T_n(x) = 1 - x + x^2 - x^3 + \cdots + (-x)^n$.

Chapter 9

Section 9.1 Preliminary Questions

1. (a) First order   (b) First order   (c) Order 3   (d) Order 2

2. Yes   **3.** Example: $y' = y^2$ **4.** Example: $y' = y^2$

5. Example: $y' + y = x$

Section 9.1 Exercises

1. (a) First order   (b) Not first order   (c) First order
(d) First order   (e) Not first order   (f) First order

3. Let $y = 4x^2$. Then $y' = 8x$ and $y' - 8x = 8x - 8x = 0$.

5. Let $y = 25e^{-2x^2}$. Then $y' = -100xe^{-2x^2}$ and
$y' + 4xy = -100xe^{-2x^2} + 4x(25e^{-2x^2}) = 0$.

7. Let $y = 4x^4 - 12x^2 + 3$. Then

$$y'' - 2xy' + 8y = (48x^2 - 24) - 2x(16x^3 - 24x) + 8(4x^4 - 12x^2 + 3)$$
$$= 48x^2 - 24 - 32x^4 + 48x^2 + 32x^4 - 96x^2 + 24 = 0$$

9. (a) Separable: $y' = \frac{9}{x}y^2$ (b) Separable: $y' = \frac{\sin x}{\sqrt{4 - x^2}}e^{3y}$
(c) Not separable   (d) Separable: $y' = (1)(9 - y^2)$

11. $y = \frac{1}{3}x^3 + 1$; $y = \frac{1}{1-x}$ **13.** (d) $y = \ln\left|\frac{1}{x} - \frac{1}{2} + e^4\right|$

15. $y = Ce^{-x^3/3}$, where C is an arbitrary constant.

17. $y = \ln\left(4t^5 + C\right)$, where C is an arbitrary constant.

19. $y = Ce^{-5x/2} + \frac{4}{5}$, where C is an arbitrary constant.

21. $y = Ce^{-\sqrt{1-x^2}}$, where C is an arbitrary constant.

23. $y = \pm\sqrt{x^2 + C}$, where C is an arbitrary constant.

25. $x = \tan(\frac{1}{2}t^2 + t + C)$, where C is an arbitrary constant.

27. $y = \sin^{-1}\left(\frac{1}{2}x^2 + C\right)$, where C is an arbitrary constant.

29. $y = C\sec t$, where C is an arbitrary constant.

31. $y = 75e^{-2x}$ **33.** $y = -\sqrt{\ln\left(x^2 + e^4\right)}$

35. $y = 2 + 2e^{x(x-2)/2}$ **37.** $y = \tan\left(x^2/2\right)$ **39.** $y = e^{1-e^{-t}}$

41. $y = \frac{et}{e^{1/t}} - 1$ **43.** $y = \sin^{-1}\left(\frac{1}{2}e^x\right)$ **45.** $a = -3, 4$

47. $t = \pm\sqrt{\pi + 4}$

49. (a) ≈ 1145 s or 19.1 min   (b) ≈ 3910 s or 65.2 min

51. $y = 8 - (8 + 0.0002215t)^{2/3}$; $t_e \approx 66000$ s or 18 hr, 20 min

55. (a) $q(t) = CV\left(1 - e^{-t/RC}\right)$

(b)
$$\lim_{t\to\infty} q(t) = \lim_{t\to\infty} CV\left(1 - e^{-t/RC}\right) = \lim_{t\to\infty} CV(1 - 0) = CV$$

(c) $q(RC) = CV\left(1 - e^{-1}\right) \approx (0.63)CV$

57. $V = (kt/3 + C)^3$, V increases roughly with the cube of time.

59. $g(x) = Ce^{(3/2)x}$, where C is an arbitrary constant; $g(x) = \frac{C}{x-1}$, where C is an arbitrary constant.

61. $y = Cx^3$ and $y = \pm\sqrt{A - \frac{x^2}{3}}$

63. (b) $v(t) = -9.8t + 100(\ln(50) - \ln(50 - 4.75t))$;
$v(10) = -98 + 100(\ln(50) - \ln(2.5)) \approx 201.573$ m/s

69. (c) $C = \dfrac{14\pi}{15B\sqrt{2g}} \cdot R^{5/2}$

Section 9.2 Preliminary Questions

1. $y(t) = 5 - ce^{4t}$ for any positive constant c

2. No   **3.** True

4. The difference in temperature between a cooling object and the ambient temperature is decreasing. Hence, the rate of cooling, which is proportional to this difference, is also decreasing in magnitude.

Section 9.2 Exercises

1. General solution: $y(t) = 10 + ce^{2t}$; solution satisfying $y(0) = 25$: $y(t) = 10 + 15e^{2t}$; solution satisfying $y(0) = 5$: $y(t) = 10 - 5e^{2t}$

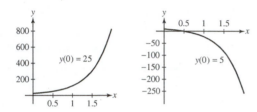

3. $y = -6 + 11e^{4x}$

5. (a) $y' = -0.02(y - 10)$ **(b)** $y = 10 + 90e^{-\frac{1}{50}t}$
(c) $100 \ln 3$ s ≈ 109.8 s

7. $\approx 5:50$ AM   **9.** ≈ 0.77 min $= 46.6$ s

11. $500 \ln \frac{3}{2}$ s ≈ 203 s $= 3$ min 23 s

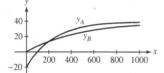

13. -58.8 m/s   **15.** -11.8 m/s

17. (a)   i. $17,563.94   **ii.** approximately 13.86 years or about 13 years 10 months
(c) $120,000   **(d)** $107,629.00   **(e)** 8%

19. $4068.73 per year

21. (a) $I(t) = \dfrac{V}{R}\left(1 - e^{-\left(\frac{R}{L}\right)t}\right)$

Section 9.3 Preliminary Questions

1. 7   **2.** $y = \pm\sqrt{1 + t}$ **3. (b)**   **4.** 20

Section 9.3 Exercises

1. 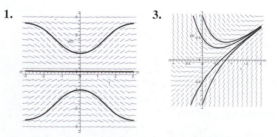   **3.**

5. (a) 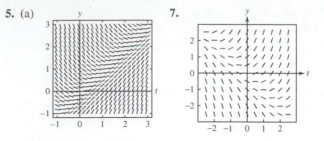   **7.**

9. For $y' = t$, y' only depends on t. The isoclines of any slope c will be the vertical lines $t = c$.

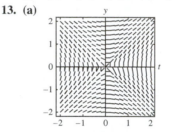

11. (i) C (ii) B (iii) F (iv) D (v) A (vi) E

13. (a)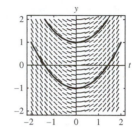

15. (a) $y_1 = 3.1$ **(b)** $y_2 = 3.231$
(c) $y_3 = 3.3919$, $y_4 = 3.58171$, $y_5 = 3.799539$, $y_6 = 4.0445851$
(d) $y(2.2) \approx 3.231$, $y(2.5) \approx 3.799539$

17. $y(0.5) \approx 1.7210$ **19.** $y(3.3) \approx 3.3364$

21. $y(2) \approx 2.8838$ **25.** $y(0.5) \approx 1.794894$

27. $y(0.25) \approx 1.094871$

Section 9.4 Preliminary Questions

1. (a) No   **(b)** Yes   **(c)** No   **(d)** Yes

2. No   **3.** Yes

Section 9.4 Exercises

1. $y = \dfrac{5}{1 - e^{-3t}/C}$ and $y = \dfrac{5}{1 + (3/2)e^{-3t}}$

3. $\lim\limits_{t \to \infty} y(t) = 2$

5. (a) $P(t) = \dfrac{2000}{1 + 3e^{-0.6t}}$ **(b)** $t = \frac{1}{0.6}\ln 3 \approx 1.83$ years

7. $k = \ln\frac{81}{31} \approx 0.96$ year^{-1}; $t = \dfrac{\ln 9}{2\ln 9 - \ln 31} \approx 2.29$ years

9. After $t = 7.6$ h, or at 3:36 PM

11. (a) $y_1(t) = \dfrac{10}{10 - 9e^{-t}}$ and $y_2(t) = \dfrac{1}{1 - 2e^{-t}}$
(b) $t = \ln\frac{9}{8}$ **(c)** $t = \ln 2$

13. (a) $A(t) = 16(1 - \frac{5}{3}e^{t/40})^2/(1 + \frac{5}{3}e^{t/40})^2$
(b) $A(10) \approx 2.1$
(c)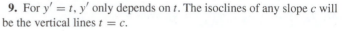

15. ≈ 943 million

17. (d) $t = -\frac{1}{k}\left(\ln y_0 - \ln(A - y_0)\right)$

Section 9.5 Preliminary Questions

1. (a) Yes **(b)** No **(c)** Yes **(d)** No

2. (b)

3. $P(x) = x^{-1}$ **4.** $P(x) = 1$

Section 9.5 Exercises

1. (c) $y = \frac{x^4}{5} + \frac{C}{x}$ **(d)** $y = \frac{x^4}{5} - \frac{1}{5x}$

5. $y = \frac{1}{2}x + \frac{C}{x}$ **7.** $y = -\frac{1}{4}x^{-1} + Cx^{1/3}$

9. $y = \frac{1}{5}x^2 + \frac{1}{3} + Cx^{-3}$ **11.** $y = -x\ln x + Cx$

13. $y = \frac{1}{2}e^x + Ce^{-x}$ **15.** $y = x\cos x + C\cos x$

17. $y = x^x + Cx^x e^{-x}$ **19.** $y = \frac{1}{5}e^{2x} - \frac{6}{5}e^{-3x}$

21. $y = \frac{\ln|x|}{x+1} - \frac{1}{x(x+1)} + \frac{5}{x+1}$ **23.** $y = -\cos x + \sin x$

25. $y = \tanh x + 3\operatorname{sech} x$

27. For $m \neq -n$: $y = \frac{1}{m+n}e^{mx} + Ce^{-nx}$; for $m = -n$:
$y = (x + C)e^{-nx}$

29. (a) $y' = 4000 - \frac{40y}{500+40t}$; $y = 1000\frac{4t^2+100t+125}{2t+25}$

(b) 40 g/L

31. 50 g/L

33. (a) $\frac{dV}{dt} = \frac{20}{1+t} - 5$ and $V(t) = 20\ln(1+t) - 5t + 100$

(b) The maximum value is $V(3) = 20\ln 4 - 15 + 100 \approx 112.726$.

(c) Estimate empty at ≈ 34 min.

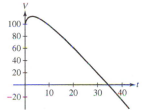

35. $I(t) = \frac{1}{10}\left(1 - e^{-20t}\right)$

37. (a) $I(t) = \frac{V}{R} - \frac{V}{R}e^{-(R/L)t}$ **(c)** Approximately 0.0184 s

39. (b) $c_1(t) = 10e^{-t/6}$

Chapter 9 Review

1. (a) No, first order **(b)** Yes, first order **(c)** No, order 3
(d) Yes, second order

3. $y = \pm\left(\frac{4}{3}t^3 + C\right)^{1/4}$, where C is an arbitrary constant.

5. $y = Cx - 1$, where C is an arbitrary constant.

7. $y = \frac{1}{2}\left(x + \frac{1}{2}\sin 2x\right) + \frac{\pi}{4}$ **9.** $y = \frac{2}{2-x^2}$

11.

13.

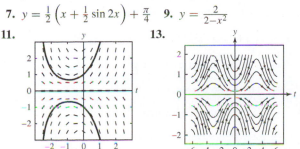

15. $y(t) = \tan t$

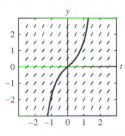

17. $y(0.1) \approx 1.1$; $y(0.2) \approx 1.209890$; $y(0.3) \approx 1.329919$

19. $y = x^2 + 2x$ **21.** $y = \frac{1}{2} + e^{-x} - \frac{11}{2}e^{-2x}$

23. $y = \frac{1}{2}\sin 2x - 2\cos x$ **25.** $y = 1 - \sqrt{t^2 + 15}$

27. $w = \tan\left(k\ln x + \frac{\pi}{4}\right)$

29. $y = -\cos x + \frac{\sin x}{x} + \frac{C}{x}$, where C is an arbitrary constant.

31. Solution satisfying $y(0) = 3$: $y(t) = 4 - e^{-2t}$; solution
satisfying $y(0) = 4$: $y(t) = 4$

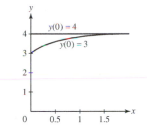

33. (a) 12

(b) ∞, if $y(0) > 12$; 12, if $y(0) = 12$; $-\infty$, if $y(0) < 12$

(c) -3

35. $400,000 - 200,000e^{0.25} \approx \$143,194.91$ **37.** $\$400,000$

39. Solutions are of the form $y = \frac{B}{A} + Ce^{-At}$ and $\lim_{t\to\infty} y = \frac{B}{A}$.

41. $\frac{-7\sqrt{10}\sqrt{y}}{240y+64,800}$: $t = 3225.88$ s or 51 min 56 s

43. 2 **45.** $t = 5\ln 441 \approx 30.45$ days

49. (a) $\frac{dc_1}{dt} = -\frac{2}{5}c_1$ **(b)** $c_1(t) = 8e^{(-2/5)t}$ g/L

Chapter 10

Section 10.1 Preliminary Questions

1. $a_4 = 12$ **2. (c)** **3.** $\lim_{n\to\infty} a_n = \sqrt{2}$ **4. (b)**

5. (a) False. Counterexample: $a_n = \cos\pi n$

(b) True **(c)** False. Counterexample: $a_n = (-1)^n$

Section 10.1 Exercises

1. (a) (iv) **(b)** (i) **(c)** (iii) **(d)** (ii)

3. $c_1 = 3, c_2 = \frac{9}{2}, c_3 = \frac{9}{2}, c_4 = \frac{27}{8}$

5. $a_1 = 2, a_2 = 5, a_3 = 47, a_4 = 4415$

7. $b_1 = 4, b_2 = 6, b_3 = 4, b_4 = 6$

9. $c_1 = 1, c_2 = \frac{3}{2}, c_3 = \frac{11}{6}, c_4 = \frac{25}{12}$

11. $b_1 = 2, b_2 = 3, b_3 = 8, b_4 = 19$

13. (a) $a_n = \frac{(-1)^{n+1}}{n^3}$ **(b)** $a_n = \frac{n+1}{n+5}$

15. $\lim_{n\to\infty} 12 = 12$ **17.** $\lim_{n\to\infty} \frac{5n-1}{12n+9} = \frac{5}{12}$

19. $\lim_{n\to\infty} (-2^{-n}) = 0$ **21.** The sequence diverges.

23. $\lim_{n\to\infty} \frac{n}{\sqrt{n^2+1}} = 1$ **25.** $\lim_{n\to\infty} \ln\left(\frac{12n+2}{-9+4n}\right) = \ln 3$

27. $\lim\limits_{n\to\infty} \sqrt{4+\frac{1}{n}} = 2$ **29.** $\lim\limits_{n\to\infty} \cos^{-1}\left(\frac{n^3}{2n^3+1}\right) = \frac{\pi}{3}$

31. **(a)** $M = 999$ **(b)** $M = 99{,}999$

35. $\lim\limits_{n\to\infty}\left(10 + \left(-\frac{1}{9}\right)^n\right) = 10$ **37.** The sequence diverges.

39. $\lim\limits_{n\to\infty} 2^{1/n} = 1$ **41.** $\lim\limits_{n\to\infty}\frac{9^n}{n!} = 0$

43. $\lim\limits_{n\to\infty}\frac{3n^2+n+2}{2n^2-3} = \frac{3}{2}$ **45.** $\lim\limits_{n\to\infty}\frac{\cos n}{n} = 0$

47. The sequence diverges. **49.** $\lim\limits_{n\to\infty}\left(2 + \frac{4}{n^2}\right)^{1/3} = 2^{1/3}$

51. $\lim\limits_{n\to\infty} \ln\left(\frac{2n+1}{3n+4}\right) = \ln\frac{2}{3}$ **53.** The sequence diverges.

55. $\lim\limits_{n\to\infty}\frac{e^n+(-3)^n}{5^n} = 0$ **57.** $\lim\limits_{n\to\infty} n\sin\frac{\pi}{n} = \pi$

59. $\lim\limits_{n\to\infty}\frac{3-4^n}{2+7\cdot4^n} = -\frac{1}{7}$ **61.** $\lim\limits_{n\to\infty}\left(1+\frac{1}{n}\right)^n = e$

63. $\lim\limits_{n\to\infty}\frac{(\ln n)^2}{n} = 0$ **65.** $\lim\limits_{n\to\infty} n(\sqrt{n^2+1} - n) = \frac{1}{2}$

67. $\lim\limits_{n\to\infty}\frac{1}{\sqrt{n^4+n^8}} = 0$ **69.** $\lim\limits_{n\to\infty}(2^n+3^n)^{1/n} = 3$ **71.** (b)

73. Any number greater than or equal to 3 is an upper bound.

75. Example: $a_n = (-1)^n$ **79.** Example: $f(x) = \sin\pi x$

87. (e) $AGM\left(1, \sqrt{2}\right) \approx 1.198$

Section 10.2 Preliminary Questions

1. The sum of an infinite series is defined as the limit of the sequence of partial sums. If the limit of this sequence does not exist, the series is said to diverge.

2. $S = \frac{1}{2}$

3. The result is negative, so the result is not valid: a series with all positive terms cannot have a negative sum. The formula is not valid because a geometric series with $|r| \geq 1$ diverges.

4. No **5.** No **6.** $N = 13$

7. No, S_N is increasing and converges to 1, so $S_N \leq 1$ for all N.

8. Example: $\sum\limits_{n=1}^{\infty}\frac{1}{n^{9/10}}$

Section 10.2 Exercises

1. **(a)** $a_n = \frac{1}{3^n}$ **(b)** $a_n = \left(\frac{5}{2}\right)^{n-1}$

(c) $a_n = (-1)^{n+1}\frac{n^n}{n!}$ **(d)** $a_n = \frac{1+\frac{(-1)^{n+1}+1}{2}}{n^2+1}$

3. $S_2 = \frac{5}{4}$, $S_4 = \frac{205}{144}$, $S_6 = \frac{5369}{3600}$

5. $S_2 = \frac{2}{3}$, $S_4 = \frac{4}{5}$, $S_6 = \frac{6}{7}$ **7.** $S_6 = 1.24992$

9. $S_{10} = 0.03535167962$, $S_{100} = 0.03539810274$, $S_{500} = 0.03539816290$, $S_{1000} = 0.03539816334$. Yes.

11. $S_3 = \frac{3}{10}$, $S_4 = \frac{1}{3}$, $S_5 = \frac{5}{14}$, $\sum\limits_{n=1}^{\infty}\left(\frac{1}{n+1} - \frac{1}{n+2}\right) = \frac{1}{2}$

13. $S_3 = \frac{3}{7}$, $S_4 = \frac{4}{9}$, $S_5 = \frac{5}{11}$, $\sum\limits_{n=1}^{\infty}\frac{1}{4n^2-1} = \frac{1}{2}$

15. $S = \frac{1}{2}$ **17.** $\lim\limits_{n\to\infty}\frac{n}{10n+12} = \frac{1}{10} \neq 0$

19. $\lim\limits_{n\to\infty}(-1)^n\left(\frac{n-1}{n}\right)$ does not exist.

21. $\lim\limits_{n\to\infty} a_n = \lim\limits_{n\to\infty}\cos\frac{1}{n+1} = 1 \neq 0$

23. $S = \frac{8}{7}$ **25.** The series diverges. **27.** $S = \frac{59{,}049}{3328}$

29. $S = \frac{1}{e-1}$ **31.** $S = \frac{35}{3}$ **33.** $S = 4$ **35.** $S = \frac{7}{15}$ **37.** $\frac{2}{9}$

39. $\frac{31}{99}$

41. The fractions $S_a = \frac{a}{9}$, where $a = 1, 2, \ldots, 8$ have repeating decimals of the form $0.aaa\ldots$

43. (b) and (c)

47. **(a)** Counterexample: $\sum\limits_{n=1}^{\infty}\left(\frac{1}{2}\right)^n = 1$.

(b) Counterexample: If $a_n = 1$, then $S_N = N$.

(c) Counterexample: $\sum\limits_{n=1}^{\infty}\frac{1}{n}$ diverges.

(d) Counterexample: $\sum\limits_{n=1}^{\infty}\cos 2\pi n \neq 1$.

49. The total area is $\frac{1}{4}$.

51. **(a)** $De^{-k} + De^{-2k} + De^{-3k} + \cdots = \frac{De^{-k}}{1-e^{-k}}$

(b) $De^{-kt} + De^{-2kt} + De^{-3kt} + \cdots = \frac{De^{-kt}}{1-e^{-kt}}$

(c) $t \geq -\frac{1}{k}\ln\left(1 - \frac{D}{S}\right)$

53. The total length of the path is $2 + \sqrt{2}$. **57.** 42 ft

Section 10.3 Preliminary Questions

1. (b)

2. A function f such that $a_n = f(n)$ must be positive, decreasing, and continuous for $x \geq 1$.

3. Convergence of p-series or integral test

4. Comparison Test

5. No; $\sum\limits_{n=1}^{\infty}\frac{1}{n}$ diverges, but since $\frac{e^{-n}}{n} < \frac{1}{n}$ for $n \geq 1$, the

Comparison Test tells us nothing about the convergence of $\sum\limits_{n=1}^{\infty}\frac{e^{-n}}{n}$.

Section 10.3 Exercises

1. $\int_1^{\infty}\frac{dx}{x^4}\,dx$ converges, so the series converges.

3. $\int_1^{\infty} x^{-1/3}\,dx = \infty$, so the series diverges.

5. $\int_{25}^{\infty}\frac{x^2}{(x^3+9)^{5/2}}\,dx$ converges, so the series converges.

7. $\int_1^{\infty}\frac{dx}{x^2+1}$ converges, so the series converges.

9. $\int_1^{\infty}\frac{dx}{x(x+1)}$ converges, so the series converges.

11. $\int_2^{\infty}\frac{1}{x(\ln x)^2}\,dx$ converges, so the series converges.

13. $\int_1^{\infty}\frac{dx}{2^{\ln x}} = \infty$, so the series diverges.

15. $\frac{1}{n^3+8n} \leq \frac{1}{n^3}$, so the series converges.

19. $\frac{1}{n2^n} \leq \left(\frac{1}{2}\right)^n$, so the series converges.

21. $\frac{1}{n^{1/3}+2^n} \leq \left(\frac{1}{2}\right)^n$, so the series converges.

23. $\frac{4}{m!+4^m} \leq 4\left(\frac{1}{4}\right)^m$, so the series converges.

25. $0 \leq \frac{\sin^2 k}{k^2} \leq \frac{1}{k^2}$, so the series converges.

27. $\frac{2}{3^n+3^{-n}} \leq 2\left(\frac{1}{3}\right)^n$, so the series converges.

29. $\frac{1}{(n+1)!} \leq \frac{1}{n^2}$, so the series converges.

31. $\frac{\ln n}{n^3} \leq \frac{1}{n^2}$ for $n \geq 1$, so the series converges.

33. $\frac{(\ln n)^{100}}{n^{1.1}} \leq \frac{1}{n^{1.09}}$ for n sufficiently large, so the series converges.

35. $\frac{n}{3^n} \leq \left(\frac{2}{3}\right)^n$ for $n \geq 1$, so the series converges.

39. The series converges.   **41.** The series diverges.

43. The series converges.   **45.** The series diverges.

47. The series converges.   **49.** The series converges.

51. The series diverges.   **53.** The series converges.

55. The series diverges.   **57.** The series converges.

59. The series diverges.   **61.** The series diverges.

63. The series diverges.   **65.** The series converges.

67. The series diverges.   **69.** The series diverges.

71. The series converges.   **73.** The series converges.

75. The series diverges.   **77.** The series converges.

79. The series converges for $a > 1$ and diverges for $a \leq 1$.

81. The series converges for $p > 1$ and diverges for $p \leq 1$.

89. $\displaystyle\sum_{n=1}^{\infty} n^{-5} \approx 1.0369540120.$

93. $\displaystyle\sum_{n=1}^{1000} \frac{1}{n^2} = 1.6439345667$ and $1 + \displaystyle\sum_{n=1}^{100} \frac{1}{n^2(n+1)} = 1.6448848903.$

The second sum is a better approximation to $\dfrac{\pi^2}{6} \approx 1.6449340668.$

Section 10.4 Preliminary Questions

1. Example: $\displaystyle\sum \frac{(-1)^n}{\sqrt[3]{n}}$ **2.** (b)   **3.** No

4. $|S - S_{100}| \leq 10^{-3}$, and S is larger than S_{100}.

Section 10.4 Exercises

3. Converges conditionally.   **5.** Converges absolutely.

7. Converges conditionally.   **9.** Converges conditionally.

11. (a)

n	S_n	n	S_n
1	1	6	0.899782407
2	0.875	7	0.902697859
3	0.912037037	8	0.900744734
4	0.896412037	9	0.902116476
5	0.904412037	10	0.901116476

13. $S_5 = 0.947$ **15.** $S_{44} = 0.06567457397$

17. Converges (by geometric series).

19. Converges (by Comparison Test).

21. Converges (by Limit Comparison Test).

23. Diverges (by Limit Comparison Test).

25. Converges (by geometric series and linearity).

27. Converges absolutely (by Integral Test).

29. Converges (by Alternating Series Test).

31. Converges (by Integral Test).   **33.** Converges conditionally.

Section 10.5 Preliminary Questions

1. $\rho = \lim\limits_{n \to \infty} \left| \dfrac{a_{n+1}}{a_n} \right|$

2. The Ratio Test is conclusive for $\displaystyle\sum_{n=1}^{\infty} \frac{1}{2^n}$ and inconclusive

for $\displaystyle\sum_{n=1}^{\infty} \frac{1}{n}$.

3. No

Section 10.5 Exercises

1. Converges absolutely.   **3.** Converges absolutely.

5. The Ratio Test is inconclusive.   **7.** Diverges.

9. Converges absolutely.   **11.** Converges absolutely.

13. Diverges.   **15.** The Ratio Test is inconclusive.

17. Converges absolutely.   **19.** Converges absolutely.

21. $\rho = \frac{1}{3} < 1$ **23.** $\rho = 2|x|$

25. $\rho = |r|$ **29.** Converges absolutely.

31. The Ratio Test is inconclusive, so the series may converge or diverge.

33. Converges absolutely.   **35.** The Ratio Test is inconclusive.

37. Converges absolutely.   **39.** Converges absolutely.

41. Converges absolutely.

43. Converges (by geometric series and linearity).

45. Diverges (by the Divergence Test).

47. Converges (by the Direct Comparison Test).

49. Diverges (by the Direct Comparison Test).

51. Converges (by the Ratio Test).

53. Converges (by the Limit Comparison Test).

55. Diverges (by p-series).   **57.** Converges (by geometric series).

59. Converges (by Limit Comparison Test).

61. Diverges (by Divergence Test).

65. (b) $\sqrt{2\pi} \approx 2.50663$

n	$\dfrac{e^n n!}{n^{n+1/2}}$
1000	2.506837
1500	2.506768
2000	2.506733
2500	2.506712
3000	2.506698

Section 10.6 Preliminary Questions

1. Yes. The series must converge for both $x = 4$ and $x = -3$.

2. (a), (c)   **3.** $R = 4$

4. $F'(x) = \displaystyle\sum_{n=1}^{\infty} n^2 x^{n-1}; R = 1$

Section 10.6 Exercises

1. $R = 2$. It does not converge at the endpoints.

3. $R = 3$ for all three series.

9. $(-1, 1)$ **11.** $[-\sqrt{2}, \sqrt{2}]$ **13.** $[-1, 1]$ **15.** $(-\infty, \infty)$

17. $(-\infty, \infty)$ **19.** $(-1, 1]$ **21.** $(-1, 1)$ **23.** $[-1, 1)$

25. $(2, 4)$ **27.** $(6, 8)$ **29.** $\left[-\frac{7}{2}, -\frac{5}{2} \right)$ **31.** $(-\infty, \infty)$

33. $\left(2 - \frac{1}{e}, 2 + \frac{1}{e} \right)$

35. $\displaystyle\sum_{n=0}^{\infty} 3^n x^n$ on the interval $\left(-\frac{1}{3}, \frac{1}{3} \right)$

37. $\displaystyle\sum_{n=0}^{\infty} \frac{x^n}{3^{n+1}}$ on the interval $(-3, 3)$

39. $\displaystyle\sum_{n=0}^{\infty} (-1)^n x^{2n}$ on the interval $(-1, 1)$

43. $\displaystyle\sum_{n=0}^{\infty} (-1)^{n+1} (x - 5)^n$ on the interval $(4, 6)$

47. (c) $S_4 = \frac{69}{640}$ and $|S - S_4| \approx 0.000386 < a_5 = \frac{1}{1920}$

49. $R = 1$ **51.** $\sum\limits_{n=1}^{\infty} \frac{n}{2^n} = 2$ **53.** $F(x) = \frac{1-x-x^2}{1-x^3}$

55. $-1 \le x \le 1$ **57.** $P(x) = \sum\limits_{n=0}^{\infty} (-1)^n \frac{x^n}{n!}$

59. N must be at least 5; $S_5 = 0.3680555556$

61. $P(x) = 1 - \frac{1}{2}x^2 - \sum\limits_{n=2}^{\infty} \frac{1\cdot3\cdot5\cdots(2n-3)}{(2n)!}x^{2n}; \ R = \infty$

Section 10.7 Preliminary Questions

1. $f(0) = 3$ and $f'''(0) = 30$

2. $f(-2) = 0$ and $f^{(4)}(-2) = 48$

3. Substitute x^2 for x in the Maclaurin series for $\sin x$.

4. $f(x) = 4 + \sum\limits_{n=1}^{\infty} \frac{(x-3)^{n+1}}{n(n+1)}$ **5.** (c)

Section 10.7 Exercises

1. $f(x) = 2 + 3x + 2x^2 + 2x^3 + \cdots$

3. $\frac{1}{1-2x} = \sum\limits_{n=0}^{\infty} 2^n x^n$ on the interval $\left(-\frac{1}{2}, \frac{1}{2}\right)$

5. $\cos 3x = \sum\limits_{n=0}^{\infty} (-1)^n \frac{9^n x^{2n}}{(2n)!}$ on the interval $(-\infty, \infty)$

7. $\sin(x^2) = \sum\limits_{n=0}^{\infty} (-1)^n \frac{x^{4n+2}}{(2n+1)!}$ on the interval $(-\infty, \infty)$

9. $\ln(1 - x^2) = -\sum\limits_{n=1}^{\infty} \frac{x^{2n}}{n}$ on the interval $(-1, 1)$

11. $\tan^{-1}(x^2) = \sum\limits_{n=0}^{\infty} (-1)^n \frac{x^{4n+2}}{2n+1}$ on the interval $[-1, 1]$

13. $e^{x-2} = \sum\limits_{n=0}^{\infty} \frac{x^n}{e^2 n!}$ on the interval $(-\infty, \infty)$

15. $\ln(1 - 5x) = -\sum\limits_{n=1}^{\infty} \frac{5^n x^n}{n}$ on the interval $\left[-\frac{1}{5}, \frac{1}{5}\right)$

17. $\sinh x = \sum\limits_{k=0}^{\infty} \frac{x^{2k+1}}{(2k+1)!}$ on the interval $(-\infty, \infty)$

19. $e^x \sin x = x + x^2 + \frac{x^3}{3} - \frac{x^5}{30} + \cdots$

21. $\frac{\sin x}{1-x} = x + x^2 + \frac{5x^3}{6} + \frac{5x^4}{6} + \cdots$

23. $(1+x)^{1/4} = 1 + \frac{1}{4}x - \frac{3}{32}x^2 + \frac{7}{128}x^3 + \cdots$

25. $e^x \tan^{-1} x = x + x^2 + \frac{1}{6}x^3 - \frac{1}{6}x^4 + \cdots$

27. $e^{\sin x} = 1 + x + \frac{1}{2}x^2 - \frac{1}{8}x^4 + \cdots$

29. $\frac{1}{x} = \sum\limits_{n=0}^{\infty} (-1)^n (x-1)^n$ on the interval $(0, 2)$

31. $\frac{1}{1-x} = \sum\limits_{n=0}^{\infty} (-1)^{n+1} \frac{(x-5)^n}{4^{n+1}}$ on the interval $(1, 9)$

33. $21 + 35(x-2) + 24(x-2)^2 + 8(x-2)^3 + (x-2)^4$ on the interval $(-\infty, \infty)$

35. $\frac{1}{x^2} = \sum\limits_{n=0}^{\infty} (-1)^n (n+1) \frac{(x-4)^n}{4^{n+2}}$ on the interval $(0, 8)$

37. $\frac{1}{1-x^2} = \sum\limits_{n=0}^{\infty} \frac{(-1)^{n+1}(2^{n+1}-1)}{2^{2n+3}} (x-3)^n$ on the interval $(1, 5)$

39. $\cos^2 x = \frac{1}{2} + \frac{1}{2} \sum\limits_{n=0}^{\infty} (-1)^n \frac{(4)^n x^{2n}}{(2n)!}$ **45.** $S_4 = 0.1822666667$

47. (a) 5 (b) $S_4 = 0.7474867725$

49. $\int_0^1 \cos(x^2)\,dx = \sum\limits_{n=0}^{\infty} \frac{(-1)^n}{(2n)!(4n+1)}; \ S_3 = 0.9045227920$

51. $\int_0^1 e^{-x^3}\,dx = \sum\limits_{n=0}^{\infty} \frac{(-1)^n}{n!(3n+1)}; \ S_5 = 0.8074461996$

53. $\int_0^x \frac{1-\cos(t)}{t}\,dt = \sum\limits_{n=1}^{\infty} (-1)^{n+1} \frac{x^{2n}}{(2n)!2n}$

55. $\int_0^x \ln(1+t^2)\,dt = \sum\limits_{n=1}^{\infty} (-1)^{n-1} \frac{x^{2n+1}}{n(2n+1)}$

57. $\frac{1}{1+2x}$ **59.** $\cos\pi = -1$ **65.** e^{x^3} **67.** $1 - 5x + \sin 5x$

69. $\frac{1}{(1-2x)(1-x)} = \sum\limits_{n=0}^{\infty} \left(2^{n+1} - 1\right) x^n$

71. $I(t) = \frac{V}{R} \sum\limits_{n=1}^{\infty} \frac{(-1)^{n+1}}{n!} \left(\frac{Rt}{L}\right)^n$

73. $f(x) = \sum\limits_{n=0}^{\infty} \frac{(-1)^n x^{6n}}{(2n)!}$ and $f^{(6)}(0) = -360$.

75. $e^{x^{20}} = 1 + x^{20} + \frac{x^{40}}{2} + \cdots$

77. No

n	Series value when $x = 2$
5	2.54297
10	−0.239933
15	41.9276
20	−764.272
25	16,595.8

83. $\lim\limits_{x\to0} \dfrac{\sin x - x + \frac{x^3}{6}}{x^5} = \dfrac{1}{120}$

85. $\lim\limits_{x\to0} \left(\dfrac{\sin(x^2)}{x^4} - \dfrac{\cos x}{x^2}\right) = \dfrac{1}{2}$

87. $S = \frac{\pi}{4} - \frac{1}{2}\ln 2$ **89.** $L \approx 28.369$

Chapter 10 Review

1. (a) $a_1^2 = 4, a_2^2 = \frac{1}{4}, a_3^2 = 0$

(b) $b_1 = \frac{1}{24}, b_2 = \frac{1}{60}, b_3 = \frac{1}{240}$

(c) $a_1 b_1 = -\frac{1}{12}, a_2 b_2 = -\frac{1}{120}, a_3 b_3 = 0$

(d) $2a_2 - 3a_1 = 5, 2a_3 - 3a_2 = \frac{3}{2}, 2a_4 - 3a_3 = \frac{1}{12}$

3. $\lim\limits_{n\to\infty} (5a_n - 2a_n^2) = 2$ **5.** $\lim\limits_{n\to\infty} e^{a_n} = e^2$

7. $\lim\limits_{n\to\infty} (-1)^n a_n$ does not exist.

9. $\lim\limits_{n\to\infty} \left(\sqrt{n+5} - \sqrt{n+2}\right) = 0$ **11.** $\lim\limits_{n\to\infty} 2^{1/n^2} = 1$

13. The sequence diverges. **15.** $\lim\limits_{n\to\infty} \tan^{-1}\left(\frac{n+2}{n+5}\right) = \frac{\pi}{4}$

17. $\lim\limits_{n\to\infty} \left(\sqrt{n^2+n} - \sqrt{n^2+1}\right) = \frac{1}{2}$

19. $\lim\limits_{m\to\infty} \left(1 + \frac{1}{m}\right)^{3m} = e^3$

21. $\lim\limits_{n\to\infty} \left(n\big(\ln(n+1) - \ln n\big)\right) = 1$

25. $\lim\limits_{n\to\infty} \dfrac{a_{n+1}}{a_n} = 3$ **27.** $S_4 = -\dfrac{11}{60}, S_7 = \dfrac{41}{630}$

29. $\sum\limits_{n=2}^{\infty} \left(\dfrac{2}{3}\right)^n = \dfrac{4}{3}$ **31.** $S = \dfrac{4}{37}$ **33.** $\sum\limits_{n=-1}^{\infty} \dfrac{2^{n+3}}{3^n} = 36$

35. $a_n = \left(\dfrac{1}{2}\right)^n + 1 - 2^n, b_n = 2^n - 1$

37. $S = \dfrac{47}{180}$ **39.** The series diverges.

41. $\int_1^{\infty} \dfrac{1}{(x+2)(\ln(x+2))^3}\, dx = \dfrac{1}{2(\ln(3))^2}$, so the series converges.

43. $\dfrac{1}{(n+1)^2} < \dfrac{1}{n^2}$, so the series converges.

45. $\sum\limits_{n=0}^{\infty} \dfrac{1}{n^{1.5}}$ converges, so the series converges.

47. $\dfrac{n}{\sqrt{n^5+2}} < \dfrac{1}{n^{3/2}}$, so the series converges.

49. $\sum\limits_{n=0}^{\infty} \left(\dfrac{10}{11}\right)^n$ converges, so the series converges.

51. Converges.

55. (b) $0.3971162690 \le S \le 0.3971172688$, so the maximum size of the error is 10^{-6}.

57. Converges absolutely.   **59.** Diverges.

61. (a) 500   (b) $K \approx \sum\limits_{n=0}^{499} \dfrac{(-1)^k}{(2k+1)^2} = 0.9159650942$

63. (a) Converges.   (b) Converges.   (c) Diverges.
(d) Converges.

65. Converges.   **67.** Converges.   **69.** Diverges.

71. Diverges.   **73.** Converges.   **75.** Converges.

77. Converges (by geometric series).

79. Converges (by geometric series).

81. Converges (by the Leibniz Test).

83. Converges (by the Leibniz Test).

85. Diverges (by the Divergence Test).

87. Converges (absolutely, by a Direct Comparison with the p-series $\sum\limits_{n=1}^{\infty} \dfrac{1}{n^{3/2}}$

89. Converges (by the Root Test).

91. Converges (by the Limit Comparison Test)

93. Converges using partial sums (the series is telescoping).

95. Diverges (by the Direct Comparison Test).

97. Converges (by the Direct Comparison Test).

99. Converges (by the Limit Comparison Test).

101. Converges on the interval $(-\infty, \infty)$.

103. Converges on the interval $[2, 4]$.   **105.** Converges at $x = 0$.

107. $\dfrac{2}{4-3x} = \dfrac{1}{2} \sum\limits_{n=0}^{\infty} \left(\dfrac{3}{4}\right)^n x^n$. The series converges on the interval $\left(\dfrac{-4}{3}, \dfrac{4}{3}\right)$.

109. (c)

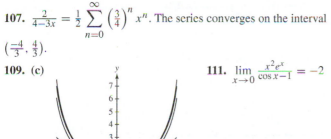

111. $\lim\limits_{x\to 0} \dfrac{x^2 e^x}{\cos x - 1} = -2$

113. $e^{4x} = \sum\limits_{n=0}^{\infty} \dfrac{4^n}{n!} x^n$

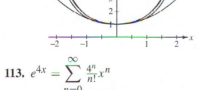

115. $x^4 = 16 + 32(x-2) + 24(x-2)^2 + 8(x-2)^3 + (x-2)^4$

117. $\sin x = \sum\limits_{n=0}^{\infty} \dfrac{(-1)^{n+1}(x-\pi)^{2n+1}}{(2n+1)!}$

119. $\dfrac{1}{1-2x} = \sum\limits_{n=0}^{\infty} \dfrac{2^n}{5^{n+1}} (x+2)^n$ **121.** $\ln\dfrac{x}{2} = \sum\limits_{n=1}^{\infty} \dfrac{(-1)^{n+1}(x-2)^n}{n2^n}$

123. $(x^2 - x)e^{x^2} = \sum\limits_{n=0}^{\infty} \left(\dfrac{x^{2n+2} - x^{2n+1}}{n!}\right)$ so $f^{(3)}(0) = -6$

125. $\dfrac{1}{1+\tan x} = 1 - x + x^2 - \dfrac{4}{3}x^3 + \cdots$, so $f^{(3)}(0) = -8$

127. $\dfrac{\pi}{2} - \dfrac{\pi^3}{2^3 3!} + \dfrac{\pi^5}{2^5 5!} - \dfrac{\pi^7}{2^7 7!} + \cdots = \sin\dfrac{\pi}{2} = 1$

Chapter 11

Section 11.1 Preliminary Questions

1. A circle of radius 3 centered at the origin

2. The center is at $(4, 5)$.   **3.** Maximum height: 4

4. Yes; No

5. (a) The line $y = x$ traversed left to right.

(b) Same path traversed left to right twice as fast.

6. $c_1(t) = (t, t^3), c_2(t) = (\sqrt[3]{t}, t), c_3(t) = (t^3, t^9)$

Section 11.1 Exercises

1. $(t = 0)(1, 9); (t = 2)(9, -3); (t = 4)(65, -39)$

3. $y = 2.5x - .000766x^2$

5. (a)   (b)   (c)   (d)

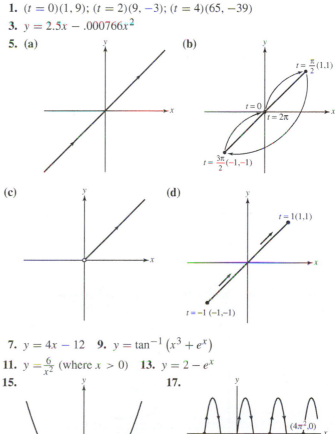

7. $y = 4x - 12$ **9.** $y = \tan^{-1}(x^3 + e^x)$

11. $y = \dfrac{6}{x^2}$ (where $x > 0$)   **13.** $y = 2 - e^x$

15.   **17.**

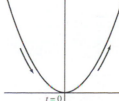

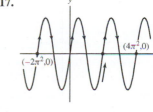

19. (a) ↔ (iv), (b) ↔ (ii), (c) ↔ (iii), (d) ↔ (i)

21. (a) $y_{max} = 100$ cm; (b) lands at $x = 2040$ cm from the origin, when $t = 20$ s

23. $c(t) = (t, 9 - 4t)$ **25.** $c(t) = \left(\frac{5+t^2}{4}, t\right)$

27. $c(t) = (-9 + 7\cos t, 4 + 7\sin t)$ **29.** $c(t) = (-4 + t, 9 + 8t)$

31. $c(t) = (3 - 8t, 1 + 3t)$ **33.** $c(t) = (1 + t, 1 + 2t)$ $(0 \le t \le 1)$

35. $c(t) = (3 + 4\cos t, 9 + 4\sin t)$ **37.** $c(t) = \left(-4 + t, -8 + t^2\right)$

39. $c(t) = (2 + t, 2 + 3t)$ **41.** $c(t) = \left(3 + t, (3 + t)^2\right)$

43. $y = \sqrt{x^2 - 1}$ $(1 \le x < \infty)$ **45.** Plot III

47.

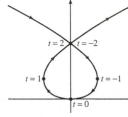

49. $\left.\dfrac{dy}{dx}\right|_{t=-4} = -\dfrac{1}{6}$ **51.** $\left.\dfrac{dy}{dx}\right|_{s=-1} = -\dfrac{3}{4}$ **53.** $-\dfrac{2}{3}$

55. $y = -\dfrac{9}{2}x + \dfrac{11}{2}$; $\dfrac{dy}{dx} = -\dfrac{9}{2}$

57. $y = x^2 + x^{-1}$; $\dfrac{dy}{dx} = 2x - \dfrac{1}{x^2}$ **59.** $(0, 0)$, $(96, 180)$

61.

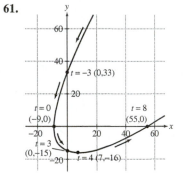

The graph is in: quadrant i for $t < -3$ or $t > 8$, quadrant ii for $-3 < t < 0$, quadrant iii for $0 < t < 3$, quadrant iv for $3 < t < 8$.

63. $(55, 0)$

65. The coordinates of P, $(R\cos\theta, r\sin\theta)$, describe an ellipse for $0 \le \theta \le 2\pi$.

69. $c(t) = (3 - 9t + 24t^2 - 16t^3, 2 + 6t^2 - 4t^3)$, $0 \le t \le 1$

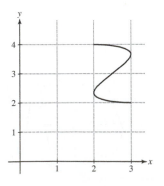

73. $y = -\sqrt{3}x + \dfrac{\sqrt{3}}{2}$

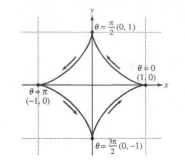

75. $((2k - 1)\pi, 2)$, $k = 0, \pm1, \pm2, \ldots$ **85.** $\left.\dfrac{d^2y}{dx^2}\right|_{t=2} = -\dfrac{21}{512}$

87. $\left.\dfrac{d^2y}{dx^2}\right|_{t=-3} = 0$ **89.** Concave up: $t > 0$ **91.** $\frac{1}{3}$ **93.** $\frac{2}{5}$

95. $\frac{2}{3}$

Section 11.2 Preliminary Questions

1. $S = \int_a^b \sqrt{x'(t)^2 + y'(t)^2}\, dt$

2. No. They are equal when the curve traced by $c(t)$ is a line segment from the initial to the final point, and $c(t)$ is a one-to-one function.

3. The speed at time t **4.** Displacement: 5; No   **5.** $L = 180$ cm

6. 4π

Section 11.2 Exercises

1. $S = 10$ **3.** $S = 800$ **5.** $S = \frac{1}{2}(65^{3/2} - 5^{3/2}) \approx 256.43$

7. $S = 3\pi$ **9.** $S = -8\left(\frac{\sqrt{2}}{2} - 1\right) \approx 2.34$

13. $S = \ln(\cosh(A))$ **15.** $\left.\dfrac{ds}{dt}\right|_{t=2} = 4\sqrt{10} \approx 12.65$ m/s

17. $\left.\dfrac{ds}{dt}\right|_{t=9} = \sqrt{41} \approx 6.4$ m/s   **19.** $\left.\dfrac{ds}{dt}\right|_{t=0} = 1$

21. $\left(\dfrac{ds}{dt}\right)_{min} \approx \sqrt{4.89} \approx 2.21$ **23.** $\dfrac{ds}{dt} = 8$

25.

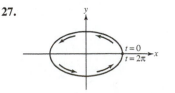

$M_{10} = 6.903734$, $M_{20} = 6.915035$, $M_{30} = 6.914949$, $M_{50} = 6.914951$

27.

$M_{10} = 25.528309$, $M_{20} = 25.526999$, $M_{30} = 25.526999$, $M_{50} = 25.526999$

29. $S = 2\pi^2 R$

39. (a)

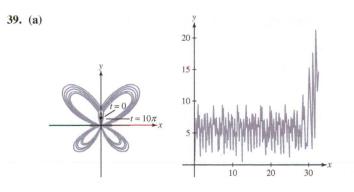

(b) $L \approx 212.096$

Section 11.3 Preliminary Questions

1. (a)   **2.** Positive: $(r,\theta) = \left(1, \frac{\pi}{2}\right)$; Negative: $(r, \theta) = \left(-1, \frac{3\pi}{2}\right)$

3. (a) Equation of the circle of radius 2 centered at the origin

(b) Equation of the circle of radius $\sqrt{2}$ centered at the origin

(c) Equation of the vertical line through the point $(2, 0)$

4. (a)

Section 11.3 Exercises

1. (A): $\left(3\sqrt{2}, \frac{3\pi}{4}\right)$; (B): $(3, \pi)$; (C):
$\left(\sqrt{5}, \pi + 0.46\right) \approx \left(\sqrt{5}, 3.60\right)$; (D): $\left(\sqrt{2}, \frac{5\pi}{4}\right)$; (E): $\left(\sqrt{2}, \frac{\pi}{4}\right)$;
(F): $\left(4, \frac{\pi}{6}\right)$; (G): $\left(4, \frac{11\pi}{6}\right)$

3. (a) $(1, 0)$ **(b)** $\left(\sqrt{12}, \frac{\pi}{6}\right)$ **(c)** $\left(\sqrt{8}, \frac{3\pi}{4}\right)$ **(d)** $\left(2, \frac{2\pi}{3}\right)$

5. (a) $\left(\frac{3\sqrt{3}}{2}, \frac{3}{2}\right)$ **(b)** $\left(-\frac{6}{\sqrt{2}}, \frac{6}{\sqrt{2}}\right)$ **(c)** $(0, 0)$ **(d)** $(0, -5)$

7. (A): $0 \le r \le 3, \pi \le \theta \le 2\pi$, (B): $0 \le r \le 3, \frac{\pi}{4} \le \theta \le \frac{\pi}{2}$, (C):
$3 \le r \le 5, \frac{3\pi}{4} \le \theta \le \pi$

9. $m = \tan \frac{3\pi}{5} \approx -3.1$ **11.** $x^2 + y^2 = 7^2$

13. $x^2 + (y - 1)^2 = 1$ **15.** $y = x - 1$ **17.** $r = \sqrt{5}$

19. $r = \tan \theta \sec \theta$ **21.** $e^r = 1$

23. (a)$\leftrightarrow$(iii), (b)$\leftrightarrow$(iv), (c)$\leftrightarrow$(i), (d)$\leftrightarrow$(ii)

25. (a) $(r, 2\pi - \theta)$ **(b)** $(r, \theta + \pi)$ **(c)** $(r, \pi - \theta)$
(d) $\left(r, \frac{\pi}{2} - \theta\right)$

27. $r \cos\left(\theta - \frac{\pi}{3}\right) = d$

29.

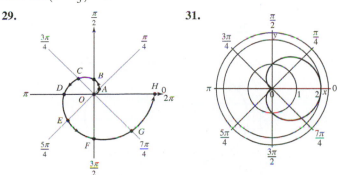

31.

33. (a) A, $\theta = 0, r = 0$; B, $\theta = \frac{\pi}{4}, r = \sin \frac{2\pi}{4} = 1$; C, $\theta = \frac{\pi}{2}$,
$r = 0$; D, $\theta = \frac{3\pi}{4}, r = \sin \frac{2 \cdot 3\pi}{4} = -1$; E, $\theta = \pi, r = 0$; F, $\theta = \frac{5\pi}{4}$,
$r = 1$; G, $\theta = \frac{3\pi}{2}, r = 0$; H, $\theta = \frac{7\pi}{4}, r = -1$; I, $\theta = 2\pi, r = 0$

(b) $0 \le \theta \le \frac{\pi}{2}$ is in the first quadrant. $\frac{\pi}{2} \le \theta \le \pi$ is in the fourth
quadrant. $\pi \le \theta \le \frac{3\pi}{2}$ is in the third quadrant. $\frac{3\pi}{2} \le \theta \le 2\pi$ is in the
second quadrant.

35.

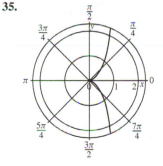

37. $\left(x - \frac{a}{2}\right)^2 + \left(y - \frac{b}{2}\right)^2 = \frac{a^2 + b^2}{4}, r = \frac{\sqrt{x^2 + y^2}}{2}$, centered at the
point $\left(\frac{a}{2}, \frac{b}{2}\right)$

39. $r^2 = \sec 2\theta$ **41.** $\left(x^2 + y^2\right)^2 = x^3 - 3y^2 x$

43. $r = 2 \sec\left(\theta - \frac{\pi}{9}\right)$ **45.** $r = 2\sqrt{10} \sec(\theta - 4.39)$

49. $r^2 = 2a^2 \cos 2\theta$

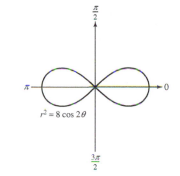

$r^2 = 8 \cos 2\theta$

53. $\theta = \frac{\pi}{2}, m = -\frac{2}{\pi}; \theta = \pi, m = \pi$

55. $\left(\frac{\sqrt{2}}{2}, \frac{\pi}{6}\right), \left(\frac{\sqrt{2}}{2}, \frac{5\pi}{6}\right), \left(\frac{\sqrt{2}}{2}, \frac{7\pi}{6}\right), \left(\frac{\sqrt{2}}{2}, \frac{11\pi}{6}\right)$

57. A: $m = 1$, B: $m = -1$, C: $m = 1$

Section 11.4 Preliminary Questions

1. (b)   **2.** Yes   **3. (c)**

Section 11.4 Exercises

1. $A = \frac{1}{2} \int_{\pi/2}^{\pi} r^2 \, d\theta = \frac{25\pi}{4}$

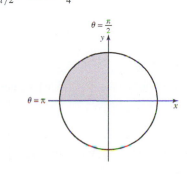

3. $A = \frac{1}{2}\int_0^\pi r^2 \, d\theta = 4\pi$ **5.** $A = \frac{\pi^5}{320} + \frac{\pi^4}{16} + \frac{\pi^3}{3}$

7. $A = \frac{3\pi}{2}$ **9.** $A = \frac{\pi}{8} \approx 0.39$

11.

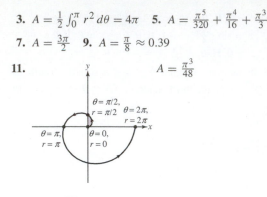

$A = \frac{\pi^3}{48}$

13. $A = \frac{\sqrt{15}}{2} + 7\cos^{-1}\left(\frac{1}{4}\right) \approx 11.163$

15. $A = \pi - \frac{3\sqrt{3}}{2} \approx 0.54$ **17.** $A = \frac{\pi}{8} - \frac{1}{4} \approx 0.14$ **19.** $A = 4\pi$

21. $A = \frac{9\pi}{2} - 4\sqrt{2}$ **23.** $S = 4\pi$

25. $L = \frac{1}{3}\left(\left(\pi^2 + 4\right)^{3/2} - 8\right) \approx 14.55$ **27.** π

29. $L = \sqrt{2}\pi/4 \approx 1.11$

31. $L = \int_0^{2\pi} \sqrt{\cos^4\theta + 4\cos^2\theta\sin^2\theta} \, d\theta \approx 5.52$

33. $\int_0^{\pi/2} \sqrt{2e^{2\theta} + 2e^\theta + 1} \, d\theta$ **35.** $\int_0^{\pi/2} \sin^2\theta\sqrt{1 + 8\cos^2\theta} \, d\theta$
37. $L \approx 6.682$ **39.** $L \approx 79.56$

Section 11.5 Preliminary Questions

1. (a) Hyperbola   **(b)** Parabola   **(c)** Ellipse

(d) Not a conic section

2. Hyperbolas   **3.** The points $(0, c)$ and $(0, -c)$

4. $\pm\frac{b}{a}$ are the slopes of the two asymptotes of the hyperbola.

Section 11.5 Exercises

1. $F_1 = \left(-\sqrt{65}, 0\right)$, $F_2 = \left(\sqrt{65}, 0\right)$. The vertices are $(9, 0)$, $(-9, 0)$, $(0, 4)$, and $(0, -4)$.

3. $F_1 = \left(\sqrt{97}, 0\right)$, $F_2 = \left(\sqrt{97}, 0\right)$. The vertices are $(4, 0)$ and $(-4, 0)$.

5. $F_1 = \left(\sqrt{65} + 3, -1\right)$, $F_2 = \left(-\sqrt{65} + 3, -1\right)$. The vertices are $(10, -1)$ and $(-4, -1)$.

7. $\frac{x^2}{6^2} + \frac{y^2}{3^2} = 1$ **9.** $\frac{(x-14)^2}{6^2} + \frac{(y+4)^2}{3^2} = 1$

11. $\frac{x^2}{5^2} + \frac{y^2}{7^2} = 1$ **13.** $\frac{x^2}{(40/3)^2} + \frac{y^2}{(50/3)^2} = 1$

15. $\left(\frac{x}{3}\right)^2 - \left(\frac{y}{4}\right)^2 = 1$ **17.** $\frac{x^2}{2^2} - \frac{y^2}{\left(2\sqrt{3}\right)^2} = 1$

19. $\left(\frac{x-2}{5}\right)^2 - \left(\frac{y}{10\sqrt{2}}\right)^2 = 1$ **21.** $y = 3x^2$

23. $y = \frac{1}{20}x^2$ **25.** $y = \frac{1}{16}x^2$ **27.** $x = \frac{1}{8}y^2$

29. Vertices: $(\pm4, 0)$, $(0, \pm2)$. Foci: $\left(\pm\sqrt{12}, 0\right)$. Centered at the origin.

31. Vertices: $(7, -5)$, $(-1, -5)$. Foci: $\left(\sqrt{65} + 3, -5\right)$, $\left(-\sqrt{65} + 3, -5\right)$. Center: $(3, -5)$. Asymptotes: $y = \frac{7}{4}x - \frac{41}{4}$ and $y = -\frac{7}{4}x + \frac{1}{4}$.

33. Vertices: $(5, 5)$, $(-7, 5)$. Foci: $\left(\sqrt{84} - 1, 5\right)$, $\left(-\sqrt{84} - 1, 5\right)$. Center: $(-1, 5)$. Asymptotes: $y = \frac{\sqrt{48}}{6}(x + 1) + 5 \approx 1.15x + 6.15$ and $y = -\frac{\sqrt{48}}{6}(x + 1) + 5 \approx -1.15x + 3.85$.

35. Vertex: $(0, 0)$. Focus: $\left(0, \frac{1}{16}\right)$.

37. Vertices: $\left(1 \pm \frac{5}{2}, \frac{1}{5}\right)$, $\left(1, \frac{1}{5} \pm 1\right)$. Foci: $\left(-\frac{\sqrt{21}}{2} + 1, \frac{1}{5}\right)$, $\left(\frac{\sqrt{21}}{2} + 1, \frac{1}{5}\right)$. Centered at $\left(1, \frac{1}{5}\right)$.

39. $D = -87$; ellipse   **41.** $D = 40$; hyperbola

47. Focus: $(0, c)$. Directrix: $y = -c$.   **49.** $A = \frac{8}{3}c^2$

51. $r = \frac{3}{2 + \cos\theta}$ **53.** $r = \frac{4}{1 + \cos\theta}$

55. Hyperbola, $e = 4$, directrix, $x = 2$

57. Ellipse, $e = \frac{3}{4}$, directrix, $x = \frac{8}{3}$ **59.** $r = \frac{-12}{5 + 6\cos\theta}$

61. $\left(\frac{x+3}{5}\right)^2 + \left(\frac{y}{4}\right)^2 = 1$ **63.** 4.5 billion miles

Chapter 11 Review

1. (a), (c)

3. $c(t) = (1 + 2\cos t, 1 + 2\sin t)$. The intersection points with the y-axis are $\left(0, 1 \pm \sqrt{3}\right)$. The intersection points with the x-axis are $\left(1 \pm \sqrt{3}, 0\right)$.

5. $c(\theta) = (\cos(\theta + \pi), \sin(\theta + \pi))$ **7.** $c(t) = (1 + 2t, 3 + 4t)$

9. $y = -\frac{x}{4} + \frac{37}{4}$ **11.** $y = -\frac{8}{(x-3)^3} + \frac{3-x}{2}$

13. $\left.\frac{dy}{dx}\right|_{t=3} = \frac{3}{14}$ **15.** $\left.\frac{dy}{dx}\right|_{t=0} = \frac{\cos 20}{e^{20}}$ **17.** $(1.41, 1.60)$

19. $c(t) = \left(-1 + 6t^2 - 4t^3, -1 + 6t - 6t^2\right)$

21. $\frac{ds}{dt} = \sqrt{3 + 2(\cos t - \sin t)}$; maximal speed: $\sqrt{3 + 2\sqrt{2}}$
23. $s = \sqrt{2}$

25.

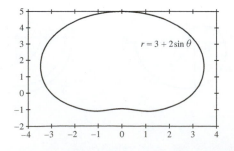

$s = 2\int_0^\pi \sqrt{\cos^2 2t + \sin^2 t} \, dt \approx 6.0972$

27. $\left(1, \frac{\pi}{6}\right)$ and $\left(3, \frac{5\pi}{4}\right)$ have rectangular coordinates $\left(\frac{\sqrt{3}}{2}, \frac{1}{2}\right)$ and $\left(-\frac{3\sqrt{2}}{2}, -\frac{3\sqrt{2}}{2}\right)$.

29. $\sqrt{x^2 + y^2} = \frac{2x}{x - y}$ **31.** $r = 3 + 2\sin\theta$

33. $A = \frac{\pi}{16}$ **35.** $e - \frac{1}{e}$

    Note: One needs to double the integral from $-\frac{\pi}{2}$ to $\frac{\pi}{2}$ in order to account for both sides of the graph.

37. $A = \frac{3\pi a^2}{2}$

39. Outer: $L \approx 36.121$, inner: $L \approx 7.5087$, difference: 28.6123

41. Ellipse. Vertices: $(\pm 3, 0)$, $(0, \pm 2)$. Foci: $(\pm\sqrt{5}, 0)$.

43. Ellipse. Vertices: $\left(\pm\frac{2}{\sqrt{5}}, 0\right)$, $\left(0, \pm\frac{4}{\sqrt{5}}\right)$. Foci: $\left(0, \pm\sqrt{\frac{12}{5}}\right)$.

45. $\left(\frac{x}{8}\right)^2 + \left(\frac{y}{\sqrt{61}}\right)^2 = 1$ **47.** $\left(\frac{x}{8}\right)^2 - \left(\frac{y}{6}\right)^2 = 1$ **49.** $x = \frac{1}{32}y^2$

51. $y = \sqrt{3}x + \left(\sqrt{3} - 5\right)$ and $y = -\sqrt{3}x + \left(-\sqrt{3} - 5\right)$

REFERENCES

The online source MacTutor History of Mathematics Archive **www-history.mcs.st-and.ac.uk** has been a valuable source of historical information.

Section 1.1

(EX 77) Adapted from *Calculus Problems for a New Century*, Robert Fraga, ed., Mathematical Association of America, Washington, DC, 1993, p. 9.

Section 1.2

(EX 25) Adapted from *Calculus Problems for a New Century*, Robert Fraga, ed., Mathematical Association of America, Washington, DC, 1993, p. 9.

Section 1.7

(EXMP 4) Adapted from B. Waits and F. Demana, "The Calculator and Computer Pre-Calculus Project," in *The Impact of Calculators on Mathematics Instruction*, University of Houston, 1994.

(EX 12) Adapted from B. Waits and F. Demana, "The Calculator and Computer Pre-Calculus Project," in *The Impact of Calculators on Mathematics Instruction*, University of Houston, 1994.

Section 2.2

(EX 63) Adapted from *Calculus Problems for a New Century*, Robert Fraga, ed., Mathematical Association of America, Washington, DC, 1993, Note 28.

Section 2.3

(EX 40) Adapted from *Calculus Problems for a New Century*, Robert Fraga, ed., Mathematical Association of America, Washington, DC, 1993, Note 28.

Chapter 2 Review

(EX 68) Adapted from *Calculus Problems for a New Century*, Robert Fraga, ed., Mathematical Association of America, Washington, DC, 1993, Note 28.

Section 3.1

(EX 77) Problem suggested by Dennis DeTurck, University of Pennsylvania.

Section 3.2

(EX 92) Problem suggested by Chris Bishop, SUNY Stony Brook.

(EX 93) Problem suggested by Chris Bishop, SUNY Stony Brook.

Section 3.4

(PQ 2) Adapted from *Calculus Problems for a New Century*, Robert Fraga, ed., Mathematical Association of America, Washington, DC, 1993, p. 25.

(EX 48) Karl J. Niklas and Brian J. Enquist, "Invariant Scaling Relationships for Interspecific Plant Biomass Production Rates and Body Size," *Proc. Natl. Acad. Sci.* 98, no. 5:2922-2927 (February 27, 2001).

Section 3.5

(EX 45) Adapted from Walter Meyer, *Falling Raindrops,* in *Applications of Calculus*, P. Straffin, ed., Mathematical Association of America, Washington, DC, 1993.

(EX 47) Adapted from a contribution by Jo Hoffacker, University of Georgia.

(EX 48–49) Adapted from a contribution by Thomas M. Smith, University of Illinois at Chicago, and Cindy S. Smith, Plainfield High School.

(EX 52, 56) Problems suggested by Chris Bishop, SUNY Stony Brook.

Section 3.10

(EX 32) Adapted from *Calculus Problems for a New Century*, Robert Fraga, ed., Mathematical Association of America, Washington, DC, 1993.

(EX 34) Problem suggested by Kay Dundas.

(EX 38, 44) Adapted from *Calculus Problems for a New Century*, Robert Fraga, ed., Mathematical Association of America, Washington, DC, 1993.

Chapter 3 Review

(EX 81, 94, 119) Problems suggested by Chris Bishop, SUNY Stony Brook.

Section 4.2

(MN p. 201) Adapted from "Stories about Maxima and Minima," V. M. Tikhomirov, AMS, (1990).

(MN p. 205) From Pierre Fermat, "On Maxima and Minima and on Tangents," translated by D.J. Struik (ed.), *A Source Book in Mathematics, 1200–1800*, Princeton University Press, Princeton, NJ, 1986.

Section 4.5

(EX 48, 79) Adapted from *Calculus Problems for a New Century*, Robert Fraga, ed., Mathematical Association of America, Washington, DC, 1993.

Section 4.6

(EX 28–29) Adapted from *Calculus Problems for a New Century*, Robert Fraga, ed., Mathematical Association of America, Washington, DC, 1993.

(EX 34) From Michael Helfgott, "Thomas Simpson and Maxima and Minima", *Convergence Magazine*, published online by the Mathematical Association of America.

(EX 42) Problem suggested by John Haverhals, Bradley University. *Source:* Illinois Agrinews.

(EX 44) Adapted from *Calculus Problems for a New Century*, Robert Fraga, ed., Mathematical Association of America, Washington, DC, 1993.

(EX 68–70) Adapted from B. Noble, *Applications of Undergraduate Mathematics in Engineering*, Macmillan, New York, 1967.

(EX 72) Adapted from Roger Johnson, "A Problem in Maxima and Minima," *American Mathematical Monthly*, 35:187-188 (1928).

Section 4.7

(EX 19) Adapted from "Do dogs know calculus?" Timothy Pennings, *The College Mathematic Journal*, Vol. 34, No. 3 (May, 2003), pp. 178–182.

(EX 38) Adapted from *Calculus for a Real and Complex World* by Frank Wattenberg, PWS Publishing, Boston, 1995.

(EX 77) Adapted from Robert J. Bumcrot, "Some Subtleties in L' Hôpital's Rule," in *A Century of Calculus*, Part II, Mathematical Association of America, Washington, DC, 1992.

Section 4.8

(EX 20) Adapted from *Calculus Problems for a New Century*, Robert Fraga, ed., Mathematical Association of America, Washington, DC, 1993, p. 52.

(EX 32–33) Adapted from E. Packel and S. Wagon, *Animating Calculus*, Springer-Verlag, New York, 1997, p. 79.

Chapter 4 Review

(EX 68) Adapted from *Calculus Problems for a New Century*, Robert Fraga, ed., Mathematical Association of America, Washington, DC, 1993.

Section 5.1

(EX 3) Problem suggested by John Polhill, Bloomsburg University.

(EX 86) Problem suggested by Chris Bishop, SUNY Stony Brook.

Section 5.5

(EX 40–41) Adapted from *Calculus Problems for a New Century*, Robert Fraga, ed., Mathematical Association of America, Washington, DC, 1993, p. 102.

(EX 42) Problem suggested by Dennis DeTurck, University of Pennsylvania.

Section 5.6

(EX 25–26) M. Newman and G. Eble, "Decline in Extinction Rates and Scale Invariance in the Fossil Record." *Paleobiology* 25:434-439 (1999).

(EX 28) From H. Flanders, R. Korfhage, and J. Price, *Calculus,* Academic Press, New York, 1970.

Section 5.7

(EX 76) Adapted from *Calculus Problems for a New Century*, Robert Fraga, ed., Mathematical Association of America, Washington, DC, 1993, p. 121.

Section 6.1

(EX 52) Adapted from Tom Farmer and Fred Gass, "Miami University: An Alternative Calculus" in *Priming the Calculus Pump*, Thomas Tucker, ed., Mathematical Association of America, Washington, DC, 1990, Note 17.

(EX 65) Adapted from *Calculus Problems for a New Century*, Robert Fraga, ed., Mathematical Association of America, Washington, DC, 1993.

Section 6.3

(EX 60, 62) Adapted from G. Alexanderson and L. Klosinski, "Some Surprising Volumes of Revolution," *Two-Year College Mathematics Journal* 6, 3:13-15 (1975).

Section 7.1

(EX 62–64, 67, 68–70, 73) Problems suggested by Brian Bradie, Christopher Newport University.

(EX 78) Adapted from *Calculus Problems for a New Century*, Robert Fraga, ed., Mathematical Association of America, Washington, DC, 1993.

(EX 87) Adapted from J. L. Borman, "A Remark on Integration by Parts," *American Mathematical Monthly* 51:32-33 (1944).

Section 7.3

(EX 52) Adapted from *Calculus Problems for a New Century*, Robert Fraga, ed., Mathematical Association of America, Washington, DC, 1993, p. 118.

Section 7.6

(EX 1–5, 9) Problems suggested by Brian Bradie, Christopher Newport University.

Section 7.7

(EX 84) Problem suggested by Chris Bishop, SUNY Stony Brook.

Section 7.9

See R. Courant and F. John, *Introduction to Calculus and Analysis*, Vol. 1, Springer-Verlag, New York, 1989.

Section 8.1

(EX 54) Adapted from G. Klambauer, *Aspects of Calculus*, Springer-Verlag, New York, 1986, Ch 6.

Section 9.1

(EX 57) Adapted from E. Batschelet, *Introduction to Mathematics for Life Scientists*, Springer-Verlag, New York, 1979.

(EX 59) Adapted from *Calculus Problems for a New Century*, Robert Fraga, ed., Mathematical Association of America, Washington, DC, 1993.

(EX 60, 65) Adapted from M. Tenenbaum and H. Pollard, *Ordinary Differential Equations*, Dover, New York, 1985.

Section 10.1

(EX 68) Adapted from G. Klambauer, *Aspects of Calculus*, Springer-Verlag, New York, 1986, p. 393.

Section 10.2

(EX 48) Adapted from *Calculus Problems for a New Century*, Robert Fraga, ed., Mathematical Association of America, Washington, DC, 1993, p. 137.

(EX 49) Adapted from *Calculus Problems for a New Century*, Robert Fraga, ed., Mathematical Association of America, Washington, DC, 1993, p. 138.

(EX 61) Adapted from George Andrews, "The Geometric Series in Calculus," *American Mathematical Monthly* 105, 1:36-40 (1998).

(EX 64) Adapted from Larry E. Knop, "Cantor's Disappearing Table," *The College Mathematics Journal* 16, 5:398-399 (1985).

Section 10.4

(EX 33) Adapted from *Calculus Problems for a New Century*, Robert Fraga, ed., Mathematical Association of America, Washington, DC, 1993, p. 145.

Section 11.2

(EX 41) Adapted from Richard Courant and Fritz John, *Differential and Integral Calculus*, Wiley-Interscience, New York, 1965.

Section 11.3

(EX 58) Adapted from *Calculus Problems for a New Century*, Robert Fraga, ed., Mathematical Association of America, Washington, DC, 1993.

Appendix D

(PROOF OF THEOREM 6) A proof without this simplifying assumption can be found in R. Courant and F. John, *Introduction to Calculus and Analysis, Vol. 1*, Springer-Verlag, New York, 1989.

INDEX

ELEMENTARY FUNCTIONS

Power Functions $f(x) = x^a$

$f(x) = x^n$, n a positive integer

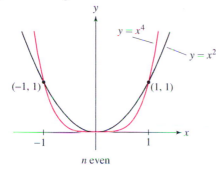

n even

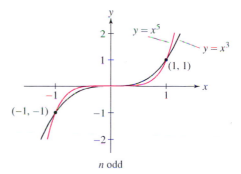

n odd

Asymptotic behavior of a polynomial function of even degree and positive leading coefficient

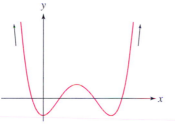

n even

Asymptotic behavior of a polynomial function of odd degree and positive leading coefficient

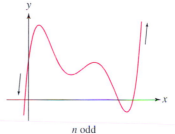

n odd

$f(x) = x^{-n} = \dfrac{1}{x^n}$

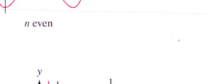

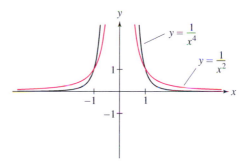

Inverse Trigonometric Functions

$\arcsin x = \sin^{-1} x = \theta$

$\Leftrightarrow \quad \sin \theta = x, \quad -\dfrac{\pi}{2} \leq \theta \leq \dfrac{\pi}{2}$

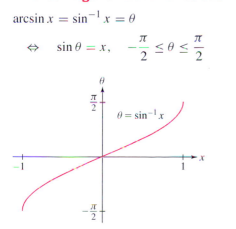

$\arccos x = \cos^{-1} x = \theta$

$\Leftrightarrow \quad \cos \theta = x, \quad 0 \leq \theta \leq \pi$

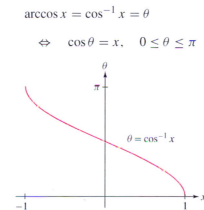

$\arctan x = \tan^{-1} x = \theta$

$\Leftrightarrow \quad \tan \theta = x, \quad -\dfrac{\pi}{2} < \theta < \dfrac{\pi}{2}$

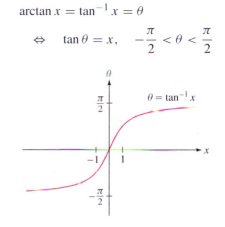

Exponential and Logarithmic Functions

$$\log_a x = y \quad \Leftrightarrow \quad a^y = x$$

$$\ln x = y \quad \Leftrightarrow \quad e^y = x$$

$$\log_a(xy) = \log_a x + \log_a y$$

$$\log_a(a^x) = x \qquad a^{\log_a x} = x$$

$$\ln(e^x) = x \qquad e^{\ln x} = x$$

$$\log_a\left(\frac{x}{y}\right) = \log_a x - \log_a y$$

$$\log_a 1 = 0 \qquad \log_a a = 1$$

$$\ln 1 = 0 \qquad \ln e = 1$$

$$\log_a(x^r) = r \log_a x$$

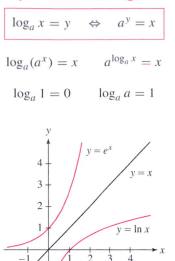

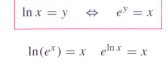

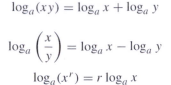

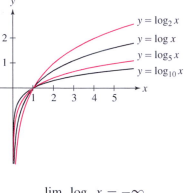

$$\lim_{x \to \infty} a^x = \infty, \quad a > 1$$

$$\lim_{x \to \infty} a^x = 0, \quad 0 < a < 1$$

$$\lim_{x \to -\infty} a^x = 0, \quad a > 1$$

$$\lim_{x \to -\infty} a^x = \infty, \quad 0 < a < 1$$

$$\lim_{x \to 0^+} \log_a x = -\infty$$

$$\lim_{x \to \infty} \log_a x = \infty$$

Hyperbolic Functions

$$\sinh x = \frac{e^x - e^{-x}}{2} \qquad \operatorname{csch} x = \frac{1}{\sinh x}$$

$$\cosh x = \frac{e^x + e^{-x}}{2} \qquad \operatorname{sech} x = \frac{1}{\cosh x}$$

$$\tanh x = \frac{\sinh x}{\cosh x} \qquad \coth x = \frac{\cosh x}{\sinh x}$$

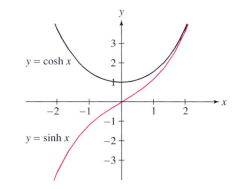

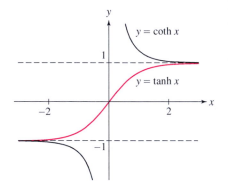

$$\sinh(x + y) = \sinh x \cosh y + \cosh x \sinh y$$

$$\cosh(x + y) = \cosh x \cosh y + \sinh x \sinh y$$

$$\sinh 2x = 2 \sinh x \cosh x$$

$$\cosh 2x = \cosh^2 x + \sinh^2 x$$

Inverse Hyperbolic Functions

$$y = \sinh^{-1} x \quad \Leftrightarrow \quad \sinh y = x$$

$$y = \cosh^{-1} x \quad \Leftrightarrow \quad \cosh y = x \text{ and } y \geq 0$$

$$y = \tanh^{-1} x \quad \Leftrightarrow \quad \tanh y = x$$

$$\sinh^{-1} x = \ln\left(x + \sqrt{x^2 + 1}\right)$$

$$\cosh^{-1} x = \ln\left(x + \sqrt{x^2 - 1}\right) \quad x > 1$$

$$\tanh^{-1} x = \frac{1}{2} \ln\left(\frac{1 + x}{1 - x}\right) \quad -1 < x < 1$$

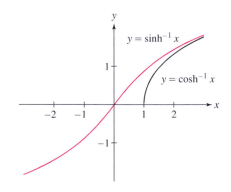

DIFFERENTIATION

Differentiation Rules

1. $\dfrac{d}{dx}(c) = 0$

2. $\dfrac{d}{dx}x = 1$

3. $\dfrac{d}{dx}(x^n) = nx^{n-1}$ (Power Rule)

4. $\dfrac{d}{dx}[cf(x)] = cf'(x)$

5. $\dfrac{d}{dx}[f(x) + g(x)] = f'(x) + g'(x)$

6. $\dfrac{d}{dx}[f(x)g(x)] = f(x)g'(x) + g(x)f'(x)$ (Product Rule)

7. $\dfrac{d}{dx}\left[\dfrac{f(x)}{g(x)}\right] = \dfrac{g(x)f'(x) - f(x)g'(x)}{[g(x)]^2}$ (Quotient Rule)

8. $\dfrac{d}{dx}f(g(x)) = f'(g(x))g'(x)$ (Chain Rule)

9. $\dfrac{d}{dx}f(x)^n = nf(x)^{n-1}f'(x)$ (General Power Rule)

10. $\dfrac{d}{dx}f(kx + b) = kf'(kx + b)$

11. $g'(x) = \dfrac{1}{f'(g(x))}$ where g is the inverse f^{-1}

12. $\dfrac{d}{dx}\ln f(x) = \dfrac{f'(x)}{f(x)}$

Trigonometric Functions

13. $\dfrac{d}{dx}\sin x = \cos x$

14. $\dfrac{d}{dx}\cos x = -\sin x$

15. $\dfrac{d}{dx}\tan x = \sec^2 x$

16. $\dfrac{d}{dx}\csc x = -\csc x \cot x$

17. $\dfrac{d}{dx}\sec x = \sec x \tan x$

18. $\dfrac{d}{dx}\cot x = -\csc^2 x$

Inverse Trigonometric Functions

19. $\dfrac{d}{dx}(\sin^{-1} x) = \dfrac{1}{\sqrt{1 - x^2}}$

20. $\dfrac{d}{dx}(\cos^{-1} x) = -\dfrac{1}{\sqrt{1 - x^2}}$

21. $\dfrac{d}{dx}(\tan^{-1} x) = \dfrac{1}{1 + x^2}$

22. $\dfrac{d}{dx}(\csc^{-1} x) = -\dfrac{1}{|x|\sqrt{x^2 - 1}}$

23. $\dfrac{d}{dx}(\sec^{-1} x) = \dfrac{1}{|x|\sqrt{x^2 - 1}}$

24. $\dfrac{d}{dx}(\cot^{-1} x) = -\dfrac{1}{1 + x^2}$

Exponential and Logarithmic Functions

25. $\dfrac{d}{dx}(e^x) = e^x$

26. $\dfrac{d}{dx}(a^x) = (\ln a)a^x$

27. $\dfrac{d}{dx}\ln |x| = \dfrac{1}{x}$

28. $\dfrac{d}{dx}(\log_a x) = \dfrac{1}{(\ln a)x}$

Hyperbolic Functions

29. $\dfrac{d}{dx}(\sinh x) = \cosh x$

30. $\dfrac{d}{dx}(\cosh x) = \sinh x$

31. $\dfrac{d}{dx}(\tanh x) = \operatorname{sech}^2 x$

32. $\dfrac{d}{dx}(\operatorname{csch} x) = -\operatorname{csch} x \coth x$

33. $\dfrac{d}{dx}(\operatorname{sech} x) = -\operatorname{sech} x \tanh x$

34. $\dfrac{d}{dx}(\coth x) = -\operatorname{csch}^2 x$

Inverse Hyperbolic Functions

35. $\dfrac{d}{dx}(\sinh^{-1} x) = \dfrac{1}{\sqrt{1 + x^2}}$

36. $\dfrac{d}{dx}(\cosh^{-1} x) = \dfrac{1}{\sqrt{x^2 - 1}}$

37. $\dfrac{d}{dx}(\tanh^{-1} x) = \dfrac{1}{1 - x^2}$

38. $\dfrac{d}{dx}(\operatorname{csch}^{-1} x) = -\dfrac{1}{|x|\sqrt{x^2 + 1}}$

39. $\dfrac{d}{dx}(\operatorname{sech}^{-1} x) = -\dfrac{1}{x\sqrt{1 - x^2}}$

40. $\dfrac{d}{dx}(\coth^{-1} x) = \dfrac{1}{1 - x^2}$

INTEGRATION

Substitution

If an integrand has the form $f(u(x))u'(x)$, then rewrite the entire integral in terms of u and its differential $du = u'(x)\,dx$:

$$\int f(u(x))u'(x)\,dx = \int f(u)\,du$$

Integration by Parts Formula

$$\int uv'\,dx = uv - \int u'v\,dx$$

TABLE OF INTEGRALS

Basic Forms

1. $\displaystyle \int u^n\,du = \frac{u^{n+1}}{n+1} + C, \quad n \neq -1$

2. $\displaystyle \int \frac{du}{u} = \ln|u| + C$

3. $\displaystyle \int e^u\,du = e^u + C$

4. $\displaystyle \int a^u\,du = \frac{a^u}{\ln a} + C$

5. $\displaystyle \int \sin u\,du = -\cos u + C$

6. $\displaystyle \int \cos u\,du = \sin u + C$

7. $\displaystyle \int \sec^2 u\,du = \tan u + C$

8. $\displaystyle \int \csc^2 u\,du = -\cot u + C$

9. $\displaystyle \int \sec u \tan u\,du = \sec u + C$

10. $\displaystyle \int \csc u \cot u\,du = -\csc u + C$

11. $\displaystyle \int \tan u\,du = \ln|\sec u| + C$

12. $\displaystyle \int \cot u\,du = \ln|\sin u| + C$

13. $\displaystyle \int \sec u\,du = \ln|\sec u + \tan u| + C$

14. $\displaystyle \int \csc u\,du = \ln|\csc u - \cot u| + C$

15. $\displaystyle \int \frac{du}{\sqrt{a^2 - u^2}} = \sin^{-1}\frac{u}{a} + C$

16. $\displaystyle \int \frac{du}{a^2 + u^2} = \frac{1}{a}\tan^{-1}\frac{u}{a} + C$

Exponential and Logarithmic Forms

17. $\displaystyle \int ue^{au}\,du = \frac{1}{a^2}(au - 1)e^{au} + C$

18. $\displaystyle \int u^n e^{au}\,du = \frac{1}{a}u^n e^{au} - \frac{n}{a}\int u^{n-1}e^{au}\,du$

19. $\displaystyle \int e^{au}\sin bu\,du = \frac{e^{au}}{a^2 + b^2}(a\sin bu - b\cos bu) + C$

20. $\displaystyle \int e^{au}\cos bu\,du = \frac{e^{au}}{a^2 + b^2}(a\cos bu + b\sin bu) + C$

21. $\displaystyle \int \ln u\,du = u\ln u - u + C$

22. $\displaystyle \int u^n \ln u\,du = \frac{u^{n+1}}{(n+1)^2}[(n+1)\ln u - 1] + C$

23. $\displaystyle \int \frac{1}{u\ln u}\,du = \ln|\ln u| + C$

Hyperbolic Forms

24. $\displaystyle \int \sinh u\,du = \cosh u + C$

25. $\displaystyle \int \cosh u\,du = \sinh u + C$

26. $\displaystyle \int \tanh u\,du = \ln\cosh u + C$

27. $\displaystyle \int \coth u\,du = \ln|\sinh u| + C$

28. $\displaystyle \int \operatorname{sech} u\,du = \tan^{-1}|\sinh u| + C$

29. $\displaystyle \int \operatorname{csch} u\,du = \ln\left|\tanh\frac{1}{2}u\right| + C$

30. $\displaystyle \int \operatorname{sech}^2 u\,du = \tanh u + C$

31. $\displaystyle \int \operatorname{csch}^2 u\,du = -\coth u + C$

32. $\displaystyle \int \operatorname{sech} u \tanh u\,du = -\operatorname{sech} u + C$

33. $\displaystyle \int \operatorname{csch} u \coth u\,du = -\operatorname{csch} u + C$

Trigonometric Forms

34. $\displaystyle \int \sin^2 u\,du = \frac{1}{2}u - \frac{1}{4}\sin 2u + C$

35. $\displaystyle \int \cos^2 u\,du = \frac{1}{2}u + \frac{1}{4}\sin 2u + C$

36. $\displaystyle \int \tan^2 u\,du = \tan u - u + C$

37. $\displaystyle \int \cot^2 u\,du = -\cot u - u + C$

38. $\displaystyle \int \sin^3 u\,du = -\frac{1}{3}(2 + \sin^2 u)\cos u + C$

39. $\displaystyle \int \cos^3 u\,du = \frac{1}{3}(2 + \cos^2 u)\sin u + C$

40. $\displaystyle \int \tan^3 u\,du = \frac{1}{2}\tan^2 u + \ln|\cos u| + C$

41. $\displaystyle \int \cot^3 u\,du = -\frac{1}{2}\cot^2 u - \ln|\sin u| + C$

42. $\displaystyle \int \sec^3 u\,du = \frac{1}{2}\sec u \tan u + \frac{1}{2}\ln|\sec u + \tan u| + C$

43. $\displaystyle\int \csc^3 u\, du = -\frac{1}{n}\csc u \cot u + \frac{1}{n}\ln|\csc u - \cot u| + C$

44. $\displaystyle\int \sin^n u\, du = -\frac{1}{n}\sin^{n-1} u \cos u + \frac{n-1}{n}\int \sin^{n-2} u\, du$

45. $\displaystyle\int \cos^n u\, du = \frac{1}{n}\cos^{n-1} u \sin u + \frac{n-1}{n}\int \cos^{n-2} u\, du$

46. $\displaystyle\int \tan^n u\, du = \frac{1}{n-1}\tan^{n-1} u - \int \tan^{n-2} u\, du$

47. $\displaystyle\int \cot^n u\, du = \frac{-1}{n-1}\cot^{n-1} u - \int \cot^{n-2} u\, du$

48. $\displaystyle\int \sec^n u\, du = \frac{1}{n-1}\tan u \sec^{n-2} u + \frac{n-2}{n-1}\int \sec^{n-2} u\, du$

49. $\displaystyle\int \csc^n u\, du = \frac{-1}{n-1}\cot u \csc^{n-2} u + \frac{n-2}{n-1}\int \csc^{n-2} u\, du$

50. $\displaystyle\int \sin au \sin bu\, du = \frac{\sin(a-b)u}{2(a-b)} - \frac{\sin(a+b)u}{2(a+b)} + C$

51. $\displaystyle\int \cos au \cos bu\, du = \frac{\sin(a-b)u}{2(a-b)} + \frac{\sin(a+b)u}{2(a+b)} + C$

52. $\displaystyle\int \sin au \cos bu\, du = -\frac{\cos(a-b)u}{2(a-b)} - \frac{\cos(a+b)u}{2(a+b)} + C$

53. $\displaystyle\int u \sin u\, du = \sin u - u \cos u + C$

54. $\displaystyle\int u \cos u\, du = \cos u + u \sin u + C$

55. $\displaystyle\int u^n \sin u\, du = -u^n \cos u + n\int u^{n-1}\cos u\, du$

56. $\displaystyle\int u^n \cos u\, du = u^n \sin u - n\int u^{n-1}\sin u\, du$

57. $\displaystyle\int \sin^n u \cos^m u\, du$

$\displaystyle = -\frac{\sin^{n-1} u \cos^{m+1} u}{n+m} + \frac{n-1}{n+m}\int \sin^{n-2} u \cos^m u\, du$

$\displaystyle = \frac{\sin^{n+1} u \cos^{m-1} u}{n+m} + \frac{m-1}{n+m}\int \sin^n u \cos^{m-2} u\, du$

Inverse Trigonometric Forms

58. $\displaystyle\int \sin^{-1} u\, du = u \sin^{-1} u + \sqrt{1-u^2} + C$

59. $\displaystyle\int \cos^{-1} u\, du = u \cos^{-1} u - \sqrt{1-u^2} + C$

60. $\displaystyle\int \tan^{-1} u\, du = u \tan^{-1} u - \frac{1}{2}\ln(1+u^2) + C$

61. $\displaystyle\int u \sin^{-1} u\, du = \frac{2u^2-1}{4}\sin^{-1} u + \frac{u\sqrt{1-u^2}}{4} + C$

62. $\displaystyle\int u \cos^{-1} u\, du = \frac{2u^2-1}{4}\cos^{-1} u - \frac{u\sqrt{1-u^2}}{4} + C$

63. $\displaystyle\int u \tan^{-1} u\, du = \frac{u^2+1}{2}\tan^{-1} u - \frac{u}{2} + C$

64. $\displaystyle\int u^n \sin^{-1} u\, du = \frac{1}{n+1}\left[u^{n+1}\sin^{-1} u - \int \frac{u^{n+1}\, du}{\sqrt{1-u^2}}\right], \quad n \neq -1$

65. $\displaystyle\int u^n \cos^{-1} u\, du = \frac{1}{n+1}\left[u^{n+1}\cos^{-1} u + \int \frac{u^{n+1}\, du}{\sqrt{1-u^2}}\right], \quad n \neq -1$

66. $\displaystyle\int u^n \tan^{-1} u\, du = \frac{1}{n+1}\left[u^{n+1}\tan^{-1} u - \int \frac{u^{n+1}\, du}{1+u^2}\right], \quad n \neq -1$

Forms Involving $\sqrt{a^2 - u^2},\ a > 0$

67. $\displaystyle\int \sqrt{a^2 - u^2}\, du = \frac{u}{2}\sqrt{a^2-u^2} + \frac{a^2}{2}\sin^{-1}\frac{u}{a} + C$

68. $\displaystyle\int u^2\sqrt{a^2 - u^2}\, du = \frac{u}{8}(2u^2-a^2)\sqrt{a^2-u^2} + \frac{a^4}{8}\sin^{-1}\frac{u}{a} + C$

69. $\displaystyle\int \frac{\sqrt{a^2-u^2}}{u}\, du = \sqrt{a^2-u^2} - a\ln\left|\frac{a+\sqrt{a^2-u^2}}{u}\right| + C$

70. $\displaystyle\int \frac{\sqrt{a^2-u^2}}{u^2}\, du = -\frac{1}{u}\sqrt{a^2-u^2} - \sin^{-1}\frac{u}{a} + C$

71. $\displaystyle\int \frac{u^2\, du}{\sqrt{a^2-u^2}} = -\frac{u}{2}\sqrt{a^2-u^2} + \frac{a^2}{2}\sin^{-1}\frac{u}{a} + C$

72. $\displaystyle\int \frac{du}{u\sqrt{a^2-u^2}} = -\frac{1}{a}\ln\left|\frac{a+\sqrt{a^2-u^2}}{u}\right| + C$

73. $\displaystyle\int \frac{du}{u^2\sqrt{a^2-u^2}} = -\frac{1}{a^2 u}\sqrt{a^2-u^2} + C$

74. $\displaystyle\int (a^2-u^2)^{3/2}\, du = -\frac{u}{8}(2u^2-5a^2)\sqrt{a^2-u^2} + \frac{3a^4}{8}\sin^{-1}\frac{u}{a} + C$

75. $\displaystyle\int \frac{du}{(a^2-u^2)^{3/2}} = \frac{u}{a^2\sqrt{a^2-u^2}} + C$

Forms Involving $\sqrt{u^2 - a^2},\ a > 0$

76. $\displaystyle\int \sqrt{u^2 - a^2}\, du = \frac{u}{2}\sqrt{u^2-a^2} - \frac{a^2}{2}\ln\left|u + \sqrt{u^2-a^2}\right| + C$

77. $\displaystyle\int u^2\sqrt{u^2 - a^2}\, du$

$\displaystyle = \frac{u}{8}(2u^2-a^2)\sqrt{u^2-a^2} - \frac{a^4}{8}\ln\left|u + \sqrt{u^2-a^2}\right| + C$

78. $\displaystyle\int \frac{\sqrt{u^2-a^2}}{u}\, du = \sqrt{u^2-a^2} - a\cos^{-1}\frac{a}{|u|} + C$

79. $\displaystyle\int \frac{\sqrt{u^2-a^2}}{u^2}\, du = -\frac{\sqrt{u^2-a^2}}{u} + \ln\left|u + \sqrt{u^2-a^2}\right| + C$

80. $\displaystyle\int \frac{du}{\sqrt{u^2-a^2}} = \ln\left|u + \sqrt{u^2-a^2}\right| + C$

81. $\displaystyle\int \frac{u^2\, du}{\sqrt{u^2-a^2}} = \frac{u}{2}\sqrt{u^2-a^2} + \frac{a^2}{2}\ln\left|u + \sqrt{u^2-a^2}\right| + C$

82. $\displaystyle\int \frac{du}{u^2\sqrt{u^2-a^2}} = \frac{\sqrt{u^2-a^2}}{a^2 u} + C$

83. $\displaystyle\int \frac{du}{(u^2-a^2)^{3/2}} = -\frac{u}{a^2\sqrt{u^2-a^2}} + C$

Forms Involving $\sqrt{a^2 + u^2},\ a > 0$

84. $\displaystyle\int \sqrt{a^2 + u^2}\, du = \frac{u}{2}\sqrt{a^2+u^2} + \frac{a^2}{2}\ln\left(u + \sqrt{a^2+u^2}\right) + C$

85. $\displaystyle\int u^2\sqrt{a^2 + u^2}\, du$

$\displaystyle = \frac{u}{8}(a^2 + 2u^2)\sqrt{a^2+u^2} - \frac{a^4}{8}\ln\left(u + \sqrt{a^2+u^2}\right) + C$

86. $\displaystyle\int \frac{\sqrt{a^2+u^2}}{u}\, du = \sqrt{a^2+u^2} - a\ln\left|\frac{a+\sqrt{a^2+u^2}}{u}\right| + C$

87. $\displaystyle\int \frac{\sqrt{a^2+u^2}}{u^2}\, du = -\frac{\sqrt{a^2+u^2}}{u} + \ln\left(u + \sqrt{a^2+u^2}\right) + C$

88. $\displaystyle\int \frac{du}{\sqrt{a^2 + u^2}} = \ln\left(u + \sqrt{a^2 + u^2}\right) + C$

89. $\displaystyle\int \frac{u^2\, du}{\sqrt{a^2 + u^2}} = \frac{u}{2}\sqrt{a^2 + u^2} - \frac{a^2}{2}\ln\left(u + \sqrt{a^2 + u^2}\right) + C$

90. $\displaystyle\int \frac{du}{u\sqrt{a^2 + u^2}} = -\frac{1}{a}\ln\left|\frac{\sqrt{a^2 + u^2} + a}{u}\right| + C$

91. $\displaystyle\int \frac{du}{u^2\sqrt{a^2 + u^2}} = -\frac{\sqrt{a^2 + u^2}}{a^2 u} + C$

92. $\displaystyle\int \frac{du}{(a^2 + u^2)^{3/2}} = \frac{u}{a^2\sqrt{a^2 + u^2}} + C$

Forms Involving $a + bu$

93. $\displaystyle\int \frac{u\, du}{a + bu} = \frac{1}{b^2}\left(a + bu - a\ln|a + bu|\right) + C$

94. $\displaystyle\int \frac{u^2\, du}{a + bu} = \frac{1}{2b^3}\left[(a + bu)^2 - 4a(a + bu) + 2a^2\ln|a + bu|\right] + C$

95. $\displaystyle\int \frac{du}{u(a + bu)} = \frac{1}{a}\ln\left|\frac{u}{a + bu}\right| + C$

96. $\displaystyle\int \frac{du}{u^2(a + bu)} = -\frac{1}{au} + \frac{b}{a^2}\ln\left|\frac{a + bu}{u}\right| + C$

97. $\displaystyle\int \frac{u\, du}{(a + bu)^2} = \frac{a}{b^2(a + bu)} + \frac{1}{b^2}\ln|a + bu| + C$

98. $\displaystyle\int \frac{du}{u(a + bu)^2} = \frac{1}{a(a + bu)} - \frac{1}{a^2}\ln\left|\frac{a + bu}{u}\right| + C$

99. $\displaystyle\int \frac{u^2\, du}{(a + bu)^2} = \frac{1}{b^3}\left(a + bu - \frac{a^2}{a + bu} - 2a\ln|a + bu|\right) + C$

100. $\displaystyle\int u\sqrt{a + bu}\, du = \frac{2}{15b^2}(3bu - 2a)(a + bu)^{3/2} + C$

101. $\displaystyle\int u^n\sqrt{a + bu}\, du$

$\displaystyle= \frac{2}{b(2n + 3)}\left[u^n(a + bu)^{3/2} - na\int u^{n-1}\sqrt{a + bu}\, du\right]$

102. $\displaystyle\int \frac{u\, du}{\sqrt{a + bu}} = \frac{2}{3b^2}(bu - 2a)\sqrt{a + bu} + C$

103. $\displaystyle\int \frac{u^n\, du}{\sqrt{a + bu}} = \frac{2u^n\sqrt{a + bu}}{b(2n + 1)} - \frac{2na}{b(2n + 1)}\int \frac{u^{n-1}\, du}{\sqrt{a + bu}}$

104. $\displaystyle\int \frac{du}{u\sqrt{a + bu}} = \frac{1}{\sqrt{a}}\ln\left|\frac{\sqrt{a + bu} - \sqrt{a}}{\sqrt{a + bu} + \sqrt{a}}\right| + C, \quad \text{if } a > 0$

$\displaystyle= \frac{2}{\sqrt{-a}}\tan^{-1}\sqrt{\frac{a + bu}{-a}} + C, \quad \text{if } a < 0$

105. $\displaystyle\int \frac{du}{u^n\sqrt{a + bu}} = -\frac{\sqrt{a + bu}}{a(n - 1)u^{n-1}} - \frac{b(2n - 3)}{2a(n - 1)}\int \frac{du}{u^{n-1}\sqrt{a + bu}}$

106. $\displaystyle\int \frac{\sqrt{a + bu}}{u}\, du = 2\sqrt{a + bu} + a\int \frac{du}{u\sqrt{a + bu}}$

107. $\displaystyle\int \frac{\sqrt{a + bu}}{u^2}\, du = -\frac{\sqrt{a + bu}}{u} + \frac{b}{2}\int \frac{du}{u\sqrt{a + bu}}$

Forms Involving $\sqrt{2au - u^2}$, $a > 0$

108. $\displaystyle\int \sqrt{2au - u^2}\, du = \frac{u - a}{2}\sqrt{2au - u^2} + \frac{a^2}{2}\cos^{-1}\left(\frac{a - u}{a}\right) + C$

109. $\displaystyle\int u\sqrt{2au - u^2}\, du$

$\displaystyle= \frac{2u^2 - au - 3a^2}{6}\sqrt{2au - u^2} + \frac{a^3}{2}\cos^{-1}\left(\frac{a - u}{a}\right) + C$

110. $\displaystyle\int \frac{du}{\sqrt{2au - u^2}} = \cos^{-1}\left(\frac{a - u}{a}\right) + C$

111. $\displaystyle\int \frac{du}{u\sqrt{2au - u^2}} = -\frac{\sqrt{2au - u^2}}{au} + C$

ESSENTIAL THEOREMS

Intermediate Value Theorem

If f is continuous on a closed interval $[a, b]$ and $f(a) \neq f(b)$, then for every value M between $f(a)$ and $f(b)$, there exists at least one value $c \in (a, b)$ such that $f(c) = M$.

Mean Value Theorem

If f is continuous on a closed interval $[a, b]$ and differentiable on (a, b), then there exists at least one value $c \in (a, b)$ such that

$$f'(c) = \frac{f(b) - f(a)}{b - a}$$

Extreme Values on a Closed Interval

If f is continuous on a closed interval $[a, b]$, then f attains both a minimum and a maximum value on $[a, b]$. Furthermore, if $c \in [a, b]$ and $f(c)$ is an extreme value (min or max), then c is either a critical point of f in (a, b) or one of the endpoints a or b.

The Fundamental Theorem of Calculus, Part I

Assume that f is continuous on $[a, b]$ and let F be an antiderivative of f on $[a, b]$. Then

$$\int_a^b f(x)\, dx = F(b) - F(a)$$

Fundamental Theorem of Calculus, Part II

Assume that f is a continuous function on $[a, b]$. Then the area function $A(x) = \displaystyle\int_a^x f(t)\, dt$ is an antiderivative of f, that is,

$$A'(x) = f(x) \quad \text{or equivalently} \quad \frac{d}{dx}\int_a^x f(t)\, dt = f(x)$$

Furthermore, $A(x)$ satisfies the initial condition $A(a) = 0$.